Mobile Robots: Navigation, Control and Sensing

Mobile Robots: Navigation, Control and Sensing

Guest Editors

Stephen Monk
David Cheneler

MDPI

Basel • Beijing • Wuhan • Barcelona • Belgrade • Novi Sad • Cluj • Manchester

Guest Editors
Stephen Monk
School of Engineering
Lancaster University
Lancaster
United Kingdom

David Cheneler
School of Engineering
Lancaster University
Lancaster
United Kingdom

Editorial Office
MDPI AG
Grosspeteranlage 5
4052 Basel, Switzerland

This is a reprint of the Special Issue, published open access by the journal *Sensors* (ISSN 1424-8220), freely accessible at: www.mdpi.com/journal/sensors/special_issues/MobileRobots.

For citation purposes, cite each article independently as indicated on the article page online and using the guide below:

Lastname, A.A.; Lastname, B.B. Article Title. *Journal Name* **Year**, *Volume Number*, Page Range.

ISBN 978-3-7258-3388-7 (Hbk)
ISBN 978-3-7258-3387-0 (PDF)
https://doi.org/10.3390/books978-3-7258-3387-0

Contents

About the Editors

Stephen Monk

Dr. Stephen D. Monk is a Senior Lecturer at Lancaster University in the areas of Nuclear and Robotic Engineering, with over 60 papers published in these areas. His previous robotics projects involved the autonomous cutting of pipework using a hydraulically actuated dual-arm system, the automation of a UR3 robotic arm to scan for radioactivity over used personal protective equipment using a semi-autonomous control and vision system, the development of a small-bore soft robot to navigate and characterize pipework within nuclear decommissioning environments, and the development of an adaptable semi-autonomous underwater decommissioning sample retrieval robot in partnership with the National Maritime Research Institute (Japan). Similarly, his nuclear projects and areas of research have included a novel neutron spectrometer for high-altitude assay, a portable neutron/gamma camera for use at ports, tritium sensing, Magnox storage pond characterization, a novel viscometer for use in uncharacterized sludge, and a silicon carbide neutron sensor developed in partnership with Kyoto University.

David Cheneler

Dr. David Cheneler is a chartered mechanical engineer, a Fellow of the IoM3, and a Senior Lecturer in Engineering at Lancaster University. He is currently the Extreme Environments theme lead at the Lancaster Intelligent, Robotic, and Autonomous Systems Centre and a part of the Control and Robotics Research Group in the School of Engineering. His expertise lies in the development of sensing, automation, and decommissioning technologies. His current research interests include, but are not limited to, the following: robotics, biomedical technologies, localization and mapping strategy development, automated advanced manufacturing such as underwater laser cutting and welding, agricultural automation, and environmental and personal condition monitoring. He currently leads a team of one PDRA and twelve Ph.D. students, several of which are projects related to the nuclear industry. He has published over 73 publications, with 40 being first/last authors, as impacts from 39 projects having been awarded GBP 3.9M in funding (GBP 1M as PI).

Preface

Navigation is one of the main challenges in robotics, utilising technologies and strategies such as sensing, positioning, mapping, approaching, tracking, formation, control, communication, human interfaces, and learning. The aim of this Special Issue is to contribute to the state-of-the-art and current applications of robot navigation. Areas of interest within this compendium include obstacle avoidance, path planning, crossing-point estimation in human–robot navigation, robust tracking control, task planning utilising large language models, kinematic analysis for control logic development, deep reinforcement learning, neural network control, the use of LIDAR in 3D mapping, optical and acoustic sensor performance, sinkhole exploration, gaze point tracking, multi-robot task allocation and path planning, harmonic maps and adaptive artificial potential fields, multi-agent deep reinforcement learning, a navigation toolbox, a fractional-order controller, a new sonar–vision system for underwater vehicles, a strategy for controlling motions of a walking robot, multi-level information fusion for robust global orientation optimization, control of a double gyros unicycle robot, improving the tracking of trajectories, optimal control tuning using real-time data, locomotion inspired by Caenorhabditis elegans, a systematic review of 2D path planning, and an autonomous surface ship decision system.

Stephen Monk and David Cheneler
Guest Editors

sensors

MDPI

Review

Obstacle Avoidance and Path Planning Methods for Autonomous Navigation of Mobile Robot

Kornél Katona *,†, Husam A. Neamah † and Péter Korondi

Department of Electrical Engineering and Mechatronics, Faculty of Engineering, University of Debrecen, 4028 Debrecen, Hungary; husam@eng.unideb.hu (H.A.N.); korondi.peter@eng.unideb.hu (P.K.)

* Correspondence: katona.kornel@eng.unideb.hu

† These authors contributed equally to this work.

Abstract: Path planning creates the shortest path from the source to the destination based on sensory information obtained from the environment. Within path planning, obstacle avoidance is a crucial task in robotics, as the autonomous operation of robots needs to reach their destination without collisions. Obstacle avoidance algorithms play a key role in robotics and autonomous vehicles. These algorithms enable robots to navigate their environment efficiently, minimizing the risk of collisions and safely avoiding obstacles. This article provides an overview of key obstacle avoidance algorithms, including classic techniques such as the Bug algorithm and Dijkstra's algorithm, and newer developments like genetic algorithms and approaches based on neural networks. It analyzes in detail the advantages, limitations, and application areas of these algorithms and highlights current research directions in obstacle avoidance robotics. This article aims to provide comprehensive insight into the current state and prospects of obstacle avoidance algorithms in robotics applications. It also mentions the use of predictive methods and deep learning strategies.

Keywords: obstacle avoidance; global path planning; local path planning; autonomous vehicles; navigation algorithms

check for updates

Citation: Katona, K.; Neamah, H.A.; Korondi, P. Obstacle Avoidance and Path Planning Methods for Autonomous Navigation of Mobile Robot. *Sensors* **2024**, *24*, 3573. https://doi.org/10.3390/s24113573

Academic Editors: David Cheneler and Stephen Monk

Received: 5 May 2024

Revised: 25 May 2024

Accepted: 29 May 2024

Published: 1 June 2024

1. Introduction

Autonomous robots are machines or devices capable of operating independently and making decisions in their environment. These robots are equipped with sensors and embedded systems to gather information about their surroundings, such as mapping, navigation, and obstacle detection. Obstacle avoidance plays a crucial role in the operation of autonomous robots, enabling them to navigate their environment efficiently and safely. Obstacle avoidance algorithms assist robots in avoiding obstacles and minimizing collisions, allowing them to reach their destination safely and accomplish their tasks. Thus, obstacle avoidance is an indispensable element of effective and reliable operation for autonomous robots. This paper explores a literature review of alternative route planning and mobile robot navigation methods. The main algorithms it considers are discussed in the following sections. Global path planning involves navigating a robot based on preexisting environmental data, which is loaded into the robot's planning system to compute a trajectory from the starting point to the destination. This method generates a complete path before the robot begins its journey, essentially optimizing the route gradually [1]. Global path planning is consciously determining the best way to move a robot from a starting point to a destination. In global route planning, the robot has already been moved from the starting location to the destination, and the robot is then released into the specified environment [2]. In contrast, local path planning involves navigating a robot in dynamic or unknown environments where the algorithm adapts to real-time obstacles and changes. This method primarily focuses on real-time obstacle avoidance using sensor-based data for safe navigation [3]. The robot typically follows the shortest, straight-line path from the start to the destination until encountering an obstacle. Upon detection, it deviates from this path while updating essential details like the new distance to the target and the point of obstacle bypass [2]. Continuous knowledge of the target's position relative to the robot is critical for accurate navigation, as depicted in Figure 1.

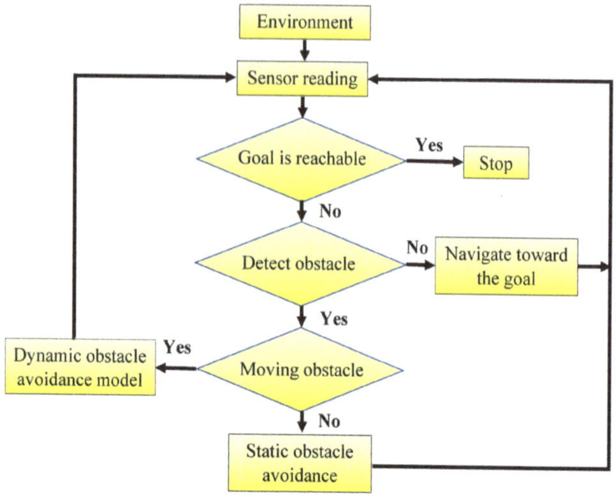

Figure 1. The obstacle avoidance procedure [4].

The diagram of the algorithms of each type included in this paper is shown in Figure 2. Another classification divides the methods into classical and heuristic algorithms (Figure 3). The classification according to classic and heuristic obstacle avoidance algorithms makes the selection and application of algorithms more transparent and manageable. Users can more easily identify which algorithm best meets the requirements of a given problem. Classical algorithms such as Dijkstra perform well for minor deterministic problems, while heuristic algorithms such as A* can be more efficient for larger and more complex issues. Moreover, the separation between global and local search algorithms is less clear. There are heuristic algorithms (such as A* or the DL-based algorithms) that have both versions of the search algorithm. This paper follows the latter classification.

One group of algorithms is called optimization methods. These mathematical procedures and algorithms aim to find the best possible solution to a given problem within the constraints available. An optimal solution is usually a combination of the values of one or more variables that maximizes or minimizes the value of the objective function while taking into account various constraints or conditions. For example, Particle Swarm Optimization, Cuckoo Search Algorithm, Artificial Bee Colony, Ant Colony Optimization, and Grey Wolf Optimization are based on such optimization techniques. These methods are called swarm (population)-based because they are inspired by animal behavior. Usually, some population of individuals (solutions) is used, and these individuals are iteratively developed and modified to find the best solution. They can effectively find optimal solutions to complex and diverse problems that traditional algorithms cannot manage with difficulty or at all.

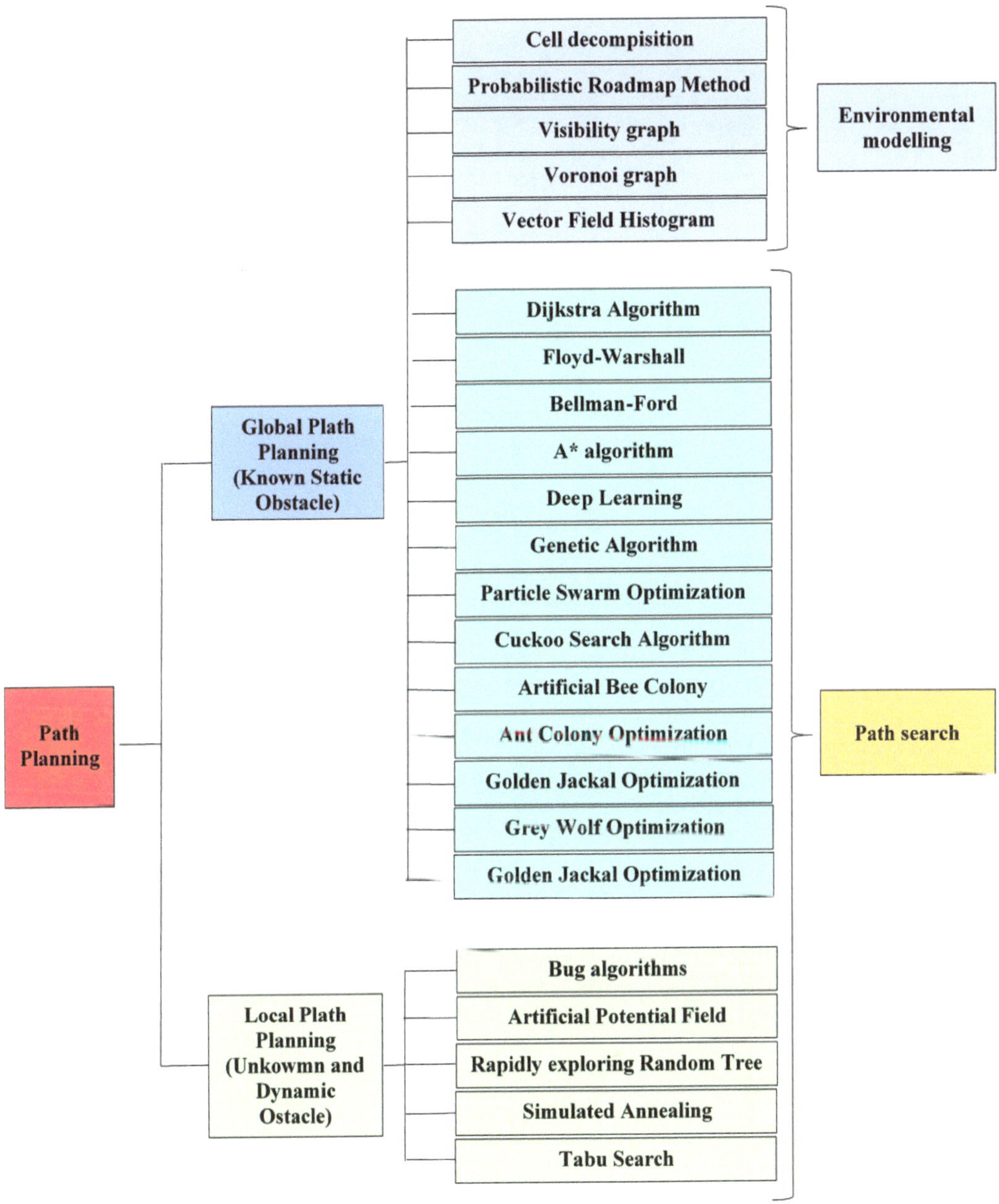

Figure 2. Diagram of the algorithms.

Section 2 discusses bright spaces and fundamental obstacle avoidance methods. Then, in the Section 3, classical avoidance algorithms such as Dijkstra, Floyd–Warshall (FW), Bellman-Ford (BF), Artificial Potential Field (APF), Bug Algorithms, Vector Field Histogram (VFH), Probabilistic Roadmap Method (PRM), Rapidly exploring Random Tree (RRT), Cell Decomposition (CD), and the Following Gap Method (FGM) are discussed. Heuristic algorithms are presented in Section 3. This includes the A* Algorithm, Fuzzy Logic (FL), Particle Swarm Optimization (PSO), Genetic Algorithm (GA), Cuckoo Search Algorithm

(CSA), Artificial Bee Colony (ABC), and Ant Colony Optimization (ACO). It also mentions using deep learning (DL) strategies and predictive methods. In this section, we discuss Artificial Neural Networks (ANNs), Model Predictive Control (MPC), and Deep Reinforcement Learning (DRL). Other algorithms are mentioned here, such as Dynamic Window Approach (DWA), Golden Jackal Optimization (GJO), or Grey Wolf Optimization (GWO). At the end of the chapter, a further so-called hybrid algorithm consists of two or more of the algorithms discussed earlier. Sliding Mode (SM) is presented in more detail among the hybrid methods. This article tries to collect and analyze most of the algorithms commonly used in practice. However, covering all currently existing methods in a single article is impossible, so this is not the aim here. This paper attempts to provide an overview of historically important and currently significant algorithms in practice as comprehensively as possible. Of course, it is impossible to discuss all possible methods (especially in the case of hybrid algorithms), but we tried to present the development directions of each algorithm. Such an extensive literature review cannot be found in other works. One of this article's most important values, after the description of the theoretical background, is the summary table of the individual algorithms, which provides a sufficient comparison based on the algorithms' main properties (e.g., convergence, calculation time).

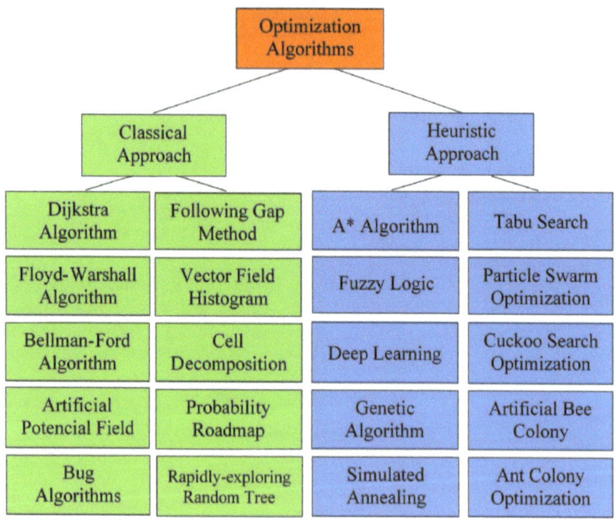

Figure 3. The classical/heuristic division of algorithms.

2. Classic Approaches

This section presents some classic approaches.

2.1. *Dijkstra Algorithm*

Dutch scientist Edsger Wybe Dijkstra introduced the Dijkstra Algorithm (DA) in 1956, which he published in 1959 [5]. The question of the shortest path between two nodes in a directed graph is solved by this method, which is one of the most commonly used techniques for mapping isolated workspace paths [6]. This method is a well-known strategy, but it is less effective when the origin and destination are farther apart. In this case, the algorithm calculates the shortest path for all nodes, even if the node is irrelevant for the optimal route. Consequently, most of the calculations may be redundant, resulting in a time-consuming process. Another factor that may contribute to the time-consuming process is the presence of long edges in the graph. In this case, the Dijkstra algorithm has to spend a considerable amount of time processing the edges [7].

To plan the shortest path in Dijkstra's algorithm, the starting position must be specified, and two heaps S and U must be introduced. The S heap records the vertices for which the shortest path is not found and the distance between the vertex and the starting point [8]. The flowchart of the Dijkstra algorithm is shown in Figure 4.

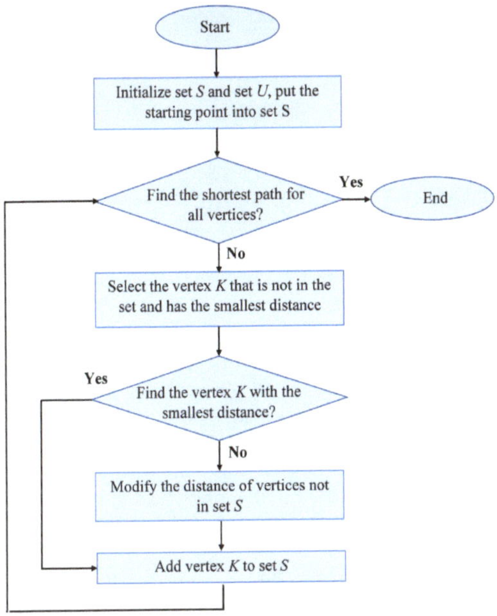

Figure 4. The flowchart of the Dijkstra.

The work in [9] used DA to define vehicle routes on toll roads. Path planning is in a localization-insecure environment based on the Dijkstra method in [10]. Dijkstra was used to determine the shortest distance between cities on the island of Java [11]. This method was later modified to handle the situation where most of the network parameters are unknown and expressed as neutrosophic values (can be true, false, and neutral simultaneously, depending on the point of view) [12]. A Dijkstra-based route planning strategy for autonomous vehicles is included in [13]. In [14], a Dijkstra algorithm is applied to unmanned aerial vehicles (UAVs). Ref. [15] presents the optimal route planning of an unmanned surface vehicle in a real-time maritime environment using the Dijkstra algorithm.

2.2. *Floyd-Warshall Algorithm (FW)*

The Floyd-Warshall algorithm can be considered dynamic programming, and it was published in 1962 by Robert Floyd. The algorithm efficiently and simultaneously finds the shortest paths between all pairs of vertices of a weighted and potentially directed graph [16,17].

The algorithm compares all possible paths for each line of all points on the graph. The graph's vertices should be numbered from 1 to n (n number of vertexes). Suppose there is also a shortest path function $f(i,j,k)$ which gives the shortest path from i to j using only the node from 1 to k as an intermediate point. The ultimate goal of using this function is to find the shortest path from each vertex i to vertex j using the intermediate node from 1 to $k+1$. This algorithm first computes the function $f(i,j,1)$ for each pair (i,j), then uses the results to compute $f(i,j,2)$ for each pair (i,j), and so on. This process continues until $k=n$, and the shortest path is found for all (i,j) pairs with the interpolation of vertices [18].

The algorithm consists of two parts: the construction of the path matrix and the state transition equation. The construction of the path matrix is based on the weight matrix of the graph to obtain the matrix D^n of the shortest path between each two points. The elements of row i and column j of the matrix D^n are the length of the shortest path from vertex i to vertex j. The state transition equation mathematically calculates the shortest distance between each point (1). The computation time is $O(n^3)$ [19].

$$d_{ij}^{k} = min\{d_{ij}^{k-1}, d_{ik}^{k-1} + d_{kj}^{k-1}\} \tag{1}$$

where the notation d_{ij}^{k} represents the shortest path from i to j that also passes through vertex k. For example, d_{ij}^{0} is the edge length between vertices i and j.

2.3. Bellman-Ford Algorithm (BF)

The Bellman-Ford algorithm is a classical method that computes the shortest paths in a weighted graph from a single source. This algorithm considers the negative-weighted edges of the graph, so it can handle graphs that contain negative-weighted cycles. These cycles generate several paths from the origin to the destination, where each cycle minimizes the shortest path length. The algorithm efficiently uses $O(n\dot{m})$ time for a graph with n vertices and m edges. The BF algorithm can handle edges with negative weights, unlike Dijkstra's algorithm, which only works with edges with positive weights. For this reason, the BF algorithm is mainly used for graphs with negative edge weights. Although its efficiency is lower than that of Dijkstra's algorithm, some problems would be impossible without negative weights. The BF algorithm is similar to Dijkstra's algorithm, but it approximates all edges instead of selecting vertices with minimum distance. This operation is performed $n-1$ times, where n is the number of vertices in the graph, and these iterations provide an exact prior in the graph [20–22].

2.4. Artificial Potential Field (APF)

The idea is that the mobile robot moves within a potential field where the robot and obstacles behave as positive charges while the target behaves as a negative charge. The mismatch between attractive and repulsive forces helps the robot to move in the environment. The attractive force attracts the robot to the target location, while the repulsive force keeps it away from each obstacle [23], as shown in Figure 5.

The final force acting on the robot is the vector sum of all repulsive and attractive forces. However, the distance determines the magnitude of the force, i.e., obstacles close to the robot will have a more significant effect. Similarly, if the robot is far from the target, its speed will be high and slow down as it approaches the target. As mentioned in the literature [24], the attractive force is the negative gradient of the attractive potential (2).

$$F_{attr} = -\nabla U_{attr} = -K_{attr}(d - d_{goal}) \tag{2}$$

where $d - d_{goal}$ is the Euclidean distance between the current position and the target, and K_{attr} is the scaling factor. The repulsive force can be calculated by adding the repulsive effect of the obstacles on the robot. This can be obtained by calculating the obstacles' distance and direction (angle) from the robot. An obstacle close to the robot has a high repulsive force. The formula described by [25] is (3)

$$U_{rep} = \sum_{i=1}^{n} U_{repi}(d) \tag{3}$$

U_{rep} negative gradient repulsive force. So (4),

$$F_{rep} = -U_{repi}(d) \tag{4}$$

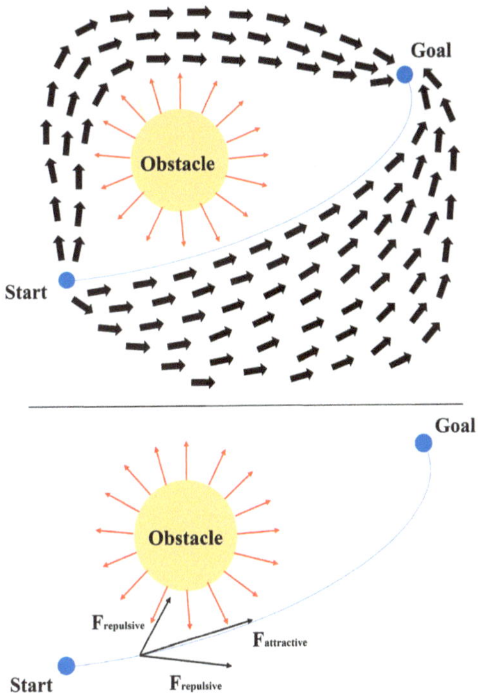

Figure 5. APF-based navigation for a mobile robot.

To avoid local minima, various methods have been devised. One such method is the left-turning potential field approach, which compels the robot to change direction when encountering a local minimum. Conversely, the virtual target point method involves strategically placing a virtual target point when the robot reaches a local minimum. During this process, the robot disregards the influence of both the target point and obstacles, enabling it to pivot and break free from the local minimum. Another critical issue with APF is its susceptibility to local minima, which can hinder the robot's progress. Symmetric and U-shaped obstacles exemplify these dead-end scenarios, leading to the robot becoming trapped. Figure 6 illustrates symmetric obstacles, where the forces exerted by the target and obstacles cancel each other out, resulting in a stalemate for the robot—a classic instance of local minima. To address this problem, significant attractor forces are temporarily applied at random locations to prevent the robot from being trapped in local minima. These measures aim to disrupt the equilibrium between attractive and repulsive forces, enabling the robot to navigate effectively [26]. In summary, while APF offers a direct path from source to destination, its susceptibility to local minima poses a significant challenge. Various strategies, such as the left-turning potential field and virtual target point methods, have been developed to mitigate this issue and ensure smoother navigation in complex environments [27].

The APF has been used in a dynamically changing, obstacle-filled environment between unmanned aerial vehicles (UAVs) [28]. Ref. [29] proposes an improved artificial potential field method for autonomous underwater vehicle (AUV) route planning. Based on an improved artificial potential field, ref. [30] introduces dynamic route planning for autonomous vehicles on icy and snowy roads. Ref. [31] discusses local path planning for multi-robot systems using an improved APF. Furthermore, ref. [32] outlines an active obstacle avoidance method for autonomous vehicles that is also based on an improved APF.

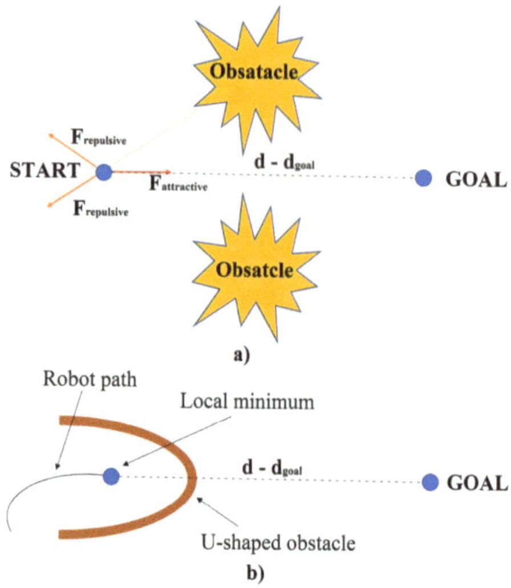

Figure 6. Dead-end scenario of the artificial potential field method: (**a**) symmetric obstacles and (**b**) U-shaped obstacle.

2.5. Bug Algorithms

Despite the presence of more efficient algorithms, Bug algorithms are still significant in robotics. These were the earliest navigation and obstacle avoidance algorithms that achieved relatively reliable results with speedy computation times. The algorithms are designed to work assuming that the robot is a single point in 2D space and that its movement is between each point. Bug algorithms are a popular type of robot navigation algorithms that provide a trajectory following an obstacle boundary in navigation scenarios with unknown obstacles, similar to the behavior of a bug [33]. The algorithm can be divided into three main variants based on their obstacle avoidance behavior, as discussed below [34,35]:

- The Bug-1 algorithm activates when the robot detects an obstacle. It starts circumnavigating the obstacle until it reaches the starting point from which it began while calculating the shortest distance from the destination to the departure point and creating a new path from the calculated departure point to the destination as it circumnavigates the obstacle. After completing full circles, it resumes circumnavigating the obstacles until it reaches the departure point, then proceeds on the newly generated path toward the destination.
- The bug-2 algorithm sets a direction from the starting position to the destination, and the robot follows it until it encounters an obstacle. Upon interruption, it follows the obstacle's edge and calculates a new direction from each new position until the new direction matches the original direction. Once reaching this position, the robot resumes following the previously generated path towards the destination.
- In contrast, the Dist-Bug algorithm relies on distances to targets and obstacles. When encountering an obstacle on the path, the robot begins following the obstacle's edge and calculates the distance between that point and the destination at each point. The point with the smallest distance to the target is called the distance point. Subsequently, the robot creates a new path along which it moves to the destination when it finds the distance point during its movement around the obstacle.

The three versions are shown in Figure 7.

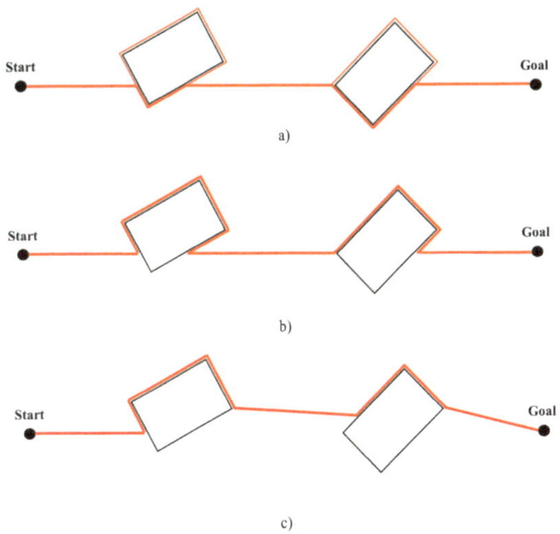

Figure 7. Obstacle avoidance with the Bug algorithms: (**a**) path of the Bug-1 algorithm, (**b**) the path of the Bug-1 algorithm, and (**c**) Dist-Bug algorithm path.

A maritime search route planning method for unmanned surface vehicles (USVs) based on the improved Bug algorithm presented here is also presented in [36].

2.6. Follow the Gap Method (FGM)

The FGM avoids obstacles by finding the gap between them. It calculates the gap angle. The minimum gap between obstacles is the threshold gap from which the robot can move. If the measured gap is larger than the threshold, the robot follows the calculated gap angle. Obstacle avoidance using the FGM is achieved in three main steps [37].

The algorithm uses sensory information to identify gaps with the largest angle and works in three steps, as follows [38]:

- The initial step involves computing the arrays of gaps. During this phase, the algorithm utilizes the current sensory data, such as information from the LIDAR sensor, to produce a gap array. This array provides details regarding the sizes of the available gaps surrounding the robot in angular form. The FGM algorithm identifies the largest gap by the conclusion of this stage.
- The FGM calculates the angle to the gap's center point using specific geometric relations.
- In the third stage, this method calculates the final heading angle, ϕ_{final}, using (5)

$$\phi_{final} = \frac{\frac{\alpha}{d_{min}}\phi_{gap-c} + \phi_{goal}}{\frac{\alpha}{d_{min}} + 1} \tag{5}$$

The weighted function described in Equation (5) comprises the angle to the center point of the widest gap ϕ_{gap-c}, the angle to the goal point ϕ_{goal}, the distance to the nearest obstacle d_{min}, and a safety parameter denoted as α. Higher values of the alpha parameter prompt the robot to maintain a safe distance from obstacles and align with the center of the safe gap. Conversely, lower values of alpha lead the robot to prioritize the goal point, potentially approaching obstacles too closely in certain scenarios.

The representation of the gaps accessible to the robot, the angle towards the midpoint of the widest gap, the angle towards the destination, and the final heading angle determined by FGM are depicted in a robot-obstacle configuration, as illustrated in Figure 8.

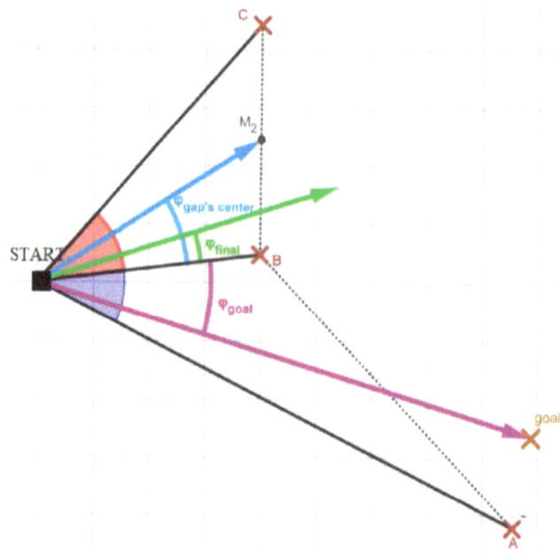

Figure 8. Robot-obstacle configuration, obstacles (A, B, and C), the midpoint of the widest angular gap (M_2), goal point (X), angle to the goal point (ϕ_{goal}), final heading angle (ϕ_{final}), and angle to the largest gap's center point ($\phi_{gap'scenter}$). [38].

So, in this procedure, the robot selects the largest gap around it and moves towards the target, taking into account the largest gap and the minimum distance from the obstacle. One of the drawbacks of this method is the lengthening of the path, which can sometimes be unnecessary. Another challenge is the subtle differences in gap sizes. This can sometimes result in the robot changing the number of selected gaps instantly, which can lead to zigzag paths [39].

In [40], the collision avoidance task is accomplished with the Follow the Gap Vector Method. A central part of the approach proposed in [2] is to identify gaps in the environment by analyzing sensor data.

2.7. Vector Field Histogram (VFH)

The algorithm initiates by generating a 2D histogram around the robot to depict obstacles. Subsequently, the 2D histogram undergoes updates with new sensor detections. It converts this 2D histogram into a 1D histogram and further into a polar histogram. Finally, the algorithm identifies the most suitable sector characterized by low polar obstacle density and computes the steering angle and velocity towards this direction. Figure 9 is from the work of [41] which illustrates the 2D histogram grid. The conversion from 2D to 1D histogram is shown in Figure 10a, and Figure 10b is a representation of the 1D polar histogram with obstacle density for a situation where the robot has three obstacles, A, B and C in its close vicinity.

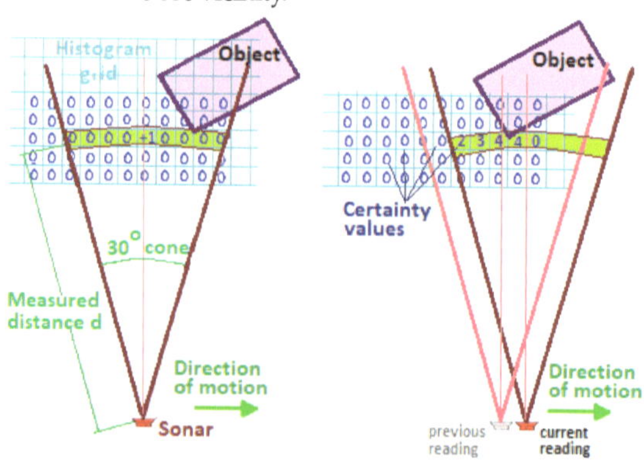

Figure 9. Structure of the 2D histogram grid map [42].

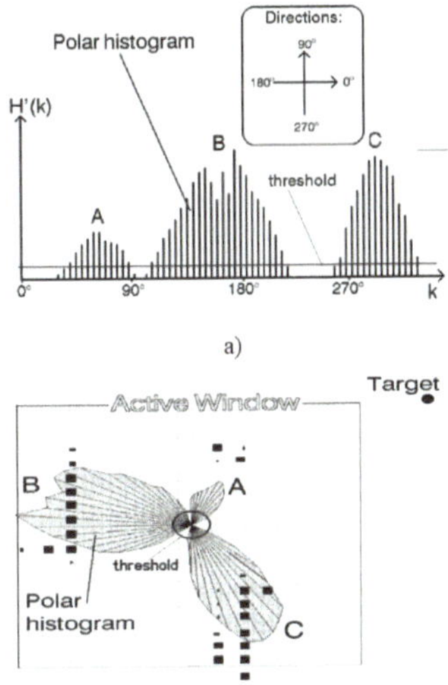

Figure 10. Representation of (**a**) the 1D histogram (**b**) the polar histogram with obstacle density for a situation where the robot has three obstacles, A, B and C in its close vicinity [42].

The first step is to sort the costs of the traversable area and then calculate the cost using the cost function based on the indicated direction of the polar histogram. The designated directions are selected from the traversable areas taking into account the robot's kinematic and dynamic characteristics. Inaccessible sectors, as determined by the robot's capabilities, are classified as impassable areas. Areas above the threshold are labeled as impassable, whereas those below the threshold are considered passable. To continue, the histogram generated in the previous step must be converted into a binary format by choosing the appropriate threshold based on the current situation. The commonly used cost function is shown as follows (6) [43]:

$$f(v) = c_1 \cdot \Delta(v, d_g) + c_2 \cdot \Delta\left(v, \frac{\Theta}{\alpha}\right) + c_3 + \cdot\Delta(v, d_{g-1}) \tag{6}$$

The candidate direction fv represents the cost value $f(v)$. c_1, c_2, and c_3 are three parameters to be determined according to the actual situation. The d_g is the target direction, d_{g-1} is the previous direction, and the orientation of the robot is $\frac{\Theta}{\alpha}$. The absolute difference between v and d_g is denoted by $\Delta(v, d_t)$. The difference between the marked direction and the orientation of the robot is denoted by $\Delta\left(v, \frac{Theta}{alpha}\right)$. The difference between v and d_{g-1} is denoted by $\Delta(v, d_{g-1})$.

To determine the robot's desired control command, this algorithm employs a two-stage data reduction process. Although this ensures accurate computation of the robot's path to the target, it necessitates additional resources, such as memory and processing power [44].

Initial tests have shown that the mobile robot can use the VHF to traverse very crowded obstacle courses at high average speeds, and can pass through narrow openings (e.g., doorways) and move through narrow corridors without oscillating [41]. In [45], an improved 3D-VFH algorithm is proposed for autonomous flight and local obstacle avoidance of multirotor UAVs in confined environments.

2.8. Cell Decomposition (CD)

The cell-by-cell technique divides the area into non-overlapping grids, called cells, and uses grids that can be connected from the initial cells to the target to move from one cell to another. This method is classified as exact, approximate, and probabilistic CD depending on the assignment of boundaries between cells. For exact CD, the resolution is lossless and the shape and size of the cells are not fixed and each element is assigned a number. In contrast, for approximate CD, the decomposition result approximates the actual map and the grid has a fixed shape and size. And the probabilistic CD is like the approximate CD, except for the cell boundaries, which do not represent a physical meaning [46]. Figure 11 shows that the CD systems can be divided into three classes.

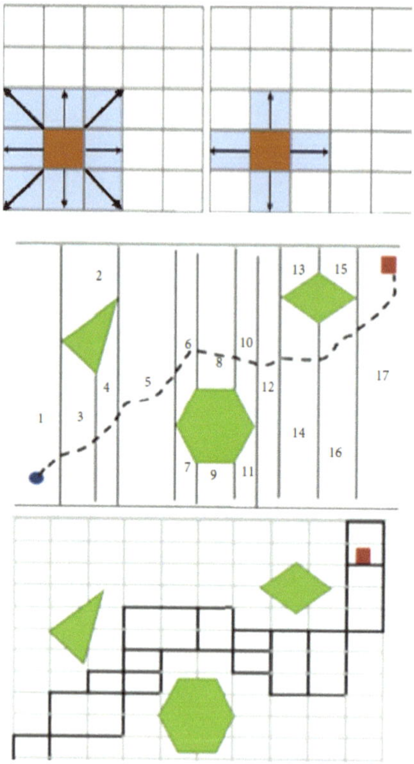

Figure 11. (**Top**): approximate CD; (**middle**): exact CD, and (**bottom**): probability CD [47].

2.9. Probabilistic Roadmap Method (PRM)

Classical methods face several drawbacks, such as high time requirements at large scales and getting bogged down in local minima, which makes them ineffective in practical scenarios. To address these limitations and increase efficiency, probabilistic algorithms have been proposed. These algorithms aim at providing practical paths for robots through static workspaces [48].

One of the most important examples is the Probabilistic Roadmap Method (PRM) [49]. It uses lines to delimit the connectivity of the robot's free areas. This includes the visibility graph and the Voronoi graph [47]. Figure 12 illustrates these two graphs.

In the visibility graph, the obstacles are represented as polygons [50], and the vertical nodes of the polygonal obstacles are connected in such a way that the path length is minimized while the lines remain close to the obstacles. In contrast, the Voronoi graph uses the two closest points of the edges of the obstacles for planning and divides the domain into subdomains. In the latter case, the robot moves farther away from the obstacles, which increases safety but results in longer paths compared with the visibility graph [51].

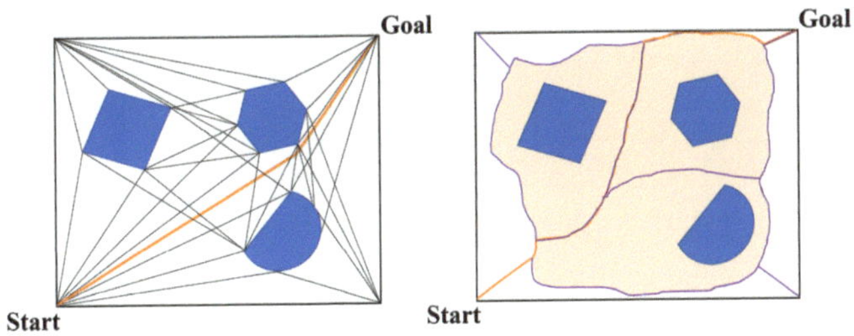

Figure 12. Visibility graph (**left**); Voronoi graph (**right**). The visibility graph is constructed based on the visibility between points, while the Voronoi graph is constructed based on the geometric relationships between areas.

To link the initial state with the goal region, PRMs explore this roadmap graph and pinpoint a sequence of states and local connections that the robot can traverse. While these algorithms can theoretically create arbitrarily accurate representations as the number of samples approaches infinity, in practice, only a handful of critical states are needed to define solution trajectories. These critical states often have significant structure, such as entries to narrow passages, but they can only be identified through exhaustive sampling [52].

These algorithms offer highly accurate representations with a theoretically infinite number of samples. In practice, however, this is only necessary in a few cases. For example, entering a bottleneck. The Voronoi graph continues to play a crucial role in the further development of different algorithms for different purposes [53]. Notably, ref. [54] presents a useful visibility Voronoi graph search algorithm for generating routes for unmanned surface vehicles. In addition, ref. [55] uses the Voronoi graph to partition agricultural areas into multiple fields, making it easier for multiple robots to perform agricultural tasks.

2.10. *Rapidly Exploring Random Tree (RRT)*

The Rapidly exploring Random Tree (RRT) method facilitates swift exploration of the configuration space [56]. Initially proposed by LaValle [57], the RRT algorithm generates a graph, termed a "tree", where nodes signify potential reachable states and edges denote transitions between states. The RRT's root denotes the initial state, with all other states reachable along the path from the root to the corresponding node. Leveraging a sampling approach, this algorithm operates effectively in complex environments, evading local minima [58]. It has proven effective in tackling nonholonomic and kinodynamic motion planning challenges. In robotics, algorithms employed to generate RRTs are versatile, allowing trajectories to incorporate turns at any angle, albeit subject to kinematic and dynamic constraints [56].

When sampling, it allows all nodes in the robot configuration space to be reached with equal probability. Based on the constraints of the algorithm, it selects a node in the random tree. On impact, it resamples and discards the previous node. If no collision occurs, the selected node is added to the random tree. If a node on the route is redundant, it is deleted; otherwise, it remains as a node in the random tree [59].

The flow chart of an RRT is shown in Figure 13:

This method does not require the modeling of space and can be used in large-scale environments. It also takes into account the objective constraints of unmanned vehicles, making it suitable for handling route planning problems in dynamic and multiobstacle environments. However, the route is randomly generated, leading to distortion. Second, the random tree has no orientation during the search process, resulting in slow convergence speed and low search efficiency [60]. Several improvements have been made to address the limitations of the algorithm. Among others, the RRT-Connect algorithm, the asymptotically optimal Rapidly Exploring Random Tree (RRT), the asymptotically optimal bidirectional Rapidly Exploring Random Tree (B-RRT), and the intelligent bidirectional RRT (IB-RRT) were born out of this necessity. Ref. [60]. The RRT* algorithm can construct an RRT whose branches converge asymptotically to the optimal solution given a given cost function. It solves feasibility problems efficiently and qualitatively concerning the cost function [56].

RRT has been used to plan the routes of ships [58], industrial robots [59] and micro aerial vehicles (MAVs) [61], among others.

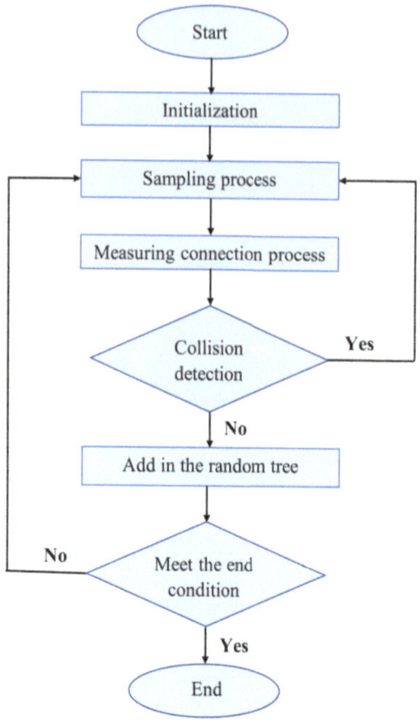

Figure 13. The RRT process.

3. Heuristic Approach

A heuristic approach is used to solve problems faster [62]. The method has proven its effectiveness and is widely used in autonomous navigation [5].

3.1. A* Algorithm

The A* algorithm is a graph search algorithm similar to Dijkstra's algorithm, developed by Hart (1968) [63] to speed up the search process of Dijkstra's algorithm. To do this, they introduced a heuristic cost function, which is the distance between the current point and the target point. Like the Dijkstra algorithm, the A* algorithm needs an environment model, e.g., a grid map. In the A* algorithm, the search area is usually divided into small squares, where each square represents a node. The algorithm can solve various routing problems with superior performance and accuracy compared with Dijkstra's algorithm. Algorithm A* solves problems by finding the path with the lowest cost (e.g., the shortest time) among all possible paths to the solution. Of these paths, it first considers those that appear to lead the fastest to the solution. The A* algorithm uses an evaluation function (7):

$$f(n) = g(n) + h(n) \tag{7}$$

The function $f(n)$ represents the cumulative cost from the starting point to the current point, extending to the target point. Meanwhile, $g(n)$ denotes the shortest cost from the initial point to the current position n, and $h(n)$ predicts the optimal path cost from the current point n to the destination, often calculated as the Manhattan distance [64]. Initially applied in port areas, Casalino used the A* algorithm [65] for local pathfinding. Guan proposed an improved version of the A* algorithm [66], which helps Unmanned Surface

Vessels (USVs) avoid static obstacles at sea and reach their destination smoothly while avoiding local minima. In addition, a collision-free trajectory planning method for space robots based on the A* algorithm has been developed in [67]. The geometric A* presented in [68] is designed for route planning of automated guided vehicles (AGVs) operating in port environments.

Despite its advantages, traditional A* does not always provide an optimal solution, as it does not take into account all feasible routes. In each iteration, A* evaluates the nodes based on their f values, which is a computationally expensive process, especially in large map search areas. Consequently, this approach can significantly slow down the speed of route planning.

3.2. Fuzzy Logic (FL)

Fuzzy logic (FL) is a technique for persuading the human intellect. FL is a uniform approximate (linguistic) method for inferring uncertain facts using uncertain rules [69]. In 1965, Lotfi A. Zadeh was the first to introduce the idea of an FL system [70]. The fuzzy sets he created are an extension of the traditional notion of a set, going beyond the Aristotelian (true–not–true; yes–no) division. The fuzzy set A is defined as follows [70]:

$$A = \{x, \mu_A(x) \mid x \in X, \mu_A(x) : X \to [0,1]\}$$

where X is the so-called reference surface, and $mu_A(x)$ is the so-called membership function, which takes values in the complete closed interval between 0 and 1. In the special case where $mu_A(x)$ takes only values 0 and 1, A reduces to a classical set. The three basic operations on fuzzy sets (intersection, union, complement) are defined as extensions of the corresponding operations on classical sets. The standard properties of sets (De Morgan, absorption, associativity, distributivity, idempotence) hold here as well. Fuzzy inference (or fuzzy reasoning) is an extension of classical inference [69]. However, Zadeh's vision was later expanded in several areas. The FL serves as a formal blueprint for representing and implementing the heuristic intelligence and observation-based methods of experts [71,72].

Figure 14 is an example of the primary FL driver used in [73]. The general architecture of a fuzzy logic controller consists of four units: IF–THEN rules, whose associated linguistic variable values can be not only true or false but can vary between the two; a fuzzy inference mechanism, which is a process to identify the output values associated with the input variables based on the fuzzy rules; an input fuzzification unit; and an output defuzzification unit. Hex Moor [74] was the first to apply the FL concept to robot path planning and obstacle avoidance. Since then, for example, the FL route planning approach has been applied in unknown environments [73]. Mobile robot routing algorithm based on FL and neural networks designed [75]. Chelsea and Kelly demonstrate FL controller for UAVs in a two-dimensional environment [6]. Then, 3D space navigation was demonstrated using FL for aerial [76] and underwater [77] robots and a Mamdani-type FL-based controller for a nonholonomic wheeled mobile robot that tracks moving obstacles [78].

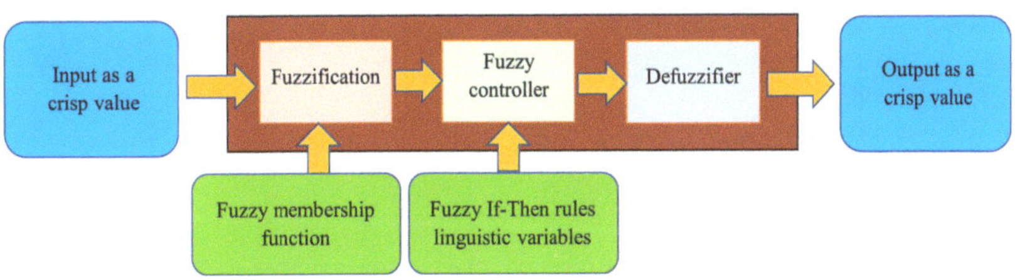

Figure 14. Basic FL controller consisting of an IF–THEN rule, an inference mechanism, an input fuzzification unit, and an output defuzzification unit.

3.3. Genetic Algorithm (GA)

A genetic algorithm is an optimization technique referring to genetics and natural selection, first introduced by Bremermann in 1958 [79]. It is based on Darwinian evolutionary theory and mimics the concept of survival of individuals best adapted to their environment. The most viable members of the population survive, while the weakest die off. The surviving members, depending on their fitness, allow the genes to be passed on to the next generation through cross-breeding, mutation, and selection. In this way, the individual fitness of the population continuously approaches the optimum. This random structure information was used to create a search algorithm that provided solutions to the problem of finding feasible pathways [79].

GAs stand for a sequence of algorithms. They randomly initialize populations with a character string and an objective function. Then, based on Darwinian evolutionary theory, they generate a new population using the three genetic operators (mutation, crossover, and selection). The new populations are created until the stopping conditions are met [50]. Such stopping conditions are a time limit, the required fitness value, and the maximum number of generations. During mutation, elements of an arbitrary string mutate with a given mutation probability. In a crossover, the elements of two strings are crossed according to a certain rule, thus creating two new strings. In selection, two strings selected by probability based on their objective function are compared based on their fitness, and the higher ranked higher-ranked one is selected to create the new population. The GA process is illustrated in Figure 15. The initial input comes from the population variables. This is followed by the encoding and decoding of chromosomes, the initialization of the population, and, the evaluation of the fitness values of the individuals within it. If the conditions are met, the optimal solution is obtained directly. Otherwise, the algorithm iterates, evolves, and selects new individuals from the population, whose fitness is re-evaluated until the condition is met. After that, the process stops.

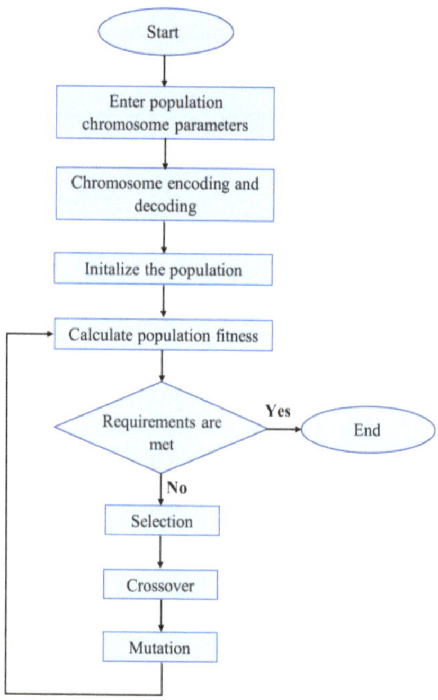

Figure 15. Process of GA [80].

GAs are used in many areas for mobile robot path planning problems, for example, for humanoid robot navigation [81], for the underwater robot navigation challenge in 3D route planning [82], and for aerial robots [83,84], as well as genetic-algorithm-based trajectory optimization for digital twin robots [85]. Work using improved genetic algorithms can be found in [86,87].

3.4. Simulated Annealing (SA) and Tabu Search (TS)

Simulated annealing and the Tabu search are approximate (heuristic) algorithms and therefore do not guarantee the optimal solution. They do not know when the optimal solution has been reached. Therefore, they need to be told when to stop. Easily designed to implement any combinatorial optimization problem, under some conditions, they converge asymptotically to the optimal solution. The same can be said for GAs [88].

3.4.1. Simulated Annealing (SA)

SA is an iterative search method based on the analogy of annealing metals. Annealing is a process in which a low-energy state of the metal is created by melting the metal and then slowly cooling it. Temperature is the control variable in the annealing process and determines how random the energy state is [88]. Consider an energy diagram with two potential barriers. A ball is randomly placed on the potential curve and can only move down the curve. The ball then has an equal chance of going to A than to B (Figure16).

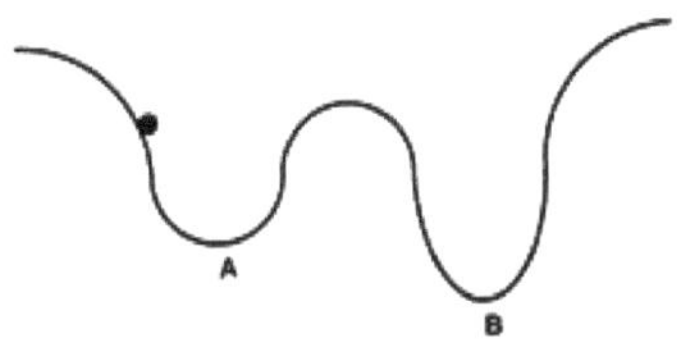

Figure 16. The potential barriers (A, B).

Upward movements can be accepted at times with a probability controlled by the parameter temperature (T). For example, if you want the ball to move from pit A to pit B with a higher probability, you have to increase its temperature. The probability of accepting upward movement decreases as T decreases. At high temperatures, the search becomes almost random, while at low temperatures it becomes almost greedy. At zero temperature, the search becomes completely greedy, i.e., it accepts only downward movements. The algorithm is based on the Metropolis procedure, which simulates the heat treatment process at a given temperature T [89]. At the beginning of the procedure, the current temperature and solution are given, as well as the time for which the heat treatment at the given temperature should be maintained. The SA algorithm should start from a high temperature. However, if the initial temperature is too high, it will only result in a loss of time. The initial temperature should be such that virtually any proposed movement is acceptable, whether upward or downward. Thereafter, the temperature will gradually decrease. The annealing time increases as the temperature decreases. The annealing process stops when the time exceeds the permissible time [90].

The main part of the algorithm consists of two circles. In the inner circle, a possible move is generated and the acceptance of the move is decided by an acceptance function. The acceptance function assigns a P_{accept} probability based on the current temperature and the cost change ΔC (8). At high temperatures, most uphill movements are likely to be accepted by the algorithm, regardless of the increase in costs. However, as temperatures fall, only downward movements are accepted. If the step is accepted, it is applied to the current path to generate the next state. The outer loop checks if the stopping condition is satisfied. Each time the inner loop completes, the temperature is updated using a function, and the stopping condition is checked again. This continues until the stop condition is met [90].

$$P_{accept} = \begin{cases} e^{-\frac{\Delta C}{T}} & \text{if} \quad \Delta C \geq 0 \\ 1 & \text{if} \quad \Delta C < 0 \end{cases} \tag{8}$$

3.4.2. Tabu Search (TS)

TS is a combinatorial optimization technique that optimizes an initial given permutation or converts it to the closest possible optimal solution, by alternating successive steps.

Using this method, it is possible to reduce the cost of a path by a series of edge swaps in a randomly generated round trip. The process continues until the path with the minimum cost is found. The selection of the best step to improve or not improve the current solution is based on the fact that good steps are more likely to reach the optimal or close to the optimal solution. The set of acceptable solutions in a given iteration forms a candidate list. The Tabu search selects the best solution from this candidate list, whose size reflects the trade-off between quality and performance. To Tabu the relocation attributes, a Tabu constraint is introduced to prevent the reversal of moves. This constraint is enforced by a Tabu list that stores the relocation attributes. The aspiration-level component allows the Tabu state to be temporarily overridden if the reversal results in a better solution than the best one achieved so far [88].

In [91], the design of minimal-cost delivery routes for goods-carrying mobile robots is developed using hybrid simulated annealing/Tabu search and approximation methods based on Tabu search algorithms, which start and end from a central warehouse while the robots serve customers. Each customer is supplied exactly once per vehicle path.

3.5. Particle Swarm Optimization (PSO)

Particle Swarm Optimization (PSO) is a nature-inspired approach that mimics the collective behavior of bird flocks, fish schools, or animal herds as they seek food, adapt to their surroundings, and interact with predators [92]. PSO draws inspiration from the foraging strategy observed in bird flocks, where individuals move towards the most favorable food sources guided by their knowledge, collective wisdom, and momentum. This behavior is emulated by the PSO algorithm through the representation of each potential solution as a particle, with personal and global best positions and inertia. Each particle maintains specific attributes such as position, velocity, and objective, striving to converge toward the global optimum over multiple iterations. The PSO process begins with the initialization of a randomly generated particle swarm, with each particle assigned a unique velocity to navigate the search space. Notably, unlike genetic algorithms, PSO assigns random weights to all potential solutions, enabling particles to explore the solution space dynamically. The algorithm's functioning revolves around the interplay between particle positions and velocities, with each particle's position updated based on its velocity conditions. Refer to Figure 17 for an illustration of this process [93,94].

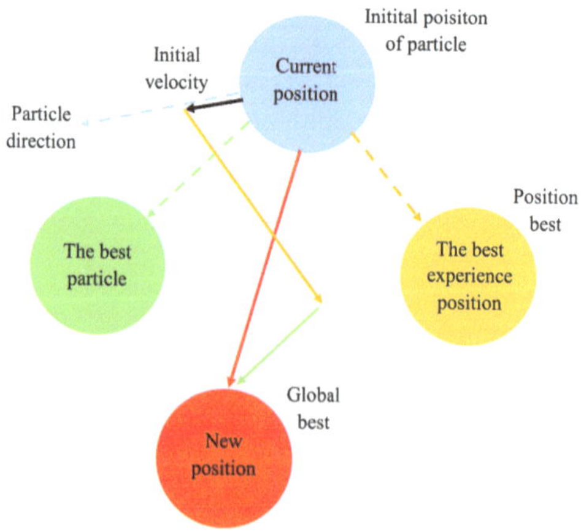

Figure 17. The basic concept of PSO.

Suppose that the search space is D-dimensional, and the ith particle of the population can be represented by a D-dimensional vector $(x_i^1, x_i^2, \dots, xi^D)^T$. The velocity of this particle can be represented by another D-dimensional vector $(V_i^1, V_i^2, \dots, V_i^D)^T$. The previously best-visited position of the ith particle is denoted by P_i, and the best particle in the swarm is denoted by P_g. The update of the particle's position is accomplished by the following two equations: Equation (9) calculates a new velocity for each particle based on its previous velocity, and (10) updates each particle's position in the search space [92,95].

$$V_{id}^{k+1} = wV_{id}^{k} + c_1 r_1 p_{id}^{k} t - x_{id} t + c_2 r_2 p_{g}^{k} t - x_{id}^{k} t$$
$$V(t+1) = wV(t) + [c_1 r_1 (P_{best} - x(t))] + [c_2 r_2 (G_{best} - x(t))] \quad (9)$$

$$x_{id}^{k+1} t + 1 = x_{id}^{k} t + v_{id}^{k+1} + 1$$
$$x(t+1) = x(t) + v(t+1) \quad (10)$$

where k is the iteration number, $d = 1, 2, 3, \ldots, D$; $i = 1, 2, 3, \ldots, N$; and N is the swarm size. w is inertia weight, which controls the momentum of the particle by weighing the contribution of the previous velocity. c_1 and c_2 are positive constants, called acceleration coefficients. Alternatively, c_1 is also called the cognitive (local or personal) weight, and c_2 is the social (or global) weight. r_1 and r_2 are random values ranging from $[0, 1]$. $V(t)$ is the velocity associated with the particle at time t, and $X(t)$ is the position of the particle at time t.

The PSO process, depicted in Figure 18, is characterized by rapid convergence but shows slower responses during particle search within a region. This limitation, due to its fixed convergence rate, can lead to localization issues [96].

PSO is widely applied in mobile robot path planning across various types, including humanoid [97], industrial, [98], wheeled [99], aerial [100], and underwater robots [101], particularly in complex three-dimensional environments.

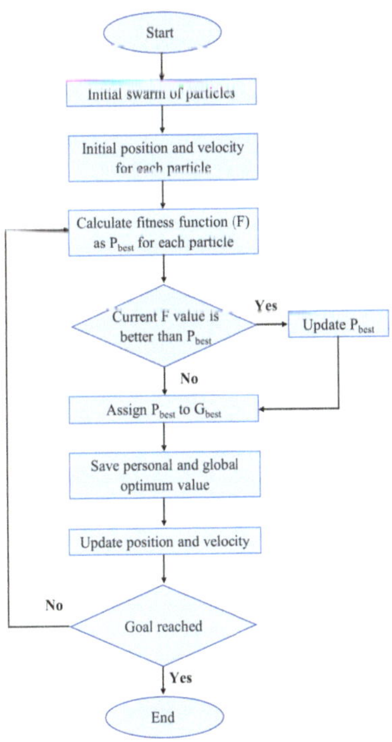

Figure 18. The PSO process.

3.6. Cuckoo Search Algorithm (CSA)

The concept of a cuckoo search is inspired by the behavior of cuckoo birds, which lay their eggs in the nests of other host birds (of different species). The cuckoo bird attempts to deposit its eggs in the nest of a host bird by removing one of the host's eggs and replacing it with one of its own, which closely resembles the host bird's eggs. Afterward, the cuckoo bird swiftly departs. The primary goal of this behavior is to safeguard its eggs from predators, as well as to ensure that its offspring have access to food and protection in the host nest. However, there is a risk that the host bird may detect the cuckoo egg and either remove it from the nest or abandon the nest to construct a new one. Consequently, the cuckoo continuously evolves its egg appearance to mimic that of the host bird's eggs, reducing the likelihood of detection. Importantly, the host bird also learns to detect foreign eggs over time, perpetuating the cycle of egg-laying and detection. Once the cuckoo successfully places its egg in the host nest, a new phase ensues. Cuckoo chicks hatch earlier than the host bird's offspring and may attempt to eject the host eggs or chicks from the nest. Additionally, cuckoo chicks compel the host mother bird to provide them with more food, potentially depriving the host chicks of sustenance altogether [102].

The interaction between the cuckoo and the host bird results in a direct conflict, as the host bird has a probability, denoted as P and ranging from 0 to 1, of detecting the cuckoo's egg. If a host bird detects cuckoo eggs in its nest, it may either discard the egg or desert the nest altogether. These fundamental occurrences form the basis of the cuckoo search algorithm. The primary features of the CSA are outlined in [103]:

In the cuckoo search algorithm, a single egg is deposited by a cuckoo in a nest selected at random, symbolizing a potential solution to an optimization problem. The nest containing the most promising eggs—representing the optimal solutions—is carried forward to subsequent iterations. The total number of available nests remains constant, and each egg laid by a cuckoo is subject to a probability (P_a) within the interval $[0, 1]$ of being detected and consequently abandoned. Consequently, during each iteration (t), a proportion (P_a) of the entire population undergoes alteration.

The efficiency of the cuckoo search algorithm is enhanced through the utilization of Levy flight instead of random walk. Numerous animals and insects exhibit the characteristic Levy flight behavior. Levy flight entails a random walk with step lengths determined by a heavy-tailed probability distribution, as depicted in (11) [104]. Levy flight outperforms random walk in this regard. Hence, we opted for the cuckoo search algorithm in this research due to its ability to achieve faster convergence rates.

$$X_i(t+1) = X_i(t) + \alpha \oplus L(\lambda) \tag{11}$$

$$\alpha = \alpha_0 \otimes (x_i(t) - x_{best}) \tag{12}$$

where $X_i(t+1)$ represents the new solution, t indicates the current generation (iteration) of the solution, α is the step-wise parameter that controls the moving step size of the cuckoo, $\oplus$ is entry-wise multiplication, and α_0 denotes the step size factor, which is usually set to 0.01. and $L(\lambda)$ is Levy exponent, which stands for a random search path, which can be expressed as

$$L(\lambda) = \frac{\varphi \times m}{|n|^{\frac{1}{\beta}}} \tag{13}$$

where m and n are two random numbers subjected to the normal distribution, β is set to 1.5. φ is defined as:

$$\varphi = \left(\frac{\Gamma(1+\beta) \times sin(\pi\frac{\beta}{2})}{\Gamma\frac{1+\beta}{2} \times \beta \times 2^{\frac{\beta-1}{2}}} \right)^{\frac{1}{\beta}} \tag{14}$$

Algorithms with high computational complexity typically demand significant resources, which may not always be feasible. The cuckoo search algorithm (CSA), however, requires only a few initial parameters, enabling efficient resolution of multimodal problems.

The CSA, depicted in Figure 19, involves three key operations: (i) Levy flight for generating new solutions, (ii) replacement of nests with superior solutions based on fitness evaluations, and (iii) greedy selection to maintain the best solutions until the goal is achieved.

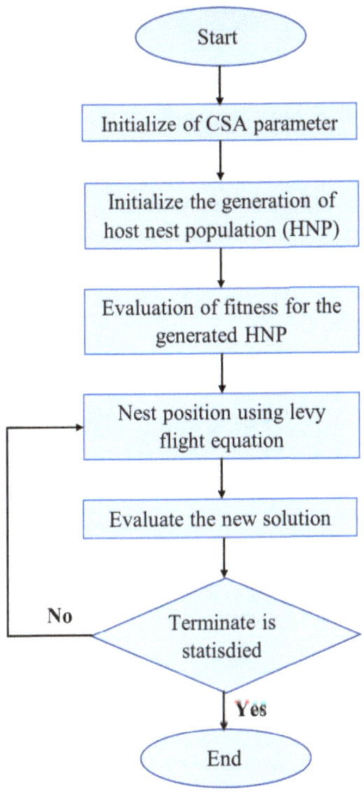

Figure 19. The CSA process.

CSA has been effectively hybridized with an adaptive neuro-fuzzy inference system for enhancing the navigation of multiple mobile robots in unknown environments [105] and applied in vehicle track design [106] and scheduling [107]. Additionally, it has been used in a novel artificial neural network approach to predict ground vibrations from mine blasting [108].

3.7. *Artificial Bee Colony (ABC)*

Karaboga developed the Artificial Bee Colony (ABC) technique, a swarm-based algorithm inspired by the foraging behaviors of bees [109]. The three rules of the ABC model are as follows: (a) Forager bees: Forager bees are sent to the food sites (the nearest colony) and inspect the quality of the food. (b) Inactive forager bees: Based on information from active forager bees, inactive bees inspect the food sources detected and assess/assess them. (c) Food sources: Forager bees that find rich food sources distribute them, while forager bees with few food sources give them up, creating a problematic situation.

The population is initialized from the set of employed and onlooker bees. Each worker is sent to the food source (x_i^j) that the bee is responsible for, and according to (15), the fitness value of each source of food is determined [110].

$$fit_i = \begin{cases} \frac{1}{1+f_i} & \text{if} \quad f_i \geq 0 \\ 1+|f_i| & \text{if} \quad f_i < 0 \end{cases} \tag{15}$$

where the objective function f_i shows the fitness value of source x_i. In addition, the failure counter, which is a limit value for each food source, is defined and initialized to zero.

Then, using (16), they try to find a better food source ($v_{i,j}$).

$$v_i^j = x_i^j + \alpha_i^j (x_i^j - x_k^j) \tag{16}$$

where $j = 1, 2, \ldots, D$. The problem dimension is defined by D. $k = 1, 2, \ldots, N$. N represents the total number of employed or onlooker bees. The value of k is not equal to i, and α_i^j is a random number generated from a uniform distribution in $[-1, 1]$.

Should the fitness value of the new position surpass that of the current one, the bee retains the newly identified food source location and disregards the previous source. The worker bee then communicates the fitness value of this new food source to the onlooker bee. The onlooker bee evaluates each food source based on the probability P_i^j and selects the optimal food source x_i. The probability evaluation of the food source is determined using Equation (17) [110].

$$P_i^j = \frac{fit_i}{\sum_{j=1}^{N} fit_j} \tag{17}$$

where fit_i is the fitness value of the solution i. This is clear from the ABC algorithm's general flow chart (depicted in Figure 20) [111].

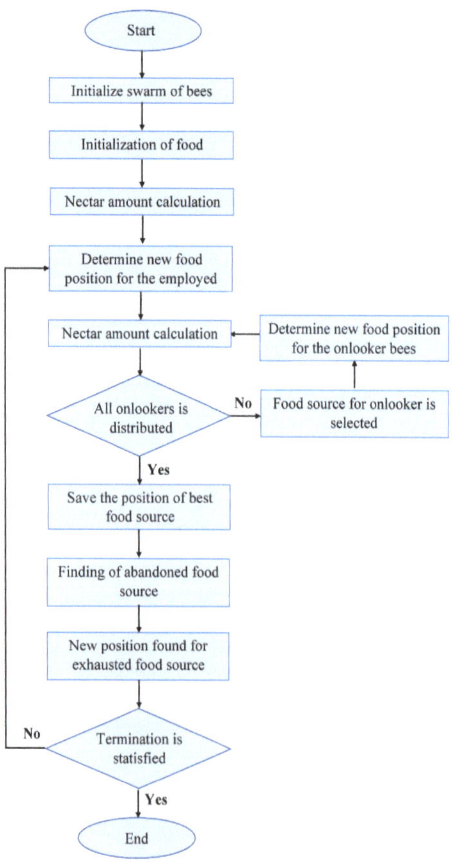

Figure 20. The ABC process.

The ABC algorithm has been used to solve many real-world problems. Ref. [112]'s applications of the ABC algorithm can be seen in many situations where MR (moving robot) systems operate in static environments [113,114]. For example, they tested a wheeled MR underwater [115], applying it to the routing problem of autonomous vehicles [116], as well as aerial robots [117]. The modified ABC algorithm was used for the Unmanned Combat Aerial Vehicle (UCAV) navigation problem [118] to plan optimal routes in a three-dimensional environment, including unmanned helicopters [119].

3.8. Ant Colony Optimization (ACO)

This algorithm is inspired by the foraging behavior and communication of ants, and it was presented by Dorigo and Maniezzo in 1991 [120]. Ants leave behind a kind of pheromone on the paths they traverse. The more ants travel along a path, the more pheromone accumulates on it, and other ants will follow stronger pheromone trails left by other ants in the area. When an ant initiates a search process in a problem, for example, searching for a route on a map, it randomly selects a route and follows it. As it progresses, the ant senses the amount of pheromones in the environment and makes decisions to modify its route based on this information. Ants prefer routes with higher pheromone concentrations. The ACO algorithm runs repeated colonies of ants and compares the results of each colony to optimize the pathways based on the amount of pheromones. In this process, the algorithm gradually converges to an optimal solution to the problem. The formulae of the ACO algorithm (19) are described in [80]:

$$P_{ij}^k(t) = \begin{cases} \frac{\tau_{ij}^{\alpha}(t) n_{ij}^{\beta}(t)}{\sum_{s\in d_k} \tau_{ij}^{\alpha}(t) n_{ij}^{\beta}(t)} & s \in d_k \\ 0 \quad \text{otherwise} \end{cases} \tag{18}$$

$$n_{ij} = \frac{1}{d_{ij}}$$

where $P_{ij}^k(t)$ is the transition probability, $\tau_{ij}^{\alpha}(t)$ represents the pheromone concentration, $n_{ij}^{\beta}(t)$ is the heuristic function, d_k is a collection of access points, and n_{ij} is a heuristic function, usually expressed as the reciprocal of the distance d_{ij} between i and j.

$$\tau_{ij}(t + \Delta t) = (1 - \rho)\tau_{ij}(t) + \Delta\tau_{ij}(t) \tag{19}$$

$$\tau_{ij} - \tau_{ij} \mid \sum_{k=1}^{M} \Delta\tau_{ij}^k \tag{20}$$

$$\Delta\tau_{ij}^k = \begin{cases} \frac{1}{d_{ij}} & \text{Ant } k \text{ pass } (i,j) \\ 0 & \text{otherwise} \end{cases} \tag{21}$$

where ρ represents the pheromone volatility coefficient, M is the total number of ants in the ant colony, and $\Delta\tau_{ij}^k$ represents the pheromone amount released by the kth ant. The ACO process is illustrated in Figure 21

Initially applied to solving the Traveling Salesman Problem (TSP) [120], the principles and mathematical models of the ACO algorithm have since been systematically studied and have undergone significant development, such as in [121] with airport AGV route optimization model based on the ant colony algorithm for optimizing Dijkstra's algorithm in urban systems. In [122], a search and rescue is presented in a maze-like environment with ant and Dijkstra algorithms. The work in [123] describes the application of odometry and Dijkstra's algorithm to warehouse mobile robot navigation and shortest path determination.

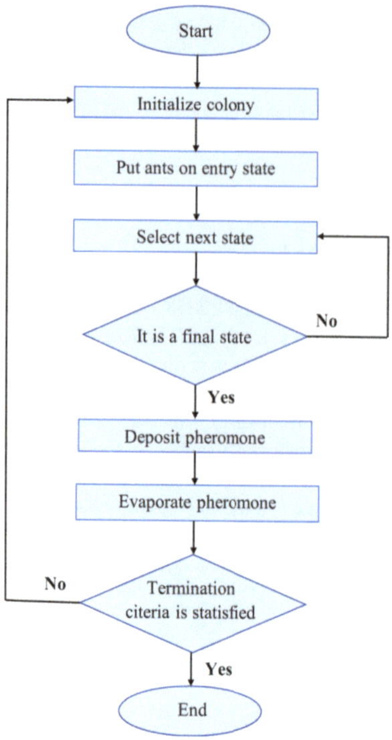

Figure 21. The ACO process.

3.9. Deep-Learning-Based Control (DL)

Machine learning (ML) is the process of using computer systems to learn and improve without their experience, explicitly programming them. Machine learning algorithms rely on recognizing patterns and rules from data and making decisions or predictions based on them. Basic machine learning techniques include supervised learning (where algorithms are trained on labeled data), unsupervised learning (where algorithms try to find structure from unlabeled data), and semisupervised learning, which uses a combination of the two methods. Deep learning is a specialized field of machine learning that uses deep neural networks to learn complex patterns and representations. Deep learning enables computer systems to learn representations of data using multilayered, hierarchical structures. These layers gradually learn higher-level features, which makes deep learning algorithms particularly effective in image recognition, speech recognition, natural language processing, and many other complex tasks. Deep learning models such as Convolutional Neural Networks (CNNs) and Recurrent Neural Networks (RNNs) have made significant breakthroughs in various application areas of artificial intelligence. The main advantage of solutions based on machine learning is that they can learn from the data, so their models already incorporate the nonlinear behavior of the control plant. This enables better performance in many control applications than classical approaches. Deep learning techniques are suitable for handling both global and local path-planning problems [124].

3.9.1. Artificial Neutral Network (ANN)

A neural network, which draws inspiration from the natural human senses, serves as an intelligent system and was originally devised for mobile robot route planning [125]. It consists of simulated networks composed of neuron-like units. These networks undergo optimization through comprehensive training on designated tasks, with the connection strengths between units being gradually adjusted over time [126]. In a neural network, the processing elements (neurons) are usually ordered topologically and interconnected in a well-defined way. The structure of the neural network plays an important role in the execution of the task. Due to the internal parallel structure of neural nets, computations can be performed in parallel, thus ensuring high processing speed. Thus, neural networks are particularly suitable for solving real-time tasks.

A general neuron structure is shown in Figure 22.

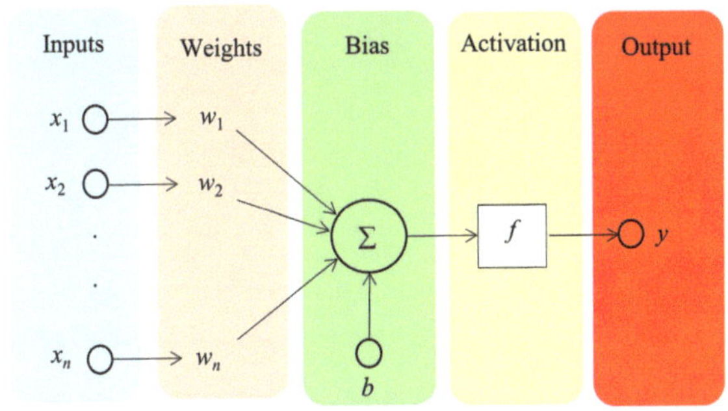

Figure 22. A general neuron structure.

Where x_i is input to the neuron, $X = [x_1, x_2, \ldots x_n]$ is the input vector (n represents the number of inputs on the neuron). b is a constant input (bias-offset value), and y is the neuron's output. w_i represents the weight factor associated with the ith input, $W = [w_1, w_2, \ldots w_n]$ is the weight vector, and f represents the activation function.

The x_i scalar inputs are summed by weighting w_i and the weighted sum is then summed to a nonlinear element. The weighted sum of the input signals, which is the input to the activation function, is called the excitation, while the output signal is called the response (activation). The f function is called the activation function. The output of a neuron can be calculated as follows:

$$y = f\left(\sum_{i=1}^{n} x_i w_i - b\right) \tag{22}$$

The weight factors determine the degree of influence on connections with neighboring neurons within a neuron's vicinity. A neural network's functionality relies on these weight factors, which encapsulate information or the processing of information during the learning phase.

Utilizing a nonlinear activation function enables the neural network to model any nonlinear function when applied to a suitable neuron. Conversely, a linear activation function leads to a linear neural network. To imbue a neural network with nonlinearity, it is imperative to incorporate at least one nonlinear activation function. Additionally, differentiation plays a crucial role, as gradient-based learning stands as the predominant method for adjusting neural network weights.

ANNs are structured into distinct layers: the input layer, where known data are fed into the model; the intermediate layers, referred to as hidden layers; and the output layer, which yields the final sought-after value. Each layer comprises various units (neurons or nodes), with each unit connected to the subsequent layer through a transfer function. Within an ANN, the output of layer $i - 1$ serves as the input for layer i. The known data enter the input layer, accompanied by a bias term. Subsequently, these data are subjected to multiplication by initial weights, followed by summation. The resulting values are then passed through functions to the subsequent layer, iterating until reaching the output layer, where the final value is derived. Transitioning from one layer to the next in an ANN involves the utilization of transfer functions [127]. A possible layout of an artificial neural network is illustrated in Figure 23.

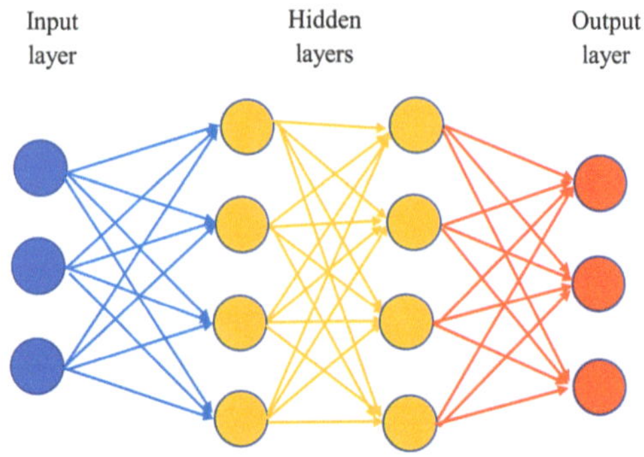

Figure 23. A possible structure of the ANN (the number of hidden layers may vary).

The operation of neural networks can typically be divided into two phases:

- Learning phase—the network stores the desired information processing procedure in some way.
- Recall phase—the stored procedure is used to execute information processing.

The main forms of learning in neural networks [128]:

- Learning with a teacher (called supervised or guided learning (also known as controlled learning)).
- Reinforcement learning.
- Learning without a teacher (unsupervised or unsupervised learning).
- Analytical learning.

Learning neural networks is nothing more than a multivariate optimization procedure based on a predefined criterion function (cost function). Various optimization techniques have been widely used for learning neural networks: gradient-based strategies [129,130], evolutionary methods, genetic algorithms [131,132], and particle swarm optimization (PSO; see later) algorithms [133].

ANN has been applied in a wide range of fields, including search optimization [134], pattern recognition [135,136], image processing [137,138], mobile robot routing [139], signal processing [140], and many more. A hybrid approach to mobile robot navigation combining an ANN and FL [141,142] was designed for a mobile robot navigation controller using a neuro-fuzzy logic system.

In [143], a single-layer approach to robot tracking control was proposed. Through experiments on a KUKA LBR4+ robotic manipulator, the feasibility of the novel ANN approximation for robot control was examined. Ref. [144] presents positioning the error compensation of an industrial robot using neural networks. Ref. [145] presents a recurrent neural network for prediction of motion paths in human robot collaborative assembly.

The ANN was extended to create the Guided Adaptive Pulse Coupled Neural Network (GAPCNN) for mobile robots [146]. The GAPCNN aims to achieve fast parameter convergence to help the robot move in both static and dynamic environments. In particular, the ANN method has been used in MATLAB for mobile robot trajectory planning problems for aerial robots [147], humanoid robots [148], underwater robots [149], and industrial robots [150].

3.9.2. Model Predictive Control (MPC)

The MPC method (Figure 24) is used to predict the behavior of the system for a given time interval and, based on the prediction, optimize the intervention signal at each time instant. As a result, it minimizes the cost function and determines the optimal control sequence. The method has the advantage of a user-friendly design process and easy implementation. It has many applications in the automotive industry, for example, it is used to solve tracking problems [151]. The path planning of autonomous vehicles can also be conducted using a predictive approach [152,153].

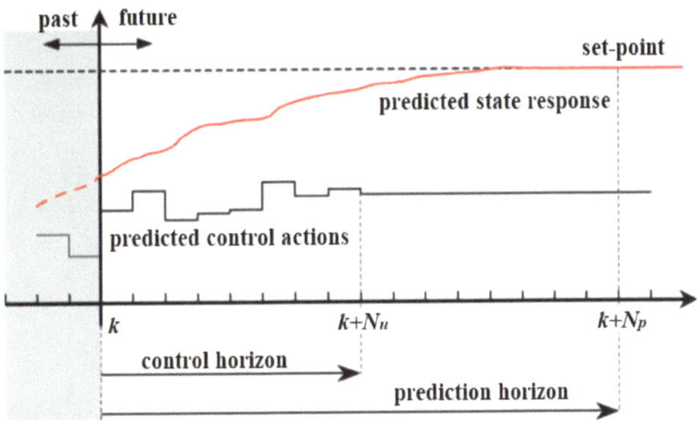

Figure 24. The general concept of the MPC.

Due to the high computational complexity of numerical optimization, it is of paramount importance to ensure real-time computability, which requires the right formulation of the problem and the choice of the appropriate procedure for its solution. The most commonly used MPC approach is based on linear models, but this also has limitations that can reduce performance. Advances in recent decades have allowed engineers to use control approaches that have a higher computational cost, such as Nonlinear Model Predictive Control (NMPC). The main drawback of an NMPC is its complexity, which can lead to high computational time. As a consequence, in most cases, only suboptimal solutions can be obtained, which may degrade the performance of the closed-loop system [154].

Ref. [155] also proposes a cooperative regulatory strategy for docking unmanned aerial vehicles (UAVs) based on MPC. The proposed strategy implements a nonlinear and a linear MPC for the coarse approach (long range) and the fine docking maneuver (short range) based on the same objective function with tailored optimization strategies. Docking is a complex, critical maneuver that requires knowledge of the flight safety of the docking route and the constraints associated with the position of the platform to be docked. In addition, nonlinear effects such as vorticity due to the close approach of the lead agent must be taken into account.

Using the MPC method, it is easier to prove the stability and performance of the system, as it does not require knowledge of the system model. Instead, a local model is used and updated every time step. In addition, other methods are available to redefine the learning characteristics compared with neural networks. For example, in [156], an MPC-based control solution is proposed where the terminal cost and the set are determined through an iterative process.

The presented algorithms do not guarantee stability, which makes their application in safety-critical systems, such as autonomous vehicles, risky. However, some solutions address this problem. For example, ref. [157] presents a control strategy based on safety settings that can modify the input signal of the system when the output of a machine learning agent may destabilize the system. Another solution is given in [158], in which a Hamilton–Jacobi reachability algorithm is exploited that can work with any machine learning-based solution. A combined approach is presented in [159], in which a classical controller is used to control the linearized system, while the machine learning-based algorithm handles the nonlinearities of the system.

B. Németh [160] incorporated machine learning into the usual model-based robust control theory framework, but emphasized it as a new tool and an additional data-driven branch. For all its learning nature, however, robust control remains, i.e., the traditional model-based solution has been extended to fit today's new approach to new types of tasks.

The method is independent of the internal structure of the learning-based control element. Hence, a control element with any structure can be incorporated in its place, providing considerable freedom in control design. For example, solutions based on neural networks, which are already well established in practice, can be implemented in the developed control solutions for reference signal training or feedback loops. However, data-based MPC-type control schemes typically have a more closed, less flexible formalism for the optimization formulated in them. Another consequence of the hierarchical structure is that the learning-based management element can be physically separated from the supervisor and robust management elements. The vehicle motion dynamics are considered in the robust control element and the learning functions in the learning-based control element. For example, in the context of automated vehicles, the supervisor-robust control dual, which has a low computational demand, can be placed on board the vehicle, while the learning-based control element can be placed on an independent platform, such as a cloud. Control solutions that rely predominantly on solving an optimization task online typically do not have this advantage. The supervisor element, which requires online computation, has significantly lower computational requirements than traditional MPC or more advanced data-based (learning) MPC solutions. This is because the supervisor performs significantly fewer tasks than the main optimization task of the MPC. In the supervisor, it is not necessary to perform an optimization over a long horizon, since the impact on future motion states is taken into account in the learning process by running on episodes or prespecified patterns.

3.9.3. Deep Reinforcement Learning (DRL)

Reinforcement learning (RL), inspired by animal psychological learning, learns optimal decision-making strategies from experience [161]. RF is a special type of ML algorithm that does not require large amounts of data for training. The RF algorithm is modeled based on reward, and several papers address the problem of autonomous vehicle control using RF methods [162,163]. Although RF-based solutions can be efficient, the stability of the closed-loop system is still an open question. A proposed solution is an RF-based algorithm combined with a robust controller [164], which achieves the stability of the algorithm by applying uncertainty models. The Deep Reinforcement Learning (DRL) model is particularly promising for solving Vehicle Routing Problems (VRPs). DRL can estimate patterns that are difficult to find with manual heuristics, especially for large-scale problems. Moreover, DRL can generate and infer routes quickly, making it extremely useful for solving time-sensitive VRPs.

The use of DRL in mobile robot navigation is a growing trend. The purpose of using the DRL algorithm in an autonomous navigation task is to find the optimal policy for guiding the robot to the target position through interaction with the environment. The advantage of DRL-based navigation is that it is map-free, has strong learning ability, and has little dependence on sensor accuracy [124].

3.10. Other Algorithms

Without wishing to be exhaustive, we briefly mention some algorithms that have recently become common.

3.10.1. Dynamic Window Approach (DWA)

The DWA can generally be classified as a heuristic method, as it does not rely on rigorous mathematical models or algorithms to solve the problem but rather on an empirical approach. This method is designed for local routing and obstacle avoidance [165]. It takes into account the robot's current speed, acceleration limits, and immediate surroundings to calculate a safe and feasible path to the destination. Creates a dynamic window based on possible velocities and angular velocities. An objective function calculates the optimal value of these pairs of velocities based on the minimum distance from the obstacles, the final bearing angle, and the velocity values of the robots. While in less complex environments the DWA can deftly avoid obstacles, its performance in extremely crowded environments may be suboptimal [166]. The DWA has the local minima and the global convergence problems [167].

Once the task creator has set one or more targets and the global route has been planned, the execution phase involves the robot scanning the surrounding environment, planning local trajectories, and moving forward. This sequence is repeated until the goal is reached. At the beginning of this flow, it samples all the speed pairs corresponding to the kinematic constraints of the robot. DWA computes the coordinates of the waypoints for each input velocity pair using iterations (23) to (25) [168].

$$x(t_n) = x(t_{n-1}) + v \cdot \Delta t \cdot \cos(\Theta(t_{n-1})) \tag{23}$$

$$y(t_n) = y(t_{n-1}) + v \cdot \Delta t \cdot \sin(\Theta(t_{n-1})) \tag{24}$$

$$\Theta(t_n) = \Theta(t_{n-1}) + \omega \cdot \Delta t \tag{25}$$

The model assumes that the robot moves a distance of $v \cdot \Delta t$ along the heading of t_{n-1} and then rotates an angle of $\omega \cdot \Delta t$, where $x(t)$ and $y(t)$ represent the coordinates, and $\Theta(t)$ represents the heading of the robot. By iterating the input velocities, this method computes the coordinates and heading of the robot from time t_0 to t_n. The computation time depends on the number of iterations. In the next step, DWA calculates the distance between each obstacle and waypoint using matrix operations. The calculation time is influenced by the number of paths and obstacle points. Subsequently, DWA swiftly determines the direction to the path's endpoint and the distance to the target. After assessing all possible speed pairs, the optimal speed command is generated. This model is extensively utilized in research involving wheeled robots [168].

3.10.2. Golden Jackal Optimization (GJO)

GJO is a metaheuristic, swarm-intelligence-based algorithm proposed by Nitish Chopra and Muhammad Mohsin Ansari, which models the cooperative hunting behavior and tactics of golden jackals in nature. Because these opportunistic animals are famous for their ability to adapt to different environments [169]. Golden jackals usually hunt with males and females. After finding the prey, they begin to move towards it cautiously. The prey is then surrounded and stalked until it stops. Finally, it is attacked and captured. Updating the position of the prey often depends on the male golden jackal. For this reason, the diversity of golden jackals is not adequate in some cases, and the search algorithm tends to fall into the local optimum [170].

GJO initiates with a randomized distribution of the first solution across the search space, as shown in Equation (26) [169]:

$$Y_0 = Y_{min} + rand(Y_{max} - Y_{min}) \tag{26}$$

where Y_{max} and Y_{min} are the maximum and minimum values of the variable Y, and $rand(Y_{max} - Y_{min})$ is a uniform random vector in the range of 0 to 1.

In [171], a hybrid-strategy-based GJO algorithm for robot path planning is presented.

3.10.3. Grey Wolf Optimization (GWO)

GWO is another type of swarm intelligence algorithm [172] that mimics the hunting strategy of wolves. It categorizes the wolves into different roles: the chief wolf, α, who leads the hunt; β, who assists the leader; δ, who scouts and guards; and the rest ω. The wolves' hunting process is generally broken down into three phases: encirclement, pursuit, and attack. During the encirclement phase, the algorithm updates positions using Equation (27):

$$X(t+1) = X_P(t) - A|C \cdot X_P(t) - X(t)| \tag{27}$$

Although GWO is efficient, it needs a unique initial population. Another drawback is its slow convergence and easily falling into a local optimum [173]. Shitu Singh [174] proposed a more advanced version using Levy's flight model to modify the population and the greedy selection method to update the path.

The Grey Wolf Optimization algorithm has been successfully applied to route planning [175]. The Golden Sine Grey Wolf Optimizer (GSGWO) has been improved from the Grey Wolf Optimizer (GWO), which provides slow convergence speed and easily falls into local optimum, especially without an obstacle-crossing function [176].

3.10.4. Gravitation Search Algorithm (GSA)

GSA is also a robust metaheuristic population-based search algorithm based on gravity rules [177]. Objects are attracted to each other by the force of gravity, and this force is responsible for the global movement of all objects towards more massive objects. The masses thus interact through gravitational force. Heavy masses, which are good solutions, move more slowly than lighter masses (bad solutions). The position of the mass corresponds to the solution of the problem. The gravitational and inertial mass of bodies is determined by a fitness function. GSA can be seen as an isolated system for masses.

Like other metaheuristic systems, GSA has parameters that greatly affect its performance. The mass j acting on mass i by mass j is the equation F_{ij} giving the gravitational force and the gravitational acceleration a_i caused by it (28) [177]:

$$F_{i,j} = G\frac{M_{aj}M_{pi}}{R^2} \tag{28}$$

$$a_i = \frac{F_{i,j}}{M_{ii}} \tag{29}$$

where M_{aj} and M_{pi} represent the active gravitational mass of particle i and passive gravitational mass of particle j, respectively, R is the distance between masses, and M_{ii} represents the inertia mass of particle i. "G(t) is the gravitational constant that decreases iteratively" [178]:

$$G(t) = G_0 e^{-\alpha\frac{t}{T}} \tag{30}$$

The gravitational constant G is the most sensitive entity in the GSA model and effectively controls the balance between the exploration and exploitation capabilities of the algorithm. α and G_0 are constant parameters that affect the performance of the algorithm. As for the tuning of the mentioned parameters, many GSA variants have been developed [178].

4. Hybrid Algorithms

As you can see, there are many other ways to avoid obstacles. Among these are many that use several classical or heuristic algorithms at the same time, which are also described in this article. These are commonly referred to as "hybrids". In this article, three such algorithms are mentioned as an addition.

4.1. New Hybrid Navigation Algorithm (NHNA)

The so-called "new hybrid navigation" algorithm consists of two independent layers, the deliberative and reactive layers. The deliberative layer plans the reference route using the A* search algorithm based on the stored preliminary information. The reactive layer takes over the reference trajectory and guides the robot autonomously along the planned route [25]. The reference path is temporary, and it can be changed by the reactive layer during movement. This layer uses the D-H error algorithm (Distance Histogram bug). It is an improved version of the bug-2 algorithm [42], which allows the robot to freely rotate at angles less than 90° to avoid obstacles. If a rotation of 90° or more is required to avoid an obstacle, the bug-2 algorithm behaves as a bug [44]. The algorithm needs prior information about the environment, which it stores as a binary grid map. The state of each grid on the map is free or occupied: free if there is no obstacle in it, and occupied if it has an obstacle. Figure 25 shows the results of [25], which shows the planned and shortest paths generated by the algorithm. Figure 26a shows the path of the robot with the Dist-Bug algorithm, while Figure 26b shows the behavior of the robot with the D-H error algorithm [25].

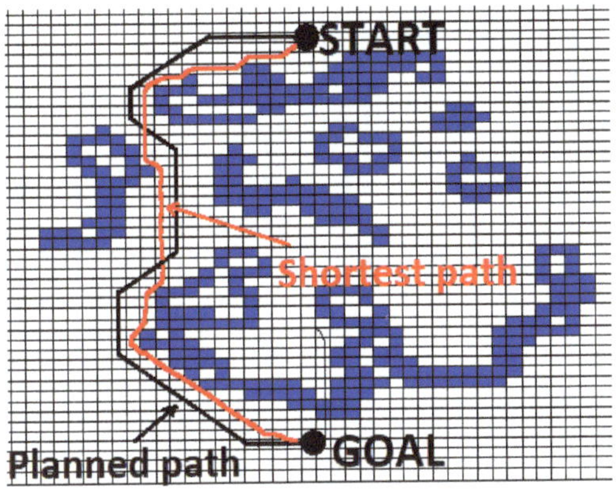

Figure 25. Raster map and robot path (with NHNA) [25].

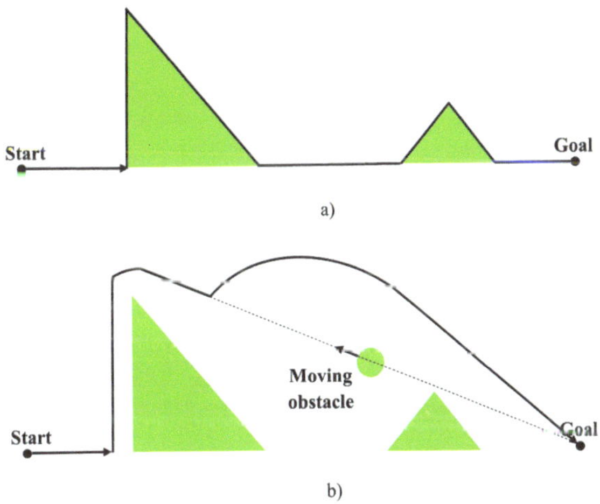

Figure 26. Obstacle avoidance strategy: (**a**) path of the Dist-Bug algorithm and (**b**) robot trajectory with the distance histogram (D-H) error algorithm.

4.2. Hybrid Navigation Algorithm with Roaming Trails (HNA)

This algorithm is designed to effectively handle environments where the robot encounters obstacles during movement. During navigation, the robot can deviate from its path to avoid obstacles using reactive navigation strategies, but is always limited within the area. By ensuring the robot moves within a convex area encompassing the target node's location, it is assured to reach the target in the presence of static obstacles by following a straight path. In certain scenarios, the mobile robot must navigate around obstacles or come to a halt when faced with an obstacle. [44]. The main difference between the hybrid navigation algorithm and NHNA is that it uses APF instead of D-H BUG in the reactive layer. NHNA did not describe any constraints on the deviation from the reference path, but HNA used the concept of roaming trails for the same purpose. Figure 27 shows the roaming traces with the preliminary map (top) and the safe trajectory of the robot on the roaming traces (bottom) [179].

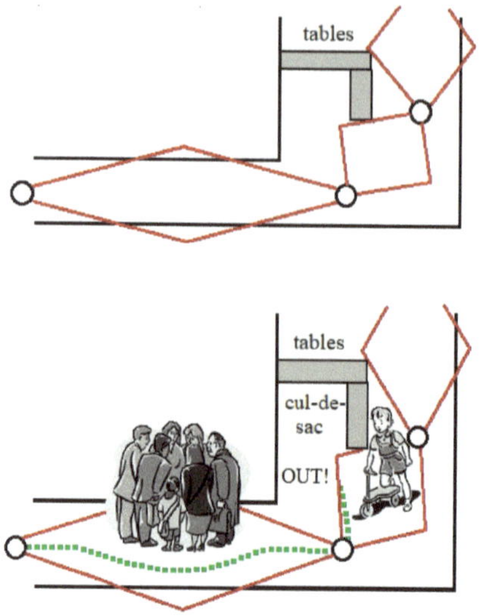

Figure 27. Route Path of the robot with roaming trails (HNA): Top: Primary map with roaming trails. Bottom: Robot trajectory (dashed line) [179].

For more than ten years, the approach has been extensively tested on robots, in particular on the autonomous robot Staffetta [179]. Staffetta is specifically designed for autonomous transport in hospitals, with a payload of 120 kg and a maximum speed of 1 m/s. The robot is equipped with sensors to detect nearby objects and touch sensors to avoid collisions. Furthermore, it is equipped with a laser-based localization system that allows regular position corrections. Based on the experimental experience gained, the second generation of the robot (Merry Porter™) has been further developed and is now independently transporting waste within the Modena Polyclinic.

4.3. Methods Based on Sliding Mode (SM)

The intelligent space learns motion control by tracking the robot's movements [180], thus being able to learn an obstacle avoidance strategy. This learning is based on a neuro-fuzzy approximation of vector-field-based obstacle avoidance. The efficiency of navigation is crucial, as the main application tasks of a mobile robot may include, for example, the guidance of visually impaired people, which requires an immediate reaction to any disturbance. Using the artificial potential field, a collision-free trajectory is guaranteed along gradient lines. The equations of motion of the robot concerning the fixed world system (x_f, y_f) can be derived as follows:

$$\begin{aligned} \dot{x} &= v_G \cos\phi \\ \dot{y} &= v_G \sin\phi \\ \dot{\phi} &= v_G / L \tan\theta \end{aligned} \tag{31}$$

where v_G denotes the velocity vector at the center of the moving platform, which is constrained along the longitudinal axis fixed to the robot due to nonholonomic kinematics. In the robot fixed coordinate system (x_R, y_R), a local harmonic potential field $\Psi(x, y)$ is generated [181]. According to Laplace's equation, this harmonic field corresponds to

$$\nabla^T \cdot \nabla\Psi(x,y) = \frac{\partial^2\Psi(x,y)}{\partial x^2} + \frac{\partial^2\Psi(x,y)}{\partial y^2} = 0 \tag{32}$$

The solution to (32) gives the potential of a singular point of power q at (0,0) in a 2D Cartesian (x, y):

$$\Psi(x,y) = q \ln \frac{1}{\sqrt{x^2+y^2}} \tag{33}$$

and the associated gradient $\rho(x,y) \in \mathbb{R}^2$:

$$\rho(x,y) = -grad\Psi(x,y) = \frac{q}{\sqrt{x^2+y^2}} \begin{pmatrix} x \\ y \end{pmatrix} \tag{34}$$

The configuration of the fundamental potential field consists of a negative unit singular point in the target and a positive singular point of magnitude $0 < q < 1$ in the middle of the obstacle, $q = \frac{R}{R+e}$, where e is the distance between the target and the center of the obstacle, and R is the radius of the circular safety zone. As circular obstacle protection zones cannot be applied directly [181], elliptical safety zones have been designed. Each obstacle has one safety ellipse, but if there are multiple obstacles, two ellipses are needed on either side of the selected route. In this case, the two potential fields must somehow be "merged" to form a single potential field. A good alternative method is to always consider only the nearest safety ellipse. However, this requires switching potential fields at the intersection of equidistant lines between ellipses. This switching, as shown in Figure 28, results in a noncontinuous gradient field.

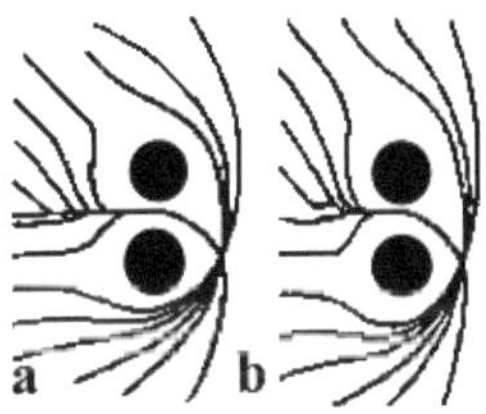

Figure 28. Gradient lines for switching noncontinuous gradients: (**a**) noncontinuous switching; (**b**) smooth switching [24]

In this case, the sliding surface can be described by the line $\sigma_{eq} = 0$. When switching between gradient lines, the scattering appears as oscillations. This effect can be reduced by smoothing the gradient lines near the equidistance line: in the boundary layer along the equidistance line between the two safety zones by spatial domain smoothing. The gradient of the resulting smooth gradient field is the weighted sum of the two gradients. The control inputs are usually the outputs of some actuator. The gradient $\rho(x,y)$ is implemented as a velocity field. Kinematics constrains robot motion from three-dimensional to two-dimensional along the velocity vector. We assume that the state variables x, y, ϕ and the kinematic parameters L and W are known. The orientation of the robot's angle ϕ must be controlled to be colinear concerning the gradient $\rho(x,y)$. So the desired orientation at the point (x, y) is [181]:

$$\phi_\rho = Atan\frac{\rho_y}{\rho_x} \quad with \quad \rho(x,y) = \begin{pmatrix} \rho_x \\ \rho_y \end{pmatrix} \in \mathbb{R}^2 \tag{35}$$

Because speed control is simple, the desired direction of movement β, $v = \beta|v|$, where β is defined by the orientation error $\triangle\phi$, as follows:

$$
\begin{aligned}
\triangle\phi = \phi_\rho - \phi + 2\pi \Rightarrow \beta = 1 \quad ha \quad & -2\pi < \phi_\rho - \phi < -\frac{3\pi}{2} \\
\triangle\phi = \phi_\rho - \phi + \pi \Rightarrow \beta = -1 \quad ha \quad & -\frac{3\pi}{2} < \phi_\rho - \phi < -\frac{\pi}{2} \\
\triangle\phi = \phi_\rho - \phi \Rightarrow \beta = 1 \quad ha \quad & -\frac{\pi}{2} < \phi_\rho - \phi < \frac{\pi}{2} \\
\triangle\phi = \phi_\rho - \phi - \pi \Rightarrow \beta = -1 \quad ha \quad & \frac{\pi}{2} < \phi_\rho - \phi < \frac{3\pi}{2} \\
\triangle\phi = \phi_\rho - \phi - 2\pi \Rightarrow \beta = 1 \quad ha \quad & \frac{3\pi}{2} < \phi_\rho - \phi < -2\pi
\end{aligned} \tag{36}
$$

The sliding surface of the orientation error is defined as follows:

$$\sigma = \beta\triangle\phi \tag{37}$$

A sliding mode along the $\sigma = 0$ surface is created, but at the same time, the direction of motion is changed, and changing the sign of β should be avoided. This can be avoided by monotonically decreasing $\triangle\phi$ by controlling the value of ϕ [181]. The Lyapunov function in this case is $V = \frac{1}{2}\sigma^T\sigma$. Differentiating this function along the trajectories of the system:

$$\sigma^T\dot{\sigma} = \sigma^T v(S\cos\theta - \frac{1}{L}) \tag{38}$$

where $S(x, y, \phi)$ describes the rate of change in the curvature of the gradient along the track lines, $\phi = \arctan SL$, and the θ is

$$\theta = \varphi + \frac{\pi}{2} sign\triangle\phi \tag{39}$$

4.4. Other Examples

Just a few more examples of hybrid algorithms are provided below:

- Ref. [182] presents a hybrid path-planning algorithm based on improved A* and an artificial potential field for unmanned surface vehicle formations.
- Researchers have also exploited GA hybridization with other approaches to MR navigation for better results in route planning problems, such as GA-PSO [183], GA-FL [184], and GA-ANN [185].
- In [186], a hybrid genetic algorithm (HGA)-based approach applied to the image denoising problem is presented. HGA provides the dynamic mutation rate and a switchable global-local search method for the mutation operator of the ordinary genetic algorithm [187].
- In [188], the dynamic modeling of the impact of polymer insulators in polluted conditions based on the HGA-PSO algorithm is presented
- Ref. [189] used a Voronoi diagram and the particle swarm optimization algorithm to achieve multirobot navigation and obstacle avoidance.
- Ref. [190] presents a UGV routing algorithm based on an improved A* with an improved artificial potential field.
- Ref. [43] used VFH*, combining the VFH+ local obstacle avoidance algorithm and the A* path planning algorithm.

5. Comparison of the Algorithms Discussed in This Paper

The advantages and disadvantages of the classical and heuristic algorithms and the convergence and computation time requirements are summarized in Tables 1–5. Convergence time and computation time are expressed in different units and scales. Data are approximate values only and may vary depending on circumstances.

Table 1. Summary table of the classical algorithms in the thesis. Part 1.

Algorithm	Advantages	Disadvantages	Convergence Time	Calculation Time	References
Dijkstra	Robust and reliable operation; Ensuring accurate route planning	Not adaptable to dynamically changing environments	Long	High	[5]
FW	Find the shortest route between all pair nodes	High memory requirements for large graphs; $O(n^3)$ running time, which is inefficient for large graphs	Medium	High	[19]
BF	Ability to handle negative weights; Detects negative cycles	Slower than Dijkstra for positive weight graphs; $O(nm)$ running time	Medium	Medium [20–22]	
APF	Simple and intuitive method; Ability to handle both static and dynamic obstacles	The robot can get stuck in local minima; Difficult to use in more complex environments	Medium	High	[23]
Bug	Simple and easy to implement algorithm; Good for avoiding static obstacles	No guarantee of the shortest route; Less effective for more complex or dynamic obstacles	Short	Low	[33,34]

Table 2. Summary table of the classical algorithms in the thesis. Part 2.

Algorithm	Advantages	Disadvantages	Convergence Time	Calculation Time	References
FGM	Very effective on narrow or fragmented gaps	Not effective on every obstacle; Difficult to use in confined maps or with large robots	Short	Low	[37,38]
VFH	Flexibility and adaptability;	Time- and computation-intensive, especially for large maps; Complex parameterization	Medium	High	[41,43]
CD	It effectively bypasses local minima to help find globally optimal solutions	High computational demand and memory requirements; Proper parameterization and fine-tuning can be critical for efficiency	High	High	[46]
PRM	Integrate sensory data and probabilistic information	High computational and memory demand; Complex parameterization and fine-tuning	Medium	High	[48,52]
RRT	Suitable for solving the route planning problem in dynamic and multi obstacle conditions; applicable to the route planning problem in high-dimensional environments	The route is randomly generated, the route is biased; The convergence speed is slow, and the search efficiency is low	High	High	[56,58,60]

Table 3. Summary table of the heuristic algorithms in the thesis. Part 1.

Algorithm	Advantages	Disadvantages	Convergence Time	Calculation Time	References
A*	Direct search; No preprocessing required	Large amount of calculation; Optimal solution not guaranteed	Medium	Medium	[63]
FL	A flexible and adaptable; React to uncertainties and foggy information	High memory requirements; the rules and parameters largely require human intervention	Medium	High	[69,75]
GA	Strong global searching ability	Slow convergence; Poor local optimization; Poor stability	Long	High	[50,79]
SA	Good for global optimization; ability to avoid local minima	Global optimum is not guaranteed; Depends on cooling schedule	Medium-Long	Medium	[88,90]
TS	Ability to avoid local minima; can be used for complex problems	High memory requirements due to the taboo list; parameter-sensitive	Medium	Medium-High	[88]
PSO	Fast search time; high convergence speed in early-stage	Slow convergence speed in later period; easy to fall into local optimum	Medium	High	[92–94]
CSA	Simple and easy to implement; efficient exploration and optimization of space	No guarantee of a global optimal solution; Less effective for more complex or large problems	Medium	Medium	[102–104]
ABC	Flexible and adaptable; finding global optimal solutions for larger systems;	Parameterization and fine-tuning is time-consuming; High memory requirements	High	High	[109,110,112]
ACO	Strong global searching ability; high efficiency; high convergence speed in later period	Slow convergence speed in early stage	High	High	[80,120]

Table 4. Summary table of the heuristic algorithms in the thesis. Part 2.

Algorithm	Advantages	Disadvantages	Convergence Time	Calculation Time	References
DWA	Fast response times in real-time applications; efficient local obstacle avoidance	Finding only local solutions; global optimum is not guaranteed	Low	Low	[165–167]
GJO	Powerful global search capability; ability to avoid local minima	Requires significant computing resources; sometimes slower convergence	Medium-Long	High	[169,170]
GWO	Powerful global search capability; handles multidimensional optimization problems well	Possible early convergence; depends on fine-tuning of parameters	Long	Medium	[172,173]
GSA	Good global search capability; robust for different types of problems	Slow convergence; requires significant computing resources	Long	High	[177,178]
ANN	Ability to learn and adapt; robust and able to handle large amounts of data	High memory requirements; during the learning phase, large data sets are needed	Long	High	[125,126]
MPC	Forward-looking optimization; ability to manage the limitations of systems	High computing resources; complex implementation	Medium-Long	High	[154]
DRL	Complex problem solving; autonomous learning; good generalization ability	High computational demand; high data demand	Long	High	[124]

Table 5. Summary table of the hybrid algorithms in the thesis.

Algorithm	Advantages	Disadvantages	Convergence Time	Calculation Time	References
NHNA	Integrate the benefits of multiple algorithms; improved accuracy and efficiency in different environments	More complex implementation; high calculation demand	Medium	Medium-High	[25]
HNA	Better route optimization; more flexibility in dealing with obstacles	Requires significant computing resources; complex parameter tuning	Medium	Medium-High	[44,179]
SM	Robust in unknown dynamic environments; fast reaction time	Sensitive to noise and discontinuities; Precise modeling required	Short-Medium	Medium	[181]

6. Discussions and Future Trends

Navigation and route planning are the central difficulties of mobile robots and have been the subject of decades of research. As a result, several methodologies have been presented and applied to the problem of route planning for mobile robots. Strategies for mobile robot optimization can be classified into deterministic or classical approaches and nondeterministic or heuristic approaches. Traditional algorithms execute a given task step by step according to predefined instructions, and their results are exact and deterministic. (One of the simplest algorithms is the Pythagorean theorem, which determines a third parameter in an identical way given two input parameters, and the term heuristic is derived from the Greek word heuresis, which means to find.) Theoretically, constructing an exact solution procedure would make it possible to calculate how to reach the goal based on the robot's current position and by analyzing all possible paths. The problem is that the situation becomes too complex above a given complexity of the environment. A heuristic algorithm does not consider all possible steps but only decides according to some logic based on a particular part of the problem space. A considerable advantage of heuristic algorithms is that they can deliver results relatively quickly for high-complexity problems with little computation. However, they have the disadvantage that the optimal solution cannot be guaranteed completely. They are helpful when the solution to a problem cannot be found within a foreseeable time by a conventional method that provides an exact solution. They can also provide an optimal or approximate solution for large problem sizes. Among the earliest developed error avoidance algorithms, they are straightforward to calibrate but time-intensive. These methods are not goal-oriented; they trace edges without considering the ultimate objective [37]. The Dijkstra algorithm is a graph search algorithm designed to find paths and determine the shortest paths [5]. The Floyd-Warshall (FW) algorithm uses a weighted and directed graph and can compute opposing weighted edges. The solutions are derived from the previous results, and multiple solutions can be generated [191]. This algorithm finds the shortest path between each pair of nodes and is particularly useful when the distance between all pair nodes of the graph needs to be determined. However, it is inefficient for large graphs due to its high memory and time requirements. The Bellman-Ford (BF) algorithm can find the shortest path from one peak to another, which is simple and does not require complex data structures to apply. The algorithm iteratively extends the search to all nodes, not just along the current shortest path. For this reason, it can be slower than Dijkstra's algorithm for positive-weight graphs but can handle graphs with negative weights, whereas Dijkstra's algorithm cannot. If there is a negative cycle in the graph, the Dijkstra algorithm would run the cycle infinitely, as this would theoretically result in an infinitely negative cost. In contrast, the BF algorithm would detect this and terminate [192]. This algorithm can handle opposing weight edges and detect negative cycles, an advantage for specific problems. Likewise, artificial potential field (APF) is a simple technique for avoiding obstacles, but robots following this principle can get stuck in so-called local minima [37,42]. This is a time-consuming algorithm, as the robot can stop before the obstacle until it moves. The Bug algorithm is also an early version of the obstacle avoidance algorithm used in robot navigation [34]. The gap tracking method (FGM) is another early obstacle avoidance algorithm used in environments where the robot must navigate narrow spaces. Still, it can not avoid U-shaped obstacles [37,193].

Fuzzy logic (FL) has been developed among the heuristic algorithms for various applications, including obstacle avoidance robotics [71,72]. Since their initial research, genetic algorithms (GAs) have been widely applied to solve various optimization problems, including obstacle avoidance algorithms [79]. Simulated annealing (SA) and Tabu search (TS) have proven to be very effective and robust in solving a wide range of problems across various applications. They are also helpful in dealing with issues where specific parameters are not known in advance. These properties are missing in all conventional optimization techniques [88]. They apply an appropriate cost function to give feedback to the algorithm on the progress of the search. The difference in principle is how and where domain-specific knowledge is used. For example, SA obtains such information mainly from the cost function. The disturbed items are selected randomly, and the acceptance or rejection of disturbances is based on the Metropolis criterion, which is a function of cost. The cooling schedule also has a significant impact on the algorithm's performance. It must be carefully tailored to the problem domain and the specific problem instance. TS differs from GA and SA because it has an explicit memory component. At each iteration, the neighborhood of the current solution is partially explored, and the move is made toward the best nontaboo solution in that neighborhood. The neighborhood function and the size and content of the Tabu list are problem-specific. Memory structures also influence the direction of the search. Particle swarm optimization (PSO) is a population-based heuristic optimization method derived from standing wave theory [48]. The cuckoo search algorithm (CSA) was introduced as an efficient and straightforward global search technique among evolutionary algorithms [194]. The Artificial Bee Colony (ABC) algorithm was developed to model the behavior of living organisms and is one of the evolutionary algorithms [109]. While machine learning requires human intervention, deep learning can learn from mistakes. Deep learning requires a more significant amount of data, which demands higher computational power. In deep learning, algorithms learn autonomously by analyzing large amounts of data. In contrast, reinforcement learning requires feedback from the agent to know what actions lead to the desired outcome. Significant developments in neural networks occurred in the 1980s and beyond and have been applied to obstacle avoidance robotics [125]. In control systems, the star of reinforcement learning solutions is now gone, replaced by data-driven MPC solutions that can provide theoretical guarantees of performance [195]. It is questionable whether a suitable fitness function can solve all our problems, not to mention the theoretical guarantees of stability or convergence. Nevertheless, it is worth using machine learning algorithms in engineering because, presumably, they will be able to solve more and more routine tasks for us. Furthermore, there is also the question of how data-driven MPC algorithms solve all control theory problems. Specifically, where is the space left for model-based robust control? The Hybrid Navigation Algorithm (HNA) with wandering trails combines different methods and techniques for optimal route planning, which has been applied to a partially known environment [179]. Recent developments have resulted in a New Hybrid Navigation Algorithm (NHNA) similar to the HNA. It is a complete algorithm that uses several approaches to achieve efficient and stable robot navigation. However, it cannot be used in unknown environments as it requires prior environmental information [25,193]. Sliding mode (SM) algorithms employ several methods and have seen significant development and application, especially in robotics and control systems [181].

Essential characteristics of algorithms are convergence time, computation time, and memory requirements. The convergence time is required for the algorithm to reach convergence, i.e., to achieve a stable or desired state. This time may vary depending on the algorithm type, the task's nature, and the initial conditions. The goal is to make the algorithm converge as fast as possible to solve the task or problem efficiently. Computation time and memory requirements are closely related. The more complex the environment in which we want to navigate the robot, the more data and more complex computations are needed to find the optimum.

Among the previously developed methods, heuristic approaches are relatively new and have significant applications in mobile robot navigation. Contemporary research increasingly focuses on optimizing algorithms through hybridization to achieve superior performance. Historically, classical methodologies were prevalent but faced limitations such as susceptibility to local minima and high computational demands. In response, researchers have shifted towards heuristic methods, particularly effective in uncertain or unknown environments. These heuristic approaches, often enhanced by hybridization with classical methods, have proven successful in complex three-dimensional workspaces, such as those encountered by underwater, unmanned aerial vehicles, and humanoid robots. This shift underscores the improved adaptability and efficiency of heuristic strategies over classical approaches in dynamic settings.

As technology advances and robotics becomes increasingly integrated into various aspects of our lives, obstacle avoidance algorithms are poised to undergo significant developments to meet the demands of emerging applications. With the proliferation of machine learning techniques, we can expect obstacle avoidance algorithms to incorporate more advanced learning-based approaches. These algorithms will be capable of adapting and improving their performance over time through experience and feedback, leading to more efficient and robust obstacle avoidance in dynamic environments. Future obstacle avoidance systems will rely on sophisticated sensor fusion techniques to integrate data from multiple sensors, such as LIDAR, cameras, radar, and ultrasonic sensors. By combining information from diverse sources, these algorithms will achieve a more comprehensive understanding of the environment, enhancing their ability to detect and avoid obstacles accurately. Future obstacle avoidance algorithms prioritize real-time adaptive planning to navigate complex and dynamic environments effectively. These algorithms will continuously analyze sensor data and adjust robot trajectories to avoid obstacles and navigate changing scenarios in real-time, unreal time, and efficient robot operation. Collaborative obstacle avoidance algorithms will become increasingly important in environments where multiple robots or autonomous vehicles operate concurrently. These algorithms will enable robots to communicate and coordinate their movements to avoid collisions and optimize path planning, leading to smoother and more efficient operations in shared spaces. Drawing inspiration from nature, future obstacle avoidance algorithms may incorporate bio-inspired principles, such as swarm intelligence or mimicry of animal behavior. These approaches could lead to innovative solutions for navigating challenging environments, leveraging the collective intelligence of swarms, or mimicking the agility and adaptability of animals in natural habitats. As robots become more prevalent, ethical considerations regarding obstacle avoidance will gain prominence. Future algorithms must balance efficiency with moral considerations, prioritizing human safety and well-being in crowded environments. Additionally, human-robot interaction will play a crucial role, with obstacle avoidance algorithms designed to anticipate and respond effectively to human intentions and behaviors. With these exciting developments, researchers and engineers can pave the way for safer, more efficient, and more adaptive robotic systems in various applications.

Author Contributions: Conceptualization, P.K.; methodology, K.K.; validation, H.A.N.; investigation, K.K.; data curation, H.A.N.; writing—original draft preparation, K.K.; writing—review and editing, K.K.; supervision, P.K.; project administration, P.K., H.A.N. All authors have read and agreed to the published version of the manuscript.

Funding: This work was supported by the Hungarian Research Fund (OTKA K143595).

Conflicts of Interest: The authors declare no conflicts of interest. The funders had no role in the design of the study; in the collection, analyses, or interpretation of data; in the writing of the manuscript; or in the decision to publish the results.

Abbreviations

The following abbreviations are used in this manuscript:

ABC	Artificial Bee Colony
ACO	Ant Colony Optimization
AGV	Automated Guided Vehicle
APF	Artificial Potential Field
ANN	Artificial Neural Network
AUV	Autonomous Underwater Vehicle
BF	Bellman–Ford Algorithm
CD	Cell Decomposition
CSA	Cuckoo Search Algorithm
DL	Deep Learning
DRL	Deep Reinforcement Learning
FGM	Follow Gap Method
FL	Fuzzy Logic
FW	Floyd–Warshall Algorithm
GA	Genetic Algorithm
HGA	Hybrid Genetic Algorithm
HNA	Hybrid Navigation Algorithm
LIDAR	Light Detection And Ranging
MAV	Micro Aerial Vehicle
MPC	Model Predictive Control
NHNA	New Hybrid Navigation Algorithm
PRM	Probabilistic Roadmap Method
PSO	Particle Swarm Optimization
RL	Reinforcement Learning
SM	Sliding Mode Method
UAV	Unmanned Aerial Vehicle
UCAV	Unmanned Combat Aerial Vehicle
USV	Unmanned Surface Vehicle
VFH	Vector Field Histogram
VPS	Vehicle Routing Problems

References

1. Sedighi, K.H.; Ashenayi, K.; Manikas, T.W.; Wainwright, R.L.; Tai, H.M. Autonomous local path planning for a mobile robot using a genetic algorithm. In Proceedings of the 2004 Congress on Evolutionary Computation (IEEE Cat. No. 04TH8753), Portland, OR, USA, 19–23 June 2004; Volume 2, pp. 1338–1345.
2. Yan, K.; Ma, B. Mapless navigation based on 2D LIDAR in complex unknown environments. *Sensors* **2020**, *20*, 5802. [CrossRef] [PubMed]
3. Vckay, E.; Aneja, M.; Deodhare, D. Solving a Path Planning Problem in a Partially Known Environment using a Swarm Algorithm. *arXiv* **2017**, arXiv:1705.03176.
4. Kamil, F.; Tang, S.; Khaksar, W.; Zulkifli, N.; Ahmad, S. A review on motion planning and obstacle avoidance approaches in dynamic environments. *Adv. Robot. Autom.* **2015**, *4*, 134–142.
5. Dijkstra, E.W. A note on two problems in connexion with graphs. In *Edsger Wybe Dijkstra: His Life, Work, and Legacy*; Association for Computing Machinery: New York, NY, USA, 2022; pp. 287–290.
6. Sabo, C.; Cohen, K. Fuzzy logic unmanned air vehicle motion planning. *Adv. Fuzzy Syst.* **2012**, *2012*, 989051. [CrossRef]
7. Gonzalez, R.; Kloetzer, M.; Mahulea, C. Comparative study of trajectories resulted from cell decomposition path planning approaches. In Proceedings of the 2017 21st International Conference on System Theory, Control and Computing (ICSTCC), Sinaia, Romania, 19–21 October 2017; pp. 49–54.
8. Liu, L.s.; Lin, J.f.; Yao, J.x.; He, D.w.; Zheng, J.s.; Huang, J.; Shi, P. Path planning for smart car based on Dijkstra algorithm and dynamic window approach. *Wirel. Commun. Mob. Comput.* **2021**, *2021*, 8881684. [CrossRef]
9. Kirono, S.; Arifianto, M.I.; Putra, R.E.; Musoleh, A.; Setiadi, R. Graph-based modeling and dijkstra algorithm for searching vehicle routes on highways. *Int. J. Mech. Eng. Technol. (IJMET)* **2018**, *9*, 1273–1280.
10. Wang, C.; Cheng, C.; Yang, D.; Pan, G.; Zhang, F. Path planning in localization uncertaining environment based on Dijkstra method. *Front. Neurorobot.* **2022**, *16*, 821991. [CrossRef] [PubMed]
11. Amaliah, B.; Fatichah, C.; Riptianingdyah, O. Finding the shortest paths among cities in Java Island using node combination based on Dijkstra algorithm. *Int. J. Smart Sens. Intell. Syst.* **2016**, *9*, 2219. [CrossRef]

12. Broumi, S.; Bakal, A.; Talea, M.; Smarandache, F.; Vladareanu, L. Applying Dijkstra algorithm for solving neutrosophic shortest path problem. In Proceedings of the 2016 International Conference on Advanced Mechatronic Systems (ICAMechS), Melbourne, Australia, 30 November–3 December 2016; pp. 412–416.
13. Chen, R.; Hu, J.; Xu, W. An RRT-Dijkstra-based path planning strategy for autonomous vehicles. *Appl. Sci.* **2022**, *12*, 11982. [CrossRef]
14. Dhulkefl, E.; Durdu, A.; Terzioğlu, H. Dijkstra Algorithm Using Uav Path Planning. *Konya J. Eng. Sci.* **2020**, *8*, 92–105. [CrossRef]
15. Singh, Y.; Sharma, S.; Sutton, R.; Hatton, D. Optimal path planning of an unmanned surface vehicle in a real-time marine environment using a dijkstra algorithm. *Mar. Navig.* **2017**, 399–402.
16. Lyu, D.; Chen, Z.; Cai, Z.; Piao, S. Robot path planning by leveraging the graph-encoded Floyd algorithm. *Future Gener. Comput. Syst.* **2021**, *122*, 204–208. [CrossRef]
17. Weisstein, E.W. Floyd-Warshall Algorithm. 2008. Available online: https://mathworld.wolfram.com/ (accessed on 28 May 2024).
18. Triana, Y.S.; Syahputri, I. Implementation floyd-warshall algorithm for the shortest path of garage. *Int. J. Innov. Sci. Res. Technol.* **2018**, *3*, 871–878.
19. Magzhan, K.; Jani, H.M. A review and evaluations of shortest path algorithms. *Int. J. Sci. Technol. Res* **2013**, *2*, 99–104.
20. Terzimehic, T.; Silajdzic, S.; Vajnberger, V.; Velagic, J.; Osmic, N. Path finding simulator for mobile robot navigation. In Proceedings of the 2011 XXIII International Symposium on Information, Communication and Automation Technologies, Sarajevo, Bosnia and Herzegovina, 27–29 October 2011; pp. 1–6.
21. AbuSalim, S.W.; Ibrahim, R.; Saringat, M.Z.; Jamel, S.; Wahab, J.A. Comparative analysis between dijkstra and bellman-ford algorithms in shortest path optimization. In *Proceedings of the IOP Conference Series: Materials Science and Engineering*; IOP Publishing: Bristol, UK, 2020; p. 012077.
22. Goldberg, A.V.; Radzik, T. *A heuristic Improvement of the Bellman-Ford Algorithm*; Stanford University, Department of Computer Science: Stanford, CA, USA, 1993.
23. Abiyev, R.; Ibrahim, D.; Erin, B. Navigation of mobile robots in the presence of obstacles. *Adv. Eng. Softw.* **2010**, *41*, 1179–1186. [CrossRef]
24. Baranyi, P.; Nagy, I.; Korondi, B.; Hashimoto, H. General guiding model for mobile robots and its complexity reduced neuro-fuzzy approximation. In Proceedings of the Ninth IEEE International Conference on Fuzzy Systems, FUZZ- IEEE 2000 (Cat. No.00CH37063), San Antonio, TX, USA, 7–10 May 2000; Volume 2, pp. 1029–1032. [CrossRef]
25. Zhu, Y.; Zhang, T.; Song, J.; Li, X. A new hybrid navigation algorithm for mobile robots in environments with incomplete knowledge. *Knowl.-Based Syst.* **2012**, *27*, 302–313. [CrossRef]
26. Sepehri, A.; Moghaddam, A.M. A motion planning algorithm for redundant manipulators using rapidly exploring randomized trees and artificial potential fields. *IEEE Access* **2021**, *9*, 26059–26070. [CrossRef]
27. Di, W.; Caihong, L.; Na, G.; Yong, S.; Tengteng, G.; Guoming, L. Local path planning of mobile robot based on artificial potential field. In Proceedings of the 2020 39th Chinese Control Conference (CCC), Shenyang, China, 27–29 July 2020; pp. 3677–3682.
28. BinKai, Q.; Mingqiu, L.; Yang, Y.; XiYang, W. Research on UAV path planning obstacle avoidance algorithm based on improved artificial potential field method. In *Proceedings of the Journal of Physics: Conference Series*; IOP Publishing: Bristol, UK, 2021; Volume 1948, p. 012060.
29. Fan, X.; Guo, Y.; Liu, H.; Wei, B.; Lyu, W. Improved artificial potential field method applied for AUV path planning. *Math. Probl. Eng.* **2020**, *2020*, 6523158. [CrossRef]
30. Duan, Y.; Yang, C.; Zhu, J.; Meng, Y.; Liu, X. Active obstacle avoidance method of autonomous vehicle based on improved artificial potential field. *Int. J. Adv. Robot. Syst.* **2022**, *19*, 17298806221115984. [CrossRef]
31. Liu, Q.; Liu, J.; Zhao, Y.; Shen, R.; Hou, L.; Zhang, Y. Local path planning for multi-robot sytems based on improved artificial potential field algorithm. In Proceedings of the 2022 IEEE 5th Advanced Information Management, Communicates, Electronic and Automation Control Conference (IMCEC), Chongqing, China, 16–18 December 2022; Volume 5, pp. 1540–1544.
32. Zhai, S.; Pei, Y. The Dynamic Path Planning of Autonomous Vehicles on Icy and Snowy Roads Based on an Improved Artificial Potential Field. *Sustainability* **2023**, *15*, 15377. [CrossRef]
33. Sivaranjani, S.; Nandesh, D.A.; Gayathri, K.; Ramanathan, R. An Investigation of Bug Algorithms for Mobile Robot Navigation and Obstacle Avoidance in Two-Dimensional Unknown Static Environments. In Proceedings of the 2021 International Conference on Communication information and Computing Technology (ICCICT), Mumbai, India, 25–27 June 2021; pp. 1–6. [CrossRef]
34. Yufka, A.; Parlaktuna, O. Performance comparison of bug algorithms for mobile robots. In Proceedings of the 5th International Advanced Technologies Symposium, Karabuk, Turkey, 13–15 May 2009; pp. 13–15.
35. Neloy, M.; Das, M.; Barua, P.; Pathak, A.; Rahat, S.U. An intelligent obstacle and edge recognition system using bug algorithm. *Am. Sci. Res. J. Eng. Technol. Sci.* **2020**, *64*, 133–143.
36. Wang, X.; Yin, Y.; Jing, Q. Maritime Search Path Planning Method of an Unmanned Surface Vehicle Based on an Improved Bug Algorithm. *J. Mar. Sci. Eng.* **2023**, *11*, 2320. [CrossRef]
37. Sezer, V.; Gokasan, M. A novel obstacle avoidance algorithm: "Follow the Gap Method". *Robot. Auton. Syst.* **2012**, *60*, 1123–1134. [CrossRef]
38. Houshyari, H.; Sezer, V. A new gap-based obstacle avoidance approach: Follow the obstacle circle method. *Robotica* **2022**, *40*, 2231–2254. [CrossRef]

39. Demir, M.; Sezer, V. Improved Follow the Gap Method for obstacle avoidance. In Proceedings of the 2017 IEEE International Conference on Advanced Intelligent Mechatronics (AIM), Munich, Germany, 3–7 July 2017; pp. 1435–1440.
40. Gul, F.; Rahiman, W.; Alhady, S.; Ali, A.; Mir, I.; Jalil, A. Meta-heuristic approach for solving multi-objective path planning for autonomous guided robot using PSO–GWO optimization algorithm with evolutionary programming. *J. Ambient Intell. Humaniz. Comput.* **2021**, *12*, 7873–7890. [CrossRef]
41. Borenstein, J.; Koren, Y. The Vector Field Histogram-Fast Obstacle Avoidance For Mobile Robots. *Robot. Autom. IEEE Trans.* **1991**, *7*, 278–288. [CrossRef]
42. Oroko, J.A.; Nyakoe, G. Obstacle avoidance and path planning schemes for autonomous navigation of a mobile robot: A review. In Proceedings of the Sustainable Research and Innovation Conference, 2–3 October 2022; pp. 314–318. Available online: https://sri.jkuat.ac.ke/jkuatsri/index.php/sri/article/view/491/422 (accessed on 4 May 2024).
43. Wu, M.; Dai, S.L.; Yang, C. Mixed reality enhanced user interactive path planning for omnidirectional mobile robot. *Appl. Sci.* **2020**, *10*, 1135. [CrossRef]
44. Alatise, M.B.; Hancke, G.P. A review on challenges of autonomous mobile robot and sensor fusion methods. *IEEE Access* **2020**, *8*, 39830–39846. [CrossRef]
45. Dong, T.; Zhang, Y.; Xiao, Q.; Huang, Y. The Control Method of Autonomous Flight Avoidance Barriers of UAVs in Confined Environments. *Sensors* **2023**, *23*, 5896. [CrossRef]
46. Debnath, S.K.; Omar, R.; Bagchi, S.; Sabudin, E.N.; Shee Kandar, M.H.A.; Foysol, K.; Chakraborty, T.K. Different cell decomposition path planning methods for unmanned air vehicles-A review. In *Proceedings of the 11th National Technical Seminar on Unmanned System Technology 2019: NUSYS'19*; Springer: Berlin/Heidelberg, Germany, 2021; pp. 99–111.
47. Patle, B.; Pandey, A.; Parhi, D.; Jagadeesh, A. A review: On path planning strategies for navigation of mobile robot. *Def. Technol.* **2019**, *15*, 582–606. [CrossRef]
48. Masehian, E.; Sedighizadeh, D. Classic and heuristic approaches in robot motion planning-a chronological review. *World Acad. Sci. Eng. Technol.* **2007**, *23*, 101–106.
49. Ab Wahab, M.N.; Nefti-Meziani, S.; Atyabi, A. A comparative review on mobile robot path planning: Classical or meta-heuristic methods? *Annu. Rev. Control* **2020**, *50*, 233–252. [CrossRef]
50. Adzhar, N.; Salleh, S.; Yusof, Y.; Ahmad, M.A. Routing problem in rectangular mesh network using shortest path based Greedy method. In *Proceedings of the Journal of Physics: Conference Series*; IOP Publishing: Bristol, UK, 2019; Volume 1358, p. 012079.
51. Gnanaprakash, M. Study on Mobile Robot Path Planning–A Review. *Int. J. Appl. Eng. Res* **2015**, *10*, 2015.
52. Ichter, B.; Schmerling, E.; Lee, T.W.E.; Faust, A. Learned critical probabilistic roadmaps for robotic motion planning. In Proceedings of the 2020 IEEE International Conference on Robotics and Automation (ICRA), Paris, France, 31 May–31 August 2020; pp. 9535–9541.
53. Huang, S.K.; Wang, W.J.; Sun, C.H. A path planning strategy for multi-robot moving with path-priority order based on a generalized Voronoi diagram. *Appl. Sci.* **2021**, *11*, 9650. [CrossRef]
54. Schoener, M.; Coyle, E.; Thompson, D. An anytime Visibility–Voronoi graph-search algorithm for generating robust and feasible unmanned surface vehicle paths. *Auton. Robot.* **2022**, *46*, 911–927. [CrossRef]
55. Kim, J.; Son, H.I. A voronoi diagram-based workspace partition for weak cooperation of multi-robot system in orchard. *IEEE Access* **2020**, *8*, 20676–20686. [CrossRef]
56. Pérez-Hurtado, I.; Martínez-del Amor, M.Á.; Zhang, G.; Neri, F.; Pérez-Jiménez, M.J. A membrane parallel rapidly-exploring random tree algorithm for robotic motion planning. *Integr. Comput.-Aided Eng.* **2020**, *27*, 121–138. [CrossRef]
57. LaValle, S. *Rapidly-Exploring Random Trees: A New Tool for Path Planning;* Research Report 9811; 1998. Available online: https://msl.cs.illinois.edu/~lavalle/papers/Lav98c.pdf (accessed on 28 May 2024).
58. Jang, D.u.; Kim, J.s. Development of Ship Route-Planning Algorithm Based on Rapidly-Exploring Random Tree (RRT*) Using Designated Space. *J. Mar. Sci. Eng.* **2022**, *10*, 1800. [CrossRef]
59. Luo, S.; Zhang, M.; Zhuang, Y.; Ma, C.; Li, Q. A survey of path planning of industrial robots based on rapidly exploring random trees. *Front. Neurorobot.* **2023**, *17*, 1268447. [CrossRef]
60. Shi, Y.; Li, Q.; Bu, S.; Yang, J.; Zhu, L. Research on intelligent vehicle path planning based on rapidly-exploring random tree. *Math. Probl. Eng.* **2020**, *2020*, 5910503. [CrossRef]
61. Löfgren, K. Rapidly-Exploring Random Trees for real-time combined Exploration andPath Planning. 2023.
62. Rachmawati, D.; Gustin, L. Analysis of Dijkstra's algorithm and A* algorithm in shortest path problem. In *Proceedings of the Journal of Physics: Conference Series*; IOP Publishing: Bristol, UK, 2020; Volume 1566, p. 012061.
63. Hart, P.E.; Nilsson, N.J.; Raphael, B. A formal basis for the heuristic determination of minimum cost paths. *IEEE Trans. Syst. Sci. Cybern.* **1968**, *4*, 100–107. [CrossRef]
64. Yao, J.; Lin, C.; Xie, X.; Wang, A.J.; Hung, C.C. Path Planning for Virtual Human Motion Using Improved A* Star Algorithm. In Proceedings of the 2010 Seventh International Conference on Information Technology: New Generations, Las Vegas, NV, USA, 12–14 April 2010; pp. 1154–1158.
65. Casalino, G.; Turetta, A.; Simetti, E. A three-layered architecture for real time path planning and obstacle avoidance for surveillance USVs operating in harbour fields. In Proceedings of the OCEANS 2009-EUROPE, Bremen, Germany, 11–14 May 2009; pp. 1–8.

66. Guan, W.; Wang, K. Autonomous collision avoidance of unmanned surface vehicles based on improved A-star and dynamic window approach algorithms. *IEEE Intell. Transp. Syst. Mag.* **2023**, *113*, 102755. [CrossRef]
67. Gao, X.; Jia, Q.; Sun, H.; Chen, G. Research on path planning for 7-DOF space manipulator to avoid obstacle based on A* algorithm. *Sens. Lett.* **2011**, *9*, 1515–1519. [CrossRef]
68. Tang, G.; Tang, C.; Claramunt, C.; Hu, X.; Zhou, P. Geometric A-star algorithm: An improved A-star algorithm for AGV path planning in a port environment. *IEEE Access* **2021**, *9*, 59196–59210. [CrossRef]
69. Tzafestas, S.G. Mobile robot control and navigation: A global overview. *J. Intell. Robot. Syst.* **2018**, *91*, 35–58. [CrossRef]
70. Zadeh, L.A. Fuzzy sets. *Inf. Control* **1965**, *8*, 338–353. [CrossRef]
71. Siegwart, R.; Nourbakhsh, I.R.; Scaramuzza, D. *Introduction to Autonomous Mobile Robots*; MIT Press: Cambridge, MA, USA, 2011.
72. Ali, M.A.; Shanono, I.H. Path planning methods for mobile robots: A systematic and bibliometric review. *ELEKTRIKA-J. Electr. Eng.* **2020**, *19*, 14–34.
73. Rafai, A.N.A.; Adzhar, N.; Jaini, N.I. A review on path planning and obstacle avoidance algorithms for autonomous mobile robots. *J. Robot.* **2022**, *2022*, 2538220. [CrossRef]
74. Vachtsevanos, G.; Hexmoor, H. A fuzzy logic approach to robotic path planning with obstacle avoidance. In Proceedings of the 1986 25th IEEE Conference on Decision and Control, Athens, Greece, 10–12 December 1986; pp. 1262–1264.
75. Zhang, Q.; Sun, J.; Xiao, G.; Tsang, E. Evolutionary algorithms refining a heuristic: A hybrid method for shared-path protections in WDM networks under SRLG constraints. *IEEE Trans. Syst. Man Cybern. Part B (Cybern.)* **2007**, *37*, 51–61. [CrossRef]
76. Abbasi, Y.; Moosavian, S.A.A.; Novinzadeh, A.B. Formation control of aerial robots using virtual structure and new fuzzy-based self-tuning synchronization. *Trans. Inst. Meas. Control* **2017**, *39*, 1906–1919. [CrossRef]
77. Xiang, X.; Yu, C.; Lapierre, L.; Zhang, J.; Zhang, Q. Survey on fuzzy-logic-based guidance and control of marine surface vehicles and underwater vehicles. *Int. J. Fuzzy Syst.* **2018**, *20*, 572–586. [CrossRef]
78. Abadi, D.N.M.; Khooban, M.H. Design of optimal Mamdani-type fuzzy controller for nonholonomic wheeled mobile robots. *J. King Saud Univ.-Eng. Sci.* **2015**, *27*, 92–100.
79. Bremermann, H.J. *The Evolution of Intelligence: The Nervous System as a Model of Its Environment*; University of Washington, Department of Mathematics: Washington, DC, USA, 1958.
80. Huang, Y.; Yu, L.; Zhang, F. A survey on puncture models and path planning algorithms of bevel-tipped flexible needles. *Heliyon* **2024**, *10*, e25002. [CrossRef]
81. Kumar, A.; Kumar, P.B.; Parhi, D.R. Intelligent navigation of humanoids in cluttered environments using regression analysis and genetic algorithm. *Arab. J. Sci. Eng.* **2018**, *43*, 7655–7678. [CrossRef]
82. Chen, J.; Zhu, H.; Zhang, L.; Sun, Y. Research on fuzzy control of path tracking for underwater vehicle based on genetic algorithm optimization. *Ocean Eng.* **2018**, *156*, 217–223. [CrossRef]
83. Roberge, V.; Tarbouchi, M.; Labonté, G. Fast genetic algorithm path planner for fixed-wing military UAV using GPU. *IEEE Trans. Aerosp. Electron. Syst.* **2018**, *54*, 2105–2117. [CrossRef]
84. Roberge, V.; Tarbouchi, M. Massively parallel hybrid algorithm on embedded graphics processing unit for unmanned aerial vehicle path planning. *Int. J. Digit. Signals Smart Syst.* **2018**, *2*, 68–93. [CrossRef]
85. Liu, X.; Jiang, D.; Tao, B.; Jiang, G.; Sun, Y.; Kong, J.; Tong, X.; Zhao, G.; Chen, B. Genetic algorithm-based trajectory optimization for digital twin robots. *Front. Bioeng. Biotechnol.* **2022**, *9*, 793782. [CrossRef]
86. Zhang, L.; Zhang, Y.; Li, Y. Path planning for indoor mobile robot based on deep learning. *Optik* **2020**, *219*, 165096. [CrossRef]
87. Li, D., Wang, L.; Cai, J.; Wang, A.; Tan, T.; Gui, J. Research on mobile robot path planning based on improved genetic algorithm. *Int. J. Model. Simul. Sci. Comput.* **2023**, *14*, 2341030. [CrossRef]
88. Youssef, H.; Sait, S.M.; Adiche, H. Evolutionary algorithms, simulated annealing and tabu search: A comparative study. *Eng. Appl. Artif. Intell.* **2001**, *14*, 167–181. [CrossRef]
89. Metropolis, N.; Rosenbluth, A.W.; Rosenbluth, M.N.; Teller, A.H.; Teller, E. Equation of state calculations by fast computing machines. *J. Chem. Phys.* **1953**, *21*, 1087–1092. [CrossRef]
90. Malek, M.; Guruswamy, M.; Pandya, M.; Owens, H. Serial and parallel simulated annealing and tabu search algorithms for the traveling salesman problem. *Ann. Oper. Res.* **1989**, *21*, 59–84. [CrossRef]
91. Osman, I.H. Metastrategy simulated annealing and tabu search algorithms for the vehicle routing problem. *Ann. Oper. Res.* **1993**, *41*, 421–451. [CrossRef]
92. Kashyap, N.; Mishra, A. A discourse on metaheuristics techniques for solving clustering and semisupervised learning models. In *Cognitive Big Data Intelligence with a Metaheuristic Approach*; Elsevier: Amsterdam, The Netherlands, 2022; pp. 1–19.
93. Ashour, A.S.; Guo, Y. Optimization-based neutrosophic set in computer-aided diagnosis. In *Optimization Theory Based on Neutrosophic and Plithogenic Sets*; Elsevier: Amsterdam, The Netherlands, 2020; pp. 405–421.
94. Shami, T.M.; El-Saleh, A.A.; Alswaitti, M.; Al-Tashi, Q.; Summakieh, M.A.; Mirjalili, S. Particle swarm optimization: A comprehensive survey. *IEEE Access* **2022**, *10*, 10031–10061. [CrossRef]
95. Kumar, A.; Pant, S.; Ram, M.; Singh, S. On solving complex reliability optimization problem using multi-objective particle swarm optimization. In *Mathematics Applied to Engineering*; Elsevier: Amsterdam, The Netherlands, 2017; pp. 115–131.
96. Zhao, X.; Ji, Y.X.; Ning, X.l. Accelerometer calibration based on improved particle swarm optimization algorithm of support vector machine. *Sens. Actuators A Phys.* **2024**, *369*, 115096. [CrossRef]

97. Kumar, P.B.; Pandey, K.K.; Sahu, C.; Chhotray, A.; Parhi, D.R. A hybridized RA-APSO approach for humanoid navigation. In Proceedings of the 2017 Nirma University International Conference on Engineering (NUiCONE), Ahmedabad, India, 23–25 November 2017; pp. 1–6.
98. Gao, M.; Ding, P.; Yang, Y. Time-optimal trajectory planning of industrial robots based on particle swarm optimization. In Proceedings of the 2015 Fifth International Conference on Instrumentation and Measurement, Computer, Communication and Control (IMCCC), Qinhuangdao, China, 18–20 September 2015; pp. 1934–1939.
99. Castillo, O.; Martinez-Marroquin, R.; Melin, P.; Valdez, F.; Soria, J. Comparative study of bio-inspired algorithms applied to the optimization of type-1 and type-2 fuzzy controllers for an autonomous mobile robot. *Inf. Sci.* **2012**, *192*, 19–38. [CrossRef]
100. Rendón, M.A.; Martins, F.F. Path following control tuning for an autonomous unmanned quadrotor using particle swarm optimization. *IFAC-PapersOnLine* **2017**, *50*, 325–330. [CrossRef]
101. He, B.; Ying, L.; Zhang, S.; Feng, X.; Yan, T.; Nian, R.; Shen, Y. Autonomous navigation based on unscented-FastSLAM using particle swarm optimization for autonomous underwater vehicles. *Measurement* **2015**, *71*, 89–101. [CrossRef]
102. Shishavan, S.T.; Gharehchopogh, F.S. An improved cuckoo search optimization algorithm with genetic algorithm for community detection in complex networks. *Multimed. Tools Appl.* **2022**, *81*, 25205–25231. [CrossRef]
103. Imran, M.; Khan, S.; Hlavacs, H.; Khan, F.A.; Anwar, S. Intrusion detection in networks using cuckoo search optimization. *Soft Comput.* **2022**, *26*, 10651–10663. [CrossRef]
104. Xiong, Y.; Zou, Z.; Cheng, J. Cuckoo search algorithm based on cloud model and its application. *Sci. Rep.* **2023**, *13*, 10098. [CrossRef] [PubMed]
105. Mohanty, P.K.; Parhi, D.R. A new hybrid optimization algorithm for multiple mobile robots navigation based on the CS-ANFIS approach. *Memetic Comput.* **2015**, *7*, 255–273. [CrossRef]
106. Xiao, L.; Hajjam-El-Hassani, A.; Dridi, M. An application of extended cuckoo search to vehicle routing problem. In Proceedings of the 2017 International Colloquium on Logistics and Supply Chain Management (LOGISTIQUA), Rabat, France, 27–28 April 2017; pp. 31–35.
107. Bibiks, K.; Hu, Y.F.; Li, J.P.; Pillai, P.; Smith, A. Improved discrete cuckoo search for the resource-constrained project scheduling problem. *Appl. Soft Comput.* **2018**, *69*, 493–503. [CrossRef]
108. Bui, X.N.; Nguyen, H.; Tran, Q.H.; Nguyen, D.A.; Bui, H.B. Predicting ground vibrations due to mine blasting using a novel artificial neural network-based cuckoo search optimization. *Nat. Resour. Res.* **2021**, *30*, 2663–2685. [CrossRef]
109. Karaboga, D. *An Idea Based on Honey Bee Swarm for Numerical Optimization*; Technical Report, Technical report-tr06; Erciyes University, Engineering Faculty, Computer Engineering Department: Kayseri, Türkiye, 2005.
110. ÖZDEMİR, D.; Dörterler, S. An adaptive search equation-based artificial bee colony algorithm for transportation energy demand forecasting. *Turk. J. Electr. Eng. Comput. Sci.* **2022**, *30*, 1251–1268. [CrossRef]
111. Ahmed, B.K.A.; Mahdi, R.D.; Mohamed, T.I.; Jaleel, R.A.; Salih, M.A.; Zahra, M.M.A. A novel secure artificial bee colony with advanced encryption standard technique for biomedical signal processing. *Period. Eng. Nat. Sci.* **2022**, *10*, 288–294. [CrossRef]
112. Kaya, E.; Gorkemli, B.; Akay, B.; Karaboga, D. A review on the studies employing artificial bee colony algorithm to solve combinatorial optimization problems. *Eng. Appl. Artif. Intell.* **2022**, *115*, 105311. [CrossRef]
113. An, D.; Mu, Y.; Wang, Y.; Li, B.; Wei, Y. Intelligent Path Planning Technologies of Underwater Vehicles: A Review. *J. Intell. Robot. Syst.* **2023**, *107*, 22. [CrossRef]
114. Liang, J.H.; Lee, C.H. Efficient collision-free path-planning of multiple mobile robots system using efficient artificial bee colony algorithm. *Adv. Eng. Softw.* **2015**, *79*, 47–56. [CrossRef]
115. Li, B.; Chiong, R.; Gong, L.g. Search-evasion path planning for submarines using the artificial bee colony algorithm. In Proceedings of the 2014 IEEE Congress on Evolutionary Computation (CEC), Beijing, China, 6–11 July 2014; pp. 528–535.
116. Bhagade, A.S.; Puranik, P.V. Artificial bee colony (ABC) algorithm for vehicle routing optimization problem. *Int. J. Soft Comput. Eng.* **2012**, *2*, 329–333.
117. Xu, C.; Duan, H.; Liu, F. Chaotic artificial bee colony approach to Uninhabited Combat Air Vehicle (UCAV) path planning. *Aerosp. Sci. Technol.* **2010**, *14*, 535–541. [CrossRef]
118. Li, B.; Gong, L.g.; Yang, W.l. An improved artificial bee colony algorithm based on balance-evolution strategy for unmanned combat aerial vehicle path planning. *Sci. World J.* **2014**, *2014*, 232704. [CrossRef] [PubMed]
119. Ding, L.; Wu, H.; Yao, Y. Chaotic artificial bee colony algorithm for system identification of a small-scale unmanned helicopter. *Int. J. Aerosp. Eng.* **2015**, *2015*, 801874. [CrossRef]
120. Dorigo, M. *Positive Feedback as a Search Strategy*; Technical report 91-16; Department of Electronics, Information and Bioengineering: Milan, Italy, 1991.
121. Zhou, Y.; Huang, N. Airport AGV path optimization model based on ant colony algorithm to optimize Dijkstra algorithm in urban systems. *Sustain. Comput. Inform. Syst.* **2022**, *35*, 100716. [CrossRef]
122. Husain, Z.; Al Zaabi, A.; Hildmann, H.; Saffre, F.; Ruta, D.; Isakovic, A. Search and rescue in a maze-like environment with ant and dijkstra algorithms. *Drones* **2022**, *6*, 273. [CrossRef]
123. Ubaidillah, A.; Sukri, H. Application of Odometry and Dijkstra Algorithm as Navigation and Shortest Path Determination System of Warehouse Mobile Robot. *J. Robot. Control (JRC)* **2023**, *4*, 413–423. [CrossRef]
124. Zhu, K.; Zhang, T. Deep reinforcement learning based mobile robot navigation: A review. *Tsinghua Sci. Technol.* **2021**, *26*, 674–691. [CrossRef]

125. Zacksenhouse, M.; DeFigueiredo, R.J.; Johnson, D.H. A neural network architecture for cue-based motion planning. In Proceedings of the 27th IEEE Conference on Decision and Control, Austin, TX, USA, 7–9 December 1988; Volume 79, p. 324327.
126. Kanwisher, N.; Khosla, M.; Dobs, K. Using artificial neural networks to ask 'why'questions of minds and brains. *Trends Neurosci.* **2023**, *46*, 240–254. [CrossRef]
127. Juan, N.P.; Valdecantos, V.N. Review of the application of Artificial Neural Networks in ocean engineering. *Ocean Eng.* **2022**, *259*, 111947. [CrossRef]
128. Kriesel, D. A Brief Introduction to Neural Networks. 2007. Available online: http://www.dkriesel.com (accessed on 28 May 2024).
129. Wang, P.; Nagrecha, K.; Vasconcelos, N. Gradient-based algorithms for machine teaching. In Proceedings of the IEEE/CVF Conference on Computer Vision and Pattern Recognition, Nashville, TN, USA, 19–25 June 2021; pp. 1387–1396.
130. Badhuk, P.; Verma, N.; Ravikrishna, R. Optimizing Chemical Reaction Mechanisms: Evaluating Parameter-Free Metaheuristic Algorithms and Gradient-Based Optimization. *Combust. Sci. Technol.* **2024**, 1–19. [CrossRef]
131. Kim, C.; Batra, R.; Chen, L.; Tran, H.; Ramprasad, R. Polymer design using genetic algorithm and machine learning. *Comput. Mater. Sci.* **2021**, *186*, 110067. [CrossRef]
132. Zhang, X.; Guo, Y.; Yang, J.; Li, D.; Wang, Y.; Zhao, R. Many-objective evolutionary algorithm based agricultural mobile robot route planning. *Comput. Electron. Agric.* **2022**, *200*, 107274. [CrossRef]
133. Wang, F.; Wang, X.; Sun, S. A reinforcement learning level-based particle swarm optimization algorithm for large-scale optimization. *Inf. Sci.* **2022**, *602*, 298–312. [CrossRef]
134. Abdolrasol, M.G.; Hussain, S.S.; Ustun, T.S.; Sarker, M.R.; Hannan, M.A.; Mohamed, R.; Ali, J.A.; Mekhilef, S.; Milad, A. Artificial neural networks based optimization techniques: A review. *Electronics* **2021**, *10*, 2689. [CrossRef]
135. Mohammad, A.S.Y.; Tahseen, A.J.A.; Sotnik, S.; Lyashenko, V. Neural Networks As A Tool For Pattern Recognition of Fasteners. *Int. J. Eng. Trends Technol.* **2021**, *69*, 151–160.
136. Kong, Q.; Cao, Y.; Iqbal, T.; Wang, Y.; Wang, W.; Plumbley, M.D. Panns: Large-scale pretrained audio neural networks for audio pattern recognition. *IEEE/ACM Trans. Audio Speech Lang. Process.* **2020**, *28*, 2880–2894. [CrossRef]
137. Tripathi, M. Analysis of convolutional neural network based image classification techniques. *J. Innov. Image Process. (JIIP)* **2021**, *3*, 100–117. [CrossRef]
138. Liu, H.; Liu, M.; Li, D.; Zheng, W.; Yin, L.; Wang, R. Recent advances in pulse-coupled neural networks with applications in image processing. *Electronics* **2022**, *11*, 3264. [CrossRef]
139. Chen, Y.; Cheng, C.; Zhang, Y.; Li, X.; Sun, L. A neural network-based navigation approach for autonomous mobile robot systems. *Appl. Sci.* **2022**, *12*, 7796. [CrossRef]
140. Hu, Y.H.; Hwang, J.N. *Handbook of Neural Network Signal Processing*; CRC Press: Boca Raton, FL, USA, 2018.
141. AbuBaker, A. A novel mobile robot navigation system using neuro-fuzzy rule based optimization technique. *Res. J. Appl. Sci. Eng. Technol.* **2012**, *4*, 2577–2583.
142. Mishra, D.K.; Thomas, A.; Kuruvilla, J.; Kalyanasundaram, P.; Prasad, K.R.; Haldorai, A. Design of mobile robot navigation controller using neuro-fuzzy logic system. *Comput. Electr. Eng.* **2022**, *101*, 108044. [CrossRef]
143. Nubert, J.; Köhler, J.; Berenz, V.; Allgöwer, F.; Trimpe, S. Safe and fast tracking on a robot manipulator: Robust mpc and neural network control. *IEEE Robot. Autom. Lett.* **2020**, *5*, 3050–3057. [CrossRef]
144. Bo, L.; Wei, T.; Zhang, C.; Fangfang, H.; Guangyu, C.; Yufei, L. Positioning error compensation of an industrial robot using neural networks and experimental study. *Chin. J. Aeronaut.* **2022**, *35*, 346–360.
145. Zhang, J.; Liu, H.; Chang, Q.; Wang, L.; Gao, R.X. Recurrent neural network for motion trajectory prediction in human-robot collaborative assembly. *CIRP Ann.* **2020**, *69*, 9–12. [CrossRef]
146. Syed, U.A.; Kunwar, F.; Iqbal, M. Guided Autowave Pulse Coupled Neural Network (GAPCNN) based real time path planning and an obstacle avoidance scheme for mobile robots. *Robot. Auton. Syst.* **2014**, *62*, 474–486. [CrossRef]
147. Zhang, C.; Hu, H.; Wang, J. An adaptive neural network approach to the tracking control of micro aerial vehicles in constrained space. *Int. J. Syst. Sci.* **2017**, *48*, 84–94. [CrossRef]
148. Sun, C.; He, W.; Ge, W.; Chang, C. Adaptive neural network control of biped robots. *IEEE Trans. Syst. Man Cybern. Syst.* **2016**, *47*, 315–326. [CrossRef]
149. Zhu, D.; Tian, C.; Sun, B.; Luo, C. Complete coverage path planning of autonomous underwater vehicle based on GBNN algorithm. *J. Intell. Robot. Syst.* **2019**, *94*, 237–249. [CrossRef]
150. Sun, C.; He, W.; Hong, J. Neural network control of a flexible robotic manipulator using the lumped spring-mass model. *IEEE Trans. Syst. Man Cybern. Syst.* **2016**, *47*, 1863–1874. [CrossRef]
151. Li, Y.; Chai, S.; Chai, R.; Liu, X. An improved model predictive control method for vehicle lateral control. In Proceedings of the 2020 39th Chinese Control Conference (CCC), Shenyang, China, 27–30 July 2020; pp. 5505–5510.
152. Dixit, S.; Montanaro, U.; Dianati, M.; Oxtoby, D.; Mizutani, T.; Mouzakitis, A.; Fallah, S. Trajectory planning for autonomous high-speed overtaking in structured environments using robust MPC. *IEEE Trans. Intell. Transp. Syst.* **2019**, *21*, 2310–2323. [CrossRef]
153. Németh, B.; Hegedűs, T.; Gáspár, P. Model predictive control design for overtaking maneuvers for multi-vehicle scenarios. In Proceedings of the 2019 18th European Control Conference (ECC), Naples, Italy, 25–28 June 2019; pp. 744–749.

154. Fényes, D. Application of Data-Driven Methods for Improving the Peformances of Lateral Vehicle Control Systems. Ph.D. Thesis, Budapest University of Technology and Economics Faculty of Transportation Engineering and Vehicle Engineering Department of Control for Transportation and Vehicle Systems, Budapest, Hungary, 2021.
155. Taner, B.; Subbarao, K. Modeling of Cooperative Robotic Systems and Predictive Control Applied to Biped Robots and UAV-UGV Docking with Task Prioritization. *Sensors* **2024**, *24*, 3189. [CrossRef] [PubMed]
156. Rosolia, U.; Zhang, X.; Borrelli, F. Robust learning model predictive control for iterative tasks: Learning from experience. In Proceedings of the 2017 IEEE 56th Annual Conference on Decision and Control (CDC), Melbourne, Australi, 12–15 December 2017; pp. 1157–1162.
157. Larsen, R.B.; Carron, A.; Zeilinger, M.N. Safe learning for distributed systems with bounded uncertainties. *IFAC-PapersOnLine* **2017**, *50*, 2536–2542. [CrossRef]
158. Fisac, J.F.; Akametalu, A.K.; Zeilinger, M.N.; Kaynama, S.; Gillula, J.; Tomlin, C.J. A general safety framework for learning-based control in uncertain robotic systems. *IEEE Trans. Autom. Control* **2018**, *64*, 2737–2752. [CrossRef]
159. Zhai, L.; Chai, T.; Ge, S.S. Stable adaptive neural network control of nonaffine nonlinear discrete-time systems and application. In Proceedings of the 2007 IEEE 22nd International Symposium on Intelligent Control, Singapore, 1–3 October 2007; pp. 602–607.
160. Németh, B.; Fényes, D.; Bede, Z.; Gáspár, P. Optimal Control Design for Traffic Flow Maximization Based on Data-Driven Modeling Method. *Energies* **2021**, *15*, 187. [CrossRef]
161. Andrew, A.M. Reinforcement learning: An introduction. *Kybernetes* **1998**, *27*, 1093–1096. [CrossRef]
162. Feher, A.; Aradi, S.; Becsi, T. Q-learning based reinforcement learning approach for lane keeping. In Proceedings of the 2018 IEEE 18th International Symposium on Computational Intelligence and Informatics (CINTI), Budapest, Hungary, 21–22 November 2018; pp. 000031–000036.
163. Xia, W.; Li, H.; Li, B. A control strategy of autonomous vehicles based on deep reinforcement learning. In Proceedings of the 2016 9th International Symposium on Computational Intelligence and Design (ISCID), Hangzhou, China, 10–11 December 2016; Volume 2, pp. 198–201.
164. Kretchmar, R.M.; Young, P.M.; Anderson, C.W.; Hittle, D.C.; Anderson, M.L.; Delnero, C.C. Robust reinforcement learning control with static and dynamic stability. *Int. J. Robust Nonlinear Control. IFAC-Affil. J.* **2001**, *11*, 1469–1500. [CrossRef]
165. Fox, D.; Burgard, W.; Thrun, S. The dynamic window approach to collision avoidance. *IEEE Robot. Autom. Mag.* **1997**, *4*, 23–33. [CrossRef]
166. Adiuku, N.; Avdelidis, N.P.; Tang, G.; Plastropoulos, A. Improved Hybrid Model for Obstacle Detection and Avoidance in Robot Operating System Framework (Rapidly Exploring Random Tree and Dynamic Windows Approach). *Sensors* **2024**, *24*, 2262. [CrossRef]
167. Hossain, T.; Habibullah, H.; Islam, R.; Padilla, R.V. Local path planning for autonomous mobile robots by integrating modified dynamic-window approach and improved follow the gap method. *J. Field Robot.* **2022**, *39*, 371–386. [CrossRef]
168. Lin, Z.; Taguchi, R. Faster Implementation of The Dynamic Window Approach Based on Non-Discrete Path Representation. *Mathematics* **2023**, *11*, 4424. [CrossRef]
169. Chopra, N.; Ansari, M.M. Golden jackal optimization: A novel nature-inspired optimizer for engineering applications. *Expert Syst. Appl.* **2022**, *198*, 116924. [CrossRef]
170. Yuan, P.; Zhang, T.; Yao, L.; Lu, Y.; Zhuang, W. A hybrid golden jackal optimization and golden sine algorithm with dynamic lens-imaging learning for global optimization problems. *Appl. Sci.* **2022**, *12*, 9709. [CrossRef]
171. Lou, T.s.; Yue, Z.p.; Jiao, Y.z.; He, Z.d. A hybrid strategy-based GJO algorithm for robot path planning. *Expert Syst. Appl.* **2024**, *238*, 121975. [CrossRef]
172. Mirjalili, S.; Mirjalili, S.M.; Lewis, A. Grey wolf optimizer. *Adv. Eng. Softw.* **2014**, *69*, 46–61. [CrossRef]
173. Faris, H.; Aljarah, I.; Al-Betar, M.A.; Mirjalili, S. Grey wolf optimizer: A review of recent variants and applications. *Neural Comput. Appl.* **2018**, *30*, 413–435. [CrossRef]
174. Singh, S.; Bansal, J.C. Mutation-driven grey wolf optimizer with modified search mechanism. *Expert Syst. Appl.* **2022**, *194*, 116450. [CrossRef]
175. Jarray, R.; Al-Dhaifallah, M.; Rezk, H.; Bouallègue, S. Parallel cooperative coevolutionary grey wolf optimizer for path planning problem of unmanned aerial vehicles. *Sensors* **2022**, *22*, 1826. [CrossRef]
176. Zhao, D.; Cai, G.; Wang, Y.; Li, X. Path Planning of Obstacle-Crossing Robot Based on Golden Sine Grey Wolf Optimizer. *Appl. Sci.* **2024**, *14*, 1129. [CrossRef]
177. Rashedi, E.; Nezamabadi-Pour, H.; Saryazdi, S. GSA: A gravitational search algorithm. *Inf. Sci.* **2009**, *179*, 2232–2248. [CrossRef]
178. Joshi, S.K. Chaos embedded opposition based learning for gravitational search algorithm. *Appl. Intell.* **2023**, *53*, 5567–5586. [CrossRef]
179. Sgorbissa, A.; Zaccaria, R. Planning and obstacle avoidance in mobile robotics. *Robot. Auton. Syst.* **2012**, *60*, 628–638. [CrossRef]
180. Morioka, K.; Lee, J.H.; Hashimoto, H. Human-following mobile robot in a distributed intelligent sensor network. *IEEE Trans. Ind. Electron.* **2004**, *51*, 229–237. [CrossRef]
181. Levant, A. Sliding order and sliding accuracy in sliding mode control. *Int. J. Control* **1993**, *58*, 1247–1263. [CrossRef]
182. Sang, H.; You, Y.; Sun, X.; Zhou, Y.; Liu, F. The hybrid path planning algorithm based on improved A* and artificial potential field for unmanned surface vehicle formations. *Ocean Eng.* **2021**, *223*, 108709. [CrossRef]

183. Wang, X.; Shi, Y.; Ding, D.; Gu, X. Double global optimum genetic algorithm–particle swarm optimization-based welding robot path planning. *Eng. Optim.* **2016**, *48*, 299–316. [CrossRef]
184. Pratihar, D.K.; Deb, K.; Ghosh, A. Fuzzy-genetic algorithms and time-optimal obstacle-free path generation for mobile robots. *Eng. Optim.* **1999**, *32*, 117–142. [CrossRef]
185. Hui, N.B.; Pratihar, D.K. A comparative study on some navigation schemes of a real robot tackling moving obstacles. *Robot. Comput.-Integr. Manuf.* **2009**, *25*, 810–828. [CrossRef]
186. de Paiva, J.L.; Toledo, C.F.; Pedrini, H. An approach based on hybrid genetic algorithm applied to image denoising problem. *Appl. Soft Comput.* **2016**, *46*, 778–791. [CrossRef]
187. Luan, P.G.; Thinh, N.T. Hybrid genetic algorithm based smooth global-path planning for a mobile robot. *Mech. Based Des. Struct. Mach.* **2023**, *51*, 1758–1774. [CrossRef]
188. Fahimi, N.; Sezavar, H.R.; Akmal, A.A.S. Dynamic modeling of flashover of polymer insulators under polluted conditions based on HGA-PSO algorithm. *Electr. Power Syst. Res.* **2022**, *205*, 107728. [CrossRef]
189. Gabbassova, Z.; Sedighizadeh, D.; Sheikhi Fini, A.; Seddighizadeh, M. Multiple robot motion planning considering shortest and safest trajectory. *Electromech. Energy Convers. Syst.* **2021**, *1*, 1–6.
190. Meng, X.; Fang, X. A UGV Path Planning Algorithm Based on Improved A* with Improved Artificial Potential Field. *Electronics* **2024**, *13*, 972. [CrossRef]
191. Hougardy, S. The Floyd–Warshall algorithm on graphs with negative cycles. *Inf. Process. Lett.* **2010**, *110*, 279–281. [CrossRef]
192. Lee, A.; Phung, A.; Swaminathan, S. Discrete Final Project: Probabilistic Shortest Paths & Robotics Navigation Applications. 2020.
193. Kovács, B.; Szayer, G.; Tajti, F.; Burdelis, M.; Korondi, P. A novel potential field method for path planning of mobile robots by adapting animal motion attributes. *Robot. Auton. Syst.* **2016**, *82*, 24–34. [CrossRef]
194. Mohanty, P.K.; Parhi, D.R. Optimal path planning for a mobile robot using cuckoo search algorithm. *J. Exp. Theor. Artif. Intell.* **2016**, *28*, 35–52. [CrossRef]
195. Berberich, J.; Köhler, J.; Müller, M.A.; Allgöwer, F. Data-driven model predictive control with stability and robustness guarantees. *IEEE Trans. Autom. Control* **2020**, *66*, 1702–1717. [CrossRef]

sensors

MDPI

Article

Crossing-Point Estimation in Human–Robot Navigation—Statistical Linearization versus Sigma-Point Transformation

Rainer Palm [1,*] and Achim J. Lilienthal [2]

1 Center for Applied Autonomous Sensor Systems (AASS), Department of Technology, Örebro University, SE-701 82 Örebro, Sweden
2 Technical University Munich (TUM), 80333 Munich, Germany; achim.j.lilienthal@tum.de
* Correspondence: rub.palm@t-online.de

Abstract: Interactions between mobile robots and human operators in common areas require a high level of safety, especially in terms of trajectory planning, obstacle avoidance and mutual cooperation. In this connection, the crossings of planned trajectories and their uncertainty based on model fluctuations, system noise and sensor noise play an outstanding role. This paper discusses the calculation of the expected areas of interactions during human–robot navigation with respect to fuzzy and noisy information. The expected crossing points of the possible trajectories are nonlinearly associated with the positions and orientations of the robots and humans. The nonlinear transformation of a noisy system input, such as the directions of the motion of humans and robots, to a system output, the expected area of intersection of their trajectories, is performed by two methods: statistical linearization and the sigma-point transformation. For both approaches, fuzzy approximations are presented and the inverse problem is discussed where the input distribution parameters are computed from the given output distribution parameters.

Keywords: human–robot interaction; Gaussian noise; sigma-point transformation; unscented Kalman filter

Citation: Palm, R.; Lilienthal, A.J. Crossing-Point Estimation in Human–Robot Navigation—Statistical Linearization versus Sigma-Point Transformation. *Sensors* **2024**, *24*, 3303. https://doi.org/10.3390/s24113303

Academic Editors: David Cheneler and Stephen Monk

Received: 20 March 2024
Revised: 16 May 2024
Accepted: 17 May 2024
Published: 22 May 2024

1. Introduction

The planning and performing of mobile robot tasks in the presence of human operators while sharing the same workspace requires a high level of stability and safety. Research activities regarding navigation, obstacle avoidance, adaptation and collaboration between robots and human agents have been widely reported [1,2]. Multiple target tracking for robots using higher control levels in a control hierarchy are discussed in [3,4]. A human-friendly interaction between robots and humans can be obtained by human-like sensor systems [5]. A prominent role in robot navigation is the trajectory-crossing problem of robots and humans [6,7] and corresponding fuzzy solutions [8]. Motivations for a fuzzy solution of the intersection problem are manifold. One point is an uncertain measurement of the position and orientation of the human agent, because of which the use of a fuzzy signal and an adequate fuzzy processing seems natural [9,10]. Another aspect is the need for decreasing the computing effort in the case of complex calculations during a very small time interval. System uncertainties and observation noise lead to uncertainties of the intersection estimations.

The objective of this work is the formulation of the crossing/intersection problem by taking into account the uncertainties in human–robot systems, including sensors and motor characteristics. An important aspect is to define permissible uncertainties in a human–robot system for a given uncertainty at a possible intersection of their trajectories. Taking into account the nonlinearities, this is performed by the differential approach and a following analysis of the regarding Gaussian distributions. This approach is compared with the sigma-point transformation, which represents a simplification of the computation and a qualitative extension of the analysis regarding the statistics of the random variables. For broader areas of possible intersections, both methods are extended to fuzzy regions together with different numbers and shapes of fuzzy sets. The most important contributions are as follows:

- An investigation of uncertainties of possible intersection areas originating from sensor noise or system uncertainties.
- A direct and inverse transformation of the error variables at the intersection areas for two input variables (orientation angles) and two output variables (intersection coordinates).
- An extension of the method from two to six input variables (two orientation angles and four position coordinates).
- An exploration of the formulations of fuzzy versions.
- A formulation of the problem by the sigma-point transformation and corresponding comparison of the two methods.

This paper deals with the one-robot one-human trajectory-crossing problem, where small uncertainties in the position and orientation may lead to high uncertainties at the intersection points. The position and orientation of the human and robot are nonlinearly coupled but can be linearized. In the following, the linear part of the nonlinear system is considered in the analysis reported for small variations in the input [11]. Then, the "direct task" is described, meaning that the parameters of the input distribution are transformed to the output distribution parameters. The "inverse task" is also solved, meaning that for the defined output distribution parameters the input parameters are calculated. In this paper, two methods are outlined:

1. The *statistical linearization*, which linearizes the nonlinearity around the operating area at the intersection. The means and standard deviations on the input parameters positions (orientations) are transformed through the linearized nonlinear system to obtain the means and standard deviations of the output parameters (the position of intersection).
2. The *sigma-point transformation*, which calculates the so-called sigma points of the input distribution, including the mean and covariance of the input. The sigma points are directly propagated through the nonlinear system [12–14] to obtain the means and covariance of the output and, with this, the standard deviations of the output (the position of intersection). The advantage of the sigma-point transformation is that it captures the first- and second-order statistics of a random variable, whereas the statistical linearization approximates a random variable only by its first order. However, the computational complexity of the extended Kalman filter (EKF, differential approach) and unscented Kalman filter (UKF, sigma-point approach) is of the same order [13].

This paper is organized as follows. Section 2 describes the related work already conducted on unscented Kalman filters in mobile robot applications. In Section 3, the general intersection problem and its analytical approach is described. Section 4 deals with the transformation/conversion of Gaussian distributions for a two-input–two-output system and for a six-input–two-output system plus the corresponding inverse and fuzzy solutions. In Section 5, the sigma-point approach plus inverse and fuzzy solutions are addressed. Section 6 presents simulations of the *statistical linearization* and the *sigma-point transformation* to show the quality of the input–output conversion of the distributions and the impact of different resolutions of fuzzy approximations on the accuracy of the random variable intersection. Finally, Section 7 concludes this paper with a discussion of the two different approaches and a comparison of the methods.

2. Related Work

The crossing problem for mobile robots has been especially dealt with by [6,7]. Both publications deal with the so-called rendezvous problem whereby the key point is the trajectory planning under time constraints, taking into account the dynamics of the contributing robots. Uncertainties of possible intersection areas that come from sensor noise or system uncertainties are not discussed deeply. A fuzzy-adaptive extended Kalman filter (FAEKF) for the real-time attitude estimation of a mobile robot is proposed in [15] where fuzzy IF–THEN rules-based adaption laws modify the noise covariance matrices of the filter. However, the use of unscented Kalman filters or sigma-point transformation has not been discussed. For the estimation of landmarks, a simultaneous localization and mapping (SLAM) method is presented by [16] where an iterated sigma-point FastSLAM (ISP-FastSLAM) algorithm is proposed to minimize statistical linearization errors through the Gaussian–Newton iteration. A further application is presented by [17] where a walking robot uses sigma-point transformation for state estimation to guarantee stability in the system's hybrid dynamics, which contains continuous and switching parts during movement. In [18], a vision-based SLAM system uses both extended Kalman filters (EKFs) and sigma-point Kalman filter (SPKF) algorithms and showed its superiority over the EKF. The pose estimation of mobile robots is discussed in [19] whereby several filter techniques like the Kalman filter (EKF), the unscented Kalman filter (UKF) and several variants of the particle filter (PF) are compared. It turns out that the UKF (also the sigma-point approach) exhibits almost the same computational cost. In [20], the inter-robot and robot–target correlations are discussed, and unscented transformation-based collaborative self-localization and a target tracking algorithm between robots are proposed. A tutorial on different approaches to exploit the structure of a system's state and measurement models to reduce the computational demand of the algorithms is presented by [21]. In this publication, the computational complexity of different state estimation algorithms is presented, showing the superiority of the sigma-point transformation algorithms.

In all these publications, the problem of obstacle avoidance and/or the crossing problem in the presence of human actors are not taken into account, because of which the present paper is a further contribution to the robot–human interaction problem.

3. Computation of Intersections

The problem can be stated as follows:

A robot and human agent move in a common area according to their tasks or intentions. To avoid collisions, possible intersections of the paths of the agents should be predicted for both the trajectory planning and on-line interactions. To accomplish this, the positions, orientations and intended movements of the robot and human should be estimated as accurately as needed.

In this connection, uncertainties and noise on the random variables' position/orientation $\mathbf{x}_R$, $\mathbf{x}_H$, ϕ_R and ϕ_H of the robot and human have a great impact on the calculation of the expected intersection position $\mathbf{x}_c$. The random variable $\mathbf{x}_c$ is calculated as the crossing point of the extension of the orientation or velocity vectors of the robot and human, which may change during motion depending on the task and current interaction. The task is to calculate the intersection and its uncertainty in the presence of the known uncertainties of the acting agent robot and human.

System noise $\mathbf{w}_R$ and $\mathbf{w}_H$ for the robot and human can be obtained from experiments. The noise $\mathbf{w}_c$ of the "virtual" intersection is composed of the nonlinear transformed noise $\mathbf{w}_R$ and $\mathbf{w}_H$ and some additional noise $\mathbf{v}_c$ that may come from uncertainties of the nonlinear computation of the intersection position $\mathbf{x}_c$ (see Figure 1). In the following, the geometrical relations are described as well as the fuzzy approximations and nonlinear transformations of the random variables $\mathbf{x}_R$, $\mathbf{x}_H$, ϕ_R and ϕ_H.

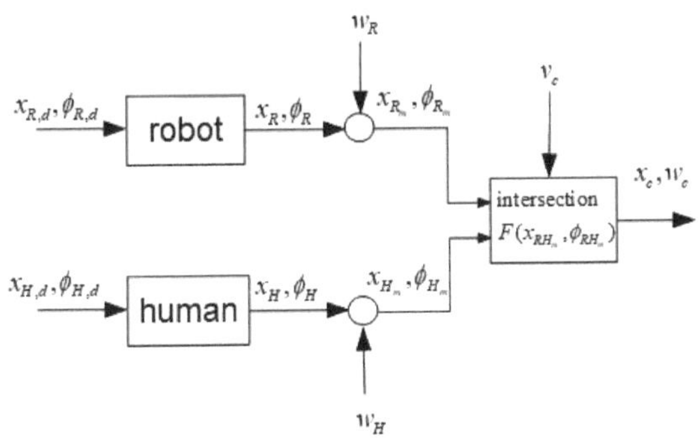

Figure 1. Intersection principle.

3.1. Geometrical Relations

Let the y-axis of the mobile coordinate frame of the robot and human be aligned with their directions of motion. Furthermore, let the orientation angles ϕ_R and ϕ_H of the robot and human be measured from the x-axis of the base frame counterclockwise. Let the intersection (x_c, y_c) of the two linear trajectories $\mathbf{x}_R(t)$ and $\mathbf{x}_H(t)$ in a plane be described by the following relations (see Figure 2):

$$\begin{aligned} x_H &= x_R + d_{RH}\cos(\phi_R + \delta_R) \\ y_H &= y_R + d_{RH}\sin(\phi_R + \delta_R) \\ x_R &= x_H + d_{RH}\cos(\phi_H + \delta_H) \\ y_R &= y_H + d_{RH}\sin(\phi_H + \delta_H) \end{aligned} \tag{1}$$

where $\mathbf{x}_H = (x_H, y_H)$ and $\mathbf{x}_R = (x_R, y_R)$ are the positions of the human and robot and ϕ_H and ϕ_R are their orientation angles, and δ_H and δ_R are the positive angles measured from the y coordinates counterclockwise. The angle at the intersection is $\tilde{\beta} = \pi - \delta_R - \delta_H$. The variables $\mathbf{x}_H, \mathbf{x}_R, \phi_R, \phi_H$ δ_H and ϕ_H δ_R; distance d_{RH}; and angle γ are assumed to be measurable. Angle γ is a bearing angle for the robot-to-human direction measured in base coordinates. If ϕ_H is not directly measurable, then it can be computed by

$$\phi_H = \arcsin((y_H - y_R)/d_{RH}) - \delta_H + \pi \tag{2}$$

The coordinates x_c and y_c of the intersection are computed straightforwardly by [8]

$$\begin{aligned} x_c &= \frac{A - B}{\tan\phi_R - \tan\phi_H} \\ y_c &= \frac{A\tan\phi_H - B\tan\phi_R}{\tan\phi_R - \tan\phi_H} \\ A &= x_R\tan\phi_R - y_R \\ B &= x_H\tan\phi_H - y_H \end{aligned} \tag{3}$$

Rewriting (3) leads to

$$\begin{aligned} x_c &= \left(x_R\frac{\tan\phi_R}{G} - y_R\frac{1}{G}\right) - \left(x_H\frac{\tan\phi_H}{G} - y_H\frac{1}{G}\right) \\ y_c &= \left(x_R\frac{\tan\phi_R\tan\phi_H}{G} - y_R\frac{\tan\phi_H}{G}\right) \\ &- \left(x_H\frac{\tan\phi_H\tan\phi_R}{G} - y_H\frac{\tan\phi_R}{G}\right) \\ G &= \tan\phi_R - \tan\phi_H \end{aligned} \tag{4}$$

After rearranging (4), we observe that $\mathbf{x}_c = (x_c, y_c)^T$ is linear in $\mathbf{x}_{RH} = (x_R, y_R, x_H, y_H)^T$

$$\mathbf{x}_c = A_{RH} \cdot \mathbf{x}_{RH} \tag{5}$$

where

$$A_{RH} = f(\phi_R, \phi_H) = \frac{1}{G} \begin{pmatrix} \tan\phi_R & -1 & -\tan\phi_H & 1 \\ \tan\phi_R \tan\phi_H & -\tan\phi_H & -\tan\phi_R \tan\phi_H & \tan\phi_R \end{pmatrix}$$

This notation is of advantage for further computations, such as the fuzzification of the intersection problem and the transformation of the error distributions.

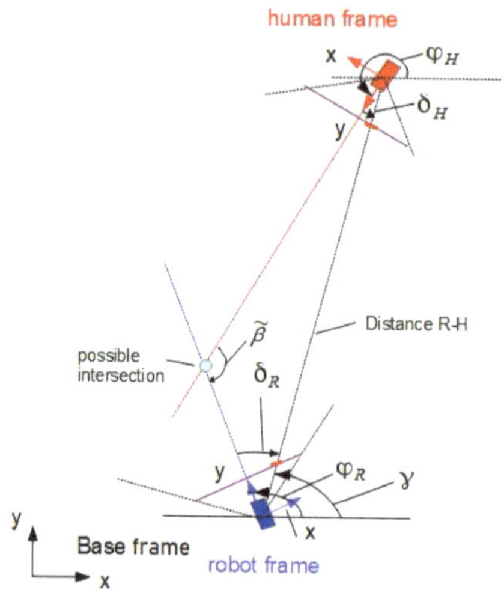

Figure 2. Human–robot scenario: geometry.

3.2. Computation of Intersections—Fuzzy Approach

The fuzzy solution presented in the following is a combination of classical analytical (crisp) methods and rule-based methods in the sense of a Takagi–Sugeno fuzzy rule base. An appropriate choice of the number of fuzzy sets and corresponding fuzzy rules depends strongly on the specific application. In the present case, fuzzy sets are used as the approximation of nonlinear functions. In the following, we introduce a fuzzy rule-based approximation of (5) with $n \times n$ fuzzy rules $R_{i,j}$

$$\begin{aligned} R_{i,j}: \quad IF \quad \phi_R &= \Phi_{Ri} \quad AND \quad \phi_H = \Phi_{Hj} \\ THEN \quad \mathbf{x}_c &= A_{RHi,j} \cdot \mathbf{x}_{RH} \end{aligned} \tag{6}$$

n—the number of fuzzy terms, Φ_{R_i} and Φ_{H_j} for ϕ_R and ϕ_H, with the result

$$\mathbf{x}_c = \sum_{i,j} w_i(\phi_R) w_j(\phi_H) \cdot A_{RHi,j} \cdot \mathbf{x}_{RH} \tag{7}$$

$i, j = 1 \ldots n, w_i(\phi_R), w_j(\phi_H) \in [0,1]$ are normalized membership functions with $\sum_i w_i(\phi_R) = 1$ and $\sum_j w_j(\phi_H) = 1$.

Let the universes of discourse for ϕ_R and ϕ_H be $\phi_R, \phi_H \in [0, 360]$. Furthermore, let these universes of discourse be divided into n partitions (for example, 6) of 60, which leads to 6×6 fuzzy rules. The corresponding membership functions are shown in Figure 3. It turns out that this resolution leads to a poor fuzzy approximation. The approximation quality can be improved by increasing the number of fuzzy sets, which however results in a quadratic increase in the number of fuzzy rules. To avoid an "explosion" of the number of fuzzy rules being computed in one time step, a set of sub-areas covering a small number of rules for each sub-area is defined. Based on the measurements of ϕ_R and ϕ_H, the appropriate sub-area is selected together with a corresponding set of rules (see Figure 4, sub-area A_R, A_H). With this, the number of rules to be activated at one time step of calculation is low, although the total number of rules can be high. At the borderlines between the sub-areas, abrupt changes may occur, which can be avoided by overlapping the sub-areas.

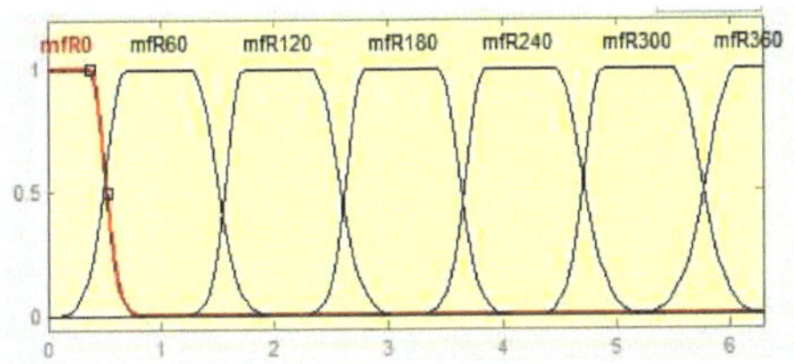

Figure 3. Membership functions for $\Delta\phi_R, \Delta\phi_H = 0 - 360°$.

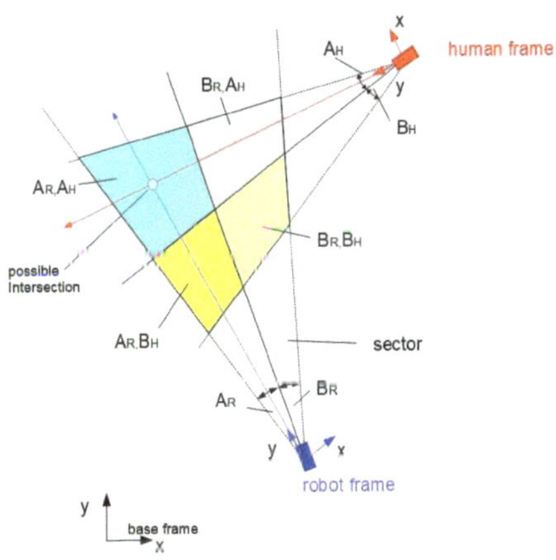

Figure 4. Fuzzy sectors.

3.3. Differential Approach

The positions and orientations of robots and humans are usually corrupted with noise originated from system uncertainties, sensor errors and motor characteristics. These uncertainties become apparent in uncertainties in the crossing/intersection areas of the trajectories of the robot and human. The analysis of uncertainty and noise at $\mathbf{x}_c$ generated by the noise at ϕ_R, ϕ_H and $\mathbf{x}_{RH} = (x_R, y_R, x_H, y_H)^T$ requires a linearization of (4) around the operating points and with this a differential strategy. Let, for simplification, only the orientation angles ϕ_R and ϕ_H be corrupted with noise. In Section 4.3, the positions $\mathbf{x}_{RH} = (x_R, y_R, x_H, y_H)^T$ are taken into account, too.

Differentiating (4) with $\mathbf{x}_{RH} = const.$ yields

$$\mathbf{dx}_c = J \cdot \mathbf{d}\text{Œ}$$
$$\mathbf{d}\text{Œ} = (d\phi_R \quad d\phi_H)^T; \quad J = \begin{pmatrix} J_{11} & J_{12} \\ J_{21} & J_{22} \end{pmatrix} \tag{8}$$

where

$$\begin{aligned} J_{11} &= \begin{pmatrix} -\tan\phi_H & 1 & \tan\phi_H & -1 \end{pmatrix} \frac{\mathbf{x}_{RH}}{G^2 \cdot \cos^2\phi_R} \\ J_{12} &= \begin{pmatrix} \tan\phi_R & -1 & -\tan\phi_R & 1 \end{pmatrix} \frac{\mathbf{x}_{RH}}{G^2 \cdot \cos^2\phi_H} \\ J_{21} &= J_{11} \cdot \tan\phi_H \\ J_{22} &= J_{12} \cdot \tan\phi_R \end{aligned}$$

The following sections deal with the accuracy of the computed intersection in the case of noisy orientation information (see Figure 5).

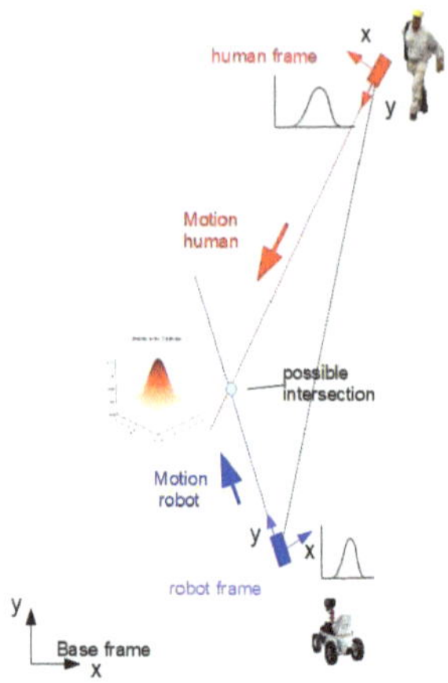

Figure 5. Intersection with noisy orientations.

4. Transformation of Gaussian Distributions

4.1. General Assumptions

Consider a nonlinear system

$$\mathbf{z} = F(\mathbf{x}) \tag{9}$$

where the random variables $\mathbf{x} = (x_1, x_2)^T$ denote the input, $\mathbf{z} = (z_1, z_2)^T$ denotes the output and F denotes a nonlinear transformation. The distribution of the uncorrelated Gaussian distributed components x_1 and x_2 is described by

$$f_{x_1,x_2} = \frac{1}{2\pi\sigma_{x_1}\sigma_{x_2}} exp(-\frac{1}{2}(\frac{e_{x_1}^2}{\sigma_{x_1}^2} + \frac{e_{x_2}^2}{\sigma_{x_2}^2})) \tag{10}$$

where $e_{x_1} = x_1 - \bar{x_1}$, with $\bar{x_1}$—the mean (x_1) and σ_{x_1}—the standard deviation x_1, and $e_{x_2} = x_2 - \bar{x_2}$, with $\bar{x_2}$—the mean (x_2) and σ_{x_2}—the standard deviation x_2.

The goal is as follows: Given the nonlinear transformation (9) and the distribution (10), compute the output signals z_1 and z_2 and their distributions together with their standard deviations and the correlation coefficient. Linear systems transform Gaussian distributions linearly such that the output signals are also Gaussian-distributed. This does not apply for nonlinear systems, but if the input standard deviation is small enough, then a local linear transfer function can be built for which the outputs are Gaussian-distributed. Suppose the input standard deviations are small with respect to the nonlinear function, then the output distribution can be written as follows:

$$f_{z_1,z_2} = \frac{1}{2\pi\sigma_{z_1}\sigma_{z_2}\sqrt{1-\rho_{z_{12}}^2}} \cdot \\ exp(-\frac{1}{2(1-\rho_{z_{12}}^2)}(\frac{e_{z_1}^2}{\sigma_{z_1}^2} + \frac{e_{z_2}^2}{\sigma_{z_2}^2} - \frac{2\rho_{z_{12}}e_{z_1}e_{z_2}}{\sigma_{z_1}\sigma_{z_2}})) \tag{11}$$

$\rho_{z_{12}}$—the correlation coefficient.

4.2. Statistical Linearization, Two Inputs–Two Outputs

Let the nonlinear transformation **F** be described by two smooth transfer functions (see block scheme Figure 6)

$$z_1 = f_1(x_1, x_2) \\ z_2 = f_2(x_1, x_2) \tag{12}$$

where $(x_1, x_2) = (\phi_R, \phi_H)$ and $(z_1, z_2) = (x_c, y_c)$.

The linearization of (12) yields

$$\mathbf{dz} = \tilde{J} \cdot \mathbf{dx} \quad or \quad \mathbf{e_z} = \tilde{J} \cdot \mathbf{e_x} \tag{13}$$

with

$$\mathbf{e_z} = (e_{z_1}, e_{z_2})^T \quad and \quad \mathbf{e_x} = (e_{x_1}, e_{x_2})^T \\ \mathbf{dz} = (dz_1, dz_2)^T \quad and \quad \mathbf{dx} = (dx_1, dx_2)^T \tag{14}$$

$$\tilde{J} = \begin{pmatrix} \partial f_1/\partial x_1, \partial f_1/\partial x_2 \\ \partial f_2/\partial x_1, \partial f_2/\partial x_2 \end{pmatrix} \tag{15}$$

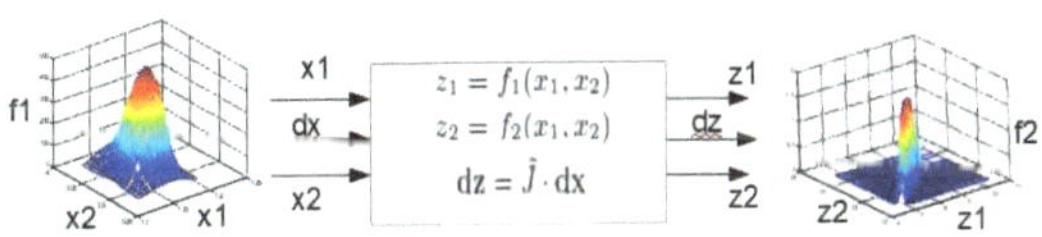

Figure 6. Differential transformation.

4.2.1. Output Distribution

To obtain the density f_{z_1,z_2} (11) of the output signal, we invert (15) and substitute the entries of $\mathbf{e_x}$ into (10). $\tilde{J}$ is invertible if it is positive definite with $|\tilde{J}| > 0$. Otherwise, there exist singularities due to different constellations of the position vector $\mathbf{x}_{RH}$ and/or the orientations ϕ_R and ϕ_H. To find all the singularities requires a further analysis, which is not the content of this paper. However, a simple heuristic leads us to some obvious situations: If $\phi_R = \phi_H$ or $\phi_R = \phi_H + \pi$, then the human and robot would move in parallel either in the same or the opposite direction. On the other hand, one may also obtain diverging trajectories with no crossing.

$$\mathbf{e_x} = J \cdot \mathbf{e_z} \tag{16}$$

with $J = \tilde{J}^{-1}$ and

$$J = \begin{pmatrix} J_{11} & J_{12} \\ J_{21} & J_{22} \end{pmatrix} = \begin{pmatrix} \mathbf{j_{xz}} \\ \mathbf{j_{yz}} \end{pmatrix} \tag{17}$$

where $\mathbf{j_{xz}} = (J_{11}, J_{12})$ and $\mathbf{j_{yz}} = (J_{21}, J_{22})$. The entries J_{ij} are the result of the inversion of $\tilde{J}$. From this substitution, we obtain

$$f_{x_1,x_2} = K_{x_1,x_2} \cdot$$
$$exp(-\frac{1}{2} \cdot \mathbf{e_z}^T \cdot (\mathbf{j_{x_1,z}}^T, \mathbf{j_{x_2,z}}^T) \cdot S_x^{-1} \cdot \begin{pmatrix} \mathbf{j_{x_1,z}} \\ \mathbf{j_{x_2,z}} \end{pmatrix} \cdot \mathbf{e_z}) \tag{18}$$

where $K_{x_1,x_2} = \frac{1}{2\pi\sigma_{x_1}\sigma_{x_2}}$ and

$$S_x^{-1} = \begin{pmatrix} \frac{1}{\sigma_{x_1}^2}, 0 \\ 0, \frac{1}{\sigma_{x_2}^2} \end{pmatrix} \tag{19}$$

The exponent of (18) is rewritten into

$$x_{po} = -\frac{1}{2} \cdot (\frac{1}{\sigma_{x_1}^2}(e_{z_1}J_{11} + e_{z_2}J_{12})^2 + \frac{1}{\sigma_{x_2}^2}(e_{z_1}J_{21} + e_{z_2}J_{22})^2) \tag{20}$$

and furthermore

$$x_{po} = -\frac{1}{2} \cdot [e_{z_1}^2(\frac{J_{11}^2}{\sigma_{x_1}^2} + \frac{J_{21}^2}{\sigma_{x_2}^2}) + e_{z_2}^2(\frac{J_{12}^2}{\sigma_{x_1}^2} + \frac{J_{22}^2}{\sigma_{x_2}^2}) + 2 \cdot e_{z_1}e_{z_2}(\frac{J_{11}J_{12}}{\sigma_{x_1}^2} + \frac{J_{21}J_{22}}{\sigma_{x_2}^2})] \tag{21}$$

Let

$$A = (\frac{J_{11}^2}{\sigma_{x_1}^2} + \frac{J_{21}^2}{\sigma_{x_2}^2}); \qquad B = (\frac{J_{12}^2}{\sigma_{x_1}^2} + \frac{J_{22}^2}{\sigma_{x_2}^2})$$
$$C = (\frac{J_{11}J_{12}}{\sigma_{x_1}^2} + \frac{J_{21}J_{22}}{\sigma_{x_2}^2}) \tag{22}$$

then a comparison of x_{po} in (21) and the exponent in (11) yields

$$\frac{1}{(1-\rho_{z_{12}}^2)}\frac{1}{\sigma_{z_1}^2} = A; \qquad \frac{1}{(1-\rho_{z_{12}}^2)}\frac{1}{\sigma_{z_2}^2} = B$$
$$\frac{-2\rho_{z_{12}}}{(1-\rho_{z_{12}}^2)}\frac{1}{\sigma_{z_1}\sigma_{z_2}} = 2C \tag{23}$$

The standard deviations σ_{z_1} and σ_{z_2} and the correlation coefficient $\rho_{z_{12}}$ yield

$$\rho_{z_{12}} = -\frac{C}{\sqrt{AB}}$$
$$\frac{1}{\sigma_{z_1}^2} = A - \frac{C^2}{B}; \qquad \frac{1}{\sigma_{z_2}^2} = B - \frac{C^2}{A} \tag{24}$$

The result is as follows: If the parameter of the input distribution and the transfer function $F(x, y)$ are known, then the output distribution parameters can be computed straightforwardly.

4.2.2. Fuzzy Solution

To save computing costs in real time, we create a TS fuzzy model that is represented by the rules R_{ij}.

$$\begin{aligned} &R_{ij}: \\ IF \quad x_1 = X_{1i} \quad AND \quad x_2 = X_{2i} \\ &THEN \quad \rho_{z_{12}} = -\frac{C_{ij}}{\sqrt{A_{ij}B_{ij}}} \\ AND \\ &\frac{1}{\sigma_{z_1}^2} = A_{ij} - \frac{C_{ij}^2}{B_{ij}}; \\ AND \\ &\frac{1}{\sigma_{z_2}^2} = B_{ij} - \frac{C_{ij}^2}{A_{ij}} \end{aligned} \tag{25}$$

where X_{1i}, X_{2i} are fuzzy terms for x_1, x_2, and A_{ij}, B_{ij}, C_{ij} are functions of the predefined variables $x_1 = x_{1i}$ and $x_2 = x_{2i}$.

From (25), we derive

$$\begin{aligned} \rho_{z_{12}} &= -\sum_{ij} w_i(x_1)w_j(x_2)\frac{C_{ij}}{\sqrt{A_{ij}B_{ij}}} \\ \frac{1}{\sigma_{z_1}^2} &= \sum_{ij} w_i(x_1)w_j(x_2)(A_{ij} - \frac{C_{ij}^2}{B_{ij}}) \\ \frac{1}{\sigma_{z_2}^2} &= \sum_{ij} w_i(x_1)w_j(x_2)(B_{ij} - \frac{C_{ij}^2}{A_{ij}}) \end{aligned} \tag{26}$$

$w_i(x_1) \in [0,1]$ and $w_j(x_2) \in [0,1]$ are the weighting functions with $\sum_i w_i(x_1) = 1$, $\sum_j w_j(x_2) = 1$.

4.2.3. Inverse Solution

The previous paragraph discussed the direct transformation task: Let the distribution parameters of the input variable be defined and find the corresponding output parameters. However, it might also be useful to solve the inverse task: Given the output parameters (standard deviation and correlation coefficient), find the corresponding input parameters. This solution of the inverse task is similar to those discussed in Section 4.2. The starting points are equations (10) and (11), which describe the distributions of the inputs and outputs, respectively. Then, we substitute (13) into (10) and rename the resulting exponent x_{po_z} into x_{po_x} and discuss the exponent x_{po_x}

$$x_{po_x} = \frac{-1}{2(1-\rho_{z_{12}}^2)}(\mathbf{e_x}^T J^T S_z^{-1} J \mathbf{e_x} - \frac{2\rho_{z_{12}} e_{z_1} e_{z_2}}{\sigma_{z_1}\sigma_{z_2}}) \tag{27}$$

with

$$S_x^{-1} = \begin{pmatrix} \frac{1}{\sigma_{z_1}^2}, 0 \\ 0, \frac{1}{\sigma_{z_2}^2} \end{pmatrix}$$

Now, comparing (27) with the exponent of (10) of the input density, we find that the mixed term in (27) must be zero, from which we obtain the correlation coefficient $\rho_{z_{12}}$ and with this the standard deviations of the inputs

$$\begin{aligned}
\rho_{z_{12}} &= (\frac{\tilde{J}_{11}\tilde{J}_{12}}{\sigma_{z_1}^2} + \frac{\tilde{J}_{21}\tilde{J}_{22}}{\sigma_{z_2}^2})\frac{\sigma_{z_1}\sigma_{z_2}}{(\tilde{J}_{11}\tilde{J}_{22} + \tilde{J}_{12}\tilde{J}_{21})} \\
\frac{1}{\sigma_x^2} &= (\frac{\tilde{J}_{11}^2}{\sigma_{z_1}^2} + \frac{\tilde{J}_{21}^2}{\sigma_{z_2}^2} - \frac{2\rho_{z_{12}}}{\sigma_{z_1}\sigma_{z_2}}\tilde{J}_{11}\tilde{J}_{21})/(1-\rho_{z_{12}}^2) \\
\frac{1}{\sigma_y^2} &= (\frac{\tilde{J}_{12}^2}{\sigma_{z_1}^2} + \frac{\tilde{J}_{22}^2}{\sigma_{z_2}^2} - \frac{2\rho_{z_{12}}}{\sigma_{z_1}\sigma_{z_2}}\tilde{J}_{12}\tilde{J}_{22})/(1-\rho_{z_{12}}^2)
\end{aligned} \tag{28}$$

The detailed development can be found in [22].

4.3. *Six Inputs–Two Outputs*

Consider again the nonlinear system

$$\mathbf{x_c} = F(\mathbf{x}) \tag{29}$$

In the previous subsections, we assumed the positions $\mathbf{x}_R$ and $\mathbf{x}_H$ not to be corrupted with noise. However, taking into account the positions to be random variables, the number of inputs is 6 so that the input vector yields $\mathbf{x} = (x_1, x_2, x_3, x_4, x_5, x_6)^T$ or $\mathbf{x} = (\phi_R, \phi_H, x_R, y_R, x_H, y_H)$ with the output vector $\mathbf{x_c} = (x_c, y_c)^T$.

Furthermore, let the uncorrelated Gaussian-distributed inputs $x_1 \ldots x_6$ be described by the 6-dim density

$$f_{x_i} = \frac{1}{(2\pi)^{6/2}|S_x|^{1/2}} exp(-\frac{1}{2}(\mathbf{e_x}^T S_x^{-1}\mathbf{e_x})) \tag{30}$$

where $\mathbf{e_x} = (e_{x1}, e_{x2}, \ldots, e_{x6})^T$; $\mathbf{e_x} = \mathbf{x} - \bar{\mathbf{x}}$, $\bar{\mathbf{x}}$—the mean($\mathbf{x}$) and S_x—the covariance matrix.

$$S_x = \begin{pmatrix} \sigma_{x_1}^2 & 0 & \ldots & 0 \\ 0 & \sigma_{x_2}^2 & \ldots & 0 \\ \ldots & \ldots & \ldots & \ldots \\ 0 & \ldots & 0 & \sigma_{x_6}^2 \end{pmatrix}$$

According to (11), the output density is described by

$$\begin{aligned}
f_{x_c,y_c} &= \frac{1}{2\pi\sigma_{x_c}\sigma_{y_c}\sqrt{1-\rho^2}} \cdot \\
exp(-\frac{1}{2(1-\rho^2)}&(\mathbf{e}_{\mathbf{x_c}}^{\mathbf{T}} S_c^{-1}\mathbf{e}_{\mathbf{x_c}} - \frac{2\rho e_{x_c} e_{y_c}}{\sigma_{x_c}\sigma_{y_c}}))
\end{aligned} \tag{31}$$

ρ—the correlation coefficient, $\mathbf{e}_{x_c} = (e_{x_c}, e_{y_c})^T$.

After some calculations [23], we find for ρ, $\frac{1}{\sigma_{x_c}^2}$ and $\frac{1}{\sigma_{y_c}^2}$

$$\begin{aligned}
\rho &= -\frac{C}{\sqrt{AD}} \\
\frac{1}{\sigma_{x_c}^2} &= A - \frac{C^2}{D}; \quad \frac{1}{\sigma_{y_c}^2} = D - \frac{C^2}{A}
\end{aligned} \tag{32}$$

with

$$\begin{aligned}
A &= \sum_{i=1}^{6} \frac{1}{\sigma_{x_i}^2} J_{i1}^2; \quad B = \sum_{i=1}^{6} \frac{1}{\sigma_{x_i}^2} J_{i1} J_{i2} \\
C &= \sum_{i=1}^{6} \frac{1}{\sigma_{x_i}^2} J_{i1} J_{i2}; \quad D = \sum_{i=1}^{6} \frac{1}{\sigma_{x_i}^2} J_{i2}^2
\end{aligned} \tag{33}$$

This is the counterpart to the 2-dim input case (24).

4.3.1. Inverse Solution

An inverse solution cannot be uniquely computed due to the undetermined character of the 6-input–2-output system. Therefore, from the required variances at the intersection position (output), the corresponding variances for the positions and orientations of the robot–human or robot–robot (input) cannot be concluded.

4.3.2. Fuzzy Approach

The steps to the fuzzy approach are very similar to those of the 2-input case:

- Define the operation points $\mathbf{x}_i = (x_1, x_2, x_3, x_4, x_5, x_6)_i^T$;
- Compute A_i, B_i and C_i at $\mathbf{x}_i = (x_1, x_2, x_3, x_4, x_5, x_6)_i^T$ from (33);
- Formulate the fuzzy rules R_i according to (25) and (26), $i = 1 \ldots n$.

The number n of rules is computed as follows:

With $l = 6$—the number of fuzzy terms and $k = 6$—the number of inputs, we obtain $n = l^k = 6^6$—the number of rules.

This number of rules is unacceptably high. To limit n to an adequate number, one has to limit the number of inputs and/or fuzzy terms to look for the most influential variables either in a heuristic or systematic way [24]. This however is not the issue to be discussed in this paper.

5. Sigma-Point Transformation

In the following, the estimation/identification of the standard deviations of possible intersection coordinates of trajectories for both the robot–robot and human–robot combinations by means of the sigma-point technique is discussed. The following method is based on the unscented Kalman filter technique where the intersections cannot be directly measured but predicted/computed only. Nevertheless, it is possible to compute the variance of the predicted events, such as possible collisions or planned rendezvous situations, by a direct propagation of statistical parameters—the sigma points—through the nonlinear geometrical relation, which is a result of the crossing of two trajectories. Let $\mathbf{x} = (x_1, x_2)^T$—the input vector and $\mathbf{x_c} = (x_{c1}, x_{c2})^T$—the output vector where for the special case $(x_1, x_2)^T = (\phi_R, \phi_H)^T$ and $(x_{c1}, x_{c2})^T = (x_c, y_c)^T$. The nonlinear relation between $\mathbf{x}$ and $\mathbf{x_c}$ is given by (34)

$$\mathbf{x_c} = \mathbf{F}(\mathbf{x}) \tag{34}$$

For the discrete case, we obtain for the state $\mathbf{x_c}$

$$\mathbf{x_c}(k) = \mathbf{F}(\mathbf{x}(k-1) + \mathbf{w}(k-1)) \tag{35}$$

and for the measured output $\mathbf{z_c}(k)$

$$\mathbf{z_c}(k) = \mathbf{h}(\mathbf{x_c})(k) + \mathbf{v}(k)) \tag{36}$$

where $\mathbf{w}$ and $\mathbf{v}$ are the system noise and measurement noise, respectively. $\mathbf{h}(\mathbf{x_c})$ is the output nonlinearity. Furthermore, let there be the following:

$\bar{\mathbf{x}}(k)$—the mean at time t_k;
$\mathbf{P}(k)$—the covariance matrix;
$\mathbf{x}_0$—the initial state with the known mean $\mu_0 = E(\mathbf{x}_0)$;
$\mathbf{P}_0(k) = E[(\mathbf{x}_0 - \mu_0)(\mathbf{x}_0 - \mu_0)^T]$.

5.1. Selection of Sigma Points

Sigma points are the selected parameters of a given error distribution of a random variable. Sigma points lie along the major eigen-axes of the covariance matrix of the random variable. The height of each sigma point (see Figure 7) represents its relative weight W^j used in the following selection procedure.

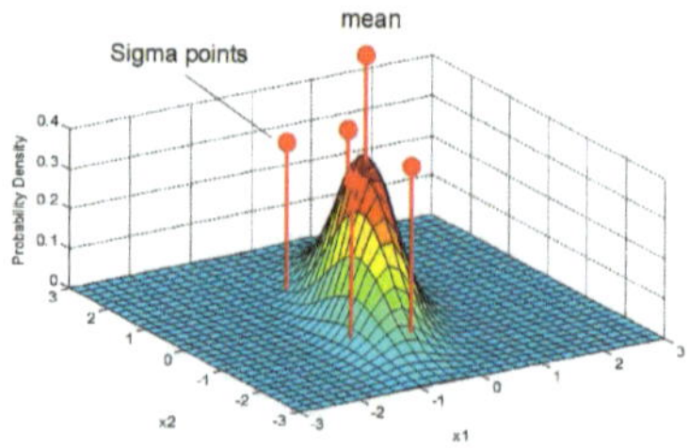

Figure 7. Sigma points for a 2-dim Gaussian random variable.

Let $\mathbf{X}(k-1)$ be a set of $2n+1$ sigma points where n is the dimension of the state space (in our example, $n = 2$).

$$\mathbf{X}(k-1) = \{(\mathbf{x}^j(k-1), W^j) | j = 0 \ldots 2n\} \tag{37}$$

Consider the following selection of sigma points

$$\begin{aligned} \mathbf{x}^0(k-1) &= \bar{\mathbf{x}}(k-1) \\ & -1 < W^0 < 1 \qquad\qquad (38) \\ W^0 &= \frac{\lambda}{n+\lambda}; \quad \lambda = \alpha^2(n+\kappa) - n \\ \mathbf{x}^i(k-1) &= \bar{\mathbf{x}}(k-1) + \sqrt{(\frac{n}{1-W^0}\mathbf{P}(k-1))}; \quad i = 1 \ldots n \\ \mathbf{x}^i(k-1) &= \bar{\mathbf{x}}(k-1) - \sqrt{(\frac{n}{1-W^0}\mathbf{P}(k-1))}; \quad i = (n+1) \ldots 2n \\ W^j &= \frac{1-W^0}{2n} \end{aligned} \tag{39}$$

under the following condition

$$\sum_{j=0}^{2n} W^j = 1 \tag{40}$$

α and κ are scaling factors. A usual choice is $\alpha = 10^{-2}$ and $\kappa = 0$. $\sqrt{\frac{n}{1-W^0}\mathbf{P}(k-1)}$ is the row/column of the matrix square root of $\frac{n}{1-W^0}\mathbf{P}$. The square root of a matrix $\mathbf{P}$ is the solution $\mathbf{S}$ for $\mathbf{P} = \mathbf{S} \cdot \mathbf{S}$, which is obtained by Cholesky factorization.

5.2. Model Forecast Step

To go on with the UKF, the following step is devoted to the model forecast. In this way, the sigma points $\mathbf{x}^j(k)$ are propagated through the nonlinear process model

$$\mathbf{x}_c^{f,j}(k) = \mathbf{F}(\mathbf{x}^j(k-1)) \tag{41}$$

where the superscript f means "forecast". From these transformed and forecasted sigma points, the mean and covariance for the forecast value of $\mathbf{x}_c(k)$ are

$$\begin{aligned} \mathbf{x}_c^f(k) &= \sum_{j=0}^{2n} W^j \mathbf{x}_c^{f,j}(k) \\ \mathbf{P}^f(k) &= \sum_{j=0}^{2n} W^j (\mathbf{x}_c^{f,j}(k) - \mathbf{x}_c^f(k))(\mathbf{x}_c^{f,j}(k) - \mathbf{x}_c^f(k))^T \end{aligned} \tag{42}$$

5.3. Measurement Update Step

In this step, the sigma points are propagated through the nonlinear observation model

$$\mathbf{z}_c^{f,j}(k) = \mathbf{h}(\mathbf{x}_c^j(k-1)) \tag{43}$$

from which we obtain the mean and covariance (innovation covariance)

$$\mathbf{z}_c^f(k-1) = \sum_{j=0}^{2n} W^j \mathbf{z}_c^{f,j}(k-1)$$
$$Cov(\tilde{\mathbf{z}}_c^f(k-1)) = \tag{44}$$
$$\sum_{j=0}^{2n} W^j (\mathbf{z}_c^{f,j}(k-1) - \mathbf{z}_c^f(k-1)) \times$$
$$(\mathbf{z}_c^{f,j}(k-1) - \mathbf{z}_c^f(k-1))^T + \mathbf{R}(k)$$

and the cross-covariance

$$Cov(\tilde{\mathbf{x}}_c^f(k), \tilde{\mathbf{z}}_c^f(k-1)) =$$
$$\sum_{j=0}^{2n} W^j (\mathbf{x}_c^{f,j}(k) - \mathbf{x}_c^f(k))(\mathbf{z}_c^{f,j}(k-1) - \mathbf{z}_c^f(k-1))^T \tag{45}$$

5.4. Data Assimilation Step

In this step, the forecast information is combined with the new information from the output $\mathbf{z}(k)$ from which we obtain, with the Kalman filter, gain $\mathbf{K}$

$$\hat{\mathbf{x}}_c(k) = \mathbf{x}_c^f(k) + \mathbf{K}(k)(\mathbf{z}_c(k) - \mathbf{z}_c^f(k-1)) \tag{46}$$

The gain $\mathbf{K}$ is given by

$$\mathbf{K}(k) = Cov(\tilde{\mathbf{x}}_c^f(k), \tilde{\mathbf{z}}_c^f(k-1)) \cdot Cov^{-1}(\tilde{\mathbf{z}}_c^f(k-1)) \tag{47}$$

and the posterior covariance is updated by

$$\mathbf{P}(k) = \mathbf{P}^f(k) - \mathbf{K}(k) \cdot Cov(\tilde{\mathbf{z}}_c^f(k-1))\mathbf{K}^T(k) \tag{48}$$

Usually, it is sufficient to compute the mean and variance for the output/state $\mathbf{x}_c$ of the nonlinear static system $\mathbf{F}(\mathbf{x})$. In this case, it is possible to stop further computing at Equation (42), meaning to rather calculate the transformed sigma points $\mathbf{x}_c^{f,j}$ and develop the specific output means and variances from (41) and (42). In this connection, it is enough to substitute the covariance matrix $\mathbf{Q}$ into (38) instead of $\mathbf{P}$. One advantage of the sigma-point approach prior to statistical linearization is the easy scalability to multi-dimensional random variables.

For the intersection problem, there are 2 cases:

1. The 2 inputs, 2 outputs (2 orientation angles and 2 crossing coordinates);
2. The 6 inputs, 2 outputs (2 orientation angles and 4 position coordinates, and 2 crossing coordinates).

For the statistical linearization (method 1), the step from the 2 inputs–2 outputs case to the (6,2)-case is computationally more costly than that for the sigma-point approach (method 2), (see Equations (20)–(24) versus Equations (37) and (40)–(42)).

5.5. Sigma Points—Fuzzy Solutions

In order to lower the computing effort, the application of the TS fuzzy interpolation may be a solution, which will be shown in the following. Having a look at the two-dimensional problem, we can see a nonlinear propagation of the input sigma points through a nonlinear function $\mathbf{F}$. Let $\mathbf{x}^j$ be the two-dimensional "input" sigma points

$$\mathbf{x}^j = (x_1^j, x_2^j)^T \tag{49}$$

or for the special case "intersection"

$$\mathbf{x}^j = (\phi_R^j, \phi_H^j)^T \tag{50}$$

The propagation through $\mathbf{F}$ leads to the "output" sigma points

$$\mathbf{x}_c^{f,j}(k) = \mathbf{F}(\mathbf{x}^j(k-1)) \tag{51}$$

or for the special case

$$\begin{aligned} \mathbf{x}_c^{f,j}(k) &= \mathbf{F}(x_1^j(k-1), x_2^j(k-1)) = \\ & \mathbf{F}(\phi_R^j(k-1), \phi_H^j(k-1)) \end{aligned} \tag{52}$$

The special nonlinear function $\mathbf{F}$ is described by (see (5))

$$\mathbf{x}_c = \mathbf{A}_{RH}(\phi_R, \phi_H) \cdot \mathbf{x}_{RH} \tag{53}$$

where $\mathbf{A}_{RH}$ is a nonlinear matrix (6) linearly combined with the position vector $\mathbf{x}_{RH} = (x_R, y_R, x_H, y_H)^T$.

A fuzzification aims at $\mathbf{A}_{RH}$:

$$\begin{aligned} \mathbf{F}^{fuzz}(\phi_R, \phi_H) = \mathbf{A}_{RH}^{fuzz} \cdot \mathbf{x}_{RH} = \\ \sum_{l_1, l_2}^{m} w^{l_1}(\phi_R) w^{l_2}(\phi_H) \cdot \mathbf{A}_{RH}(\phi_R^{l_1}, \phi_H^{l_2}) \cdot \mathbf{x}_{RH} \end{aligned} \tag{54}$$

Applied to the sigma points (ϕ_R^j, ϕ_H^j), we obtain a TS fuzzy model described by the following rules R_{l_1,l_2}

$$\begin{aligned} & R_{l_1,l_2}: \\ & IF \quad \phi_R^j = \Phi_{Rl_1}^j \quad AND \quad \phi_H^j = \Phi_{Hl_2}^j \\ & THEN \quad \mathbf{x}_c^{f,j} = \mathbf{A}_{RH}(\phi_R^{l_1,j}, \phi_H^{l_2,j}) \cdot \mathbf{x}_{RH} \end{aligned} \tag{55}$$

where $\Phi_{Rl_1}^j, \Phi_{Hl_2}^j$ are fuzzy terms for ϕ_R^j, ϕ_H^j; the matrices $\mathbf{A}_{RH}$ are functions of the predefined variables ϕ_R^j and ϕ_H^j. This set of rules leads to the result

$$\begin{aligned} \mathbf{x}_c^{f,j} = \mathbf{F}^{fuzz}(\phi_R^j, \phi_H^j) = \\ \sum_{l_1, l_2}^{m} w^{l_1}(\phi_R^j) w^{l_2}(\phi_H^j) \cdot \mathbf{A}_{RH}(\phi_R^{l_1,j}, \phi_H^{l_2,j}) \cdot \mathbf{x}_{RH} \end{aligned} \tag{56}$$

$w^{l_1}(\phi_R^j) \in [0,1]$ and $w^{l_2}(\phi_H^j) \in [0,1]$ are weighting functions with $\sum_{l_1} w^{l_1} = 1, \sum_{l_2} w^{l_2} = 1$. The advantage of this approach is that the $l_1 \times l_2$ matrices $\mathbf{A}_{RH}^{l_1,l_2,j} = \mathbf{A}_{RH}(\phi_R^{l_1,j}, \phi_H^{l_2,j})$ can be computed off-line. Then, the calculation of the mean and covariance matrix is obtained by

$$
\begin{aligned}
\mathbf{x}_c^f(k) &= \sum_{j=0}^{2n} \mathbf{W}^j \mathbf{x}_c^{f,j}(k) \\
\mathbf{P}^f(k) &= \sum_{j=0}^{2n} \mathbf{W}^j \tilde{\mathbf{x}}_c^{f,j}(k)(\tilde{\mathbf{x}}_c^{f,j}(k))^T \\
\tilde{\mathbf{x}}_c^{f,j} &= \mathbf{x}_c^{f,j} - \mathbf{x}_c^f
\end{aligned}
\tag{57}
$$

From the covariance $\mathbf{P}^f$, the variances $\sigma_{cxx}, \sigma_{cyy}, \sigma_{cxy}$ can be obtained

$$
\begin{aligned}
\sigma_{cxx} &= E((x_c^f - \bar{x}_c^f)^2)) \\
\sigma_{cyy} &= E((y_c^f - \bar{y}_c^f)^2)) \\
\sigma_{cxy} = \sigma_{cyx} &= E((x_c^f - \bar{x}_c^f) \cdot (y_c^f - \bar{y}_c^f))
\end{aligned}
\tag{58}
$$

5.6. Inverse Solution

The inverse solution for the sigma-point approach is much easier to obtain than that for the statistical linearization method. Starting from Equation (34), we build the inverse function

$$
\mathbf{x} = \mathbf{F}^{-1}(\mathbf{x}_c) \tag{59}
$$

on the condition that $\mathbf{F}^{-1}$ exists. Then, the covariance matrix $\mathbf{P}$ is defined in correspondence to the required variances $\sigma_{cxx}, \sigma_{cyy}$ and σ_{cxy}. The following steps correspond to Equations (34)–(42). The position vector $\mathbf{x}_{RH}$ is assumed to be known. The inversion of $\mathbf{F}$ requires a linearization of $\mathbf{x}_{RH}$ and a starting point to obtain a stable convergence to the inverse $\mathbf{F}^{-1}$. The result is the mean $\mathbf{x}$ and the covariance $\mathbf{Q}$ at the input. A reliable inversion is only possible for the 2-input–2-output case.

5.7. Six-Inputs–Two-Outputs

This case works exactly as the 2-input–2-output case along with Equations (34)–(42) due to the fact that the computation of the sigma points (38)–(40) and the propagation through the nonlinearity $\mathbf{F}$ automatically include the input and output dimensions.

6. Simulation Results

The following simulations show the results of the uncertainties of the predicted intersections based on statistical linearization and sigma-point transformation. For both methods, identical parameters are employed for comparison reasons (see Figure 2). The position/orientation of the robot and human are given by the following:

$\mathbf{x}_R = (x_R, y_R)^T = (2,0)^T$m;
$\mathbf{x}_H = (x_H, y_H)^T = (4,10)^T$m;
$\phi_R = 1.78$ rad = 102°;
$\phi_H = 3.69$ rad = 212°.
ϕ_R and ϕ_H are corrupted by Gaussian noise with standard deviations (std) of $\sigma_{\phi_R} = \sigma_{x_1} = 0.02$ rad, (= 1.1°), and $\sigma_{\phi_H} = \sigma_{x_2} = 0.02$ rad, (=1.1°) .

6.1. Statistical Linearization

Table 1 shows a comparison of the non-fuzzy method with the fuzzy approach using sectors of 60°, 30°, 15°, 7.5° of the unit circle for the orientations of the robot and human. The notations in Table 2 are as follows: σ_{xc}—std-computed, σ_{xm}—std-measured, etc. As expected, we see that higher resolutions lead to a better match between the fuzzy and analytical approach. Furthermore, the match between the measured and calculated values depends on the form of membership functions (MFS). For example, low input standard deviations (0.02 rad) show a better match for Gaussian membership functions, and higher input standard deviations (0.05 rad = 2.9°) require Gaussian bell-shaped membership functions, which comes from different smoothing effects (see columns 4 and 5 in Table 2).

A comparison of the control surfaces and corresponding measurements x_{cm}, y_{cm} (black and red dots) is depicted in Figures 8–10. Figure 8 shows the control surface of x_c and y_c for the non-fuzzy case (4). The control surfaces of the fuzzy approximations (7) for the 30° and 7.5° sectors are shown in Figures 9 and 10. The resolution 30° (Figure 9) shows a very high deviation compared to the non-fuzzy approach (Figure 8), which decreases further down to the resolution 7.5° (Figure 10). This explains the high differences between the measured and computed standard deviations and correlation coefficients, in particular for sector sizes of 30° and higher.

Table 1. Standard deviations and fuzzy and non-fuzzy results.

Input Std	0.02 Gauss, Bell Shaped (GB)				Gauss	0.05 GB
sector size/ °	60°	30°	15°	**7.5°**	**7.5°**	**7.5°**
non-fuzz σ_{xc}	0.143	0.140	0.138	**0.125**	**0.144**	**0.366**
fuzz σ_{xc}	0.220	0.184	0.140	**0.126**	**0.144**	**0.367**
non-fuzz σ_{xm}	0.160	0.144	0.138	**0.126**	**0.142**	**0.368**
fuzz σ_{xm}	0.555	0.224	0.061	**0.225**	**0.164**	**0.381**
non-fuzz σ_{y_c}	0.128	0.132	0.123	**0.114**	**0.124**	**0.303**
fuzz σ_{y_c}	0.092	0.087	0.120	**0.112**	**0.122**	**0.299**
non-fuzz σ_{y_m}	0.134	0.120	0.123	**0.113**	**0.129**	**0.310**
fuzz σ_{y_m}	0.599	0.171	0.034	**0.154**	**0.139**	**0.325**
non-fuzz ρ_{xy_c}	0.576	0.541	0.588	**0.561**	**0.623**	**0.669**
fuzz ρ_{xy_c}	−0.263	0.272	0.478	**0.506**	**0.592**	**0.592**
non-fuzz ρ_{xy_m}	0.572	0.459	0.586	**0.549**	**0.660**	**0.667**
fuzz ρ_{xy_m}	0.380	0.575	0.990	**0.711**	**0.635**	**0.592**

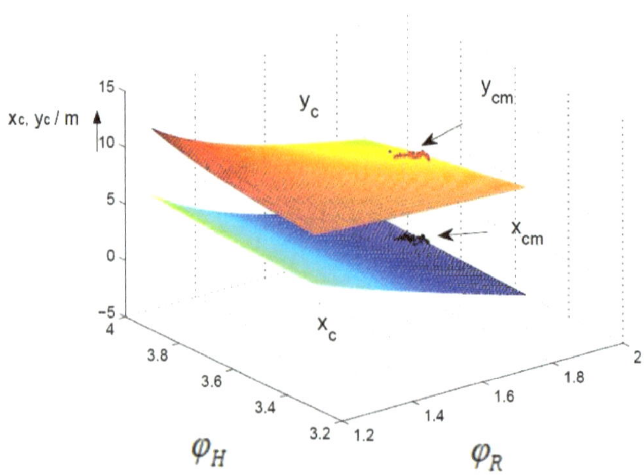

Figure 8. Control surface non-fuzzy, units of ϕ_R and ϕ_H in rad.

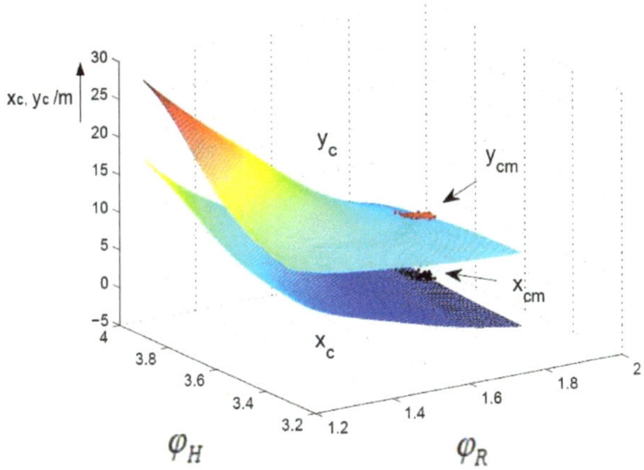

Figure 9. Control surface fuzzy, 30°, units of ϕ_R and ϕ_H in rad.

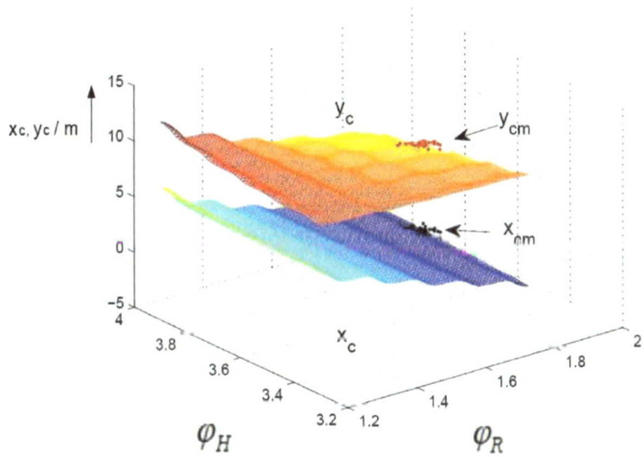

Figure 10. Control surface fuzzy, 7.5°, units of ϕ_R and ϕ_H in rad.

6.2. *Sigma-Point Method*

Two-inputs–two-outputs:

The simulation of the sigma-point method is based on a Matlab implementation of an unscented Kalman filter by [25]. The first example deals with the 2-inputs–2-outputs case in which only the orientations are taken into account, but the disturbances of the positions of the robot and human are not part of the sigma-point calculation. A comparison between the computed and measured covariance shows a very good match. The same holds for the standard deviations $\sigma_{xc}, \sigma_{y_c}$. A comparison with the statistical linearization shows a good match as well (see Table 2, rows 1 and 2).

A view at the sigma points presents the following results: Figure 11 shows the two-dimensional distribution of the orientation angles (ϕ_R, ϕ_H) and the corresponding sigma points $s_1, \ldots, s_5$ where s_1 denotes the mean value. Figure 12 shows the two-dimensional distribution of the intersection coordinates (x_c, y_c) with the sigma points $S_1, \ldots, S_5$. S_1 denotes the mean value and $S_1, \ldots, S_5$ are distributed in such a way that the s_i are transformed into S_i, $i = 1 \ldots 5$. From both figures, an optimal selection of both $s_1, \ldots, s_5$ and $S_1, \ldots, S_5$ can be observed, which results in a good match of the computed and measured standard deviations σ_{x_c}.

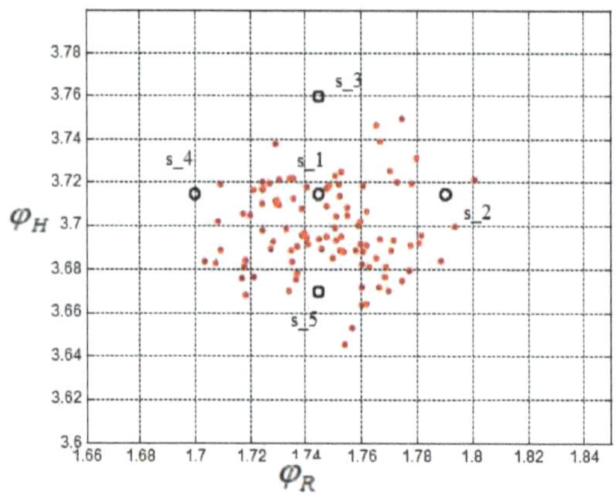

Figure 11. *Sigma points*, input, units of ϕ_R and ϕ_H in rad.

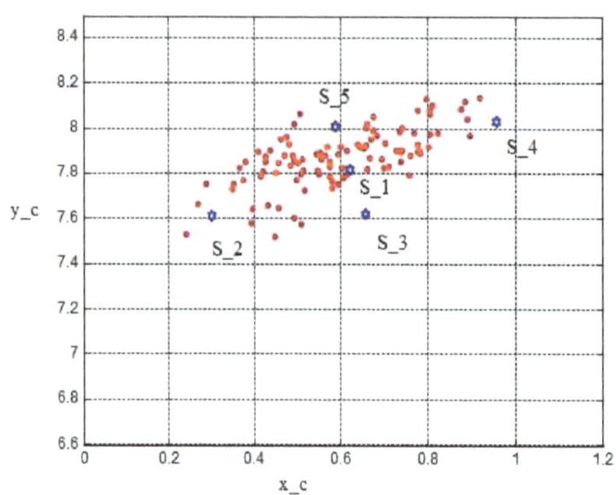

Figure 12. *Sigma points*, output.

Six-inputs–two-outputs:

The 6-inputs–2-outputs example shows that the additional consideration of 4 input position coordinates with $\sigma_{\mathbf{x}_R} = 0.02$ leads to similar results both for the computed and measured covariances and between the sigma-point method and statistical linearization (see $P(7,7) = \sigma_{x_c}^2$, $P(8,8) = \sigma_{y_c}^2$ and $covar(7,7) = \sigma_{x_m}^2$, $covar(8,8) = \sigma_{y_m}^2$, and $\sigma_{x_c}^2$—computed, and $\sigma_{x_m}^2$—the measured variation). Table 2 shows the covariance submatrix considering the output positions only.

Computed covariance:

$$P = 10^{-1} \times \begin{pmatrix} 0.004 & -0.000 & -0.000 & 0.000 & -0.000 & -0.000 & -0.030 & -0.018 \\ -0.000 & 0.004 & 0.000 & -0.000 & -0.000 & -0.000 & 0.003 & -0.017 \\ -0.000 & 0.000 & 0.004 & 0.000 & -0.000 & -0.000 & 0.004 & 0.002 \\ 0.000 & -0.000 & 0.000 & 0.004 & -0.000 & -0.000 & 0.001 & 0.000 \\ -0.000 & -0.000 & -0.000 & -0.000 & 0.004 & 0.000 & 0.000 & -0.002 \\ -0.000 & -0.000 & -0.000 & -0.000 & 0.000 & 0.004 & -0.001 & 0.004 \\ -0.030 & 0.003 & 0.004 & 0.001 & 0.000 & -0.001 & \mathbf{0.235} & \mathbf{0.127} \\ -0.018 & -0.017 & 0.002 & 0.000 & -0.002 & 0.004 & \mathbf{0.127} & \mathbf{0.165} \end{pmatrix} \quad (60)$$

$$\sigma_{xc} = 0.153, \quad \sigma_{y_c} = 0.122$$

Measured covariance:

$$
cover = 10^{-1} \times \begin{pmatrix} 0.004 & 0.000 & 0.000 & 0.000 & 0.000 & -0.000 & -0.028 & -0.020 \\ 0.000 & 0.004 & 0.000 & 0.001 & 0.000 & -0.000 & 0.000 & -0.020 \\ 0.000 & 0.000 & 0.004 & -0.000 & 0.001 & -0.001 & 0.003 & 0.001 \\ 0.000 & 0.001 & -0.000 & 0.004 & -0.000 & -0.000 & -0.000 & -0.003 \\ 0.000 & 0.000 & 0.001 & -0.000 & 0.005 & -0.000 & -0.001 & -0.006 \\ -0.000 & -0.000 & -0.001 & -0.000 & -0.000 & 0.005 & -0.000 & 0.005 \\ -0.028 & 0.000 & 0.003 & -0.000 & -0.001 & -0.000 & \mathbf{0.213} & \mathbf{0.131} \\ -0.020 & -0.020 & 0.001 & -0.003 & -0.006 & 0.005 & \mathbf{0.131} & \mathbf{0.182} \end{pmatrix} \quad (61)
$$

$$
\sigma_{xc} = 0.145, \quad \sigma_{y_c} = 0.134
$$

Two-inputs–two-outputs, direct and inverse solution

The next example shows the computation of the direct and inverse cases. In the direct case, we obtain again similar values between the computed and measured covariances and, with this, the standard deviations. The results of the inverse solution lead to similar values of the original inputs (orientations $x_1 = \phi_R, x_2 = \phi_H$) (see Table 2). The simulations of the fuzzy versions showed the same similarities and can therefore be left out here.

Table 2. Covariances, standard deviations—computed and measured.

Outputs	Covariance, Computed	Covariance, Measured	σ_{xc}, Comp/Meas	σ_{y_c}, Comp/Meas
2 inputs	$P = \begin{pmatrix} 0.0213 & 0.0114 \\ 0.0114 & 0.0159 \end{pmatrix}$	$covar = \begin{pmatrix} 0.0264 & 0.0146 \\ 0.0146 & 0.0166 \end{pmatrix}$	0.145/0.144	0.126/0.134
2 inputs, stat. lin.	-	-	0.144/0.142	0.124/0.129
6 inputs	$P = \begin{pmatrix} 0.0235 & 0.0127 \\ 0.0127 & 0.0165 \end{pmatrix}$	$covar = \begin{pmatrix} 0.0213 & 0.0131 \\ 0.0131 & 0.0182 \end{pmatrix}$	0.135/0.145	0.122/0.134
Direct solution	$P = \begin{pmatrix} 0.0234 & 0.0133 \\ 0.0133 & 0.0151 \end{pmatrix}$	$covar = \begin{pmatrix} 0.0264 & 0.0146 \\ 0.0146 & 0.0166 \end{pmatrix}$	0.152/0.162	0.128/128
Inverse solution	$P = 10^{-3} \times \begin{pmatrix} 0.4666 & 0.0522 \\ 0.0522 & 0.4744 \end{pmatrix}$	$covar = 10^{-3} \times \begin{pmatrix} 0.4841 & 0.0190 \\ -0.0190 & 0.396 \end{pmatrix}$	0.0215/0.0220	0.0217/0.0190

Two-inputs–two-outputs, moving robot–human

The next example deals with the robot and human in motion. Figure 13 shows the positions and orientations of the robot and human at selected time steps $t_1 \ldots t_5$ and the development of the corresponding intersections $\mathbf{x}_c$.

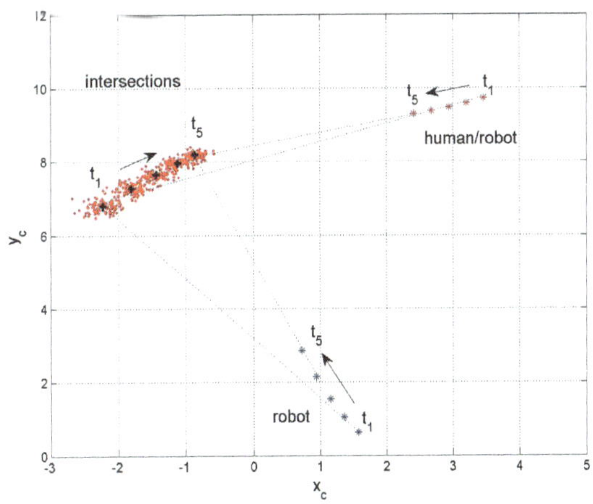

Figure 13. Moving robot and human.

Figure 14 shows the corresponding time plot. The time steps $t_1 \dots t_5$ are taken at 0.58 s, $\dots$, 4.58 s with a time distance of 1 s, which is 25 time steps of 0.04 s each. The robot and human start at
$\mathbf{x}_R = (x_R, y_R)^T = (2,0)^T \mathrm{m}$
$\mathbf{x}_H = (x_H, y_H)^T = (4,10)^T \mathrm{m}$
with the velocities
$\dot{x}_R(k) = -0.21\ \mathrm{m/s}$;
$\dot{y}_R(1) = +0.24\ \mathrm{m/s}$;
$\dot{x}_H(k) = -0.26\ \mathrm{m/s}$;
$\dot{y}_H(1) = -0.24\ \mathrm{m/s}$.
k is the time step.

The x components of the velocities $\dot{x}_R(k)$ and $\dot{x}_H(k)$ stay constant during the whole simulation.

The y components change their velocities with constant factors

$$\begin{aligned} \dot{y}_R(k+1) &= K_R \cdot \dot{y}_R(k) \\ \dot{y}_H(k+1) &= K_H \cdot \dot{y}_H(k) \end{aligned}$$

where $K_R = 1.2$ and $K_H = 0.9$. The orientation angles start with the following:
$\phi_R = 1.78$ rad;
$\phi_H = 3.69$ rad.

They change their values every second according to the direction of motion.

From both plots, one observes an expected decrease in the output standard deviations for a mutual decrease in their distances to the specific intersection and a good match between the computed and measured values $\mathbf{x}_c$ (see Table 3). With the information about the distance of the robot and the standard deviation from and at the expected intersection, respectively, it becomes possible to plan either an avoidance strategy or mutual cooperation between the robot and human.

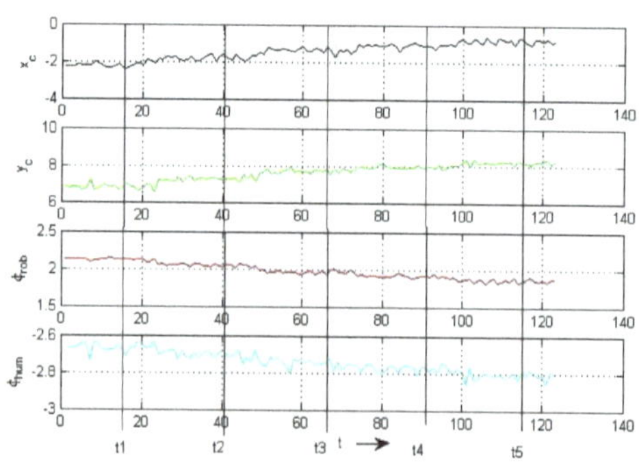

Figure 14. Time plot, robot and human.

Table 3. Covariances, standard deviations—computed and measured, moving robot–human.

Outputs	Covariance, Computed	Covariance, Measured	σ_{xc}, Comp/Meas	σ_{y_c}, Comp/Meas
t_1	$P = \begin{pmatrix} 0.0220 & 0.0017 \\ 0.0017 & 0.0163 \end{pmatrix}$	$covar = \begin{pmatrix} 0.0246 & -0.0002 \\ -0.0002 & 0.0202 \end{pmatrix}$	0.148/0.156	0.127/0.142
t_2	$P = \begin{pmatrix} 0.0198 & 0.0023 \\ 0.0023 & 0.0138 \end{pmatrix}$	$covar = \begin{pmatrix} 0.0222 & 0.0018 \\ 0.0018 & 0.0153 \end{pmatrix}$	0.140/0.148	0.117/0.123
t_3	$P = \begin{pmatrix} 0.0168 & 0.0030 \\ 0.0030 & 0.0107 \end{pmatrix}$	$covar = \begin{pmatrix} 0.0140 & 0.0040 \\ 0.0040 & 0.0088 \end{pmatrix}$	0.129/0.118	0.103/0.093
t_4	$P = \begin{pmatrix} 0.0151 & 0.0029 \\ 0.0029 & 0.0083 \end{pmatrix}$	$covar = \begin{pmatrix} 0.0127 & 0.0014 \\ 0.0014 & 0.0073 \end{pmatrix}$	0.122/0.112	0.091/0.085
t_5	$P = \begin{pmatrix} 0.0125 & 0.0023 \\ 0.0023 & 0.0061 \end{pmatrix}$	$covar = \begin{pmatrix} 0.0102 & 0.0030 \\ 0.0030 & 0.0056 \end{pmatrix}$	0.112/0.101	0.078/0.074

7. Summary and Conclusions

The content of this work is the prediction of encounter situations of mobile robots and human agents in shared areas by analyzing planned/intended trajectories in the presence of uncertainties and system and observation noise. In this context, the problem of intersections of trajectories with respect to system uncertainties and Gaussian noise of the position and orientation of the agents involved is discussed. The problem is addressed by two methods: the statistical linearization of distributions and the sigma-point transformation of the distribution parameters. The positions and orientations of the robot and human are corrupted with Gaussian noise represented by the parameters' mean and standard deviation. The goal is to calculate the mean and standard deviation/variation at the intersection via the nonlinear relation between the positions/orientations of the robot and human, on the one hand, and the position of the intersection of their intended trajectories, on the other hand.

This analysis is realized by the statistical linearization of the nonlinear relation between the statistics of the robot and human (input) and the statistics of the intersection (output). The output results are the mean and standard deviation of the intersection as functions of the input parameters' mean and standard deviation of the positions and orientations of robot and human. This work is first carried out for two-input–two-output relations (two orientations of the robot–human and two intersection coordinates) and then for six inputs–two outputs (two orientations and four position coordinates of the robot–human and two intersection coordinates). These cases were extended to their fuzzy versions by different Takagi–Sugeno (TS) fuzzy approximations and compared with the non-fuzzy case. Up to a certain resolution, the approximation works as accurately as the original non-fuzzy version. For the two-input–two-output case, an inverse solution is derived, except for the six-input–two-output case because of the undetermined nature of the differential input–output relation.

The sigma-point transformation aims at transforming/propagating distribution parameters—the sigma points—directly through nonlinearities. The transformed sigma points are then converted into the distribution parameters' mean and covariance matrix. The sigma-point transformation is closely connected to the unscented Kalman filter, which is used in the example of the robot and human in motion. The specialty of the example is a computed virtual system output ("observation")—the intersection of two intended trajectories—where the corresponding output uncertainty is a sum of the transformed position/orientation noise and the computational uncertainty from the fuzzy approximation. In total, the comparison between the computed and measured covariances shows a very good match and the comparison with the statistical linearization shows good coincidences as well. Both the sigma-point transformation and the differential statistical linearization scales for more than two variables linearly. Their computational complexity is in the same order [13]. However, if the model is nonlinear, then the differential linearization (EKF) serves as the first-order or second-order approximating estimator. If the system is highly nonlinear, the EKF may diverge and the sigma-point approach produces typically better results. In summary, a prediction of the accuracy of human–robot trajectories using the methods presented in this work increases the performance of human–robot collaboration and human safety. In future work, this method can be used for robot–human scenarios in factory workshops and for robots working in complicated environments like rescue operations in cooperation with human operators.

Author Contributions: R.P. developed the methods and implemented the simulation programs. A.J.L. conceived the project and gave advice and support. All authors have read and agreed to the published version of the manuscript.

Funding: This work has received funding from the European Union's Horizon 2020 research and innovation programme under grant agreement No. 101017274 (DARKO).

Institutional Review Board Statement: Not applicable.

Informed Consent Statement: Not applicable.

Data Availability Statement: Data are contained within the article.

Conflicts of Interest: The authors declare that they have no competing interests.

References

1. Khatib, O. Real-time obstacle avoidance for manipulators and mobile robots. In Proceedings of the IEEE International Conference on Robotics and Automation, St. Louis, MI, USA, 25–28 March 1985; p. 500505.
2. Firl, J. Probabilistic Maneuver Recognition in Traffic Scenarios. Ph.D. Thesis, KIT Karlsruhe Institute of Technology, Karlsruhe, Germany, 2014.
3. Luo, W.; Xing, J.; Milan, A.; Zhang, X.; Liu, W.; Zhao, X.; Kim, T. Multiple object tracking: A literature review. *Artif. Intell.* **2021**, *293*, 103448. [CrossRef]
4. Chen, J.; Wang, C.; Chou, C. Multiple target tracking in occlusion area with interacting object models in urban environments. *Robot. Auton. Syst.* **2018**, *103*, 68–82. [CrossRef]
5. Kassner, M.; Patera, W.; Bulling, A. Pupil: An open source platform for pervasive eye tracking and mobile gaze-based interaction. In Proceedings of the 2014 ACM International Joint Conference on Pervasive and Ubiquitous Computing, Seattle, DC, USA, 13–17 September 2014; pp. 1151–1160.
6. Bruce, J.; Wawer, J.; Vaughan, R. Human-robot rendezvous by co-operative trajectory signals. In Proceedings of the 10th ACM/IEEE International Conference on Human-Robot Interaction Workshop on Human-Robot Teaming, Portland, OR, USA, 2–5 March 2015; pp. 1–2.
7. Kunwar, F.; Benhabib, B. Advanced predictive guidance navigation for mobile robots: A novel strategy for rendezvous in dynamic settings. *Intern. J. Smart Sens. Intell. Syst.* **2008**, *1*, 858–890. [CrossRef]
8. Palm, R.; Lilienthal, A. Fuzzy logic and control in human-robot systems: Geometrical and kinematic considerations. In Proceedings of the WCCI 2018: 2018 IEEE International Conference on Fuzzy Systems (FUZZ-IEEE), Rio de Janeiro, Brazil, 8–13 July 2018; pp. 827–834.
9. Palm, R.; Driankov, D. Fuzzy inputs. In *Fuzzy Sets and Systems—Special Issue on Modern Fuzzy Control*; IEEE: Piscataway, NJ, USA, 1994; pp. 315–335.
10. Foulloy, L.; Galichet, S. Fuzzy control with fuzzy inputs. *IEEE Trans. Fuzzy Syst.* **2003**, *11*, 437–449. [CrossRef]
11. Banelli, P. Non-linear transformations of gaussians and gaussian-mixtures with implications on estimation and information theory. *IEEE Trans. Inf. Theory* **2013**.
12. Julier, S.J.; Uhlmann, J.K. Unscented filtering and nonlinear estimation. *Proc. IEEE* **2004**, *92*, 401–422. [CrossRef]
13. Merwe, R.v.d.; Wan, E.; Julier, S.J. Sigma-point kalman filters for nonlinear estimation and sensor-fusion: Applications to integrated navigation. In Proceedings of the AIAA 2004, Guidance, Navigation and Control Conference, Portland, OR, USA, 28 June–1 July 2004.
14. Terejanu, G.A. *Unscented Kalman Filter Tutorial*; Department of Computer Science and Engineering University at Buffalo: Buffalo, NY, USA, 2011.
15. Odry, Á.; Kecskes, I.; Sarcevic, P.; Vizvari, Z.; Toth, A.; Odry, P. A novel fuzzy-adaptive extended kalman filter for real-time attitude estimation of mobile robots. *Sensors* **2020**, *20*, 803. [CrossRef] [PubMed]
16. Song, Y.; Song, Y.; Li, Q. Robust iterated sigma point fastslam algorithm for mobile robot simultaneous localization and mapping. *Chin. J. Mech. Eng.* **2011**, *24*, 693. [CrossRef]
17. Bittler, J.; Bhounsule, P.A. Hybrid unscented kalman filter: Application to the simplest walker. In Proceedings of the 3rd Modeling, Estimation and Control Conference MECC, Lake Tahoe, NV, USA, 2–5 October 2023.
18. Alireza, S.D.; Shahri, M. Vision based simultaneous localization and mapping using sigma point kalman filter. In Proceedings of the 2011 IEEE International Symposium on Robotic and Sensors Environments (ROSE), Montreal, QC, Canada, 17–18 September 2011.
19. Xue, Z.; Schwartz, H. A comparison of several nonlinear filters for mobile robot pose estimation. In Proceedings of the 2013 IEEE International Conference on Mechatronics and Automation (ICMA), Kagawa, Japan, 4–7 August 2013.
20. Lyu, Y.; Pan, Q.; Lv, J. Unscented transformation-based multi-robot collaborative self-localization and distributed target tracking. *Appl. Sci.* **2019**, *9*, 903. [CrossRef]
21. Raitoharju, M.; Piche, R. On computational complexity reduction methods for kalman filter extensions. *IEEE Aerosp. Electron. Syst. Mag.* **2019**, *34*, 2–19. [CrossRef]
22. Palm, R.; Lilienthal, A. Uncertainty and fuzzy modeling in human-robot navigation. In Proceedings of the 11th Joint International Computer Conference (IJCCI 2019), Vienna, Austria, 17–19 September 2019; pp. 296–305.
23. Palm, R.; Lilienthal, A. Fuzzy geometric approach to collision estimation under gaussian noise in human-robot interaction. In Proceedings of the Computational Intelligence: 11th International Joint Conference, IJCCI 2019, Vienna, Austria, 17–19 September 2019; Springer: Cham, Switzerland; pp. 191–221.

24. Schaefer, J.; Strimmer, K. A shrinkage to large scale covariance matrix estimation and implications for functional genomics. *Stat. Appl. Genet. Mol. Biol.* **2005**, 4, 32. [CrossRef] [PubMed]
25. Cao, Y. Learning the Unscented Kalman Filter. 2021. Available online: https://www.mathworks.com/matlabcentral/fileexchange/18217-learning-the-unscented-kalman-filter (accessed on 14 May 2024).

sensors

MDPI

Article

Robust Tracking Control of Wheeled Mobile Robot Based on Differential Flatness and Sliding Active Disturbance Rejection Control: Simulations and Experiments

Amine Abadi [1,*], Amani Ayeb [2], Moussa Labbadi [3], David Fofi [1], Toufik Bakir [1] and Hassen Mekki [4]

1 Laboratory ImViA EA 7535, University of Bourgogne, 21000 Dijon, France; david.fofi@u-bourgogne.fr (D.F.); toufik.bakir@u-bourgogne.fr (T.B.)
2 National Institute of Applied Science and Technology, Physics and Instrumentation Department, Tunis 1080, Tunisia; amani.ayeb@insat.ucar.tn
3 LIS UMR CNRS 7020, Aix-Marseille University, 13013 Marseille, France; moussa.labbadi@lis-lab.fr
4 NOCCS Laboratory, National School of Engineering of Sousse, University of Sousse, Sousse 4054, Tunisia; mekki.hassen@gmail.com
* Correspondence: amine.abadi@u-bourgogne.fr

Abstract: This paper proposes a robust tracking control method for wheeled mobile robot (WMR) against uncertainties, including wind disturbances and slipping. Through the application of the differential flatness methodology, the under-actuated WMR model is transformed into a linear canonical form, simplifying the design of a stabilizing feedback controller. To handle uncertainties from wheel slip and wind disturbances, the proposed feedback controller uses sliding mode control (SMC). However, increased uncertainties lead to chattering in the SMC approach due to higher control inputs. To mitigate this, a boundary layer around the switching surface is introduced, implementing a continuous control law to reduce chattering. Although increasing the boundary layer thickness reduces chattering, it may compromise the robustness achieved by SMC. To address this challenge, an active disturbance rejection control (ADRC) is integrated with boundary layer sliding mode control. ADRC estimates lumped uncertainties via an extended state observer and eliminates them within the feedback loop. This combined feedback control method aims to achieve practical control and robust tracking performance. Stability properties of the closed-loop system are established using the Lyapunov theory. Finally, simulations and experimental results are conducted to compare and evaluate the efficiency of the proposed robust tracking controller against other existing control methods.

Keywords: differential flatness; sliding mode control; active disturbance rejection control; extended state observer; wheeled mobile robot

Citation: Abadi, A.; Ayeb, A.; Labbadi, M.; Fofi, D.; Bakir, T.; Mekki, H. Robust Tracking Control of Wheeled Mobile Robot Based on Differential Flatness and Sliding Active Disturbance Rejection Control: Simulations and Experiments. *Sensors* **2024**, *24*, 2849. https://doi.org/10.3390/s24092849

Academic Editors: David Cheneler and Stephen Monk

Received: 22 March 2024
Revised: 17 April 2024
Accepted: 25 April 2024
Published: 29 April 2024

1. Introduction

The domain of robotics finds mobile robots to be particularly intriguing, attracting considerable fascination and study. Designed to operate in dynamic settings, be it indoors or outdoors, these robots demonstrate the capacity to navigate autonomously or with minimal human input. Central to their functionality is their mobility, achieved through diverse locomotion methods, such as wheels, tracks, or legs. This mobility empowers them to traverse diverse terrains, overcoming obstacles encountered during their journeys. Recently, mobile robots have been used in various domains, including civilian, industrial, and military, to carry out diverse tasks such as surveillance [1], transportation [2], agricultural operations [3], and exploration [4]. Given the broad application spectrum and critical nature of tasks involving mobile robots, there exists an urgent need to develop performance tracking controllers to execute proposed missions with exceptional accuracy. However, achieving this objective remains a significant challenge due to the inherent under-actuation and nonlinearity in WMRs, constrained by nonholonomic limitations. Consequently, researchers have directed their efforts towards investigating the control of mobile robotic systems.

In the past few decades, substantial progress has been made in the field of tracking control for wheeled mobile robots (WMR) through the application of nonlinear control theory [5–7]. Among these control methodologies, linearization controllers, such as the flatness controller [8], have risen as a popular approach that can significantly simplify the controller design process. The flatness property is a technique used to define the dynamic behavior of nonlinear underactuated models by identifying a set of core system variables known as flat outputs. This perspective has significant implications for control systems, as will be demonstrated. The first step in flatness control involves generating a desired realizable trajectory that implicitly incorporates the system model. Following that, the nonlinear WMR model can be linearized, resulting in the canonical Brunovsky form [9,10]. This special form simplifies the concept of a feedback controller capable of achieving exact trajectory tracking. In fact, controlling a linear system is easier than controlling an underactuated nonlinear system, and this feature has encouraged researchers to use the properties of flatness in several application domains, such as the control of hydraulic systems [11], exoskeleton robots [12], microgrid [13], underwater robot [14], and quadrotor [15,16].

Numerous research studies on WMR have utilized the concept of flatness control. Abadi [17] introduced an approach for optimal path planning for WMR using the collocation method, flatness control, and spline curves. This method effectively reduces the time needed to compute optimal robot trajectories during navigation, which is crucial for real-world applications. Kaniche [18] proposed a flatness visual servoing control for WMR subjected to disturbances. Salah [19] developed an approach to generate the upper coverage trajectory of a mobile robot by leveraging flatness. Yakovlev [20] combined flatness control with predictive control to enable safe navigation of a WMR among static and dynamic obstacles.

There is always is a difference between the mathematical model describing the movement of WMR and reality. This difference is due to environmental phenomena neglected during modeling, such as wind, slipping, etc. The question that arises is how flatness control applied to WMR can ensure the accurate tracking of a desired trajectory despite the presence of uncertainties. To resolve this problem, a robust feedback controller must be combined with flatness, taking into account the impact of uncertainties to the model. Up to the present, there have been limited methods in the literature concerning the robustness issues of flatness systems. Among these approaches, the sliding mode control (SMC) has been successfully utilized in a variety of systems [21–23].

SMC is a robust control technique used to manage dynamic systems in the presence of uncertainties and disturbances. At its core, SMC aims to drive the system state onto a designated sliding surface within the state space. Once on this surface, the system's behavior is constrained, allowing for effective regulation. SMC achieves this through discontinuous control actions, known as switching control, which dynamically alternate between different control laws. This switching mechanism ensures that the system remains on the sliding surface, enhancing robustness against external influences. Despite its effectiveness, SMC is associated with a phenomenon called chattering [24], characterized by rapid switching between control actions near the sliding surface. While chattering can theoretically improve tracking accuracy, it can lead to practical issues such as mechanical wear and high-frequency oscillations. To resolve this problem, numerous approaches have been suggested in the existing literature, such as high-order SMC [25], boundary layer [26], and active adaptive continuous nonsingular terminal sliding mode algorithm [27]. A frequently utilized approach for mitigating the chattering phenomenon involves incorporating the boundary layer technique within SMC. This entails replacing the sign function with a smooth function. However, this strategy presents its own set of challenges. Firstly, there exists a trade-off between the size of the boundary layer and the performance of SMC, which impacts the effectiveness of chattering reduction. Secondly, the robustness and accuracy of the system may not always be guaranteed within the boundary layer. Additionally, beyond addressing the chattering issue, achieving precise control of a robotic system necessitates knowledge about its state, typically obtained through real instruments, which can incur high costs and complicate the system's structure. Moreover, in many instances, directly measuring certain system parameters may be impractical. To overcome these limitations, one potential solution involves implementing software sensors or observers, commonly referred to as virtual sensors. Therefore, to tackle both the reduced robustness resulting from the boundary layer approach and the challenge of state estimation, we propose a novel robust feedback controller that integrates the boundary layer sliding method with a disturbance observer.

In recent times, disturbance observers have emerged as potent tools for handling consolidated uncertainties, closely tied to disturbance-observer-based control. Within the domain of nonlinear disturbance observers, two notable approaches stand out: the uncertainty and disturbance estimator (UDE) [28] and the active disturbance rejection control [29] based on the extended state observer (ESO). In the UDE, only the disturbance is estimated, though in general, the observer equations depend on system states and inputs. Thus, a state observer is necessary unless all states are measurable. The idea of the ESO is to extend the original state vector by the disturbance vector and possibly, some of its time-derivatives, and then design a state observer for the extended system. ESO distinguishes itself by incorporating a dynamic model of disturbances or uncertainties into its estimation methodology, enabling it to identify and mitigate uncertainties not explicitly accounted for in the system model. The design of ESO is distinguished by its minimal dependence on system data and its freedom from the traditional system model, which simplifies its implementation process. Furthermore, several other types of disturbance observers are available, such as the nonlinear extended state observer (NLESO) [30], the adaptive extended state observer (AESO) [31], and the extended high-gain observer (EHGO) [32]. EHGO, part of the ESO family, stands out in two key aspects: it does not necessitate slow variations in disturbances, and it estimates a matched disturbance term originating from model uncertainty and external disturbances. Given the advantages offered by ESO, considerable research effort has been devoted to developing advanced controls for robotic systems.

In Ref. [33], Xie introduced a controller that integrates the backstepping technique with ESO to improve tracking performance for underwater robots. Additionally, Qi [34] improved the bandwidth of ESO to achieve more accurate disturbance estimation. Subsequently, they utilized a simple feedback controller to ensure attitude stabilization over a 3D hovering quadrotor system. In the work by Aole [35], an improved ADRC methodology for controlling lower limb exoskeletons is presented. The proposed approach integrates Linear ESO with a tracking differentiator, nonlinear state error feedback, and a proportional controller. Simulation results demonstrate the effectiveness of the suggested ADRC in efficiently regulating the hip and knee movements of the robot in the presence of disturbances. Hu [36] integrated a predictive control technique with ESO for unmanned underwater vehicles, offering a solution to concurrently handle external disturbances and system measurement noises. Based on this observation, the main contributions of our research can be summarized as follows:

1. The kinematic model for WMR is structured in a standard format that systematically tackles underactuation and transforms nonmatching disturbances into matching ones through a flatness-based approach;
2. The designated trajectory is feasible in practice because of the concept of differential flatness, which equates differential flatness with controllability, ensuring its physical achievability;
3. Continuous sliding mode control (SMC) is employed to eliminate chattering, an essential necessity for the efficient application of control in real-world scenarios;
4. SMC is integrated with ESO for the uncertain kinematic WMR model. This strategy seeks to improve the practicality and resilience of the tracking controller by reducing

chattering through boundary layer SMC and estimating the lumped disturbance affecting the WMRs via ESO, which is then employed as a feedforward compensation;

5. The proposed control method was compared with several other control methods, including traditional flatness control, backstepping tracking control flatness-based sliding control, and flatness active disturbance rejection control and backstepping sliding active disturbance rejection control. These comparisons were validated through simulations conducted in Matlab/Simulink and experiments carried out on the TurtleBot WMR.

The structure of the remaining sections of this article is outlined as follows. Section 2 provides a thorough overview of the flatness control technique for WMR. Section 3 elaborates on the concept of flatness-based sliding mode tracking control of WMRs. The proposed robust tracking controller is delineated in Section 4. Sections 5 and 6 present and discuss the results of simulations and experiments. Finally, Section 7 concludes the paper by summarizing the key findings and suggesting potential future directions.

2. Flatness-Based Tracking Control

In our study, we analyzed a differential two-wheeled mobile robot (see Figure 1) that consists of two independent active wheels and a third passive wheel (a standard freewheel). This robotic system is widely regarded as an effective trade-off between control ease and the degrees of freedom that enable the robot to meet mobility requirements. The configuration of the mobile robot with wheels can be described by the vector $q_r = [x, y, \theta]$. In this notation, x and y represent the coordinates of the robot's center position in the stationary frame (O, X, Y), while θ represents the orientation angle of the robot. The state equation of the WMR kinematic model, neglecting uncertainties, is represented as follows:

$$\begin{aligned} \dot{x} &= cos(\theta)v \\ \dot{y} &= sin(\theta)v \\ \dot{\theta} &= w \end{aligned} \tag{1}$$

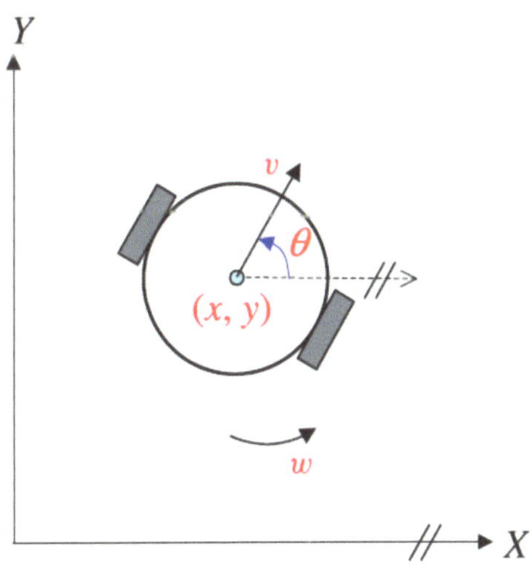

Figure 1. Two-wheeled mobile robot.

The robot's translational and rotational velocities are denoted by v and w, respectively. The angular velocities of the right and left wheels (w_r and w_l) can be defined as functions of the robot's translational and rotational velocities as follows:

$$v = (\frac{w_r + w_l}{2})r \tag{2}$$

$$w = (\frac{w_r - w_l}{2b})r \tag{3}$$

where the variables r and $2b$ represent the radius and distance between the wheels, respectively. The nonholonomic limitation is defined as follows, based on the nonslip requirement:

$$\dot{x}sin\theta - \dot{y}cos\theta = 0 \tag{4}$$

The accuracy of the tracking will be guaranteed through the flatness property, which involves describing all system states and inputs, as well as their finite time derivatives, within the framework of a flat output. Considering the following nonlinear system:

$$\dot{x} = f(x, u) \tag{5}$$

where $x \in \mathbb{R}^n$ and $u \in \mathbb{R}^m$ represent the state and the input vector.

The nonlinear system (5) is differentially flat if there exists an output λ in the following form:

$$\lambda = \xi(x, u, \dot{u}, \ldots, u^{(c)}) \in \mathbb{R}^m \tag{6}$$

such that the state and the input can be expressed as follows:

$$x = \kappa_1(\lambda, \dot{\lambda}, \ddot{\lambda} \ldots, \lambda^{(a)}) \tag{7}$$

$$u = \kappa_2(\lambda, \dot{\lambda}, \ddot{\lambda} \ldots, \lambda^{(a+1)}) \tag{8}$$

where a and c are finite multi-indices, and ξ, κ_1, and κ_2 are smooth vector functions of the output vector λ and its derivatives. By introducing the functions κ_1 and κ_2, this flat output is composed of a set of variables that enable the parameterization of all other system variables: the state, the command, and also the output λ. Indeed, if the output of the system is defined by a relation of the form $\lambda = \Xi(x, u, \dot{u}, \ldots, u^{(p)})$, then necessarily, the quantities described in Equations (7) and (8) make it possible to affirm that there exists an integer c such that:

$$f = \Xi(\lambda, \dot{\lambda}, \ddot{\lambda} \ldots, \lambda^{(c)}) \tag{9}$$

The flat output combines all unconstrained variables of the system since the components of λ are differentially independent. Alternatively, based on Equation (9), we can argue that the flat output λ solely relies on the state and the command. This would make it an endogenous variable of the system, in contrast to the state of an observer, which would be an example of an exogenous variable of the observed system. In addition, Lie–Bäcklund's notion of differential equivalence [8] shows that the number of components of λ is the same as the number of components of the control:

$$dim\lambda = dimu \tag{10}$$

This fundamental characteristic allows us to determine the requisite number of independent variables needed in a model to establish its flatness. A key benefit of the flatness property lies in its facilitation of various transformations, such as diffeomorphism and feedback linearization. These transformations enable the conversion of a nonlinear system into a controllable linear system, where the flat outputs represent the state vector.

Several studies in the literature, including Ref. [37], have shown that the WMR kinematic modeling can be defined as a differentially flat model, where the positional coordinates denoted as $\lambda = [\lambda_{11}, \lambda_{21}]^T = [x, y]^T$ serve as the flat outputs. Therefore, the entire set of state and control components pertaining to the WMR system are expressed using the flat variable λ and its derivatives, as demonstrated below:

$$\theta = arctan\frac{\dot{\lambda}_{21}}{\dot{\lambda}_{11}} \tag{11}$$

$$v = \sqrt{\dot{\lambda}_{11}^2 + \dot{\lambda}_{21}^2} \tag{12}$$

$$w = \frac{\dot{\lambda}_{11}\ddot{\lambda}_{21} - \ddot{\lambda}_{11}\dot{\lambda}_{21}}{\dot{\lambda}_{11}^2 + \dot{\lambda}_{21}^2} \tag{13}$$

The differentially flat nature of the WMR's kinematic model has been demonstrated in the literature by various researchers [37]. This implies that all the states and controls of the kinematic WMR model can be expressed as functions of λ and its derivatives. However, the noninvertible relationship between the control input vectors w and v and the highest derivatives of the flat output limits the development of static feedback linearization for the nonlinear WMR. To address this constraint, we incorporate the control input v into the kinematic model defined by Equation (1) by treating it as an additional state. As a result, we obtain a revised system that can be defined as follows:

$$\begin{aligned} \dot{x} &= cos(\theta)v \\ \dot{y} &= sin(\theta)v \\ \dot{v} &= u_{r1} \\ \dot{\theta} &= u_{r2} \end{aligned} \tag{14}$$

The state and control inputs of the modified system defined by Equation (14) are represented by $X_r = [x, y, v, \theta]^T$ and $u_{r1} = \dot{v}$ and $u_{r2} = w$. In order to establish a bijective relationship between the inputs u_{r1}, u_{r2}, and higher-order derivatives of $\lambda_{11} = x$, $\lambda_{21} = y$, we apply successive differentiations to the flat outputs until at least one of the input variables appears in the resulting expressions, as illustrated below:

$$\begin{bmatrix} \ddot{\lambda}_{11} \\ \ddot{\lambda}_{21} \end{bmatrix} = B_{rob} \begin{bmatrix} u_{r1} \\ u_{r2} \end{bmatrix} \tag{15}$$

where B_{rob} is described as follows:

$$B_{rob} = \begin{bmatrix} cos(\theta) & -vsin(\theta) \\ sin(\theta) & vcos(\theta) \end{bmatrix} \tag{16}$$

The matrix B_{rob} is not singular if $v \neq 0$. In this case, we can define the control as follows:

$$\begin{bmatrix} u_{r1} \\ u_{r2} \end{bmatrix} = B_{rob}^{-1} \begin{bmatrix} \ddot{\lambda}_{11} \\ \ddot{\lambda}_{21} \end{bmatrix} \tag{17}$$

To arrive at the linearized system, referred to as the Burnovsky Form (BF), we can substitute the control input (17) into Equation (15). This substitution yields the following modified expression:

$$(BF_1)\begin{cases} \dot{\lambda}_{11} = \lambda_{12} \\ \dot{\lambda}_{12} = v_1 \\ Y_1 = \lambda_{11} = x \end{cases} \qquad (BF_2)\begin{cases} \dot{\lambda}_{21} = \lambda_{22} \\ \dot{\lambda}_{22} = v_2 \\ Y_2 = \lambda_{21} = y \end{cases} \tag{18}$$

where v_1 and v_2 represent a suitable feedback controller defined as follows:

$$v_1 = \ddot{\lambda}_{xd} - \sigma_{x2}(\lambda_{12} - \dot{\lambda}_{xd}) - \sigma_{x1}(\lambda_{11} - \lambda_{xd}) \tag{19}$$

$$v_2 = \ddot{\lambda}_{yd} - \sigma_{y2}(\lambda_{22} - \dot{\lambda}_{yd}) - \sigma_{y1}(\lambda_{21} - \lambda_{yd}) \tag{20}$$

where λ_{xd} and λ_{yd} denote the desired trajectories for the flat output λ_{11} and λ_{21}, respectively. Meanwhile, the controller gains are represented by σ_{x1}, σ_{x2}, σ_{y1}, and σ_{y2}. The polynomial of the Burnovsky system (18) can be defined as follows:

$$s^2 + \sigma_{x2}s + \sigma_{x1} = s^2 + 2m_x\epsilon_{xc} + \epsilon_{xc}^2 \tag{21}$$

$$s^2 + \sigma_{y2}s + \sigma_{y1} = s^2 + 2m_y\epsilon_{yc} + \epsilon_{yc}^2 \tag{22}$$

where the parameters m_x and m_y are the damping coefficients, and ϵ_{xc} and ϵ_{yc} are the frequencies in Equations (21) and (22). We can calculate the controller gain as follows:

$$\sigma_{x1} = \epsilon_{xc}^2, \sigma_{x2} = 2m_x\epsilon_{xc}, \sigma_{y1} = \epsilon_{yc}^2, \sigma_{y2} = 2m_y\epsilon_{yc} \tag{23}$$

By integrating the feedback law, as described in Equations (19) and (20), into the system (17), we can express the flatness-based tracking control (FBTC) utilized for the mobile robot in the following manner:

$$\begin{bmatrix} u_{FBTCx} \\ u_{FBTCy} \end{bmatrix} = B_{rob}^{-1} \begin{bmatrix} \ddot{\lambda}_{xd} - \sigma_{x2}\dot{e}_1 - \sigma_{x1}e_1 \\ \ddot{\lambda}_{yd} - \sigma_{y2}\dot{e}_2 - \sigma_{y1}e_2 \end{bmatrix} \tag{24}$$

where $e_1 = \lambda_{11} - \lambda_{xd}$ and $e_2 = \lambda_{21} - \lambda_{yd}$.

In ideal conditions where uncertainties such as wind and wheel slip are negligible in the kinematic model of the WMR, the control input defined by Equation (24) can achieve satisfactory tracking performance for the desired trajectory. However, it is practically impossible to have a model that accurately represents the real-world movement of the robot in all environmental conditions. As a result, the following section will focus on developing a robust tracking control for a WMR kinematic model that is subject to uncertainties.

3. Flatness-Based Sliding Tracking Control

In order to account for real-world conditions, we consider uncertainties such as slippage and external environmental disturbances when describing the kinematic model of WMR (Figure 2). As a result, the model is defined differently, as shown below.

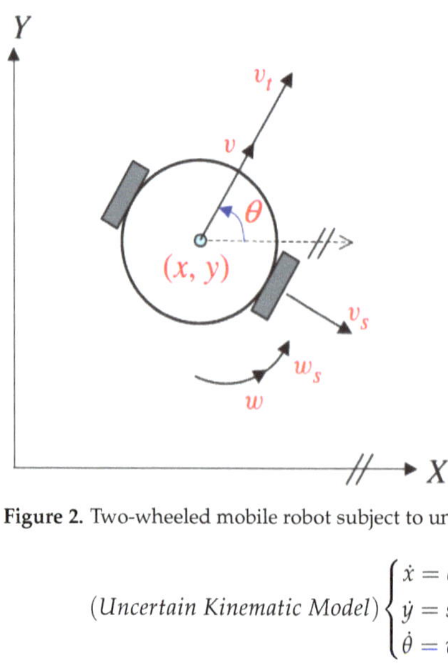

Figure 2. Two-wheeled mobile robot subject to uncertainties.

$$(Uncertain\ Kinematic\ Model) \begin{cases} \dot{x} = cos(\theta)v + v_t cos(\theta) + v_s sin(\theta) + p_x \\ \dot{y} = sin(\theta)v + v_t sin(\theta) - v_s cos(\theta) + p_y \\ \dot{\theta} = w + w_s \end{cases} \tag{25}$$

where the variables p_x and p_y represent the external environmental disturbances, indicating the potential influences from the surrounding conditions. On the other hand, v_t and v_s represent the slip velocities, where v_t denotes the slip velocity along the forward direction and v_s represents the slip velocity normal to it. Additionally, w_s denotes the angular slip velocity. According to [37], it is assumed that the slippage phenomenon can be defined and bounded as follows:

$$v_t(t) = v_s(t) = w_s(t) = \kappa_1 v(t) \tag{26}$$

$$||v_t|| \leq \varepsilon_1 ||v||,\ ||v_s|| \leq \varepsilon_2 ||v||, ||w_s|| \leq \varepsilon_3 \tag{27}$$

where $\kappa_1, \varepsilon_1, \varepsilon_2$ and ε_3 are positive constants.

Assuming that λ_{xd} and λ_{yd} are the reference trajectories for λ_{11} and λ_{21}, respectively, we can define the error dynamics as $e_i = \lambda_{i1} - \lambda_{id}$ for $i = 1, 2$. To achieve convergence of the tracking error e_i to zero in the presence of uncertainties, we employ a sliding mode control approach that relies on the principles of the flatness law. By incorporating this control strategy, we aim to ensure robust and accurate tracking performance even in the face of system uncertainties. The design of the sliding mode control involves two essential stages: the choice of the sliding surface and the development of the control law. These steps play a crucial role in establishing an effective and stable sliding mode control strategy. The selection of the sliding surface determines the desired system behavior and convergence properties, while the design of the control law focuses on generating control signals that guide the system towards the desired sliding surface and ensure its maintenance on that surface. In the context of the tracking example for the WMR, we make use of the sliding variable $\sigma_r = [s_x, s_y]^T$ to represent the tracking error. To define the sliding surface, we consider the desired tracking behavior and express it as follows, taking into account the specific requirements of the system:

$$s_x = \dot{e}_1 + \beta_1 e_1 \tag{28}$$

$$s_y = \dot{e}_2 + \beta_2 e_2 \tag{29}$$

where the gains β_1 and β_2 can be selected using pole-placement techniques to ensure the asymptotic convergence of the tracking errors $e_1 = \lambda_{11} - \lambda_{xd}$ and $e_2 = \lambda_{21} - \lambda_{yd}$ to zero. In this tracking example, the sliding variable $\sigma_r = [s_x, s_y]^T$ is chosen as the tracking error. Therefore, the sliding surface for the WMR can be defined as follows:

$$\dot{e}_1 + \beta_1 e_1 = 0 \tag{30}$$

$$\dot{e}_2 + \beta_2 e_2 = 0 \tag{31}$$

As suggested by Mauledoux [38], to guarantee that the sliding surface $\sigma_r = 0$ is attractive, we can enforce the dynamics of σ_r as follows:

$$\dot{\sigma}_r = -k_i sgn(\sigma_r) \tag{32}$$

where the standard signum function is denoted by sgn, and k_i $(i = 1, 2)$ is a constant. One approach to proving the error dynamics stability is to analyze the following Lyapunov function:

$$V_s = \frac{1}{2}\sigma_r^2 \tag{33}$$

The derivative of V_s is defined as follows:

$$\dot{V}_s = \sigma_r \dot{\sigma}_r \tag{34}$$

We can conclude that V_s is a positive function and its derivative $\dot{V}_s$ is negative or zero. Hence, the system exhibits asymptotic Lyapunov stability. Using Equations (28), (29) and (32) we obtain:

$$-k_1 sign(s_x) = \ddot{e}_1 + \beta_1 \dot{e}_1 \quad (35)$$

$$-k_2 sign(s_y) = \ddot{e}_2 + \beta_2 \dot{e}_2 \quad (36)$$

As a result, by using Equations (35) and (36), we can obtain:

$$\ddot{\lambda}_{11} = \ddot{\lambda}_{xd} - \beta_1 \dot{e}_1 - k_1 sgn(s_x) \quad (37)$$

$$\ddot{\lambda}_{21} = \ddot{\lambda}_{yd} - \beta_2 \dot{e}_2 - k_2 sgn(s_y) \quad (38)$$

Substituting $\ddot{\lambda}_{11}$ and $\ddot{\lambda}_{21}$ with their new expressions defined by Equations (35) and (36) in the control defined by (17), the flatness-based sliding mode tracking controller (FSMC) applied to WMR is defined as follows:

$$\begin{bmatrix} u_{FSMCx} \\ u_{FSMCy} \end{bmatrix} = B_{rob}^{-1} \begin{bmatrix} \ddot{\lambda}_{xd} - \beta_1 \dot{e}_1 - k_1 sgn(s_x) \\ \ddot{\lambda}_{yd} - \beta_2 \dot{e}_2 - k_2 sgn(s_y) \end{bmatrix} \quad (39)$$

The FSMC defined by Equation (39) contains a discontinuous control term due to the function $sgn(\sigma)$. Although selecting sufficiently large values for k_1 and k_2 can achieve convergence to sliding variable in limited time and provide robustness against perturbations, it also causes the phenomenon of chattering. Thus, to avoid this problem, the function $sgn(\sigma)$ can be replaced by the function Sat defined as follows:

$$Sat(\sigma_r) \begin{cases} \frac{\sigma_r}{a_s} & if \quad |\sigma_r| \leq a_s \\ sgn(\sigma_r) & if \quad |\sigma_r| > a_s \end{cases} \quad (40)$$

where a_s is the width of the threshold of the saturation function.

The thickness of the boundary layer, denoted as a_s, within the saturation function stands as a pivotal parameter influencing the efficacy of the sliding mode controller. As the value of a_s increases, the approximation diverges more from the ideal sgn function, resulting in enhanced reduction of chattering. However, this improvement comes at the cost of diminished robustness. Conversely, if the value of the parameter a_s is reduced, the change of the control signal will be too frequent, which leads to inevitable chatter of the control signal. Therefore, a variable-thickness boundary layer a_s is tailored to strike a balance between mitigating chattering and upholding system robustness amid uncertainties. In the upcoming section, the FSMC described by Equation (39) will be integrated with active disturbance rejection control to enhance the robustness lost by the Sat function and maintain the advantage of reducing chattering.

4. Proposed Robust Tracking Controller

In this section, we introduce a novel cascade control strategy that utilizes a combination of flatness property, active disturbance rejection control (ADRC), and boundary layer sliding mode control to solve the problem of reduced robustness obtained when replacing the function sgn by the function sat in the FSMC defined by Equation (39). Given the uncertain kinematic model (25), we can obtain the following relationship by differentiating λ_{11} and λ_{21} until the input terms u_1 and u_2 become evident:

$$\begin{bmatrix} \ddot{\lambda}_{11} \\ \ddot{\lambda}_{21} \end{bmatrix} = B_{rob} \begin{bmatrix} u_{r1} \\ u_{r2} \end{bmatrix} + C_{rob} + D_{rob} \begin{bmatrix} u_{r1} \\ u_{r2} \end{bmatrix} \quad (41)$$

where C_{rob} and D_{rob} are defined as follows:

$$C_{rob} = \begin{bmatrix} cos(\theta)(v_s w_s + \dot{v}_t) + sin(\theta)(\dot{v}_s - v w_s - v_t w_s) + \dot{p}_x \\ sin(\theta)(v_s w_s + \dot{v}_t) - cos(\theta)(\dot{v}_s - v w_s - v_t w_s) + \dot{p}_y \end{bmatrix}, D_{rob} = \begin{bmatrix} 0 & -v_t sin(\theta) + v_s cos(\theta) \\ 0 & v_t cos(\theta) + v_s sin(\theta) \end{bmatrix} \tag{42}$$

By utilizing the control input described in Equation (17) on system (41), we achieve:

$$\ddot{\lambda} = v + \delta \tag{43}$$

where $\ddot{\lambda} = [\ddot{\lambda}_{11}, \ddot{\lambda}_{21}]^T$,$v = [v_1, v_2]^T$ and $\delta = [\delta_1, \delta_2]^T = d_{rob} B_{rob}^{-1} v + C_{rob}$.

Rewriting Equation (43) in terms of two linear integrator systems subject to perturbation yields the following expressions:

$$MBF_1 \begin{cases} \dot{\lambda}_{11} = \lambda_{12} \\ \dot{\lambda}_{12} = v_1 + \delta_1 \\ Y_1 = \lambda_{11} \end{cases} \quad MBF_2 \begin{cases} \dot{\lambda}_{21} = \lambda_{22} \\ \dot{\lambda}_{22} = v_2 + \delta_2 \\ Y_2 = \lambda_{21} \end{cases} \tag{44}$$

Consider Δ_1 and Δ_2 as the differentials of δ_1 and δ_2 with respect to time t, respectively. We assume that both δ_i and Δ_i $(i = 1, 2)$ are bounded. In practical applications, determining the actual values of the lumped disturbances δ_1 and δ_2 that affect the system is considered a challenging problem. Hence, an observer is required to estimate these values.

4.1. ESO Design

The extended state observer (ESO) plays a vital role in system control by simultaneously estimating the system states and uncertainties. This capability enables the ESO to effectively reject or compensate for disturbances, enhancing the system's robustness and performance. The ESO takes into account all factors that affect the system and treats parameter uncertainties and external perturbations as a single observed disturbance. The ESO is named as such because it estimates uncertainties as an extended state. Its benefits include not being reliant on the mathematical model of the system, as well as having a straightforward implementation and demonstrating good performance. Consider $\lambda_{13} = \delta_1$, $\alpha_{23} = \delta_2$ as an extended state for system (44). The latter can be expressed as follows:

$$\begin{cases} \dot{\lambda}_{11} = \lambda_{12} \\ \dot{\lambda}_{12} = \lambda_{13} + v_1 \\ \dot{\lambda}_{13} = \Delta_1 \\ Y_1 = \lambda_{11} \end{cases} \quad \begin{cases} \dot{\lambda}_{21} = \lambda_{22} \\ \dot{\lambda}_{22} = \lambda_{23} + v_2 \\ \dot{\lambda}_{23} = \Delta_2 \\ Y_1 = \lambda_{21} \end{cases} \tag{45}$$

We can express systems (45) in matrix form as follows:

$$\begin{cases} \dot{\lambda}_1 = A_x \lambda_1 + B_x v_1 + E_x \Delta_1 \\ Y_1 = C_x \lambda_1 \end{cases} \tag{46}$$

$$\begin{cases} \dot{\lambda}_2 = A_y \lambda_2 + B_y v_2 + E_y \Delta_2 \\ Y_2 = C_y \lambda_2 \end{cases} \tag{47}$$

where $\lambda_1 = [\lambda_{11}, \lambda_{12}, \lambda_{13}]^T$, $\lambda_2 = [\lambda_{21}, \lambda_{22}, \lambda_{23}]^T$, $A_x = A_y = \begin{bmatrix} 0 & 1 & 0 \\ 0 & 0 & 1 \\ 0 & 0 & 0 \end{bmatrix}$, $B_x = B_y = \begin{bmatrix} 0 \\ 1 \\ 0 \end{bmatrix}$, $C_x = C_y = \begin{bmatrix} 1 & 0 & 0 \end{bmatrix}$, $E_x = E_y = \begin{bmatrix} 0 & 0 & 1 \end{bmatrix}^T$. The expression for the Extended State Observer (ESO) corresponding to each extended system (46) and (47) can be given as follows:

$$\dot{\hat{\lambda}}_1 = A_x \hat{\lambda}_1 + B_x v_x + \alpha_{gx} C_x (\lambda_1 - \hat{\lambda}_1) \tag{48}$$

$$\dot{\hat{\lambda}}_2 = A_y\hat{\lambda}2 + B_y v_y + \alpha_{gy} C_y(\lambda_2 - \hat{\lambda}_2) \tag{49}$$

where $\alpha_{gx} = [\alpha_{11}, \alpha_{12}, \alpha_{13}]^T$, $\alpha_{gy} = [\alpha_{21}, \alpha_{22}, \alpha_{23}]^T$. To determine the observer gains α_{ij} $(i = 1,2,3)$, $(j = 1,2,3)$, we can adopt the methodology proposed by Gao [39] outlined in the following manner:

$$s^3 + \alpha_{11}s^2 + \alpha_{12}s + \alpha_{13} = (s + \gamma_{xo})^3 \tag{50}$$

$$s^3 + \alpha_{21}s^2 + \alpha_{22}s + \alpha_{23} = (s + \gamma_{yo})^3 \tag{51}$$

The choice of γ_{xo} and γ_{yo} is made to ensure that Equations (50) and (51) form Hurwitz polynomials with respect to the complex variable. The observer gain can be formulated as a function of the ESO bandwidth by utilizing Equations (50) and (51), as demonstrated below:

$$\begin{aligned} \alpha_{11} = 3\gamma_{xo}, \alpha_{12} = 3\gamma_{xo}^2, \alpha_{13} = \gamma_{xo}^3 \\ \alpha_{21} = 3\gamma_{yo}, \alpha_{22} = 3\gamma_{yo}^2, \alpha_{23} = \gamma_{yo}^3. \end{aligned} \tag{52}$$

The observer error associated with each ESO can be defined by employing Equations (46)–(49) as follows:

$$\dot{\hat{e}}_x = \dot{\lambda}_1 - \dot{\hat{\lambda}}_1 = (A_x - \alpha_{gx}C_x)\hat{e}_x + E_x\Delta_1 \tag{53}$$

$$\dot{\hat{e}}_y = \dot{\lambda}_2 - \dot{\hat{\lambda}}_2 = (A_y - \alpha_{gy}C_y)\hat{e}_y + E_y\Delta_2 \tag{54}$$

It is possible to express Equations (53) and (54) in matrix form as shown below:

$$\dot{\hat{e}} = \hat{H}\hat{e} + E_d \tag{55}$$

where $\hat{e} = [\hat{e}_x, \dot{\hat{e}}_x, \ddot{\hat{e}}_x, \hat{e}_y, \dot{\hat{e}}_y, \ddot{\hat{e}}_y]^T$, $\hat{H} = \begin{bmatrix} \hat{H}_1 & 0_3 \\ 0_3 & \hat{}H_2 \end{bmatrix}$, $\hat{H}_1 = \begin{bmatrix} -\alpha_{11} & 1 & 0 \\ -\alpha_{12} & 0 & 1 \\ -\alpha_{13} & 0 & 0 \end{bmatrix}$, $\hat{H}_2 = \begin{bmatrix} -\alpha_{21} & 1 & 0 \\ -\alpha_{22} & 0 & 1 \\ -\alpha_{23} & 0 & 0 \end{bmatrix}$, $E_d = \begin{bmatrix} 0 & 0 & \Delta_1 & 0 & 0 & \Delta_2 \end{bmatrix}^T$.

Lemma 1. *In Equation (55), the boundedness of* $\lim_{t\to\infty} \hat{e}(t)$ *can be guaranteed if at least one of the following two conditions is satisfied:*

- $\delta_i < n_1, i = 1,2$ *for all time t;*
- $\Delta_i < n_2, i = 1,2$ *for all time t.*

Asymptotic stability of the estimated error dynamics can be achieved when the values of δ_i, $i = 1,2$, *are either directly obtained or assumed to be constant, leading to* $\Delta_i = 0, i = 1,2$. *In this scenario, the positive constants* n_1 *and* n_2 *play a vital role in ensuring the system's stability. Lemma 1, as stated in Zhang et al. [40] , establishes that the roots of the matrix* $\hat{H}$ *in Equation (55) reside in the left half plane. This result is ensured by the nonnegativity of the bandwidths* γ_{xo} *and* γ_{yo}. *Consequently, it can be deduced that the estimated error dynamics described by Equations (53) and (54) are asymptotically stable.*

4.2. New Robust Feedback Controller

The feedback controller presented in Equations (19) and (20) relies on state measurements, but except for λ_{11} and λ_{21}, the remaining states cannot be accurately measured. To solve this problem, the state estimation obtained through the two ESOs defined in Equations (48) and (49) are used instead. Furthermore, in order to simplify the compensation of the lumped disturbances δ_1 and δ_2, they are replaced by their approximations,

$\hat{\delta}_1$ and $\hat{\delta}_2$. By incorporating the results of the extended state observers (ESOs), a robust feedback controller can be developed in the following manner:

$$v_{SADRCx} = \ddot{\lambda}_{xd} - \beta_1 \dot{\hat{e}}_1 - k_1 sat(\hat{s}_x) - \hat{\delta}_1 \quad (56)$$

$$v_{SADRCy} = \ddot{\lambda}_{yd} - \beta_2 \dot{\hat{e}}_2 - k_2 sat(\hat{s}_y) - \hat{\delta}_2 \quad (57)$$

according to the sliding mode active disturbance rejection control feedback given in Equations (56) and (57), we can obtain the new robust tracking controller named Flatness-Sliding-Active-Disturbance-Rejection Control (FSADRC), defined as follows:

$$\begin{bmatrix} u_{FSADRCx} \\ u_{FSADRCy} \end{bmatrix} = B_r^{-1} \begin{bmatrix} \ddot{\lambda}_{xd} - \beta_1 \dot{\hat{e}}_1 - k_1 sat(\hat{s}_x) - \hat{\delta}_1 \\ \ddot{\lambda}_{yd} - \beta_2 \dot{\hat{e}}_2 - k_2 sat(\hat{s}_y) - \hat{\delta}_2 \end{bmatrix} \quad (58)$$

where $\hat{e}_{r1} = \hat{\lambda}_{11} - \lambda_{xd}$ and $\hat{e}_{r2} = \hat{\lambda}_{21} - \lambda_{yd}$. The schematic diagram presented in Figure 3 illustrates the principle of trajectory tracking control for a mobile robot.

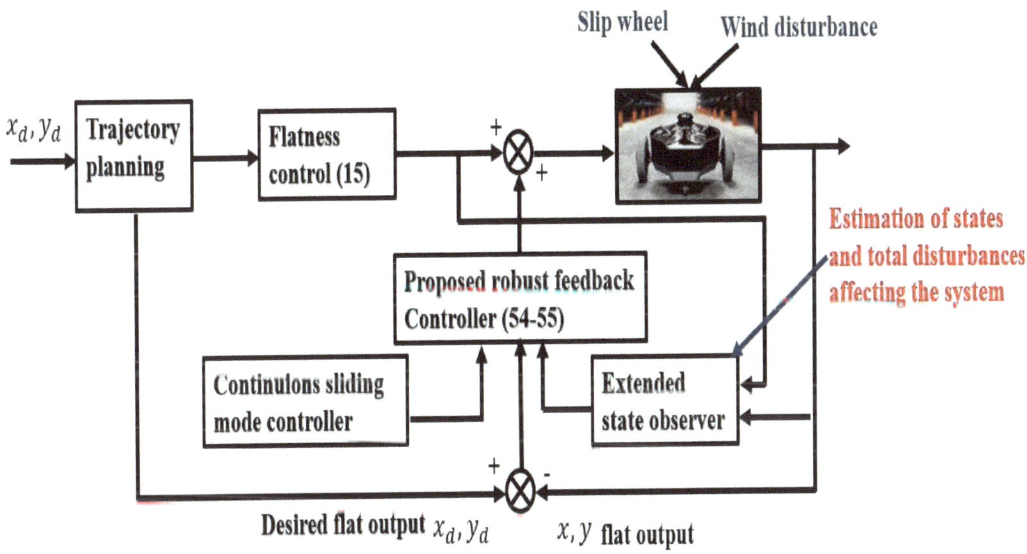

Figure 3. Mobile robot trajectory tracking control principle scheme.

4.3. Stability Analysis of the Closed-Loop System

This section will address the stability analysis of the tracking error systems for x and y, utilizing the estimation error defined by Equations (53) and (54). In order to prove the stability of the error dynamics of position x, Lyapunov's function is chosen as follows:

$$V_{sx} = \frac{1}{2} s_x^2 \quad (59)$$

where $s_x = \dot{e}_1 + \beta_1 e_1 = \dot{\lambda}_{11} - \dot{\lambda}_{xd} + \beta_1(\lambda_{11} - \lambda_{xd}), \lambda_{11} = x, \lambda_{xd} = x_d$.

We can define the derivative of the Lyapunov function V_{sx} as follows:

$$\dot{V}_{sx} = s_x \dot{s}_x = s_x(\ddot{\lambda}_{11} - \ddot{\lambda}_{xd} + \beta_1(\dot{\lambda}_{11} - \dot{\lambda}_{xd})) \quad (60)$$

When replacing $\ddot{\lambda}_{11}$ by its Equation (44) defined by $\dot{\lambda}_{11} = v_1 + \delta_1$, we obtain:

$$\dot{V}_{sx} = s_x \dot{s}_x = s_x(v_1 + \delta_1 - \ddot{\lambda}_{xd} + \beta_1(\dot{\lambda}_{11} - \dot{\lambda}_{xd})) \quad (61)$$

When v_1 represents the feedback controller, substituting it with the proposed robust feedback tracking control, denoted as v_{SADRCx} defined by Equation (56), yields:

$$\begin{aligned} \dot{V}_{sx} &= s_x\dot{s}_x = s_x(\ddot{\lambda}_{xd} - \beta_1(\hat{\dot{\lambda}}_{11} - \dot{\lambda}_{xd}) - k_1 sat(\hat{s}_x) - \hat{\delta}_1 + \delta_1 - \ddot{\lambda}_{xd} + \beta_1(\dot{\lambda}_{11} - \dot{\lambda}_{xd})) \\ \dot{V}_{sx} &= s_x\dot{s}_x = s_x(\beta_1(\dot{\lambda}_{11} - \hat{\dot{\lambda}}_{11}) + \delta_1 - \hat{\delta}_1 - k_1 sat(\hat{s}_x)) \end{aligned} \tag{62}$$

where s_x is defined as follows:

$$Sat(s_x) = \begin{cases} \frac{s_x}{a_{sx}} & if \quad |s_x| \le a_{sx} \\ sgn(s_x) & if \quad |s_x| > a_{sx} \end{cases} \tag{63}$$

Concerning the stability and boundedness of the ESO defined by Equation (53), it can be achieved by choosing α_{gx} in such a way that the eigenvalues of $A_x - \alpha_{gx}C_x$ are negative, indicating poles in the left-hand plane, and ensuring that uncertainty is bounded. As a result, the error $\hat{e}x \to 0$. This implies that $\hat{\lambda}11 \to \lambda_{11}$, $\hat{\delta}_1 \to \delta_1$, and $\hat{s}_x \to s_x$. In this scenario, the Lyapunov function defined by Equation (62) is formulated as follows:

$$\dot{V}_{sx} = -s_x(k_1 sat(s_x)) \tag{64}$$

Since $Sat(s_x)$, defined by Equation (63), is divided into two segments, the proof process will be analyzed in two cases. In the first scenario, when the saturation function is defined as described by:

$$Sat(s_x) = \frac{s_x}{a_{sx}} \tag{65}$$

Moreover, the Lyapunov function is defined as follows:

$$\dot{V}_{sx} = -\frac{k}{a_{sx}}(s_x^2) \le 0 \tag{66}$$

Alternatively, when the saturation function is given by:

$$Sat(s_x) = sgn(s_x) \tag{67}$$

the Lyapunov function takes the form:

$$\dot{V}_{sx} = -k_1 s_x sgn(s_x) \le 0 \tag{68}$$

Thus, based on Equations (66) and (68), it can be concluded that the Lyapunov function $\dot{V}_{sx}$ is negative regardless of the definition of the function $Sat(s_x)$. As a result, the tracking error of the position x is stable. Similarly, the same conclusions about the stability of the closed-loop system y can be drawn.

5. Simulation Results

This section presents simulation tests to validate the efficacy and superiority of the suggested controller, flatness sliding active disturbance rejection control (FSADRC), as defined by Equation (58). The proposed control is evaluated against flatness sliding mode control (FSMC), represented by Equation (39), and flatness-based tracking control (FBTC), as defined in Equation (24), using computer simulation results. The parameters of the WMR are $r = 0.1$ m, $b = 0.15$ m. To enhance the observation and comparison of the simulation results, we have chosen two types of reference trajectories: a circular path and a Bézier curve. Additionally, we also consider two different scenarios of perturbation. The controller design parameters of FBTC, FSMC, and FSADRC are chosen as $m_x = m_y = 1$, $\epsilon_{xc} = \epsilon_{yc} = 2$, $\beta_1 = \beta_2 = 5$, and $k_1 = k_2 = 10$. As suggested by Gao [39], it is advisable to select the observer bandwidth to be sufficiently higher than the controller bandwidth. This ensures that the observer dynamics remain faster than the system dynamics, enabling effective disturbance estimation and compensation. In our case, we have chosen observer bandwidths of $\gamma_{xo} = \gamma_{yo} = 6$ rad/s to fulfill this requirement and ensure robust performance of the control system. To ensure that the sliding mode control system achieves both satisfactory dynamic and steady-state performance, and to prevent chatter in the control signal, the cut and dry method is frequently employed to establish the thickness of the boundary layer. Specifically, in this case, $a_{sx} = a_{sy} = 0.3$ is chosen.

5.1. First Scenario

In this simulation, we consider that slip velocities v_t and v_s can be up to 30% of the forward speed. Thus, $\kappa_1 = 0.3$. In addition, the WMR is subjected to constant wind perturbation defined as follows:

$$p_x = p_y = 3\ \mathrm{m/s}, w_s = 0.5\ \mathrm{rad/s} \tag{69}$$

The reference trajectory considered in this scenario is a circle, which is defined by the following equation:

$$x_r = cos(t), y_r = sin(t) \tag{70}$$

The performance of the uncertain WMR systems under different control strategies, namely FBTC, FSMC, and FSADRC, is depicted in Figure 4. Figure 5 shows the results of the estimated lumped disturbance affecting the x and y channels obtained using the extended state observer (ESO). Figure 6 illustrates the control input applied to the wheeled mobile robot under the conditions of the first scenario. The simulation results indicate that the uncertainty caused by slow wind perturbation and slip decreases the tracking performance in trajectory following, rendering FBTC ineffective as a controller. On the other hand, both FSMC and FSADRC demonstrate robustness in handling the overall disturbance affecting the WMR model. These controllers exhibit the ability to mitigate disturbances and successfully maintain the desired trajectory of the WMR system. Consequently, it can be inferred that controllers that disregard uncertain models, despite being feedback controllers, may exhibit unsatisfactory performance. The fundamental distinction between the FSMC and FSADRC controllers lies in their design methodologies and approaches. FSMC relies on finely-tuned gains to achieve disturbance rejection, which can lead to chattering due to the relatively high gain values. In contrast, FSADRC combines the advantages of the boundary layer method to minimize chattering and an ESO to estimate and eliminate lumped disturbance.

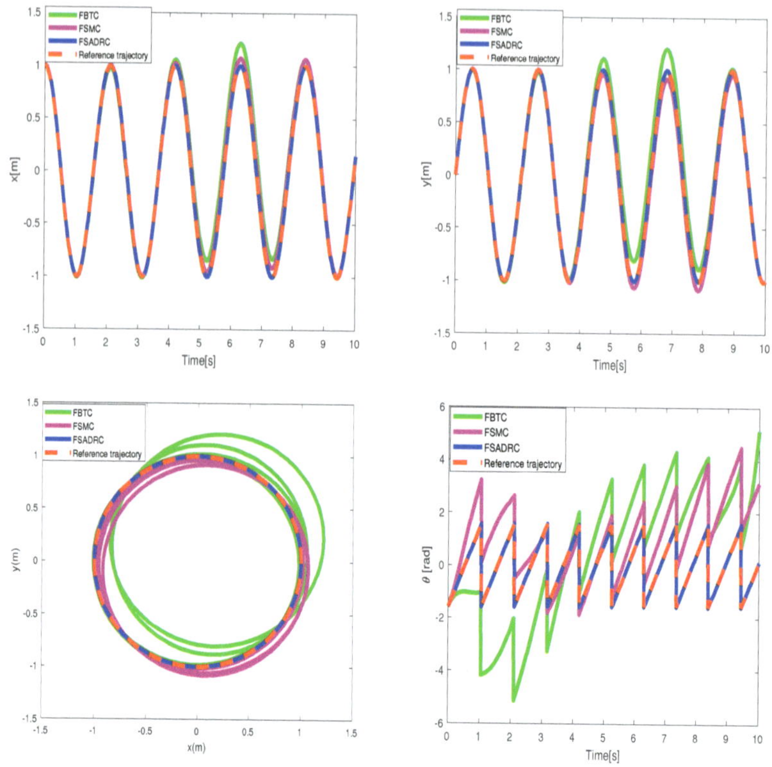

Figure 4. Simulation tracking results of the wheeled mobile robot under the conditions of the first scenario.

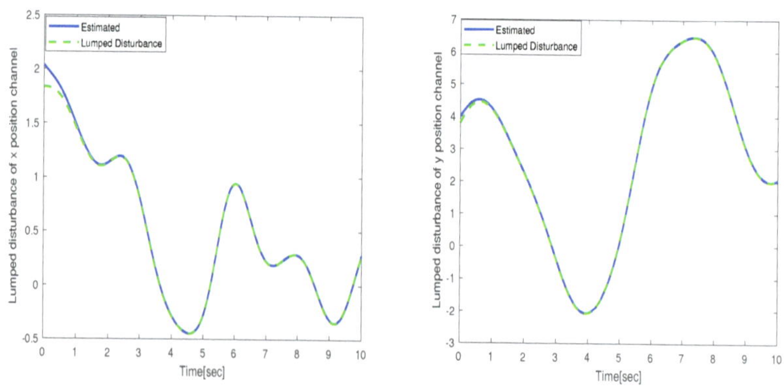

Figure 5. Lumped disturbance affecting the x and y position channels in the context of the first scenario.

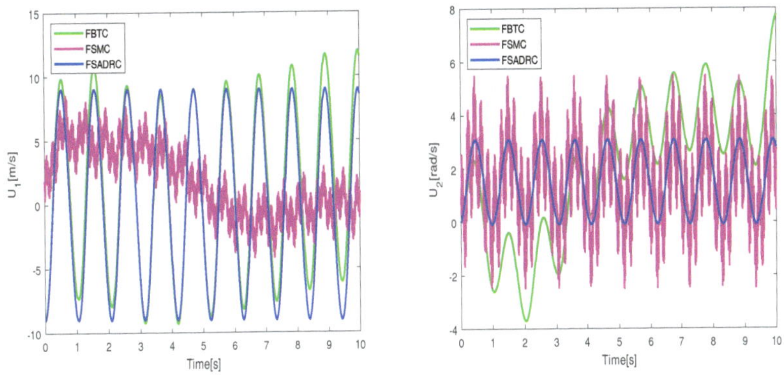

Figure 6. Control input applied to the wheeled mobile robot under the conditions of the first scenario.

5.2. Second Scenario

The objective of this simulation is to create and follow a trajectory for a robot, starting from an initial state where $x(0) = y(0) = 0$, and reaching a final state specified by $x(10) = 3.5$ and $y(10) = 5$. This trajectory must navigate through a room containing obstacles, while also considering time-varying wind disturbances and slipping. The desired trajectory should meet the following criteria: minimizing energy consumption, maneuvering around static obstacles, and adhering to the specified state constraints as follows:

$$0\,m \leq \lambda_{xd} \leq 4\,m, 0\,m \leq \lambda_{yd} \leq 6\,m \tag{71}$$

The optimal trajectory generation method proposed in [17] offers a solution to obtain the desired trajectory by solving a nonlinear optimization problem. By integrating the principles of flatness, the collocation method, and B-spline functions, this method efficiently generates trajectories while guaranteeing constraint satisfaction. To ensure consistency in the simulation results, the parameters for all three controllers remain unchanged from the previous simulations. Considering an uncertain initial condition of $\hat{x}(0) = 1$ and $\hat{y}(0) = 1$ for the wheeled mobile robot (WMR), we further specify that the slip velocities v_t and v_s can potentially reach up to 50% to 70%. In addition, we take into account the influence of sinusoidal wind disturbances. In contrast to the initial scenario, the disturbance signals consist of combinations of multi-frequency sinusoidal signals representing time-varying disturbances, particularly wind, defined as follows:

$$p_x = p_y = 1.5 + 2.5sin(4t) + 4.5cos(2t), w_s = 1.5 + 3cos(2t) \tag{72}$$

The simulation results regarding trajectory tracking performance of the second scenario are depicted in Figure 7. Based on these figures, it can be observed that the WMR system experiences significant divergence from the desired trajectory when affected by slippage and external disturbances, rendering FBTC ineffective as a controller. The FSMC controller's intervention through the sliding mode's discontinuous term eliminates uncertainty effects and maintains the stability of the closed-loop control. However, as shown in Figure 8, the presence of chattering in the FSMC control signals negatively impacts the system's behavior. Hence, it can be inferred that while FSMC is a robust control approach, its practical applicability is quite restricted. Therefore, developing a control approach capable of mitigating the chattering effect while maintaining the robustness advantage provided by SMC is necessary. The results of the lumped disturbance estimation for this simulation are illustrated in Figure 9. According to the simulation findings, the mobile robot satisfactory trajectory tracking performance when confronted with model disturbances and uncertain initial conditions while employing the FSADRC controller. Of greater significance, the proposed control methodology achieves superior tracking of the desired trajectory, devoid of the chattering phenomenon, and enhances tracking performance against aggressive disturbances.

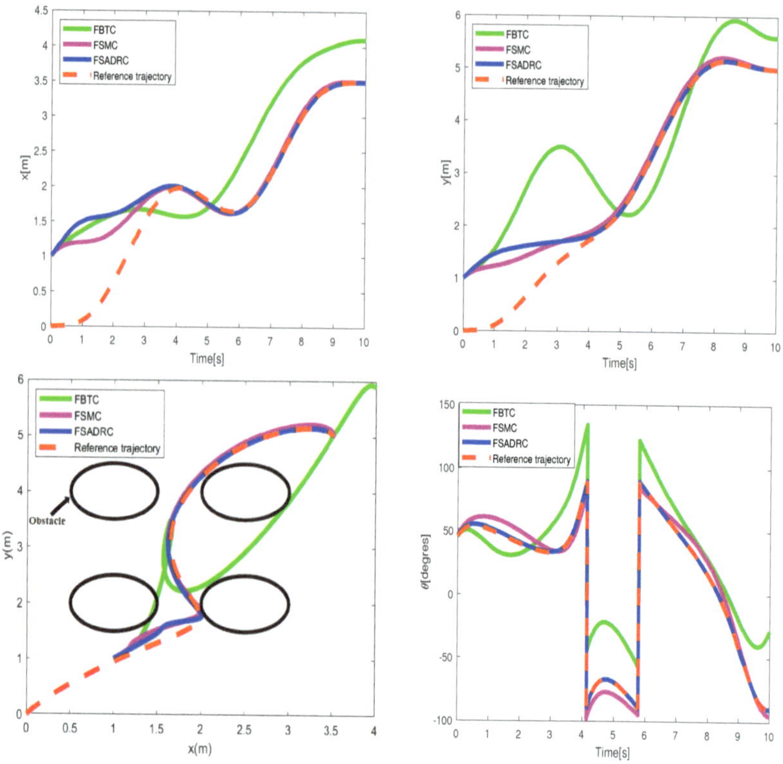

Figure 7. Simulation tracking results of the wheeled mobile robot in the conditions of the second scenario.

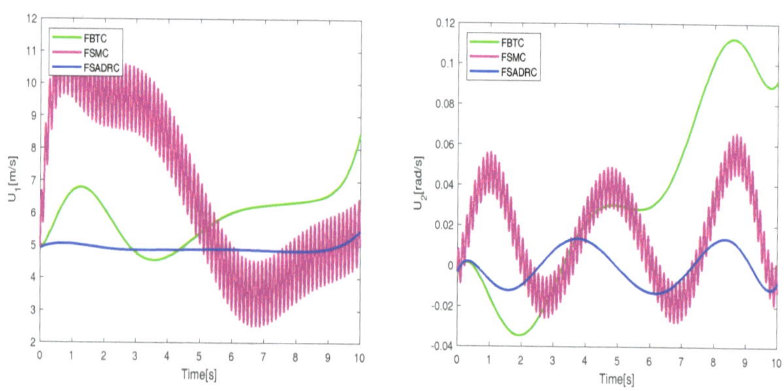

Figure 8. Control input applied to the wheeled mobile robot under the conditions of the second scenario.

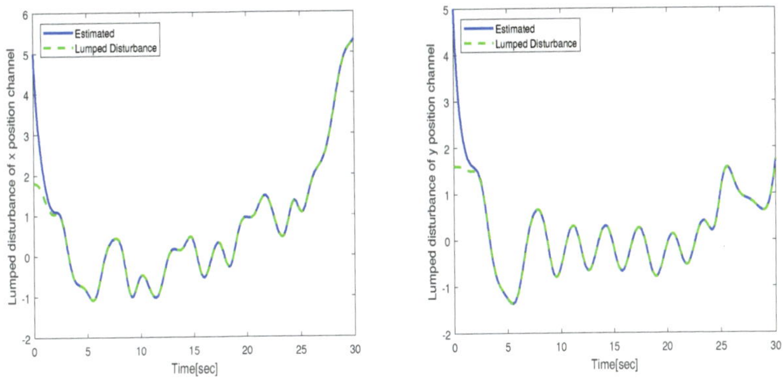

Figure 9. Lumped disturbance affecting the x and y position channels in the context of the second scenario.

6. Tracking the Experimental Results of a Wheeled Mobile Robot

This section outlines experiments conducted with the TurtleBot3, a Wheeled Mobile Robot (WMR), to evaluate a proposed methodology. The TurtleBot provides a cost-effective platform for researchers to explore and validate control algorithms without requiring expensive robotic systems. Its compatibility with the Robot Operating System (ROS) enhances its functionalities, offering resources for algorithm development and experimentation. With LiDAR, IMU, and wheel encoders onboard, the TurtleBot3 provides precise environmental feedback, facilitating algorithm optimization. Researchers can augment the system with additional sensors or hardware components to evaluate various control algorithms across diverse scenarios. To facilitate the observation and comparison of experimental results, we have selected two types of reference trajectories: an eight-shaped path and a Bézier curve. Additionally, we have considered two different scenarios of perturbation: the first involves slowly time-varying disturbances, while the second entails aggressive time-varying disturbances. For further validation, the performance of the proposed control method is compared with other state-of-the-art control techniques such as backstepping tracking control (BTC) [41], flatness active disturbance rejection control (FADRC) introduced in [42], flatness-based tracking control (FBTC) as defined by Equation (39), and backstepping sliding active disturbance rejection control (BSADRC) [43]. The controller design parameters selected for the experimental results are identical to those chosen for the simulation results.

6.1. First Experiment with Slowly Time-Varying Disturbances

In this experiment, eight shapes were chosen for the reference trajectory, as outlined below:

$$x_r = 2cos(t), y_r = -2sin(t) \quad (73)$$

To replicate real-world navigation conditions for the WMR, high-speed fans are utilized in the laboratory to simulate windy environments. Additionally, a stick is employed to disturb the castors of the WMR, creating slipping incidents, thus adding further realism to the testing environment. Figure 10 illustrates the real-time tracking of the eight-shaped reference trajectory of the WMR using the proposed control method described in this paper.

Figure 10. Real-time trajectory tracking experiment.

Simulation of the experiment under identical conditions reveal tracking trajectories in Figure 11. Figure 12 illustrates lumped disturbance estimation, while Figure 13 displays control torques. Based on the experimental results shown in Figure 11, it is evident that FADRC, FSADRC, and BSADRC methods excel at tracking trajectories even in the face of genuine uncertainty. Conversely, the FBTC and BTC methods demonstrate significant shortcomings when it comes to handling uncertainties. To assess the superiority of the proposed control, we will conduct a thorough study in the subsequent section. This study will include a quantitative analysis of the controllers under more severe disturbance conditions.

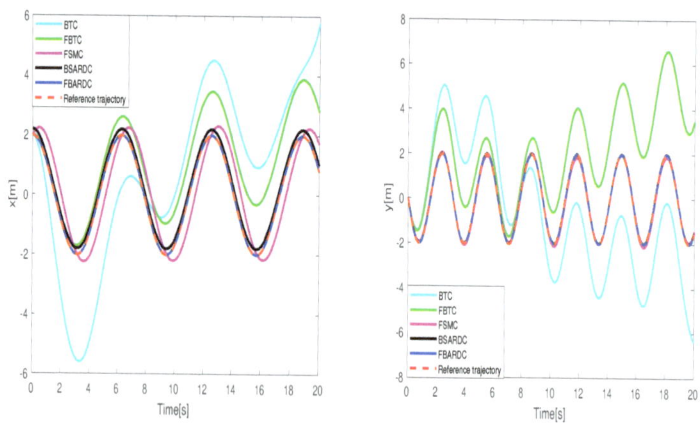

Figure 11. *Cont.*

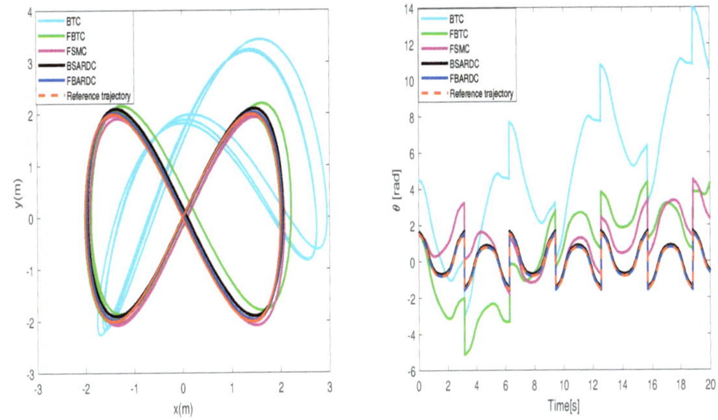

Figure 11. Results of the wheeled mobile robot's tracking under the conditions of the first experiment scenario.

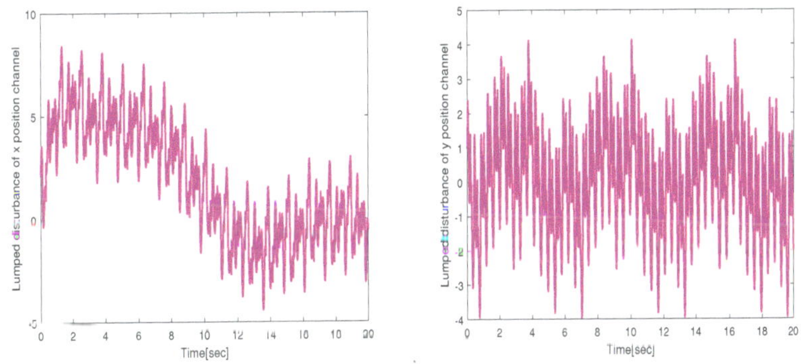

Figure 12. Estimation values of the lumped disturbances under the conditions of the first experiment scenario.

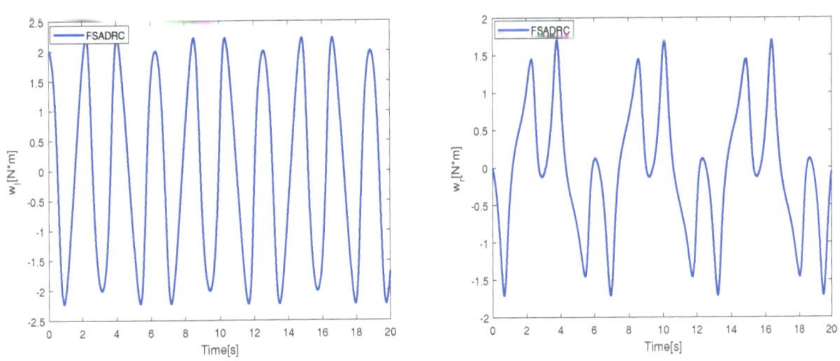

Figure 13. Control torques applied to the right and left wheels to track the eight-shaped reference trajectory.

6.2. Second Experiment with Aggressive Time-Varying Disturbances

In this experiment, we intensify the frequency of disturbance variation generated by the industrial ventilator and subject the robot to aggressive impacts with a stick to assess the effectiveness of the proposed controller. Additionally, we adopt the eighth-order Bézier curve as a reference trajectory for both the x and y positions, defined as follows:

$$\begin{aligned} \lambda_{xd} = x_r = P_{x0}(1-t)^8 + 8P_{x1}(1-t)^7 t + 28P_{x2}(1-t)^6 t^2 P + 56P_{x3}(1-t)^5 t^3 + \ldots \\ 70P_{x4}(1-t)^4 t^4 + 56P_{x5}(1-t)^3 t^5 + 28P_{x6}(1-t)^2 t^6 + 8P_{x7}(1-t)t^7 + P_{x8}t^8. \\ \lambda_{yd} = y_r = P_{y0}(1-t)^8 + 8P_{y1}(1-t)^7 t + 28P_{y2}(1-t)^6 t^2 P + 56P_{y3}(1-t)^5 t^3 + \ldots \\ 70P_{y4}(1-t)^4 t^4 + 56P_{y5}(1-t)^3 t^5 + 28P_{y6}(1-t)^2 t^6 + 8P_{y7}(1-t)t^7 + P_{y8}t^8. \end{aligned} \tag{74}$$

where P_{xj}, P_{yj}, and $j = 0 \ldots 8$ represent the control parameters of the reference trajectory. These parameters may vary depending on several factors, including the robot's initial position, the desired final position, and constraints such as obstacle avoidance. As an example, we select control parameters that allow the WMR to transition from its initial state $q_r(0) = [0,0,0]^T$ to the desired final state $q_r(20) = [2,2,0]^T$. The tracking experiment results of the WMR under aggressive time-varying disturbances are depicted in Figure 14. In Figure 15, the lumped disturbance affecting the WMR within the context of the second experimental scenario is displayed. Similarly, Figure 16 illustrates the proposed control input applied to the wheeled mobile robot within the same context.

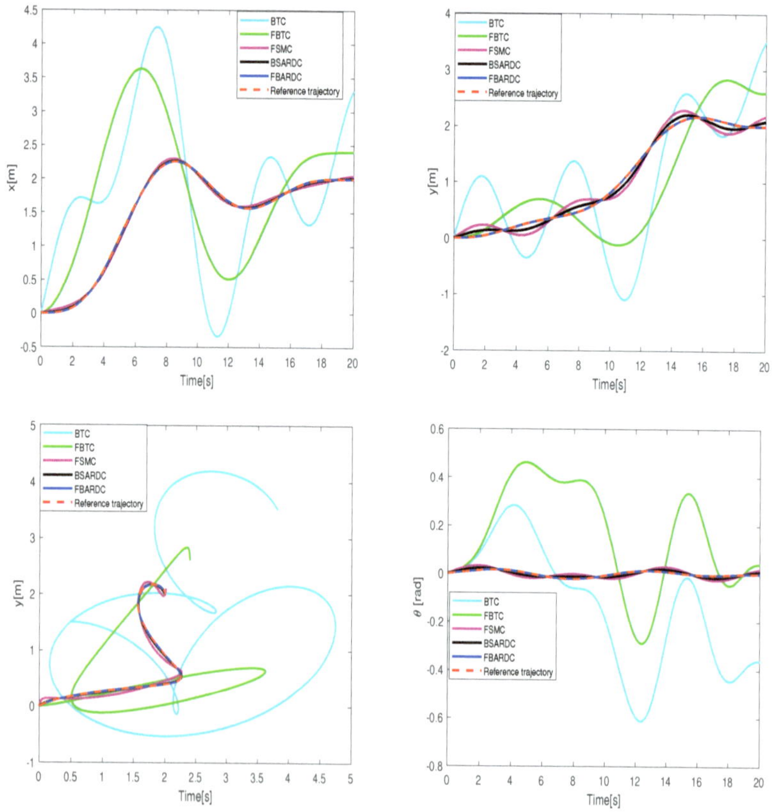

Figure 14. Results of the wheeled mobile robot's tracking under the conditions of the second experiment scenario.

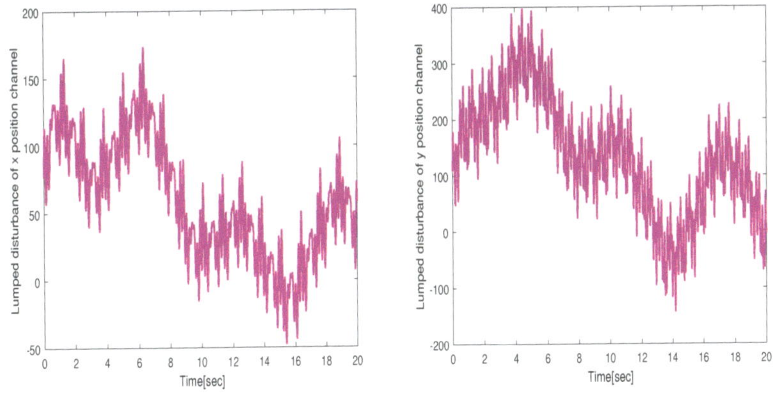

Figure 15. Estimated values of the lumped disturbances under the conditions of the second experiment scenario.

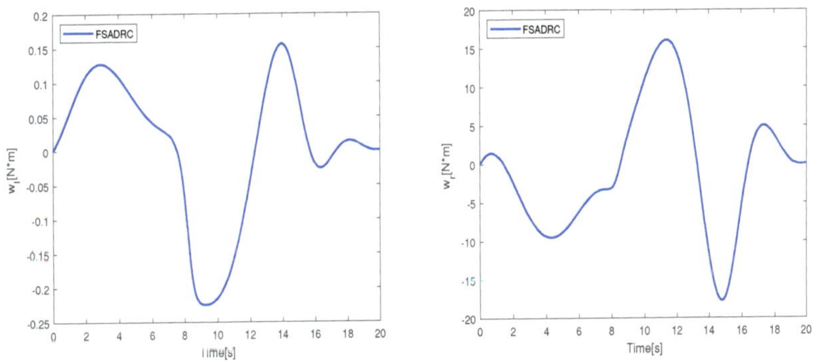

Figure 16. Control torques applied to the right and left wheels to track the Bézier reference trajectory.

To quantitatively assess the tracking performance of the WMR, we employed the integral absolute error (IAE) and the control effort performance index as comparison metrics. The IAE is computed for each of the control strategies in the following manner:

$$\text{IAE}_i = \int_0^{t_f} |e_i(t)|dt. \quad e_i(t) = \lambda_i(t) - \lambda_{id}(t), \tag{75}$$

where t_f is the total simulation duration and $i = 1,2$, represents the position in the x and y direction, respectively. The control effort is given as follows:

$$\text{P}_{\text{avg}} = \frac{1}{N}\sum_{k=1}^{N} u^2(k) \tag{76}$$

where N indicates the total count of samples. The associated key performance indicators IAE and P_{avg} for both strategies are provided in Table 1.

Table 1. Performance indexes IAE and P_{avg}.

Index	BTC	FBTC	FADRC	FSADRC	BSADRC
IAE	5.5351	4.2654	0.07	0.0127	0.02
P_{avg}	2.5351	0.261	0.1266	0.13	1.253

Examining the data in Table 1, it is evident that the FSADRC controller outperforms the BTC, FBTC, FADRC, and FSMC methods in terms of tracking performance. Although its tracking performance is nearly comparable to that of the BSADRC, the FSADRC requires minimal effort to accomplish its task compared to the BSADRC. This characteristic is particularly crucial in contexts where energy resources are limited, such as in mobile or autonomous applications. The enhanced efficiency of the FSADRC over the BSADRC is explained by the advantage of flatness control, which simplifies controller design by transforming the nonlinear system into a linear one. This feature makes all control based on the concept of flatness less complex than control based on backstepping. Ultimately, the experiment and table findings show that the disturbance rejection function simplifies the system model by addressing real-time modeling uncertainties. Consequently, the FSADRC method relies less on an exact analytical model description, treating unknown dynamics as internal disturbances compensated for by the rejection function. This enhances the robustness of FSADRC, which also incorporates the boundary layer technique to alleviate chattering effects.

7. Conclusions

This paper aims to introduce a robust control methodology for uncertain wheeled mobile robots (WMR). By employing flatness-based control, the nonlinear kinematic model of the WMR undergoes transformation into a canonical form, enabling the implementation of a robust feedback controller that incorporates boundary layer sliding mode control and extended state observer techniques. Simulation results conducted under various scenarios of uncertainties illustrate the effectiveness of FSADRC in enhancing the trajectory tracking performance of the WMR when compared to BTC, FBTC, and FADRC, even amid variations in slipping and external wind disturbances. Furthermore, within the same context, FSADRC demonstrates comparable efficiency to BSADRC in terms of trajectory tracking, while exhibiting an advantage in effort usage due to its flatness property. The smooth operation of FSADRC, coupled with its resilience against parameter variations and external disturbances, renders it a practical choice for real-world applications. Moreover, experimental findings using the TurtleBot3 validate the efficacy of the proposed FSADRC in real-world navigational tasks. In future studies, the application of FSADRC will extend to other robotic systems, such as quadrotors and arm manipulators, to assess its effectiveness and explore its potential for broader deployment.

Author Contributions: Methodology, A.A. (Amine Abadi); Validation, A.A. (Amine Abadi) and A.A. (Amani Ayeb); Resources, D.F.; Writing—original draft, A.A. (Amine Abadi); Writing—review & editing, M.L. and H.M.; Visualization, D.F.; Project administration, T.B.; Funding acquisition, T.B. All authors have read and agreed to the published version of the manuscript.

Funding: This research received no external funding.

Institutional Review Board Statement: Not applicable.

Informed Consent Statement: Not applicable.

Data Availability Statement: Data are contained within the article.

Conflicts of Interest: The authors declare no conflicts of interest.

Abbreviations

The following abbreviations are used in this manuscript:

WMR	Wheeled mobile robot
ADRC	Active disturbance rejection control
ESO	Extended state observer
BTC	Backstepping tracking control
FBTC	Flatness-based tracking control
FSMC	Flatness sliding mode control

FSADRC	Flatness sliding active disturbance rejection control
BSADRC	Backstepping sliding active disturbance rejection control

References

1. Al, A.S.M.A.O.; Al-Qassa, A.; Nasser, A.R.; Alkhayyat, A.; Humaidi, A.J.; Ibraheem, I.K. Embedded design and implementation of mobile robot for surveillance applications. *Indones. J. Sci. Technol.* **2021**, *6*, 427–440. [CrossRef]
2. Ebel, H.; Rosenfelder, M.; Eberhard, P. Cooperative object transportation with differential-drive mobile robots: Control and experimentation. *Robot. Auton. Syst.* **2024**, *173*, 104612. [CrossRef]
3. Yépez-Ponce, D.F.; Salcedo, J.V.; Rosero-Montalvo, P.D.; Sanchis, J. Mobile robotics in smart farming: Current trends and applications. *Front. Artif. Intell.* **2023**, *6*, 1213330. [CrossRef]
4. Garaffa, L.C.; Basso, M.; Konzen, A.A.; de Freitas, E.P. Reinforcement learning for mobile robotics exploration: A survey. *IEEE Trans. Neural Netw. Learn. Syst.* **2021**, *34*, 3796–3810. [CrossRef]
5. Zangina, U.; Buyamin, S.; Abidin, M.S.Z.; Azimi, M.S.; Hasan, H.S. Non-linear PID controller for trajectory tracking of a differential drive mobile robot. *J. Mech. Eng. Res. Dev.* **2020**, *43*, 255–269. [CrossRef]
6. Benchouche, W.; Mellah, R.; Bennouna, M.S. The Impact of the dynamic model in feedback linearization trajectory tracking of a mobile robot. *Period. Polytech. Electr. Eng. Comput. Sci.* **2021**, *65*, 329–343. [CrossRef]
7. Yousuf, B.M.; Saboor Khan, A.; Munir Khan, S. Dynamic modeling and tracking for nonholonomic mobile robot using PID and backstepping. *Adv. Control. Appl. Eng. Ind. Syst.* **2021**, *3*, e71. [CrossRef]
8. Fliess, M.; Lévine, J.; Martin, P.; Rouchon, P. A lie-backlund approach to equivalence and flatness of nonlinear systems. *IEEE Trans. Autom. Control.* **1999**, *44*, 922–937. [CrossRef]
9. Rigatos, G.; Zervos, N.; Siano, P.; Wira, P.; Abbaszadeh, M. Flatness-based control for steam-turbine power generation units using a disturbance observer. *IET Electr. Power Appl.* **2021**, *15*, 1013–1028. [CrossRef]
10. Rigatos, G.; Hamida, M.A.; Abbaszadeh, M.; Siano, P. Flatness-based disturbance observer for condition monitoring of marine power generation units. *Proc. Inst. Mech. Eng. Part I J. Syst. Control.* **2023**, *237*, 1620–1634. [CrossRef]
11. Jing, C.; Xu, H.; Song, X.; Lu, B. Adaptive extended state observer-based flatness nonlinear output control for torque tracking of electrohydraulic loading system. *Trans. Inst. Meas. Control* **2018**, *40*, 2999–3009. [CrossRef]
12. Rigatos, G.; Wira, P.; Abbaszadeh, M.; Pomares, J. Flatness-based control in successive loops for industrial and mobile robots. In Proceedings of the IECON 2022–48th Annual Conference of the IEEE Industrial Electronics Society, Brussels, Belgium, 17–20 October 2022; pp. 1–6. [CrossRef]
13. Lin-Shi, X.; Simon, T.; Trégouët, J.F.; Morel, H. Flatness-Based Control of an m-Branch Power Flow Controller for Meshed DC Microgrids. In Proceedings of the IEEE 1st International Power Electronics and Application Symposium (PEAS), Shanghai, China, 13–15 November 2021; pp. 1–6. [CrossRef]
14. Rigatos, G.; Abbaszadeh, M.; Pomares, J.; Wira, P.; Cuccurullo, G. Flatness-based control in successive loops for robotic manipulators and autonomous vehicles. *Int. J. Syst. Sci.* **2024**, *55*, 954–979. [CrossRef]
15. Yu, X.; Zhou, X.; Guo, K.; Jia, J.; Guo, L.; Zhang, Y. Safety flight control for a quadrotor UAV using differential flatness and dual-loop observers. *IEEE Trans. Ind. Electron.* **2021**, *69*, 13326–13336. [CrossRef]
16. Abadi, A.; El Amraoui, A.; Mekki, H.; Ramdani, N. Guaranteed trajectory tracking control based on interval observer for quadrotors. *Int. J. Control* **2020**, *93*, 2743–2759. [CrossRef]
17. Abadi, A.; Mekki, H.; Brahim, A.B.H.; El Amraoui, A.; Ramdani, N. Optimal trajectory generation and flatness tracking control for a mobile robot. In Proceedings of the 18th International Conference on Sciences and Techniques of Automatic Control and Computer Engineering (STA), Monastir, Tunisia, 21–23 December 2017; pp. 223–228. [CrossRef]
18. Kaaniche, K.; El-Hamrawy, O.; Rashid, N.; Albekairi, M.; Mekki, H. Mobile Robot Control Based on 3D Visual Servoing: A New Approach Combining Pose Estimation by Neural Network and Differential Flatness. *Appl. Sci.* **2022**, *12*, 6167. [CrossRef]
19. Nasr, S.; Mekki, H.; Bouallegue, K. A multi-scroll chaotic system for a higher coverage path planning of a mobile robot using flatness controller. *Chaos Solitons Fractals* **2022**, *118*, 366–375. [CrossRef]
20. Yakovlev, K,S.; Andreychuk, A.; Belinskaya, J.; Makarov, D. Safe interval path planning and flatness-based control for navigation of a mobile robot among static and dynamic obstacles. *Autom. Remote Control* **2022**, *83*, 903–918. [CrossRef]
21. Khalesi, R.; Yousefi, M.; Pishkenari, H.N.; Vossoughi, G. Robust independent and simultaneous position control of multiple magnetic microrobots by sliding mode controller. *Mechatronics* **2022**, *84*, 102776. [CrossRef]
22. Lian, S.; Meng, W.; Shao, K.; Zheng, J.; Zhu, S.; Li, H. Full attitude control of a quadrotor using fast nonsingular terminal sliding mode with angular velocity planning. *IEEE Trans. Ind. Electron.* **2022**, *70*, 3975–3984. [CrossRef]
23. Li, J.; Wang, J.; Peng, H.; Hu, Y.; Su, H. Fuzzy-torque approximation-enhanced sliding mode control for lateral stability of mobile robot. *IEEE Trans. Syst. Man Cybern.* **2021**, *52*, 2491–2500. [CrossRef]
24. Utkin, V.; Lee, H. Chattering problem in sliding mode control systems. In Proceedings of the International Workshop on Variable Structure Systems, Alghero, Sardinia, 5–7 June 2006; pp. 346–350. [CrossRef]
25. Utkin, V. Discussion aspects of high-order sliding mode control. *IEEE Trans. Autom. Control* **2015**, *61*, 829–833. [CrossRef]
26. Boiko, I.M. Chattering in sliding mode control systems with boundary layer approximation of discontinuous control. *Int. J. Syst. Sci.* **2015**, *44*, 1126–1133. [CrossRef]

27. Guo, R.; Ding, Y.; Yue, X. Active adaptive continuous nonsingular terminal sliding mode controller for hypersonic vehicle. *Aerosp. Sci. Technol.* **2023**, *137*, 108279. [CrossRef]
28. Belguedri, M.; Benrabah, A.; Khoucha, F.; Benbouzid, M.; Benmansour, K. An improved uncertainty and disturbance estimator-based speed control for grid-connected pumping kite wind generator. *Control Eng. Pract.* **2024**, *143*, 105795. [CrossRef]
29. Han, J. From PID to active disturbance rejection control. *IEEE Trans. Ind. Electron.* **2023**, *56*, 900–906. [CrossRef]
30. Liu, L.; Wang, D.; Peng, Z. State recovery and disturbance estimation of unmanned surface vehicles based on nonlinear extended state observers. *Ocean Eng.* **2019**, *171*, 625–632. [CrossRef]
31. Zhao, L.; Li, Z.; Li, H.; Liu, B. Backstepping integral sliding mode control for pneumatic manipulators via adaptive extended state observers. *ISA Trans.* **2024**, *144*, 374–384. [CrossRef] [PubMed]
32. Chowdhury, D.; Al-Nadawi, Y.K.; Tan, X. Dynamic inversion-based hysteresis compensation using extended high-gain observer. *Automatica* **2022**, *135*, 109977. [CrossRef]
33. Xie, T.; Li, Y.; Jiang, Y.; An, L.; Wu, H. Backstepping active disturbance rejection control for trajectory tracking of underactuated autonomous underwater vehicles with position error constraint. *Int. J. Adv. Robot. Syst.* **2020**, *17*. [CrossRef]
34. Qi, G.; Hu, J.; Li, L.; Li, K. Integral Compensation Function Observer and Its Application to Disturbance-Rejection Control of QUAV Attitude. *IEEE Trans. Cybern.* **2024**, 1–12. [CrossRef]
35. Aole, S.; Elamvazuthi, I.; Waghmare, L.; Patre, B.; Meriaudeau, F. Improved active disturbance rejection control for trajectory tracking control of lower limb robotic rehabilitation exoskeleton. *Sensors* **2020**, *20*, 3681. [CrossRef] [PubMed]
36. Hu, Y.; Li, B.; Jiang, B.; Han, J.; Wen, C.Y. Disturbance Observer-Based Model Predictive Control for an Unmanned Underwater Vehicle. *J. Mar. Sci. Eng.* **2024**, *12*, 94. [CrossRef]
37. Ryu, J.C.; Agrawal, S.K. Differential flatness-based robust control of mobile robots in the presence of slip. *Int. Robot. Res.* **2011**, *30*, 463–475. [CrossRef]
38. Mauledoux, M.; Mejia-Ruda, E.; Aviles Sanchez, O.; Dutra, M.S.; Rojas Arias, A. Design of Sliding Mode Based Differential Flatness Control of Leg-wheel Hybrid Robot. *Appl. Mech. Mater.* **2011**, *835*, 681–686. [CrossRef]
39. Gao, Z. Scaling and bandwidth-parameterization based controller tuning. In Proceedings of the 2003 American Control Conference, Denver, CO, USA, 4–6 June 2003; pp. 4989–4996. [CrossRef]
40. Zhang, Y.; Jiang, Z.; Yang, H.; Cheng, J.; Zhang, W. High-order extended state observer-enhanced control for a hypersonic flight vehicle with parameter uncertainty and external disturbance. *Proc. Inst. Mech. Eng. Part G J. Aerosp. Eng.* **2015**, *229*, 2481–2496. [CrossRef]
41. Rudra, S.; Barai, R.K.; Maitra, M. Design and implementation of a block-backstepping based tracking control for nonholonomic wheeled mobile robot. *Int. J. Robust Nonlinear Control* **2016**, *26*, 3018–3035. [CrossRef]
42. Abadi, A.; El Amraoui, A.; Mekki, H.; Ramdani, N. Flatness-Based Active Disturbance Rejection Control For a Wheeled Mobile Robot Subject To Slips and External Environmental Disturbances. *IFAC-PapersOnLine* **2020**, *53*, 9571–9576. [CrossRef]
43. Dou, J.X.; Kong, X.X.; Wen, B.C. Backstepping sliding mode active disturbance rejection control of quadrotor attitude and its stability. *J. Northeast. Univ. (Nat. Sci.)* **2016**, *37*, 1415. [CrossRef]

sensors

MDPI

Article

Enhancing Robot Task Planning and Execution through Multi-Layer Large Language Models

Zhirong Luan [1,*], Yujun Lai [1], Rundong Huang [1], Shuanghao Bai [2], Yuedi Zhang [2], Haoran Zhang [2] and Qian Wang [1]

[1] School of Electrical Engineering, Xi'an University of Technology, Xi'an 710000, China; 2221920082@stu.xaut.edu.cn (Y.L.); 3211712276@stu.xaut.edu.cn (R.H.); wangqian77@xaut.edu.cn (Q.W.)
[2] College of Artificial Intelligence, Xi'an Jiaotong University, Xi'an 710000, China; baishuanghao@stu.xjtu.edu.cn (S.B.); zyd993@stu.xjtu.edu.cn (Y.Z.); zhr2001@stu.xjtu.edu.cn (H.Z.)
* Correspondence: luanzhirong@xaut.edu.cn

Abstract: Large language models have found utility in the domain of robot task planning and task decomposition. Nevertheless, the direct application of these models for instructing robots in task execution is not without its challenges. Limitations arise in handling more intricate tasks, encountering difficulties in effective interaction with the environment, and facing constraints in the practical executability of machine control instructions directly generated by such models. In response to these challenges, this research advocates for the implementation of a multi-layer large language model to augment a robot's proficiency in handling complex tasks. The proposed model facilitates a meticulous layer-by-layer decomposition of tasks through the integration of multiple large language models, with the overarching goal of enhancing the accuracy of task planning. Within the task decomposition process, a visual language model is introduced as a sensor for environment perception. The outcomes of this perception process are subsequently assimilated into the large language model, thereby amalgamating the task objectives with environmental information. This integration, in turn, results in the generation of robot motion planning tailored to the specific characteristics of the current environment. Furthermore, to enhance the executability of task planning outputs from the large language model, a semantic alignment method is introduced. This method aligns task planning descriptions with the functional requirements of robot motion, thereby refining the overall compatibility and coherence of the generated instructions. To validate the efficacy of the proposed approach, an experimental platform is established utilizing an intelligent unmanned vehicle. This platform serves as a means to empirically verify the proficiency of the multi-layer large language model in addressing the intricate challenges associated with both robot task planning and execution.

Keywords: robots; large language models; natural language; semantic alignment method

Citation: Luan, Z.; Lai, Y.; Huang, R.; Bai, S.; Zhang, Y.; Zhang, H.; Wang, Q. Enhancing Robot Task Planning and Execution through Multi-Layer Large Language Models. *Sensors* **2024**, *24*, 1687. https://doi.org/10.3390/s24051687

Academic Editors: David Cheneler and Stephen Monk

Received: 30 January 2024
Revised: 22 February 2024
Accepted: 1 March 2024
Published: 6 March 2024

1. Introduction

Grounded in experiential learning and knowledge accumulation, humans demonstrate a remarkable ability to comprehend intricate tasks through simple communication. Large language models (LLMs), when subjected to extensive and diverse datasets during training, possess the capability to emulate human-like language understanding and the discernment of human intentions. Exploiting the inherent augmentation capability within large language models enables the decomposition of tasks into multiple subtasks of reduced complexity [1]. This distinctive feature can be harnessed for task planning within robotic systems, ultimately leading to more efficient and seamless human–robot interactions.

The current state of research on robotic task planning, grounded in large language models, remains at its nascent stage. Prevailing studies have predominantly concentrated on tasks characterized by low complexity, such as robotic arm trajectory planning and robotic object handling. While these investigations have contributed significantly to establishing a theoretical framework for large-model-based robot control, they fall short in addressing tasks of elevated complexity. Illustratively, consider the task wherein Bob requests a drink of water from Sam. Sam translates this request into a series of small tasks, encompassing finding a cup, locating a water source, filling the cup, returning to Bob's location, and passing him the cup. The significance of hydration is often underestimated, and the subdivision of such high-complexity tasks into smaller, manageable components is pivotal. Each subtask should be designed to be straightforward, executed through muscle memory. However, prevailing research has predominantly fixated on smaller and more basic tasks. To navigate the intricacies of more complex tasks [2], the employment of a large language model for robotic task planning on a macro level becomes imperative. Furthermore, the expansion into more intricate tasks necessitates the robot's ability to adeptly handle and integrate complex environmental information into the task planning process.

In tackling this challenge, our investigation reveals that the direct generation of a robot control code using a large language model (LLM) is impractical, leading to considerable latency and errors. A comprehensive examination of cognitive processes underscores that the precision of outcomes produced by a step-by-step model exceeds that of directly generated results [3]. Consequently, it is advisable for the large language model to transition to a step-by-step mode for optimizing the effectiveness of robot motion planning.

Recognizing the superior aptitude of large language models (LLMs) in understanding semantic-level information and delivering accurate feedback, this paper introduces a multi-layer task decomposition architecture employing large language models. The initial step involves breaking down a complex task into a sequence of low-complexity tasks, resembling a common-sense-like progression, aimed at mitigating the overall complexity and execution difficulty of the task [4]. However, these task sequences remain impractical for direct execution by the robot due to the absence of essential environmental information.

To address this limitation, a visual language model is constructed to sense the physical attributes of environmental information. This model interacts with the information through the large language model, thereby acquiring localized environmental information [5]. Subsequently, a subsequent round of task decomposition ensues, generating fine-grained tasks by amalgamating the acquired environment information with low-complexity tasks. To enable effective robot control, alignment between the decomposed tasks and robot commands is achieved at the semantic level. This alignment ensures that the output of the LLM corresponds seamlessly with the semantic requirements of the robot task commands, achieved through feature vector alignment [6]. Consequently, the large language model can output tasks at the semantic level that precisely control the robot to execute the corresponding actions. The methodology outlined in Figure 1 provides a comprehensive overview of the proposed approach in this paper.

We devised a two-dimensional computable space through the implementation of the "heat map algorithm". This methodology involves the mapping of visual information onto a 2D computable space, offering insights into the relative positions of objects. The resultant mapping guides the robot's movements by generating dense, point-like trajectories within the heat map. The real-time generation of this image-level mapping allows for dynamic trajectory adjustments in response to changes in the environment.

It is pertinent to highlight that our approach incorporates the sensing of environmental information through a visual language model, which subsequently feeds this information into a large language model [7]. This collaborative interaction facilitates the generation of environment-specific policies by the large language model, diverging from the reliance on pre-trained policies derived from extensive robot data. As a result, our methodology achieves zero-sample robot control within an open instruction set. The integration of the heat map algorithm into a planning framework, encompassing multiple large language models and visual language models, empowers the robot to comprehend abstract semantic information [8]. This integration not only facilitates accurate task execution but also enables free-form natural language control of robots for tasks of heightened complexity. Concurrently, we substantiate the efficacy of our approach in natural language understanding and the guidance of robot behavior.

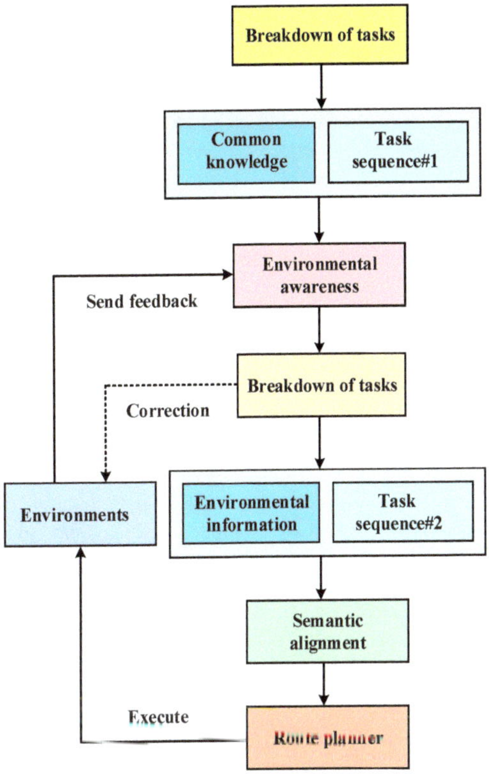

Figure 1. Overview of our approach.

Our contributions can be succinctly summarized as follows:

(1) Multi-Layer Task Decomposition: Our architecture uses large language models to guide robot behaviors through natural language, enhancing control in complex tasks.

(2) Integration of Environmental Perception: We employ a visual language model to input environmental information into the large language model, enabling task customization.

(3) Semantic Alignment for Task Control: Using semantic similarity methods, we align natural language descriptions with robot control instructions.

(4) Heat Map Navigation Algorithm: Our novel algorithm generates motion trajectories in a 2D space, guiding realistic robot behaviors.

The remainder of the paper is structured as follows: Section 2 provides an overview of related work; Section 3 delineates the architectural design and methodological principles; and Section 4 expounds upon the experimental methodology and presents an analysis of the results, while Sections 5 and 6 delve into discussions and summarize the methodology elucidated in this paper.

2. Related Work

Natural Language Interaction: Natural languages have been extensively researched for instruction extraction and robot control, where the language is able to make constraints and give behavioral specifications for robot behavior. Tellex et al. [9] described core aspects of language use in robots, including understanding natural language requests, using language to drive learning about the physical world, and engaging in collaborative dialogue with humans. These linguistic specifications can be used to reason about intermediate processes in natural language [10]. Micheli et al. [11] introduced a two-stage process and enhanced the performance of model training by interacting with the environment. Previous work has used classical methods for task sequence extraction, such as lexical analysis and formal logic to disassemble tasks. Thomason et al. [12] designed a mobile robotic dialogue agent that understands human commands through semantic analysis. More often than not, the focus of existing research has shifted from online to offline control of robot motion, with the help of local arithmetic enhancements, capable of executing local end-to-end behavioral patterns [13,14]. Brown et al. [15] trained GPT-3 and demonstrated that large language models can greatly improve the correctness of zero-sample recognition. A great deal of work has revolved around giving robot data the form of building robot datasets through natural language annotation. Model learning as we know it with imitation learning to reinforcement learning, all of these methods require a large amount of data in the form of natural language to generate a model with the robot's data, and the control of the robot can only be realized by interacting with a large amount of data Jiang et al. [16] argued that the combinatorial nature of language is crucial for learning different sub-skills and systematically generalizing them to new ones. The model of controlling a predictor for behavioral interaction with a robot through linguistic commands is closer to our ideas. Sharma et al. [17] optimized the predictor's model through supervised learning while generating collision-free trajectories in planar computable space. Huang et al. [18] used the code generation and language interaction capabilities of a pre-trained large language model to introduce the knowledge of the large language model into a three-dimensional computational space to instruct a machine to perform precision actions. In contrast, our work focuses on using the semantic understanding capability of the large language model to align instructions at the semantic level, solving the problem of difficulty in matching the task output from the large language model with the robot control instructions.

Language Models for Robotics: The use of large amounts of robotic data to train language models for solving real-world problems in the physical world is a popular field. A large amount of work has focused on the ability of models to understand natural language and interact with it. Zeng et al. [19] showed that such pre-trained micromodels have generic knowledge and that such models are capable of storing different forms of knowledge from a variety of domains. As a carrier of generic knowledge, the large language model needs to be combined with local scenario information to generate specialized knowledge, and a large amount of work has focused on combining environmental information with the macrolanguage model to enable the model to interact with the environment. Liang et al. [20] demonstrated that the large language model is capable of generating policy code from document strings, and they proposed a hierarchical code generation approach to enable the generation of complex code. Huang et al. [21] implemented robot behaviors by constructing action sequences using the general common sense of the large language model. After obtaining the environment information, the large language model is able to understand the environment information, but still lacks the ability to act, and executing the ability to act requires the large language model to invoke robot motion control commands, which often requires the provision of pre-generated libraries of action commands. Wu et al. [22] found that the large language model has excellent summarization and inductive capabilities and that the large language model summarizes user preferences, generates corresponding motion strategies, and invokes the mechanical base and robotic arm to perform the task of item summarization. In contrast, our focus is on improving the correctness of the large language model in controlling the robot's behavior so that the robot can understand complex semantic information and perform more complex tasks.

3. Method

First, we constructed a multi-layer large language model task decomposition architecture, describing in detail the design details and working principles of the architecture (Section 3.1). We then intervened in the process of fine-grained task decomposition by means of a semantic similarity approach to align the sequence of subtasks of the task decomposition with the atomic tasks we set out to perform (Section 3.2). We then demonstrated the generation of trajectories in 2D space to guide realistic robot behavior by sensing environmental information through visual and linguistic models (Section 3.3).

3.1. Multi-Layer Large Language Model Architecture for Task Decomposition

In our architectural framework, we incorporated two large language models, as depicted in Figure 2. The first large language model is responsible for comprehending human instructions and subsequently generating an executable coarse-grained plan for the robot. This process involves task decomposition at the level of common sense. However, it is essential to note that the task sequences produced by the first large language model may not precisely reflect the robot's behaviors [23]. This discrepancy arises from the inherent limitation of common sense, as it lacks environmental information. Consequently, the task sequences resulting from coarse-grained task decomposition exhibit a deficiency in incorporating the capability for interaction with the physical world environment.

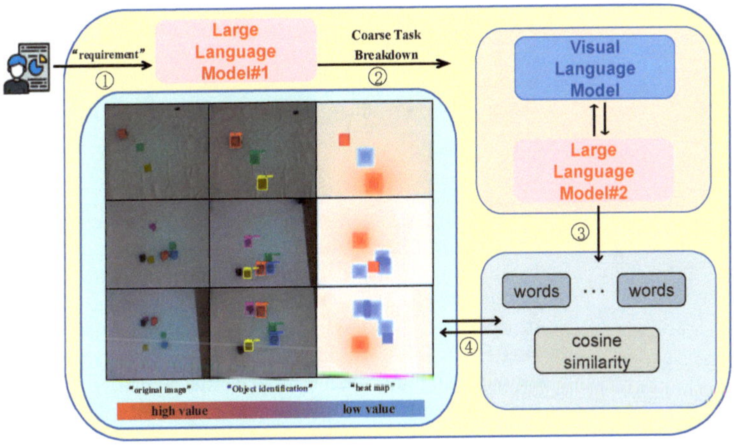

Figure 2. Overall structure and various functional modules.

To enhance the adaptability of the generalized knowledge within the large language model across diverse environments, the outcomes of the coarse-grained task decomposition were input into the subsequent functional module. This functional module comprises a large language model and a visual model. The visual model is equipped with an extensive repository of pre-trained a priori knowledge, enabling it to discern the categories of items within an image and thereby acquire information about the items in the environment [24]. It is worth mentioning that both LLMs used in this paper were based on ChatGPT 3.5. Additionally, the VLM, which interacts with the environment, is the OWL-ViT model developed by Google and released as open source. We utilized the APIs of ChatGPT 3.5 and OWL-ViT separately to facilitate information exchange among the various models. The large language model interacts with the visual model to extract environmental information perceived by the latter. Subsequently, it engages in a refined task decomposition process, incorporating the environment information. This iterative decomposition yields more precise fine-grained tasks, aligning with the specific nuances of the environment [25]. The generation of a sequence of fine-grained tasks is achieved by organizing these tasks based on the general knowledge embedded in the large language model.

While these tasks provide a precise linguistic description of the robot's motion, they encounter a challenge in controlling the robot's movement due to a misalignment issue between the generated task sequences and the robot's motion instructions. To facilitate the invocation of robot motion by the large language model, we employed a semantic similarity evaluation method to align task sequences with instruction sequences at the vector level. In this process, the text is first vectorized to represent tasks and instructions in a numerical format [26]. Subsequently, vector normalization is applied to mitigate the influence of text length on the vectors. Finally, at the vector level, tasks and instructions are aligned, ensuring semantic consistency. This alignment process allowed the fine-grained tasks generated by the large language model to be effectively mapped to the robot control instructions, overcoming the challenge of controlling the robot's motion.

Our approach involved transmitting the video captured by the camera to the visual model for processing, enabling the robot to execute appropriate behaviors upon receiving a command, such as navigating to single or multiple target locations [27]. Based on the recognition results derived from the visual information, a two-dimensional planar heat map is generated. In this heat map, the target location is characterized by a high heat value, while the remaining objects exhibited lower heat values. These heat values diverged toward the periphery, forming a comprehensive heat map. Specifically, in this experiment, the robot is represented in the heat map as solid red and blue blocks. Areas with high heat values are depicted as blocks with a red gradient, while non-target locations are represented by blocks with a blue gradient. Utilizing the principles of a greedy algorithm, we can calculate a trajectory to the target location with the highest heat value. The proposed thermal map is designed for real-time updating, ensuring prompt responsiveness to changes in environmental information [28]. Consequently, the navigation method based on the thermal map can quickly adapt and generate corresponding navigation instructions as the environmental context evolves.

3.2. Semantic Similarity-Based Alignment of Task Descriptions with Robot Control Instructions

Controlling robot motion through natural language strategies poses a significant challenge due to potential discrepancies between the task planning output from the large language model and the corresponding motion control functions of the robot. The content of the task planning output by the large language model exhibits variations in specific descriptions, introducing ambiguity in the understanding of the specific actions the robot needs to perform. To address this challenge, we employed semantic similarity, emphasizing semantic alignment rather than textual similarity [29], to correlate the task planning with the robot motion control instructions. Recognizing that some fine-grained tasks may require further decomposition to reach a machine-executable level, we proposed a cyclic semantic alignment method. This method aims to iteratively refine the alignment process, enhancing the correspondence between the nuanced task descriptions and the robot's motion control instructions.

In executable task planning, the task description output from the large language model and the robot control instruction are essentially two semantically similar texts, and we usually used cosine similarity to judge when measuring the similarity of the two texts. The semantic similarity can be computed by first embedding the text in the feature space, and then performing the similarity computation in the feature space [30]. Specifically, this paper used cosine similarity to compute semantic similarity.

We vectorized the text, assuming that the task output from the large language model is text A and the robot control command is text B. We first represented them as vectors $\mathbf{v}_A$ and $\mathbf{v}_B$, which are vectorized using the word embedding method, and these vectors represent the position of the text in the vector space.

$$\begin{aligned} \mathbf{v}_A &= (w_{A1}, w_{A2}, \ldots, w_{An}) \\ \mathbf{v}_B &= (w_{B1}, w_{B2}, \ldots, w_{Bm}) \end{aligned} \quad (1)$$

where n and m denote the size of the vocabulary in texts A and B, respectively, and w_{Ai} and w_{Bi} are the corresponding vocabulary weights.

In order to remove the effect of text length, the text vector can be normalized. The normalized vectors are denoted as $\mathbf{u}_A$ and $\mathbf{u}_B$:

$$\begin{aligned} \mathbf{u}_A &= \frac{\mathbf{v}_A}{\|\mathbf{v}_A\|} \\ \mathbf{u}_B &= \frac{\mathbf{v}_B}{\|\mathbf{v}_B\|} \end{aligned} \quad (2)$$

In Equation (2), $\mathbf{u}_A$ and $\mathbf{u}_B$ represent the normalized text vectors, while $\| \mathbf{v}_A \|$ and $\| \mathbf{v}_B \|$ represent the norms of vectors $\mathbf{v}_A$ and $\mathbf{v}_B$, respectively. The normalization operation essentially involves dividing each element in the vector by the vector's norm.

The effect of doing this is that, regardless of the original length of the text vector, the normalized text vectors all have unit lengths. As a result, when computing distances or comparing similarities between text vectors, they are not affected by the length of the text, thus allowing for better comparison of text similarities.

Cosine similarity is measured by calculating the dot product of two vectors and dividing by the product of the norms of the two vectors. The cosine similarity formula is as follows:

$$\text{Similarity}(\mathbf{u}_A, \mathbf{u}_B) = \frac{\mathbf{u}_A \cdot \mathbf{u}_B}{\| \mathbf{u}_A \| \cdot \| \mathbf{u}_B \|} \tag{3}$$

$$\text{Similarity}(\mathbf{u}_A, \mathbf{u}_B) = \frac{\sum\limits_{i=1}^{n} w_{Ai} \cdot w_{Bi}}{\sqrt{\sum_{i=1}^{n} w_{Ai}^2} \cdot \sqrt{\sum_{i=1}^{m} w_{Bi}^2}} \tag{4}$$

With text similarity matching, we were able to perform alignment between tasks and instructions. The detailed procedure is in Algorithm 1.

In the process of generating task plans, one situation that may occur is that the tasks generated by the large language model in conjunction with the environmental information are incorrect, and this error results in the semantics of the tasks not being able to be aligned with the semantics of the commands, leading to the inability to invoke the robot control commands [31]. For text that cannot be aligned at the semantic level, we fed the task output from the large language model back into the model for a new round of decomposition to re-generate the task sequence, and this process continued until the latest coarse-grained task is fed into the model [32].

Algorithm 1: Semantic Information Vector Space Alignment Methods
Cosine similarity vector alignment (outline)

1. **Input:** *text message*
2. **quantitative:**
3. $\mathbf{v}_A = (w_{A1}, w_{A2}, \ldots, w_{An})$, $\mathbf{v}_B = (w_{B1}, w_{B2}, \ldots, w_{Bm})$ //vectorization and location information
4. **normalization:**
5. $\mathbf{u}_A = \frac{\mathbf{v}_A}{\|\mathbf{v}_A\|}$, $\mathbf{u}_B = \frac{\mathbf{v}_B}{\|\mathbf{v}_B\|}$ //vector normalization
6. **cosine similarity:**
7. **Similarity**$(\mathbf{u}_A, \mathbf{u}_B) = \frac{\mathbf{u}_A \cdot \mathbf{u}_B}{\|\mathbf{u}_A\| \cdot \|\mathbf{u}_B\|}$ //Text Similarity Determination
8. **Similarity**$(\mathbf{u}_A, \mathbf{u}_B) = \frac{\sum_{i=1}^{n} w_{Ai} \cdot w_{Bi}}{\sqrt{\sum_{i=1}^{n} w_{Ai}^2} \cdot \sqrt{\sum_{i=1}^{m} w_{Bi}^2}}$ //Alignment of text A and text B
9. **Determining text similarity:** (−1 to 1) The closer to 1 the vectors are, the more similar they are.

3.3. Robot Heatmap Navigation Algorithm for Open Environment Awareness

Based on Sections 3.1 and 3.2, we modeled the behavior of a robot controlled by an open natural language ℓ, e.g., by having the robot go first to location A and then to location B. However, generating a robot trajectory based on ℓ is very difficult because the information in ℓ is too granular and lacks the detailed process of the task and, at the same time, lacks information about the environment [32]. Considering that the environment in which the robot works is not static, it is necessary to allow the robot to consider real-time environmental information when performing tasks. Assuming that the large language model decomposes ℓ into several subtasks $(\ell_1, \ell_2, \ldots, \ell_n)$, at this stage of task generation, we concentrated on the real-time environment to create detailed tasks ℓ_i to generate motion control commands that the robot can execute [33]. The above approach decomposes complex tasks into subtasks of lower complexity and senses the environment for each of the low-complexity tasks. The core problem of this subsection is how to make full use of the environment information to generate the motion trajectory $\tau_i^{\mathbf{r}}$ for the robot for each task ℓ_i. In this paper, the real robots that execute the motion trajectory $\tau_i^{\mathbf{r}}$ are multiple McNamee-wheeled unmanned vehicles, and with reference to the work of voxposer [18], we combined the environment information with the trajectory generation problem and summarized the problem as follows:

$$\min_{\tau_i^{\mathbf{r}}}\{\mathcal{F}_{task}(\mathbf{T}_i,\ell_i)+\mathcal{F}_{control}(\tau_i^{\mathbf{r}})\} \quad \text{subject to} \quad \mathcal{C}(\mathbf{T}_i) \tag{5}$$

where $\mathbf{T}_i$ is an environmental sensing information, $\tau_i^{\mathbf{r}} \subseteq \mathbf{T}_i$ is the trajectory of the unmanned cart in the dynamic environment, $\mathcal{C}(\mathbf{T}_i)$ denotes the constraints of the unmanned cart in the dynamic environment, $\mathcal{F}_{task}(\mathbf{T}_i,\ell_i)$ denotes the completion of the corresponding task within the confines of the dynamic environment, and $\mathcal{F}_{control}(\tau_i^{\mathbf{r}})$ specifies the control cost desired by the shortest path or the least task execution time.

It is very difficult to compute $\mathcal{F}_{task}(\mathbf{T}_i,\ell_i)$ based on open natural language, on the one hand because of the problem of difficult alignment between the natural language and the robot task ℓ_i, and on the other hand, because of the lack of dynamic environment information and real-time robot position. In this regard, we provided a 2D computable space describing the relative position information of objects in a dynamic environment $\mathbf{V} \in \mathbb{R}^{w\times h}$. We called it a calorific heatmap. It reflects objects in the environment that we are interested in or not interested in [34]; for objects of interest, we defined a high heat value for them, and objects that are not of interest are reflected in the heat map as low heat values. It directs the movement of objects with high heat values in the environment, creating trajectory curves between the robot and the objects of interest. The heat map assigns heat values to various objects in the surroundings. Using sub-tasks defined by the large language model, the task objective is labeled with a high heat value, attracting the robot toward the target area. Objects not of interest have low heat values on the heat map, repelling the robot and guiding it away from non-target areas.

We denoted the high calorific heat target as $\mathbf{e}$ and the robot trajectory as $\tau^{\mathbf{e}}$. For the subtask ℓ_i in $\mathcal{F}_{task}(\mathbf{T}_i,\ell_i)$, we can numericalize the task in the two-dimensional space $\mathbf{V} \in \mathbb{R}^{w\times h}$ by means of a calorific heatmap. The corresponding task $\mathcal{F}_{task}$ in the environment can be approximated by the continuous accumulation of $\mathbf{e}$ in the two-dimensional space $\mathbf{V} \in \mathbb{R}^{w\times h}$. The formula is as follows:

$$\mathcal{F}_{task} = -\sum_{j=1}^{|\tau_i^{\mathbf{e}}|}\mathbf{V}\left(p_j^{\mathbf{e}}\right), \ \textbf{where } p_j^{\mathbf{e}} \in \mathbb{N}^2 \text{ is } \mathbf{e} \text{ discrete position } (x,y) \text{ in step } j. \tag{6}$$

Large language models (LLMs) exhibit the capability to adapt their output in response to contextual information. We can influence the LLM to generate content aligning with our preferences through a prompting approach. In this study, prompt engineering is implemented through question-and-answer pairs, comprising questions and results, along with related objects and corresponding questions [35]. Illustrated in Figure 3, for the robot's shortest path-planning problem, we integrated information about objects in the environment. Using a second LLM, we decomposed the problem into a fine-grained task sequence. The resulting task sequences incorporated environmental information, empowering the robot to execute tasks in a real-world setting.

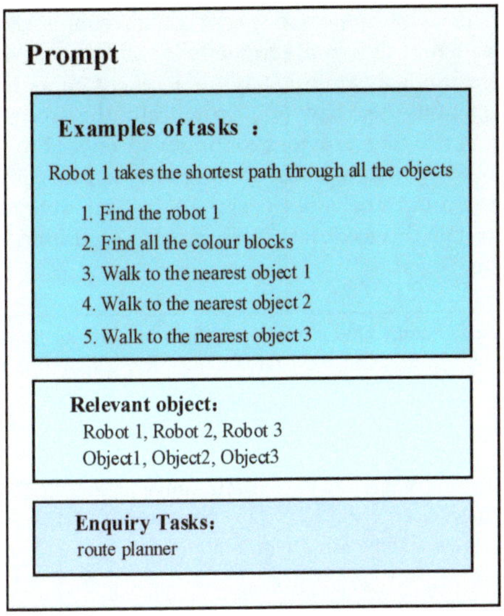

Figure 3. Example of a prompt engineering.

The prompt-engineered large language model is capable of recognizing objects of interest and understanding the relative spatial information in order to generate motion strategies [36]. Specifically, it can (1) perceive the environmental information by calling the visual model Application Programming Interface (API) to obtain the relative position information of the scene objects; (2) generate the task for the specialized scene based on the perceived environmental information combined with the generic knowledge of the first large language model; and (3) input the tasks into the heat map module through code form to generate the robot motion trajectory $\boldsymbol{\tau}^{\mathbf{e}}$ corresponding to each step of the task ℓ_i. The heat map $\mathbf{V}_i^t = \mathbf{heatmap}(\mathbf{o}^t, \ell_i)$ can be further obtained, where $\mathbf{o}^t$ is the camera observation at the moment t and ℓ_i is the task being executed.

In order to be able to generate a smooth trajectory for the robot to travel to the target area, we represented each step of the task as a mathematical problem $\mathcal{F}_{task}(\mathbf{T}_i, \ell_i)$. The motion trajectory can now be planned through the problem defined in Equation (1). The heat map reveals item properties and their positions in the scene. Objects of interest have high heat values, drawing the robot toward them, while non-target items are seen as obstacles with low heat values, pushing the robot away. All positions in the heat map have computable heat values. Our goal is to create a trajectory with the highest heat value, capable of reaching specified subtask locations. We defined the path's "heat value" as the sum of heat values of all nodes on the path.

$$R(P) = \sum_{i=1}^{k} H(v_i) \tag{7}$$

where k denotes the path length and $H(v)$ denotes the heat value of node v. We wished to find a path P that maximises the value of $R(P)$. A greedy idea was used to select the next node v_i in each step such that $H(v_i)$ is maximal, which can be expressed as:

$$v_i = \mathrm{argmax}_{v \in N(v_{i-1})} H(v) \tag{8}$$

where $N(v_{i-1})$ denotes the set of neighbouring nodes of node v_{i-1}. In this way, we can find a path $P = \{v_1, v_2, \ldots, v_k\}$, where v_1 is the starting point, and v_i chosen at each step makes $H(v_i)$ maximal until the node with the highest calorific value is reached. We provided dense rewards in the space to generate a planning path. During robot operation, because 2D visual information does not contain complete spatial information, it provides a positional relationship between the robot and the environment relative to each other. The robot continuously approximates our generated motion trajectory through this real-time feedback of the heat map [37]. Specifically, the process of updating the environment information in real time provides continuous feedback to control the robot motion nodes so that the robot's motion profile continuously approaches our generated path trajectory, and when the robot's behavior is shifted, it is possible to replan the real robot motion trajectory through this feedback. Please refer to Algorithm 2 for more specific details about the algorithm.

Algorithm 2: Dynamic navigation algorithms for calorific heat maps
Environmental Interaction and Mathematical Representation (outline)

1. **Input:** ***natural language representation task*** ℓ
2. **Breakdown of tasks:**
3. $(\ell_1, \ell_2, \ldots, \ell_n)$
4. **The problem is reduced to the optimization equation**
5. $\min_{\tau_i^r} \{\mathcal{F}_{task}(\mathrm{T}_i, \ell_i) + \mathcal{F}_{control}(\tau_i^r)\}$ **subject to** $\mathcal{C}(\mathrm{T}_i)$
6. **Constructing a mathematical representation of** $\mathcal{F}_{task}$
7. $\mathcal{F}_{task} = -\sum_{j=1}^{|\tau_i^e|} \mathrm{V}\left(p_j^e\right)$, **where** $p_j^e \in \mathbb{N}^2$
8. **Construct a two-dimensional space:** $\mathrm{V} \in \mathbb{R}^{w \times h}$
9. **Constructing calorific heat maps:**
10. $\mathrm{V}_i^t = \mathrm{heatmap}(\mathrm{o}^t, \ell_i)$
11. **Define a high calorific value path:**
12. $R(P) = \sum_{i=1}^{k} H(v_i)$//We define the heat value of a path as the sum of the heat values of all nodes on the path.
13. **Path generation:**
14. $v_i = argmax_{v \in N(v_{i-1})} H(v)$//$N(v_{i-1})$ denotes the set of neighbouring nodes of node v_{i-1}
15. **Output:** $P = \{v_1, v_2, \ldots, v_k\}$ //Path with the highest calorific value

4. Experiments and Analyses

We first discuss the design and execution of the holistic experiment and implement the holistic experiment in a real-world environment (Section 4.1). In order to make the holistic task planning more interpretable, we added an intermediate process of task planning, which allows humans to intervene in the intermediate process if the results do not meet human expectations (Section 4.2). We conducted semantic similarity-based feedback experiments and discussed the optimal number of loops (Section 4.3). We presented some failed cases from the experiment and conducted discussions and analyses on these cases (Section 4.4).

4.1. Experimental Design

In this experiment, we constructed a real-world experimental platform to support the theory proposed in this paper, and the large language model and the visual model were deployed in the server to extract the generic knowledge by calling API [38]. In this experiment, we used three unmanned carts and a drone to achieve the motion control of the robots through the Ros system. Specifically, the drone provides visual information; the visual information is transmitted back to the server to perceive the environment information through the visual model; and the large language model combines the environment information with the generic knowledge to generate the robot motion scheme to cope with the environment [39] and then generates the optimized trajectory through the heat map algorithm to guide the robots to complete the corresponding tasks as shown in Figure 4.

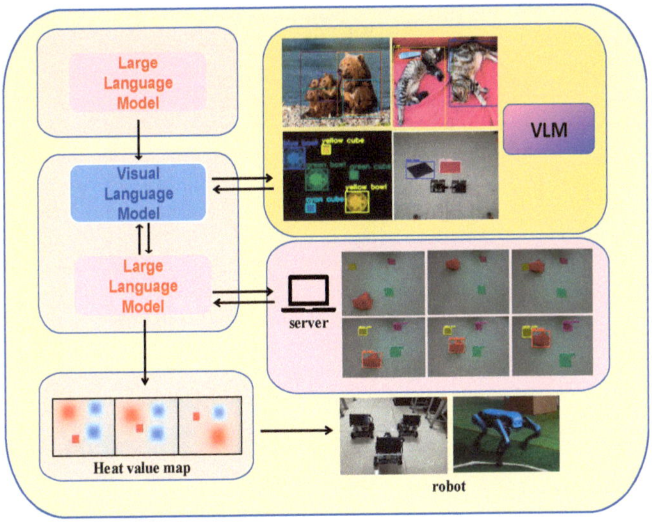

Figure 4. Overall structure of the experiment. In the figure we have used VLM to recognize real images and determined three different categories of target objects, robots, and obstacles, and mapped these categories of objects differently so that they are represented differently in the heat map.

In Figure 5, we present a detailed design of the experiment, illustrating the intermediate process from the task given in natural language to the generation of heat maps depicting the movement trajectories of robotic; As depicted in Figure 5, the video is captured by the camera mounted on the UAV, providing a global view. Extracting environment information by intercepting images from the video allows us to discern details about the items in the surroundings. The large-scale language model receives a natural language task from a human and, in conjunction with the information extracted by the visual model, decomposes the task and collaboratively generates a sequence of subtasks [40]. This process breaks down the complex task into steps executable by the robot and performs semantic similarity matching in vector space to align with the robot instructions. Trajectories for each task step are generated in a heat map, enabling the robot to approach the target using relative position information provided by these trajectories. Continuous correction of offset through visual information feedback ensures successful execution of the navigation task. For clarity, we used color blocks to cover the unmanned vehicle throughout the experiment, facilitating observation. In the task illustrated in Figure 5, the objective is for the vehicle to initially reach the area where the yellow color block is located and subsequently reach the area where the green color block is situated. To enhance visual representation of the experimental flow, the vehicle is covered with red color blocks in this particular task.

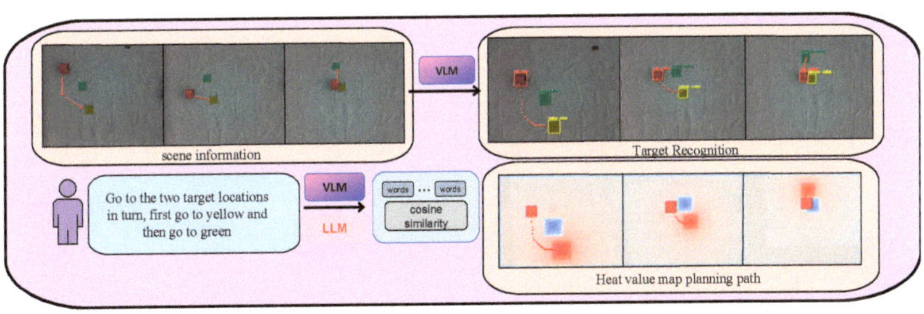

Figure 5. Experimental design that describes the process of generating calorimetric map motion trajectories from image information.

Figure 6 displays five tasks of varying difficulty levels designed by us. We described these five tasks using natural language and employed the framework proposed in this paper to enable the robot to accomplish navigation tasks. It is worth mentioning that all five tasks are types of robot navigation tasks. We progressively increased the complexity of navigation tasks from simple to complex. These tasks required the LLMs to understand natural language and make accurate judgments. In the navigation tasks, we introduced obstacle detection conditions and multi-robot multi-target navigation, covering various difficulty levels of the experimental scenarios. We (1) had the robot arrive at two specified target locations in succession; (2) had the robot arrive at three specified target locations in succession; (3) planned the shortest path through the human-specified locations, focusing on judging whether the strategy generated by the large language model is consistent with the robot's execution; (4) determined whether there is a human-specified marker in the field and travelled to a specified location A if there is, and to a specified location B if there is not; and (5) designed a task rich in complex semantic information that instructs three unmanned vehicles to perform multi-robot, multi-objective navigation tasks that incorporate and temporally sequence task judgment conditions. Specifically, unmanned vehicle A and unmanned vehicle B each travel to a different human-specified location, and unmanned vehicle C judges that vehicle A has arrived at the specified location before starting to travel to a new location. It is worth noting that all of the above tasks can be interacted with real robots using natural language, and dynamic feedback can be provided in real time to reduce interference caused by changes in environmental information. We successfully completed the above five experiments in a real environment, and the images in Figure 6 were all captured during the actual experimental process. The experiments show that the task decomposition of the large language model, with the addition of environmental awareness and semantic alignment, can control the robots to perform tasks of higher complexity.

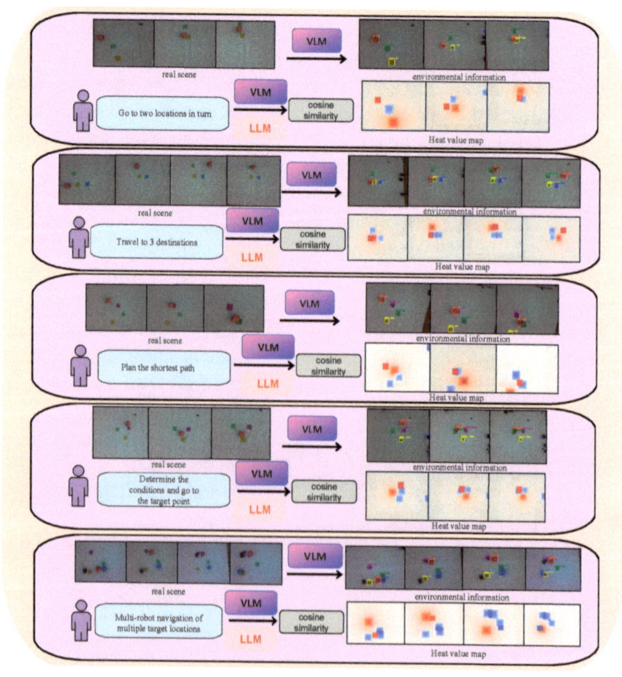

Figure 6. The 5 experiments are designed to include tasks of varying complexity. The flow of the 5 experiments is described, including our process of target recognition from raw image information to finally generating a heat map.

4.2. Intermediate Process of Task Planning

We used Figure 7 to visually illustrate the output formats of the two LLMs. This enables controllability of each module's output, allowing for manual intervention in a specific module to generate results that align with our expectations. Introducing an intermediate step of task planning can enhance the interpretability of the overall task planning process. Simultaneously, this intermediate step opens up the opportunity for human intervention in the task planning process, particularly when the generated results deviate from the intended human task execution process [40]. Human intervention is facilitated across three dimensions: coarse-grained task decomposition, fine-grained task decomposition, and motion control instructions [41]. Foreseeably, by controlling the robot through a human-in-the-loop model built upon the foundation laid in this paper, we can enhance the reliability and safety of the robot's control.

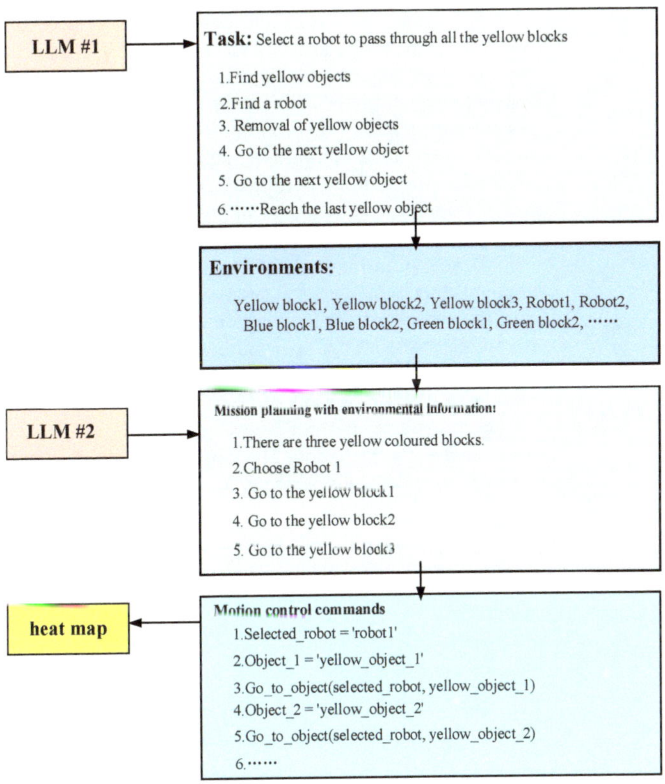

Figure 7. Example of the intermediate process of task planning.

Figure 7 illustrates an example of the intermediate task planning process, wherein the given problem involves selecting a robot to traverse all yellow objects. The task is decomposed, and a detailed intermediate process is displayed. Initially, the first large language model (LLM) conducts a coarse-grained decomposition of the problem, generating a sequence of tasks devoid of environmental information. These tasks involve finding the yellow objects and locating the robot to execute the tasks, followed by sequential movements of the robots to the positions of the yellow objects. In this process, lacking environmental information, these decomposed tasks are answered based on the information contained in the question. The second LLM utilizes environmental information and incorporates the fine-grained task breakdown from sequences produced by the first LLM. This integration results in new task sequences enriched with scene information. It is evident that the LLM can generate more realistic task sequences by incorporating environmental information. Subsequently, by aligning with corresponding motion control commands, a motion trajectory is generated in the heat map to guide the robot in executing the task.

4.3. Feedback Experiments Based on Semantic Similarity

The output of semantically similar results by the large language model indicates its ability to comprehend the problem and generate relatively accurate results, which are already acceptable at the semantic level [42]. We facilitated the mapping of the results generated by the large language model to robot control commands through vector alignment. It is noteworthy that when alignment is conducted at the semantic level, the large language model is proficient in decomposing semantically correct tasks [43]. Consequently, these tasks can be successfully mapped to corresponding motion commands through semantic-level alignment.

We conducted a feedback experiment based on semantic similarity and explored the optimal number of iterations. Throughout the experimentation process, we observed that the results generated by the large language model are not consistently optimal. This inconsistency stems from the fact that the process of prompting the large language model does not encompass all working conditions [44]. Consequently, when describing certain complex tasks with detailed natural language, the large language model may face challenges in understanding, leading to difficulties in comprehending intricate task nuances [45]. As a result of the understanding deviation in the problem formulation process, the output results of the large language model may occasionally deviate from our anticipated direction [46].

As shown in Figure 8, we provide feedback on the tasks that posed challenges in aligning at the semantic level. Subsequently, we reintroduced the erroneous task sequences into the second macro-language model for a new round of task decomposition. This iterative process serves to semantically bias the macro-language model toward our cue words by incorporating feedback information about the erroneous tasks [47]. Given that the cue words encompass robot motion instructions, this bias encourages the macro-language model to generate natural language that aligns with the correct instructions enriched with semantic information. This approach enhances the likelihood of outputting correct command results at the semantic level.

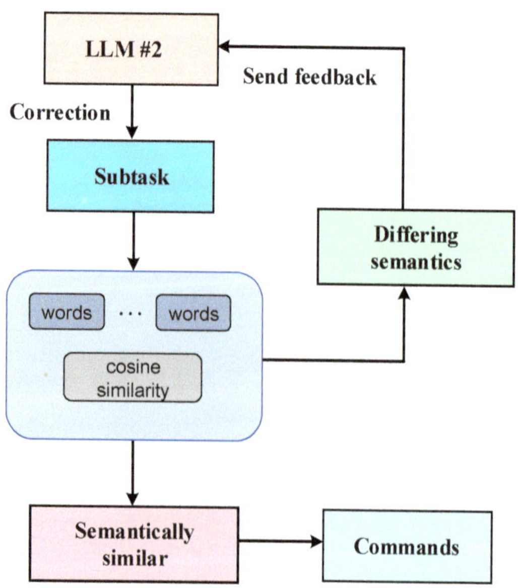

Figure 8. Idea of the cyclic semantic alignment method.

To illustrate this process in detail, as shown in Figure 9, we present an example of prompts for the second LLM. This segment indicates that when the LLM receives feedback indicating semantic mismatch, the next output will prefer the format of prompt words we designed. This pattern is advantageous for the LLM to generate outputs biased toward our desired expectations.

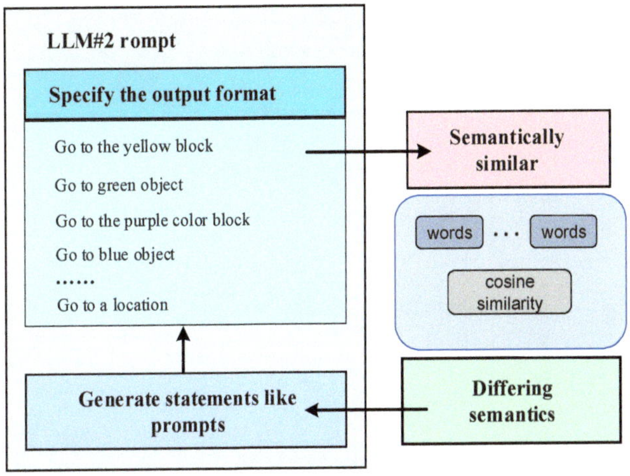

Figure 9. Example of a hint for LLM#2 in a cyclic semantic alignment approach.

As indicated in Table 1, the feedback proves advantageous in improving the success rate of task execution. By informing the large language model about the inaccuracies in task decompositions through feedback, it effectively reduces information uncertainty and steers the decomposition preference toward our cue words [48,49]. The experimental results reveal that the correctness rate reaches 78% after the third feedback. From the fourth feedback onwards, the increase in task success becomes slow. We considered three or four iterations as the optimal number of feedback attempts, demonstrating high system efficiency.

Table 1. In the middle of the two steps of task decomposition and vector alignment, different numbers of feedback attempts were set, and five tasks were performed to compare mandated execution success rates.

Task	Feedback 0	Feedback 1	Feedback 2	Feedback 3	Feedback 4	Feedback 5
Task 1: Travel to two target sites	4/10	7/10	8/10	9/10	10/10	10/10
Task 2: Travel to three target sites	3/10	6/10	8/10	8/10	8/10	8/10
Task 3: Planning the shortest route	3/10	6/10	7/10	8/10	8/10	8/10
Task 4: Self-determination of target location	4/10	5/10	7/10	7/10	7/10	7/10
Task 5: Multi-robot to multi-objective tasks	2/10	5/10	6/10	7/10	7/10	7/10
total	32%	58%	72%	78%	80%	80%

4.4. Discussion and Analysis of Failed Cases

During the experiment, there may be reasons leading to the failure of the experiment, specifically when the robot fails to execute the correct tasks. We analyzed these failed cases, which resulted from either the LLMs misinterpreting natural language or the VLM detecting targets incorrectly during the environmental perception process. Because the motion command is aimed at directing the robot to a specific location, which is relatively simple natural language, errors in semantic similarity modules often occur due to incorrect semantic information provided by the LLMs and VLM or incorrect target detections. Therefore, we focused on discussing the failures caused by these two modules, the LLMs and VLM.

When transmitting natural language to the LLMs, it is essential for the LLMs to correctly understand the semantics of the natural language and generate the correct strategy. However, the LLMs cannot always produce accurate results. As shown in Figure 10, when given a task like "Select a robot to pass through all yellow-colored blocks," the correct task decomposition logic should be for the robot to pass through blocks it has not reached yet. However, the LLMs occasionally generate confused task decomposition processes, such as generating a task to go to a yellow-colored block, which is obviously incorrect. Such logical errors in the task decomposition process can result in the robot failing to execute the task successfully.

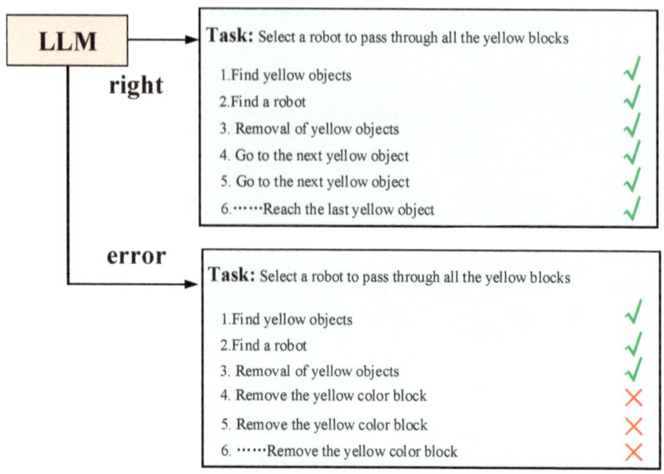

Figure 10. Example of LLM-generated policy error.

Another scenario is when the VLM makes errors during environmental perception, as shown in Figure 11. The VLM may misidentify targets in the presence of changes in light, leading the LLMs to receive incorrect environmental information, resulting in the failure of the robot to execute tasks. Alternatively, the VLM may identify irrelevant items in the scene that users do not want it to recognize, causing interference in the experimental process and potentially generating incorrect results in the heat map.

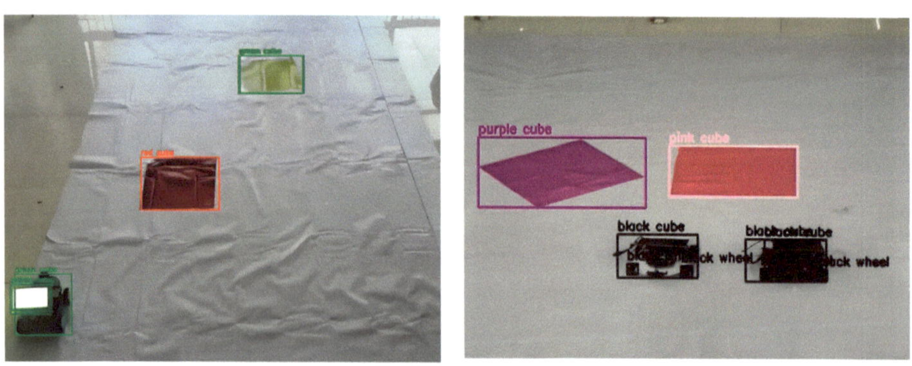

(**a**) (**b**)

Figure 11. (**a**) VLM identifies both yellow-colored blocks and black-colored cars as green objects when there are changes in light brightness. (**b**) VLM will recognize the objects along with the attached blocks, such as identifying black wheel.

5. Discussion

In this work, we achieved semantic-level alignment between the output of the large language model and the robot control commands and experimentally validated that the approach is capable of controlling a robot using natural language for tasks of higher complexity [50]. The method proposed in this paper enables easier human–robot interaction to accomplish corresponding tasks in the real world, without requiring extensive expertise or skills. Humans can describe task requirements in natural language, and using the framework designed in this paper, the tasks can be understood at the semantic level and translated into task sequences that the robot can execute, thereby driving the robot to achieve task goals. Importantly, this method, through the form of multi-layer task decomposition and interaction with the environment, enables robots to understand and complete tasks with complex semantic information. Compared to traditional natural language-controlled robot technologies that can only handle relatively simple tasks, this method enables robots to perform more complex and abstract tasks.

This work has some limitations. First, this experiment relies on the environment perception module to obtain environment information, which will be limited by the perception module when the visual language model analyzes the object properties. Second, this work performs a numerical analysis in a two-dimensional computable space, which does not provide the robot with high-precision environmental information and constrains the robot from performing more detailed tasks [51]. The third point is that this work designs a feedback mechanism during the task decomposition process, although it solves the problem of aligning the task decomposition with the robot control instructions. But it does not guarantee that the results generated by the large language model do not have logic problems [52]. The robot displacement caused by such logic errors can only be adjusted by the real-time feedback of visual information, and the system will make the robot re-plan the correct trajectory through subsequent task sequences. This process increases the time for task execution.

We can enhance the specificity of the model output through fine tuning. Specifically, we can use the self-instruct method [53] to generate corpora for each model in the multi-layered large model architecture. We can then fine-tune the respective models using these corpora to improve the accuracy of task decomposition.

We can utilize additional environmental perception modules, which will help to better transform the open environment into a computable space for guiding precise robot actions. Because we are currently only using cameras, in the future, we can enhance the complexity of numerical space by adding different sensors, such as LiDAR and depth cameras, and fusing multimodal sensor information.

In future work, firstly, we can add the multi-environmental sensing module, which can obtain more three-dimensional and rich environmental information through multi-modal environmental sensing. Secondly, we can construct more complex numerical spaces to optimize the motion control strategy so that the robot can perform more delicate tasks. Finally, we will also design prompt words with more semantic information and use prompt engineering to reduce the number of cycles in the task decomposition process and optimize the model system.

6. Conclusions

In this research, we present a novel methodology that integrates a large language model (LLM) with a visual language model (VLM) and a calorific heat map. This approach facilitated a multi-layer decomposition of tasks, thereby elevating the precision of task decomposition. Additionally, we introduced an intermediate task planning process to bolster the reliability of robot control. To ensure alignment between LLM outputs and motion control instructions, a vector alignment method is employed. Through rigorous testing and evaluation in a real-world robot scenario, our findings substantiate that the proposed methodology enhances the LLM's proficiency in comprehending intricate real-world tasks. Furthermore, it amplifies the likelihood of aligning the LLM outputs with motion control commands.

Author Contributions: Conceptualization, Z.L.; formal analysis, Z.L.; project administration, Z.L.; validation, Y.L. and R.H.; algorithm construction, S.B.; investigation, Y.Z. and H.Z.; data curation, Q.W.; writing—original draft, Z.L. and Y.L. All authors have read and agreed to the published version of the manuscript.

Funding: This research is supported by the National Natural Science Foundation of China (NSFC), grant number No. U21A20485.

Institutional Review Board Statement: Not applicable.

Informed Consent Statement: Not applicable.

Data Availability Statement: Data are contained within the article.

Conflicts of Interest: The authors declare no conflicts of interest.

References

1. Chowdhery, A.; Narang, S.; Devlin, J.; Bosma, M.; Mishra, G.; Roberts, A.; Fiedel, N. Palm: Scaling language modeling with pathways. *J. Mach. Learn. Res.* **2023**, *24*, 1–113.
2. Toussaint, M.; Harris, J.; Ha, J.S.; Driess, D.; Hönig, W. Sequence-of-Constraints MPC: Reactive Timing-Optimal Control of Sequential Manipulation. In Proceedings of the 2022 IEEE/RSJ International Conference on Intelligent Robots and Systems (IROS), Kyoto, Japan, 23–27 October 2022; pp. 13753–13760.
3. Liu, W.; Paxton, C.; Hermans, T.; Fox, D. Structformer: Learning spatial structure for language-guided semantic rearrangement of novel objects. In Proceedings of the 2022 International Conference on Robotics and Automation (ICRA), Philadelphia, PA, USA, 23–27 May 2022; pp. 6322–6329.
4. Thomason, J.; Padmakumar, A.; Sinapov, J.; Walker, N.; Jiang, Y.; Yedidsion, H.; Mooney, R. Jointly improving parsing and perception for natural language commands through human-robot dialog. *J. Artif. Intell. Res.* **2020**, *67*, 327–374. [CrossRef]
5. Jang, E.; Irpan, A.; Khansari, M.; Kappler, D.; Ebert, F.; Lynch, C.; Finn, C. Bc-z: Zero-shot task generalization with robotic imitation learning. In Proceedings of the Conference on Robot Learning, PMLR, Auckland, New Zealand, 14–18 December 2022; pp. 991–1002.
6. Brohan, A.; Brown, N.; Carbajal, J.; Chebotar, Y.; Dabis, J.; Finn, C.; Zitkovich, B. Rt-1: Robotics transformer for real-world control at scale. *arXiv* **2022**, arXiv:2212.06817.
7. Andreas, J.; Klein, D.; Levine, S. Learning with latent language. *arXiv* **2017**, arXiv:1711.00482.
8. Bommasani, R.; Hudson, D.A.; Adeli, E.; Altman, R.; Arora, S.; von Arx, S.; Liang, P. On the opportunities and risks of foundation models. *arXiv* **2021**, arXiv:2108.07258.
9. Tellex, S.; Gopalan, N.; Kress-Gazit, H.; Matuszek, C. Robots that use language. *Annu. Rev. Control Robot. Auton. Syst.* **2020**, *3*, 25–55. [CrossRef]
10. Huang, W.; Abbeel, P.; Pathak, D.; Mordatch, I. Language models as zero-shot planners: Extracting actionable knowledge for embodied agents. In Proceedings of the International Conference on Machine Learning, PMLR, Baltimore, MD, USA, 17–23 July 2022; pp. 9118–9147.
11. Micheli, V.; Fleuret, F. Language models are few-shot butlers. *arXiv* **2021**, arXiv:2104.07972.
12. Thomason, J.; Zhang, S.; Mooney, R.J.; Stone, P. Learning to interpret natural language commands through human-robot dialog. In Proceedings of the Twenty-Fourth International Joint Conference on Artificial Intelligence, Buenos Aires, Argentina, 25–31 July 2015.
13. Shwartz, V.; West, P.; Bras, R.L.; Bhagavatula, C.; Choi, Y. Unsupervised commonsense question answering with self-talk. *arXiv* **2020**, arXiv:2004.05483.
14. Blukis, V.; Knepper, R.A.; Artzi, Y. Few-shot object grounding and mapping for natural language robot instruction following. *arXiv* **2020**, arXiv:2011.07384.
15. Brown, T.; Mann, B.; Ryder, N.; Subbiah, M.; Kaplan, J.D.; Dhariwal, P.; Amodei, D. Language models are few-shot learners. *Adv. Neural Inf. Process. Syst.* **2020**, *33*, 1877–1901.
16. Jiang, Y.; Gu, S.S.; Murphy, K.P.; Finn, C. Language as an abstraction for hierarchical deep reinforcement learning. *Adv. Neural Inf. Process. Syst.* **2019**, *32*. Available online: https://proceedings.neurips.cc/paper_files/paper/2019/file/0af787945872196b42c9f73ead2565c8-Paper.pdf (accessed on 2 February 2024).
17. Sharma, P.; Sundaralingam, B.; Blukis, V.; Paxton, C.; Hermans, T.; Torralba, A.; Fox, D. Correcting robot plans with natural language feedback. *arXiv* **2022**, arXiv:2204.05186.
18. Huang, W.; Wang, C.; Zhang, R.; Li, Y.; Wu, J.; Fei-Fei, L. Voxposer: Composable 3d heat maps for robotic manipulation with language models. *arXiv* **2023**, arXiv:2307.05973.

19. Zeng, A.; Attarian, M.; Ichter, B.; Choromanski, K.; Wong, A.; Welker, S.; Florence, P. Socratic models: Composing zero-shot multimodal reasoning with language. *arXiv* **2022**, arXiv:2204.00598.
20. Liang, J.; Huang, W.; Xia, F.; Xu, P.; Hausman, K.; Ichter, B.; Zeng, A. Code as policies: Language model programs for embodied control. In Proceedings of the 2023 IEEE International Conference on Robotics and Automation (ICRA), London, UK, 29 May–2 June 2023; pp. 9493–9500.
21. Huang, W.; Xia, F.; Shah, D.; Driess, D.; Zeng, A.; Lu, Y.; Ichter, B. Grounded decoding: Guiding text generation with grounded models for robot control. *arXiv* **2023**, arXiv:2303.00855.
22. Wu, J.; Antonova, R.; Kan, A.; Lepert, M.; Zeng, A.; Song, S.; Funkhouser, T. Tidybot: Personalized robot assistance with large language models. *arXiv* **2023**, arXiv:2305.05658.
23. Radford, A.; Kim, J.W.; Hallacy, C.; Ramesh, A.; Goh, G.; Agarwal, S.; Sutskever, I. Learning transferable visual models from natural language supervision. In Proceedings of the International Conference on Machine Learning, PMLR, Virtual, 18–24 July 2021; pp. 8748–8763.
24. Gu, X.; Lin, T.Y.; Kuo, W.; Cui, Y. Open-vocabulary object detection via vision and language knowledge distillation. *arXiv* **2021**, arXiv:2104.13921.
25. Kamath, A.; Singh, M.; LeCun, Y.; Synnaeve, G.; Misra, I.; Carion, N. Mdetr-modulated detection for end-to-end multi-modal understanding. In Proceedings of the IEEE/CVF International Conference on Computer Vision, Montreal, BC, Canada, 11–17 October 2021; pp. 1780–1790.
26. Minderer, M.; Gritsenko, A.; Stone, A.; Neumann, M.; Weissenborn, D.; Dosovitskiy, A.; Houlsby, N. Simple open-vocabulary object detection. In Proceedings of the European Conference on Computer Vision, Tel Aviv, Israel, 23–27 October 2022; Springer Nature: Cham, Switzerland, 2022; pp. 728–755.
27. Zellers, R.; Holtzman, A.; Peters, M.E.; Mottaghi, R.; Kembhavi, A.; Farhadi, A.; Choi, Y. PIGLeT: Language grounding through neuro-symbolic interaction in a 3D world. *arXiv* **2021**, arXiv:2106.00188.
28. Zellers, R.; Lu, X.; Hessel, J.; Yu, Y.; Park, J.S.; Cao, J.; Choi, Y. Merlot: Multimodal neural script knowledge models. *Adv. Neural Inf. Process. Syst.* **2021**, *34*, 23634–23651.
29. Shah, D.; Osiński, B.; Levine, S. Lm-nav: Robotic navigation with large pre-trained models of language, vision, and action. In Proceedings of the Conference on Robot Learning, PMLR, Atlanta, GA, USA, 6–9 November 2023; pp. 492–504.
30. Cui, Y.; Karamcheti, S.; Palleti, R.; Shivakumar, N.; Liang, P.; Sadigh, D. No, to the Right: Online Language Corrections for Robotic Manipulation via Shared Autonomy. In Proceedings of the 2023 ACM/IEEE International Conference on Human-Robot Interaction, Stockholm, Sweden, 13–16 March 2023; pp. 93–101.
31. Stone, A.; Xiao, T.; Lu, Y.; Gopalakrishnan, K.; Lee, K.H.; Vuong, Q.; Hausman, K. Open-world object manipulation using pre-trained vision-language models. *arXiv* **2023**, arXiv:2303.00905.
32. Ma, Y.J.; Liang, W.; Som, V.; Kumar, V.; Zhang, A.; Bastani, O.; Jayaraman, D. LIV: Language-Image Representations and Rewards for Robotic Control. *arXiv* **2023**, arXiv:2306.00958.
33. Nair, S.; Mitchell, E.; Chen, K.; Savarese, S.; Finn, C. Learning language-conditioned robot behavior from offline data and crowd-sourced annotation. In Proceedings of the Conference on Robot Learning, PMLR, Auckland, New Zealand, 14–18 December 2022; pp. 1303–1315.
34. Singh, I.; Blukis, V.; Mousavian, A.; Goyal, A.; Xu, D.; Tremblay, J.; Garg, A. Progprompt: Generating situated robot task plans using large language models. In Proceedings of the 2023 IEEE International Conference on Robotics and Automation (ICRA), London, UK, 29 May–2 June 2023, pp. 11523–11530
35. Raman, S.S.; Cohen, V.; Rosen, E.; Idrees, I.; Paulius, D.; Tellex, S. Planning with large language models via corrective re-prompting. In Proceedings of the NeurIPS 2022 Foundation Models for Decision Making Workshop, New Orleans, LA, USA, 3 December 2022.
36. Liu, B.; Jiang, Y.; Zhang, X.; Liu, Q.; Zhang, S.; Biswas, J.; Stone, P. Llm+ p: Empowering large language models with optimal planning proficiency. *arXiv* **2023**, arXiv:2304.11477.
37. Vemprala, S.; Bonatti, R.; Bucker, A.; Kapoor, A. Chatgpt for robotics: Design principles and model abilities. *Microsoft Auton. Syst. Robot. Res.* **2023**, *2*, 20.
38. Lin, K.; Agia, C.; Migimatsu, T.; Pavone, M.; Bohg, J. Text2motion: From natural language instructions to feasible plans. *arXiv* **2023**, arXiv:2303.12153. [CrossRef]
39. Driess, D.; Xia, F.; Sajjadi, M.S.; Lynch, C.; Chowdhery, A.; Ichter, B.; Florence, P. Palm-e: An embodied multimodal language model. *arXiv* **2023**, arXiv:2303.03378.
40. Yuan, H.; Zhang, C.; Wang, H.; Xie, F.; Cai, P.; Dong, H.; Lu, Z. Plan4mc: Skill reinforcement learning and planning for open-world minecraft tasks. *arXiv* **2023**, arXiv:2303.16563.
41. Xie, Y.; Yu, C.; Zhu, T.; Bai, J.; Gong, Z.; Soh, H. Translating natural language to planning goals with large-language models. *arXiv* **2023**, arXiv:2302.05128.
42. Lu, Y.; Lu, P.; Chen, Z.; Zhu, W.; Wang, X.E.; Wang, W.Y. Multimodal Procedural Planning via Dual Text-Image Prompting. *arXiv* **2023**, arXiv:2305.01795.
43. Kwon, M.; Xie, S.M.; Bullard, K.; Sadigh, D. Reward design with language models. *arXiv* **2023**, arXiv:2303.00001.
44. Du, Y.; Watkins, O.; Wang, Z.; Colas, C.; Darrell, T.; Abbeel, P.; Andreas, J. Guiding pretraining in reinforcement learning with large language models. *arXiv* **2023**, arXiv:2302.06692.

45. Hu, H.; Sadigh, D. Language instructed reinforcement learning for human-ai coordination. *arXiv* **2023**, arXiv:2304.07297.
46. Bahl, S.; Mendonca, R.; Chen, L.; Jain, U.; Pathak, D. Affordances from Human Videos as a Versatile Representation for Robotics. In Proceedings of the IEEE/CVF Conference on Computer Vision and Pattern Recognition, Vancouver, BC, Canada, 17–24 June 2023; pp. 13778–13790.
47. Patel, D.; Eghbalzadeh, H.; Kamra, N.; Iuzzolino, M.L.; Jain, U.; Desai, R. Pretrained Language Models as Visual Planners for Human Assistance. *arXiv* **2023**, arXiv:2304.09179.
48. Wang, G.; Xie, Y.; Jiang, Y.; Mandlekar, A.; Xiao, C.; Zhu, Y.; Anandkumar, A. Voyager: An open-ended embodied agent with large language models. *arXiv* **2023**, arXiv:2305.16291.
49. Tam, A.; Rabinowitz, N.; Lampinen, A.; Roy, N.A.; Chan, S.; Strouse, D.J.; Hill, F. Semantic exploration from language abstractions and pretrained representations. *Adv. Neural Inf. Process. Syst.* **2022**, *35*, 25377–25389.
50. Mu, J.; Zhong, V.; Raileanu, R.; Jiang, M.; Goodman, N.; Rocktäschel, T.; Grefenstette, E. Improving intrinsic exploration with language abstractions. *Adv. Neural Inf. Process. Syst.* **2022**, *35*, 33947–33960.
51. Li, S.; Puig, X.; Paxton, C.; Du, Y.; Wang, C.; Fan, L.; Zhu, Y. Pre-trained language models for interactive decision-making. *Adv. Neural Inf. Process. Syst.* **2022**, *35*, 31199–31212.
52. Wang, Y.; Kordi, Y.; Mishra, S.; Liu, A.; Smith, N.A.; Khashabi, D.; Hajishirzi, H. Self-instruct: Aligning language model with self generated instructions. *arXiv* **2022**, arXiv:2212.10560.
53. Yang, R.; Song, L.; Li, Y.; Zhao, S.; Ge, Y.; Li, X.; Shan, Y. Gpt4tools: Teaching large language model to use tools via self-instruction. *arXiv* **2023**, arXiv:2305.18752.

sensors

MDPI

Article

Kinematic Analysis and Application to Control Logic Development for RHex Robot Locomotion

Piotr Burzyński [1], Ewa Pawłuszewicz [1], Leszek Ambroziak [1] and Suryansh Sharma [2,*]

[1] Department of Industrial Process Automation, Faculty of Mechanical Engineering, Bialystok University of Technology, Wiejska 45C, 15-351 Bialystok, Poland; p.burzynski@doktoranci.pb.edu.pl (P.B.); e.pawluszewicz@pb.edu.pl (E.P.); l.ambroziak@pb.edu.pl (L.A.)

[2] Networked Systems Group, Delft University of Technology, Building 28, Van Mourik Broekmanweg 6, 2628 XE Delft, The Netherlands

* Correspondence: suryansh.sharma@tudelft.nl

Abstract: This study explores the kinematic model of the popular RHex hexapod robots which have garnered considerable interest for their locomotion capabilities. We study the influence of tripod trajectory parameters on the RHex robot's movement, aiming to craft a precise kinematic model that enhances walking mechanisms. This model serves as a cornerstone for refining robot control strategies, enabling tailored performance enhancements or specific motion patterns. Validation conducted on a bespoke test bed confirms the model's efficacy in predicting spatial movements, albeit with minor deviations due to motor load variations and control system dynamics. In particular, the derived kinematic framework offers valuable insights for advancing control logic, particularly navigating in flat terrains, thereby broadening the RHex robot's application spectrum.

Keywords: C-legged hexapod; mobile robot; walking robot; kinematics modeling; simulation

Citation: Burzyński, P.; Pawłuszewicz, E.; Ambroziak, L.; Sharma, S. Kinematic Analysis and Application to Control Logic Development for RHex Robot Locomotion. *Sensors* **2024**, *24*, 1636. https://doi.org/10.3390/s24051636

Academic Editors: David Cheneler and Stephen Monk

Received: 8 February 2024
Revised: 27 February 2024
Accepted: 28 February 2024
Published: 2 March 2024

1. Introduction

The domain of mobile robotics has witnessed a notable evolution in recent years, driven by technological advancements and an ever-evolving array of applications across various contexts. This evolution has been paralleled by substantial research endeavors aimed at mitigating electronic costs [1], harnessing the augmented computational capabilities of microchips, and revolutionizing intelligent and adaptable manufacturing processes [2]. Considerable endeavors extend to the refinement of sophisticated control strategies [3] and the enhancement of autonomous navigation capabilities, alongside the development of measurement techniques and sensors resilient to noise interference [4,5]. Furthermore, the integration of a broad spectrum of artificial intelligence methodologies, encompassing machine learning and neuromorphic control systems, is being pursued at multiple tiers within the robot's architecture [6,7]. Mobile robots, classified according to their operational domains: ground (Unmanned Ground Vehicles, UGVs), aerial (Unmanned Aerial Vehicles, UAVs), aquatic—submersible (Autonomous Underwater Vehicles, AUVs), and surface-based (Unmanned Surface Vehicles, USVs) [8], have extended their applicability beyond traditional settings, adeptly adapting to intricate terrains and even modifying their morphology and locomotion to optimize efficiency [9]. The 1980s marked a turning point with the introduction of dynamic locomotion in robots, significantly advanced by research at Tokyo University and MIT's LegLab [10,11]. The field further evolved with Honda's P2 humanoid in the 1990s, demonstrating greater versatility and leading to broader commercial and research interest. Recent decades have focused on enhancing the dynamic walking and running capabilities of legged robots, with ongoing challenges in improving efficiency, speed, and robustness [12,13].

Within the realm of land-based robotics, the distinction among wheeled, tracked, and legged configurations underscores a trade-off between velocity and terrain adaptability.

Wheeled robots exhibit superior speed and energy efficiency on well-defined terrains, yet encounter challenges in rugged, obstacle-laden environments where legged robots, with their capacity to navigate discrete footholds and utilize additional appendages for stability, excel [14]. Despite the inherent complexities associated with their intricate mechanics [15], propulsion [16], and control systems [17], legged robots offer unparalleled versatility in natural and unstructured environments, drawing inspiration from various animal locomotion mechanisms to achieve dynamic stability and mobility [18,19]. Single-legged robots are inspired by saltatorial animals (animals that locomote by jumping) [20], two-legged robots by humanoids [21], four legs on quadrupeds [22], and more than four legs are inspired by insects [23–25].

This paper delves into the hexapod configuration, particularly focusing on the RHex robot, which embodies a fusion of stability and adaptability. Hexapod robots, celebrated for their static stability and versatility, exemplify the ongoing research pursuit to harmonize mobility, efficiency, and intricacy [26]. Through an examination of the RHex robot, this study contributes to a nuanced comprehension of legged robotics, proposing innovations in leg design and control methodologies to enhance its functionality across diverse terrains, including aquatic environments, as demonstrated by the flapped-paddle amphibian variant-FLHex [27]. Our exploration of the kinematics and control mechanisms of the RHex robot endeavors to push the boundaries of attainable feats in legged robotics, thereby establishing a new standard for adaptability and performance in mobile robotic applications.

The RHex hexapod robot introduces an innovative departure from the traditional multi-segmented, multi-degree-of-freedom leg configuration commonly found in legged robots, opting instead for a singular C-shaped leg with a solitary degree of freedom per leg [23,28]. This design ingeniously strikes a balance between the intricacies of legged locomotion and the efficiency of wheeled mobility, endowing the RHex with the capability to traverse uneven terrains with remarkable stability and resilience. The robot's legs, capable of high-speed synchronized rotations, facilitate a broad spectrum of mobility tasks, encompassing navigating slopes [29] and stairs [30,31], overcoming obstacles [27], and executing intricate maneuvers such as flipping [32].

Among the array of gaits employed by RHex robots, the alternating tripod gait emerges as particularly notable for its efficiency and dynamic stability, drawing inspiration from the locomotive patterns observed in insects such as cockroaches and beetles. This gait organizes the robot's legs into two tripods—front and rear legs on one side paired with the middle leg on the opposite side—ensuring continuous ground contact for one tripod while the other repositions for the subsequent step [33]. This methodology not only facilitates efficient forward movement but also enhances the robot's dynamic stability, as depicted by the support triangle illustrated in Figure 1B. The utilization of the alternating tripod gait ensures the continued placement of the robot's center of gravity inside the support triangle, as demonstrated in Figure 2.

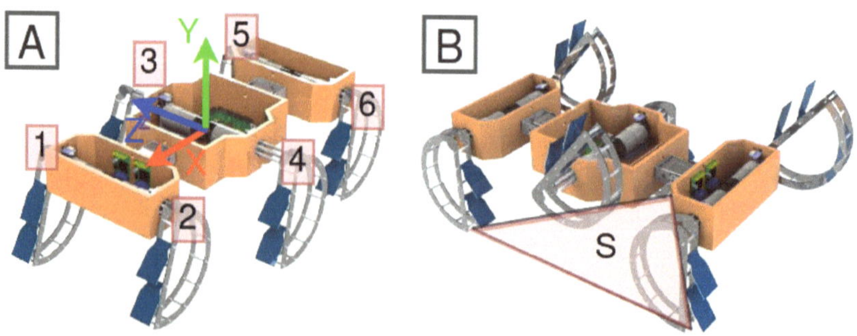

Figure 1. A flapped-paddle amphibian variant of the RHex robot, FLHex (Video [34]) showing (**A**) center of mass and coordinate axes of the robot and (**B**) alternating tripod pairs where tripod legs that are in contact with ground define a support triangle S.

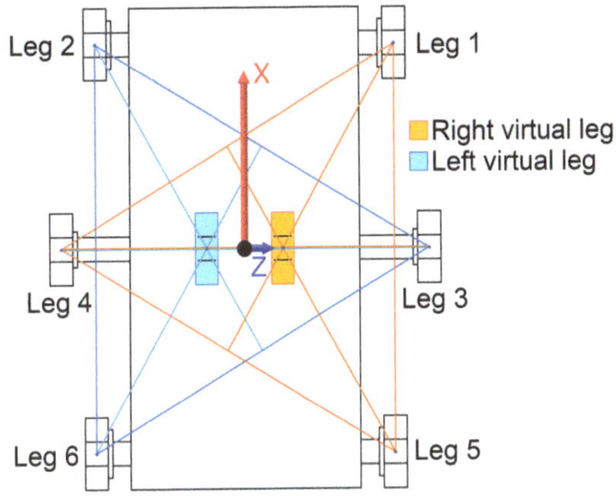

Figure 2. FLHex robot tripods acting as virtual legs.

Further simplifying the complex design of the RHex, the concept of 'virtual legs' combines the physical legs within each tripod, effectively transforming the robot's structure into a body flanked by two composite legs. This abstraction plays a pivotal role in the development of control algorithms and mathematical models for the RHex, streamlining the approach to robot dynamics and control [35]. The adaptability and efficiency demonstrated by the RHex's gait, validated through practical implementations and captured in video demonstrations [34], underscore the advanced design of the robot and its potential for diverse applications.

The principal objective of this investigation is to elucidate the dynamics governing the motion of the RHex robot, particularly when employing the alternating tripod gait—a gait of considerable significance and prevalence among RHex robots. We undertake an in-depth exploration into the development of a kinematic model specifically tailored to this gait, detailing the impact of various parameters on the robot's spatial displacement for a predetermined number of steps. This model serves as a fundamental tool for refining the control and navigation of the RHex robot, offering insights into its operational capabilities.

The structure of this manuscript is organized to facilitate the comprehension of our findings and methodologies. Section 2 introduces an incremental model that focuses on a single C-shaped leg, mirroring the crawling gait of the RHex robot. Subsequently, Section 3 delves into the temporal facets of the leg trajectories integral to the alternating tripod gait. In Section 4, we extend the model to incorporate dual C-shaped legs, reflecting the walking gait of the RHex robot. Section 5 is devoted to the empirical validation of our model, drawing upon data derived from rigorous testing conducted on a specially constructed experimental setup. The ensuing analysis in Section 6 leverages the model to dissect how the intricacies of tripod trajectories and the nuances of the robot's design influence a single gait cycle. This analysis culminates in the formulation of a comprehensive set of design principles aimed at optimizing the control of RHex robots. The manuscript concludes in Section 7, wherein we delineate our findings and discuss their implications for the field.

2. The Kinematics of a Single RHex Robot Leg

Let us start by considering a simplified system of one of the legs of RHex robots shown in Figure 3. The square represents the chassis of the robot, connected to the end of the robot's leg (C-shaped curve) at its point of rotation located in the middle of the square (Point A in Figure 3). The leg shown as a circular arc with radius r and central angle $180^\circ + \alpha$, where α describes the elongated part of the leg, with $\alpha \in [0^\circ, 90^\circ]$. In most RHex robots, the legs are in the shape of a semicircle ($\alpha = 0^\circ$) [1]. However, there are RHex robot designs with an extended leg ($\alpha > 0^\circ$) or even a fully circular leg that is used to increase the smoothness of the robot's movement [35]. The chassis has only two degrees of freedom and it can displace only in the X-axis and Y-axis directions. The position of the robot's leg is described by the angle $\theta \in [-180°, 180°]$ between the leg's diameter at its current position and the leg's diameter in the upright position of the leg (see Figure 3a). The system describing the absolute position of the leg is shown in Figure 3b.

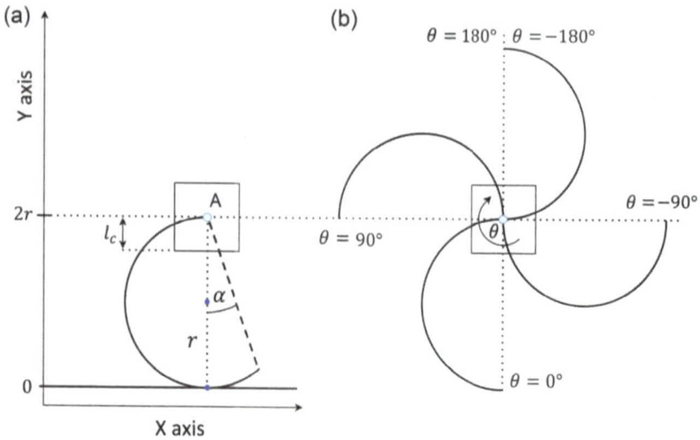

Figure 3. RHex robot C-leg kinematics (**a**) when positioned in the most upright position, (**b**) with varying value of θ as the leg rotates.

The maximum position on the Y-axis corresponds to the leg's most upright position, equivalent to the leg's diameter. The minimum position on the Y-axis is the distance from the robot's leg rotational joint (pivot) to the bottom of the chassis represented by l_c. When the Y-axis position of the pivot is l_c, the robot's chassis is in contact with the ground, the legs rotate in the air, and the system becomes stationary, as illustrated in Figure 4. The system, consisting of a body equipped with a single rotating C-shaped leg, initiates movement when the leg makes contact with the ground at position θ_{start} and returns to a stationary state when the leg breaks contact with the ground at position θ_{end}. These positions, θ_{start} and θ_{end}, are depicted in Figure 4. We see their dependence on the design parameters of the robot and its leg. The value of θ_{start} is determined by r, l_c and the value of θ_{end} is influenced by r, l_c, and α. These can be calculated using the equations shown below.

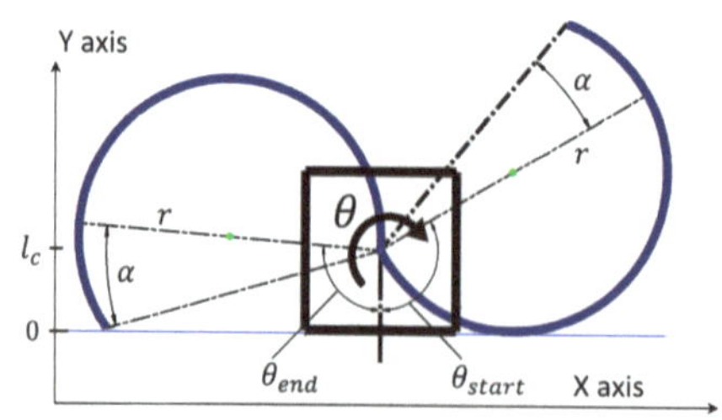

Figure 4. The robot leg positions at θ_{start}, and θ_{end} showing when the leg-ground contact changes.

For θ_{start}

$$1 + \cos(\theta_{\text{start}}) = \frac{l_c}{r} \tag{1}$$

$$\cos\left(\frac{\theta_{\text{start}}}{2}\right) = \pm\sqrt{\frac{l_c}{2r}} \tag{2}$$

where $\pm$ becomes $-$ due to the leg being located in the II or III quadrant of the coordinate system (see Figure 3b), so

$$\cos\left(\frac{\theta_{\text{start}}}{2}\right) = -\sqrt{\frac{l_c}{2r}} \tag{3}$$

$$\theta_{\text{start}} = 90^\circ - 2\arccos\left(\sqrt{\frac{l_c}{2r}}\right) \tag{4}$$

Similarly, for θ_{end} it holds

$$l_c = 2r\cos(\alpha)\cos(\theta_{\text{end}} - \alpha) \tag{5}$$

$$\theta_{\text{end}} = \alpha + \arccos\left(\frac{l_c}{2r\cos\alpha}\right), \alpha \neq \pm 90^\circ \tag{6}$$

The variation of θ_{start} and θ_{end} with varying r, l_c and α are shown in Figure 5. It can be seen that changing any of these parameters has a significant impact on the leg surface that will participate in movement.

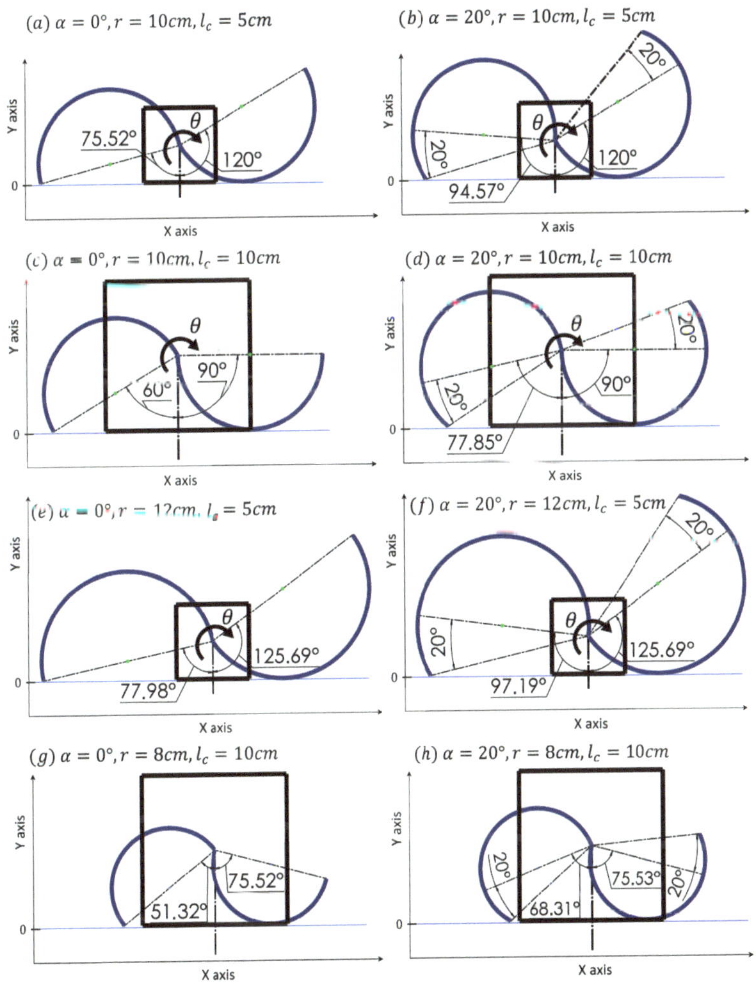

Figure 5. The influence of RHex leg parameters α, r, l_c on θ_{start} and θ_{end}.

We consider the crawling gait where the legs alter their positions in a cyclic manner over time, denoted by $\theta(t)$, by rotating at a constant velocity. This rotation propels the robot's body along the X-axis (parallel to the ground in the direction the robot traverses) and the Y-axis (perpendicular to the ground, aligned with the direction of gravitational pull). The locomotion of the body with a single C-shaped leg through one complete revolution of the leg is depicted for $\alpha = 0°$ and $\alpha \in [0, 90°]$ in Figure 6.

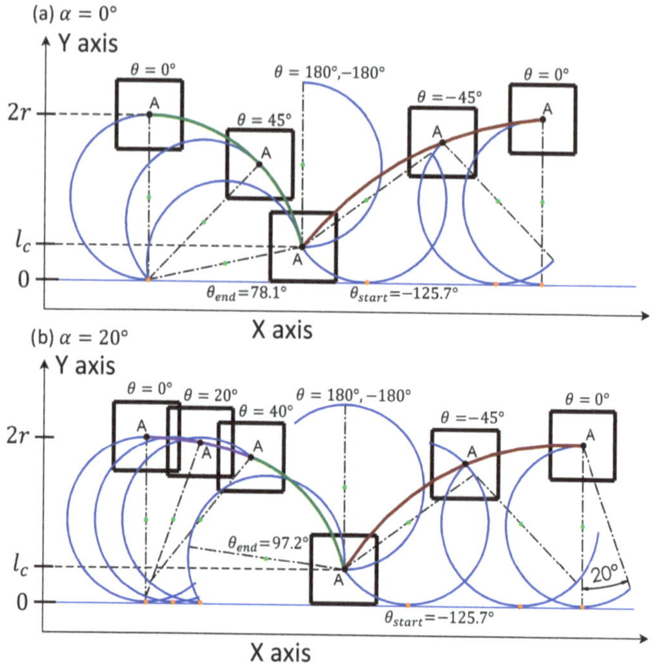

Figure 6. Framed motion of a single RHex robot leg for (**a**) $\alpha = 0°$ and (**b**) $\alpha = 20°$.

In the first case (see Figure 6a for $\alpha = 0°$), the system starts at the position where the contact point of the leg with the ground is located at the tip of the leg and the system pivots around the contact point until the chassis is in contact with the ground at leg position θ_{end}. In that pivoting movement for $\theta \in [0°, \theta_{\text{end}}]$, the position $x(\theta)$ of the robot in the X-axis and $y(\theta)$ in the Y-axis are described by:

$$x(\theta) = 2r \sin\theta(t_n) \tag{7}$$

$$y(\theta) = 2r \cos\theta(t_n) \tag{8}$$

where time $t_n = nT, n \in \mathbb{N}$, resulting from the sampling process with the period T and $t_0 = 0, \theta \in [0, \theta_{end}], x(0°) = 0, y(0°) = 2r$. The position of the center of mass (Point A in Figure 3a) as a discrete signal can be described as

$$\begin{bmatrix} x(t_n) \\ y(t_n) \end{bmatrix} = \begin{bmatrix} x(t_n - t_{n\text{-}1}) + 2r\sin\theta(t_n) - 2r\sin\theta(t_n - t_{n\text{-}1}) \\ 2r\cos\theta(t_n) \end{bmatrix} \tag{9}$$

The pivoting motion ends when the chassis of the robot comes into contact with the ground and the robot's leg detaches from the ground (as the leg rotates, see Point A in Figure 6a). The leg is then rotated over the robot body and leans on the ground in front of the chassis. It then starts another motion of the center of mass. When the robot leg is detached from the ground for $\theta \in [-180°, \theta_{\text{start}}] \cup [\theta_{\text{end}}, 180°]$, the system becomes stationary in the so-called aerial phase. After making ground contact following the aerial phase, further rotation of the leg drives the system into a second ground phase with an ascending type of motion that corresponds to a cycloid (contact point of the leg with the ground is moving through the surface of the C-shaped leg towards its tip). Therefore, positions $x(\theta)$ and $y(\theta)$ can be described as

$$x(\theta) = r\Big(\theta(t_n) + 180^\circ - \sin\left(\theta(t_n) + 180^\circ\right)\Big) = r\Big(\theta(t_n) + 180^\circ + \sin\theta(t_n)\Big) \tag{10}$$

$$y(\theta) = r\Big(1 - \cos\left(\theta(t_n) + 180^\circ\right)\Big) = r\Big(1 + \cos\theta(t_n)\Big) \tag{11}$$

In that ascending movement for $\theta \in [\theta_{\text{start}}, 0^\circ]$, the position $x(\theta)$ of the robot in the X-axis and $y(\theta)$ in Y-axis are described as a discrete signal with a sampling period T using

$$\begin{bmatrix} x(t_n) \\ y(t_n) \end{bmatrix} = \begin{bmatrix} x(t_n - t_{n\text{-}1}) + r(\theta(t_n) - \theta(t_n - t_{n\text{-}1}) + \sin\theta(t_n) - \sin\theta(t_n - t_{n\text{-}1})) \\ r + r\cos\theta(t_n) \end{bmatrix} \tag{12}$$

At the end of that phase ($\theta = 0^\circ$), the whole process is repeated. Thus, one rotation of the leg in the single-leg system for $\alpha = 0^\circ$ can be divided into an aerial phase for $\theta \in [-180^\circ, \theta_{start}] \cup [\theta_{end}, 180^\circ]$ and a ground phase for $\theta \in [\theta_{start}, \theta_{end}]$. The ground phase can be further divided into a descending (pivoting) and ascending (cycloid) motion where the transition between them occurs at $\theta = 0^\circ$.

For the second case with $\alpha > 0^\circ$ (see Figure 6b), the sequence is slightly changed. At the start it is not the tip of the leg that makes contact with the ground but it is the leg surface. The rotation of the leg causes a descending movement that corresponds to a cycloid. For $\theta = 2\alpha$, the leg-ground contact point reaches the leg's tip moving the system in a pivoting style of motion similar to when $\alpha = 0^\circ$ but the pivoting is with a smaller radius due to the enlarged part of the leg. After this point, the rest of the movement is identical to the first case. The descending cycloidal movement at the start and ascending cycloidal movement at the end phase of the cycle are fragments of the same cycloid.

By comparing both cases, one can observe that the basic types of the movements in both cases are the same and the only difference lies in the ranges of leg position at which specific types of motion occur. To be more specific, in both cases, the transition between the pivoting and cycloid type of motion in the ground phase occurs at $\theta = 2\alpha$. By combining the obtained information, the X and Y axis crawling gait displacement of the center of mass of the RHex robot (after discretization with the sampling period T) can be described using

$$\begin{bmatrix} x(t_n) \\ y(t_n) \end{bmatrix} = \begin{cases} \begin{bmatrix} x(t_n - t_{n\text{-}1}) + r(\theta(t_n) - \theta(t_n - t_{n\text{-}1}) + \sin\theta(t_n) - \sin\theta(t_n - t_{n\text{-}1})) \\ r + r\cos\theta(t_n - t_{n\text{-}1}) \end{bmatrix} & \text{if } \theta(t_n) \in [\theta_{start}, 2\alpha) \\ \begin{bmatrix} x(t_n - t_{n\text{-}1}) + 2r\cos\alpha\Big(\sin(\theta(t_n) - \alpha) - \sin\big(\theta(t_n - t_{n\text{-}1}) - \alpha\big)\Big) \\ 2r\cos\alpha\cos(\theta(t_n) - \alpha) \end{bmatrix} & \text{if } \theta(t_n) \in [2\alpha, \theta_{end}] \\ \begin{bmatrix} x(t_n - t_{n\text{-}1}) \\ l_c \end{bmatrix} & \text{if } \theta(t_n) \in (-180^\circ, \theta_{start}) \cup (\theta_{end}, 180^\circ] \end{cases} \tag{13}$$

3. RHex Tripods Motion Profile For Walking/Running Scenario

For the RHex robot, various locomotion modes such as walking, running, turning, and climbing are achieved through the employment of predetermined periodic leg position setpoint functions. These functions are synchronized for each leg within one tripod (comprising legs 1-4-5 or 2-3-6, as illustrated in Figure 1A) and an alternated version for the opposite tripod. This coordination is widely recognized as the tripod gait, characterized by the robot maintaining a minimum of three points of contact with the ground at any given time. These contact points form a support triangle S (Figure 1B), which invariably encompasses the projection of the RHex robot's center of mass (Figure 2), ensuring both dynamic and static stability. The rotation of both tripods is unidirectional, following a specific cyclic pattern. Within a single walking cycle, every leg of the RHex robot completes a full rotation, encompassing slow and fast swing phases. The slow swing phase facilitates the execution of a step, whereas the fast swing phase repositions the leg in preparation for the subsequent step. This alternating tripod stepping mechanism culminates in a stable walking pattern for the RHex robot. The alteration in position $\theta(t)$ of the RHex robot's tripods (as depicted in Figure 1) during the slow swing phase induces a displacement akin to that described in Section 2. A notable distinction, however, lies in the variable rotation speed of the legs and the implementation of alternating motion profiles, which precludes any chassis–ground contact.

In the tripod gait, the legs of each corresponding tripod adjust their positions according to a cyclic time function $\theta(t)$, with a single cycle depicted in Figure 7. These leg position trajectories, often referred to as the 'Buehler clock' or 'motion profiles of tripods', define the robot's kinematic behavior [33]. The trajectory $\theta(t)$ is characterized by several parameters: the period of the motion profiles t_c, the duty factor of each tripod within a cycle t_s, the angle covered during the slow swing phase ϕ_s, with $\phi_s \in [0, 180°]$, and the motion profile offsets ϕ_o. Typically, $t_s \in (0, t_c]$, but for faster movement, a duty factor in the range $t_s \in \left[\frac{t_c}{2}, t_c\right]$ is advised. Each tripod undergoes slow and fast swing phases within a cycle, spanning angles ϕ_s and $360° - \phi_s$, respectively, to complete a full rotation.

The optimal walking gait of the RHex robot can be attained through precise control of the parameters t_c, t_s, ϕ_s, and ϕ_o. Manipulating these values allows for the adjustment of the distance covered in a single walking cycle, modulation of the robot's body turbulence along the Y-axis (as shown in Figure 1), and the timing of the double support phase (t_d shown in Figure 7), where all six legs potentially make simultaneous ground contact during the slow swing phase. The extent of the double support phase t_d is contingent upon the duty factors of the two tripods. A scenario with $t_s = \frac{t_c}{2}$ eliminates the double support phase entirely ($t_d = 0$). The implementation of double support is particularly beneficial under conditions of heightened load on the leg drive motor or when enhanced stability is necessary, such as during transport of a payload by the robot. Nonetheless, prolonging the dual support phase inversely affects the robot's locomotive speed, a detail further explored later in this study. The motion profile offset, denoted by ϕ_o, adjusts the trajectory relative to the vertical (Figure 7) and is typically set to 0° in most applications.

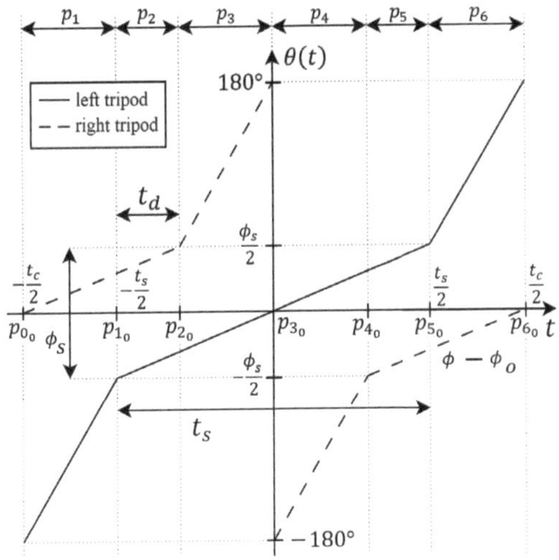

Figure 7. The motion profiles and essential parameters for the left and right tripods in a single walking cycle.

Based on motion profiles presented in Figure 7 the rotation speed in fast swing (in the aerial phase) can be calculated as

$$\dot{\theta}_F(t) = \frac{360^\circ - \phi_s}{t_c - t_s} \tag{14}$$

And is the for the rotation speed in slow swing (ground phase):

$$\dot{\theta}_S(t) = \frac{\phi_s}{t_s} \tag{15}$$

Thus, the rotation speed in time $\dot{\theta}_L(t)$ for the left and rotation speed in time $\dot{\theta}_R(t)$ for the right tripod in one cycle of the tripod gait can be presented, respectively, as:

$$\dot{\theta}_R(t) = \begin{cases} \dot{\theta}_S(t), & t \in [p_{0_0}, p_{2_0}] \cup [p_{4_0}, p_{6_0}] \\ \dot{\theta}_F(t), & t \in (p_{2_0}, p_{4_0}) \end{cases} \tag{16}$$

$$\dot{\theta}_L(t) = \begin{cases} \dot{\theta}_S(t), & t \in (p_{1_0}, p_{5_0}) \\ \dot{\theta}_F(t), & t \in [p_{0_0}, p_{1_0}] \cup [p_{5_0}, p_{6_0}] \end{cases} \tag{17}$$

where p_{i_0}, $i = 1, \ldots, 6$ are time stamps of the p_i phase end and p_{i+1} phase start (see Figure 7).

In Table 1, a single Buehler clock cycle is segmented into six distinct phases, labeled as $p_i, i = 1, \ldots, 6$ in Figure 7. The duration of phases p_1, p_3, p_4, p_6 is set at $\frac{t_c - t_s}{2}$, whereas phases p_2, p_5 span a timeframe of $t_s - \frac{t_c}{2}$. This configuration establishes the temporal markers $p_{i_0}, i = 1, \ldots, 6$ that signify the conclusion of phase p_i and the commencement of phase p_{i+1} within a single walking cycle. Observations from a recorded RHex robot tripod gait, as documented in [34], reveal variations in the rotation speed of the tripods between successive movement phases. Specifically, the fast swing phases are attributed to the aerial phase, while the slow swing phases correlate with the ground contact phase of the tripod's motion. Additionally, Table 1 delineates the type of movement along the Y-axis for the corresponding phases as witnessed in the recordings.

Table 1. Motion phases $p_i, i = 1,2,\ldots,6$ of the RHex robot in tripod alternating gait for walking cycle $(n_j, j = 0,1,\ldots)$.

Phase p_i of Motion	Left Tripod Motion Type	Right Tripod Motion Type	Tripod Responsible for Movement	Robot Movement in Y Axis	Time Stamp of Phase End p_{i0} for Each Walking Cycle
p_1	Fast swing	Slow swing	Right	Descending	$n_j t_c + (t_c - t_s)/2$
p_2	Slow swing	Slow swing	Transition from right to left	From descending to ascending	$n_j t_c + t_s/2$
p_3	Slow swing	Fast swing	Left	Ascending	$n_j t_c + t_c/2$
p_4	Slow swing	Fast swing	Left	Descending	$(n_j + 1)t_c - t_s/2$
p_5	Slow swing	Slow swing	Transition from left to right	From descending to ascending	$n_j t_c + (t_c + t_s)/2$
p_6	Fast swing	Slow swing	Right	Ascending	$(n_j + 1)t_c$

By analyzing Figure 7 and Table 1, the movement phases can be categorized into pairs $\{p_1; p_6\}$, $\{p_3; p_4\}$, and $\{p_2; p_5\}$. Within these pairs, both the rotation speed of the specific tripods and the duration of the phases are identical. The pairs $\{p_1; p_6\}$ and $\{p_3; p_4\}$ facilitate the slow swing for the right and left tripods, respectively, with the primary distinction being the tripod that is active during each phase. Conversely, phases p_2 and p_5 correspond to the double support phases, during which both tripods engage in a slow swing and may simultaneously make contact with the ground.

Note that the left and right tripod in tripod gait rotate in the same direction (see Figure 7) to cause a forward displacement of the RHex robot. The direction of rotation of the tripods can be reversed to achieve a backward motion. However, it is not as optimal and stable as forward running and can be harmful to the leg drive motor because of sudden load increase when the leg starts to touch the ground.

4. RHex Incremental Kinematic Model For Walking in Flat Terrain

To obtain an incremental kinematic model of the RHex robot for walking gaits in flat terrain that uses alternating tripod motion profiles, some simplifying assumptions have been made:

- The leg has no mass—the RHex robot's legs are a very small fraction of the total mass of the robot. Therefore, assuming a massless leg will not largely impact the motion mechanics of the system.
- No bending of the leg—in this analysis the C-shaped leg is considered to be a rigid body, despite the potential for leg deformation under load that can improve the robot's mobility by functioning as a form of suspension. This aspect is particularly relevant for running gaits, where the RHex robot's vertical (Y-axis) motion may exhibit distinctive characteristics. For instance, at high leg rotation speeds, the system may predominantly engage in ascending motion phases, where the combined effects of momentum and gravitational forces enable the robot to execute a series of jumps, minimizing the descending motion phases. However, in walking gaits, where the forces involved are considerably lower, the compliance of the legs does not significantly alter the system's fundamental motion patterns. By modeling the legs as rigid bodies, the robot's movement can be simplified to a combination of pivoting and cycloidal motions, which will be further explored as representing the foundational movement patterns of the RHex robot.
- No slipping between the robot's legs and the ground.
- No change in the system mass.
- $\phi_s \in (0°, 360°], t_s \in [\frac{t_c}{2}, t_c]$.

Within the framework of the tripod gait, the RHex robot can be conceptually simplified to a two-degree-of-freedom rigid body, outfitted with two semicircular legs that share a common axis of rotation. Each leg in this model epitomizes one half of the robot's bipartite tripod mechanism. This reduction is justified by the dynamic stability inherent to the tripod gait, which, during locomotion across flat terrains, restricts the robot's body displacement to the X and Y axes, as illustrated in Figure 1. The synchronized movement of the legs forming each tripod (effectively acting as a 'virtual leg') consistently maintains three points of contact with the ground, effectively nullifying any rotational movement of the body. Consequently, it is reasonable to posit that the center of mass displacement in the actual RHex robot, when employing the tripod gait, mirrors that of its simplified counterpart.

We develop a simplified model as depicted in Figure 8a, that serves as a partial representation of the RHex robot. This model comprises a square chassis, symbolizing the robot's body, and a pair of C-shaped legs, each originating from a different tripod. The initial leg positions correspond to those outlined in the motion profiles (refer to Figure 7). The leg's pivot point—where it attaches to the motor—is situated at Point A in Figure 8a. It's crucial to note that for the model to be applicable, all legs of the RHex robot must share a common Y-axis level at their pivoting points. The blue dot in the figure denotes the contact point where the leg meets the ground, represented by a solid black line at zero meters elevation. The X-axis, running horizontally and parallel to the ground, signifies the direction of the robot's forward and backward traversal, while the Y-axis, perpendicular to the ground, aligns with the gravitational pull. These axes align with those presented in Figure 1.

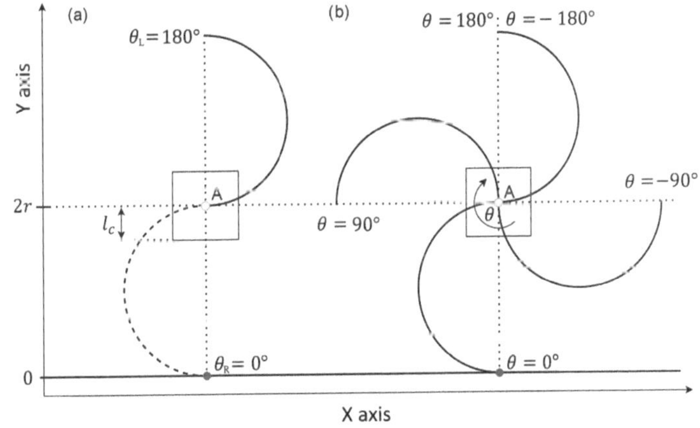

Figure 8. Kinematics of the simplified RHex robot with C-shaped leg for $\alpha = 0°$ showing the (**a**) initial position of the legs in each cycle and (**b**) the position θ of the legs.

As the leg rotates clockwise (as shown in Figure 8b), the ground contact point shifts, prompting the system to displace. The construction attributes of the RHex robot, such as the leg radius r, the distance l_c from the leg's pivot Point A to the chassis bottom, and the leg extension α, are described using the same parameters introduced in Section 2. The range $\theta(t_n) \in [\theta_{start}, \theta_{end}]$ within which the robot's leg maintains ground contact is defined in a manner analogous to that in Section 2.

Initial positions of the right and left leg (each represent corresponding tripod of the RHex robot) in the simplified system can be described using:

$$\theta_R(t_0) = 0 \tag{18}$$

$$\theta_L(t_0) = -180° \tag{19}$$

Using the initial position of the legs, the initial position of the robot's center of mass is determined as:

$$x(t_0) = x_0 = 0 \tag{20}$$

$$y(t_0) = y_0 = 2r \tag{21}$$

To prevent the robot's chassis from contacting the ground during the tripod gait, alternating motion profiles are employed. The trajectories for $t \in (0, \frac{t_c}{2})$ and $t \in (\frac{t_c}{2}, t_c)$, as illustrated in Figure 7, mirror each other irrespective of the specific tripod in action, rendering them as odd functions. Consequently, the robot's movement is characterized by a series of half-cycle displacements. Furthermore, as elucidated in Section 2, the locomotion of the RHex robot when operating with a single leg encompasses both pivoting and cycloidal motions, as delineated by Equations (9) and (12), respectively, [1]. Building on these observations, it can be inferred that the movement of a bipedal configuration in the RHex robot consists of a cyclic pattern of ascending and descending motions. To gain a clearer understanding of the RHex robot's displacement during a half-cycle of the tripod gait, a detailed visualization is provided in Figure 9. For the initial leg positions depicted in Figure 8a, the legs adjust their positions following the trajectory outlined in Figure 7, with parameters set to $r = 5$ cm, $\alpha = 0°$, $\phi_s = 90°$ and $t_s = \frac{t_c}{2}$.

Initially, the system pivots on the right leg while the left leg rotates freely in the air. At a certain instance, both legs momentarily make contact with the ground, as depicted in the visualization at the lowest central position. Subsequently, the right leg loses ground contact, and the system's progression is driven by the left leg's motion, albeit in a cycloidal fashion. By the conclusion of the half-cycle's visualization, the system reverts to a state akin to the initial condition, albeit with the left and right legs' positions interchanged. This sequence recurs twice within a single tripod gait cycle, culminating with the legs reverting to their original positions ($\theta_L(t = t_c) = 180°$ and $\theta_R(t = t_c) = 360°$). The determining factor of which tripod is engaged with the ground, thereby facilitating displacement, hinges on the greater value of $d(t)$—the distance from the leg's pivot point (illustrated by the purple dot in Figure 9) to the leg's furthest extremity toward the ground along the Y-axis. For each tripod, $d(t)$ is contingent upon the current position of its 'virtual leg' $\theta(t)$, and thus, is time-dependent. A leg is considered in contact with the ground when its $d(t)$ is equal to or surpasses that of the alternate virtual leg. The distance $d(t_n)$ resulting from the sampling process with the period T for each virtual leg is defined as follows:

$$d(t_n) = \begin{cases} r - r\cos\theta(t_n) & \text{if } \theta(t_n) \in [\theta_{\text{start}}, 2\alpha] \\ 2r\cos(\alpha)\arccos(\theta(t_n) - \alpha) & \text{if } \theta(t_n) \in (2\alpha, \theta_{\text{end}}] \\ l_c & \text{if } \theta(t_n) \in (-180°, \theta_{\text{start}}) \cup (\theta_{\text{end}}, 180°] \end{cases} \tag{22}$$

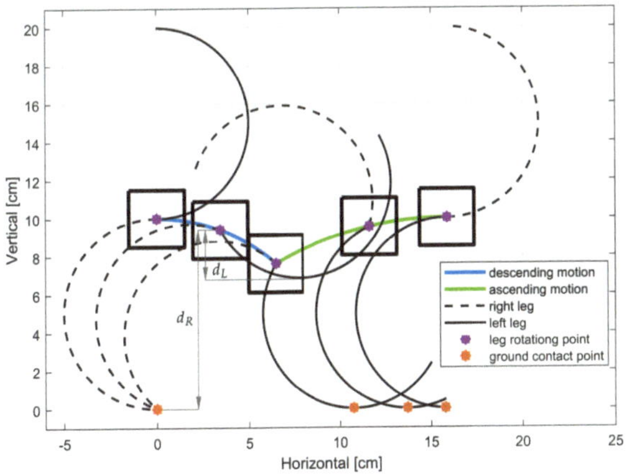

Figure 9. Tripod gait half cycle visualization for $r = 5$ cm, $\alpha = 0°$, $\phi_s = 90°$ and $t_s = \frac{t_c}{2}$.

Designation of the tripod responsible for the movement and its current position at any given time is derived from comparing the distance $d(t_n)$ of both legs as

$$\theta_G(t_n) = \begin{cases} \theta_L(t_n), & \text{if } d_R(t_n) \leqslant d_L(t_n) \\ \theta_R(t_n), & \text{if } d_R(t_n) > d_L(t_n) \end{cases} \tag{23}$$

where $\theta_L(t_n)$ is the current position of the legs in the left tripod with distance $d_L(t_n)$, $\theta_R(t_n)$ is the current position of the legs in the right tripod with distance $d_R(t_n)$, see Figure 9. By combining Equation (13) of one-legged RHex robot system with Equation (23), the X and Y displacement of the center of mass of the RHex robot (after discretization with sampling period T) for tripod gait can be described as

$$\begin{bmatrix} x(t_n) \\ y(t_n) \end{bmatrix} = \begin{cases} \begin{bmatrix} x(t_n\text{-}t_{n\text{-}1})+r\left(\theta_G(t_n)\text{-}\theta_G(t_n\text{-}t_{n\text{-}1})+\sin\theta_G(t_n)\text{-}\sin\theta_G(t_n\text{-}t_{n\text{-}1})\right) \\ r+r\cos\theta_G(t_n) \end{bmatrix} & \text{if } \theta_G(t_n) \in [\theta_{\text{start}}, 2\alpha) \\ \begin{bmatrix} x(t_n\text{-}t_{n\text{-}1})+2r\cos\alpha\left(\sin(\theta_G(t_n)\text{-}\alpha)\text{-}\sin(\theta_G(t_n\text{-}t_{n\text{-}1})\text{-}\alpha)\right) \\ 2r\cos\alpha\cos(\theta_G(t_n)\text{-}\alpha) \end{bmatrix} & \text{if } \theta_G(t_n) \in [2\alpha, \theta_{\text{end}}] \\ \begin{bmatrix} x(t_n\text{-}t_{n\text{-}1}) \\ l_c \end{bmatrix} \quad \text{if } \theta_G(t_n) \in (-180°, \theta_{\text{start}}) \cup (\theta_{\text{end}}, 180°] \end{cases} \tag{24}$$

The developed model featuring $\alpha = 0°$, variable t_s, r, and ϕ_s was simulated to evaluate the impact of these parameters on the robot's locomotion. The results are presented in Figure 10. Consistent with the model's premises, the robot's displacement embodies the movement types previously delineated. Notably, the displacement along the X-axis and Y-axis is directly influenced by the leg's radius, highlighting that even minor adjustments to the leg's radius can significantly affect the robot's operational range—a crucial factor in defining its potential applications.

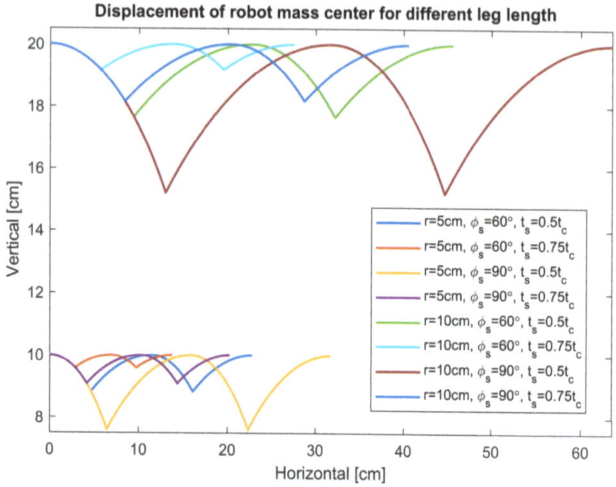

Figure 10. Comparison of robot displacement in X and Y axis for different r, t_S and ϕ_S.

To alter the robot's walking, adjustments to other parameters are necessary. For instance, augmenting t_S within the motion profile diminishes the X-axis displacement while concurrently reducing the oscillation amplitude of the robot's virtual center of mass along the Y-axis. This reduction is particularly advantageous when employing optical sensors. Similar effects are observed with a decrease in ϕ_S. Therefore, by simultaneously increasing t_S and decreasing ϕ_S, comparable walking can be achieved through diverse motion profiles. This interdependency offers valuable insights for designing varied gaits tailored to specific tasks such as running or load-bearing, where the duration of double support phases may necessitate adjustment.

5. Experimental Validation of the RHex Walking Model

To corroborate the kinematic model presented in the preceding section, the creation of an experimental test bed congruent with the model's premises was imperative. A critical aspect of this setup was the constriction of the system's degrees of freedom exclusively in the X and Y axes. This limitation was essential to ensure no extraneous resistance was introduced, thereby allowing for an accurate emulation of the motion of the mobile robot's center of mass.

For the validation of the kinematic model, the experimental test bed depicted in Figure 11 was meticulously designed and fabricated. The test bed incorporates a dual set of linear guideway blocks and rails, commonly found in CNC machinery, configured to permit motion along the X- and Y-axes while constraining movement and rotation across other axes. To ensure minimal resistance and friction, lubricants and bearings were integrated within the blocks. A subsystem comprising a pair of RHex robot legs, representing the mobile robot, was mounted onto this bespoke structure. To faithfully replicate the robot's control mechanisms, components identical to those utilized in the FLHex robot, as documented in [27], were employed. This setup includes an Arduino Mega 2560 (Arduino.cc Corp.) microcontroller and a Pololu VNH5019 (Pololu Corp., Las Vegas, NV, USA) motor driver for control, a high-torque 12V DC motor Pololu 37Dx70L (Pololu Corp., Las Vegas, NV, USA) series with a 50:1 gearbox for actuation, and a quadrature magnetic encoder for position sensing. The leg position control system of the RHex robot is governed by a fractional-order PID (FOPID) controller with optimally derived coefficients [36] and aims to maintain the leg positions in tight alignment with the predefined motion profiles through a series of carefully executed steps.

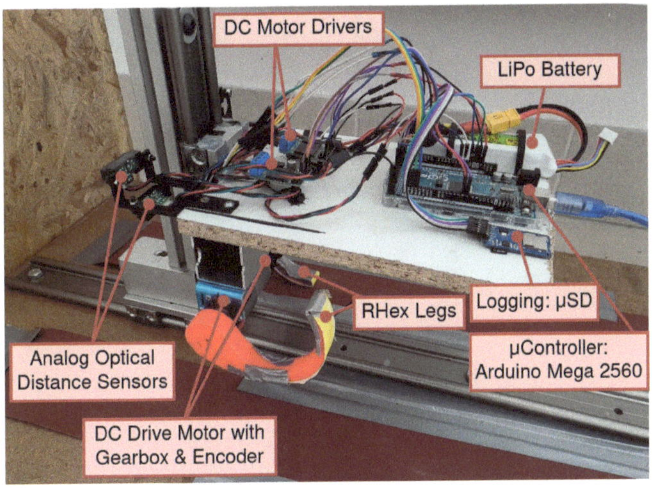

Figure 11. Experimental test bed fabricated to emulate and validate the kinematic model.

The leg center positions during rotation were estimated using analog optical distance sensors: a Sharp GP2Y0A41SK0F (Sharp Corp., Osaka, Japan) for the Y-axis and a Sharp GP2Y0A21YK0F (Sharp Corp., Osaka, Japan) for the X-axis. To mitigate sensor noise, a quadratic regression was applied over a 100-sample window. Both sensors were interfaced with an Arduino Mega, which was collecting and filtering this data. Arduino board was receiving FOPID controller outputs from a laptop running MATLAB 2022b where horizontal and vertical legs positions were calculated and visualized. For these experiments, the leg radius was set at 5 cm, designed to meet the model's stipulated requirements. As shown in Figure 11, the motor shaft and leg end form a semicircle. A sandpaper was used as a walking surface to prevent slippage. Additionally, the leg was engineered to minimize bending.

Experimental results are juxtaposed with model predictions for legs with parameters $r = 5$ cm, $\phi_s = 60^\circ$ in Figure 12, and with parameter $\phi_s = 90^\circ$ in Figure 13. The results are with varying t_s. Each figure delineates the horizontal and vertical displacements alongside the corresponding leg positions over time. Notably, the most significant discrepancies were observed in height changes, although these deviations were minimal relative to the leg's size. These differences could partially result from friction between the guideway block and rail or errors in optical distance measurement. Some degree of unavoidable leg bending may also contribute to these discrepancies. Crucially, the experimental travel distances align with the model's predictions, affirming the kinematic model's applicability to the walking/running control logic of the robot and its overall accuracy.

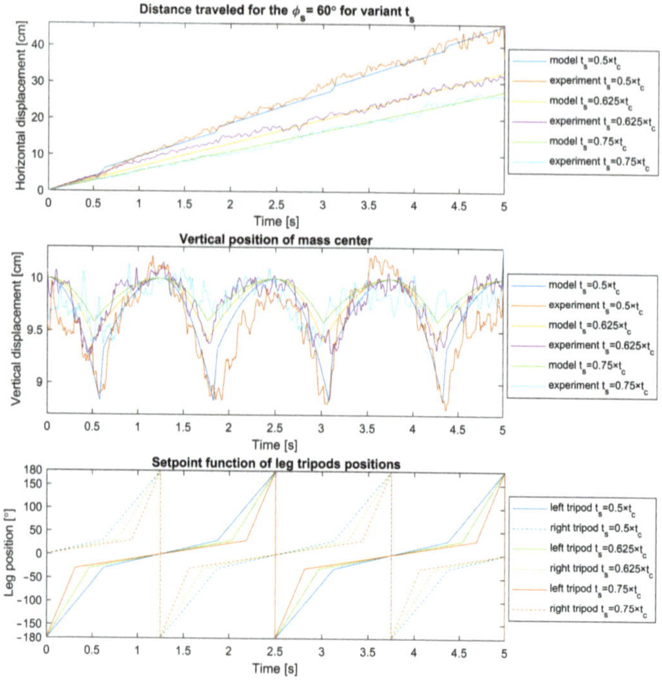

Figure 12. Experimental validation of model for $t_c = 2.5$ s, $\phi_o = 0°$, $\phi_s = 60°$ with varying t_s.

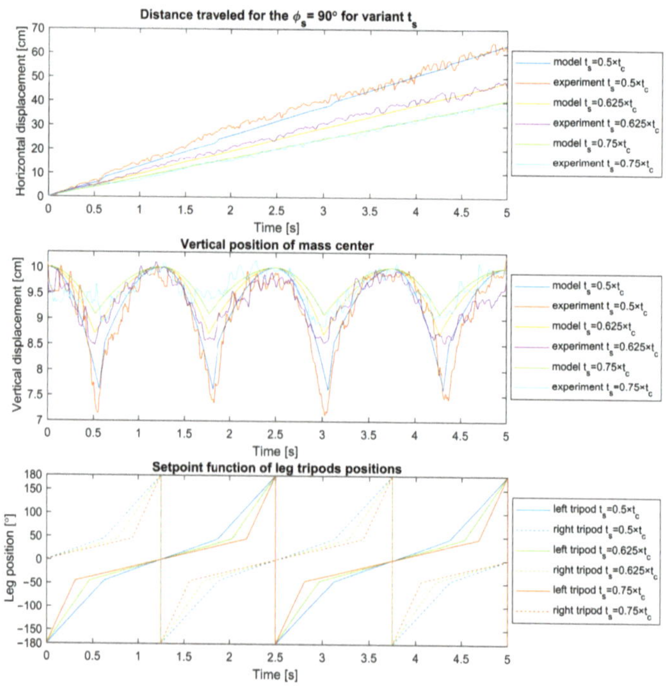

Figure 13. Experimental validation of model for $t_c = 2.5$ s, $\phi_o = 0°$, $\phi_s = 90°$ with varying t_s.

6. Results and Discussion

A significant observation in both the experimental tests and kinematic model prediction for $t_s = 0.5t_c$ is the abrupt increase in both horizontal and vertical displacements observed immediately following the transition between tripod sets during ground contact (shifting from the declining phase of one tripod's step to the ascending phase of the other). This transition results in an augmented displacement for the robot over a larger number of steps to some degree.

To delve deeper into the cause of this phenomenon, further visualizations were conducted for a leg with parameters $r = 5$ cm, $\alpha = 0°$, $t_c = 2.5$ s, and varying t_s. The visualizations are showcased in Figures 14–16 and depict the robot's displacement during a single step cycle in the walking gait for $t_s = 0.5t_c$, $t_s = 0.625t_c$, and $t_s = 0.75t_c$, respectively, with the leg positions visualized at equal time intervals of $t = 0.03125t_c$. Notably, Figure 14 illustrates that around the 1.8-second, the points are significantly more spaced out compared to other instances, indicating a higher velocity during these periods as evidenced in Figures 12 and 13. Upon examination of the visualized leg positions, it becomes apparent that in this scenario, the tripod designated for the aerial fast swing phase inadvertently makes ground contact, while the other tripod, which is supposed to propel the system through its motion and maintain ground contact, is detached. Such an occurrence is undesirable, as it prevents the leg from executing its intended function during that movement phase. Furthermore, an increased rotation speed of the leg upon ground contact may pose a risk of damage or accelerated wear to the robot's drive components.

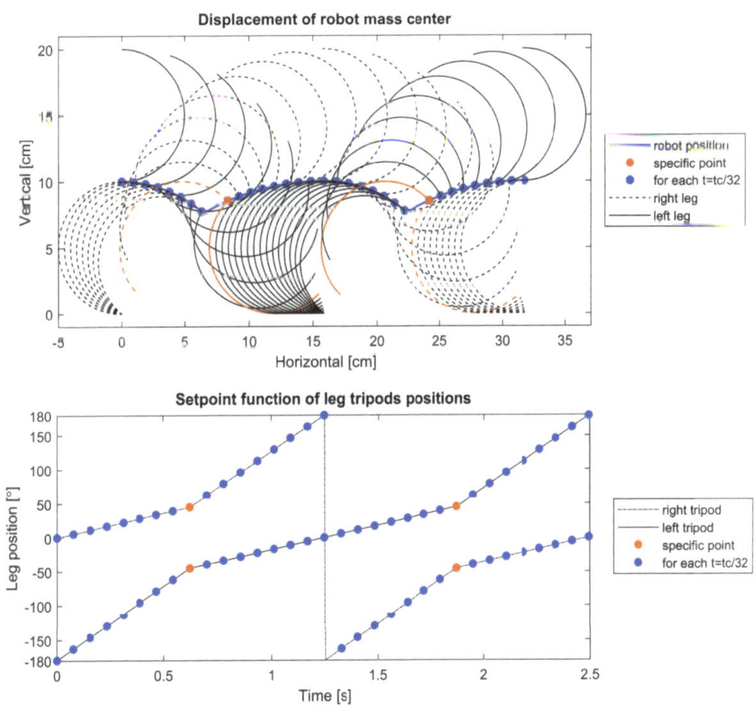

Figure 14. Displacement of the robot for $t_s = 0.5t_c$.

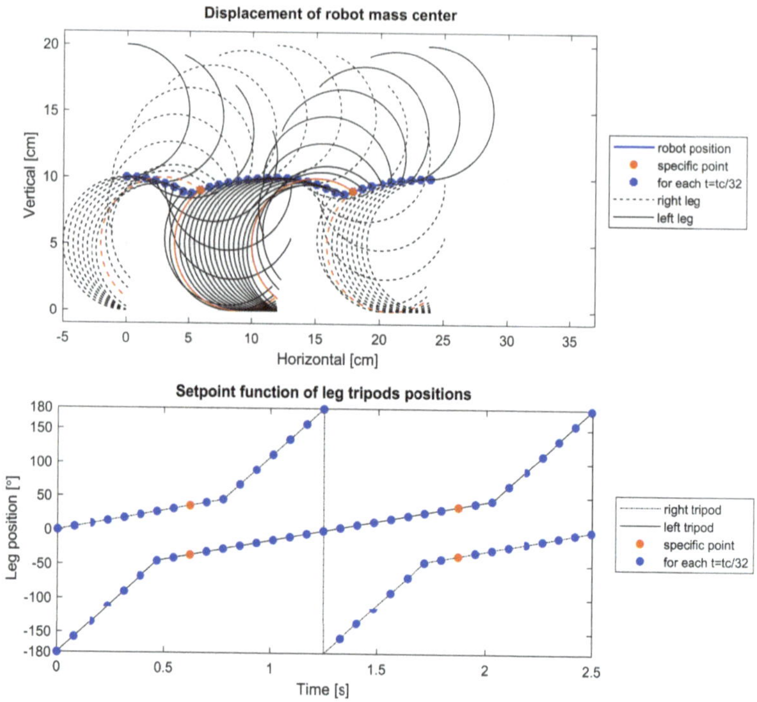

Figure 15. Displacement of the robot for $t_s = 0.625t_c$.

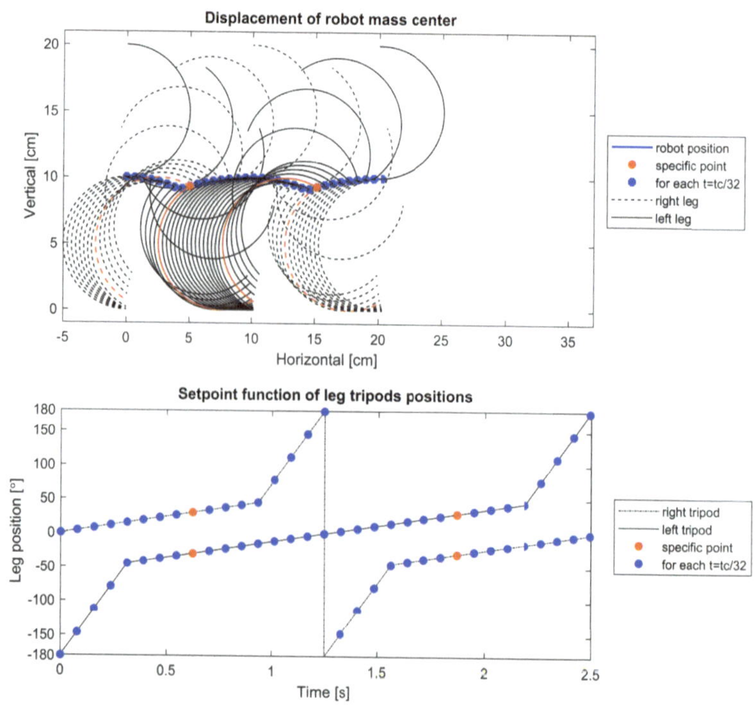

Figure 16. Displacement of the robot for $t_s = 0.75t_c$.

To ascertain which aspects of the robot's design or motion profiles might lead to the aforementioned undesirable occurrences, additional simulations were conducted. The initial focus was on specific parameters, setting $\alpha = 0°$ while varying r, t_s, and ϕ_s. The robot's displacement along the X-axis during a single walking gait cycle served as the criterion for identifying instances of the undesired event, as such occurrences would typically manifest as a noticeable increase in displacement. The findings are documented in Figure 17. In all scenarios with $\alpha = 0°$, variations in r or ϕ_s resulted in a linear alteration of the X-axis displacement per cycle. Conversely, a non-linear response was observed for t_s, particularly when $t_s < 0.585t_c$, where the X-axis displacement significantly increased, confirming the presence of the undesired event upon reviewing the leg positions in these instances.

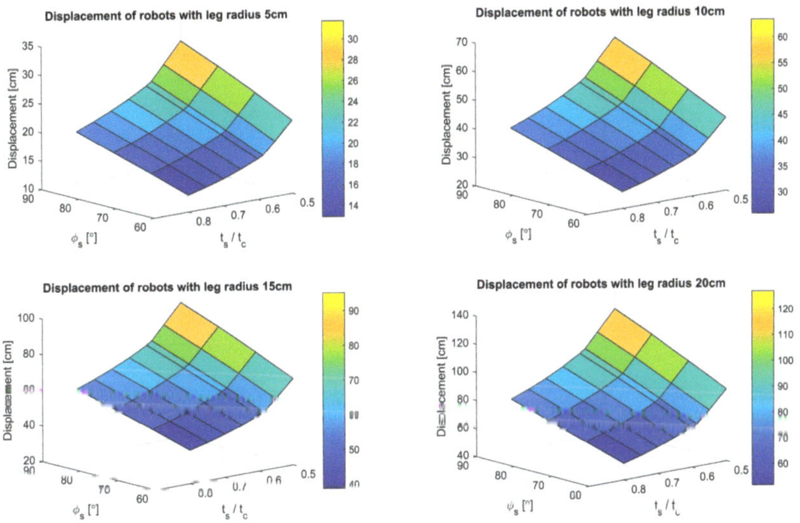

Figure 17. Comparison of the influence of t_s and ϕ_s on the distance traveled by the robot for different leg radius.

However, the situation grows more intricate with $\alpha > 0°$, revealing that the likelihood of the undesired event is influenced by a combination of α, t_s, and ϕ_s. To circumvent this event, reference to a supplementary graph, illustrated in Figure 18, is recommended. Utilizing this graph involves selecting α and t_s values such that their corresponding point on the graph resides on or above the line designated for the chosen ϕ_s within the motion profile.

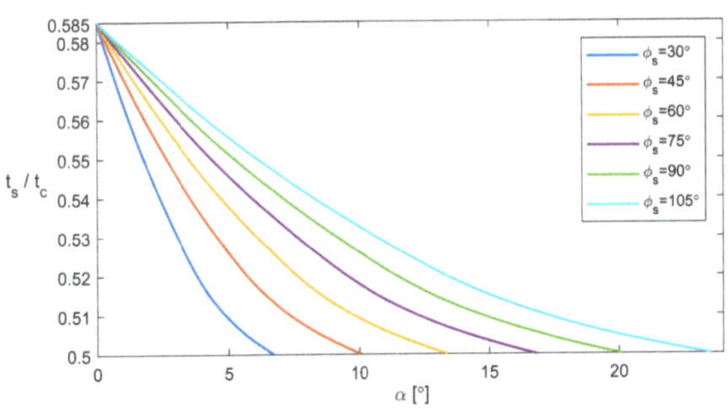

Figure 18. Minimal value of α for specified dependency of $\frac{t_s}{t_c}$ where legs in aerial phase do not make contact with the ground.

7. Conclusions

In this study, we have comprehensively delineated the gait of RHex-type robots, highlighting the pivotal parameters influencing their locomotion. The investigative efforts culminated in the formulation of a kinematic iterative model, tailored for the gait control of such hexapod robots. This model's fidelity was substantiated through rigorous experimental validations conducted on a specially designed test bench. Comparative analyses between the model and experimental outcomes revealed the manifestation of specific undesired phenomena under certain conditions dictated by the robot's leg design and motion profile parameters. Crucially, these insights facilitated the creation of a heuristic graph, poised to guide the optimization of the RHex robot's running gaits in forthcoming control strategies. By judiciously adjusting the gait parameters, it becomes feasible to tailor the kinematics to suit varying double support duration, catering to the robot's immediate operational requirements. Nonetheless, it is imperative to acknowledge a fundamental compromise: enhancing the robot's velocity, particularly in running gaits, invariably introduces increased oscillations along the Y-axis. This phenomenon could potentially compromise the accuracy of concurrent measurements, underscoring a critical consideration in the pursuit of elevated movement speeds. The forthcoming phase of this research will be dedicated to a comprehensive analysis of the RHex robot's locomotion across terrains of heterogeneous characteristics, encompassing surfaces such as sand and the transitional zones from shorelines to aquatic environments. A focal point of the investigation will be the exploration of how the robot's leg material properties influence the incidence of slippage between the leg and the terrain, thereby affecting the robot's dynamic performance. Particular attention will be given to the impact of the flexibility and texture of the robot's C-shaped legs, which play a pivotal role in its movement, on its interaction with diverse ground conditions. This in-depth examination aims to elucidate the intricate relationship between the robot's structural design and its adaptability to complex environmental challenges. The authors have also initiated the integration of a neuromorphic walking controller into the robotic framework described within the article.

Author Contributions: Conceptualization, P.B., L.A. and S.S.; methodology, P.B. and L.A.; software, P.B. and L.A.; validation, P.B. and L.A.; formal analysis, P.B., L.A., and E.P.; investigation, P.B. and L.A.; resources, P.B., L.A., and E.P.; data curation, P.B. and L.A.; writing—original draft preparation, P.B., L.A., E.P., and S.S.; writing–review and editing, P.B., L.A., E.P., and S.S.; visualization, P.B., L.A. and S.S.; supervision, P.B., L.A., and E.P.; project administration, P.B. and L.A.; funding acquisition, P.B. and L.A.; All authors have read and agreed to the published version of the manuscript.

Funding: This article was supported by statutory funds of the Department of Mechanical Engineering (WZ/WM-IIM/4/2023).

Institutional Review Board Statement: Not applicable.

Informed Consent Statement: Not applicable.

Data Availability Statement: Data are contained within the article.

Conflicts of Interest: The authors declare no conflicts of interest.

References

1. Vina, A.; Barrientos, A. C-Legged Hexapod Robot Design Guidelines Based on Energy Analysis. *Appl. Sci.* **2021**, *11*, 2513. [CrossRef]
2. Li, Y.; Ge, S.; Dai, S.; Zhao, L.; Yan, X.; Zheng, Y.; Shi, Y. Kinematic modeling of a combined system of multiple mecanum-wheeled robots with velocity compensation. *Sensors* **2020**, *20*, 75. [CrossRef] [PubMed]
3. Yang, C.; Lu, W.; Xia, Y. Positioning Accuracy Analysis of Industrial Robots Based on Non-Probabilistic Time-Dependent Reliability. *IEEE Trans. Reliab.* **2023**. [CrossRef]
4. Yang, C.; Liang, K.; Zhang, X.; Geng, X. Sensor placement algorithm for structural health monitoring with redundancy elimination model based on sub-clustering strategy. *Mech. Syst. Signal Process.* **2019**, *124*, 369–387. [CrossRef]
5. Yang, C.; Xia, Y. Interval Pareto front-based multi-objective robust optimization for sensor placement in structural modal identification. *Reliab. Eng. Syst. Saf.* **2024**, *242*, 109703. [CrossRef]

6. Kreiser, R.; Pienroj, P.; Renner, A.; Sandamirskaya, Y. Pose Estimation and Map Formation with Spiking Neural Networks: Towards Neuromorphic SLAM. In Proceedings of the 2018 IEEE/RSJ International Conference on Intelligent Robots and Systems (IROS), Madrid, Spain, 1–5 October 2018.
7. Blum, H.; Dietmüller, A.; Milde, M.; Conradt, J.; Indiveri, G.; Sandamirskaya, Y. A neuromorphic controller for a robotic vehicle equipped with a dynamic vision sensor. In Proceedings of the Robotics: Science and Systems XIII, Cambridge, MA, USA, 12–16 July 2017.
8. Jahn, U.; Heß, D.; Stampa, M.; Sutorma, A.; Röhrig, C.; Schulz, P.; Wolff, C. A taxonomy for mobile robots: Types, applications, capabilities, implementations, requirements, and challenges. *Robotics* **2020**, *9*, 109. [CrossRef]
9. Russo, M.; Ceccarelli, M. A Survey on Mechanical Solutions for Hybrid Mobile Robots. *Robotics* **2020**, *9*, 32. [CrossRef]
10. Wu, Y.; Li, D.; Dong, X.; Wang, X.; Zheng, C.; Zhu, A.; Zhang, Z.; Zhou, X.; Zhang, Y. Design and Experimental Evaluation of a Hexapod Mobile Robot with C-Shaped Legs. In Proceedings of the 2023 8th IEEE International Conference on Advanced Robotics and Mechatronics, ICARM 2023, Sanya, China, 8–10 July 2023; pp. 336–341. [CrossRef]
11. Kouame Yann Olivier, A.; Biradar, R.C.; Karthik, R.; Devanagavi, G.D. Design of a Robot for carrying out research on hybrid robot's mobility: Case of a mecanum wheel-legged robot. In Proceedings of the 2023 5th International Conference on Electrical, Computer and Communication Technologies, ICECCT 2023, Erode, India, 22–24 February 2023. [CrossRef]
12. Mahkam, N.; Yilmaz, T.B.; Özcan, O. Smooth and inclined surface locomotion and obstacle scaling of a c-legged miniature modular robot. In Proceedings of the 2021 IEEE 4th International Conference on Soft Robotics, RoboSoft 2021, New Haven, CT, USA, 12–16 April 2021; pp. 9–14. [CrossRef]
13. Xu, K.; Lu, Y.; Shi, L.; Li, J.; Wang, S.; Lei, T. Whole-body stability control with high contact redundancy for wheel-legged hexapod robot driving over rough terrain. *Mech. Mach. Theory* **2023**, *181*, 105199. [CrossRef]
14. Trojnacki, M.; Dąbek, P. Mechanical Properties of Modern Wheeled Mobile Robots. *J. Autom. Mob. Robot. Intell. Syst.* **2019**, *13*, 3–13. [CrossRef]
15. Barros, R.J.; Silva Filho, J.L.; Neto, J.V.; Nascimento, T.P. An Open-Design Warehouse Mobile Robot. In Proceedings of the 2020 Latin American Robotics Symposium, 2020 Brazilian Symposium on Robotics and 2020 Workshop on Robotics in Education, LARS-SBR-WRE 2020, Natal, Brazil, 9–13 November 2020. [CrossRef]
16. Yang, K.; Rong, X.; Zhou, L.; Li, Y. Modeling and analysis on energy consumption of hydraulic quadruped robot for Optimal Trot motion control. *Appl. Sci.* **2019**, *9*, 1771. [CrossRef]
17. Chen, Z.; Wang, S.; Wang, J.; Xu, K.; Lei, T.; Zhang, H.; Wang, X.; Liu, D.; Si, J. Control strategy of stable walking for a hexapod wheel legged robot. *ISA Trans.* **2021**, *108*, 367–380. [CrossRef] [PubMed]
18. Bruzzone, L.; Quaglia, G. Review article: Locomotion systems for ground mobile robots in unstructured environments. *Mech. Sci.* **2012**, *3*, 49–62. [CrossRef]
19. Bruzzone, L.; Nodehi, S.E.; Fanghella, P. Tracked Locomotion Systems for Ground Mobile Robots: A Review. *Machines* **2022**, *10*, 648. [CrossRef]
20. Haldane, D.W.; Yim, J.K.; Fearing, R.S. Repetitive extreme-acceleration (14-g) spatial jumping with Salto-1P. In Proceedings of the IEEE International Conference on Intelligent Robots and Systems, Vancouver, BC, Canada, 24–28 September 2017; pp. 3345–3351. [CrossRef]
21. Patel, N.A.; Pradhan, S.N.; Shah, K.D. Two legged robot design, simulation and realization. In Proceedings of the 4th International Conference on Autonomous Robots and Agents, Wellington, New Zealand, 10–12 February 2009; pp. 426–429. [CrossRef]
22. Dat, T.T.K.; Phuc, T.T. A study on locomotions of quadruped robot. *Lect. Notes Electr. Eng.* **2014**, *282*, 595–604. [CrossRef]
23. Saranli, U.; Buehler, M.; Koditschek, D.E. RHex: A simple and highly mobile hexapod robot. *Int. J. Robot. Res.* **2001**, *20*, 616–631. [CrossRef]
24. Summary of the RHex Robot Platform. Available online: https://www.rhex.web.tr/ (accessed on 15 January 2024).
25. Ma, J.; Qiu, G.; Guo, W.; Li, P.; Ma, G. Design, Analysis and Experiments of Hexapod Robot with Six-Link Legs for High Dynamic Locomotion. *Micromachines* **2022**, *13*, 1404. [CrossRef]
26. Krishna, A.; Nandanan, K.; Pradeep Kumar, S.S.; Srihari, K.S.; Sivraj, P. Design and fabrication of a hexapod robot. In Proceedings of the 2014 International Conference on Embedded Systems (ICES), Coimbatore, India, 3–5 July 2014; pp. 225–230. [CrossRef]
27. Burzynski, P.; Simha, A.; Kotta, Ü.; Pawluszewicz, E.; Sastry, S. FLHex: A flapped-paddle hexapod for all-terrain amphibious locomotion. *Bull. Pol. Acad. Sci. Tech. Sci.* **2021**, *69*, 139007. [CrossRef]
28. Altendorfer, R.; Moore, N.; Komsuoglu, H.; Buehler, M.; Brown, H.B.; Mcmordie, D.; Saranli, U.; Full, R.; Koditschek, D.E. RHex: A biologically inspired hexapod runner. *Auton. Robot.* **2001**, *11*, 207–213. [CrossRef]
29. Martone, M.; Pavlov, C.; Zeloof, A.; Bahl, V. Enhancing the vertical mobility of a robot hexapod using microspines. *arXiv* **2019**, arXiv:1906.04811. [CrossRef]
30. Moore, E.Z.; Campbell, D.; Grimminger, F.; Buehler, M. Reliable stair climbing in the simple hexapod 'RHex'. In Proceedings of the IEEE International Conference on Robotics and Automation, Washington, DC, USA, 11–15 May 2002; Volume 3, pp. 2222–2227. [CrossRef]
31. Campbell, D.; Buehler, M. Stair descent in the simple hexapod 'RHex'. In Proceedings of the IEEE International Conference on Robotics and Automation, Taipei, Taiwan, 14–19 September 2003; Volume 1, pp. 1380–1385. [CrossRef]
32. Saranli, U.; Rizzi, A.A.; Koditschek, D.E. Model-Based Dynamic Self-Righting Maneuvers for a Hexapedal Robot. *Int. J. Robot. Res.* **2004**, *23*, 903–918. [CrossRef]

33. Saranli, U.; Buehler, M.; Koditschek, D.E. Design, modeling and preliminary control of a compliant hexapod robot. In Proceedings of the IEEE International Conference on Robotics and Automation, San Francisco, CA, USA, 24–28 April 2000; Volume 3, pp. 2589–2596. [CrossRef]
34. FLHex: A Flapped-Paddle Hexapod—YouTube. Available online: https://www.youtube.com/watch?v=Ux1AlOFUUco (accessed on 15 January 2024).
35. De León, J.; Cebolla, R.; Barrientos, A. A sensor fusion method for pose estimation of c-legged robots. *Sensors* **2020**, *20*, 6741. [CrossRef] [PubMed]
36. Burzynski, P. Discrete fractional order PID controller in case of the FLHex robot leg position control system. In Proceedings of the 2022 26th International Conference on Methods and Models in Automation and Robotics (MMAR), Międzyzdroje, Poland, 22–25 August 2022. [CrossRef]

sensors

MDPI

Article

Path Following for Autonomous Mobile Robots with Deep Reinforcement Learning

Yu Cao , Kan Ni, Takahiro Kawaguchi and Seiji Hashimoto *

Program of Intelligence and Control, Cluster of Electronics and Mechanical Engineering, School of Science and Technology, Gunma University, 1-5-1 Tenjin-cho, Kiryu 376-8515, Japan; t202d602@gunma-u.ac.jp (Y.C.); t202d003@gunma-u.ac.jp (K.N.); kawaguchi@gunma-u.ac.jp (T.K.)
* Correspondence: hashimotos@gunma-u.ac.jp; Tel.: +81-0277-30-1741

Abstract: Autonomous mobile robots have become integral to daily life, providing crucial services across diverse domains. This paper focuses on path following, a fundamental technology and critical element in achieving autonomous mobility. Existing methods predominantly address tracking through steering control, neglecting velocity control or relying on path-specific reference velocities, thereby constraining their generality. In this paper, we propose a novel approach that integrates the conventional pure pursuit algorithm with deep reinforcement learning for a nonholonomic mobile robot. Our methodology employs pure pursuit for steering control and utilizes the soft actor-critic algorithm to train a velocity control strategy within randomly generated path environments. Through simulation and experimental validation, our approach exhibits notable advancements in path convergence and adaptive velocity adjustments to accommodate paths with varying curvatures. Furthermore, this method holds the potential for broader applicability to vehicles adhering to nonholonomic constraints beyond the specific model examined in this paper. In summary, our study contributes to the progression of autonomous mobility by harmonizing conventional algorithms with cutting-edge deep reinforcement learning techniques, enhancing the robustness of path following.

Keywords: autonomous mobile robot; path following; velocity control; deep reinforcement learning; soft actor-critic

check for updates

Citation: Cao, Y.; Ni, K.; Kawaguchi, T.; Hashimoto, S. Path Following for Autonomous Mobile Robots with Deep Reinforcement Learning. *Sensors* **2024**, *24*, 561. https://doi.org/10.3390/s24020561

Academic Editors: David Cheneler and Stephen Monk

Received: 21 December 2023
Revised: 10 January 2024
Accepted: 15 January 2024
Published: 16 January 2024

1. Introduction

Autonomous Mobile Robots (AMRs) refer to robotic systems designed to exhibit minimal or no human intervention in their movement [1]. These robots are engineered to autonomously follow a predefined path, whether in indoor or outdoor environments. AMRs are increasingly appreciated for their expanding applications across diverse domains, including logistics transportation [2], security surveillance [3], and robotic cleaning services [4]. Within the customer service industry, an escalating number of AMRs are being deployed to enhance customer experiences by providing daily life conveniences [5,6]. Moreover, the agricultural industry is increasingly expressing interest in AMRs, driven by their ability to address issues such as labor shortages, natural phenomena, and economic challenges that have the potential to significantly diminish opportunities in farming [7].

The challenge of following a predetermined path has long been a focal point in the control engineering community. Path following involves a vehicle navigating a globally defined geometric path with loose time constraints and can be divided into control theory-based methods and geometric methods [8]. Control theory-based methods, such as Proportional–Integral–Derivative (PID) controllers, face challenges in finding optimal parameters [9]. Fuzzy controllers rely on expert experience or prior knowledge [10], while model predictive controllers require consideration of computational costs and precise modeling for reliable results [11]. In comparison to control theory-based methods, geometric methods have become more popular due to their simplicity, robustness, and suitability for real-time control. The Pure Pursuit (PP) controller, proposed as the earliest geometric approach for path following, fits a circle through the vehicle's current position to a point on the path ahead of the vehicle by a look-ahead distance [8,12]. It was first discussed in [13] and later formally elaborated in [14], where the PP strategy and its applications were introduced. The straightforward nature of this strategy has contributed to its popularity in various applications. Notably, the PP controller has been employed in two vehicles during the DARPA Grand Challenge [15] and three vehicles in the DARPA Urban Challenge [16]. However, this approach assumes that the vehicle is operating at a constant speed and the path is free of curvature, leading to degraded performance on curved paths [12,17]. Additionally, especially when the vehicle deviates from the path, and the distance between the vehicle and the path exceeds the look-ahead distance, there is no corresponding control law.

In recent years, Reinforcement Learning (RL) has achieved remarkable success in various fields, particularly in robotics, garnering increased attention and widespread recognition [18]. RL is a machine learning method that addresses the challenge of enabling a decision-making agent to learn optimal actions within an environment. The introduction of Deep Neural Networks (DNNs) into RL, owing to their outstanding ability to approximate nonlinear functions and extract relevant features from raw inputs, has given rise to the advent of Deep Reinforcement Learning (DRL). This approach excels in tasks such as defeating the world champion in the game of Go [19] and mastering intricate robotics manipulation tasks [20]. Naturally, DRL has found applications in path following. Liu et al. [21] introduced a multiple kernel feature learning framework for value function approximation, addressing the challenges of feature representation and online learning ability in RL. Their simulation results demonstrated better performance in tracking precision and smoothness. Chen et al. [22] proposed a steering control approach that combines PID control and PP control. In this setup, RL is employed to learn the weights of the two controllers, balancing the trade-off between smooth control and tracking error. Subsequently, they extended their work by updating the constant speed control to a new speed adaptation method using fuzzy logic [23]. This modification allows the original approach to be applicable not only to low-speed urban environments, but also to high-speed scenarios, reaching speeds of up to 80 km/h. Chen et al. [24] presented a hybrid approach combining DRL and PP control. Similarly, the steering output is a combination of the outputs from both, treating DRL as a compensatory mechanism for PP control. The previously mentioned methods primarily emphasize steering control, with many adopting constant speed control, which lacks generality. Moreover, in certain instances, manual design of reference speeds is adopted, demanding additional optimization efforts.

In this paper, our focus is on exploring the combination of PP control and DRL to address the challenges in path following. PP is responsible for steering control, while DRL, specifically employing the Soft Actor-Critic (SAC) algorithm, takes charge of velocity control, creating a complementary relationship between the two. We believe that the implementation of PP, being an easily deployable method, can compensate for the shortcomings of the algorithm itself through adaptive velocity adjustment, resulting in improved path convergence. Simulation and experimental results, validated using a nonholonomic mobile robot, demonstrate the ease of training for our proposed approach and its superiority in tracking paths with varying curvatures.

2. Problem Formulation

In this section, we introduce the problem of path following for nonholonomic mobile robots in a planar environment.

2.1. Kinematics Modeling

Nonholonomic mobile robots constitute a category of mobile robots with constrained mobility, unlike typical holonomic mobile robots that can move freely in a plane. Steering in nonholonomic mobile robots is accomplished by independently controlling the speed of the wheels on each side of the vehicle. When the speeds of the wheels on both sides are not equal, the vehicle will turn [25]. Basic differential steering robots are equipped with two driven wheels and a front and rear caster for added stability, as illustrated in Figure 1. In the context of a mobile robot situated on a 2D plane with a defined global Cartesian coordinate system $\{O\}$, the robot possesses three degrees of freedom represented by its posture,

$$\mathbf{x} = \begin{bmatrix} x \\ y \\ \psi \end{bmatrix} \tag{1}$$

where (x, y) represents the robot's current position in the global coordinate system, and ψ represents the heading angle, measured counterclockwise from the x-axis.

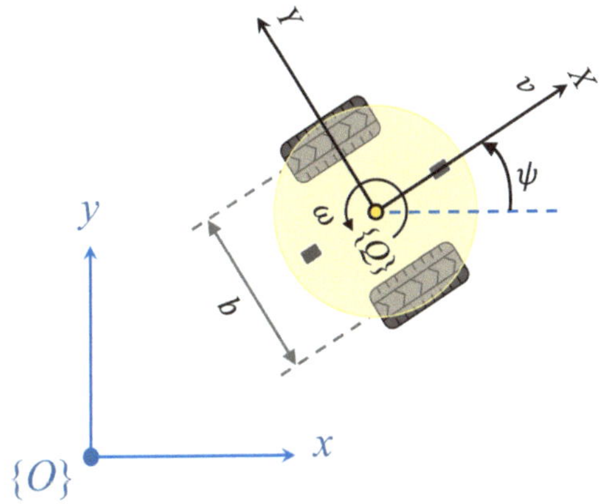

Figure 1. A two-wheeled independently driven nonholonomic mobile robot with the definition of the global coordinate frame $\{O\}$ and the body coordinate frame $\{Q\}$.

The mobile robot's motion is controlled by its linear velocity v and rotational velocity ω, as their positive directions are defined in Figure 1. The mobile robot's kinematics is then defined as follows [25–27]:

$$\dot{\mathbf{x}} = \begin{bmatrix} \dot{x} \\ \dot{y} \\ \dot{\psi} \end{bmatrix} = \begin{bmatrix} \cos\psi & 0 \\ \sin\psi & 0 \\ 0 & 1 \end{bmatrix} \mathbf{u} \tag{2}$$

where $\mathbf{u} = [v, \omega]^T \in \mathcal{U} \subset \mathbb{R}^2$ define input constraints. Path following of such a nonholonomic wheeled mobile robot involves the design of algorithms to generate reference commands for $\mathbf{u}$.

Let v_{max} represent the maximum linear velocity achievable by the mobile robot. It is crucial to account for the condition that, even when the robot simultaneously moves forward at a linear velocity v and rotates at an angular velocity ω, the velocity of the outer wheel should not exceed the maximum allowed velocity [27]. Therefore, the following constraint is applied:

$$\hat{v} + \frac{b}{2}\hat{\omega} \leq v_{max} \tag{3}$$

where $\hat{v}$ and $\hat{\omega}$ represent the set maximum velocities in practical use, and b denotes the wheelbase, which is the distance between the centers of the wheels. The experimental robot in the paper is tested to move at a maximum velocity slightly exceeding 0.5 m/s. Through the constraint outlined in Equation (3), the maximum values are determined and presented in Table 1.

Table 1. Mobile robot parameters.

Symbol	Description	Value
$\hat{v}$	Set maximum linear velocity	0.4 m/s
$\hat{\omega}$	Set maximum angular velocity	1.0 rad/s
b	Wheelbase	0.172 m

2.2. Path Following

The objective of path following is to design a controller, such that the mobile robot follows an arc-length parametrized reference path [11],

$$\mathcal{P} = \{\mathbf{p} \in \mathbb{R}^2 | \mathbf{p} = \mathbf{p}_r(\lambda), \forall \lambda \geq 0\}. \tag{4}$$

For any given parameter λ, a local reference coordinate system $\{R\}$ centered at $\mathbf{p}_r(\lambda)$ can be defined, denoted by the subscript r. The relative angle δ_r between the global coordinate system $\{O\}$ and the local reference coordinate system $\{R\}$ can be calculated with Equation (5).

$$\delta_r(\lambda) = \text{atan2}(y_r^{'}(\lambda), x_r^{'}(\lambda)) \tag{5}$$

where the function atan2 used here is the four-quadrant version of arctan, which calculates the angle between the positive x-axis and the robot position $[x_r, y_r]^T$ in the Cartesian plane, with a positive counterclockwise direction. $x_r^{'}$ and $y_r^{'}$ are the first order derivatives. Moreover, it is clear that the parametrized reference path must exhibit continuous differentiability.

Considering the robot's posture at time t as $[x(t), y(t), \psi(t)]$, the error in path following, commonly referred to as the cross-track error, is determined by Equation (6), which is a cross product between two vectors [12]. The control objective is to guarantee that the cross-track error converges such that $\lim_{t \to \infty} e_p(t) = 0$.

$$e_p(t) = d_y \hat{t}_x - d_x \hat{t}_y \tag{6}$$

where $\mathbf{d} = (d_x, d_y)$ is the tracking error vector and $\hat{\mathbf{t}} = (\hat{t}_x, \hat{t}_y)$ is the unit tangent vector to the reference path at $\lambda(t)$, as defined in Equations (7) and (8), respectively.

$$\mathbf{d} = (x(t), y(t)) - (x_r(\lambda(t)), y_r(\lambda(t))) \tag{7}$$

$$\hat{\mathbf{t}} = \frac{(x_r^{'}(\lambda(t)), y_r^{'}(\lambda(t)))}{||(x_r^{'}(\lambda(t)), y_r^{'}(\lambda(t)))||} \tag{8}$$

The orientation error $\psi_e(t)$ between the robot and the reference path at time t is determined by Equation (9), which is frequently incorporated and is typically treated as a secondary objective, or used to assist in the elimination of the cross-track error. It indicates moving towards or away from the direction of the path.

$$\begin{aligned} \psi_e(t) &= \psi(t) - \delta_r(\lambda(t)) \\ \psi_e(t) &= \text{atan2}(\sin(\psi_e(t)), \cos(\psi_e(t))) \end{aligned} \tag{9}$$

where $\psi_e(t)$ is normalized within the range of $[-\pi, \pi]$. While the trigonometric operations remain unchanged, constraining the range to this specific interval aids in reducing the observation space. A graphic representation of the path following errors is illustrated in Figure 2.

To determine the point along the reference path for calculating the cross-track error, the point nearest to the robot is selected [12,28]. It leads to an optimization problem of finding the parameter λ that minimizes the distance between the robot's position and the reference path. The optimization problem can be expressed as in Equation (10), with a preference for the squared Euclidean distance due to its equivalence with the original optimization problem and computational convenience.

$$\lambda(t) = \arg\min_{\lambda} \|(x(t), y(t)) - (x_r(\lambda), y_r(\lambda))\|^2 \quad (10)$$

A common approach for updating the path variable λ involves iteratively computing the value that minimizes the distance between the robot and the reference path. This process can be accomplished through the application of the conjugate gradient method, which is highly efficient when solving quadratic convex optimization problems, often converging to the optimal solution within a finite number of steps [29]. Furthermore, the feature of guaranteeing only a local optimum serves to prevent abrupt jumps in the path parameter, ensuring stability in the optimization process. Further implementation details can be found in [29,30].

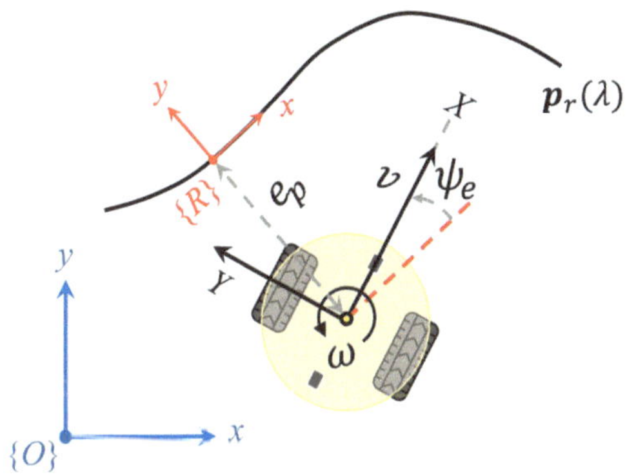

Figure 2. Schematic representation of cross-track error e_p and heading error ψ_e with respect to the reference path $\mathbf{p}_r(\lambda)$.

In the end, the proposed path-following control system is illustrated in Figure 3. In this system, the path following algorithm utilizes both path information and model states as inputs to compute errors. Subsequently, the commands for linear velocity and angular velocity are generated by the DRL and PP, respectively. The actual current velocities undergo saturation processing and are related through transfer functions as inputs to the kinematics model. Here, we consider the relationship between the velocity command and velocity as an identity transform to simplify the upcoming analysis. This consideration is also made to fulfill the Markov property, where the future evolution of a process depends solely on the present state and not on past history.

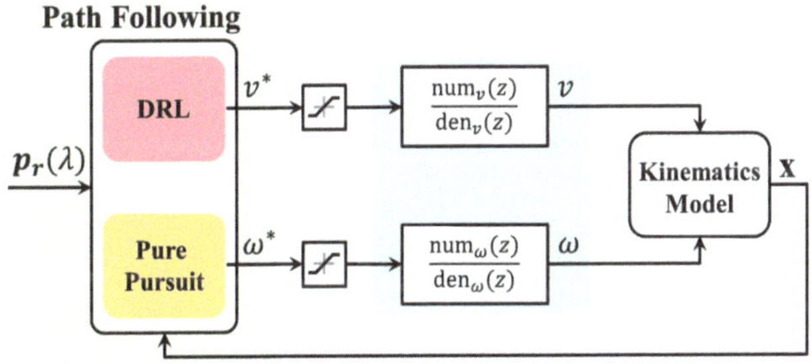

Figure 3. Separate longitudinal and latitudinal control structure of the proposed path-following control system.

3. Design and Implementation

In this section, we introduce our proposal for path following based on DRL. Firstly, we present a variant of PP utilized in steering control. The aim of this variant is to better align with our proposal and address an inherent issue in PP. Following that, we delve into the application of SAC and the design of a training environment with the objective of minimizing cross-track error while encouraging the maximization of linear velocity. Finally, we discuss implementation details.

3.1. Pure Pursuit Steering Control

The original pure pursuit algorithm fits a circle between the robot's position and a look-ahead point on the reference path, assuming that the robot moves along this trajectory. The conventional selection for the look-ahead point is a point on the reference path such that $\|(x,y) - (x_r(\lambda), y_r(\lambda))\| = L$, representing a distance L from the robot's current position. Since there are potentially multiple points fulfilling this criterion, the one with the highest value of the parameter λ is selected. The main issue with this selection method is that when the robot deviates from the path by more than the distance L, the control law is not defined [12]. Consequently, the failure to stay within the distance L results in the optimization problem's failure.

In this paper, a variant is proposed with the primary objective of addressing the aforementioned issues. Leveraging an arc-length parameterized path, an enhancement can be achieved by choosing a point situated at an arc length of d forward along the reference path from the point closest to the robot's current position relative to the path, denoted as $[x_r(\lambda), y_r(\lambda)]$. This newly selected point, denoted as $[x_r(\lambda + d), y_r(\lambda + d)]$, is then assigned as the look-ahead point, as illustrated in Figure 4. The look-ahead distance is subsequently calculated as $L = \|(x,y) - (x_r(\lambda + d), y_r(\lambda + d))\|$. Another advantage accompanying such a choice is that we only need to solve the optimization problem once, specifically for the nearest point.

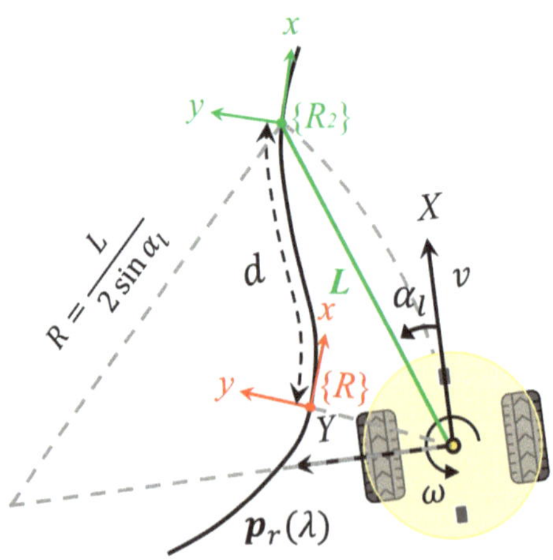

Figure 4. Geometry of the variant pure pursuit algorithm, which ensures defined control at arbitrary positions.

Eventually, the commanded heading rate ω^* for a robot traveling at linear velocity v is defined as per Equation (11).

$$\omega^* = \frac{2v \sin \alpha_l}{L} \tag{11}$$

where the look-ahead angle α_l is given by

$$\alpha_l = \arctan(\frac{y_r(\lambda + d) - y}{x_r(\lambda + d) - x}) - \psi. \tag{12}$$

3.2. Soft Actor-Critic in Velocity Control

A Markov Decision Process (MDP) is characterized by a sequential decision process that is fully observable, operates in a stochastic environment, and possesses a transition model adhering to the Markov property [31]. The MDP can be concisely represented as a tuple $(\mathcal{S}, \mathcal{A}, p, r)$, where $\mathcal{S}$ represents the set of all states called the state space; $\mathcal{A}$ denotes the actions available to the agent called the action space; $p : \mathcal{S} \times \mathcal{S} \times \mathcal{A} \to [0, 1]$ denotes the transition probability $p(s'|\mathbf{s}, \mathbf{a})$, representing the likelihood that action $\mathbf{a}$ in state $\mathbf{s}$ will result in state $\mathbf{s}'$; and $r : \mathcal{S} \times \mathcal{A}$ represents the immediate reward received after transitioning from state $\mathbf{s}$ to state $\mathbf{s}'$ as a result of action $\mathbf{a}$.

Detailed knowledge and the algorithm of SAC can be referenced from [32,33]. Here, we introduce only the essential components used in the proposed system. SAC is a RL algorithm built upon the Maximum Entropy RL (MERL) framework, which generalizes the objective of standard RL by introducing a regularization term. This regularization term ensures that the optimal policy π^* maximizes both expected return and entropy simultaneously as follows:

$$\pi^* = \arg\max_{\pi} \sum_t \mathbb{E}_{(\mathbf{s}_t,\, \mathbf{a}_t) \sim \rho^\pi} \left[r(\mathbf{s}_t, \mathbf{a}_t) + \alpha \mathcal{H}(\pi(\cdot|\mathbf{s}_t)) \right] \tag{13}$$

where ρ^π represents the trajectory of state-action pairs that the agent encounters under the control policy π, and $\mathcal{H}(\pi(\cdot|\mathbf{s}_t))$ is the entropy associated with the parameter α. This parameter acts as the temperature, influencing the balance between the entropy term and the reward.

The soft Q-function is formulated to evaluate state-action pairs, as outlined in Equation (14).

$$Q(\mathbf{s}_t, \mathbf{a}_t) = r(\mathbf{s}_t, \mathbf{a}_t) + \gamma \mathbb{E}_{\mathbf{s}_{t+1} \sim p}[V(\mathbf{s}_{t+1})] \tag{14}$$

where γ represents the discount rate for preventing the infinitely large return, and the soft state-value function $V(\mathbf{s}_t)$ based on MERL is defined by

$$V(\mathbf{s}_t) = \mathbb{E}_{\mathbf{a}_t \sim \pi}[Q(\mathbf{s}_t, \mathbf{a}_t) - \alpha \log \pi(\mathbf{a}_t|\mathbf{s}_t)] \tag{15}$$

The optimization is performed for function approximators of both the soft Q-function and the policy. The soft Q-function is parameterized by $\theta \in \mathbb{R}^n$, representing a vector of n parameters, and can be effectively modeled using a DNN. The optimization of the soft Q-function is achieved by employing a policy evaluation algorithm, such as Temporal-Difference (TD) learning. The parameters of the soft Q-function can be optimized by minimizing the mean squared loss given by Equation (16). The loss is approximated using state-action pairs stored in the experience replay buffer, denoted by $\mathcal{D}$.

$$L_Q(\theta) = \mathbb{E}_{\mathbf{s}_t, \mathbf{a}_t \sim \mathcal{D}}[(Q_\theta(\mathbf{s}_t, \mathbf{a}_t) - y_t)^2] \tag{16}$$

where y_t given in Equation (17) is the TD target that the soft Q-function is updating towards. The update makes use of a target soft Q-function with parameters $\bar{\theta}$, where $i = 1, 2$ denotes the double Q-function approximators, referred to as the clipped double-Q trick [34].

$$y_t = r(\mathbf{s}_t, \mathbf{a}_t) + \gamma \mathbb{E}_{\mathbf{s}_{t+1} \sim p}[\min_{i=1,2} Q_{\bar{\theta}_i}(\mathbf{s}_{t+1}, \mathbf{a}_{t+1}) - \alpha \log \pi(\mathbf{a}_{t+1}|\mathbf{s}_{t+1})] \tag{17}$$

Similarly, the policy, parameterized by $\phi \in \mathbb{R}^m$, is modeled as a Gaussian with a mean and a standard deviation determined by a DNN. The objective for updating the policy parameters is defined by maximizing the expected return and the entropy, as depicted in Equation (18). Due to the challenges in directly sampling latent action and computing gradients, the policy is reparameterized such that $\mathbf{a}_t = f_\phi(\xi_t; \mathbf{s}_t) = \tanh(\mu_\phi(\mathbf{s}_t) + \sigma_\phi(\mathbf{s}_t) \odot \xi_t)$ within a finite bound. Here, μ_ϕ and σ_ϕ represent the mean and standard deviation of the action, respectively, and $\xi_t \sim \mathcal{N}(0, I)$ is an input noise vector. The reparameterized sample is thus differentiable.

$$J_\pi(\phi) = \mathbb{E}_{\mathbf{s}_t \sim \mathcal{D},\, \xi_t \sim \mathcal{N}}[\min_{i=1,2} Q_{\theta_i}(\mathbf{s}_t, f_\phi(\xi_t; \mathbf{s}_t)) - \alpha \log \pi_\phi(f_\phi(\xi_t; \mathbf{s}_t)|\mathbf{s}_t)] \tag{18}$$

where π_ϕ is implicitly defined in relation to f_ϕ.

In practice, the temperature α is adjusted automatically to constrain the average entropy of the policy, allowing for variability in entropy at different states. The objective of discovering a stochastic policy with maximal expected return, satisfying a minimum expected entropy constraint, is presented in Equation (19).

$$J(\alpha) = \mathbb{E}_{\mathbf{a}_t \sim \pi_t}[-\alpha \log \pi_t(\mathbf{a}_t|\mathbf{s}_t) - \alpha \bar{\mathcal{H}}] \tag{19}$$

where the entropy target $\bar{\mathcal{H}}$ is set to be the negative of the action space dimension.

The overall architecture of the SAC-based controller is depicted in Figure 5. It employs a pair of neural networks—one specifically designed for learning the policy and the other for learning the value function. To mitigate bias in value estimation, target networks are introduced. Following the optimization of the networks, the parameters of the target networks undergo an update using a soft update strategy, denoted by Equation (20). Specifically, a fraction of the updated network parameters is blended with the target network parameters.

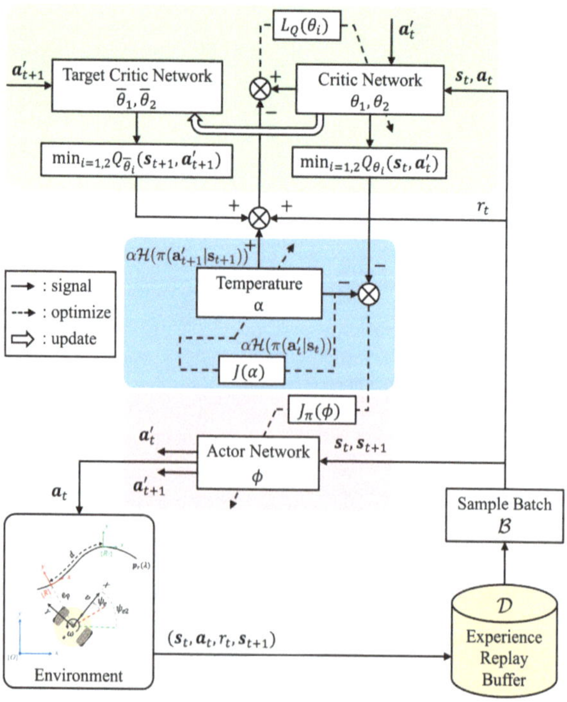

Figure 5. The optimization process of SAC-based controller at each time-step. In order to explicitly distinguish between different sources, $\mathbf{a}'$ represents an action that is resampled from the current policy, rather than being drawn from prior experiences.

$$\bar{\theta} \leftarrow \tau\theta + (1-\tau)\bar{\theta} \tag{20}$$

where parameter τ indicates how fast the update is carried on and the update is performed at each step after optimizing the online critic networks.

Moreover, an experience replay buffer $\mathcal{D}$ for storing and replaying samples, effectively reducing sample correlation, enhancing sample efficiency, and improving the learning capability of the algorithm is integrated. Through interaction with the training environment, the mobile robot, acting as the agent, takes an action $\mathbf{a}_t$ based on the current policy according to observed states $\mathbf{s}_t$, receives immediate rewards r_t, and transitions to the next state $\mathbf{s}_{t+1}$. This experience is stored in the replay buffer. During each optimization step, a mini-batch sample $\mathcal{B}$ is randomly drawn from the buffer to approximate the required expected values. The detailed description of the designed training environment is provided below.

3.2.1. Observation Space and Action Space

The observation $\mathbf{s}$, which serves as the input to the velocity controller for path following, is designed as follows:

$$\mathbf{s} = \{e_p, \psi_e, v, \omega, \psi_{e2}\} \tag{21}$$

where e_p is the cross-track error, $\psi_e \in [-\pi, \pi]$ is the normalized orientation error between the path and the mobile robot, v and ω are the current linear velocity and rotational velocity of the robot, respectively. ψ_{e2}, selected from the look-ahead point as discussed in the previous section, functions as an augmented observation that provides information about the curvature of the path in the future.

A graphical explanation is presented in Figure 6. Through trial and adjustment, the arc-length divergence parameter d, which serves both steering control and the observations in this paper, is set to 0.2 m. In configuring this parameter, our primary consideration is on selecting a look-ahead point that ensures the robot quickly regains the path. A small value of d makes the robot approach the path rapidly, but it may result in overshooting and oscillations along the reference path. Conversely, a large d reduces oscillations but might increase cross track errors, especially around corners.

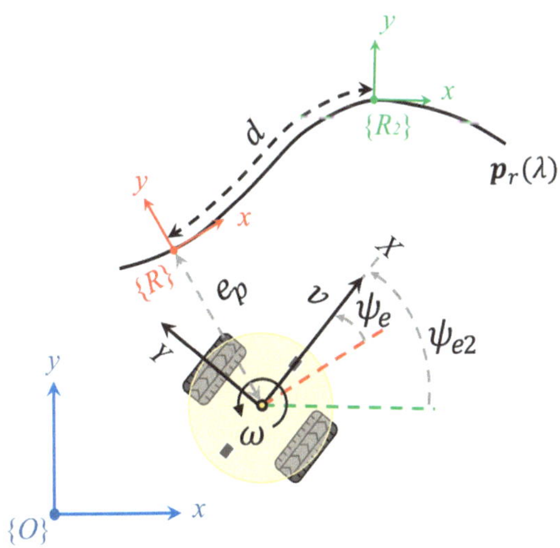

Figure 6. Observations concerning a predetermined reference path.

The action, denoted as $\mathbf{a}(t) = \pi(\mathbf{s}(t)|\phi)$, represents the rate of linear velocity $\dot{v}$ normalized by the velocity itself. This selection is adopted to mitigate undesired rapid changes. Adopting an incremental control input for the robot makes it easier to achieve smooth motions without the need for additional rewards or penalties for excessive velocity changes. Furthermore, constraining the velocity rate within a specified range, i.e., $\dot{v} \in [\dot{v}_{min}, \dot{v}_{max}]$, provides a better stability. The linear velocity command $v^*(t)$ after saturation operation for the mobile robot at each time step is calculated using Equation (22).

$$\begin{aligned} \mathbf{a}(t) &= \tanh(\mu_\phi(\mathbf{s}(t)) + \sigma_\phi(\mathbf{s}(t)) \odot \xi(t)) \\ \dot{v}^*(t) &= k\mathbf{a}(t) + b \\ v^*(t) &= \mathrm{clip}(v(t) + \dot{v}^*(t)\Delta t, v_{min}, v_{max}) \end{aligned} \tag{22}$$

where $k = (\dot{v}_{max} - \dot{v}_{min})/2$ and $b = (\dot{v}_{max} + \dot{v}_{min})/2$ are the scale and bias, respectively, to recover the normalized action to the range of the desired action. The range of linear velocity rate is designed as $[-0.5, 0.3]$ $\mathrm{m/s^2}$. The exploration space for deceleration is slightly larger than that for acceleration, addressing situations requiring urgent braking.

3.2.2. Reward Function

The reward function is designed to penalize the robot when it deviates from the path, while rewarding the robot's velocity as much as possible, as depicted in Equation (23).

$$r(t) = -k_1|e_p(t)| + k_2 v(t)\left(1 - \frac{1}{e_{tol}}|e_p(t)|\right) - k_3 F(t) \tag{23}$$

where k_1, k_2, and k_3 are positive constants that define the importance of each term. e_{tol} is the tolerance for cross-track error within which the robot receives positive velocity rewards. Based on intuitive considerations, it is desirable for the robot to decrease its velocity when deviating from the path to prevent further error expansion. To achieve this, a segmented penalty approach is also introduced. When the cross-track error exceeds a critical threshold, the penalty on velocity increases accordingly. This design ensures that the policy receives velocity rewards only when the cross-track error is within the critical threshold.

The reward function with the first two terms is visualized in Figure 7, with the range of $[-e_{tol}, e_{tol}]$. Within this range, the reward at each step ranges from a maximum value of 1, indicating perfect tracking of the path at maximum speed, to a minimum value of -1. Due to improvements in the pure pursuit algorithm, the mobile robot can consistently track the point ahead of itself on the path at any lateral distance. In this scenario, it is sufficient to solely investigate the reward associated with the robot traveling along the path.

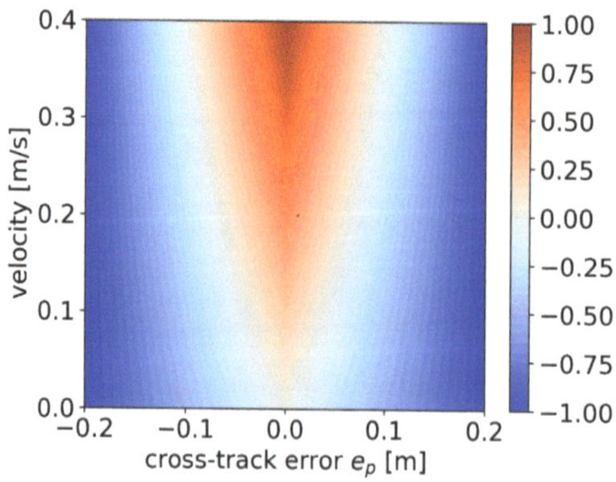

Figure 7. Reward function for path following within the e_p range of $[-0.2, 0.2]$ m.

Additionally, it has been observed in experiments that the agent may choose to discontinue forward movement at challenging turns to avoid potential penalties. To address this situation, a flag F defined as

$$F(t) = \begin{cases} 1 & \text{if } v(t) < \epsilon, \\ 0 & \text{otherwise.} \end{cases}$$

indicating a stationary state has been introduced, where ϵ is an extremely small value such as 1×10^{-6} for numerical stability. The parameters for the reward function are designed as follows: $k_1 = 5.0$, $k_2 = 2.5$, $k_3 = 0.2$, and $e_{tol} = 0.2$ m.

3.2.3. Environment and Details

The training environment encompasses the kinematics of the robot itself and a reference path. Ensuring the adaptability of the policy to various challenges is crucial, as it cultivates the ability to handle generalized scenarios, thereby reducing the risk of overfitting to specific paths. We employed a stochastic path generation algorithm proposed in our previous work [35], randomly generating a reference path for the robot to follow at the beginning of each episode. In this paper, the parameters of the stochastic path generation algorithm is defined with $N_w = 5$, $L_{min} = 0.5$ m, and $L_{max} = 2.0$ m. Straight paths with a generation probability of 0.1 are also introduced into the training. The straight path is achieved by setting $N_w = 2$ and $L_w = 2.5$ m. The example of randomly generated curved paths is illustrated in Figure 8.

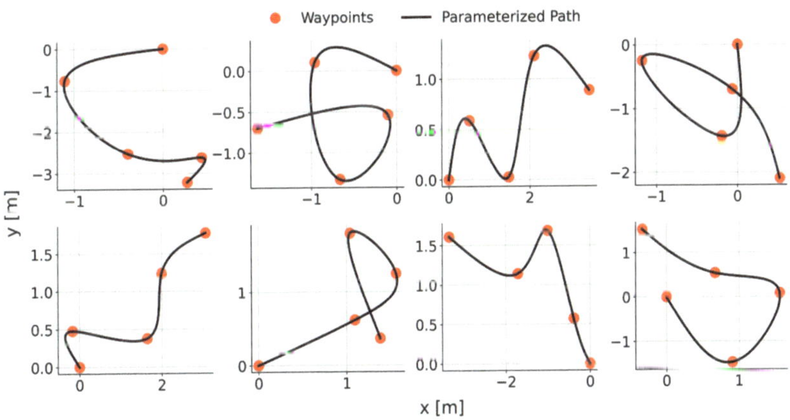

Figure 8. Example of randomly generated reference paths.

After the generation of a reference path, the initial posture of the robot is randomly sampled from a uniform distribution, with a position error range of $[-0.1, 0.1]$ meters for both the global x-axis and y-axis, and a heading error range of $[-0.0873, 0.08723]$ radians with respect to the reference path. A warm-up strategy is also implemented to gather completely random experiences. During the initial training phase, the agent takes random actions uniformly sampled from the action space. Each episode terminates when the robot reaches the endpoint, or when reaching 400 time steps. The summarized training parameters are presented in Table 2.

The actor and critic neural networks are both structured with two hidden layers. Each layer is equipped with Rectified Linear Unit (ReLU) activation function, featuring 256 neurons in both hidden layers. The actor's final layer outputs the mean $\mu_\phi(\mathbf{s}_t)$ and standard deviation $\sigma_\phi(\mathbf{s}_t)$ of a distribution, facilitating the sampling of a valid action. Subsequently, the action undergoes a tanh transformation to confine its range, as outlined in Equation (22). In the critic network, the action and state are concatenated to form an input. For the optimization of neural networks, the Adam optimizer [36] is utilized with a minibatch size of 256. Subsequently, training is conducted five times separately, allowing for an assessment of the algorithm's effectiveness and stability. Each training utilizes a distinct random seed to control factors such as path generation parameters and the initial posture of the robot. This approach ensures the reproducibility of the experiments. Hyperparameters are summarized in Table 3.

Table 2. Hyperparameters in training the SAC for path following.

Description	Value
Sampling period	0.05 s
Target soft update rate	0.005
Discount factor	0.99
Entropy target	-1
Maximum time steps per episode	4×10^2
Warm-up time steps	5×10^3
Maximum time steps	5×10^5
Experience replay buffer size	5×10^5

Table 3. Hyperparameters of both actor and critic networks for path following.

Description	Value
Number of hidden layers	2
Number of neurons per layers	256
Activation function	ReLU
Optimizer	Adam
Learning rate	3×10^{-4}
Minibatch size	256

4. Results and Analysis

In this section, we discuss the achievements attained by the path following control of the proposed method. We first analyze the training process and then introduce two evaluation criteria, failure rate and completion rate, to assess the advantages of the proposed method over PP control in the majority of samples. Finally, we test and analyze the superiority of the proposed method over PP control in both simulation and experimental environments.

4.1. Training Process

The learning curves of average return and average velocity for the path following problem are depicted in Figure 9. In the initial stage, where the average velocity consistently increases, the learned policy exhibits a relatively high average velocity but with lower rewards obtained. This suggests that the learned policy prioritizes speed improvement while neglecting the reduction in the cross-tracking error. Additionally, the initial stage of the five trials demonstrates a remarkably high level of consistency, with minimal standard deviation (noted by the absence of shaded regions in the figures). Subsequently, the average velocity noticeably decreases, while rewards, on the contrary, increase. This indicates that the policy starts learning how to decelerate to handle curves, especially challenging bends, resulting in a certain degree of over-deceleration.

After progressing halfway through the learning process, a more suitable policy is identified that balances both velocity and cross-track error. For all five trials, the overall training tends to stabilize. Moreover, compared to other OpenAI Gym RL tasks [37] that often require steps in the order of tens of millions [32,33], our approach converges with fewer steps, confirming the ease and stability of algorithm convergence.

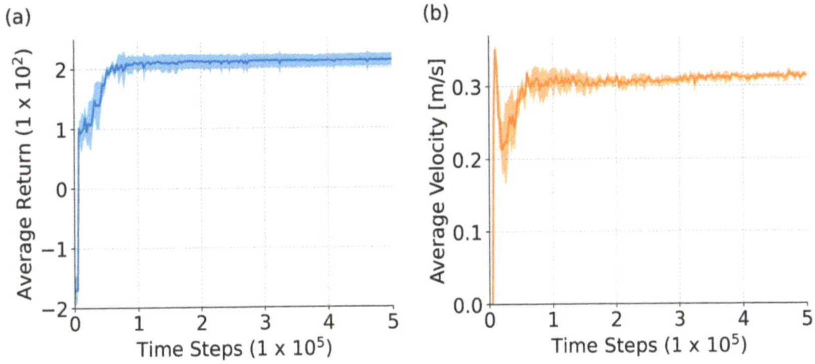

Figure 9. Learning curves for path following over 5 trials: (**a**) average return; (**b**) average velocity. The solid line represents the mean, while the shaded area corresponds to the confidence interval represented by the standard deviation.

4.2. Simulation Results

4.2.1. Quantitative Evaluation

The performance of both the PP control and the proposed method is assessed across 1000 paths generated using the same algorithm employed during training. Performance is evaluated by checking whether the cross-track error surpasses a predefined threshold within a specified number of time steps, set at 400 steps, aligning with the duration used for training the SAC-based controller per episode. In each path-following task, if this threshold is exceeded, the task is marked as a failure, and the execution of the task is terminated. The failure rate is then defined as the proportion of failures out of the total 1000 tasks.

$$\text{failure rate} = \frac{N_{failure}\ [\text{sample}]}{N_{total}\ [\text{sample}]} \tag{24}$$

where $N_{failure}$ represents the number of failed samples. Additionally, the overall completion of the path is evaluated at that specific moment.

With the path parameterized by arc length, the completion rate is defined as the ratio of the path parameter at which the robot concludes the task to the parameter of the path's endpoint, as expressed in Equation (25). This parameterization allows for a meaningful measure of how much of the path has been covered when the robot finishes its trajectory.

$$\text{completion rate} = \frac{\lambda_n\ [\text{m}]}{\lambda_{end}\ [\text{m}]} \tag{25}$$

where λ_n denotes the arc length parameter of the nearest point as in Equation (10), while λ_{end} represents the arc length parameter of the path's endpoint.

Firstly, the performance of the PP control is evaluated. The results of failure rate and completion rate for three different thresholds of cross-track error (0.1, 0.2, and 0.3 m) are summarized in Tables 4 and 5, respectively. From the perspective of failure rate, it is certain that a higher threshold leads to a lower failure rate for any given velocity. Within the same threshold criteria, as reference velocity increases, the failure rate also increases, primarily due to poor performance at high velocities in turns. Conversely, from the perspective of completion rate, lower velocities may result in minimal failure in path following, but the overall completion rate is not high. Increasing velocity is associated with an improvement in completion rate, but beyond a certain velocity, the completion rate decreases due to premature failures caused by excessive velocity. Moreover, the higher the velocity, the greater the variation in completion rates across different paths.

Table 4. Failure rates of the PP control at various reference velocities for three cross-track error thresholds.

Velocity [m/s]	Threshold [m]		
	0.1	**0.2**	**0.3**
0.10	0.000	0.000	0.000
0.15	0.075	0.000	0.000
0.20	0.276	0.053	0.002
0.25	0.486	0.257	0.043
0.30	0.606	0.431	0.238
0.35	0.698	0.562	0.394
0.40	0.767	0.643	0.516

Table 5. Completion rates of the PP control at various reference velocities for three cross-track error thresholds. For instance, the value 0.400 ± 0.077 indicates the mean result with a standard deviation range, derived from 1000 tasks.

Velocity [m/s]	Threshold [m]		
	0.1	**0.2**	**0.3**
0.10	0.400 ± 0.077	0.400 ± 0.077	0.400 ± 0.077
0.15	0.578 ± 0.132	0.597 ± 0.112	0.597 ± 0.112
0.20	0.678 ± 0.215	0.758 ± 0.150	0.776 ± 0.122
0.25	0.699 ± 0.279	0.803 ± 0.234	0.893 ± 0.130
0.30	0.671 ± 0.312	0.773 ± 0.287	0.865 ± 0.237
0.35	0.619 ± 0.314	0.710 ± 0.307	0.802 ± 0.281
0.40	0.571 ± 0.307	0.662 ± 0.312	0.739 ± 0.300

Secondly, the five trained policies are evaluated on the same set of 1000 paths. The trained policies exhibit significant improvements in both the failure rate and completion rate. At the most stringent threshold of 0.1 m, there is a low failure rate. Despite these failures, the completion rate reaches as high as 0.873, representing the average result of the five policies. Beyond the 0.2 m threshold, the absence of failures and the path completion rate approaching 1 imply that, after training the velocity control ensures a reduction in cross-track error while maximizing velocity in regions with low curvature. In other words, it allows for higher velocities when possible and slows down where necessary. The reason for not reaching 1 is that some randomly generated paths have a substantial arc length, and they cannot be fully completed within the 400-step limit. At the same time, it is evident that the performance among the five policies is quite similar, indirectly indicating the stability of the learning process and outcomes. More detailed results are summarized in Tables 6 and 7.

Table 6. Failure rates of proposed SAC-based path following control for three cross-track error thresholds.

Method	Threshold [m]		
	0.1	**0.2**	**0.3**
Policy 1	0.155	0.004	0.000
Policy 2	0.114	0.005	0.000
Policy 3	0.157	0.005	0.000
Policy 4	0.118	0.005	0.000
Policy 5	0.262	0.008	0.000

Table 7. Completion rates of proposed SAC-based path following control for three cross-track error thresholds. For instance, the value 0.883 ± 0.284 indicates the mean result with a standard deviation range, derived from 1000 tasks.

Method	Threshold [m]		
	0.1	**0.2**	**0.3**
Policy 1	0.883 ± 0.284	0.984 ± 0.070	0.987 ± 0.070
Policy 2	0.891 ± 0.285	0.972 ± 0.124	0.975 ± 0.124
Policy 3	0.880 ± 0.286	0.981 ± 0.080	0.985 ± 0.080
Policy 4	0.892 ± 0.277	0.975 ± 0.099	0.978 ± 0.099
Policy 5	0.817 ± 0.326	0.968 ± 0.140	0.971 ± 0.140

In conclusion, the results above confirm that under constant-speed control, PP control is insufficient to handle diverse path scenarios. However, the outcomes of our proposed method demonstrate that without a path-specific designed reference velocity, an adaptive velocity control strategy is learned and greatly enhances path following performance under the two criteria we examined.

4.2.2. Path Convergence and Adaptive Velocity

The test path is an eight-shaped curve that is widely used for testing path following algorithm, as defined in Equation (26). It includes straight segments as well as curves with varying degrees of curvature, representing commonly encountered scenarios in real-world applications.

$$\begin{cases} x &= a\sin(\lambda) \\ y &= a\sin(\lambda)\cos(\lambda) \end{cases} \tag{26}$$

where a is a constant that determines the size and shape of the curve, set as 1.0. Notably, despite the provided parameterization equation for the path here, it is still necessary to undergo arc-length parameterization, as referenced in [38]. Moreover, based on the algorithm for generating random paths and multiple tests, as examined in Figure 8, this specific path is highly unlikely to occur in the training environment.

The trajectory and cross-track error results for tracking the eight-shaped curve using PP control and the proposed SAC-based control are illustrated in Figure 10, with velocity comparisons presented in Figure 11. As a comparison, the results of the PP control are based on a velocity command of 0.4 m/s. Since our proposed method aims to maximize velocity, and considering that we can view this specific path as a kind of randomly generated trajectory, we are uncertain about how to set a reference velocity for PP control in this particular scenario. The initial posture is set to a randomly generated value of $[0.009, -0.044, 0.736]$ for both methods.

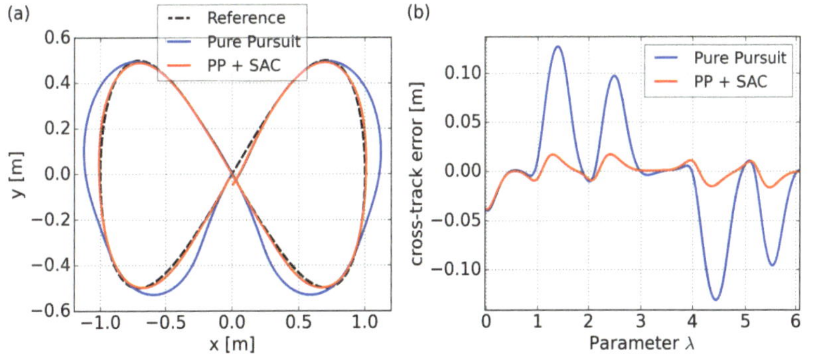

Figure 10. Path following comparison of the eight-shaped path in simulation: (**a**) trajectories results; (**b**) cross-track error results.

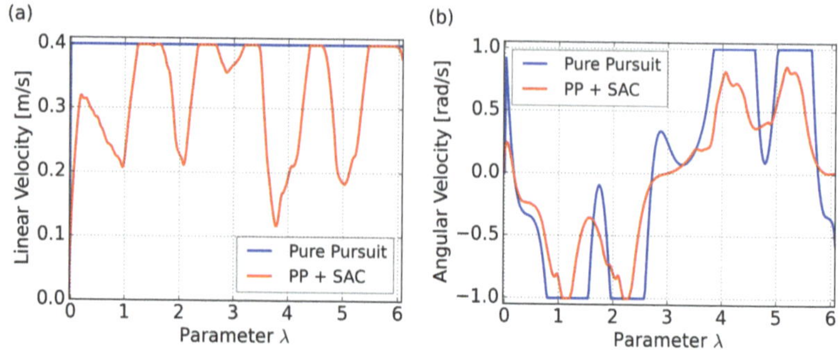

Figure 11. Velocity comparison for the eight-shaped path in simulation: (**a**) linear velocity results; (**b**) angular velocity results.

It is noteworthy that, due to the differing time consumption of the two methods, utilizing time as the horizontal axis for comparing cross-track errors and velocity changes may not yield a clear comparison. The use of λ as the horizontal axis allows for a more accurate comparison of the two methods at the same path curvature. Subsequent graphical comparisons will follow the same principle.

The results of the PP control indicates poor performance in curved sections, due to excessively high velocity and saturated angular velocities, resulting in insufficient turning. On the other hand, our proposed method, specifically policy 4 in the figure, can achieve minimal cross-track error due to adaptive velocity adjustment. This significantly improves the success rate of path following and prevents entering the saturation area of angular velocity. Specific comparative results are summarized in Table 8, where $\bar{e_p}$ represents the root mean squared cross-track error, $|e_p|_{max}$ represents the maximum absolute error occurred, and $\bar{v}$ represents the average velocity. Both path convergence and velocity performance are consistently demonstrated across the five trained policies. A more intuitive visualization is presented in a trajectory scatter plot plotted using velocity magnitude, as shown in Figure 12.

Table 8. Results for one lap of the eight-shaped path in simulation.

Method	$\bar{e}_p$ [m]	$\lvert e_p\rvert_{max}$ [m]	$\bar{v}$ [m/s]
Pure Pursuit	0.0593	0.1311	0.4000
Policy 1	0.0117	0.0385	0.2868
Policy 2	0.0118	0.0385	0.2793
Policy 3	0.0119	0.0384	0.2819
Policy 4	0.0115	0.0385	0.2688
Policy 5	0.0121	0.0385	0.2958

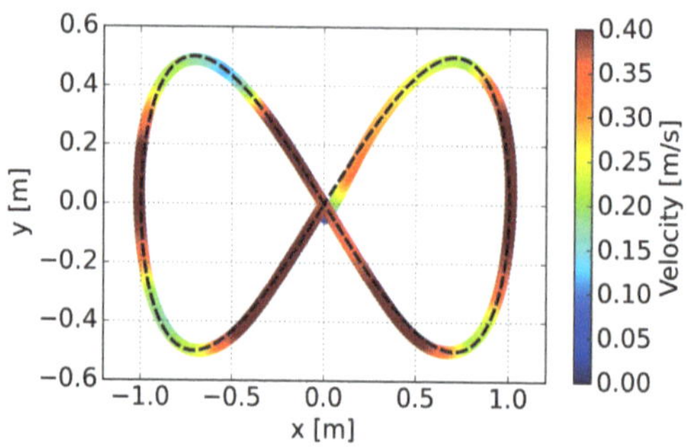

Figure 12. Linear velocities along the trajectory for the eight-shaped path in simulation.

It is evident that the proposed method can adjust velocity in a smooth way, decelerating before entering a curve, accelerating when exiting a curve rapidly, and maintaining maximum velocity on relatively straight segments. Smooth variations in linear velocity are also evident in this representation.

4.3. *Experimental Results*

4.3.1. Experimental Setup

The experimental nonholonomic wheeled mobile robot, as depicted in Figure 13, features the placement of the active wheel at the center of the chassis, with an additional omnidirectional wheel at both the front and rear. The wheelbase, measured and calibrated to 0.172 m, corresponds precisely to the parameter employed in simulation. The overall dimensions of the robot measure 216 × 216 × 171 mm.

At the top layer of the robot, a 360 Degree Laser Scanner, utilizing Laser imaging, Detection, and Ranging (LiDAR), is configured for subsequent mapping and localization functions. This setup serves the purpose of perceiving the robot's posture. The middle layer comprises the main control Raspberry Pi 4 Model B, motor control module, and battery. Although an RGB camera is also configured, it was not utilized in this paper.

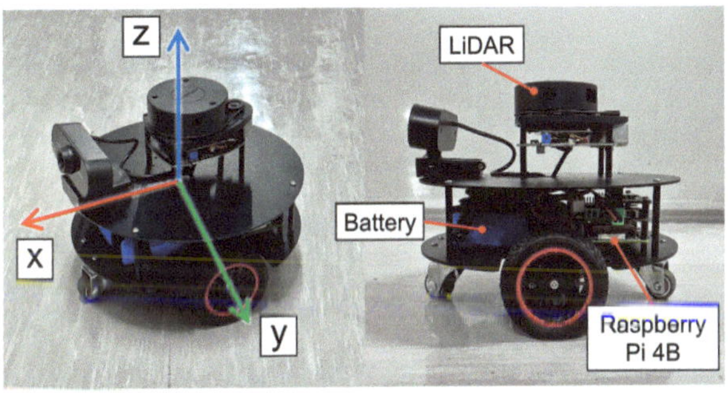

Figure 13. Configuration of the experimental nonholonomic wheeled mobile robot.

The Raspberry Pi operates on the Ubuntu 20.04 system to support the execution of the Robot Operating System (ROS), a set of software libraries and tools designed to facilitate the development of robot applications. This content specifically utilizes the ROS Noetic version. Figure 14 illustrates the robot's connectivity and communication. This necessitates that the laptop and the Raspberry Pi system be within the same local area network for effective communication, with the ROS facilitating the connection and interaction.

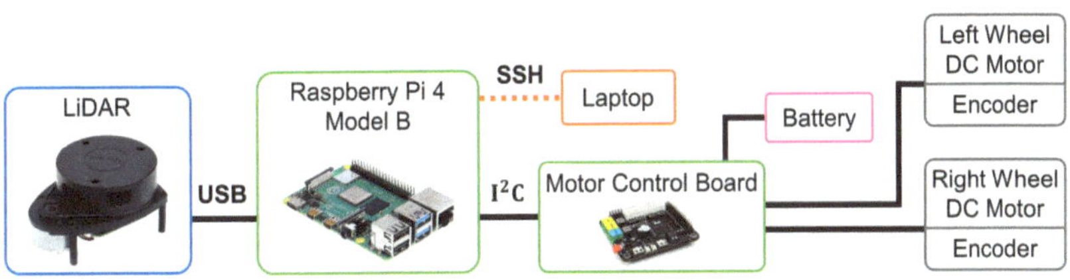

Figure 14. Diagram of the robot's connectivity and communication.

The primary focus of this paper is path following; therefore, the methods for obtaining the real-world posture of the robot are briefly summarized. Given that a robot equipped with LiDAR has already been configured, the experiments are conducted indoors. The mapping component utilizes Gmapping [39], a Simultaneous Localization and Mapping (SLAM) algorithm based on 2D laser range data for constructing 2D grid maps. After obtaining the map, during the execution of the path following algorithm, the robot's posture is determined using Adaptive Monte Carlo Localization (AMCL) [40], which employs a particle filter to track the robot's posture against a known map.

Due to the fact that the policy for velocity control is trained using PyTorch modules, in order to use it on the Raspberry Pi, we have converted it to the Open Neural Network Exchange (ONNX) format. In one control cycle, the measured and computed observations, obtained through the robot's posture via AMCL, are input into the ONNX model. The resulting velocity commands are then published at a frequency of 20 Hz. Table 9 summarizes the parameter count, Floating-Point Operations (FLOPs), and the inference time during actual operation of our model. The results highlight the remarkable lightweight nature and computational efficiency of our model, perfectly fulfilling our control objectives.

Table 9. Computational cost in real-time, where the inference time is an average of 100 inferences.

Parameters	FLOPs	Inference Time
67,842	67,328	97.8 μs

4.3.2. Path Convergence and Adaptive Velocity

The trained policies were found to be capable of tracking paths with small cross-track errors in experimental testing, similar to simulation results. However, the actual velocity was observed to be lower in reality compared to simulation. We attribute this difference in velocity tracking to delays in input–output caused by communication issues inherent in real-world robots. This phenomenon has also been identified and discussed in [41,42]. In this paper, we employ the method of modifying the control dynamics of the robot by scaling the output of the policy as used in [42] for its simplicity. After testing, an appropriate scaling factor in experiment was determined to be 2.2.

The trajectory and cross-track error results for tracking the eight-shaped path are shown in Figure 15, demonstrating consistent results with the simulation. Specifically, the results in the figure represent the outcomes of policy 2. The acceleration of the policy in the experiment appears more cautious, attributed to the transfer relationship and delay between input and output of motor wheels. Setting aside this aspect, the consistency of the policies has been validated, and a noticeable deceleration is observed in the curved segments. Similarly, the proposed method's angular velocity avoids the shortcomings of PP control, preventing steering from entering the saturation region. The results of the five experiments are summarized in Table 10. The experimental results for the five policies consistently outperform traditional PP control in both RMSE and maximum error. However, due to the slower velocity on straight segments in the middle portion compared to the simulated speed (see Figure 16), there is a slight overall decrease in average velocity. Video S1, which is an experimental recording captured from a bird's-eye view perspective, can be found in the supplementary materials for reference.

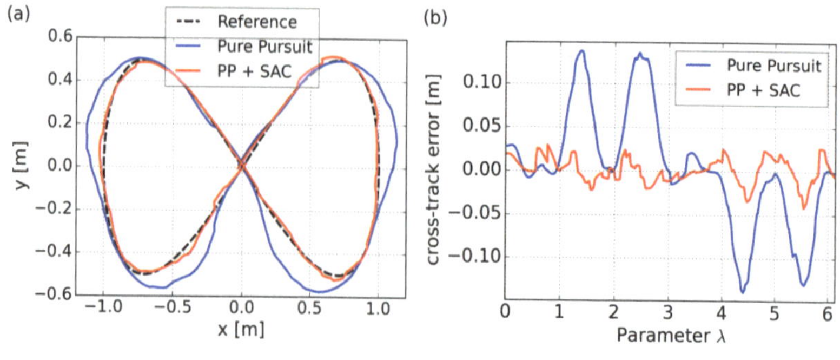

Figure 15. Path following comparison of the eight-shaped path in experiment: (**a**) trajectories results; (**b**) cross-track error results.

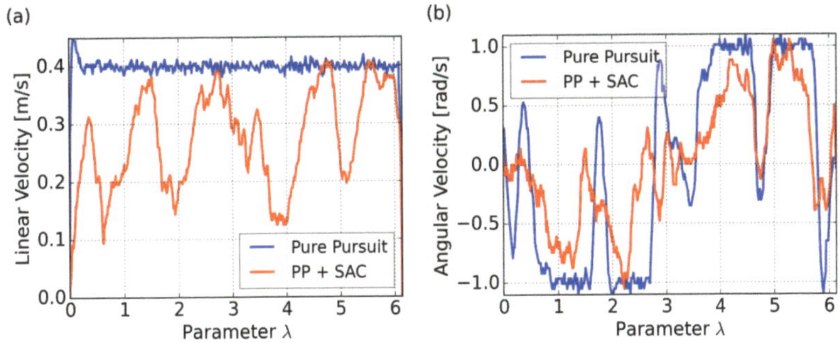

Figure 16. Velocity comparison for the eight-shaped path in experiment: (**a**) linear velocity results; (**b**) angular velocity results.

Table 10. Results for one lap of the eight-shaped path in experiment.

Method	$\bar{e}_p$ [m]	$\lvert e_p \rvert_{max}$ [m]	$\bar{v}$ [m/s]
Pure Pursuit	0.0674	0.1386	0.3949
Policy 1	0.0207	0.0493	0.2317
Policy 2	0.0146	0.0410	0.2257
Policy 3	0.0108	0.0314	0.2444
Policy 4	0.0134	0.0376	0.2097
Policy 5	0.0117	0.0368	0.2090

Velocities along the trajectory in the experiment are visualized in Figure 17. Similar performance to the simulation is confirmed, with deceleration observed before entering a curve and rapid acceleration when exiting the curve. This behavior ensures reduction in cross-track error and success in steering.

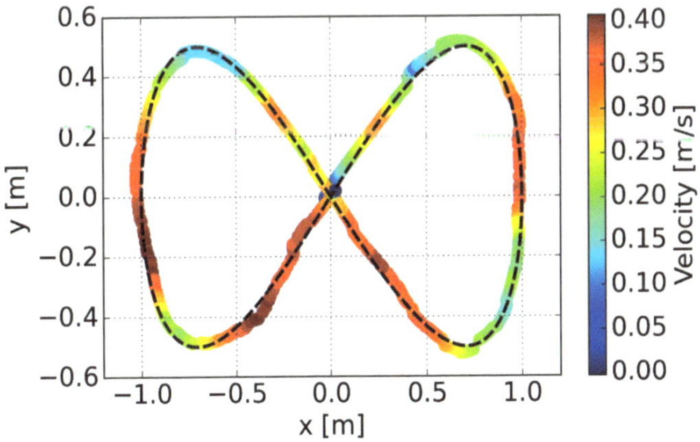

Figure 17. Linear velocities along the trajectory for the eight-shaped path in experiment.

5. Conclusions

In this paper, we propose a path-following control method that combines traditional steering control with DRL. Through interactive learning in a stochastic environment, the proposed method is demonstrated to have learned an adaptive velocity control strategy capable of addressing various path scenarios. It is quantitatively evaluated using two criteria: failure rate and completion rate. The results show a significant out-performance in both criteria compared to constant speed control under PP control. Notably, our method demonstrates robustness across diverse paths without the need for repeated design of reference velocities. Both simulation and experimental tests on an eight-shaped path confirm the learned control strategy's ability to reduce cross-track error and achieve smooth velocity adjustments. This is manifested by deceleration before entering a curve, rapid acceleration when exiting a curve, and maintaining maximum velocity on relatively straight segments. The entire approach underscores the powerful capabilities of DRL in addressing path-following challenges.

Despite the achievements we have made, we discovered that directly applying the trained policies in experiments, while successful in path-following tasks with minimal cross-track error, exhibits differences in velocity tracking compared to simulation. This discrepancy is primarily attributed to the additional uncertainties in the control dynamics of the real-world robot, including input–output delays caused by communication. Recent research has delved into RL in delayed environments, suggesting the potential to enhance velocity tracking performance and bridge the gap between simulation and the real world [41]. On the other hand, safety considerations, which are often crucial in practical robot scenarios, were not taken into account when formulating the problem. Exceeding a certain threshold of cross-track error may be deemed unsafe. Incorporating control barrier functions [43] into the problem has been proven as an effective and successful approach when safety control becomes a necessary consideration [44,45]. We will address the aforementioned shortcomings in future work.

In conclusion, our proposal contributes to the advancement of autonomous mobility by integrating conventional algorithm with state-of-the-art DRL techniques, thereby enhancing the robustness of path following. Moreover, we believe its potential for broader application to vehicles constrained by nonholonomic principles, extending beyond the specific model studied in this paper.

Supplementary Materials: The following supporting information can be downloaded at: https://www.mdpi.com/article/10.3390/s24020561/s1, Video S1: Path following comparison of the eight-shaped path in experiment.

Author Contributions: Conceptualization, Y.C. and K.N.; methodology, Y.C. and T.K.; software, Y.C. and K.N.; validation, K.N. and T.K.; formal analysis, Y.C. and T.K.; investigation, K.N.; resources, T.K. and S.H.; data curation, S.H.; writing—original draft preparation, Y.C.; writing—review and editing, Y.C.; visualization, Y.C. and K.N.; supervision, T.K. and S.H.; project administration, S.H.; funding acquisition, S.H. All authors have read and agreed to the published version of the manuscript.

Funding: This research received no external funding.

Institutional Review Board Statement: Not applicable.

Informed Consent Statement: Not applicable.

Data Availability Statement: The data that support the findings of this study are available from the author, Seiji Hashimoto, upon reasonable request.

Conflicts of Interest: The authors declare no conflict of interest.

Abbreviations

The following abbreviations are used in this manuscript:

AMR	Autonomous Mobile Robot
PID	Proportional-Integral-Derivative
PP	Pure Pursuit
RL	Reinforcement Learning
DNN	Deep Neural Network
DRL	Deep Reinforcement Learning
SAC	Soft Actor-Critic
MDP	Markov Decision Process
MERL	Maximum Entropy Reinforcement Learning

TD	Temporal-Difference
ReLU	Rectified Linear Unit
LiDAR	Laser imaging, Detection, and Ranging
ROS	Robot Operating System
SLAM	Simultaneous Localization and Mapping
AMCL	Adaptive Monte Carlo Localization
ONNX	Open Neural Network Exchange
FLOP	Floating-Point Operation

References

1. Alatise, M.B.; Hancke, G.P. A review on challenges of autonomous mobile robot and sensor fusion methods. *IEEE Access* **2020**, *8*, 39830–39846. [CrossRef]
2. Li, G.; Lin, R.; Li, M.; Sun, R.; Piao, S. A master-slave separate parallel intelligent mobile robot used for autonomous pallet transportation. *Appl. Sci.* **2019**, *9*, 368. [CrossRef]
3. Hancke, G.P.; Markantonakis, K.; Mayes, K.E. Security challenges for user-oriented RFID applications within the Internet of thing. *J. Internet Technol.* **2010**, *11*, 307–313.
4. Jin, X.; Ray, A. Navigation of autonomous vehicles for oil spill cleaning in dynamic and uncertain environments. *Int. J. Control* **2014**, *87*, 787–801. [CrossRef]
5. El Hafi, L.; Isobe, S.; Tabuchi, Y.; Katsumata, Y.; Nakamura, H.; Fukui, T.; Matsuo, T.; Garcia Ricardez, G.; Yamamoto, M.; Taniguchi, A.; et al. System for augmented human–robot interaction through mixed reality and robot training by non-experts in customer service environments. *Adv. Robot.* **2020**, *34*, 157–172. [CrossRef]
6. Yu, Q.; Yuan, C.; Fu, Z.; Zhao, Y. An autonomous restaurant service robot with high positioning accuracy. *Ind. Robot. Int. J.* **2012**, *39*, 271–281. [CrossRef]
7. Roshanianfard, A.; Noguchi, N.; Okamoto, H.; Ishii, K. A review of autonomous agricultural vehicles (The experience of Hokkaido University). *J. Terramech.* **2020**, *91*, 155–183. [CrossRef]
8. Shan, Y.; Yang, W.; Chen, C.; Zhou, J.; Zheng, L.; Li, B. CF-pursuit: A pursuit method with a clothoid fitting and a fuzzy controller for autonomous vehicles. *Int. J. Adv. Robot. Syst.* **2015**, *12*, 134. [CrossRef]
9. Zhao, P.; Chen, J.; Song, Y.; Tao, X.; Xu, T.; Mei, T. Design of a control system for an autonomous vehicle based on adaptive-pid. *Int. J. Adv. Robot. Syst.* **2012**, *9*, 44. [CrossRef]
10. Huang, S.; Lin, G. Parallel auto-parking of a model vehicle using a self-organizing fuzzy controller. *Proc. Inst. Mech. Eng. Part J. Automob. Eng.* **2010**, *224*, 997–1012. [CrossRef]
11. Faulwasser, T.; Kern, B.; Findeisen, R. Model predictive path-following for constrained nonlinear systems. In Proceedings of the 48h IEEE Conference on Decision and Control (CDC) Held Jointly with 2009 28th Chinese Control Conference, Shanghai, China, 15–18 December 2009; pp. 8642–8647. [CrossRef]
12. Paden, B.; Čáp, M.; Yong, S.Z.; Yershov, D.; Frazzoli, E. A Survey of Motion Planning and Control Techniques for Self-Driving Urban Vehicles. *IEEE Trans. Intell. Veh.* **2016**, *1*, 33–55. [CrossRef]
13. Wallace, R.S.; Stentz, A.; Thorpe, C.E.; Moravec, H.P.; Whittaker, W.; Kanade, T. First Results in Robot Road-Following. In Proceedings of the 9th International Joint Conference on Artificial Intelligence, San Francisco, CA, USA, 18–23 August 1985; Volume 2, pp. 1089–1095.
14. Amidi, O.; Thorpe, C.E. Integrated mobile robot control. In Proceedings of the Mobile Robots V. SPIE, Boston, MA, USA, 1 March 1991; Volume 1388, pp. 504–523. [CrossRef]
15. Buehler, M.; Iagnemma, K.; Singh, S. *The 2005 DARPA Grand Challenge: The Great Robot Race*; Springer: Berlin, Germany, 2007; Volume 36.
16. Buehler, M.; Iagnemma, K.; Singh, S. *The DARPA Urban Challenge: Autonomous Vehicles in City Traffic*; Springer: Berlin, Germany, 2009; Volume 56.
17. Coulter, R.C. *Implementation of the Pure Pursuit Path Tracking Algorithm*; Technical Report; Carnegie-Mellon UNIV Robotics INST: Pittsburgh, PA, USA, 1992.
18. Li, Y. Deep reinforcement learning: An overview. *arXiv* **2017**, arXiv:1701.07274.
19. Silver, D.; Huang, A.; Maddison, C.J.; Guez, A.; Sifre, L.; Van Den Driessche, G.; Schrittwieser, J.; Antonoglou, I.; Panneershelvam, V.; Lanctot, M.; et al. Mastering the game of Go with deep neural networks and tree search. *Nature* **2016**, *529*, 484–489. [CrossRef] [PubMed]
20. Levine, S.; Finn, C.; Darrell, T.; Abbeel, P. End-to-end training of deep visuomotor policies. *J. Mach. Learn. Res.* **2016**, *17*, 1334–1373.
21. Liu, J.; Huang, Z.; Xu, X.; Zhang, X.; Sun, S.; Li, D. Multi-kernel online reinforcement learning for path tracking control of intelligent vehicles. *IEEE Trans. Syst. Man, Cybern. Syst.* **2020**, *51*, 6962–6975. [CrossRef]
22. Chen, L.; Chen, Y.; Yao, X.; Shan, Y.; Chen, L. An adaptive path tracking controller based on reinforcement learning with urban driving application. In Proceedings of the 2019 IEEE Intelligent Vehicles Symposium (IV), Paris, France, 9–12 June 2019; pp. 2411–2416. [CrossRef]
23. Shan, Y.; Zheng, B.; Chen, L.; Chen, L.; Chen, D. A reinforcement learning-based adaptive path tracking approach for autonomous driving. *IEEE Trans. Veh. Technol.* **2020**, *69*, 10581–10595. [CrossRef]

24. Chen, I.M.; Chan, C.Y. Deep reinforcement learning based path tracking controller for autonomous vehicle. *Inst. Mech. Eng. Part J. Automob. Eng.* **2021**, *235*, 541–551. [CrossRef]
25. González, R.; Rodríguez, F.; Guzmán, J.L. Autonomous tracked robots in planar off-road conditions. In *Modeling, Localization, and Motion Control*; Springer: Berlin, Germany, 2014.
26. Corke, P.I.; Jachimczyk, W.; Pillat, R. *Robotics, Vision and Control: Fundamental Algorithms in MATLAB*; Springer: Berlin, Germany, 2011; Volume 73.
27. Kanayama, Y.; Kimura, Y.; Miyazaki, F.; Noguchi, T. A stable tracking control method for an autonomous mobile robot. In Proceedings of the IEEE International Conference on Robotics and Automation, Cincinnati, OH, USA, 13–18 May 1990; pp. 384–389. [CrossRef]
28. Martinsen, A.B.; Lekkas, A.M. Curved path following with deep reinforcement learning: Results from three vessel models. In Proceedings of the OCEANS 2018 MTS/IEEE Charleston, Charleston, SC, USA, 22–25 October 2018; pp. 1–8. [CrossRef]
29. Boyd, S.P.; Vandenberghe, L. *Convex Optimization*; Cambridge University Press: Cambridge, UK, 2004.
30. Hager, W.W.; Zhang, H. A survey of nonlinear conjugate gradient methods. *Pac. J. Optim.* **2006**, 2, 35–58.
31. Puterman, M.L. Markov decision processes. In *Handbooks in Operations Research and Management Science*; North Holland: Amsterdam, The Netherlands, 1990; Volume 2, pp. 331–434.
32. Haarnoja, T.; Zhou, A.; Abbeel, P.; Levine, S. Soft actor-critic: Off-policy maximum entropy deep reinforcement learning with a stochastic actor. In Proceedings of the 35th International Conference on Machine Learning. PMLR, Stockholmsmässan, Stockholm, Sweden, 10–15 July 2018; Volume 80, pp. 1861–1870.
33. Haarnoja, T.; Zhou, A.; Hartikainen, K.; Tucker, G.; Ha, S.; Tan, J.; Kumar, V.; Zhu, H.; Gupta, A.; Abbeel, P.; et al. Soft actor-critic algorithms and applications. *arXiv* **2018**, arXiv:1812.05905.
34. Fujimoto, S.; Hoof, H.; Meger, D. Addressing function approximation error in actor-critic methods. In Proceedings of the 35th International Conference on Machine Learning, PMLR, Stockholmsmässan, Stockholm, Sweden, 10–15 July 2018; pp. 1587–1596.
35. Cao, Y.; Ni, K.; Jiang, X.; Kuroiwa, T.; Zhang, H.; Kawaguchi, T.; Hashimoto, S.; Jiang, W. Path following for Autonomous Ground Vehicle Using DDPG Algorithm: A Reinforcement Learning Approach. *Appl. Sci.* **2023**, *13*, 6847. [CrossRef]
36. Kingma, D.P.; Ba, J. Adam: A method for stochastic optimization. *arXiv* **2014**, arXiv:1412.6980.
37. Brockman, G.; Cheung, V.; Pettersson, L.; Schneider, J.; Schulman, J.; Tang, J.; Zaremba, W. Openai gym. *arXiv* **2016**, arXiv:1606.01540.
38. Wang, H.; Kearney, J.; Atkinson, K. Arc-length parameterized spline curves for real-time simulation. In Proceedings of the 5th International Conference on Curves and Surfaces, Saint-Malo, France, 27 June–3 July 2002; Volume 387396.
39. Gmapping Package. Available online: http://wiki.ros.org/gmapping (accessed on 18 November 2023).
40. Adaptive Monte Carlo Localization Package. Available online: http://wiki.ros.org/amcl (accessed on 18 November 2023).
41. Ramstedt, S.; Bouteiller, Y.; Beltrame, G.; Pal, C.; Binas, J. Reinforcement learning with random delays. *arXiv* **2020**, arXiv:2010.02966.
42. Rubí, B.; Morcego, B.; Pérez, R. Deep reinforcement learning for quadrotor path following with adaptive velocity. *Auton. Robot.* **2021**, *45*, 119–134. [CrossRef]
43. Ames, A.D.; Xu, X.; Grizzle, J.W.; Tabuada, P. Control barrier function based quadratic programs for safety critical systems. *IEEE Trans. Autom. Control* **2016**, *62*, 3861–3876. [CrossRef]
44. Molnar, T.G.; Ames, A.D. Safety-Critical Control with Bounded Inputs via Reduced Order Models. *arXiv* **2023**, arXiv:2303.03247.
45. Janwani, N.C.; Daş, E.; Touma, T.; Wei, S.X.; Molnar, T.G.; Burdick, J.W. A learning-based framework for safe human-robot collaboration with multiple backup control barrier functions. *arXiv* **2023**, arXiv:2310.05865.

sensors

MDPI

Article

Dynamic Output Feedback and Neural Network Control of a Non-Holonomic Mobile Robot

Manuel Cardona † and **Fernando E. Serrano** *,†

Research Department, Universidad Don Bosco, San Salvador 1874, El Salvador; manuel.cardona@udb.edu.sv
* Correspondence: serranofer@eclipso.eu
† These authors contributed equally to this work.

Abstract: This paper presents the design and synthesis of a dynamic output feedback neural network controller for a non-holonomic mobile robot. First, the dynamic model of a non-holonomic mobile robot is presented, in which these constraints are considered for the mathematical derivation of a feasible representation of this kind of robot. Then, two control strategies are provided based on kinematic control for this kind of robot. The first control strategy is based on driftless control; this means that considering that the velocity vector of the mobile robot is orthogonal to its restriction, a dynamic output feedback and neural network controller is designed so that the control action would be zero only when the velocity of the mobile robot is zero. The Lyapunov stability theorem is implemented in order to find a suitable control law. Then, another control strategy is designed for trajectory-tracking purposes, in which similar to the driftless controller, a kinematic control scheme is provided that is suitable to implement in more sophisticated hardware. In both control strategies, a dynamic control law is provided along with a feedforward neural network controller, so in this way, by the Lyapunov theory, the stability and convergence to the origin of the mobile robot position coordinates are ensured. Finally, two numerical experiments are presented in order to validate the theoretical results synthesized in this research study. Discussions and conclusions are provided in order to analyze the results found in this research study.

Keywords: mobile robot; non-holonomy; driftless control

check for updates

Citation: Cardona, M.; Serrano, F.E. Dynamic Output Feedback and Neural Network Control of a Non-Holonomic Mobile Robot. *Sensors* **2023**, *23*, 6875. https://doi.org/10.3390/s23156875

Academic Editors: Stephen Monk and David Cheneler

Received: 2 June 2023
Revised: 7 July 2023
Accepted: 20 July 2023
Published: 3 August 2023

1. Introduction

Mobile robots have been widely implemented since their introduction several decades ago due to the vast applications of these kinds of robots. It is important to mention also that mobile robots have been implemented for several tasks in military and civil missions. For these reasons, it is important to design and synthesize several kinds of control strategies for these robots, taking into consideration the imperatives for trajectory-tracking, path following, leader–follower missions, etc. Due to the simplicity of the mobile robot dynamics and their implementation in hardware platforms, it is important to remark that kinematic control is abundant for these kinds of robots. The control strategies for robots are diverse; among these control strategies are the strategies based on robust control, sliding mode control, fuzzy control, and neural control, among others. Restriction in the dynamics of mobile robots which are found commonly are of non-holonomic type. These kinds of restrictions are based considering the kinematic and dynamics properties of the mobile robot. That is why these restrictions provide a way to develop driftless control strategies in order to provide an easy way to implement these control approaches.

Taking into consideration that the dynamic modeling of mobile robots is very important to this research study, it is crucial to mention the following research papers in which this topic is considered. So, for example, in papers like [1], the dynamic modeling with its uncertainties is shown. Then, in [2], a non-holonomic wheeled mobile robot with unknown dynamics is represented mathematically for control purposes. Then, in [3], a non-holonomic mobile robot is modeled with unknown dynamics. In [4], a remarkable paper which is very important for this research study, the researchers analyze the dynamics of a lightweight mobile robot for longitudinal motion. Meanwhile, in [5], the dynamic modeling and sliding mode control of a tractor mobile robot is presented. Then, in [6], a complete book chapter about the dynamic modeling of mobile robots is presented considering the longitudinal and lateral slip. All these research studies consider the dynamic modeling of a mobile robot. However, it is important to mention that while in general, the dynamics of mobile robots is quite simple, sometimes, in order to obtain the mathematical representation of a mobile robot's dynamics, it is important to consider unmodeled and unknown dynamics. The studies the literature have yet to clarify that the dynamic controllers and neural controllers are sufficiently robust for kinds of mobile robots.

Holonomic constraints are crucial to mention in this research study, taking into consideration that these kinds of constraints, especially the non-holonomic constraints, are found in mobile robots. These constraints are related to the velocity vector field of the mobile robot, and the mathematical explanations of these kind of constraints are mentioned in the content of this research paper. For these reasons, it is important to mention the following research papers found in the literature. For example, in papers like [7], non-holonomic constraints for geometric control theory are explained. Meanwhile, in [8], an interesting research paper about the singularities of holonomic and non-holonomic robotic systems is presented. In [9], the researchers studied holonomic and non-holonomic deformations in the AB equations, which are useful for atmospheric fluid modeling. Then, in [10], the dynamical invariant-based quantum gates are presented. In [11,12], researchers presented the model predictive control of a holonomic mobile robot and an adaptive robust controller for mechanical systems with non-holonomic trajectories, respectively.

A mobile robot's dynamic model constraints allow us a way to obtain efficient control strategies by the driftless control method. The driftless control method consists of obtaining a zero control effort only if the velocity of the mobile robot is equal to zero. In the literature, there are different driftless control techniques for different kinds of robots or mechanisms, so for example, in [13], the calculation of the control effort of two input driftless control systems is presented. Then, in [14], the switched driftless control of a kind of non-holonomic system is shown. Meanwhile, in [15], a driftless oscillation control for a nonlinear systems is provided. Then, in [16], the researchers studied the controllability of a driftless nonlinear time-delayed system. In [17,18], the involutive flows of a nonlinear driftless control system and the asymptotic control for wheeled mobile robot with driftless constraints are evinced, respectively.

One of the most important theoretical fundamentals for this research study is the design and implementation of dynamic output feedback controllers, taking into consideration that a hybrid control strategy based on dynamic output feedback control is implemented for these kinds of mobile robots. In papers like [19], a dynamic output feedback control for a switched affine system based on $L-\infty$ control is presented. It is important to mention [20,21], in which the dynamic output feedback control of a networked control system is presented and a mixed dynamic output feedback control for an active suspension system with actuator saturation and time delays are presented, respectively. It is important to mention that in the second paper, a hybrid control strategy with dynamic output feedback and fuzzy type-2 controllers is implemented, taking into consideration that this control strategy provides an optimal framework for the present research study. In [22], the dynamic output feedback control of a Luré system is proposed. Then, in [23,24], the consensus of a linear multi-agent system by reduced order dynamic output feedback and the robust stabilization for an uncertain singular Markovian-jump system via dynamic output feedback control is achieved.

It is important to mention that neural control is a suitable control strategy taking into consideration that a hybrid control strategy for a mobile non-holonomic robot is presented. For example, in [25], a neural fault-tolerant controller with input constraints for an output manipulator with output constraints is presented. Meanwhile, in [26], indirect neural control for an unmanned surface vessel is presented considering injection and deception attacks. Then, in [27,28], a quasi-optimal neural control for solar thermal systems and neural-based fixed optimal control for the attitude tracking of a space vehicle with output constraints are evinced, respectively. Finally, in [29,30], a space manipulator neural output constrained control for a space manipulator using a Lyapunov tan-barrier functional and the neural network control of nuclear plants are evinced, respectively.

It is also important to mention two phenomena found in practice in mobile robotics regarding the control, stabilization and trajectory tracking of mobile robots in the presence of input saturation and time delays. It is important to consider the following references regarding the trajectory tracking control of mobile robots with input saturation. For example, in [31], the adaptive stabilization and control of mobile robots with input saturation is shown. This paper is important for this present research study considering that non-holonomic constraints are included in the development of the control strategy. In this paper, input saturation is not considered because of the design of driftless and non-driftless control approaches, which are novel. In [32], the input saturation is considered for the control allocation of a mobile robot. Then, in [33], the visual tracking of a mobile robot is performed with the saturated inputs of velocity and acceleration. In [34], the robust control tracking design of a mobile robot with input saturation is presented.

The time-delay phenomenon is found in many kinds of linear and nonlinear dynamic systems. Mobile robots are not the exception in which time delays are found. It is important to mention the following references regarding this topic. For example, in [35], the control synchronization of mobile robots with input time delays is evinced. Meanwhile, in [36], the control of mobile robots with time delays is presented. Then, in [37], a predictive control for mobile robots with time delays is presented. Finally, in [38], the control of mobile robots with time-varying delays and noise attenuation is shown.

It is good to clarify that all the numerical simulations were performed in GNU Octave 4.2.2. The implementation of multibody dynamics commercial software or a real-time experimental setup will be implemented as a future direction of this research study. Despite this, it is important to mention the following references regarding the multibody dynamics simulation implementing commercial software. It is found in references like [39], where an 8×8 vehicle is simulated by the implementation of multibody dynamics commercial software. Meanwhile, in [40], a railway vehicle is simulated by the implementation of the previous mentioned commercial software. Finally, in [41,42], the researchers explained other multibody analyses for different kinds of mechanisms or vehicles.

This paper presents the implementation of a hybrid control strategy which consists of a neural and dynamic output feedback controller for a non-holonomic mobile robot. We show two control strategies: the first one is a driftless control system based on the dynamic output feedback and neural controller, and the second one is a full controller based also on neural and feedback output control. The dynamic output feedback and neural control are designed implementing a Lyapunov functional suitable to obtain the control law in both cases. The neural network implemented in this research study is a feedforward neural network in order to facilitate the controller design. The two control strategies ensure the stability for trajectory tracking purposes and the convergence to zero for the error variable, which comprise the difference between the desired trajectory and the measured trajectory. It is important to mention that two numerical experiments are provided in order to validate the theoretical results of this research study. Discussion and conclusions are provided at the end of this research study.

It is important to remark that the main contribution of this research study is that non-holonomy is one of the complexities found in some kinds of mobile and robotics systems. So, for this reason, this research study provides two control approaches related to non-holonomy in mobile robotic systems. The first approach consists of a driftless controller that is simple and easy to implement in specific in available hardware in many research laboratories in the world. The driftless control strategy provides the easiness of being implemented in mobile robotics due to its compactness and low computational effort.

In addition, the driftless control strategy provides the important characteristic that when the control input is zero, the mobile robot's velocity is zero. Meanwhile, the non driftless control strategy is more adequate when a robust and compact strategy is necessary for the trajectory tracking of the mobile robot. The neural networks are tuned offline, so the implementation of this control strategy is straightforward due to the implementation in real-time hardware only requiring matricial and vectorial operations, making the controller adequate for a more viable control strategy for these kinds of mobile robots.

It is important to mention that the neural controller component of this hybrid control strategy comprises a standard feedforward neural network. The difference between the neural network implemented in this research study and the neural network used for comparison purposes is that the latest one is tuned and implemented standalone in comparison with the dynamic and neural network controller. It is important to mention that the neural network implemented for comparison purposes is marginally stable in comparison with the neural network which comprises the proposed hybrid control strategy, which is asymptotically stable.

The strengths of the proposed control strategy rely on the improvements of the performance in comparison that other control strategies found in the literature, such as PID control, sliding mode control and neural network control when these strategies are implemented standalone in comparison to when these strategies are implemented as a hybrid control strategy. It is important to mention that the theoretical results are validated numerically, corroborating that this control strategy surpasses other control strategies found in the literature. It is worthwhile to mention that one advantage of the proposed control strategy is that the neural network controller is tuned offline in order to obtain the optimal performance; meanwhile, as long as the neural network controller component meets the stability results, the closed-loop stability is ensured. It is worthwhile to mention that the dynamic controller part along with the neural controller part improves the closed-loop performance in comparison with other control techniques. Finally, it is observed that a novelty of this research study is that the proposed controller is designed for driftless and non-driftless control.

2. Related Work

In this section, some related work that is worthwhile to mention in this research study is presented. This literature review consists of the following items related to non-holonomic mobile robots and their control:

- Kinematics of mobile robots.
- Non-holonomic mobile robots.
- Dynamic output feedback of mobile robots.
- Neural control of mobile robots.
- Miscellaneous control strategies for mobile robots.

It is important to mention the following references related to the kinematics of mobile robots because they are crucial for this research study. For example, in [43], a kinematic Lyapunov-based controller is presented for mobile robots. Then, in [44], a kinematic-based control strategy for a spherical mobile robot driven by a 2D pendulum is shown. Meanwhile, in [45], a singularity free kinematic model of a degenerated mobile robot is obtained. Meanwhile, in the following references [46–48], the kinematic control and two kinematic models of mobile robots are presented.

In this section, it is important to mention the following research study taking into consideration that a kinematic and dynamic model design for non-holonomic mobile robots is important for this research study. Among these research studies, we found the following. In [49], the distance-based control of a non-holomic mobile robot swarm is presented. Meanwhile, in [50], the motion and force control of mobile robots is presented by means of a fuzzy wavelet neural network controller. Then, in [51], the distributed formation control for a swarm of mobile robots is presented, taking into consideration the velocity constraints and iterative learning. Meanwhile, in [11,52], a model predictive path following the control strategy for holonomic mobile robots and the real-time identification of different types of non-holonomic mobile robots are presented, respectively. In [53], the trajectory tracking of a non-holonomic mobile robot is presented by means of sliding mode control with disturbances.

Dynamic output feedback control has been implemented specifically for mobile robots. For this reason, it is important to mention the following control strategies, which are found in the literature considering the importance that they have for this research study. In [54,55], the researchers mentioned the cooperative output control of a mobile flexible manipulator and also the distributed output feedback control of non-holonomic mobile robots with only the leader's position measurement. Then, in [56,57], an output trajectory tracking of mobile robots and an adaptive tracking control by means of output feedback for mobile robots are presented, respectively. Then, in [58], an output tracking of a non-holonomic mobile robot with fractional order visual feedback is evinced. In [59], a finite-time output feedback tracking control of a non-holonomic mobile robot is presented.

It is important to mention in this research study some neural robot control strategies which are found in the literature. So, for example, in papers like [60], a vision approach with deep neural networks to control autonomous mobile robots is presented. Meanwhile, in [61], the kinematics of a cable-driven parallel robot is achieved by the implementation of neural networks. Then, in [62], the trajectory tracking of a self-balancing robot by adaptive neural networks is performed. Then, in [63], a fault diagnosis for the harmonic reducer of industrial robots is achieved by means of neural networks. Other examples of the implementation of neural networks are found in [64,65] in which in the first case, the integrated consensus control of multi-robots using neural model predictive control is outlined. Meanwhile, in the second reference, a neural control for a robotic manipulator with an input deadzone is presented.

To finalize this section, the following references are related to miscellaneous robot control strategies related in any other way to mobile robot control. In papers like [66,67], model predictive and model control are implemented for the control of different kinds of robots. Then, in [68,69], two control strategies are presented for the trajectory tracking of mobile robots. Finally, in [11,70], model predictive and adaptive full state constrained tracking control for mobile robots are presented, respectively.

3. Notation

In this section, the notations used in this research study are presented in order to facilitate the paper's readiness and clarify the theoretical results obtained in this research study. All the operators used in this research study are evinced in this section in order to elucidate the mathematical background used.

1. $\langle .,. \rangle$ is the inner product defined in a Hilbert space.
2. $[f,g] = \frac{\partial g}{\partial x} f - \frac{\partial f}{\partial x} g$ is the Lie bracket.
3. $\frac{\partial f}{\partial x} = \begin{pmatrix} \frac{\partial f_1}{\partial x_1} & \cdots & \frac{\partial f_1}{\partial x_n} \\ \vdots & \ddots & \vdots \\ \frac{\partial f_n}{\partial x_1} & \cdots & \frac{\partial f_n}{\partial x_n} \end{pmatrix}$ is the Jacobian of a vector field.
4. $||.||$ is the 2-norm defined in a Euclidean space.

4. Problem Formulation

In this section, the kinematic model of the mobile robot is presented. It is important to mention that only a generic mobile robot is implemented in this research study. The intention is to drive the robot according to a pre-specified trajectory. The kinematics of the mobile robot is given by the following equations:

$$\begin{aligned} sin(\theta)\dot{x} - cos(\theta)\dot{y} &= 0 \\ sin(\theta+\phi)\dot{x} - cos(\theta+\phi)\dot{y} &= 0 \end{aligned} \tag{1}$$

As can be noticed in Figure 1, the coordinates and mobile robot length are defined as shown in the figure. In order to facilitate the mathematical tractability of this kinematic model in order to obtain the proposed control strategies, the following equivalent kinematic model is obtained, as shown in [71]. Consider the following change of variable $q = [x, y, \theta, \phi]^T$, so (1) can be obtained as:

$$\begin{aligned} \langle \omega_1, \dot{q} \rangle &= \begin{bmatrix} sin(\theta) & cos(\theta) & 0 & 0 \end{bmatrix} \dot{q} = 0 \\ \langle \omega_2, \dot{q} \rangle &= \begin{bmatrix} sin(\theta+\phi) & -cos(\theta+\phi) & -dcos(\phi) & 0 \end{bmatrix} \dot{q} = 0 \end{aligned} \tag{2}$$

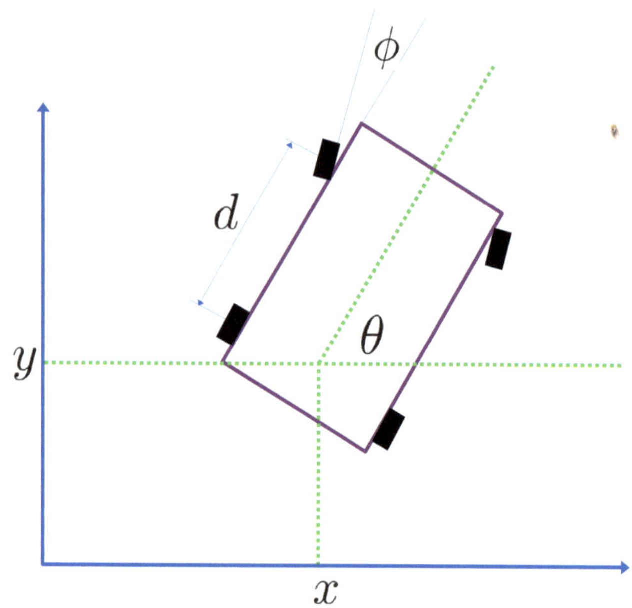

Figure 1. Schematic drawing of the mobile robot used in this research study.

In order to verify the holonomy of the previous mentioned system, the following definition is needed [71]:

Definition 1. *Consider two vector fields given by $f \in \mathbb{R}^n$ and $g \in \mathbb{R}^n$, so the Lie brackets between these two vectors are given by:*

$$[f, g] = \frac{\partial g}{\partial x} f - \frac{\partial f}{\partial x} g \tag{3}$$

In which $x = [x_1, x_2, \ldots, x_n]^T$ in which:

$$\frac{\partial f}{\partial x} = \begin{pmatrix} \frac{\partial f_1}{\partial x_1} & \cdots & \frac{\partial f_1}{\partial x_n} \\ \vdots & \ddots & \vdots \\ \frac{\partial f_n}{\partial x_1} & \cdots & \frac{\partial f_n}{\partial x_n} \end{pmatrix} \tag{4}$$

So, by defining the following vector fields:

$$g_1 = \begin{pmatrix} 0 \\ 0 \\ 0 \\ 1 \end{pmatrix} \qquad g_2 = \begin{pmatrix} -cos(\phi) \\ -sin(\phi) \\ -\frac{1}{d}\frac{sin(\theta)}{\cos(\theta)} \\ 0 \end{pmatrix} \tag{5}$$

Somehow, the Lie bracket of g_1 and g_2 is given by:

$$[g_1, g_2] = \begin{pmatrix} -sin(\phi) \\ -cos(\phi) \\ -\frac{1}{d}cos(\theta)sin(\phi)cotan(\phi) \end{pmatrix} \tag{6}$$

So, the previous Lie bracket does not span the set of g_1 and g_2 and the system is non-holonomic [71].

5. Control Strategies Definitions

In this section, we define the two control strategies proposed in this research study. This section is divided into the following subsections in order to evince the main theoretical results:

- Neural controller definition.
- Driftless control strategy.
- Non-Driftless control strategy.

5.1. Neural Controller Structure

The neural controller structure consists of the following components:

$$\tilde{y}_j = \sum_{i=1}^{n} (w_{ji}\sigma(p + \theta_i) + \theta_j) \tag{7}$$

In which w_{ji} represents the hidden unit weights of the neural network, θ_i represents the input weights, θ_j represents the bias of the neural network, and $\sigma(.)$ is the activation function, which in this case is a sigmoidal function, and $p = \sum_{r=1}^{k} q_r$ for the inputs q_r and the j output. The neural network (7) can be written in vector matrix form as follow:

$$\begin{aligned} \tilde{y} &= \sum_{i=1}^{n} \underbrace{\begin{bmatrix} w_{1i} \\ w_{2i} \end{bmatrix}}_{w_i} \sigma(p + \theta_i) + \underbrace{\begin{bmatrix} \theta_1 \\ \theta_2 \end{bmatrix}}_{\theta} \\ \tilde{y} &= \sum_{i=1}^{n} w_i\sigma(p + \theta_i) + \theta \end{aligned} \tag{8}$$

In which $w_i \in \mathbb{R}^2$ is the hidden unit weights and $\theta \in \mathbb{R}^2$ is the bias vector.

5.2. Driftless Control of the Mobile Robot

For the dynamic neural network control of the mobile robot, consider the following dynamic neural controller:

$$\dot{x}_c = K_1 e + K_2 x_c + \sum_{i=1}^{n} w_i\sigma(p + \theta_i) + \theta \tag{9}$$

In which $x_c \in \mathbb{R}^2$ is the controller variable, $K_1 \in \mathbb{R}^{2\times 2}$ and $K_2 \in \mathbb{R}^{2\times 2}$ are the gain matrices and p is the neural network controller input. In order to establish the driftless control system, the following scheme is implemented:

$$\dot{q} = g_1 u_1 + g_2 u_2 = \underbrace{\begin{bmatrix} g_1 & g_2 \end{bmatrix}}_{G} \underbrace{\begin{bmatrix} u_1 \\ u_2 \end{bmatrix}}_{u} \tag{10}$$

In the following theorem, we define the driftless control law for the mobile robot:

Theorem 1. *The driftless dynamic system (10) is stabilized by the following control law:*

$$\begin{aligned} U &= -G^{-1}\frac{q}{\alpha\|q\|^2}\eta x_c^T K_1 e - G^{-1}\frac{q}{\alpha\|q\|^2}\eta x_c^T K_2 x_c \\ &- \frac{q}{\alpha\|q\|^2}\eta x_c^T\left[\sum_i^n w_i\sigma(p+\theta_i)+\theta\right] - G^{-1}q \end{aligned} \tag{11}$$

in which $\eta \in \mathbb{R}^+$ and $\alpha\eta \in \mathbb{R}^+$ are positive gain constants and $e = q_d - q$ is the error variable in which q_d is the desired trajectory of the mobile robot.

Proof. Consider the following Lyapunov function:

$$V = \frac{\eta}{2}x_c^T x_c + \frac{\alpha}{2}q^T q \tag{12}$$

Obtaining the first time derivative of the previous Lyapunov function yields:

$$\dot{V} = \eta x_c^T \dot{x}_c + \alpha q^T \dot{q} \tag{13}$$

Now, making the required substitution in the previous equation yields:

$$\begin{aligned} \dot{V} &= \eta x_c^T K_1 e + \eta x_c^T K_2 x_c \\ &+ \eta x_c^T\left[\sum_i^n w_i\sigma(p+\theta_i)+\theta\right] + \alpha q^T G U \end{aligned} \tag{14}$$

So, making the required substitutions of (11) into the previous equation yields:

$$\dot{V} = -\alpha q^T q < 0 \tag{15}$$

and the proof is completed. □

5.3. Non Driftless Control of the Mobile Robot

To achieve this, it is necessary to define the dynamic model of the mobile robot in the following way:

$$\underbrace{\begin{bmatrix} sin(\theta) & -cos(\theta) & 0 & 0 \\ sin(\theta+\phi) & -cos(\theta+\phi) & -dcos(\phi) & 0 \end{bmatrix}}_{G}\underbrace{\begin{bmatrix} \dot{x} \\ \dot{y} \\ \dot{\theta} \\ \dot{\phi} \end{bmatrix}}_{\dot{q}} = \underbrace{\begin{bmatrix} u_1 \\ u_2 \end{bmatrix}}_{U} \tag{16}$$

in which $U \in \mathbb{R}^2$ is the control input. Now, consider the following neural dynamic controller:

$$\dot{x}_c = K_1 e + K_2 x_c + \sum_{i=1}^n w_i\sigma(p+\theta_i) + \theta \tag{17}$$

In which $K_1 \in \mathbb{R}^4$ and $K_2 \in \mathbb{R}^4$ are appropriate gain matrices. Meanwhile $\dot{e} = \dot{q}_d - \dot{q}$ is the error dynamics and $q_d \in \mathbb{R}^4$ is the desired trajectory of the mobile robot. The following theorem evinces how the dynamics of the mobile robot can be stabilized.

Theorem 2. *The dynamic system of the mobile robot (16) is stabilized by the following control law as shown in:*

$$
\begin{aligned}
U &= G\dot{q}_d - G\frac{e}{\beta\|e\|^2}\rho x_c^T K_1 e - G\frac{e}{\beta\|e\|^2}\rho x_c^T K_2 x_c \\
&- G\frac{e}{\beta\|e\|^2}\rho x_c^T\left[\sum_{i=1}^{n} w_i\sigma(p+\theta_i)+\theta\right] + Ge
\end{aligned}
\tag{18}
$$

In which $\rho \in \mathbb{R}^+$ and $\beta \in \mathbb{R}^+$ are appropriate control parameters.

Proof. Consider the following Lyapunov functional:

$$
V = \frac{\rho}{2}x_c^T x_c + \frac{\beta}{2}e^T e \tag{19}
$$

So, taking the derivative of the previous Lyapunov function and making the appropriate substitutions yields:

$$
\begin{aligned}
\dot{V} &= \rho x_c^T\left[K_1 e + K_2 x_c + \sum_{i=1}^{n} w_i\sigma(p+\theta_i)+\theta\right] \\
&+ \beta e^T\left[\dot{q}_d - G^{-1}U\right]
\end{aligned}
\tag{20}
$$

Now by substituting (18) into the previous equation yields:

$$
\dot{V} = -\beta e^T e < 0 \tag{21}
$$

So the system is globally stable, and the proof is complete. □

6. Numerical Experiments

In this section, two numerical experiments are performed to test and validate the theoretical results found in this research study. The numerical experiments conducted in this research study are intended to verify the following performance indicators:

- Minimization of the tracking error.
- Speed of response of the controller.
- Improvement in comparison with other control strategies.

The experiments performed in this research study are the following:

- Driftless control strategy.
- Non-driftless control strategy.

The conditions in which these experiments are performed is basically $d = 0.5$ m taking into consideration that these control strategies are intended for kinematic control purposes.

6.1. Numerical Experiment 1

For this experiment, the following gain constants are implemented: $\eta = 1 \times 10^{-6}$, $\alpha = 1 \times 10^{-6}$. Now, consider the neural network parameters defined as:

$$
\begin{aligned}
W_i &= \begin{bmatrix} 0.084825 & 0.087038 & 0.073776 & 0.098989 \end{bmatrix} \\
\theta_i &= \begin{bmatrix} 0.071581 & 0.096176 & 0.026009 & 0.086780 \\ 0.048502 & 0.041201 & 0.083732 & 0.015706 \\ 0.044176 & 0.037104 & 0.033042 & 0.024069 \\ 0.081635 & 0.010157 & 0.035023 & 0.089209 \end{bmatrix} \\
\theta &= \begin{bmatrix} 0.080926 \\ 0.066993 \\ 0.044356 \\ 0.046736 \end{bmatrix}
\end{aligned}
\tag{22}
$$

The experiment consists of driving the position variables x and y to the origin starting from an initial condition in order to obtain the maximum accuracy until the final desired value must be reached to obtain the maximum performance.

In Figures 2 and 3 present the evolution in time of the mobile robot when it is driven from the initial condition to the origin in finite time. It is important to notice that the action of the controller drives these variables to the origin in approximately 1 s proving that the controller is effective despite the conditions in which the experiment is performed.

Meanwhile, Figures 4 and 5 show the evolution in time of the angles of the mobile robot in order to drive the position variables x and y to the desired final value in finite time. The action of the neural controller and the dynamic surface controller are demonstrated to be fast and accurate in order to follow a predefined trajectory.

Meanwhile, Figure 6 shows the trajectory of the mobile robot in 3D. It is corroborated in this figure how the trajectory is completed considering not only the position of the mobile robot but also the orientation of the mobile robot.

Finally, Figures 7 and 8 show the evolution in time of the control inputs U_1 and U_2. It is verified that a small control input is necessary to drive the position and orientation of the mobile robot in finite time.

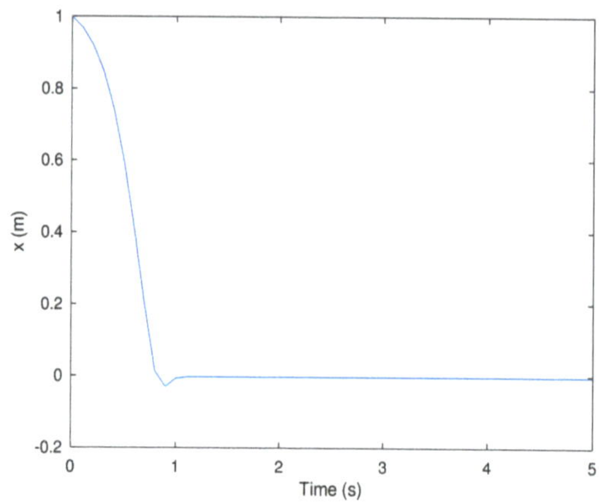

Figure 2. Position of the mobile robot in the x frame.

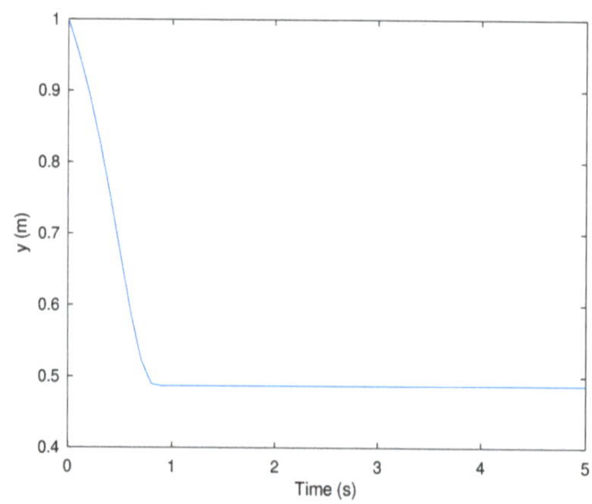

Figure 3. Position of the mobile robot in the y frame.

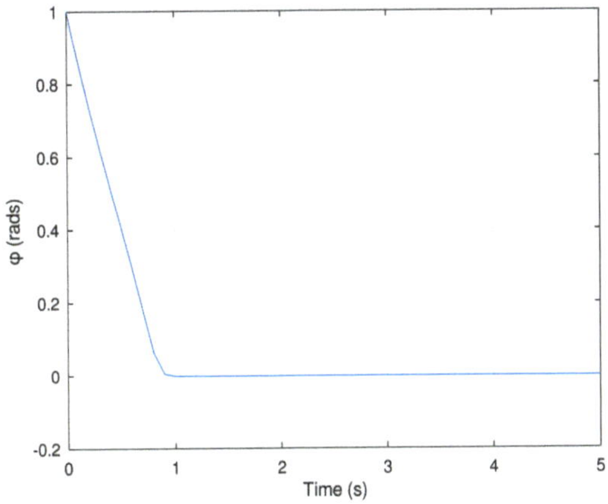

Figure 4. Evolution in time of the variable ϕ.

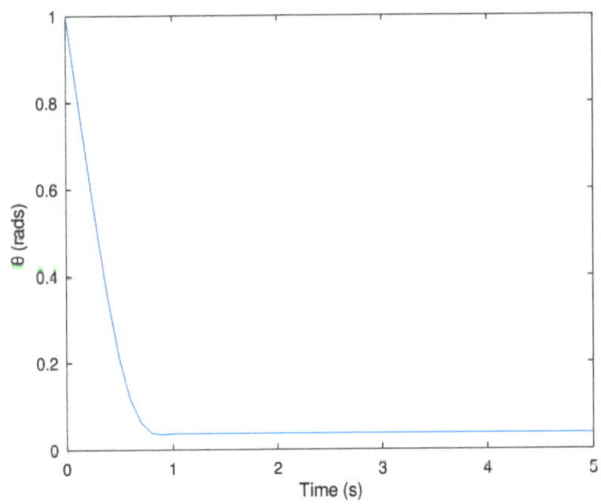

Figure 5. Evolution in time of the variable θ.

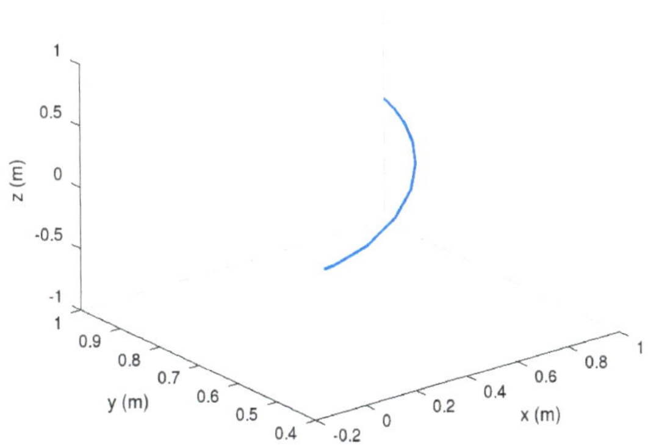

Figure 6. Evolution in time of the mobile robot trajectory.

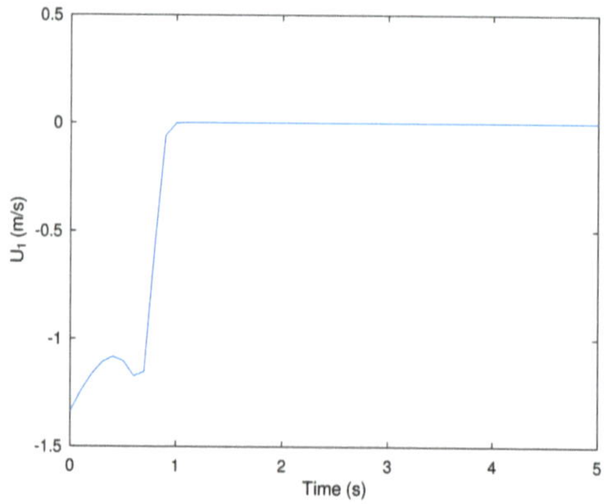

Figure 7. Evolution in time of the variable U_1.

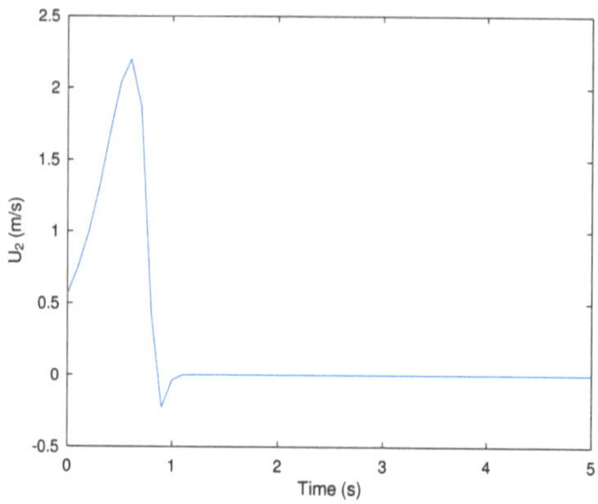

Figure 8. Evolution in time of the variable U_2.

6.2. Numerical Experiment 2

In this numerical experiment, we tested and validated the theoretical results regarding the synthesis of a non-driftless control of a mobile robot. This strategy, similar to the first experiment, consists of designing a neural-dynamic controller for trajectory-tracking purposes. In this experiment, the proposed control strategy is compared with the following strategies:

- Neural controller.
- Neural proportional-derivative PD controller.

The simulation parameters are the following; $\rho = 0.1$, $\beta = 0.1$, $K_1 = 1 \times 10^{-1}$, $K_2 = 1 \times 10^{-8}$. Meanwhile, the neural controller component has the following weights:

$$\begin{aligned} W_i &= \begin{bmatrix} 1.1557 \times 10^{-2} & 7.6815 \times 10^{-3} & 7.8527 \times 10^{-2} & 7.4791 \times 10^{-2} \end{bmatrix} \\ \theta_i &= \begin{bmatrix} 0.083379 & 0.02538 & 0.051005 & 0.092401 \\ 0.018102 & 0.090383 & 0.044346 & 0.067841 \\ 0.063258 & 0.062726 & 0.079339 & 0.090871 \\ 0.0421146 & 0.01078 & 0.006508 & 0.0038533 \end{bmatrix} \\ \theta &= \begin{bmatrix} 0.044756 \\ 0.083565 \\ 0.041632 \\ 0.076618 \end{bmatrix} \end{aligned} \tag{23}$$

Figures 9 and 10 show the evolution in time of the position variables x and y of the mobile robot. It is evinced that the controller synthetized with the proposed control strategy is more accurate in comparison with the neural and neural PD controllers. The proposed control strategy is more accurate in comparison with the strategies used as a comparative benchmark. The reason is because of the addition of a dynamic controller component that makes this control strategy more accurate and faster than the other control strategies.

Meanwhile, Figures 11 and 12 corroborate how the error variables reach the origin in finite time. As evinced in the previous figures, it is verified that the error variables yielded by the proposed control strategies reach the origin faster and more accurately than the neural and neural PD control strategies.

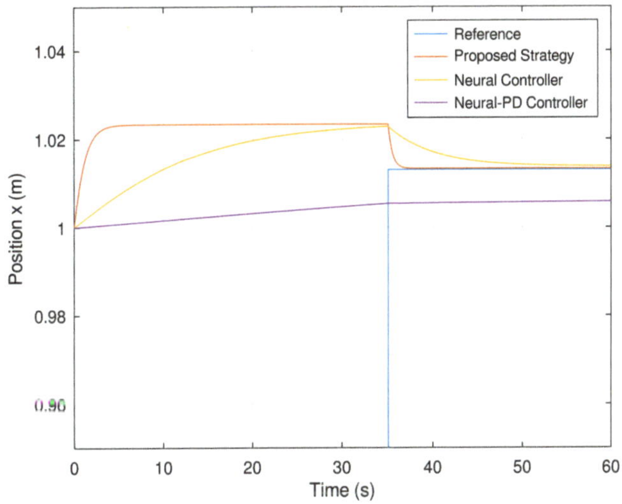

Figure 9. Position of the mobile robot in the x frame.

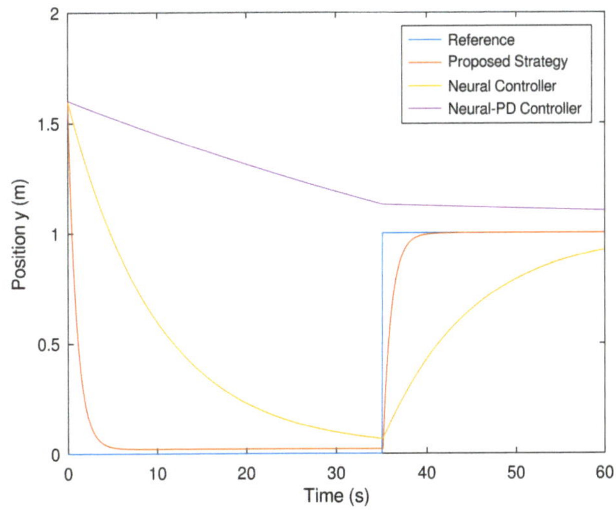

Figure 10. Position of the mobile robot in the y frame.

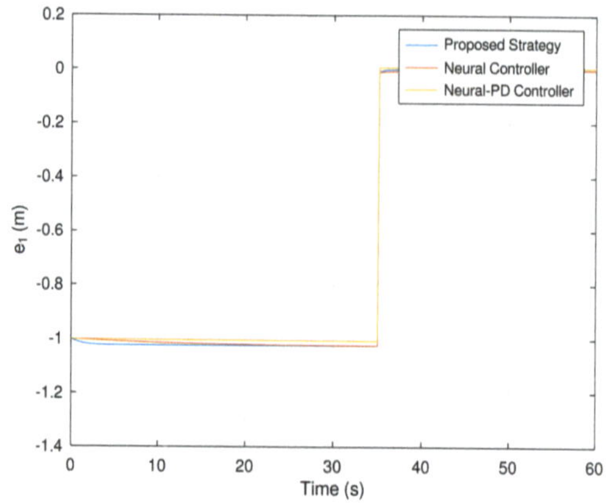

Figure 11. Evolution in time of the error variable e_1.

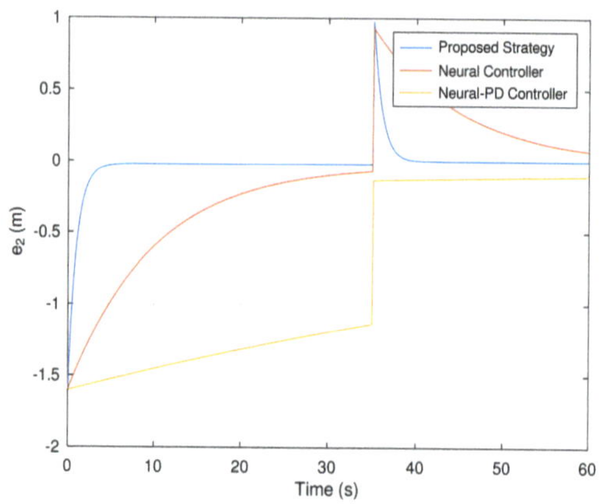

Figure 12. Evolution in time of the error variable e_2.

Then, Figure 13 presents the trajectory of the mobile robot in 3D. It is evinced that the trajectory tracked by proposed controller is significantly better than the neural and neural PD controllers.

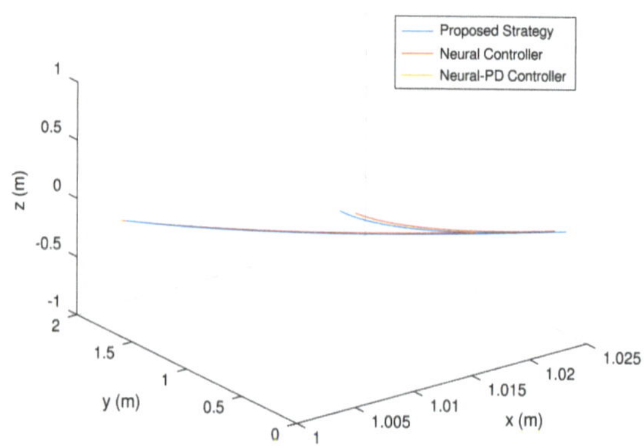

Figure 13. Evolution in time of the mobile robot trajectory.

Finally, Figures 14 and 15 show the evolution in time of the control efforts for the variables U_1 and U_2. It is important to notice that the control effort for both control inputs is significantly smaller in comparison with the control effort generated by the neural and neural PD control strategies. This is an advantage when the controller is implemented in a real experimental setup. If the control effort is smaller, then unwanted effects like saturation are avoided.

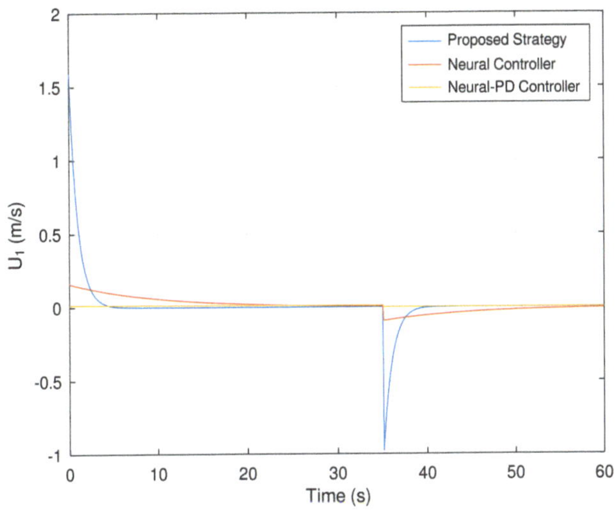

Figure 14. Evolution in time of the variable U_1.

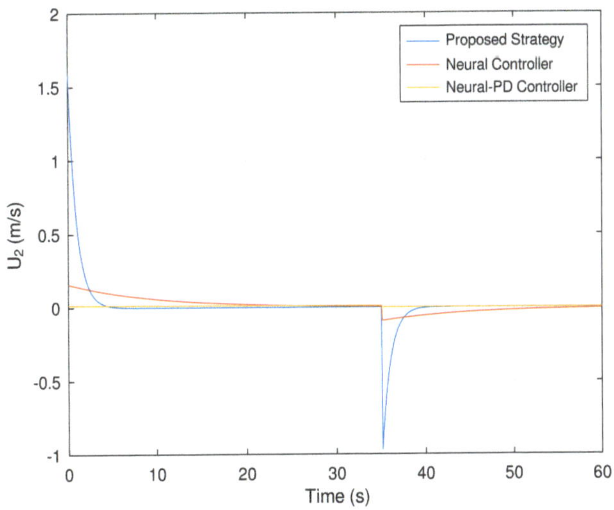

Figure 15. Evolution in time of the variable U_2.

7. Discussion

According to the theoretical and experimental results of this research paper, it is important to mention and discuss some important results and findings obtained. First, considering that the driftless control strategies with neural networks have not been reported extensively in the literature, in this research study, we proposed a combined neural network and dynamic controller for the trajectory tracking of mobile robots. We verified that the driftless control strategy is less costly in terms of control effort in comparison with other control strategies. In addition, the driftless control strategy presented in this research paper is useful for non-holonomic dynamics, which in this case is a mobile robot.

This paper demonstrates that the driftless control strategy provides an efficient mobile robot navigation strategy that is fast, reliable and accurate. It is worthwhile to mention that the neural network component of the proposed strategy provides an adequate control strategy that can be implemented and tuned relatively easily. The weights of the neural network do not need to be tuned offline by training methods like the Newton or Gaussian method, but it is important to remark that the neural networks used in this research study can be tuned by different optimization algorithms with less computational effort. The stable trajectory tracking of the mobile robot by implementing the neural network controller component ensured a precise stabilization by meeting the appropriate weight requirements, so several training and optimization algorithms can be implemented.

The dynamic controller part ensures that the stability of the driftless controller must be accurate and fast by meeting the adequate requirements of closed loop global stability. The controller was synthesized by selecting an appropriate Lyapunov function and obtaining the adequate control law. It is important to remark that the controller was carefully designed in order to be implemented in hardware easily, so the proposed control approach was relatively simple, considering also that the dynamics of the mobile robot is relatively simple but recognizing the complexity of non-holonomy of the mobile robot dynamics. This control strategy provides a smaller computational effort taking into consideration that the proposed controller is compact and the neural network controller only requires matricial and vectorial operations.

Meanwhile, the non-driftless controller is also reliable, accurate and fast, taking into consideration that this dynamic controller is more robust but requires slightly higher control effort than the driftless control system, but it is even more adequate for trajectory tracking in comparison with the driftless control system. It is important to notice that depending on the application of the mobile robot, one of these two control strategies is suitable. For this reason, in this research study, both strategies are investigated, taking into consideration that the control effort in some cases is required to be smaller to avoid some effects like saturation that can produce unwanted effects. In cases where the required hardware is available, the non-driftless control strategy is more adequate, so the driftless control strategy is suitable when the adequate hardware is not available.

8. Conclusions

This research paper presents a driftless and non-driftless control strategy based on neural dynamic controllers for the trajectory tracking of a mobile robot. Considering the non-holonomic characteristics of the mobile robot dynamics, an appropriate, relatively simple and implementable control strategy is provided in order for these results would be implementable in real-time hardware. The Lyapunov stability theorem is implemented for the design of the two control approaches in order to obtain a globally closed-loop stable system. Numerical examples validate the theoretical results obtained in this research study, proving that the theoretical results yield an appropriate performance of the mobile robot in order to track a predefined trajectory.

As a future direction, the next steps are to design a robust controller for the mobile robot considering different types of uncertainty for dynamic and kinematic control. Considering that uncertainties are found in real systems in practice, the dynamic modeling and design of a robust control strategy will be proposed. Besides, the consideration of disturbances and the implementation in a real experimental setup will be proposed in the future. The design of a novel disturbance rejection control will be also considered.

Author Contributions: Conceptualization, F.E.S. and M.C.; methodology, F.E.S.; software, F.E.S.; validation, M.C.; formal analysis, F.E.S.; investigation, F.E.S. and M.C.; resources, M.C.; data curation, F.E.S.; writing—original draft preparation, F.E.S. and M.C.; writing—review and editing, F.E.S. and M.C.; visualization, M.C.; supervision, M.C.; project administration, M.C.; funding acquisition, M.C. All authors have read and agreed to the published version of the manuscript.

Funding: This research was funded by the Universidad Don Bosco El Salvador, grant number 2023-VC-IV-1921.

Institutional Review Board Statement: Not applicable.

Informed Consent Statement: Not applicable.

Data Availability Statement: No data was used in this research study.

Conflicts of Interest: The authors declare no conflict of interest.

References

1. Zhao, L.; Li, J.; Li, H.; Liu, B. Double-loop tracking control for a wheeled mobile robot with unmodeled dynamics along right angle roads. *ISA Trans.* **2022**, *136*, 525–534. [CrossRef] [PubMed]
2. Wu, Y.; Wang, Y.; Fang, H. Full-state constrained neural control and learning for the nonholonomic wheeled mobile robot with unknown dynamics. *ISA Trans.* **2022**, *125*, 22–30. [CrossRef] [PubMed]
3. Yoo, S.J.; Park, B.S. Quantized feedback control strategy for tracking performance guarantee of nonholonomic mobile robots with uncertain nonlinear dynamics. *Appl. Math. Comput.* **2021**, *407*, 126349. [CrossRef]
4. Trojnacki, M.; Dąbek, P. Studies of dynamics of a lightweight wheeled mobile robot during longitudinal motion on soft ground. *Mech. Res. Commun.* **2017**, *82*, 36–42. [CrossRef]
5. Alipour, K.; Robat, A.B.; Tarvirdizadeh, B. Dynamics modeling and sliding mode control of tractor-trailer wheeled mobile robots subject to wheels slip. *Mech. Mach. Theory* **2019**, *138*, 16–37. [CrossRef]
6. Tzafestas, S.G. 3-Mobile Robot Dynamics. In *Introduction to Mobile Robot Control*; Tzafestas, S.G., Ed.; Elsevier: Oxford, UK, 2014; pp. 69–99.
7. Gover, A.R.; Slovák, J. Non-holonomic equations for the normal extremals in geometric control theory. *J. Geom. Phys.* **2022**, *171*, 104395. [CrossRef]
8. Tchoń, K.; Ratajczak, J. Singularities of holonomic and non-holonomic robotic systems: A normal form approach. *J. Frankl. Inst.* **2021**, *358*, 7698–7713. [CrossRef]
9. Abhinav, K.; Mukherjee, I.; Guha, P. Non-holonomic and quasi-integrable deformations of the AB equations. *Phys. D Nonlinear Phenom.* **2022**, *433*, 133186. [CrossRef]
10. Li, Y.; Xin, T.; Qiu, C.; Li, K.; Liu, G.; Li, J.; Wan, Y.; Lu, D. Dynamical-invariant-based holonomic quantum gates: Theory and experiment. *Fundam. Res.* **2022**, *3*, 229–236. [CrossRef]
11. Cenerini, J.; Mehrez, M.W.; Woo Han, J.; Jeon, S.; Melek, W. Model Predictive Path Following Control without terminal constraints for holonomic mobile robots. *Control Eng. Pract.* **2023**, *132*, 105406. [CrossRef]
12. Chen, X.; Liang, W.; Zhao, H.; Al Mamun, A. Adaptive robust controller using intelligent uncertainty observer for mechanical systems under non-holonomic reference trajectories. *ISA Trans.* **2022**, *122*, 79–87. [CrossRef]
13. LI, S.J.; Responder, W. Describing and Calculating Flat Outputs of Two-input Driftless Control Systems. *IFAC Proc. Vol.* **2010**, *43*, 683–688. [CrossRef]
14. Ishikawa, M. Switched Feedback Control for a class of First-order Nonholonomic Driftless Systems. *IFAC Proc. Vol.* **2008**, *41*, 4761–4766. [CrossRef]
15. Zuyev, A.; Grushkovskaya, V. Obstacle Avoidance Problem for Driftless Nonlinear Systems with Oscillating Controls. *IFAC-PapersOnLine* **2017**, *50*, 10476–10481. [CrossRef]
16. Califano, C.; Li, S.; Moog, C.H. Controllability of driftless nonlinear time-delay systems. *Syst. Control Lett.* **2013**, *62*, 294–301. [CrossRef]
17. Altafini, C. Involutive flows and discretization errors for nonlinear driftless control systems. *Syst. Control Lett.* **2017**, *110*, 29–37. [CrossRef]
18. Shim, H.S.; Sung, Y.G. Asymptotic control for wheeled mobile robots with driftless constraints. *Robot. Auton. Syst.* **2003**, *43*, 29–37. [CrossRef]
19. Xie, H.; Zong, G.; Yang, D.; Chen, Y.; Shi, K. Dynamic output feedback L-Infinity control of switched affine systems: An event-triggered mechanism. *Nonlinear Anal. Hybrid Syst.* **2023**, *47*, 101278. [CrossRef]
20. Zhang, D.; Zhang, L.; Yu, Z.; Li, H.; Shu, L. Dynamic output feedback control for networked control systems: A sum-based discrete event-triggered approach. *IFAC-PapersOnLine* **2022**, *55*, 61–66. [CrossRef]
21. Xie, Z.; Wang, D.; Wong, P.K.; Li, W.; Zhao, J. Dynamic-output-feedback based interval type-2 fuzzy control for nonlinear active suspension systems with actuator saturation and delay. *Inf. Sci.* **2022**, *607*, 1174–1194. [CrossRef]

22. Bertolin, A.L.; Oliveira, R.C.; Valmorbida, G.; Peres, P.L. Dynamic output-feedback control of continuous-time Lur'e systems using Zames-Falb multipliers by means of an LMI-based algorithm. *IFAC-PapersOnLine* **2022**, *55*, 109–114. [CrossRef]
23. Silva, B.M.; Ishihara, J.Y.; Tognetti, E.S. LMI-based consensus of linear multi-agent systems by reduced-order dynamic output feedback. *ISA Trans.* **2022**, *129*, 121–129. [CrossRef]
24. Chen, W.; Gao, F.; Xu, S.; Li, Y.; Chu, Y. Robust stabilization for uncertain singular Markovian jump systems via dynamic output-feedback control. *Syst. Control Lett.* **2023**, *171*, 105433. [CrossRef]
25. Yao, Q. Fixed-time neural adaptive fault-tolerant control for space manipulator under output constraints. *Acta Astronaut.* **2023**, *203*, 483–494. [CrossRef]
26. Wu, C.; Zhu, G.; Lu, J. Indirect adaptive neural tracking control of USVs under injection and deception attacks. *Ocean Eng.* **2023**, *270*, 113641. [CrossRef]
27. Friese, J.; Brandt, N.; Schulte, A.; Kirches, C.; Tegethoff, W.; Köhler, J. Quasi-optimal control of a solar thermal system via neural networks. *Energy AI* **2023**, *12*, 100232. [CrossRef]
28. Alsaade, F.W.; Yao, Q.; Al-zahrani, M.S.; Alzahrani, A.S.; Jahanshahi, H. Neural-based fixed-time attitude tracking control for space vehicle subject to constrained outputs. *Adv. Space Res.* **2022**, *71*, 3588–3599. [CrossRef]
29. Jahanshahi, H.; Yao, Q.; Ijaz Khan, M.; Moroz, I. Unified neural output-constrained control for space manipulator using tan-type barrier Lyapunov function. *Adv. Space Res.* **2022**, *71*, 3712–3722. [CrossRef]
30. Zhou, G.; Tan, D. Review of nuclear power plant control research: Neural network-based methods. *Ann. Nucl. Energy* **2023**, *181*, 109513. [CrossRef]
31. Huang, J.; Wen, C.; Wang, W.; Jiang, Z.P. Adaptive stabilization and tracking control of a nonholonomic mobile robot with input saturation and disturbance. *Syst. Control Lett.* **2013**, *62*, 234–241. [CrossRef]
32. Alves, J.G.; Lizarralde, F.; Monteiro, J.C. Control Allocation for Wheeled Mobile Robots Subject 500 to Input Saturation. *IFAC-PapersOnLine* **2020**, *53*, 3904–3909. [CrossRef]
33. Wang, R.; Zhang, X.; Fang, Y. Visual tracking of mobile robots with both velocity and acceleration saturation constraints. *Mech. Syst. Signal Process.* **2021**, *150*, 107274. [CrossRef]
34. Martínez, E.A.; Ríos, H.; Mera, M. Robust tracking control design for Unicycle Mobile Robots with input saturation. *Control Eng. Pract.* **2021**, *107*, 104676. [CrossRef]
35. Velasco-Villa, M.; Castro-Linares, R.; Rosales-Hernández, F.; del Muro-Cuéllar, B.; Hernández-Pérez, M. Discrete-time synchronization strategy for input time-delay mobile robots. *J. Frankl. Inst.* **2013**, *350*, 2911–2935. [CrossRef]
36. Koumboulis, F.N.; Kouvakas, N.D. Mobile robots in singular time-delay form—Modeling and control. *J. Frankl. Inst.* **2016**, *353*, 160–179. [CrossRef]
37. Kojima, K.; Oguchi, T.; Alvarez-Aguirre, A.; Nijmeijer, H. Predictor-based Tracking Control of A Mobile Robot with Time-delays. *IFAC Proc. Vol.* **2010**, *43*, 167–172. [CrossRef]
38. Santos, J.; Conceição, A.; Santos, T.; Araújo, H. Remote control of an omnidirectional mobile robot with time-varying delay and noise attenuation. *Mechatronics* **2018**, *52*, 7–21. [CrossRef]
39. Gorelov, V.; Komissarov, A.; Miroshnichenko, A. 8 × 8 Wheeled Vehicle Modeling in a Multibody Dynamics Simulation Software. *Procedia Eng.* **2015**, *129*, 300–307. [CrossRef]
40. Tang, Z.; Yuan, X.; Xie, X.; Jiang, J.; Zhang, J. Implementing railway vehicle dynamics simulation in general-purpose multibody simulation software packages. *Adv. Eng. Softw.* **2019**, *131*, 153–165. [CrossRef]
41. Wang, H.; Chen, J.; Feng, Z.; Li, Y.; Deng, C.; Chang, Z. Dynamics analysis of underwater glider based on fluid-multibody coupling model. *Ocean Eng.* **2023**, *278*, 114330. [CrossRef]
42. Gan, J.; Zhou, Z.; Yu, A.; Ellis, D.; Attwood, R.; Chen, W. Co-simulation of multibody dynamics and discrete element method for hydraulic excavators. *Powder Technol.* **2023**, *414*, 118001. [CrossRef]
43. Panahandeh, P.; Alipour, K.; Tarvirdizadeh, B.; Hadi, A. A kinematic Lyapunov-based controller to posture stabilization of wheeled mobile robots. *Mech. Syst. Signal Process.* **2019**, *134*, 106319. [CrossRef]
44. Li, W.; Zhan, Q. Kinematics-based four-state trajectory tracking control of a spherical mobile robot driven by a 2-DOF pendulum. *Chin. J. Aeronaut.* **2019**, *32*, 1530–1540. [CrossRef]
45. LiBretto, M.; Qiu, Y.; Kim, E.; Pluckter, K.; Yuk, N.S.; Ueda, J. Singularity-free solutions for inverse kinematics of degenerate mobile robots. *Mech. Mach. Theory* **2020**, *153*, 103988. [CrossRef]
46. Zhao, L.; Jin, J.; Gong, J. Robust zeroing neural network for fixed-time kinematic control of wheeled mobile robot in noise-polluted environment. *Math. Comput. Simul.* **2021**, *185*, 289–307. [CrossRef]
47. Jilek, T.; Burian, F.; Kriz, V. Kinematic Models for Odometry of a Six-Wheeled Mobile Robot. *IFAC-PapersOnLine* **2016**, *49*, 305–310. [CrossRef]
48. Jiang, H.; Xu, G.; Zeng, W.; Gao, F. Design and kinematic modeling of a passively-actively transformable mobile robot. *Mech. Mach. Theory* **2019**, *142*, 103591. [CrossRef]
49. Hernández-León, P.; Dávila, J.; Salazar, S.; Ping, X. Distance-Based Formation Maneuvering of Non-Holonomic Wheeled Mobile Robot Multi-Agent System. *IFAC-PapersOnLine* **2020**, *53*, 5665–5670. [CrossRef]
50. Wang, Y.; Mai, T.; Mao, J. Adaptive motion/force control strategy for non-holonomic mobile manipulator robot using recurrent fuzzy wavelet neural networks. *Eng. Appl. Artif. Intell.* **2014**, *34*, 137–153. [CrossRef]

51. Hou, R.; Cui, L.; Bu, X.; Yang, J. Distributed formation control for multiple non-holonomic wheeled mobile robots with velocity constraint by using improved data-driven iterative learning. *Appl. Math. Comput.* **2021**, *395*, 125829. [CrossRef]
52. Ma, Y.; Zheng, G.; Perruquetti, W. Real-time Identification of different types of non-holonomic mobile robots. *IFAC Proc. Vol.* **2013**, *46*, 791–796. [CrossRef]
53. Goswami, N.K.; Padhy, P.K. Sliding mode controller design for trajectory tracking of a non-holonomic mobile robot with disturbance. *Comput. Electr. Eng.* **2018**, *72*, 307–323. [CrossRef]
54. Zhang, S.; Wu, Y.; He, X. Cooperative output feedback control of a mobile dual flexible manipulator. *J. Frankl. Inst.* **2021**, *358*, 6941–6961. [CrossRef]
55. Zou, Y.; Deng, C.; Dong, L.; Ding, L.; Lu, M. Distributed output feedback consensus tracking control of multiple nonholonomic mobile robots with only position information of leader. *Appl. Math. Comput.* **2022**, *422*, 126962. [CrossRef]
56. Andreev, A.S.; Peregudova, O.A. On Output Feedback Trajectory Tracking Control of an Omni-Mobile Robot**This work was financially supported by the Ministry of Education and Science of Russia under Grant [9.5994.2017/BP] and Russian Foundation for Basic Research under Grant [19-01-00791]. *IFAC-PapersOnLine* **2019**, *52*, 37–42. [CrossRef]
57. Huang, J.; Wen, C.; Wang, W.; Jiang, Z.P. Adaptive output feedback tracking control of a nonholonomic mobile robot. *Automatica* **2014**, *50*, 821–831. [CrossRef]
58. Chen, H.; Chen, Y.; Chen, W.; Yang, F. Output Tracking of Nonholonomic Mobile Robots with a Model-free Fractional-order Visual Feedback. *IFAC-PapersOnLine* **2016**, *49*, 736–741. [CrossRef]
59. Wu, D.; Cheng, Y.; Du, H.; Zhu, W.; Zhu, M. Finite-time output feedback tracking control for a nonholonomic wheeled mobile robot. *Aerosp. Sci. Technol.* **2018**, *78*, 574–579. [CrossRef]
60. Sleaman, W.K.; Hameed, A.A.; Jamil, A. Monocular vision with deep neural networks for autonomous mobile robots navigation. *Optik* **2023**, *272*, 170162. [CrossRef]
61. Chawla, I.; Pathak, P.; Notash, L.; Samantaray, A.; Li, Q.; Sharma, U. Inverse and Forward Kineto-Static Solution of a Large-Scale Cable-Driven Parallel Robot using Neural Networks. *Mech. Mach. Theory* **2023**, *179*, 105107. [CrossRef]
62. Gandarilla, I.; Montoya-Cháirez, J.; Santibáñez, V.; Aguilar-Avelar, C.; Moreno-Valenzuela, J. Trajectory tracking control of a self-balancing robot via adaptive neural networks. *Eng. Sci. Technol. Int. J.* **2022**, *35*, 101259. [CrossRef]
63. He, Y.; Chen, J.; Zhou, X.; Huang, S. In-situ fault diagnosis for the harmonic reducer of industrial robots via multi-scale mixed convolutional neural networks. *J. Manuf. Syst.* **2023**, *66*, 233–247. [CrossRef]
64. Xiao, H.; Chen, C.P.; Lai, G.; Yu, D.; Zhang, Y. Integrated nonholonomic multi-robot consensus tracking formation using neural network-optimized distributed model predictive control strategy. *Neurocomputing* **2023**, *518*, 282–293. [CrossRef]
65. Wu, Y.; Niu, W.; Kong, L.; Yu, X.; He, W. Fixed-time neural network control of a robotic manipulator with input deadzone. *ISA Trans.* **2022**, *135*, 449–461. [CrossRef] [PubMed]
66. Mai, T.; Tran, H. An adaptive robust backstepping improved control scheme for mobile manipulators robot. *ISA Trans.* **2023**, 137, 446–456. [CrossRef]
67. Raikwar, S.; Fehrmann, J.; Herlitzius, T. Navigation and control development for a four-wheel-steered mobile orchard robot using model-based design. *Comput. Electron. Agric.* **2022**, *202*, 107410. [CrossRef]
68. Rosenfelder, M.; Ebel, H.; Eberhard, P. Cooperative distributed nonlinear model predictive control of a formation of differentially-driven mobile robots. *Robot. Auton. Syst.* **2022**, *150*, 103993. [CrossRef]
69. Zhang, J.; Li, S.; Meng, H.; Li, Z.; Sun, Z. Variable gain based composite trajectory tracking control for 4-wheel skid-steering mobile robots with unknown disturbances. *Control Eng. Pract.* **2023**, *132*, 105428. [CrossRef]
70. Luo, X.; Mu, D.; Wang, Z.; Ning, P.; Hua, C. Adaptive full-state constrained tracking control for mobile robotic system with unknown dead-zone input. *Neurocomputing* **2023**, *524*, 31–42. [CrossRef]
71. Spong, M.W.; Hutchinson, S.; Vidyasagar, M. *Robot Modeling and Control*; John Wiley and Sons: Hoboken, NJ, USA, 2006.

sensors

MDPI

Article

A LiDAR-Camera-Inertial-GNSS Apparatus for 3D Multimodal Dataset Collection in Woodland Scenarios

Mário P. Cristóvão [1,*], David Portugal [1,2], Afonso E. Carvalho [1,2] and João Filipe Ferreira [1,3]

1 Institute of Systems and Robotics, Department of Electrical Engineering and Computers, University of Coimbra, 3030-290 Coimbra, Portugal
2 Department of Electrical Engineering and Computers, University of Coimbra, 3030-290 Coimbra, Portugal
3 Computational Intelligence and Applications Research Group, Department of Computer Science, School of Science and Technology, Nottingham Trent University, Nottingham NG11 8NS, UK
* Correspondence: mario.cristovao@isr.uc.pt

Abstract: Forestry operations have become of great importance for a sustainable environment in the past few decades due to the increasing toll induced by rural abandonment and climate change. Robotics presents a promising solution to this problem; however, gathering the necessary data for developing and testing algorithms can be challenging. This work proposes a portable multi-sensor apparatus to collect relevant data generated by several onboard sensors. The system incorporates Laser Imaging, Detection and Ranging (LiDAR), two stereo depth cameras and a dedicated inertial measurement unit (IMU) to obtain environmental data, which are coupled with an Android app that extracts Global Navigation Satellite System (GNSS) information from a cell phone. Acquired data can then be used for a myriad of perception-based applications, such as localization and mapping, flammable material identification, traversability analysis, path planning and/or semantic segmentation toward (semi-)automated forestry actuation. The modular architecture proposed is built on Robot Operating System (ROS) and Docker to facilitate data collection and the upgradability of the system. We validate the apparatus' effectiveness in collecting datasets and its flexibility by carrying out a case study for Simultaneous Localization and Mapping (SLAM) in a challenging woodland environment, thus allowing us to compare fundamentally different methods with the multimodal system proposed.

Keywords: multi-sensor apparatus; multimodal dataset collection; forestry robotics; LiDAR; inertial measurement unit; depth cameras; GNSS

check for updates

Citation: Cristóvão, M.P.; Portugal, D.; Carvalho, A.E.; Ferreira, J.F. A LiDAR-Camera-Inertial-GNSS Apparatus for 3D Multimodal Dataset Collection in Woodland Scenarios. *Sensors* **2023**, *23*, 6676. https://doi.org/10.3390/s23156676

Academic Editors: David Cheneler and Stephen Monk

Received: 13 June 2023
Revised: 21 July 2023
Accepted: 25 July 2023
Published: 26 July 2023

1. Introduction

Forest and woodland maintenance are crucial and challenging tasks that require monitoring forests, planting trees, and removing invasive species. These tasks can be physically demanding and time-consuming for human workers, posing significant safety risks. While autonomous robots have the potential to revolutionize forestry maintenance, the existing technology has limitations that prevent widespread adoption. One of the most important of these tasks, landscape maintenance, has become particularly relevant as forest fires have become increasingly prevalent in recent decades.

Current forestry robots [1,2] often lack the flexibility needed to easily navigate through complex and dynamic forest environments, making data acquisition slow and cumbersome. Forest areas are characterized by various obstacles, such as dense vegetation, uneven terrains, and dynamic changes due to growth and decay. Conventional forestry robots may struggle to maneuver through these challenging conditions, leading to limited coverage and incomplete data acquisition [3]. The lack of flexibility in conventional forestry robots also hampers their ability to access hard-to-reach areas.

To overcome existing limitations, this paper proposes the development of a lightweight and portable LiDAR-Camera-Inertial-GNSS apparatus with an onboard computer for acquiring datasets in forests and woodlands. The apparatus, illustrated in Figure 1, collects multiple sensor modalities such as accelerometer, gyroscope and magnetometer data from an Inertial Measurement Unit (IMU), RGB and 3D Depth information from two cameras, 3D Light Detection and Ranging (LiDAR) scans, and Global Navigation Satellite System (GNSS) information to create detailed and accurate datasets of forest environments.

The apparatus aims to solve the critical challenge of data acquisition, facilitating the planning, testing and deployment of autonomous robots for forestry maintenance, and it has key potential benefits, including improving the safety and efficiency of acquiring datasets and reducing costs associated with deploying an automated vehicle in the field. This paper also presents an experimental evaluation of the system in a real forest environment, where Simultaneous Localization and Mapping (SLAM) implementations have been tested as a case study, demonstrating the feasibility, flexibility and potential of the system proposed.

By proposing a lightweight and portable multi-sensor apparatus, this study provides significant contributions to the field of forestry robotics, such as a publicly available ready-to-use dataset [4] that enables researchers to analyze forest environments in depth, and developing and testing perception-based methods with real-world data. In addition to the apparatus design description and important lessons learned, the architecture developed to record and store datasets also represents an important contribution, bringing a novel, modular, and user-friendly architecture for acquiring multisensory outdoor datasets, allowing different sensor configurations to be used with minor adjustments. Moreover, we also contribute with an in-house developed Android App for easily exposing smartphone GNSS data with ROS for use in Robotics [5]. As such, this work has the potential to pave the way for a more widespread use of autonomous robots in forests and woodland scenarios.

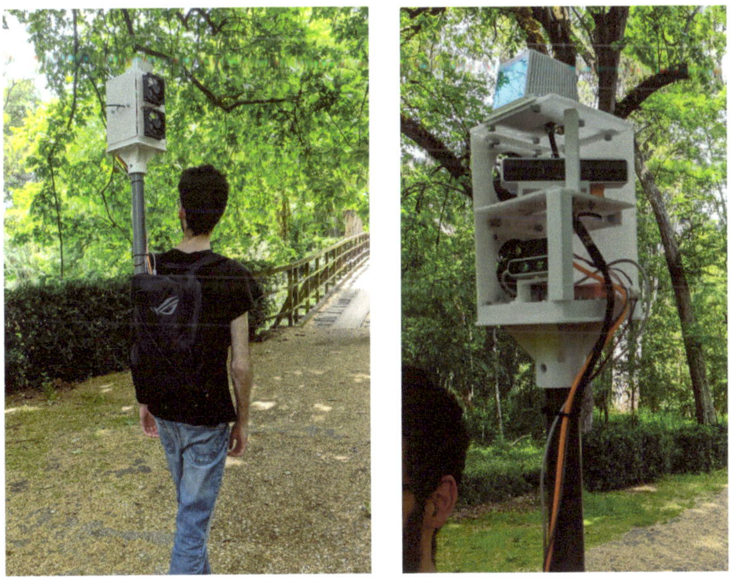

(**a**) (**b**)

Figure 1. Illustration of the LiDAR-Camera-Inertial-GNSS apparatus proposed for dataset collection. (**a**) Operator carrying the apparatus backpack. (**b**) Close-up view of the sensors.

2. Use Case Scenario

Simultaneous Localization And Mapping, commonly referred to as SLAM, is the action of progressively building a map of an environment perceived by a moving entity (e.g., a robot) while persistently localizing in that map as the entity moves through space [6,7]. SLAM algorithms play a crucial role in enabling autonomous robots to navigate and perceive unknown environments in real time. However, the complex and dynamic nature of forest environments presents particularly significant challenges, making it essential to evaluate and refine SLAM methods offline using realistic datasets.

Below, we formulate a potential use case scenario to clarify our motivations and demonstrate how the collection of datasets can be used to test methods for artificial perception in Forestry Robotics as well as for localization and mapping in particular:

In this use case, a research engineer undertakes a data collection mission in a remote forest area using an in-house developed multi-sensor apparatus. The mission aims to gather a diverse range of multimodal data, such as LiDAR, IMU, Depth and RGB cameras, and GNSS, to support research efforts.

He follows a systematic circular route through the dense vegetation of the forest while equipped with the multi-sensor apparatus worn as a backpack. The LiDAR sensor accurately measures the three-dimensional structure of trees and vegetation, while the IMU tracks the backpack's orientation, including magnetic heading. Depth cameras and RGB images provide visual details of the forest, and GPS records positioning data throughout the entire journey.

Upon completing the route and returning to the starting point, he concludes the data collection process and disconnects the apparatus. Before sharing the data, he transfers the acquired dataset to his laptop as a ROS bag file and performs initial assessments as well as functional checks to ensure the quality of the dataset.

The shared dataset becomes a valuable resource for the Forestry Robotics community. For instance, a PhD student utilizes the collected data to test and refine SLAM algorithms, aiming to improve the precision and efficiency of robots in mapping and navigating forest environments, and contributing to the progress of forestry management and conservation practices by harnessing technological advancements.

3. Background and Related Work

In this section, we start with an analysis of previous multi-sensor apparatuses designed for dataset collection, including those specifically tailored for SLAM and forestry applications, followed by a review of seminal work on 3D SLAM for a better understanding of existing state-of-the-art methods.

In the past, a few portable and light sensing apparatuses designed to collect data from the environment around us have already been proposed. To the best of our knowledge, Oveland et al. [8] was the first to develop a portable apparatus to be used in a forest environment. They compared different methodologies to study the Diameter at Breast Height (DBH), an important feature in forestry inventory, reaching the conclusion that a portable apparatus with multiple sensors, such as LiDAR and an IMU, is a viable alternative to perform forest inventory. In Proudman et al.'s work [9], a portable system was designed for estimating the DBH of trees in forestry applications. However, their choice of using a metal stick instead of a backpack introduces the issue of user fatigue. This design may lead to excessive variations in stick position, resulting in unintelligible and uncontrolled movements, which can negatively impact data collection. While their system had the benefit of a built-in display for real-time data visualization, the design limitation raises concerns about the accuracy and consistency of the measurements. On the other hand, Su et al. [10] and Xiao et al. [10] developed an accurate backpack system with two orthogonally positioned LiDARs. Their backpack design overcomes the user fatigue issue associated with the metal stick approach, allowing for more stable movements during data acquisition. Since Su et al. aimed to measure DBH, their paper lacks effective metrics to assess the precision of localization. In contrast, Xiao et al. present the Relative Translation Error metric, introduced in [11], but they lack RGB-D and stereo information. Jelavic et al. developed a system for forestry-harvesting missions [2]. Their system aims to acquire a dataset to generate an *a priori* map of the deployment location for an autonomous harvesting excavator. While their implementation shares similarities with the present study by providing metrics to evaluate the precision of the SLAM algorithm used, it falls short in incorporating RGB-D and GNSS information. Sier et. al [12] designed a very complete apparatus on top of a cart wheel with the objective of comparing the performance of six different LiDARs in GNSS denied environments. LiDARs with both spinning and solid-state technologies were considered as well as a stereo fish-eye camera. The authors compare different state-of-the-art SLAM methods with different LiDAR configurations to assess the most appropriate combination. For forest environments, the authors concluded that the most robust combinations are the FAST-LIO [13] implementation using the more precise Ouster spinning LiDARs and the Livox Horizon using a LIO-based SLAM design for the Livox Horizon. In a recent study conducted by Faitli et al. [14], a new measurement setup was developed to collect LiDAR and IMU data for localization and mapping using a LIO-SAM-based method. While the system design shares similarities with the previous work of Proudman et al. [9], Faitli et al. focused more on evaluating the performance of their SLAM algorithm specifically designed for forest environments. However, it is important to note that their dataset did not include RGB information, which restricts the potential applications of their dataset. Another recent study by Li et al. [15] presented a new sensing kit that collected LiDAR-IMU datasets in multiple GNSS-denied scenarios, including a forest environment. Instead of using a backpack or a handheld design, the authors chose to develop a helmet that integrated the sensors, such as LiDAR, IMU, and GNSS, while storing the rest of the hardware in a backpack. According to the authors, the motion characteristics of the helmet approach were similar to those found in the handheld counterpart, involving quick shifting and shaking, whereas the backpack design only accounted for quick shifting. However, the level of fatigue that would result from supporting a 1.5 Kg load on top of the operator's head remains uncertain. Once again, the datasets produced in this study lacked RGB-D information, which hinders the potential use cases of the collected datasets.

One of the most popular implementations of SLAM is Real-Time Appearance-Based Mapping (RTAB-Map) [16]. RTAB-Map is a graph-based SLAM system that relies on an image loop closure detector, offering several options for the back-end, namely GTSAM (default) [17], g2o [18] and TORO [19]. The loop closure detector uses a bag-of-words approach to determine the likelihood that a new image was taken from a previous or a new location. It can estimate odometry from IMU and wheel encoders, but it also supports Visual and LiDAR odometry as optional odometry sources. When executing loop closure, RTAB-Map reuses the features that were previously matched in Visual or LiDAR Odometry, which improves the overall performance. RTAB-Map can generate both 2D and 3D Occupancy grids.

Several LiDAR-based methods derive from LiDAR Odometry And Mapping, which is commonly known as LOAM. Although LOAM can create highly accurate maps, it usually performs poorly in places with few landmarks, such as long corridors. LeGO-LOAM [20] adds two additional modules to the LOAM technique: point cloud segmentation and loop closure. These extra components allow an improvement in computing performance and drift reduction over long distances but does not improve the results when used in a featureless environment. LeGO-LOAM uses the naive ICP algorithm to perform loop closure, but a more robust approach based on a point cloud descriptor is implemented in SC-LeGO-LOAM [20,21]. To help improve the performance in a low features environment, researchers have been recently adding an IMU to similar systems in a tightly coupled approach (see [13,22,23]), giving rise to the term LiDAR Inertial Odometry (LIO). For instance, the LIO-SAM [24] approach proposes a tightly coupled LiDAR framework atop of a factor graph. The implementation considers four different factors, namely IMU preintegration, LiDAR odometry, GPS and a loop closure factor, making it ideal for multi-sensor fusion and global optimization. Lately, several methods based on similar principles have been proposed [25–32].

Cartographer is Google's implementation to solve the SLAM problem [33]. It is also a LiDAR-based graph SLAM divided into two main components: local SLAM (the front end) and global SLAM (the back-end). This approach takes input of a range-finding sensor, e.g., a LiDAR, and applies a band-pass filter to the input data. IMU can also be used to help figure out the robot rotation and to provide information on gravity direction, which is used in the 3D variant.

In order to evaluate SLAM systems, vital metrics such as Relative Translation Error (RTE) and Absolute Position Error (APE) were introduced in [11]. RTE measures the accuracy of estimating the relative translation between two positions. If the two positions are taken from the same location, a lower RTE implies a more precise localization estimate. On the other hand, APE quantifies the accuracy of absolute position estimation by comparing the estimated positions with ground truth values. A smaller APE indicates better localization accuracy.

The systems reviewed in this section provide valuable insights into the challenges and opportunities in the development of a portable apparatus for forestry applications. Table 1 presents a comparison between the reviewed systems and the SLAM algorithms tested using the data collected with each of these frameworks. Some systems focus on the forest application inventory, and other applications are particularly focused on determining the DBH of trees. Key metrics are missing in some works, making it difficult to effectively evaluate their performance for SLAM. The majority of the apparatuses reviewed in this study lack RGB-colored images of the environment. RGB information plays a crucial role in various artificial perception algorithms and methods, and its absence limits the potential use cases for both the apparatus and the dataset it generates. It is imperative to include RGB information in recorded datasets to enable a wider range of use cases. Furthermore, these systems are typically expensive, which is primarily due to the high costs associated with the prevalent LiDAR technology incorporated in them.

Table 1. Comparison table between previously mentioned systems.

Work	IMU	LiDAR	GNSS	Stereo	RGB-D	SLAM Method	Error (m)	Forestry	Structure	[1] Cost (USD)
Oveland et al. [8], 2018	✓	✓	✓	—	—	GeoSLAM (proprietary) [34]	N/A	✓	Backpack	$13,500
Proudman et al. [9], 2021	✓	✓	✓	—	✓	Factor-Graph LIO [35] + Elevation Mapping [36]	0.11 (RTE)	✓	Handheld	$19,000
Su et al. [10], 2021	✓	✓	—	—	—	Custom LiDAR-SLAM inspired by [37,38]	N/A	✓	Backpack	$9000
Jelavic et al. [2], 2021	✓	✓	—	✓	—	Cartographer [33]	0.41 (APE)	✓	Handheld	$10,500
Xiao et al. [39], 2022	✓	✓	✓	—	—	LIO-SAM [24]	0.03 (RTE)	—	Backpack	$10,000
Sier et al. [12], 2022	✓	✓	✓	✓	—	FAST-LIO [13] LIO-Livox [40]	0.05 (APE)	✓	Wheeled Cart	$43,000
Faitli et al. [14], 2023	✓	✓	✓	—	—	LIO-SAM-based [24]	0.02-0.16 (APE)	✓	Handheld	$34,000
Li et al. [15], 2023	✓	✓	✓	—	—	FAST-LIO [13] LOAM [41] LIO-LIVOX [40]	0.13-0.35 (APE)	✓	Helmet	$41,000
Our solution	✓	✓	✓	✓	✓	RTAB-Map [16] Cartographer [33]	0.09-0.28 (RTE)	✓	Backpack	$4000

[1] The estimated costs outlined represent a lower bound value for the examined systems.

An estimate of the overall value of the above-mentioned works has been included in Table 1. While some costs are provided by the authors, most are estimated based on the current unit price of the sensors that make up the various apparatus systems.

By building upon the lessons learned from previous works, we have built an apparatus that combines multiple sensor modalities into a single lightweight, inexpensive and portable backpack system. The integration of multiple sensors allows for comprehensive data acquisition, which in turn enables a detailed and accurate perception of the forest environments. The apparatus presented in this study overcomes the limitations of prior systems by providing a wider range of sensory information. Additionally, the collected datasets are made publicly available, including high accuracy pose estimation off the shelf, with applicability potential beyond localization and mapping algorithms, as we demonstrate later on.

4. System Description

Among the reviewed approaches, the use of a wheel cart system, as demonstrated in [12], offers the most ergonomic solution. However, this approach significantly hampers the maneuverability of the apparatus. Forest environments present several challenges and obstacles such as uneven terrain, rocky surfaces, and dense vegetation, making wheeled vehicles generally impractical for navigation. On the other hand, mounting the entire system on a metal rod held firmly in the operator's hands greatly enhances maneuverability, facilitating precise pointing and feature capture. Nonetheless, this approach comes with its limitations. The human operator may experience fatigue over time, resulting in reduced stability when holding the system and ultimately affecting the quality of the dataset.

The backpack method strikes a balance between ergonomics and maneuverability. It offers easier portability and improved stabilization during extended distances compared to the rod approach while remaining highly adaptable to uneven terrains. Taking into account the unique demands of forestry scenarios, the backpack method emerges as the most practical compromise, enabling operators to navigate through the challenging environments while maintaining stability and minimizing fatigue. Therefore, as seen in Figure 1, we decided on a backpack-based design for the apparatus proposed.

Our objective emphasizes the importance of collecting datasets that encompass a diverse range of sensory information from different sensor types. In mapping applications, accurate depth sensors play a crucial role, with LiDAR being widely recognized as the predominant sensor in this field. By incorporating an RGB-D camera, the datasets we generate become more versatile and can be utilized for various applications, such as segmentation and fuel identification algorithms. Unfortunately, this important component lacks in most of the reviewed literature. The inclusion of an IMU and GNSS information in the datasets enables a more efficient exploration and testing of sensor fusion algorithms and localization methods. By encompassing these different sensor modalities, our datasets become comprehensive resources for advancing research and development in various domains.

Alongside the sensor possibilities, there are additional requirements that must be fulfilled to ensure a competent working solution. Firstly, it is critical that the system can run continuously for a minimum of two hours so that longer expeditions and/or multiple consecutive datasets can be collected without recharging its batteries. Additionally, the software should be well integrated using a commonly used middleware for Robotics applications such as ROS (Robot Operating System), and the system should be able to endure high outdoor temperatures to allow working under most weather conditions. This includes considering appropriate cooling solutions to avoid the sensors to go beyond their maximum operating temperature. Other important requirements include modularity for adding, swapping, or removing sensors and processing nodes and real-time notification of sensor malfunctions during startup and runtime. These characteristics are mainly achieved through ROS and Docker—further explained later on—and highly improve the system's robustness and potential for adoption in different scenarios. In addition, the apparatus should be affordable (below $4500), and its weight should be kept to a minimum so as not to cause discomfort to the user. Naturally, multiple sensors and a small-factor onboard computer must be incorporated to achieve the proposed goals, and all sensors must be kept fixed in the apparatus structure with a well-known geometrical relationship between them at all times.

As shown in Figure 2a, the system comprises the following components: a Xsens MTi IMU, a Mid-70 Livox LiDAR, a Mynt Eye S1030 stereo camera, and an Intel Realsense D435i RGB-D camera. The LiDAR publishes point clouds at 50 Hz with precise measurements and a maximum range of 90 m. However, the LiDAR's limited circular Field of View (FoV) of 70.4° restricts the amount of information it can capture. To complement the LiDAR data, the Mynt Eye provides point clouds with a larger horizontal FoV of 146°. The Intel Realsense D435i camera not only provides additional depth information but also serves as the single source of RGB and infrared information, which is useful for identifying relevant forest entities, such as flammable material [3]. The system also contains an onboard computer, the Udoo Bolt V3, that is responsible for receiving and recording data from every sensor. The onboard computer is equipped with a high-speed M.2 NVMe Solid-State Drive (SSD) to allow the simultaneous recording of high volumes of data acquired by the different sensors. The entire system is powered by a 14.8 V Turnigy battery with 10,000 mAh, which can provide approximately 4 h of continuous operation. For further clarification, a diagram showing how the various modules are physically interconnected and powered is presented in Figure 2b.

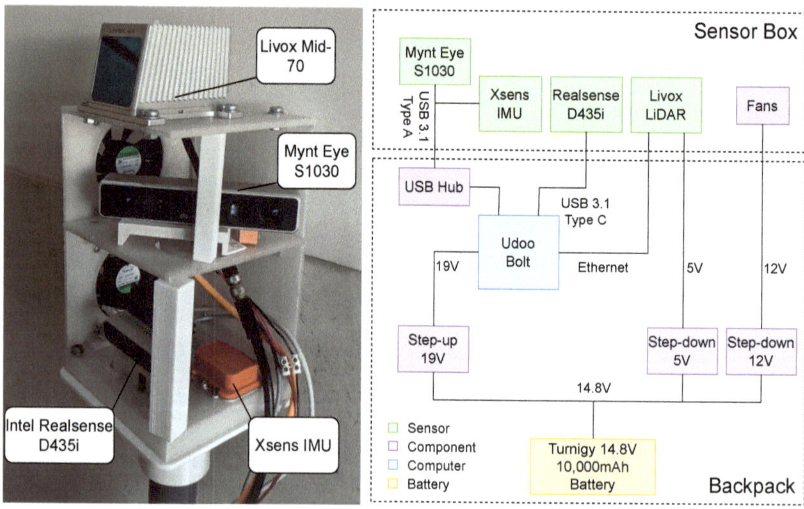

(a) (b)

Figure 2. A closer look at the multiple sensors incorporated in the system and their connectivity and power management. (**a**) Sensor framework. (**b**) Physical system block diagram.

To ensure durability and functionality, the physical structure of the apparatus was divided into two distinct components: "sensor box" and "backpack". The sensor box houses the sensors, while the backpack accommodates the computer, battery, and voltage regulators (cf. Figure 2b). The system is intended for use in outdoor environments, where ambient temperatures can reach over 35 °C for extended periods of time, and sensors experience increased heating. In such conditions, the use of some plastic materials such as Polylactic Acid (PLA) are not suitable due to their potential to deform or degrade. Given that the host computer can reach high temperatures, the structure inside the backpack is made of Acrylonitrile Butadiene Styrene (ABS), which can withstand temperatures of about 80 °C without significant degradation [42]. The sensor box, on the other hand, is built using polyethylene terephthalate glycol (PETG), which is a material that has a glass transition temperature at around 75 °C [43] but offers more adequate ultraviolet resistance, which is important considering that this is the most exposed part of the apparatus. To mitigate potential thermal issues in the sensor box, several fans are set up to blow fresh air into the cameras, which tend to heat up after long periods of operation. A heat sink is also attached to back of the Realsense D435i camera to improve heat dissipation. These measures not only help maintain optimal performance but also ensure that the temperature of the sensors remains within safe limits, minimizing any potential safety risks to the operator. Moreover, to facilitate future upgrades, the Livox LiDAR is mounted on top, allowing for easy replacement with a LiDAR with a larger horizontal FoV in the mid-term future. Since the primary purpose of the Mynt Eye is to increase the FoV for mapping, the mount that holds it in place was designed to allow easy rotation around the yaw axis. This enables users to adjust the extent to which the Mynt Eye's data overlaps with the FoV of the LiDAR and the D435i as needed in their specific application.

Sensor poses are known in a common frame of reference from the Computer-Aided Design (CAD) of the system with the exception of the yaw angle of the Mynt Eye due to its adjustable nature. Sensor registration is completed manually using the transformations provided by CAD. Once the Mynt Eye is physically set by the user in its final orientation, its pose is passed as an input parameter of the system by visually comparing the 3D intersection of the different sensor point clouds. From our experience, this procedure yields appropriate results in general. Yet in the future, we intend to work on more precise and automated extrinsic calibration of the apparatus' sensors.

We use a software platform for packaging and running applications in isolated containers, since the sensor drivers run in different, not fully cross-compatible versions of ROS. Therefore, the project's complete architecture for the dataset recording process, as depicted in Figure 3, is designed around Docker. It includes containers for the different ROS versions needed as well as a container running a ROS bridge server, which exposes port 9090 on the host computer to receive GNSS information from an in house developed Android Sensor ROS application (see [5] for more information). Additionally, another container is responsible for writing the dataset into the `rosbag` format, and a debugging node runs on a System Diagnostic container, recording a separate `rosbag` dataset with various useful monitoring information, such as sensor acquisition frequencies and CPU temperature. The use of Docker also enables easy replication of the architecture for different apparatuses with sensor configurations specific to each use case. The codebase for the complete system architecture can be accessed at https://github.com/Forestry-Robotics-UC/fruc_dataset_apparatus, accessed on 12 June 2023.

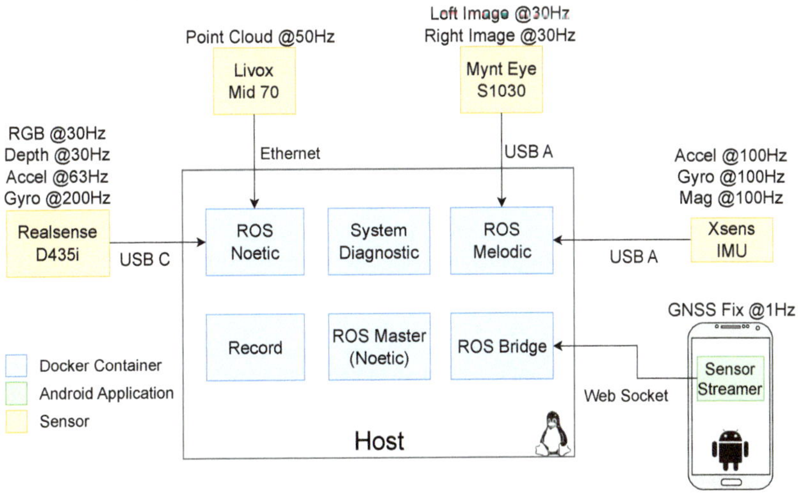

Figure 3. Complete system architecture, highlighting the different sensors, devices and Docker containers.

5. Experimental Procedure

A dataset [4] was collected at the Choupal National Woods (40°13′13.3″ N; 8°26′38.1″ W) in Coimbra, Portugal, with the specific aim of evaluating the apparatus' effectiveness in challenging outdoor conditions. The dataset was collected on a sunny day, where the user performed two circular loop laps, amounting to a distance of approximately 800 m (as depicted in Figure 4) with a duration of 15 min 33 s. Throughout the experiment, the smartphone collecting GNSS information was kept in a fixed position in relation to the apparatus, with the android application running in the foreground with the phone's screen actively on. The phone was connected to a network carrier, and the cellular data plan was activated to improve the Assisted GNSS estimation. The woodland environment featured a diverse range of visual elements, including tree trunks, trees, bushes, and leaves, providing rich features for evaluation.

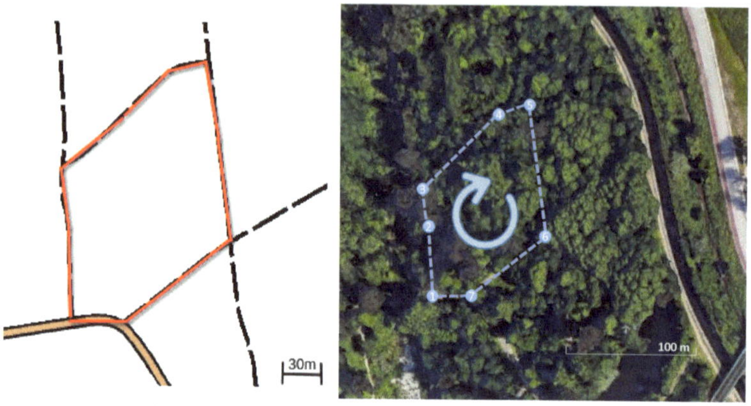

(**a**) (**b**)

Figure 4. Aerial/top views of the navigated path at the experimental site in the Choupal National Woods. (**a**) Route map [44]. (**b**) Satellite view (Google Earth).

The dataset entails a diverse range of ROS sensor data acquired at different frequencies. The stored data includes IMU readings, i.e., 3D orientation, angular velocity and linear acceleration; magnetic field readings and its internal temperature values (all at 99.26 Hz). The Livox LiDAR provides 3D point clouds at 49.88 Hz, while the smartphone provides Assisted GNSS (A-GNSS) fix data (latitude, longitude, etc.) at 1.05 Hz. Left and right monochromatic stereo images are obtained by the Mynt Eye camera at 19.82 Hz, while the Intel Realsense acquires RGB and depth images at 29.68 Hz, together with accelerometer (63.33 Hz) and gyroscope (197.90 Hz) measurements from the internal IMU. The data captured offer a comprehensive view of the environment, enabling extensive analysis and facilitating research and development for numerous applications.

In order to assess the quality of the collected dataset, we employ different SLAM methods to obtain reliable localization data. Localization plays a critical role in various applications that involve navigation, such as map building and transversability analysis. It serves as the foundation for perception-based algorithms to operate effectively, and therefore, it is paramount to provide localization information alongside raw data in a dataset to increase the potential use cases. In addition, SLAM algorithms can also serve as a valuable means to evaluate the quality of the dataset. For instance, if an RGB feature-matching algorithm in SLAM performs well and successfully maps and recognizes loop closures, it suggests that the images captured by the RGB-D camera possess sufficient quality for other algorithms like metric-semantic mapping. The same logic can be applied for the LiDAR scans. The utilization of various types of SLAM methods and the subsequent evaluation of their performance enables us to gauge the dataset's overall quality and its suitability for a wide range of applications. This showcases the flexibility and usability potential of the dataset provided. For this, we focus on two prominent, distinct and proven open-source ROS-based SLAM algorithms: Cartographer and RTAB-Map. These are compared and evaluated based on a decoupled multimodal architecture, as illustrated in Figure 5.

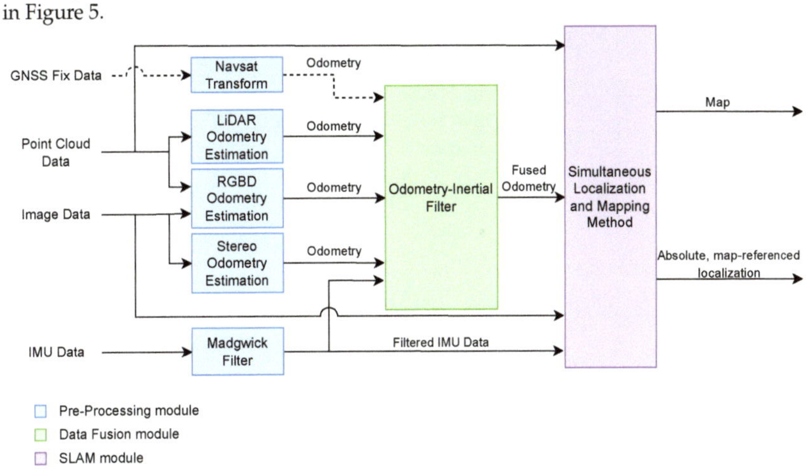

Figure 5. High-level diagram of the process to acquire a map of the environment and an absolute, map-referenced localization using the data from the available sensors. Dashed arrows represent optional connections.

Relevant raw data collected with the diverse sensors have been processed using dedicated ROS nodes to convert sensory information into odometry motion estimates. LiDAR odometry from the Livox Mid 70 is computed using the `livox_mapping` [45] ROS package, which extracts feature points from point clouds to obtain relative poses via frame matching and is especially tailored for the Livox LiDAR series.

Odometry estimation from the Intel Realsense D435i's RGB-D images is obtained using `rgbd_odometry`, which is part of the RTAB-Map ROS package. The node computes visual features and depth information from depth images; then, it applies feature correspondences between images and Random Sample Consensus (RANSAC) to extract the most likely transformation between consecutive images. For stereo images from the Mynt Eye S1030 camera, we use the `stereo_odometry` node, which is also included in the RTAB-Map ROS package. The node computes visual features extracted from the left images with their depth information computed by finding the same features on the right images. Then, it also uses feature correspondences and a RANSAC approach to extract the most likely transformation between the consecutive left images.

GNSS Fix data can also be converted into an odometry input for late fusion with the remaining estimates of relative pose. For this, one can use the `navsat_transform_node` included in the `robot_localization` ROS package. In our architecture, we set this input as optional, given the limited capabilities of GNSS in the woodland scenario where the experiments were performed due to tall trees and large canopies that can obstruct the line of sight between the GNSS receiver and the satellites, and introduce multipath interference, thus leading to a degradation of GNSS positioning.

The odometry estimates derived are then fused with the IMU measurements. For this, the angular velocities, linear accelerations, and magnetic readings from the XSens MTi IMU are firstly fused using a Madgwick Filter (see [46] and `imu_filter_madgwick` ROS package). This provides a 3D orientation estimate that is passed on as an additional input together with the odometry estimates into the data-fusion module to provide a 6D fused odometry input (3D position and 3D orientation) to the SLAM method.

The fusion node uses an Unscented Kalman Filter implementation from [47], which is available in the `robot_localization` ROS package. The resulting fused odometry, along with sensor point clouds and filtered IMU measurements feed either the Cartographer or the RTAB-Map SLAM methods to generate a map and an absolute map-referenced localization.

It is important to note, however, that the dataset collected as described above is unlabeled and also lacks ground-truth reference localization information. Therefore, a set of quantitative and qualitative metrics need to be defined for algorithm evaluation and/or training. For SLAM, in particular, since in this experiment, the user's start pose matches their final pose, the quantitative RTE and Relative Angular Error (RAE) metrics can be computed as:

$$\mathrm{RTE} = \sqrt{(x_{final} - x_{start})^2 + (y_{final} - y_{start})^2 + (z_{final} - z_{start})^2}, \quad (1)$$

and

$$\mathrm{RAE} = \sqrt{(\phi_{final} - \phi_{start})^2 + (\theta_{final} - \theta_{start})^2 + (\psi_{final} - \psi_{start})^2}, \quad (2)$$

where ϕ, θ, and ψ represent the Euler angles (roll, pitch, and yaw respectively).

To assess the alignment and consistency of the SLAM algorithm outputs with the real-world environment, a qualitative metric was employed. This metric involves overlaying the localization onto a map of the Choupal National Woods, which was generated by O-Solutions [44] through the human practice of cartography. This visual comparison provides clear insights into the alignment and consistency of the SLAM algorithms' outputs with respect to the real-world environment.

6. Results and Discussion

The experimental evaluation results reveal distinct characteristics of the two SLAM methods tested. Cartographer and RTAB-Map employ different sensor modalities to estimate their respective localizations. RTAB-Map utilizes Odometry, IMU, and RGB-D information, while Cartographer leverages Odometry, IMU, and LiDAR data. The versatility of the recorded dataset and the apparatus becomes evident in supporting these diverse sensor configurations, showcasing its adaptability and effectiveness to accommodate SLAM approaches with distinct assumptions.

Figures 6 and 7 illustrate the localization results of adopting the SLAM techniques along the navigated path followed by the user while carrying the apparatus on his back. It can be observed that both methods are able to continuously localize the system with different levels of success when applied to the data collected.

Since the methods are based on GraphSLAM, special attention is placed on loop closure identification and its impact on the overall results. RTAB-Map uses visual features for loop closure, while Cartographer is supported by LiDAR-based loop closure. Both approaches successfully identify loop closures along the path (see Figures 6 and 7), which helps to significantly reduce the RTE and RAE for the two methods, as shown in Table 2. This indicates that the dataset has enough distinct features from multiple modalities for the SLAM implementations to recognize previously visited locations and enhance the overall mapping accuracy. However, the RTAB-Map with loop closure exhibits a noticeable negative variation along the Z-axis, despite the dataset being collected in a relatively flat terrain. This is not observed when RTAB-Map operates without loop closure, as illustrated in Figure 7c. This suggests that the Z-axis variation issue is primarily related to the back-end graph optimization of the SLAM algorithm. In woodland and forest scenarios, visual similarities and locations that look alike are common, which may cause false positives in loop closure. This can also be caused by sensor noise and/or lighting changes, and it is more likely to occur if visual features are used, as in the case of RTAB-Map. Moreover, we made use of the default graph optimization approach GTSAM [48] in RTAB-Map. Yet, it also supports TORO [19] and g2o [18]. We hypothesize that further tuning of the back-end parameters could eventually enhance the Z-axis variation and the overall trajectory accuracy for RTAB-Map. In contrast, Cartographer demonstrates higher localization precision, with the back-end successfully optimizing large sparse pose graphs with the Ceres Solver [49] while keeping the trajectory on leveled ground (refer to Figure 7c). As a result, it achieves superior overall localization performance as shown in Figure 8 when compared to RTAB-Map, overlaying accurately on top of the reference route map from [44], as illustrated in Figure 8a.

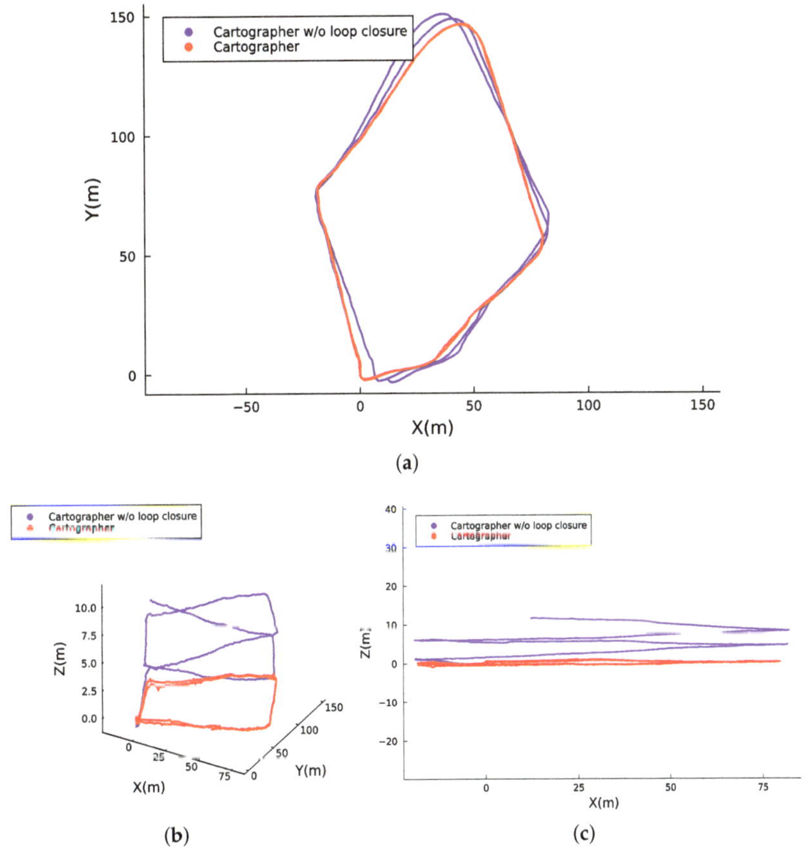

Figure 6. Absolute map referenced localization of the traveled path computed with and without loop closure by Cartographer. (**a**) Top view. (**b**) 3D isometric perspective. (**c**) Side view.

Table 2. Relative Pose Error (Translation and Rotation) of the SLAM methods tested at the same start and end pose.

Method	RTE (m)	RAE (rad)
RTAB-Map	0.089	0.25
RTAB-Map without loop closure	77.10	1.18
Cartographer	0.28	0.069
Cartographer without loop closure	17.10	0.21

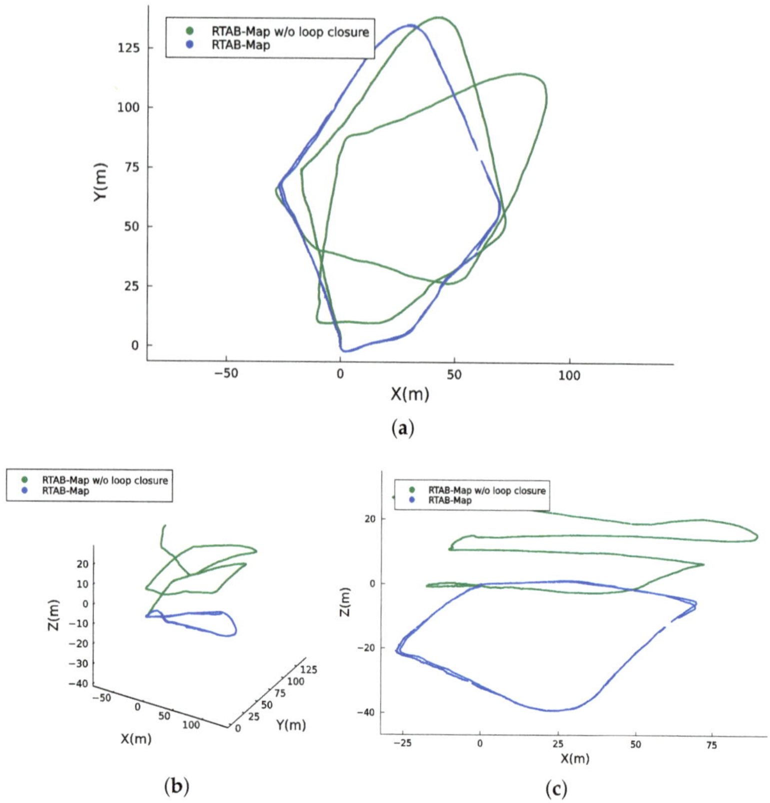

Figure 7. Absolute map referenced localization of the traveled path computed with and without loop closure by RTAB-Map. (**a**) Top view. (**b**) 3D isometric perspective. (**c**) Side view.

The A-GNSS data recorded from the smartphone did not yield satisfactory results in terms of accuracy and trajectory representation, as illustrated in Figure 9. While the two loops followed by the user are discernible, the overall shape of the trajectory is not consistent with the navigated path. The woodland environment, with its tall trees and dense canopies, poses significant challenges for acquiring reliable GNSS data. The obstructed visibility of satellites in such an environment hampers the quality and reliability of the A-GNSS measurements. Even though the proposed architecture supports GNSS input and this is provided in the dataset for further study, recognizing the sub-optimal nature of these measurements, we have decided not to include them in the sensor fusion node. This decision avoids the potential degradation of localization performance in both SLAM algorithms, as the incorporation of unreliable data could introduce errors and inconsistencies into the fusion process.

The maps built when executing the two SLAM methods are significantly different. RTAB-Map supports the generation of 3D point clouds [50] or a 3D octomap [51] of the environment, while Cartographer produces a 2D occupancy grid [52]. To allow for a side-by-side comparison, one can use the localization yielded by the SLAM method together with a mapping framework that reconstructs a comprehensive 3D RGB map utilizing the RGB-D information contained in the dataset. In this work, we employ these SLAM methods alongside the UFOMap 3D mapping framework [53] to build consistent and optimized 3D colored octree maps of the environment, as depicted in Figure 10. Comparing the generated maps with an image (Figure 10a) captured from a similar vantage point, it is evident that both maps exhibit consistency. The distinct features of the path are clearly discernible amidst the surrounding vegetation, including small bushes and grass, highlighting the accuracy and fidelity of the reconstructed map.

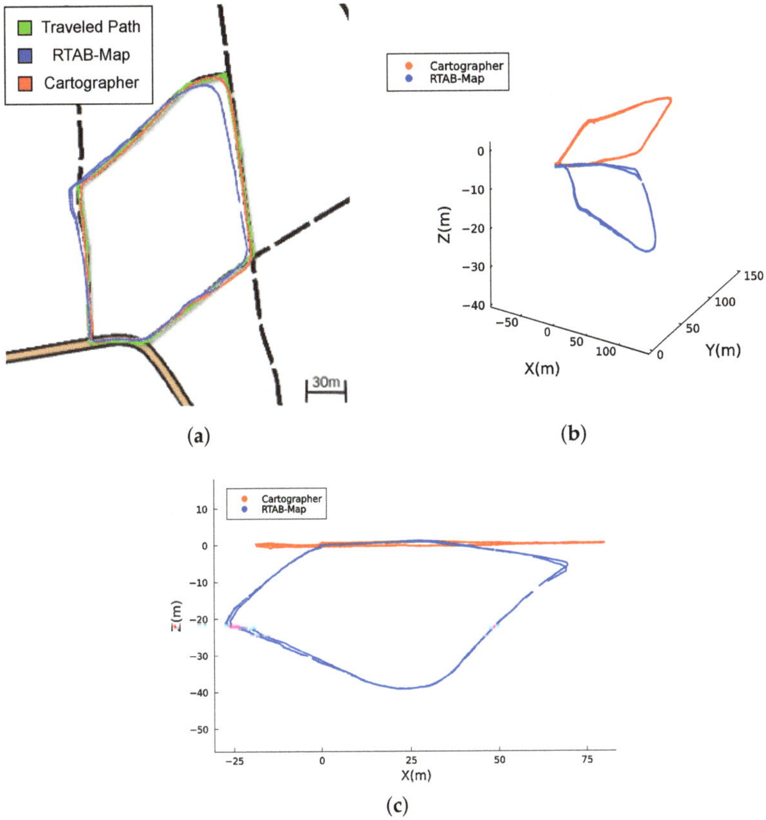

Figure 8. Absolute map referenced localization of the traveled path computed with loop closure by Cartographer and RTAB-Map. (**a**) Overlay of RTAB-Map and Cartographer localization with the route map of Figure 4a. (**b**) 3D isometric perspective. (**c**) Side view.

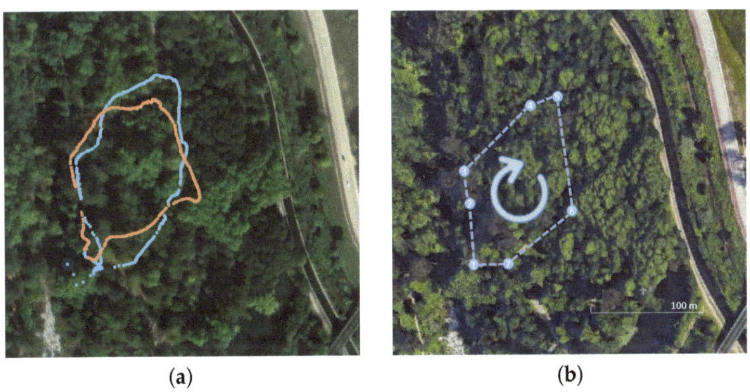

Figure 9. Assisted GNSS (A-GNSS) positioning from a smartphone integrated in the apparatus. (**a**) A-GNSS data during first (blue) and second (orange) laps of the navigated path (Google Maps). (**b**) Satellite view (Google Earth) of the navigated path. Replicated from Figure 4b for comparison.

Figure 10. Octree maps generated by using each method's localization with loop closure correction and the RGB-D outputs as inputs to the UFOMap package [53]. (**a**) Reference RGB image. (**b**) Scene reconstruction using Cartographer. (**c**) Scene reconstruction using RTAB-Map.

In the above discussion, we have used the RGB-D camera data to build the 3D colored octree map of the environment. However, another representation of the environment can be accomplished by registering the colorless point clouds obtained from the 3D LiDAR sensor using the localization data derived from the SLAM methods. This is illustrated in Figure 11. This approach allows for the inclusion of more distant features in the map due to the extended range of the LiDAR. The resulting map is consistent and provides clear identification of salient features such as trees and the borders of the path. When examining the map from a close-up of a 3rd person view of the apparatus shown in Figure 11a, individual trees are reconstructed with a high level of detail, showcasing the dataset's ability to capture fine details. Figure 11b,c provide isometric perspectives on the complete maps generated by the Cartographer and RTAB-Map localizations, respectively. Although both maps demonstrate consistency, it is evident that the Cartographer map exhibits significantly sharper details as the features on Figure 11c appear more blurred. A video of the data acquired with the generation of the maps using the dataset is available at https://youtu.be/9EXIwiExvWs, accessed on 16 June 2023. This detailed and precise representation of the environment enables researchers to gain valuable insights into the environment, facilitating tasks such as path planning and traversability analysis, and further analysis of the forest landscape, for instance through semantic segmentation or metric-semantic mapping.

As a final, bonus example going beyond our use case scenario, Figure 12 shows the mechanical effort-based traversability technique proposed by Carvalho et al. [54] running with the data provided in our dataset. It uses point clouds to infer terrain gradient and the location of obstacles in space, and from there, it generates a global 2D costmap with mechanical effort information to guide the agent from one place to another in a way that minimizes the mechanical effort it is subject to and potentially its energy/fuel consumption.

The rich and accurate data extraction procedure also highlights the utility of acquiring multimodal datasets with the proposed apparatus, enabling a deeper understanding of the forest's structural characteristics through robust perception capabilities.

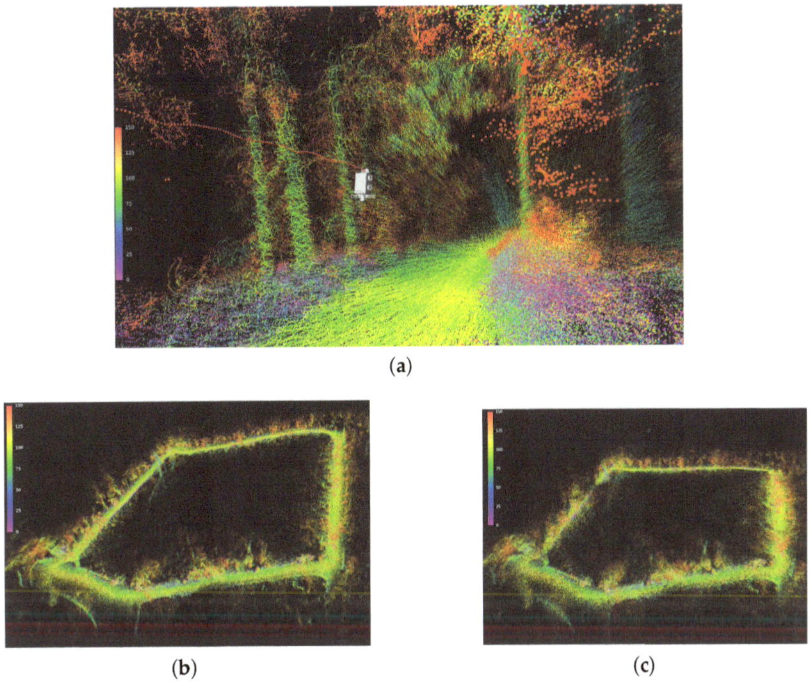

Figure 11. Livox LiDAR point clouds registered into a map using the 6D localization computed by Cartographer and RTAB-Map, with both methods using loop closure correction. In this representation, colors represent the LiDAR intensity, i.e., the strength of the backscattered echo at each point. (**a**) Close-up view of registered point clouds while traversing the navigated path with localization extracted from Cartographer. (**b**) 3D reconstruction of travelled path in isometric perspective using Cartographer localization. (**c**) 3D reconstruction of travelled path in isometric perspective using RTAB-Map localization.

Figure 12. Mechanical effort costmap generated by the method presented in [54] with data from our dataset, in which lighter values in the grayscale represent easier to traverse areas and vice versa.

7. Conclusions

In this study, we propose the development of a portable, lightweight and inexpensive apparatus for collecting multisensory data considering the requirements for forest and woodland environments while also allowing for the collection of datasets in any type of environment. Through experimental evaluation, providing insights into the performance of state-of-the-art SLAM implementations on the collected data, we demonstrate the versatility, feasibility and potential of the proposed approach in facilitating the planning, testing, and deployment of autonomous robots for forestry maintenance.

The dataset generated by the multi-sensor apparatus, openly available in [4], presents a contribution to the field of forestry robotics, as it provides the bulk of data required by researchers to analyze forest environments in depth, obtain an *a priori* map for robot operations, and label and train segmentation algorithms. The novel architecture developed for recording and storing the dataset provides a modular and user-friendly solution for acquiring extensive and dense datasets, seamlessly integrating into diverse platforms with various sensor combinations. Moreover, we also contribute to the community with an Android mobile app implementation, available in [5], which delivers GNSS/A-GNSS data for ROS systems out-of-the-box.

This opens up new possibilities for a more widespread adoption of autonomous robots in the field, improving the efficiency of data acquisition and reducing costs associated with automated vehicle deployment. Our study lays the foundation for future research and development in autonomous forestry maintenance, ultimately leading to safer and more efficient practices in forestry management.

Looking ahead, we plan to integrate a GNSS Real-Time Kinematic (RTK) station in our multi-sensor apparatus to facilitate reliable comparison between localization and/or SLAM algorithms. GNSS-RTK can deliver absolute gold standard positioning with centimeter-level precision, which is particularly valuable for localization-dependent algorithms, enabling more precise and refined results in these areas of research. Future datasets will be collected in various forest environments, with a particular focus on locations that have significant terrain variations. The current dataset lacks annotated images, but this limitation can be turned into an opportunity for users to apply their domain knowledge and expertise in annotating the images according to their specific needs, making the dataset more adaptable.

Author Contributions: Conceptualization, M.P.C., D.P. and A.E.C.; methodology, M.P.C., D.P. and J.F.F.; software, M.P.C.; validation, M.P.C.; formal analysis, M.P.C., D.P. and J.F.F.; investigation, M.P.C.; resources, D.P. and J.F.F.; data curation, M.P.C.; writing—original draft preparation, M.P.C.; writing—review and editing, M.P.C., D.P., A.E.C. and J.F.F.; visualization, M.P.C.; supervision, D.P., A.E.C. and J.F.F.; project administration, D.P. and J.F.F.; funding acquisition, D.P. and J.F.F. All authors have read and agreed to the published version of the manuscript.

Funding: This work was funded by the Programa Operacional Regional do Centro, Portugal 2020, European Union FEDER, research project Semi-Autonomous Robotic System For Forest Cleaning and Fire Prevention (SafeForest, ref. CENTRO-01-0247-FEDER-045931), co-funded by the Agência Nacional de Inovação within the CMU Portugal Large Scale Collaborative Projects program. Support for this research was also provided by the Fundação para a Ciência e a Tecnologia (Portuguese Foundation for Science and Technology) through the Carnegie Mellon Portugal Affiliated Ph.D. Program under Grant PRT/BD/153920/2022, and through the Scientific Employment Stimulus 5th Edition, under contract 2022.05726.CEECIND.

Institutional Review Board Statement: Not applicable.

Informed Consent Statement: Not applicable.

Data Availability Statement: The data presented in this study are openly available in the Zenodo repository. DOI: 10.5281/zenodo.8139205.

Acknowledgments: The authors thank M. Eduarda Andrada for her insights and feedback, and they gratefully acknowledge Paulo Peixoto, Rui P. Rocha, Jorge Lobo and Mahmoud Tavakoli for conceding some of the materials used in the apparatus.

Conflicts of Interest: The authors declare no conflict of interest.

Abbreviations

The following abbreviations are used in this manuscript:

A-GNSS	Assisted Global Navigation Satellite System
ABS	Acrylonitrile Butadiene Styrene
APE	Absolute Position Error
CAD	Computer-Aided Design
CPU	Central Processing Unit
DBH	Diameter at Breast Height
FoV	Field of View
g2o	General Graph Optimization
GNSS	Global Navigation Satellite System
GPS	Global Positioning System
GraphSLAM	Graph-based Simultaneous Localization and Mapping
GTSAM	Georgia Tech Smoothing and Mapping
ICP	Iterative Closest Point
IMU	Inertial Measurement Unit
LIO	LiDAR-Inertial Odometry
LIO-SAM	LiDAR-Inertial Odometry with Smoothing and Mapping
LiDAR	Light Detection And Ranging
LeGO-LOAM	Lightweight and Ground-Optimized LiDAR Odometry and Mapping
LOAM	LiDAR Odometry and Mapping
PETG	Polyethylene Terephthalate Glycol
RANSAC	Random Sample Consensus
RGB-D	Red, Green, Blue and Depth channels/sensor
ROS	Robot Operating System
RTAB-Map	Real-Time Appearance-Based Mapping
RAE	Relative Angular Error
RTE	Relative Translation Error
RTK	Real-Time Kinematics
SC-LeGO-LOAM	Scan Context Lightweight and Ground-Optimized LOAM
SLAM	Simultaneous Localization And Mapping
SSD	Solid-State Drive
TORO	Tree-based netwORk Optimizer
UFOMap	Unknown Free Occupied Map

References

1. Portugal, D.; Andrada, M.E.; Araújo, A.G.; Couceiro, M.S.; Ferreira, J.F. ROS Integration of an Instrumented Bobcat T190 for the SEMFIRE Project. *Robot. Oper. Syst. Ros Complet. Ref.* **2021**, *6*, 87–119.
2. Jelavic, E.; Jud, D.; Egli, P.; Hutter, M. Robotic Precision Harvesting: Mapping, Localization, Planning and Control for a Legged Tree Harvester. *Field Robot.* **2022**, *2*, 1386–1431. [CrossRef]
3. Ferreira, J.F.; Portugal, D.; Andrada, M.E.; Machado, P.; Rocha, R.P.; Peixoto, P. Sensing and Artificial Perception for Robots in Precision Forestry—A Survey. *Forests* 2023, *in press*.
4. Cristóvão, M. FRUC Multiple Sensor Forest Dataset Including Absolute, Map-Referenced Localization, 2023. Available online: https://zenodo.org/record/8139205 (accessed on 13 June 2023).
5. Cristóvão, M. ROS Streaming Sensors (Android App). 2022. Available online: https://github.com/mjpc13/SensorStreamer (accessed on 13 June 2023).
6. Thrun, S.; Burgard, W.; Fox, D. *Probabilistic Robotics*; Intelligent Robotics and Autonomous Agents; MIT Press: Cambridge, MA, USA, 2005; pp. I–XX, 1–647.

7. Santos, J.M.; Couceiro, M.S.; Portugal, D.; Rocha, R.P. Fusing sonars and LRF data to perform SLAM in reduced visibility scenarios. In Proceedings of the 2014 IEEE International Conference on Autonomous Robot Systems and Competitions (ICARSC), Espinho, Portugal, 14–15 May 2014; pp. 116–121.
8. Oveland, I.; Hauglin, M.; Giannetti, F.; Schipper Kjørsvik, N.; Gobakken, T. Comparing Three Different Ground Based Laser Scanning Methods for Tree Stem Detection. *Remote Sens.* **2018**, *10*, 538. [CrossRef]
9. Proudman, A.; Ramezani, M.; Fallon, M. Online Estimation of Diameter at Breast Height (DBH) of Forest Trees Using a Handheld LiDAR. In Proceedings of the 2021 European Conference on Mobile Robots (ECMR), Bonn, Germany, 31 August–3 September 2021; pp. 1–7.
10. Su, Y.; Guo, Q.; Jin, S.; Guan, H.; Sun, X.; Ma, Q.; Hu, T.; Wang, R.; Li, Y. The Development and Evaluation of a Backpack LiDAR System for Accurate and Efficient Forest Inventory. *IEEE Geosci. Remote Sens. Lett.* **2021**, *18*, 1660–1664. . [CrossRef]
11. Sturm, J.; Engelhard, N.; Endres, F.; Burgard, W.; Cremers, D. A benchmark for the evaluation of RGB-D SLAM systems. In Proceedings of the 2012 IEEE/RSJ International Conference on Intelligent Robots and Systems, Vilamoura-Algarve, Portugal, 7–12 October 2012; pp. 573–580. [CrossRef]
12. Sier, H.; Qingqing, L.; Xianjia, Y.; Queralta, J.P.; Zou, Z.; Westerlund, T. A Benchmark for Multi-Modal Lidar SLAM with Ground Truth in GNSS-Denied Environments. *arXiv* **2022**, arXiv:2210.00812.
13. Xu, W.; Zhang, F. FAST-LIO: A Fast, Robust LiDAR-Inertial Odometry Package by Tightly-Coupled Iterated Kalman Filter. *IEEE Robot. Autom. Lett.* **2021**, *6*, 3317–3324. . [CrossRef]
14. Faitli, T.; Hakala, T.; Kaartinen, H.; Hyyppä, J.; Kukko, A. Real-Time Lidar-Inertial Positioning and Mapping for Forestry Automation. *Int. Arch. Photogramm. Remote Sens. Spat. Inf. Sci.* **2023**, *XLVIII-1/W1-2023* , 145–150. . [CrossRef]
15. Li, J.; Wu, W.; Yang, B.; Zou, X.; Yang, Y.; Zhao, X.; Dong, Z. WHU-Helmet: A Helmet-Based Multisensor SLAM Dataset for the Evaluation of Real-Time 3-D Mapping in Large-Scale GNSS-Denied Environments. *IEEE Trans. Geosci. Remote Sens.* **2023**, *61*, 1–16. [CrossRef]
16. Labbé, M.; Michaud, F. RTAB-Map as an open-source lidar and visual simultaneous localization and mapping library for large-scale and long-term online operation. *J. Field Robot.* **2019**, *36*, 416–446. [CrossRef]
17. Dellaert, F.; Contributors, G. Borglab/gtsam. 2022. Available online: https://zenodo.org/record/7582634 (accessed on 13 June 2023).
18. Kummerle, R.; Grisetti, G.; Strasdat, H.; Konolige, K.; Burgard, W. G^2o: A general framework for graph optimization. In Proceedings of the 2011 IEEE International Conference on Robotics and Automation, Shanghai, China, 9–13 May 2011; pp. 3607–3613. [CrossRef]
19. Grisetti, G.; Stachniss, C.; Burgard, W. Nonlinear Constraint Network Optimization for Efficient Map Learning. *IEEE Trans. Intell. Transp. Syst.* **2009**, *10*, 428–439. [CrossRef]
20. Shan, T.; Englot, B. LeGO-LOAM: Lightweight and Ground-Optimized Lidar Odometry and Mapping on Variable Terrain. In Proceedings of the IEEE/RSJ International Conference on Intelligent Robots and Systems (IROS), Madrid, Spain, 1–5 October 2018; pp. 4758–4765.
21. Kim, G.; Kim, A. Scan Context: Egocentric Spatial Descriptor for Place Recognition Within 3D Point Cloud Map. In Proceedings of the IEEE/RSJ International Conference on Intelligent Robots and Systems (IROS), Madrid, Spain, 1–5 October 2018. [CrossRef]
22. Ye, H.; Chen, Y.; Liu, M. Tightly Coupled 3D Lidar Inertial Odometry and Mapping. In Proceedings of the 2019 IEEE International Conference on Robotics and Automation (ICRA), Montreal, QC, Canada, 20–24 May 2019.
23. Xu, W.; Cai, Y.; He, D.; Lin, J.; Zhang, F. FAST-LIO2: Fast Direct LiDAR-inertial Odometry. *arXiv* **2021**, arXiv:2107.06829.
24. Shan, T.; Englot, B.; Meyers, D.; Wang, W.; Ratti, C.; Daniela, R. LIO-SAM: Tightly-coupled Lidar Inertial Odometry via Smoothing and Mapping. In Proceedings of the IEEE/RSJ International Conference on Intelligent Robots and Systems (IROS), Las Vegas, NV, USA, 24 October 2020–24 January 2021; pp. 5135–5142.
25. Reinke, A.; Palieri, M.; Morrell, B.; Chang, Y.; Ebadi, K.; Carlone, L.; Agha-Mohammadi, A.A. LOCUS 2.0: Robust and Computationally Efficient Lidar Odometry for Real-Time 3D Mapping. *IEEE Robot. Autom. Lett.* **2022**, *7*, 9043–9050. [CrossRef]
26. Lin, J.; Zhang, F. R 3 LIVE: A Robust, Real-time, RGB-colored, LiDAR-Inertial-Visual tightly-coupled state Estimation and mapping package. In Proceedings of the 2022 International Conference on Robotics and Automation (ICRA), Philadelphia, PA, USA, 23–27 May 2022; pp. 10672–10678.
27. Yin, H.; Li, S.; Tao, Y.; Guo, J.; Huang, B. Dynam-SLAM: An Accurate, Robust Stereo Visual-Inertial SLAM Method in Dynamic Environments. *IEEE Trans. Robot.* **2022**, *39*, 289–308. [CrossRef]
28. Wang, Y.; Ma, H. mVIL-Fusion: Monocular Visual-Inertial-LiDAR Simultaneous Localization and Mapping in Challenging Environments. *IEEE Robot. Autom. Lett.* **2022**, *8*, 504–511. [CrossRef]
29. Yuan, Z.; Wang, Q.; Cheng, K.; Hao, T.; Yang, X. SDV-LOAM: Semi-Direct Visual-LiDAR Odometry and Mapping. *IEEE Trans. Pattern Anal. Mach. Intell.* **2023** . [CrossRef]
30. He, D.; Xu, W.; Chen, N.; Kong, F.; Yuan, C.; Zhang, F. Point-LIO: Robust High-Bandwidth Light Detection and Ranging Inertial Odometry. *Adv. Intell. Syst. 5* **2023**, 2200459. [CrossRef]
31. Vizzo, I.; Guadagnino, T.; Mersch, B.; Wiesmann, L.; Behley, J.; Stachniss, C. KISS-ICP: In Defense of Point-to-Point ICP Simple, Accurate, and Robust Registration If Done the Right Way. *IEEE Robot. Autom. Lett.* **2023**, *8*, 1029–1036. [CrossRef]
32. Karfakis, P.; Couceiro, M.S.; Portugal, D. NR5G-SAM: A SLAM Framework for Field Robot Applications based on 5G New Radio. *Sensors* **2023**, *23*, 5354. [CrossRef]

33. Hess, W.; Kohler, D.; Rapp, H.; Andor, D. Real-Time Loop Closure in 2D LIDAR SLAM. In Proceedings of the 2016 IEEE International Conference on Robotics and Automation (ICRA), Stockholm, Sweden, 16–21 May 2016; pp. 1271–1278.
34. FARO Technologies, Inc. GeoSLAM with ZEB1 Handheld SLAM Scanner. 2018. Available online: https://geoslam.com (accessed on 13 June 2023).
35. Wisth, D.; Camurri, M.; Fallon, M. Robust legged robot state estimation using factor graph optimization. *IEEE Robot. Autom. Lett.* **2019**, *4*, 4507–4514. [CrossRef]
36. Fankhauser, P.; Bloesch, M.; Hutter, M. Probabilistic terrain mapping for mobile robots with uncertain localization. *IEEE Robot. Autom. Lett.* **2018**, *3*, 3019–3026. [CrossRef]
37. Nuchter, A.; Lingemann, K.; Hertzberg, J.; Surmann, H. 6D SLAM with approximate data association. In Proceedings of the ICAR'05. Proceedings., 12th International Conference on Advanced Robotics, Seattle, WA, USA, 18–20 July 2005; pp. 242–249.
38. Li, Y.; Olson, E.B. Extracting general-purpose features from LIDAR data. In Proceedings of the 2010 IEEE International Conference on Robotics and Automation, Anchorage, Alaska, 3–8 May 2010; pp. 1388–1393.
39. Xiao, K.; Yu, W.; Liu, W.; Qu, F.; Ma, Z. High-Precision SLAM Based on the Tight Coupling of Dual Lidar Inertial Odometry for Multi-Scene Applications. *Appl. Sci.* **2022**, *12*, 939. [CrossRef]
40. Livox. LIO-Livox (A Robust LiDAR-Inertial Odometry for Livox LiDAR). 2021. Available online: https://github.com/uuumxx/lio-livox (accessed on 13 June 2023).
41. Zhang, J.; Singh, S. LOAM : Lidar Odometry and Mapping in real-time. In Proceedings of the Robotics: Science and Systems Conference (RSS), Berkeley, CA, USA, 12–16 July 2014; pp. 109–111.
42. Tiganis, B.; Burn, L.; Davis, P.; Hill, A. Thermal degradation of acrylonitrile–butadiene–styrene (ABS) blends. *Polym. Degrad. Stab.* **2002**, *76*, 425–434. [CrossRef]
43. Latko-Durałek, P.; Dydek, K.; Boczkowska, A. Thermal, Rheological and Mechanical Properties of PETG/rPETG Blends. *J. Polym. Environment* **2019**, *27*, 2600–2606. [CrossRef]
44. Rafael Miguel. O-Solutions—Choupal. 2019. Available online: https://o-solutions.pt/wp-content/uploads/2020/11/17_Choupal-Coimbra-2019_10_05_RM.png (accessed on 13 June 2023).
45. Livox. Livox Mapping. 2020. Available online: https://github.com/Livox-SDK/livox_mapping (accessed on 13 June 2023).
46. Madgwick, S.O.; Harrison, A.J.; Vaidyanathan, R. Estimation of IMU and MARG orientation using a gradient descent algorithm. In Proceedings of the 2011 IEEE international conference on rehabilitation robotics, Zurich, Switzerland, 29 June–1 July 2011; pp. 1–7.
47. Moore, T.; Stouch, D. A generalized extended kalman filter implementation for the robot operating system. In Proceedings of the Intelligent Autonomous Systems 13: Proceedings of the 13th International Conference IAS-13, Padova, Italy, 15–18 July 2014; Springer: Berlin/Heidelberg, Germany, 2016; pp. 335–348.
48. Dellaert, F. *Factor Graphs and GTSAM: A Hands-on Introduction*; Technical Report; Georgia Institute of Technology: Atlanta, GA, USA, 2012.
49. Agarwal, S.; Mierle, K.; Team, T.C.S. Ceres Solver. 2022. Available online: https://github.com/ceres-solver/ceres-solver (accessed on 13 June 2023)
50. Rusu, R.B.; Cousins, S. 3d is here: Point cloud library (pcl). In Proceedings of the 2011 IEEE International Conference on Robotics and Automation, Shanghai, China, 9–13 May 2011; pp. 1–4.
51. Hornung, A.; Wurm, K.M.; Bennewitz, M.; Stachniss, C.; Burgard, W. OctoMap: An efficient probabilistic 3D mapping framework based on octrees. *Auton. Robot.* **2013**, *34*, 189–206. [CrossRef]
52. Elfes, A. Using occupancy grids for mobile robot perception and navigation. *Computer* **1989**, *22*, 46–57. [CrossRef]
53. Duberg, D.; Jensfelt, P. UFOMap: An Efficient Probabilistic 3D Mapping Framework That Embraces the Unknown. *IEEE Robot. Autom. Lett.* **2020**, *5*, 6411–6418. [CrossRef]
54. Carvalho, A.E.; Ferreira, J.F.; Portugal, D. 3D Traversability Analysis and Path Planning Based on Mechanical Effort for UGVs in Forest Environments. *Robot. Auton. Syst.* **2023**. (Under Review).

sensors

Article

Commercial Optical and Acoustic Sensor Performances under Varying Turbidity, Illumination, and Target Distances

Fredrik Fogh Sørensen [†], Christian Mai [†], Ole Marius Olsen, Jesper Liniger and Simon Pedersen [*]

AAU Energy, Aalborg University, Niels Bohrs Vej 8, 6700 Esbjerg, Denmark; ffso@energy.aau.dk (F.F.S.); chrimai@energy.aau.dk (C.M.); omo@energy.aau.dk (O.M.O.); jel@et.aau.dk (J.L.)

* Correspondence: spe@energy.aau.dk

† These authors contributed equally to this work.

Abstract: Acoustic and optical sensing modalities represent two of the primary sensing methods within underwater environments, and both have been researched extensively in previous works. Acoustic sensing is the premier method due to its high transmissivity in water and its relative immunity to environmental factors such as water clarity. Optical sensing is, however, valuable for many operational and inspection tasks and is readily understood by human operators. In this work, we quantify and compare the operational characteristics and environmental effects of turbidity and illumination on two commercial-off-the-shelf sensors and an additional augmented optical method, including: a high-frequency, forward-looking inspection sonar, a stereo camera with built-in stereo depth estimation, and color imaging, where a laser has been added for distance triangulation. The sensors have been compared in a controlled underwater environment with known target objects to ascertain quantitative operation performance, and it is shown that optical stereo depth estimation and laser triangulation operate satisfactorily at low and medium turbidites up to a distance of approximately one meter, with an error below 2 cm and 12 cm, respectively; acoustic measurements are almost completely unaffected up to two meters under high turbidity, with an error below 5 cm. Moreover, the stereo vision algorithm is slightly more robust than laser-line triangulation across turbidity and lighting conditions. Future work will concern the improvement of the stereo reconstruction and laser triangulation by algorithm enhancement and the fusion of the two sensing modalities.

Keywords: sensor testing and evaluation; multiple-sensor systems; imaging sensors; acoustic sensors; sonar measurements; stereo vision; laser triangulation; illumination; turbidity

Citation: Sørensen, F.F.; Mai, C.; Olsen, O.M.; Liniger, J.; Pedersen, S. Commercial Optical and Acoustic Sensor Performances under Varying Turbidity, Illumination, and Target Distances. *Sensors* **2023**, *23*, 6575. https://doi.org/10.3390/s23146575

Academic Editors: David Cheneler and Stephen Monk

Received: 16 June 2023
Revised: 7 July 2023
Accepted: 18 July 2023
Published: 21 July 2023

1. Introduction

Just as in the case above water [1–3], a large variety of motivating applications and solution algorithms exist for the use of sensor information in many operational contexts, such as localization and inspection, including 2D/3D reconstruction of underwater objects and scenes [4]. Acoustic sensing is the premier sensing modality used in underwater environments due to the high speed of sound and low attenuation in water [5]. Simultaneously, many underwater sensing tasks such as inspection are advantageously performed using optical cameras because they deliver high sensing resolution and are easily interpreted by operators [6]. However, optical sensing is considerably affected by turbidity, attenuation, and lighting (both natural sunlight and artificial illumination), factors which do not significantly affect acoustic methods [7,8]. Hence, the sensing modalities have complementary advantages; combined sensing solutions lead to a robust solution which is often required for use in automated solutions, as noted in [9,10].

Given these complementary sensing effects, it is desirable to quantify the effects of environmental influences such as turbidity on sensing performance to elucidate the operational limitations for each sensing modality.

The contribution of this work is to reproduce and expand on previous works concerning the effect of environmental turbidity and lighting on target reconstruction by the precise control of target distance using a 3D servo-driven gantry; the recording of simultaneous stereo, color image, laser-triangulation, and acoustic imaging in the same controlled experiment; and quantitative evaluation of sensor noise and accuracy by conversion to real-world-unit point clouds for each sensor.

The hypotheses are that optical sensing accuracy will be negatively affected as a function of increasing turbidity; that an optimum illumination level that provides the best performance exists; and that it will break down when exceeding a certain turbidity and target distance; contrarily, the acoustic sensor should be negligibly affected by these environmental parameters.

The rest of the paper is organized as follows: firstly, related works are outlined; secondly, the materials and methods applied in the experiments are described, including the chosen commercial sensors and experimental facility; thirdly, the results from the sensor's raw measurements are evaluated for their operating limits and accuracy, with examples of measurements additionally illustrated; finally, the discussion summarizes the qualitative and quantitative behavior of the sensors.

2. Related Work

Previous investigations have focused on different objectives: for example, the reconstruction of undistorted and clear visual images from subsea images for the purposes of presentation to operators and the reconstruction of 2D/3D objects for the purposes of object detection, segmentation, classification, and structural damage detection [4] have been studied. Both qualitative and quantitative investigations of this nature have been performed in recent years.

In O'Byrne et al. [11], an image repository was created with various target objects under varying turbidites using a setup with two waterproof cameras to test stereo reconstruction algorithms. Some algorithms for 3D reconstruction and damage detection were demonstrated on this dataset in O'Byrne et al. [12–14]. Just as the case above water, structured light can be added to the scene to aid in reconstruction, demonstrated by Aykin et al. [15], Bruno et al. [16].

In Mai et al. [17], the fidelity was evaluated for high-frequency sonar, stereo vision, and time-of-flight (ToF) cameras of determining distance to and shape of a target object, with a focus on the comparison of sensor accuracy and noise. It was shown that stereo vision delivers the highest measurement fidelity, followed by the ToF camera; finally, sonar has the lowest measurement fidelity. A ToF camera was also investigated in Risholm et al. [18], wherein the camera used a range-gated strategy to successfully reduce backscatter from turbidity, in this case, to monitor fish in turbid environments.

An example of using optical and acoustic sensing modalities together is shown in Roman et al. [19], where a high-frequency sonar, stereo imaging, and a laser triangulation method were compared for archeological 3D measurements in the Aegean Sea. It was shown that the sensing modalities all provide useable fidelity in the given environment; however, turbidity and other environmental influences were not measured. In Yang et al. [20], the emphasis was on examining sharpness and color reproduction under varying turbidity and lighting conditions using a monocular color camera. A ColorChecker and SFR chart were used to estimate the image quality and color reproduction.

More recently, in Scott and Marburg [21], the quantitative effects of turbidity on various stereo reconstruction methods showed that stereo vision depth estimation is possible with usable robustness under low (17NTU) and medium (20NTU) turbidity conditions. Apart from inspection tasks, visual sensors can also be used for concurrent localization, such as those described in Concha et al. [22], where localization and dense mapping are demonstrated from a monocular camera sequence.

3. Materials and Methods

To perform the experiments, a commercial sensor was selected to embody each of the sensing modalities; then, these sensors were mounted in a rigid aluminum frame to fix the extrinsics between the sensors themselves and the target objects. First, we describe the selected sensors and their specifications; then, we describe the experimental setup, including the data acquisition and the selected target objects used in the performance evaluation; and finally, we describe the experimental procedure.

3.1. Sensors

For each sensing modality, a commercial-off-the-shelf (COTS) sensor was selected based on the maximum sensing distance which was used in the experiments, 2 m, while maintaining a high sensing fidelity under the given distance range. The stereo and color camera modalities were both embodied by the Intel D435i camera [23], and the acoustic modality was embodied by the BluePrint subsea M3000d sonar [24].

3.1.1. Stereo and Color Camera

A COTS stereo camera, the Intel D435i [23], embodied the optical sensing modality. This camera was chosen based on having a minimum 2 megapixel resolution color imager as well as on-board stereo imaging; in particular, it had built-in stereo depth estimation processing (to reduce the need for external computation in an end-use application). The stereo camera sensor specifications are given in Table 1. For the Intel D435i, the color imaging sensor was the OmniVision OV2740, while the stereo imaging sensors were OmniVision OV9282s. Since the stereo depth estimation is a built-in function of the camera, the main stereo-sensing specifications are listed in Table 2.

Table 1. Intel D435i imaging sensor manufacturer specifications in air.

Parameter	Stereo Imager	Color Imager
Resolution	1280 px × 800 px	1920 px × 1080 px
Shutter type	Global shutter	Rolling shutter
Data format	10-bit RAW	10-bit RAW RGB
Horizontal FOV	$91 \pm 1°$	$69 \pm 1°$
Vertical FOV	$66 \pm 1°$	$42 \pm 1°$
Diagonal FOV	$101 \pm 1°$	$77 \pm 1°$

Table 2. Intel D435i stereo depth estimation manufacturer specifications at a 2 m distance, recommended settings, in air.

Parameter	Value
Resolution	848 px × 480 px
Frame rate (max)	90 FPS
Data format	16-bit (1 mm/LSB)
Horizontal FOV	$86 \pm 3°$
Vertical FOV	$57 \pm 3°$
Diagonal FOV	$94 \pm 3°$
Min. distance	195 mm
Depth accuracy	$\leq$2%
RMS error	$\leq$2%
Temporal noise	$\leq$1%
Fill rate	$\geq$99%

3.1.2. Forward-Looking Imaging Sonar

The acoustic sensing modality was similarly embodied by a COTS forward-looking imaging sonar, the BluePrint subsea M3000d [24]; this sonar was selected for its small range resolution <1 cm and small angular resolution <1° with a suitable minimum distance of ≤0.1 m and a maximum distance of ≥1 m. The forward-looking imaging sonar sensing specifications are given in Table 3. For the purposes of this work, the sonar was used exclusively in the high-frequency mode shown on the right.

Table 3. Oculus m3000d sonar manufacturer specifications, * indicates range-dependent specification.

Common parameters		
Update rate max *	40 Hz	
Number of beams (max)	512	
Range (min)	0.1 m	
Vertical aperture	20°	
Mode parameters	**Low-frequency mode**	**High-frequency mode**
Operating frequency	1.2 MHz	3.0 MHz
Range (max)	30 m	5 m
Range resolution *	2.5 mm	2 mm
Horizontal aperture	130°	40°
Angular resolution	0.6°	0.4°
Beam separation	0.25°	0.1°

3.1.3. Laser-Line Augmentation

The laser specifications are given in Table 4. The laser was fitted with a line-generation lens immediately after the focusing lens and was mounted in a waterproof enclosure with a flat port acrylic window. The laser was focused at approximately 2 m and was mounted to be within view of the color camera at both the minimum and maximum test distances. The closest observable distance was determined by the intersection of the laser plane with the lower plane of the camera field-of-view (FOV), and the maximum distance was determined by the intersection with the upper plane of the FOV, as shown in Figure 1.

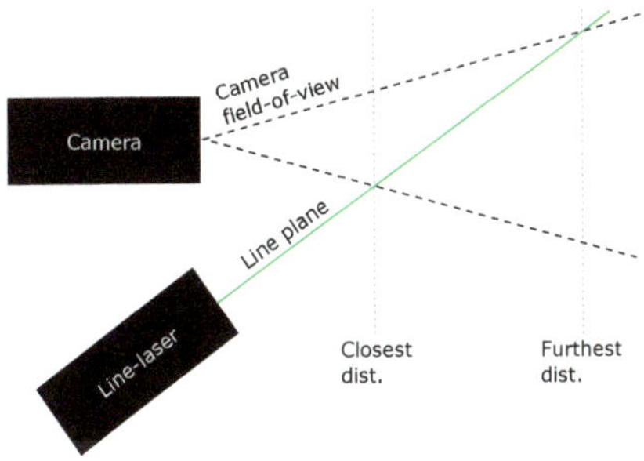

Figure 1. Laser (green) and camera field-of-view geometry (dashed lines).

Table 4. Laser-line specifications.

Parameter	Value
Laser type	Diode laser
Laser wavelength	532 nm
Beam class	Class 3B

3.1.4. Turbidity Sensor

The turbidity sensor was an optical nephelometric sensor, model Aanderaa Turbidity Sensor 4296 [25]. The sensor was mounted to measure the turbidity in the forward direction towards the target into an unoccluded volume to avoid reflections from the pool's interior surfaces and the water surface. The turbidity sensor's main specifications are given in Table 5.

Table 5. Aanderaa Turbidity Sensor 4296 [25].

Parameter	Value
Range	0 FTU to 25 FTU
Resolution	0.1%
Accuracy	±3% of range

3.2. Experimental Setup

The experimental setup consisted of three overall parts: the sensors being tested, the test pool filled with test medium, and the test targets mounted on a 3D gantry (traverse). To ensure the extrinsics were fixed between the sensors, they were mounted on a rigid frame made of aluminum profiles, shown in Figure 2.

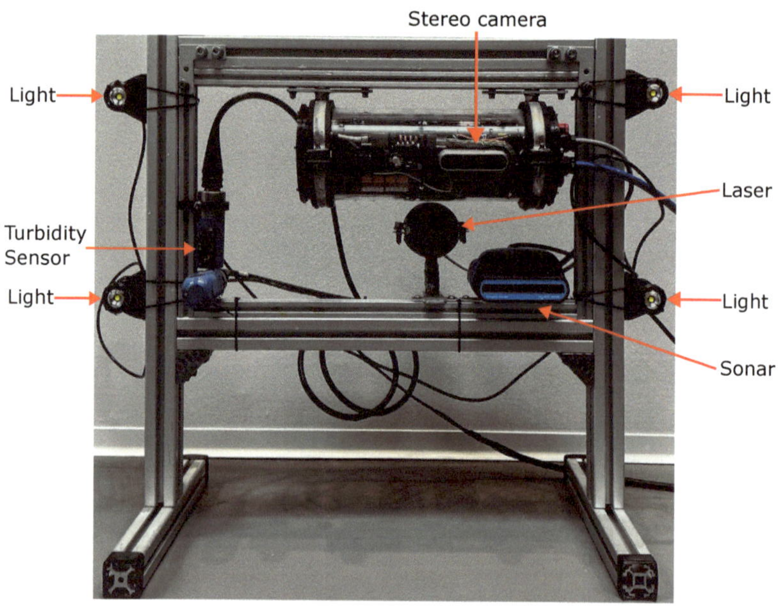

Figure 2. Image of the sensor frame. Upper right: stereo camera, lower right: sonar, center: laser, lower left: turbidity sensor; exterior of frame: 4 pcs. LED lights.

The test pool was filled with tap water, and Kaolin [26] was used to control the turbidity. The test targets were mounted on a 3D gantry which allowed them to be moved with respect to the sensor frame, such that the distance between the target and the sensor reference planes could be varied. The complete experimental setup is shown in Figure 3. To prevent disturbances from external light sources, the experiments were conducted in a laser-safety-rated laboratory where external lighting could be reduced to near zero levels.

Figure 3. Image of the experimental setup. Left: target object on the gantry. Right: sensor frame with lights. Lower right: mixer. Pool dimensions: 180 cm by 260 cm.

3.3. Target Objects

Two target objects were used during the experimental measurements: an ISO 12233:2017 edge spatial frequency response chart (eSFR chart) [27] used for image quality analysis—this eSFR target was printed in a 16:9 format and was printed with near-infrared and visible reflective inkjet technology—and a metal cylinder that resembles part of an offshore structure. The eSFR chart was glued to an aluminum sandwich backing plate and is shown in Figure 4a; the metal cylinder is shown in Figure 4b.

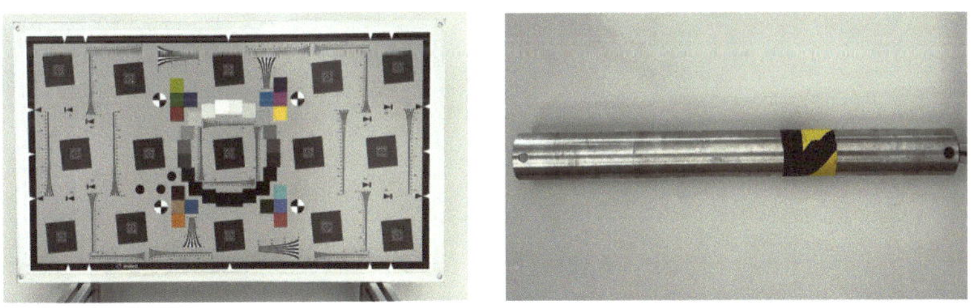

(**a**) (**b**)

Figure 4. Pictures of target objects, eSFR, and metal cylinder. (**a**) eSFR target object [27]. (**b**) Cylinder target object (vertical during experiments). Material: aluminium, diameter: 10 cm, wall thickness: 5 mm.

3.4. Data Acquisition

The data acquisition was performed using the Robot Operating System (ROS) Noetic built on Ubuntu 20.04, running on an NVIDIA Jetson Xavier NX, which is located within the stereo camera submersible enclosure. The Xavier NX was connected through serial communication (RS232) to the turbidity and conductivity sensors, by USB 3.1 to the stereo camera, and by gigabit ethernet to a switch outside the experimental tank. The Xavier NX and sensors were powered using power-over-ethernet (PoE) from the switch, apart from the sonar, which had a separate power supply and ethernet connection. The interconnection between the sensor components and the data capture equipment are shown in brief in Figure 5. See also Figure 2 for the physical layout of the sensors.

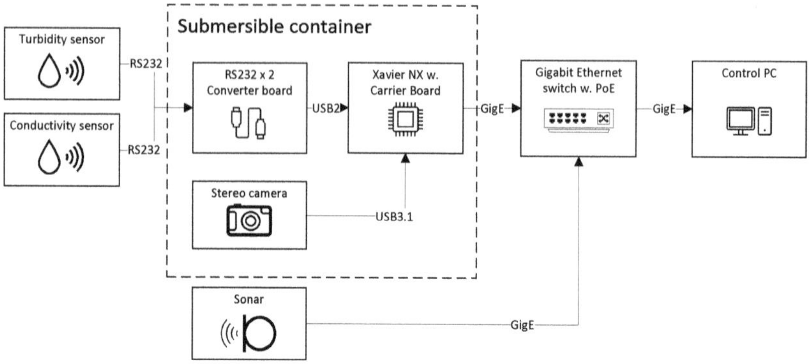

Figure 5. Block diagram of the data acquisition setup. The inner box denotes components in the transparent submersible container, wherein the stereo camera is mounted.

3.5. Experimental Parameters

The experiments were conducted at a set of turbidities, target distances, and illumination settings. Table 6a lists the desired and achieved turbidities for the experiment series, including the standard deviation as given by fluctuations in the turbidity sensor measurement. Table 6b lists the desired and achieved distances for the experiment series, including the measurement uncertainty; note that when transitioning to/from the far distances, the sensor frame was moved within the pool and the target distance was re-initialized using an external laser distance meter. The used lighting levels are shown in Table 6c.

Table 6. Experimental conditions.

(**a**) Experimental turbidities.

Desired Turbidity	Average Measured Turbidity with Standard Deviation
0 FTU	0.31 ± 0.02 FTU
1 FTU	1.03 ± 0.11 FTU
2 FTU	2.11 ± 0.22 FTU
6 FTU	5.99 ± 0.66 FTU

(**b**) Experimental target distances.

Set	Target Distances with Uncertainty
Close	43, 48, 53, 58, 63 cm $\pm$ 1.5‰
Medium	73, 83, 93, 103 cm $\pm$ 1.5‰
Far	140, 170, 200 cm $\pm$ 1.5‰

(**c**) Experimental lighting settings.

Light Settings
25, 50, 75, 100%

3.6. Experimental Procedure

The experiments were conducted using a repetitive procedure which is also illustrated in Figure 6. The inner loop corresponds to light level variations; the intermediate loop corresponds to distance variations; and the outer loop corresponds to turbidity variations.

The procedure was designed to have the least experimental disturbances during variable changes since light changes cause no physical movements, whereas the control of Kaolin content is additive in nature.

1. The sensor frame is placed within the test pool.
2. The test target is reset, and the base distance is measured using a laser distance meter.
3. Measurements are performed at each distance:
   (a) Measurements are performed at each light level:
      i. Light level is set at the selected percentage: see Table 6c
      ii. The experiment is allowed to settle for 10 s.
      iii. Sensor data are recorded in ROSbag format; then, point (i) is repeated.
   (b) The distance is changed by control of the gantry: see Table 6b; then, point (a) is repeated
4. Kaolin is added until the desired turbidity is reached—see Table 6a—as measured by the turbidity sensor; then, point 3 is repeated.

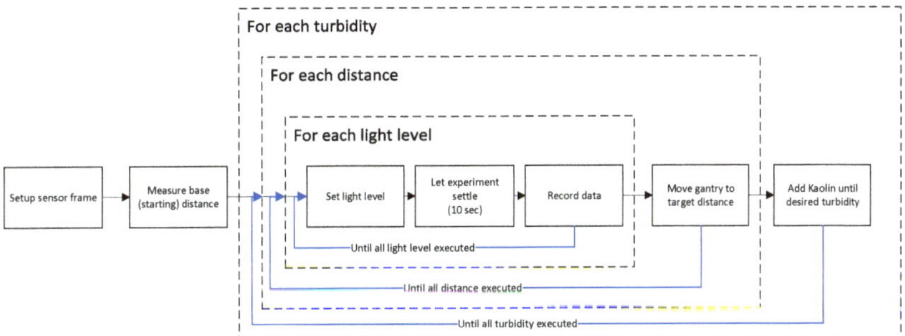

Figure 6. Block diagram of the experimental procedure. Black lines denote experimental flow, blue lines denote experimental repetitions.

4. Results

Using the ROSbags generated through the experiments, the performance of three sensing modalities has been evaluated: stereo depth estimation based on the built-in algorithm of the Intel camera—see Appendix A.3, laser triangulation implemented through the color camera and the MATLAB triangulation algorithm—see Appendix A.1; and the high-frequency imaging sonar—see Appendix A.2. For all modalities, the measurement accuracy has been analyzed through MATLAB, as described in the Appendix A.

4.1. Illumination Effects

The light level naturally influences the results for the optical methods, influencing both stereo depth estimation and laser-line triangulation. By review of the sensor measurements, it is evident that for both visual methods, the optimal light level in the experiments is 50%, with an example illustrated in Figure 7b. Less illumination, 25%, results in less clear features for stereo estimation and increased laser glare, shown in Figure 7a, while illumination levels of 75% to 100%, shown in Figure 7c,d, results in reduced contrast for the laser as well as increased backscatter, which reduces visual features in the resulting images.

(a) (b) (c) (d)

Figure 7. Color images of eSFR target at 1.03 m under varying light levels. (**a**) RGB image at light level of 25%, (**b**) RGB image at light level of 50%, (**c**) RGB image at light level of 75%, (**d**) RGB image at light level of 100%.

4.2. Stereo Depth Estimation

For the stereo camera, the performance has been evaluated for a rectangular region of interest (ROI) in the central 20% of the depth image frame, as illustrated in Figure 8. To determine the operational limits, the cut-off for valid distance measurements has been set at 50% valid pixels within the ROI, i.e., a pixel fill rate of >50% is considered as valid. The measurement accuracy as analyzed with Appendix A.3 is shown in Table 7 and Figures 9–11 while an example of the depth image is illustrated in Figure 8. Note how the background of the pool is still estimated at 0.3 FTU, Figure 8a, but begins to disappear at 2.1 FTU, Figure 8b, while the target remains valid in both cases. For the cylinder geometry estimation, the results show a very high deviation, which most likely stems from an insufficient quality of stereo intrinsic calibration, since it is evident that the eSFR plate behind the cylinder is also heavily distorted, as illustrated in Figure 12.

Table 7. Stereo depth accuracy and operation limits; dash "-" denotes no valid measurement. Δ denotes mean deviation from ground truth distance, shown with ±standard devation.

Target Dist.	Δ at 0.3 FTU	Δ at 1.0 FTU	Δ at 1.4 FTU	Δ at 2.1 FTU	Δ at 6.0 FTU
43.00 cm ± 1.5‰	11.30 (1.16) cm	11.80 (2.12) cm	11.76 (1.38) cm	11.58 (1.00) cm	12.12 (0.68) cm
48.00 cm ± 1.5‰	12.18 (1.27) cm	12.62 (1.17) cm	12.48 (0.93) cm	12.22 (0.90) cm	12.72 (0.91) cm
53.00 cm ± 1.5‰	12.42 (1.82) cm	12.78 (1.09) cm	12.70 (1.03) cm	12.64 (1.08) cm	13.34 (1.38) cm
58.00 cm ± 1.5‰	13.16 (1.41) cm	13.10 (1.33) cm	13.22 (1.11) cm	13.24 (1.30) cm	13.74 (1.60) cm
63.00 cm ± 1.5‰	13.36 (1.43) cm	13.74 (1.58) cm	13.64 (1.34) cm	13.64 (1.37) cm	13.92 (2.13) cm
73.00 cm ± 1.5‰	13.22 (11.08) cm	13.84 (1.87) cm	14.44 (2.65) cm	14.06 (1.77) cm	-
83.00 cm ± 1.5‰	12.58 (11.68) cm	13.64 (2.48) cm	13.54 (2.28) cm	13.44 (4.15) cm	-
93.00 cm ± 1.5‰	13.00 (2.77) cm	13.00 (2.81) cm	13.56 (3.06) cm	12.80 (2.74) cm	-
103.00 cm ± 1.5‰	10.38 (3.02) cm	11.12 (3.11) cm	10.40 (3.28) cm	11.14 (3.04) cm	-
140.00 cm ± 1.5‰	1.22 (6.47) cm	3.52 (5.76) cm	−0.32 (7.01) cm	0.46 (9.62) cm	-
170.00 cm ± 1.5‰	−10.98 (8.41) cm	−10.58 (9.34) cm	−13.82 (14.72) cm	−15.52 (19.36) cm	-
200.00 cm ± 1.5‰	−36.20 (23.68) cm	−33.30 (14.78) cm	-	-	-

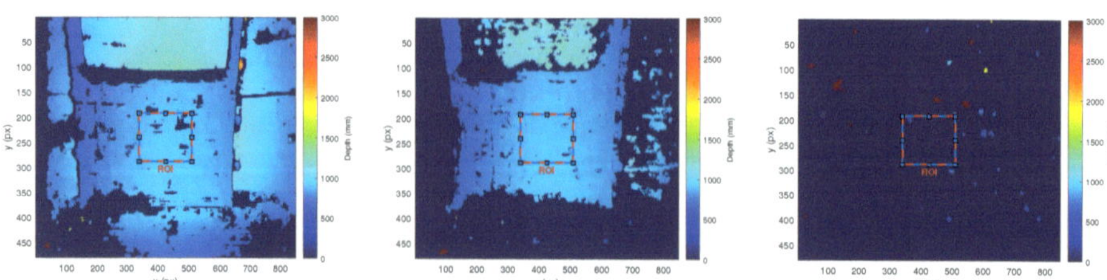

(a) (b) (c)

Figure 8. Depth images of eSFR target at 1.03 m. (**a**) Depth image of eSFR target at 0.3 FTU, (**b**) Depth image of eSFR target at 2.1 FTU, (**c**) Depth image of eSFR target at 6.0 FTU.

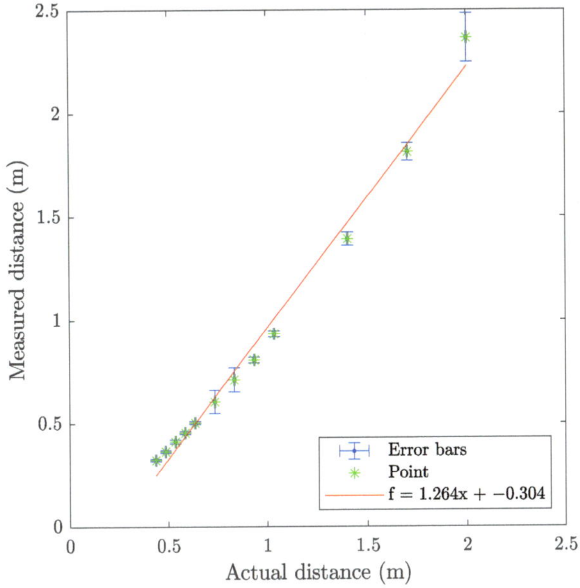

Figure 9. Graph showing the measuring accuracy at 0.3 FTU for stereo depth estimation.

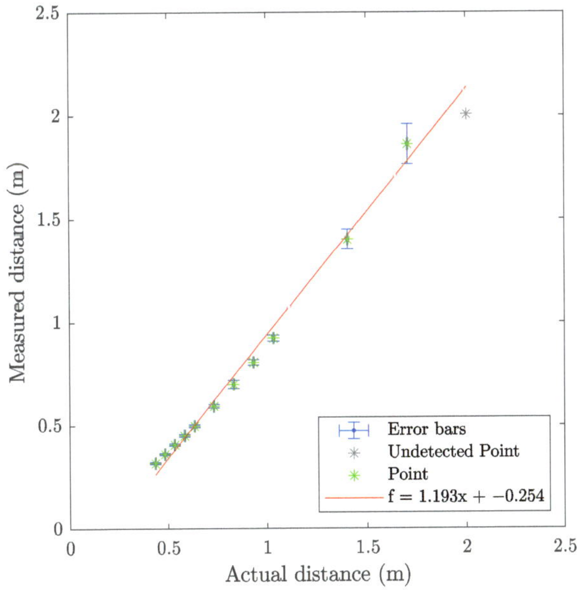

Figure 10. Graph showing the measuring accuracy at 2.1 FTU for imaging sonar.

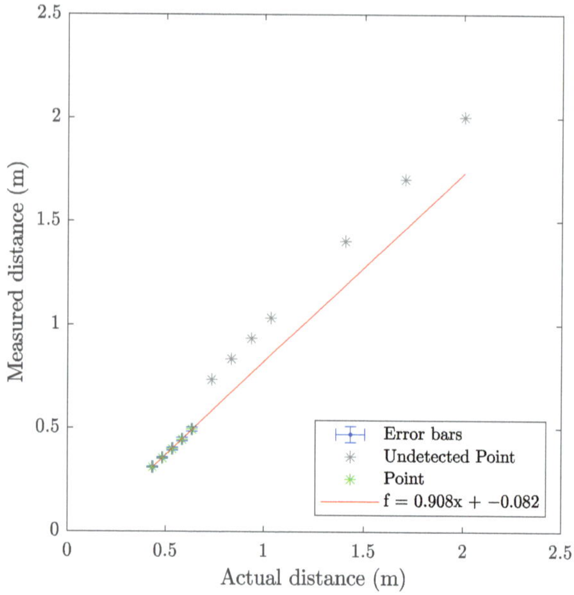

Figure 11. Graph showing the measuring accuracy at 6.0 FTU for imaging sonar.

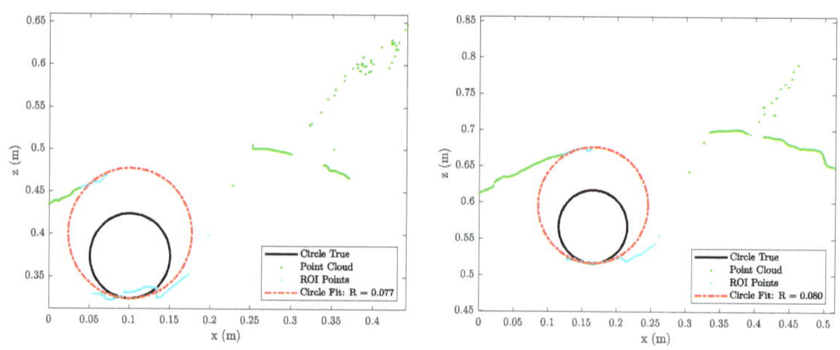

(**a**) (**b**)

Figure 12. Circlefits of cylinder target at 0.3 FTU. (**a**) Circlefit of cylinder target at 0.64 m to eSFR target. (**b**) Circlefit of cylinder target at 0.84 m to eSFR target.

4.3. *Laser Triangulation*

The laser triangulation is performed by detecting the laser-line and projection as described in Appendix A.1, with examples shown on Figure 13 and results shown in Table 8 and Figures 14–16. The laser triangulation has an accuracy of a single centimeter up to a range of about 50 cm, increasing to an error of 3 cm at a range of 200 cm. The behavior of the deviation over distance seems to indicate some remaining uncompensated error in the camera intrinsics calibration since the error is non-monotonic with respect to the target distance. The sensing functions up to a distance of 103 cm for turbidities of $\leq$2.1 FTU—see Figures 14 and 15—and drops to 43 cm at 6 FTU: only very close range sensing is possible at this high turbidity; see Figures 13c and 16. The laser-line is naturally much easier to detect due to the improved contrast at low turbidities, which is evident from Figure 13a,b. For the cylindrical target, the geometric reproduction accuracy is shown in Figure 17, where the detected circle has a radius close to the actual value of 5 cm; the main outliers stem from the specular reflection along the long axis of the cylinder. The deviation is increased as the distance to the cylinder target is increased, as shown in Figure 17c. Overall, the fidelity of the geometric reproduction is satisfactory at close distances.

Table 8. Laser triangulation accuracy and operation limits; dash "-" denotes no valid measurement. Δ denotes mean deviation from ground truth distance, shown with ±standard devation.

Target Dist.	Δ at 0.3 FTU	Δ at 1.0 FTU	Δ at 2.1 FTU	Δ at 6.0 FTU
43.00 cm ± 1.5‰	0.59 (0.70) cm	0.96 (0.73) cm	0.54 (0.30) cm	0.90 (0.22) cm
48.00 cm ± 1.5‰	0.33 (0.64) cm	0.49 (0.38) cm	0.61 (0.53) cm	-
53.00 cm ± 1.5‰	0.29 (0.89) cm	0.27 (0.42) cm	0.79 (1.19) cm	-
58.00 cm ± 1.5‰	0.20 (1.11) cm	0.28 (0.65) cm	0.21 (0.44) cm	-
63.00 cm ± 1.5‰	−0.05 (1.28) cm	−0.09 (0.40) cm	0.01 (0.84) cm	-
73.00 cm ± 1.5‰	−0.51 (1.30) cm	−0.62 (0.42) cm	−0.38 (3.11) cm	-
83.00 cm ± 1.5‰	−0.87 (1.14) cm	−1.00 (0.63) cm	−0.62 (0.49) cm	-
93.00 cm ± 1.5‰	−1.50 (0.73) cm	−1.40 (0.52) cm	−0.73 (0.71) cm	-
103.00 cm ± 1.5‰	−1.21 (2.00) cm	−1.54 (0.90) cm	−1.39 (0.98) cm	-
140.00 cm ± 1.5‰	0.42 (2.61) cm	-	-	-
170.00 cm ± 1.5‰	1.24 (2.79) cm	-	-	-
200.00 cm ± 1.5‰	2.64 (3.41) cm	-	-	-

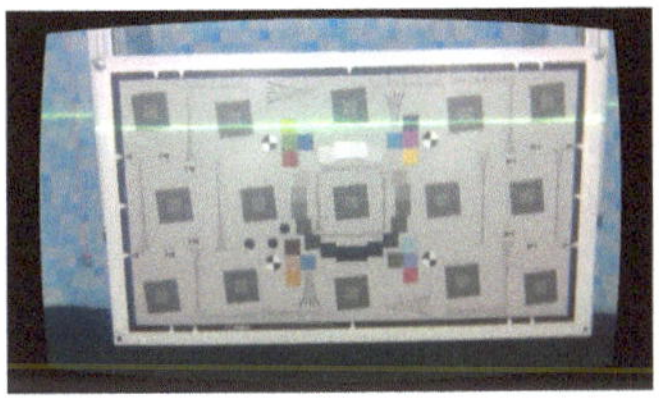

(**a**) Image of eSFR target at 0.3 FTU.

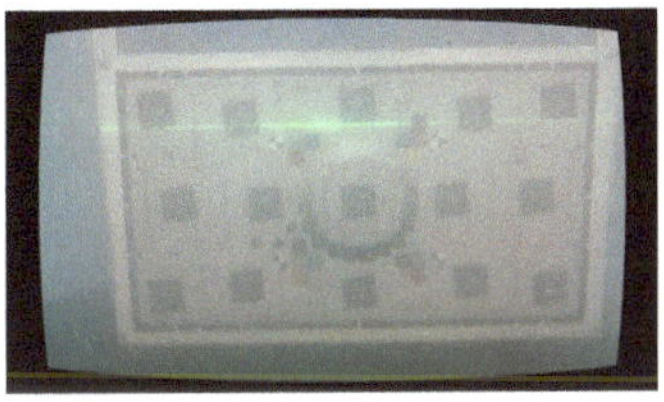

(**b**) Image of eSFR target at 2.1 FTU.

(**c**) Image of eSFR target at 6.0 FTU.

Figure 13. Images of eSFR target at 1.03 m.

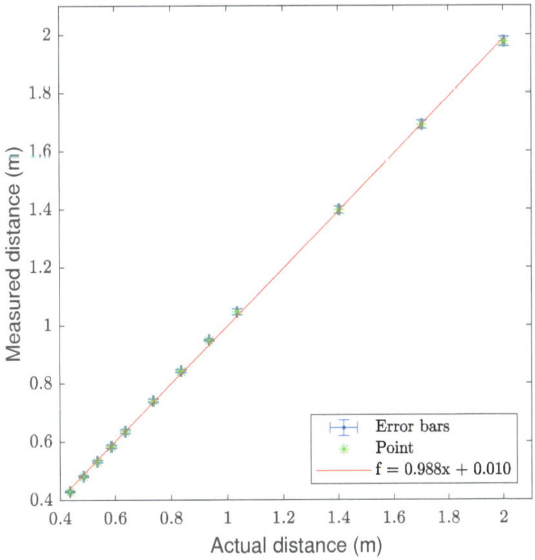

Figure 14. Graph showing the measuring accuracy at 0.3 FTU for laser triangulation.

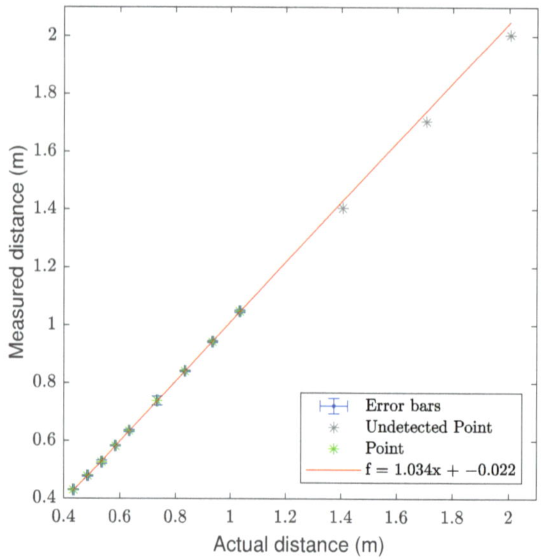

Figure 15. Graph showing the measuring accuracy at 2.1 FTU for laser triangulation.

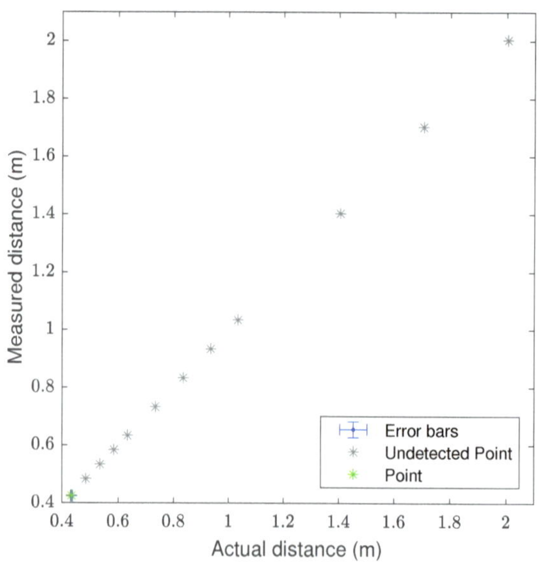

Figure 16. Graph showing the measuring accuracy at 6.0 FTU for laser triangulation.

4.4. Acoustic (Sonar)

The sonar data have been processed using the program described in Appendix A.2, with examples shown on Figure 18 and results summarized in Table 9 and Figures 19–21. The sonar target object distances show excellent linearity >0.98% and consistent monotonic error for all turbidities. Of particular note in the resulting images is the specular acoustic artifact arising at close distances, which creates a radial high-intensity echo tangential to the plane of the target object. The cylindrical target information is illustrated with a binarized image in Figure 22, where it is clear that the cylinder is detected; however, there is a substantial amount of noise at the front and rear boundaries of the cylinder.

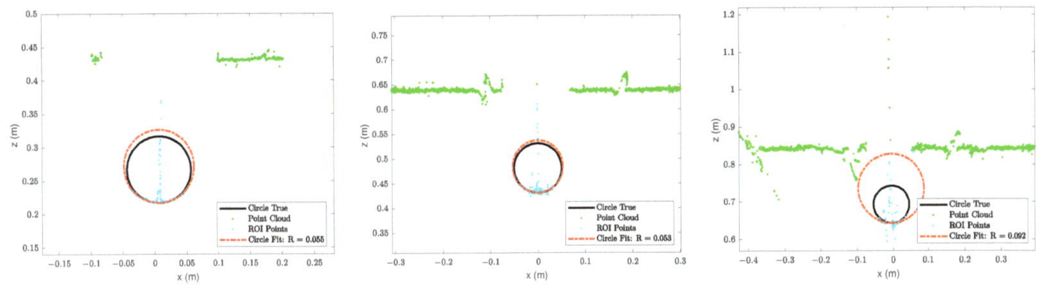

(**a**) (**b**) (**c**)

Figure 17. Circlefits of cylinder target at 0.3 FTU. (**a**) Circlefit of cylinder target at 0.44 m to eSFR target. (**b**) Circlefit of cylinder target at 0.64 m to eSFR target. (**c**) Circlefit of cylinder target at 0.84 m to eSFR target.

Table 9. Imaging sonar accuracy and operation limits. Δ denotes mean deviation from ground truth distance, shown with ±standard devation.

Target Dist.	Δ at 0.3 FTU	Δ at 1.0 FTU	Δ at 1.4 FTU	Δ at 2.1 FTU	Δ at 6.0 FTU
43.00 cm ± 1.5‰	1.61 (2.18) cm	2.06 (2.10) cm	1.92 (2.41) cm	1.81 (2.05) cm	2.16 (2.41) cm
48.00 cm ± 1.5‰	1.80 (2.30) cm	2.36 (2.33) cm	1.81 (2.57) cm	2.10 (2.25) cm	1.97 (2.28) cm
53.00 cm ± 1.5‰	2.07 (2.41) cm	2.26 (2.30) cm	1.86 (2.82) cm	2.08 (2.17) cm	2.33 (2.28) cm
58.00 cm ± 1.5‰	2.18 (2.74) cm	2.40 (2.43) cm	2.14 (2.86) cm	2.29 (2.53) cm	2.62 (2.46) cm
63.00 cm ± 1.5‰	2.14 (2.55) cm	2.70 (2.10) cm	2.42 (2.90) cm	2.59 (2.34) cm	2.67 (2.00) cm
73.00 cm ± 1.5‰	2.21 (3.26) cm	2.74 (2.68) cm	2.51 (3.16) cm	2.47 (3.51) cm	3.08 (2.47) cm
83.00 cm ± 1.5‰	2.68 (3.92) cm	2.63 (3.13) cm	2.76 (4.60) cm	2.44 (3.94) cm	3.50 (3.82) cm
93.00 cm ± 1.5‰	2.51 (4.73) cm	2.83 (4.14) cm	3.24 (5.17) cm	2.49 (4.28) cm	2.97 (4.29) cm
103.00 cm ± 1.5‰	2.74 (5.13) cm	2.88 (4.47) cm	2.94 (5.54) cm	2.53 (4.70) cm	3.38 (4.57) cm
140.00 cm ± 1.5‰	5.94 (6.58) cm	4.25 (4.58) cm	2.23 (6.06) cm	2.42 (6.02) cm	3.06 (6.24) cm
170.00 cm ± 1.5‰	2.90 (5.68) cm	3.84 (4.28) cm	2.50 (5.58) cm	2.75 (5.32) cm	3.59 (5.52) cm
200.00 cm ± 1.5‰	3.06 (4.03) cm	4.05 (4.19) cm	2.38 (3.53) cm	2.81 (3.57) cm	3.57 (3.26) cm

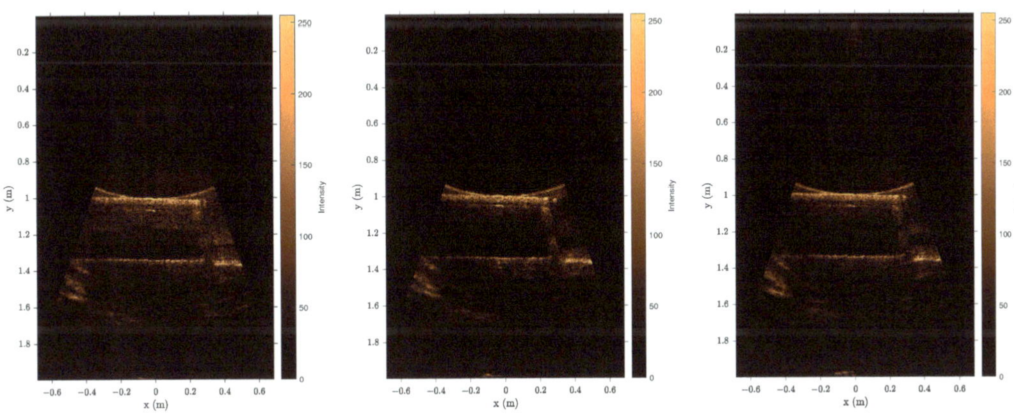

(**a**) (**b**) (**c**)

Figure 18. Sonar images of eSFR target at 1.03 m. (**a**) Sonar image of eSFR target at 0.3 FTU. (**b**) Sonar image of eSFR target at 2.1 FTU. (**c**) Sonar image of eSFR target at 6.0 FTU.

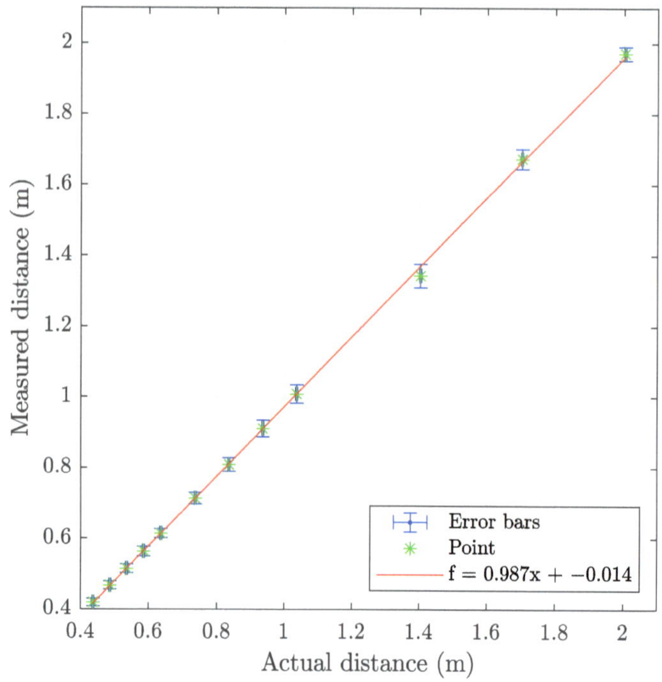

Figure 19. Graph showing the measuring accuracy at 0.3 FTU for imaging sonar.

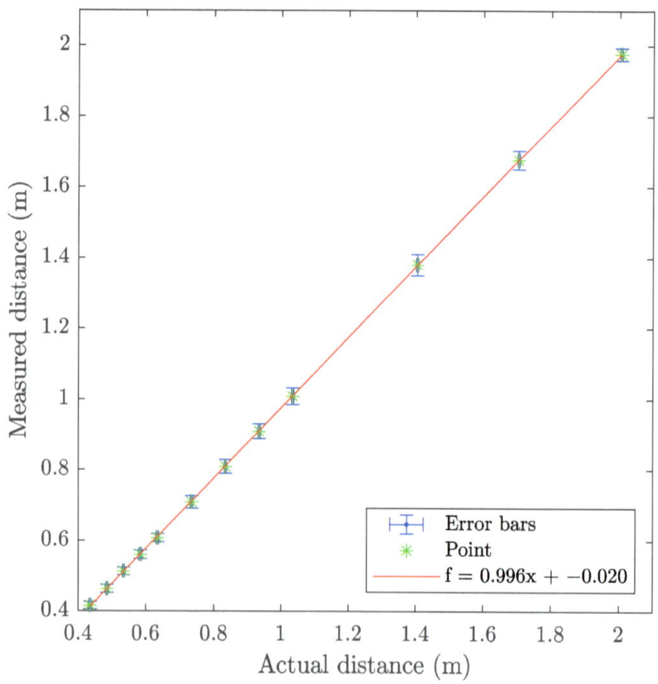

Figure 20. Graph showing the measuring accuracy at 2.1 FTU for imaging sonar.

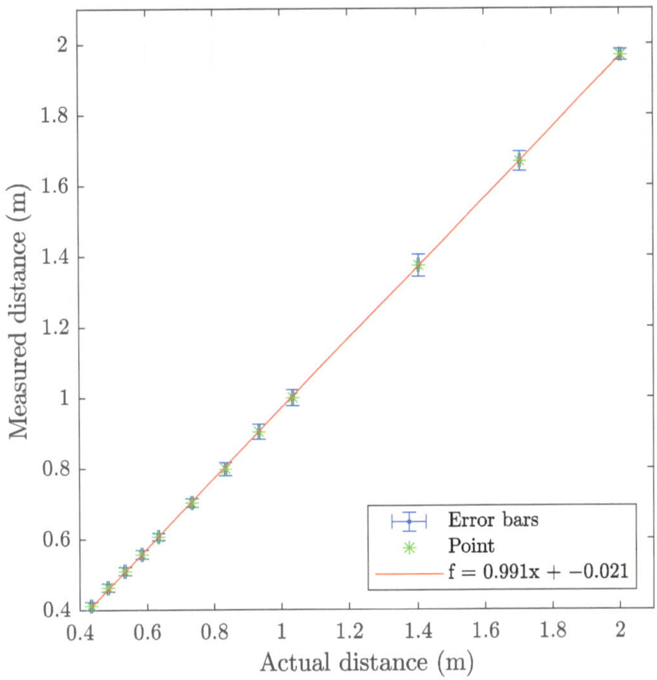

Figure 21. Graph showing the measuring accuracy at 6.0 FTU for imaging sonar.

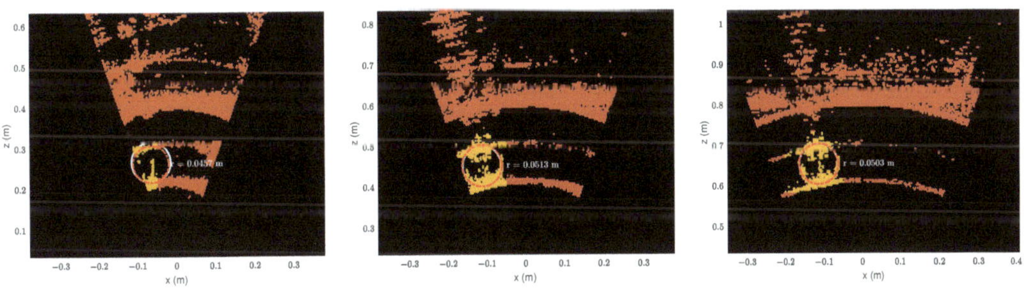

(**a**) (**b**) (**c**)

Figure 22. Circlefit of cylinder target at 0.3 FTU. (**a**) Circlefit of cylinder target at 0.44 m to eSFR target. (**b**) Circlefit of cylinder target at 0.64 m to eSFR target. (**c**) Circlefit of cylinder target at 0.84 m to eSFR target.

5. Discussion

In general, acoustic sensing is mostly stable across operating conditions; however, stereo vision and laser-line triangulation can also operate successfully under low and medium turbidity conditions: 0.3 FTU to 2.1 FTU at ranges of up to 100 cm. For laser triangulation, the accuracy is relatively constant in the range of 0.3 FTU to 2.1 FTU, with a total maximum mode deviation of 1.54 (0.90) cm at a range of 103 cm. Stereo depth estimation suffers from some non-linearity and increased deviation up to 36 cm—though, it is lower, ≤15 cm, at a distance below 140 cm; this most likely related to an insufficient quality of the intrinsic parameter calibration in particular, which warrants further work. At 6 FTU, the operating range is severely limited for the optical methods (laser-line and stereo depth estimation) but still usable up to distances ≤43 cm for laser distance measurements and up to ≤63 cm for stereo depth estimations. For all of the sensors, it is possible to detect and estimate the cylinder targets' geometry within 10% of the actual dimensions at distances closer than 63 cm. However, accuracy is substantially worse at longer distances. The operating depth for the considered approaches is generally limited by the manufacturer constraints for the commercial sensors as noted in their specifications; for the laser triangulation, the operating depth is additionally limited by the amount of ambient light; operating very close to the surface would not be possible. For the stereo camera, a large distortion still remains after the execution of the built-in calibration procedures: this would most likely need to be further corrected in a real application, depending on the particular application requirements. For acoustic sensing, other environmental parameters such as salinity or suspended particulate matter of large sizes may be more interesting to investigate since these are more likely to affect performance and operating limits with respect to the target distance. In summary, this work entails that these optical methods are usable even under relatively high turbidities if they are used for operations where only short-range measurements are needed; the useful operating range increases with decreasing turbidites, up until a maximum experimental distance of 200 cm. Contrarily, ranging using the acoustic sensor is, for the purpose of detecting used target objects under the given distances and environmental effects, unaffected, even at the highest turbidity and target distance.

6. Concluding Remarks

The experimental evaluation confirms the hypotheses that these optical methods provide great spatial details of the target objects and that increased turbidity affects their accuracy negatively. However, even at some substantial turbidity levels, i.e., 2.1 FTU, they still provide reasonable target object information at close ranges. Conversely, the sonar is not affected to a notable degree by turbidities of up to 6 FTU, but it provides the least amount of spatial information. In summary, this warrants investigation of sensor fusion where the complementary advantages of the different modalities can be fully exploited. Other future work includes the possible improvement of the laser-line distance measurement algorithm to improve the operating range and rejection of specular reflections. Alternatively, a modulated or rotational laser approach can also be investigated. Improvement of the stereo camera calibration to lower the distortion or external processing of the stereo camera information can be studied in order to ascertain whether larger operating conditions are achievable with other algorithms; this can be performed in extension to or completely replace the built-in stereo depth estimation. The addition of other environmental influences, such as salinity and suspended particulate matter, may lead to additional effects worth investigating, particularly for acoustic sensing.

Author Contributions: Conceptualization, C.M., F.F.S., J.L. and S.P.; methodology, C.M. and F.F.S.; software, F.F.S. and C.M.; validation, F.F.S. and C.M.; formal analysis, F.F.S. and C.M.; investigation, O.M.O., F.F.S. and C.M.; resources, J.L. and S.P.; data curation, F.F.S.; writing—original draft preparation, C.M.; writing—review and editing, O.M.O., F.F.S., J.L. and S.P.; visualization, F.F.S. and C.M.; supervision, J.L. and S.P.; project administration, J.L. and S.P.; funding acquisition, J.L. and S.P. All authors have read and agreed to the published version of the manuscript.

Funding: The research is part of the ACOMAR project, which is funded by the Energy Technology Development and Demonstration Program (EUDP), journal number 64020-1093.

Data Availability Statement: The data presented in this study are available on request from the authors.

Acknowledgments: The authors would like to thank the support from the Energy Technology Development and Demonstration Program (EUDP) via the "ACOMAR—Auto Compact Marine Growth Remover" project (J.No. 64020-1093). Thanks also go to our project partners SubC Partner, Sihm Højtryk, Mati2ilt, EIVA, Total E&P Denmark, and Siemens Gamesa Renewable Energy, and our colleagues from Aalborg University for many valuable discussions and technical support.

Conflicts of Interest: The authors declare no conflict of interest.

Appendix A. Data Processing Code

For each of the sensors, MATLAB code has been used to process the detected targets as described in the following sections. The data from the experiments are represented in the data tables "ColorTable**FTU.mat", "SonarTable**FTU.mat", and "StereoTable**FTU.mat" as referenced in the code, for the color images, sonar, and stereo data, respectively. The data are available on request in both the raw ROSbag format and the MATLAB mat files.

Appendix A.1. Laser-Line Triangulation MATLAB Code

First, the projective tomography, denoted *laserTform*, was derived. This was achieved by detecting the laser-line on a distinct set of calibration images. For these images, the laser-line points were projected into 2D world coordinates $[X, Y]$ using the camera intrinsics, and the corresponding Z distance to the laser-line was set using a separate laser-distance meter measurement, thus yielding a set of points $[X, Y, Z]$ which represent the laser-line position in real-world coordinates. From these world-coordinate points and the corresponding image plane coordinates, the projective homography was calculated. Subsequently, in other test images, the detected laser-line can be transformed into world-coordinates using the 2D projective homography to yield the target distances, as shown in the code Listing A1.

Listing A1. Code for laser processing.

```
% Load data for each turbidity
ColorTable{1} = {"ColorTable00FTU.mat"; 0.312};
ColorTable{2} = {"ColorTable05FTU.mat"; 0.520};
ColorTable{3} = {"ColorTable15FTU.mat"; 1.034};
ColorTable{4} = {"ColorTable10FTU.mat"; 1.404};
ColorTable{5} = {"ColorTable20FTU.mat"; 2.107};
ColorTable{6} = {"ColorTable60FTU.mat"; 5.988};

% Load the RGB camera calibration parameters.
load("intelCameraParams.mat")
load("laserTForm_projective_V2.mat")

numOfPics = 5; % Choose the number of pictures to average over.
for bb = 1:numel(ColorTable)
    % Clear terminal and not specified variables
    clc
    close all
    clearvars -except ColorTable intelCameraParams laserTform numOfPics bb

    % Load data for specified turbidity level
    load(ColorTable{bb}{1})

    % Get ground truth distances for dataset
    Dists = unique(dataTableColor.TrueDistance)'/1000;
    numOfDists = numel(Dists);

    for iii = 1:numOfDists
        for ii = 1:numOfPics
            % Load image and undistort it.
            im_raw = loadImageData(dataTableColor,ii,"eSFR",Dists(iii),50,"color");
```

```
                im_undist = undistortImage(im_raw, intelCameraParams);
                im_rgb = imrotate(im_undist,180);

                % Process color image and generate point cloud from line laser
                sigma = 700;
                RGB = imflatfield(im_rgb,sigma);
                [maxline,centroidline,j] = getLaserLine(RGB,false);
                test_imagePoints(:,:) = [j.',centroidline];
                test_imagePoints(any(isnan(test_imagePoints), 2), :) = [];
                worldPointsTest = transformPointsForward(laserTform, test_imagePoints);
                worldPointsTest = [worldPointsTest(:,1) zeros(size(test_imagePoints,1),1) worldPointsTest(:,2)];
                PC_Laser = pointCloud(worldPointsTest);
                % Extract 400 points closes to the center at the ground truth distance
                [indX,~] = findNearestNeighbors(PC_Laser,[0.0,0.0,Dists(iii)],400);
                NearstPoints = PC_Laser.Location(indX,:);

                pcLaser(iii,ii).X = PC_Laser.Location(:,1);
                pcLaser(iii,ii).Y = PC_Laser.Location(:,2);
                pcLaser(iii,ii).Z = PC_Laser.Location(:,3);
                pcLaser(iii,ii).NearstPoints = PC_Laser.Location(indX,3);

                % Check if points are found and calculate mode and std for image
                if(size(pcLaser(iii,ii).NearstPoints,1)>0)
                    mode_laser(iii,ii) = mode(pcLaser(iii,ii).NearstPoints);
                    std_laser(iii,ii) = std(pcLaser(iii,ii).NearstPoints);
                else
                    mode_laser(iii,ii) = NaN;
                    std_laser(iii,ii) = NaN;
                end
            end
            actual_laser(iii) = Dists(iii); % Same as Dists if all distances are used.
        end
        modes_laser = mean(sonar_mode,2)';
        stds_laser = mean(sonar_std,2)';
end
```

Appendix A.2. Sonar Target Detection MATLAB Code

For the sonar data, the sonar image was first binarized with a threshold of 0.3 (normalized). Then, the sonar points were projected into 2D coordinates, followed by projection into 3D. The 100 closest points within 10% of the ground truth target distance to the origin were found and used to calculate the mode and std.dev. of the target distance measurement, as shown in Listing A2.

Listing A2. Code for sonar processing.

```
% Load data for each turbidity
SonarTable{1} = {"SonarTable00FTU.mat"; 0.312};
SonarTable{2} = {"SonarTable05FTU.mat"; 0.520};
SonarTable{3} = {"SonarTable10FTU.mat"; 1.034};
SonarTable{4} = {"SonarTable15FTU.mat"; 1.404};
SonarTable{5} = {"SonarTable20FTU.mat"; 2.107};
SonarTable{6} = {"SonarTable60FTU.mat"; 5.988};

numOfPics = 5; % Choose the number of pictures to average over.
for bb = 1:numel(SonarTable)
    % Clear terminal and not specified variables
    clc
    close all
    clearvars -except SonarTable numOfPics bb

    % Load data for specified turbidity level
    load(SonarTable{bb}{1})

    % Get ground truth distances for dataset
    Dists = unique(dataTableSonar.TrueDistance)'/1000;
    numOfDists = numel(Dists);

    for iii = 1:numOfDists
        for ii = 1:numOfPics

            %Load image
            [sonar_data,dist] = loadImageData(dataTableSonar, ii,"eSFR",Dists(iii),60,"sonar");

            % Process sonar image and generate point cloud
            sonar_drawn_sonar = sonar_data.Image_rectified;
            BW = imbinarize(sonar_drawn_sonar,0.3);
            sonar_binary = BW(:,:,1);
            xWorldLimits = [-sin(deg2rad(130/2)) sin(deg2rad(130/2))];
            yWorldLimits = [0 2];
            RA = imref2d(size(sonar_binary),xWorldLimits, yWorldLimits);
            aa = 1;
            for y = 1:size(sonar_binary, 1)    % for number of rows of the image
                for x = 1:size(sonar_binary, 2)    % for number of columns of the image
                  if(sonar_binary(y,x))
                      [xWorld, yWorld] = intrinsicToWorld(sonar_data.SpacialReference,x,y);
                      xyzPoints(aa,:) = [xWorld, yWorld,0];
                      aa = aa + 1;
                  end
                end
```

```
            end
            PC_sonar{1} = pointCloud(xyzPoints);

            % Define region of interest (ROI) and extract points with ROI
            roi = [-0.65 0.3 dist*0.95 dist*1.05 -0.2 0.2];
            sampleIndices = findPointsInROI(PC_sonar{1},roi);
            PC_sonar_cropped{1} = select(PC_sonar{1}, sampleIndices);

            % Calculate mode and std for image
            sonar_mode(iii,ii) = mode(PC_sonar_cropped{1}.Location(:,2));
            sonar_std(iii,ii) = std(PC_sonar_cropped{1}.Location(:,2));
        end
        actual_sonar(iii) = dist; % Same as Dists if all distances are used.
    end
    modes_sonar = mean(sonar_mode,2)';
    stds_sonar = mean(sonar_std,2)';
end
```

Appendix A.3. Stereo Depth MATLAB Code

For the stereo depth images, a centered ROI corresponding to 20% of the image width and height was cropped for evaluation. The points of the central ROI was transformed to a point cloud, and the Z-axis was extracted to calculate the distance measurement mode and standard deviation. The code for used for evaluation is given in Listing A3.

Listing A3. Code for stereo processing.

```
% Load data for each turbidity
DepthTable{1} = {"DepthTable00FTU.mat"; 0.312};
DepthTable{2} = {"DepthTable05FTU.mat"; 0.520};
DepthTable{3} = {"DepthTable10FTU.mat"; 1.034};
DepthTable{4} = {"DepthTable15FTU.mat"; 1.404};
DepthTable{5} = {"DepthTable20FTU.mat"; 2.107};
DepthTable{6} = {"DepthTable60FTU.mat"; 5.988};

% Load the depth camera intrinsics.
load("depthIntrinsics.mat")

numOfPics = 5; % Choose the number of pictures to average over.

for bb = 1:numel(DepthTable)
    % Clear terminal and not specified variables
    clc
    close all
    clearvars -except DepthTable intrinsics numOfPics bb

    % Load data for specified turbidity level
    load(DepthTable{bb}{1})

    % Get ground truth distances for dataset
```

```
    Dists = unique(dataTableDepth.TrueDistance)'/1000;
    numOfDists = numel(Dists);

    for iii = 1:numOfDists
        for ii = 1:numOfPics
            [depthImage,dist] = loadImageData(dataTableDepth,
                ii,target,Dists(iii),50,"depth");
            % Full Image Point Cloud
            ptCloud = pcfromdepth(depthImage,depthScaleFactor,
                intrinsics);

            % ROI Point Cloud
            ROI = [848*0.4,480*0.4,848*0.2,480*0.2];
            h = images.roi.Rectangle(gca,'Position',ROI,'
                StripeColor','r');
            depthImageROI = double(imcrop(depthImage,h.
                Position));
            depthImageROI(depthImageROI == 0) = NaN;
            ptCloudROI = pcfromdepth(depthImageROI,
                depthScaleFactor,intrinsics);

            % Get depth measurement of point cloud
            B = ptCloudROI.Location(:,:,3)';
            B = double(B(:)');

            % Check if valid points are found and coverage is
                more than 50%
            % and calculate mode and std for image
            if(mean(B,"omitnan")<0 || sum(isnan(B))>numel(B)
                *0.5)
                depth_mode(iii,ii) = NaN;
                depth_std(iii,ii) = NaN;
            else
                depth_mode(iii,ii) = mode(B);
                depth_std(iii,ii) = std(B,"omitnan");
            end
        end
        actual_depth(iii) = dist; % Same as Dists if all
            distances are used.
    end
    modes_depth = mean(depth_mode,2)';
    stds_depth = mean(depth_std,2)';
end
```

References

1. Sergiyenko, O. *Optoelectronic Devices in Robotic Systems*; Springer Nature: Berlin/Heidelberg, Germany, 2022. [CrossRef]
2. Chen, F.; Brown, G.M.; Song, M. Overview of 3-D shape measurement using optical methods. *Opt. Eng.* **2000**, *39*, 10–22. [CrossRef]
3. Cyganek, B.; Siebert, J.P. *An Introduction to 3D Computer Vision Techniques and Algorithms*; John Wiley & Sons: Hoboken, NJ, USA, 2009; pp. 1–483. [CrossRef]
4. Massot-Campos, M.; Oliver-Codina, G. Optical sensors and methods for underwater 3D reconstruction. *Sensors* **2015**, *15*, 31525–31557. [CrossRef] [PubMed]
5. Hovem, J.M. Underwater acoustics: Propagation, devices and systems. *J. Electroceramics* **2007**, *19*, 339–347. [CrossRef]

6. Liniger, J.; Jensen, A.L.; Pedersen, S.; Sørensen, H.; Mai, C. On the Autonomous Inspection and Classification of Marine Growth on Subsea Structures. In Proceedings of the IEEE OCEANS 2022 Conference, Chennai, India, 21–24 February 2022.
7. Jian, M.; Liu, X.; Luo, H.; Lu, X.; Yu, H.; Dong, J. Underwater image processing and analysis: A review. *Signal Process. Image Commun.* **2021**, *91*, 116088. [CrossRef]
8. Foote, K.G. Using a sonar in a different environment from that of its calibration: Effects of changes in salinity and temperature. In Proceedings of the OCEANS 2018 MTS/IEEE Charleston, Charleston, SC, USA, 22–25 October 2018; pp. 1–5. [CrossRef]
9. Pedersen, S.; Liniger, J.; Sørensen, F.F.; Schmidt, K.; Von Benzon, M.; Klemmensen, S.S. Stabilization of a ROV in three-dimensional space using an underwater acoustic positioning system. *IFAC-PapersOnLine* **2019**, *52*, 117–122. [CrossRef]
10. von Benzon, M.; Sorensen, F.; Liniger, J.; Pedersen, S.; Klemmensen, S.; Schmidt, K. Integral Sliding Mode control for a marine growth removing ROV with water jet disturbance. In Proceedings of the 2021 European Control Conference (ECC), Delft, The Netherlands, 29 June–2 July 2021. [CrossRef]
11. O'Byrne, M.; Schoefs, F.; Pakrashi, V.; Ghosh, B. An underwater lighting and turbidity image repository for analysing the performance of image-based non-destructive techniques. *Struct. Infrastruct. Eng.* **2018**, *14*, 104–123. [CrossRef]
12. O'Byrne, M.; Ghosh, B.; Pakrashi, V.; Schoefs, F. Effects of turbidity and lighting on the performance of an image processing based damage detection technique. In Proceedings of the Safety, Reliability, Risk and Life-Cycle Performance of Structures and Infrastructures—Proceedings of the 11th International Conference on Structural Safety and Reliability, ICOSSAR 2013, New York, NY, USA, 16–20 June 2013 ; pp. 2645–2650. [CrossRef]
13. O'Byrne, M.; Pakrashi, D.; Schoefs, D.; Ghosh, D. A Comparison of Image based 3D Recovery Methods For Underwater Inspections. In Proceedings of the 7th European Workshop on Structural Health Monitoring, Nantes, France, 8–11 July 2014; pp. 671–678.
14. O'Byrne, M.; Pakrashi, V.; Schoefs, F.; Ghosh, B. A Stereo-Matching Technique for Recovering 3D Information from Underwater Inspection Imagery. *Comput.-Aided Civ. Infrastruct. Eng.* **2018**, *33*, 193–208. [CrossRef]
15. Aykin, M.D.; Negahdaripour, S.; Wang, Y. Calibration and 3D reconstruction of underwater objects with non-single-view projection model by structured light stereo imaging. *Appl. Opt.* **2016**, *55*, 6564–6575. [CrossRef]
16. Bruno, F.; Bianco, G.; Muzzupappa, M.; Barone, S.; Razionale, A.V. Experimentation of structured light and stereo vision for underwater 3D reconstruction. *ISPRS J. Photogramm. Remote Sens.* **2011**, *66*, 508–518. [CrossRef]
17. Mai, C.; Liniger, J.; Jensen, A.L.; Sørensen, H.; Pedersen, S. Experimental investigation of non-contact 3D sensors for marine-growth cleaning operations. In Proceedings of the 5th IEEE International Conference on Image Processing, Applications and Systems (IPAS 2022), Genova, Italy, 5–7 December 2022.
18. Risholm, P.; Thorstensen, J.; Thielemann, J.T.; Softley, C.; Yates, C.; Abrosimov, I.; Tschudi, J.; Alexander, J.; Haugholt, K.H.; Kaspersen, K. Real-time super-resolved 3D in turbid water using a fast range-gated CMOS camera. *Appl. Opt.* **2018**, 57, 3927–3937. [CrossRef] [PubMed]
19. Roman, C.; Inglis, G.; Rutter, J. Application of structured light imaging for high resolution mapping of underwater archaeological sites. In Proceedings of the OCEANS'10 IEEE Sydney, OCEANSSYD 2010, Sydney, Australia, 24–27 May 2010; pp. 1–9. [CrossRef]
20. Yang, M.; Yin, G.; Wang, H.; Dong, J.; Xie, Z.; Zheng, B. A Underwater Sequence Image Dataset for Sharpness and Color Analysis. *Sensors* **2022**, *22*, 3550. [CrossRef] [PubMed]
21. Scott, M.; Marburg, A. Quantifying the Degradation of Optical Algorithms in Increasingly Turbid Mediums. In Proceedings of the Oceans Conference Record (IEEE), San Diego, CA, USA, 20–23 September 2021. [CrossRef]
22. Concha, A.; Drews, P., Jr.; Campos, M.; Civera, J. Real-time localization and dense mapping in underwater environments from a monocular sequence. In Proceedings of the MTS/IEEE OCEANS 2015—Genova: Discovering Sustainable Ocean Energy for a New World, Genova, Italy, 18–21 May 2015. [CrossRef]
23. Intel Corporation. *Depth Camera D435i—Intel® RealSense™ Depth and Tracking Cameras*; Intel Corporation: Santa Clara, CA, USA, 2022.
24. Blueprint Design Engineering Ltd. *Oculus m-Series*; Blueprint Design Engineering Ltd.: Ulverston, UK, 2022 .
25. Aanderaa Data Instruments AS. *Aanderaa Optical Turbidity Sensor*; Aanderaa Data Instruments AS: Nesttun, Norway, 2022.
26. Sigma-Aldrich. *Kaolin Tested According to Ph Eur 1332-58-7*; Sigma-Aldrich: St. Louis, MO, USA, 2022.
27. Imatest LLC. *ISO 12233:2017 Edge SFR (eSFR) Inkjet Chart*; Imatest LLC: Boulder, CO, USA, 2022.

sensors

MDPI

Article

RUDE-AL: Roped UGV Deployment Algorithm of an MCDPR for Sinkhole Exploration

David Orbea *, Christyan Cruz Ulloa, Jaime Del Cerro and Antonio Barrientos *

Centro de Automática y Robótica (UPM-CSIC), Universidad Politécnica de Madrid—Consejo Superior de Investigaciones Científicas, 28006 Madrid, Spain; christyan.cruz.ulloa@upm.es (C.C.U.); j.cerro@upm.es (J.D.C.)
* Correspondence: david.orbea@upm.es (D.O.); antonio.barrientos@upm.es (A.B.)

Abstract: The presence of sinkholes has been widely studied due to their potential risk to infrastructure and to the lives of inhabitants and rescuers in urban disaster areas, which is generally addressed in geotechnics and geophysics. In recent years, robotics has gained importance for the inspection and assessment of areas of potential risk for sinkhole formation, as well as for environmental exploration and post-disaster assistance. From the mobile robotics approach, this paper proposes RUDE-AL (Roped UGV DEployment ALgorithm), a methodology for deploying a Mobile Cable-Driven Parallel Robot (MCDPR) composed of four mobile robots and a cable-driven parallel robot (CDPR) for sinkhole exploration tasks and assistance to potential trapped victims. The deployment of the fleet is organized with node-edge formation during the mission's first stage, positioning itself around the area of interest and acting as anchors for the subsequent release of the cable robot. One of the relevant issues considered in this work is the selection of target points for mobile robots (anchors) considering the constraints of a roped fleet, avoiding the collision of the cables with positive obstacles through a fitting function that maximizes the area covered of the zone to explore and minimizes the cost of the route distance performed by the fleet using genetic algorithms, generating feasible target routes for each mobile robot with a configurable balance between the parameters of the fitness function. The main results show a robust method whose adjustment function is affected by the number of positive obstacles near the area of interest and the shape characteristics of the sinkhole.

Keywords: genetic algorithms; multi-robot system; CDPR; MCDPR; ROS; roped fleet; navigation

Citation: Orbea, D.; Cruz Ulloa, C.; Del Cerro, J.; Barrientos, A. RUDE-AL: Roped UGV Deployment Algorithm of an MCDPR for Sinkhole Exploration. *Sensors* **2023**, *23*, 6487. https://doi.org/10.3390/s23146487

Academic Editors: David Cheneler and Stephen Monk

Received: 30 May 2023
Revised: 12 July 2023
Accepted: 15 July 2023
Published: 18 July 2023

1. Introduction

Sinkholes are hollows in the ground surface formed by the dissolution of limestone or geological features, considered closed cavities drained due to subsoil dissolution in karst rocks [1]. This land subsidence and collapse represents a major hazard that substantially impacts economic and human losses [2], as well as in nearby infrastructures [3]. In the last decade, several karst collapses have taken place, causing the collapse of roads and buildings in urban areas , causing safety risks to residents [4]. Different natural storms have caused sinkholes such as in Guatemala (Guatemala City—2010) [5], USA (Florida—2004) [6].

Pressure wireless sensor network (WSN) technologies and neural network learning databases use UAVs and thermal cameras for sinkhole detection and monitoring [7]. There are also approaches for the use of robotic total stations (RTS) together with total stations (TS) to calculate the horizontal and vertical displacement of the earth for sinkhole detection [8], opening the line of integration of mobile robots and their application for monitoring, exploration, and assistance for search and rescue (SAR) tasks.

The success of search and rescue (SAR) missions depends on the performance of robotic platforms individually [9]. However, search and exploration tasks can be enhanced through cooperative systems using multiple unmanned ground vehicles (UGVs) [10]. There are several types of cooperative systems used in urban search and rescue (USAR) [11], wilderness search and rescue (WiSAR) [12] and air-sea rescue (ASR) operations [13].

Regarding deployment of robot fleets in disaster environments, [14] details of thirty-four official deployments have been reported and analyzed, where UGVs have been used in twenty-two incidents. In contrast, UAVs and unmanned aquatic vehicles (UMVs) in eight incidents [15], mainly at the subway level in mining environments, building collapse, earthquakes, and hurricanes, due to their potential to save lives by offering faster response times [16], support in hazardous environments [17], real-time monitoring [18], and area mapping [19].

The most important robotic challenges in subway environments are the limitation of wireless communications and detection task such as toxic gas analysis. Robots used for WiSAR tasks must be prepared for difficult weather conditions and interfere minimally with environmental data acquisition [20], so robots with mobile base cables with robust situational awareness are currently proposed for autonomous navigation, where environmental monitoring sensors are mounted for gas discrimination [21], air quality sensing, or gas concentration [22], with which information assistance is provided to rescue personnel for hazardous area detection.

In terms of SAR tasks using mobile robots, these systems can be considered cooperative Cable-Driven Parallel Robots (CDPR), where each robot represents an anchor point for the wires. This configuration is a major challenge for base localization in the environment, obstacle avoidance, and adaptive control of payload position and orientation [23].

This work is part of a Robotics and Cybernetics Group (ROBCIB) of the Polytechnic University of Madrid (UPM) project, which main objective is to perform the deployment of a Mobile Cable-Driven Parallel Robot (MCDPR) inside a sinkhole automatically, considering mobile robots as mobile bases that will release a parallel cable-driven robot in the region of interest. Therefore, challenges are related to the deployment of the mobile robots (mobile bases) and relative localization of the CDPR. Accordingly, the strategy is focused on two stages: first, the traversability of a path for the mobile bases from the base station to the surroundings of the sinkhole, considering that the mobile bases of the MCDPR will be attached by ropes; then, the release of the CDPR, avoiding contact with positive obstacles. Regarding the relative localization challenge, in order to get a good precision of portable localization systems (radar, infrared, UWB), using trilateration methods, a first approach presents the use of a fleet of four mobile robots and a CDPR carried on top of one of the mobile platforms, applying two physical configurations: node-edge fleet and roped individual.

This paper's main contribution is RUDE-AL (Roped UGV Deployment Algorithm), an algorithm for the autonomous selection of anchoring points of a mobile robot fleet for sinkhole exploration through computer vision techniques to filter positive and negative obstacles, and genetic algorithms for solving optimization problems.

The mission comprises two stages: the first one consists of finding a feasible fleet trajectory in a node-edge configuration with four nodes and three edges; the second corresponds to the deployment of each robot to its corresponding vertex.

For this purpose, multiple tests were carried out on 2D images representing 2.5D maps, which were generated manually (using image edition tools) and automatically (images containing splines generated using Bézier curves), evaluating the behavior of the algorithm to produce feasible roped navigation paths for environments with different physical characteristics.

Implementing a heuristic method (genetic algorithms) considerably reduces computational time, allowing to adjust the weights of the variables to be optimized through a weighted fitness function for multi-objective optimization. The algorithm is applied to the navigation of a fleet of four simulated robots in Gazebo.

The work is made up of the following sections. Section 2 presents the state of the art and related works. Section 3 details the problem formulation and the proposal. Section 4 describes the generation of the different environments for testing and description of the performed experiments. Section 5 shows the results obtained by the target point selection algorithm. Finally, Section 6 presents the conclusions obtained.

2. Related Work

2.1. Mobile CDPRs

CDPRs are emerging as an attractive proposition due to their advantages in covering large volumes with fast movements while maintaining a balance between light weight and high durability [24].

CDPR-related research has been reported since 1984, originally for underwater applications. The RoboCrane project [25], developed under the Defense Advanced Research Project Agency (DARPA), extended CDPR applications to land, sea, air, and space. CDPRs show high load-to-weight ratio, while relying on stable configurations, flexibility, and maneuverability over rough terrain surfaces.

An example of mobile CDPRs are the so-called extended-crane systems, which represent a combination between a cable robot and a conventional crane, where the robotic configuration is intended to separate positioning and balancing tasks. Another example of a reconfigurable structure for cable robots is found in agricultural applications, adding mobile pillars to transport winches while controlling the position of the mobile platform.

In terms of CDPR applications, the IPAnema family of robots [26] is used for industrial inspection, handling, and assembly tasks, heavy lifting, CoGiro [27], motion simulators, CableRobot [28], and recreation of underwater environments [29]; in logistics warehousing tasks, CABLAR [30] and FASTKIT [31], and in search and rescue operations, the MARIONET family of robots [32].

At conceptual level, the MoPick prototype [33] presents an approach to parallel cable robots with mobile bases (MCDPR) for pick and place tasks. A parallel cable robot for multiple mobile cranes (CPRMCs) is proposed by [34], detailing the design and a multilateration-based localization algorithm, whose global planning is performed with a grid-based artificial potential field method. It also uses sensors for cooperative obstacle avoidance, integrating autonomous level control for the platform, together with a co-simulation by using Matlab and Labview.

A CDPR with three mobile cranes for search and rescue operations is shown by [35], where each mobile base consists of a reconfigurable telescopic boom that can rotate (Figure 1a). The cable is mounted from the tip of the telescopic arm to the end effector. Although the cranes are fixed, the system can be reconfigured to increase the working space and keep the system in static equilibrium. The objective of this study is to compare the stress distribution and the size of the working space when applying payloads.

Considering strategies for motion planning of a fleet of mobile robots for deployment of a cable robot, ref. [36] proposes a modular CDPR carried by a rover for inspection and light manipulation tasks on celestial bodies (Figure 1b), whose applications focus on solar panel inspection and maintenance, as well as lava cave exploration (Figure 2).

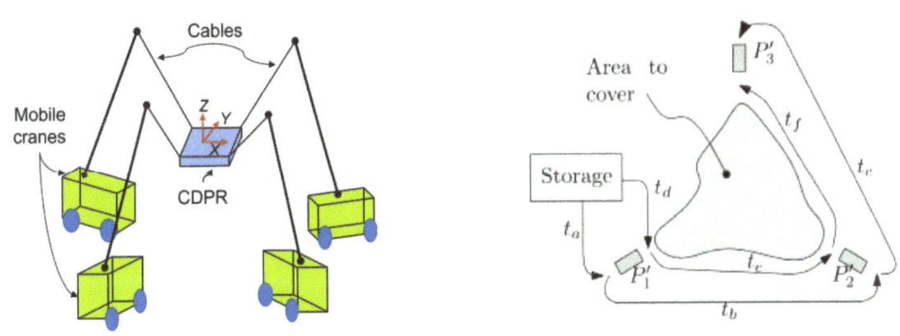

(a) (b)

Figure 1. MCDPR system configurations and deployment. (**a**) Schematic of CDPR with mobile cranes. Source: Authors. (**b**) Deployment procedure of a rover. Obtained from [36].

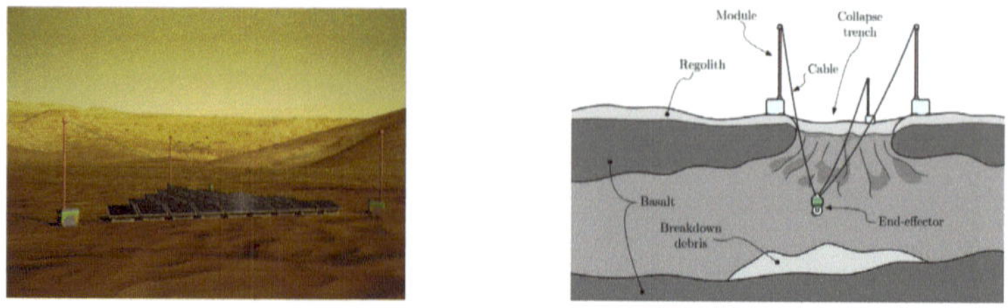

(**a**) (**b**)

Figure 2. Applications of modular CDPR. (**a**) Maintenance of large ground solar array. Obtained from [36]. (**b**) Exploration of lava tubes on celestial bodies. Obtained from [36].

In the related works described in Table 1, no strategies for the implementation of a self-contained portable CDPR have been found.

Table 1. Comparison of state-of-the-art CDPR prototypes.

Reference	Robot Name	Mobile Bases	Application	Environment	Physical Implementation	Number of Robotic Platforms	Degrees of Freedom
[25]	Robocrane	YES	Research	Indoor/Outdoor	YES	3	6
[26]	IPAnema	NO	Industry	Indoor	YES	1	6
[27]	CoGiro	NO	Research	Indoor	YES	1	6
[28]	CableRobot	NO	Research	Indoor	YES	1	6
[29]	-	NO	Research	Indoor	YES	1	6
[30]	CABLAR	NO	Logistics	Indoor	YES	1	6
[31]	FASTKIT	YES	Logistics	Indoor	YES	2	6
[32]	MARIONET-CRANE	YES	Rescue	Outdoor	YES	1	6
[33]	MoPick	YES	Logistics	Indoor	YES	4	3
[34]	-	YES	Research	Outdoor	NO	4	4
[35]	-	YES	Research	Outdoor	NO	3	3
[36]	-	YES	Research	Outdoor	NO	3	3

2.2. *Multi-Objective Optimization*

State-of-the-art path planning methods are commonly formulated based on a single-objective optimization for distance cost [37]. However, more factors need to be considered in real-world applications, which turns the study problem into a multi-objective optimization method applicable for multi-robot fleet path planning [38]. Common approaches use methods such as summing the weights of each objective as a function, known as the scalarized approach or weighted sum method [39], whose importance lies in correctly deciding the weight coefficients based on empirical checks.

The use of multi-objective genetic algorithms has been widely studied in building design [40], balancing of operations in hydroelectric reservoirs [41], sizing of microgrid systems [42], and sustainable mechanization allocation for spraying and harvesting systems [43]. It has been shown relevant in robotics for trajectory optimization [44,45], controller design [46], industrial robotic arm design [47], and cloud robotic platform service scheduling [48], and for multi-robot trajectory planning for area coverage [49].

On the application of evolutionary algorithms (EA) for target selection in multi-robot systems, ref. [50] proposes the use of an evolutionary algorithm with Indirect Representation and Extended Nearest Neighbor (IREANN) with a simple mutation operator for GTSPC (Generalized Travelling Salesman Problem with Coverage). The trajectory optimization for MDCPR of the MoPick platform using direct transcription optimization method where the optimization task adds the CDPR constraints in planning, and direct transcription increases the confidence of the data, is studied by [51].

3. Methodology

3.1. *Problem Statement*

As explained in Section 1, the strategy proposed to perform the deployment of the MCDPR must assure the traversability of a path for the mobile bases from a start point to the surrounds of the sinkhole, considering that the robots are attached by ropes that cannot pass through positive obstacles. After that, using optimization techniques, a feasible configuration that minimizes the fleet path distance cost and maximizes the area of the region of interest (ROI) to be covered must be found.

For the conceptualization of the problem, the starting point is a 2D map of free space, positive and negative obstacles. Figure 3a shows positive obstacles (gray) which block the traversability of mobile robots and ropes, while negative obstacles (black) block the traversability of mobile robots, but allow the passage of ropes. The figure is the actual image used as an input for the proposed algorithm.

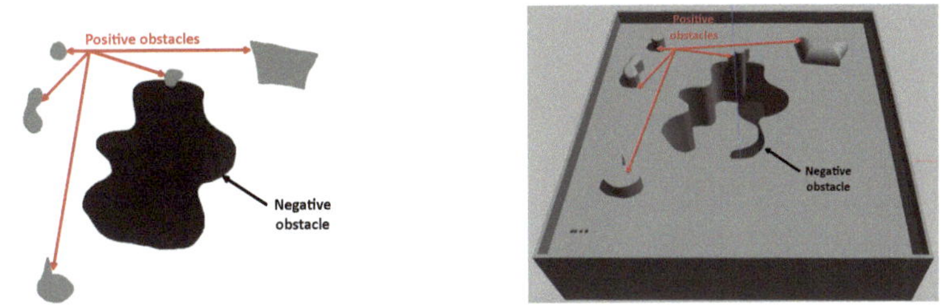

(a) (b)

Figure 3. Map representation of environment. (**a**) A 2.5D map representation. (**b**) A 3D map representation. Source: Authors.

Figure 3b shows a Gazebo representation of a structured environment, where the undercut corresponds to a negative obstacle.

Physically, the self-contained CDPR is on top of one of the mobile robots (Figure 4). However, the scope of the proposed algorithm is in two dimensions, so the corded fleet as seen from the top plane is four robots with one node from which the cables are projected.

Figure 4. Physical configuration of the fleet. Source: Authors.

Before performing path planning, the possibility of finding nearby positive obstacles that affect feasible target points for fleet route planning or deployment route must be considered. For these cases, a range of up to two nearby positive obstacles affecting the planning is defined, dividing into zones of interest where the user selects which area to explore.

Once the feasible candidates are available, the fleet route planning is performed for each candidate point, calculating the distance cost of each route. Then, candidate point ranges are defined for each mobile robot, where for each combination of candidate points the area covered is calculated. The distance cost of the minimum fleet route is minimized and the area covered is maximized for each combination of candidate points using a weighted fit function.

Through the optimization of the fitness function, the target points are obtained, with which the deployment routes between them are planned (Figure 5a); subsequently, the feasible deployment routes are evaluated, avoiding collisions of the ropes with positive obstacles near the sinkhole to be explored (Figure 5b). Finally, the mission is executed in the simulator, adding a 2D display of the roped fleet.

(**a**) (**b**)

Figure 5. Deployment procedure, (**a**) definition of fleet path (red) in node-edge configuration and deployment path (blue) in individual configuration, (**b**) feasible deployment configuration with roped restrictions. Source: Authors.

3.2. RUDE-AL Algorithm

The main contribution of this article is the Roped UGV fleet DEployment ALgorithm (RUDE-AL), designed to provide a feasible goal selection based on the minimization of distance cost of fleet deployment and maximization of covered area, considering rope contact restrictions for negative and positive obstacles. It has been divided into two phases and three stages:

- Phase 1. Node-edge fleet configuration
  - Map, obstacle, and start selection for offline planning
  - Definition, sorting, and choosing of feasible goal candidates based on area coverage and distance cost of fleet deployment
- Phase 2. Roped individual configuration
  - Goal local planning and establishment of deployment configuration based on rope restrictions

The flow diagram of the process of RUDE-AL is shown in Figure 6.

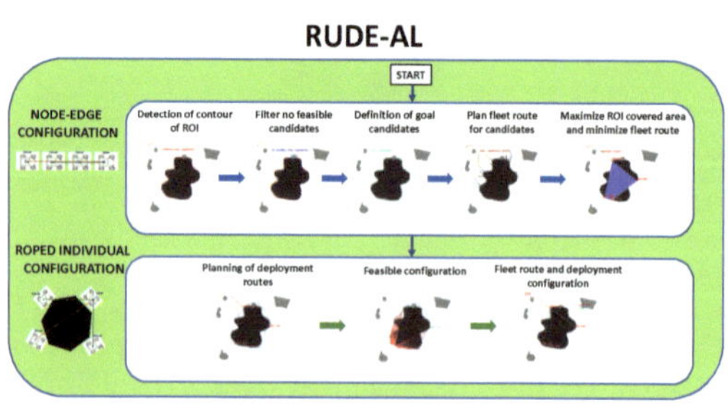

Figure 6. RUDE-AL algorithm procedure. Source: Authors.

3.2.1. Phase 1. Node-Edge Fleet Configuration

In this phase, the four robots are formed in a worm configuration consisting of four nodes and three edges (Figure 7). For navigation, the strategy used is to assign the front robot as the leader, and the others as followers. Each robot is assigned the fleet route, and minimum distance constraints are added to avoid collisions between them.

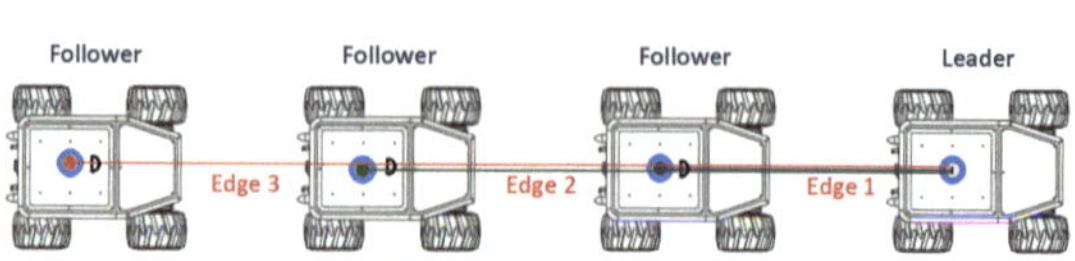

Figure 7. Node-edge configuration with cable robot on the leader robot. Source: Authors.

Map, Obstacle, and Start Selection for Offline Planning

In this stage, computer vision techniques are used to generate two-dimensional maps of positive and negative obstacles, separating them into individual masks for individual processing. Table 2 shows the associated color coding, as well as land navigation conditions and roped fleet restrictions for traversability.

Table 2. RGB color definition used in 2D maps for RUDE-AL. Source: Authors.

Tipo	RGB Color	UGVs Navigation Allowed	Cable Cross Allowed
Free space	(255, 255, 255)	YES	YES
Positive obstacle	(136, 138, 133)	NO	NO
Negative obstacle	(0, 0, 0)	NO	YES

OpenCV tools are used for contour detection, obstacle inflation (erode and dilate tools), and graphical representation of the algorithm. The starting point of the mission and the negative obstacle to be explored are manually selected. The existence of positive obstacles near the undercut that affect the roped deployment process is evaluated, and a group of candidate points is automatically obtained. Figure 8 shows the candidate point selection process for maps with several positive obstacles and one negative obstacle.

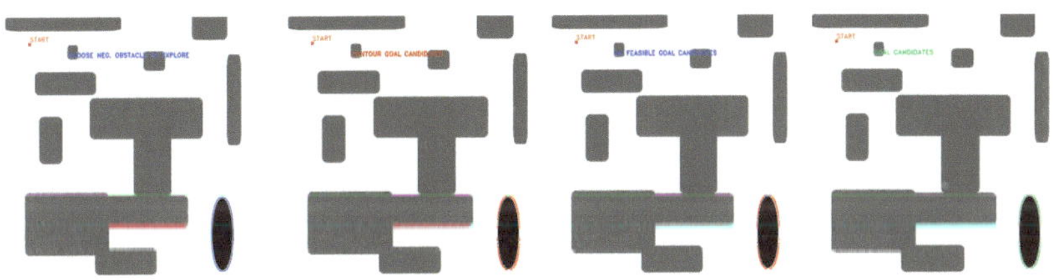

(a) (b) (c) (d)

Figure 8. First stage candidate points with no nearby positive obstacles, (**a**) obstacle selection, (**b**) contour candidates, (**c**) evaluation of feasible and non-feasible candidates (red feasible, blue no feasible, (**d**) first stage candidate points (green). Source: Authors.

Figure 9 shows the process for selecting candidate points with a positive obstacle close to the zone of interest.

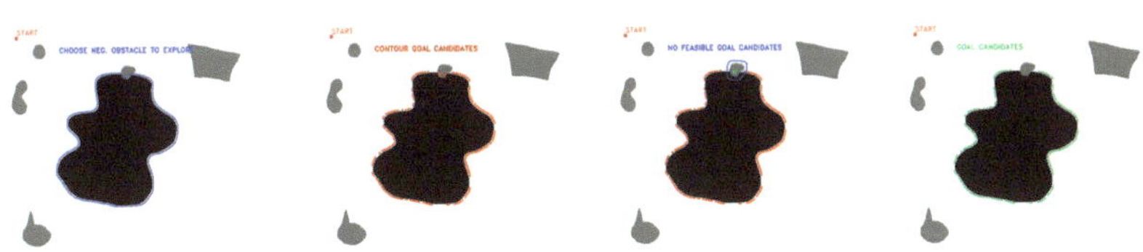

(a) (b) (c) (d)

Figure 9. First stage candidate points with one nearby positive obstacles, (**a**) obstacle selection, (**b**) contour candidates, (**c**) evaluation of feasible and non-feasible candidates (red feasible, blue no feasible, (**d**) first stage candidate points (green). Source: Authors.

Figure 10 shows the selection of candidate points when there are two positive obstacles close to the area of interest.

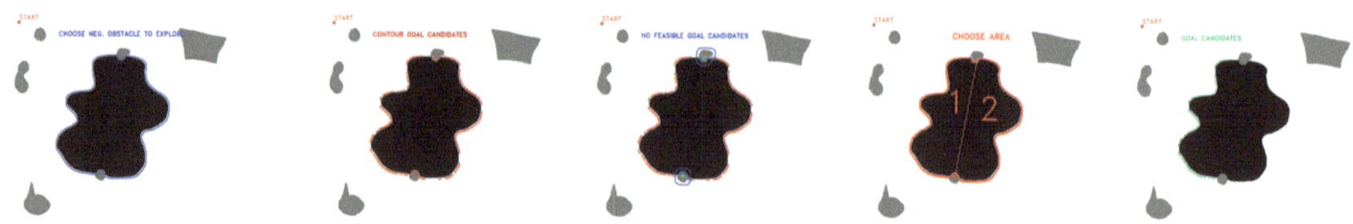

(a) (b) (c) (d) (e)

Figure 10. First stage candidate points with two nearby positive obstacles, (**a**) obstacle selection, (**b**) contour candidates, (**c**) evaluation of feasible and non-feasible candidates (red feasible, blue no feasible), (**d**) selection of available areas to explore, (**e**) first stage candidate points (green). Source: Authors.

Then, path planning is performed with a multi-query planner, in this case, a probabilistic roadmap (PRM), in order to obtain the candidate fleet routes from the starting point to the candidate target points of the first stage (Figure 11).

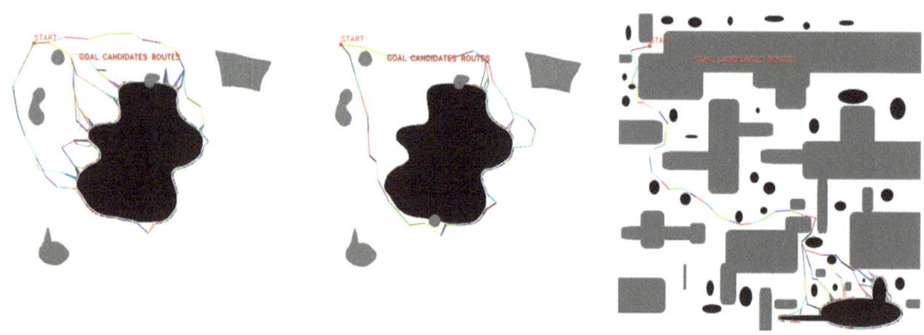

Figure 11. Multiple goal path planning for different scenarios. Source: Authors.

Algorithm 1 shows the pseudocode implemented for this stage, using as input variables: the required map, which is an RGB image in png format with a resolution of 1000 × 1000 pixels, whose colors correspond to the requirements of Table 2, and the value of the starting point start. The following functions explain the processes:

- Erode (line 12), Dilate (line 13), Range (line 14), Contours (line 16) and Centroids (line 18): are OpenCV functions used to inflate regions of images, define masks, and get characteristics of contours.
- Remove_dup (line 28): is a function to delete the multiple occurrences of an object in a list.
- Line(line 35): is a function that returns the slope and the constant "b" for a y = mx + b line between two input points
- Check_click (line 51): is a function that determines where the user clicks and returns the selected zone to explore (options are 1 or 2).
- Route (line 51) is a function that performs the prm path planning between two points. Returns the feasible path between them.

Algorithm 1 Fleet Planning

```
Input: map,start
Output: c_point, c_path
//init_candidates: set of initial candidate points
//c_point: set of candidate points
//c_path: set of candidate paths
//index_roi: index of the negative obstacle to explore
//index_pos_obst: list of index of positive obstacles close to ROI
//m: slope of line between two points
//b: constant of line between two points
//y_sup: bool that defines if "y" coordinate of a point is greater that the one described by equation constants
//upper_c: candidate points greater that line defined between two near positive obstacles
//lower_c: candidate points lower that line defined between two near positive obstacles
map ← gray(map)
erodedn_m ← Erode(map)
erodedp_m ← Dilate(map)
black_m ← Range(erodedn_m, 0, 10)
gray_m ← Range(erodedp_m, 50, 150)
neg_contours ← Contours(black_m)
pos_contours ← Contours(gray_m)
pos_centroids ← Centroids(pos_contours)
c_point ← neg_contours[index_roi]
for j ∈ pos_centroids do
    for j ∈ neg_contours do
        if gray_m[c_point[i]] == 1 then
            Delete from c_point this neg_contours [i]
            Append in index_pos_obst this j
        end if
    end for
end for
index_pos_obst ← Remove_dup[index_pos_obst]
if length(index_pos_obst) == 1 then
    for i ∈ c_point do
        c_path = Route(start, i)
    end for
end if
if length(index_pos_obst) == 2 then
    m, b ← Line(pos_centroid[index_pos_obst[0]], pos_centroid[index_pos_obst[1]])
    for i ∈ c_point do
        y_sup ← check_line(c_point, m, b)
        if y_sup == true then
            Append in upper_c this c_point[i]
        else
            Append in lower_c this c_point[i]
        end if
    end for
    zone=Check_click()
    if zone == 1 then
        c_point = upper_c
    else
        c_point = lower_c
    end if
    for i ∈ c_point do
        c_path[i] = Route(start, i)
    end for
end if
return c_point, c_path
```

Definition, Sorting, and Choice of Feasible Goal Candidates Based on Area Coverage and Distance Cost of Fleet Deployment

At this stage, the distance cost of the fleet route is minimized and the area covered by the polygon formed by the combination of four points corresponding to the position of the mobile robots is maximized. The candidate points are defined as c_1, c_2, c_3, c_4. Initially, to reduce the number of combinations to be processed, a minimum distance criterion is applied between subsequent points, defining minimum and maximum indices for each point in the general list of candidates, as shown in Figure 12, from which different vectors are obtained for each candidate point.

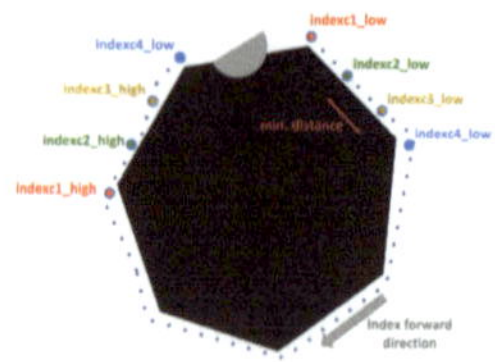

Figure 12. Definition of minimum and maximum index for each candidate point. Source: Authors.

The calculation of the area covered by the combination of the candidate points is made with the sum of the areas of the triangles that form the quadrilateral 1-2-3-4, shown in Figure 13, with the Equation (1):

$$a_{cov} = \frac{1}{2}|\vec{a} \times \vec{b}| + \frac{1}{2}|\vec{b} \times \vec{c}| \tag{1}$$

where:

- a_{cov}: covered area by points
- $\vec{a}$: vector between points 1–4
- $\vec{b}$: vector between points 1–3
- $\vec{c}$: vector between points 1–2

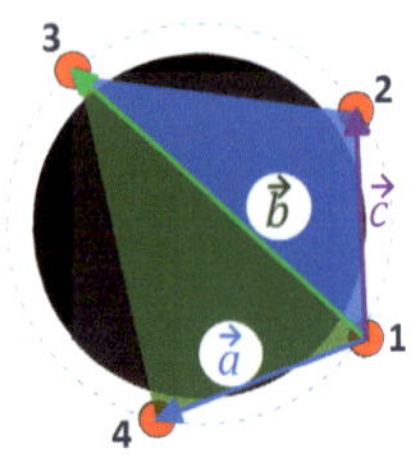

Figure 13. Vectors to calculate covered area. Source: Authors.

For the calculation of the distance cost, in each fleet route associated to each candidate point (Figure 11), the Equation (2) is used:

$$c_{route} = \sum_{i=1}^{i=n} d(p_i, p_{i-1}) \tag{2}$$

where:

- c_{route}: accumulated distance cost of route
- d: euclidean distance between points

The cost values associated to each candidate route are normalized with respect to the maximum distance cost. Since the aim is to maximize the fitness function, and the area covered calculated with respect to the area of the obstacle obtained with OpenCV, a weighted fitness function is defined using the scalarized approach (weighted sum method) with the Equation (3):

$$f_{fitness} = \frac{w_{area} * a_{candidate}}{a_{obstacle}} + w_{route} * max(inv\_route\_cost_{candidates}) \tag{3}$$

where:

- $f_{fitness}$: fitness function
- w_{area}: weight applied for area coverage
- $a_{candidate}$: weight applied for path distance cost
- $a_{obstacle}$: covered by 4-candidate points combination

- w_{route}: area of obstacle obtained from contours
- $inv\_route\_cost_{candidates}$: vector of inverse normalized route cost for candidate points

To estimate candidates that maximize the fitness function while reducing the computational cost, genetic algorithms, implemented with the PyGAD package [52], are used. To use the package, it is necessary to adapt the problem to the structure of a genetic algorithm.

A combination of four candidate points is sought, whose coordinates are defined with values of type "integer", where each candidate point is contained in a pool of feasible solutions, and has an associated fleet path cost value. For the application of PyGAD to the problem, the indices of the lists of each candidate point are used as solutions, having as objective the maximization of the $f_{fitness}$ function, delimiting the maximum values of mutation range to the length of the vector of each feasible candidate. The data type is of type "int". For the generation of the instance, the parameters described in Table 3 are set.

Table 3. Parameters defined in PyGAD instance.

Parameter	Definition	Value
num_generations	Number of generations	1200
mutation_num_genes	Number of genes por instance random_mutation()	4
num_parents_mating	Number of solutions to be selected as parents	2
sol_per_pop	Number of solutions in the population	70
num_genes	Number of genes in each solution	4
fitness_function	Fitness function	$f_{fitness}$
init_range_low	Lower value of random range where initial population is selected.	0
init_range_high	Upper value of random range where initial population is selected.	len(candidate1)/2
crossover_type	Type of crossover operation.	"single_point"
random_mutation_min_val	Start value of the range from which a random value is selected to be added to the gene.	1
random_mutation_max_val	End value of the range from which a random value is selected to be added to the gene.	100
mutation_type	Type of mutation operation	"random"
gene_space	Specify the posible values for each gene in order to restrict the gene values.	[range (0, len(candidate1)), range (0, len(candidate2)), range (0, len(candidate3)), range (0, len(candidate4))]
gene_type	Gene type (numeric data type)	int

The value of w_{area} is 2, and w_{route} is 3, prioritizing the minimization of the fleet route distance cost, which represents the process with the highest energy cost due to the relationship with the number of robots in the fleet.

Algorithm 2 takes as input variables the candidate points c_{point}, and the paths of each candidate c_{path}. The following functions explain the process:

- Get_area (line 13) is a function that gets the area of the contour made by a list of points. The output is the area of the contour.
- Route_cost (line 14) is a function that calculates the accumulated individual distance cost for a list of paths.
- Index_range (line 15) is a function that defines the lower and upper limit indexes for each candidate point according to a minimum distance between points. The output is a range of index for each of the four candidate points.
- Get_candidates (line 16) is a function that defines independent lists of candidate points of a main list according to the given limits. The output is four lists of points.

- Fitness_function (line 17) is a function that iterates testing different combinations according to Ga_instance parameters. It includes the Area_calc() in order to calculate area for every iteration different combination of points.
- Area_calc (line 17) is a function that calculates the area described between four points according to cross product and sum of areas described in Equation (1).
- Ga_instance (line 18) is pygad instance to configure the genetic algorithm that include the parameters described in Table 2. The instance must run using Ga_instance.run. The output of Ga_instance.best_solution() is the best combination of four points according to the fitness function.

Algorithm 2 Points Selection

Input: c_{point}, c_{path}
Output: sol_points,fleet_path

1: *//roi_area: area on negative obstacle in* $pixels^2$
2: *//*c_{point}*: set of candidate points*
3: *//*c_{path}*: set of candidate paths*
4: *//range_c_points: 2x4 matrix of range of points for candidates 1, 2, 3, 4*
5: *//c1: list of candidate 1 points*
6: *//c2: list of candidate 2 points*
7: *//c3: list of candidate 3 points*
8: *//c4: list of candidate 4 points*
9: *//sol_points: solution of four points*
10: *//fit_function: fitness function used in pygad instance*
11: *//sol_paths: fleet paths for solution points*
12: *//path_cost: list of distance costs related to the input* c_{path}
13: $roi\_area \leftarrow Get\_area(c_{point})$
14: path_cost=Route_cost(c_{path})
15: range_c1, range_c2, range_c3, range_c4=Index_range(c_{point})
16: c1, c2, c3, c4=Get_candidates(range_c_points, c_{point})
17: fit_function=fitness_function(c1,c2,c3,c4, roi_area, path_cost, Area_calc())
18: ga_instance(fit_function, range_c_points)
19: sol_points=ga_instance.best_solution()
20: **for** $i \in sol\_points$ **do**
21:     sol_paths[i]=c_{path}[sol_points[i]]
22: **end for**
23: fleet_path=min(sol_paths)
24: **return** sol_points,fleet_path

3.2.2. Phase 2. Roped Individual Configuration

In this phase, the robots are freed from the constraints of the worm configuration, where each mobile robot is attached by a rope to the parallel cable robot (Figure 14). In this phase, it is important to note that the lead robot will not always carry the wire robot, as the robot to carry will be defined by obtaining the feasible deployment configuration. Each robot in the fleet will have its own route, added to the fleet route from the previous phase.

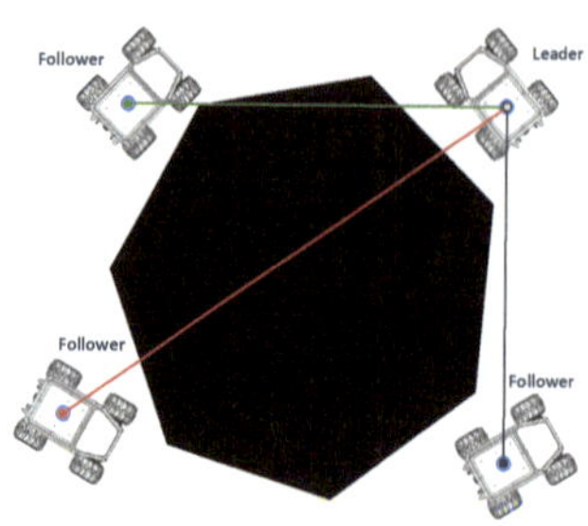

Figure 14. Roped individual configuration. Source: Authors.

Goal Local Planning and Establishment of Deployment Configuration Based on Rope Restrictions

After having selected the best feasible candidate points, a route planning is performed between the selected points to obtain the deployment routes and propose a feasible deployment configuration considering the roped fleet constraints (Table 2). By having one fleet route and several deployment routes, different combinations of possible routes can be generated, shown in Figure 15.

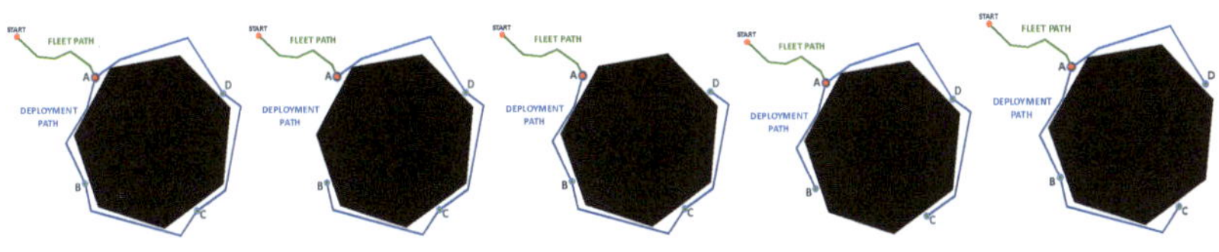

(a) (b) (c) (d) (e)

Figure 15. Candidate deployment configurations, (**a**) available deployment paths, (**b**) deployment path without A-B route, (**c**) deployment path without A-D route, (**d**) deployment path without B-C route, (**e**) deployment path without C-D route. Source: Authors.

The point corresponding to the fleet robot carrying the cable robot is also taken into account since, in addition to generating a feasible deployment path for UGV navigation, corded fleet constraints for positive obstacles must be considered; see Figure 16.

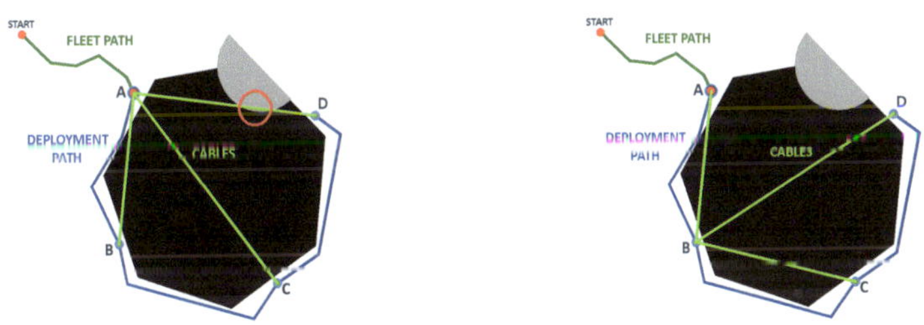

(a) (b)

Figure 16. Cable robot position, (**a**) At point A (no roped feasible configuration because of collision), (**b**) At point B (roped feasible configuration). Source: Authors.

In addition to the previous considerations, it must be evaluated that at each instant of the deployment, there is a rope connecting each mobile robot with the cable robot, where contact with positive obstacles must be avoided; see Figure 17.

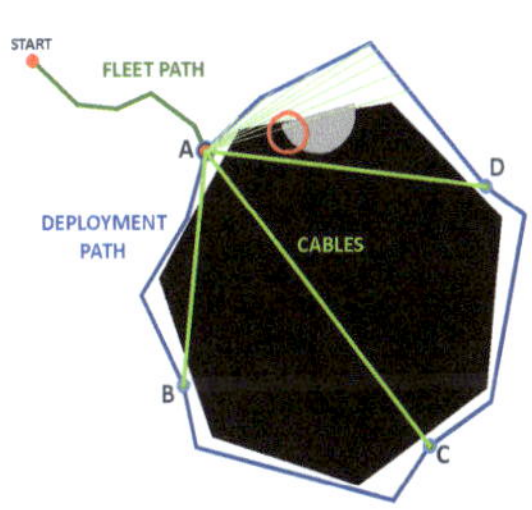

Figure 17. Evaluation of rope contact with positive obstacles at interpolated path for follower robot at A-D candidate route. Source: Authors.

Table 4 details the conditions considered to obtain feasible paths and feasible configurations for possible situations using different start positions.

Table 4. Feasible configuration for deployment path.

Point	1st Feasible Path	2nd Feasible Path	3rd Feasible Path	Feasible Configuration
A	R_1	R_2	R_3	A $\rightarrow$(R_1,R_2,R_3)
	R_1	R_2	R_4	A $\rightarrow$(R_1,R_2,R_4)
	R_1	R_3	R_4	A $\rightarrow$(R_1,R_3,R_4)
	R_1	R_3	R_4	A $\rightarrow$(R_2,R_3,R_4)
B	R_1	R_2	R_3	B $\rightarrow$(R_1,R_2,R_3)
	R_1	R_2	R_4	B $\rightarrow$(R_1,R_2,R_4)
	R_1	R_3	R_4	B $\rightarrow$(R_1,R_3,R_4)
	R_1	R_3	R_4	B $\rightarrow$(R_2,R_3,R_4)
C	R_1	R_2	R_3	C $\rightarrow$(R_1,R_2,R_3)
	R_1	R_2	R_4	C $\rightarrow$(R_1,R_2,R_4)
	R_1	R_3	R_4	C $\rightarrow$(R_1,R_3,R_4)
	R_1	R_3	R_4	C $\rightarrow$(R_2,R_3,R_4)
D	R_1	R_2	R_3	D $\rightarrow$(R_1,R_2,R_3)
	R_1	R_2	R_4	D $\rightarrow$(R_1,R_2,R_4)
	R_1	R_3	R_4	D $\rightarrow$(R_1,R_3,R_4)
	R_1	R_3	R_4	D $\rightarrow$(R_2,R_3,R_4)

Finally, the routes of each robot are defined according to the possible configurations for fleet deployment (Figure 18), using the criteria defined in Table 5.

Table 5. Definition of individual robot paths for feasible deployment path configurations.

Principal Node Point	Feasible Configuration	Robot 1 Path	Robot 2 Path	Robot 3 Path	Robot 4 Path
A	type 1	FP+ADCB	FP + ADC	FP + AD	FP
	type 2	FP + ABCD	FP + ABC	FP + AB	FP
	type 3	FP + ADC	FP + AD	FP + AB	FP
	type 4	FP + ABC	FP + AB	FP + AD	FP
B	type 1	FP + BADC	FP + BAD	FP + BA	FP
	type 2	FP + BCDA	FP + BCD	FP + BC	FP
	type 3	FP + BAD	FP + BA	FP + BC	FP
	type 4	FP + BCD	FP + BC	FP + BA	FP
C	type 1	FP + CBAD	FP + CBA	FP + CB	FP
	type 2	FP + CDAB	FP + CDA	FP + CD	FP
	type 3	FP + CBA	FP + CB	FP + CD	FP
	type 4	FP + CDA	FP + CD	FP + CB	FP
D	type 1	FP + DCBA	FP + DCB	FP + DC	FP
	type 2	FP + DABC	FP + DAB	FP + DA	FP
	type 3	FP + DCB	FP + DC	FP + DA	FP
	type 4	FP + DAB	FP + DA	FP + DC	FP

The criterion for selecting robot routes is to apply the longest route to the leader robot, progressively descending to the shortest route to follower robot 3, in type 1 and type 2 configurations. For type 3 and type 4 configurations, the path to the longest boundary node is assigned to the leader robot, the shortest leg of the path from the leader robot to follower robot 1, the path to the shortest boundary node to follower robot 2, and the shortest path to follower robot 3.

The pseudocode is detailed in Algorithm 3. The functions that help to understand this section are:

- Route (line: 10) is a function that performs the PRM path planning between two points. It returns the feasible path between them.
- Route_check (line: 14) is a function that creates an interpolated line between input point and every interpolated point of the input route, and checks if there are collisions between the interpolated line (rope) and a positive obstacle. Output is a Boolean true if there is collision, and false if there is no collisions.

- GetRobotPaths (line: 35) is a function in which input is feasibility of each path of the roped configuration deployment check, and it returns the path for each robot according to Table 5.

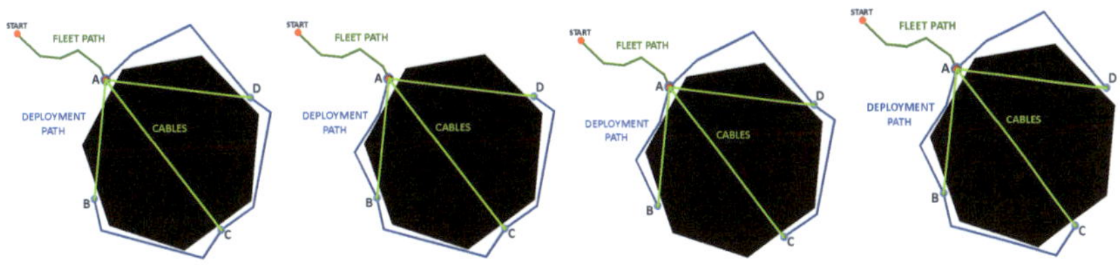

(a) (b) (c) (d)

Figure 18. Feasible deployment path configuration using point A as principal node. (**a**) Type 1. (**b**) Type 2. (**c**) Type 3. (**d**) Type 4. Source: Authors.

Algorithm 3 Roped deployment

```
    Input: sol_points, fleet_path, map
    Output: fleet_path, r1_path, r2_path, r3_path, r4_path
1:  //init_candidates: set of initial candidate points
2:  //check_path: is a Boolean that indicates if the path is feasible to navigate with roped fleet conditions (true=no valid,
    false=valid)
3:  //p1_check: is a list that contains boolean values related to a feasible path between point 1 and a route
4:  //p2_check: is a list that contains boolean values related to a feasible path between point 2 and a route
5:  //p3_check: is a list that contains boolean values related to a feasible path between point 3 and a route
6:  //p4_check: is a list that contains boolean values related to a feasible path between point 4 and a route
7:  map ← gray(map)
8:  gray_m ← Range(map, 50, 150)
9:  for i ∈ (sol_points − 1) do
10:     deployment_paths[i]=Route(sol_points[i],sol_points[i+1])
11: end for
12: for i ∈ (sol_points) do
13:     for j ∈ deployment_paths do
14:         check_path=Route_check(sol_points[i],deployment_paths[j])
15:         if check_path == true then
16:             if i == 1 then
17:                 Append in p1_check this check_path
18:                 break
19:             end if
20:             if i == 2 then
21:                 Append in p2_check this check_path
22:                 break
23:             end if
24:             if i == 3 then
25:                 Append in p3_check this check_path
26:                 break
27:             end if
28:             if i == 4 then
29:                 Append in p4_check this check_path
30:                 break
31:             end if
32:         end if
33:     end for
34: end for
35: robot1_path, robot2_path, robot3_path, robot4_path=GetRobotPaths(p1_check, p2_check, p3_check, p4_check)
36: return robot1_path, robot2_path, robot3_path, robot4_path
```

4. Experiments

In order to test the performance of the algorithm on environments with different characteristics, two groups of environments are defined. The first, GLOBAL TEST, is intended to test different combinations of environments with various positive and negative obstacles, to test the robustness of the fleet in node-edge configuration; this group consists of manual generation of maps, using combinations associated with the characteristics of the fleet environment. The second group, SHAPE TEST, modifies the shape features of the region of interest to be explored, in order to test the robustness of the algorithm and identify the mission characteristics (shape of ROI, number of positive obstacles near the ROI, start point) that can affect the performance.

For GLOBAL TEST, the experiments are performed on six maps with the characteristics shown in Table 6.

Table 6. Characteristics of GLOBAL TEST environment maps.

Map	Number of Positive Obstacles	Number of Negative Obstacles	Number of Positive Obstacles around ROI	Shape of Negative Obstacle
1	0	1	0	Irregular
2	4	1	1	Irregular
3	5	1	2	Irregular
4	8	1	0	Ellipse
5	12	6	0	Ellipse
6	17	34	0	Irregular ellipse

For each map, feasible fleet paths are obtained just once to get the feasible fleet candidate path cost, because the main goal is evaluating the performance of algorithm, not to be a PRM path planner.

For GLOBAL TEST, the maps shown in Figure 19 are used with the characteristics described in Table 6.

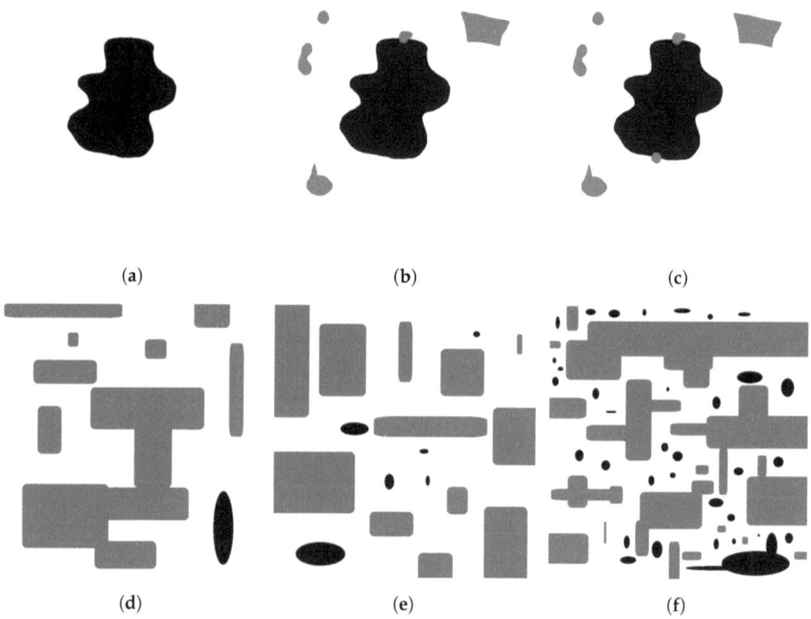

Figure 19. Manually generated maps for GLOBAL TEST, (**a**) map 1, (**b**) map 2, (**c**) map 3, (**d**) map 4, (**e**) map 5, (**f**) map 6. Source: Authors.

For SHAPE TEST, thirty types of maps are automatically generated for each number of positive obstacles around the area of interest, varying the shape of the sinkhole. These maps are generated through a point connection script using cubic Bezier curves [53]. The criteria for autogeneration of the maps are explained in Table 7.

Table 7. Characteristics for automatic generated maps.

Maps	Number of Positive Obstacles around ROI	Cp_{rad}	Smoothness	N_{random}	Scale
1–10	0–2	0.2	0.05	6	500
11–20	0–2	0.3	0.08	7	250
21–30	0–2	0.1	0.1	8	300

Where:

- Maps: The range of maps that are auto-generated with the indicated characteristics.
- N_{random}: number of random points to connect to generate the Bezier curve.
- Cp_{rad}: radius around the points where control points are. A larger radius means a sharper feature.
- Smoothness: Parameter to define the smoothness of the curve.
- Scale: X and Y pixel rectangle size where the random points will be generated.

The total number of generated maps is 90. Tests are performed by planning the routes from four start points ([100,100], [100,900], [800,100], [800,800]), with a total of 360 tests of the algorithm.

The 1000 × 1000 pixel resolution maps have a scale of 0.1 m per pixel, thus, the size range of the generated sinkholes goes from 20 m to 50 m approximately. The reference mobile robot size is the SummitXL (0.7 m length, 0.6 m width, 0.45 m height).

Some of the used maps are shown in Figure 20.

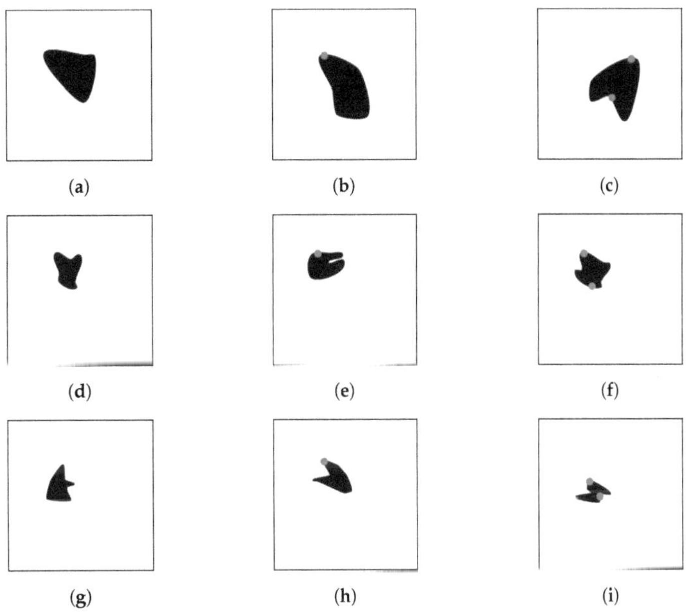

Figure 20. Automatic generated maps for SHAPE TEST, (**a**) maps 1–10 with 0 nearby positive obstacles, (**b**) maps 1–10 with 1 nearby positive obstacles, (**c**) maps 1-10 with 2 nearby positive obstacles, (**d**) maps 11–20 with 0 nearby positive obstacles, (**e**) maps 11–20 with 1 nearby positive obstacles, (**f**) maps 11–20 with 2 nearby positive obstacles, (**g**) maps 21–30 with 0 nearby positive obstacles, (**h**) maps 21–30 with 1 nearby positive obstacles, (**i**) maps 21–30 with 2 nearby positive obstacles. Source: Authors.

All experiments were carried on Intel Core i7-9750H PC with 16GB RAM, under the ROS Noetic operating system and Python 3.8, on a computer running Ubuntu 20.04. For simulation, Gazebo is used through Robotnik Stack for Summit-XL [54], using mobile platforms of four wheels with differential locomotion.

5. Results

This chapter presents the results of the experiments. Table 8 shows the summary of the results obtained, detailing the exceptions for SHAPE TEST, which are explained in the following sections. The results are encouraging, allowing us to identify the circumstances of the map that caused incorrect results.

Table 8. Summary of results.

Experiment	Number of Maps	Number of Experiments	Number of Successful Experiments	Successful Rate
Global TEST	6	24	24	100%
SHAPE TEST (feasible paths)	90	360	357	99.16%
SHAPE TEST (feasible paths with workspace limitation)	90	360	352	97.78%
SHAPE TEST (feasible paths with collision risk)	90	360	350	97.2%
SHAPE TEST (feasible paths with exceptions)	90	360	354	98.33%
SHAPE TEST	90	360	333	92.5%

Tests are performed from the starting point [100,100], obtaining a feasible fleet route, deployment routes, and cable layout. Figures 21–26.

Figure 21. GLOBAL TEST procedure for Map 1. Source: Authors.

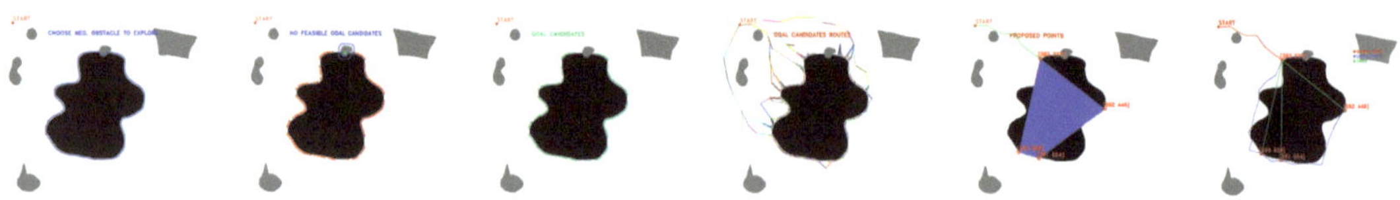

Figure 22. GLOBAL TEST procedure for Map 2. Source: Authors.

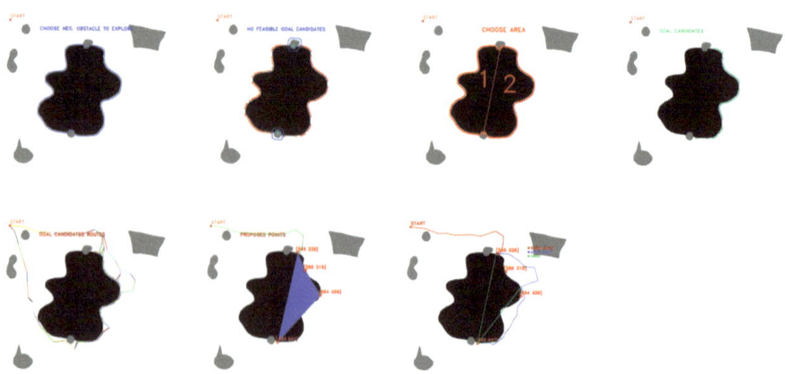

Figure 23. GLOBAL TEST procedure for Map 3. Source: Authors.

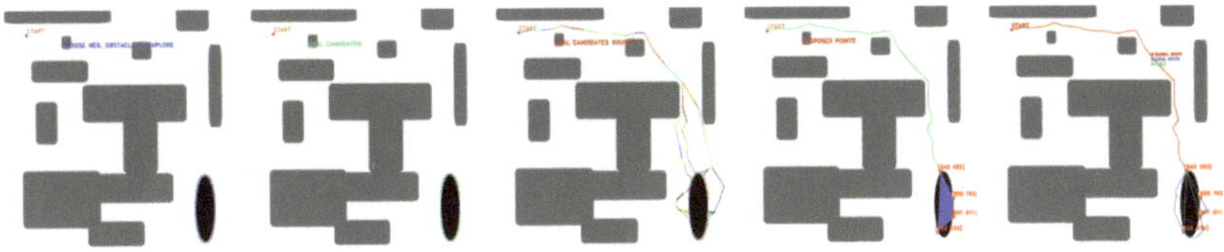

Figure 24. GLOBAL TEST procedure for Map 4. Source: Authors.

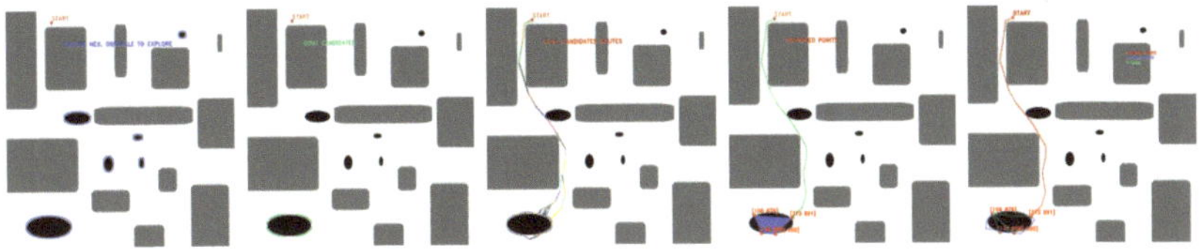

Figure 25. GLOBAL TEST procedure for Map 5. Source: Authors.

Figure 26. GLOBAL TEST procedure for Map 6. Source: Authors.

In GLOBAL TEST experiments, the algorithm finds feasible solutions for 100% of the cases.

In SHAPE TEST experiments, maps with the features in Table 7 are tested. Figure 27 shows some of the results obtained by the algorithm in the experiments performed in the SHAPE TEST environment.

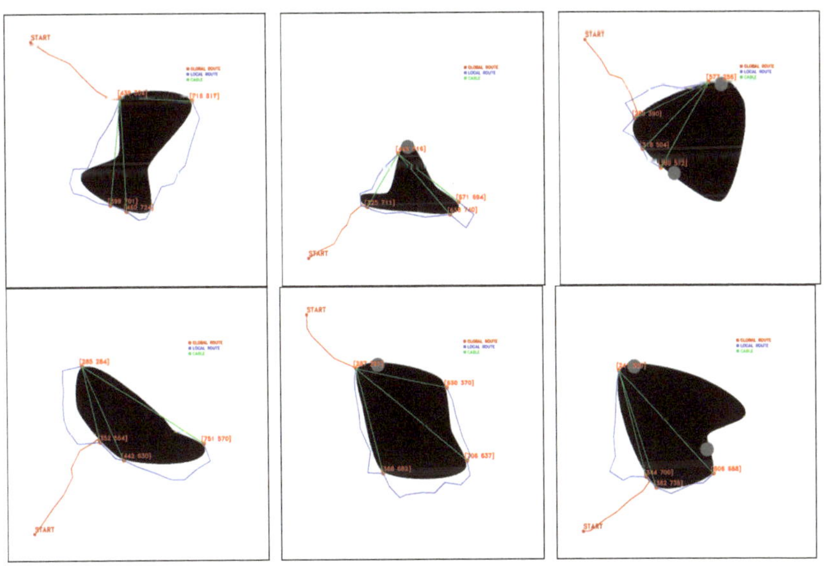

Figure 27. Feasible fleet and deployment paths tested in SHAPE TEST generated environment. Source: Authors.

For the 360 tests, the algorithm returns feasible routes in 357, representing 99.16% efficiency for fleet route and deployment route generation.

Since the scope of the paper is delimited to point selection and feasible route generation for a fleet of corded robots, the workspace of the effector mounted on one of the robots is not considered. However, post-fleet navigation issues are discussed, through caveats defined in two groups: limitations of the corded robot workspace, and risk of collision of the cords in the release of the end effector.

Selected points representing a limitation to the workspace are obtained eight times (Figure 28), i.e., 2.22%, while the risk of cable collision with positive obstacles occurs 10 times (Figure 29), 2.78% of the time.

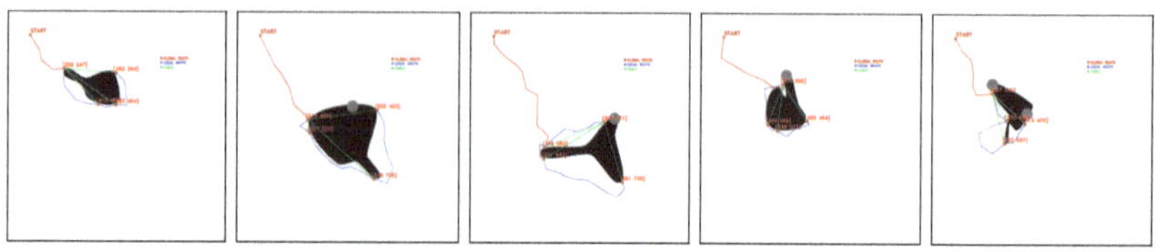

Figure 28. Selected points for feasible fleet and deployment paths that limit the release of the CDPR. Source: Authors.

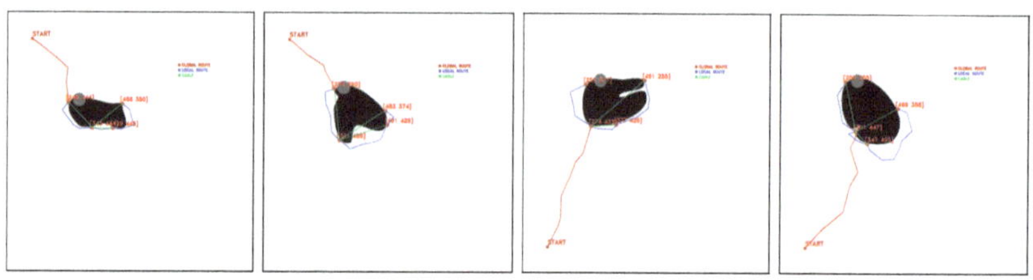

Figure 29. Selected points for feasible fleet and deployment paths close to hit positive obstacles during release of cable drive robot. Source: Authors.

There are also exceptional cases where navigation and deployment of the fleet is feasible, but the release of the effector would present an error. These cases occur in six occasions (1.67%), shown in Figure 30.

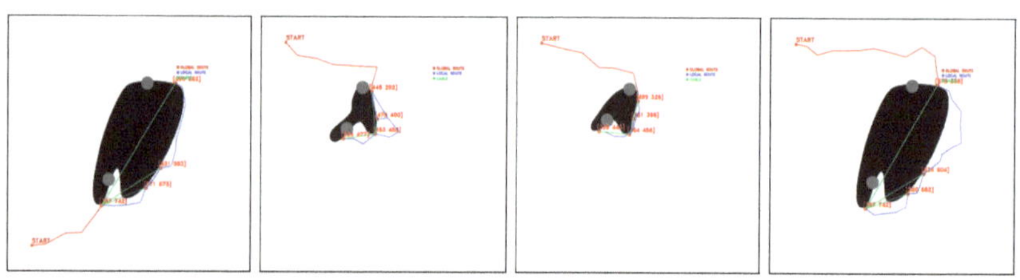

Figure 30. Selected feasible points that present problems during the release of the cable driven robot. Source: Authors.

5.1. Fitness

In order to analyze statistical significance, box and whisker plots associated with the relationships between the fit variables (fitness, weighted fleet route distance cost, weighted area covered) and map characteristics (number of positive obstacles in the ROI, shape characteristics for map generation, starting points) are presented.

Figure 31 shows the box and whisker plots of fitness and map characteristics. The one-way analysis of variance (ANOVA) shows that there are significant differences with $\alpha = 0.05$ of fitness vs. number of positive obstacles in the ROI (F = 305.53, $p = 0$), fitness vs. shape (F = 4.04, $p = 0.0184$), and that there are no significant differences of fitness vs. starting point data (F = 1.08, $p = 0.3588$).

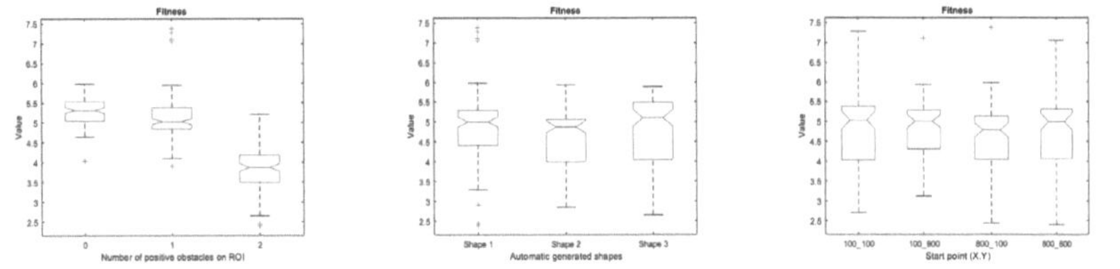

(a) (b) (c)

Figure 31. Box and whiskers diagrams for fitness and its variables, (**a**) fitness vs. number of positive obstacles on ROI, (**b**) fitness vs. automatic generated shapes, (**c**) fitness vs. start points. Source: Authors.

Figure 32 collects the plots for the weighted cost of the area covered and the map features. Analysis of variance indicates significant differences with $\alpha = 0.05$ of area weighted cost vs. number of positive obstacles in the ROI (F = 266.65, $p = 0$), area weighted cost vs. shape (F = 3.4, $p = 0.0345$), and that there are no significant differences of area weighted cost vs. start point data (F = 0.07, $p = 0.9748$).

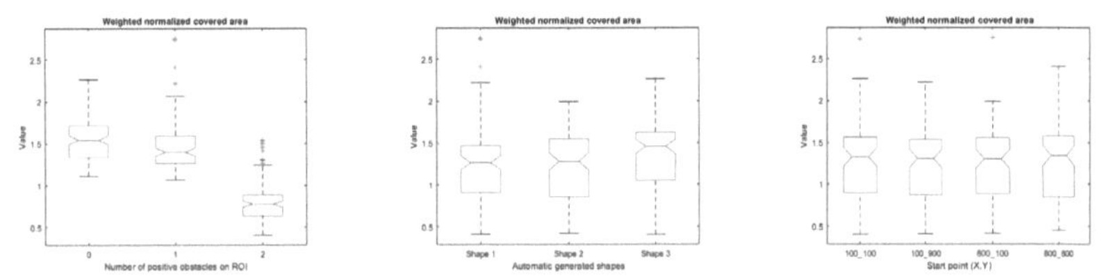

(a) (b) (c)

Figure 32. Box and whiskers diagrams for weighted normalized covered area and its variables, (**a**) weighted normalized covered area vs. number of positive obstacles on ROI, (**b**) weighted normalized covered area vs. automatic generated shapes, (**c**) weighted normalized covered area vs. start points. Source: Authors.

Figure 33 points out the plots for the weighted fleet route distance cost and map features. Analysis of variance reveals significant differences with $\alpha = 0.05$ of weighted fleet route distance cost vs. number of positive obstacles in the ROI (F = 9.64, $p = 0$), weighted fleet route distance cost vs. shape (F = 5.98, $p = 0.0028$), and that there are no significant differences of fleet route distance vs. start point data (F = 1.64, $p = 0.1791$).

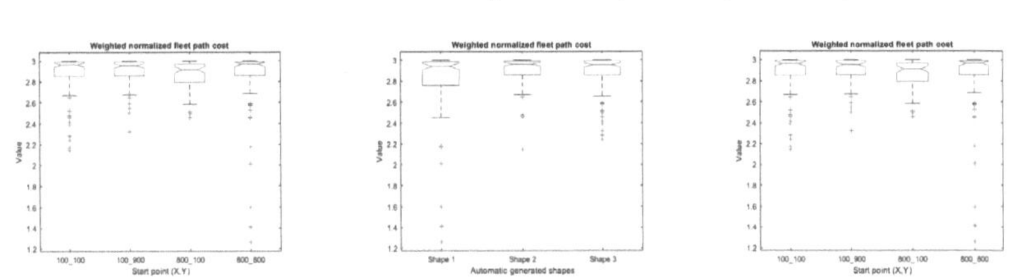

(a) (b) (c)

Figure 33. Box and whiskers diagrams for weighted normalized fleet path cost and its variables, (**a**) weighted normalized fleet path cost vs. number of positive obstacles on ROI, (**b**) weighted normalized fleet path cost vs. automatic generated shapes, (**c**) weighted normalized fleet path cost vs. start points. Source: Authors.

In summary, it can be said that the fitness value obtained as a result of the algorithm is mainly affected by the number of positive obstacles close to the ROI, and secondarily, by the shape of the land depression to be explored. The starting point does not affect in a relevant way the fitness function value.

5.2. Algorithm Complexity

The complexity of an algorithm is usually calculated based on the Big-O notation, divided into time complexity, referring to the execution time of the algorithm, and space complexity, the amount of memory used by an algorithm [55]. In this section, the temporal complexity of the genetic algorithm is discussed. The amount of data processed by an algorithm is represented by N. If the algorithm does not depend on N, it has a constant complexity, represented by the notation O(1). If the algorithm depends on N, the complexity is represented as a function of this variable with the notations O(N), O(N^2), O(log N), O(N log N), O(2^N), O(N!). Figure 34 shows the evolution of the execution time of the genetic algorithm as a function of the number of samples. Input range goes from 76 to 806, with a maximum time of 16.55 s, and an average time of 9.48 s. A linear regression is performed to estimate the fit to the measured time data, obtaining the Equation (4):

$$time = 0.0101 * n_{inputs} + 5.8094 \quad (4)$$

With a value of R^2 of 0.5523, the algorithm has a proportional behaviour, understood in the Big-O notation as O(n), considered a fair complexity.

Figure 34. Execution time of GA algorithm. Source: Authors.

5.3. Test on Gazebo Simulator

For the fleet navigation simulation, the Summit XL robot ROS package is adapted to use four robots. The 3D environments are developed by approximating the splines of the generated maps and exporting to a compatible format for import into the Gazebo environment. A script is implemented to track the trajectories of each robot, and a viewer is set up parallel to the execution of the mission. Figure 35 shows the operation of the algorithm and the navigation of the robot fleet. The size scale used is 0.1 m per pixel for the map, so the sinkhole is approximately 50 m long, 30 m wide, and 20 m deep.

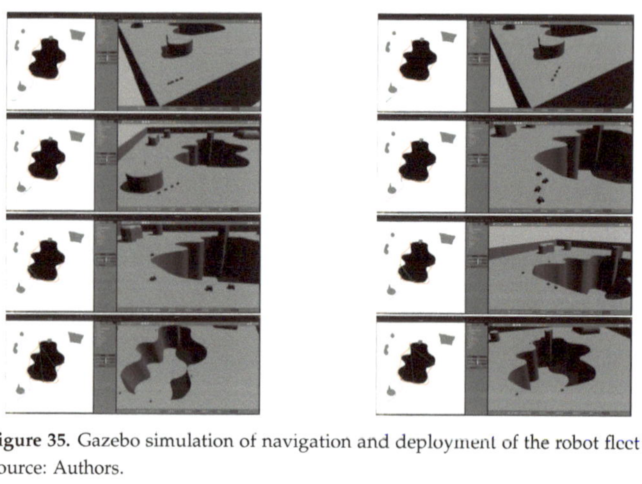

Figure 35. Gazebo simulation of navigation and deployment of the robot fleet around a sinkhole. Source: Authors.

6. Conclusions

In this work, the use of a fleet of UGV land mobile robots is proposed for the deployment of a mobile cable-driven parallel robot to perform the exploration of sinkholes.

For the positioning of the mobile bases of the MCDPR, fleet route planning and corded deployment are performed using maps with different characteristics, generated either manually or automatically.

Several experiments have been performed by varying the fleet route environment and the features of the sinkhole to explore. The results of the experiments show the robustness of the algorithm for the generation of feasible routes, in addition to corroborating that both the shape of the sinkhole and the number of positive obstacles in the ROI region of interest present significant differences for the cost function fit. The validity of the routes in a simulation model in Gazebo has been also evaluated.

The use of evolutionary algorithms considerably reduces the calculation time of the algorithm. However, parameter settings should be tailored for each task, so it is important to initially consider the scope and scalability of the proposed solution.

Future work should focus on optimization of fleet routes, smoothing of planned routes to improve navigation, consideration of physical characteristics of ropes, and integration into real roped robotic fleets. For real-world scenarios, navigation can be improved with the addition of local planning techniques to avoid dynamic obstacles. Simulation can focus on testing the algorithm on rope capable simulators considering the dynamical restrictions of the real roped robotic fleet. In addition, the workspace of the released CDPR as part of the proposed weighted fitness function should be analyzed.

Author Contributions: Conceptualization, A.B., D.O. and J.D.C.; methodology, A.B., D.O. and C.C.U.; software, D.O. and C.C.U.; validation, D.O. and C.C.U.; formal analysis, D.O., C.C.U. and A.B.; investigation, D.O., C.C.U., A.B. and J.D.C.; resources, A.B. and J.D.C.; data curation, D.O. and C.C.U.; writing—original draft preparation, D.O., C.C.U. and J.D.C.; writing—review and editing, D.O., C.C.U. and A.B.; visualization, C.C.U., A.B. and J.D.C.; supervision, C.C.U., A.B. and J.D.C., project administration, A.B. and J.D.C.; funding acquisition, A.B. and J.D.C. All authors have read and agreed to the published version of the manuscript.

Funding: This research has been made possible thanks to the financing of RoboCity2030-DIH-CM, Madrid Robotics Digital Innovation Hub, S2018/NMT-4331, funded by "Programas de Actividades I+D en la Comunidad Madrid" and cofunded by Structural Funds of the EU and TASAR (Team of Advanced Search And Rescue Robots) (PID2019-105808RB-I00), funded by MCIN/ AEI /10.13039/501100011033. This research was developed in Centro de Automática y Robótica—Universidad Politécnica de Madrid—Consejo Superior de Investigaciones Científicas (CAR UPM-CSIC).

Institutional Review Board Statement: Not applicable.

Informed Consent Statement: Not applicable.

Data Availability Statement: A video description of the article can be found on Appendix A. The reported data can be found on Appendix B.

Conflicts of Interest: The authors declare no conflict of interest.

Abbreviations

The following abbreviations are used in this manuscript:

ANOVA	Analysis of Variance
ASR	Air Sea Rescue
CDPR	Cable Driven Parallel Robot
CPRMC	Cable Parallel robot for multiple mobile cranes
DARPA	Defense Advanced Research Project Agency
EA	Evolutionary Algorithms

GA	Genetic Algorithms
MCDPR	Mobile Cable-Driven Parallel Robot
PRM	Probabilistic Roadmap
ROI	Region of Interest
RTS	Robotic Total Station
RUDE-AL	Roped UGV fleet Deployment ALgorithm
SAR	Search and Rescue
TS	Total Station
UGV	Unmanned Ground Vehicle
USAR	Urban Search and Rescue
WiSAR	Wilderness SAR

Appendix A

Link to the video RUDE-AL: Roped UGV deployment algorithm of a MCDPR for sinkhole exploration: (accesed on 30 May 2023). https://youtu.be/2yYPXLVhq2I.

Appendix B

The obtained results can be found on https://github.com/davidorbea92/rude_al (accesed on 30 May 2023).

References

1. Maleki, M.; Salman, M.; Sahebi, S.; Szilard, V. GIS based sinkhole susceptibility mapping using the best worst method. *Spat. Inf. Res.* **2023**. [CrossRef]
2. La Rosa, A.; Pagli, C.; Molli, G.; Casu, F.; De Luca, C.; Pieroni, A.; D'Amato Avanzi, G. Growth of a sinkhole in a seismic zone of the northern Apennines (Italy). *Nat. Hazards Earth Syst. Sci.* **2018**, *18*, 2355–2366. [CrossRef]
3. Montgomery, J.; Jackson, D.; Kiernan, M.; Anderson, J.B.; Ginn, S. *Final Report for ALDOT Project 930-945 Use of Geophysical Methods for Sinkhole Exploration*; Auburn University: Auburn, AL, USA, 2020. pp. 1–120.
4. Liu, H.; Ge, J.; Wang, Y.; Li, J.; Ding, K.; Zhang, Z.; Guo, Z.; Li, W.; Lan, J. Multi-UAV Optimal Mission Assignment and Path Planning for Disaster Rescue Using Adaptive Genetic Algorithm and Improved Artificial Bee Colony Method. *Actuators* **2022**, *11*, 4. [CrossRef]
5. Hermosilla, R.G. The Guatemala City sinkhole collapses. *Carbonates Evaporites* **2012**, *27*, 103–107. [CrossRef]
6. English, S.; Heo, J.; Won, J. Investigation of sinkhole formation with human influence: A case study fromwink sink in Winkler county, Texas. *Sustainability* **2020**, *12*, 3537. [CrossRef]
7. Ali, H.; Choi, J.H. A review of underground pipeline leakage and sinkhole monitoring methods based on wireless sensor networking. *Sustainability* **2019**, *11*, 4007. [CrossRef]
8. Gutiérrez, F.; Benito-Calvo, A.; Carbonel, D.; Desir, G.; Sevil, J.; Guerrero, J.; Martínez-Fernández, A.; Karamplaglidis, T.; García-Arnay, Á.; Fabregat, I. Review on sinkhole monitoring and performance of remediation measures by high-precision leveling and terrestrial laser scanner in the salt karst of the Ebro Valley, Spain. *Eng. Geol.* **2019**, *248*, 283–308. [CrossRef]
9. Munoz-Ceballos, N.D.; Suarez-Rivera, G. Performance criteria for evaluating mobile robot navigation algorithms: A review. *Rev. Iberoam. Autom. Inform. Ind.* **2022**, *19*, 132–143. [CrossRef]
10. Kashino, Z.; Nejat, G.; Benhabib, B. Aerial Wilderness Search and Rescue with Ground Support. *J. Intell. Robot. Syst. Theory Appl.* **2020**, *99*, 147–163. [CrossRef]
11. Reardon, C.; Fink, J. Air-ground robot team surveillance of complex 3D environments. In Proceedings of the SSRR 2016—International Symposium on Safety, Security and Rescue Robotics, Lausanne, Switzerland, 23–27 October 2016; pp. 320–327. [CrossRef]
12. Delmerico, J.; Mueggler, E.; Nitsch, J.; Scaramuzza, D. Active autonomous aerial exploration for ground robot path planning. *IEEE Robot. Autom. Lett.* **2017**, *2*, 664–671. [CrossRef]
13. Xiao, X.; Dufek, J.; Woodbury, T.; Murphy, R. UAV assisted USV visual navigation for marine mass casualty incident response. In Proceedings of the IEEE International Conference on Intelligent Robots and Systems, Vancouver, BC, Canada, 24–28 September 2017; pp. 6105–6110. [CrossRef]
14. Adrian, R.; García, G.; Arias-Montiel, M. *Prototipo Virtual de un Robot Móvil Multi-Terreno Para Aplicaciones de Búsqueda y Rescate Ballbot Lego NXT Control View Project Materiales de Construcción View Project*; Technical Report October; Universidad Tecnologica de la Mixteca: Huajuapan de León, Mexico, 2016.
15. Murphy, R.R.; Nardi, D.; Erkmen, A.M.; Fiorini, P. Search and Rescue Robotics. In *Springer Handbook of Robotics*; Springer: Berlin/Heidelberg, Germany, 2008. [CrossRef]
16. Mehmood, S.; Ahmed, S.; Kristensen, A.S.; Ahsan, D. Multi criteria decision analysis (MCDA) of unmanned aerial vehicles (UAVS) as a part of standard response to emergencies. *Int. J. Innov. Technol. Explor. Eng.* **2018**, *8*, 79–85.

17. Sung, Y. Multi-Robot Coordination for Hazardous Environmental Monitoring. Ph.D. Thesis, Virginia Polytechnic Institute and State University, Blacksburg, VA, USA, 2019.
18. Merino, L.; Caballero, F.; Martínez-de Dios, J.R.; Ollero, A. Cooperative fire detection using unmanned aerial vehicles. *Proc. IEEE Int. Conf. Robot. Autom.* **2005**, *2005*, 1884–1889. [CrossRef]
19. Brenner, S.; Gelfert, S.; Rust, H. New Approach in 3D Mapping and Localization for Search and Rescue Missions. In Proceedings of the CERC, Karlsruhe, Germany, 22–23 September 2017; pp. 105–111.
20. Peña Queralta, J.; Taipalmaa, J.; Pullinen, B.C.; Katha Sarker, V.; Gia, T.N.; Tenhunen, H.; Gabbouj, M.; Raitoharju, J.; Westerlund, T. Collaborative Multi-Robot Systems for Search and Rescue: Coordination and Perception. 2020; pp. 1–28. Available online: http://xxx.lanl.gov/abs/2008.12610 (accessed on 29 May 2023).
21. Fan, H.; Hernandez Bennetts, V.; Schaffernicht, E.; Lilienthal, A.J. Towards gas discrimination and mapping in emergency response scenarios using a mobile robot with an electronic nose. *Sensors* **2019**, *19*, 685. [CrossRef] [PubMed]
22. Zhao, J.; Gao, J.; Zhao, F.; Liu, Y. A search-and-rescue robot system for remotely sensing the underground coal mine environment. *Sensors* **2017**, *17*, 2426. [CrossRef] [PubMed]
23. Qian, S.; Zi, B.; Shang, W.W.; Xu, Q.S. A review on cable-driven parallel robots. *Chin. J. Mech. Eng. (Engl. Ed.)* **2018**, *31*, 66. [CrossRef]
24. Surdilovic, D.; Radojicic, J.; Bremer, N. Efficient Calibration of Cable-Driven Parallel Robots with Variable Structure. *Mech. Mach. Sci.* **2015**, *32*, 113–128. [CrossRef]
25. Bostelman, R.V.; Albus, J.S.; Dagalakis, N.G.; Jacoff, A. Robocrane project: An advanced concept for large scale manufacturing. In Proceedings of the Association for Unmanned Vehicles Systems International, Orlando, FL, USA, 1 July 1996.
26. Pott, A.; Mütherich, H.; Kraus, W.; Schmidt, V.; Miermeister, P.; Verl, A. IPAnema: A family of cable-driven parallel robots for industrial applications. *Mech. Mach. Sci.* **2013**, *12*, 119–134. [CrossRef]
27. El-ghazaly, G.; Gouttefarde, M.; Creuze, V. Adaptive Terminal Sliding Mode Control of a Manipulator: CoGiRo. In *CableCon: Cable-Driven Parallel Robots*; Springer: Cham, Switzerland , 2014; pp.179–200.
28. Miermeister, P.; Lächele, M.; Boss, R.; Masone, C.; Schenk, C.; Tesch, J.; Kerger, M.; Teufel, H.; Pott, A.; Bülthoff, H.H. The CableRobot simulator large scale motion platform based on Cable Robot technology. In Proceedings of the IEEE International Conference on Intelligent Robots and Systems, Daejeon, Republic of Korea, 9–14 October 2016; pp. 3024–3029. [CrossRef]
29. Rodriguez-Barroso, A.; Saltaren, R. Cable-Driven Robot to Simulate the Buoyancy Force for Improving the Performance of Underwater Robots. In *Cable-Driven Parallel Robots*; Springer International Publishing: Cham, Switzerland, 2021; Volume 104, pp. 413–425. [CrossRef]
30. Gouttefarde, M.; Lamaury, J.; Reichert, C.; Bruckmann, T. A Versatile Tension Distribution Algorithm for n-DOF Parallel Robots Driven by n + 2 Cables. *IEEE Trans. Robot.* **2015**, *31*, 1444–1457. [CrossRef]
31. Pedemonte, N.; Rasheed, T.; Marquez-Gamez, D.; Long, P.; Hocquard, É.; Babin, F.; Fouché, C.; Caverot, G.; Girin, A.; Caro, S. FASTKIT: A Mobile Cable-Driven Parallel Robot for Logistics. *Springer Tracts Adv. Robot.* **2020**, *132*, 141–163. [CrossRef]
32. Merlet, J.P. MARIONET, a family of modular wire-driven parallel robots. In *Advances in Robot Kinematics: Motion in Man and Machine*; Springer: Dordrecht, The Netherlands, 2010. [CrossRef]
33. Rasheed, T. Collaborative Mobile Cable-Driven Parallel Robots. Ph.D. Thesis, L'ÉCole Centrale De Nantes, Nantes, France, 2020.
34. Zi, B.; Lin, J.; Qian, S. Localization, obstacle avoidance planning and control of a cooperative cable parallel robot for multiple mobile cranes. *Robot. Comput.-Integr. Manuf.* **2015**, *34*, 105–123. [CrossRef]
35. Tan, H.; Nurahmi, L.; Pramujati, B.; Caro, S. On the Reconfiguration of Cable-Driven Parallel Robots with Multiple Mobile Cranes. In Proceedings of the 2020 5th International Conference on Robotics and Automation Engineering (ICRAE), Singapore, 20–22 November 2020; Volume 1, pp. 126–130. [CrossRef]
36. Seriani, S.; Gallina, P.; Wedler, A. A modular cable robot for inspection and light manipulation on celestial bodies. *Acta Astronaut.* **2016**, *123*, 145–153. [CrossRef]
37. Aguilar, W.G.; Morales, S.; Ruiz, H.; Abad, V. RRT* GL Based Optimal Path Planning for Real-Time Navigation of UAVs. In *Lecture Notes in Computer Science*; Springer: Cham, Switzerland, 2017; pp. 585–595. [CrossRef]
38. Zhu, Z.; Xiao, J.; Li, J.Q.; Wang, F.; Zhang, Q. Global path planning of wheeled robots using multi-objective memetic algorithms. *Integr. Comput.-Aided Eng.* **2015**, *22*, 387–404. [CrossRef]
39. Cui, Y.; Geng, Z.; Zhu, Q.; Han, Y. Review: Multi-objective optimization methods and application in energy saving. *Energy* **2017**, *125*, 681–704. [CrossRef]
40. Yu, W.; Li, B.; Jia, H.; Zhang, M.; Wang, D. Application of multi-objective genetic algorithm to optimize energy efficiency and thermal comfort in building design. *Energy Build.* **2015**, *88*, 135–143. [CrossRef]
41. Feng, Z.; Niu, W.; Cheng, C. Optimization of hydropower reservoirs operation balancing generation benefit and ecological requirement with parallel multi-objective genetic algorithm. *Energy* **2018**, *153*, 706–718. [CrossRef]
42. Ramli, M.A.; Bouchekara, H.R.; Alghamdi, A.S. Optimal sizing of PV/wind/diesel hybrid microgrid system using multi-objective self-adaptive differential evolution algorithm. *Renew. Energy* **2018**, *121*, 400–411. [CrossRef]
43. Hormozi, M.A.; Zaki Dizaji, H.; Bahrami, H.; Sharifyazdi, M.; Monjezi, N. Multi-objective optimization of allocating sustainable mechanization for spraying and harvesting systems in paddy fields. *Iran. J. Biosyst. Eng.* **2023**. [CrossRef]
44. Xue, Y. Mobile Robot Path planning with a non-dominated sorting genetic algorithm. *Appl. Sci.* **2018**, *8*, 2253. [CrossRef]

45. Xue, Y.; Sun, J.Q. Solving the path planning problem in mobile robotics with the multi-objective evolutionary algorithm. *Appl. Sci.* **2018**, *8*, 1425. [CrossRef]
46. Pham, V.T.; Stefek, A.; Krivanek, V.; Nguyen, T.S. Design of a Saving-Energy Fuzzy Logic Controller for a Differential Drive Robot Based on an Optimization. *Appl. Sci.* **2023**, *13*, 997. [CrossRef]
47. Kouritem, S.A.; Abouheaf, M.I.; Nahas, N.; Hassan, M. A multi-objective optimization design of industrial robot arms. *Alex. Eng. J.* **2022**, *61*, 12847–12867. [CrossRef]
48. Yin, L.; Liu, J.; Zhou, F.; Gao, M.; Li, M. Cost-based hierarchy genetic algorithm for service scheduling in robot cloud platform. *J. Cloud Comput.* **2023**, *12*, 35. [CrossRef]
49. Le, A.V.; Parween, R.; Mohan, R.E.; Nhan, N.H.K.; Abdulkader, R.E. Optimization complete area coverage by reconfigurable htrihex tiling robot. *Sensors* **2020**, *20*, 3170. [CrossRef] [PubMed]
50. Kulich, M.; Kubalík, J.; Přeučil, L. An integrated approach to goal selection in mobile robot exploration. *Sensors* **2019**, *19*, 1400. [CrossRef]
51. Rasheed, T.; Long, P.; Roos, A.S.; Caro, S. Optimization based Trajectory Planning of Mobile Cable-Driven Parallel Robots. In Proceedings of the IEEE International Conference on Intelligent Robots and Systems, Macau, China, 3–8 November 2019; pp. 6788–6793. [CrossRef]
52. Gad, A.F. PyGAD: An Intuitive Genetic Algorithm Python Library. 2021. Available online: http://xxx.lanl.gov/abs/2106.06158 (accessed on 29 May 2023).
53. Vailland, G.; Gouranton, V.; Babel, M. Cubic Bézier Local Path Planner for Non-holonomic Feasible and Comfortable Path Generation. In Proceedings of the IEEE International Conference on Robotics and Automation, Xi'an, China, 30 May–5 June 2021; pp. 7894–7900. [CrossRef]
54. Bermúdez Salguero, D.J. Métodos de Ayuda a la Navegación en Exteriores Para Robot Summit en Entorno ROS. Ph.D. Thesis, Universidad de Sevilla Escuela Técnica Superior de Ingeniería, Sevilla, Spain, 2022.
55. Abdiansah, A.; Wardoyo, R. Time Complexity Analysis of Support Vector Machines (SVM) in LibSVM. *Int. J. Comput. Appl.* **2015**, *128*, 28–34. [CrossRef]

sensors

Article

Gaze Point Tracking Based on a Robotic Body–Head–Eye Coordination Method

Xingyang Feng [1,†], Qingbin Wang [2,†], Hua Cong [1], Yu Zhang [1] and Mianhao Qiu [1,*]

1 Army Academy of Armored Forces, Beijing 100072, China; m17695720631@163.com (X.F.); 18911025632@163.com (H.C.); zhangyuzgy2018@163.com (Y.Z.)
2 Research Center of Precision Sensing and Control, Institute of Automation, Chinese Academy of Sciences, Beijing 100190, China; maokang94@163.com
* Correspondence: 15801000128@163.com
† These authors contributed equally to this work.

Abstract: When the magnitude of a gaze is too large, human beings change the orientation of their head or body to assist their eyes in tracking targets because saccade alone is insufficient to keep a target at the center region of the retina. To make a robot gaze at targets rapidly and stably (as a human does), it is necessary to design a body–head–eye coordinated motion control strategy. A robot system equipped with eyes and a head is designed in this paper. Gaze point tracking problems are divided into two sub-problems: in situ gaze point tracking and approaching gaze point tracking. In the in situ gaze tracking state, the desired positions of the eye, head and body are calculated on the basis of minimizing resource consumption and maximizing stability. In the approaching gaze point tracking state, the robot is expected to approach the object at a zero angle. In the process of tracking, the three-dimensional (3D) coordinates of the object are obtained by the bionic eye and then converted to the head coordinate system and the mobile robot coordinate system. The desired positions of the head, eyes and body are obtained according to the object's 3D coordinates. Then, using sophisticated motor control methods, the head, eyes and body are controlled to the desired position. This method avoids the complex process of adjusting control parameters and does not require the design of complex control algorithms. Based on this strategy, in situ gaze point tracking and approaching gaze point tracking experiments are performed by the robot. The experimental results show that body–head–eye coordination gaze point tracking based on the 3D coordinates of an object is feasible. This paper provides a new method that differs from the traditional two-dimensional image-based method for robotic body–head–eye gaze point tracking.

Keywords: bionic eyes; gaze point tracking; gaze point approaching; body–eye–head coordination; 3D coordinates

check for updates

Citation: Feng, X.; Wang, Q.; Cong, H.; Zhang, Y.; Qiu, M. Gaze Point Tracking Based on a Robotic Body–Head–Eye Coordination Method. *Sensors* **2023**, *23*, 6299. https://doi.org/10.3390/s23146299

Academic Editors: David Cheneler and Stephen Monk

Received: 1 June 2023
Revised: 29 June 2023
Accepted: 6 July 2023
Published: 11 July 2023

1. Introduction

When the magnitude of a gaze is too large, human beings change the orientation of their head or body to assist their eyes in tracking targets because saccade alone is insufficient to keep a target at the center region of the retina. Studies on body–head–eye coordination gaze point tracking are still rare because the body–head–eye coordination mechanism of humans is prohibitively complex. Multiple researchers have investigated the eye–head coordination mechanism, binocular coordination mechanism and bionic eye movement control. In addition, researchers have validated the eye–head coordination models on eye–head systems. This work is significant for the development of intelligent robots for human–robot interaction. However, most of these methods are based on the principle of neurology, and their further developments and applications may be limited by people's understanding of human processes. However, binocular coordination based on the 3D coordinates of an object is simple and practical, as verified by our previous paper [1].

When the fixation point transfers greatly, the head and eyes should move in coordination to accurately shift the gaze to the target. Multiple studies have built models of eye–head coordination based on the physiological characteristics of humans. For example, Kardamakis A A et al. [2] researched eye–head movement and gaze shifting. The best balance between eye movement speed and the duration time was sought, and the optimal control method was used to minimize the loss of motion. Freedman E G et al. [3] studied the physiological mechanism of coordinated eye–head movement. However, they did not establish an engineering model. Nakashima et al. [4] proposed a method for gaze prediction that combines information on the head direction with a saliency map. In another study [5], the authors presented a robotic head for social robots to attend to scene saliency with bio-inspired saccadic behaviors. The scene saliency was determined by measuring low-level static scene information, motion, and prior object knowledge. Law et al. [6] described a biologically constrained architecture for developmental learning of eye–head gaze control on an iCub robot. They also identified stages in the development of infant gaze control and proposed a framework of artificial constraints to shape the learning of the robot in a similar manner. Other studies have investigated the mechanisms of eye–head movement for robots and achieved satisfactory performance [7,8].

Some application studies based on coordinated eye–head movement have been carried out in addition to the mechanism research. For example, Kuang et al. [9] developed a method for egocentric distance estimation based on the parallax that emerges during compensatory head–eye movements. This method was tested in a robotic platform equipped with an anthropomorphic neck and two binocular pan–tilt units. Reference [10]'s model is capable of reaching static targets posed at a starting distance of 1.2 m in approximately 250 control steps. Hülse et al. [11] introduced a computational framework that integrates robotic active vision and reaching. Essential elements of this framework are sensorimotor mappings that link three different computational domains relating to visual data, gaze control and reaching.

Some researchers have applied the combined movement of the eyes, head and body in mobile robots. In one study [12], large reorientations of the line of sight, involving combined rotations of the eyes, head, trunk and lower extremities, were executed either as fast single-step or as slow multiple-step gaze transfers. Daye et al. [13] proposed a novel approach for the control of linked systems with feedback loops for each part. The proximal parts had separate goals. In addition, an efficient and robust human tracker for a humanoid robot was implemented and experimentally evaluated in another study [14].

On the one hand, human eyes can obtain three-dimensional (3D) information from objects. This 3D information is useful for humans to make decisions. Human can shift their gaze stably and approach a target using the 3D information of the object. When the human gaze shifts to a moving target, the eyes first rotate to the target, and then the head and even the body rotate if the target leaves the sight of the eyes [15]. Therefore, the eyes, head and body move in coordination to shift the gaze to the target with minimal energy expenditure. On the other hand, when a human approaches a target, the eyes, head and body rotate to face the target and the body moves toward the target. The two movements are typically executed with the eyes, head and body acting in conjunction. A robot that can execute these two functions will be more intelligent. Such a robot would need to exploit the smooth pursuit of eyes [16], coordinated eye–head movement [17], target detection and the combined movement of the eyes, head and robot body to carry out these two functions. Studies have achieved many positive results in these aspects.

Mobile robots can track and locate objects according to 3D information. Some special cameras such as deep cameras and 3D lasers have been applied to obtain the 3D information of the environment and target. In one study [18], a nonholonomic underactuated robot with bounded control was described that travels within a 3D region. A single sensor provided the value of an unknown scalar field at the current location of the robot. Nefti-Meziani S et al. [19] presented the implementation of a stereo-vision system integrated in a humanoid robot. The low cost of the vision system is one of the main aims, avoiding expensive investment in hardware when used in robotics for 3D perception. Namavari A et al. [20] presented an automatic system for the gauging and digitalization of 3D indoor environments. The configuration consisted of an autonomous mobile robot, a reliable 3D laser rangefinder and three elaborated software modules.

The main forms of motion of bionic eyes include saccade [1], smooth pursuit, vergence [21], vestibule–ocular reflex (VOR) [22] and optokinetic reflex (OKR) [23]. Saccade and smooth pursuit are the two most important functions of the human eye. Saccade is used to move eyes voluntarily from one point to another by rapid jumping, while smooth pursuit can be applied to track moving targets. In addition, binocular coordination and eye–head coordination are of high importance to realize object tracking and gaze control.

It is of great significance for robots to be able change their fixation point quickly. In control models, the saccade control system should be implemented using a position servo controller to change and keep the target at the center region of the retina with minimum time consumption. Researchers have been studying the implementation of saccade on robots over the last twenty years. For example, in 1997, Bruske et al. [24] incorporated saccadic control into a binocular vision system by using the feedback error learning (FEL) strategy. In 2013, Wang et al. [25] designed an active vision system that can imitate saccade and other eye movements. The saccadic movements were implemented with an open-loop controller, which ensures faster saccadic eye movements than a closed-loop controller can accommodate. In 2015, Antonelli et al. [26] achieved saccadic movements on a robot head by using a model called recurrent architecture (RA). In this model, the cerebellum is regarded as an adaptive element used to learn an internal model, while the brainstem is regarded as a fixed-inverse model. The experimental results on the robot showed that this model is more accurate and less sensitive to the choice of the inverse model relative to the FEL model.

The smooth pursuit system acts as a velocity servo controller to rotate eyes at the same angular rate as the target while keeping them oriented toward the desired position or in the desired region. In Robinson's model of smooth pursuit [27], the input is the velocity of the target's image across the retina. The velocity deviation is taken as the major stimulus to pursue and is transformed into an eye velocity command. Based on Robinson's model, Brown [28] added a smooth predictor to accommodate time delays. Deno et al. [29] applied a dynamic neural network, which unified two apparently disparate models of smooth pursuit and dynamic element organization to the smooth pursuit system. The dynamic neural network can compensate for delays from the sensory input to the motor response. Lunghi et al. [30] introduced a neural adaptive predictor that was previously trained to accomplish smooth pursuit. This model can explain a human's ability to compensate for the 130 ms physiological delay when they follow external targets with their eyes. Lee et al. [31] applied a bilateral OCS model on a robot head and established rudimentary prediction mechanisms for both slow and fast phases. Avni et al. [32] presented a framework for visual scanning and target tracking with a set of independent pan–tilt cameras based on model predictive control (MPC). In another study [33], the authors implemented smooth pursuit eye movement with prediction and learning in addition to solving the problem of time delays in the visual pathways. In addition, some saccade and smooth pursuit models have been validated on bionic eye systems [34–37]. Santini F et al. [34] showed that the oculomotor strategies by which humans scan visual scenes produce parallaxes that provide an accurate estimation of distance. Other studies have realized the coordinated control of eye and arm movements through configuration and training [35]. Song Y et al. [36] proposed a binocular control model, which was derived from a neural pathway, for smooth pursuit. In their smooth pursuit experiments, the maximum retinal error was less than 2.2°, which is sufficient to keep a target in the field of view accurately. An autonomous mobile manipulation system was developed in the form of a modified image-based visual servo (IBVS) controller in a study [37].

The above-mentioned work is significant for the development of intelligent robots. However, there are some shortcomings. First, most of the existing methods are based on the principle of neurology, and further developments and applications may be limited by people's understanding aimed at humans. Second, only two-dimensional (2D) image information is applied when gaze shifts to targets are implemented, while 3D information is ignored. Third, the studies of smooth pursuit [16], eye–head coordination [17], gaze shift and approach are independent and have not been integrated. Fourth, bionic eyes are different from human eyes; for example, some of them are two eyes that are fixed without movement or move with only 1 DOF, whereas some of them use special cameras or a single camera. Fifth, the movements of bionic eyes and heads are performed separately, without coordination.

To overcome the shortcomings mentioned above to a certain extent, a novel control method that implements the gaze shift and approach of a robot according to 3D coordinates is proposed in this paper. A robot system equipped with bionic eyes, a head and a mobile robot is designed to help nurses deliver medicine in hospitals. In this system, both the pan and each eye have 2 DOF (namely, tilt and pan [38]), and the mobile robot can rotate and move forward over the ground. When the robot gaze shifts to the target, the 3D coordinates of the target are acquired by the bionic eyes and transferred to the eye coordination system, head coordination system and robot coordination system. The desired position of the eye, head and robot are calculated based on the 3D information of the target. Then, the eye, head and mobile robot are driven to the desired positions. When the robot approaches the target, the eye, head and mobile robot first rotate to the target and then move to the target. This method allows the robot to achieve the above-mentioned functions with minimal resource consumption and can separate the control of the eye, head and mobile robot, which can improve the interactions between robots, human beings and the environment.

The rest of the paper is organized as follows. In Section 2, the robot system platform is introduced, and the control system is presented. In Section 3, the desired position is discussed and calculated. Robot pose control is described in Section 4. The experimental results are given and discussed in Section 5; finally, conclusions are drawn in Section 6.

2. Platform and Control System

To study the gaze point tracking of the robot, this paper designs a robot experiment platform including the eye–head subsystem and the mobile robot subsystem.

2.1. Robot Platform

The physical object of the robot is shown in Figure 1. With the mobile robot as a carrier, a head with two degrees of freedom is fixed on the mobile robot, and the horizontal and vertical rotations of the head are controlled by M_{hu} and M_{hd}, respectively. The bionic eye system is fixed to the head. The mobile robot is driven by two wheels, each of which is individually controlled by a servo motor. The angle and displacement of the robot platform can be determined by controlling the distance and speed of each wheel's movement. The output shaft of each stepper motor of the head and eye is equipped with a rotary encoder to detect the position of the motor. Using the frequency multiplication technique, the resolution of the rotary encoder is 0.036°. The purpose of using a rotary encoder is to prevent the effects of lost motor motion on the 3D coordinate calculations. The movement of each motor is limited by a limit switch. The initial positioning of the eye system is based on the visual positioning plate [39].

The robot system includes two eyes and one mobile robot. To simulate the eyes and the head, six DOFs are designed in this system. The left eye's pan and tilt are controlled by motors M_{lu} and M_{ld}, respectively. The right eye's pan and tilt are controlled by motors M_{ru} and M_{rd}, respectively. The head's pan and tilt are controlled by motors M_{hu} and M_{hd}, respectively. The mobile robot has two driving wheels and can perform rotation and forward movement. When the mobile robot needs to rotate, two wheels are set to turn the same amount in different directions. When the mobile robot needs to go forward, two wheels are set to turn the same amount in the same direction.

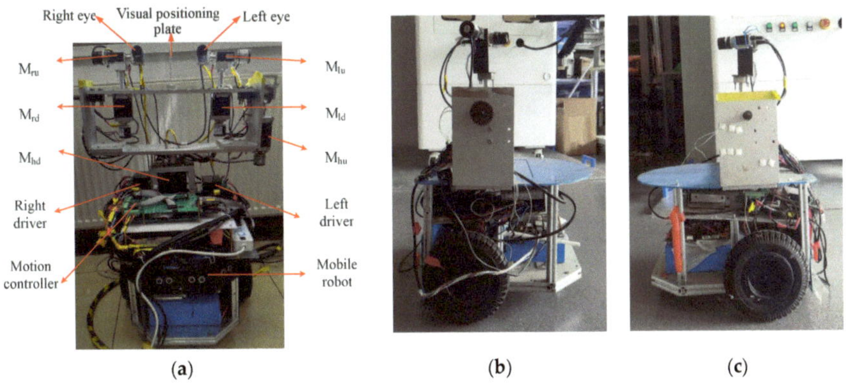

Figure 1. Physical implementation of the robot system. (**a**) The front side. (**b**) The left side. (**c**) The right side.

A diagram of the robot system's organization is shown in Figure 2. The host computer and the mobile robot motion controller, the head motion controller and the eye motion controller all communicate through the serial ports. For satisfactory communication quality and stability, the baud rate of serial communication is 9600 bps. The camera communicates with the host computer via a GigE Gigabit Network. The camera's native resolution is 1600 × 1200 pixels. To increase the calculation speed, the system uses an image downsampled to 400 × 300 pixels.

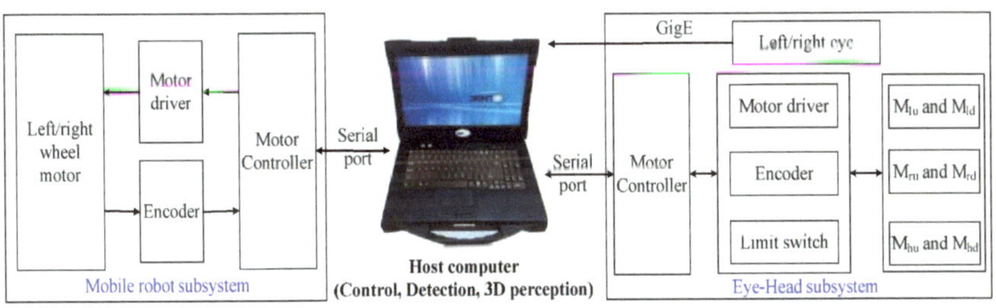

Figure 2. Robot system's organization diagram.

2.2. Control System

Figure 3 shows the control block diagram of the gaze point tracking of the mobile robot. First, based on binocular stereo-vision perception, the binocular pose and the left and right images are used to calculate the 3D coordinates of the target [40], and the coordinates of the target in the eye coordinate system are converted to the head and mobile robot coordinate system. Then, the desired poses of the eyes, head and mobile robot are calculated according to the 3D coordinates of the target. Finally, according to the desired pose, the motor is controlled to move to the desired position, and the change in the position of the motor is converted into changes in the eyes, head and mobile robot.

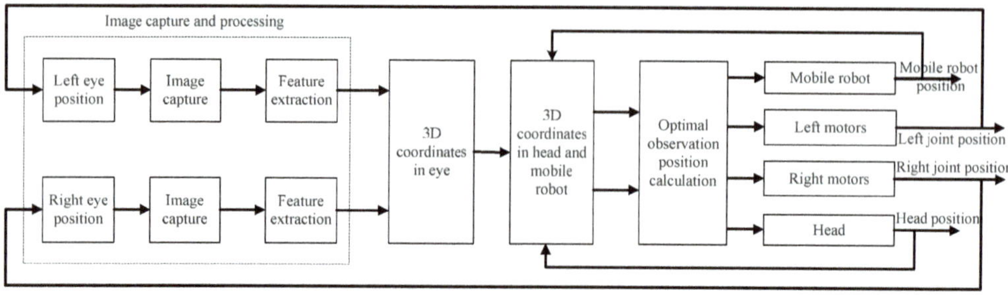

Figure 3. Block diagram of the gaze point tracking control system.

The tracking and approaching motion control problem based on the target 3D coordinates [1] is equivalent to solving the index J minimization problem of Equation (1), where f_i is the current state vector of the joint pose of the eye, head and mobile robot and f_q is the desired state vector:

$$J = \left\| f_i - f_q \right\| \tag{1}$$

where J is the indicator function.

Figure 4a shows the definition of each coordinate system of the robot. The coordinate system of the eye is $O_eX_eY_eZ_e$, which coincides with the left motion module's base coordinate system at the initial position. The head coordinate system is $O_hX_hY_hZ_h$, and the coordinates $\boldsymbol{P}_h$ (x_h, y_h, z_h) of the point $\boldsymbol{P}$ in the head coordinate system can be calculated using the coordinates $\boldsymbol{P}_e$ (x_e, y_e, z_e) in the eye coordinate system. The definitions of d_x and d_y are shown in Figure 4b. The robot coordinate system $O_wX_wY_wZ_w$ coincides with the head coordinate system of the initial position. In the bionic eye system, the axis of rotation of the robot approximately coincides with Y_w.

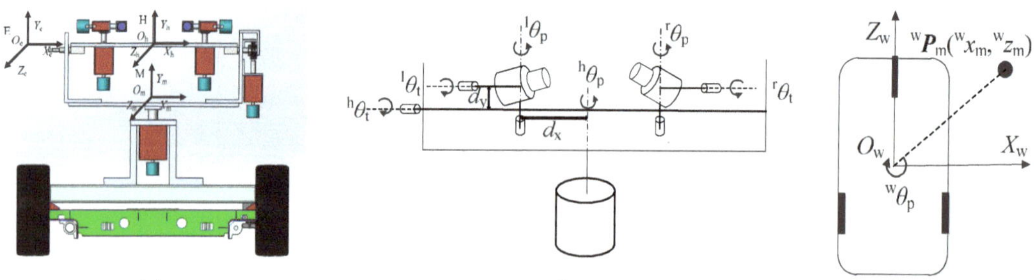

Figure 4. Robot coordinates system and system parameter definition, (**a**) coordinate system definition, (**b**) eye–head system parameters and (**c**) mobile robot parameters.

Figure 4b,c show the definition of each system parameter. ${}^{l}\theta_p$ and ${}^{l}\theta_t$ are the pan and tilt of the left eye, respectively. ${}^{r}\theta_p$ and ${}^{r}\theta_t$ are the pan and tilt of the right eye, respectively. ${}^{h}\theta_p$ and ${}^{h}\theta_t$ are the pan and tilt of the head, respectively. The angle of the robot that rotates around the Y_w axis is ${}^{w}\theta_p$. The robot can not only rotate around Y_w but can also shift in the $X_wO_wZ_w$ plane. When the robot moves, the robot coordinate system at time i is the base coordinate system, and the position of the robot at time $i + 1$ relative to the base coordinate system is ${}^{w}\boldsymbol{P}_m$ (${}^{w}x_m$, ${}^{w}z_m$). When the robot performs gaze point tracking or approaches the target, the 3D coordinates of the target are first calculated at time i, and then the desired posture f_q of each part of the robot at time $i + 1$ is calculated according to the 3D coordinates of the target. When the current pose f_i of the robot system is equal to the desired pose, the robot maintains the current pose; when not equal, the system controls the various parts of the robot to move to the desired pose. The current pose vector of the robot system is $f_i = ({}^{w}x_{mi}, {}^{w}z_{mi}, {}^{w}\theta_{pi}, {}^{h}\theta_{pi}, {}^{h}\theta_{ti}, {}^{l}\theta_{pi}, {}^{l}\theta_{ti}, {}^{r}\theta_{pi}, {}^{r}\theta_{ti})$, and the desired pose is $f_q = ({}^{w}x_{mq}, {}^{w}z_{mq}, {}^{w}\theta_{pq}, {}^{h}\theta_{pq}, {}^{h}\theta_{tq}, {}^{l}\theta_{pq}, {}^{l}\theta_{tq}, {}^{r}\theta_{pq}, {}^{r}\theta_{tq})$. When performing in situ gaze point tracking, the robot performs only pure rotation and does not move forward. When the robot approaches the target, it first turns to the target and then moves straight toward the target. Therefore, the definition of f_q in the two tasks is different. Let ${}^{g}f_q$ be the desired pose when the gaze point is tracked and ${}^{a}f_q$ be the desired pose of the robot when approaching the target.

After analyzing the control system, we found that the most important step in solving this control problem is to determine the desired pose.

3. Desired Pose Calculation

When performing in situ gaze point tracking, the robot performs only pure rotation and does not move forward. When the robot approaches the target, it first turns to the target and then moves straight toward the target. Therefore, the calculation of the desired pose can be divided into two sub-problems: (1) desired pose calculation for in situ gaze point tracking and (2) desired pose calculation for approaching gaze point tracking.

The optimal observation position is used for the accurate acquisition of 3D coordinates. The 3D coordinate accuracy is related to the baseline, time difference and image distortion. In the bionic eye platform, the baseline is changed with the changes in the cameras' positions because the optical center is not coincident with the center of rotation. The 3D coordinate error of the target is smaller when the baseline of the two cameras is longer. Therefore, it is necessary to keep the baseline unchanged. On the other hand, there is a time difference caused by unstick synchronization between image acquisition and camera position acquisition. In addition, it is necessary to keep the target in the center areas of the two camera images to obtain accurate 3D coordinates of the target.

3.1. Optimal Observation Position of Eyes

In the desired pose of the robot, the most important aspect is the expected pose of the bionic eye [40]. Following the definition of this parameter, the calculation of the desired pose of the robot system is greatly simplified; thus, we present an engineering definition here of the desired pose of the bionic eye.

As shown in Figure 5, ${}^{l}\boldsymbol{m}_i$ (${}^{l}u_i$, ${}^{l}v_i$) and ${}^{r}({}^{r}u_i, {}^{r}v_i)$ are the image coordinates of point ${}^{e}\boldsymbol{P}$ in the camera at time i. ${}^{l}\boldsymbol{m}_o$ and ${}^{r}\boldsymbol{m}_o$ are the image centers of the left and right cameras, respectively. ${}^{l}\boldsymbol{P}$ is the vertical point of ${}^{e}\boldsymbol{P}$ along the line ${}^{l}O_c{}^{l}Z_c$, and ${}^{r}\boldsymbol{P}$ is the vertical point of ${}^{e}\boldsymbol{P}$ along the line ${}^{r}O_c{}^{r}Z_c$. ${}^{l}\Delta m$ is the distance between ${}^{l}\boldsymbol{m}$ and ${}^{l}\boldsymbol{m}_o$. ${}^{r}\Delta m$ is the distance between ${}^{r}\boldsymbol{m}$ and ${}^{r}\boldsymbol{m}_o$. D_b is the baseline length. The pan angles of the left and right cameras in the optimal observation position are ${}^{l}\theta_p$ and ${}^{r}\theta_p$, respectively. The tilt angles of the left and right cameras in the optimal observation position are ${}^{l}\theta_t$ and ${}^{r}\theta_t$, respectively. $\boldsymbol{P}_{ob}$ (${}^{l}\theta_p$, ${}^{l}\theta_t$, ${}^{r}\theta_p$, ${}^{r}\theta_t$) is the optimal observation position.

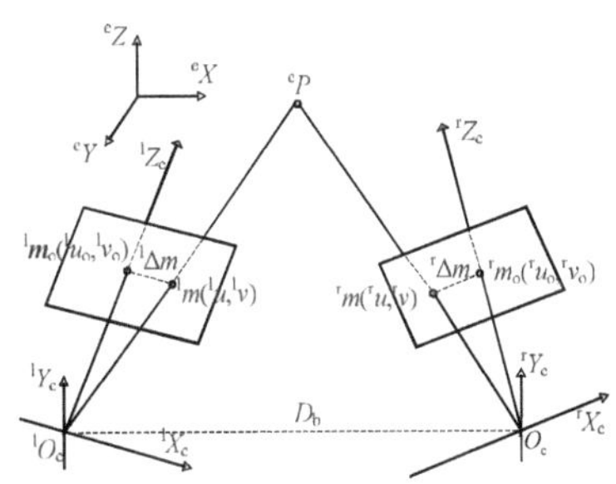

Figure 5. Schematic of the relationship between a Cartesian point and its image point.

When the two eyeballs of the bionic eye move relative to each other, the 3D coordinates of the target obtained by the bionic eye produces a large error. To characterize this error, we give a detailed analysis of its origins in Appendix A. Through analysis, we obtain the following conclusions to reduce the measurement error of the bionic eye:

(1) Make the length of D_b long enough, and maintain as much length as possible during the movement;

(2) Try to observe the target closer to the target so that the depth error is as small as possible;

(3) During the movement of the bionic eye, control the two cameras so that they move at the same angular velocity;

(4) Try to keep the target symmetrical, and make ${}^{l}\Delta m$ and ${}^{r}\Delta m$ as equal as possible in the left and right camera images.

Based on these four methods, the motion strategy of the motor is designed, and the measurement accuracy of the target's 3D information can be effectively improved.

According to the conclusion, we can define a definition of the optimal observed pose of the bionic eye to reduce the measurement error.

The optimal observation position needed to meet the conditions is listed in Equation (2). When the target is very close to the eyes, the target's optimal observation position cannot be obtained because the image position of the target can be kept at the image center region. It is challenging to obtain the optimal solution of the observation position based on Equation (12). However, a suboptimal solution can be obtained by using a simplified calculation method. First, ${}^{l}\theta_t$ and ${}^{r}\theta_t$ are calculated in the case that ${}^{l}\theta_t$ and ${}^{r}\theta_t$ are equal to zero; then, ${}^{l}\theta_t$ and ${}^{r}\theta_t$ are calculated while ${}^{l}\theta_t$ and ${}^{r}\theta_t$ are kept equal to the calculated value. Trial-and-error methods can be used to obtain the optimal solution when the suboptimal solution is obtained.

$$\begin{cases} {}^{l}\theta_{pq} = {}^{r}\theta_{pq} = \theta_p \\ {}^{l}\theta_{tq} = {}^{r}\theta_{tq} = \theta_t \\ {}^{l}\Delta \boldsymbol{m} = -{}^{r}\Delta \boldsymbol{m} \end{cases} \tag{2}$$

where

$${}^{l}\Delta \boldsymbol{m} = \begin{pmatrix} {}^{l}\Delta u \\ {}^{l}\Delta v \end{pmatrix} = \begin{pmatrix} {}^{l}u_i - {}^{l}u_0 \\ {}^{l}v_i - {}^{l}v_0 \end{pmatrix} \tag{3}$$

$${}^{r}\Delta \boldsymbol{m} = \begin{pmatrix} {}^{r}\Delta u \\ {}^{r}\Delta v \end{pmatrix} = \begin{pmatrix} {}^{r}u_i - {}^{r}u_0 \\ {}^{r}v_i - {}^{r}v_0 \end{pmatrix} \tag{4}$$

3.2. Desired Pose Calculation for In Situ Gaze Point Tracking

When the range of target motion is large and the desired posture of the eyeball exceeds its reachable posture, the head and mobile robot move to keep the target in the center region of the image. In robotic systems, eye movements tend to consume the least amount of resources and do not have much impact on the stability of the head and mobile robot during exercise. Head rotation consumes more resources than the eyeball but consumes fewer resources than trunk rotation. At the same time, the rotation of the head affects the stability of the eyeball but does not have much impact on the stability of the trunk. Mobile robot rotation consumes the most resources and has a large impact on the stability of the head and eyeball. When tracking the target, one needs only to keep the target in the center region of the binocular image. Therefore, when performing gaze point tracking, the movement mechanism of the head, eyes and mobile robot are designed with the principle of minimal resource consumption and maximum system stability. When the eyeball can perceive the 3D coordinates of the target in the reachable and optimal viewing posture, only the eye is rotated; otherwise, the head is rotated. The head also has an attainable range of poses. When the desired pose exceeds this range, the mobile robot needs to be turned so that the bionic eye always perceives the 3D coordinates of the target in the optimal viewing position. Let ${}^{h}\gamma_p$ and ${}^{h}\gamma_t$ be the angles between the head and the gaze point in the $X_hO_hZ_h$ and $Y_hO_hZ_h$ planes, respectively. The range of binocular rotation in the horizontal direction is $[-{}^{e}\theta_{pmax}, {}^{e}\theta_{pmax}]$, and the range of binocular rotation in the vertical direction is $[-{}^{e}\theta_{tmax}, {}^{e}\theta_{tmax}]$. The range of head rotation in the horizontal direction is $[-{}^{h}\theta_{pmax},$

${}^{\mathrm{h}}\theta_{\mathrm{pmax}}]$, and the range of head rotation in the vertical direction is $[-{}^{\mathrm{h}}\theta_{\mathrm{tmax}}, {}^{\mathrm{h}}\theta_{\mathrm{tmax}}]$. For the convenience of calculation, the angles between the head and the fixation point in the horizontal direction and the vertical direction are designated as $[-{}^{\mathrm{h}}\gamma_{\mathrm{pmax}}, {}^{\mathrm{h}}\gamma_{\mathrm{pmax}}]$ and $[-{}^{\mathrm{h}}\gamma_{\mathrm{tmax}}, {}^{\mathrm{h}}\gamma_{\mathrm{tmax}}]$, respectively. When the angle between the head and the target exceeds a set threshold, the head needs to be rotated to the ${}^{\mathrm{h}}\theta'_{\mathrm{p}}$ and ${}^{\mathrm{h}}\theta'_{\mathrm{t}}$ positions in the horizontal and vertical directions, respectively. When ${}^{\mathrm{h}}\theta'_{\mathrm{p}}$ exceeds the angle that the head can attain, the angle at which the mobile robot needs to be compensated is ${}^{\mathrm{w}}\theta_{\mathrm{p}}$. In the in situ gaze point tracking task, the cart does not need to translate in the $X_{\mathrm{w}}O_{\mathrm{w}}Z_{\mathrm{w}}$ plane, so $x_{\mathrm{w}} = 0$, and $z_{\mathrm{w}} = 0$. Furthermore, according to the definition of the optimal observation pose of the bionic eye, the conditions that ${}^{\mathrm{g}}f_{\mathrm{q}}$ should satisfy are

$$
{}^{\mathrm{g}}f_{\mathrm{q}} = \begin{pmatrix}
{}^{\mathrm{w}}x_{\mathrm{mq}} = 0 \\
{}^{\mathrm{w}}z_{\mathrm{mq}} = 0 \\
{}^{w}\theta_{\mathrm{pq}} = \left\{\theta \mid \ \left|\theta\right| \leq 2\pi, {}^{\mathrm{h}}\theta_{\mathrm{pq}} + \theta = {}^{\mathrm{h}}\theta'_{\mathrm{p}}\right\} \\
{}^{\mathrm{h}}\theta_{\mathrm{pq}} = \{\theta \mid \ |\theta| \leq {}^{\mathrm{h}}\theta_{\mathrm{pmax}}, \left|{}_{\mathrm{h}}\gamma_{\mathrm{p}}\right| \leq^{\mathrm{h}} \gamma_{\mathrm{pmax}}\} \\
{}^{\mathrm{h}}\theta_{\mathrm{tq}} = \{\theta \mid \ |\theta| \leq {}^{\mathrm{h}}\theta_{\mathrm{tmax}}, \left|{}_{\mathrm{h}}\gamma_{\mathrm{t}}\right| \leq^{\mathrm{h}} \gamma_{\mathrm{tmax}}\} \\
{}^{\mathrm{l}}\theta_{\mathrm{pq}} = {}^{\mathrm{r}}\theta_{\mathrm{pq}} = \{\theta \mid \ |\theta| \leq {}^{\mathrm{e}}\theta_{\mathrm{pmax}}, {}^{\mathrm{l}}\Delta m_{\mathrm{l}} = -\Delta m_{\mathrm{r}}\} \\
{}^{\mathrm{l}}\theta_{\mathrm{tq}} = {}^{\mathrm{r}}\theta_{\mathrm{tq}} = \{\theta \mid \ |\theta| \leq {}^{\mathrm{e}}\theta_{\mathrm{tmax}}, \Delta m_{\mathrm{l}} = -\Delta m_{\mathrm{r}}\}
\end{pmatrix} \tag{5}
$$

The desired pose needs to be calculated based on the 3D coordinates of the target. Therefore, to obtain the desired pose, it is necessary to acquire the 3D coordinates of the target according to the current pose of the robot.

3.2.1. Three-Dimensional Coordinate Calculation

The mechanical structure and coordinate settings of the system are shown in Figure 6a. The principle of binocular stereoscopic 3D perception is shown in Figure 6b. E is the eye coordinate system, $\mathrm{E_l}$ is the left motion module's end coordinate system, $\mathrm{E_r}$ is the right motion module's end coordinate system, $\mathrm{B_l}$ is the left motion module's base coordinate system, $\mathrm{B_r}$ is the right motion module's base coordinate system, $\mathrm{C_l}$ is the left camera coordinate system and $\mathrm{C_r}$ is the right camera coordinate system. In the initial position, $\mathrm{E_l}$ coincides with $\mathrm{B_l}$, and $\mathrm{E_r}$ overlaps with $\mathrm{B_r}$. When the binocular system moves, the base coordinate system does not change. ${}^{\mathrm{l}}\boldsymbol{T}$ represents the transformation matrix of the eye coordinate system E to the left motion module's base coordinate system $\mathrm{B_l}$, ${}^{\mathrm{r}}\boldsymbol{T}$ represents the transformation matrix of E to $\mathrm{B_r}$, ${}^{\mathrm{l}}\boldsymbol{T}_{\mathrm{e}}$ represents the transformation matrix of $\mathrm{B_l}$ to $\mathrm{E_l}$, ${}^{\mathrm{r}}\boldsymbol{T}_{\mathrm{e}}$ represents the transformation matrix of $\mathrm{B_r}$ to $\mathrm{E_r}$ and ${}^{\mathrm{l}}\boldsymbol{T}_{\mathrm{m}}$ represents the leftward motion. The module end coordinate system corresponds to the transformation matrix of the left camera coordinate system, and ${}^{\mathrm{r}}\boldsymbol{T}_{\mathrm{m}}$ represents the transformation matrix of the right motion module's end coordinate system to the right camera coordinate system. ${}^{\mathrm{l}}\boldsymbol{T}_{\mathrm{r}}$ represents the transformation matrix of the right camera coordinate system to the left camera coordinate system at the initial position.

The origin ${}^{\mathrm{l}}O_{\mathrm{c}}$ of $\mathrm{C_l}$ lies at the optical center of the left camera, the ${}^{\mathrm{l}}Z_{\mathrm{c}}$ axis points in the direction of the object parallel to the optical axis of the camera, the ${}^{\mathrm{l}}X_{\mathrm{c}}$ axis points horizontally to the right along the image plane and the ${}^{\mathrm{l}}Y_{\mathrm{c}}$ axis points vertically downward along the image plane. The origin ${}^{\mathrm{r}}O_{\mathrm{c}}$ of $\mathrm{C_r}$ lies at the optical center of the right camera, ${}^{\mathrm{r}}Z_{\mathrm{c}}$ is aligned with the direction of the object parallel to the optical axis of the camera, ${}^{\mathrm{r}}X_{\mathrm{c}}$ points horizontally to the right along the image plane and ${}^{\mathrm{r}}Y_{\mathrm{c}}$ points vertically downward along the image plane. $\mathrm{E_l}$'s origin ${}^{\mathrm{l}}O_{\mathrm{e}}$ is set at the intersection of the two rotation axes of the left motion module, ${}^{\mathrm{l}}Z_{\mathrm{e}}$ is perpendicular to the two rotation axes and points to the front of the platform, ${}^{\mathrm{l}}X_{\mathrm{e}}$ coincides with the vertical rotation axis and ${}^{\mathrm{l}}Y_{\mathrm{e}}$ coincides with the horizontal rotation axis. Similarly, the origin ${}^{\mathrm{r}}O_{\mathrm{e}}$ of the coordinate system $\mathrm{E_r}$ is set at the intersection of the two rotation axes of the right motion module, ${}^{\mathrm{r}}Z_{\mathrm{e}}$ is perpendicular to the two rotation axes and points toward the front of the platform, ${}^{\mathrm{r}}X_{\mathrm{e}}$ coincides with the vertical rotation axis and ${}^{\mathrm{r}}Y_{\mathrm{e}}$ coincides with the horizontal rotation axis.

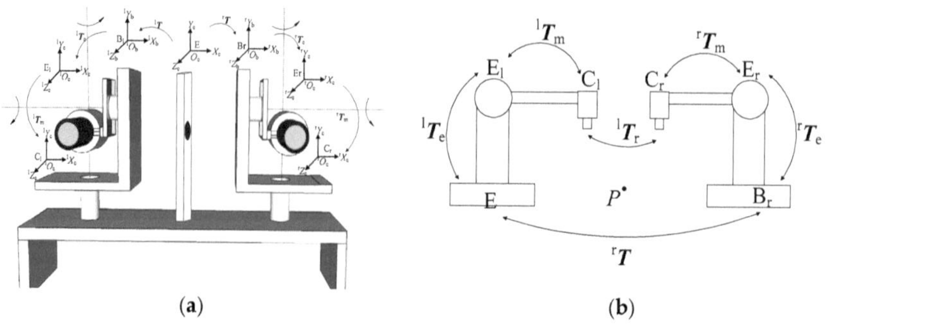

Figure 6. (**a**) Mechanical structure and coordinate systems of the bionic eye platform and (**b**) binocular 3D perception principle of bionic eyes.

The left motion module's base coordinates system B_l coincides with the eye coordinate system E; thus, ${}^{l}\boldsymbol{T}$ consists of an identity matrix. To calculate the 3D coordinates of the feature points in real time from the camera pose, it is necessary to calculate ${}^{r}\boldsymbol{T}$. At the initial position of the system, the external parameters ${}^{l}\boldsymbol{T}_r$ of the left and right cameras are calibrated offline, as are the hand–eye parameters of the left–right motion module to the camera coordinate system.

When the system is in its initial configuration, the coordinates of point $\boldsymbol{P}$ in the eye coordinate system are $\boldsymbol{P}_e$ (x_e, y_e, z_e). Its coordinates in B_l are ${}^{l}\boldsymbol{P}_e$ (${}^{l}x_e$, ${}^{l}y_e$, ${}^{l}z_e$), and its coordinates ${}^{l}\boldsymbol{P}_c$ (${}^{l}x_c$, ${}^{l}y_c$, ${}^{l}z_c$) in C_l are

$$ {}^{l}\boldsymbol{P}_c = {}^{l}\boldsymbol{T}_m^{-1}\,{}^{l}\boldsymbol{P}_e \tag{6} $$

The coordinates ${}^{r}\boldsymbol{P}_e$ (${}^{r}x_e$, ${}^{r}y_e$, ${}^{r}z_e$) of point P in B_r are

$$ {}^{r}\boldsymbol{P}_e = {}^{r}\boldsymbol{T}\boldsymbol{P}_e \tag{7} $$

The coordinates ${}^{r}\boldsymbol{P}_c$ (${}^{r}x_c$, ${}^{r}y_c$, ${}^{r}z_c$) of point $\boldsymbol{P}$ in C_r are

$$ {}^{r}\boldsymbol{P}_c = {}^{r}\boldsymbol{T}_m^{-1}\,{}^{r}\boldsymbol{T}\boldsymbol{P}_e \tag{8} $$

The point in C_r is transformed into C_l:

$$ {}^{l}\boldsymbol{P}_c = {}^{l}\boldsymbol{T}_r\,{}^{r}\boldsymbol{T}_m^{-1}\,{}^{r}\boldsymbol{T}\boldsymbol{P}_e \tag{9} $$

Based on the Equations (6) and (9), ${}^{r}\boldsymbol{T}$ is available:

$$ {}^{r}\boldsymbol{T} = {}^{r}\boldsymbol{T}_m\,{}^{l}\boldsymbol{T}_r^{-1}\,{}^{l}\boldsymbol{T}_m^{-1} \tag{10} $$

During the movement of the system, when the left motion module rotates by ${}^{l}\theta_p$ and ${}^{l}\theta_t$ in the horizontal and vertical directions, respectively, the transformation relationship between B_l and E_l is

$$ {}^{l}\boldsymbol{T}_e = \begin{pmatrix} \mathrm{Rot}(Y, {}^{l}\theta_p)\mathrm{Rot}(X, {}^{l}\theta_t) & 0 \\ 0 & 1 \end{pmatrix} \tag{11} $$

The coordinates of point $\boldsymbol{P}$ in C_l are

$$ {}^{l}\boldsymbol{P}_e = {}^{l}\boldsymbol{T}_m^{-1}\,{}^{l}\boldsymbol{T}_e\boldsymbol{P}_w = {}^{l}\boldsymbol{T}_d\boldsymbol{P}_e \tag{12} $$

Assume that

$$^{l}T_{d}=\begin{pmatrix} ^{l}n_x & ^{l}o_x & ^{l}a_x & ^{l}p_x \\ ^{l}n_y & ^{l}o_y & ^{l}a_y & ^{l}p_y \\ ^{l}n_z & ^{l}o_z & ^{l}a_z & ^{l}p_z \\ 0 & 0 & 0 & 1 \end{pmatrix} \tag{13}$$

The point $^{l}P_{1c}$ ($^{l}x_{1c}$, $^{l}y_{1c}$) at which line $P^{l}O_c$ intersects $^{l}Z_c = 1$ is

$$\begin{pmatrix} ^{l}x_{1c} \\ ^{l}y_{1c} \end{pmatrix}=\begin{pmatrix} \frac{^{l}n_x x_e+^{l}o_x y_e+^{l}a_x z_e+^{l}p_x}{^{l}n_z x_e+^{l}o_z y_e+^{l}a_z z_e+^{l}p_z} \\ \frac{^{l}n_y x_e+^{l}o_y y_e+^{l}a_y z_e+^{l}p_y}{^{l}n_z x_e+^{l}o_z y_e+^{l}a_z z_e+^{l}p_z} \end{pmatrix} \tag{14}$$

The image coordinates of $^{l}P_{1c}$ in the left camera are m_l (u_l, v_l), ($^{l}x_{1c}$, $^{l}y_{1c}$) and (u_l, v_l) and can be converted by the parameters of the camera. According to the camera's internal parameter model, the following can be obtained:

$$\begin{pmatrix} ^{l}x_{1c} \\ ^{l}y_{1c} \\ 1 \end{pmatrix}=^{l}M_{in}^{-1}\begin{pmatrix} u_l \\ v_l \\ 1 \end{pmatrix} \tag{15}$$

where $^{l}M_{in}$ is the internal parameter matrix of the left camera. The value of ($^{l}x_{1c}$, $^{l}y_{1c}$) can be obtained by the image coordinates of $^{l}P_{1c}$, and the parameters of the left camera can be obtained by substituting (15) into (14):

$$\begin{cases} (^{l}n_x-^{l}x_{1c}{}^{l}n_z)x_e+(^{l}o_x-^{l}x_{1c}{}^{l}o_z)y_e+(^{l}a_x-^{l}x_{1c}{}^{l}a_z)z_e+^{l}p_x-^{l}x_{1c}{}^{l}p_z=0 \\ (^{l}n_y-^{l}y_{1c}{}^{l}n_z)x_e+(^{l}o_y-^{l}y_{1c}{}^{l}o_z)y_e+(^{l}a_y-^{l}y_{1c}{}^{l}a_z)z_e+^{l}p_y-^{l}y_{1c}{}^{l}p_z=0 \end{cases} \tag{16}$$

During the motion of the system, when the right motion module rotates through $^{r}\theta_p$ and $^{r}\theta_t$ in the horizontal and vertical directions, respectively, the transformation relationship between B_r and E_r is

$$^{r}T_e=\begin{pmatrix} \mathrm{Rot}(Y,^{r}\theta_p)\mathrm{Rot}(X,^{r}\theta_t) & 0 \\ 0 & 1 \end{pmatrix} \tag{17}$$

The coordinates of point P in C_r are

$$^{r}P_e=^{r}T_m^{-1r}T_e{}^{r}TP_e=^{r}T_dP_e \tag{18}$$

Assume that

$$^{r}T_{d}=\begin{pmatrix} ^{r}n_x & ^{r}o_x & ^{r}a_x & ^{r}p_x \\ ^{r}n_y & ^{r}o_y & ^{r}a_y & ^{r}p_y \\ ^{r}n_z & ^{r}o_z & ^{r}a_z & ^{r}p_z \\ 0 & 0 & 0 & 1 \end{pmatrix} \tag{19}$$

The point $^{l}P_{1c}$ ($^{r}x_{1c}$, $^{r}y_{1c}$) at which line $P^{r}O_c$ intersects $^{r}Z_c = 1$ is

$$\begin{pmatrix} ^{r}x_{1c} \\ ^{r}y_{1c} \end{pmatrix}=\begin{pmatrix} \frac{^{r}n_x x_e+^{r}o_x y_e+^{r}a_x z_e+^{r}p_x}{^{r}n_z x_e+^{r}o_z y_e+^{r}a_z z_e+^{r}p_z} \\ \frac{^{r}n_y x_e+^{r}o_y y_e+^{r}a_y z_e+^{r}p_y}{^{r}n_z x_e+^{r}o_z y_e+^{r}a_z z_e+^{r}p_z} \end{pmatrix} \tag{20}$$

The image coordinates of $^{r}P_{1c}$ in the camera, namely, m_r (u_r, v_r), ($^{r}x_{1c}$, $^{r}y_{1c}$) and (u_r, v_r), can be converted using the parameters of the camera. According to the camera's internal parameter model, the following can be obtained:

$$\begin{pmatrix} ^{r}x_{1c} \\ ^{r}y_{1c} \\ 1 \end{pmatrix}=^{l}M_{in}^{-1}\begin{pmatrix} u_r \\ v_r \\ 1 \end{pmatrix} \tag{21}$$

where ${}^{\mathrm{r}}\boldsymbol{M}_{\mathrm{in}}$ is the inner parameter matrix of the right camera. The value of $({}^{\mathrm{r}}x_{1\mathrm{c}}, {}^{\mathrm{r}}y_{1\mathrm{c}})$ can be obtained by the image coordinates of ${}^{\mathrm{r}}\boldsymbol{P}_{1\mathrm{c}}$ and the parameters in the camera, and the following can be obtained by substituting (21) into (20):

$$\begin{cases} ({}^{\mathrm{r}}n_x - {}^{\mathrm{r}}x_{1\mathrm{c}}{}^{\mathrm{r}}n_z)x_{\mathrm{e}} + ({}^{\mathrm{r}}o_x - {}^{\mathrm{r}}x_{1\mathrm{c}}{}^{\mathrm{r}}o_z)y_{\mathrm{e}} + ({}^{\mathrm{r}}a_x - {}^{\mathrm{r}}x_{1\mathrm{c}}{}^{\mathrm{r}}a_z)z_{\mathrm{e}} + {}^{\mathrm{r}}p_x - {}^{\mathrm{r}}x_{1\mathrm{c}}{}^{\mathrm{r}}p_z = 0 \\ ({}^{\mathrm{r}}n_y - {}^{\mathrm{r}}y_{1\mathrm{c}}{}^{\mathrm{r}}n_z)x_{\mathrm{e}} + ({}^{\mathrm{r}}o_y - {}^{\mathrm{r}}y_{1\mathrm{c}}{}^{\mathrm{r}}o_z)y_{\mathrm{e}} + ({}^{\mathrm{r}}a_y - {}^{\mathrm{r}}y_{1\mathrm{c}}{}^{\mathrm{r}}a_z)z_{\mathrm{e}} + {}^{\mathrm{r}}p_y - {}^{\mathrm{r}}y_{1\mathrm{c}}{}^{\mathrm{r}}p_z = 0 \end{cases} \tag{22}$$

Four equations can be obtained from Equations (16) and (22) for x_{e}, y_{e} and z_{e}, and the 3D coordinates of point $\boldsymbol{P}_{\mathrm{e}}$ can be calculated by the least squares method.

The 3D coordinates $\boldsymbol{P}_{\mathrm{h}}$ $(x_{\mathrm{h}}, y_{\mathrm{h}}, z_{\mathrm{h}})$ in the head coordinate system can be obtained by Equation (23). d_{x} and d_{y} are illustrated in Figure 4.

$$\begin{pmatrix} x_{\mathrm{h}} \\ y_{\mathrm{h}} \\ z_{\mathrm{h}} \end{pmatrix} = \begin{pmatrix} x_{\mathrm{e}} - d_{\mathrm{x}} \\ y_{\mathrm{e}} - d_{\mathrm{y}} \\ z_{\mathrm{e}} \end{pmatrix} \tag{23}$$

Let the angles at which the current moment of the head rotate relative to the initial position be ${}^{\mathrm{h}}\theta_{\mathrm{p}i}$ and ${}^{\mathrm{h}}\theta_{\mathrm{t}i}$; the coordinates of the target in the robot coordinate system are

$$\begin{pmatrix} {}^{\mathrm{w}}x_{\mathrm{m}} \\ {}^{\mathrm{w}}y_{\mathrm{m}} \\ {}^{\mathrm{w}}z_{\mathrm{m}} \\ 1 \end{pmatrix} = \begin{pmatrix} \mathrm{Rot}(X, {}^{\mathrm{h}}\theta_{\mathrm{t}i})\mathrm{Rot}(Y, {}^{\mathrm{h}}\theta_{\mathrm{p}i}) & 0 \\ 0 & 1 \end{pmatrix}^{-1} \begin{pmatrix} x_{\mathrm{h}} \\ y_{\mathrm{h}} \\ z_{\mathrm{h}} \\ 1 \end{pmatrix} \tag{24}$$

According to the 3D coordinates of the target in the head coordinate system, the angle between the target and Z_{h} in the horizontal direction and the vertical direction can be obtained as follows:

$${}^{\mathrm{h}}\gamma_{\mathrm{p}} = \arctan(\frac{x_{\mathrm{h}}}{z_{\mathrm{h}}}) \tag{25}$$

$${}^{\mathrm{h}}\gamma_{\mathrm{t}} = \arctan(\frac{y_{\mathrm{h}}}{z_{\mathrm{h}}}) \tag{26}$$

When ${}^{\mathrm{h}}\gamma_{\mathrm{p}}$ and ${}^{\mathrm{h}}\gamma_{\mathrm{t}}$ exceed a set threshold, the head needs to rotate. To leave a certain margin for the rotation of the eyeball and for the convenience of calculation, the angles required for the head to rotate in the horizontal direction and the vertical direction are calculated by the principle shown in Figure 7a,b, respectively. Figure 7a shows the calculation principle of the horizontal direction angle when the target's x coordinates of the head coordinate system is greater than zero. After the head is rotated to ${}^{\mathrm{h}}\theta'_{\mathrm{p}}$, the target point is on the ${}^{\mathrm{l}}Z_{\mathrm{e}}$ axis of the left motion module end coordinate system, and the left motion module reaches the maximum rotatable threshold ${}^{\mathrm{e}}\theta_{\mathrm{pmax}}$. Figure 7b shows the calculation principle of the vertical direction when the target's y coordinates of the head coordinate system are greater than d_y. After the head is rotated to ${}^{\mathrm{h}}\theta'_{\mathrm{t}}$, the target point is on the Z_{e} axis of the eye coordinate system, and the eye reaches the maximum threshold ${}^{\mathrm{e}}\theta_{\mathrm{tmax}}$ that can be rotated.

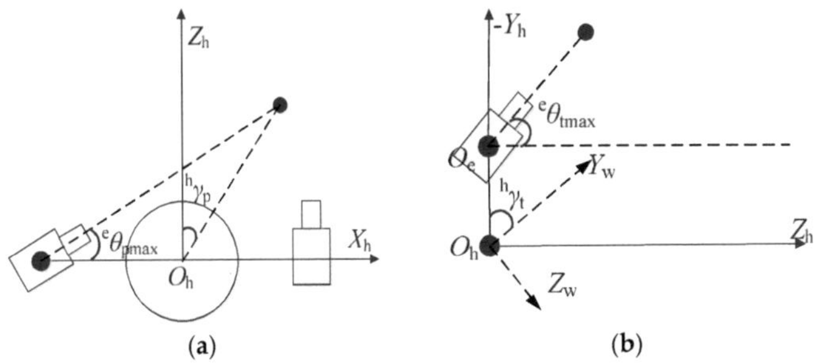

Figure 7. Principle of head rotation calculation in fixation point tracking: (**a**) horizontal rotation angle and (**b**) vertical rotation angle.

3.2.2. Horizontal Rotation Angle Calculation

Let the current angle of the head in the horizontal direction be ${}^{h}\theta_{pi}$. When the head is rotated in the horizontal direction to ${}^{h}\theta'_{p}$, the 3D coordinates of the target in the new head coordinate system are

$$\begin{pmatrix} x'_h \\ y'_h \\ z'_h \\ 1 \end{pmatrix} = \begin{pmatrix} \cos({}^{h}\theta'_{p} - {}^{h}\theta_{pi}) & 0 & -\sin({}^{h}\theta'_{p} - {}^{h}\theta_{pi}) & 0 \\ 0 & 1 & 0 & 0 \\ \sin({}^{h}\theta'_{p} - {}^{h}\theta_{pi}) & 0 & \cos({}^{h}\theta'_{p} - {}^{h}\theta_{pi}) & 0 \\ 0 & 0 & 0 & 1 \end{pmatrix} \begin{pmatrix} x_h \\ y_h \\ z_h \\ 1 \end{pmatrix} \quad (27)$$

Therefore,

$$\begin{pmatrix} x'_h \\ y'_h \\ z'_h \end{pmatrix} = \begin{pmatrix} x_h \cos({}^{h}\theta'_{p} - {}^{h}\theta_{pi}) - z_h \sin({}^{h}\theta'_{p} - {}^{h}\theta_{pi}) \\ y_h \\ x_h \sin({}^{h}\theta'_{p} - {}^{h}\theta_{pi}) + z_h \cos({}^{h}\theta'_{p} - {}^{h}\theta_{pi}) \end{pmatrix} \quad (28)$$

The coordinates of the target in the new eye coordinate system are

$$\begin{pmatrix} {}^{e}x'_h \\ {}^{e}y'_h \\ {}^{e}z'_h \end{pmatrix} = \begin{pmatrix} d_x + x_h \cos({}^{h}\theta'_{p} - {}^{h}\theta_{pi}) - z_h \sin({}^{h}\theta'_{p} - {}^{h}\theta_{pi}) \\ y_h + d_y \\ x_h \sin({}^{h}\theta'_{p} - {}^{h}\theta_{pi}) + z_h \cos({}^{h}\theta'_{p} - {}^{h}\theta_{pi}) \end{pmatrix} \quad (29)$$

After turning, the left motion module reaches the maximum threshold ${}^{e}\theta_{pmax}$ that can be rotated, so that

$$\tan({}^{e}\theta_{pmax}) = \frac{{}^{e}z'_h}{{}^{e}x'_h} = \frac{x_h \sin({}^{h}\theta'_{p} - {}^{h}\theta_{pi}) + z_h \cos({}^{h}\theta'_{p} - {}^{h}\theta_{pi})}{d_x + x_h \cos({}^{h}\theta'_{p} - {}^{h}\theta_{pi}) - z_h \sin({}^{h}\theta'_{p} - {}^{h}\theta_{pi})} \quad (30)$$

Simplifying Equation (30), we have

$$\sin({}^{h}\theta'_{p} - {}^{h}\theta_{pi}) = \frac{d_x \tan({}^{e}\theta_{qmax})}{x_h + z_h \tan({}^{e}\theta_{qmax})} + \frac{x_h \tan({}^{e}\theta_{qmax}) - z_h}{x_h + z_h \tan({}^{e}\theta_{qmax})} \cos({}^{h}\theta'_{p} - {}^{h}\theta_{pi}) \quad (31)$$

Assume that

$$\begin{cases} k_1 = \frac{x_h \tan({}^{e}\theta_{qmax}) - z_h}{x_h + z_h \tan({}^{e}\theta_{qmax})} \\ k_2 = \frac{d_x \tan({}^{e}\theta_{qmax})}{x_h + z_h \tan({}^{e}\theta_{qmax})} \end{cases} \quad (32)$$

According to the triangular relationship,

$$[k_1 \cos({}^{h}\theta'_{p} - {}^{h}\theta_{pi}) + k_2]^2 + \cos^2({}^{h}\theta'_{p} - {}^{h}\theta_{pi}) = 1 \quad (33)$$

The solution of Equation (33) is

$$\cos({}^{\mathrm{h}}\theta'_{\mathrm{p}} -{}^{\mathrm{h}} \theta_{\mathrm{p}i}) = \frac{-k_1k_2 \pm \sqrt{k_1^2 - k_2^2 + 1}}{k_1^2 + 1} \tag{34}$$

Therefore,

$${}^{\mathrm{h}}\theta'_{\mathrm{p}} ={}^{\mathrm{h}} \theta_{\mathrm{p}i} + \arccos(\frac{-k_1k_2 \pm \sqrt{k_1^2 - k_2^2 + 1}}{k_1^2 + 1}) \tag{35}$$

Equation (35) has two solutions; therefore, we choose the solution in which the deviation e of Equation (36) is minimized:

$$e = \left| \tan({}^{\mathrm{e}}\theta_{\mathrm{qmax}}) - \frac{x_{\mathrm{h}} \sin({}^{\mathrm{h}}\theta'_{\mathrm{p}} -{}^{\mathrm{h}} \theta_{\mathrm{p}i}) + z_{\mathrm{h}} \cos({}^{\mathrm{h}}\theta'_{\mathrm{p}} -{}^{\mathrm{h}} \theta_{\mathrm{p}i})}{d_x + x_{\mathrm{h}} \cos({}^{\mathrm{h}}\theta'_{\mathrm{p}} -{}^{\mathrm{h}} \theta_{\mathrm{p}i}) - z_{\mathrm{h}} \sin({}^{\mathrm{h}}\theta'_{\mathrm{p}} -{}^{\mathrm{h}} \theta_{\mathrm{p}i})} \right| \tag{36}$$

When the obtained ${}^{\mathrm{h}}\theta'_p$ is outside of the range $[-{}^{\mathrm{h}}\theta_{\mathrm{pmax}}, {}^{\mathrm{h}}\theta_{\mathrm{pmax}}]$, the value of ${}^{\mathrm{h}}\theta_{\mathrm{pq}}$ is

$${}^{\mathrm{h}}\theta_{\mathrm{pq}} = \begin{cases} {}^{\mathrm{h}}\theta_{\mathrm{pmax}} & , \ {}^{\mathrm{h}}\theta'_{\mathrm{p}} \geq{}^{\mathrm{h}} \theta_{\mathrm{pmax}} \\ -{}^{\mathrm{h}}\theta_{\mathrm{pmax}} & , \ {}^{\mathrm{h}}\theta'_{\mathrm{p}} \leq -{}^{\mathrm{h}}\theta_{\mathrm{pmax}} \\ {}^{\mathrm{h}}\theta'_{\mathrm{p}} & , \ \text{else} \end{cases} \tag{37}$$

Finally, one can obtain the ${}^{\mathrm{w}}\theta_{\mathrm{pq}}$ value:

$${}^{\mathrm{w}}\theta_{\mathrm{pq}} = \begin{cases} {}^{\mathrm{h}}\theta'_{\mathrm{p}} -{}^{\mathrm{h}} \theta_{\mathrm{pmax}} & , \ {}^{\mathrm{h}}\theta'_{\mathrm{p}} >{}^{\mathrm{h}} \theta_{\mathrm{pmax}} \\ {}^{\mathrm{h}}\theta'_{\mathrm{p}} +{}^{\mathrm{h}} \theta_{\mathrm{pmax}} & , \ {}^{\mathrm{h}}\theta'_{\mathrm{p}} < -{}^{\mathrm{h}}\theta_{\mathrm{pmax}} \\ 0 & , \ \text{else} \end{cases} \tag{38}$$

Based on the same principle, when the x coordinate of the target in the head coordinate system is less than 0, the coordinates of the target in the right motion module base coordinate system after the rotation are

$$\begin{pmatrix} {}^{\mathrm{r}}x'_{\mathrm{e}} \\ {}^{\mathrm{r}}y'_{\mathrm{e}} \\ {}^{\mathrm{r}}z'_{\mathrm{e}} \\ 1 \end{pmatrix} = \begin{pmatrix} x_{\mathrm{h}} \cos({}^{\mathrm{h}}\theta'_{\mathrm{p}} -{}^{\mathrm{h}} \theta_{\mathrm{p}i}) - z_{\mathrm{h}} \sin({}^{\mathrm{h}}\theta'_{\mathrm{p}} -{}^{\mathrm{h}} \theta_{\mathrm{p}i}) - d_x \\ y_{\mathrm{h}} + d_y \\ x_{\mathrm{h}} \sin({}^{\mathrm{h}}\theta'_{\mathrm{p}} -{}^{\mathrm{h}} \theta_{\mathrm{p}i}) + z_{\mathrm{h}} \cos({}^{\mathrm{h}}\theta'_{\mathrm{p}} -{}^{\mathrm{h}} \theta_{\mathrm{p}i}) \\ 1 \end{pmatrix} \tag{39}$$

After turning, the right motion module reaches $-{}^{\mathrm{e}}\theta_{\mathrm{pmax}}$, and the following can be obtained:

$$\tan(-{}^{\mathrm{e}}\theta_{\mathrm{qmax}}) = \frac{{}^{\mathrm{r}}z'_{\mathrm{e}}}{{}^{\mathrm{r}}x'_{\mathrm{e}}} = \frac{x_{\mathrm{h}} \sin({}^{\mathrm{h}}\theta'_{\mathrm{p}} -{}^{\mathrm{h}} \theta_{\mathrm{p}i}) + z_{\mathrm{h}} \cos({}^{\mathrm{h}}\theta'_{\mathrm{p}} -{}^{\mathrm{h}} \theta_{\mathrm{p}i})}{x_{\mathrm{h}} \cos({}^{\mathrm{h}}\theta'_{\mathrm{p}} -{}^{\mathrm{h}} \theta_{\mathrm{p}i}) - z_{\mathrm{h}} \sin({}^{\mathrm{h}}\theta'_{\mathrm{p}} -{}^{\mathrm{h}} \theta_{\mathrm{p}i}) - d_x} \tag{40}$$

We simplify Equation (40) as follows:

$$\sin({}^{\mathrm{h}}\theta'_{\mathrm{p}} -{}^{\mathrm{h}} \theta_{\mathrm{p}i}) = \frac{d_x \tan({}^{\mathrm{e}}\theta_{\mathrm{qmax}})}{x_{\mathrm{h}} - z_{\mathrm{h}} \tan({}^{\mathrm{e}}\theta_{\mathrm{qmax}})} - \frac{x_{\mathrm{h}} \tan({}^{\mathrm{e}}\theta_{\mathrm{qmax}}) + z_{\mathrm{h}}}{x_{\mathrm{h}} - z_{\mathrm{h}} \tan({}^{\mathrm{e}}\theta_{\mathrm{qmax}})} \cos({}^{\mathrm{h}}\theta'_{\mathrm{p}} -{}^{\mathrm{h}} \theta_{\mathrm{p}i}) \tag{41}$$

Let

$$\begin{cases} k'_1 = -\frac{x_{\mathrm{h}} \tan({}^{\mathrm{e}}\theta_{\mathrm{qmax}}) + z_{\mathrm{h}}}{x_{\mathrm{h}} - z_{\mathrm{h}} \tan({}^{\mathrm{e}}\theta_{\mathrm{qmax}})} \\ k'_2 = \frac{d_x \tan({}^{\mathrm{e}}\theta_{\mathrm{qmax}})}{x_{\mathrm{h}} - z_{\mathrm{h}} \tan({}^{\mathrm{e}}\theta_{\mathrm{qmax}})} \end{cases} \tag{42}$$

The same two solutions are available:

$$ {}^{\mathrm{h}}\theta'_{\mathrm{p}} = {}^{\mathrm{h}}\theta_{\mathrm{p}i} + \arccos\left(\frac{-k'_1k'_2 \pm \sqrt{(k'_1)^2 - (k'_2)^2 + 1}}{(k'_1)^2 + 1}\right) \tag{43} $$

Select the solution in which the deviation e of Equation (44) is minimized:

$$ e = \left| -\tan({}^{\mathrm{e}}\theta_{\mathrm{qmax}}) - \frac{x_{\mathrm{h}}\sin({}^{\mathrm{h}}\theta'_{\mathrm{p}} - {}^{\mathrm{h}}\theta_{\mathrm{p}i}) + z_{\mathrm{h}}\cos({}^{\mathrm{h}}\theta'_{\mathrm{p}} - {}^{\mathrm{h}}\theta_{\mathrm{p}i})}{x_{\mathrm{h}}\cos({}^{\mathrm{h}}\theta'_{\mathrm{p}} - {}^{\mathrm{h}}\theta_{\mathrm{p}i}) - z_{\mathrm{h}}\sin({}^{\mathrm{h}}\theta'_{\mathrm{p}} - {}^{\mathrm{h}}\theta_{\mathrm{p}i}) - d_x} \right| \tag{44} $$

Using Equations (37) and (38), ${}^{\mathrm{h}}\theta_{\mathrm{pq}}$ and ${}^{\mathrm{w}}\theta_{\mathrm{pq}}$ can be obtained.

3.2.3. Vertical Rotation Angle Calculation

When the target's y coordinate in the head coordinate system is greater than d_y, the current angle of the head in the vertical direction is ${}^{\mathrm{h}}\theta_{\mathrm{t}i}$, and when the head is rotated in the vertical direction to ${}^{\mathrm{h}}\theta'_{\mathrm{t}}$, the target is in the new head coordinate system. The 3D coordinates are

$$ \begin{pmatrix} x'_{\mathrm{h}} \\ y'_{\mathrm{h}} \\ z'_{\mathrm{h}} \\ 1 \end{pmatrix} = \begin{pmatrix} 1 & 0 & 0 & 0 \\ 0 & \cos({}^{\mathrm{h}}\theta'_{\mathrm{t}} - {}^{\mathrm{h}}\theta_{\mathrm{t}i}) & \sin({}^{\mathrm{h}}\theta'_{\mathrm{t}} - {}^{\mathrm{h}}\theta_{\mathrm{t}i}) & 0 \\ 0 & -\sin({}^{\mathrm{h}}\theta'_{\mathrm{t}} - {}^{\mathrm{h}}\theta_{\mathrm{t}i}) & \cos({}^{\mathrm{h}}\theta'_{\mathrm{t}} - {}^{\mathrm{h}}\theta_{\mathrm{t}i}) & 0 \\ 0 & 0 & 0 & 1 \end{pmatrix} \begin{pmatrix} x_{\mathrm{h}} \\ y_{\mathrm{h}} \\ z_{\mathrm{h}} \\ 1 \end{pmatrix} \tag{45} $$

Therefore,

$$ \begin{pmatrix} x'_{\mathrm{h}} \\ y'_{\mathrm{h}} \\ z'_{\mathrm{h}} \end{pmatrix} = \begin{pmatrix} x_{\mathrm{h}} \\ y_{\mathrm{h}}\cos({}^{\mathrm{h}}\theta'_{\mathrm{t}} - {}^{\mathrm{h}}\theta_{\mathrm{t}i}) + z_{\mathrm{h}}\sin({}^{\mathrm{h}}\theta'_{\mathrm{t}} - {}^{\mathrm{h}}\theta_{\mathrm{t}i}) \\ z_{\mathrm{h}}\cos({}^{\mathrm{h}}\theta'_{\mathrm{t}} - {}^{\mathrm{h}}\theta_{\mathrm{t}i}) - y_{\mathrm{h}}\sin({}^{\mathrm{h}}\theta'_{\mathrm{t}} - {}^{\mathrm{h}}\theta_{\mathrm{t}i}) \end{pmatrix} \tag{46} $$

Using Equation (29), the coordinates of the eye coordinate system after the rotation of the target can be calculated:

$$ \begin{pmatrix} {}^{\mathrm{e}}x'_{\mathrm{h}} \\ {}^{\mathrm{e}}y'_{\mathrm{h}} \\ {}^{\mathrm{e}}z'_{\mathrm{h}} \end{pmatrix} = \begin{pmatrix} d_x + x_{\mathrm{h}} \\ y_{\mathrm{h}}\cos({}^{\mathrm{h}}\theta'_{\mathrm{t}} - {}^{\mathrm{h}}\theta_{\mathrm{t}i}) + z_{\mathrm{h}}\sin({}^{\mathrm{h}}\theta'_{\mathrm{t}} - {}^{\mathrm{h}}\theta_{\mathrm{t}i}) + d_y \\ z_{\mathrm{h}}\cos({}^{\mathrm{h}}\theta'_{\mathrm{t}} - {}^{\mathrm{h}}\theta_{\mathrm{t}i}) - y_{\mathrm{h}}\sin({}^{\mathrm{h}}\theta'_{\mathrm{t}} - {}^{\mathrm{h}}\theta_{\mathrm{t}i}) \end{pmatrix} \tag{47} $$

After rotation, the left and right motion modules reach the rotatable maximum value ${}^{\mathrm{e}}\theta_{\mathrm{tmax}}$ in the vertical direction, so that

$$ \tan({}^{\mathrm{e}}\theta_{\mathrm{tmax}}) = \frac{{}^{\mathrm{e}}y'_{\mathrm{h}}}{{}^{\mathrm{e}}z'_{\mathrm{h}}} = \frac{y_{\mathrm{h}}\cos({}^{\mathrm{h}}\theta'_{\mathrm{t}} - {}^{\mathrm{h}}\theta_{\mathrm{t}i}) + z_{\mathrm{h}}\sin({}^{\mathrm{h}}\theta'_{\mathrm{t}} - {}^{\mathrm{h}}\theta_{\mathrm{t}i}) + d_y}{z_{\mathrm{h}}\cos({}^{\mathrm{h}}\theta'_{\mathrm{t}} - {}^{\mathrm{h}}\theta_{\mathrm{t}i}) - y_{\mathrm{h}}\sin({}^{\mathrm{h}}\theta'_{\mathrm{t}} - {}^{\mathrm{h}}\theta_{\mathrm{t}i})} \tag{48} $$

Simplifying Equation (48), we obtain

$$ \sin({}^{\mathrm{h}}\theta'_{\mathrm{t}} - {}^{\mathrm{h}}\theta_{\mathrm{t}i}) = \frac{z_{\mathrm{h}}\tan({}^{\mathrm{e}}\theta_{\mathrm{tmax}}) - y_{\mathrm{h}}}{z_{\mathrm{h}} + y_{\mathrm{h}}\tan({}^{\mathrm{e}}\theta_{\mathrm{tmax}})}\cos({}^{\mathrm{h}}\theta'_{\mathrm{t}} - {}^{\mathrm{h}}\theta_{\mathrm{t}i}) - \frac{d_y}{z_{\mathrm{h}} + y_{\mathrm{h}}\tan({}^{\mathrm{e}}\theta_{\mathrm{tmax}})} \tag{49} $$

Let

$$ \begin{cases} k_1 = \frac{z_{\mathrm{h}}\tan({}^{\mathrm{e}}\theta_{\mathrm{tmax}}) - y_{\mathrm{h}}}{z_{\mathrm{h}} + y_{\mathrm{h}}\tan({}^{\mathrm{e}}\theta_{\mathrm{tmax}})} \\ k_2 = -\frac{d_y}{z_{\mathrm{h}} + y_{\mathrm{h}}\tan({}^{\mathrm{e}}\theta_{\mathrm{tmax}})} \end{cases} \tag{50} $$

Therefore,

$$ {}^{\mathrm{h}}\theta'_{\mathrm{t}} = {}^{\mathrm{h}}\theta_{\mathrm{t}i} + \arccos\left(\frac{-k_1k_2 \pm \sqrt{k_1^2 - k_2^2 + 1}}{k_1^2 + 1}\right) \tag{51} $$

Equation (51) has two solutions; therefore, we choose the solution in which the deviation e of Equation (52) is minimized:

$$e = \left| \tan({}^{e}\theta_{tmax}) - \frac{y_h \cos({}^{h}\theta'_t - {}^{h}\theta_{ti}) + z_h \sin({}^{h}\theta'_t - {}^{h}\theta_{ti}) + d_y}{z_h \cos({}^{h}\theta'_t - {}^{h}\theta_{ti}) - y_h \sin({}^{h}\theta'_t - {}^{h}\theta_{ti})} \right| \tag{52}$$

Similarly, when the target's y coordinates in the head coordinate system are less than d_y, we have

$$\tan(-{}^{e}\theta_{tmax}) = \frac{{}^{e}y'_h}{{}^{e}z'_h} = \frac{y_h \cos({}^{h}\theta'_t - {}^{h}\theta_{ti}) + z_h \sin({}^{h}\theta'_t - {}^{h}\theta_{ti}) + d_y}{z_h \cos({}^{h}\theta'_t - {}^{h}\theta_{ti}) - y_h \sin({}^{h}\theta'_t - {}^{h}\theta_{ti})} \tag{53}$$

$$\sin({}^{h}\theta'_t - {}^{h}\theta_{ti}) = -\frac{z_h \tan({}^{e}\theta_{tmax}) + y_h}{z_h - y_h \tan({}^{e}\theta_{tmax})} \cos({}^{h}\theta'_t - {}^{h}\theta_{ti}) - \frac{d_y}{z_h - y_h \tan({}^{e}\theta_{tmax})} \tag{54}$$

$$\begin{cases} k_1 = -\frac{z_h \tan({}^{e}\theta_{tmax}) + y_h}{z_h - y_h \tan({}^{e}\theta_{tmax})} \\ k_2 = -\frac{d_y}{z_h - y_h \tan({}^{e}\theta_{tmax})} \end{cases} \tag{55}$$

$$e = \left| -\tan({}^{e}\theta_{tmax}) - \frac{y_h \cos({}^{h}\theta'_t - {}^{h}\theta_{ti}) + z_h \sin({}^{h}\theta'_t - {}^{h}\theta_{ti}) + d_y}{z_h \cos({}^{h}\theta'_t - {}^{h}\theta_{ti}) - y_h \sin({}^{h}\theta'_t - {}^{h}\theta_{ti})} \right| \tag{56}$$

When the obtained ${}^{h}\theta'_t$ is outside of the range $[-{}^{h}\theta_{tmax}, {}^{h}\theta_{tmax}]$, the value of ${}^{h}\theta_{tq}$ is

$${}^{h}\theta_{tq} = \begin{cases} {}^{h}\theta_{tmax} & , \ {}^{h}\theta'_t \geq {}^{h}\theta_{tmax} \\ -{}^{h}\theta_{tmax} & , \ {}^{h}\theta'_t \leq -{}^{h}\theta_{tmax} \\ {}^{h}\theta'_t & , \ \text{else} \end{cases} \tag{57}$$

After obtaining ${}^{h}\theta_{pq}$, ${}^{h}\theta_{tq}$ and ${}^{w}\theta_{pq}$, $\mathbf{P}'_e(x'_e, y'_e, z'_e)$ are the coordinates of the target in the eye coordinate system after the mobile robot and the head are rotated:

$$\begin{pmatrix} x'_e \\ y'_e \\ z'_e \\ 1 \end{pmatrix} = \begin{pmatrix} \text{Rot}(X, {}^{h}\theta_{tq})\text{Rot}(Y, {}^{h}\theta_{pq}) & 0 \\ 0 & 1 \end{pmatrix} \begin{pmatrix} \text{Rot}(Y, {}^{w}\theta_{pq}) & 0 \\ 0 & 1 \end{pmatrix} \begin{pmatrix} x_w \\ y_w \\ z_w \\ 1 \end{pmatrix} + \begin{pmatrix} d_x \\ d_y \\ 0 \\ 0 \end{pmatrix} \tag{58}$$

The desired observation pose of the eye, characterized by ${}^{l}\theta_{tq}$, ${}^{l}\theta_{pq}$, ${}^{r}\theta_{tq}$ and ${}^{r}\theta_{pq}$, can be obtained using the method described in the following section.

3.2.4. Calculation of the Desired Observation Poses of the Eye

According to Formula (2), ${}^{l}\theta_{tq} = {}^{r}\theta_{tq} = \theta_t$, and ${}^{l}\theta_{pq} = {}^{r}\theta_{pq} = \theta_p$.

The inverse of the hand–eye matrix of the left camera and left motion module end coordinate system is

$${}^{l}\boldsymbol{T}_m^{-1} = \begin{pmatrix} {}^{l}n_x & {}^{l}o_x & {}^{l}a_x & {}^{l}p_x \\ {}^{l}n_y & {}^{l}o_y & {}^{l}a_y & {}^{l}p_y \\ {}^{l}n_z & {}^{l}o_z & {}^{l}a_z & {}^{l}p_z \\ 0 & 0 & 0 & 1 \end{pmatrix} \tag{59}$$

The coordinate ${}^{l}\boldsymbol{P}_c$ $({}^{l}x_c, {}^{l}y_c, {}^{l}z_c)$ of $\mathbf{P}'_e(x'_e, y'_e, z'_e)$ in the left camera coordinate system satisfies the following relationship:

$$\begin{pmatrix} {}^{l}\boldsymbol{P}_c \\ 1 \end{pmatrix} = {}^{l}\boldsymbol{T}_m^{-1} \begin{pmatrix} \text{Rot}(X, -{}^{l}\theta_t) & 0 \\ 0 & 1 \end{pmatrix} \begin{pmatrix} \text{Rot}(Y, -{}^{l}\theta_p) & 0 \\ 0 & 1 \end{pmatrix} \begin{pmatrix} \boldsymbol{P}'_e \\ 1 \end{pmatrix} \tag{60}$$

According to the small hole imaging model, the imaging coordinates of the $\mathbf{P}'_e(x'_e, y'_e, z'_e)$ point in the left camera are

$$\begin{pmatrix} {}^{l}u \\ {}^{l}v \\ 1 \end{pmatrix} = {}^{l}M_{in}{}^{l}P_{lc} = \begin{pmatrix} {}^{l}k_x & 0 & {}^{l}u_0 \\ 0 & {}^{l}k_y & {}^{l}v_0 \\ 0 & 0 & 1 \end{pmatrix} \begin{pmatrix} {}^{l}x_c/{}^{l}z_c \\ {}^{l}y_c/{}^{l}z_c \\ 1 \end{pmatrix} = \begin{pmatrix} {}^{l}k_x{}^{l}x_c/{}^{l}z_c + {}^{l}u_0 \\ {}^{l}k_y{}^{l}y_c/{}^{l}z_c + {}^{l}v_0 \\ 1 \end{pmatrix} \tag{61}$$

Substituting Equation (61) into Equation (2), we obtain

$$\begin{pmatrix} {}^{l}\Delta u \\ {}^{l}\Delta v \end{pmatrix} = \begin{pmatrix} {}^{l}k_x{}^{l}x_c/{}^{l}z_c \\ {}^{l}k_y{}^{l}y_c/{}^{l}z_c \\ 1 \end{pmatrix} \tag{62}$$

Based on the same principle, the coordinate ${}^{r}\boldsymbol{P}_c$ (${}^{r}x_c$, ${}^{r}y_c$, ${}^{r}z_c$) of $\mathbf{P}'_e(x'_e, y'_e, z'_e)$ in the right camera coordinate system is

$$\begin{pmatrix} {}^{r}\boldsymbol{P}_c \\ 1 \end{pmatrix} = {}^{r}\boldsymbol{T}_m^{-1} \begin{pmatrix} \mathrm{Rot}(X, -{}^{r}\theta_t) & 0 \\ 0 & 1 \end{pmatrix} \begin{pmatrix} \mathrm{Rot}(Y, -{}^{r}\theta_p) & 0 \\ 0 & 1 \end{pmatrix} {}^{r}\boldsymbol{T}_e \begin{pmatrix} \boldsymbol{P}'_e \\ 1 \end{pmatrix} \tag{63}$$

The imaging coordinates of point $\mathbf{P}'_e(x'_e, y'_e, z'_e)$ in the right camera are

$$\begin{pmatrix} {}^{r}u \\ {}^{r}v \\ 1 \end{pmatrix} = {}^{r}M_{in}{}^{r}P_{lc} = \begin{pmatrix} {}^{r}k_x & 0 & {}^{r}u_0 \\ 0 & {}^{r}k_y & {}^{r}v_0 \\ 0 & 0 & 1 \end{pmatrix} \begin{pmatrix} {}^{r}x_c/{}^{r}z_c \\ {}^{r}y_c/{}^{r}z_c \\ 1 \end{pmatrix} = \begin{pmatrix} {}^{r}k_x{}^{r}x_c/{}^{r}z_c + {}^{r}u_0 \\ {}^{r}k_y{}^{r}y_c/{}^{r}z_c + {}^{r}v_0 \\ 1 \end{pmatrix} \tag{64}$$

$$\begin{pmatrix} {}^{r}\Delta u \\ {}^{r}\Delta v \end{pmatrix} = \begin{pmatrix} {}^{r}k_x{}^{r}x_c/{}^{r}z_c \\ {}^{r}k_y{}^{r}y_c/{}^{r}z_c \\ 1 \end{pmatrix} \tag{65}$$

By Equations (2), (62) and (65), two equations related to θ_t and θ_p (see Appendix C for the complete equations) can be obtained. It is challenging to calculate the values of θ_t and θ_p directly from these two equations, however. To obtain a solution, we consider a suboptimal observation pose and use this pose as the initial value; then, we use the trial-and-error method to obtain the optimal observation pose. When θ_t is calculated, let $\theta_p = 0$; the solution of θ_t can then be obtained by $\Delta v_l = -\Delta v_r$. When θ_p is calculated, the solution of θ_p is solved by $\Delta u_l = -\Delta u_r$. The solution $\boldsymbol{P}_{ob}$ (θ_t, θ_t, θ_p, θ_p) is a suboptimal observed pose. Based on the suboptimal observation pose, the trial-and-error method can be used to obtain the optimal solution with the smallest error. The range of θ_t is $[-\theta_{tmax}, \theta_{tmax}]$. The range of θ_p is $[-\theta_{pmax}, \theta_{pmax}]$.

According to Equations (60) and (63), let θ_p be equal to 0 to obtain

$$\begin{pmatrix} {}^{l}\mathbf{P}_c \\ 1 \end{pmatrix} = ({}^{l}\boldsymbol{T}_m)^{-1} \begin{pmatrix} \mathrm{Rot}(X, -\theta_t) & 0 \\ 0 & 1 \end{pmatrix} {}^{l}\boldsymbol{T}_e \begin{pmatrix} \mathbf{P}'_e \\ 1 \end{pmatrix} \tag{66}$$

The following result is also available:

$$\begin{pmatrix} {}^{r}\mathbf{P}_c \\ 1 \end{pmatrix} = ({}^{r}\boldsymbol{T}_m)^{-1} \begin{pmatrix} \mathrm{Rot}(X, -\theta_t) & 0 \\ 0 & 1 \end{pmatrix} {}^{r}\boldsymbol{T}_e \begin{pmatrix} \mathbf{P}'_e \\ 1 \end{pmatrix} \tag{67}$$

The base coordinate system of the left motion module is the world coordinate system. Therefore, ${}^{l}\boldsymbol{T}_w$ is a unit matrix. To simplify the calculation, we have

$$\begin{pmatrix} {}^{r}\boldsymbol{P}_e \\ 1 \end{pmatrix} = \begin{pmatrix} {}^{r}x'_e \\ {}^{r}y'_e \\ {}^{r}z'_e \\ 1 \end{pmatrix} = {}^{r}\boldsymbol{T}_e \begin{pmatrix} \mathbf{P}'_e \\ 1 \end{pmatrix} \tag{68}$$

According to the calculation principle of Section 3.2.1, we have the following:

$$\begin{cases} \Delta u_{\mathrm{l}} = {}^{\mathrm{l}}f_x \frac{({}^{\mathrm{l}}o_x y'_{\mathrm{e}} + {}^{\mathrm{l}}a_x z'_{\mathrm{e}})\cos\theta_{\mathrm{t}} + ({}^{\mathrm{l}}o_x z'_{\mathrm{e}} - {}^{\mathrm{l}}a_x y'_{\mathrm{e}})\sin\theta_{\mathrm{t}} + ({}^{\mathrm{l}}p_x + {}^{\mathrm{l}}n_x x'_{\mathrm{e}})}{({}^{\mathrm{l}}o_z y'_{\mathrm{e}} + {}^{\mathrm{l}}a_z z'_{\mathrm{e}})\cos\theta_{\mathrm{t}} + ({}^{\mathrm{l}}o_z z'_{\mathrm{e}} - {}^{\mathrm{l}}a_z y'_{\mathrm{e}})\sin\theta_{\mathrm{t}} + ({}^{\mathrm{l}}p_z + {}^{\mathrm{l}}n_z x'_{\mathrm{e}})} \\ \Delta v_{\mathrm{l}} = {}^{\mathrm{l}}f_y \frac{({}^{\mathrm{l}}o_y y'_{\mathrm{e}} + {}^{\mathrm{l}}a_y z'_{\mathrm{e}})\cos\theta_{\mathrm{t}} + ({}^{\mathrm{l}}o_y z'_{\mathrm{e}} - {}^{\mathrm{l}}a_y y'_{\mathrm{e}})\sin\theta_{\mathrm{t}} + ({}^{\mathrm{l}}p_y + {}^{\mathrm{l}}n_y x'_{\mathrm{e}})}{({}^{\mathrm{l}}o_z y'_{\mathrm{e}} + {}^{\mathrm{l}}a_z z'_{\mathrm{e}})\cos\theta_{\mathrm{t}} + ({}^{\mathrm{l}}o_z z'_{\mathrm{e}} - {}^{\mathrm{l}}a_z y'_{\mathrm{e}})\sin\theta_{\mathrm{t}} + ({}^{\mathrm{l}}p_z + {}^{\mathrm{l}}n_z x'_{\mathrm{e}})} \end{cases} \tag{69}$$

$$\begin{cases} \Delta u_{\mathrm{r}} = {}^{\mathrm{r}}f_x \frac{({}^{\mathrm{r}}o_x {}^{\mathrm{r}}y'_{\mathrm{e}} + {}^{\mathrm{r}}a_x {}^{\mathrm{r}}z'_{\mathrm{e}})\cos\theta_{\mathrm{t}} + ({}^{\mathrm{r}}o_x {}^{\mathrm{r}}z'_{\mathrm{e}} - {}^{\mathrm{r}}a_x {}^{\mathrm{r}}y'_{\mathrm{e}})\sin\theta_{\mathrm{t}} + ({}^{\mathrm{r}}p_x + {}^{\mathrm{r}}n_x {}^{\mathrm{r}}x'_{\mathrm{e}})}{({}^{\mathrm{r}}o_z {}^{\mathrm{r}}y'_{\mathrm{e}} + {}^{\mathrm{r}}a_z {}^{\mathrm{r}}z'_{\mathrm{e}})\cos\theta_{\mathrm{t}} + ({}^{\mathrm{r}}o_z z'_{\mathrm{e}} - {}^{\mathrm{r}}a_z {}^{\mathrm{r}}y'_{\mathrm{e}})\sin\theta_{\mathrm{t}} + ({}^{\mathrm{r}}p_z + {}^{\mathrm{r}}n_z {}^{\mathrm{r}}x'_{\mathrm{e}})} \\ \Delta v_{\mathrm{r}} = {}^{\mathrm{r}}f_y \frac{({}^{\mathrm{r}}o_y {}^{\mathrm{r}}y'_{\mathrm{e}} + {}^{\mathrm{r}}a_y {}^{\mathrm{r}}z'_{\mathrm{e}})\cos\theta_{\mathrm{t}} + ({}^{\mathrm{r}}o_y {}^{\mathrm{r}}z'_{\mathrm{e}} - {}^{\mathrm{r}}a_y {}^{\mathrm{r}}y'_{\mathrm{e}})\sin\theta_{\mathrm{t}} + ({}^{\mathrm{r}}p_y + {}^{\mathrm{r}}n_y {}^{\mathrm{r}}x'_{\mathrm{e}})}{({}^{\mathrm{r}}o_z {}^{\mathrm{r}}y'_{\mathrm{e}} + {}^{\mathrm{r}}a_z {}^{\mathrm{r}}z'_{\mathrm{e}})\cos\theta_{\mathrm{t}} + ({}^{\mathrm{r}}o_z {}^{\mathrm{r}}z'_{\mathrm{e}} - {}^{\mathrm{r}}a_z {}^{\mathrm{r}}y'_{\mathrm{e}})\sin\theta_{\mathrm{t}} + ({}^{\mathrm{r}}p_z + {}^{\mathrm{r}}n_z {}^{\mathrm{r}}x'_{\mathrm{e}})} \end{cases} \tag{70}$$

Assume the following:

$$E_{\mathrm{sv}} = |\Delta v_{\mathrm{l}}| + |\Delta v_{\mathrm{r}}| \tag{71}$$

The solution to θ_{t} that keeps the target at the center of the two cameras needs to satisfy the following conditions:

$$\begin{cases} \Delta v_{\mathrm{l}} + \Delta v_{\mathrm{r}} = 0 \\ -\theta_{\mathrm{tmax}} \leq \theta_{\mathrm{t}} \leq \theta_{\mathrm{tmax}} \\ \theta_{\mathrm{t}} = \mathrm{argmin}(E_{\mathrm{sv}}) \end{cases} \tag{72}$$

Substituting the second equation of Equations (69) and (70) into Equation (72) and solving the equation, we have

$$k_1 \cos^2\theta_{\mathrm{t}} + k_2 \sin^2\theta_{\mathrm{t}} + k_3 \sin\theta_{\mathrm{t}}\cos\theta_{\mathrm{t}} + k_4\cos\theta_{\mathrm{t}} + k_5\sin\theta_{\mathrm{t}} + k_6 = 0 \tag{73}$$

where k_1, k_2, k_3, k_4, k_5 are

$$k_1 = {}^{\mathrm{l}}f_y({}^{\mathrm{l}}o_y y'_{\mathrm{e}} + {}^{\mathrm{l}}a_y z'_{\mathrm{e}})({}^{\mathrm{r}}o_z {}^{\mathrm{r}}y'_{\mathrm{e}} + {}^{\mathrm{r}}a_z {}^{\mathrm{r}}z'_{\mathrm{e}}) + {}^{\mathrm{r}}f_y({}^{\mathrm{l}}o_z y'_{\mathrm{e}} + {}^{\mathrm{l}}a_z z'_{\mathrm{e}})({}^{\mathrm{r}}o_y {}^{\mathrm{r}}y'_{\mathrm{e}} + {}^{\mathrm{r}}a_y {}^{\mathrm{r}}z'_{\mathrm{e}}) \tag{74}$$

$$k_2 = {}^{\mathrm{l}}f_y({}^{\mathrm{l}}o_y z'_{\mathrm{e}} - {}^{\mathrm{l}}a_y y'_{\mathrm{e}})({}^{\mathrm{r}}o_z {}^{\mathrm{r}}z'_{\mathrm{e}} - {}^{\mathrm{r}}a_z {}^{\mathrm{r}}y'_{\mathrm{e}}) + {}^{\mathrm{r}}f_y({}^{\mathrm{l}}o_z z'_{\mathrm{e}} - {}^{\mathrm{l}}a_z y'_{\mathrm{e}})({}^{\mathrm{r}}o_y {}^{\mathrm{r}}z'_{\mathrm{e}} - {}^{\mathrm{r}}a_y {}^{\mathrm{r}}y'_{\mathrm{e}}) \tag{75}$$

$$\begin{aligned} k_3 = {}&{}^{\mathrm{l}}f_y({}^{\mathrm{l}}o_y y'_{\mathrm{e}} + {}^{\mathrm{l}}a_y z'_{\mathrm{e}})({}^{\mathrm{r}}o_z {}^{\mathrm{r}}z'_{\mathrm{e}} - {}^{\mathrm{r}}a_z {}^{\mathrm{r}}y'_{\mathrm{e}}) + {}^{\mathrm{l}}f_y({}^{\mathrm{l}}o_y z'_{\mathrm{e}} - {}^{\mathrm{l}}a_y y'_{\mathrm{e}})({}^{\mathrm{r}}o_z {}^{\mathrm{r}}y'_{\mathrm{e}} + {}^{\mathrm{r}}a_z {}^{\mathrm{r}}z'_{\mathrm{e}}) \\ &+ {}^{\mathrm{r}}f_y({}^{\mathrm{l}}o_z y'_{\mathrm{e}} + {}^{\mathrm{l}}a_z z'_{\mathrm{e}})({}^{\mathrm{r}}o_y {}^{\mathrm{r}}z'_{\mathrm{e}} - {}^{\mathrm{r}}a_y {}^{\mathrm{r}}y'_{\mathrm{e}}) + {}^{\mathrm{r}}f_y({}^{\mathrm{l}}o_z z'_{\mathrm{e}} - {}^{\mathrm{l}}a_z y'_{\mathrm{e}})({}^{\mathrm{r}}o_y {}^{\mathrm{r}}y'_{\mathrm{e}} + {}^{\mathrm{r}}a_y {}^{\mathrm{r}}z'_{\mathrm{e}}) \end{aligned} \tag{76}$$

$$\begin{aligned} k_4 = {}&{}^{\mathrm{l}}f_y({}^{\mathrm{l}}o_y y'_{\mathrm{e}} + {}^{\mathrm{l}}a_y z'_{\mathrm{e}})({}^{\mathrm{r}}p_z + {}^{\mathrm{r}}n_z {}^{\mathrm{r}}x'_{\mathrm{e}}) + {}^{\mathrm{l}}f_y({}^{\mathrm{l}}p_y + {}^{\mathrm{l}}n_y x'_{\mathrm{e}})({}^{\mathrm{r}}o_z {}^{\mathrm{r}}y'_{\mathrm{e}} + {}^{\mathrm{r}}a_z {}^{\mathrm{r}}z'_{\mathrm{e}}) \\ &+ {}^{\mathrm{r}}f_y({}^{\mathrm{l}}o_z y'_{\mathrm{e}} + {}^{\mathrm{l}}a_z z'_{\mathrm{e}})({}^{\mathrm{r}}p_y + {}^{\mathrm{r}}n_y {}^{\mathrm{r}}x'_{\mathrm{e}}) + {}^{\mathrm{r}}f_y({}^{\mathrm{l}}p_z + {}^{\mathrm{l}}n_z x'_{\mathrm{e}})({}^{\mathrm{r}}o_y {}^{\mathrm{r}}y'_{\mathrm{e}} + {}^{\mathrm{r}}a_y {}^{\mathrm{r}}z'_{\mathrm{e}}) \end{aligned} \tag{77}$$

$$\begin{aligned} k_5 = {}&{}^{\mathrm{l}}f_y({}^{\mathrm{l}}o_y z'_{\mathrm{e}} - {}^{\mathrm{l}}a_y y'_{\mathrm{e}})({}^{\mathrm{r}}p_z + {}^{\mathrm{r}}n_z {}^{\mathrm{r}}x'_{\mathrm{e}}) + {}^{\mathrm{l}}f_y({}^{\mathrm{l}}p_y + {}^{\mathrm{l}}n_y x'_{\mathrm{e}})({}^{\mathrm{r}}o_z {}^{\mathrm{r}}z'_{\mathrm{e}} - {}^{\mathrm{r}}a_z {}^{\mathrm{r}}y'_{\mathrm{e}}) \\ &+ {}^{\mathrm{r}}f_y({}^{\mathrm{l}}o_z z'_{\mathrm{e}} - {}^{\mathrm{l}}a_z y'_{\mathrm{e}})({}^{\mathrm{r}}p_y + {}^{\mathrm{r}}n_y {}^{\mathrm{r}}x'_{\mathrm{e}}) + {}^{\mathrm{r}}f_y({}^{\mathrm{l}}p_z + {}^{\mathrm{l}}n_z x'_{\mathrm{e}})({}^{\mathrm{r}}o_y {}^{\mathrm{r}}z'_{\mathrm{e}} - {}^{\mathrm{r}}a_y {}^{\mathrm{r}}y'_{\mathrm{e}}) \end{aligned} \tag{78}$$

$$k_6 = {}^{\mathrm{l}}f_y({}^{\mathrm{l}}p_y + {}^{\mathrm{l}}n_y x'_{\mathrm{e}})({}^{\mathrm{r}}p_z + {}^{\mathrm{r}}n_z {}^{\mathrm{r}}x'_{\mathrm{e}}) + {}^{\mathrm{r}}f_y({}^{\mathrm{l}}p_z + {}^{\mathrm{l}}n_z x'_{\mathrm{e}})({}^{\mathrm{r}}p_y + {}^{\mathrm{r}}n_y {}^{\mathrm{r}}x'_{\mathrm{e}}) \tag{79}$$

According to the triangle relationship, we have

$$\cos^2\theta_{\mathrm{t}} + \sin^2\theta_{\mathrm{t}} = 1 \tag{80}$$

Replacing $\cos\theta_{\mathrm{t}}$ in Equation (73) with $\sin\theta_{\mathrm{t}}$, we obtain the following:

$$k'_1 \sin^4\theta_{\mathrm{t}} + k'_2 \sin^3\theta_{\mathrm{t}} + k'_3 \sin^2\theta_{\mathrm{t}} + k'_4 \sin\theta_{\mathrm{t}} + k'_5 = 0 \tag{81}$$

where k_1, k_2, k_3, k_4, k_5 are

$$k'_1 = (k_2 - k_1)^2 + k_3^2 \tag{82}$$

$$k'_2 = 2(k_2 - k_1)k_5 + 2k_3k_4 \tag{83}$$

$$k'_3 = 2(k_2 - k_1)k_6 + k_5^2 + k_4^2 - k_3^2 \tag{84}$$

$$k'_4 = 2k_5k_6 - 2k_3k_4 \tag{85}$$

$$k'_5 = k_6^2 - k_4^2 \tag{86}$$

Four solutions can be obtained using Equation (81). The optimal solution is a real number, and the most suitable solution can be selected by the condition of Equation (72).

After θ_t is obtained, θ_p can be solved based on the obtained θ_t.

According to Equations (60) and (63), $\overline{\theta}_t$ is the solution obtained in Section 3.2.2, so that

$$\begin{pmatrix} {}^{l}\mathbf{P}_c \\ 1 \end{pmatrix} = \left({}^{l}\boldsymbol{T}_m\right)^{-1} \begin{pmatrix} \mathrm{Rot}(X, -\overline{\theta}_t) & 0 \\ 0 & 1 \end{pmatrix} \begin{pmatrix} \mathrm{Rot}(Y, -{}^{l}\theta_p) & 0 \\ 0 & 1 \end{pmatrix} {}^{l}\boldsymbol{T}_e \begin{pmatrix} \mathbf{P}'_e \\ 1 \end{pmatrix} \tag{87}$$

The following result is also available:

$$\begin{pmatrix} {}^{r}\mathbf{P}_c \\ 1 \end{pmatrix} = \left({}^{r}\boldsymbol{T}_m\right)^{-1} \begin{pmatrix} \mathrm{Rot}(X, -\overline{\theta}_t) & 0 \\ 0 & 1 \end{pmatrix} \begin{pmatrix} \mathrm{Rot}(Y, -{}^{l}\theta_p) & 0 \\ 0 & 1 \end{pmatrix} {}^{r}\boldsymbol{T}_e \begin{pmatrix} \mathbf{P}'_e \\ 1 \end{pmatrix} \tag{88}$$

Since $\overline{\theta}_t$ is known, for convenience of calculation, we set

$${}^{l}\boldsymbol{T}'_m = \left({}^{l}\boldsymbol{T}_m\right)^{-1} \begin{pmatrix} \mathrm{Rot}(X, -\overline{\theta}_t) & 0 \\ 0 & 1 \end{pmatrix} = \begin{pmatrix} {}^{l}n'_x & {}^{l}o'_x & {}^{l}a'_x & {}^{l}p'_x \\ {}^{l}n'_y & {}^{l}o'_y & {}^{l}a'_y & {}^{l}p'_y \\ {}^{l}n'_z & {}^{l}o'_z & {}^{l}a'_z & {}^{l}p'_z \\ 0 & 0 & 0 & 1 \end{pmatrix} \tag{89}$$

$${}^{r}\boldsymbol{T}'_m = \left({}^{r}\boldsymbol{T}_m\right)^{-1} \begin{pmatrix} \mathrm{Rot}(X, -\overline{\theta}_t) & 0 \\ 0 & 1 \end{pmatrix} = \begin{pmatrix} {}^{r}n'_x & {}^{r}o'_x & {}^{r}a'_x & {}^{r}p'_x \\ {}^{r}n'_y & {}^{r}o'_y & {}^{r}a'_y & {}^{r}p'_y \\ {}^{r}n'_z & {}^{r}o'_z & {}^{r}a'_z & {}^{r}p'_z \\ 0 & 0 & 0 & 1 \end{pmatrix} \tag{90}$$

The following results are obtained:

$$\begin{cases} \Delta u_l = {}^{l}f_x \dfrac{({}^{l}n'_x x'_e + {}^{l}a'_x z'_e)\cos\theta_p + ({}^{l}a'_x x'_e - {}^{l}n'_x z'_e)\sin\theta_p + ({}^{l}p'_x + {}^{l}o'_x y'_e)}{({}^{l}n'_z x'_e + {}^{l}a'_z z'_e)\cos\theta_p + ({}^{l}a'_z x'_e - {}^{l}n'_z z'_e)\sin\theta_p + ({}^{l}p'_z + {}^{l}o'_z y'_e)} \\ \Delta v_l = {}^{l}f_y \dfrac{({}^{l}n'_y x'_e + {}^{l}a'_y z'_e)\cos\theta_p + ({}^{l}a'_y x'_e - {}^{l}n'_y z'_e)\sin\theta_p + ({}^{l}p'_y + {}^{l}o'_y y'_e)}{({}^{l}n'_z x'_e + {}^{l}a'_z z'_e)\cos\theta_p + ({}^{l}a'_z x'_e - {}^{l}n'_z z'_e)\sin\theta_p + ({}^{l}p'_z + {}^{l}o'_z y'_e)} \end{cases} \tag{91}$$

$$\begin{cases} \Delta u_r = {}^{r}f_x \dfrac{({}^{r}n'_x {}^{r}x'_e + {}^{r}a'_x {}^{r}z'_e)\cos\theta_p + ({}^{r}a'_x {}^{r}x'_e - {}^{r}n'_x {}^{r}z'_e)\sin\theta_p + ({}^{r}p'_x + {}^{r}o'_x {}^{r}y'_e)}{({}^{r}n'_z {}^{r}x'_e + {}^{r}a'_z {}^{r}z'_e)\cos\theta_p + ({}^{r}a'_z {}^{r}x'_e - {}^{r}n'_z {}^{r}z'_e)\sin\theta_p + ({}^{r}p'_z + {}^{r}o'_z {}^{r}y'_e)} \\ \Delta v_r = {}^{r}f_y \dfrac{({}^{r}n'_y {}^{r}x'_e + {}^{r}a'_y {}^{r}z'_e)\cos\theta_p + ({}^{r}a'_y {}^{r}x'_e - {}^{r}n'_y {}^{r}z'_e)\sin\theta_p + ({}^{r}p'_y + {}^{r}o'_y {}^{r}y'_e)}{({}^{r}n'_z {}^{r}x'_e + {}^{r}a'_z {}^{r}z'_e)\cos\theta_p + ({}^{r}a'_z {}^{r}x'_e - {}^{r}n'_z {}^{r}z'_e)\sin\theta_p + ({}^{r}p'_z + {}^{r}o'_z {}^{r}y'_e)} \end{cases} \tag{92}$$

Assume that

$$E_{su} = |\Delta u_l| + |\Delta u_r| \tag{93}$$

The solution to θ_p that keeps the target at the center of the two cameras needs to satisfy the following conditions:

$$\begin{cases} \Delta u_l + \Delta u_r = 0 \\ -\theta_{pmax} \le \theta_p \le \theta_{pmax} \\ \theta_p = \mathrm{argmin}(E_{su}) \end{cases} \tag{94}$$

Substituting the second equation of Equations (91) and (92) into Equation (94) and solving the available equation, we obtain

$$k_1 \cos^2\theta_p + k_2 \sin^2\theta_p + k_3 \sin\theta_p \cos\theta_p + k_4 \cos\theta_p + k_5 \sin\theta_p + k_6 = 0 \tag{95}$$

where

$$k_1 = {}^{l}f_x({}^{l}n'_x x'_e + {}^{l}a'_x z'_e)({}^{r}n'_z {}^{r}x'_e + {}^{r}a'_z {}^{r}z'_e) + {}^{r}f_x({}^{r}n'_x {}^{r}x'_e + {}^{r}a'_x {}^{r}z'_e)({}^{l}n'_z x'_e + {}^{l}a'_z z'_e) \tag{96}$$

$$k_2 = {}^{\mathrm{l}}f_x({}^{\mathrm{l}}n'_x z'_{\mathrm{e}} - {}^{\mathrm{l}}a'_x x'_{\mathrm{e}})({}^{\mathrm{r}}n'_z {}^{\mathrm{r}}z'_{\mathrm{e}} - {}^{\mathrm{r}}a'_z {}^{\mathrm{r}}x'_{\mathrm{e}}) + {}^{\mathrm{r}}f_x({}^{\mathrm{r}}n'_x {}^{\mathrm{r}}z'_{\mathrm{e}} - {}^{\mathrm{r}}a'_x {}^{\mathrm{r}}x'_{\mathrm{e}})({}^{\mathrm{l}}n'_z z'_{\mathrm{e}} - {}^{\mathrm{l}}a'_z x'_{\mathrm{e}}) \tag{97}$$

$$\begin{aligned} k_3 = {} & {}^{\mathrm{l}}f_x({}^{\mathrm{l}}n'_x x'_{\mathrm{e}} + {}^{\mathrm{l}}a'_x z'_{\mathrm{e}})({}^{\mathrm{r}}n'_z {}^{\mathrm{r}}z'_{\mathrm{e}} - {}^{\mathrm{r}}a'_z {}^{\mathrm{r}}x'_{\mathrm{e}}) + {}^{\mathrm{l}}f_x({}^{\mathrm{l}}n'_x z'_{\mathrm{e}} - {}^{\mathrm{l}}a'_x x'_{\mathrm{e}})({}^{\mathrm{r}}n'_z {}^{\mathrm{r}}x'_{\mathrm{e}} + {}^{\mathrm{r}}a'_z {}^{\mathrm{r}}z'_{\mathrm{e}}) \\ & + {}^{\mathrm{r}}f_x({}^{\mathrm{r}}n'_x {}^{\mathrm{r}}z'_{\mathrm{e}} - {}^{\mathrm{r}}a'_x {}^{\mathrm{r}}x'_{\mathrm{e}})({}^{\mathrm{l}}n'_z x'_{\mathrm{e}} + {}^{\mathrm{l}}a'_z z'_{\mathrm{e}}) + {}^{\mathrm{r}}f_x({}^{\mathrm{r}}n'_x {}^{\mathrm{r}}x'_{\mathrm{e}} + {}^{\mathrm{r}}a'_x {}^{\mathrm{r}}z'_{\mathrm{e}})({}^{\mathrm{l}}n'_z z'_{\mathrm{e}} - {}^{\mathrm{l}}a'_z x'_{\mathrm{e}}) \end{aligned} \tag{98}$$

$$\begin{aligned} k_4 = {} & {}^{\mathrm{l}}f_x({}^{\mathrm{l}}n'_x x'_{\mathrm{e}} + {}^{\mathrm{l}}a'_x z'_{\mathrm{e}})({}^{\mathrm{r}}o'_z {}^{\mathrm{r}}y'_{\mathrm{e}} + {}^{\mathrm{r}}p'_z) + {}^{\mathrm{l}}f_x({}^{\mathrm{l}}o'_x y'_{\mathrm{e}} + {}^{\mathrm{l}}p'_x)({}^{\mathrm{r}}n'_z {}^{\mathrm{r}}x'_{\mathrm{e}} + {}^{\mathrm{r}}a'_z {}^{\mathrm{r}}z'_{\mathrm{e}}) \\ & + {}^{\mathrm{r}}f_x({}^{\mathrm{r}}o'_x {}^{\mathrm{r}}y'_{\mathrm{e}} + {}^{\mathrm{r}}p'_x)({}^{\mathrm{l}}n'_z x'_{\mathrm{e}} + {}^{\mathrm{l}}a'_z z'_{\mathrm{e}}) + {}^{\mathrm{r}}f_x({}^{\mathrm{r}}n'_x {}^{\mathrm{r}}x'_{\mathrm{e}} + {}^{\mathrm{r}}a'_x {}^{\mathrm{r}}z'_{\mathrm{e}})({}^{\mathrm{l}}o'_z y'_{\mathrm{e}} + {}^{\mathrm{l}}p'_z) \end{aligned} \tag{99}$$

$$\begin{aligned} k_5 = {} & {}^{\mathrm{l}}f_x({}^{\mathrm{l}}n'_x z'_{\mathrm{e}} - {}^{\mathrm{l}}a'_x x'_{\mathrm{e}})({}^{\mathrm{r}}o'_z {}^{\mathrm{r}}y'_{\mathrm{e}} + {}^{\mathrm{r}}p'_z) + {}^{\mathrm{l}}f_x({}^{\mathrm{l}}o'_x y'_{\mathrm{e}} + {}^{\mathrm{l}}p'_x)({}^{\mathrm{r}}n'_z {}^{\mathrm{r}}z'_{\mathrm{e}} - {}^{\mathrm{r}}a'_z {}^{\mathrm{r}}x'_{\mathrm{e}}) \\ & + {}^{\mathrm{r}}f_x({}^{\mathrm{r}}o'_x {}^{\mathrm{r}}y'_{\mathrm{e}} + {}^{\mathrm{r}}p'_x)({}^{\mathrm{l}}n'_z z'_{\mathrm{e}} - {}^{\mathrm{l}}a'_z x'_{\mathrm{e}}) + {}^{\mathrm{r}}f_x({}^{\mathrm{r}}n'_x {}^{\mathrm{r}}z'_{\mathrm{e}} - {}^{\mathrm{r}}a'_x {}^{\mathrm{r}}x'_{\mathrm{e}})({}^{\mathrm{l}}o'_z y'_{\mathrm{e}} + {}^{\mathrm{l}}p'_z) \end{aligned} \tag{100}$$

$$k_6 = {}^{\mathrm{l}}f_x({}^{\mathrm{l}}o'_x y'_{\mathrm{e}} + {}^{\mathrm{l}}p'_x)({}^{\mathrm{r}}o'_z {}^{\mathrm{r}}y'_{\mathrm{e}} + {}^{\mathrm{r}}p'_z) + {}^{\mathrm{r}}f_x({}^{\mathrm{r}}o'_x {}^{\mathrm{r}}y'_{\mathrm{e}} + {}^{\mathrm{r}}p'_x)({}^{\mathrm{l}}o'_z y'_{\mathrm{e}} + {}^{\mathrm{l}}p'_z) \tag{101}$$

Replacing $\cos\theta_{\mathrm{p}}$ in Equation (73) with $\sin\theta_{\mathrm{p}}$, we obtain

$$k'_1 \sin^4 \theta_{\mathrm{p}} + k'_2 \sin^3 \theta_{\mathrm{p}} + k'_3 \sin^2 \theta_{\mathrm{p}} + k'_4 \sin \theta_{\mathrm{p}} + k'_5 = 0 \tag{102}$$

where

$$k'_1 = (k_2 - k_1)^2 + (k_3)^2 \tag{103}$$

$$k'_2 = 2(k_2 - k_1)k_5 + 2k_3k_4 \tag{104}$$

$$k'_3 = 2(k_2 - k_1)k_6 + (k_5)^2 + (k_4)^2 - (k_3)^2 \tag{105}$$

$$k'_4 = 2k_5k_6 - 2k_3k_4 \tag{106}$$

$$k'_5 = (k_6)^2 - (k_4)^2 \tag{107}$$

Four solutions can be obtained using Equation (102). The optimal solution must be a real number, and the most suitable solution can be selected using the condition of Equation (94). For the case where the four solutions cannot satisfy Equation (94), the position of the target is beyond the position that the bionic eye can reach. In this case, compensation is required through the head or torso. $\overline{\theta}_{\mathrm{t}}$ and $\overline{\theta}_{\mathrm{p}}$ obtained at this time are suboptimal solutions close to the optimal solution. θ_{t} and θ_{p} are the optimal solutions.

Through the above steps, the desired observation pose can be calculated. The calculation steps of ${}^{\mathrm{g}}f_{\mathrm{q}}$ can be summarized by the flow chart shown in Figure 8.

3.3. Desired Pose Calculation for Approaching Gaze Point Tracking

The mobile robot approaches the target in two steps: the first step is that the robot and the head rotate in the horizontal direction until the robot and the head are facing the target, and the second step is that the robot moves straight toward the target. The desired position of the approaching motion should satisfy the following conditions: (1) the target should be on the Z axis of the robot and the head coordinate system, (2) the distance between the target and the robot should be less than the set threshold D_{T} and (3) the eye should be in the optimal observation position. ${}^{\mathrm{a}}f_{\mathrm{q}}$ can be defined as

$$
{}^{a}f_{q} = \begin{pmatrix} {}^{w}x_{mq} = 0 \\ {}^{w}z_{mq} = \{z|\ \ 0 < {}^{w}z_{m} - z \leq D_{T}\} \\ {}^{w}\theta_{pq} = \{\theta|\ \ |\theta| \leq 2\pi, {}^{w}\gamma_{p} = 0\} \\ {}^{h}\theta_{pq} = 0 \\ {}^{h}\theta_{tq} = \{\theta|\ \ |\theta| \leq {}^{h}\theta_{tmax}, |{}_{h}\gamma_{t}| \leq {}^{h}\gamma_{tmax}\} \\ {}^{l}\theta_{pq} = {}^{r}\theta_{pq} = \{\theta|\ \ |\theta| \leq {}^{e}\theta_{pmax}, \Delta m_{l} = -\Delta m_{r}\} \\ {}^{l}\theta_{tq} = {}^{r}\theta_{tq} = \{\theta|\ \ |\theta| \leq {}^{e}\theta_{tmax}, \Delta m_{l} = -\Delta m_{r}\} \end{pmatrix} \tag{108}
$$

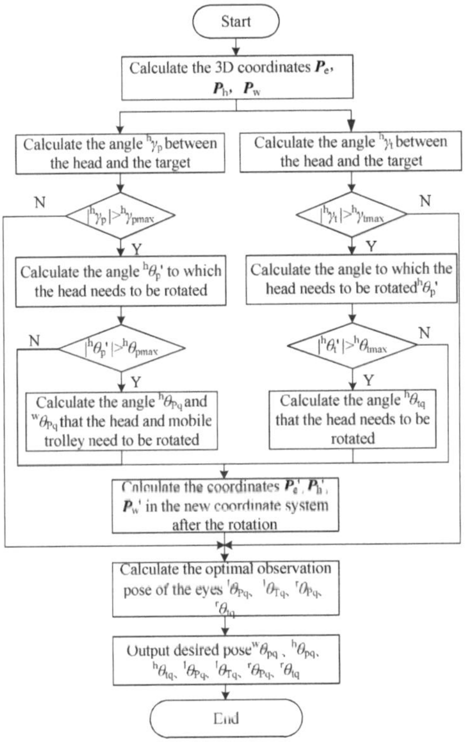

Figure 8. Steps for calculating the desired pose of the fixation point.

The desired rotation angle ${}^{w}\theta_{pq}$ of the moving robot is the same as the angle ${}^{b}\gamma_{p}$ between the robot and the target and can be obtained by

$$
{}^{w}\theta_{pq} = {}^{w}\gamma_{p} = \arctan(\frac{{}^{w}z_{m}}{{}^{w}x_{m}}) \tag{109}
$$

${}^{h}\theta_{tq}$ can be obtained using the method described in Section 3.2. The optimal observation pose described in Section 3.2.4 can be used to obtain ${}^{l}\theta_{tq}$, ${}^{l}\theta_{pq}$, ${}^{r}\theta_{tq}$ and ${}^{r}\theta_{pq}$.

4. Robot Pose Control

After obtaining the desired pose of the robot system, the control block diagram shown in Figure 9 is used to control the robot to move to the desired pose.

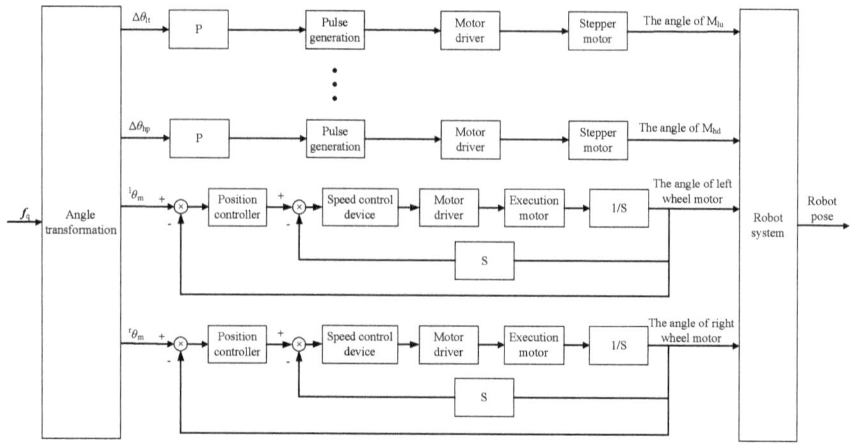

Figure 9. Robot pose control block diagram.

The desired pose is converted to the desired position of the motor. $\Delta\theta_{\mathrm{lt}}$, $\Delta\theta_{\mathrm{lp}}$, $\Delta\theta_{\mathrm{rt}}$, $\Delta\theta_{\mathrm{rp}}$, $\Delta\theta_{\mathrm{ht}}$ and $\Delta\theta_{\mathrm{hp}}$ are deviations of the desired angle from the current angle of motor M_{lu}, motor M_{ld}, motor M_{ru}, motor M_{rd}, motor M_{hu} and motor M_{hd}, respectively. ${}^{\mathrm{l}}\theta_{\mathrm{m}}$ and ${}^{\mathrm{r}}\theta_{\mathrm{m}}$ are the angles at which each wheel of the moving robot needs to be rotated. During the in situ gaze point tracking process, the moving robot performs only the rotation of the original position, and the angle of the robot movement can be calculated according to the desired angle of the robot. When the robot rotates, the two wheels move in opposite directions at the same speed. Let the distance between the two wheels of the moving robot be D_{r}; when the robot rotates around an angle ${}^{\mathrm{w}}\theta_{\mathrm{pq}}$, the distance that each wheel needs to move is

$$S = {}^{\mathrm{w}}\theta_{\mathrm{pq}}\frac{D_{\mathrm{r}}}{2} \tag{110}$$

The diameter of each wheel is d_{w}, and the angle of rotation of each wheel is (where counterclockwise is positive)

$${}^{\mathrm{r}}\theta_m = -{}^{\mathrm{l}}\theta_m = \frac{2S}{d_{\mathrm{w}}} \tag{111}$$

In the process of approaching the target, the moving robot follows a straight line, and the angle of rotation of each wheel is

$${}^{\mathrm{r}}\theta_m = {}^{\mathrm{l}}\theta_m = \frac{2^{\mathrm{w}}z_{\mathrm{mq}}}{d_{\mathrm{w}}} \tag{112}$$

The movement of the moving robot is achieved by controlling the rotation of each wheel. Each wheel is equipped with a DC brushless motor, and a DSP2000 controller is used to control the movement of the DC brushless motor. Position servo control is implemented in the DSP2000 controller.

In the robot system, the weight of the camera and lens is approximately 80 g, the weight of the camera and the fixed mechanical parts is approximately 50 g and the motor that controls the vertical rotation of the camera (rotating around the horizontal axis of rotation) and the corresponding encoder weighs approximately 250 g. The mechanical parts of the fixed vertical rotating motor and encoder weigh approximately 100 g. The radius of the rotation of the camera in the vertical direction is approximately 1 cm, and the rotation in the horizontal direction (rotation about the vertical axis of rotation) has a radius of approximately 2 cm. Therefore, when the gravitational acceleration is 9.8 $\mathrm{m/s^2}$, the torque required for the vertical rotating electric machine is approximately 0.013 N·m, and the torque required for the horizontal rotating electric machine is approximately 0.043 N·m. The vertical rotating motor uses a 28BYG5401 stepping motor with a holding torque of 0.1 N·m and a positioning torque of 0.008 N·m. The driver is HSM20403A. The horizontal rotating motor is a 57BYGH301 stepping motor with a holding torque of 1.5 N·m, a positioning torque of 0.07 N·m and drive model HSM20504A. The four stepping motors of the eye have a step angle of 1.8° and are all subdivided by 25, so the actual step angle of each motor is 0.072°, and the minimum pulse width that the driver can receive is 2.5 μs. The stepper motor has a maximum angular velocity of 200°/s.

The head vertical rotary motor uses a 57BYGH401 stepper motor with a holding torque of 2.2 N·m, a positioning torque of 0.098 N·m and drive model HSM20504A. The head horizontal rotary motor is an 86BYG350B three-phase AC stepping motor with a holding torque of 5 N·m, a positioning torque of 0.3 N-m and an HSM30860M driver. The step angle of the head motor after subdivision is also 0.072°. The head vertical motor has a load of approximately 5 kg and a radius of rotation of less than 1 cm. The head horizontal rotary motor has a load of approximately 9.5 kg and a radius of rotation of approximately 5 cm. In the experiment, we found that the maximum horizontal pulse frequency that the head horizontal rotary motor can receive is 0.6 Kpps. Its maximum angular velocity is 43.2°/s.

5. Experiments and Discussion

Using the robot platform introduced in Section 2, experiments on in situ gaze point tracking and approaching gaze point tracking were performed

Each camera has a resolution of 400 × 300 pixels. The directions of rotation are [−45°, 45°]. The range of rotation of the head is [−30°, 30°]. d_x and d_y are 150 mm and 200 mm, respectively. The internal and external parameters, distortion parameters, initial position parameters and left- and right-hand–eye parameters of the dual purpose method are calibrated as follows:

$$ {}^{\mathrm{l}}\boldsymbol{M}_{\mathrm{in}} = \begin{pmatrix} 341.58 & 0 & 201.6 \\ 0 & 341.97 & 147.62 \\ 0 & 0 & 1 \end{pmatrix} \tag{113} $$

$$ K_{\mathrm{l}} = \begin{pmatrix} -0.1905 & 0.2171 & -0.0018 & -0.0005 & -0.0823 \end{pmatrix} \tag{114} $$

$$ {}^{\mathrm{l}}T_m = \begin{pmatrix} 1.0 & 0.0078 & -0.0022 & 58.4172 \\ 0.0001 & 0.9954 & 0.0959 & 3.6042 \\ 0.0013 & 0.0959 & 0.9954 & 51.9366 \\ 0 & 0 & 0 & 1 \end{pmatrix} \tag{115} $$

$$ {}^{\mathrm{r}}\boldsymbol{M}_{\mathrm{in}} = \begin{pmatrix} 335.13 & 0 & 184.32 \\ 0 & 335.5 & 141.26 \\ 0 & 0 & 1 \end{pmatrix} \tag{116} $$

$$ K_{\mathrm{r}} = \begin{pmatrix} -0.1861 & 0.1987 & -0.004 & -0.0011 & -0.0739 \end{pmatrix} \tag{117} $$

$$ {}^{\mathrm{r}}T_m = \begin{pmatrix} 0.9999 & -0.0086 & -0.0125 & -45.0147 \\ 0.0190 & -0.9969 & -0.0782 & -24.5528 \\ 0.0097 & -0.0784 & 0.9970 & 42.9270 \\ 0 & 0 & 0 & 1 \end{pmatrix} \tag{118} $$

$$ {}^{\mathrm{l}}T_r = \begin{pmatrix} 0.9998 & -0.0099 & -0.0193 & 189.5922 \\ 0.0095 & 0.9997 & -0.0215 & -0.0426 \\ 0.0195 & 0.0213 & 0.9996 & 8.9671 \\ 0 & 0 & 0 & 1 \end{pmatrix} \tag{119} $$

The experimental in situ gaze point tracking scene is shown in Figure 10, with a checkerboard target used as the target. For in situ gaze point tracking, the target is held by a person. In the approaching target gaze tracking experiment, the target is fixed in front of the robot.

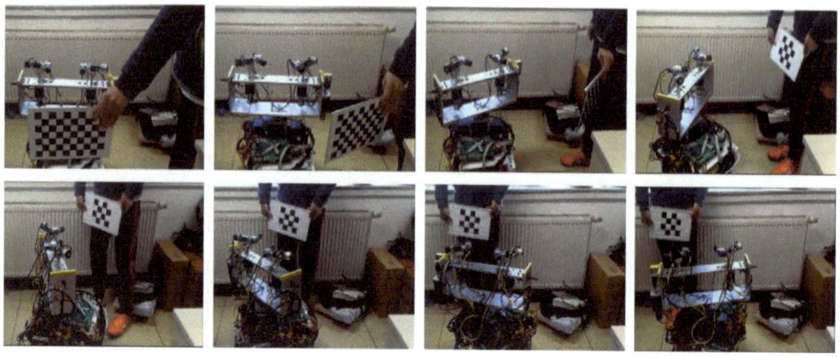

Figure 10. Experimental in situ gaze point tracking scene.

5.1. In Situ Gaze Point Tracking Experiment

In the in situ gaze experiment, the target moves at a low speed within a certain range, and the robot combines the movement of the eye, the head and the mobile robot so that the binocular vision can always perceive the 3D coordinates of the target at the optimal observation posture. This experiment prompts the robot to find the target and gaze at it. In the gaze point tracking process, binocular stereo vision is used to calculate the 3D coordinates of the target in the eye coordinate system in real time. Through the positional relationship between the eye and the head, the coordinate system of the target in the eye can be converted to the head coordinate system. Similarly, the 3D coordinates of the target in the robot coordinate system can be obtained. Through the 3D coordinates, the desired poses of the eyes, head and mobile robot are calculated according to the method proposed in this paper. Then, the camera is controlled to the desired position by the stepping motor; after reaching the desired position, the image and the motor position information are collected again, and the 3D coordinates of the target are calculated.

In the experiment, the angles between the head and the target, ${}^{\mathrm{h}}\gamma_{\mathrm{pmax}}$ and ${}^{\mathrm{h}}\gamma_{\mathrm{pmax}}$, are each 30°. The method described in Section 3 is used to calculate the desired pose of each joint of the robot based on the 3D coordinates of the target. In the experiment, the actual coordinate position and desired coordinate position of the target in the binocular image space, the actual position and desired position of the eye and head motor, the angle between the head and the robot and the target, and the target in the robot coordinate system are stored. Figure 11a,b show the u and v coordinates of the target on the left image, respectively, and Figure 11c,d show the u and v coordinates of the target on the right image, respectively. The desired image coordinates are recalculated based on the optimal observation position. Figure 11e–h show the positions of the tilt motor ($\mathrm{M_{lu}}$) of the left eye, the pan motor ($\mathrm{M_{ld}}$) of the left eye, the tilt motor ($\mathrm{M_{ru}}$) of the right eye and the pan motor ($\mathrm{M_{rd}}$) of the right eye, respectively. Figure 11i shows the positions of the pan motor ($\mathrm{M_{hd}}$) of the head. Since the target moves in the vertical direction with small amplitude, the motor $\mathrm{M_{hu}}$ does not rotate, and the case is similar to the motion principle of the motor $\mathrm{M_{hd}}$, so the motor position of the head only provides the result of the motor $\mathrm{M_{hd}}$. Figure 11j shows the angle deviation and rotation. In this figure, T-h is the angle between the head and target, T-r is the angle between the robot and target, R-r is the angle of the robot rotation from the origin location and T-o is the angle of the target to the origin location. Figure 11k shows the coordinates (${}^{w}x$, ${}^{w}z$) of the target in the world coordinate system. Figure 11l shows the coordinates (${}^{o}x$, ${}^{o}z$) of the target in the world coordinate system of the origin location.

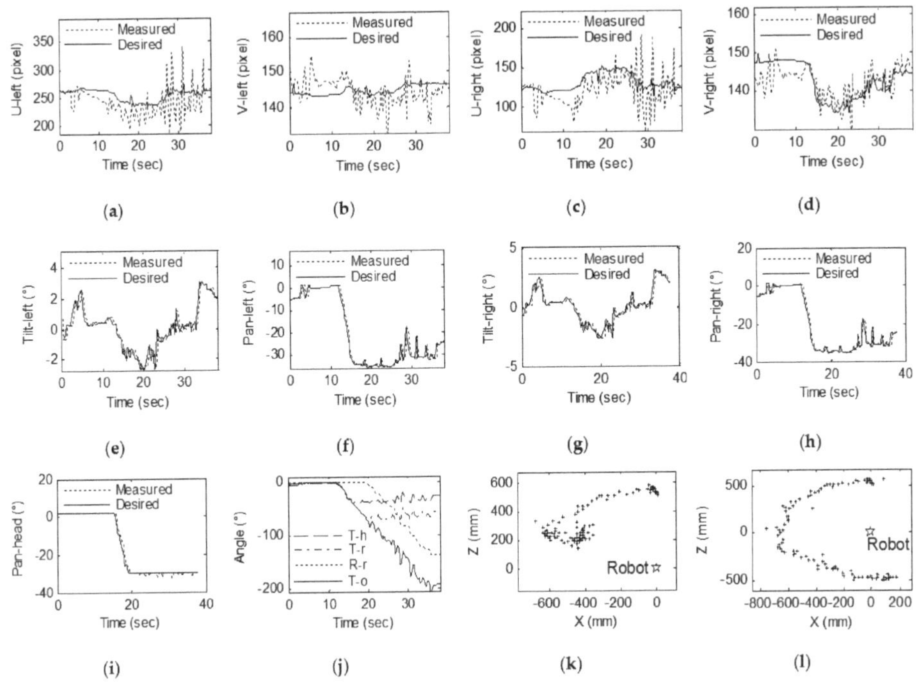

Figure 11. Experimental results of gaze shifting to the target: (**a**) U coordinates of the target on the left image. (**b**) V coordinates of the target on the left image. (**c**) U coordinates of the target on the right image. (**d**) V coordinates of the target on the right image. (**e**) Left camera tilt. (**f**) Left camera pan. (**g**) Right camera tilt. (**h**) Right camera pan. (**i**) Head pan. (**j**) Angle deviation and rotation. (**k**) Coordinates (^{w}x, ^{w}z) of the target in the world coordinate system. (**l**) Coordinates (^{n}x, ^{n}z) of the target in the world coordinate system based on the origin location. The "+" in the subfigures (**k**,**l**) represents the position of the target in the coordinate system, and the "☆" represents the position of the robot in the coordinate system.

As shown in Figure 11, the image coordinate of the target is substantially within ±40 pixels in the central region of the left and right images in the x direction. These coordinates are kept within ±10 pixels of the center region of the left and right images in the y direction. Throughout the experiment, the target was rotated approximately 200° around the robot. The robot moved approximately 140°, the head rotated 30° and the target could be kept in the center region of the binocular images. The motor position curve shows that the motor's operating position can track the desired position very well. The angle variation curve shows that the angle between the target and the head and the robot changes and that the robot turning angles are suitably consistent. The coordinates of the target shown in Figure 11 in the robot coordinate system and the coordinates of the target in the initial position of the world coordinate system are very close to the actual position change in the target's position.

Through the above analysis, we can determine the following: (1) It is feasible to realize gaze point tracking of a robot based on 3D coordinates. (2) Using the movement of the head, eyes and mobile robot used in this paper, it is possible to achieve gaze point tracking of the target while ensuring minimum resource consumption.

5.2. Approaching Gaze Point Tracking Experiment

The approaching gaze point tracking experimental scene is shown in Figure 12.

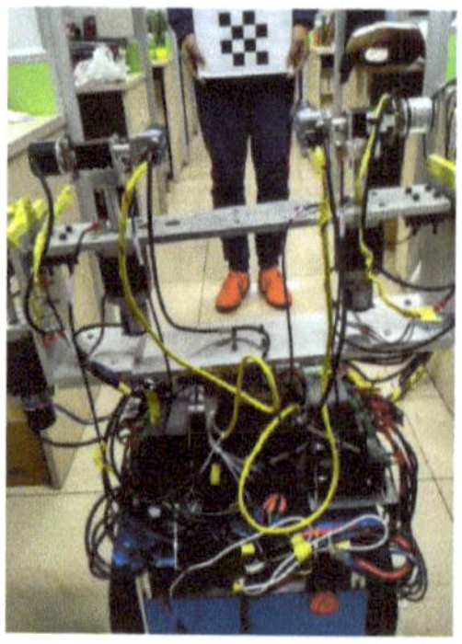

Figure 12. Experimental approaching gaze point tracking scene.

The robot approaches the target without obstacles and reaches the area in which the robot can operate on the target. The target can be grasped or carefully observed. In the approaching gaze experiment, a target is fixed at a position 2.2 m from the robot, and when the robot moves to a position where the distance from the target to robot is 0.6 m, the motion is stopped, and the maximum speed of the moving robot is 1 m/s. The experiment realizes the approaching movement to the target in two steps: first, the head, the eye and the moving robot chassis are rotated so that the head and the moving robot are facing the target, and the head observes the target in the optimal observation posture; second, the movement is controlled. The robot moves linearly in the target's direction. During the movement, the angles of the head and the eye are fine-tuned, and the 3D coordinates of the target are detected in real time until the z coordinate of the target in the robot coordinate system is less than the threshold set to stop the motion.

Figure 13 shows the results of the approaching gaze point tracking experiment. Figure 13a,b show the u and v coordinates of the target on the left image, respectively, and Figure 13c,d show the u and v coordinates of the target on the right image, respectively. The desired image coordinates are recalculated based on the optimal observation position. Figure 13e–h show the positions of the tilt motor (M_{lu}) of the left eye, the pan motor (M_{ld}) of the left eye, the tilt motor (M_{ru}) of the right eye and the pan motor (M_{rd}) of the right eye, respectively. Figure 13i shows the positions of the pan motor (M_{hd}) of the head. Figure 13j shows the angle deviation and rotation. T-h is the angle between the head and the target. T-r is the angle between the robot and the target, R-r is the angle of the robot's rotation from the origin location and T-o is the angle of the target to the origin location. Figure 13k shows the coordinates (^{w}x, ^{w}z) of the target in the world coordinate system. Figure 13l shows the robot's forward distance and the distance between the target and the robot.

The change in the image's coordinate curve indicates that the coordinates of the target in the left and right images move from the initial position to the central region of the image and stabilize in the center region of the image during the approach process. In the process of turning towards the target in the first step, the target coordinates in the image fluctuate because the head motor rotates a large amount and is accompanied by a certain vibration during the rotation, which can be avoided by using a system with better stability. The variety curve of the motor position in Figure 13 shows that the motion of the motor can track the target well with the desired pose, and the prediction of the 3D coordinates is not used during the tracking process, so this prediction is accompanied by a cycle lag. The changes in angle in Figure 13 show that the robot system achieves the task of steering toward the target in the first few control cycles and then moves toward the target at a stable angle. Figure 13a shows the change in the coordinates of the target in the robot coordinate system. When the robot rotates, fluctuations arise around the measured x coordinate, mainly due to the measurement error caused by the shaking of the system. The experimental results in Figure 13b show that the robot's movement toward the target is very consistent. During the approach process, the target can be kept within ±50 pixels of the desired position in the horizontal direction of the image while being within ±20 pixels of the desired position in the vertical direction of the image. The eye motor achieves fast tracking of the target in 1.5 s. The angle between the target and the head is reduced from 20° to 0°, and the angle between the target and the robot is reduced from 35° to 0°. The robot then over-turns. At 34°, the target changes by 34° from the initial position.

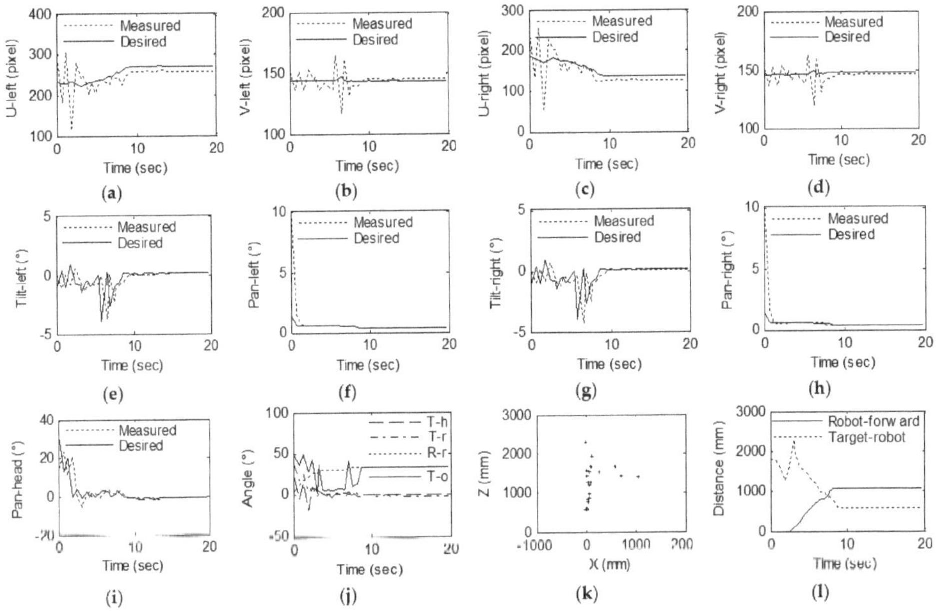

Figure 13. Experimental results of gaze shifting to the target: (**a**) U coordinates of the target on the left image. (**b**) V coordinates of the target on the left image. (**c**) U coordinates of the target on the right image. (**d**) V coordinates of the target on the right image. (**e**) Left camera tilt. (**f**) Left camera pan. (**g**) Right camera tilt. (**h**) Right camera pan. (**i**) Head pan. (**j**) Angular deviation and rotation. (**k**) Coordinates $({}^{w}x, {}^{w}z)$ of the target in the world coordinate system. (**l**) Robot forward distance and the distance between the target and robot.

Through the above analysis, it can be found that by using the combination of the head, the eye and the trunk in the present method, the approach toward the target can be achieved while ensuring that the robot is gazing at the target.

6. Conclusions

This study achieved gaze point tracking based on the 3D coordinates of the target. First, a robot experiment platform was designed. Based on the bionic eye experiment platform, a head with two degrees of freedom was added, using the mobile robot as a carrier.

Based on the characteristics of the robot platform, this paper proposed a method of gaze point tracking. To achieve in situ gaze point tracking, the combination of the eyes, head and trunk is designed based on the principles of minimum resource consumption and maximum system stability. Eye rotation consumes the least amount of resources and has minimal impact on the stability of the overall system during the exercise. The head rotation consumes more resources than the eyeball but fewer than the trunk rotation. At the same time, the rotation of the head affects the stability of the eyeball but only minimally affects the stability of the entire robotic system. The resources consumed by the rotation of the trunk generally predominate, and the rotation of the trunk tends to affect the stability of the head and the eye. Therefore, when the eye can observe the target in the optimal observation posture, only the eye is rotated; otherwise, the head is rotated, and when the angle at which the head needs to move exceeds its threshold, the mobile robot rotates. When approaching gaze point tracking is performed, the robot and head first face the target and then move straight toward the vicinity of the target. Based on the proposed gaze point tracking method, this paper provides an expected pose calculation method for the horizontal rotation angle and the vertical rotation angle.

Based on the experimental robot platform, a series of experiments was performed, and the effectiveness of the gaze point tracking method was verified. In our future works, a practical task of delivering medicine in a hospital and more detailed comparative experiments, as well as discussions with other similar studies, will be implemented.

Author Contributions: Conceptualization, Q.W. and M.Q.; Methodology, X.F.; Software, X.F.; Validation, X.F.; Formal analysis, H.C.; Investigation, M.Q.; Data curation, Q.W. and Y.Z.; Writing—original draft, X.F. and Q.W.; Writing—review & editing, M.Q.; Project administration, H.C. All authors have read and agreed to the published version of the manuscript.

Funding: This research received no external funding.

Institutional Review Board Statement: Not applicable.

Informed Consent Statement: Not applicable.

Data Availability Statement: No new data was created.

Conflicts of Interest: The authors declare no conflict of interest.

Appendix A. Measurement Error Analysis of Binocular Stereo Vision

It is assumed that the angle at which the left motion module rotates in the horizontal direction with respect to the initial position is ${}^{l}\theta_{p}$ and that the angle at which the right motion module rotates in the horizontal direction with respect to the initial position is ${}^{r}\theta_{p}$. When using the bionic eye platform for 3D coordinate calculation, we find that when ${}^{l}\theta_{p} = {}^{r}\theta_{p}$, the measured three-dimensional coordinate ratio is higher than ${}^{l}\theta_{p} > 0$ and ${}^{r}\theta_{p} < 0$, which is more accurate, as shown in Figure A1a,b. In this case, the measurement accuracy is higher than that in the case shown in Figure A1c. This chapter analyzes the measurement error of three-dimensional coordinates, explains the reason for this phenomenon and proposes a method to improve the accuracy of three-dimensional coordinate measurement.

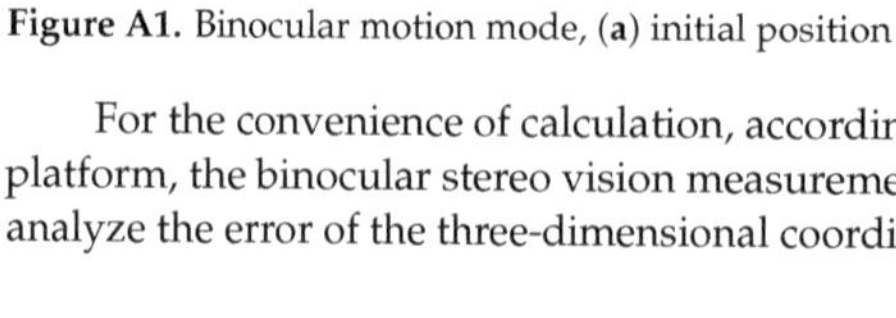

Figure A1. Binocular motion mode, (**a**) initial position ${}^{l}\theta_{p} = {}^{r}\theta_{p}$, (**b**) ${}^{l}\theta_{p} = {}^{r}\theta_{p}$, (**c**) ${}^{l}\theta_{p} > 0$ and ${}^{r}\theta_{p} < 0$.

For the convenience of calculation, according to the characteristics of the bionic eye platform, the binocular stereo vision measurement model shown in Figure A2 is used to analyze the error of the three-dimensional coordinate measurement.

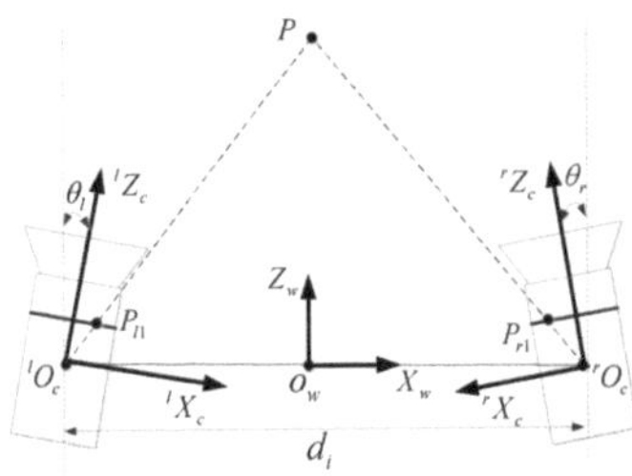

Figure A2. The principle of the vision system of bionic eyes.

Appendix A.1. Vision System of Bionic Eyes

Two cameras are used to imitate human eyes and the principle of the vision system of bionic eyes is shown in Figure 1. We suppose that the optical axes of two cameras used to imitate human eyes are coplanar. As shown in Figure 1, $O_wX_wZ_w$ is the world coordinate system, the X_w axis is along the baseline of two cameras and the Z_w axis is in the plane $\prod$, which consists of the two cameras' optical axes. ${}^lO_c{}^lX_c{}^lZ_c$ is the coordinate system of the left camera, where ${}^lZ_c$ is along the optical axis of the left camera and ${}^lX_c$ axis is in the plane $\prod$. ${}^rO_c{}^rX_c{}^rZ_c$ is the coordinate system of the right camera, ${}^rZ_c$ is along the optical axis of the right camera and ${}^rX_c$ axis is in the plane $\prod$. The two cameras can move cooperatively to imitate the movement of human eyes.

Appendix A.2. Depth Measurement Model

In Figure A2, the position vector of point ${}^lO_c$ in $O_wX_wZ_w$ is ${}^w\boldsymbol{O}_l = [-d_i/2, 0]^{\mathrm{T}}$ and the position vector of point ${}^rO_c$ in $O_wX_wZ_w$ is ${}^w\boldsymbol{O}_r = [d_i/2, 0]^{\mathrm{T}}$. d_i is the length of the baseline. Let ${}^l\boldsymbol{P} = [x_l, z_l]^{\mathrm{T}}$ and ${}^r\boldsymbol{P} = [x_r, z_r]^{\mathrm{T}}$ be the position vectors of the object point P in the left and right camera coordinate systems, respectively. Let ${}^w\boldsymbol{P} = [x_w, z_w]^{\mathrm{T}}$ be the position vector of point P in $O_wX_wZ_w$; then, ${}^l\boldsymbol{P}$ can be obtained as follows:

$$ {}^l\boldsymbol{P} = {}^l\boldsymbol{R}_w({}^w\boldsymbol{P} - {}^w\boldsymbol{O}_l) \tag{A1} $$

where ${}^l\boldsymbol{R}_w$ is the rotation transformation matrix from the world coordinate system to the left camera coordinate system. ${}^l\boldsymbol{R}_w$ can be expressed as

$$ {}^l\boldsymbol{R}_w = \begin{bmatrix} \cos\theta_l & \sin\theta_l \\ -\sin\theta_l & \cos\theta_l \end{bmatrix} \tag{A2} $$

where θ_l is the rotation angle which is defined as the angle between camera optical axis and the Z_w axis.

Based on the same principle, we can obtain

$$ {}^r\boldsymbol{P} = {}^r\boldsymbol{R}_w({}^w\boldsymbol{P} - {}^w\boldsymbol{O}_r) \tag{A3} $$

where

$$ {}^r\boldsymbol{R}_w = \begin{bmatrix} \cos\theta_r & \sin\theta_r \\ -\sin\theta_r & \cos\theta_r \end{bmatrix} \tag{A4} $$

As shown in Figure 1, P_{l1} is the intersection point of the line ${}^lO_cP$ and the normalized image plane of the left camera, and P_{r1} is the intersection point of the line ${}^rO_cP$ and the normalized image plane of the right camera. From the geometric relationship as shown in Figure 1, the position vector of point P_{l1} in ${}^lO_c{}^lX_c{}^lZ_c$ can be expressed as ${}^l\boldsymbol{P}_{l1} = [x_l/z_l, 1]^{\mathrm{T}}$, and the position vector of point P_{r1} in ${}^rO_c{}^rX_c{}^rZ_c$ can be expressed as ${}^r\boldsymbol{P}_{r1} = [x_r/z_r, 1]$. Let ${}^w\boldsymbol{P}_{l1} = [{}^wx_{l1}, {}^wz_{l1}]^{\mathrm{T}}$ be the position vector of point P_{l1} in $O_wX_wZ_w$ and ${}^w\boldsymbol{P}_{r1} = [{}^wx_{r1}, {}^wz_{r1}]^{\mathrm{T}}$

be the position vector of point P_{r1} in $O_wX_wZ_w$. ${}^w\boldsymbol{P}_{l1}$ and ${}^w\boldsymbol{P}_{r1}$ can be calculated in (A5) and (A6) by coordinate transformation according to (A1) and (A3).

$$ {}^w\boldsymbol{P}_{l1} = \left({}^l\boldsymbol{R}_w\right)^{-1} {}^l\boldsymbol{P}_{l1} + {}^w\boldsymbol{O}_l \tag{A5} $$

$$ {}^w\boldsymbol{P}_{r1} = ({}^r\boldsymbol{R}_w)^{-1} {}^r\boldsymbol{P}_{r1} + {}^w\boldsymbol{O}_r \tag{A6} $$

From Equations (99) and (100), we can obtain

$$ {}^w\boldsymbol{P}_{l1} = \begin{bmatrix} {}^w x_{l1} \\ {}^w z_{l1} \end{bmatrix} = \begin{bmatrix} \frac{x_l}{z_l}\cos\theta_l - \sin\theta_l - \frac{d_i}{2} \\ \frac{x_l}{z_l}\sin\theta_l + \cos\theta_l \end{bmatrix} \tag{A7} $$

$$ {}^w\boldsymbol{P}_{r1} = \begin{bmatrix} {}^w x_{r1} \\ {}^w z_{r1} \end{bmatrix} = \begin{bmatrix} \frac{x_r}{z_r}\cos\theta_r - \sin\theta_r + \frac{d_i}{2} \\ \frac{x_r}{z_r}\sin\theta_r + \cos\theta_r \end{bmatrix} \tag{A8} $$

According to ${}^w\boldsymbol{O}_l = [-d_i/2, 0]^{\mathrm{T}}$ and Equation (A7), the line ${}^lO_cP$ can be expressed as Equation (A9) in the world coordinate system $O_wX_wZ_w$.

$$ z = \frac{{}^w z_{l1}}{\frac{d_i}{2} + {}^w x_{l1}} x + \frac{{}^w z_{r1} \times \frac{d_i}{2}}{\frac{d_i}{2} + {}^w x_{l1}} \tag{A9} $$

Based on ${}^w\boldsymbol{O}_r = [d_i/2, 0]^{\mathrm{T}}$ and Equation (A8), the line ${}^rO_cP$ in the world coordinate system $O_wX_wZ_w$ can be expressed as follows:

$$ z = -\frac{{}^w z_{r1}}{\frac{d_i}{2} - {}^w x_{r1}} x + \frac{{}^w z_{r1} \times \frac{d_i}{2}}{\frac{d_i}{2} - {}^w x_{l1}} \tag{A10} $$

The intersection point of the lines ${}^rO_cP$ and ${}^lO_cP$ is the point P, so the depth z_w of point P can be obtained by Equations (A9) and (A10).

$$ z_w = \frac{-d_i {}^w z_{r1} {}^w z_{r1}}{{}^w x_{r1} {}^w z_{l1} - {}^w z_{l1}\frac{d_i}{2} - {}^w z_{r1}\frac{d_i}{2} - {}^w x_{l1} {}^w z_{r1}} \tag{A11} $$

Appendix A.3. Measurement Error Analysis

In the real situation, there are the eyes' rotation angle errors (usually caused by time difference between image acquisition and motor angle acquisition when the two cameras move, the stepper motor's clearance error and the encoder's resolution) and image errors (usually caused by image distortion, image resolution and image feature extraction error). Let Δm_l and Δm_r be the image errors of the two cameras, respectively, then ${}^l\boldsymbol{P}_{l1} = [x_l/z_l, 1]^{\mathrm{T}}$ and ${}^r\boldsymbol{P}_{r1} = [x_r/z_r, 1]$ can be revised as ${}^l\boldsymbol{P}'_{l1} = [x_l/z_l + \Delta m_l, 1]^{\mathrm{T}}$ and ${}^r\boldsymbol{P}'_{r1} = [x_r/z_r + \Delta m_r, 1]$. Let $\Delta\theta_l$ and $\Delta\theta_r$ be the errors of two cameras' rotation angles, respectively. Therefore, we can rewrite Equations (A7) and (A8) as follows:

$$ {}^w\boldsymbol{P}'_{l1} = \begin{bmatrix} {}^w x'_{l1} \\ {}^w z'_{l1} \end{bmatrix} = \begin{bmatrix} \left(\frac{x_l}{z_l} + \Delta m_l\right)\cos(\theta_l + \Delta\theta_l) - \sin(\theta_l + \Delta\theta_l) - \frac{d_i}{2} \\ \left(\frac{x_l}{z_l} + \Delta m_l\right)\sin(\theta_l + \Delta\theta_l) + \cos(\theta_l + \Delta\theta_l) \end{bmatrix} \tag{A12} $$

$$ {}^w\boldsymbol{P}'_{r1} = \begin{bmatrix} {}^w x'_{r1} \\ {}^w z'_{r1} \end{bmatrix} = \begin{bmatrix} \left(\frac{x_r}{z_r} + \Delta m_r\right)\cos(\theta_r + \Delta\theta_r) - \sin(\theta_r + \Delta\theta_r) + \frac{d_i}{2} \\ \left(\frac{x_r}{z_r} + \Delta m_r\right)\sin(\theta_r + \Delta\theta_r) + \cos(\theta_r + \Delta\theta_r) \end{bmatrix} \tag{A13} $$

Based on the same principle, the revised depth z'_w of point P can be obtained as follows:

$$z'_w = \frac{-d_i {}^w z'_{l1} {}^w z'_{r1}}{{}^w x'_{r1} {}^w z'_{l1} - {}^w z'_{l1} \frac{d_i}{2} - {}^w z'_{r1} \frac{d_i}{2} - {}^w x'_{l1} {}^w z'_{r1}} \tag{A14}$$

According to the cooperative movement pattern of human eyes, the absolute values of θ_l and θ_r are restricted to a limited range and assumed to be equal as follows:

$$-\theta_l = \theta_r = \theta \;\; s.t. \;\; 0 \le \theta < \frac{\pi}{2} \tag{A15}$$

Since Δm_r, Δm_l, $\Delta\theta_l$ and $\Delta\theta_r$ are usually close to 0, the simplified expression of the error Δz between the actual value z_w and measurement value z'_w can be derived from (A7), (A8) and (A11)–(A14).

$$\Delta z = z'_w - z_w \approx -z_w \frac{z_w(\Delta m_l \cos^2\theta - \Delta m_r \cos^2\theta - \Delta\theta_l + \Delta\theta_r)}{d_i + z_w(\Delta m_l \cos^2\theta - \Delta m_r \cos^2\theta - \Delta\theta_l + \Delta\theta_r)} \tag{A16}$$

In addition, the relative error of z_w is

$$\Delta z_r = \frac{\Delta z}{z_w} \approx -\frac{z_w(\Delta m_l \cos^2\theta - \Delta m_r \cos^2\theta - \Delta\theta_l + \Delta\theta_r)}{d_i + z_w(\Delta m_l \cos^2\theta - \Delta m_r \cos^2\theta - \Delta\theta_l + \Delta\theta_r)} \tag{A17}$$

Let

$$\varepsilon = \Delta m_l \cos^2\theta - \Delta m_r \cos^2\theta - \Delta\theta_l + \Delta\theta_r \tag{A18}$$

In practice, ε has very small value and $z_w\varepsilon << d_i$, so Δz_r can be simplified as follows:

$$\Delta z_r \approx -\frac{z_w\varepsilon}{d_i} \tag{A19}$$

From Equation (A19), it can be known that relative error of z_w is proportional to $z_w\varepsilon$ and inversely proportional to d_i. Thus, we can adopt the following strategies to reduce the error of depth:

(1) Keep d_i long enough and constant when the bionic eyes move.

(2) Observe the target as close as possible since the depth error is smaller when bionic eyes observe target in a close distance.

(3) Control the two cameras of the bionic eyes with the same angular velocity during the process of the eyes' movement. In this way, $\Delta\theta^l$ and $\Delta\theta^r$ will be approximately equal to each other, and ε can be reduced.

(4) Keep the target on the Z_w axis if possible, so that Δm_l and Δm_r are close to each other.

These strategies can be used to design effective motion control methods so that bionic eyes can perceive the target's 3D information accurately.

Appendix B. Proof of Equation (100)

Let

$$a = -d_i {}^w z'_{l1} {}^w z'_{r1} \tag{A20}$$

$$b = {}^w x'_{r1} {}^w z'_{l1} - {}^w z'_{l1} \frac{d_i}{2} - {}^w z'_{r1} \frac{d_i}{2} - {}^w x'_{l1} {}^w z'_{r1} \tag{A21}$$

Then, z'_w in Equation (A14) can be expressed as

$$z'_w = \frac{a}{b} \tag{A22}$$

From Equations (A7) and (A8), we can obtain

$$ {}^{w}z'_{l1} = (\frac{x_l}{z_l} + \Delta m_l)\sin(\theta_l + \Delta\theta_l) + \cos(\theta_l + \Delta\theta_l) \tag{A23} $$

$$ {}^{w}z'_{r1} = (\frac{x_r}{z_r} + \Delta m_r)\sin(\theta_r + \Delta\theta_r) + \cos(\theta_r + \Delta\theta_r) \tag{A24} $$

From (A10), (A23) and (A24), we can obtain

$$ \begin{aligned} a = \; & -d_i \tfrac{x_r}{z_r}\tfrac{x_l}{z_l}[\sin\theta_r\cos\Delta\theta_r + \cos\theta_r\sin\Delta\theta_r][\sin\theta_l\cos\Delta\theta_l + \cos\theta_l\sin\Delta\theta_l] \\ & -d_i \tfrac{x_r}{z_r}[\sin\theta_r\cos\Delta\theta_r + \cos\theta_r\sin\Delta\theta_r][\sin\theta_l\cos\Delta\theta_l + \cos\theta_l\sin\Delta\theta_l]\Delta m_l \\ & -d_i \tfrac{x_r}{z_r}[\sin\theta_r\cos\Delta\theta_r + \cos\theta_r\sin\Delta\theta_r][\cos\theta_l\cos\Delta\theta_l - \sin\theta_l\sin\Delta\theta_l] \\ & -d_i \tfrac{x_l}{z_l}[\sin\theta_r\cos\Delta\theta_r + \cos\theta_r\sin\Delta\theta_r][\sin\theta_l\cos\Delta\theta_l + \cos\theta_l\sin\Delta\theta_l]\Delta m_r \\ & -d_i[\sin\theta_r\cos\Delta\theta_r + \cos\theta_r\sin\Delta\theta_r][\sin\theta_l\cos\Delta\theta_l + \cos\theta_l\sin\Delta\theta_l]\Delta m_l\Delta m_r \\ & -d_i[\sin\theta_r\cos\Delta\theta_r + \cos\theta_r\sin\Delta\theta_r][\cos\theta_l\cos\Delta\theta_l - \sin\theta_l\sin\Delta\theta_l]\Delta m_r \\ & -d_i \tfrac{x_l}{z_l}[\sin\theta_l\cos\Delta\theta_l + \cos\theta_l\sin\Delta\theta_l][\cos\theta_r\cos\Delta\theta_r - \sin\theta_r\sin\Delta\theta_r] \\ & -d_i[\sin\theta_l\cos\Delta\theta_l + \cos\theta_l\sin\Delta\theta_l][\cos\theta_r\cos\Delta\theta_r - \sin\theta_r\sin\Delta\theta_r]\Delta m_l \\ & -d_i[\cos\theta_r\cos\Delta\theta_r - \sin\theta_r\sin\Delta\theta_r][\cos\theta_l\cos\Delta\theta_l - \sin\theta_l\sin\Delta\theta_l] \end{aligned} \tag{A25} $$

Δm_r, Δm_l, $\Delta\theta_l$ and $\Delta\theta_r$ are usually close to 0, so

$$ \begin{cases} \cos\Delta\theta_r \approx 1 \\ \cos\Delta\theta_l \approx 1 \\ \sin\Delta\theta_r\sin\Delta\theta_l \approx 0 \\ \Delta m_l\Delta m_r \approx 0 \\ \Delta m_l\sin\Delta\theta_l \approx 0 \\ \Delta m_r\sin\Delta\theta_l \approx 0 \\ \Delta m_l\sin\Delta\theta_r \approx 0 \\ \Delta m_r\sin\Delta\theta_r \approx 0 \end{cases} \tag{A26} $$

From (A25) and (A26), we can obtain

$$ \begin{aligned} a \approx \; & d_i \tfrac{1}{z_r z_l}[-x_l x_r\sin\theta_l\sin\theta_r - x_r z_l\cos\theta_l\sin\theta_r - x_l z_r\sin\theta_l\cos\theta_r - z_r z_l\cos\theta_r\cos\theta_l \\ & + \sin\Delta\theta_l(-x_l x_r\sin\theta_r\cos\theta_l + x_r z_l\sin\theta_r\sin\theta_l - x_l z_r\cos\theta_r\cos\theta_l + z_r z_l\cos\theta_r\sin\theta_l) \\ & + \sin\Delta\theta_r(x_l z_r\sin\theta_l\sin\theta_r + z_r z_l\cos\theta_l\sin\theta_r - x_l x_r\sin\theta_l\cos\theta_r - x_r z_l\cos\theta_l\cos\theta_r) \\ & -\Delta m_l(x_r z_l\sin\theta_l\sin\theta_{rl} + z_r z_l\sin\theta_l\cos\theta_r) - \Delta m_r(x_l z_r\sin\theta_r\sin\theta_l - z_r z_l\sin\theta_r\cos\theta_l)] \end{aligned} \tag{A27} $$

From (A7), (A8) and (A19), we can obtain

$$ \begin{cases} x_l = (x_w + \frac{d_i}{2})\cos\theta - z_w\sin\theta \\ z_l = -(x_w + \frac{d_i}{2})\sin\theta + z_w\cos\theta \\ x_r = (x_w - \frac{d_i}{2})\cos\theta + z_w\sin\theta \\ z_r = (x_w - \frac{d_i}{2})\sin\theta + z_w\cos\theta \end{cases} \tag{A28} $$

Equation (A19) can be derived from (A19), (A27) and (A28):

$$ \begin{aligned} a \approx \; & z_w^2 \frac{-d_i + [\frac{d_i}{2}\sin 2\theta + \frac{d_i(x_w + \frac{d_i}{2})\sin^2\theta}{z_w}]\Delta m_l}{z_l z_r} + z_w^2 \frac{[\frac{d_i(x_w - \frac{d_i}{2})\sin^2\theta}{z_w} - \frac{d_i}{2}\sin 2\theta]\Delta m_r}{z_l z_r} \\ & + z_w^2 \frac{\frac{-d_i(x_w + \frac{d_i}{2})\sin\Delta\theta_l}{z_w} + \frac{d_i(x_w - \frac{d_i}{2})\sin\Delta\theta_r}{z_w}}{z_l z_r} \end{aligned} \tag{A29} $$

Δm_r, Δm_l, $\Delta\theta_l$ and $\Delta\theta_r$ are usually close to 0, $x_w \ll z_w$, and $d_i \ll z_w$. Thus,

$$ a \approx z_w^2 \frac{-d_i}{z_l z_r} \tag{A30} $$

From (A21), (A23) and (A24), we can obtain

$$
\begin{aligned}
b = \; & \tfrac{x_l}{z_l}\tfrac{x_r}{z_r}(\sin\theta_l\cos\Delta\theta_l + \cos\theta_l\sin\Delta\theta_l)(\cos\theta_r\cos\Delta\theta_r - \sin\theta_r\sin\Delta\theta_r) \\
& + \tfrac{x_l}{z_l}(\sin\theta_l\cos\Delta\theta_l + \cos\theta_l\sin\Delta\theta_l)(\cos\theta_r\cos\Delta\theta_r - \sin\theta_r\sin\Delta\theta_r)\Delta m_r \\
& - \tfrac{x_l}{z_l}(\sin\theta_l\cos\Delta\theta_l + \cos\theta_l\sin\Delta\theta_l)(\sin\theta_r\cos\Delta\theta_r + \cos\theta_r\sin\Delta\theta_r) \\
& + \tfrac{x_r}{z_r}(\sin\theta_l\cos\Delta\theta_l + \cos\theta_l\sin\Delta\theta_l)(\cos\theta_r\cos\Delta\theta_r - \sin\theta_r\sin\Delta\theta_r)\Delta m_l \\
& + (\sin\theta_l\cos\Delta\theta_l + \cos\theta_l\sin\Delta\theta_l)(\cos\theta_r\cos\Delta\theta_r - \sin\theta_r\sin\Delta\theta_r)\Delta m_l\Delta m_r \\
& - (\sin\theta_l\cos\Delta\theta_l + \cos\theta_l\sin\Delta\theta_l)(\sin\theta_r\cos\Delta\theta_r + \cos\theta_r\sin\Delta\theta_r)\Delta m_l \\
& + \tfrac{x_r}{z_r}(\cos\theta_l\cos\Delta\theta_l - \sin\theta_l\sin\Delta\theta_l)(\cos\theta_r\cos\Delta\theta_r - \sin\theta_r\sin\Delta\theta_r) \\
& + (\cos\theta_l\cos\Delta\theta_l - \sin\theta_l\sin\Delta\theta_l)(\cos\theta_r\cos\Delta\theta_r - \sin\theta_r\sin\Delta\theta_r)\Delta m_r \\
& - (\cos\theta_l\cos\Delta\theta_l - \sin\theta_l\sin\Delta\theta_l)(\sin\theta_r\cos\Delta\theta_r + \cos\theta_r\sin\Delta\theta_r) \\
& - \tfrac{x_r}{z_r}\tfrac{x_l}{z_l}(\sin\theta_r\cos\Delta\theta_r + \cos\theta_r\sin\Delta\theta_r)(\cos\theta_l\cos\Delta\theta_l - \sin\theta_l\sin\Delta\theta_l) \\
& - \tfrac{x_r}{z_r}(\sin\theta_r\cos\Delta\theta_r + \cos\theta_r\sin\Delta\theta_r)(\cos\theta_l\cos\Delta\theta_l - \sin\theta_l\sin\Delta\theta_l)\Delta m_l \\
& + \tfrac{x_r}{z_r}(\sin\theta_r\cos\Delta\theta_r + \cos\theta_r\sin\Delta\theta_r)(\sin\theta_l\cos\Delta\theta_l + \cos\theta_l\sin\Delta\theta_l) \\
& - \tfrac{x_l}{z_l}(\sin\theta_r\cos\Delta\theta_r + \cos\theta_r\sin\Delta\theta_r)(\cos\theta_l\cos\Delta\theta_l - \sin\theta_l\sin\Delta\theta_l)\Delta m_r \\
& - (\sin\theta_r\cos\Delta\theta_r + \cos\theta_r\sin\Delta\theta_r)(\cos\theta_l\cos\Delta\theta_l - \sin\theta_l\sin\Delta\theta_l)\Delta m_l\Delta m_r \\
& + (\sin\theta_r\cos\Delta\theta_r + \cos\theta_r\sin\Delta\theta_r)(\sin\theta_l\cos\Delta\theta_l + \cos\theta_l\sin\Delta\theta_l)\Delta m_r \\
& - \tfrac{x_l}{z_l}(\cos\theta_r\cos\Delta\theta_r - \sin\theta_r\sin\Delta\theta_r)(\cos\theta_l\cos\Delta\theta_l - \sin\theta_l\sin\Delta\theta_l) \\
& - (\cos\theta_r\cos\Delta\theta_r - \sin\theta_r\sin\Delta\theta_r)(\cos\theta_l\cos\Delta\theta_l - \sin\theta_l\sin\Delta\theta_l)\Delta m_l \\
& + (\cos\theta_r\cos\Delta\theta_r - \sin\theta_r\sin\Delta\theta_r)(\sin\theta_l\cos\Delta\theta_l + \cos\theta_l\sin\Delta\theta_l)
\end{aligned}
\tag{A31}
$$

From (A26) and (A31), we can obtain:

$$
\begin{aligned}
b \approx \; & \tfrac{1}{z_l z_r}[x_l x_r\sin\theta_l\cos\theta_r - x_l z_r\sin\theta_l\sin\theta_r + z_l x_r\cos\theta_l\cos\theta_r - z_l z_r\cos\theta_l\sin\theta_r \\
& - x_l x_r\sin\theta_r\cos\theta_l + z_l z_r\sin\theta_l\cos\theta_r - x_l z_r\cos\theta_l\cos\theta_r + z_l x_r\sin\theta_l\sin\theta_r \\
& + \sin\Delta\theta_l(x_l x_r\cos\theta_l\cos\theta_r - x_l z_r\cos\theta_l\sin\theta_r - z_l x_r\cos\theta_r\sin\theta_l + x_l z_r\sin\theta_l\cos\theta_r \\
& + z_l z_r\cos\theta_r\cos\theta_l + z_l z_r\sin\theta_l\sin\theta_r + x_l x_r\sin\theta_l\sin\theta_r + z_l x_r\sin\theta_r\cos\theta_l) \\
& + \sin\Delta\theta_r(-x_l x_r\sin\theta_l\sin\theta_r - x_l z_r\sin\theta_l\cos\theta_r - z_l x_r\cos\theta_l\sin\theta_r - z_l z_r\cos\theta_l\cos\theta_r \\
& \quad x_l x_r\cos\theta_l\cos\theta_r + z_l x_r\sin\theta_l\cos\theta_r + x_l z_r\cos\theta_l\sin\theta_r - z_l z_r\sin\theta_l\sin\theta_r) \\
& + \Delta m_l(z_l x_r\sin\theta_l\cos\theta_r - z_l z_r\sin\theta_l\sin\theta_r - z_l x_r\cos\theta_l\sin\theta_r - z_l z_r\cos\theta_l\cos\theta_r) \\
& + \Delta m_r(x_l z_r\sin\theta_l\cos\theta_r + z_l z_r\cos\theta_l\cos\theta_r - x_l z_r\cos\theta_l\sin\theta_r + z_l z_r\sin\theta_l\sin\theta_r)]
\end{aligned}
\tag{A32}
$$

Equation (A33) can be derived by (A19), (A28) and (A32):

$$
\begin{aligned}
b \approx \; & z_w\frac{-d_i - \Delta m_l[2x_w\sin\theta\cos\theta + \frac{(x_w^2 - \frac{d_i^2}{4})}{z_w}\sin^2\theta + z_w\cos^2\theta]}{z_l z_r} \\
& + z_w\frac{\Delta m_r[z_w\cos^2\theta + \frac{(x_w^2 - \frac{d_i^2}{4})}{z_w}\sin^2\theta - 2x_w\sin\theta\cos\theta]}{z_l z_r} \\
& + z_w\frac{\sin\Delta\theta_l[z_w + \frac{x_w^2 - \frac{d_i^2}{4}}{z_w}] - \sin\Delta\theta_r[z_w + \frac{x_w^2 - \frac{d_i^2}{4}}{z_w}]}{z_l z_r}
\end{aligned}
\tag{A33}
$$

Δm_r, Δm_l, $\Delta\theta_l$ and $\Delta\theta_r$ are usually close to 0, $x_w \ll z_w$, and $d_i \ll z_w$. Thus,

$$
b \approx z_w\frac{-d_i + z_w(\Delta m_r\cos^2\theta - \Delta m_l\cos^2\theta + \sin\Delta\theta_l - \sin\Delta\theta_r)}{z_l z_r}
\tag{A34}
$$

From (A22), (A29) and (A34), we can obtain

$$
z'_w \approx \frac{z_w d_i}{d_i - z_w(\Delta m_r\cos^2\theta - \Delta m_l\cos^2\theta + \sin\Delta\theta_l - \sin\Delta\theta_r)}
\tag{A35}
$$

So,

$$
\Delta z = z'_w - z_w \;\approx\; -z_w\frac{z_w(\Delta m_l\cos^2\theta - \Delta m_r\cos^2\theta - \sin\Delta\theta_l + \sin\Delta\theta_r)}{d_i + z_w(\Delta m_l\cos^2\theta - \Delta m_r\cos^2\theta - \sin\Delta\theta_l + \sin\Delta\theta_r)}
\tag{A36}
$$

$\Delta\theta_l$ and $\Delta\theta_r$ are usually close to 0, so

$$\Delta z \approx -z_w \frac{z_w(\Delta m_l \cos^2\theta - \Delta m_r \cos^2\theta - \Delta\theta_l + \Delta\theta_r)}{d_i + z_w(\Delta m_l \cos^2\theta - \Delta m_r \cos^2\theta - \Delta\theta_l + \Delta\theta_r)} \quad \text{(A37)}$$

The proof is completed.

Appendix C. Two Equations Related to θ_t and θ_p

Substituting Equation (59) into Equation (60), we can obtain:

$$\begin{pmatrix} {}^1x_c \\ {}^1y_c \\ {}^1z_c \\ 1 \end{pmatrix} = \begin{pmatrix} {}^1n_x & {}^1o_x & {}^1a_x & {}^1p_x \\ {}^1n_y & {}^1o_y & {}^1a_y & {}^1p_y \\ {}^1n_z & {}^1o_z & {}^1a_z & {}^1p_z \\ 0 & 0 & 0 & 1 \end{pmatrix} \begin{pmatrix} 1 & 0 & 0 & 0 \\ 0 & \cos^1\theta_t & \sin^1\theta_t & 0 \\ 0 & -\sin^1\theta_t & \cos^1\theta_t & 0 \\ 0 & 0 & 0 & 1 \end{pmatrix} \begin{pmatrix} \cos^1\theta_p & 0 & -\sin^1\theta_p & 0 \\ 0 & 1 & 0 & 0 \\ \sin^1\theta_p & 0 & \cos^1\theta_p & 0 \\ 0 & 0 & 0 & 1 \end{pmatrix} \begin{pmatrix} x_w \\ y_w \\ z_w \\ 1 \end{pmatrix} \quad \text{(A38)}$$

Equation (A38) can be factored out:

$$\begin{pmatrix} {}^1x_c \\ {}^1y_c \\ {}^1z_c \end{pmatrix} = \begin{pmatrix} {}^1n_x x_w \cos{}^1\theta_p - {}^1n_x z_w \sin{}^1\theta_p + {}^1o_x x_w \sin{}^1\theta_p \sin{}^1\theta_t \\ +{}^1o_x y_w \cos{}^1\theta_t + {}^1o_x z_w \cos{}^1\theta_p \sin{}^1\theta_t + {}^1a_x x_w \sin{}^1\theta_p \cos{}^1\theta_t \\ -{}^1a_x y_w \sin{}^1\theta_t + {}^1a_x z_w \cos{}^1\theta_p \cos{}^1\theta_t + {}^1p_x \\ {}^1n_y x_w \cos{}^1\theta_p - {}^1n_y z_w \sin{}^1\theta_p + {}^1o_y x_w \sin{}^1\theta_p \sin{}^1\theta_t \\ +{}^1o_y y_w \cos{}^1\theta_t + {}^1o_y z_w \cos{}^1\theta_p \sin{}^1\theta_t + {}^1a_y x_w \sin{}^1\theta_p \cos{}^1\theta_t \\ -{}^1a_y y_w \sin{}^1\theta_t + {}^1a_y z_w \cos{}^1\theta_p \cos{}^1\theta_t + {}^1p_y \\ {}^1n_z x_w \cos{}^1\theta_p - {}^1n_z z_w \sin{}^1\theta_p + {}^1o_z x_w \sin{}^1\theta_p \sin{}^1\theta_t \\ +{}^1o_z y_w \cos{}^1\theta_t + {}^1o_z z_w \cos{}^1\theta_p \sin{}^1\theta_t + {}^1a_z x_w \sin{}^1\theta_p \cos{}^1\theta_t \\ -{}^1a_z y_w \sin{}^1\theta_t + {}^1a_z z_w \cos{}^1\theta_p \cos{}^1\theta_t + {}^1p_z \end{pmatrix} \quad \text{(A39)}$$

Substituting Equation (A39) into Equation (62), we can obtain:

$$\begin{pmatrix} \Delta u_1 \\ \Delta v_1 \end{pmatrix} = \begin{pmatrix} \dfrac{\begin{matrix} ({}^1k_x {}^1n_x x_w \cos{}^1\theta_p - {}^1k_x {}^1n_x z_w \sin{}^1\theta_p + {}^1k_x {}^1o_x x_w \sin{}^1\theta_p \sin{}^1\theta_t \\ +{}^1k_x {}^1o_x y_w \cos{}^1\theta_t + {}^1k_x {}^1o_x z_w \cos{}^1\theta_p \sin{}^1\theta_t + {}^1k_x {}^1a_x x_w \sin{}^1\theta_p \cos{}^1\theta_t \\ -{}^1k_x {}^1a_x y_w \sin{}^1\theta_t + {}^1k_x {}^1a_x z_w \cos{}^1\theta_p \cos{}^1\theta_t + {}^1k_x {}^1p_x) \end{matrix}}{\begin{matrix} ({}^1n_z x_w \cos{}^1\theta_p - {}^1n_z z_w \sin{}^1\theta_p + {}^1o_z x_w \sin{}^1\theta_p \sin{}^1\theta_t \\ +{}^1o_z y_w \cos{}^1\theta_t + {}^1o_z z_w \cos{}^1\theta_p \sin{}^1\theta_t + {}^1a_z x_w \sin{}^1\theta_p \cos{}^1\theta_t \\ -{}^1a_z y_w \sin{}^1\theta_t + {}^1a_z z_w \cos{}^1\theta_p \cos{}^1\theta_t + {}^1p_z) \end{matrix}} \\ \dfrac{\begin{matrix} ({}^1k_y {}^1n_y x_w \cos{}^1\theta_p - {}^1k_y {}^1n_y z_w \sin{}^1\theta_p + {}^1k_y {}^1o_y x_w \sin{}^1\theta_p \sin{}^1\theta_t \\ +{}^1k_y {}^1o_y y_w \cos{}^1\theta_t + {}^1k_y {}^1o_y z_w \cos{}^1\theta_p \sin{}^1\theta_t + {}^1k_y {}^1a_y x_w \sin{}^1\theta_p \cos{}^1\theta_t \\ -{}^1k_y {}^1a_y y_w \sin{}^1\theta_t + {}^1k_y {}^1a_y z_w \cos{}^1\theta_p \cos{}^1\theta_t + {}^1k_y {}^1p_y) \end{matrix}}{\begin{matrix} ({}^1n_z x_w \cos{}^1\theta_p - {}^1n_z z_w \sin{}^1\theta_p + {}^1o_z x_w \sin{}^1\theta_p \sin{}^1\theta_t \\ +{}^1o_z y_w \cos{}^1\theta_t + {}^1o_z z_w \cos{}^1\theta_p \sin{}^1\theta_t + {}^1a_z x_w \sin{}^1\theta_p \cos{}^1\theta_t \\ -{}^1a_z y_w \sin{}^1\theta_t + {}^1a_z z_w \cos{}^1\theta_p \cos{}^1\theta_t + {}^1p_z) \end{matrix}} \end{pmatrix} \quad \text{(A40)}$$

Based on the same principle, substituting into each matrix and factoring the value of Δm_r, we can obtain

$$\begin{pmatrix}\Delta u_{\mathrm{r}}\\ \Delta v_{\mathrm{r}}\end{pmatrix}=\begin{pmatrix}\dfrac{\begin{array}{c}({}^{\mathrm{r}}k_x{}^{\mathrm{r}}n_x{}^{\mathrm{r}}x_{\mathrm{w}}\cos{}^{\mathrm{r}}\theta_{\mathrm{p}}-{}^{\mathrm{r}}k_x{}^{\mathrm{r}}n_x{}^{\mathrm{r}}z_{\mathrm{w}}\sin{}^{\mathrm{r}}\theta_{\mathrm{p}}+{}^{\mathrm{r}}k_x{}^{\mathrm{r}}o_x{}^{\mathrm{r}}x_{\mathrm{w}}\sin{}^{\mathrm{r}}\theta_{\mathrm{p}}\sin{}^{\mathrm{r}}\theta_{\mathrm{t}}\\ +{}^{\mathrm{r}}k_x{}^{\mathrm{r}}o_x{}^{\mathrm{r}}y_{\mathrm{w}}\cos{}^{\mathrm{r}}\theta_{\mathrm{t}}+{}^{\mathrm{r}}k_x{}^{\mathrm{r}}o_x{}^{\mathrm{r}}z_{\mathrm{w}}\cos{}^{\mathrm{r}}\theta_{\mathrm{p}}\sin{}^{\mathrm{r}}\theta_{\mathrm{t}}+{}^{\mathrm{r}}k_x{}^{\mathrm{r}}a_x{}^{\mathrm{r}}x_{\mathrm{w}}\sin{}^{\mathrm{r}}\theta_{\mathrm{p}}\cos{}^{\mathrm{r}}\theta_{\mathrm{t}}\\ -{}^{\mathrm{r}}k_x{}^{\mathrm{r}}a_x{}^{\mathrm{r}}y_{\mathrm{w}}\sin{}^{\mathrm{r}}\theta_{\mathrm{t}}+{}^{\mathrm{r}}k_x{}^{\mathrm{r}}a_x{}^{\mathrm{r}}z_{\mathrm{w}}\cos{}^{\mathrm{r}}\theta_{\mathrm{p}}\cos{}^{\mathrm{r}}\theta_{\mathrm{t}}+{}^{\mathrm{r}}k_x{}^{\mathrm{r}}p_x)\end{array}}{\begin{array}{c}({}^{\mathrm{r}}n_z{}^{\mathrm{r}}x_{\mathrm{w}}\cos{}^{\mathrm{r}}\theta_{\mathrm{p}}-{}^{\mathrm{r}}n_z{}^{\mathrm{r}}z_{\mathrm{w}}\sin{}^{\mathrm{r}}\theta_{\mathrm{p}}+{}^{\mathrm{r}}o_z{}^{\mathrm{r}}x_{\mathrm{w}}\sin{}^{\mathrm{r}}\theta_{\mathrm{p}}\sin{}^{\mathrm{r}}\theta_{\mathrm{t}}\\ +{}^{\mathrm{r}}o_z{}^{\mathrm{r}}y_{\mathrm{w}}\cos{}^{\mathrm{r}}\theta_{\mathrm{t}}+{}^{\mathrm{r}}o_z{}^{\mathrm{r}}z_{\mathrm{w}}\cos{}^{\mathrm{r}}\theta_{\mathrm{p}}\sin{}^{\mathrm{r}}\theta_{\mathrm{t}}+{}^{\mathrm{r}}a_z{}^{\mathrm{r}}x_{\mathrm{w}}\sin{}^{\mathrm{r}}\theta_{\mathrm{p}}\cos{}^{\mathrm{r}}\theta_{\mathrm{t}}\\ -{}^{\mathrm{r}}a_z{}^{\mathrm{r}}y_{\mathrm{w}}\sin{}^{\mathrm{r}}\theta_{\mathrm{t}}+{}^{\mathrm{r}}a_z{}^{\mathrm{r}}z_{\mathrm{w}}\cos{}^{\mathrm{r}}\theta_{\mathrm{p}}\cos{}^{\mathrm{r}}\theta_{\mathrm{t}}+{}^{\mathrm{r}}p_z)\end{array}}\\ \dfrac{\begin{array}{c}({}^{\mathrm{r}}k_y{}^{\mathrm{r}}n_y{}^{\mathrm{r}}x_{\mathrm{w}}\cos{}^{\mathrm{r}}\theta_{\mathrm{p}}-{}^{\mathrm{r}}k_y{}^{\mathrm{r}}n_y{}^{\mathrm{r}}z_{\mathrm{w}}\sin{}^{\mathrm{r}}\theta_{\mathrm{p}}+{}^{\mathrm{r}}k_y{}^{\mathrm{r}}o_y{}^{\mathrm{r}}x_{\mathrm{w}}\sin{}^{\mathrm{r}}\theta_{\mathrm{p}}\sin{}^{\mathrm{r}}\theta_{\mathrm{t}}\\ +{}^{\mathrm{r}}k_y{}^{\mathrm{r}}o_y{}^{\mathrm{r}}y_{\mathrm{w}}\cos{}^{\mathrm{r}}\theta_{\mathrm{t}}+{}^{\mathrm{r}}k_y{}^{\mathrm{r}}o_y{}^{\mathrm{r}}z_{\mathrm{w}}\cos{}^{\mathrm{r}}\theta_{\mathrm{p}}\sin{}^{\mathrm{r}}\theta_{\mathrm{t}}+{}^{\mathrm{r}}k_y{}^{\mathrm{r}}a_y{}^{\mathrm{r}}x_{\mathrm{w}}\sin{}^{\mathrm{r}}\theta_{\mathrm{p}}\cos{}^{\mathrm{r}}\theta_{\mathrm{t}}\\ -{}^{\mathrm{r}}k_y{}^{\mathrm{r}}a_y{}^{\mathrm{r}}y_{\mathrm{w}}\sin{}^{\mathrm{r}}\theta_{\mathrm{t}}+{}^{\mathrm{r}}k_y{}^{\mathrm{r}}a_y{}^{\mathrm{r}}z_{\mathrm{w}}\cos{}^{\mathrm{r}}\theta_{\mathrm{p}}\cos{}^{\mathrm{r}}\theta_{\mathrm{t}}+{}^{\mathrm{r}}k_y{}^{\mathrm{r}}p_y)\end{array}}{\begin{array}{c}({}^{\mathrm{r}}n_z{}^{\mathrm{r}}x_{\mathrm{w}}\cos{}^{\mathrm{r}}\theta_{\mathrm{p}}-{}^{\mathrm{r}}n_z{}^{\mathrm{r}}z_{\mathrm{w}}\sin{}^{\mathrm{r}}\theta_{\mathrm{p}}+{}^{\mathrm{r}}o_z{}^{\mathrm{r}}x_{\mathrm{w}}\sin{}^{\mathrm{r}}\theta_{\mathrm{p}}\sin{}^{\mathrm{r}}\theta_{\mathrm{t}}\\ +{}^{\mathrm{r}}o_z{}^{\mathrm{r}}y_{\mathrm{w}}\cos{}^{\mathrm{r}}\theta_{\mathrm{t}}+{}^{\mathrm{r}}o_z{}^{\mathrm{r}}z_{\mathrm{w}}\cos{}^{\mathrm{r}}\theta_{\mathrm{p}}\sin{}^{\mathrm{r}}\theta_{\mathrm{t}}+{}^{\mathrm{r}}a_z{}^{\mathrm{r}}x_{\mathrm{w}}\sin{}^{\mathrm{r}}\theta_{\mathrm{p}}\cos{}^{\mathrm{r}}\theta_{\mathrm{t}}\\ -{}^{\mathrm{r}}a_z{}^{\mathrm{r}}y_{\mathrm{w}}\sin{}^{\mathrm{r}}\theta_{\mathrm{t}}+{}^{\mathrm{r}}a_z{}^{\mathrm{r}}z_{\mathrm{w}}\cos{}^{\mathrm{r}}\theta_{\mathrm{p}}\cos{}^{\mathrm{r}}\theta_{\mathrm{t}}+{}^{\mathrm{r}}p_z)\end{array}}\end{pmatrix}\tag{A41}$$

By Equations (2), (A40) and (A41), Equation (A42) related to θ_{t} and θ_{p} can be obtained. It can be found from Equation (A32) that both θ_{t} and θ_{p} appear in the form of a trigonometric function, and it is difficult to obtain values of θ_{t} and θ_{p} directly from these two equations. In order to obtain the solution available in the project, we firstly obtain a sub-optimal observation pose and then use the sub-optimal observation pose as the initial value. We finally use the trial and error method to obtain the optimal observation pose.

$$\left\{\begin{array}{l}({}^{\mathrm{l}}k_x{}^{\mathrm{l}}n_x x_{\mathrm{w}}\cos\theta_{\mathrm{p}}-{}^{\mathrm{l}}k_x{}^{\mathrm{l}}n_x z_{\mathrm{w}}\sin\theta_{\mathrm{p}}+{}^{\mathrm{l}}k_x{}^{\mathrm{l}}o_x x_{\mathrm{w}}\sin\theta_{\mathrm{p}}\sin\theta_{\mathrm{t}}\\ +{}^{\mathrm{l}}k_x{}^{\mathrm{l}}o_x y_{\mathrm{w}}\cos\theta_{\mathrm{t}}+{}^{\mathrm{l}}k_x{}^{\mathrm{l}}o_x z_{\mathrm{w}}\cos\theta_{\mathrm{p}}\sin\theta_{\mathrm{t}}+-{}^{\mathrm{l}}k_x{}^{\mathrm{l}}a_x y_{\mathrm{w}}\sin\theta_{\mathrm{t}}\\ {}^{\mathrm{l}}k_x{}^{\mathrm{l}}a_x x_{\mathrm{w}}\sin\theta_{\mathrm{p}}\cos\theta_{\mathrm{t}}+{}^{\mathrm{l}}k_x{}^{\mathrm{l}}a_x z_{\mathrm{w}}\cos\theta_{\mathrm{p}}\cos\theta_{\mathrm{t}}+{}^{\mathrm{l}}k_x{}^{\mathrm{l}}p_x)\\ ({}^{\mathrm{r}}n_z{}^{\mathrm{r}}x_{\mathrm{w}}\cos\theta_{\mathrm{p}}-{}^{\mathrm{r}}n_z{}^{\mathrm{r}}z_{\mathrm{w}}\sin\theta_{\mathrm{p}}+{}^{\mathrm{r}}o_z{}^{\mathrm{r}}x_{\mathrm{w}}\sin\theta_{\mathrm{p}}\sin\theta_{\mathrm{t}}+{}^{\mathrm{r}}p_z\\ +{}^{\mathrm{r}}o_z{}^{\mathrm{r}}y_{\mathrm{w}}\cos\theta_{\mathrm{t}}+{}^{\mathrm{r}}o_z{}^{\mathrm{r}}z_{\mathrm{w}}\cos\theta_{\mathrm{p}}\sin\theta_{\mathrm{t}}+{}^{\mathrm{r}}a_z{}^{\mathrm{r}}x_{\mathrm{w}}\sin\theta_{\mathrm{p}}\cos\theta_{\mathrm{t}}\\ -{}^{\mathrm{r}}a_z{}^{\mathrm{r}}y_{\mathrm{w}}\sin\theta_{\mathrm{t}}+{}^{\mathrm{r}}a_z{}^{\mathrm{r}}z_{\mathrm{w}}\cos\theta_{\mathrm{p}}\cos\theta_{\mathrm{t}})=({}^{\mathrm{r}}k_x{}^{\mathrm{r}}n_x{}^{\mathrm{r}}x_{\mathrm{w}}\cos\theta_{\mathrm{p}}\\ -{}^{\mathrm{r}}k_x{}^{\mathrm{l}}n_x{}^{\mathrm{r}}z_{\mathrm{w}}\sin\theta_{\mathrm{p}}+{}^{\mathrm{r}}k_x{}^{\mathrm{r}}o_x{}^{\mathrm{r}}x_{\mathrm{w}}\sin\theta_{\mathrm{p}}\sin\theta_{\mathrm{t}}+{}^{\mathrm{r}}k_x{}^{\mathrm{r}}o_x{}^{\mathrm{r}}y_{\mathrm{w}}\cos\theta_{\mathrm{t}}\\ +{}^{\mathrm{r}}k_x{}^{\mathrm{r}}o_x{}^{\mathrm{r}}z_{\mathrm{w}}\cos\theta_{\mathrm{p}}\sin\theta_{\mathrm{t}}+{}^{\mathrm{r}}k_x{}^{\mathrm{r}}a_x{}^{\mathrm{r}}x_{\mathrm{w}}\sin\theta_{\mathrm{p}}\cos\theta_{\mathrm{t}}+{}^{\mathrm{r}}k_x{}^{\mathrm{r}}p_x\\ -{}^{\mathrm{r}}k_x{}^{\mathrm{r}}a_x{}^{\mathrm{r}}y_{\mathrm{w}}\sin\theta_{\mathrm{t}}+{}^{\mathrm{r}}k_x{}^{\mathrm{r}}a_x{}^{\mathrm{r}}z_{\mathrm{w}}\cos\theta_{\mathrm{p}}\cos\theta_{\mathrm{t}})({}^{\mathrm{l}}n_z x_{\mathrm{w}}\cos\theta_{\mathrm{p}}\\ -{}^{\mathrm{l}}n_z z_{\mathrm{w}}\sin\theta_{\mathrm{p}}+{}^{\mathrm{l}}o_z x_{\mathrm{w}}\sin\theta_{\mathrm{p}}\sin\theta_{\mathrm{t}}+{}^{\mathrm{l}}o_z y_{\mathrm{w}}\cos\theta_{\mathrm{t}}+{}^{\mathrm{l}}o_z z_{\mathrm{w}}\cos\theta_{\mathrm{p}}\sin\theta_{\mathrm{t}}\\ +{}^{\mathrm{l}}a_z x_{\mathrm{w}}\sin\theta_{\mathrm{p}}\cos\theta_{\mathrm{t}}-{}^{\mathrm{l}}a_z y_{\mathrm{w}}\sin\theta_{\mathrm{t}}+{}^{\mathrm{l}}a_z z_{\mathrm{w}}\cos\theta_{\mathrm{p}}\cos\theta_{\mathrm{t}}+{}^{\mathrm{l}}p_z)\\ ({}^{\mathrm{l}}k_y{}^{\mathrm{l}}n_y x_{\mathrm{w}}\cos\theta_{\mathrm{p}}-{}^{\mathrm{l}}k_y{}^{\mathrm{l}}n_y z_{\mathrm{w}}\sin\theta_{\mathrm{p}}+{}^{\mathrm{l}}k_y{}^{\mathrm{l}}o_y x_{\mathrm{w}}\sin\theta_{\mathrm{p}}\sin\theta_{\mathrm{t}}\\ +{}^{\mathrm{l}}k_y{}^{\mathrm{l}}o_y y_{\mathrm{w}}\cos\theta_{\mathrm{t}}+{}^{\mathrm{l}}k_y{}^{\mathrm{l}}o_y z_{\mathrm{w}}\cos\theta_{\mathrm{p}}\sin\theta_{\mathrm{t}}-{}^{\mathrm{l}}k_y{}^{\mathrm{l}}a_y y_{\mathrm{w}}\sin\theta_{\mathrm{t}}\\ +{}^{\mathrm{l}}k_y{}^{\mathrm{l}}a_y x_{\mathrm{w}}\sin\theta_{\mathrm{p}}\cos\theta_{\mathrm{t}}+{}^{\mathrm{l}}k_y{}^{\mathrm{l}}a_y z_{\mathrm{w}}\cos\theta_{\mathrm{p}}\cos\theta_{\mathrm{t}}+{}^{\mathrm{l}}k_y{}^{\mathrm{l}}p_y)\\ ({}^{\mathrm{r}}n_z{}^{\mathrm{r}}x_{\mathrm{w}}\cos\theta_{\mathrm{p}}-{}^{\mathrm{r}}n_z{}^{\mathrm{r}}z_{\mathrm{w}}\sin\theta_{\mathrm{p}}+{}^{\mathrm{r}}o_z{}^{\mathrm{r}}x_{\mathrm{w}}\sin\theta_{\mathrm{p}}\sin\theta_{\mathrm{t}}+{}^{\mathrm{r}}p_z\\ +{}^{\mathrm{r}}o_z{}^{\mathrm{r}}y_{\mathrm{w}}\cos\theta_{\mathrm{t}}+{}^{\mathrm{r}}o_z{}^{\mathrm{r}}z_{\mathrm{w}}\cos\theta_{\mathrm{p}}\sin\theta_{\mathrm{t}}+{}^{\mathrm{r}}a_z{}^{\mathrm{r}}x_{\mathrm{w}}\sin\theta_{\mathrm{p}}\cos\theta_{\mathrm{t}}\\ -{}^{\mathrm{r}}a_z{}^{\mathrm{r}}y_{\mathrm{w}}\sin\theta_{\mathrm{t}}+{}^{\mathrm{r}}a_z{}^{\mathrm{r}}z_{\mathrm{w}}\cos\theta_{\mathrm{p}}\cos\theta_{\mathrm{t}})=({}^{\mathrm{r}}k_y{}^{\mathrm{r}}n_y{}^{\mathrm{r}}x_{\mathrm{w}}\cos\theta_{\mathrm{p}}\\ -{}^{\mathrm{r}}k_y{}^{\mathrm{r}}n_y{}^{\mathrm{r}}z_{\mathrm{w}}\sin\theta_{\mathrm{p}}+{}^{\mathrm{r}}k_y{}^{\mathrm{r}}o_y{}^{\mathrm{r}}x_{\mathrm{w}}\sin\theta_{\mathrm{p}}\sin\theta_{\mathrm{t}}+{}^{\mathrm{r}}k_y{}^{\mathrm{r}}o_y{}^{\mathrm{r}}y_{\mathrm{w}}\cos\theta_{\mathrm{t}}\\ +{}^{\mathrm{r}}k_y{}^{\mathrm{r}}o_y{}^{\mathrm{r}}z_{\mathrm{w}}\cos\theta_{\mathrm{p}}\sin\theta_{\mathrm{t}}+{}^{\mathrm{r}}k_y{}^{\mathrm{r}}a_y{}^{\mathrm{r}}x_{\mathrm{w}}\sin\theta_{\mathrm{p}}\cos\theta_{\mathrm{t}}+{}^{\mathrm{r}}k_y{}^{\mathrm{r}}p_y\\ -{}^{\mathrm{r}}k_y{}^{\mathrm{r}}a_y{}^{\mathrm{r}}y_{\mathrm{w}}\sin\theta_{\mathrm{t}}+{}^{\mathrm{r}}k_y{}^{\mathrm{r}}a_y{}^{\mathrm{r}}z_{\mathrm{w}}\cos\theta_{\mathrm{p}}\cos\theta_{\mathrm{t}})({}^{\mathrm{l}}n_z x_{\mathrm{w}}\cos\theta_{\mathrm{p}}\\ -{}^{\mathrm{l}}n_z z_{\mathrm{w}}\sin\theta_{\mathrm{p}}+{}^{\mathrm{l}}o_z x_{\mathrm{w}}\sin\theta_{\mathrm{p}}\sin\theta_{\mathrm{t}}+{}^{\mathrm{l}}o_z y_{\mathrm{w}}\cos\theta_{\mathrm{t}}+{}^{\mathrm{l}}o_z z_{\mathrm{w}}\cos\theta_{\mathrm{p}}\sin\theta_{\mathrm{t}}\\ +{}^{\mathrm{l}}a_z x_{\mathrm{w}}\sin\theta_{\mathrm{p}}\cos\theta_{\mathrm{t}}-{}^{\mathrm{l}}a_z y_{\mathrm{w}}\sin\theta_{\mathrm{t}}+{}^{\mathrm{l}}a_z z_{\mathrm{w}}\cos\theta_{\mathrm{p}}\cos\theta_{\mathrm{t}}+{}^{\mathrm{l}}p_z)\end{array}\right.\tag{A42}$$

References

1. Wang, Q.; Zou, W.; Xu, D.; Zhu, Z. Motion control in saccade and smooth pursuit for bionic eye based on three-dimensional coordinates. *J. Bionic Eng.* **2017**, *14*, 336–347. [CrossRef]
2. Kardamakis, A.A.; Moschovakis, A.K. Optimal control of gaze shifts. *J. Neurosci.* **2009**, *29*, 7723–7730. [CrossRef] [PubMed]
3. Freedman, E.G.; Sparks, D.L. Coordination of the eyes and head: Movement kinematics. *Exp. Brain Res.* **2000**, *131*, 22–32. [CrossRef] [PubMed]
4. Nakashima, R.; Fang, Y.; Hatori, Y.; Hiratani, A.; Matsumiya, K.; Kuriki, I.; Shioiri, S. Saliency-based gaze prediction based on head direction. *Vis. Res.* **2015**, *117*, 59–66. [CrossRef] [PubMed]
5. He, H.; Ge, S.S.; Zhang, Z. A saliency-driven robotic head with bio-inspired saccadic behaviors for social robotics. *Auton. Robot.* **2014**, *36*, 225–240. [CrossRef]
6. Law, J.; Shaw, P.; Lee, M. A biologically constrained architecture for developmental learning of eye–head gaze control on a humanoid robot. *Auton. Robot.* **2013**, *35*, 77–92. [CrossRef]

7. Wijayasinghe, I.B.; Aulisa, E.; Buttner, U.; Ghosh, B.K.; Glasauer, S.; Kremmyda, O. Potential and optimal target fixating control of the human head/eye complex. *IEEE Trans. Control Syst. Technol.* **2015**, *23*, 796–804. [CrossRef]
8. Ghosh, B.K.; Wijayasinghe, I.B.; Kahagalage, S.D. A geometric approach to head/eye control. *IEEE Access* **2014**, *2*, 316–332. [CrossRef]
9. Kuang, X.; Gibson, M.; Shi, B.E.; Rucci, M. Active vision during coordinated head/eye movements in a humanoid robot. *IEEE Trans. Robot.* **2012**, *28*, 1423–1430. [CrossRef]
10. Vannucci, L.; Cauli, N.; Falotico, E.; Bernardino, A.; Laschi, C. Adaptive visual pursuit involving eye-head coordination and prediction of the target motion. In Proceedings of the IEEE-RAS International Conference on Humanoid Robots, Madrid, Spain, 18–20 November 2014; pp. 541–546.
11. Huelse, M.; McBride, S.; Law, J.; Lee, M. Integration of active vision and reaching from a developmental robotics perspective. *IEEE Trans. Auton. Ment. Dev.* **2010**, *2*, 355–367. [CrossRef]
12. Anastasopoulos, D.; Naushahi, J.; Sklavos, S.; Bronstein, A.M. Fast gaze reorientations by combined movements of the eye, head, trunk and lower extremities. *Exp. Brain Res.* **2015**, *233*, 1639–1650. [CrossRef]
13. Daye, P.M.; Optican, L.M.; Blohm, G.; Lefèvre, P. Hierarchical control of two-dimensional gaze saccades. *J. Comput. Neurosci.* **2014**, *36*, 355–382. [CrossRef] [PubMed]
14. Rajruangrabin, J.; Popa, D.O. Robot head motion control with an emphasis on realism of neck–eye coordination during object tracking. *J. Intell. Robot. Syst.* **2011**, *63*, 163–190. [CrossRef]
15. Schulze, L.; Renneberg, B.; Lobmaier, J.S. Gaze perception in social anxiety and social anxiety disorder. *Front. Hum. Neurosci.* **2013**, *7*, 1–5. [CrossRef] [PubMed]
16. Liu, Y.; Zhu, D.; Peng, J.; Wang, X.; Wang, L.; Chen, L.; Li, J.; Zhang, X. Real-time robust stereo visual SLAM system based on bionic eyes. *IEEE Trans. Med. Robot. Bionics* **2020**, *2*, 391–398. [CrossRef]
17. Guitton, D. Control of eye–head coordination during orienting gaze shifts. *Trends Neurosci.* **1993**, *15*, 174–179. [CrossRef]
18. Matveev, A.S.; Hoy, M.C.; Savkin, A.V. 3D environmental extremum seeking navigation of a nonholonomic mobile robot. *Automatica* **2014**, *50*, 1802–1815. [CrossRef]
19. Nefti-Meziani, S.; Manzoor, U.; Davis, S.; Pupala, S.K. 3D perception from binocular vision for a low cost humanoid robot NAO. *Robot. Auton. Syst.* **2015**, *68*, 129–139. [CrossRef]
20. Surmann, H.; Nüchter, A.; Hertzberg, J. An autonomous mobile robot with a 3D laser range finder for 3D exploration and digitalization of indoor environments. *Robot. Auton. Syst.* **2003**, *45*, 181–198. [CrossRef]
21. Song, W.; Minami, M.; Shen, L.Y.; Zhang, Y.N. Bionic tracking method by hand & eye-vergence visual servoing. *Adv. Manuf.* **2016**, *4*, 157–166.
22. Li, H.Y.; Luo, J.; Huang, C.J.; Huang, Q.Z.; Xie, S.R. Design and control of 3-DoF spherical parallel mechanism robot eyes inspired by the binocular vestibule-ocular reflex. *J. Intell. Robot. Syst.* **2015**, *78*, 425–441. [CrossRef]
23. Masseck, O.A.; Hoffmann, K.P. Comparative neurobiology of the optokinetic reflex. *Ann. N. Y. Acad. Sci.* **2009**, *1164*, 430–439. [CrossRef] [PubMed]
24. Bruske, J.; Hansen, M.; Riehn, L.; Sommer, G. Biologically inspired calibration-free adaptive saccade control of a binocular camera-head. *Biol. Cybern.* **1997**, *77*, 433–446. [CrossRef]
25. Wang, X.; Van De Weem, J.; Jonker, P. An advanced active vision system imitating human eye movements. In Proceedings of the 2013 16th International Conference on Advanced Robotics, Montevideo, Uruguay, 25–29 November 2013; pp. 5–10.
26. Antonelli, M.; Duran, A.J.; Chinellato, E.; Pobil, A.P. Adaptive saccade controller inspired by the primates' cerebellum. In Proceedings of the IEEE International Conference on Robotics and Automation, Seattle, WA, USA, 26–30 May 2015; pp. 5048–5053.
27. Robinson, D.A.; Gordon, J.L.; Gordon, S.E. A model of the smooth pursuit eye movement system. *Biol. Cybern.* **1986**, *55*, 43–57. [CrossRef] [PubMed]
28. Brown, C. Gaze controls with interactions and delays. *IEEE Trans. Syst. Man Cybern.* **1990**, *20*, 518–527. [CrossRef]
29. Deno, D.C.; Keller, E.L.; Crandall, W.F. Dynamical neural network organization of the visual pursuit system. *IEEE Trans. Biomed. Eng.* **1989**, *36*, 85–92. [CrossRef] [PubMed]
30. Lunghi, F.; Lazzari, S.; Magenes, G. Neural adaptive predictor for visual tracking system. In Proceedings of the 20th Annual International Conference of the IEEE Engineering in Medicine and Biology Society, Hong Kong, China, 1 November 1998; Volume 20, pp. 1389–1392.
31. Lee, W.J.; Galiana, H.L. An internally switched model of ocular tracking with prediction. *IEEE Trans. Neural Syst. Rehabil. Eng.* **2005**, *13*, 186–193. [CrossRef]
32. Avni, O.; Borrelli, F.; Katzir, G.; Rivlin, E.; Rotstein, H. Scanning and tracking with independent cameras-a biologically motivated approach based on model predictive control. *Auton. Robot.* **2008**, *24*, 285–302. [CrossRef]
33. Zhang, M.; Ma, X.; Qin, B.; Wang, G.; Guo, Y.; Xu, Z.; Wang, Y.; Li, Y. Information fusion control with time delay for smooth pursuit eye movement. *Physiol. Rep.* **2016**, *4*, e12775. [CrossRef]
34. Santini, F.; Rucci, M. Active estimation of distance in a robotic system that replicates human eye movement. *Robot. Auton. Syst.* **2007**, *55*, 107–121. [CrossRef]
35. Chinellato, E.; Antonelli, M.; Grzyb, B.J.; Del Pobil, A.P. Implicit sensorimotor mapping of the peripersonal space by gazing and reaching. *IEEE Trans. Auton. Ment. Dev.* **2011**, *3*, 43–53. [CrossRef]

36. Song, Y.; Zhang, X. An active binocular integrated system for intelligent robot vision. In Proceedings of the IEEE International Conference on Intelligence and Security Informatics, Washington, DC, USA, 11–14 June 2012; pp. 48–53.
37. Wang, Y.; Zhang, G.; Lang, H.; Zuo, B.; De Silva, C.W. A modified image-based visual servo controller with hybrid camera configuration for robust robotic grasping. *Robot. Auton. Syst.* **2014**, *62*, 1398–1407. [CrossRef]
38. Lee, Y.C.; Lan, C.C.; Chu, C.Y.; Lai, C.M.; Chen, Y.J. A pan-tilt orienting mechanism with parallel axes of flexural actuation. *IEEE-ASME Trans. Mechatron.* **2013**, *18*, 1100–1112. [CrossRef]
39. Wang, Q.; Zou, W.; Zhang, F.; Xu, D. Binocular initial location and extrinsic parameters real-time calculation for bionic eye system. In Proceedings of the 11th World Congress on Intelligent Control and Automation, Shenyang, China, 29 June–4 July 2014; pp. 74–80.
40. Fan, D.; Liu, Y.Y.; Chen, X.P.; Meng, F.; Liu, X.L.; Ullah, Z.; Cheng, W.; Liu, Y.H.; Huang, Q. Eye gaze based 3D triangulation for robotic bionic eyes. *Sensors* **2020**, *20*, 5271. [CrossRef] [PubMed]

sensors

MDPI

Article

A Convex Optimization Approach to Multi-Robot Task Allocation and Path Planning

Tingjun Lei [1,†], Pradeep Chintam [1,†], Chaomin Luo [1,*], Lantao Liu [2] and Gene Eu Jan [3,4]

1 Department of Electrical and Computer Engineering, Mississippi State University, Mississippi State, MS 39762, USA
2 Department of Intelligent Systems Engineering, Indiana University, Bloomington, IN 47408, USA
3 Department of Electrical Engineering, National Taipei University, New Taipei City 23741, Taiwan
4 Tainan National University of the Arts, Tainan City 72045, Taiwan
* Correspondence: chaomin.luo@ece.msstate.edu
† These authors contributed equally to this work.

Abstract: In real-world applications, multiple robots need to be dynamically deployed to their appropriate locations as teams while the distance cost between robots and goals is minimized, which is known to be an NP-hard problem. In this paper, a new framework of team-based multi-robot task allocation and path planning is developed for robot exploration missions through a convex optimization-based distance optimal model. A new distance optimal model is proposed to minimize the traveled distance between robots and their goals. The proposed framework fuses task decomposition, allocation, local sub-task allocation, and path planning. To begin, multiple robots are firstly divided and clustered into a variety of teams considering interrelation and dependencies of robots, and task decomposition. Secondly, the teams with various arbitrary shape enclosing intercorrelative robots are approximated and relaxed into circles, which are mathematically formulated to convex optimization problems to minimize the distance between teams, as well as between a robot and their goals. Once the robot teams are deployed into their appropriate locations, the robot locations are further refined by a graph-based Delaunay triangulation method. Thirdly, in the team, a self-organizing map-based neural network (SOMNN) paradigm is developed to complete the dynamical sub-task allocation and path planning, in which the robots are dynamically assigned to their nearby goals locally. Simulation and comparison studies demonstrate the proposed hybrid multi-robot task allocation and path planning framework is effective and efficient.

Keywords: multi-robot deployment; convex optimization; task allocation; SOM neural networks; path planning; task decomposition

Citation: Lei, T.; Chintam, P.; Luo, C.; Liu, L.; Jan, G.E. A Convex Optimization Approach to Multi-Robot Task Allocation and Path Planning. *Sensors* **2023**, *23*, 5103. https://doi.org/10.3390/s23115103

Academic Editors: David Cheneler and Stephen Monk

Received: 17 March 2023
Revised: 13 May 2023
Accepted: 22 May 2023
Published: 26 May 2023

1. Introduction

Multi-robot deployment is an essential issue in robotics field, which requires mobile robots to be deployed in the workspace to cooperatively fulfill tasks [1–5]. Dynamic deployment problems in combinatorial optimization consist of finding an optimal solution in some objectives, such as timing, workload, distance traveled, energy, and deadlines [6–13]. In real-world applications, such as mine detection, environmental exploration, and rescue mission, many robots need to be dynamically deployed to goals (targets) while total distance cost among robots, and between robots and goals is minimized [14–20].

1.1. Related Works

Although there have been many studies on multi-robot deployment via optimization models, very few existing deployment algorithms focus on distance constraints. Previous research on multi-robot deployment and task allocation may be categorized into various methods, such as distributed-based [21–24], linear programming [24,25], graph-based [26–28], market-based [22,29,30], conflict-based search method [31], neural networks [32,33], etc.

Sung et al. [21] addressed a distributed approach to integrate multi-robot deployment and multi-target tracking. A local algorithm is adopted to achieving performance close to the optimal algorithms with limited communication with assistance of distributed approach. For a heterogeneous multi-robot system, where tasks form disjoint groups and where there are restrictions on the number of tasks, a robot may accomplish (both within the overall mission and within each task group), Luo et al. [34] provided a provably-good distributed task allocation methods. Their goal of task allocation is to maximize (minimize) the total payout (cost) of the robots, whereby each robot receives a payment (or incurs a cost) for completing a job. Basil et al. [35] proposed a modular robot to be a morphologically flexible, autonomous kinematic machines with hundreds or even millions of modules that work together to exhibit intelligent behavior. The self-reconfiguration process, which seeks to identify a series of reconfiguration activities to convert robots from an initial form to a target one, may be enhanced by clustering the modules in modular robots. Luo et al. [24] extended distributed algorithms by taking deadline into account in multi-robot deployment. Purohit et al. [26] presented a spanning tree method associated with self-localizing capability in the graph. However, robot teams are not yet successfully clustered and assigned to multiple tasks effectively. To efficiently and securely explore and recon a given region with a large number of robots, Li et al. [36] introduced an enhanced genetic algorithm (IGA) to tackle the job assignment issue of a multi-robot system. Searching the numerous identical sections of the specified region is a subtask that must be completed in order to solve a challenge. Qin et al. [37] formally described the challenge of completing dynamic tasks with several robots by modeling their interactions as a series of state transitions, or behavior trees. Through the use of a unique priority system, a framework-associated distributed algorithms for inter-robot communication, negotiation, and agreement protocols was provided. Bai et al. [38] investigated the multi-robot task allocation issue, in which a team of geographically-dispersed robots must effectively move a number of packages from their originating locations to their respective destinations within a certain amount of time. A market-based approach is suggested by Rossi et al. [29] for simultaneous task subdivision and allocation in heterogeneous multi-robot systems. García et al. [30] implemented a behavior-based architecture with many layers allowing the market-based method to achieve varying degrees of coordination. However, as the number of robots in the team or the complexity of the problem increases, market-based approaches suffer from scalability and dynamics issues, which tend to hinder these processes, especially when this happens in real time.

Neural network (NN) methods have been broadly applied to multi-robot path planning and task allocation. Luo and Yang [33] developed a neural dynamics model to assign multiple robots to environmental exploration collaboratively. Multiple robots cooperate to achieve a common sweeping goal effectively. However, energy consumption of multi-robot system has not been considered in this paper. Later, Luo et al. [32] extended their research by addressing the computational complexity and energy efficiency of multi-robot system with navigation [39]. The energy consumption of multiple tasks for an arbitrary number of robots is considered, in which a bio-inspired neural networks model for multi-robot navigation applied to cleaning robots is developed. In this model, multiple robots are assigned to complete terrain coverage task cooperatively, extendable to unknown exploration environments [39]. Distance cost optimization is considered in the recent research. For instance, Lee et al. [31] suggested a master–slave based multi-robot deployment with time and distance minimization consideration; however, the distance is not globally optimized. Some researchers combined a couple of models to take advantage of various benefits. For instance, Michael et al. [22] effectively integrated distributed algorithm and market-based method for multi-robot deployment. Motes et al. [40] concatenated path-finding method and conflict-based search method to multi-robot deployment and navigation with inter-team conflict avoidance.

The centralization or decentralization of task allocation methods is another essential characteristic. Using the well-known Kiva warehouse robot as an example [41], both task allocation and path planning are conducted centrally; however, this may not be feasible if the multi-robot system operates in an environment that lacks robust sensing and computational capabilities. Clustering is a potential middle alternative between centralized and decentralized methods that aims to balance performance and computational load [42]. Martin et al. [43] proposed an algorithm to group the players into balanced clusters, applied randomized methods to large problems to relieve the computational load, and assessed feasibility in a large scenario and contrasted with a genetic approach.

In general, market/auction-based approaches can be solved decentrally [44]. However, the problem structure must be straightforward enough so that each agent can act as a bidder and bid on tasks; this can be challenging when some tasks require the cooperation of multiple agents. Optimization-based methods permit more complex problem structures but can be challenging to solve decentralized. Bo et al. [45] proposed a stochastic programming framework, which optimizes the decomposition, allocation, and scheduling of tasks for a group of agents. The framework enables teams of mobile robots in different locations to perform different tasks. However, the scalability and dynamics issues arise as the number of robots in the team or the complexity of the problem increases. Therefore, a framework based on optimization is proposed in this paper for decentralized task allocation that is appropriate for complex problem models.

1.2. Proposed Framework and Original Contributions

In this paper, a two-stage team-based decentralized deployment framework through convex optimization model is developed. It aims for multiple robots scattered in arbitrary space to minimize distance cost among robots, and between robots and goals so as to reach all the task locations effectively shown in Figure 1.

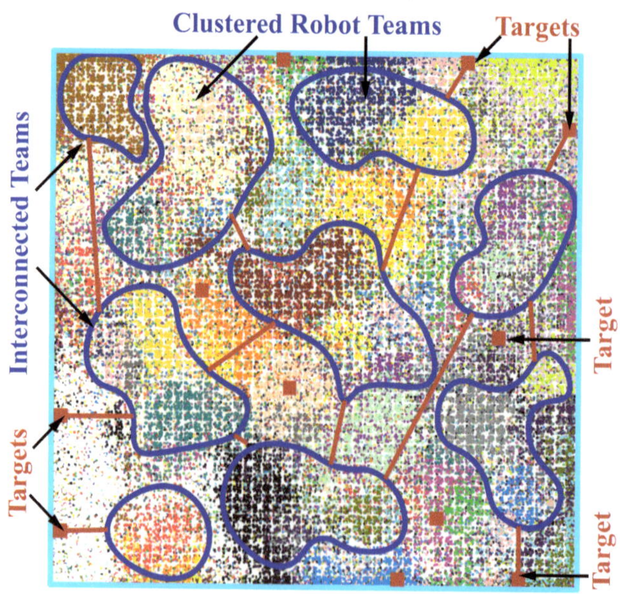

Figure 1. Illustration of an exploration environment with robot teams (swarms) (adapted from Eklavya 2019 [46]). In this multi-robot deployment problem, a *known* environment is given with $\mathcal{N}$ target locations, in which some locations are located within the workspace (such as warehouses), but other locations are in the edges of the workspace. There are varying numbers of robots in each team.

In the first stage, multiple robots in workspace are classified and clustered into a variety of teams enclosing correlative robots as *task decomposition*. Robot teams in arbitrary shape are represented as circles, in order to formulate it into a convex optimization model.

The relative locations of teams are obtained through this convex optimization model. Then, the locations of circles are refined in a refinement stage via a Delaunay triangulation method to clean up the overlaps of teams. In the second stage, within the team, a self-organizing map (SOM) neural networks (NN) paradigm is considered to complete the dynamical sub-task allocation and path planning by dynamically assigning them to their nearby goals locally. The proposed framework couples task decomposition, deployment, local sub-task allocation and path planning to support cases where the optimal solution depends on robot interrelation, dependencies, and availability, as well as inter-team conflict avoidance.

The contributions of this paper are summarized as follows:

(1) A two-stage convex-optimization-based framework is proposed for decentralized multi-robot task allocation and task decomposition.
(2) The first stage of the proposed convex-optimization-based framework is designed to determine the relative locations of the robot teams. By obtaining data regarding the robots and the targets, the robots are categorized and aggregated into multiple teams with respect to the total distance cost.
(3) The second stage of the proposed convex optimization-based framework aims to locally assign teams of robots to final goals. The local self-organizing map based neural network (SOMNN) method is developed to subtask allocations and robot path planning.
(4) A Delaunay triangulation (DT) method is employed to refine the team locations and, thus, connect the two stages.

The overall workflow of the proposed multi-robot task allocation framework is illustrated in Figure 2.

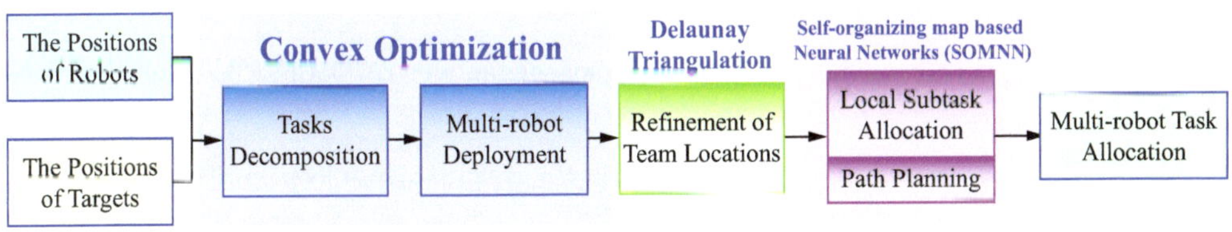

Figure 2. The overall workflow of team-based decentralized deployment framework through convex optimization models.

The rest of this paper is organized as follows. In Section 2, the problem statement and formulation is presented. In Section 3, the distance optimal-based convex optimization model for multi-robot deployment is proposed. Section 4 presents the SOMNN sub-task allocation and path planning approach. Numerical experiments, simulations, and comparison studies are presented in Section 5. Several important properties of the presented framework are summarized in Section 6.

2. Problem Statement and Formulation

In this multi-robot deployment problem, a known environment is given with $\mathcal{N}$ target locations, in which some locations are located within the workspace (such as warehouses), but other locations are in the edges of the workspace (see Figure 1). There are m robots to be deployed that are classified into g swarms (groups). In every swarm, there are k robots (k is a variable). Robots are initially located within different irregular shapes as swarms (see Figure 1). Robots moves as a swarm while maintaining a minimized total distance cost among robots shown in Figure 3. Interrelated robots are classified into the same swarm to be deployed in a nearby terrain. Some robots are enclosed in one swarm in collaboration with the robots in other swarms. Robots with relatively high connections are arranged close to one another for connectivity. The distance between pairs of robots with high connection is to be minimized. Likewise, the distance between robots and target locations is to be minimized.

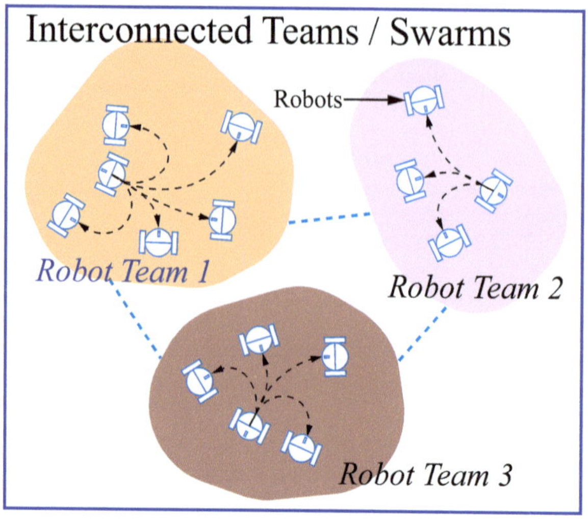

Figure 3. Illustration of multi-robot deployment with robot teams. Robots move as a swarm while minimizing the total distance cost between them. The classification of interrelated robots into the relevant swarm for deployment on adjacent terrain. Some robots are contained within one swarm in conjunction with robots from other swarms.

The relative locations of the robot swarms in the known environment are provided by an attractor–repeller convex optimization algorithm. The robots enclosed in swarms are deployed into their relative locations in the workspace. Given the relative locations of the robots, a planar graph and relative location matrix are obtained by a Delaunay triangulation approach, to enforce no overlap between any two circles, used for the next step. The robots contained in their swarms are locally assigned to their locations. The robots in a group are required to have the shortest possible distance from the fixed target location on the edge of the working environment.

3. Convex Optimization Model

The proposed framework couples task decomposition, deployment, local subtask allocation and path planning to support scenarios where the optimal solution depends on robot interrelation, dependencies, and availability, as well as inter-team conflict avoidance. We introduce an efficient technique that addresses the deployment problem of a team of heterogeneous robots. For multiple scattered tasks in arbitrary space, the objective to be solved is to minimize distance for robots to reach all the task locations.

3.1. Convex Optimization Algorithms

Convex optimization algorithms to minimize the total distance cost are described in this section. There are several definitions of the proposed algorithms.

Definition 1. *Team of robots.*

The deployment of robot swarms is described as follows:

- Each swarm of robots is labeled 1, 2, ..., M, represented as a circle with radius r_i, $i = 1, 2, ..., M$. The radius r_i is determined by the number of robots in this swarm.
- The location of each swarm 1, 2, ..., M is given by the coordinates of its center depicted as (x_i, y_i).
- The non-negative cost per unit distance between swarms i and j is denoted by c_{ij}, which is equivalent to the weight between swarms.

- The distance measured from center to center of swarms i and j by Euclidean distance (L_2 norm) is represented by d_{ij}, that is, $d_{ij}=\sqrt{(x_i-x_j)^2+(y_i-y_j)^2}$.

Therefore, the multi-robot deployment problem may be mathematically formulated as weighted distance minimization model among swarms as follows.

$$\begin{aligned} \min_{(x_i,y_j)} \quad & \sum_{1\le i<j\le N} c_{ij}d_{ij} \\ \text{s.t.} \quad & r_i+r_j-d_{ij}\le 0, \quad \forall \ 1\le i<j\le M \end{aligned} \tag{1}$$

The circles containing robots are deployed as close as possible, but they should not overlap. The objective function $\sum_{1\le i<j\le M} c_{ij}d_{ij}$ attempts to make distance d_{ij} as low as possible, which attracts pairs of circles i and j towards each other so as to function as an attractor. The constraints, $r_i+r_j\le d_{ij}(\forall \ 1\le i<j\le M)$, push any pair of circles away from each other with no overlapping [47]. Swarms are approximated by circles whose radii are proportional to the amount of encircled robots. The constraint to prevent circles from overlapping has the mathematical form: $(x_i-x_j)^2+(y_i-y_j)^2\ge(r_i+r_j)^2$, where (x_i,y_i) and (x_j,y_j) denote the coordinates of the centers of two circles i and j, whereas r_i and r_j represent their corresponding radii, respectively. The distance between circles is minimized, while they have no overlaps as constraints. This problem may be formulated as a non-linear optimization model with its superiority of convexity, inspired by a dynamic spring system [48] as follows.

$$\min_{(x_i,y_j),w_F,h_F} \quad \sum_{i,j\in M}\frac{1}{2}c_{ij}d_{ij}^2 \ + \sum_{i,j\in M}\max\left\{0,\omega_{ij}(r_i+r_j-d_{ij})\right\} \tag{2}$$

s.t.

$$\begin{aligned} & \frac{1}{2}w_F\ge x_i+r_i \ \text{ and } \ \frac{1}{2}w_F\ge r_i-x_i, \text{ for all } i\in M, \\ & \frac{1}{2}h_F\ge y_i+r_i \ \text{ and } \ \frac{1}{2}h_F\ge r_i-y_i, \text{ for all } i\in M, \\ & w_F^{up}\ge w_F\ge w_F^{low}, \\ & h_F^{up}\ge h_F\ge h_F^{low}, \end{aligned} \tag{3}$$

where ω_{ij} is a constant, $\omega_{ij}>0$. $\omega_{ij}(r_i+r_j-d_{ij})$ is penalized in the total energy function to generate a repulsive force. d_{ij} and r_i, r_j of circles are defined above. If $d_{ij}<r_i+r_j$, then circles i and j overlap. $d_{ij}\ge r_i+r_j$ prevents circles i and j from overlapping each other.

In order to formulate convex optimization model, in which circles enclosing robots are formulated to have no overlaps, repulsive force is introduced. In addition to this attractive force, we consider move this *max* function by introducing repulsive force. Afterwards, robots as swarms are deployed into their appropriate potions. Target distance concept employed in this model was initially introduced by Anjos and Vannelli [49]. Let each swarm of robots i be represented by a circle with radius r_i, where r_i is proportional to $\sqrt{a_i}$, the square root of the area of circle i, measured by the amount of robots encircled. Following [49], the target distance for each pair of circles i,j is defined as $t_{ij}:=\alpha(r_i+r_j)^2$, where $\alpha>0$ is a parameter. To prevent circles from overlapping, the target distance is enforced by introducing a penalty term $f(\frac{D_{ij}}{t_{ij}})$, where $f(z)=\frac{1}{z}-1$ for $z>0$, and $D_{ij}=(x_i-x_j)^2+(y_i-y_j)^2$. The objective function is thus given by

$$\sum_{1\le i<j\le n} c_{ij}D_{ij}+f(\frac{D_{ij}}{t_{ij}}). \tag{4}$$

The repeller term (i.e., the penalty function) enforced by holding a positive value, functions as a repulsive force, if $r_i + r_j > d_{ij}$, to hinder the circles from overlapping.

The attractor in the objective function used to apply an attractive force to the two circles, makes the two swarms of robots (circles) move closer together (see Figure 4). The repeller term disappears or becomes a negative value implying that there is no any overlap between circles, if $r_i + r_j \leq d_{ij}$. An attractive force is enforced to the two circles through the attractor in the objective function enforces, if $D_{ij} \geq t_{ij}$. In this case, there is no any overlap between circles and the repeller term is zero or slightly negative. Conversely, the repeller term is positive, which tends to positive infinity as D_{ij} tends to zero so as to prevent the circles from overlapping fully, if $D_{ij} < t_{ij}$. There is no force between circles i and j exactly if $D_{ij}^2 = t_{ij}/c_{ij}$. The generalized target distance T_{ij} is defined to contribute to this optimization model.

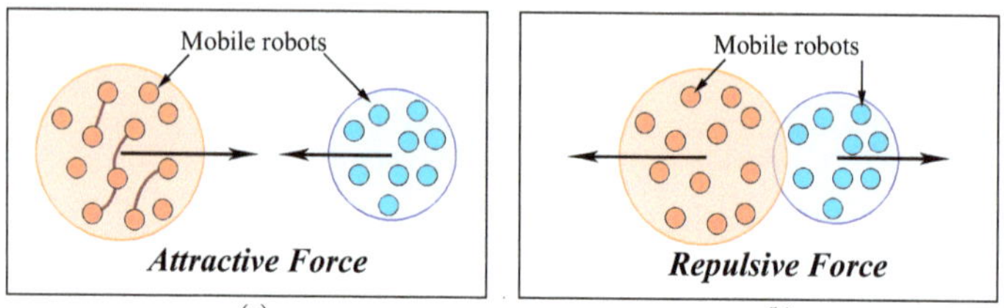

(a) (b)

Figure 4. Two circles with attractive and repulsive forces. (**a**) Two disconnected circles with attractive force, $r_i + r_j \leq d_{ij}$; (**b**) Two connected circles with repulsive force, $r_i + r_j > d_{ij}$.

The model aims to ensure that $\frac{D_{ij}}{t_{ij}} = 1$ at optimality, so choosing $\alpha < 1$ sets a target value t_{ij} that allows some overlap of the respective circles, which means that the non-overlap requirement is relaxed. In practice, by properly adjusting the value of α we achieve a reasonable separation between all pairs of circles. Let M and P denote the set of mobile teams (circles) and the set of targets (goals), respectively. Target distances are applied only for pairs of mobile robot teams (circles). The complete attractor-repeller (AR) model is

$$\min_{(x_i,y_i),i\in M,w_F,h_F} \sum_{i,j\in M\cup P} c_{ij}D_{ij} + \sum_{i,j\in M} f\left(\frac{D_{ij}}{t_{ij}}\right) \tag{5}$$

s.t.

$$\begin{aligned} & x_i + r_i \leq \frac{1}{2}w_F \text{ and } r_i - x_i \leq \frac{1}{2}w_F, \text{ for all } i \in M, \\ & y_i + r_i \leq \frac{1}{2}h_F \text{ and } r_i - y_i \leq \frac{1}{2}h_F, \text{ for all } i \in M, \\ & w_F^{low} \leq w_F \leq w_F^{up}, \\ & h_F^{low} \leq h_F \leq h_F^{up}, \end{aligned}$$

where (x_i, y_i) are the coordinates of the centre of circle i as previously defined; w_F, h_F are the width and height of the workspace; and w_F^{low}, w_F^{up}, h_F^{low}, and h_F^{up} are the lower and upper bounds on the width and height, respectively. The first two sets of constraints require that all the circles be entirely contained in the workspace, and the remaining two pairs of inequalities bound the width and height of the workspace of robots. In particular, for certain robot environment, we set $w_F^{low} = w_F^{up} = \bar{w}_F$ and $h_F^{low} = h_F^{up} = \bar{h}_F$, where $\bar{w}_F$ and $\bar{h}_F$ are the width and height of the workspace to be explored by robots.

Definition 2. *Generalized target distance*

$$T_{ij} := \sqrt{\frac{t_{ij}}{c_{ij} + \varepsilon}}, \tag{6}$$

where $\varepsilon > 0$ *is sufficiently small to ensure that* $D_{ij} \approx \sqrt{\frac{t_{ij}}{c_{ij}}}$ *if* $D_{ij} \approx T_{ij}$.

In real-world applications, the distances D_{ij} between the circles should be inversely proportional to c_{ij} representing the weights on the wire-length, and should be proportional to the relative size of the teams through the value of t_{ij}. Hence, a generalized target distance, T_{ij}, is defined such that $D_{ij} \approx T_{ij}$ at optimality. Using T_{ij}, a convex version of the AR model may be described in the following section with the following term.

$$F_{ij}(x_i, x_j, y_i, y_j) := \begin{cases} c_{ij}z + \frac{t_{ij}}{z} - 1, & z \geq T_{ij} \\ 2\sqrt{c_{ij}t_{ij}} - 1, & 0 \leq z < T_{ij} \end{cases} \tag{7}$$

with $z = (x_i - x_j)^2 + (y_i - y_j)^2$. It is clear that this problem is convex, and that by construction F_{ij} attains its minimum value whenever the locations of circles i and j satisfy $D_{ij} \leq T_{ij}$. This includes the case where $D_{ij} = 0$, i.e., both circles completely overlap. The idea is to add to the objective function a term of the form $-\ln(D_{ij}/T_{ij})$ for each pair i, j of circles. Hence, the model solved in the first stage of our method is

$$\min_{(x_i, y_i), w_F, h_F} \sum_{1 \leq i < j \leq n} F_{ij}(x_i, x_j, y_i, y_j) - \beta K \ln(\frac{D_{ij}}{T_{ij}}) \tag{8}$$

s.t.

$$\begin{aligned} & x_i + r_i \leq \frac{1}{2}w_F \ \text{ and } r_i - x_i \leq \frac{1}{2}w_F, \ \text{for all modules } i, \\ & y_i + r_i \leq \frac{1}{2}h_F \ \text{ and } r_i - y_i \leq \frac{1}{2}h_F, \ \text{for all modules } i, \\ & w_F^{low} \leq w_F \leq w_F^{up}, \\ & h_F^{low} \leq h_F \leq h_F^{up}, \end{aligned}$$

where β is a parameter selected empirically. K is chosen to reflect the weights of all the pairs of mobile circles (teams) in the objective function by $K = \sum_{i<j} c_{ij}$.

The topological relationships between terms are obtained in this first stage. Without the term $-\beta K \ln(D_{ij}/T_{ij})$ in Equation (8), this problem is convex. By solving it, the solution of the first stage provides relative locations within the workspace for all the robot teams (circles) represented by circles. The relative locations of the robot swarms in the known environment are provided by an attractor–repeller convex optimization algorithm shown in Figure 5. The team-based convex optimization algorithm for robot deployment is summarized in Algorithm 1.

Algorithm 1: Team-based convex optimization algorithm for robot deployment

Input: Initial configuration (q_s), Goal configuration (q_g), and Required cost c_{ij}, d_{ij}
Output: *Path* (or sequence of nodes from q_s to q_g), Robots' final locations (L_{goal})
Begin
1: Classify m robots into k teams $\mathcal{T}_i, i \in [1, k]$;
2: Approximate teams $\mathcal{T}_i$ to k circles at center of $\mathcal{T}_j(x_j, y_j), j \in [1, k]$;
3: Solve the convex optimization model (Equation (8)) having $C_j(x_j, y_j), j \in [1, k]$;
4: Obtain relative locations of teams. New circle locations $\tilde{C}_j(\tilde{x}_j, \tilde{y}_j), j \in [1, k]$;
5: Apply SOMNN to subtask allocation to reach g goals $\mathbf{G}(\mathbf{x}_l, \mathbf{y}_l), l \in [1, g]$;
End

3.2. Refinement of Team Locations

The circles are allowed to overlap in the convex optimization stage and the precise locations of team have not been determined. The solution of the first stage provides relative locations for all the circles enclosing robots. In this paper, we consider a relative location matrix to encode relative locations. Using this technique the non-overlap constraints that are originally disjunctive, non-linear, and non-convex can be linearized and easily enforced in the second stage model. We consider the Delaunay triangulation (DT) method for two main reasons, (i) it spreads out circles thus teams in the workspace and (ii) it transforms the relative location graph into a planar graph. In this framework, robot deployment resolved by the convex optimization with relative locations in Figure 5 is refined by the proposed DT method shown in Figure 6.

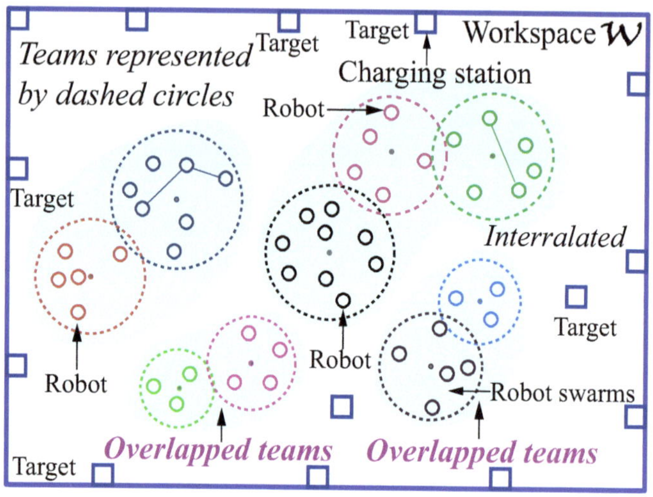

Figure 5. Robot teams (swarms) deployment initially resolved by the proposed convex optimization model.

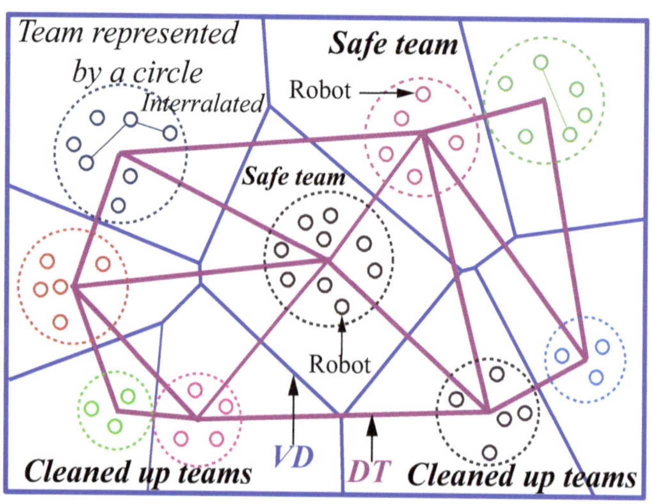

Figure 6. Delaunay triangulation used to linearize and refine the overlapped circles/teams.

4. SOM Neural Networks for Sub-Task Allocation and Path Planning

In this stage, a swarm of robots are assigned locally as sub-tasks to their goals through self-organizing map based neural networks (SOMNN) method in static or dynamic environments. The principal technical superiority of this SOMNN model is well established, given that the robot trajectory planning is fused with the sub-task allocation in every team and their goals. Once the relative locations of teams with robots are defined, the robots move to their goals in the dynamical environments. In our research, there are k teams, each of which has a swarm of robots. There are α robots in one team $\mathcal{T}_1(\alpha)$ in a workspace, which are assigned to η goals with pre-defined locations, such as recharging pile for autonomous electric vehicles. In this paper, SOM-based neural network model is made up of two layers of neurons (nodes). The first layer configured as the input layer consists of two neurons (u_i, v_i) representing coordinates of the ith goal $G_i(u_i, v_i)$. The second layer as the output layer contains $\alpha \times \eta$ neurons, denoted as $r_1^1, r_1^2, \ldots, r_1^\eta, r_2^1, r_2^2, \ldots, r_2^\eta, r_\alpha^1, r_\alpha^2, \ldots, r_\alpha^\eta$, representing the locations of the α robots and their trajectories. All locations of goals form input dataset. Each neuron in the output layer is fully connected to the neurons in the input layer [50] (see Figure 7).

$$\begin{aligned} \mathcal{N}_{\dot{\alpha}}^{\dot{\eta}} \Longleftarrow \dot{\mathcal{D}}_{\dot{\alpha}}^{\dot{\eta}} = \min\{D_\alpha^\eta|_i, i = 1, \ldots, \alpha; \dot{\alpha} = 1, \ldots, \alpha \\ \dot{\eta} = 1, \ldots, \eta; \text{ and } \{\dot{\alpha}, \dot{\eta}\} \in \varpi\} \end{aligned} \tag{9}$$

$$D_\alpha^\eta|_i = \mathcal{D}(\mathcal{G}_i, \mathfrak{r}_{\dot{\alpha}}^{\dot{\eta}})(1 + \mathcal{P}) \tag{10}$$

$$\mathcal{D}(\mathcal{G}_i, \mathfrak{r}_{\dot{\alpha}}^{\dot{\eta}}) = \left|\mathcal{G}_i - \mathfrak{r}_{\dot{\alpha}}^{\dot{\eta}}\right| = \sqrt{\left(x_i - w_{\dot{\alpha}}^{\dot{\eta}}(x)\right)^2 + \left(y_i - w_{\dot{\alpha}}^{\dot{\eta}}(y)\right)^2} \tag{11}$$

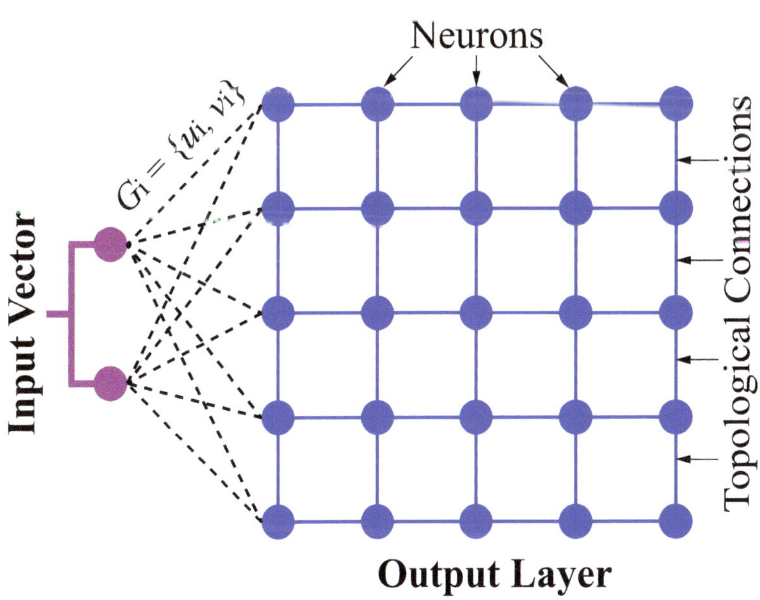

Figure 7. SOM-based NN architecture used for subtask allocation.

A weight vector connecting the two input nodes to output nodes is defined as $r_{\dot{\alpha}}^{\dot{\eta}} = \langle w_{\dot{\alpha}}^{\dot{\eta}}(x), w_{\dot{\alpha}}^{\dot{\eta}}(y)\rangle$, where $\dot{\alpha} = 1, 2, \cdots, \alpha$; $\dot{\eta} = 1, 2, \cdots, \eta$. The weight vectors of the neurons for each robot are initialized based on the coordinates of the initial robot location. Therefore,

α robots move to the η goals while following planned path, due to this SOM neural network algorithm. A winner $\mathcal{N}_{\dot{\alpha}}^{\dot{\eta}}$ is determined in light of the following criterion.

$$\mathcal{N}_{\dot{\alpha}}^{\dot{\eta}} \Longleftarrow \dot{\mathcal{D}}_{\dot{\alpha}}^{\dot{\eta}} = \min\Big\{ D_{\dot{\alpha}}^{\eta}|_i, i = 1,\ldots,\alpha; \dot{\alpha} = 1,\ldots,\alpha \\ \dot{\eta} = 1,\ldots,\eta; \text{ and } \{\dot{\alpha},\dot{\eta}\} \in \varpi\Big\}, \tag{12}$$

$$D_{\dot{\alpha}}^{\eta}|_i = \mathcal{D}(\mathcal{G}_i, \mathbf{r}_{\dot{\alpha}}^{\dot{\eta}})(1+\mathcal{P}), \tag{13}$$

where $\mathcal{N}_{\dot{\alpha}}^{\dot{\eta}}$ is the $\dot{\alpha}$-th neuron of the $\dot{\eta}$ th in the group of goals. ϖ denotes the set of output neurons on the $\dot{\eta}$ th in the group of goals, which has not yet been a winner. The weighted distance, $\dot{\mathcal{D}}_{\dot{\alpha}}^{\eta}$, is minimum of $D_{\dot{\alpha}}^{\eta}|_i$ described as follows.

$$\mathcal{D}(\mathcal{G}_i, \mathbf{r}_{\dot{\alpha}}^{\dot{\eta}}) = \left|\mathcal{G}_i - \mathbf{r}_{\dot{\alpha}}^{\dot{\eta}}\right| = \sqrt{\left(x_i - w_{\dot{\alpha}}^{\dot{\eta}}(x)\right)^2 + \left(y_i - w_{\dot{\alpha}}^{\dot{\eta}}(y)\right)^2} \tag{14}$$

where $\mathcal{P}$ is a parameter determining the equitable distribution of sub-task of workload for robots. $|\cdot|$ represents the Euclidean distance. The parameter $\mathcal{P}$ is expressed as

$$\mathcal{P} = \frac{\ell_{\dot{\alpha}} - \tilde{\mathcal{V}}_A}{1+\tilde{\mathcal{V}}_A}, \tag{15}$$

where $\ell_{\dot{\alpha}}$ is the trajectory length of the $\dot{\alpha}$ th robot ($\dot{\alpha} = 1, 2, \cdots, \alpha$). $\tilde{\mathcal{V}}_A$ is the average trajectory length of the robots, given by the following Equation (16).

$$\tilde{\mathcal{V}}_A = \frac{1}{\alpha}\sum_{\dot{\alpha}=1}^{\alpha} \ell_{\dot{\alpha}} \tag{16}$$

The winner implies the neuron in the group of the output neurons with a lower workload of that location, and the neuron with the minimum distance toward the input data. The SOM NN is updated by modifying its weigh vectors $r_{\dot{\alpha}}^{\dot{\eta}} = \langle w_{\dot{\alpha}}^{\dot{\eta}}(x), w_{\dot{\alpha}}^{\dot{\eta}}(y)\rangle$ ($\dot{\alpha} = 1, 2, \cdots, \alpha$; $\dot{\eta} = 1, 2, \cdots, \eta$), by the following rule (17), until the weight vectors remain unchanged.

$$\mathbf{r}_{\dot{\alpha}}^{\dot{\eta}}(t+1) \\ = \begin{cases} \mathbf{G}_i, & \text{if } D_{\dot{\alpha}}^{\eta}|_i < \mu\psi \\ \mathbf{r}_{\dot{\alpha}}^{\dot{\eta}}(t) + \vartheta f\left(d_{\dot{\alpha}}^{\dot{\eta}}\right)\left(\mathbf{G}_i - \mathbf{r}_{\dot{\alpha}}^{\dot{\eta}}(t)\right), & \text{otherwise} \end{cases} \tag{17}$$

where ϑ is the learning rate. μ is a little constant, usually smaller than 0.5. ψ is the minimum distance between any two neurons of the goal locations. $f(d_{\dot{\alpha}}^{\dot{\eta}})$ is the neighborhood function, defined as

$$f\left(d_{\dot{\alpha}}^{\dot{\eta}}\right) = \begin{cases} e^{-(d_{\dot{\alpha}}^{\dot{\eta}})^2/G^2}, & \text{if } d_{\dot{\alpha}}^{\dot{\eta}} < \Lambda\eta \\ 0, & \text{otherwise} \end{cases} \tag{18}$$

where Λ is a small constant denoting the range of neighborhood, normally less than 0.4. $d_{\dot{\alpha}}^{\dot{\eta}}$ is the distance measured between the neuron and the winner neuron, $\mathcal{N}_{\dot{\alpha}}^{\dot{\eta}}$, from the group of the output neurons. The gain constant ψ with an initial value of 10 is given by $\mathbf{\Delta}(t+1) = (1-\psi)\mathbf{\Delta}(t)$, where ψ is the gain changing rate usually determined empirically. Usually, the range of ψ is between 0.001 and 0.05. Smaller ψ leads to shorter total trajectory and longer computational time. t is the number of iterations. The implementation of the proposed multi-robot deployment method is described in Algorithm 2.

Algorithm 2: Multi-robot deployment algorithm with distance minimization

Input : Initialized 2D Cartesian workspace $\mathcal{W}$;
Coordinates of m robots, and q targets;
Required c_{ij} and d_{ij}.
Output: Robot location deployed.
Multi-robot deployment with distance minimization;
Multi-robot_Deployment ();
repeat
for each robot $\mathcal{G}(x,y) \Leftarrow R(x_i,y_j) \in \mathcal{R}(x,y)_{group}$;
for $\phi \leftarrow 1$ **to** m **do**
// m robots to be deployed
1. Classify m robots to k teams according to their interrelation
$\mathcal{T}_1(\alpha) \leftarrow r_1^1(x_1,y_1), r_1^2(x_2,y_2), \ldots, r_1^\alpha(x_\alpha,y_\alpha)$;
// α robots are enclosed in Team 1
$\mathcal{T}_2(\beta) \leftarrow r_2^1(x_1,y_1), r_2^2(x_2,y_2), \ldots, r_2^\beta(x_\beta,y_\beta)$;
// β robots are enclosed in Team 2 ...
$\mathcal{T}_k(\sigma) \leftarrow r_k^1(x_1,y_1), r_k^2(x_2,y_2), \ldots, r_k^\sigma(x_\sigma,y_\sigma)$;
// σ robots are enclosed in Team k
$k \longleftarrow \alpha \bigcup \beta \bigcup \ldots \bigcup \sigma$;
$\exists\ (r_1^1) \bigcup (r_1^2) \bigcup \ldots \bigcup (r_1^\alpha) \in \mathcal{T}_1(\alpha)$;
$\exists\ (r_2^1) \bigcup (r_2^2) \bigcup \ldots \bigcup (r_2^\beta) \in \mathcal{T}_2(\beta)$;
$\exists\ (r_k^1) \bigcup (r_k^2) \bigcup \ldots \bigcup (r_k^\sigma) \in \mathcal{T}_k(\sigma)$;
2. Approximate teams $\mathcal{T}_i$ to k circles
$\mathcal{T}_1(\alpha) \leftarrow \langle\ \{r_1^1\} \bigcup \{r_1^2\} \bigcup \ldots \bigcup \{r_1^\alpha\}\ \rangle$;
$\mathcal{T}_2(\beta) \leftarrow \langle\ \{r_2^1\} \bigcup \{r_2^2\} \bigcup \ldots \bigcup \{r_2^\beta\}\ \rangle$;
$\mathcal{T}_k(\sigma) \leftarrow \langle\ \{r_k^1\} \bigcup \{r_k^2\} \bigcup \ldots \bigcup \{r_k^\sigma\}\ \rangle$;
until
$r_{\bar{k}}^{\bar{m}}|_\phi \in \emptyset\ (\bar{k} = 1,2,\ldots,k; \bar{m} = 1,2,\ldots,m)$;
3. Solve the convex optimization model (Equation (8))
$\mathcal{C}_1(\alpha) = \{x_1,y_1\}, \mathcal{C}_2(\beta) = \{x_2,y_2\}, \ldots, \mathcal{C}_k(\sigma) = \{x_k,y_k\}$;
// Relative locations of teams are found 4. Refinement of robot swarms by DT method
$\tilde{\mathcal{C}_1}(\alpha) = \{\tilde{x_1},\tilde{y_1}\}, \tilde{\mathcal{C}_2}(\beta) = \{\tilde{x_2},\tilde{y_2}\}, \ldots, \tilde{\mathcal{C}_k}(\sigma) = \{\tilde{x_k},\tilde{y_k}\}$;
// Robot teams are spread out 5. SOM NN to sub-task allocation
if $\exists(\bar{m},\bar{k}) \in \mathcal{P}_r(i,j)/\mathcal{P}_{EntireWS}, \forall 1 \le \bar{m} \le m, 1 \le \bar{\le} k$
s.t. $\xi(x,y) \le \xi(x_c,y_c)$ **then**
$G_\alpha^\eta|_\phi \leftarrow (x_i,y_\phi)$;
// Mark it as goals discovered

until $\mathcal{S}(x,y)_{cov} + \bigcup_{l=1}^{n_{ob}} \mathcal{OB} = \mathcal{P}(x,y)_{EntireWS}$;
return $P(x_\phi,y_\phi), \xi(x_c\phi,y_\phi) \in G_\alpha^\eta|_\phi$

5. Simulated Experiments and Comparison Studies

In this section, simulation and comparison studies are conducted to validate our proposed framework.

5.1. Numerical Experiments by Convex Optimization

Numerical experiments are carried out to validate the convex optimization model. In the first set of experiments, we focus on the multi-robot deployment problem through the convex optimization model with equal number of robots and goals, summarized in Table 1. It aims to experimentally verify the correctness of the results obtained from the proposed algorithm. In this experiment, multiple robots are enclosed in their teams before they are deployed to their goals. It assumes that number of robots is equal to the goals, except in

Team 2, 50 robots will be assigned to two goals. The total distance the robots traveled is calculated. The first set of experimental results are illustrated in Figure 8.

Table 1. The first set of numerical experiments with robot teams.

Teams	# of Robots	# of Teams	# of Goals	Length
T_1	73	9	73	38.43
T_2	50	10	2	42.01
T_3	45	11	45	13.18
T_4	42	33	42	6.03
T_5	75	49	22	68.46

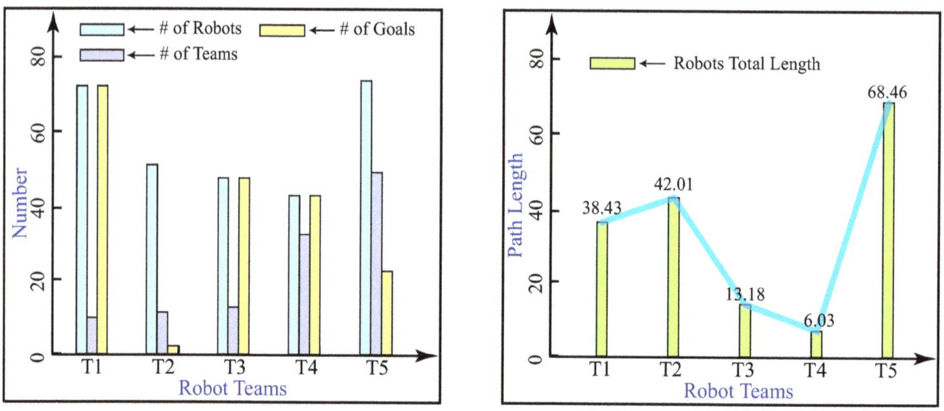

Figure 8. The first set of numerical experimental results.

In the second set of experiments, multi-robot deployment through the proposed convex optimization model is obtained with different numbers of robots and goals. The results are summarized in Table 2. The second set of experimental results are illustrated in Figure 9.

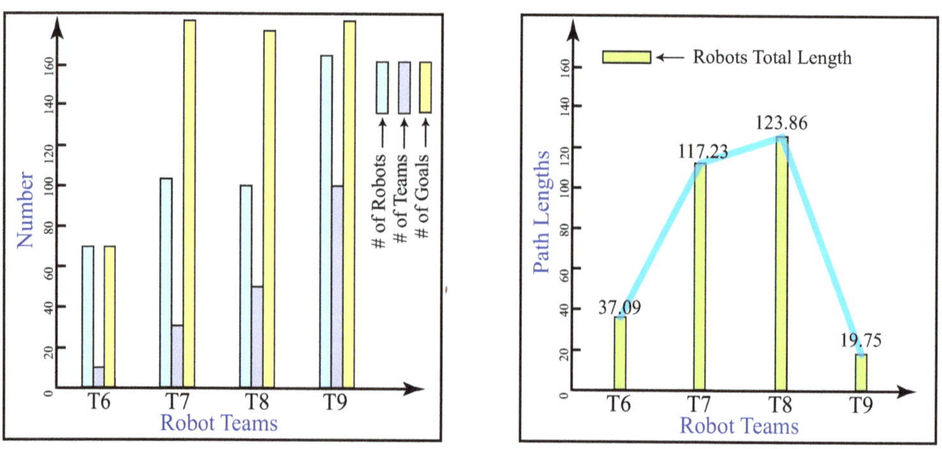

Figure 9. The second set of numerical experimental results.

Table 2. The second set of numerical experiments with robot teams.

Teams	# of Robots	# of Teams	# of Goals	Length
T_6	69	10	69	37.09
T_7	106	30	212	117.23
T_8	100	50	209	123.86
T_9	167	100	334	19.75

For lack of space, we present our numerical experiments concerning some maps and scenarios, but our results are representative for a broad range of environments. In the next section, we will consider two maps obtained from the publicly available papers.

5.2. Evaluation Using Standard Environments

In this section, complex robot allocation experiments in standard environments with multiple constraints are conducted to further validate the robustness and effectiveness of our proposed convex optimization framework. Standard environments, such as apte, xerox, hp, ami33, and ami49, provided by Microelectronics Center of North Carolina (MCNC) are used in our analyses.

In these scenarios, robots are clustered into predetermined teams. For instance, in the apte environment, 287 robots are distributed into nine teams (Table 3). They are required to perform complex task allocation under 97 constraints. Constraints are limitations or rules that need to be respected by the optimization framework. Several constraints are listed in Table 4. The first constraint is to ensure that nine robots must reach goal location 37. These nine robots must be one from each team $T_1, T_2, \ldots, T_9$, respectively, as outlined in Table 4. The second constraint is to ensure that eight robots reach goal location 55, and each of these eight robots must be one from each team $T_1, T_2, \ldots, T_8$, respectively. Similarly, the third constraint is that two robots must reach goal location 17, and these two robots must be from groups T_1 and T_2. Some robots in the teams are required to reach specific goals. In the apte standard environment, 97 such constraints must be respected. The initial apte environment setting of robots and goals is shown in Figure 10. In this stage, the robots are clustered into nine teams that are assigned to goals located on the boundary. In such scenario, our framework identified an optimized solution on an average in 0.69 s where the average length of all robots path is 425.09 m.

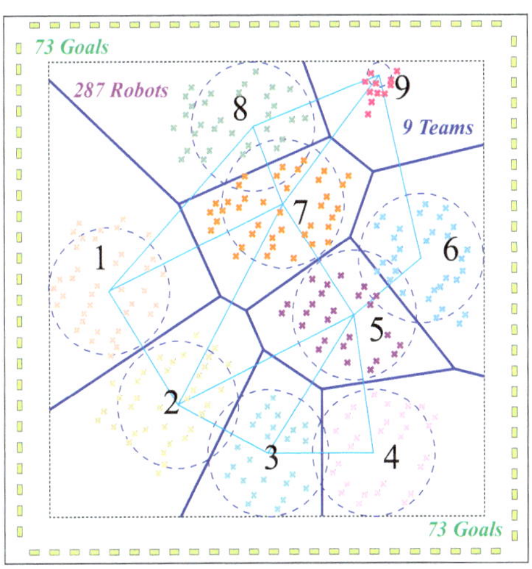

Figure 10. The initial settings of apte standard environment with 287 robots in nine teams. Robots are assigned to goals located on the boundary.

Table 3. The average execution time and path length of the proposed convex optimization-based framework for the standard MCNC environments.

Standard Environment	Goals	Teams	Robots	Constraints	Workspace (m^2)	Avg_time (s)	Avg_length (m)
apte	73	9	287	97	46.56	0.69	425.09
xerox	2	10	698	203	19.35	1.12	411
hp	45	11	309	83	8.30	1.17	154.84
ami33	42	33	522	123	1.16	14.16	65.31
ami49	22	49	953	408	35.4	9.96	699

Table 4. The constraints in the apte standard environment. 287 robots from nine teams are assigned to 73 goals.

# of Constraints	To # of Goal	# of Robots	From # of Teams
1	37	9	T_1,T_2,T_3,T_4,T_5,T_6,T_7,T_8,T_9
2	55	8	T_1,T_2,T_3,T_4,T_5,T_6,T_7,T_8
3	17	2	T_1,T_2
4	19	2	T_1,T_2
5	16	2	T_1,T_2
6	15	2	T_1,T_2
7	2	8	T_1,T_2,T_3,T_4,T_5,T_6,T_7,T_8
8	14	2	T_1,T_5
9	13	2	T_1,T_5
...	...	...	...

Similarly, in the xerox standard environment with 203 constraints, 698 robots were distributed into 10 teams and were tasked to reach two goals. On average, the optimized solution in this scenario was discovered in 1.12 s where the average path length of all robots is 411 m. Likewise, our framework was able to identify optimal solutions in hp, ami33, and ami49 environments in an average of 1.17, 14.16, and 9.96 s, respectively. As presented in Table 3, the average path lengths for all environments are 154.84, 65.31, and 699 m, respectively. From Table 3, it can also be observed that though number of constraints are more in xerox environment compared to apte, the average length discovered by our framework is smaller in the xerox environment than that of apte. Similar observations can be made from the results for other environments. This is attributed to the fact that not all constraints in different environments are similar. Therefore, the execution time taken by the model and the path length required by the robots to reach, respectively, goals vary significantly based on the environments.

5.3. Simulations and Comparison Studies

Two test environmental maps are obtained using a SLAM (Simultaneous Localization and Mapping) algorithm through a publicly available dataset [26]. We used these two building maps to test our framework by assuming the correlative robots using nodes as targets. In this building test scenario, we demonstrate our framework is applicable for the seven-robot deployment. As there are seven nodes corresponding to seven targets, seven robots will be assigned to these seven rooms illustrated in Figure 11a. In this scenario, the deployment aims to deploy at least one robot in each of nodes (rooms). Robots are assigned to seven rooms with minimized travel distance cost, which is formulated to the developed convex optimization model. As this test case is straightforward with only seven targets, the relative locations of solution is obtained from the convex optimization model before they are refined and spread out. Finally, these seven robots are deployed to seven rooms with total distance minimized.

A more complicated indoor room environment is depicted in Figure 11b. In this test scenario, 22 rooms are supposed to be deployed by 22 robots, in which one goal is reached by at least one robot. Our model initially selects Nodes 7, 9, 10, and 13 as the main teams. Based on the interconnection depicted in Figure 11b, a team consists of several robots to be deployed. In this case, Node 7 consists of Robots r_1, r_2, r_3, r_4,r_5, r_6, r_{11}, and r_9. Node 10 contains Robots r_{20}, r_{21}, and r_{22}. Node 13 contains Robots r_9, r_{14}, and r_{15}. Node 9 contains Robots r_{12}, r_{16}, r_{17}, and r_{19}. Therefore, our developed convex optimization model creates relative locations of Nodes (Teams) 7, 9, 10, and 13. The total distance cost of among these nodes/teams is minimized, where the 22-robot swarms are assigned their final goals in this building.

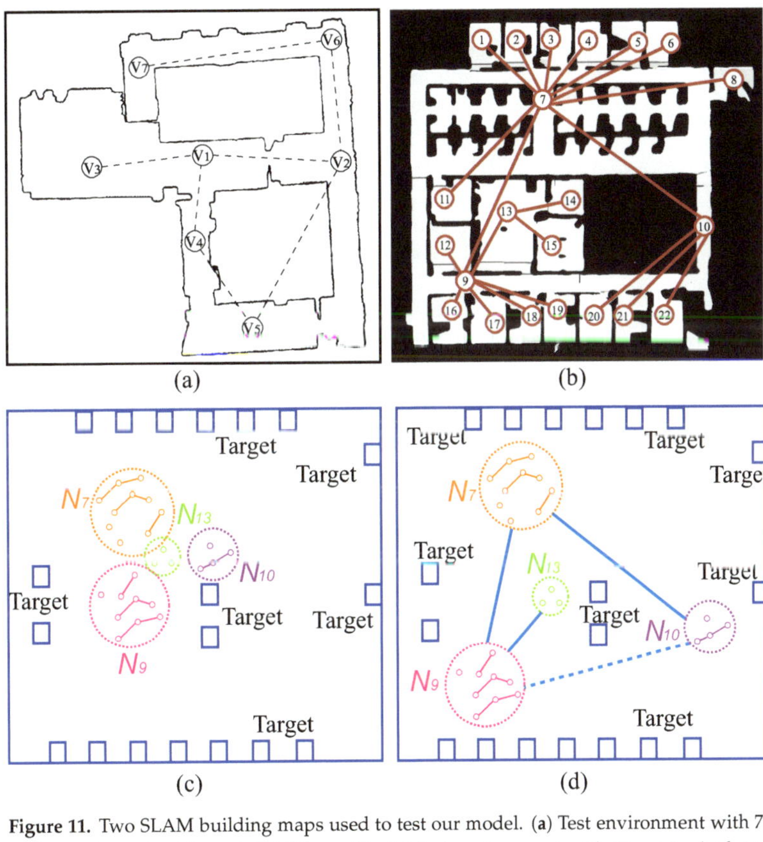

Figure 11. Two SLAM building maps used to test our model. (**a**) Test environment with 7 rooms (redrawn from Purohit et al. [26] Figure 2). (**b**) Test environment with 22 rooms (redrawn from Purohit et al. [26] Figure 3). (**c**) Formation of 4 teams through proposed convex optimization model. (**d**) Spread out teams via Delaunay triangulation (DT).

After applying for the refinement of teams, we obtain their locations of nodes. In each team, it consists of a few robots, such as Node 7 consists of Robots r_1, r_2, r_3, r_4, r_5, r_6, and r_{11}. The next stage, these seven robots driven by the SMNN model are navigated to seven rooms. It is clear that node as team containing robots could be the team member of another node. For instance, Node 9 is a member but it is contained in Node 7. This is beneficial of the developed convex optimization able to solve this sort of optimization problem.

In this test scenario of simulation, these 22 robots are clustered into four teams in task decomposition according to their interconnection of robots and correlation of teams. There are four teams, $\mathcal{N}_7$ with 10 robots; $\mathcal{N}_9$ with 9 robots; $\mathcal{N}_{10}$ with 4 robots; and $\mathcal{N}_{13}$ with 3 robots. First stage with our convex optimization model creates relative locations of four teams, in which the total distance cost is minimized shown in Figure 11c. Once the relative locations of these four teams (circles) are determined, a DT method ia applied to spread out them to their appropriate locations shown in Figure 11d. The correlation of teams are shown in Figure 11d. For instance, $\mathcal{N}_7$ is assumed to connect with $\mathcal{N}_9$ and $\mathcal{N}_{10}$. Robots within their teams are located in vicinity of their targets (rooms). The SOMNN model is used to assign members of robots to their targets (rooms). Therefore, the robots are deployed to their final targets—rooms illustrated in Figure 11b.

Our framework is compared with Hungarian algorithm [51]. Assume that m robots are assigned to k tasks located at certain goals. The interrelation of robots are assumed. Our framework clusters these robots into teams before they are assigned to goals. Hungarian algorithm directly assigns robots to their tasks and goals. We simulate 6, 11, 30, 33, and 49 robots to be deployed to same number of goals. The results show that, in terms of total path length, our proposed model is 10.14%, 14.26%, 8.69%, 3.96%, and 18.22% lower than the Hungarian algorithm for 6, 11, 30, 33, and 49 targets, respectively. The distance robots traveled to goals as a measure is compared with Hungarian algorithm and illustrated in Figure 12.

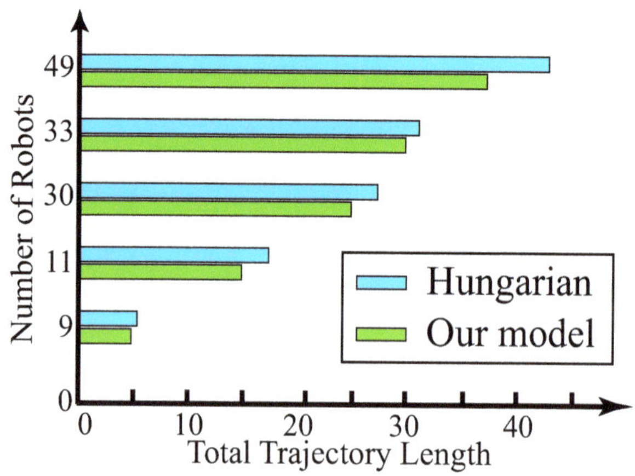

Figure 12. Comparison with Hungarian algorithm in total distance.

6. Conclusions

A new framework of team-based multi-robot deployment is proposed through convex optimization model. Multiple robots are deployed to their appropriate locations while the total distance cost is minimized, by the convex optimization mission and robot path planning. The SOMNN paradigm is used to dynamically fulfill the sub-task allocation task. The proposed model couples task decomposition, deployment, local subtask allocation and path planning to support cases where the optimal solution depends on robot interrelation, dependencies, and availability, as well as inter-team conflict avoidance. Simulation and comparison studies validate the effectiveness of our proposed framework. This framework will be implemented on actual robots in the near future work.

Author Contributions: Conceptualization, T.L. and C.L.; methodology, T.L., P.C. and C.L.; software, T.L. and P.C.; validation, T.L. and P.C.; formal analysis, T.L., P.C. and C.L.; investigation, T.L., P.C. and C.L.; resources, C.L.; data curation, T.L. and P.C.; writing—original draft preparation, C.L., T.L. and P.C.; writing—review and editing, P.C., C.L., L.L. and G.E.J.; supervision, C.L., L.L. and G.E.J.; project administration, C.L. All authors have read and agreed to the published version of the manuscript.

Funding: This research was supported by the Mississippi Space Grant Consortium under NASA EPSCoR RID grant.

Institutional Review Board Statement: Not applicable.

Informed Consent Statement: Not applicable.

Data Availability Statement: Not applicable.

Acknowledgments: The authors would like to thank the editor-in-chief, the associate editor, and the anonymous reviewers for their valuable comments.

Conflicts of Interest: The authors declare no conflict of interest.

References

1. Liu, L.; Luo, C.; Shen, F. Multi-agent formation control with target tracking and navigation. In Proceedings of the 2017 IEEE International Conference on Information and Automation (ICIA), Macao, China, 18–20 July 2017; pp. 98–103.
2. Jiménez, A.C.; García-Díaz, V.; Bolaños, S. A decentralized framework for multi-agent robotic systems. *Sensors* **2018**, *18*, 417. [CrossRef] [PubMed]
3. Patil, A.; Bae, J.; Park, M. An algorithm for task allocation and planning for a heterogeneous multi-robot system to minimize the last task completion time. *Sensors* **2022**, *22*, 5637. [CrossRef] [PubMed]
4. Zhang, Q.; Luo, R.; Zhao, D.; Luo, C.; Qian, D. Model-free reinforcement learning based lateral control for lane keeping. In Proceedings of the 2019 International Joint Conference on Neural Networks (IJCNN), Budapest, Hungary, 14–19 July 2019; pp. 1–7.
5. Wang, J.; Luo, C. Automatic wall defect detection using an autonomous robot: A focus on data collection. In Proceedings of the ASCE International Conference on Computing in Civil Engineering 2019: Data, Sensing, and Analytics, Atlanta, GA, USA, 17–19 June 2019; pp. 312–319.
6. Poskart, B.; Iskierka, G.; Krot, K.; Burduk, R.; Gwizdal, P.; Gola, A. Multi-Parameter Predictive Model of Mobile Robot's Battery Discharge for Intelligent Mission Planning in Multi-Robot Systems. *Sensors* **2022**, *22*, 9861. [CrossRef] [PubMed]
7. Lei, T.; Sellers, T.; Luo, C.; Zhang, L. A bio-inspired neural network approach to robot navigation and mapping with nature-inspired algorithms. In Proceedings of the Advances in Swarm Intelligence: 13th International Conference, ICSI 2022, Xi'an, China, 15–19 July 2022; pp. 3–16.
8. Luo, C.; Yang, S.X.; Meng, M.Q.-H. Neurodynamics based complete coverage navigation with real-time map building in unknown environments. In Proceedings of the 2006 IEEE/RSJ International Conference on Intelligent Robots and Systems, Beijing, China, 9–15 October 2006; pp. 4228–4233.
9. Lei, T.; Luo, C.; Jan, G.E.; Fung, K. Variable speed robot navigation by an ACO approach. In Proceedings of the Advances in Swarm Intelligence: 10th International Conference, ICSI 2019, Chiang Mai, Thailand, 26–30 July 2019; pp. 232–242.
10. Jan, G.E.; Luo, C.; Hung, L.P.; Shih, S.T. A computationally efficient complete area coverage algorithm for intelligent mobile robot navigation. In Proceedings of the 2014 International Joint Conference on Neural Networks (IJCNN), Beijing, China, 6–11 July 2014; pp. 961–966.
11. Jayaraman, E.; Lei, T.; Rahimi, S.; Cheng, S.; Luo, C. Immune system algorithms to environmental exploration of robot navigation and mapping. In Proceedings of the Advances in Swarm Intelligence: 12th International Conference, ICSI 2021, Qingdao, China, 17–21 July 2021; pp. 73–84.
12. Lei, T.; Li, G.; Luo, C.; Zhang, L.; Liu, L.; Gates, R. An informative planning-based multi-layer robot navigation system as applied in a poultry barn. *Intell. Robot.* **2022**, *2*, 313–332. [CrossRef]
13. Wang, S.; Wang, Y.; Li, D.; Zhao, Q. Distributed Relative Localization Algorithms for Multi-Robot Networks: A Survey. *Sensors* **2023**, *23*, 2399. [CrossRef]
14. Lei, T.; Sellers, T.; Rahimi, S.; Cheng, S.; Luo, C. A nature-inspired algorithm to adaptively safe navigation of a COVID-19 disinfection robot. In Proceedings of the Intelligent Robotics and Applications: 14th International Conference, ICIRA 2021, Yantai, China, 22–25 October 2021; pp. 123–134.
15. Tardós, J.; Aragues, R.; Sagüés, C.; Rubio, C. Simultaneous deployment and tracking multi-robot strategies with connectivity maintenance. *Sensors* **2018**, *18*, 927. [CrossRef]
16. Lei, T.; Luo, C.; Jan, G.E.; Bi, Z. Deep learning-based complete coverage path planning with re-joint and obstacle fusion paradigm. *Front. Robot. AI* **2022**, *9*, 843816. [CrossRef]
17. Romeh, A.E.; Mirjalili, S. Multi-Robot Exploration of Unknown Space Using Combined Meta-Heuristic Salp Swarm Algorithm and Deterministic Coordinated Multi-Robot Exploration. *Sensors* **2023**, *23*, 2156. [CrossRef]
18. Lei, T.; Chintam, P.; Carruth, D.W.; Jan, G.E.; Luo, C. Human-Autonomy Teaming-Based Robot Informative Path Planning and Mapping Algorithms with Tree Search Mechanism. In Proceedings of the 2022 IEEE 3rd International Conference on Human–Machine Systems (ICHMS), Orlando, FL, USA, 17–19 November 2022; pp. 1–6.
19. Chu, Z.; Wang, F.; Lei, T.; Luo, C. Path planning based on deep reinforcement learning for autonomous underwater vehicles under ocean current disturbance. *IEEE Trans. Intell. Veh.* **2023**, *8*, 108–120. [CrossRef]

20. Lei, T.; Luo, C.; Sellers, T.; Rahimi, S. A bat-pigeon algorithm to crack detection-enabled autonomous vehicle navigation and mapping. *Intell. Syst. Applic.* **2021**, *12*, 200053. [CrossRef]
21. Sung, Y.; Budhiraja, A.K.; Williams, R.K.; Tokekar, P. Distributed simultaneous action and target assignment for multi-robot multi-target tracking. In Proceedings of the 2018 IEEE International conference on robotics and automation (ICRA), Brisbane, Australia, 21–25 May 2018; pp. 3724–3729.
22. Michael, N.; Zavlanos, M.M.; Kumar, V.; Pappas, G.J. Distributed multi-robot task assignment and formation control. In Proceedings of the 2008 IEEE International Conference on Robotics and Automation, Pasadena, CA, USA, 19–23 May 2008; pp. 128–133.
23. Lei, T.; Chintam, P.; Luo, C.; Rahimi, S. Multi-Robot Directed Coverage Path Planning in Row-based Environments. In Proceedings of the 2022 IEEE Fifth International Conference on Artificial Intelligence and Knowledge Engineering (AIKE), Laguna Hills, CA, USA, 19–21 September 2022; pp. 114–121.
24. Luo, L.; Chakraborty, N.; Sycara, K. Distributed algorithm design for multi-robot task assignment with deadlines for tasks. In Proceedings of the 2013 IEEE International Conference on Robotics and Automation, Karlsruhe, Germany, 6–10 May 2013; pp. 3007–3013.
25. Lei, T.; Luo, C.; Sellers, T.; Wang, Y.; Liu, L. Multitask allocation framework with spatial dislocation collision avoidance for multiple aerial robots. *IEEE Trans. Aerosp. Electron. Syst.* **2022**, *58*, 5129–5140. [CrossRef]
26. Purohit, A.; Zhang, P.; Sadler, B.M.; Carpin, S. Deployment of swarms of micro-aerial vehicles: From theory to practice. In Proceedings of the 2014 IEEE International Conference on Robotics and Automation (ICRA), Hong Kong, China, 31 May–7 June 2014; pp. 5408–5413.
27. Lei, T.; Luo, C.; Ball, J.E.; Rahimi, S. A graph-based ant-like approach to optimal path planning. In Proceedings of the 2020 IEEE congress on evolutionary computation (CEC), Glasgow, UK, 19–24 July 2020; pp. 1–6.
28. Sellers, T.; Lei, T.; Luo, C.; Jan, G.E.; Ma, J. A node selection algorithm to graph-based multi-waypoint optimization navigation and mapping. *Intell. Robot.* **2022**, *2*, 333–354. [CrossRef]
29. Rossi, C.; Aldama, L.; Barrientos, A. Simultaneous task subdivision and allocation using negotiations in multi-robot systems. *Int. J. Adv. Robot. Syst.* **2015**, *12*, 16. [CrossRef]
30. García, P.; Caamaño, P.; Duro, R.J.; Bellas, F. Scalable task assignment for heterogeneous multi-robot teams. *Int. J. Adv. Robot. Syst.* **2013**, *10*, 105. [CrossRef]
31. Lee, H.; Jeon, J.; Lee, B.H. An efficient cooperative deployment of robots for multiple tasks. In Proceedings of the 2015 IEEE International Conference on Robotics and Automation (ICRA), Seattle, WA, USA, 26–30 May 2015; pp. 5419–5425.
32. Luo, C.; Yang, S.X.; Li, X.; Meng, M.Q.-H. Neural-dynamics-driven complete area coverage navigation through cooperation of multiple mobile robots. *IEEE Trans. Ind. Electron.* **2016**, *64*, 750–760. [CrossRef]
33. Luo, C.; Yang, S.X. A real-time cooperative sweeping strategy for multiple cleaning robots. In Proceedings of the IEEE Internatinal Symposium on Intelligent Control, Vancouver, BC, Canada, 30 October 2002; pp. 660–665.
34. Luo, L.; Chakraborty, N.; Sycara, K. Provably-good distributed algorithm for constrained multi-robot task assignment for grouped tasks. *IEEE Trans. Robot.* **2014**, *31*, 19–30. [CrossRef]
35. Bassil, J.; Makhoul, A.; Piranda, B.; Bourgeois, J. Distributed Size-Constrained Clustering Algorithm for Modular Robot-Based Programmable Matter. *ACM Trans. Auton. Adapt. Syst.* **2023**, *18*, 1. [CrossRef]
36. Li, S.; Xu, X.; Zuo, L. Task assignment of multi-robot systems based on improved genetic algorithms. In Proceedings of the 2015 IEEE International Conference on Mechatronics and Automation (ICMA), Beijing, China, 2–5 August 2015; pp. 1430–1435.
37. Yang, Q.; Luo, Z.; Song, W.; Parasuraman, R. Self-reactive planning of multi-robots with dynamic task assignments. In Proceedings of the 2019 International Symposium on Multi-Robot and Multi-Agent Systems (MRS), New Brunswick, NJ, USA, 22–23 August 2019; pp. 89–91.
38. Bai, X.; Fielbaum, A.; Kronmüller, M.; Knoedler, L.; Alonso-Mora, J. Group-based distributed auction algorithms for multi-robot task assignment. *IEEE Trans. Autom. Sci. Eng.* **2022**, *20*, 1292–1303. [CrossRef]
39. Luo, C.; Yang, S.X. A bioinspired neural network for real-time concurrent map building and complete coverage robot navigation in unknown environments. *IEEE Trans. Neural Netw.* **2008**, *19*, 1279–1298. [CrossRef]
40. Motes, J.; Sandström, R.; Lee, H.; Thomas, S.; Amato, N.M. Multi-robot task and motion planning with subtask dependencies. *IEEE Robot. Autom. Lett.* **2020**, *5*, 3338–3345. [CrossRef]
41. Wurman, P.R.; D'Andrea, R.; Mountz, M. Coordinating hundreds of cooperative, autonomous vehicles in warehouses. *AI Mag.* **2008**, *29*, 9.
42. Dutta, A.; Ufimtsev, V.; Asaithambi, A.; Czarnecki, E. Coalition formation for multi-robot task allocation via correlation clustering. *Cybern. Syst.* **2019**, *50*, 711–728. [CrossRef]
43. Martin, J.G.; Muros, F.J.; Maestre, J.M.; Camacho, E.F. Multi-robot task allocation clustering based on game theory. *Robot. Auton. Syst.* **2023**, *161*, 104314. [CrossRef]
44. Harman, H.; Sklar, E.I. A practical application of market-based mechanisms for allocating harvesting tasks. In Proceedings of the 19th International Conference on Practical Applications of Agents and Multi-Agent Systems (PAAMS 2021), Salamanca, Spain, 6–8 October 2021; pp. 114–126.
45. Fu, B.; Smith, W.; Rizzo, D.M.; Castanier, M.; Ghaffari, M.; Barton, K. Robust task scheduling for heterogeneous robot teams under capability uncertainty. *IEEE Trans. Robot.* **2022**, *39*, 1087–1105. [CrossRef]

46. Sarkar, E. Artificial Neural Networks: Kohonen Self-Organising Maps. Ph.D. Thesis, University of Liverpool, Liverpool, UK, 2018.
47. Wang, A.; Gounaris, C.E. On tackling reverse convex constraints for non-overlapping of unequal circles. *J. Glob. Optim.* **2021**, *80*, 357–385. [CrossRef]
48. Castillo, I.; Sim, T. A spring-embedding approach for the facility layout problem. *J. Oper. Res. Soc.* **2004**, *55*, 73–81. [CrossRef]
49. Anjos, M.F.; Vannelli, A. A new mathematical-programming framework for facility-layout design. *INFORMS J. Comput.* **2006**, *18*, 111–118. [CrossRef]
50. Yuan, X.; Yang, S.X. Multirobot-based nanoassembly planning with automated path generation. *IEEE/ASME Trans. Mechatronics* **2007**, *12*, 352–356. [CrossRef]
51. Kuhn, H.W. The Hungarian method for the assignment problem. *Nav. Res. Logist. Q.* **1955**, *2*, 83–97. [CrossRef]

sensors

MDPI

Article

Robot Navigation in Complex Workspaces Employing Harmonic Maps and Adaptive Artificial Potential Fields

Panagiotis Vlantis [1], Charalampos P. Bechlioulis [1,*] and Kostas J. Kyriakopoulos [2]

[1] Department of Electrical and Computer Engineering, University of Patras, 26504 Patras, Greece; panvlantis@outlook.com

[2] School of Mechanical Engineering, National Technical University of Athens, 15780 Athens, Greece; kkyria@mail.ntua.gr

* Correspondence: chmpechl@upatras.gr

Abstract: In this work, we address the single robot navigation problem within a planar and arbitrarily connected workspace. In particular, we present an algorithm that transforms any static, compact, planar workspace of arbitrary connectedness and shape to a disk, where the navigation problem can be easily solved. Our solution benefits from the fact that it only requires a fine representation of the workspace boundary (i.e., a set of points), which is easily obtained in practice via SLAM. The proposed transformation, combined with a workspace decomposition strategy that reduces the computational complexity, has been exhaustively tested and has shown excellent performance in complex workspaces. A motion control scheme is also provided for the class of non-holonomic robots with unicycle kinematics, which are commonly used in most industrial applications. Moreover, the tuning of the underlying control parameters is rather straightforward as it affects only the shape of the resulted trajectories and not the critical specifications of collision avoidance and convergence to the goal position. Finally, we validate the efficacy of the proposed navigation strategy via extensive simulations and experimental studies.

Keywords: motion and path planning; collision avoidance; autonomous vehicle navigation; artificial potential fields

check for updates

Citation: Vlantis, P.; Bechlioulis, C.P.; Kyriakopoulos, K.J. Robot Navigation in Complex Workspaces Employing Harmonic Maps and Adaptive Artificial Potential Fields. *Sensors* **2023**, *23*, 4464. https://doi.org/10.3390/s23094464

Academic Editors: Stephen Monk and David Cheneler

Received: 17 April 2023
Revised: 29 April 2023
Accepted: 2 May 2023
Published: 3 May 2023

1. Introduction

The navigation of autonomous robots in cluttered environments is a widely studied topic in the field of robotics. Popular methodologies that have been employed in the related literature to address it include, but are not limited to, *configuration space decomposition approaches* [1,2]; *probabilistic sampling methods* such as rapidly exploring random trees [3,4], probabilistic roadmaps [5,6] and manifold samples [7,8]; and *optimal control strategies* such as receding horizon control [9,10] and path homotopy invariants [11,12]. Apart from the aforementioned discrete methods regarding the workspace and/or the decision domain, Artificial Potential Fields (APFs) that were originally introduced in [13] generally provide a simpler means of encoding collision avoidance specifications, with their negated gradient functioning as a reference motion direction that drives the robot towards the desired goal configuration. As shown in [14], despite their intuitive nature, this class of controllers suffers unavoidably from the presence of unwanted equilibria induced by the workspace's topology, whose region of attraction may not be trivial. In their seminal work [15], Rimon and Koditschek presented a family of APFs called Navigation Functions (NFs) for point and sphere worlds, as well as a constructive transformation for mapping workspaces cluttered by sequences of star-shaped obstacles into such worlds. However, certain design parameters require tedious tuning to eliminate unwanted local minima and render the transformation a diffeomorphism. In practice, this solution suffers by the fact that the allowable values of the design parameters may cause both the potential and the corresponding transformation to vary too abruptly close to the obstacles (the issue of "disappearing valleys" [15]), thus pushing the trajectories of the robot very close to them. Density functions for remedying such drawbacks or adjustable NFs for relaxing some generally conservative requirements are presented in [16,17]. Additionally, attempts to extend the NF framework directly to non-sphere worlds can be found in [18,19]. Finally, a novel approach based on power diagrams which can be used for designing tune-free vector fields for navigation within convex workspaces is also presented in [20].

Artificial Harmonic Potential Fields (AHPFs) constitute an interesting subclass of APFs, since they are free of unwanted local minima by construction. However, no simple method exists for constructing safe (with respect to obstacle avoidance), harmonic potentials even for simple workspaces. AHPFs suitable for navigation in realistic environments were originally utilized in [21], where computationally expensive numerical techniques were employed to solve the associated Dirichlet and Neumann problems. Several extensions of the aforementioned methodology followed [22,23], addressing issues such as numerical precision and computation, dynamic environments, etc. The panel method was also employed in [24–26] to build harmonic potentials to coordinate the motion of single and multiple robots in polygonal environments. In [27,28], well-known closed-form solutions of the uncompressed fluid flow around simple geometries was used in order to safely drive a robot among moving obstacles. Harmonic potential fields have also been used in [29,30] to address the Simultaneous Localization and Mapping problem (SLAM) by coordinating the robot motion in unknown environments. Moreover, a methodology based on the evaluation of the harmonic potential field's streamlines was used in [31,32] for mapping a multiply connected workspace to a disk, collapsing inner obstacles to line segments or arcs. In a recent work [33], the problem of designing closed form harmonic potentials in sphere worlds was addressed by the introduction of a diffeomorphism [34], which allows mapping such workspaces to the euclidean plane with some of its points removed. Finally, extensions of this work addressing topologically complex three-dimensional workspaces or multi-robot scenarios by introducing appropriate constructive workspace transformations can be found in [35,36], respectively.

1.1 Contributions

We address the navigation problem for a robot operating within a static, compact, planar workspace of arbitrary connectedness and shape by designing a control law that safely drives the robot to a given goal position from almost any initial feasible configuration. The goal of this work is twofold. **(A) To cope with the topological complexity of the workspace,** we employed numerical techniques in order to build a transformation that maps the workspace onto a punctured disk and delved into the respective construction in detail. We remark that, although the transformation constructed using this method is an approximation of a harmonic map ideal for navigation, our solution benefits from the fact that it only needs a sufficiently fine polygonal workspace description that can be easily acquired in practice (e.g., through SLAM), contrary to [15,34,36] that require an explicit representation of the workspace boundaries (i.e., as the level sets of sufficiently smooth functions). Moreover, unlike the solutions proposed in [15,36], our approach does not require the decomposition of the workspace obstacles into sequences of simpler overlapping shapes and computes the desired transformation in one step. **(B) To steer the robot to its desired configuration,** we employed a control law based on closed-form AHPFs coupled with adaptive laws for their parameters to eliminate the necessity of explicitly defined local activation neighborhoods around the workspace boundaries for ensuring collision avoidance. Our approach is reactive (closed loop) since it selects the velocity of the robot based on the positions of the robot, the desired goal and the workspace boundary. As such, it is more robust against position measurement errors than other open loop approaches such as configuration space decomposition approaches [2] or probabilistic sampling methods such as rapidly exploring random trees [4], probabilistic roadmaps [6] and manifold samples [8], where an open loop path is initially extracted and executed by a trajectory tracking controller. In this way, even small position errors risk the safe execution of the calculated plan. We remark that our overall control scheme only requires solving a computationally expensive problem once for a given static workspace, independent of the robot's initial and goal configurations, in contrast to the solutions presented in [21,22]. Finally, we adapt our methodology to the class of differential drive robots, which are commonly encountered in real-world applications and propose an algorithm that decomposes the overall workspace into small neighbouring subsets to render the problem of addressing large workspaces tractable. An overview of the proposed methodology's pros and cons compared to alternative transformations and potential fields can be seen in Tables 1 and 2, respectively.

Preliminary results were included in our conference paper [37]. We have to stress though that the algorithmic calculation of the harmonic map is given in the present work, along with a rigorous formulation of the panel method. A modification of the adaptive laws for the parameters of the underlying potential field is also introduced to simplify the tuning process by eliminating the necessity of heuristically defined local activation neighborhoods around the workspace boundaries for ensuring collision avoidance. Moreover, an extension for tackling the navigation problem under unicycle kinematics is also provided. Finally, new comparative simulation results are provided to highlight the strong points of the proposed method with respect to other related works, accompanied by an experiment employing an actual robot navigating within a complex office workspace.

Table 1. Comparison between the Harmonic Transformation (HM) proposed in this work and the (i) Star-to-Sphere Transformation (SST) [15], (ii) Multi-Agent Navigation Transformation (MANT) [36] and (iii) the Navigation Transformation (NT) [34]. Although HMs require global knowledge of the workspace's geometry to be constructed, HMs are infinitely differentiable and require the domain to be represented by closed polygonal curves (which can be easily obtained using SLAM methodologies), unlike the alternatives that require the domain boundaries to be represented as sets of sufficiently differentiable implicit equations.

	Geometry Representation	**Global**	**Analytic**
HM	Points on the boundary	Yes	Yes
SST	Trees of Stars	Yes	Yes
NT	C^2-manifolds	No	No
MANT	Trees of C^2-manifolds	No	No

Table 2. Comparison of Adaptive Harmonic Potential Fields (proposed herein) with common alternatives, specifically Rimon–Koditchek Navigation Functions (RKNF) [15], Harmonic Navigation Functions [33] and approximate Harmonic Potential Fields obtained using numerical techniques [21]. Unlike RKNFs that require tuning for ensuring convergence to the goal from almost all initial configurations and HNFs that require tuning for guaranteeing collision avoidance with the workspace boundaries, the proposed control law enjoys both properties by design.

	Convergence	**Collision Avoidance**	**Computational Cost**
AHPF	By design	By design	Cheap
HPF	By design	By design	Expensive
RKNF	Requires tuning	By design	Cheap
HNF	By design	Requires tuning	Cheap

1.2. Preliminaries

We use $\mathcal{D}_r(x)$ to denote an open disk with radius $r > 0$ centered at $x \in \mathbb{R}^2$. Additionally, $\mathcal{D}$ and $\partial\mathcal{D}$ denote the closed disk and circle with unit radii centered at the origin of $\mathbb{R}^2$, respectively. Furthermore, let $\mathfrak{I}_N \triangleq \{1, 2, \ldots, N\}$ and $\mathfrak{I}_N^\star \triangleq \{0\} \bigcup \mathfrak{I}_N$. Given sets $A, B \subseteq \mathbb{R}^n$, we use $\mathrm{cl}(A)$, ∂A, $\mathrm{int}(A)$ and $\overline{A}$ to denote the closure, boundary, interior and complement of A with respect to $\mathbb{R}^n$, respectively, and $A \setminus B$ to denote the complement of B with respect to A. Furthermore, we use $\mathbf{0}_N$ and $\mathbf{1}_N$ to denote the all-zeros and all-ones column vectors of length N, respectively, and $\mathbf{0}_{N\times M}$ to denote the $N \times M$ zero matrix. We also define $\mathbf{1}^k_{N\times M}$, $k \in \mathfrak{I}_M$ as the $N \times M$ matrix whose k-th column is equal to $\mathbf{1}_N$ and every other column is equal to $\mathbf{0}_N$. Given a vector function $f(x)$, we use $\nabla_x f$ to denote its Jacobian matrix. Furthermore, given an arc C, we use $|C|$ to denote its length. We will also say that a set A is attractive (repulsive) under a potential function ψ when there exists a point $p_0 \notin \mathrm{cl}(A)$ such that if we initialize at p_0 and move along the negated gradient of ψ, we will converge (not converge) to ∂A. Finally, a potential function ψ is called harmonic if it satisfies the Laplace equation, i.e., $\nabla^2\psi = 0$, where ∇^2 denotes the Laplacian operator. An important property of harmonic functions is the principle of superposition, which follows from the linearity of the Laplace equation. Moreover, the extrema of a non-constant harmonic function occur on the boundary of the domain of definition, thus excluding any local minima/maxima within it (a desirable property for motion planning).

2. Problem Formulation

We consider a robot operating within a compact workspace $\mathcal{W} \subset \mathbb{R}^2$ bounded by a single outer and a finite set of inner disjoint Jordan curves (a Jordan curve is a non-self-intersecting continuous planar closed curve), which correspond to the boundaries of static obstacles. It is assumed that $\mathcal{W}$ can be written as:

$$\mathcal{W} = \overline{\mathcal{W}_0} \setminus \bigcup_{i\in\mathfrak{I}_N} \mathcal{W}_i$$

where $\mathcal{W}_i$, $i \in \mathfrak{I}^\star_N$ denote regions of $\mathbb{R}^2$ that the robot cannot occupy (see left subplot in Figure 1). Particularly, the complement of $\mathcal{W}_0$ is considered to be a bounded, simply connected region that may also include a strict subset of its own boundary (this corresponds to cases when we wish to place the robot's goal configuration on some part of the workspace outer boundary which is not physically occupied by an actual obstacle, e.g., the door of a compartment (refer to Section 5.2 for more details)) and $\mathcal{W}_1, \mathcal{W}_2, \ldots, \mathcal{W}_N$ are assumed to be closed, simply connected compact sets that are contained in $\overline{\mathcal{W}_0}$ and are pairwise disjoint. Let $p = [x, y]^T \in \mathbb{R}^2$ denote the robot's position and assume that the robot's motion is described by the single integrator model:

$$\dot{p} = u \tag{1}$$

where $u \in \mathbb{R}^2$ is the corresponding control input vector.

Problem 1. *Our goal is to design a control law to successfully drive a robot with kinematics* (1) *towards a given goal configuration $p_d \in \mathcal{W}$ from almost any feasible initial configuration $p_{\text{init}} \in \mathcal{W}$, while ensuring collision avoidance, i.e., $p(t) \in \mathcal{W}$ for all $t \geq 0$.*

Remark 1. *The results presented in this work can be readily employed for the navigation of disk robots with radius $R > 0$ by appropriately augmenting the workspace boundaries with the robot's size.*

3. Harmonic Maps for Planar Navigation

In this section, we present a methodology that maps the robot's workspace onto a punctured unit disk, over which the robot's control law is designed. Particularly, our goal is to construct a transformation, $T : \mathrm{cl}(\mathcal{W}) \mapsto \mathcal{D}$, from the closure of the robot's configuration space $\mathrm{cl}(\mathcal{W})$ onto the unit disk $\mathcal{D}$ with the following properties:

1. $T(\cdot)$ maps the outer boundary $\partial\mathcal{W}_0$ to the unit circle $\partial\mathcal{D}$;
2. $T(\cdot)$ maps the boundary $\partial\mathcal{W}_i, i \in \mathfrak{I}_N$ of each obstacle to a distinct point $q_i = [\mathrm{u}_i, \mathrm{v}_i]^T \in \mathrm{int}(\mathcal{D})$;
3. $T(\cdot)$ is a diffeomorphism for all $p \in \mathrm{int}(\mathcal{W})$.

To that end, we compute a transformation $\tilde{T}(p) = [\tilde{u}(p), \tilde{v}(p)]^T$, with $\tilde{u}(p)$ and $\tilde{v}(p)$ being harmonic functions with respect to p, by approximating an ideal harmonic map T that meets the aforementioned properties and the existence of which was proven in Theorem 2 of [38], accompanied by sufficient conditions that render it a diffeomorphism as outlined in the corresponding proof.

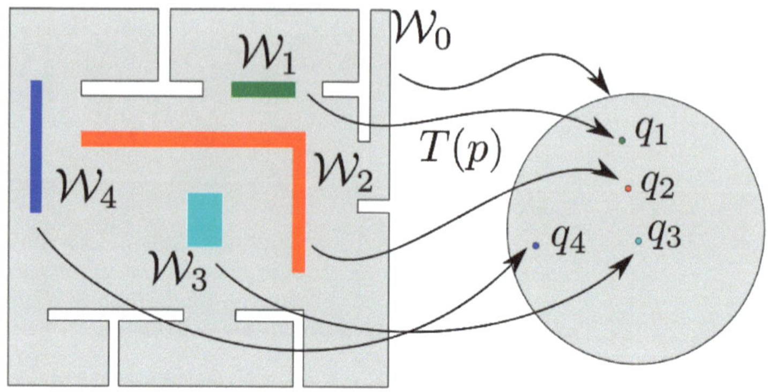

Figure 1. Transformation of a workspace onto a punctured disk.

Particularly, this theorem implies that given an orientation-preserving, weak homeomorphism $T_\partial : \partial\mathcal{W}_0 \to \partial\mathcal{D}$ (such a transformation can be easily obtained for any given planar Jordan curve C by (1) arbitrarily selecting a point p_o on C, (2) defining $\ell(p), \forall p \in C$ as the length of the arc $\widehat{p_o p}$, assuming one travels from p_o to p on C while having the curve's interior to its left and (3) choosing $T_\partial(p) = [\cos(2\pi\ell(p)/L), \sin(2\pi\ell(p)/L)]^T$, where $L = |C|$) from the workspace outer boundary $\partial\mathcal{W}_0$ to the boundary of the unit disk, the harmonic map T that satisfies the conditions:

$$T(p) = T_\partial(p) \triangleq [\bar{u}(p), \bar{v}(p)]^T, \quad \forall p \in \partial\mathcal{W}_0, \tag{2}$$

$$\int_{\partial\mathcal{W}_i} \frac{\partial T}{\partial n_p} ds = 0, \quad \forall i \in \mathfrak{I}_N \tag{3}$$

with n_p denoting the unit vector that is normal to the boundary at the point $p \in \partial\mathcal{W}_i, i \in \mathfrak{I}_N$, is a diffeomorphism that maps $\mathrm{cl}(\mathcal{W})$ to the target set and collapses each inner obstacle $\mathcal{W}_i$ onto a distinct point q_i within its interior (see Figure 1). Note that the coordinates of q_i, i.e., the images of the internal obstacles, are not explicitly specified but are computed as part of the solution, such that the aforementioned constraints are satisfied.

Given that closed-form solutions to the aforementioned problem are generally not available for non-trivial domains, in this work, we employed numerical techniques and particularly the Panel Method [24,39,40] (similar formulations can be obtained by employing other numerical techniques such as the Boundary Element Method (BEM), the Finite Element Method (FEM) or the Finite Differences Method (FDM)) in order to construct a harmonic map $\tilde{T}$ that sufficiently approximates T. As such, by subdividing separately the workspace's outer and inner boundaries into $\tilde{M}_0, \tilde{M}_1, \ldots, \tilde{M}_N$ number of elements (see Figure 2), we define the components of $\tilde{T}(p) = [\tilde{u}(p), \tilde{v}(p)]^T$ as follows:

$$\tilde{u}(p) = \sum_{i=0}^{N} \sum_{j=1}^{\tilde{M}_i} \sum_{l=1}^{\tilde{M}_C} \tilde{C}^x_{ijl} \tilde{H}_{ijl}(p)$$

$$\tilde{v}(p) = \sum_{i=0}^{N} \sum_{j=1}^{\tilde{M}_i} \sum_{l=1}^{\tilde{M}_C} \tilde{C}^y_{ijl} \tilde{H}_{ijl}(p)$$

$$\tilde{H}_{ijl}(p) = \int_{\tilde{E}_{ij}} \tilde{G}_{ijl}(s) \ln(\|p - \tilde{p}_{i,j}(s)\|) ds \tag{4}$$

where $\tilde{M}_C$ is the number of control parameters per element, $\tilde{E}_{ij}$ denotes the j-th element of the i-th boundary's approximation, $\tilde{p}_{i,j}(s) : [0, |\tilde{E}_{ij}|] \mapsto \tilde{E}_{ij}$ is a bijective parameterization of $\tilde{E}_{ij}$, $\tilde{G}_{ijl} : [0, |\tilde{E}_{ij}|] \mapsto \mathbb{R}$ is the shape function corresponding to the l-th control parameter of $\tilde{E}_{ij}$ and $\tilde{C}^x_{ijl}, \tilde{C}^y_{ijl} \in \mathbb{R}$ are control parameters that need to be appropriately selected so that $\tilde{T}$ satisfies properties 1–3 for all $l \in \mathfrak{I}_{\tilde{M}_C}$, $j \in \mathfrak{I}_{\tilde{M}_i}$ and $i \in \mathfrak{I}^\star_N$. It is worth noting that for common choices of $\tilde{G}_{ijl}$ (e.g., constant or linear shape functions) and simple types of $\tilde{E}_{ij}$ (e.g., line segments), the integral in (4) can be easily evaluated to obtain a closed-form expression for $\tilde{H}_{ijl}$. As an illustration, for a line segment element $\tilde{E}_{ij}$ with two control parameters (i.e., $\tilde{M}_C = 2$), a typical choice for linear shape functions (see Figure 2) is $\tilde{G}_{ij1}(s) = s/|\tilde{E}_{ij}|$, $\tilde{G}_{ij2}(s) = \left(1 - s/|\tilde{E}_{ij}|\right)$ and $\tilde{p}_{i,j}(s) = \tilde{p}_{i,j,A} + \left(\tilde{p}_{i,j,B} - \tilde{p}_{i,j,A}\right)s/|\tilde{E}_{ij}|$ for the corresponding parameterization, where $\tilde{p}_{i,j,A}, \tilde{p}_{i,j,B}$ are the element's end-points. To obtain the unknown control parameters as well as the images of the workspace's inner obstacles, one needs to solve two independent linear systems of equations:

$$\tilde{A}\tilde{X} = \tilde{B}_x, \quad \tilde{A}\tilde{Y} = \tilde{B}_y \tag{5}$$

for the unknown vectors:

$$\tilde{X} = \left[\tilde{C}^x_{0,1,1}, \cdots, \tilde{C}^x_{1,1,1}, \cdots, \tilde{C}^x_{N,\tilde{M}_N,\tilde{M}_C}, \mathrm{u}_1, \cdots, \mathrm{u}_N\right]^T$$
$$\tilde{Y} = \left[\tilde{C}^y_{0,1,1}, \cdots, \tilde{C}^y_{1,1,1}, \cdots, \tilde{C}^y_{N,\tilde{M}_N,\tilde{M}_C}, \mathrm{v}_1, \cdots, \mathrm{v}_N\right]^T.$$

The matrix $\tilde{A}$ and the right hand side vectors $\tilde{B}_x$ and $\tilde{B}_y$ are constructed by selecting a set of $\sum_{i \in \mathfrak{I}^\star_N} \tilde{m}_i$ arbitrary points $\tilde{p}^\star_{i,j}$ (a typical strategy is to select the points $\tilde{p}^\star_{i,j}$ uniformly on the outer and inner boundaries of the given domain) such that a) $\tilde{p}^\star_{i,j} \in \partial\mathcal{W}_i$ for all $j \in \mathfrak{I}_{\tilde{m}_i}$ and $i \in \mathfrak{I}^\star_N$ and b) $\sum_{i \in \mathfrak{I}^\star_N} \tilde{m}_i = \tilde{M}_C \sum_{i \in \mathfrak{I}^\star_N} \tilde{M}_i$, and evaluating (2) and (3) at those points as follows:

$$\tilde{A} = \begin{bmatrix} \tilde{A}_0, & \mathbf{0}_{\tilde{m}_0 \times N} \\ \tilde{A}_1, & -\mathbf{1}^1_{\tilde{m}_1 \times N} \\ \vdots & \vdots \\ \tilde{A}_N, & -\mathbf{1}^N_{\tilde{m}_N \times N} \\ \tilde{A}_\dagger, & \mathbf{0}_{N \times N} \end{bmatrix}, \; \tilde{B}_x = \begin{bmatrix} \tilde{B}_{x,0} \\ \mathbf{0}_{\tilde{m}_1} \\ \vdots \\ \mathbf{0}_{\tilde{m}_N} \\ \mathbf{0}_N \end{bmatrix}, \; \tilde{B}_y = \begin{bmatrix} \tilde{B}_{y,0} \\ \mathbf{0}_{\tilde{m}_1} \\ \vdots \\ \mathbf{0}_{\tilde{m}_N} \\ \mathbf{0}_N \end{bmatrix}$$

$$\tilde{A}_k = \begin{bmatrix} \tilde{H}_{0,1,1}(\tilde{p}^\star_{k,1}) & \cdots & \tilde{H}_{N,\tilde{M}_N,\tilde{M}_C}(\tilde{p}^\star_{k,1}) \\ \tilde{H}_{0,1,1}(\tilde{p}^\star_{k,2}) & \cdots & \tilde{H}_{N,\tilde{M}_N,\tilde{M}_C}(\tilde{p}^\star_{k,2}) \\ \vdots & \vdots & \vdots \\ \tilde{H}_{0,1,1}(\tilde{p}^\star_{k,\tilde{m}_k}) & \cdots & \tilde{H}_{N,\tilde{M}_N,\tilde{M}_C}(\tilde{p}^\star_{k,\tilde{m}_k}) \end{bmatrix}, \; \forall k \in \mathfrak{I}^\star_N$$

$$\tilde{A}_\dagger = \begin{bmatrix} \sum_{k=1}^{\tilde{m}_1} \frac{\partial \tilde{H}_{0,1,1}}{\partial n_{0,1}}(\tilde{p}^\star_{1,k}) & \cdots & \sum_{k=1}^{\tilde{m}_1} \frac{\partial \tilde{H}_{N,\tilde{M}_N,\tilde{M}_C}}{\partial n_{N,\tilde{M}_N}}(\tilde{p}^\star_{1,k}) \\ \sum_{k=1}^{\tilde{m}_2} \frac{\partial \tilde{H}_{0,1,1}}{\partial n_{0,1}}(\tilde{p}^\star_{2,k}) & \cdots & \sum_{k=1}^{\tilde{m}_2} \frac{\partial \tilde{H}_{N,\tilde{M}_N,\tilde{M}_C}}{\partial n_{N,\tilde{M}_N}}(\tilde{p}^\star_{2,k}) \\ \vdots & \vdots & \vdots \\ \sum_{k=1}^{\tilde{m}_N} \frac{\partial \tilde{H}_{0,1,1}}{\partial n_{0,1}}(\tilde{p}^\star_{N,k}) & \cdots & \sum_{k=1}^{\tilde{m}_N} \frac{\partial \tilde{H}_{N,\tilde{M}_N,\tilde{M}_C}}{\partial n_{N,\tilde{M}_N}}(\tilde{p}^\star_{N,k}) \end{bmatrix}$$

$$\tilde{B}_{x,0} = \begin{bmatrix} \bar{u}(\tilde{p}^{\star}_{0,1}) \\ \bar{u}(\tilde{p}^{\star}_{0,2}) \\ \vdots \\ \bar{u}(\tilde{p}^{\star}_{0,\tilde{m}_0}) \end{bmatrix} \quad \tilde{B}_{y,0} = \begin{bmatrix} \bar{v}(\tilde{p}^{\star}_{0,1}) \\ \bar{v}(\tilde{p}^{\star}_{0,2}) \\ \vdots \\ \bar{v}(\tilde{p}^{\star}_{0,\tilde{m}_0}) \end{bmatrix}.$$

Notice that by discretizing the workspace boundaries into a large number of sufficiently small elements, the overall approximation error between the solution $\tilde{T}$ of the aforementioned linear problem and the exact transformation T can be rendered arbitrarily small (see [39,40]). However, the complexity of constructing the mapping is of order $O(\bar{M}^3)$, where $\bar{M}$ denotes the number of total elements of the mapping (i.e., the complexity of the solution of the dense system of linear Equation (5)). Nevertheless, the construction of the transformation, which is the main computational bottleneck, is performed only once at the beginning. Additionally, apart from the straightforward user-defined homeomorphism T_∂ on the workspace boundary, no tedious trial and error tuning is needed to extract the diffeomorphic transformation $\tilde{T}$, in contrast to other related works such as the Star-to-Sphere Transformation (SST) [15], the Multi-Agent Navigation Transformation (MANT) [36] and the Navigation Transformation (NT) [34].

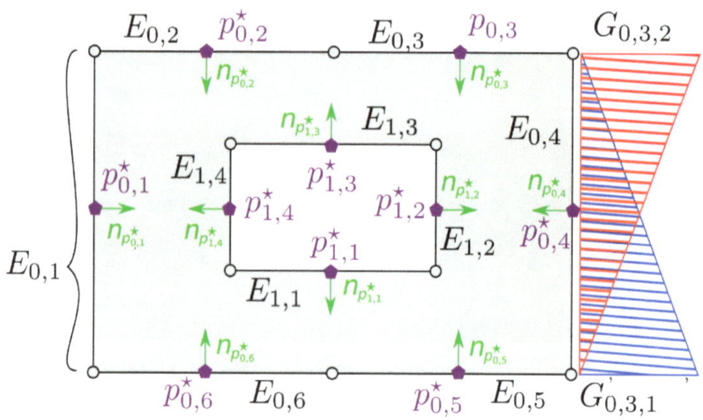

Figure 2. Discretization of a given domain's boundary using line segment elements. By convention, the outer boundary is considered to be clockwise oriented, whereas inner boundaries are counterclockwise oriented. The normal direction of each element is depicted using green colored vectors. Furthermore, the values of the two linear shape functions $\tilde{G}_{0,3,1}$ and $\tilde{G}_{0,3,2}$ are plotted along the associated element $\tilde{E}_{0,3}$.

4. Control Design

To address Problem 1, we equip the robot with the aforementioned transformation $q = T(p)$ from the closure of its configuration space $\mathcal{W}$ onto the unit disk $\mathcal{D}$ and an artificial potential $\psi(q,k)$ augmented with an adaptive control law $\dot{k} = f_k(q,k)$ for its parameters $k = [k_d, k_1, k_2, \ldots, k_N]^T$. The robot velocity control law is calculated as follows:

$$u = -\mathcal{K}_u s(q,k) J^{-1}(p) \nabla_q \psi(q,k) \tag{6}$$

where $J(p)$ denotes the Jacobian matrix of $T(p)$, $s(q,k) \geq 0$ is a continuously differentiable gain function given by:

$$s(q,k) = \gamma \sigma_p \left(\frac{1 - \|q\|}{\epsilon_p} \right) + (1-\gamma)\sigma_v \left(\frac{(\nabla_q \psi)^T q}{\epsilon_v + \|\nabla_q \psi\| \|q\|} \right) \tag{7}$$

with

$$\sigma_p(x) = \begin{cases} x^2(3-2x), & \text{if } x \le 1 \\ 1, & \text{if } x > 1 \end{cases},$$

$$\sigma_v(x) = \begin{cases} x^2, & \text{if } x \ge 0 \\ 0, & \text{if } x < 0 \end{cases} \tag{8}$$

and $\mathcal{K}_u$, γ, ϵ_p and ϵ_v are scalar constants such that $\mathcal{K}_u, \epsilon_v > 0$ and $\gamma, \epsilon_p \in (0,1)$. More specifically, $s(q,k)$ consists of two individual terms, with the first vanishing as the robot approaches the workspace's outer boundary (and its distance from the unit circle is less than ϵ_p) and the second vanishing when the robot's velocity points away from the disk's center. The scalar parameter γ can be used for adjusting the contribution of each respective term of $s(q,k)$. Finally, ψ is a harmonic artificial potential field defined on the image $T(\mathcal{W})$ of the workspace $\mathcal{W}$ and whose negated gradient $-\nabla_q \psi(q,k)$ defines the direction of the robot's motion in the real workspace $\mathcal{W}$ via the inverse Jacobian $J^{-1}(p)$. By design, the resultant vector field precludes collisions between the robot and the workspace's inner obstacles and renders the goal configuration almost globally attractive except for a set of measure zero initial configurations. However, since $\mathcal{W}_0$ may not be repulsive under ψ for an arbitrary, fixed selection of k, we also introduce the adaptive law $f_k(q,k)$ which, along with $s(q,k)$, guarantees forward invariance of the workspace without compromising the convergence and stability properties of the overall system. The following subsections elaborate on each component of the proposed control law individually.

4.1. Artificial Harmonic Potential Fields

We construct an artificial harmonic potential field on the disk space $\mathcal{D}$ employing point sources placed at the desired configuration $q_d = T(p_d)$ as well as at the points $q_i = T(\partial \mathcal{W}_i)$, $\forall i \in \mathfrak{I}_N$ that correspond to the inner obstacles, as follows:

$$\phi(q,k) = k_d \ln\left(\frac{\|q-q_d\|}{2}\right) - \sum_{i=1}^{N} k_i \ln\left(\frac{\|q-q_i\|}{2}\right) \tag{9}$$

where $k_d > 0$ and $k_i \ge 0$ denote harmonic source strengths which vary according to adaptive laws that are presented later. An interesting property of the above potential field, which stems from the maximum principle for harmonic functions, is that, for fixed k, the only minima of ϕ are located at q_d and, possibly, at infinity. As a direct consequence of this property, the Hessian $\nabla_q^2 \phi$ computed at a non-degenerate critical point of ϕ in our domain's interior has one positive and one negative eigenvalue with the same magnitude, e.g., λ and $-\lambda$ with $\lambda > 0$.

Next, we define a reference potential ψ based on ϕ, which is given by:

$$\psi(q,k) = \frac{1 + \tanh(\phi(q,k)/w_\phi)}{2} \tag{10}$$

where w_ϕ is a positive scaling constant. Note that ψ maps the extended real line to the closed interval $[0,1]$. As $\tanh(\phi/w_\phi)$ is a strictly increasing function, the only critical points of ψ are the ones inherited from ϕ with their indices preserved. Furthermore, the gradient of ψ with respect to q, given by

$$\nabla_q \psi = \frac{1 - \left(\tanh(\phi/w_\phi)\right)^2}{2w_\phi} \nabla_q \phi, \tag{11}$$

is well defined and bounded for all $q \in \mathcal{D}$.

If the workspace was radially unbounded, selecting k fixed with $k_d > \sum_{i=1}^{N} k_i$ would render the potential field (10) sufficient for navigation. The author of [33] addresses bounded workspaces that are diffeomorphic to sphere worlds by simply mapping the outer bounding circle to infinity. In this work, we would like to be able to place q_d on regions of $\partial\mathcal{D}$ that are not physically occupied by obstacles (such as passages to other compartments, see, for example, Section 5.2); thus, we cannot follow the same procedure since that would render the effect of the sole attractor on the robot null. Instead, we design appropriate adaptive laws for the parameters k of ϕ to render the outer boundary repulsive and establish the forward completeness of the proposed scheme at all times.

Before proceeding with the definition of the adaptive law, we first state two propositions that will be used in the subsequent analysis, the proofs of which can be found in the Appendix A.

Proposition 1. *Let $k_d > 0$ and $q' \in \partial\mathcal{D} \setminus \{q_d\}$. There exists $k' > 0$ such that if $k_i < k'$, $\forall i \in \mathfrak{I}_N$, then q' is repulsive under ψ.*

Proposition 2. *If k_i are non-negative and bounded, there exists $k'_d > 0$ such that ψ is Morse for all $k_d \geq k'_d$.*

4.2. Adaptive Laws

We now present the adaptive law $\dot{k} = f_k(q, k)$ that updates the parameters of the potential field ψ. Its primary goal is to render (a) the workspace outer boundary repulsive and (b) any critical point of ϕ in the vicinity of the robot non-degenerate, a property that will be used later in the analysis. In particular, we consider f_k of the form:

$$\begin{aligned} \dot{k}_d &= \xi_1(\lambda + \|\nabla_q \phi\|; \epsilon_1) \\ \dot{k}_i &= \left(\overline{k}_i - k_i\right) w_i \ell_i g_i - \mathcal{K}_k k_i h_i w_0 \left(g_0 + \xi_1(s; \epsilon_2)\right), \ \forall i \in \mathfrak{I}_N \end{aligned} \tag{12}$$

where w_i and g_i, $i \in \mathfrak{I}_N^\star$, as well as h_i, $i \in \mathfrak{I}_N$, are functions to be defined later, $\overline{k}_i$, $i \in \mathfrak{I}_N$ are desired upper bounds for k_i, λ denotes the non-negative eigenvalue of $\nabla_q^2 \phi$, $\mathcal{K}_k$ is a positive control gain and ϵ_1 and ϵ_2 are small positive constants. The continuously differentiable switch $\xi_1(x; \epsilon)$ and functions $\ell_i(q)$ are, respectively, given by:

$$\xi_1(x; \epsilon) = 1 - \sigma_p(x/\epsilon)$$

$$\ell_i(q) = -\frac{\mathcal{K}_u s(q, k)}{\ln\left(\frac{\|q - q_i\|}{2}\right)}. \tag{13}$$

According to Proposition 1, our first requirement can be accomplished by designing f_k to reduce k_i as the robot approaches $\partial\mathcal{D}$. To do so without compromising the inherent inner obstacle collision avoidance properties of ϕ, we need to also ensure that each k_i does not vanish within some neighborhood of q_i for all $i \in \mathfrak{I}_N$. To that end, firstly we define g_i, employing the smoothly vanishing function defined in (8) to serve as pseudo-metrics of the alignment between the robot's velocity and the directions towards the goal and inner obstacles, respectively, given by:

$$g_i = \sigma_v(\overline{g}_i), \quad \forall i \in \mathfrak{I}_N^\star \tag{14}$$

with

$$\begin{aligned} \overline{g}_0 &= \frac{1}{4}\left(\alpha \|\nabla_q \psi\| \|q - q_d\| - (\nabla_q \psi)^T (q - q_d)\right) \\ \overline{g}_i &= \frac{1}{2}(\nabla_q \psi)^T (q - q_i), \quad \forall i \in \mathfrak{I}_N \end{aligned}$$

where $\alpha \in (0, 1]$ is a fixed constant that is used for selecting the desired alignment between the robot's motion and the direction to the goal. We also define the accompanying weights

w_i as follows to ensure that only one term of (12) dominates as the robot approaches a particular boundary of $\mathcal{W}$:

$$w_0 = \frac{\xi_2(\overline{w}_0; \epsilon_3)}{\overline{w}_0 + \sum_{j=1}^{N} (\bar{k}_j \overline{w}_j)} \qquad (15)$$
$$w_i = \frac{\overline{w}_i}{\overline{w}_0 + \sum_{j=1}^{N} (\bar{k}_j \overline{w}_j)}, \quad \forall i \in \mathfrak{I}_N$$

with

$$\overline{w}_i = \overline{r}_i / (r_i + \overline{r}_i), \quad \forall i \in \mathfrak{I}_N^{\star},$$

$$r_0 = (1 - \|q\|)^2, \quad r_i = \|q - q_i\|^2, \forall i \in \mathfrak{I}_N$$

$$\overline{r}_i = \sqrt[m]{\sum_{j \neq i} (r_j)^m}, \quad \forall i \in \mathfrak{I}_N \qquad (16)$$

$$\xi_2(x; \epsilon) = \begin{cases} 0, & \text{if } x < \epsilon \\ \left(\frac{x-\epsilon}{1-\epsilon}\right)^2 \left(3 - 2\frac{x-\epsilon}{1-\epsilon}\right), & \text{if } \epsilon \leq x \leq 1 \\ 1, & \text{otherwise} \end{cases}$$

for a scalar constant $\epsilon_3 \in (0, 1)$ in (15) and some integer $m < -1$ in (16) that serves as a smooth under-approximation of $\min_{j \neq i}(r_j), i \in \mathfrak{I}_N$. Finally, the weights h_i, $i \in \mathfrak{I}_N$ are defined as follows:

$$h_i = 1 + \frac{\sigma_v(\overline{h}_i)}{1 + \sum_{j \in \mathfrak{I}_N} \sigma_v(\overline{h}_j)}$$

with

$$\overline{h}_i = k_i \left(1 - (\tanh(\phi / w_\phi))^2\right)^2 \left(\frac{q_d - q}{\|q_d - q\|^2}\right)^T \frac{q_i - q}{\|q_i - q\|^2}$$

whose purpose is to accelerate the decay of those k_i that contribute the most to the component of $\nabla_q \psi$ that pushes the robot toward the workspace's outer boundary.

Regarding the second requirement, as shown in Proposition 1, selecting a k_d above a certain threshold is sufficient to render ϕ free of degenerate equilibria. On the other hand, for a given $\bar{k}_i$, increasing k_d steers the robot closer to the workspace's inner obstacles. Nevertheless, since the robot may never actually enter the vicinity of a degenerate equilibrium, instead of setting k_d sufficiently large a priori, the adaptive law for the parameter k_d is introduced to increase k_d only when it is actually needed, thus alleviating the aforementioned shortcoming.

4.3. Stability Analysis

Let us consider the overall system:

$$\dot{z} = f_z(z) \qquad (17)$$

where $z = (q, k)$ and $f_z(z) = (f_q, f_k)$ with $f_q = Ju$. Furthermore, let Ω denote the image of $\mathcal{W}$ under T, i.e., $\Omega = T(\mathcal{W})$. Note that Ω consists of $\text{int}(\mathcal{D})$, possibly with a subset of $\partial\mathcal{D}$, with the points q_i removed. In this section, we elaborate on the stability properties of (17) under the proposed control scheme (6) and (12). First, we formalize the safety properties of the closed-loop system dynamics, which guarantee that our robot does not collide with any obstacle.

Proposition 3. *The workspace $\mathcal{W}$ is invariant under the dynamics* (17) *with control laws* (6) *and* (12), *i.e., $p(t) \in \mathcal{W}$ for all $t \geq 0$.*

Proof. For the proof, refer to the Appendix A. □

Having eliminated the possibility of the robot colliding with the workspace's boundaries, we proceed by showing that all critical points of ψ, where (17) may converge to, are either non-degenerate saddles or q_d. Additionally, we show that the latter is a stable equilibrium.

Proposition 4. *The artificial potential ψ decreases along the trajectories of the closed-loop system and its time derivative vanishes only at its critical points. Additionally, the preimage of q_d is a set of stable equilibria of* (1).

Proof. For the proof, refer to the Appendix A. □

Proposition 5. *Let $z^\star = (q^\star, k^\star)$ be a critical point of the closed-loop system dynamics with $q^\star \in \Omega \setminus \{q_d\}$. Then, $q^\star$ is a non-degenerate saddle point of ψ.*

Proof. For the proof, refer to the Appendix A. □

Finally, we conclude this section with the main theoretical findings.

Theorem 1. *System* (1) *under the control law* (6) *and* (12) *converges safely to p_d, for almost all initial configurations, thus addressing successfully Problem 1.*

Proof. For the proof, refer to the Appendix A. □

Remark 2. *Owing to the adaptive laws (12) that modify the harmonic source strengths online to secure the safety and convergence properties at all times, the selection of the fixed control parameters in the proposed scheme, i.e., $\mathcal{K}_u$, γ, ϵ_p, ϵ_v, w_ϕ, $\mathcal{K}_k$, ϵ_1, ϵ_2, α and ϵ_3, is straightforward as it affects only the trajectory evolution within the workspace and not the aforementioned critical properties. Consequently, their values should be set freely as opposed to NFs, where the selection of the main parameters severely affects the convergence properties of the adopted scheme and cannot be conducted constructively for generic workspaces of arbitrary topology.*

5. Extensions

In this section, we present certain extensions of the proposed approach to (a) address the safe navigation problem for unicycle robots which are frequently encountered in many application domains and (b) tackle computational complexity issues that affect the numerical computation of the harmonic map presented in Section 3 as the size of the workspace increases.

5.1. Unicycle Robot Kinematics

In this subsection, we consider robots whose motion is subjected to Pfaffian constraints of the form:

$$\begin{aligned} \dot{p} &= n_\theta v \\ \dot{\theta} &= \omega \end{aligned} \tag{18}$$

where $\theta \in [0, 2\pi)$ denotes the robot's orientation, $n_\theta = [\cos(\theta), \sin(\theta)]^T$, and $v, \omega \in \mathbb{R}$ are control inputs corresponding to the robot's linear and angular velocities, respectively. First, let us define the robot's kinematics in the image of the configuration space via the proposed transformation as follows:

$$\begin{aligned} \dot{q} &= n_{\hat{\theta}} \hat{v} \\ \dot{\hat{\theta}} &= \hat{\omega} \end{aligned} . \tag{19}$$

Note that the orientations θ and $\hat{\theta}$ are related by:

$$n_{\hat{\theta}} = \frac{J_p n_\theta}{\|J_p n_\theta\|}$$

where $J_p = J(p)$. To safely drive the robot to its goal configuration, we consider the following control laws:

$$\begin{aligned} \hat{v} &= -\mathcal{K}_v s_v(q,\hat{\theta},k)\left(n_{\hat{\theta}}\right)^T \nabla_q \psi(q,k) \\ \hat{\omega} &= -\mathcal{K}_\omega \left(n_{\hat{\theta}}^{\perp}\right)^T \nabla_q \psi(q,k) \end{aligned} \tag{20}$$

with $\mathcal{K}_v, \mathcal{K}_\omega \in \mathbb{R}$ positive constant gains, $n_{\hat{\theta}}^{\perp} = [-\sin(\theta), \cos(\theta)]^T$ and

$$s_v(q,\hat{\theta},k) = \gamma\sigma_p\left(\frac{1-\|q\|}{\epsilon_p}\right) + (1-\gamma)\sigma_v\left(\frac{(n_{\hat{\theta}}{}^T\nabla_q\psi)n_{\hat{\theta}}{}^T q}{\epsilon_v + |n_{\hat{\theta}}{}^T\nabla_q\psi|\|q\|}\right).$$

Additionally, we need to employ a modified version of the adaptive law for the potential field parameters, which is obtained by substituting s with s_v in (12) and (13) and $\overline{g}_i$, $i \in \mathfrak{I}_N^\star$, with

$$\begin{aligned} \overline{g}_{v,0} &= \frac{1}{4}\left(\alpha\left|n_{\hat{\theta}}{}^T\nabla_q\psi\right|\|q-q_d\| - \left(n_{\hat{\theta}}{}^T\nabla_q\psi\right)n_{\hat{\theta}}{}^T(q-q_d)\right) \\ \overline{g}_{v,i} &= \frac{1}{2}\left(n_{\hat{\theta}}{}^T\nabla_q\psi\right)n_{\hat{\theta}}{}^T(q-q_i), \quad \forall i \in \mathfrak{I}_N \end{aligned}$$

respectively, in (14). Finally, by expressing the aforementioned control laws to the robot's actual configuration space, we obtain:

$$\begin{aligned} v &= \nu\hat{v} \\ \omega &= \omega_{dq} + \omega_{d\hat{\theta}} \end{aligned} \tag{21}$$

where ω_{dq} and $\omega_{d\hat{\theta}}$ are terms corresponding to angular velocities induced by translational and rotational motion of the robot in the workspace's image, respectively, given by:

$$\begin{aligned} \omega_{dq} &= -\hat{v}\nu^2\left(\left(J_p\begin{bmatrix} n_\theta & \nu n_{\hat{\theta}}^{\perp}\end{bmatrix}\right)^{-1}\frac{\partial}{\partial n_\theta}J_p n_0\right)^T\begin{bmatrix}0\\1\end{bmatrix} \\ \omega_{d\hat{\theta}} &= \hat{\omega}\left(\left(J_p\begin{bmatrix} n_0 & \nu n_{\hat{\theta}}^{\perp}\end{bmatrix}\right)^{-1} n_{\hat{\theta}}^{\perp}\right)^T\begin{bmatrix}0\\1\end{bmatrix} \end{aligned}$$

with $\nu = \|J_p^{-1}n_{\hat{\theta}}\|$ and $\frac{\partial}{\partial n_\theta}J_p$ denoting the directional derivative of J_p along n_0.

The stability properties of the aforementioned closed-loop system dynamics are formalized below.

Theorem 2. *The workspace $\mathcal{W}$ is invariant under the dynamics of* (18) *equipped with the proposed control law. Additionally, the robot will asymptotically converge either to an interior critical point of ϕ or to the pre-image of q_d, which is stable.*

Proof. For the proof, refer to the Appendix A. □

Remark 3. *The result of Theorem 2 is weaker compared to that of Theorem 1, since there is no guarantee that the set of configurations which converge to a critical point of ϕ (other than the pre-image of q_d) has Lebesgue measure zero.*

5.2. Atlas of Harmonic Maps

As the size of the workspace increases, the problem of computing the transformation T grows in complexity as well, because the resources required by commonly employed numerical techniques that can solve the problem presented in Section 3 are polynomial in the number of elements used for representing $\mathcal{W}$. Alternatively, to cope with large workspaces efficiently, we propose instead the construction of an atlas $\mathcal{A} \triangleq \{(\mathcal{P}_i, T_i) \mid i \in$

$\mathfrak{I}_{N_A}\}$ obtained by separating the workspace $\mathcal{W}$ into N_A overlapping subsets $\mathcal{P}_i \subset \mathcal{W}$, such that $\bigcup_{i\in\mathfrak{I}_{N_A}} \mathcal{P}_i = \mathcal{W}$ and constructing a separate harmonic map T_i for each $\mathcal{P}_i$ (see Figure 3).

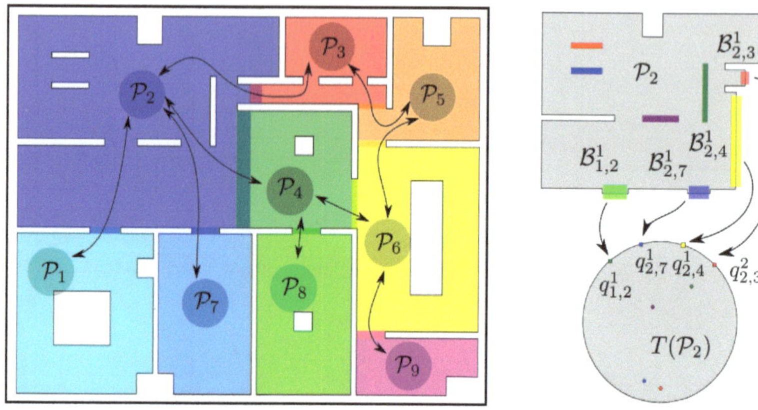

Figure 3. The partition of a complex workspace into overlapping subsets along with the corresponding graph and the tranformation T_2 of the second partition $\mathcal{P}_2$.

This essentially allows us to solve many small (and computationally less intensive) problems instead of a large one, thus reducing the overall resources required for addressing a given workspace. Therefore, given such a partitioning of $\mathcal{W}$, we define the graph $\mathcal{G} = (\mathcal{V}, \mathcal{E})$, where $\mathcal{V} = \{\mathcal{P}_i \mid i \in \mathfrak{I}_{N_A}\}$ denotes the set of corresponding nodes (workspace partitions) and $\mathcal{E}$ denotes the set of edges between the elements of $\mathcal{V}$, with each edge indicating a feasible transition from one partition to another, i.e., $(i,j) \in \mathcal{E}$ if and only if $\left(\mathrm{cl}(\mathcal{P}_i) \cap \mathrm{cl}\left(\mathcal{P}_j\right)\right) \neq \emptyset$. Note that $\mathcal{G}$ is undirected by definition, i.e., $(i,j) \in \mathcal{E}$ only if $(j,i) \in \mathcal{E}$. Additionally, since the workspace is connected, $\mathcal{G}$ should also be connected. Thus, for a given atlas $\mathcal{A}$, an initial configuration p_{init} and a final configuration p_d, we can employ standard graph search algorithms to obtain a sequence of indices $\mathcal{S} = \{s_1, s_2, \dots s_n\}$ corresponding to partitions that the robot can tranverse to reach its goal. (In general, more than one such sequence of partitions may exist connecting the initial and the final configurations. However, the selection of one that corresponds to some sort of "optimal" path is beyond the scope of this work.) Additionally, note that since the partitioning of $\mathcal{W}$ does not need to be fine, the size of $\mathcal{G}$ will generally be small, rendering the cost of finding $\mathcal{S}$ negligible.

We now concentrate on how the transition between two consecutive elements of $\mathcal{S}$ is implemented. Let $\mathcal{C}_{i,j} \triangleq \mathrm{cl}(\mathcal{P}_i) \cap \mathrm{cl}\left(\mathcal{P}_j\right)$ denote the common region of $\mathrm{cl}(\mathcal{P}_i)$ and $\mathrm{cl}\left(\mathcal{P}_j\right)$ and let $\mathcal{B}_{i,j} \triangleq \partial\mathcal{P}_i \cap \mathcal{P}_j$ denote the set of points on the boundary of $\mathcal{P}_i$ that also belong to $\mathcal{P}_j$ and are not occupied by obstacles for all $i \in \mathfrak{I}_{N_A}$ and all j such that $(i,j) \in \mathcal{E}$. Without loss of generality, we assume that $\mathcal{A}$ is constructed such that the sets $\mathcal{B}_{\ell,i} \cap \mathcal{B}_{\ell,j}$ are either empty or consist of isolated points. We note that in order to successfully complete the transition between two consecutive nodes $\mathcal{P}_i$ and $\mathcal{P}_j$ of $\mathcal{S}$, it suffices for the robot to reach any single point of $\mathcal{B}_{i,j}$ from $\mathcal{P}_i$. We also observe that each $\mathcal{B}_{i,j}$ may consist of one or more disjoint components $\mathcal{B}^{\ell}_{i,j}$, $\ell \in \mathcal{L}(i,j)$, with $\mathcal{L}(i,j)$ being some valid indexing of those. By exploiting the fact that Theorem 2 [38] imposes a weak homeomorphism requirement on T_i, we can construct each T_i such that each disjoint subset of $\partial\mathcal{P}_i$ collapses into a separate point, i.e., $T_i(\mathcal{B}^{\ell}_{i,j}) = q^{\ell}_{i,j} \in \partial\mathcal{D}$ (see Figure 3), which, in turn, implies that selecting $q^{\ell}_{i,j}$ as an intermediate goal configuration suffices to render the entire $\mathcal{B}^{\ell}_{i,j}$ attractive. Building upon this fact, for each consecutive pair of $\mathcal{P}_i$ and $\mathcal{P}_j$ in $\mathcal{S}$, we (arbitrarily) select a $\mathcal{B}^{\ell}_{i,j}$ and we construct a transformation $T_i : \mathcal{P}_i \mapsto \mathcal{D}$, with $q^{[i]} = T_i(p)$, and artificial potential field $\phi_i(q^{[i]}, k^{[i]})$ with goal configuration $q^{[i]}_d = q^{\ell}_{i,j}$. Additionally, to smooth the transition between consecutive partitions, when they overlap, we propose the following modified control law for the robot:

$$u = u^{[i]} + \eta_{c,i,j} \cdot \eta_{t,i,j} \cdot u^{[j]}, \quad \forall p \in \mathcal{C}_{i,j} \tag{22}$$

where $u^{[i]}$ and $u^{[j]}$ denote the control inputs as defined in (6) and evaluated using ψ_i, T_i and ψ_j, T_j, respectively; the function $\eta_{t,i,j} : \mathcal{C}_{i,j} \mapsto [0,1]$ is any smooth bump function such that

$$\eta_{t,i,j}(p) = \begin{cases} 0, & \text{if } p \in \mathcal{B}_{j,i} \\ 1, & \text{if } p \in \mathcal{B}_{i,j} \end{cases}$$

and

$$\eta_{c,i,j}(p, k^{[i]}, k^{[j]}) = \begin{cases} \frac{(\zeta_{i,j})^2}{\epsilon_4 + (\zeta_{i,j})^2}, & \text{if } \zeta_{i,j} \geq 0 \\ 0, & \text{if } \zeta_{i,j} < 0 \end{cases}$$

with $\zeta_{i,j} = \left(\nabla_p \psi_i\right)^T \cdot \left(\nabla_p \psi_j\right)$; and $\epsilon_4 > 0$ is a fixed parameter. What this modification essentially does is incrementally add an extra component, with the direction of $\nabla_p \psi_j$, to the robot's velocity when that component is cosine similar (two vectors u and v are cosine similar if their inner product is positive) with $\nabla_p \psi_i$. We note that $\eta_{c,i,j} \to 1$ and $\eta_{t,i,j} \to 1$ as the robot approaches the boundary of the corresponding partition. We also remark that once the robot has completed its transition to $\mathcal{P}_j$, we do not concern ourselves with $u^{[i]}$ anymore, i.e., $u = u^{[j]}$ even if p returns to $\mathcal{C}_{i,j}$. The overall scheme employed for navigating a holonomic robot to its goal configuration using an altas constructed as described above can be found in Algorithm 1.

Regarding the stability analysis of the modified system, by following the same procedure as in Section 4.3 and by virtue of $\eta_{c,i,j}$, it is trivial to verify the following statement.

Theorem 3. *System* (1) *equipped with Algorithm 1 converges safely to a given goal configuration* $p_d \in \mathcal{W}$ *from almost all initial configurations* $p_{\text{init}} \in \mathcal{W}$.

Proof. For the proof, refer to the Appendix A. □

Algorithm 1 Altas-based motion planning scheme for a holonomic robot

---

Require: $\mathcal{A}, p_{\text{init}}, p_d$
  $\mathcal{S} \leftarrow$ FINDPATHTOGOAL($\mathcal{G}, p_{\text{init}}, p_d$)
  Initialize $k^{[s]}$ for all $s \in \mathcal{S}$.
  for all i in $\mathfrak{I}_{n-1}$ **do**
    $\mathrm{s}, \mathrm{s}' \leftarrow \mathrm{s}_i, \mathrm{s}_{i+1}$
    Select (arbitrary) ℓ such that $\ell \in \mathcal{L}(\mathrm{s}, \mathrm{s}')$.
    Place goal configuration of ψ_{s} at $q^{\ell}_{\mathrm{s},\mathrm{s}'}$.
  end for
  Place goal configuration of ψ_{s_n} at $T_{\mathrm{s}_n}(p_d)$.
  $\ell \leftarrow 1$
  loop
    if $\ell = n$ or $p \in \mathcal{P}_{\mathrm{s}_\ell} \setminus \mathcal{P}_{\mathrm{s}_{\ell+1}}$ **then**
      Update p using (6) and $k^{[\mathrm{s}_\ell]}$ using (12).
    else if $p \in \mathcal{C}_{\mathrm{s}_\ell, \mathrm{s}_{\ell+1}}$ **then**
      Update p using (22) with $i = \mathrm{s}_\ell$ and $j = \mathrm{s}_{\ell+1}$.
      Update $k^{[\mathrm{s}_\ell]}$ and $k^{[\mathrm{s}_\ell]}$ using (12).
    else
      $\ell \leftarrow \ell + 1$
    end if
  end loop

---

6. Simulations and Experimental Results

In order to demonstrate the efficacy of the proposed control scheme, we have conducted various simulation and experimental studies, the results of which are presented in this section. The algorithm that computes the harmonic transformation and its Jacobian was implemented in C++, while the proposed control protocols were implemented in Python. Code implementations can be accessed at https://github.com/maxchaos/hntf2d (accessed on 16 April 2023). All simulations were carried out on a PC with an Intel i5 processor operating at 2.2 Ghz, with 4 GB RAM and running a GNU/Linux operating system. For more details regarding both simulations and experiments, the reader may refer to the accompanying video material at https://youtu.be/I6WUS81iDh4 (accessed on 16 April 2023).

6.1. Simulations—Full Workspace Transformation

In the first case study, a single transformation of the entire 8 m × 5 m workspace (see Figure 3) was constructed and the robot was instructed to navigate to various goal configurations starting from the same initial position. The initial configuration and the parameters of our controller were selected such as to better demonstrate the guaranteed collision avoidance properties of our scheme. Particularly, the initial values for the parameters of the adaptive law were selected as $k_d = 20$, $k_i = 1$ and $\bar{k}_i = 20$ for all $i \in \mathfrak{I}_{10}$. The values of the remaining parameters were $\mathcal{K}_u = 100$, $w_\phi = 20$, $\mathcal{K}_k = 100$, $\alpha = 1$, $\epsilon_p = 0.025$, $\epsilon_v = 0.1$, $\gamma = 0.7$, $\epsilon_1 = 0.01$, $\epsilon_2 = 0.1$ and $\epsilon_3 = 0.1$. The goal configurations and the trajectories executed by the robot, both in the real and transformed workspace, are illustrated in Figure 4.

The simulations were conducted using the Euler method with 10 ms steps. Regarding the computational complexity of the control scheme, the construction of the harmonic transformation for this large workspace that was carried out offline once required 5.4 s to complete for a sufficient approximation of the workspace boundary with 3680 segments. Finally, the online computation of the transformation $T(p)$ and its Jacobian $J(p)$ required an average of 6.0 ms per step.

6.2. Simulations—Atlas of Harmonic Maps

In this case study, we decomposed the aforementioned workspace into separate partitions (see Figure 3) and constructed a harmonic transformation T_i for each one (we adopted the door of each room as the common boundary between neighboring partitions). The robot was initialized at the same position as the previous study and it was instructed to navigate towards the same set of individual goal configurations. The initial values selected for the parameters of the adaptive law were $k^{[i]} = N^{[i]} + 3$, $k_j^{[i]} = 1$ and $\bar{k}_j^{[i]} = k^{[i]}$ for all $j \in \mathfrak{I}_{N^{[i]}}$ and $i \in \mathfrak{I}_{N_A}$, where $N^{[i]}$ denotes the amount of obstacles inside the corresponding partition. All remaining control parameters were selected as in Section 6.1. The trajectories of the robot are depicted in Figure 5. The time spent to construct the corresponding harmonic transformations varied from $0.019s$ to $0.211s$ (depending on the amount of elements required for sufficiently approximating each room, ranging between 320 and 1000 segments) and was significantly much less than the full map construction of the previous case (5.4 s). Additionally, the online computation of $T_i(p)$ and $J_i(p)$ in each of these rooms required an average time between 1.0 ms and 2.2 ms per step, respectively. Finally, it should be noted that in this case, the workspace inner obstacles were mapped to points further away from the boundaries of the partitions, which is an interesting result as it alleviates possible numerical issues that may arise in the computation of the transformation near the obstacles (the condition number of the Jacobian of the transformation is improved). It should be stressed that the length of the paths in the second case was less (improvement of 0.5 m on average), owing to the fact that the robot gets closer to the workspace boundary since the individual transformations in each room obtain a better conditioned Jacobian (condition number 0.212 against 0.093) and thus are more fine than the first approach, where a transformation is built for the whole workspace.

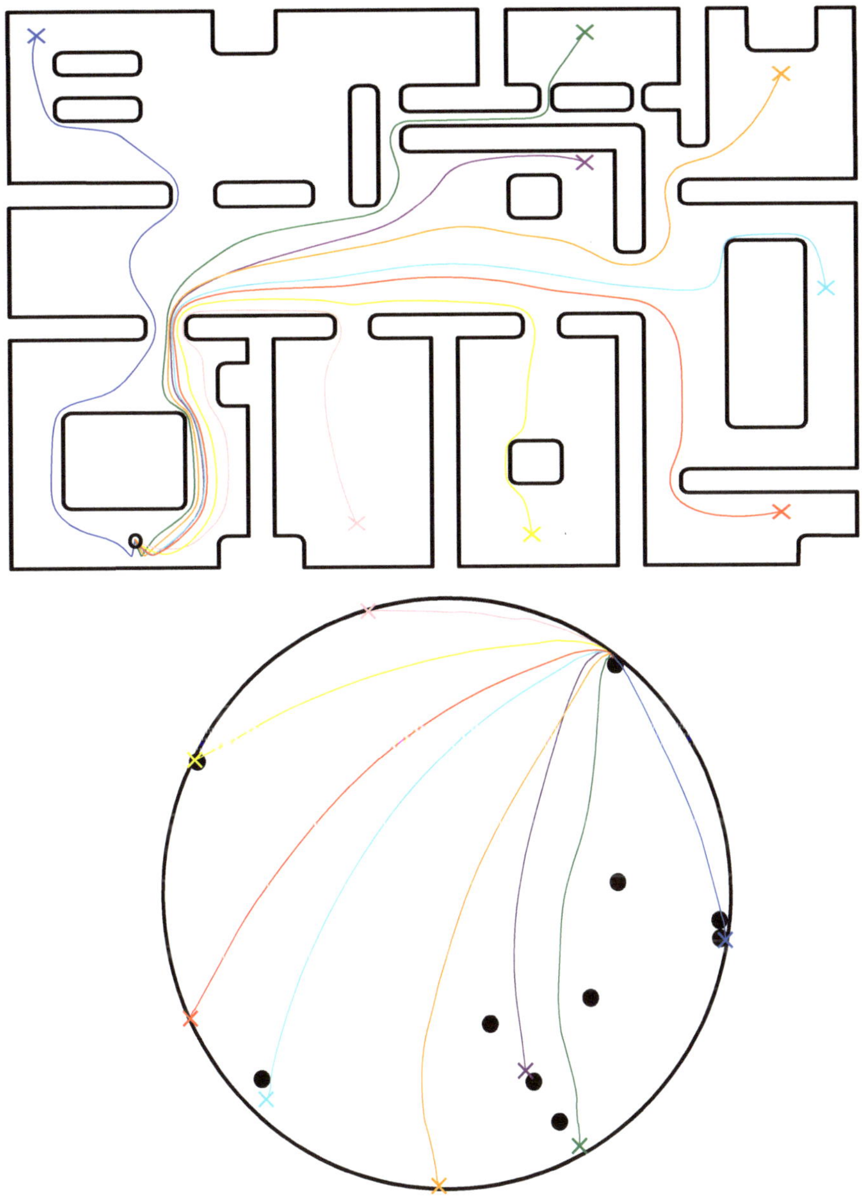

Figure 4. Robot trajectories illustrated with various colors in both the actual workspace (**upper plot**) and their image (**bottom plot**) using a full workspace transformation. The robot starts from an arbitrary location at the bottom left part of the workspace (black circle) and navigates to different goal configurations (colored crosses). Note that the ten black dots in the lower plot correspond to the points where the internal obstacles of the actual workspace have been transformed.

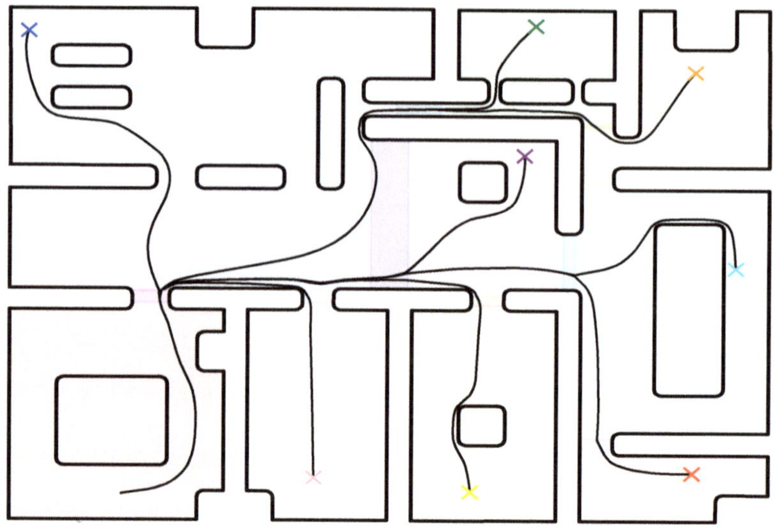

Figure 5. The resulting robot trajectories for various goal configurations depicted with colored crosses, using an atlas of the workspace.

6.3. Comparative Study—Workspace Transformation

In this subsection, we provide a comparative study of the harmonic map presented in this work against readily available workspace transformation methods employed in the motion planning literature. Particularly, we consider four 4 m $\times$ 4 m compact workspaces, each associated with a pair of initial and goal positions, and construct appropriate transformations for each one by employing the methodology presented in this work (HM), as well as (i) the Star-to-Sphere Transformation (SST) [15], (ii) the Multi-Agent Navigation Transformation (MANT) [36] and (iii) the Navigation Transformation (NT) [34] (with the aforementioned Star-to-Sphere transformation serving as the underlying map). The trajectories of the robot executed while tracing the line segment connecting the initial and goal configurations in the images of each domain can be seen in Figure 6. We note that manual tuning of the compared transformations was necessary in order to render each a diffeomorphism but without making them too steep around the obstacles. Furthermore, the domain boundaries considered here had to be sufficiently smooth in order for methodologies such as MANT to be applicable. Finally, we remark that the trajectories corresponding to the proposed transformation are, in general, less abrupt compared to the rest, a property attributed to the fact that our approach is global as opposed to the other transformations, i.e., the distortion caused by each obstacle is not limited to some narrow neighborhood around it. The total length, maximum curvature and distance from the obstacles of each executed trajectory can be seen in Tables 3–5, respectively. We can see from these values that the actual trajectories yielded using harmonic maps are among the shorter and smoother ones, although they tend to approach the obstacles more than the rest.

6.4. Comparative Study—Control Law

In this subsection, we provide a comparative study of our control scheme against other motion planning methodologies.

6.4.1. APF-Based Schemes

To demonstrate the efficacy of the proposed control scheme compared to other APF-based schemes, we considered the 12 m $\times$ 16 m workspace depicted in Figure 7, for which we constructed a harmonic map as described in Section 3. Next, we equipped a holonomic robot with three alternative control laws and instructed it to visit four distinct goal positions using these controllers, starting each time from a fixed initial configuration. Particularly, we considered a conventional navigation function-based controller (NF) [15] augmented by [17], for the selection of its notorious parameter, and a harmonic navigation function-based controller (HNF) [33], in addition to our adaptive control scheme (AHNF) described in Section 4. We note that all three control laws considered here make use of the same underlying harmonic map T constructed as described above in order to drive the robot to its instructed goal positions. The trajectories executed by the robot can be seen in Figure 7. We remark that, in general, our approach steers the robot away from inner obstacles that lie between its initial and goal configurations, unlike "greedy" schemes such as the conventional NF-based controller, while keeping the traced paths shorter compared to HNFs with fixed source weights, a property attributed to the proposed adaptive laws (12) which penalize misalignment between the robot's velocity and the direction towards the goal configuration.

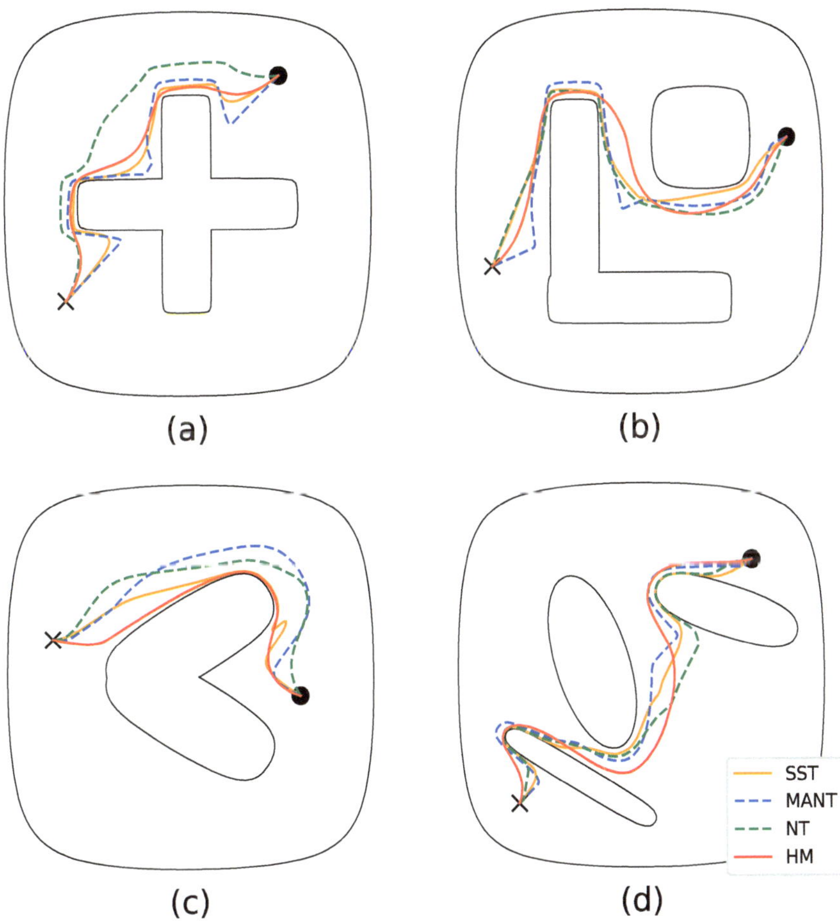

Figure 6. Robot trajectories from initial configuration (black circle) straight to the desired configuration (black cross) by employing various domain transformation methods in each workspace (**a**–**d**).

Table 3. Trajectory lengths (m) executed by employing the four alternative transformations in each workspace displayed in Figure 6.

	a	b	c	d
SST	3.63	4.18	2.12	4.09
MANT	4.26	4.69	2.34	4.45
NT	3.35	4.30	2.18	4.22
HM	3.19	4.21	2.05	4.32

Table 4. Maximum value of curvature (m^{-1}) associated with each trajectory displayed in Figure 6.

	a	b	c	d
SST	4.22	5.43	86.93	2.16
MANT	1.23	1.47	13.56	2.06
NT	66.97	25.23	14.89	6.92
HM	2.47	2.49	14.76	2.77

Table 5. Minimum distance (m) between each robot trajectory and the corresponding workspace boundaries displayed on Figure 6.

	a	b	c	d
SST	0.0303	0.0283	0.0159	0.0063
MANT	0.0644	0.1253	0.1870	0.0648
NT	0.1386	0.0506	0.0915	0.0058
HM	0.0335	0.0377	0.0103	0.0181

The total length and distance from the obstacles of each executed trajectory can be seen in Tables 6 and 7, respectively. First, we have to stress that the length trajectory corresponds to the travelled path towards the goal configuration and thus needs to be small, whereas the minimum distance to the workspace boundary refers to the closest point of the trajectory to the workspace boundary and thus needs to be large to have a safe trajectory. Consequently, note from Table 6 that the NF scheme yielded shorter path lengths than the proposed method in two cases (blue and yellow); nevertheless, such paths approach closer to the workspace boundary as indicated in Table 7, thus resulting in more risky paths. On the other hand, the Adaptive Harmonic Potential Field yields a good trade-off between path length and minimum distance to the boundary, since it achieves the shortest paths for two cases without compromising them, as is the case with the NF. On the other hand, the HPF tend to travel around the obstacle closer to the outer workspace boundary and hence exhibit more safe trajectories but they are significantly longer than the other two schemes.

6.4.2. Sampling-Based Scheme

To compare the control scheme proposed in this work against sampling-based methods, we considered a holonomic point-sized robot positioned inside a 6 m $\times$ 8 m compact workspace and a desired goal configuration. To complete this task, we employed two different controllers, namely the one proposed in this work and an admissible planner based on an improved probabilistic roadmap method (PRM) [6]. The trajectories executed by the robot using our control law as well as two of the trajectories generated by the PRM-based planner can be seen in Figure 8. The construction of the associated transformation took 31 s to complete for a given boundary approximation made of 7842 elements, whereas the PRM-based planner required approximately 24 s on average over 10 successful runs to yield a solution (we have to stress that we ran 14 trials to get 10 solutions, since four runs did not complete as they exceeded the 500 s calculation time), using the same boundary approximation for collision checking. The robot trajectories exhibited similar lengths in both algorithms (22.5 m for our method against 21.8 m on average), although no path optimization was employed in our case. Additionally, the proposed scheme resulted in a smoother robot trajectory (based on the resulting sequence of points in both cases, we calculated the minimum curvature radius as 0.23 m for our method against 0.12 m on average for the PRM method). On the other hand, note that our approach solves the motion planning problem for any pair of initial and final configurations within the workspace, whereas the sampling-based scheme considers only one go-to problem. Thus, a different initial or final configuration would require a new solution with the PRM method. On the contrary, the proposed transformation needs to be calculated only once to solve the motion planning problem for any pair of initial and final configurations. Finally, it should be noted that for a narrower corridor in Figure 8, the sampling-based approach failed to derive a solution with a reasonable execution time (no solution was calculated within 500 s), since the probability of sampling connected points within this snaky passage reduces drastically. On the contrary, the proposed transformation took 38 s to complete for the same number of elements (i.e., 7842 elements).

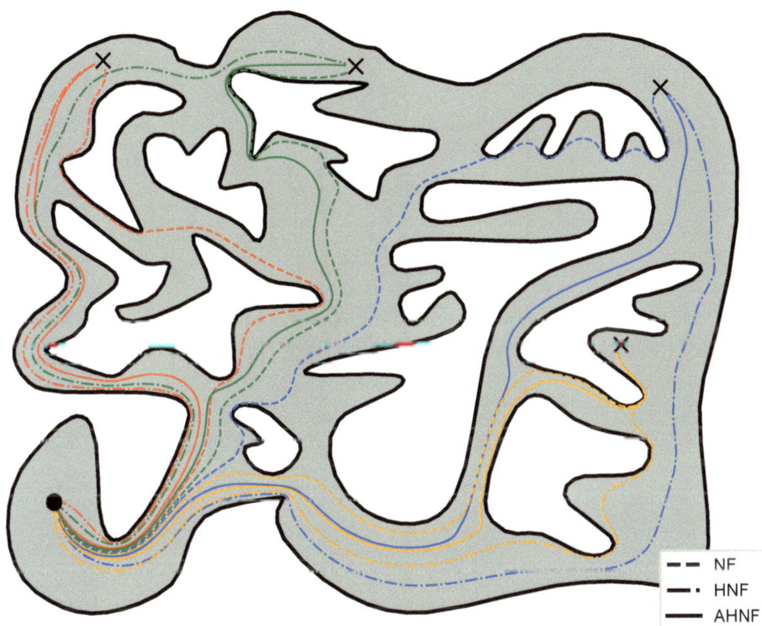

Figure 7. Trajectories of the robot navigating to four distinct goal configurations (black crosses) with red, green, yellow and blue color starting from the same initial position (black circle) while using various alternative APF-based controllers.

Table 6. Length of trajectories (m) executed by each controller (Rimon–Koditchek Navigation Functions, Harmonic Navigation Functions and Adaptive Harmonic Potential Fields) in Figure 7.

	Red	Green	Blue	Yellow
NF	19.781	20.427	**22.090**	**18.397**
HNF	18.224	22.538	26.959	20.062
AHNF	**17.874**	**19.419**	23.364	18.595

Table 7. Minimum distance (m) between the corresponding workspace boundaries and each trajectory displayed in Figure 7.

	Red	Green	Blue	Yellow
NF	0.1158	0.0102	0.1210	0.1103
HNF	**0.3347**	**0.2135**	**0.2591**	**0.2166**
AHPF	0.1310	0.0352	0.2043	0.1854

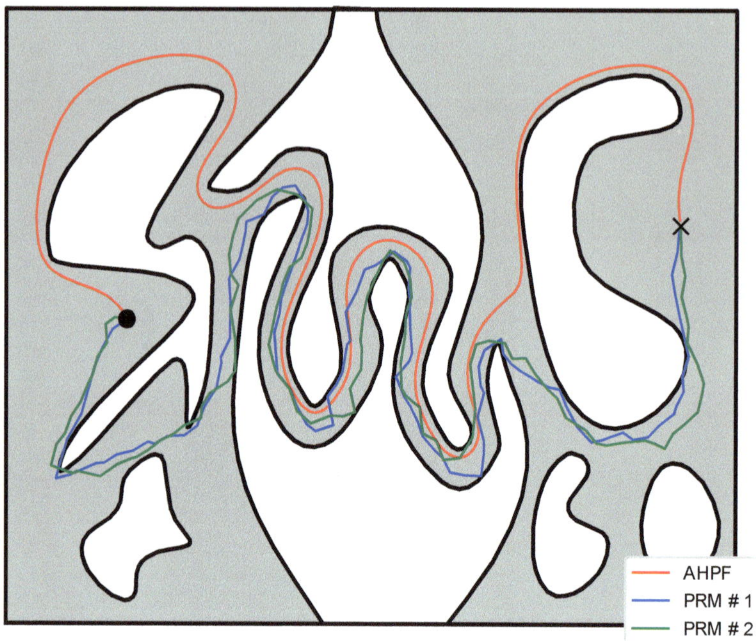

Figure 8. Trajectories of the robot navigating to its goal configuration (black cross) generated using the proposed control law and a PRM-based planner.

6.5. Experiments

In order to verify the results presented in Section 5.1, real experiments were conducted on a non-holonomic robotic platform (Robotnik Summit-XL) operating within the 10 m $\times$ 25 m compact workspace that is depicted in Figures 9 and 10. The boundaries of the workspace were obtained using readily available SLAM algorithms and were later augmented with the robot's shape (approximated by a disk). The workspace was partitioned into six overlapping subsets and the robot was instructed to visit three different goal configurations, each located in a different room. An off-the-shelf localization algorithm was employed for estimating the robot's position and orientation using its on-board sensors (laser scanners and RBG-D cameras), providing feedback at approximately 5 Hz to the robot's linear and angular commanded velocities. The construction of the associated transformations over the six subsets of the workspace took from 1.3 s for the simple and smaller partitions with 800 elements to 3.1 s for the more complex ones employing 1500 elements. On the other hand, the evaluation of the mapping as well as its Jacobian took less than 6 ms on average, which was satisfactory given the low position update rate. Note that our algorithm successfully managed to drive the robot safely (the minimum distance to the workspace boundary was 0.15 m when passing through the doors) to its specified goal configurations, as one can verify from the trajectories (see Figure 9, Figure 11 and the accompanying video material). However, an issue that needs to be pointed out is the oscillating behavior that the robot exhibited in the configuration space's image, particularly in subsets p1 and p2 as depicted in Figure 11. Such behavior is attributed both to (a) the relative slow update of the robot's pose estimation and (b) the inversion of the Jacobian which is ill-conditioned close to extremely narrow passages of the domain. Nevertheless, such shortcomings can be alleviated by a better choice of partitions, e.g., by partitioning the domain into more subsets with less complex shapes. As a future research direction, we shall investigate whether the condition number of the Jacobian of the transformation is a fine criterion, since the condition number is usually used to measure how sensitive a function is to changes or errors in the input, and the output error results from an error in the input via the Jacobian.

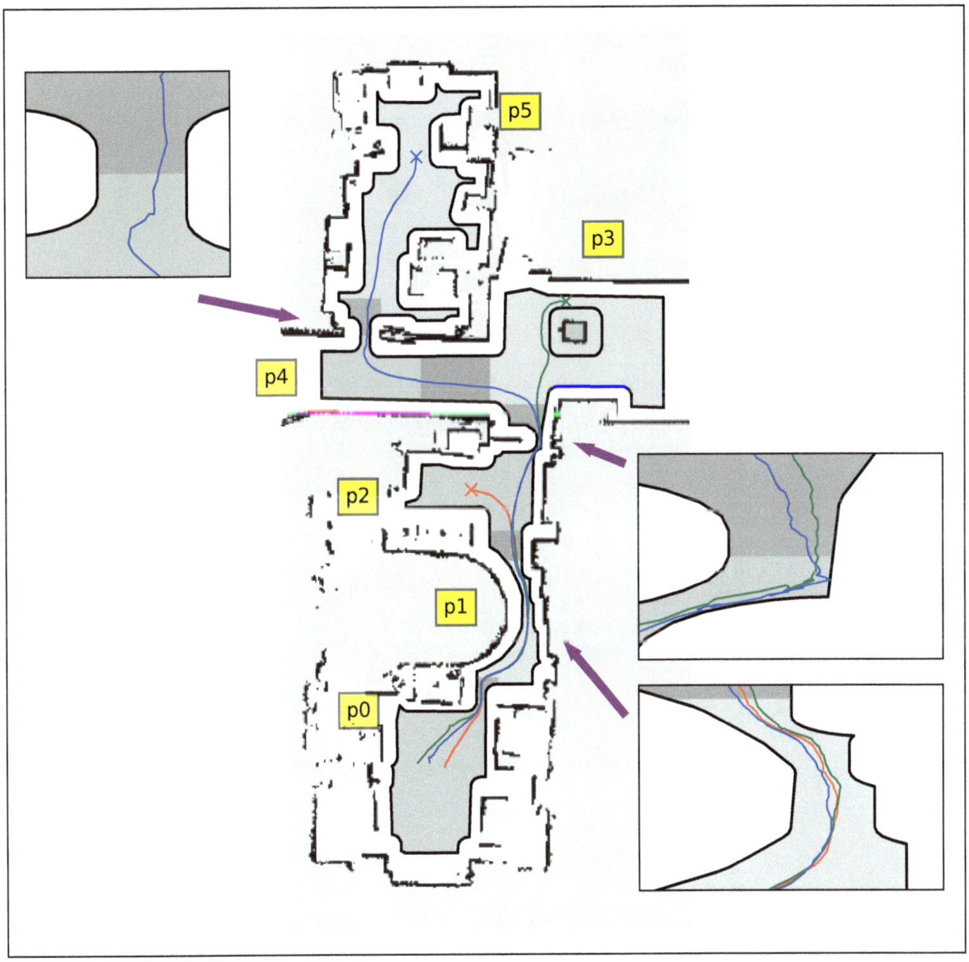

Figure 9. Trajectories of the unicycle-like robot in the real workspace, with each color (red, green and blue) corresponding to a different goal configuration. Dark gray regions indicate areas where partitions overlap.

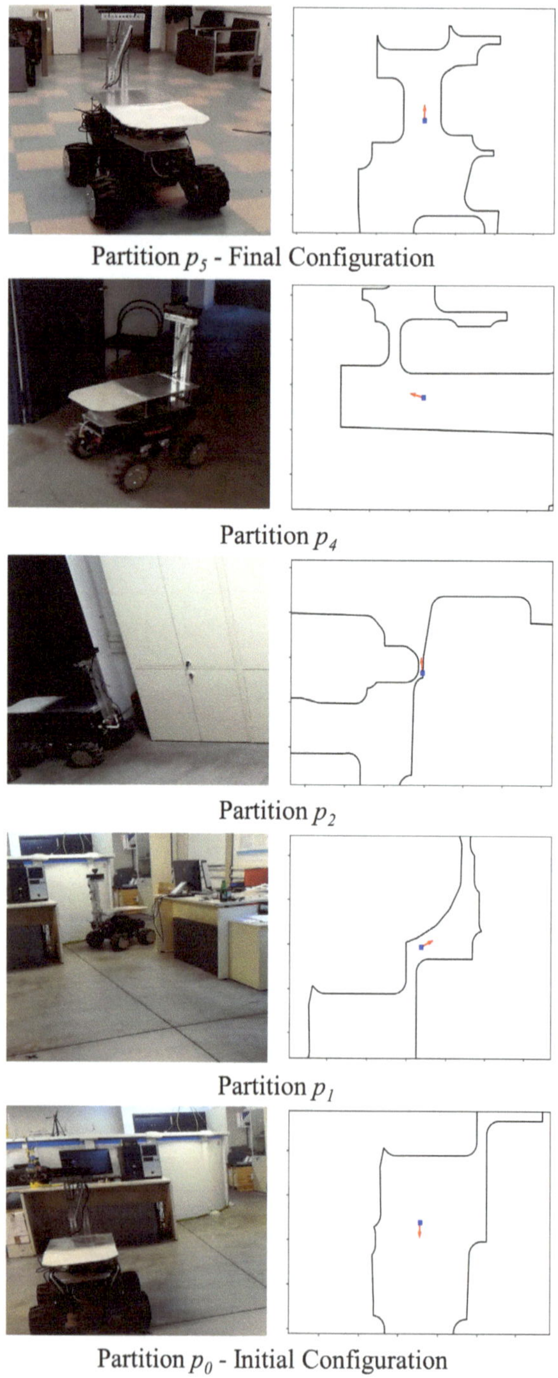

Figure 10. Snapshots of the robot navigating through the workspace. The figures on the right illustrate the position and orientation of the robot with respect to the workspace map.

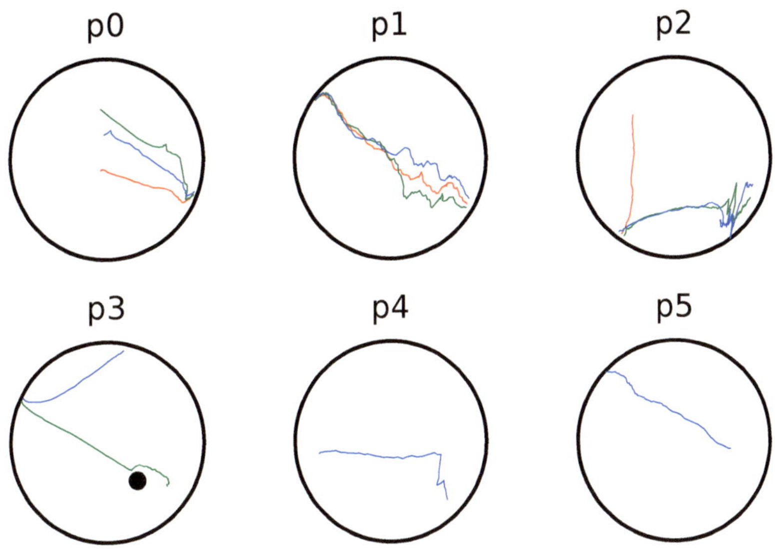

Figure 11. Robot image trajectories within each partition of the workspace for three experiments in red, green and blue color.

7. Conclusions and Future Work

In this work, we employed harmonic map theory to devise a transformation of complex workspaces directly to point worlds that are appropriate for robot navigation. Subsequently, we presented a novel motion planning control scheme based on closed-form harmonic potential fields equipped with appropriate adaptive laws for their parameters, which can safely navigate a robot to its goal state from almost all initial configurations. Additionally, we extended our approach to accommodate the navigation problem of non-holonomic robots and kept the numeric computations tractable for large workspaces.

Regarding future directions, our aim is first to increase the applicability of the proposed navigation framework by addressing partially known dynamic workspaces, which is far from being straightforward. To remedy the issue of calculation time in this case, we shall adopt a sensitivity analysis approach so that we do not solve the whole problem from scratch, but find how the solution deviates when a small change in the workspace occurs. In this way, we envision a reasonable calculation time (except from the first calculation) that would result in an almost real-time calculation of the transformation and thus allow us to consider even moving obstacles in dynamic environments. However, critical issues have to be studied concerning cases where the workspace changes topologically (e.g., in the case of antagonistically moving obstacles) and this results in significant changes in the transformation. In the same vein, switching in the transformation output might raise practical issues such as chattering that have to be carefully considered. Note that the aforementioned research direction could also serve as a first step towards the solution of the multi-robot motion planning problem, where for each robot all other robots should be considered as moving obstacles, operating antagonistically to achieve their goal configurations. Finally, another challenging research direction concerns the extension to 3D workspaces. Unfortunately, the harmonic maps have been studied only for 2D workspaces, since they rely heavily on complex analyses. Nevertheless, we propose to decompose the 3D motion planning problem into several 2D sub-problems, where the proposed solution works, and then combine them (e.g., decompose the motion along the z-axis and on the x-y plane).

Author Contributions: Methodology, P.V. and C.P.B.; Validation, P.V.; Formal analysis, P.V. and C.P.B.; Writing—original draft, P.V.;Writing—review & editing, C.P.B. and K.J.K.; Supervision, C.P.B. and K.J.K. All authors have read and agreed to the published version of the manuscript.

Funding: The publication fees of this manuscript have been financed by the Research Council of the University of Patras.

Institutional Review Board Statement: Not applicable.

Informed Consent Statement: Not applicable.

Data Availability Statement: Not applicable.

Conflicts of Interest: The authors declare no conflict of interest.

Appendix A

Appendix A.1. Proof of Proposition 1

By construction, it holds that $1 - (\tanh(\phi/w_\phi))^2 > 0$ for all $q \in \partial\mathcal{D} \setminus \{q_d\}$. The gradient of ϕ with respect to q is given by

$$\nabla_q \phi = k_d \frac{q - q_d}{\|q - q_d\|^2} - \sum_{i=1}^{N} k_i \frac{q - q_i}{\|q - q_i\|^2}. \tag{A1}$$

Computing the inner product of $\nabla_q \phi$ and q yields:

$$\begin{aligned} (\nabla_q \phi)^T q &= k_d \frac{\|q\|^2 - q_d^T q}{\|q - q_d\|^2} - \sum_{i=1}^{N} k_i \frac{(q - q_i)^T q}{\|q - q_i\|^2} \\ &\geq k_d \frac{1 - q_d^T q}{\|q - q_d\|^2} - \max_i(k_i) \sum_{i=1}^{N} \frac{1}{\|q - q_i\|}. \end{aligned} \tag{A2}$$

Given that all q_i lie within int($\mathcal{D}$), the second term on the right-hand side of (A2) is finite for all $q \in \partial\mathcal{D}$. Similarly, the first term on the right-hand side of (A2) is positive for all $q \neq q_d$. Let $q' \in \partial\mathcal{D} \setminus \{q_d\}$. Additionally, the continuity of $(1 - q_d^T q)/\|q - q_d\|^2$ and $(1 - \tanh(\phi/w_\phi)^2)/(2w_\phi)$ implies that there exists a closed neighborhood $\mathcal{F}(q')$ of q', not containing q_d, where both are positive. Hence, selecting

$$k' = k_d \min_{q \in \mathcal{F}(q')} \left(\frac{1 - q_d^T q}{\|q - q_d\|^2} \frac{1}{\sum_{i=1}^{N} \frac{1}{\|q - q_i\|}} \right)$$

ensures that $(\nabla_q \phi)^T q > 0$ for all $q \in \mathcal{F}(q')$. Moreover, computing the derivative of $d = 1 - \|q\|^2$ with respect to time for all $q \in \mathcal{F}(q')$ and assuming $k_i < k'$, $\forall i \in \mathfrak{I}_N$ yields $\dot{d} = 2\mathcal{K}_u s \nabla_q \psi^T q > 0$; thus, the distance from the workspace boundary increases, which concludes the proof.

Appendix A.2. Proof of Proposition 2

Similarly to the proof of Proposition 3 in [33], we proceed by defining $\tilde{q}_d \triangleq q - q_d$, $\tilde{q}_i \triangleq q - q_i$ for all $i \in \mathfrak{I}_N$. Let also $\hat{q}_d \triangleq \tilde{q}_d/\|\tilde{q}_d\|$ and $\hat{q}_i \triangleq \tilde{q}_i/\|\tilde{q}_i\|$. Accordingly, the Hessian of ϕ can be computed by:

$$\nabla_q^2 \phi = \frac{k_d}{\|\tilde{q}_d\|^2} \left(\mathcal{I}_2 - 2\hat{q}_d \hat{q}_d^T \right) - \sum_{i \in \mathfrak{I}_N} \frac{k_i}{\|\tilde{q}_i\|^2} \left(\mathcal{I}_2 - 2\hat{q}_i \hat{q}_i^T \right). \tag{A3}$$

Note that at a critical point of ϕ it holds that:

$$\begin{aligned} k_d \frac{\hat{q}_d}{\|\tilde{q}_d\|} &= \sum_{i\in\mathfrak{I}_N} k_i \frac{\hat{q}_i}{\|\tilde{q}_i\|} \implies \\ \frac{k_d^2}{\|\tilde{q}_d\|^2}\hat{q}_d\hat{q}_d^T &= \sum_{i\in\mathfrak{I}_N} \frac{k_i^2}{\|\tilde{q}_i\|^2}\hat{q}_i\hat{q}_i^T + \\ &\sum_{i\in\mathfrak{I}_N}\sum_{j\in\mathfrak{I}_N\setminus\{i\}} \frac{k_i k_j}{\|\tilde{q}_i\|\|\tilde{q}_j\|}\left(\hat{q}_i\hat{q}_j^T + \hat{q}_j\hat{q}_i^T\right). \end{aligned} \tag{A4}$$

Substituting (A4) into (A3) and re-arranging the terms yields:

$$\begin{aligned} \nabla_p^2\phi = &\left(\frac{k_d}{\|\tilde{q}_d\|^2} - \sum_{i\in\mathfrak{I}_N}\frac{k_i}{\|\tilde{q}_i\|^2}\right)\mathcal{I} + 2\Bigg(\sum_{i\in\mathfrak{I}_N}\frac{k_i(k_d-k_i)}{k_d}\frac{1}{\|\tilde{q}_i\|^2}\hat{q}_i\hat{q}_i^T - \\ &\frac{1}{k_d}\sum_{i\in\mathfrak{I}_N}\sum_{j\in\mathfrak{I}_N\setminus\{i\}}\frac{k_ik_j}{\|\tilde{q}_i\|\|\tilde{q}_j\|}\left(\hat{q}_i\hat{q}_j^T + \hat{q}_j\hat{q}_i^T\right)\Bigg). \end{aligned}$$

Next, we argue that for any given set of radii $\rho_i > 0$ such that $\mathcal{D}_{\rho_i}(q_i)$, $i \in \mathfrak{I}_N$ are disjoint disks that lie entirely within our domain, there exists $k_d' > 0$ such that no critical point of ϕ exists within $\mathcal{D}\setminus\bigcup_{i\in\mathfrak{I}_N}\mathcal{D}_{\rho_i}(q_i)$ for all $k_d > k_d'$. This implies that, by choosing a sufficiently large k_d, each critical point of ϕ belongs to a single $\mathcal{D}_{\rho_i}(q_i)$. Let $q^\star$ be a critical point and $\ell = \operatorname{argmin}_{i\in\mathfrak{I}_N}\|q^\star - q_i\|$. To show that $\nabla_q^2\phi(q^\star)$ is not degenerate, it suffices to show that its eigenvalue $\lambda(q^\star)$ is positive. We recall that λ is lower bounded by the quadratic form $\hat{x}^T\nabla_q^2\phi\hat{x}$ for all $\|\hat{x}\| = 1$. By considering the direction of $\tilde{q}_\ell$ and after some tedious calculations, we obtain:

$$\begin{aligned} \hat{q}_\ell^T\nabla_q^2\phi(q^\star)\hat{q}_\ell = &\frac{k_d}{\|\tilde{q}_d\|^2} - \sum_{i\in\mathfrak{I}_N\setminus\{\ell\}}\frac{k_i}{\|\tilde{q}_i\|^2} \\ &+ \frac{k_\ell}{\|\tilde{q}_\ell\|^2}\left(\frac{k_d - 2k_\ell}{k_d} - 4\frac{\|\tilde{q}_\ell\|}{k_d}\sum_{i\in\mathfrak{I}_N\setminus\{\ell\}}\frac{k_i}{\|\tilde{q}_i\|}2(\hat{q}_\ell^T\hat{q}_i)\right) \\ &+ 2\sum_{i\in\mathfrak{I}_N\setminus\{\ell\}}\frac{k_i(k_d-k_i)}{k_d}\frac{1}{\|\tilde{q}_i\|^2}\left(\hat{q}_\ell^T\hat{q}_i\right)^2 \\ &- \frac{2}{k_d}\sum_{i\in\mathfrak{I}_N\setminus\{\ell\}}\sum_{j\in\mathfrak{I}_N\setminus\{i,\ell\}}\frac{k_ik_j}{\|\tilde{q}_i\|\|\tilde{q}_j\|}2(\hat{q}_\ell^T\hat{q}_i)(\hat{q}_\ell^T\hat{q}_j). \end{aligned} \tag{A5}$$

The first right-hand side term of (A5) is strictly positive. Since all k_i are bounded and non-negative, choosing a sufficiently large k_d renders the second and third right-hand side terms non-negative. Furthermore, note that the fourth and fifth right-hand side terms are bounded for all $q^\star \in \mathcal{D}_{\rho_\ell}(q_\ell)$. Thus, by choosing a sufficiently large k_d, the first three terms of (A5) can be made dominant, thus rendering $\hat{q}_\ell^T\nabla_q^2\phi\hat{q}_\ell$ positive at $q^\star$, which concludes the proof.

Appendix A.3. Proof of Proposition 3

Firstly, we will show that the robot cannot escape through the workspace's outer boundary. Let us assume that $q \to q' \in \partial\mathcal{D}\setminus\{q_d\}$. Then, $\dot{q} \to 0$ by virtue of (7), since $s(q,k) = 0$ for all $\|q\| = 1$ with $\left(\nabla_q\phi\right)^T q \le 0$. Additionally, $w_0 \to 1$ and $w_i \to 0$, for all $i \in \mathfrak{I}_N$. Thus, $\dot{k}_i < 0$ holds within a neighborhood of $\partial\mathcal{D}$, while $k_i > 0$, which implies that $k_i \to 0$ for all $i \in \mathfrak{I}_N$. Moreover, Proposition 1 dictates that there exists $k' > 0$ for which any point in $\partial\mathcal{D}\setminus\{q_d\}$ is repulsive under ψ. Since (12) dictates that all k_i become less than k' in finite time, this contradicts our supposition.

Next, we consider collision avoidance between the robot and the inner obstacles. Let us assume that the robot approaches obstacle i. By construction, $w_i \to 1$ while $\nabla_q \psi \to 0$ and $\overline{w}_j \to 0$ for all $j \in \mathfrak{I}_N^\star \setminus \{i\}$. Note that there exists a neighborhood $\mathcal{N}_i$ of q_i such that $w_0 = 0$ for all $q \in \mathcal{N}_i$ due to continuity of $\overline{w}_0$ and $\xi_2(\overline{w}_0; \epsilon_3)$. Additionally, since the robot is assumed to approach q_i, $\dot{q}^T(q - q_i)$ cannot be identically zero inside $\mathcal{N}_i$. As such, as long as $k_i < \overline{k}_i$, $\dot{k}_i \geq 0$ inside $\mathcal{N}_i$ without $\dot{k}_i = 0$ for all $q \in \mathcal{N}_i$. This implies that $k_i \not\to 0$ as $q \to q_i$, thus rendering q_i a local maximum of ψ. Thus, there exists a neighborhood of q_i inside which $(\nabla_q \psi)^T (q - q_i) > 0$, which contradicts our assumption.

Appendix A.4. Proof of Proposition 4

Let $V \triangleq \psi(q, k)$, as defined in (10), be a candidate Lyapunov function, which is non-negative and vanishes only when $q = q_d$. Differentiating V along the system's trajectories yields:

$$\dot{V} = \frac{1 - \tanh{(\phi/w_\phi)}^2}{2w_\phi} \left((\nabla_q \phi)^T \dot{q} + \ln\left(\frac{\|q - q_d\|}{2}\right) \dot{k}_d - \sum_{i \in \mathfrak{I}_N} \ln\left(\frac{\|q - q_i\|}{2}\right) \dot{k}_i \right). \tag{A6}$$

Given that $\dot{q} = J\dot{p}$, the first term of (A6) can be further expanded as follows:

$$(\nabla_q \phi)^T \dot{q} = -\mathcal{K}_u s \frac{1 - \tanh{(\phi/w_\phi)}^2}{2w_\phi} \|J^T \nabla_q \phi\|^2, \tag{A7}$$

which is non-positive for all $q \in \Omega$ and becomes zero only at the critical points of ψ. The second term of (A6) is non-positive since $\dot{k}_d \geq 0$ by construction and invariance of $\mathcal{W}$ (see Proposition 3) implies $\|q - q_d\| \leq 2$ which, in turn, implies $\ln\left(\frac{\|q-q_d\|}{2}\right) \leq 0$. Similarly, the sign of each term of the sum is determined solely by the sign of the corresponding $\dot{k}_i$. Substituting (12) yields:

$$-\sum_{i \in \mathfrak{I}_N} \ln\left(\frac{\|q - q_i\|}{2}\right) \dot{k}_i \leq -\sum_{i \in \mathfrak{I}_N} \ln\left(\frac{\|q - q_i\|}{2}\right) \left(\overline{k}_i - k_i\right) w_i \ell_i g_i. \tag{A8}$$

Given that $g_i \leq \|\nabla_q \psi\|^2$ and $\sum_{i \in \mathfrak{I}_N} \overline{k}_i w_i \leq 1$ by construction, expanding ℓ_i into the right-hand side of (A8) leads to:

$$\begin{aligned} -\sum_{i \in \mathfrak{I}_N} \ln\left(\frac{\|q - q_i\|}{2}\right) \left(\overline{k}_i - k_i\right) w_i \ell_i g_i &\leq \mathcal{K}_u s \|\nabla_q \psi\|^2 \sum_{i \in \mathfrak{I}_N} \left(\overline{k}_i - k_i\right) w_i \\ &\leq \mathcal{K}_u s \frac{1 - \tanh{(\phi/w_\phi)}^2}{2w_\phi} \|\nabla_q \phi\|^2. \end{aligned} \tag{A9}$$

Thus, (A6) is non-positive. Therefore, by invoking Lyapunov's Stability Theorem (Theorem 3.1 [41]) we may conclude that q_d is stable. Finally, LaSalle's Theorem (Theorem 3.4 [41]) dictates that the system will converge to the largest invariant set, which, in our case, consists of the critical points of ψ, thus concluding the proof.

Appendix A.5. Proof of Proposition 5

At the critical point $z^\star$ of system (17), the Hessian $\nabla_q^2 \phi$ of ϕ is non-degenerate, since otherwise $\dot{k}_d \neq 0$. Additionally, $q^\star \in \Omega \setminus \{q_d\}$ implies that $1 - \left(\tanh{(\phi/w_\phi)}\right)^2 \neq 0$. These two facts mean that $\nabla_q^2 \psi$ has two eigenvalues at $z^\star$, namely λ and $-\lambda$, with $\lambda > 0$. Computing the Jacobian of f_q with respect to q at $z^\star$ yields:

$$\begin{aligned} \nabla_q f_q &= -\mathcal{K}_u (\nabla_q \psi)^T \nabla_q s - \mathcal{K}_u s \nabla_q^2 \psi \\ &= -\mathcal{K}_u s \nabla_q^2 \psi \end{aligned}$$

since $\nabla_q \psi(q^\star) = 0$. Furthermore, by construction of the adaptive law (12), the Jacobian of f_k with respect to z at $z^\star$ is $\mathbf{0}_{(1+N)\times(3+N)}$. Thus, linearization of the system f_z at $z^\star$ yields

$$\nabla_z f_z(z^\star) = -\mathcal{K}_u s \frac{1-\tanh\left(\phi/w_\phi\right)^2}{2w_\phi} \begin{bmatrix} \nabla_q^2 \phi & \frac{\partial \phi}{\partial k_d} & \frac{\partial \phi}{\partial k_1} & \cdots & \frac{\partial \phi}{\partial k_N} \\ & \mathbf{0}_{(1+N)\times(3+N)} & & & \end{bmatrix}.$$

Since the top-left block $\nabla_q^2 \phi$ is invertible at $z^\star$, using the well-known property of block matrix determinants, we can see that $\nabla_{z^\star} f_z$ has two non-zero eigenvalues, particularly the eigenvalues of $\nabla_q^2 \psi$ and a zero eigenvalue with multiplicity $1+N$. Thus, $\nabla_z f_z(z^\star)$ has exactly one positive eigenvalue, rendering $z^\star$ a saddle point of (17) (Theorem 3.7 [41]).

Appendix A.6. Proof of Theorem 1

In Proposition 4, we have proven that $\dot{\psi} < 0$ for all $q \in \Omega \setminus \{q_d\}$, except for the critical points of ϕ that lie in it. Lasalle's Invariance Theorem (Theorem 3.4 [41]) dictates that system (17) will converge to either (a) the desired configuration q_d, (b) the obstacles q_i or (c) a critical point $z^\star = (q^\star, k^\star)$ with $q^\star \in \Omega \setminus \{q_d\}$. We know from Proposition 3 that the critical points of case (b) are repulsive; therefore, no trajectory of the system may converge to them. Regarding the critical point $z^\star$ corresponding to case (c), Proposition 5 dictates that it must be a non-isolated, degenerate equilibrium of the whole of system (17), since $\nabla_z f_z$ has one positive, one negative and several zero eigenvalues. Let $\bar{k}_d$ be the upper bound of k_d that the closed-loop system can possibly attain, as indicated by Proposition 2. In order to prove that the set of initial conditions leading to these points has zero Lebesgue measure, we will study the properties of the gradient-like system (by definition, a gradient-like system is a pair of a scalar cost functions and a dynamical system for which each non-equilibrium initial condition moves the state towards a new one whose cost is less than that of the initial state) $(\psi(z), F_{z,\tau}(z))$ in the domain $\mathcal{S}_z$, where the scalar potential $\psi(z)$ is treated as a function to be minimized, the map $F_{z,\tau}(z) : \mathcal{S}_z \mapsto \mathbb{R}^{N+3}$ is given by

$$F_{z,\tau}(z(t)) \triangleq z(t+\tau) = z(t) + \int_t^{t+\tau} f_z(z(s))ds$$

for any $\tau > 0$ and $\mathcal{S}_z \triangleq \mathcal{D} \times [1, \bar{k}_d] \times [0, \bar{k}_1] \times \ldots [0, \bar{k}_N]$. Note that $\mathcal{S}_z$ is convex and closed. Additionally, the map $F_{z,\tau}(z)$ is a locally Lipschitz diffeomorphism in $\mathcal{S}_z$ and $\mathcal{S}_z$ is forward invariant under $F_{z,\tau}(z)$ (by virtue of Proposition 3 and design of adaptive law (12)) for all $\tau > 0$. Furthermore, the unwanted equilibria of $F_{z,\tau}$ are strict saddles. Thus, following similar arguments as the proof of Theorem 3 in [42], we conclude that the set of all initial conditions that converge to these saddles has zero Lebesgue measure, which implies that almost every trajectory of the system converges to q_d, i.e., the only stable equilibrium of (17), thus completing the proof.

Appendix A.7. Proof of Theorem 2

We begin by noting that, by virtue of (21), we only need to study the trajectories of (19) in the workspace's image, since that motion is traced exactly by our robot. Considering the first part of the Theorem 2, we note that by following the same arguments as in the proof of Proposition 3, we may conclude that the robot cannot escape throught the workspace's outer boundary. Likewise, assuming that $q \to q_i$ for some $i \in \mathfrak{I}_N$ implies that $({n_\theta}^T J^T \nabla_q \psi)^T {n_{\hat{\theta}}}^T (q - q_i)$ cannot be identically zero in a neighborhood of q_i. As such, since $\dot{k}_i \geq 0$ in the neighborhood of q_i, k_i cannot vanish as the robot approaches q_i, which contradicts our original supposition.

To prove the second part of the Theorem 2, first we show that the only equilibria of the closed-loop system coincide with the critical points of ψ. Assuming that $s_v \neq 0$, it is readily seen that both inner products in (20) vanish simultaneously only when $\nabla_q \psi = 0$. Considering now the case when $s_v \neq 0$, we note that this can only happen when $q \in \partial\mathcal{D}$ and $n_{\hat{\theta}}$ is tangent to $\partial\mathcal{D}$. For $\hat{\omega}$ to also vanish when $s_v \neq 0$, the gradient $\nabla_q \psi$ should also be tangent to $\partial\mathcal{D}$. Recalling that the adaptive laws for k ensure that $\nabla_q \psi$ will eventually point inwards, we conclude that no equilibria other than the critical points of ψ exist.

Next, we consider ψ as a lyapunov candidate function, whose derivative along the systems trajectories is given by (A6) (note that ψ does not depend on θ). Substituting (20) into the first term of (A6) yields:

$$(\nabla_q \phi)^T \dot{q} = -\mathcal{K}_v s_v \frac{1 - \tanh{(\phi/w_\phi)}^2}{2w_\phi} \left((n_{\hat{\theta}})^T \nabla_q \phi\right)^2. \tag{A10}$$

Regarding the remaining terms of (A6), given that $g_{v,i} \le \left(n_{\hat{\theta}}{}^T \nabla_q \psi\right)^2$, one can readily verify that:

$$-\sum_{i \in \mathfrak{I}_N} \ln\left(\frac{\|q - q_i\|}{2}\right) \dot{k}_i \le -\mathcal{K}_v s_v \frac{1 - \tanh{(\phi/w_\phi)}^2}{2w_\phi} \left((n_{\hat{\theta}})^T \nabla_q \phi\right)^2. \tag{A11}$$

Thus, invoking Lyapunov's Stability Theorem (Theorem 3.1 [41]) and LaSalle's Theorem (Theorem 3.4 [41]) concludes the proof similarly to Proposition 4.

Appendix A.8. Proof of Theorem 3

Regarding the robot's safety under the closed-loop system, we note that when $p \in \mathcal{P}_{s_\ell} \setminus \mathcal{P}_{s_{\ell+1}}$ or $p \in \mathcal{P}_{s_n}$ for all $\ell < n$, the individual control laws render every point on the corresponding partition boundaries repulsive. When $p \in \mathcal{C}_{s_\ell, s_{\ell+1}}$, we note that, by construction, both $u^{[s_\ell]}$ and $u^{[s_{\ell+1}]}$ vanish when the robot approaches any point of $\partial\mathcal{P}_{s_\ell} \cap \partial\mathcal{P}_{s_{\ell+1}}$, preventing the robot from escaping. Additionally, the adaptive laws of each individual potential field will eventually render both $\nabla_p \psi_{s_\ell}$ and $\nabla_p \psi_{s_{\ell+1}}$ inward-looking with respect to $\mathcal{W}$, rendering $\partial\mathcal{P}_{s_\ell} \cap \partial\mathcal{P}_{s_{\ell+1}}$ repulsive.

While $p \in \mathcal{P}_{s_\ell}$, we consider $V \triangleq \psi_{s_\ell}$ as a Lyapunov function candidate and we examine its time derivative along the system's trajectories when $p \in \mathcal{C}_{s_\ell, s_{\ell+1}}$:

$$\begin{aligned} \dot{V} &= \left(\nabla_p \psi_{s_\ell}\right)^T \dot{p} + \left(\nabla_{k^{[s_\ell]}} \psi_{s_\ell}\right)^T \dot{k}^{[s_\ell]} \\ &= \left(\nabla_p \psi_{s_\ell}\right)^T \left(u^{[s_\ell]} + \eta_{c,s_\ell,s_{\ell+1}} \eta_{t,s_\ell,s_{\ell+1}} u^{[s_{\ell+1}]}\right) + \left(\nabla_{\dot{k}^{[s_\ell]}} \psi_{s_\ell}\right)^T k^{[s_\ell]} \\ &= \left(\nabla_p \psi_{s_\ell}\right)^T u^{[s_\ell]} + \left(\nabla_{\dot{k}^{[s_\ell]}} \psi_{s_\ell}\right)^T k^{[s_\ell]} + \eta_{c,s_\ell,s_{\ell+1}} \eta_{t,s_\ell,s_{\ell+1}} \left(\nabla_p \psi_{s_\ell}\right)^T u^{[s_{\ell+1}]}. \end{aligned} \tag{A12}$$

We recall that the first two right-hand side terms of (A12) are non-positive, as shown in Proposition 4. Likewise, the last term is rendered non-positive by virtue of $\eta_{c,s_\ell,s_{\ell+1}}$. Additionally, we note that the equilibrium points of the system in $p \in \mathcal{C}_{s_\ell, s_{\ell+1}}$ correspond only to critical points of ψ_{s_ℓ}. By virtue of $\eta_{c,s_\ell,s_{\ell+1}}$, which vanishes at a critical point of ψ_{s_ℓ}, along with its derivative, one can easily verify that the Jacobian of (22) is equal to the one of $u^{[s_\ell]}$, whose properties were studied in Proposition 2. Finally, following a similar procedure as in the proof of Theorem 1, we conclude that the system will converge to the specified goal configuration for almost all initial configurations.

References

1. Canny, J. *The Complexity of Robot Motion Planning*; MIT Press: Cambridge, MA, USA, 1988.
2. Mesesan, G.; Roa, M.A.; Icer, E.; Althoff, M. Hierarchical Path Planner Using Workspace Decomposition and Parallel Task-Space RRTs. In Proceedings of the IEEE International Conference on Intelligent Robots and Systems, Madrid, Spain, 1–5 October 2018; pp. 6524–6531.
3. Karaman, S.; Frazzoli, E. Sampling-based algorithms for optimal motion planning. *Int. J. Robot. Res.* **2011**, *30*, 846–894. [CrossRef]
4. Chi, W.; Wang, C.; Wang, J.; Meng, M.Q. Risk-DTRRT-Based Optimal Motion Planning Algorithm for Mobile Robots. *IEEE Trans. Autom. Sci. Eng.* **2019**, *16*, 1271–1288. [CrossRef]
5. Kavraki, L.E.; Svestka, P.; Latombe, J..; Overmars, M.H. Probabilistic roadmaps for path planning in high-dimensional configuration spaces. *IEEE Trans. Robot. Autom.* **1996**, *12*, 566–580. [CrossRef]

6. Chen, G.; Luo, N.; Liu, D.; Zhao, Z.; Liang, C. Path planning for manipulators based on an improved probabilistic roadmap method. *Robot. Comput.-Integr. Manuf.* **2021**, *72*, 102196. [CrossRef]
7. Salzman, O.; Hemmer, M.; Halperin, D. On the Power of Manifold Samples in Exploring Configuration Spaces and the Dimensionality of Narrow Passages. *IEEE Trans. Autom. Sci. Eng.* **2015**, *12*, 529–538. [CrossRef]
8. Kingston, Z.; Moll, M.; Kavraki, L.E. Sampling-Based Methods for Motion Planning with Constraints. *Annu. Rev. Control Robot. Auton. Syst.* **2018**, *1*, 159–185. [CrossRef]
9. Ogren, P.; Leonard, N.E. A convergent dynamic window approach to obstacle avoidance. *IEEE Trans. Robot.* **2005**, *21*, 188–195. [CrossRef]
10. Wen, J.; Zhang, X.; Gao, H.; Yuan, J.; Fang, Y. E^{3}MoP: Efficient Motion Planning Based on Heuristic-Guided Motion Primitives Pruning and Path Optimization With Sparse-Banded Structure. *IEEE Trans. Autom. Sci. Eng.* **2021**, *19*, 2762–2775. [CrossRef]
11. Bhattacharya, S.; Ghrist, R. Path Homotopy Invariants and their Application to Optimal Trajectory Planning. *Ann. Math. Artif. Intell.* **2018**, *84*, 139–160. [CrossRef]
12. Diaz-Arango, G.; Vázquez-Leal, H.; Hernandez-Martinez, L.; Pascual, M.T.S.; Sandoval-Hernandez, M. Homotopy Path Planning for Terrestrial Robots Using Spherical Algorithm. *IEEE Trans. Autom. Sci. Eng.* **2018**, *15*, 567–585. [CrossRef]
13. Khatib, O. Real-time obstacle avoidance for manipulators and mobile robots. In Proceedings of the IEEE International Conference on Robotics and Automation, St. Louis, MO, USA, 25–28 March 1985; pp. 500–505.
14. Koditschek, D. Exact robot navigation by means of potential functions: Some topological considerations. In Proceedings of the IEEE International Conference on Robotics and Automation, Raleigh, NC, USA, 31 March–3 April 1987; pp. 1–6.
15. Rimon, E.; Koditschek, D. Exact Robot Navigation Using Artificial Potential Functions. *IEEE Trans. Robot. Autom.* **1992**, *8*, 501–518. [CrossRef]
16. Loizou, S.G.; Jadbabaie, A. Density Functions for Navigation-Function-Based Systems. *IEEE Trans. Autom. Control* **2008**, *53*, 612–617. [CrossRef]
17. Filippidis, I.; Kyriakopoulos, K.J. Adjustable navigation functions for unknown sphere worlds. In Proceedings of the IEEE Conference on Decision and Control and European Control Conference, Orlando, FL, USA, 12–15 December 2011; pp. 4276–4281.
18. Filippidis, I.F.; Kyriakopoulos, K.J. Navigation Functions for everywhere partially sufficiently curved worlds. In Proceedings of the IEEE International Conference on Robotics and Automation, St Paul, MN, USA, 14–18 May 2012; pp. 2115–2120.
19. Paternain, S.; Koditschek, D.E.; Ribeiro, A. Navigation Functions for Convex Potentials in a Space With Convex Obstacles. *IEEE Trans. Autom. Control* **2018**, *63*, 2944–2959. [CrossRef]
20. Arslan, O.; Koditschek, D.E. Sensor-based reactive navigation in unknown convex sphere worlds. *Int. J. Robot. Res.* **2019**, *38*, 196–223. [CrossRef]
21. Connolly, C.; B Burns, J.; Weiss, R. Path Planning Using Laplace's Equation. In Proceedings of the IEEE International Conference on Robotics and Automation, Cincinnati, OH, USA, 13–18 May 1990; pp. 2102–2106.
22. Mantegh, I.; Jenkin, M.R.M.; Goldenberg, A.A. Path Planning for Autonomous Mobile Robots Using the Boundary Integral Equation Method. *J. Intell. Robot. Syst.* **2010**, *59*, 191–220. [CrossRef]
23. Golan, Y.; Edelman, S.; Shapiro, A.; Rimon, E. Online Robot Navigation Using Continuously Updated Artificial Temperature Gradients. *IEEE Robot. Autom. Lett.* **2017**, *2*, 1280–1287. [CrossRef]
24. Kim, J.O.; Khosla, P.K. Real-time obstacle avoidance using harmonic potential functions. *IEEE Trans. Robot. Autom.* **1992**, *8*, 338–349. [CrossRef]
25. Merheb, A.; Gazi, V.; Sezer-Uzol, N. Implementation Studies of Robot Swarm Navigation Using Potential Functions and Panel Methods. *IEEE/ASME Trans. Mechatron.* **2016**, *21*, 2556–2567. [CrossRef]
26. Velagić, J.; Vuković, L.; Ibrahimović, B. Mobile Robot Motion Framework Based on Enhanced Robust Panel Method. *Int. J. Control Autom. Syst.* **2020**, *18*, 1264–1276. [CrossRef]
27. Feder, H.J.S.; Slotine, J.J.E. Real-time path planning using harmonic potentials in dynamic environments. In Proceedings of the IEEE International Conference on Robotics and Automation, Albuquerque, NM, USA, 20–25 April 1997; pp. 874–881.
28. Marchidan, A.; Bakolas, E. A local reactive steering law for 2D collision avoidance with curvature constraints and constant speed. *Robot. Auton. Syst.* **2022**, *155*, 104156. [CrossRef]
29. e Silva, E.P.; Engel, P.M.; Trevisan, M.; Idiart, M.A. Exploration Method Using Harmonic Functions. *Robot. Auton. Syst.* **2002**, *40*, 25–42. [CrossRef]
30. Maffei, R.; Souza, M.P.; Mantelli, M.; Pittol, D.; Kolberg, M.; Jorge, V.A.M. Exploration of 3D terrains using potential fields with elevation-based local distortions. In Proceedings of the IEEE International Conference on Robotics and Automation, Paris, France, 31 May–31 August 2020; pp. 4239–4244.
31. Voruganti, H.; Dasgupta, B.; Hommel, U. A novel potential field based domain mapping method. In Proceedings of the 10th WSEAS Conference on Computers (ICCOMPP06), Athens, Greece, 13 July 2006; pp. 655–661.
32. Gautam Pradeepkumar, B.; Gondegaon, S.; Voruganti, H.K. Domain Mapping for Path Planning and Mesh Generation. In Proceedings of the International Conference on Theoretical, Applied, Computational and Experimental Mechanics, Kharagpur, India, 29–31 December 2014.
33. Loizou, S.G. Closed form Navigation Functions based on harmonic potentials. In Proceedings of the IEEE Conference on Decision and Control and European Control Conference, Orlando, FL, USA, 12–15 December 2011; pp. 6361–6366.
34. Loizou, S.G. The Navigation Transformation. *IEEE Trans. Robot.* **2017**, *33*, 1516–1523. [CrossRef]

35. Loizou, S.G. Navigation functions in topologically complex 3-D workspaces. In Proceedings of the American Control Conference, Montreal, QC, Canada, 27–29 June 2012; pp. 4861–4866.
36. Loizou, S.G. The Multi-Agent Navigation Transformation: Tuning-Free Multi-Robot Navigation. In Proceedings of the Robotics: Science and Systems, Berkeley, CA, USA, 14–16 July 2014; pp. 1–9.
37. Vlantis, P.; Vrohidis, C.; Bechlioulis, C.P.; Kyriakopoulos, K.J. Robot Navigation in Complex Workspaces Using Harmonic Maps. In Proceedings of the IEEE International Conference on Robotics and Automation, Brisbane, Australia, 21–25 May 2018; pp. 1726–1731.
38. Duren, P.; Hengartner, W. Harmonic mappings of multiply connected domains. *Pac. J. Math.* **1997**, *180*, 201–220. [CrossRef]
39. Biringen, S.; Chow, C.Y. *An Introduction to Computational Fluid Mechanics by Example*; John Wiley & Sons, Inc.: Hoboken, NJ, USA, 2011.
40. Kuethe, A.M.; Chow, C.Y.; Fung, Y.C. *Foundations of Aerodynamics: Bases of Aerodynamics Design*, 5th ed.; John Wiley & Sons, Inc.: Hoboken, NJ, USA, 1997.
41. Khalil, H.K. *Nonlinear Systems*; Prentice-Hall: Hoboken, NJ, USA, 1996.
42. Panageas, I.; Piliouras, G. Gradient Descent Only Converges to Minimizers: Non-Isolated Critical Points and Invariant Regions. *arXiv* **2016**, arXiv:1605.00405.

sensors

MDPI

Review

Multi-Agent Deep Reinforcement Learning for Multi-Robot Applications: A Survey

James Orr and Ayan Dutta *

School of Computing, University of North Florida, Jacksonville, FL 32224, USA
* Correspondence: a.dutta@unf.edu

Abstract: Deep reinforcement learning has produced many success stories in recent years. Some example fields in which these successes have taken place include mathematics, games, health care, and robotics. In this paper, we are especially interested in multi-agent deep reinforcement learning, where multiple agents present in the environment not only learn from their own experiences but also from each other and its applications in multi-robot systems. In many real-world scenarios, one robot might not be enough to complete the given task on its own, and, therefore, we might need to deploy multiple robots who work together towards a common global objective of finishing the task. Although multi-agent deep reinforcement learning and its applications in multi-robot systems are of tremendous significance from theoretical and applied standpoints, the latest survey in this domain dates to 2004 albeit for traditional learning applications as deep reinforcement learning was not invented. We classify the reviewed papers in our survey primarily based on their multi-robot applications. Our survey also discusses a few challenges that the current research in this domain faces and provides a potential list of future applications involving multi-robot systems that can benefit from advances in multi-agent deep reinforcement learning.

Keywords: deep reinforcement learning; multi-robot systems; multi-agent learning; survey

check for updates

Citation: Orr, J.; Dutta, A. Multi-Agent Deep Reinforcement Learning for Multi-Robot Applications: A Survey. *Sensors* **2023**, *23*, 3625. https://doi.org/10.3390/s23073625

Academic Editors: David Cheneler and Stephen Monk

Received: 23 January 2023
Revised: 22 March 2023
Accepted: 28 March 2023
Published: 30 March 2023

1. Introduction

In a multi-robot application, several robots are usually deployed in the same environment [1–3]. Over time, they interact with each other via radio communication, for example, and coordinate to complete a task. Application areas include precision agriculture, space exploration, and ocean monitoring, among others. However, in all such real-world applications, many situations might arise that have not been thought of before deployment, and, therefore, the robots must need to plan online based on their past experiences. Reinforcement learning (RL) is one computing principle that we can use to tackle such dynamic and non-deterministic scenarios. Its primary foundation is trial and error—in a single-agent setting, the agent takes an action in a particular state of the environment, receives a corresponding reward, and transitions to a new state [4]. Over time, the agent learns which state–action pairs are worth re-experiencing based on the received rewards and which ones are not [5]. However, the number of state–action pairs becomes intractable, even for smallish computational problems. This has led to the technique known as deep reinforcement learning (DRL), where the expected utilities of the state–action pairs are approximated using deep neural networks [6]. Such deep networks can have hundreds of hidden layers [7]. Deep reinforcement learning has recently been used in finding a faster matrix multiplication solution [8], for drug discovery [9], to beat humans in Go [10], play Atari [6], and for routing in communication networks [11], among others. Robotics is no different—DRL has been used in applications ranging from path planning [12] and coverage [13] to locomotion learning [14] and manipulation [15].

Going one step further, if we introduce multiple agents to the environment, this increases the complexity [16]. Now, the agents not only need to learn from their own observations in the environment but also be mindful of other agents' transitions. This essentially means that one agent's reward may now be influenced by the actions of other agents, and this might lead to a non-stationary system. Although inherently more difficult, the use of multiple robots and, consequently, a multi-agent reinforcement learning framework for the robots is significant [17]. Such learning multi-robot systems (MRS) may be used for precision agriculture [18], underwater exploration [19], search and rescue [20], and space missions [21]. Robots' onboard sensors play a significant role in such applications. For example, the state space of the robots might include the current discovered map of the environment, which could be created by the robots' laser scanners [22]. The state might also include locations and velocities, for which the robot might need sensory information from GPS or an overhead camera [23]. Furthermore, vision systems, such as regular or multi-spectral cameras, can be used by the robots to observe the current state of the environment, and data collected by such cameras can be used for robot-to-robot coordination [24]. Therefore, designing deep reinforcement learning algorithms, potentially lightweight and sample-efficient, that will properly utilize such sensory information, is not only of interest to the artificial intelligence research community but to robotics as well. However, the last survey that reviewed the relevant multi-robot system application papers that use multi-agent reinforcement learning techniques was conducted by Yang and Gu in 2004 [17]. Note that the entire sub-field of *DRL was not invented until 2015* [6].

In this paper, we fill this significant void by reviewing and documenting relevant MRS papers that specifically use multi-agent deep reinforcement learning (MADRL). Since today's robotic applications can have a large state space and, potentially, large action spaces, we believe that reviewing only the DRL-based approaches, and not the classic RL frameworks, is of interest to the relevant communities. The primary contribution of this article is that, to the best of our knowledge, this is the *only* study that surveys multi-robot applications via multi-agent deep reinforcement learning technologies. This survey provides a foundation for future researchers to build upon in order to develop state-of-the-art multi-robot solutions, for applications ranging from task allocation and swarm behavior modeling to path planning and object transportation. An illustration of this is shown in Figure 1.

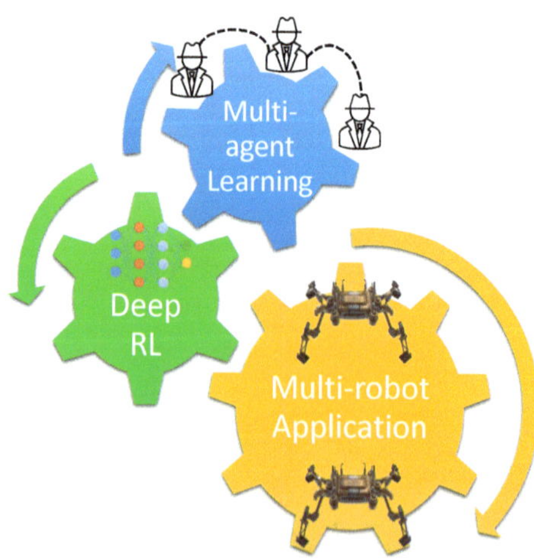

Figure 1. The main contribution of this article is that we have reviewed the latest multi-robot application papers that use multi-agent learning techniques via deep reinforcement learning. Readers will be able to find out how these three concepts are used together in the discussed studies and this survey will provide them with insight into possible future developments in this field, which, in turn, will advance the state-of-the-art.

We first provide a brief technical background and introduce terminologies necessary to understand some of the concepts and algorithms described in the reviewed papers

(Section 2). In Section 3, we categorize the multi-robot applications into (1) coverage, (2) path planning, (3) swarm behavior, (4) task allocation, (5) information collection, (6) pursuit–evasion, (7) object transportation, and (8) construction. We identify and discuss a list of crucial challenges that, in our opinion, the current studies in the literature face in Section 4, and then, finally, we conclude.

2. Background

In this section, we provide technical backgrounds on relevant computing principles.

2.1. MDP and Q-Learning

Let S and A denote the set of all states and actions available to an agent. Let R: $S \times A \to \mathbb{R}$ denote a reward function that gives the agent a virtual reward for taking action $a \in A$ in state $s \in S$. Let T denote the transition function. In a deterministic world, T: $S \times A \to S$, i.e., the actions of the agent is deterministic, whereas in a stochastic world, these actions might be probabilistic—T: $S \times A \to prob(S)$. We can use a Markov Decision Process (MDP) to model such a stochastic environment, which is defined as a tuple $\langle S, A, T, R \rangle$. The objective is to find a (optimal) policy π: $S \to A$ that maximizes the expected cumulative reward. To give higher preference to the immediate rewards than to the future ones, we discount the future reward values. The sum of the discounted rewards is called value. Therefore, to solve an MDP, we will maximize the expected value (V) over all possible sequences of states. Thus, the expected utility in a state $s \in S$ can be recursively defined as follows:

$$V(s) = R(s,a) + \gamma \max_{a' \in A} \sum_{s'} P(s'|s,a)V(s') \tag{1}$$

The above is called the Bellman equation, where $P(s'|s,a)$ is the probability of transitioning into s' from s by taking an action a. We can use value or policy iteration algorithms to solve an MDP.

However, in a situation where the R and T functions are unknown, the agent will have to try out different actions in every state to know which states are good and what action it should take in a particular state to maximize its utility. This leads to the idea of reinforcement learning (RL) where the agent will execute a in state s of the environment, and will receive a reward signal R from the environment as a result. Over time, the agent will learn the optimal policy based on this interaction between the agent and the environment [25]. An illustration is shown in Figure 2. In a *model-based* RL, the agent learns an empirical MDP by using estimated transition and reward functions. Note that these functions are approximated by interacting with the environment as mentioned earlier. Next, similar to an MDP, value or policy iteration algorithm can be employed to solve this empirical MDP model. In a *model-free* RL, the agent does not have access to T and R. This is true for numerous robotic applications in the real world. Therefore, most of the robotics papers we review in this survey use model-free RL techniques. This is also true for RL algorithms in general.

The goal of RL is to find a policy that maximizes the expected reward of the agent. Temporal difference learning is one of the most popular approach in model-free RL to learn learn the optimal utility values of each state. Q-learning is one such model-free RL technique, where the Q-value of a state–action pair (s,a) indicate the expected usefulness of that pair, which is updated as follows.

$$Q(s,a) = (1-\alpha)Q(s,a) + \alpha(R(s,a) + \gamma \max_{a' \in A} Q(s',a')) \tag{2}$$

α is the learning rate that weighs the new observations against the old. It is off-policy learning and converges to an optimal policy π^* following

$$\pi^*(s) = \arg\max_{a \in A} Q(s,a) \tag{3}$$

An excellent overview of classic RL applications in robotics can be found in [26]. Keeping track of Q-values for all possible state–action pairs in such an RL setting becomes infeasible with, for example, a million such combinations. In recent years, artificial neural networks have been used to approximate the optimal Q-values instead of storing the values in a table. This has given birth to the domain of *deep* reinforcement learning.

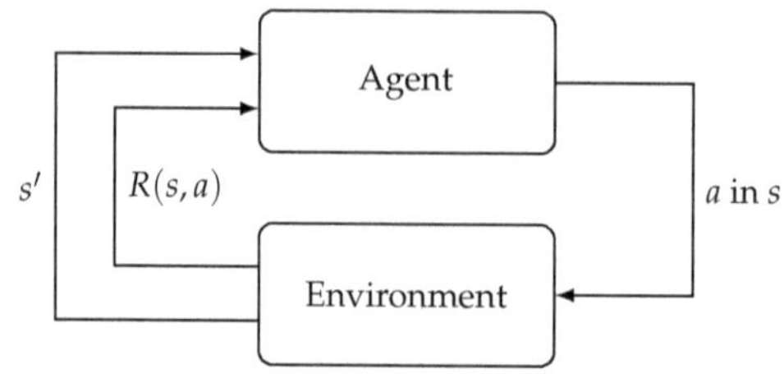

Figure 2. Illustration of Reinforcement Learning.

2.2. Multi-Agent Q-Learning

Assuming that the state space S is shared among n agents N and that there exists a common transition function T, an MDP for N is represented by the following tuple $\langle N, S, \mathbf{A}, \mathbf{O}, T, \mathbf{R} \rangle$, where the joint action space is denoted by $\mathbf{A} \leftarrow A_1 \times A_2 \cdots \times A_n$; the joint reward is denoted by $\mathbf{R} \leftarrow R_1 \times R_2 \cdots \times R_n$; and $\mathbf{O}$ denotes the joint observation of the agents, $\mathbf{O} \leftarrow O_1 \times O_2 \cdots \times O_n$. As there is more than one agent present, the action of one agent can potentially affect the reward and the consequent actions of the other agents. Therefore, the goal is to find a joint policy $\boldsymbol{\pi}^*$. However, due to the non-stationary environment and, consequently, the removal of the Markov property, convergence cannot be guaranteed unlike the single-agent setting [27]. One of the earliest approaches to learning a joint policy for two competitive agents is due to Littman [28]. It was modeled as a zero-sum two-player stochastic game (SG). It is also known as Markov Game in game theory. In SG, the goal is to find the Nash equilibrium, assuming the R and T functions are known. In a Nash equilibrium, the agents (or the players) will not have any incentive to change their adopted strategies. We slightly abuse the notation here and denote the strategy of agent N_i with π_i. Therefore, in a Nash equilibrium, the following is true

$$V_i^{\pi_i^*, \pi_{-i}^*}(s) \geq V_i^{\pi_i \pi_{-i}^*}(s), \forall \pi_i \tag{4}$$

where $V(s)$ denotes the value of state $s \in S$ to the i-th agent and π_{-i} is the strategy of the other players. Here, we assume the agents to be rational, and, therefore, all the agents always follow their optimal strategies. This general SG setting can now be used to solve multi-agent reinforcement learning (MARL) problems.

In a cooperative setting, the agents have a common goal in mind. Most of the studies in the robotics literature that use MARL use such a cooperative setting. In this case, the agents have the same reward function, R. Given this, all the agents in N will have the same value function, and, consequently, the same Q-function. The Nash equilibrium will be the optimal solution for this problem. Two main types of learning frameworks are prevalent—independent and joint learners. In an independent learning scenario, each agent ignores the presence of other agents in the environment and considers their influence as noise. The biggest advantage is that each agent/robot can implement its own RL algorithm and there is no need for coordination and, consequently, a joint policy calculation [16,27]. Independent classic Q-learners have shown promising results in AI [29,30], as well as in robotics [31,32]. On the other hand, the joint learners aim to learn the joint optimal policy from $\mathbf{O}$ and $\mathbf{A}$. Typically, an explicit coordination, potentially via communication in an MRS, is in place and the agents learn a better joint policy compared to the independent learners [16,27]. However, the complexity increases exponentially with the number of agents causing these not to scale very well. The joint Q-learning algorithms are also popular in robotics [24,33,34], as well as in general AI [28,35]. A comprehensive survey for

MARL techniques can be found in [27,36]. The authors in [36] also discuss the application domains for MARL, which includes multi-robot teams. A specific relevant example that is discussed is multi-robot object transportation.

2.3. (Multi-Agent) Deep Q-Learning

As the state and the action spaces increase in size, maintaining a table for the Q-values for all possible state–action pairs might be infeasible. To tackle this challenge, Mnih et al. [6] have proposed a neural network-based approach to approximate the Q-values directly from the sensory inputs. This has given birth of 'deep' Q-learning, as the Q-values of the state–action pairs are updated using a deep neural network.

2.3.1. Q-Networks

In their seminal paper, Mnih et al. [6] have proposed DQN—a convolutional neural network (CNN) to approximate the Q-values for a single agent. This is called the Q-network, which is parameterized by θ. The current state s_t is passed as an input to the network that outputs the Q-values for all the possible actions. An action is chosen next based on the highest Q-value, i.e.,

$$a^* = \arg\max_{a \in A} Q(s_t, a) \tag{5}$$

To ensure that the agent explores the state space enough, a^* is chosen with probability ϵ and the agent takes a random action with $(1 - \epsilon)$ probability. Due to this action, the state transitions to s_{t+1}. To avoid instability, a *target* network is maintained—it is identical to the Q network, but the parameter set θ is periodically copied to the parameters of this target network, θ^-. The state transitions are maintained in an experience replay buffer $\mathcal{D}$. Mini-batches from $\mathcal{D}$ are selected and target Q-values are predicted. θ is regressed toward the target values by finding the gradient descent of the following temporal loss function

$$\mathcal{L} = \mathbb{E}[(y_t - Q(s_t, a_t))^2] \tag{6}$$

$$y_t = R + \gamma Q(s_{t+1}, \arg\max Q(s_{t+1}, a_{t+1})) \tag{7}$$

One of the most popular extensions of DQN is Double DQN (DDQN) [37], which reduces overestimation in Q-learning. DDQN uses the Q-network for action selection following the ϵ-greedy policy, as mentioned above, but uses the target network for the evaluation of the state–action values. DQN and DDQN are extremely popular in robotics [13,38–41]. A visual working procedure of the generic DQN algorithm is presented in Figure 3.

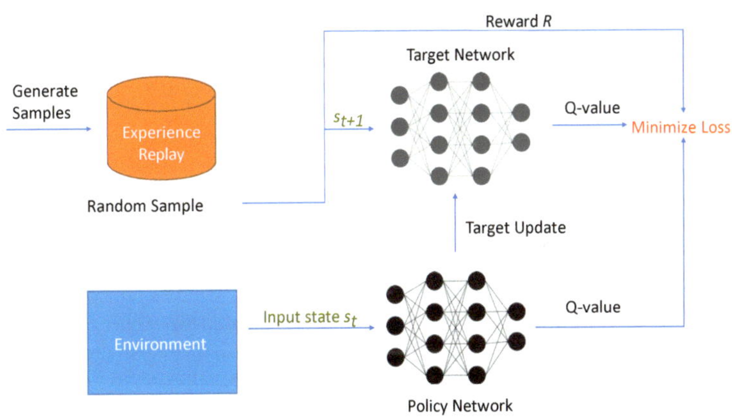

Figure 3. An illustration of the DQN architecture with a target network and an experience replay.

2.3.2. Policy Optimization Techniques

In policy optimization methods, the neural network outputs the probability distribution of these actions instead of outputting the Q-values of the available actions. Instead of using something like the ϵ-greedy strategy to derive a policy from the Q-values, the actions with higher probability outputs from the network will have higher chances of being selected. Let us denote a θ-parameterized policy by π_θ. The objective is to maximize the cumulative discounted reward

$$J(\theta) = \mathbb{E}[\mathcal{R}|\pi_\theta] \tag{8}$$

where $\mathcal{R}$ is the finite-horizon discounted cumulative reward. By optimizing the parameter set θ, e.g., by following the gradient of the policy, we aim to maximize the expected reward. Similar to the Q-networks, the learning happens in episodes. In general, the parameters in episode $i+1$, θ_{i+1}, will be an optimized version of θ_i as the following standard gradient ascent formula

$$\theta_{i+1} = \theta_i + \alpha \frac{\delta J(\theta_i)}{\delta \theta_i}. \tag{9}$$

In the vanilla form, similar to the Q-networks, the mean square error between the value of the policy (usually approximated using a neural network) and the reward-to-go (i.e., the sum of rewards received after every state transition so far) is calculated and the approximate value function parameters are regressed. Some of the popular policy optimization techniques include Deep Deterministic Policy Gradient (DDPG) [42], Proximal Policy Optimization (PPO) [43], Trust Region Policy Optimization (TRPO) [44], and Asynchronous Advantage Actor–Critic (A3C) [45], among others. Among these, DDPG is one of the most widely used for multi-robot applications [46–50]. It learns a Q-function similar to DQN and uses that to learn a policy. The policy DDPG learns is deterministic and the objective of this is to find actions that maximize the Q-values. As the action space A is assumed to be continuous, the Q-function is differentiable. To optimize θ and update the policy, we perform one-step policy ascent as follows:

$$\max_\theta \mathbb{E}_{s \in \mathcal{D}}[Q(s, \pi_\theta(s))]. \tag{10}$$

DDPG uses a sophisticated technique called actor–critic to achieve the successful combination of these two types of deep Q-learning. The actor essentially represents the policy and the critic represents the value network, respectively. The actor is updated towards the target and the critic is regressed by minimizing the error with the target [51]. The difference between the expected state value and the Q-value for an action a is called the *advantage*. One of the most popular algorithms that uses such an actor–critic framework is A3C [45]. In this algorithm, parallel actors explore the state space via different trajectories making the algorithm asynchronous; therefore, it does not require maintaining an experience replay. Another popular algorithm in the multi-robot domain is PPO, potentially because of its relatively simple implementation [43]. PPO-clip and PPO-penalty are its two primary variants that are used in robotics [52–57].

2.3.3. Extensions to Multi-Agent

As described earlier, in independent learning frameworks, any of the previously mentioned deep RL techniques, such as DQN, DDPG, A3C, or PPO, can be implemented on each agent. Note that no coordination mechanism is needed to be implemented for this [16,27,58].

For multi-agent DQN, a common experience memory can be used, which will combine the transitions of all the agents, and, consequently, they will learn from their global experiences while virtually emulating a stationary environment. Each agent can have its own network that will lead it to take an action from its Q-values [59]. Yang et al. [60] have proposed a mean field Q-learning algorithm for large-scale multi-agent learning applications. A mean-field formulation essentially brings down the complexity of an n-agent learning problem to a 2-agent learning problem by creating a virtual mean agent from the other $(n-1)$ agents in the environment. In [61], the authors have introduced the multi-agent extension of DDPG (MADDPG). Here, the actor remains decentralized, but the critic is centralized. Therefore, the critic needs information on the actions, observations, and target policies of all of the agents to evaluate the quality of the joint actions. Figure 4 shows an illustration of this process. Yu et al. [62] have proposed a multi-agent extension of PPO in cooperative settings (MAPPO). Similar to MADDPG, it uses centralized training with a decentralized execution strategy.

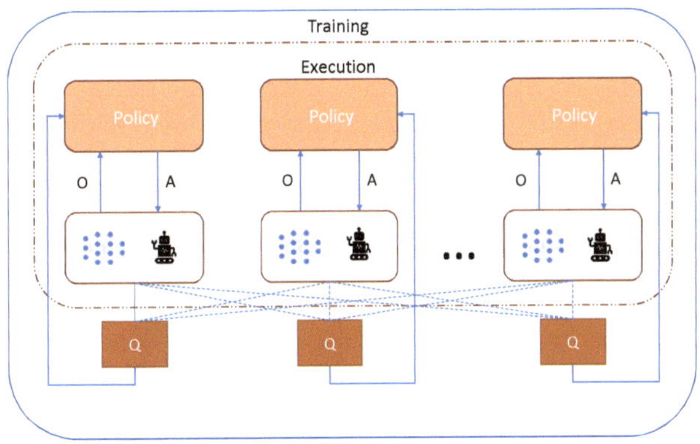

Figure 4. Illustration of multi-agent DDPG (MADDPG) [61].

Another approach to extending a single-agent DRL algorithm to a multi agent setting is to model it as a centralized RL, where all the information from agents is input together. This might create an infeasibly large state and action space for the joint agent. To alleviate this, researchers have looked into how to find each agent's contribution to the joint reward. This is named Value Function Factorization. VDN [63] is one such algorithm for cooperative settings where the joint Q-value is the addition of the local Q-values of the agents. A summary of the main types of RL algorithms used in multi-robot applications is presented in Table 1. The reader is referred to [64,65] for recent comprehensive surveys on state-of-the-art MADRL techniques and challenges. Furthermore, Oroojlooy and Hajinezhad [66] have recently published a survey paper reviewing the state-of-the-art MADRL algorithms specifically for cooperative multi-agent systems. As in most of the scenarios, the robots in an MRS work together towards solving a common problem, we believe that the survey in [66] would be a valuable asset for the robotics community.

Table 1. Types of deep RL algorithms used in the surveyed papers are listed. If a popular algorithm is used as a foundation, the algorithm's name is also mentioned within parentheses.

Q-Networks	Policy Gradients		
	DDPG	PPO	Other
[23,24,39,40,59,67–92] (QMIX), [38] (DDQN), [93] (DDQN), [94] (DDQN), [95] (DQN), [96] (DQN)	[46–50,97–106],	[22,52–57,107–128]	[86,88,129–140] (TRPO), [141] (TRPO), [81] (TRPO), [142] (TD3), [143] (SAC), [144] (SAC)

3. Multi-Robot System Applications of Multi-Agent Deep Reinforcement Learning

A summary of the discussed multi-robot applications is presented in Figure 5.

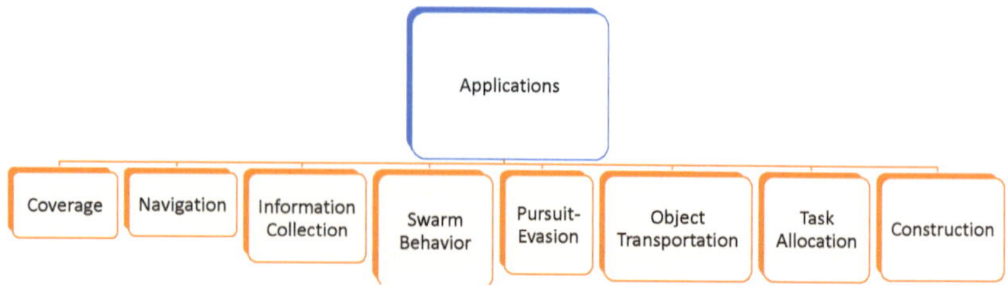

Figure 5. Multi-robot system applications that primarily use MADRL techniques.

3.1. Coverage and Exploration

The goal of an MRS in a coverage path planning (CPP) application is that every point in the environment is visited by at least one robot while some constraints are satisfied (e.g., no collision among the robots) and user-defined criteria are optimized (e.g., minimizing the travel time) [145]. CPP is one of the most popular topics in robotics. For multi-robot coverage, several popular algorithms exist even with performance guarantees and worst-case time bounds [146–149]. In exploration, however, the objective might not be the same as the multi-robot CPP problem. It is assumed that the sensor radius $r > 0$, and, therefore, the robots do not need to visit all the points on the plane. For example, the robots might be equipped with magnetic, acoustic, or infrared sensors in ground and aerial applications whereas a group of underwater vehicles might be equipped with water temperature and current measuring sensors. The robots will need GPS for outdoor localization. Such exploration can be used for mapping and searching applications among others [150–152]. Constraints such as maintaining wireless connectivity for robots with limited communication ranges might be present [153]. Inter-robot communication can be achieved via ZigBee or Wi-Fi. An example is shown in Figure 6.

Mou et al. [68] studied area coverage problems and proposed deep reinforcement learning for UAV swarms to efficiently cover irregular three-dimensional terrain. The basis of their UAV swarm structure is with the leader and the follower UAVs. The authors implement an observation history model based on convolutional neural networks and a mean embedding method to address limited communication. Li et al. [69] proposed the use of DDQN to train individual agents in a simulated grid-world environment. Then during the decision-making stage, where previously trained agents are placed in a test environment, the authors use their proposed multi-robot deduction method, which has foundations in Monte Carlo Tree Search. Zhou et al. [154] have developed a multi-robot coverage path planning mechanism that incorporates four different modules: (1) a map module, (2) a communication module, (3) a motion control module, and (4) a path generation module. They implement an actor–critic framework and natural gradient for updating the network. Up to three robots have been used in a simulation for testing the proposed coverage technique in a grid world. The two cornerstones of the study by Hu et al. [155] are (1) Voronoi partitioning-based area assignment to the robots and (2) the proposed DDPG-based DRL technique for the robots to have a collision-avoidance policy and evade objects in the field. The control of the robots is provided by the underlying neural network. The authors use a Prioritised Experience Replay (PER) [156] to store human demonstrations. The simulation was performed within Gazebo [157] and three Turtlebot3 Waffle Pi mobile robots were used to explore an unknown room during validation. Bromo [53], in his thesis, used MADRL on a team of UAVs using a modified version of PPO to map an area. During training, the policy function is shared among the robots and updated based on their current paths.

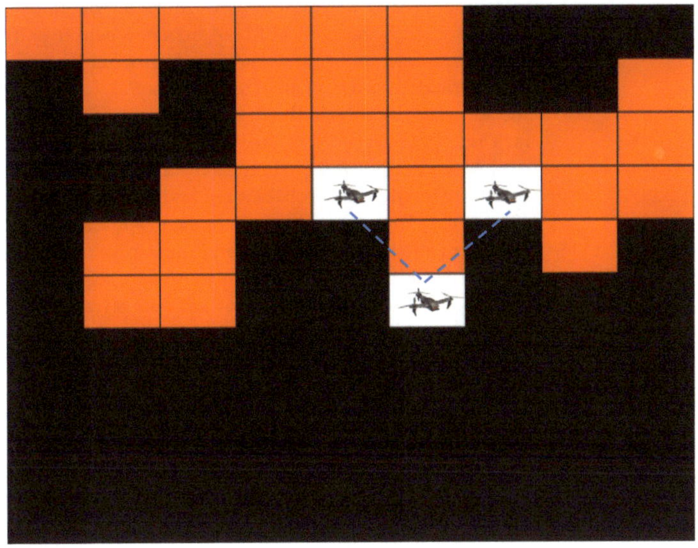

Figure 6. A multi-robot coverage scenario under continuous connectivity: the black, orange, and white cells represent the unvisited, previously visited, and the current cells of robots, respectively. The dotted lines represent the available communication channels among the robots. The goal is to cover the entire environment while maintaining a connected communication network throughout the mission.

For multi-UAV coverage, Tolstaya et al. [110] use graph neural networks (GNNs) [158] as a way for the robots to learn the environment through the abstractions of nodes and edges. GNNs have been successfully used in various coordination problems for multi-robot systems, and more recently, Graph Convolution Networks (GCNs) have been used [158]. The authors in this paper use "behavior cloning" as a heuristic to train the GNN on robots' previous experiences. In order for the individual UAVs to learn information distant from their position, they use up to 19 graph operation layers. PPO is the base DRL algorithm in this paper. Aydemir and Cetin [159] proposed a distributed system for the multi-UAV coverage in partially observable environments using DRL. Only the nearby robots share their state information and observations with each other. Blumenkamp et al. [111] developed a framework for decentralized coordination of an MRS. The RL aspect of their system uses GNNs and PPO. The agents train and develop a policy within a simulated environment and then the physical implementation of the policy with the robots occurs in a test environment. The authors also compare centralized control and communication levels to decentralized decision-making.

Similarly, Zhang et al. [160] have also proposed to employ graph neural networks for multi-robot exploration. The authors emphasize the "coarse-to-fine" exploration method of the robots, where the graph representation of the state space to be explored is explored in "hops" of greater detail. Simulation experiments involved up to 100 robots. Exploration can also be used for searching for a target asset. Liu et al. [84] have proposed a novel algorithm for cooperative search missions with a group of unmanned surface vehicles. Their algorithm makes use of two modules based on a divide-and-conquer architecture: an environmental sense module that utilizes sensing information and a policy module that is responsible for the optimal policy of the robots. Gao and Zhang [161] study a cooperative search problem while using MADRL as the solution method. The authors use independent learners on the robots to find the Nash equilibrium solution with the incomplete information available to the robots. Setyawan et al. [101] also use MADRL for multi-robot search and exploration. Unlike the previously mentioned studies, the authors have adopted a hierarchical RL approach, where they break down an abstraction of the global problem space into smaller sub-problem levels in order for the robot system to more efficiently learn in an actor–critic style. The lowest level in this order decides the robots' motor actions in the field. Sheng et al. [162] propose a novel probability density factorized multi-agent DRL method for solving the multi-robot reliable search problem. According to this study, when each robot follows its own policy to maximize its own reliability metric (e.g., probability of finding the target), the global reliability metric is also maximized. The authors implement the proposed technique on multiple simulated search environments including offices and museums, as well as on real robots. Another study in a similar application domain is done by Xia et al. [127]. Specifically, the authors have used MADRL for the multi-agent multi-target hunting problem. The authors make use of a feature embedding block to extract features from the agents' observations. The neural network architecture uses fully connected layers and a Gated Recurrent Unit (GRU) [163]. Simulation experiments included up to 24 robots and 12 targets. Caccavale et al. [96] proposed a DRL framework for a multi-robot system to clean and sanitize a railway station by coordinating the robots' efforts for maximum coverage. Their approach is decentralized where each robot runs its own CNN and the foundation of their technique is DQN. Note that the robots learn to cooperate online while taking the presence of the passengers into account.

Not only with the ground and aerial vehicles, but MADRL has also been used for ocean monitoring with a team of floating buoys as well. Kouzehgar et al. [105] proposed two area coverage approaches for such monitoring: (1) swarm-based (i.e., the robots follow simple swarming rules [164]) and (2) coverage-range-based (i.e., the robots with fixed sensing radius). The swarm-based model was trained using MADDPG and the latter model MARL was trained using a modified (consisting of eliminating reward sharing, collective reward, sensing their own share of the reward function, and independence based on individual reward) MADDPG algorithm.

Communication is one of the most important methods of coordination among a group of robots. More often than not, when and with whom the communication will happen is pre-defined. However, if the robots are non-cooperative, such an assumption does not work. Blumenkamp and Prorok [118] propose a learning model based on reinforcement learning that allows individual, potentially non-cooperative, agents to manipulate communication policies while the robots share a differentiable communication channel. The authors use GNN with PPO in their method. The proposed technique has also been successfully employed for multi-robot path planning. Along a similar path, Liang et al. [165] proposed the use of DRL to learn a high-level communication strategy. The authors presume the environment to be partially observable and they take a hierarchical learning approach. The implemented application is a cooperative patrolling field with moving targets. Meng and Kan [102] also put multi-robot communication at the forefront of their study while tackling the coverage problem. The goal of the robots has to cover an entire environment while maintaining connectivity in the team, e.g., via a tree topology. The authors use a modified version of MADDPG to solve the stated problem.

MADRL has also been used for sensor coverage, alongside area coverage [166]. In sensor-based coverage, the objective is to cover all the points in an environment with a sensor footprint. An example of this is communication coverage, where the goal of a team of UAVs is to provide Wi-Fi access to all the locations in a particular region. This might be extremely valuable after losing communication in a natural disaster, for example. The authors in [167] presented a solution for UAV coverage using mean field games [168]. This study was targeted toward UAVs that provide network coverage when network availability is down due to natural disasters. The authors constructed the Hamilton–Jacobi–Bellman [169] and Fokker–Planck–Kolmogorov [170] equations via mean field games. Their proposed neural network-based learning method is a modification of TRPO [44] and named mean-field trust region policy optimization (MFTRPO). Liu et al. [104] proposed a coverage method to have a system of UAVs cover an area and provide communication connectivity while maintaining energy efficiency and fairness of coverage. The authors utilize an actor–critic-based DDPG algorithm. Simulation experiments were carried out with up to 10 UAVs. Similar to these, Nemer et al. [171] proposed a DDPG-based MADRL framework for multi-UAV systems to provide better coverage, efficiency, and fairness for network coverage of an area. One of the key differentiating factors of this paper is that the authors also model energy-efficient controls of the UAVs to reduce the overall energy consumption by them during the mission. For a similar communication coverage application, Liu et al. [172] proposed that the UAVs have their own actor-critic networks for a fully-distributed control framework to maximize temporal mean coverage reward.

3.2. Path Planning and Navigation

In multi-robot path planning (or path finding), each robot is given a unique start and a goal location. Their objective is to plan a set of joint paths from the start to the goal, such that some pre-defined criteria, such as time and/or distance, are optimized and the robots avoid colliding with each other while following the paths. An illustration is presented in Figure 7. Planning such paths optimally has been proven to be NP-complete [173]. Like A^* [174], which is used for single-agent path planning in a discreet space, M^* [175] can be used for an MRS. Unfortunately, M^* lacks scalability. There exist numerous heuristic solutions for such multi-robot planning that scale well [176–179]. Overhead cameras and GPS can be used to localize the robots in indoor and outdoor applications, respectively. In GPS and communication-denied environments, vision systems can be used as a proxy [180]. Recently, researchers have started looking into deep reinforcement learning solutions to solve this notoriously difficult problem.

Figure 7. An illustration of multi-robot path planning with 4 e-puck robots, where the starting positions are shown and each robot's goal position is its orthogonal robot's starting location. The boxes represent the obstacles in the environment.

One of the most popular works that use MADRL for collision avoidance is due to Long et al. [22]. They propose a decentralized method using PPO while using CNNs to train the robots, which use their onboard sensors to detect obstacles. Up to 100 robots were trained and tested via simulation. Lin et al. [109] proposed a novel approach for centralized training and decentralized execution for a team of robots that need to concurrently reach a destination while avoiding objects in the environment. The authors implement their method using CNNs and PPO as well. The learned policy maps LiDAR measurements to the controls of the robots. Bae et al. [72] also use CNNs to train multiple robots to plan paths. The environment is treated as an image where the CNN extracts the features from the environment, and the robots share the network parameters.

Fan et al. [107] have proposed a DRL model technique using the policy gradient method to train the robots to avoid collisions with each other while navigating in an environment. The authors use LiDAR data for training, and, during testing, this drives the decision-making process to avoid collisions. The authors then transfer the learned policy to physical robots for real-world feasibility testing. The simulation included up to 100 robots with the objective of avoiding collisions with each other, static objects, and, finally, pedestrians. It builds on their previous work from 2018 [131]. Wang et al. [181] also use CNNs for multi-robot collision avoidance and coordination. The authors also use a recurrent module, namely Long Short Term Memory (LSTM) [182] to memorize the actions of the robots to smooth the trajectories. The authors have shown that the combined use of CNN and LSTM can produce smoother paths for the robots in a continuous domain.

Yang et al. [71] use a priori knowledge to augment the DDQN algorithm to improve the learning efficiency in multi-robot path planning. To avoid random exploration at the beginning of the learning process, the authors have used A^* [174] paths for single robots in static environments. This provides better preliminary Q-values to the networks, and, thus, the overall learning process converges relatively quickly. Wang and Deng [39] propose a novel neural network structure for task assignment and path planning where one network processes a top–down view of the environment and another network processes the first-person view of the robot. The foundation of the algorithm is also based on DQN. Na et al. [49] have used MADRL for collision avoidance among autonomous vehicles via modeling virtual pheromones inspired by nature. The authors also used a similar pheromone-based technique, along with a modified version of PPO in [55] for the same objective. Ourari et al. [123] also used a biologically-inspired method (specifically from the behavior of flocks of starlings) for multi-robot collision avoidance while a DRL method, namely PPO, is at its foundation. Their method is executed in a distributed manner and each robot incorporates information from k-nearest neighbors.

For multi-robot target assignment and navigation, Han, Chen, and Hao [117] proposed to train the policy in a simulated environment using randomization to decrease the performance transfer from simulation to the real world. The architecture they developed utilized communication amongst the robots to share experiences. They also developed a training algorithm for navigation policy, target allocation, and collision avoidance. It uses PPO as a foundation. Moon et al. [38] used MADRL for the coordination of multiple UAVs that track first responders in an emergency response situation. One of the key ideas behind their method is the inclusion of the Cramér–Rao lower bound into the learning process. The intent of the authors was to use the DRL-based UAV control algorithm to accurately track the target(s) of the UAV system. They used DDQN as their foundation technique.

Marchesini and Farinelli [74] extended their work in [75] by incorporating an Evolutionary Policy Search (EPS) for multi-robot navigation. It had two main components: navigation (reaching a target) and avoiding collisions. They extended their prior work [75] (using DDQN and LSTM at its core) by incorporating the EPS, which integrated randomization and genetic learning into the MARL technique to enhance the ability for the policy to explore and help the robots learn to navigate better.

Lin et al. [112] developed a novel deep reinforcement learning approach for coordinating the movements of an MRS such that the geometric center of the robots reached a target destination while maintaining a connected communication graph throughout the mission. Similarly, Li et al. [183] proposed a DRL method for multi-robot navigation while maintaining connectivity among the robots. The presented technique used constrained policy optimization [184] and behavior cloning. Real-world experiments with five ground robots show the efficacy of the proposed method. Maintaining such connectivity has previously been studied in an information collection application [185] applied to precision agriculture, albeit from a combinatorial optimization perspective [18].

On the other hand, Huang et al. [167] proposed a deep Q-learning method for maintaining connectivity between leader and follower robots. Interestingly, the authors do not use CNNs, instead, they rely only on dense fully connected layers in their network. Similar to these, Challita et al. [167] developed a novel DRL framework for UAVs to learn an optimal joint path while maintaining cellular connectivity. Their main contribution is founded in game theory. The authors used an Echo State Network (ESN), a type of recurrent neural network. In a similar setting, the authors' other work [186] studied minimizing interference from the cellular network using MADRL. Wang et al. [187] proposed to incorporate environmental spatiotemporal information. The proposed method used a global path planning algorithm with reinforcement learning at the local level via DDQN combined with an LSTM module. Choi et al. [92] also used a recurrent module, namely GRU along with CNN for the multi-agent path planning problem in an autonomous warehouse setting. The base of their work was the popular QMIX [188] algorithm, a form of value function factorization algorithm similar to VDN [63]. Another study of multi-robot path planning for warehouse production scenarios was carried out by Li and Guo [128]. They proposed a supervised DRL approach efficient path planning and collision avoidance. More specifically, using imitation learning and PPO, Li and Guo aimed to increase the learning performance of the vehicles in object transportation tasks.

Yao et al. [115] developed a map-based deep reinforcement learning approach for multi-robot collision avoidance, where the robots do not communicate with one another for coordination. The authors used an egocentric map as the basis of information that the robots use to avoid collisions. Three robots have been used for real-world implementations. Similar to this, Chen et al. [189] also did not rely on inter-robot communication for multi-robot coordinated path planning and collision avoidance while also navigating around pedestrians. Chen et al. [94]'s study on multi-robot path planning also considered non-communicating and decentralized agents using DDQN. Simulation experiments involved up to 96 robots.

Tan et al. [116] have developed a novel algorithm, called DeepMNavigate that uses local and global map information, PPO, and CNNs for navigation and collision avoidance learning. Their algorithm also makes use of multi-staged training for robots. Simulation experiments involved up to 90 robots. Chen et al. [190] proposed a method of using DRL in order for robots to learn human social patterns to better avoid collisions. As human behaviors are difficult to model mathematically, the authors noted that social rules usually emerge from local interactions, which drives the formulation of the problem. Chen et al. [87] proposed a novel DRL framework using hot-supervised contrastive loss (via supervised contrastive learning) combined with DRL loss for pathfinding. The robots do not use communication. They also incorporated a self-attention mechanism in the training. Their network structure used CNNs with DQN while up to 64 agents have been used for testing the approach in simulation. Navigation control using MADRL was also studied in [88], where the authors showed that the robots could recover from reaching a dead end. Alon and Zhou [135] have proposed a multi-critic architecture that also included multiple value networks. Path planning is also important in delivering products to the correct destinations. Ding et al. [91] have proposed a DQN-based MADRL technique for this specific application while combining it with a classic search technique, namely the Warshall–Floyd algorithm.

Transfer learning and federated deep learning have also been used for multi-robot path planning. In transfer learning, the assumption that the training and the testing data come from the same domain does not need to hold, which makes it attractive in many real-world scenarios, including robotics [191]. The objective here is to *transfer* the learning from one or more source domains to a potentially different target domain. Wen et al. [133] developed two novel reinforcement learning frameworks that extend the PPO algorithm and incorporate transfer learning via meta-learning for path planning. The robots learn policies in the source environments and obtain their policies following the proposed training algorithm. Next, this learning is then transferred to target environments, which might have more complex obstacle configurations. This increases the efficiency of finding the solutions in the target environments. The authors used LSTM in their neural network for memorizing the history of robot actions. In federated deep learning, training data might still be limited similar to the transfer learning applications. In this case, each agent has its own training data instead of using data shared by a central observer [192]. For example, each robot might have access to a portion of the environment, and they are not allowed to share the local images with each other, where the objective is still to train a high-quality global model. Luo et al. [193] have employed such a federated deep RL technique for multi-robot communication. The authors, in this paper, avoid blockages in communication signals due to large obstacles while avoiding inter-robot collisions. It has been shown that the proposed semi-distributed optimization technique is 86% more efficient than a central RL technique. Another federated learning-based path planning technique can be found in [130]. To reduce the volume of exchanged data between a central server and an individual robot, the proposed technique only shares the weights and biases of the networks from each agent. This might be significant in scenarios where the communication bandwidth is limited. The authors show that the presented technique in their paper offers higher robustness than a centralized training model.

PRIMAL is a multi-agent path-finding framework that uses MADRL and is proposed by Sartoretti et al. [194]. PRIMAL used the A3C [45] algorithm and an LSTM module. It also makes use of imitation learning whereby each agent can be given a copy of the centrally trained policy by an expert [195]. One of the highlights of this paper is that the proposed technique could scale up to 1024 robots albeit in simulation. $PRIMAL_2$ [196] is the advanced version of PRIMAL and was proposed by Damani et al. in 2021. It also uses A3C as its predecessor, offers real-time path re-planning, and scales up to 2048 robots—double that which PRIMAL could do.

Curriculum learning [197] has also been used for multi-robot path planning in [198], where the path planning is modeled as a lesson, going from easy to hard difficulty levels. An end-to-end MADRL system for multi-UAV collision avoidance using PPO has been proposed by Wang et al. [57]. Asayesh et al. [137] proposed a novel module for safety control of a system of robots to avoid collisions. The authors use LSTM and a Variational Auto-Encoder [199]. Li [200] has proposed using a lightweight decentralized learning framework for multi-agent collision avoidance by using only a two-layer neural network. Thumiger and Deghat [56] used PPO with an LSTM module for multi-UAV decentralized collision avoidance. Along the same line, Han et al. [54] used GRUs and their proposed reward function used reciprocal velocity obstacle for distributed collision avoidance.

For collaborative motion planning with multiple manipulators, Zhao et al. [108] proposed a PPO-based technique. The manipulators learned from their own experiences, and then, a common policy was updated while the arms continued to learn from individual experiences. This created differences in accuracy or actuator ability among the manipulators. Similarly, Gu et al. [50] proposed a method for asynchronous training of manipulator arms using DDPG and Normalized Advantage Function (NAF). Real-world experiments were carried out with two manipulators. Prianto et al. [143] proposed the use of the Soft Actor–Critic (SAC) algorithm [14] due to its efficiency in exploring large state spaces for path planning with a multi-arm manipulator system, i.e., each arm has its own unique start and goal configurations. Unlike the previous works in this domain, the authors used Hindsight Experience Replay (HER) [201] for sample-efficient training. On the other hand, Cao et al. [144] proposed a DRL framework for a multi-arm manipulator to track trajectories. Similarly to [143], Cao et al. also used SAC as their base algorithm. The main distinguishing factor of this study is that the multiple manipulator arms were capturing a non-cooperative object. Results show that the dual-arm manipulator can capture a rotating object in space with variable rotating speeds. An illustration of such a dual-arm manipulation application is shown in Figure 8.

Everett et al. [202] have proposed to use LSTM and extend their previous DRL algorithm [189] for multi-robot path planning to enhance the ability of the robots to avoid collisions. Semnani et al. [203] proposed an extension of the work proposed in [202] by using a new reward function for multi-agent motion planning in three-dimensional dense spaces. They used a hybrid control framework by combining DRL and force-based motion planning. Khan et al. [136] have proposed using GCN and a DRL algorithm called Graph Policy Gradients [134] for unlabeled motion planning of a system of robots. The multi-robot system must find the goal assignments while optimizing their trajectories.

Song et al. [90] designed a new actor–critic algorithm and a method for extracting the state features via a local-and-global attention module for a more robust MADRL method with an increasing number of agents present in the environment. The simulated experiments used dynamic environments with simulated pedestrians. Zhang et al. [204] proposed a method for using a place-timed Petri net and DRL for the multi-vehicle path planning problem. They used a curriculum-based DRL model. Huang et al. [205] proposed a vision-based decentralized policy for path planning. The authors use Soft Actor–Critic with auto encoders [206] as their deep RL technique for training a multi-UAV system. The 3D images captured by the UAVs and their inertial measurement values were used as inputs, whereas the control commands were rejected by the neural network. Simulation experiments with up to 14 UAVs were performed within the Airsim simulator. Jeon et al. [207] proposed to use MADRL to improve the energy efficiency of coordinating multiple UAVs within a logistic delivery service. The authors show that their model performs better in terms of consumed energy while delivering similar numbers of goods.

Figure 8. An illustration of dual-arm manipulation is shown using a Baxter robot where each arm might act as an RL agent.

MADRL has also found its way into coordinating multiple autonomous vehicles. The authors in [208] provide a solution to the "double merge" scenario for autonomous driving cars that consists of three primary contributions in this field: (1) the variance of the gradient estimate can be minimized without Markovian assumptions, (2) trajectory planning with hard constraints to maintain the safety of the maneuver [209], and (3) introduction of a hierarchical temporal abstraction [25] that they call an "Option Graph" to reduce the effective horizon which ultimately reduces the variance of the gradient estimation [210,211]. Similar to this, Liang et al. [212] have modeled the cooperative lane changing problem among autonomous cars as a multi-agent cooperation problem and solved it via MADRL. Specifically, the authors have used a hierarchical DRL method that breaks down the problem into "high-level option selection" and "low-level control" of the agent. Real-world experiments were performed using a robotic test track with four robots, where two of them performed the cooperative lane change.

Finally, Sivanathan et al. [119] proposed a decentralized motion planning framework and a Unity-based simulator specifically for a multi-robot system that uses DRL. The simulator can handle both independent learners and common policies. The simulator was tested with up to four cooperative non-holonomic robots that shared limited information. PPO was used as the base algorithm to train the policies.

3.3. Swarm Behavior Modeling

Navigation of a swarm of robots through a complex environment is one of the most researched topics in swarm robotics. To have a stable formation, each robot should be aware of the positions of the nearby robots. A swarm consisting of miniature robots might not have a sophisticated set of sensors available. For example, a compass can be used to know the heading of the robot. Additionally, range and bearing sensors can also be available [213,214]. Infrared sensors can be used for communication in such a swarm system [215]. Inspired by swarms of birds or schools of fish, robots usually follow three simple rules to maintain such formations: cohesion, collision avoidance, and velocity alignment [164]. It is no surprise that multi-agent deep reinforcement learning techniques have been extensively employed to mimic such swarm behaviors and solve similar problems. An illustration of forming a circle with a swarm of five e-puck robots is presented in Figure 9.

Figure 9. An illustration of pattern (circle) formation with five e-puck robots which are guarding a box in the center.

Zhu et al. [216] proposed a novel algorithm for multi-robot flocking. The algorithm builds on MADDPG and uses PER. Results from three robots show that the proposed algorithm improves over the standard MADDPG. Similarly, Salimi and Pasquier [106] have proposed the use of DDPG with centralized training and a decentralized execution mechanism to train the flocking policy for a system of UAVs. Such flocking with UAVs might be challenging due to complex kinematics. The authors show that the UAVs reach the flocking formation using a leader–follower technique without any parameter tuning. Lan et al. [217] developed a control scheme for the cooperative behavior of a swarm. The basis of their control scheme is pulled from joint multi-agent reinforcement learning theory, where the robots not only share state information, but also a performance index designed by the authors. Notably, the convergence of the policy and the value networks is theoretically guaranteed. Following the above-mentioned works, Kheawkhem and Khuankrue [99] also proposed using MADDPG to solve the multi-agent flocking control problem.

Qiu et al. [100] used MADRL to improve sample efficiency, reduce overfitting, and allow better performance, even when agents had little or "bad" sample data in a flocking application. The main idea was to train a swarm offline with demonstration data for pre-training. The presented method is based on MADDPG. Not only for coverage as described earlier, but GNNs are also popular in general for coordination in a swarm system, especially in spatial domains. For example, Kortvelesy and Prorok [218] developed a framework, called ModGNN, which aimed to provide a generalized, neural network framework, that can be applied to varying multi-robot applications. The architecture is modular in nature. They tested the framework for a UAV flocking application with 32 simulated robots.

Yan et al. [219] studied flocking in a swarm of fixed-wing UAVs operating in a continuous space. Similar studies on flocking can also be found in more recent papers from these authors [83,220]. Similar to Yan et al.'s body of work, Wang et al. [142] proposed a TD3-based [221] solution for a similar application—flocking with fixed-wing UAVs where the authors test the method with up to 30 simulated UAVs. Not strictly for swarms, Lyu et al. [47]

addressed the multi-agent flocking control problem specifically for a multi-vehicle system using DDPG with centralized training and decentralized execution. Notably, the authors take connectivity preservation into account while designing their reward function—the maximum distance could not go beyond the communication range and the minimum distance was kept at d_s, a physically safe distance between two vehicles. Interestingly, the mission waypoints are pre-defined in this paper. Bezcioglu et al. [48] also study flocking in a swarm system using DDPG and CNN, and tested it with up to 100 robots. The authors have used bio-inspired self-organizing dynamics for the joint motion of the robots.

Wang et al. [113] used MADRL to organize a swarm in specific patterns using auto-encoders [222] to learn compressed versions of the states and they tested the presented solution with up to 20 robots. Li et al. [46] proposed using a policy gradient method, namely MADDPG, with an actor–critic structure for circle formation control with a swarm of quad-rotors. Although circle formation is a popular application [223–226], this is one of the few studies that employed MADRL techniques. Sadhukhan and Selmic [121] have used PPO in order to train a multi-robot system to navigate through narrow spaces and reform into a designated formation. They used two reward schemes (one individual to the agents and one depending on the contributions to the team) and the system was centrally trained. In [125], Sadhukhan and Selmic extended their prior works by proposing a bearing-based reward function for training the swarm system, which utilizes a single policy shared among the robots.

Chen et al. [97] have developed an improved DDPG to enhance the ability of a robot to learn human intuition-style navigation without using a map. Furthermore, they create a parallel version of DDPG to extend their algorithm to a multi-robot application. Thereby, providing the robots with a method of sharing information/experiences in order to maintain formation, navigate an indoor environment, and avoid collisions. Qamar et al. [138] proposed novel reward functions and an island policy-based optimization framework for multiple target tracking using a swarm system. Along a similar line, Ma et al. [98] developed a DDPG-based algorithm for multi-robot formation control around a target, particularly in a circle around a designated object. The algorithm allows the robots to independently control their actions using local teammates' information.

Recently, Zhang et al. [124] have also proposed a target encirclement solution that uses a decentralized DRL technique. The main contribution of their work is the use of three relational graphs among the robots and other entities in the system designed using a graph attention network [227]. In their simulation experiments, the authors use six robots encircling two targets. Similarly, Khan et al. [134] have used a graph representation of the robot formation and proposed using graph convolutional neural networks [158,228,229] to extract features, i.e., local features of robot formations, for policy learning. Simulation policies were trained on three robots and then the policy is transferred to over 100 robots for testing. The robots are initialized to certain positions and are to form a specific formation while reaching an end goal.

Zhou et al. [230] recognized the problem of computational complexity with existing MADRL methods for multi-UAV multi-target tracking while proposing a decentralized solution. Their proposed solution has its root in the reciprocal altruism mechanism of cooperation theory [231]. The experience replay is shared among the UAVs in this work. Zhou et al. [139] also study target tracking with a swarm of UAVs. Not only do they learn to track a target, but the robots also learn to communicate better (i.e., the content of the message) for such tracking following the proposed policy gradient technique.

Yasuda and Ohkura [78] used a shared replay memory along with DQN to accelerate the training process for the swarm with regard to path planning. By using more robots contributing their individual experiences to the replay memory, the swarm system was able to learn the joint policy faster. Communication is an important aspect of swarm systems. Usually, researchers use pre-defined communication protocols for coordination among the swarm robots. Hüttenrauch et al. [140] proposed a histogram-based communication protocol for swarm coordination, where the robots use DRL to learn decentralized policies using TRPO [44]. An example task is graph building formation, where the robots aim to cover a certain area through coordination. Another considered task is establishing a communication link between the robots and connecting two points on a map. Along the same line, in 2019, Hüttenrauch et al. [141] used TRPO again to find MADRL-based solutions for rendezvous and pursuit–evasion in a swarm system. The main contribution of their work is the incorporation of Mean Embedding [232] into the DRL method they use to simplify the state information each agent obtains from other agents. Up to 100 robots were used in simulation experiments.

3.4. Pursuit-Evasion

In a pursuit–evasion game, usually, multiple pursuers try to capture potentially multiple evaders. When all the evaders are captured or a given maximum time elapses, the game finishes [233–235]. For a detailed taxonomy of such problems, the reader is referred to [233]. Some of the sensors that the robots might use in this application include sonar, LiDAR, and 3D cameras, among others. A unified model to analyze data from a suit of sensors can also be used [236]. An illustration is shown in Figure 10.

Egorov [59] proposed a solution for the classic pursuit–evasion problem [233] using an extension of single-agent DQN, called multi-agent DQN (MADQN). The state space is represented as a four-channel image consisting of a map, opponent location(s), ally location(s), and a self-channel. Yu et al. [40] proposed the use of a decentralized training method for pursuit evasion where each agent learns its policy individually and used limited communication with other agents during the training process. This is unlike traditional MADRL techniques where the training is centralized. The execution of the policy for each agent is also decentralized.

Figure 10. An illustration of multi-robot pursuit–evasion scenario where UAVs 1 and 2 have captured and "grounded" UAV 3, which is an evader robot.

Wang et al. [23] proposed to extend a MARL algorithm called cooperative double Q-learning (Co-DQL) for the multi-UAV pursuit–evasion problem. The foundation of Co-DQL is Q-networks with multi-layer perceptrons. Unlike traditional applications where the evader might move around randomly, in this paper, the authors assume that the target also learns to move intelligently up to a certain degree via RL. In [237], the authors consider a setup with one superior evader and multiple pursuers. They use a centralized critic model, where the actors are distributed. Unlike traditional broadcasting techniques, the authors smartly use a leader–follower line topology network for inter-robot communication that reduces the communication cost drastically. Although not strictly pursuit-evasion, Zhang et al. [76] use MADRL for coordinated territory defense, which is modeled as a game where two defender robots coordinate to block an intruder from entering a target zone.

Gupta et al. [81] argue that instead of using a centralized multi-agent DRL framework, where the model learns joint actions from joint states and observations, a more sophisticated parameter-sharing approach can be used. A drawback of the centralized learning system is that the complexity grows exponentially with the number of agents. The authors use TRPO as their base algorithm and the policy is trained with the experiences of all agents simultaneously via parameter sharing. The multi-agent scenarios they use for testing the quality of the proposed solution are pursuit–evasion and a multi-walker system with bipedal walkers.

3.5. Information Collection

The objective of information gathering about an ambient phenomenon (e.g., temperature monitoring or weed mapping) using a group of mobile robots is to explore parts of an unknown environment, such that uncertainty about the unseen locations is minimized. Relevant sensors for information gathering include RGB, Normalized Difference Vegetation Index (NDVI), or multi-spectral cameras, and thermal and humidity sensors, among others. This is unlike coverage, where the goal is to visit all the locations. There are two main reasons for this: (1) information (e.g., temperature measurements) in nearby points are highly correlated, and, therefore, the robots do not need to go to all the locations within a neighborhood [238]; and (2) the robot might not have enough battery power to cover the entire environment. This is especially true in precision agriculture, where the fields are usually too large to cover [18]. An illustration is shown in Figure 11, where the robots are tasked with collecting information from their unique sub-regions, and through communication, they will need to learn the underlying model.

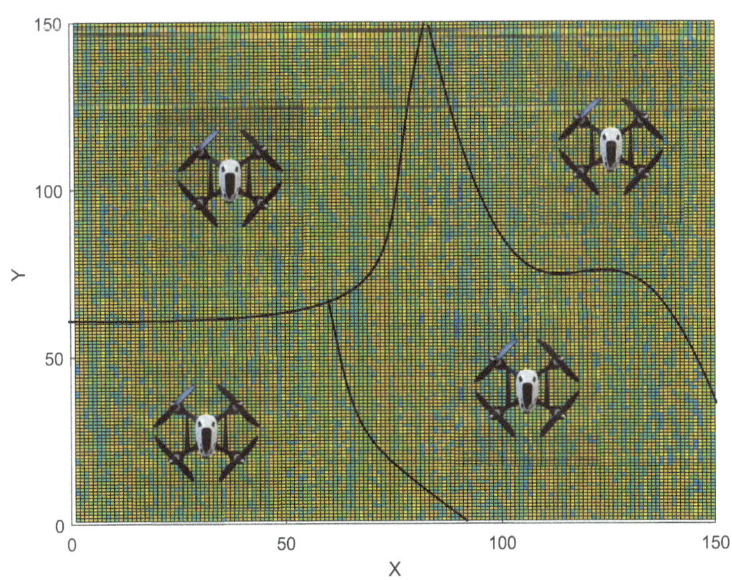

Figure 11. An illustration of multi-robot information collection. The environment is divided into four sub-regions and the underlying heatmap represents measures of soil acidity in parts of the USA [239].

Viseras and Garcia [240] have developed a novel DRL algorithm based on the popular A3C [45] algorithm. They also provide a model-based version of their original algorithm for gathering information, which uses CNNs. Said et al. [241] have proposed a mean field-based DRL technique that uses an LSTM module—a type of recurrent neural network for multi-robot information collection about an unknown ambient phenomenon. The robots are battery-powered with limited travel ranges. Recently, Wei and Zheng [67] also used MADRL for multi-robot informative path planning. They develop two strategies for cooperative learning: (1) independent Q-learning with credit assignment [4], and (2) sequential rollout using a GRU. Along the same line, Viseras et al. [85] have proposed using a MADRL framework for a multi-robot team to monitor a wildfire front. The two main components in this framework are (1) individually-trained Q-learning robots and (2) value decomposition networks. The authors have used up to 9 UAVs for testing the efficiency of their presented work.

3.6. *Task Allocation*

Multi-robot task allocation (MRTA) is a combinatorial optimization problem. Given a set of n robots and m tasks, the goal is to allocate the robots to the tasks such that a given utility function is optimized. Now, if multiple robots need to form a team to complete a single task, then it is a single-task, multi-robot allocation problem. On the other hand, if one robot can offer its services to multiple tasks, then it is called a single-robot, multi-task allocation problem. The robots might be connected to a central server via Wi-Fi, e.g., in a warehouse setting, and can receive information about tasks and other robots. Similarly, communication can happen with other robots via this central server using Wi-Fi as well. Overhead cameras or tracking systems can be used for robot localization in such a scenario. Comprehensive reviews about such MRTA concepts and solutions can be found in [242,243]. An example task allocation scenario is presented in Figure 12.

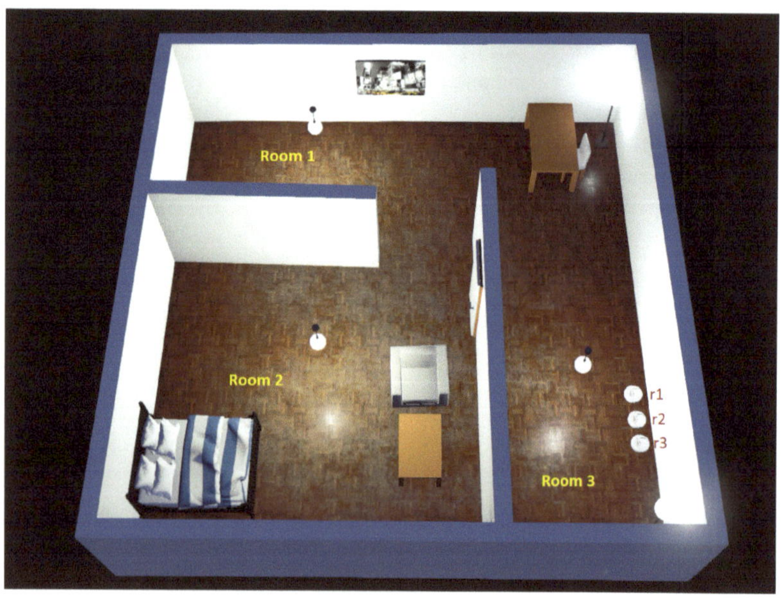

Figure 12. An illustration of multi-robot task allocation: there are 3 iRobot Roombas ($r1$–$r3$) and 3 rooms to clean. In a one-to-one matching scenario, the objective would be to assign one Roomba to a certain room. However, as room 2 is larger in size, two robots might be needed to clean it, whereas the third robot ($r3$) might be assigned to rooms 1 and 3.

Elfakharany and Ismail [132] developed a novel multi-robot task allocation and navigation method. This is the first work to propose a MADRL method to tackle task allocation, as well as the navigation problem. They use PPO with actor–critic. Their centralized training and decentralized execution method uses CNNs. Paul et al. [129] proposed to use DRL for multi-robot task allocation. They proposed a neural network architecture that they called a Capsule Attention-based Mechanism, which contains a Graph Capsule Convolutional Neural Network (GCapCN) [244] and a Multi-head Attention mechanism (MHA) [245,246]. The underlying architecture is a GNN. The task graph is encoded using GCapCN and combined with the context, which contains information on the robot, time, and neighboring robots. This information is then decoded with the MHA. Although not strictly task assignment, MADRL has been used for forming teams of heterogeneous agents (such as ambulance and fire brigade in a rescue operation) to complete a given task by Goyal [86]. Goyal has applied this technique for training a team of fire brigades to collaboratively extinguish a fire in a city within the Robocup Rescue Simulator.

Devin et al. [247] developed a novel method of compartmentalizing a trained deep reinforcement learning model into task-specific and robot-specific components. Due to this, the policies can be transferred between robots and/or tasks. Park et al. [114] propose a PPO-based DRL technique for task allocation. Their solution is tested with single-task, multi-robot, and time-extended assignments. They use an encoder–decoder architecture to represent robots and tasks, where a cross-attention layer is used to derive the relative importance of the tasks for the robots.

Scheduling tasks is another important aspect of task planning. Wang and Gombolay [82] used GNNs and imitation learning for a multi-robot system to learn a policy for task scheduling. The proposed model is based on graph attention networks [227]. The scheduling policy is first learned using a Q-network with two fully-connected layers. Imitation learning is then used to train the network from an expert dataset that contains schedules from other solutions. On the other hand, Johnson et al. [93] study the problem of dynamic flexible job shop scheduling, where an assembly line of robots must dynamically change tasks for a new job series over time. The robots learn to coordinate their actions in the assembly line. Agrawal et al. [52] performed a case study on a DRL approach to handling a homogeneous multi-robot system that can communicate while operating in an industry setting. PPO is used as the foundation algorithm. The objective of this work is to train the robots to work with each other to increase throughput and minimize the travel distances to the allocated tasks while taking the current states of the robots and the machines on the floor into account.

One of the most recent studies on deep RL-based MRTA is due to [89], which aims to use DRL for the parallelization of processing tasks for MRTA. The authors base their method on Branching Dueling Q-Network [248] with respect to multi-robot search and rescue tasks. In such a network, multiple branches of a network shares a common decision-making module where each branch handles one action dimension. This helps to reduce the curse of dimensionality in the action space. In total, 20 robots have been used within a simulation to test the feasibility of the proposed technique.

A very different and interesting task assignment application in defense systems is studied by Liu et al. [120]. The authors presented a DRL framework for multi-agent task allocation for weapon target assignment in air defense systems. They use PPO-clip along with a multi-head attention mechanism for task assignments of a (army) general and multiple narrow agents. The neural network architecture uses fully connected layers and a GRU. The major aim of this work is to increase processing efficiency and solution speed of the multi-agent task assignment problem at a large scale. Simulation experiments were carried out in a virtual digital battlefield. The experimental setup includes offensive forces and defensive forces. The defensive forces have places to protect and need to make real-time task allocation decisions for defense purposes. The defensive forces are tested with 12 and the offensive forces are tested with a total of 32 agents.

3.7. Object Transportation

To transport an object using two or more cooperative mobile robots, the goal is to design a strategy where the robots' actions are highly coordinated. Communication among the robots may or may not be possible. The robots can use depth cameras or laser scanners for avoiding obstacles. On the other hand, an optic-flow sensor can be used to determine if the pushing force from the robot has resulted in any object movement or not [249]. A force-torque sensor can be used on the robot to measure the force amount placed on the object. For a comprehensive review of this topic, please refer to [250]. An illustration is shown in Figure 13.

Zhang et al. [73] have used a modified version of DQN that controls each robot individually without a centralized controller or a decision maker. To quantitatively measure how well the robots are working together, they use the absolute error of estimated state–action values. The main idea is to use DQN to have homogeneous robots carry a rod to a target location. Each robot acts independently with neither leading nor following. Niwa et al. [251] proposed a MADRL-based solution to the cooperative transportation and obstacle removal problem. The basis of their solution is to use MARL to train individual robots' decentralized policies in a virtual environment. The policies are trained using MADDPG [61]. The authors then use the trained policies on real teams of robots to validate the effectiveness. The robots are supposed to push a target object to a final waypoint while moving a physical barrier out of the way to accomplish the task. Manko et al. [77] used CNN-based DRL architecture for multi-robot collaborative transportation where the objective is to carry an object from the start to the goal location. Eoh and Park [79] proposed a curriculum-based deep reinforcement learning method for training robots to cooperatively transport an object. In a curriculum-based RL, past experiences are organized and sorted to improve training efficiency [252]. In this paper, a region-based curriculum starts by training robots in a smaller area, before transitioning to a larger area and a single-robot to multi-robot curriculum begins by training a single robot to move an object, then transferring that learned policy to multiple robots for multi-robot transportation.

Figure 13. An illustration of multi-robot object transportation is presented where 2 iRobot Create robots carry a cardboard box and plan to go through a door in front of them.

3.8. Collective Construction

In a collective construction setup, multiple cooperative mobile robots are required. The robots might have heterogeneous properties [253]. The robots can follow simple rules and only rely on local information [254]. In the popular TERMES project from Harvard University [254], a large number of simple robots collect, carry, and place building blocks to develop a user-specified 3D structure. The robots might use onboard vision systems to access the progress in construction. A force sensor-equipped gripper can be used for holding the materials. Furthermore, a distance sensor, e.g., sonar can be used for maintaining a safe distance from the construction as well as other robots [255].

Sartoretti et al. [256] developed a framework using A3C to train robots to coordinate the construction of a user-defined structure. The proposed neural network architecture includes CNNs and an LSTM module. Each robot runs its own copy of the policy without communicating with other agents during testing.

A summary of the state and action spaces and reward functions used in some of the papers reviewed in this article are listed in Table 2.

Table 2. Examples of state and action spaces and reward functions used in prior studies.

Refs.	State	Action	Reward
[59]	Map of the environment and robots' locations with 4 channels	Discreet	Based on the locations of the robots
[67]	Robot locations and budgets	Discreet	Based on collected sensor data
[68]	Position of the leader UAVs, the coverage map, and the connection network	Discreet	Based on the overall coverage and connectivity of the Leader UAVs.
[69]	Map with obstacles and the coverage area	Discreet	Based on the robot reaching a coverage region within its task area.
[24]	Map with covered area	Discreet	Based on robot coverage.
[71]	Map of the environment	Discreet	Based on the robot movements and reaching the target without collisions.
[70]	Map with robots' positions	Discreet	Based on the herding pattern.
[72]	Map with robots' positions and target locations	Discreet	Distance from the goal and collision status.
[39]	Map with robots' positions and target locations	Discreet	Distance from the goal and collision status.
[40]	Pursuer and evader positions	Discreet	Collision status and time to capture the predator.
[73]	The map, locations, and orientations of the robots, and the objects the robots are connected to	Discreet	Based on the position of the object and the robots hitting the boundaries.
[74]	The map, and the locations of the robots	Discreet	Based on distance from the target and collisions.
[76]	The regions of the robots, positions of the defender and the attacker UAVs and the intruder	Discreet	Based on distance.
[77]	The distance from the MRS center to the goal, the difference in orientation of the direction of MRS to the goal, and the distance between the robots	Discreet	Based on the distance to the goal, orientation to the goal, proximity of obstacles, and the distance between the robots.
[78]	Sensor input information that includes distance to other robots and the target landmarks	Discreet	Based on becoming closer to the target landmark.
[79]	Spatial information on the robots, the object, and the goal	Discreet	Based on the object reaching the goal while avoiding collisions.
[80]	Robot position and velocity	Discreet	Based on the robots being within sensing range of one another.
[38]	The positions and speed of the first responders and UAVs	Discreet	Based on the Cramér–Rao lower bound (CRLB) for the whole system.
[93]	The agents' positions, types, and remaining jobs	Discreet	Based on minimizing the makespan.
[83]	The position and direction of the leader and the followers	Discreet	Based on the distance from followers to leaders and collision status.
[84]	Information on the target, other agents, maps, and collisions	Discreet	Based on finding targets and avoiding obstacles.
[85]	The robot's position, position relative to other robots, and angle and direction of the robots	Discreet	Based on covering a location on fire.

Table 2. *Cont.*

Ref.	State	Action	Reward
[23]	The positions, velocities, and distances between UAVs	Discreet	Determined by the distance from the target and the evader being reached by the pursuer.
[87]	Consists of static obstacle locations and the locations of other agents	Discreet	Based on the robots' movements toward the goal while avoiding collisions.
[94]	Contains the sensor data for the location of the target relative to the robot and the last action done by the robot	Discreet	Determined by the robot reaching the goal, reducing the number of direction changes, and avoiding collisions.
[95]	A map that includes the agent's locations, empty cells, obstacle cells, and the location of the tasks	Discreet	Determined by laying pieces of flooring in the installation area.
[110]	Map of the environment represented with waypoints, locations of the UAVs, and points of interest	Discreet	Based on the coverage of the team of robots.
[114]	The positions and tasks of the robots, the state of the robot	Discreet	Based on minimizing the number of timesteps in an episode.
[96]	Map of the area to be sterilized and the positions of the agents, the cleaning priority, size, and area of the cleaning zone	Discreet	Based on the agents cleaning priority areas for sanitation.
[86]	Temperature and "fieryness" of a building, location of the robots, water in the tanks, and busy or idle status	Discreet	Based on keeping the fires to a minimum "fieryness" level.
[53]	Sensory information on obstacles	Discreet	Based on the UAV's coverage of the area.
[52]	Robots' positions and velocities and the machine status	Discreet	Determined by robots completing machine jobs to meet the throughput goal, and their motions while avoiding collisions.
[107]	Includes the laser readings of the robots, the goal position, and the robot's velocity	Continuous	Based on the smooth movements of the robots while avoiding collisions.
[22]	Includes the laser readings of the robots, the goal position, and the robot's velocity	Continuous	Based on the time to reach the target while avoiding collisions.
[108]	An environment that includes the coordinates of the manipulator arm gripper	Continuous	Based on reaching the target object.
[109]	Laser measurements of the robots and their velocities	Continuous	Based on the centroid of the robot team reaching the goal.
[117]	The state of the robot, other robots, obstacles, and the target position	Continuous	Based on the robots' relative distance from the target location.
[97]	Sensed LiDAR data	Continuous	Based on the robot approaching and arriving at the target, avoiding collisions and the formation of the robots.
[132]	Consists of the goal positions, the robots' positions, past observations	Continuous	Based on the robot moving towards the goal in the shortest amount of time.
[133]	Contains laser data, speeds and positions of the robots, and the target position	Continuous	Based on arriving at the target, avoiding collisions, and relative position to other robots.
[113]	The position information of other robots (three consecutive frames)	Continuous	Based on time for formation, collisions, and the formation progress.
[115]	The most recent three frames of the map, local goals that include positions and directions	Continuous	Based on minimizing the arrival time of each robot while avoiding collisions.
[116]	Map of the environment, robot positions and velocities, and laser scans	Continuous	Based on the arrival time of the robot to the destination, avoiding collisions, and smoothness of travel.
[104]	The coverage score and coverage state for each point of interest and the energy consumption of each UAV	Continuous	Defined by coverage score, connectivity fairness, and energy consumption.
[134]	Robot's relative position to the goal and its velocity	Continuous	Based on the robots having collisions.
[88]	Robot motion parameters, relative distance and orientation to the goal, and their laser scanner data	Discreet/ Continuous	Determined by reaching the goal without timing out and avoiding collisions.

4. Challenges and Discussion

Although we find that a plethora of studies have used multi-agent deep reinforcement learning techniques in recent years, a number of challenges remain before we can expect wide adaptation of them in academia as well as commercially. One of the biggest challenges that we identify is scalability. Most of the papers reviewed in this article do not scale beyond tens of robots. This limits real-world adaptation. Although this is an issue with multi-robot systems in general, the data-hungry nature of most of today's DRL techniques makes the situation worse. In the future, the research community needs to come up with lightweight techniques that potentially are inspired by nature, such as swarming in biology or particle physics while making necessary changes to the underlying RL technique to fit these appropriately.

The second drawback we found in most of the studies is the lack of resources to make them reproducible. One of the overarching goals of academic research is that researchers across the world should be able to reproduce the results reported in one paper and propose a novel technique that potentially advances the field. In the current setup, most papers employing MADRL use their own (simulation) environments for their robots, which makes it extremely difficult for others to reproduce the results. As a community, we need to come up with an accepted set of benchmarks and/or simulators that the majority of the researchers can use for method design and experiments, which, in turn, will advance the field.

The next challenge is to transfer the learned models to real robots and real-world applications. We find that most experiments in the literature are conducted virtually, i.e., in simulation, rather than with physical robots. This leads to a gap in understanding the feasibility. This corroborates the finding by Liang et al. [257]. Unless we can readily use the learned models on real robots in real-world situations, we might not be able to widely adopt such techniques. It is tied up with the previously-mentioned issue of scalability. Additionally, in the deployment phase, the algorithms need to be lightweight while considering the bandwidth limitation for communication among the robots.

Software plays a significant role in developing and testing novel techniques in any robotic domain and applications of MADRL are no different. Here, we discuss some software that are popularly used for testing the feasibility of the proposed techniques in simulation.

- VMAS. Vectorized Multi-Agent Simulator for Collective Robot Learning (VMAS) is an open-source software for multi-robot application benchmarking [258]. Some applications that are part of the software include swarm behaviors, such as flocking and dispersion, as well as object transportation and multi-robot football. Note that it is a 2D physics simulator powered by PyTorch [259].
- MultiRoboLearn: Similar to VMAS, this is an open-source framework for multi-robot deep reinforcement learning applications [260]. The authors aim to unify the simulation and the real-world experiments with multiple robots via this presented software tool, which is accomplished by integrating ROS into the simulator. Mostly multi-robot navigation scenarios were tested. It would be interesting to extend this software to other multi-robot applications, especially where the robots might be static.
- MARLlib: Although not strictly built for robots, Multi-Agent RLlib (MARLlib) [261] is a multi-agent DRL software framework that is built upon Ray [262] and its toolkit RLlib [263]. This is a rich open-source software that follows Open AI Gym standards and provides frameworks for not only cooperative tasks, but for competitive multi-robot applications as well. Currently, ten environments are supported by MARLlib among which the grid world environment might be the most relevant one to the multi-robot researchers. Many baseline algorithms including the ones that are highly popular among roboticists, e.g., DDPG, PPO, and TRPO are available as baselines. The authors also show that this software is much more versatile than some of the existing ones including [264,265].

Not only these specialized ones, but other traditional robot simulators, such as Webots [266], V-rep [267], and Gazebo [157], can also be used for training and testing multiple robots. These established software platforms provide close-to-reality simulation models for many popular robot platforms. This is especially useful for robotics researchers as we have seen in this survey that MADRL applications range from aerial and ground robots to underwater robots and manipulators. Table 3 summarizes the main types of robots that have been used for MADRL applications. Not only software development, but another challenge is training data. As most of the state-of-the-art algorithms rely on massive amounts of training data, it is not always easy to train a robot with sufficient data. Dasari et al. [268] have created an open-source database for sharing robotic experiences. It contains 15 million video frames from 7 different robot manipulators including Baxter, Sawyer, Kuka, and Fetch arms. Researchers can use this dataset for efficient training while adding new experiences from their experiments to the dataset itself.

Table 3. Types of robots in the reviewed papers. If the type is not specified in the paper, it is not listed here.

Aerial Robots	Ground Robots	Manipulators
[23,24,38,46,53,56,57,68,76,83,85,91,104,106,110,110, 120,123,134,135,138,139,141,142,161,167,171,172,186, 205,207,218–220,230,240,269]	[22,39,49,52,54,55,59,70,71,73–75,77–79,82,87,90,92,97,107,109,111,112,115,117,120,124,128, 130,132,133,137,155,162,183,189,190,194,202,212,251]	[50,93,108,131,143,144]

As we have seen, sensors play a major role in creating the perception about the environment, as well as aiding the robots with communication capabilities. The robots might need to collect multi-modal sensor data and fuse them for better perceptions. These sensory observations about the environment can then be used as state inputs to the deep neural networks. Modern sensors have high sampling rates, e.g., standard LiDAR samples over 1 million data points per second. Without state-of-art learning mechanisms, it would have been almost infeasible to process and extract meaningful information from such large amounts of data (for tasks such as target recognition, classification, and semantic feature analysis, among others).

Although many robotic applications are utilizing the progress in multi-agent reinforcement learning, we have not seen any paper on modular self-reconfigurable robotics (MSRs) [270–273] where MADRL has been utilized. We believe that the field of modular robots can benefit from these developments especially given the fact that MSRs can change their shapes and the new shape might not have been pre-defined. Therefore, its control is undefined as well and it might need to learn to move around and complete tasks on-the-fly using techniques, such as MADRL, where each module acts as an intelligent agent.

On the other hand, we have found MADRL-based solutions for manipulation and motion separately. The next question that should be answered is how one can simultaneously learn those two actions where they might affect each other. For example, in a scenario, where multiple UAVs are learning to maintain a formation while manipulating an object with their onboard manipulators. This task would potentially require the robots to learn two actions simultaneously. The research question then would be how to best model the agents, their goals, and the rewards in this complex scenario.

5. Conclusions

In this paper, we have reviewed state-of-the-art studies that use multi-agent deep reinforcement learning techniques for multi-robot system applications. The types of such applications range from exploration and path planning to manipulation and object transportation. The types of robots that have been used encompass ground, aerial, and underwater applications. Although most applications involve mobile robots, we reviewed a few papers that use non-mobile (manipulator) robots as well. Most of the reviewed papers have used convolutional neural networks, potentially combining them with fully connected layers, recurrent layers, and/or graph neural networks. It is worth investigating such reinforcement learning techniques for robotics as they have the potential to learn high-level causal relationships among the robots, as well as between the robots and their environment, which might have been extremely difficult to model using a non-learning approach. As better hardware is available on a smaller scale and at a lower price, we expect to see significant growth in novel multi-robot system applications that use multi-agent reinforcement learning techniques. Furthermore, with the progress of the field of artificial intelligence in general, we expect that more studies will have theoretical underpinnings along with their showcased empirical advancements. Although a number of challenges remain to be solved, we are perhaps not too far away from seeing autonomous robots tightly integrated into our daily lives.

Author Contributions: Conceptualization, J.O. and A.D.; methodology, J.O. and A.D.; investigation, J.O. and A.D.; resources, J.O. and A.D.; data curation, J.O. and A.D.; writing—original draft preparation, J.O. and A.D.; writing—review and editing, J.O. and A.D.; visualization, A.D.; supervision, A.D.; project administration, A.D. All authors have read and agreed to the published version of the manuscript.

Funding: This research received no external funding.

Institutional Review Board Statement: Not applicable.

Informed Consent Statement: Not applicable.

Data Availability Statement: Not applicable.

Conflicts of Interest: The authors declare no conflict of interest.

References

1. Arai, T.; Pagello, E.; Parker, L.E. Advances in multi-robot systems. *IEEE Trans. Robot. Autom.* **2002**, *18*, 655–661. [CrossRef]
2. Gautam, A.; Mohan, S. A review of research in multi-robot systems. In Proceedings of the 7th IEEE International Conference on Industrial and Information Systems (ICIIS), Chennai, India, 6–9 August 2012; pp. 1–5.
3. Rizk, Y.; Awad, M.; Tunstel, E.W. Cooperative heterogeneous multi-robot systems: A survey. *ACM Comput. Surv. (CSUR)* **2019**, *52*, 1–31. [CrossRef]
4. Sutton, R.S.; Barto, A.G. *Reinforcement Learning: An introduction*; MIT Press: Cambridge, MA, USA, 2018.
5. Watkins, C.J.; Dayan, P. Q-learning. *Mach. Learn.* **1992**, *8*, 279–292. [CrossRef]
6. Mnih, V.; Kavukcuoglu, K.; Silver, D.; Rusu, A.A.; Veness, J.; Bellemare, M.G.; Graves, A.; Riedmiller, M.; Fidjeland, A.K.; Ostrovski, G.; et al. Human-level control through deep reinforcement learning. *Nature* **2015**, *518*, 529–533. [CrossRef]
7. LeCun, Y.; Bengio, Y.; Hinton, G. Deep learning. *Nature* **2015**, *521*, 436–444. [CrossRef]
8. Fawzi, A.; Balog, M.; Huang, A.; Hubert, T.; Romera-Paredes, B.; Barekatain, M.; Novikov, A.; Ruiz, F.J.R.; Schrittwieser, J.; Swirszcz, G.; et al. Discovering faster matrix multiplication algorithms with reinforcement learning. *Nature* **2022**, *610*, 47–53. [CrossRef]
9. Popova, M.; Isayev, O.; Tropsha, A. Deep reinforcement learning for de novo drug design. *Sci. Adv.* **2018**, *4*, eaap7885. [CrossRef]
10. Silver, D.; Huang, A.; Maddison, C.J.; Guez, A.; Sifre, L.; Van Den Driessche, G.; Schrittwieser, J.; Antonoglou, I.; Panneershelvam, V.; Lanctot, M.; et al. Mastering the game of Go with deep neural networks and tree search. *Nature* **2016**, *529*, 484–489. [CrossRef]
11. Mammeri, Z. Reinforcement learning based routing in networks: Review and classification of approaches. *IEEE Access* **2019**, *7*, 55916–55950. [CrossRef]
12. Panov, A.I.; Yakovlev, K.S.; Suvorov, R. Grid path planning with deep reinforcement learning: Preliminary results. *Procedia Comput. Sci.* **2018**, *123*, 347–353. [CrossRef]
13. Theile, M.; Bayerlein, H.; Nai, R.; Gesbert, D.; Caccamo, M. UAV coverage path planning under varying power constraints using deep reinforcement learning. In Proceedings of the 2020 IEEE/RSJ International Conference on Intelligent Robots and Systems (IROS), Las Vegas, NV, USA, 24 October 2020–24 January 2021; pp. 1444–1449.
14. Haarnoja, T.; Zhou, A.; Abbeel, P.; Levine, S. Soft actor-critic: Off-policy maximum entropy deep reinforcement learning with a stochastic actor. In Proceedings of the International Conference on Machine Learning, PMLR, Stockholm, Sweden, 10–15 July 2018; pp. 1861–1870.
15. Nguyen, H.; La, H. Review of deep reinforcement learning for robot manipulation. In Proceedings of the 3rd IEEE International Conference on Robotic Computing (IRC), Naples, Italy, 25–27 February 2019; pp. 590–595.
16. Busoniu, L.; Babuska, R.; De Schutter, B. A comprehensive survey of multiagent reinforcement learning. *IEEE Trans. Syst. Man Cybern. Part C (Appl. Rev.)* **2008**, *38*, 156–172. [CrossRef]
17. Yang, E.; Gu, D. *Multiagent Reinforcement Learning for Multi-Robot Systems: A Survey*; Technical Report of the Department of Computer Science; 2004.
18. Dutta, A.; Roy, S.; Kreidl, O.P.; Bölöni, L. Multi-Robot Information Gathering for Precision Agriculture: Current State, Scope, and Challenges. *IEEE Access* **2021**, *9*, 161416–161430. [CrossRef]
19. Zhou, Z.; Liu, J.; Yu, J. A survey of underwater multi-robot systems. *IEEE/CAA J. Autom. Sin.* **2021**, *9*, 1–18. [CrossRef]

20. Queralta, J.P.; Taipalmaa, J.; Pullinen, B.C.; Sarker, V.K.; Gia, T.N.; Tenhunen, H.; Gabbouj, M.; Raitoharju, J.; Westerlund, T. Collaborative multi-robot search and rescue: Planning, coordination, perception, and active vision. *IEEE Access* **2020**, *8*, 191617–191643. [CrossRef]
21. Yliniemi, L.; Agogino, A.K.; Tumer, K. Multirobot coordination for space exploration. *AI Mag.* **2014**, *35*, 61–74. [CrossRef]
22. Long, P.; Fan, T.; Liao, X.; Liu, W.; Zhang, H.; Pan, J. Towards optimally decentralized multi-robot collision avoidance via deep reinforcement learning. In Proceedings of the 2018 IEEE International Conference on Robotics and Automation (ICRA), Brisbane, QLD, Australia, 21–25 May 2018; pp. 6252–6259.
23. Wang, X.; Xuan, S.; Ke, L. Cooperatively pursuing a target unmanned aerial vehicle by multiple unmanned aerial vehicles based on multiagent reinforcement learning. *Adv. Control Appl. Eng. Ind. Syst.* **2020**, *2*, e27. [CrossRef]
24. Pham, H.X.; La, H.M.; Feil-Seifer, D.; Nefian, A. Cooperative and distributed reinforcement learning of drones for field coverage. *arXiv* **2018**, arXiv:1803.07250.
25. Sutton, R.S.; Precup, D.; Singh, S. Between MDPs and semi-MDPs: A framework for temporal abstraction in reinforcement learning. *Artif. Intell.* **1999**, *112*, 181–211. [CrossRef]
26. Kober, J.; Bagnell, J.A.; Peters, J. Reinforcement learning in robotics: A survey. *Int. J. Robot. Res.* **2013**, *32*, 1238–1274. [CrossRef]
27. Bloembergen, D.; Tuyls, K.; Hennes, D.; Kaisers, M. Evolutionary dynamics of multi-agent learning: A survey. *J. Artif. Intell. Res.* **2015**, *53*, 659–697. [CrossRef]
28. Littman, M.L. Markov games as a framework for multi-agent reinforcement learning. In *Machine Learning Proceedings 1994*; Elsevier: Amsterdam, The Netherlands, 1994; pp. 157–163.
29. Bowling, M.; Veloso, M. Multiagent learning using a variable learning rate. *Artif. Intell.* **2002**, *136*, 215–250. [CrossRef]
30. Kaisers, M.; Tuyls, K. Frequency adjusted multi-agent Q-learning. In Proceedings of the 9th International Conference on Autonomous Agents and Multiagent Systems, Toronto, ON, Canada, 10–14 May 2010; Volume 1, pp. 309–316.
31. Dutta, A.; Dasgupta, P.; Nelson, C. Adaptive locomotion learning in modular self-reconfigurable robots: A game theoretic approach. In Proceedings of the 2017 IEEE/RSJ International Conference on Intelligent Robots and Systems (IROS), Vancouver, BC, Canada, 24–28 September 2017; pp. 3556–3561.
32. Matignon, L.; Laurent, G.J.; Le Fort-Piat, N. Hysteretic q-learning: An algorithm for decentralized reinforcement learning in cooperative multi-agent teams. In Proceedings of the 2007 IEEE/RSJ International Conference on Intelligent Robots and Systems, San Diego, CA, USA, 29 October–2 November 2007; pp. 64–69.
33. Dutta, A.; Dasgupta, P.; Nelson, C. Distributed adaptive locomotion learning in modred modular self-reconfigurable robot. In *Distributed Autonomous Robotic Systems*; Springer: Berlin/Heidelberg, Germany, 2018; pp. 345–357.
34. Sadhu, A.K.; Konar, A. Improving the speed of convergence of multi-agent Q-learning for cooperative task-planning by a robot-team. *Robot. Auton. Syst.* **2017**, *92*, 66–80. [CrossRef]
35. Hu, J.; Wellman, M.P. Nash Q-learning for general-sum stochastic games. *J. Mach. Learn. Res.* **2003**, *4*, 1039–1069.
36. Buşoniu, L.; Babuška, R.; Schutter, B.D. Multi-agent reinforcement learning: An overview. In *Innovations in Multi-Agent Systems and Applications—1*; Springer: Berlin/Heidelberg, Germany, 2010; pp. 183–221.
37. Van Hasselt, H.; Guez, A.; Silver, D. Deep reinforcement learning with double q-learning. In Proceedings of the AAAI Conference on Artificial Intelligence, Phoenix, AZ, USA, 12–17 February 2016; Volume 30.
38. Moon, J.; Papaioannou, S.; Laoudias, C.; Kolios, P.; Kim, S. Deep reinforcement learning multi-UAV trajectory control for target tracking. *IEEE Internet Things J.* **2021**, *8*, 15441–15455. [CrossRef]
39. Wang, D.; Deng, H. Multirobot coordination with deep reinforcement learning in complex environments. *Expert Syst. Appl.* **2021**, *180*, 115128. [CrossRef]
40. Yu, C.; Dong, Y.; Li, Y.; Chen, Y. Distributed multi-agent deep reinforcement learning for cooperative multi-robot pursuit. *J. Eng.* **2020**, *2020*, 499–504. [CrossRef]
41. Zellner, A.; Dutta, A.; Kulbaka, I.; Sharma, G. Deep Recurrent Q-learning for Energy-constrained Coverage with a Mobile Robot. *arXiv* **2022**, arXiv:2210.00327.
42. Lillicrap, T.P.; Hunt, J.J.; Pritzel, A.; Heess, N.; Erez, T.; Tassa, Y.; Silver, D.; Wierstra, D. Continuous control with deep reinforcement learning. *arXiv* **2015**, arXiv:1509.02971.
43. Schulman, J.; Wolski, F.; Dhariwal, P.; Radford, A.; Klimov, O. Proximal policy optimization algorithms. *arXiv* **2017**, arXiv:1707.06347.
44. Schulman, J.; Levine, S.; Abbeel, P.; Jordan, M.; Moritz, P. Trust region policy optimization. In Proceedings of the International Conference on Machine Learning, PMLR, Lille, France, 6–11 July 2015; pp. 1889–1897.
45. Mnih, V.; Badia, A.P.; Mirza, M.; Graves, A.; Lillicrap, T.; Harley, T.; Silver, D.; Kavukcuoglu, K. Asynchronous methods for deep reinforcement learning. In Proceedings of the International Conference on Machine Learning, PMLR, New York, NY, USA, 20–22 June 2016; pp. 1928–1937.
46. Li, B.; Li, S.; Wang, C.; Fan, R.; Shao, J.; Xie, G. Distributed Circle Formation Control for Quadrotors Based on Multi-agent Deep Reinforcement Learning. In Proceedings of the 2021 IEEE China Automation Congress (CAC), Beijing, China, 22–24 October 2021; pp. 4750–4755.
47. Xu, Z.; Lyu, Y.; Pan, Q.; Hu, J.; Zhao, C.; Liu, S. Multi-vehicle flocking control with deep deterministic policy gradient method. In Proceedings of the 14th IEEE International Conference on Control and Automation (ICCA), Anchorage, AK, USA, 2–15 June 2018; pp. 306–311.

48. Bezcioglu, M.B.; Lennox, B.; Arvin, F. Self-Organised Swarm Flocking with Deep Reinforcement Learning. In Proceedings of the 7th IEEE International Conference on Automation, Robotics and Applications (ICARA), Prague, Czech Republic, 4–6 February 2021; pp. 226–230.
49. Na, S.; Niu, H.; Lennox, B.; Arvin, F. Bio-Inspired Collision Avoidance in Swarm Systems via Deep Reinforcement Learning. *IEEE Trans. Veh. Technol.* **2022**, *71*, 2511–2526. [CrossRef]
50. Gu, S.; Holly, E.; Lillicrap, T.; Levine, S. Deep reinforcement learning for robotic manipulation with asynchronous off-policy updates. In Proceedings of the 2017 IEEE International Conference on Robotics and Automation (ICRA), Singapore, 29 May–3 June 2017; pp. 3389–3396.
51. Li, Y. Deep reinforcement learning: An overview. *arXiv* **2017**, arXiv:1701.07274.
52. Agrawal, A.; Won, S.J.; Sharma, T.; Deshpande, M.; McComb, C. A multi-agent reinforcement learning framework for intelligent manufacturing with autonomous mobile robots. *Proc. Des. Soc.* **2021**, *1*, 161–170. [CrossRef]
53. Bromo, C. Reinforcement Learning Based Strategic Exploration Algorithm for UAVs Fleets. Ph.D. Thesis, Politecnico di Torino, Turin, Italy, 2022.
54. Han, R.; Chen, S.; Wang, S.; Zhang, Z.; Gao, R.; Hao, Q.; Pan, J. Reinforcement Learned Distributed Multi-Robot Navigation With Reciprocal Velocity Obstacle Shaped Rewards. *IEEE Robot. Autom. Lett.* **2022**, *7*, 5896–5903. [CrossRef]
55. Na, S.; Niu, H.; Lennox, B.; Arvin, F. Universal artificial pheromone framework with deep reinforcement learning for robotic systems. In Proceedings of the 6th IEEE International Conference on Control and Robotics Engineering (ICCRE), Beijing, China, 16–18 April 2021; pp. 28–32.
56. Thumiger, N.; Deghat, M. A Multi-Agent Deep Reinforcement Learning Approach for Practical Decentralized UAV Collision Avoidance. *IEEE Control Syst. Lett.* **2021**, *6*, 2174–2179. [CrossRef]
57. Wang, G.; Liu, Z.; Xiao, K.; Xu, Y.; Yang, L.; Wang, X. Collision Detection and Avoidance for Multi-UAV based on Deep Reinforcement Learning. In Proceedings of the 40th IEEE Chinese Control Conference (CCC), Shanghai, China, 26–28 July 2021; pp. 7783–7789.
58. Tampuu, A.; Matiisen, T.; Kodelja, D.; Kuzovkin, I.; Korjus, K.; Aru, J.; Aru, J.; Vicente, R. Multiagent cooperation and competition with deep reinforcement learning. *PLoS ONE* **2017**, *12*, e0172395. [CrossRef] [PubMed]
59. Egorov, M. Multi-agent deep reinforcement learning. In *CS231n: Convolutional Neural Networks for Visual Recognition*; 2016; pp. 1–8.
60. Yang, Y.; Luo, R.; Li, M.; Zhou, M.; Zhang, W.; Wang, J. Mean field multi-agent reinforcement learning. In Proceedings of the International Conference on Machine Learning, PMLR, Playa Blanca, Spain, 9–11 April 2018; pp. 5571–5580.
61. Lowe, R.; Wu, Y.I.; Tamar, A.; Harb, J.; Pieter Abbeel, O.; Mordatch, I. Multi-agent actor-critic for mixed cooperative-competitive environments. *Adv. Neural Inf. Process. Syst.* **2017**, *30*, 6382–6393.
62. Yu, C.; Velu, A.; Vinitsky, E.; Wang, Y.; Bayen, A.; Wu, Y. The surprising effectiveness of ppo in cooperative, multi-agent games. *arXiv* **2021**, arXiv:2103.01955.
63. Sunehag, P.; Lever, G.; Gruslys, A.; Czarnecki, W.M.; Zambaldi, V.; Jaderberg, M.; Lanctot, M.; Sonnerat, N.; Leibo, J.Z.; Tuyls, K.; et al. Value-decomposition networks for cooperative multi-agent learning. *arXiv* **2017**, arXiv:1706.05296.
64. Du, W.; Ding, S. A survey on multi-agent deep reinforcement learning: From the perspective of challenges and applications. *Artif. Intell. Rev.* **2021**, *54*, 3215–3238. [CrossRef]
65. Nguyen, T.T.; Nguyen, N.D.; Nahavandi, S. Deep reinforcement learning for multiagent systems: A review of challenges, solutions, and applications. *IEEE Trans. Cybern.* **2020**, *50*, 3826–3839. [CrossRef]
66. OroojlooyJadid, A.; Hajinezhad, D. A review of cooperative multi-agent deep reinforcement learning. *arXiv* **2019**, arXiv:1908.03963.
67. Wei, Y.; Zheng, R. Multi-Robot Path Planning for Mobile Sensing through Deep Reinforcement Learning. In Proceedings of the INFOCOM 2021-IEEE Conference on Computer Communications, Vancouver, BC, Canada, 10–13 May 2021; pp. 1–10.
68. Mou, Z.; Zhang, Y.; Gao, F.; Wang, H.; Zhang, T.; Han, Z. Deep reinforcement learning based three-dimensional area coverage with UAV swarm. *IEEE J. Sel. Areas Commun.* **2021**, *39*, 3160–3176. [CrossRef]
69. Li, W.; Zhao, T.; Dian, S. Multirobot Coverage Path Planning Based on Deep Q-Network in Unknown Environment. *J. Robot.* **2022**, *2022*, 6825902. [CrossRef]
70. Kakish, Z.; Elamvazhuthi, K.; Berman, S. Using reinforcement learning to herd a robotic swarm to a target distribution. In *Proceedings of the International Symposium Distributed Autonomous Robotic Systems*; Springer: Berlin/Heidelberg, Germany, 2021; pp. 401–414.
71. Yang, Y.; Juntao, L.; Lingling, P. Multi-robot path planning based on a deep reinforcement learning DQN algorithm. *CAAI Trans. Intell. Technol.* **2020**, *5*, 177–183. [CrossRef]
72. Bae, H.; Kim, G.; Kim, J.; Qian, D.; Lee, S. Multi-robot path planning method using reinforcement learning. *Appl. Sci.* **2019**, *9*, 3057. [CrossRef]
73. Zhang, L.; Sun, Y.; Barth, A.; Ma, O. Decentralized control of multi-robot system in cooperative object transportation using deep reinforcement learning. *IEEE Access* **2020**, *8*, 184109–184119. [CrossRef]
74. Marchesini, E.; Farinelli, A. Enhancing deep reinforcement learning approaches for multi-robot navigation via single-robot evolutionary policy search. In Proceedings of the 2022 IEEE International Conference on Robotics and Automation (ICRA), Philadelphia, PA, USA, 23–27 May 2022; pp. 5525–5531.

75. Marchesini, E.; Farinelli, A. Centralizing state-values in dueling networks for multi-robot reinforcement learning mapless navigation. In Proceedings of the 2021 IEEE/RSJ International Conference on Intelligent Robots and Systems (IROS), Prague, Czech Republic, 27 September–1 October 2021; pp. 4583–4588.
76. Zhang, H.; Li, D.; He, Y. Multi-robot cooperation strategy in game environment using deep reinforcement learning. In Proceedings of the 2018 IEEE International Conference on Robotics and Biomimetics (ROBIO), Kuala Lumpur, Malaysia, 12–15 December 2018; pp. 886–891.
77. Manko, S.V.; Diane, S.A.; Krivoshatskiy, A.E.; Margolin, I.D.; Slepynina, E.A. Adaptive control of a multi-robot system for transportation of large-sized objects based on reinforcement learning. In Proceedings of the 2018 IEEE Conference of Russian Young Researchers in Electrical and Electronic Engineering (EIConRus), Moscow and St. Petersburg, Russia, 29 January–1 February 2018; pp. 923–927.
78. Yasuda, T.; Ohkura, K. Collective behavior acquisition of real robotic swarms using deep reinforcement learning. In Proceedings of the 2nd IEEE International Conference on Robotic Computing (IRC), Laguna Hills, CA, USA, 31 January–2 February 2018; pp. 179–180.
79. Eoh, G.; Park, T.H. Cooperative object transportation using curriculum-based deep reinforcement learning. *Sensors* **2021**, *21*, 4780. [CrossRef]
80. Huang, W.; Wang, Y.; Yi, X. Deep q-learning to preserve connectivity in multi-robot systems. In Proceedings of the 9th International Conference on Signal Processing Systems, ICSPS 2017, Auckland, New Zealand, 27–30 November 2017; pp. 45–50.
81. Gupta, J.K.; Egorov, M.; Kochenderfer, M. Cooperative multi-agent control using deep reinforcement learning. In *Proceedings of the International Conference on Autonomous Agents and Multiagent Systems*; Springer: Berlin/Heidelberg, Germany, 2017; pp. 66–83.
82. Wang, Z.; Gombolay, M. Learning scheduling policies for multi-robot coordination with graph attention networks. *IEEE Robot. Autom. Lett.* **2020**, *5*, 4509–4516. [CrossRef]
83. Yan, C.; Wang, C.; Xiang, X.; Lan, Z.; Jiang, Y. Deep reinforcement learning of collision-free flocking policies for multiple fixed-wing uavs using local situation maps. *IEEE Trans. Ind. Inform.* **2021**, *18*, 1260–1270. [CrossRef]
84. Liu, Y.; Peng, Y.; Wang, M.; Xie, J.; Zhou, R. Multi-usv system cooperative underwater target search based on reinforcement learning and probability map. *Math. Probl. Eng.* **2020**, *2020*, 7842768. [CrossRef]
85. Viseras, A.; Meissner, M.; Marchal, J. Wildfire front monitoring with multiple uavs using deep q-learning. *IEEE Access* **2021**. [CrossRef]
86. Goyal, A. Multi-Agent Deep Reinforcement Learning for Robocup Rescue Simulator. Ph.D. Thesis, The University of Texas, Austin, TX, USA, 2020.
87. Chen, L.; Wang, Y.; Mo, Y.; Miao, Z.; Wang, H.; Feng, M.; Wang, S. Multi-Agent Path Finding Using Deep Reinforcement Learning Coupled With Hot Supervision Contrastive Loss. *IEEE Trans. Ind. Electron.* **2022**, *70*, 7032–7040. [CrossRef]
88. Jestel, C.; Surmann, H.; Stenzel, J.; Urbann, O.; Brehler, M. Obtaining Robust Control and Navigation Policies for Multi-robot Navigation via Deep Reinforcement Learning. In Proceedings of the 7th IEEE International Conference on Automation, Robotics and Applications (ICARA), Prague, Czech Republic, 4–6 February 2021; pp. 48–54.
89. Gautier, P.; Laurent, J.; Diguet, J.P. Deep Q-Learning-Based Dynamic Management of a Robotic Cluster. *IEEE Trans. Autom. Sci. Eng.* **2022**, 1–13. [CrossRef]
90. Song, C.; He, Z.; Dong, L. A Local-and-Global Attention Reinforcement Learning Algorithm for Multiagent Cooperative Navigation. *IEEE Trans. Neural Netw. Learn. Syst.* **2022**, 1–11. [CrossRef] [PubMed]
91. Ding, S.; Aoyama, H.; Lin, D. Combining Multiagent Reinforcement Learning and Search Method for Drone Delivery on a Non-grid Graph. In Proceedings of the International Conference on Practical Applications of Agents and Multi-Agent Systems; Springer: Berlin/Heidelberg, Germany, 2022; pp. 112–126.
92. Choi, H.B.; Kim, J.B.; Ji, C.H.; Ihsan, U.; Han, Y.H.; Oh, S.W.; Kim, K.H.; Pyo, C.S. MARL-based Optimal Route Control in Multi-AGV Warehouses. In Proceedings of the 2022 IEEE International Conference on Artificial Intelligence in Information and Communication (ICAIIC), Jeju Island, Republic of Korea, 21–24 February 2022; pp. 333–338.
93. Johnson, D.; Chen, G.; Lu, Y. Multi-Agent Reinforcement Learning for Real-Time Dynamic Production Scheduling in a Robot Assembly Cell. *IEEE Robot. Autom. Lett.* **2022**, *7*, 7684–7691. [CrossRef]
94. Chen, L.; Zhao, Y.; Zhao, H.; Zheng, B. Non-communication decentralized multi-robot collision avoidance in grid map workspace with double deep Q-network. *Sensors* **2021**, *21*, 841. [CrossRef]
95. Miyashita, Y.; Sugawara, T. Analysis of coordinated behavior structures with multi-agent deep reinforcement learning. *Appl. Intell.* **2021**, *51*, 1069–1085. [CrossRef]
96. Caccavale, R.; Calà, V.; Ermini, M.; Finzi, A.; Lippiello, V.; Tavano, F. Multi-robot Sanitization of Railway Stations Based on Deep Q-Learning. In Proceedings of the 8th Italian Workshop on AI and Robotics (AIRO), Online, 30 November 2021.
97. Chen, W.; Zhou, S.; Pan, Z.; Zheng, H.; Liu, Y. Mapless collaborative navigation for a multi-robot system based on the deep reinforcement learning. *Appl. Sci.* **2019**, *9*, 4198. [CrossRef]
98. Ma, J.; Lu, H.; Xiao, J.; Zeng, Z.; Zheng, Z. Multi-robot target encirclement control with collision avoidance via deep reinforcement learning. *J. Intell. Robot. Syst.* **2020**, *99*, 371–386. [CrossRef]
99. Kheawkhem, P.; Khuankrue, I. Study on Deep Reinforcement Learning for Mobile Robots Flocking Control in Certainty Situations. In Proceedings of the 19th IEEE International Conference on Electrical Engineering/Electronics, Computer, Telecommunications and Information Technology (ECTI-CON), Prachuap Khiri Khan, Thailand, 24–27 May 2022; pp. 1–4.

100. Qiu, Y.; Zhan, Y.; Jin, Y.; Wang, J.; Zhang, X. Sample-Efficient Multi-Agent Reinforcement Learning with Demonstrations for Flocking Control. *arXiv* **2022**, *arXiv:2209.08351*.
101. Setyawan, G.E.; Hartono, P.; Sawada, H. Cooperative Multi-Robot Hierarchical Reinforcement Learning. *Int. J. Adv. Comput. Sci. Appl.* **2022**, *13*, 35–44. [CrossRef]
102. Meng, S.; Kan, Z. Deep reinforcement learning-based effective coverage control with connectivity constraints. *IEEE Control Syst. Lett.* **2021**, *6*, 283–288. [CrossRef]
103. Hamed, O.; Hamlich, M. Hybrid Formation Control for Multi-Robot Hunters Based on Multi-Agent Deep Deterministic Policy Gradient. *Mendel* **2021**, *27*, 23–29. [CrossRef]
104. Liu, C.H.; Chen, Z.; Tang, J.; Xu, J.; Piao, C. Energy-efficient UAV control for effective and fair communication coverage: A deep reinforcement learning approach. *IEEE J. Sel. Areas Commun.* **2018**, *36*, 2059–2070. [CrossRef]
105. Kouzehgar, M.; Meghjani, M.; Bouffanais, R. Multi-agent reinforcement learning for dynamic ocean monitoring by a swarm of buoys. In Proceedings of the Global Oceans 2020: Singapore–US Gulf Coast, IEEE, Biloxi, MS, USA, 5–30 October 2020; pp. 1–8.
106. Salimi, M.; Pasquier, P. Deep Reinforcement Learning for Flocking Control of UAVs in Complex Environments. In Proceedings of the 6th IEEE International Conference on Robotics and Automation Engineering (ICRAE), Guangzhou, China, 19–22 November 2021; pp. 344–352.
107. Fan, T.; Long, P.; Liu, W.; Pan, J. Distributed multi-robot collision avoidance via deep reinforcement learning for navigation in complex scenarios. *Int. J. Robot. Res.* **2020**, *39*, 856–892. [CrossRef]
108. Zhao, W.; Queralta, J.P.; Qingqing, L.; Westerlund, T. Towards closing the sim-to-real gap in collaborative multi-robot deep reinforcement learning. In Proceedings of the 5th IEEE International Conference on Robotics and Automation Engineering (ICRAE), Singapore, 20–22 November 2020; pp. 7–12.
109. Lin, J.; Yang, X.; Zheng, P.; Cheng, H. End-to-end decentralized multi-robot navigation in unknown complex environments via deep reinforcement learning. In Proceedings of the 2019 IEEE International Conference on Mechatronics and Automation (ICMA), Tianjin, China, 4–7 August 2019; pp. 2493–2500.
110. Tolstaya, E.; Paulos, J.; Kumar, V.; Ribeiro, A. Multi-robot coverage and exploration using spatial graph neural networks. In Proceedings of the 2021 IEEE/RSJ International Conference on Intelligent Robots and Systems (IROS), Prague, Czech Republic, 27 September–1 October 2021; pp. 8944–8950.
111. Blumenkamp, J.; Morad, S.; Gielis, J.; Li, Q.; Prorok, A. A framework for real-world multi-robot systems running decentralized GNN-based policies. In Proceedings of the 2022 IEEE International Conference on Robotics and Automation (ICRA), Philadelphia, PA, USA, 23–27 May 2022; pp. 8772–8778.
112. Lin, J.; Yang, X.; Zheng, P.; Cheng, H. Connectivity guaranteed multi-robot navigation via deep reinforcement learning. In Proceedings of the Conference on Robot Learning, PMLR, Virtual, 16–18 November 2020; pp. 661–670.
113. Wang, J.; Cao, J.; Stojmenovic, M.; Zhao, M.; Chen, J.; Jiang, S. Pattern-rl: Multi-robot cooperative pattern formation via deep reinforcement learning. In Proceedings of the 18th IEEE International Conference On Machine Learning And Applications (ICMLA), Boca Raton, FL, USA, 16–19 December 2019; pp. 210–215.
114. Park, B.; Kang, C.; Choi, J. Cooperative Multi-Robot Task Allocation with Reinforcement Learning. *Appl. Sci.* **2021**, *12*, 272. [CrossRef]
115. Yao, S.; Chen, G.; Pan, L.; Ma, J.; Ji, J.; Chen, X. Multi-robot collision avoidance with map-based deep reinforcement learning. In Proceedings of the 32nd IEEE International Conference on Tools with Artificial Intelligence (ICTAI), Baltimore, MD, USA, 9–11 November 2020; pp. 532–539.
116. Tan, Q.; Fan, T.; Pan, J.; Manocha, D. DeepMNavigate: Deep reinforced multi-robot navigation unifying local & global collision avoidance. In Proceedings of the 2020 IEEE/RSJ International Conference on Intelligent Robots and Systems (IROS), Las Vegas, NV, USA, 24 October–24 January 2020; pp. 6952–6959.
117. Han, R.; Chen, S.; Hao, Q. Cooperative multi-robot navigation in dynamic environment with deep reinforcement learning. In Proceedings of the 2020 IEEE International Conference on Robotics and Automation (ICRA), Paris, France, 31 May–31 August 2020; pp. 448–454.
118. Blumenkamp, J.; Prorok, A. The emergence of adversarial communication in multi-agent reinforcement learning. *arXiv* **2020**, arXiv:2008.02616.
119. Sivanathan, K.; Vinayagam, B.; Samak, T.; Samak, C. Decentralized motion planning for multi-robot navigation using deep reinforcement learning. In Proceedings of the 3rd IEEE International Conference on Intelligent Sustainable Systems (ICISS), Thoothukudi, India, 3–5 December 2020; pp. 709–716.
120. Liu, J.y.; Wang, G.; Fu, Q.; Yue, S.h.; Wang, S.y. Task assignment in ground-to-air confrontation based on multiagent deep reinforcement learning. *Def. Technol.* **2022**, *19*, 210–219. [CrossRef]
121. Sadhukhan, P.; Selmic, R.R. Multi-agent formation control with obstacle avoidance using proximal policy optimization. In Proceedings of the 2021 IEEE International Conference on Systems, Man, and Cybernetics (SMC), Melbourne, Australia, 17–20 October 2021; pp. 2694–2699.
122. Sadhukhan, P. Proximal Policy Optimization for Formation Control and Obstacle Avoidance in Multi-Agent Systems. Ph.D. Thesis, Concordia University, Montreal, QC, Canada, 2021.
123. Ourari, R.; Cui, K.; Koeppl, H. Decentralized swarm collision avoidance for quadrotors via end-to-end reinforcement learning. *arXiv* **2021**, arXiv:2104.14912.

124. Zhang, T.; Liu, Z.; Pu, Z.; Yi, J. Multi-Target Encirclement with Collision Avoidance via Deep Reinforcement Learning using Relational Graphs. In Proceedings of the 2022 IEEE International Conference on Robotics and Automation (ICRA), Philadelphia, PA, USA, 23–27 May 2022; pp. 8794–8800.
125. Sadhukhan, P.; Selmic, R.R. Proximal policy optimization for formation navigation and obstacle avoidance. *Int. J. Intell. Robot. Appl.* **2022**, *6*, 746–759. [CrossRef]
126. Allen, R.E.; Gupta, J.K.; Pena, J.; Zhou, Y.; Bear, J.W.; Kochenderfer, M.J. Health-Informed Policy Gradients for Multi-Agent Reinforcement Learning. *arXiv* **2019**, arXiv:1908.01022.
127. Xia, J.; Luo, Y.; Liu, Z.; Zhang, Y.; Shi, H.; Liu, Z. Cooperative multi-target hunting by unmanned surface vehicles based on multi-agent reinforcement learning. *Defence Technol.* 2022, *in press*. [CrossRef]
128. Li, S.; Guo, W. Supervised Reinforcement Learning for ULV Path Planning in Complex Warehouse Environment. *Wirel. Commun. Mob. Comput.* **2022**, *2022*, 4384954. [CrossRef]
129. Paull, S.; Ghassemi, P.; Chowdhury, S. Learning Scalable Policies over Graphs for Multi-Robot Task Allocation using Capsule Attention Networks. In Proceedings of the 2022 IEEE International Conference on Robotics and Automation (ICRA), Philadelphia, PA, USA, 23–27 May 2022; pp. 8815–8822.
130. Na, S.; Krajník, T.; Lennox, B.; Arvin, F. Federated Reinforcement Learning for Collective Navigation of Robotic Swarms. *arXiv* **2022**, arXiv:2202.01141.
131. Fan, T.; Long, P.; Liu, W.; Pan, J. Fully distributed multi-robot collision avoidance via deep reinforcement learning for safe and efficient navigation in complex scenarios. *arXiv* **2018**, arXiv:1808.03841.
132. Elfakharany, A.; Ismail, Z.H. End-to-end deep reinforcement learning for decentralized task allocation and navigation for a multi-robot system. *Appl. Sci.* **2021**, *11*, 2895. [CrossRef]
133. Wen, S.; Wen, Z.; Zhang, D.; Zhang, H.; Wang, T. A multi-robot path-planning algorithm for autonomous navigation using meta-reinforcement learning based on transfer learning. *Appl. Soft Comput.* **2021**, *110*, 107605. [CrossRef]
134. Khan, A.; Tolstaya, E.; Ribeiro, A.; Kumar, V. Graph policy gradients for large scale robot control. In Proceedings of the Conference on Robot Learning, PMLR, Virtual, 16–18 November 2020; pp. 823–834.
135. Alon, Y.; Zhou, H. Multi-agent reinforcement learning for unmanned aerial vehicle coordination by multi-critic policy gradient optimization. *arXiv* **2020**, arXiv:2012.15472.
136. Khan, A.; Kumar, V.; Ribeiro, A. Graph policy gradients for large scale unlabeled motion planning with constraints. *arXiv* **2019**, arXiv:1909.10704.
137. Asayesh, S.; Chen, M.; Mehrandezh, M.; Gupta, K. Least-restrictive multi-agent collision avoidance via deep meta reinforcement learning and optimal control. *arXiv* **2021**, arXiv:2106.00936.
138. Qamar, S.; Khan, S.H.; Arshad, M.A.; Qamar, M.; Gwak, J.; Khan, A. Autonomous Drone Swarm Navigation and Multi-target Tracking with Island Policy-based Optimization Framework. *IEEE Access* **2022**, *10*, 91073–91091. [CrossRef]
139. Zhou, W.; Li, J.; Zhang, Q. Joint Communication and Action Learning in Multi-Target Tracking of UAV Swarms with Deep Reinforcement Learning. *Drones* **2022**, *6*, 339. [CrossRef]
140. Hüttenrauch, M.; Šošić, A.; Neumann, G. Local communication protocols for learning complex swarm behaviors with deep reinforcement learning. In *Proceedings of the International Conference on Swarm Intelligence*; Springer: Berlin/Heidelberg, Germany, 2018; pp. 71–83.
141. Hüttenrauch, M.; Adrian, S.; Neumann, G. Deep reinforcement learning for swarm systems. *J. Mach. Learn. Res.* **2019**, *20*, 1–31.
142. Wang, W.; Wang, L.; Wu, J.; Tao, X.; Wu, H. Oracle-Guided Deep Reinforcement Learning for Large-Scale Multi-UAVs Flocking and Navigation. *IEEE Trans. Veh. Technol.* **2022**, *71*, 10280–10292. [CrossRef]
143. Prianto, E.; Kim, M.; Park, J.H.; Bae, J.H.; Kim, J.S. Path planning for multi-arm manipulators using deep reinforcement learning: Soft actor–critic with hindsight experience replay. *Sensors* **2020**, *20*, 5911. [CrossRef]
144. Cao, Y.; Wang, S.; Zheng, X.; Ma, W.; Xie, X.; Liu, L. Reinforcement Learning with Prior Policy Guidance for Motion Planning of Dual-Arm Free-Floating Space Robot. *arXiv* **2022**, arXiv:2209.01434.
145. Galceran, E.; Carreras, M. A survey on coverage path planning for robotics. *Robot. Auton. Syst.* **2013**, *61*, 1258–1276. [CrossRef]
146. Agmon, N.; Hazon, N.; Kaminka, G.A. Constructing spanning trees for efficient multi-robot coverage. In Proceedings of the IEEE International Conference on Robotics and Automation, ICRA 2006, Orlando, FL, USA, 15–19 May 2006; pp. 1698–1703.
147. Kapoutsis, A.C.; Chatzichristofis, S.A.; Kosmatopoulos, E.B. DARP: Divide areas algorithm for optimal multi-robot coverage path planning. *J. Intell. Robot. Syst.* **2017**, *86*, 663–680. [CrossRef]
148. Rekleitis, I.; New, A.P.; Rankin, E.S.; Choset, H. Efficient boustrophedon multi-robot coverage: An algorithmic approach. *Ann. Math. Artif. Intell.* **2008**, *52*, 109–142. [CrossRef]
149. Zheng, X.; Jain, S.; Koenig, S.; Kempe, D. Multi-robot forest coverage. In Proceedings of the 2005 IEEE/RSJ International Conference on Intelligent Robots and Systems, Edmonton, AB, Canada, 2–6 August 2005; pp. 3852–3857.
150. Marjovi, A.; Nunes, J.G.; Marques, L.; De Almeida, A. Multi-robot exploration and fire searching. In Proceedings of the 2009 IEEE/RSJ International Conference on Intelligent Robots and Systems, St. Louis, MO, USA, 10–15 October 2009; pp. 1929–1934.
151. Nieto-Granda, C.; Rogers III, J.G.; Christensen, H.I. Coordination strategies for multi-robot exploration and mapping. *Int. J. Robot. Res.* **2014**, *33*, 519–533. [CrossRef]

152. Simmons, R.; Apfelbaum, D.; Burgard, W.; Fox, D.; Moors, M.; Thrun, S.; Younes, H. Coordination for multi-robot exploration and mapping. In Proceedings of the 17th National Conference on Artificial Intelligence (AAAI-00), Austin, TX, USA, 30 July–3 August 2000; pp. 852–858.
153. Rooker, M.N.; Birk, A. Multi-robot exploration under the constraints of wireless networking. *Control Eng. Pract.* **2007**, *15*, 435–445. [CrossRef]
154. Zhou, X.; Liu, X.; Wang, X.; Wu, S.; Sun, M. Multi-Robot Coverage Path Planning based on Deep Reinforcement Learning. In Proceedings of the 24th IEEE International Conference on Computational Science and Engineering (CSE), Shenyang, China, 20–22 October 2021; pp. 35–42.
155. Hu, J.; Niu, H.; Carrasco, J.; Lennox, B.; Arvin, F. Voronoi-based multi-robot autonomous exploration in unknown environments via deep reinforcement learning. *IEEE Trans. Veh. Technol.* **2020**, *69*, 14413–14423. [CrossRef]
156. Schaul, T.; Quan, J.; Antonoglou, I.; Silver, D. Prioritized experience replay. *arXiv* **2015**, arXiv:1511.05952.
157. Koenig, N.; Howard, A. Design and use paradigms for gazebo, an open-source multi-robot simulator. In Proceedings of the 2004 IEEE/RSJ International Conference on Intelligent Robots and Systems (IROS)(IEEE Cat. No. 04CH37566), Sendai, Japan, 28 September–2 October 2004; Volume 3, pp. 2149–2154.
158. Gama, F.; Marques, A.G.; Leus, G.; Ribeiro, A. Convolutional neural network architectures for signals supported on graphs. *IEEE Trans. Signal Process.* **2018**, *67*, 1034–1049. [CrossRef]
159. Aydemir, F.; Cetin, A. Multi-Agent Dynamic Area Coverage Based on Reinforcement Learning with Connected Agents. *Comput. Syst. Sci. Eng.* **2023**, *45*, 215–230. [CrossRef]
160. Zhang, H.; Cheng, J.; Zhang, L.; Li, Y.; Zhang, W. H2GNN: Hierarchical-Hops Graph Neural Networks for Multi-Robot Exploration in Unknown Environments. *IEEE Robot. Autom. Lett.* **2022**, *7*, 3435–3442. [CrossRef]
161. Gao, M.; Zhang, X. Cooperative Search Method for Multiple UAVs Based on Deep Reinforcement Learning. *Sensors* **2022**, *22*, 6737. [CrossRef]
162. Sheng, W.; Guo, H.; Yau, W.Y.; Zhou, Y. PD-FAC: Probability Density Factorized Multi-Agent Distributional Reinforcement Learning for Multi-Robot Reliable Search. *IEEE Robot. Autom. Lett.* **2022**, *7*, 8869–8876. [CrossRef]
163. Chung, J.; Gulcehre, C.; Cho, K.; Bengio, Y. Empirical evaluation of gated recurrent neural networks on sequence modeling. *arXiv* **2014**, arXiv:1412.3555.
164. Reynolds, C.W. Flocks, herds and schools: A distributed behavioral model. In Proceedings of the 14th Annual Conference on Computer Graphics and Interactive Techniques, Anaheim, CA, USA, 27–31 July 1987; pp. 25–34.
165. Liang, Z.; Cao, J.; Lin, W.; Chen, J.; Xu, H. Hierarchical Deep Reinforcement Learning for Multi-robot Cooperation in Partially Observable Environment. In Proceedings of the 3rd IEEE International Conference on Cognitive Machine Intelligence (CogMI), Atlanta, GA, USA, 13–15 December 2021; pp. 272–281.
166. Acar, E.U.; Choset, H.; Lee, J.Y. Sensor-based coverage with extended range detectors. *IEEE Trans. Robot.* **2006**, *22*, 189–198. [CrossRef]
167. Chen, D.; Qi, Q.; Zhuang, Z.; Wang, J.; Liao, J.; Han, Z. Mean field deep reinforcement learning for fair and efficient UAV control. *IEEE Internet Things J.* **2020**, *8*, 813–828. [CrossRef]
168. Zhang, Y.; Yang, C.; Li, J.; Han, Z. Distributed interference-aware traffic offloading and power control in ultra-dense networks: Mean field game with dominating player. *IEEE Trans. Veh. Technol.* **2019**, *68*, 8814–8826. [CrossRef]
169. Guéant, O.; Lasry, J.M.; Lions, P.L. Mean field games and applications. In *Paris Princeton Lectures on Mathematical Finance 2010*; Springer: Berlin/Heidelberg, Germany, 2011; pp. 205–266.
170. Kadanoff, L.P.; Martin, P.C. Statistical physics: Statics, dynamics, and renormalization. *Phys. Today* **2001**, *54*, 54–55.
171. Nemer, I.A.; Sheltami, T.R.; Belhaiza, S.; Mahmoud, A.S. Energy-Efficient UAV Movement Control for Fair Communication Coverage: A Deep Reinforcement Learning Approach. *Sensors* **2022**, *22*, 1919. [CrossRef] [PubMed]
172. Liu, C.H.; Ma, X.; Gao, X.; Tang, J. Distributed energy-efficient multi-UAV navigation for long-term communication coverage by deep reinforcement learning. *IEEE Trans. Mob. Comput.* **2019**, *19*, 1274–1285. [CrossRef]
173. Surynek, P. An optimization variant of multi-robot path planning is intractable. In Proceedings of the AAAI Conference on Artificial Intelligence, Atlanta, GA, USA, 11–15 July 2010; Volume 24, pp. 1261–1263.
174. Hart, P.E.; Nilsson, N.J.; Raphael, B. A formal basis for the heuristic determination of minimum cost paths. *IEEE Trans. Syst. Sci. Cybern.* **1968**, *4*, 100–107. [CrossRef]
175. Wagner, G.; Choset, H. Subdimensional expansion for multirobot path planning. *Artif. Intell.* **2015**, *219*, 1–24. [CrossRef]
176. Bennewitz, M.; Burgard, W.; Thrun, S. Finding and optimizing solvable priority schemes for decoupled path planning techniques for teams of mobile robots. *Robot. Auton. Syst.* **2002**, *41*, 89–99. [CrossRef]
177. Dutta, A.; Dasgupta, P. Bipartite graph matching-based coordination mechanism for multi-robot path planning under communication constraints. In Proceedings of the 2017 IEEE International Conference on Robotics and Automation (ICRA), Singapore, 29 May–3 June 2017; pp. 857–862.
178. Kimmel, A.; Bekris, K. Decentralized multi-agent path selection using minimal information. In *Distributed Autonomous Robotic Systems*; Springer: Berlin/Heidelberg, Germany, 2016; pp. 341–356.
179. Yu, J.; LaValle, S.M. Multi-agent path planning and network flow. In *Algorithmic Foundations of Robotics X*; Springer: Berlin/Heidelberg, Germany, 2013; pp. 157–173.

180. Xu, Y.; Wei, Y.; Wang, D.; Jiang, K.; Deng, H. Multi-UAV Path Planning in GPS and Communication Denial Environment. *Sensors* **2023**, *23*, 2997. [CrossRef]
181. Wang, D.; Deng, H.; Pan, Z. Mrcdrl: Multi-robot coordination with deep reinforcement learning. *Neurocomputing* **2020**, *406*, 68–76. [CrossRef]
182. Hochreiter, S.; Schmidhuber, J. Long short-term memory. *Neural Comput.* **1997**, *9*, 1735–1780. [CrossRef] [PubMed]
183. Li, M.; Jie, Y.; Kong, Y.; Cheng, H. Decentralized Global Connectivity Maintenance for Multi-Robot Navigation: A Reinforcement Learning Approach. In Proceedings of the 2022 IEEE International Conference on Robotics and Automation (ICRA), Philadelphia, PA, USA, 23–27 May 2022; pp. 8801–8807.
184. Achiam, J.; Held, D.; Tamar, A.; Abbeel, P. Constrained policy optimization. In Proceedings of the International Conference on Machine Learning, PMLR, Sydney, Australia, 6–11 August 2017; pp. 22–31.
185. Dutta, A.; Ghosh, A.; Kreidl, O.P. Multi-robot informative path planning with continuous connectivity constraints. In Proceedings of the 2019 IEEE International Conference on Robotics and Automation (ICRA), Montreal, QC, Canada, 20–24 May 2019; pp. 3245–3251.
186. Challita, U.; Saad, W.; Bettstetter, C. Interference management for cellular-connected UAVs: A deep reinforcement learning approach. *IEEE Trans. Wirel. Commun.* **2019**, *18*, 2125–2140. [CrossRef]
187. Wang, B.; Liu, Z.; Li, Q.; Prorok, A. Mobile robot path planning in dynamic environments through globally guided reinforcement learning. *IEEE Robot. Autom. Lett.* **2020**, *5*, 6932–6939. [CrossRef]
188. Rashid, T.; Samvelyan, M.; Schroeder, C.; Farquhar, G.; Foerster, J.; Whiteson, S. Qmix: Monotonic value function factorisation for deep multi-agent reinforcement learning. In Proceedings of the International Conference on Machine Learning, PMLR, Playa Blanca, Spain, 9–11 April 2018; pp. 4295–4304.
189. Chen, Y.F.; Liu, M.; Everett, M.; How, J.P. Decentralized non-communicating multiagent collision avoidance with deep reinforcement learning. In Proceedings of the 2017 IEEE International Conference on Robotics and Automation (ICRA), Singapore, 29 May–3 June 2017; pp. 285–292.
190. Chen, Y.F.; Everett, M.; Liu, M.; How, J.P. Socially aware motion planning with deep reinforcement learning. In Proceedings of the 2017 IEEE/RSJ International Conference on Intelligent Robots and Systems (IROS), Vancouver, BC, Canada, 24–28 September 2017; pp. 1343–1350.
191. Weiss, K.; Khoshgoftaar, T.M.; Wang, D. A survey of transfer learning. *J. Big Data* **2016**, *3*, 1–40. [CrossRef]
192. Konečný, J.; McMahan, B.; Ramage, D. Federated optimization: Distributed optimization beyond the datacenter. *arXiv* **2015**, arXiv:1511.03575.
193. Luo, R.; Ni, W.; Tian, H.; Cheng, J. Federated Deep Reinforcement Learning for RIS-Assisted Indoor Multi-Robot Communication Systems. *IEEE Trans. Veh. Technol.* **2022**, *71*, 12321–12326. [CrossRef]
194. Sartoretti, G.; Kerr, J.; Shi, Y.; Wagner, G.; Kumar, T.S.; Koenig, S.; Choset, H. Primal: Pathfinding via reinforcement and imitation multi-agent learning. *IEEE Robot. Autom. Lett.* **2019**, *4*, 2378–2385. [CrossRef]
195. Ross, S.; Gordon, G.; Bagnell, D. A reduction of imitation learning and structured prediction to no-regret online learning. In Proceedings of the 14th International Conference on Artificial Intelligence and Statistics. JMLR Workshop and Conference Proceedings, Fort Lauderdale, FL, USA, 11–13 April 2011; pp. 627–635.
196. Damani, M.; Luo, Z.; Wenzel, E.; Sartoretti, G. PRIMAL _2: Pathfinding via reinforcement and imitation multi-agent learning-lifelong. *IEEE Robot. Autom. Lett.* **2021**, *6*, 2666–2673. [CrossRef]
197. Bengio, Y.; Louradour, J.; Collobert, R.; Weston, J. Curriculum learning. In Proceedings of the 26th Annual International Conference on Machine Learning, Montreal, QC, Canada, 14–18 June 2009; pp. 41–48.
198. Sun, L.; Yan, J.; Qin, W. Path planning for multiple agents in an unknown environment using soft actor critic and curriculum learning. *Comput. Animat. Virtual Worlds* **2023**, *34*, e2113. [CrossRef]
199. Pu, Y.; Gan, Z.; Henao, R.; Yuan, X.; Li, C.; Stevens, A.; Carin, L. Variational autoencoder for deep learning of images, labels and captions. In Proceedings of the 30th International Conference on Neural Information Processing Systems, NIPS'16, Barcelona, Spain, 5–10 December 2016; Volume 29.
200. Li, H. Decentralized Multi-Agent Collision Avoidance and Reinforcement Learning. Ph.D. Thesis, The Ohio State University, Columbus, OH, USA, 2021.
201. Andrychowicz, M.; Wolski, F.; Ray, A.; Schneider, J.; Fong, R.; Welinder, P.; McGrew, B.; Tobin, J.; Pieter Abbeel, O.; Zaremba, W. Hindsight experience replay. *Adv. Neural Inf. Process. Syst.* **2017**, *30*, 5048–5058.
202. Everett, M.; Chen, Y.F.; How, J.P. Motion planning among dynamic, decision-making agents with deep reinforcement learning. In Proceedings of the 2018 IEEE/RSJ International Conference on Intelligent Robots and Systems (IROS), Madrid, Spain, 1–5 October 2018; pp. 3052–3059.
203. Semnani, S.H.; Liu, H.; Everett, M.; De Ruiter, A.; How, J.P. Multi-agent motion planning for dense and dynamic environments via deep reinforcement learning. *IEEE Robot. Autom. Lett.* **2020**, *5*, 3221–3226. [CrossRef]
204. Zhang, H.; Luo, J.; Lin, X.; Tan, K.; Pan, C. Dispatching and Path Planning of Automated Guided Vehicles based on Petri Nets and Deep Reinforcement Learning. In Proceedings of the 2021 IEEE International Conference on Networking, Sensing and Control (ICNSC), Xiamen, China, 3–5 December 2021; Volume 1, pp. 1–6.
205. Huang, H.; Zhu, G.; Fan, Z.; Zhai, H.; Cai, Y.; Shi, Z.; Dong, Z.; Hao, Z. Vision-based Distributed Multi-UAV Collision Avoidance via Deep Reinforcement Learning for Navigation. *arXiv* **2022**, arXiv:2203.02650.

206. Yarats, D.; Zhang, A.; Kostrikov, I.; Amos, B.; Pineau, J.; Fergus, R. Improving sample efficiency in model-free reinforcement learning from images. In Proceedings of the AAAI Conference on Artificial Intelligence, Virtual, 2–9 February 2021; Volume 35, pp. 10674–10681.
207. Jeon, S.; Lee, H.; Kaliappan, V.K.; Nguyen, T.A.; Jo, H.; Cho, H.; Min, D. Multiagent Reinforcement Learning Based on Fusion-Multiactor-Attention-Critic for Multiple-Unmanned-Aerial-Vehicle Navigation Control. *Energies* **2022**, *15*, 7426. [CrossRef]
208. Shalev-Shwartz, S.; Shammah, S.; Shashua, A. Safe, multi-agent, reinforcement learning for autonomous driving. *arXiv* **2016**, arXiv:1610.03295.
209. Ammar, H.B.; Tutunov, R.; Eaton, E. Safe policy search for lifelong reinforcement learning with sublinear regret. In Proceedings of the International Conference on Machine Learning, PMLR, Lille, France, 6–11 July 2015; pp. 2361–2369.
210. Schulman, J.; Moritz, P.; Levine, S.; Jordan, M.; Abbeel, P. High-dimensional continuous control using generalized advantage estimation. *arXiv* **2015**, arXiv:1506.02438.
211. Taskar, B.; Chatalbashev, V.; Koller, D.; Guestrin, C. Learning structured prediction models: A large margin approach. In Proceedings of the 22nd International Conference on Machine Learning, Bonn, Germany, 7–11 August 2005; pp. 896–903.
212. Liang, Z.; Cao, J.; Jiang, S.; Saxena, D.; Xu, H. Hierarchical Reinforcement Learning with Opponent Modeling for Distributed Multi-agent Cooperation. *arXiv* **2022**, arXiv:2206.12718.
213. Farrow, N.; Klingner, J.; Reishus, D.; Correll, N. Miniature six-channel range and bearing system: Algorithm, analysis and experimental validation. In Proceedings of the 2014 IEEE International Conference on Robotics and Automation (ICRA), Hong Kong, China, 31 May–7 June 2014; pp. 6180–6185.
214. Shiell, N.; Vardy, A. A bearing-only pattern formation algorithm for swarm robotics. In Proceedings of the Swarm Intelligence: 10th International Conference, ANTS 2016, Brussels, Belgium, 7–9 September 2016; pp. 3–14.
215. Rubenstein, M.; Cornejo, A.; Nagpal, R. Programmable self-assembly in a thousand-robot swarm. *Science* **2014**, *345*, 795–799. [CrossRef]
216. Zhu, P.; Dai, W.; Yao, W.; Ma, J.; Zeng, Z.; Lu, H. Multi-robot flocking control based on deep reinforcement learning. *IEEE Access* **2020**, *8*, 150397–150406. [CrossRef]
217. Lan, X.; Liu, Y.; Zhao, Z. Cooperative control for swarming systems based on reinforcement learning in unknown dynamic environment. *Neurocomputing* **2020**, *410*, 410–418. [CrossRef]
218. Kortvelesy, R.; Prorok, A. ModGNN: Expert policy approximation in multi-agent systems with a modular graph neural network architecture. In Proceedings of the 2021 IEEE International Conference on Robotics and Automation (ICRA), Xi'an, China, 30 May–5 June 2021; pp. 9161–9167.
219. Yan, C.; Xiang, X.; Wang, C. Fixed-Wing UAVs flocking in continuous spaces: A deep reinforcement learning approach. *Robot. Auton. Syst.* **2020**, *131*, 103594. [CrossRef]
220. Yan, C.; Xiang, X.; Wang, C.; Lan, Z. Flocking and Collision Avoidance for a Dynamic Squad of Fixed-Wing UAVs Using Deep Reinforcement Learning. In Proceedings of the 2021 IEEE/RSJ International Conference on Intelligent Robots and Systems (IROS), Prague, Czech Republic, 27 September–1 October 2021; pp. 4738–4744.
221. Fujimoto, S.; Hoof, H.; Meger, D. Addressing function approximation error in actor-critic methods. In Proceedings of the International Conference on Machine Learning, PMLR, Stockholm, Sweden, 10–15 July 2018; pp. 1587–1596.
222. Ng, A. Sparse Autoencoder. CS294A Lecture Notes. 2011; Volume 72, pp. 1–19.
223. Bhagat, S.; Das, B.; Chakraborty, A.; Mukhopadhyaya, K. k-Circle Formation and k-epf by Asynchronous Robots. *Algorithms* **2021**, *14*, 62. [CrossRef]
224. Datta, S.; Dutta, A.; Gan Chaudhuri, S.; Mukhopadhyaya, K. Circle formation by asynchronous transparent fat robots. In *Proceedings of the International Conference on Distributed Computing and Internet Technology*; Springer: Berlin/Heidelberg, Germany, 2013; pp. 195–207.
225. Dutta, A.; Gan Chaudhuri, S.; Datta, S.; Mukhopadhyaya, K. Circle formation by asynchronous fat robots with limited visibility. In *Proceedings of the International Conference on Distributed Computing and Internet Technology*; Springer: Berlin/Heidelberg, Germany, 2012; pp. 83–93.
226. Flocchini, P.; Prencipe, G.; Santoro, N.; Viglietta, G. Distributed computing by mobile robots: Uniform circle formation. *Distrib. Comput.* **2017**, *30*, 413–457. [CrossRef]
227. Veličković, P.; Cucurull, G.; Casanova, A.; Romero, A.; Lio, P.; Bengio, Y. Graph attention networks. *arXiv* **2017**, arXiv:1710.10903.
228. Kipf, T.N.; Welling, M. Semi-supervised classification with graph convolutional networks. *arXiv* **2016**, arXiv:1609.02907.
229. Wu, Z.; Pan, S.; Chen, F.; Long, G.; Zhang, C.; Philip, S.Y. A comprehensive survey on graph neural networks. *IEEE Trans. Neural Netw. Learn. Syst.* **2020**, *32*, 4–24. [CrossRef] [PubMed]
230. Wenhong, Z.; Jie, L.; Zhihong, L.; Lincheng, S. Improving multi-target cooperative tracking guidance for UAV swarms using multi-agent reinforcement learning. *Chin. J. Aeronaut.* **2022**, *35*, 100–112.
231. Nowak, M.A. Five rules for the evolution of cooperation. *Science* **2006**, *314*, 1560–1563. [CrossRef]
232. Smola, A.; Gretton, A.; Song, L.; Schölkopf, B. A Hilbert space embedding for distributions. In *Proceedings of the International Conference on Algorithmic Learning Theory*; Springer: Berlin/Heidelberg, Germany, 2007; pp. 13–31.
233. Chung, T.H.; Hollinger, G.A.; Isler, V. Search and pursuit-evasion in mobile robotics. *Auton. Robot.* **2011**, *31*, 299–316. [CrossRef]
234. Sincák, D. Multi–robot control system for pursuit–evasion problem. *J. Electr. Eng* **2009**, *60*, 143–148.

235. Stiffler, N.M.; O'Kane, J.M. A sampling-based algorithm for multi-robot visibility-based pursuit-evasion. In Proceedings of the 2014 IEEE/RSJ International Conference on Intelligent Robots and Systems, Chicago, IL, USA, 14–18 September 2014; pp. 1782–1789.
236. Oh, S.; Schenato, L.; Chen, P.; Sastry, S. Tracking and coordination of multiple agents using sensor networks: System design, algorithms and experiments. *Proc. IEEE* **2007**, *95*, 234–254. [CrossRef]
237. Wang, Y.; Dong, L.; Sun, C. Cooperative control for multi-player pursuit-evasion games with reinforcement learning. *Neurocomputing* **2020**, *412*, 101–114. [CrossRef]
238. Tokekar, P.; Vander Hook, J.; Mulla, D.; Isler, V. Sensor planning for a symbiotic UAV and UGV system for precision agriculture. *IEEE Trans. Robot.* **2016**, *32*, 1498–1511. [CrossRef]
239. Batjes, N.H.; Ribeiro, E.; Van Oostrum, A.; Leenaars, J.; Hengl, T.; Mendes de Jesus, J. WoSIS: Providing standardised soil profile data for the world. *Earth Syst. Sci. Data* **2017**, *9*, 1–14. [CrossRef]
240. Viseras, A.; Garcia, R. DeepIG: Multi-robot information gathering with deep reinforcement learning. *IEEE Robot. Autom. Lett.* **2019**, *4*, 3059–3066. [CrossRef]
241. Said, T.; Wolbert, J.; Khodadadeh, S.; Dutta, A.; Kreidl, O.P.; Bölöni, L.; Roy, S. Multi-robot information sampling using deep mean field reinforcement learning. In Proceedings of the 2021 IEEE International Conference on Systems, Man, and Cybernetics (SMC), Melbourne, Australia, 17–20 October 2021; pp. 1215–1220.
242. Khamis, A.; Hussein, A.; Elmogy, A. Multi-robot task allocation: A review of the state-of-the-art. *Coop. Robot. Sens. Netw.* **2015**, *2015*, 31–51.
243. Korsah, G.A.; Stentz, A.; Dias, M.B. A comprehensive taxonomy for multi-robot task allocation. *Int. J. Robot. Res.* **2013**, *32*, 1495–1512. [CrossRef]
244. Verma, S.; Zhang, Z.L. Graph capsule convolutional neural networks. *arXiv* **2018**, arXiv:1805.08090.
245. Kool, W.; Van Hoof, H.; Welling, M. Attention, learn to solve routing problems! *arXiv* **2018**, arXiv:1803.08475.
246. Vaswani, A.; Shazeer, N.; Parmar, N.; Uszkoreit, J.; Jones, L.; Gomez, A.N.; Kaiser, Ł.; Polosukhin, I. Attention is all you need. *Adv. Neural Inf. Process. Syst.* **2017**, *30*, 5998–6008.
247. Devin, C.; Gupta, A.; Darrell, T.; Abbeel, P.; Levine, S. Learning modular neural network policies for multi-task and multi-robot transfer. In Proceedings of the 2017 IEEE International Conference on Robotics and Automation (ICRA), Singapore, 29 May–3 June 2017; pp. 2169–2176.
248. Tavakoli, A.; Pardo, F.; Kormushev, P. Action branching architectures for deep reinforcement learning. In Proceedings of the AAAI Conference on Artificial Intelligence, New Orleans, LO, USA, 2–7 February 2018; Volume 32.
249. Alkilabi, M.H.M.; Narayan, A.; Tuci, E. Cooperative object transport with a swarm of e-puck robots: Robustness and scalability of evolved collective strategies. *Swarm Intell.* **2017**, *11*, 185–209. [CrossRef]
250. Tuci, E.; Alkilabi, M.H.; Akanyeti, O. Cooperative object transport in multi-robot systems: A review of the state-of-the-art. *Front. Robot. AI* **2018**, *5*, 59. [CrossRef] [PubMed]
251. Niwa, T.; Shibata, K.; Jimbo, T. Multi-agent Reinforcement Learning and Individuality Analysis for Cooperative Transportation with Obstacle Removal. In *Proceedings of the International Symposium Distributed Autonomous Robotic Systems*; Springer: Berlin/Heidelberg, Germany, 2021; pp. 202–213.
252. Narvekar, S.; Peng, B.; Leonetti, M.; Sinapov, J.; Taylor, M.E.; Stone, P. Curriculum Learning for Reinforcement Learning Domains: A Framework and Survey. *J. Mach. Learn. Res.* **2020**, *21*, 181:1–181:50.
253. Stroupe, A.; Huntsberger, T.; Okon, A.; Aghazarian, H.; Robinson, M. Behavior-based multi-robot collaboration for autonomous construction tasks. In Proceedings of the 2005 IEEE/RSJ International Conference on Intelligent Robots and Systems, Edmonton, AB, Canada, 2–6 August 2005; pp. 1495–1500.
254. Werfel, J.K.; Petersen, K.; Nagpal, R. Distributed multi-robot algorithms for the TERMES 3D collective construction system. In Proceedings of the Robotics: Science and Systems VII, Institute of Electrical and Electronics Engineers, Los Angeles, CA, USA, 25–30 September 2011.
255. Werfel, J.; Petersen, K.; Nagpal, R. Designing collective behavior in a termite-inspired robot construction team. *Science* **2014**, *343*, 754–758. [CrossRef] [PubMed]
256. Sartoretti, G.; Wu, Y.; Paivine, W.; Kumar, T.; Koenig, S.; Choset, H. Distributed reinforcement learning for multi-robot decentralized collective construction. In *Distributed Autonomous Robotic Systems*; Springer: Berlin/Heidelberg, Germany, 2019; pp. 35–49.
257. Liang, Z.; Cao, J.; Jiang, S.; Saxena, D.; Chen, J.; Xu, H. From Multi-agent to Multi-robot: A Scalable Training and Evaluation Platform for Multi-robot Reinforcement Learning. *arXiv* **2022**, arXiv:2206.09590.
258. Bettini, M.; Kortvelesy, R.; Blumenkamp, J.; Prorok, A. VMAS: A Vectorized Multi-Agent Simulator for Collective Robot Learning. *arXiv* **2022**, arXiv:2207.03530.
259. Paszke, A.; Gross, S.; Massa, F.; Lerer, A.; Bradbury, J.; Chanan, G.; Killeen, T.; Lin, Z.; Gimelshein, N.; Antiga, L.; et al. Pytorch: An imperative style, high-performance deep learning library. *Adv. Neural Inf. Process. Syst.* **2019**, *32*, 8024–8035.
260. Chen, J.; Deng, F.; Gao, Y.; Hu, J.; Guo, X.; Liang, G.; Lam, T.L. MultiRoboLearn: An open-source Framework for Multi-robot Deep Reinforcement Learning. *arXiv* **2022**, arXiv:2209.13760.
261. Hu, S.; Zhong, Y.; Gao, M.; Wang, W.; Dong, H.; Li, Z.; Liang, X.; Chang, X.; Yang, Y. MARLlib: Extending RLlib for Multi-agent Reinforcement Learning. *arXiv* **2022**, arXiv:2210.13708.

262. Moritz, P.; Nishihara, R.; Wang, S.; Tumanov, A.; Liaw, R.; Liang, E.; Elibol, M.; Yang, Z.; Paul, W.; Jordan, M.I.; et al. Ray: A distributed framework for emerging {AI} applications. In Proceedings of the 13th USENIX Symposium on Operating Systems Design and Implementation (OSDI 18), Carlsbad, CA, USA, 8–10 October 2018; pp. 561–577.
263. Liang, E.; Liaw, R.; Nishihara, R.; Moritz, P.; Fox, R.; Gonzalez, J.; Goldberg, K.; Stoica, I. Ray rllib: A composable and scalable reinforcement learning library. *arXiv* **2017**, arXiv:1712.09381.
264. Hu, J.; Jiang, S.; Harding, S.A.; Wu, H.; Liao, S.w. Rethinking the implementation tricks and monotonicity constraint in cooperative multi-agent reinforcement learning. *arXiv* **2021**, arXiv:2102.03479 .
265. Zhou, M.; Wan, Z.; Wang, H.; Wen, M.; Wu, R.; Wen, Y.; Yang, Y.; Zhang, W.; Wang, J. Malib: A parallel framework for population-based multi-agent reinforcement learning. *arXiv* **2021**, arXiv:2106.07551.
266. Michel, O. Cyberbotics Ltd. Webots™: Professional mobile robot simulation. *Int. J. Adv. Robot. Syst.* **2004**, *1*, 5. [CrossRef]
267. Rohmer, E.; Singh, S.P.; Freese, M. V-REP: A versatile and scalable robot simulation framework. In Proceedings of the 2013 IEEE/RSJ International Conference on Intelligent Robots and Systems, Tokyo, Japan, 3–7 November 2013; pp. 1321–1326.
268. Dasari, S.; Ebert, F.; Tian, S.; Nair, S.; Bucher, B.; Schmeckpeper, K.; Singh, S.; Levine, S.; Finn, C. Robonet: Large-scale multi-robot learning. *arXiv* **2019**, arXiv:1910.11215.
269. Challita, U.; Saad, W.; Bettstetter, C. Deep reinforcement learning for interference-aware path planning of cellular-connected UAVs. In Proceedings of the 2018 IEEE International Conference on Communications (ICC), Kansas City, MO, USA, 20–24 May 2018; pp. 1–7.
270. Baca, J.; Hossain, S.; Dasgupta, P.; Nelson, C.A.; Dutta, A. Modred: Hardware design and reconfiguration planning for a high dexterity modular self-reconfigurable robot for extra-terrestrial exploration. *Robot. Auton. Syst.* **2014**, *62*, 1002–1015. [CrossRef]
271. Chennareddy, S.; Agrawal, A.; Karuppiah, A. Modular self-reconfigurable robotic systems: A survey on hardware architectures. *J. Robot.* **2017**, *2017*.
272. Tan, N.; Hayat, A.A.; Elara, M.R.; Wood, K.L. A framework for taxonomy and evaluation of self-reconfigurable robotic systems. *IEEE Access* **2020**, *8*, 13969–13986. [CrossRef]
273. Yim, M.; Shen, W.M.; Salemi, B.; Rus, D.; Moll, M.; Lipson, H.; Klavins, E.; Chirikjian, G.S. Modular self-reconfigurable robot systems [grand challenges of robotics]. *IEEE Robot. Autom. Mag.* **2007**, *14*, 43–52. [CrossRef]

sensors

Article

Navigation with Polytopes: A Toolbox for Optimal Path Planning with Polytope Maps and B-spline Curves

Ngoc Thinh Nguyen [1,*], Pranav Tej Gangavarapu [1], Niklas Fin Kompe [1], Georg Schildbach [2] and Floris Ernst [1]

1 Institute for Robotics and Cognitive Systems, University of Lübeck, 23562 Lübeck, Germany
2 Institute for Electrical Engineering in Medicine, University of Lübeck, 23562 Lübeck, Germany
* Correspondence: nguyen@rob.uni-luebeck.de

Abstract: To deal with the problem of optimal path planning in 2D space, this paper introduces a new toolbox named "Navigation with Polytopes" and explains the algorithms behind it. The toolbox allows one to create a polytopic map from a standard grid map, search for an optimal corridor, and plan a safe B-spline reference path used for mobile robot navigation. Specifically, the B-spline path is converted into its equivalent Bézier representation via a novel calculation method in order to reduce the conservativeness of the constrained path planning problem. The conversion can handle the differences between the curve intervals and allows for efficient computation. Furthermore, two different constraint formulations used for enforcing a B-spline path to stay within the sequence of connected polytopes are proposed, one with a guaranteed solution. The toolbox was extensively validated through simulations and experiments.

Keywords: path planner; B-spline; Bézier; polytopes; optimization; navigation with polytopes toolbox

check for updates

Citation: Nguyen, N.T.; Gangavarapu, P.T.; Kompe, N.F.; Schildbach, G.; Ernst, F. Navigation with Polytopes: A Toolbox for Optimal Path Planning with Polytope Maps and B-spline Curves. *Sensors* **2023**, *23*, 3532. https://doi.org/10.3390/s23073532

Academic Editors: Stephen Monk and David Cheneler

Received: 19 February 2023
Revised: 12 March 2023
Accepted: 23 March 2023
Published: 28 March 2023

1. Introduction

Motion planning is an important component of the technology stack for enabling the autonomous navigation of unmanned vehicles [1]. It involves the computation of an admissible path or trajectory from the current position/configuration of the robot to a target area/point on a given map with obstacles. The difficulty of a motion planning task depends on the particular setup and problem formulation. It may involve complications such as kinodynamic constraints, uncertainties, and dynamic obstacles. Almost all approaches currently used in robotics involve a spatial discretization of the given map, called a *grid map* or *occupancy grid*. Important planning methods comprise graph search algorithms, such as Dijkstra, A*, and variants thereof, and sampling-based methods such as rapidly exploring random trees (RRT) [2–4]. These have been successfully applied in various works, such as for the development of a hybrid path planning algorithm and a bio-inspired control for an omni-directional mobile robot [5], or the control of a nonholonomic vehicle in tight environments [6]. They have an important drawback, however, in that the complexity of the planning problem increases rapidly with the dimensions of the map as well as the resolution of the grid map. Moreover, the grid map is an artificial construct that may complicate the path planning problem, e.g., over large empty areas or for kinodynamic constraints, and lead to unsafe or conservative results.

For this reason, this paper touts the idea of continuous motion planning and makes several contributions toward turning it into a competitive alternative. Several algorithms are proposed for efficient continuous motion planning, including the *generation of a polytope map* and a *spline-based planner*. They are described in detail in this paper and a ready-to-use implementation is provided as a Python-based toolbox, called *Navigation with Polytopes*.

In previous work, continuous motion planners using spline-based interpolations have been combined with the standard discrete frameworks [7–9]. In [7], the movement of a system between two exact discrete moments was studied, which relaxed some of the stringent requirements for optimal controller design in drones. In [8,9], a standard grid map, obtained with existing mapping tools such as *gmapping* [10], was transformed into a *polytope map*, in which the feasible area was decomposed into a finite number of (convex) polytopes, called *feasible polytopes*. The polytope map allows the computation of B-spline paths that completely stay inside the feasible area and, hence, the free space of the grid map. B-splines have been chosen for their *local convexity*: Each interval is bounded by the convex hull of the local control points [8,11,12]. This leads to the simple rule that the *B-spline control boundary*, i.e., the convex hull of the B-spline control point, must be fully contained inside the feasible area of the polytope map [13–15]. The rule has been applied widely in the literature to solve different types of motion planning problems. For example, in [16], the authors generalize the methods for motion planning with B-spline curves for constrained flatness systems. Reference [17] proposed a path planner using a B-spline curve with an obstacle avoidance property for heavy mining vehicles while [18] introduced the solution for the same problem but for Maritime autonomous surface ships. In [19], the authors further ensured the constraints on the B-spline path's curvature for autonomous cars.

The approach of using B-spline parametrization, however, is conservative, as illustrated in Figure 1, which has been obtained with the *Navigation with Polytopes* toolbox (https://gitlab.rob.uni-luebeck.de/robPublic/navigation_with_polytopes, accessed on 19 February 2023). Here, the green area depicts the control boundary for the second-degree B-spline curve in red, with a portion of it highlighted in yellow. Even though the entire curve does not leave the feasible area, the control boundary is not fully contained in the feasible area. In other words, this path, despite being safe, cannot be represented with a feasible B-spline. Generally speaking, the control boundaries for B-spline control points are relatively large compared to the area covered by the curve itself.

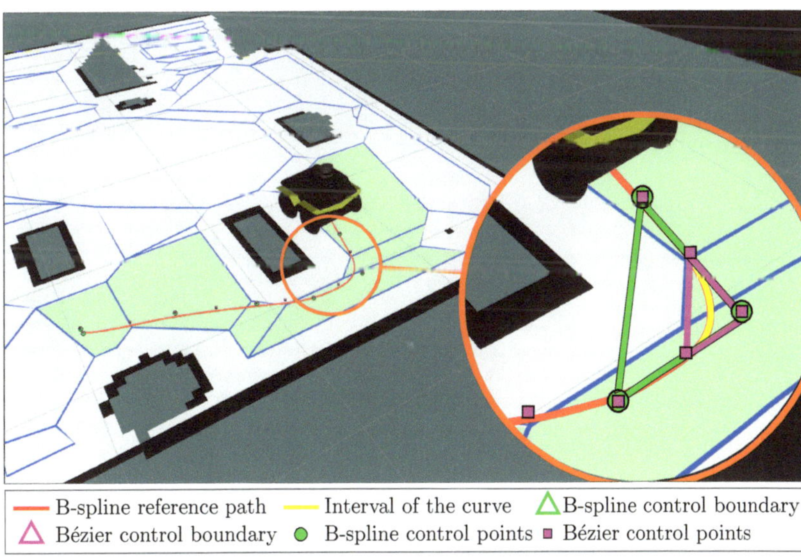

Figure 1. B-spline path planned within a polytope map with the *Navigation with Polytopes* (https://gitlab.rob.uni-luebeck.de/robPublic/navigation_with_polytopes, accessed on 19 February 2023) toolbox.

This *conservatism* can be reduced based on prior work in the computer-aided design (CAD) community regarding the conversion between B-splines and equivalent Bézier curves [20,21]. As shown in Figure 1 with the magenta triangle, the control boundary of the corresponding Bézier curve is fully contained inside the feasible region of the polytopic map. In fact, for the same curve, the control boundary of the Bézier curve (magenta triangle) is only one-quarter the size of the original B-spline control boundary (green triangle). Thus, the usage of equivalent Bézier control points allows the design of the geometrical constraints for the computed curve to be more flexible.

A new constraint formulation for a B-spline path to stay within the feasible region of the polytope map was derived in [8]. However, the calculation of the B-spline-to-Bézier conversion parameters is not easy and is usually inefficient to compute via recursive functions, due to the original recursive formulation of the B-spline curve. For example, in [20], the authors can only derive a calculation of the conversion parameters within some middle intervals of a B-spline curve when having a sufficiently large number of control points while neglecting the rest (for more details, see Remark 1).

Sharing the line of research with the existing works [8,9] and serving as their extensions, this paper concentrates on applying two efficient tools: the polytope map of the surrounding environment and the equivalent Bézier format of a B-spline curve to solve the path planning problem for mobile robots. In particular, the following novelties are presented compared to the current state of the literature:

1. A complete procedure to construct the polytope map from a standard occupancy grid map and seek an appropriate corridor (sequence of connected polytopes), leading to the destination.
2. A new algorithm to calculate the B-spline-to-Bézier conversion matrix of a uniform B-spline curve: It takes into account the differences between each interval of the whole curve and the dependencies on the total number of control points as well as the degree of the curve.
3. New path planning constraints for a B-spline path to stay inside a sequence of connected polytopes in 2D. The equivalent Bézier representation is introduced in two variants:
    (a) Constraints that use the minimal number of control points [8];
    (b) Constraints that guarantee the existence of a valid path by providing an algebraic solution [9].
4. *Navigation with Polytopes* (https://gitlab.rob.uni-luebeck.de/robPublic/navigation_with_polytopes, accessed on 19 February 2023) toolbox: It comes as a complete Python package and serves as a framework for direct and quick implementation of existing polytope-based navigation control techniques on a realistic grid map of the environment with ROS (robot operating system) compatibility (c.f. Figure 1). It provides the following features:
    (a) Construction of a polytope map from a standard grid map with consideration of the robot's dimension and possible noises.
    (b) Search for a sequence of connected polytopes (i.e., a polytopic corridor) connecting two given points with minimal distance.
    (c) Optimal B-spline path planning algorithm using the B-spline-to-Bézier conversion with multiple choices of algorithms [8,9].
    (d) Library for calculating and storing the B-spline-to-Bézier conversion matrix.

The remainder of the paper is organized as follows. The path planning problem and relevant details are formulated in Section 2. Next, Section 3 introduces the process of constructing a polytope map from a grid map. Section 4 introduces the notions of B-splines and its equivalent Bézier representation as well as the calculation of the B-spline-to-Bézier conversion matrix. Different path-planning constraint formulations are detailed in Section 5. Then, Section 6 introduces the *Navigation with Polytopes* toolbox. The results of the validation process using simulations and experiments are presented in Section 7 and further discussed in Section 8. Finally, Section 9 presents the conclusions and remarks on future work.

2. Problem Description

This paper addresses the problem of planning a 2D optimal reference path for a mobile robot to navigate between two points given the standard occupancy grid map of the surrounding environment. More specifically, the principal tool in our work is the polytope map, which describes the safety region with non-overlapping convex polytopes. It was created from the grid map via a decomposition algorithm. Within the polytope map, an appropriate sequence of connected polytopes connecting the two end-points was selected by using a graph-search algorithm. The sequence is denoted as follows:

$$\mathcal{S} \triangleq S_1 \cup S_2 \cup \cdots \cup S_q, \tag{1}$$

where $\{S_1, \ldots, S_q\}$ is an ordered list of $q \geq 2$ connected polytopes. Any pair of two consecutively connected polytopes (S_i, S_{i+1}) share a common edge denoted by E_i:

$$E_i = S_i \cap S_{i+1}. \tag{2}$$

It is also assumed that the starting and ending poses (P_s, P_f) belong to the first and last polytopes, respectively:

$$P_s \in S_1,\ P_f \in S_q. \tag{3}$$

This allows for safe travel from P_s to P_f by staying inside the set $\mathcal{S}$. Given the sequence polytopes, a smooth geometric path $p(t)$ (with t being the curve variable, which can represent the path length, pseudo-time increment, etc.) was generated:

$$p(t) : [t_s, t_f] \to \mathbb{R}^2, \tag{4}$$

which is required to satisfy the end-point constraints as well as the safety condition

$$p(t_s) = P_s, \quad p(t_f) = P_f, \tag{5}$$

$$p(t) \in \mathcal{S},\ \forall t \in [t_s, t_f]. \tag{6}$$

In this work, the geometrical properties of B-spline curves are exploited (i.e., endpoint interpolation and local convexity) in order to generate a reference B-spline path satisfying the aforementioned constraints (5) and (6). Furthermore, the equivalent Bézier representation of a B-spline curve was used to reduce the conservativeness of the path planning problem. The whole planning process will be detailed sequentially throughout the rest of the paper, while the next section begins with the construction of the polytope map from a grid map.

3. Polytope Map

This section focuses on modeling the free space environment by describing it as a continuous polytope map. Contrary to the discrete-based occupancy grid representation, the polytope map is a continuous representation of the environment. It is defined as a list of connected 2D convex polytopes within the free space of an environment. A general convention of each polytope involves a list of ordered vertices.

3.1. Construction of Polytope Map from an Occupancy Grid Map

This section presents an algorithm for the conversion of a standard grid map into a polytope map. The grid map can either be a binary map or a ternary representation, which is a common map used in ROS for standard navigation purposes. For example, Figure 2a shows an occupancy grid map of a simulation environment provided by ROBOTIS for the TurtleBot3 mobile robot [22]. The map is obtained by using the ROS package *gmapping* [10].

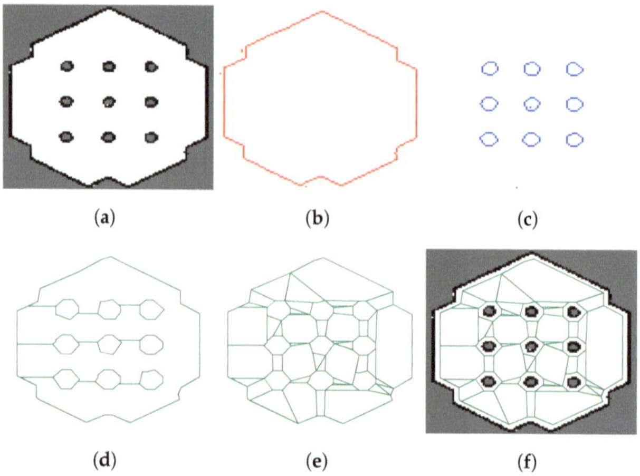

Figure 2. Illustration of the procedure for creating a polytope map from a standard grid map. (**a**) Occupancy grid map; (**b**) free space boundary extraction; (**c**) obstacle boundaries extraction; (**d**) free region with holes; (**e**) partition of free region into connected polytopes; (**f**) polytope map versus occupancy grid map.

Below, one can find the Python process, which is used to construct the polytope map from a standard occupancy grid map (with a corresponding illustration on the aforementioned grid map of TurtleBot3):

1. Extract the outer boundary of the complete map using the function *findContours* with the option RETR_EXTERNAL of the OpenCV toolbox (https://opencv.org/, accessed on 19 February 2023) as shown in Figure 2b.
2. Extract the boundaries for all of the obstacles by using the same function *findContours* with the option RETR_LIST as shown in Figure 2c.
3. Simplify the contours obtained using the RDP (Ramer–Douglas–Peucker) algorithm (https://github.com/biran0079/crdp, accessed on 19 February 2023) with two parameters $\varepsilon_{rdp,o}$ for the outer boundary and $\varepsilon_{rdp,i}$ for inner obstacles [23].
4. Shrink the outer boundary and enlarge the obstacles by a safety offset o_p by using the Gdspy toolbox (https://github.com/heitzmann/gdspy, accessed on 19 February 2023) and apply the Boolean operation to remove obstacles from the outer boundary polytope, as shown in Figure 2d.
5. Partition the obstacle-free polytope (possibly with holes) into connected polytopes by using Mark Bayazit's algorithm (https://github.com/wsilva32/poly_decomp.py, accessed on 19 February 2023), as shown in Figure 2e.

The result of the entire procedure is the polytope map shown in Figure 2f , where it is overlaid with the original grid map. It can be seen that the free space in the environment has shrunk far from the occupied cells (i.e., obstacles) and is divided into smaller and connected polytopes. In comparison with the usage of the configuration space map in safe navigation [2], the proposed approach is slightly simpler, i.e., it simply applies an offset with the safety distance o_p to all objects within the map. In contrast to this, the configuration space method requires calculating the Minkowski sums of the robot's shape and the objects.

3.2. Finding of Appropriate Sequence of Polytopes for Navigation

After obtaining a polytope map, the next step is to find a sequence of connected polytopes (i.e., defined as an ordered list of a finite number of polytopes), which forms a corridor connecting the given initial point to the final goal. Among the sequences, two consecutive polytopes share a common edge (c.f. Figures 3 and 4). In order to find that sequence, the first step is to represent the polytope map as a graph, as shown in Figure 3, in which each polytope is a node. Two nodes are considered "connected to each other" when they share a common edge (e.g., the red edge between polytopes A and B). The connection also evaluates the distance between the two polytopes by using the Euclidean distance between their center points. Then a graph search can be performed in order to obtain the shortest sequence connecting two polytopes, which contain the start and end poses. The complete process of finding such a sequence is as follows:

1. Each pair of polytopes is examined to find out if they share a common edge. If yes, then they are recognized as a connected pair.
2. From the information, an adjacency graph is created (c.f Figure 3b), which presents all polytopes as nodes and their connections to other polytopes.
3. Then a weighted graph is created from the adjacency graph by adding the distances between the center points of any pairs of connected polytopes.
4. Next, there is a search for the starting and ending polytopes by checking which polytopes contain the points (P_s, P_f).
5. A graph search algorithm can then be implemented on the weighted graph to obtain the sequence of polytopes $\mathcal{S} \triangleq S_1 \cup S_2 \cup \cdots \cup S_q$ with minimal travel distance.

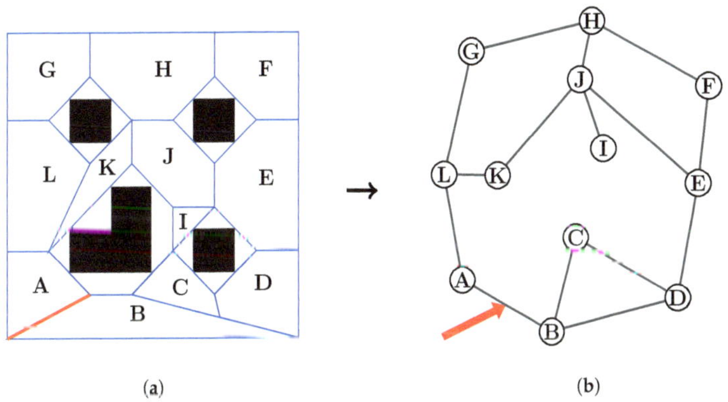

Figure 3. Illustration of a polytope map of an environment with four obstacles (black) and its graph representation. (**a**) Sample polytope map; (**b**) graph representation of the polytope map.

3.3. Transition Zone and Extended Polytope

As an intermediate step toward the full navigation task between (P_s, P_f), consider the problem of computing a path between two connected polytopes of the sequence $\mathcal{S}$ (6). In order to avoid collisions with obstacles, a so-called *transition zone* is introduced, which is a subset of the second polytope and whose union with the first polytope is convex (cf. Figure 4). Thus, a robot can travel safely from the first polytope to the second one by adding a transit at the transition zone.

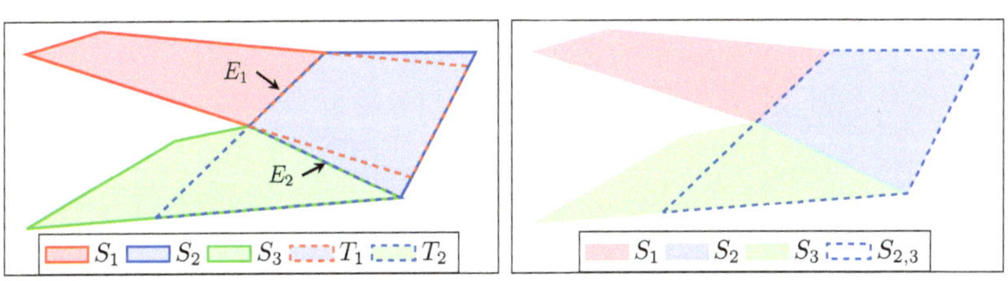

Figure 4. Illustration of connected polytopes, transition zones, and extended polytopes, according to Definitions 1 and 2.

Definition 1 (Transition zone [8,9]). *The transition zone T_i is defined for two connected polytopes S_i and S_{i+1} from* (2) *as*

$$T_i = S_{i+1} \cap (S_i | E_i), \tag{7}$$

in which E_i is the common edge as defined in (2) *and the operation $(S_i|E_i)$ gives the (possibly unbounded) polytope formed by the half-space representation of S_i without the constraint corresponding to the edge E_i.*

Definition 2 (Extended polytope [8,9]). *$S_{i,i+1}$ is defined as the extension of the polytope S_i toward the polytope S_{i+1}:*

$$S_{i,i+1} = S_i \cup T_i, \tag{8}$$

with T_i the transition zone defined as in (7).

For consistency, the last extended polytope is also the last polytope, i.e., $S_{q,q+1} \triangleq S_q$. Any extended polytope $S_{i,i+1}$ as defined in (8) is convex and the transition zone can also be achieved from the corresponding extended polytopes:

$$T_i = S_{i,i+1} \cap S_{i+1} = S_{i,i+1} \cap S_{i+1,i+2}. \tag{9}$$

This section presents the search for the sequence of connected polytopes leading to the goal. The next section introduces an interesting path parametrization, which is called the B-spline curve, whose geometrical properties allow us to control its shape via intuitive tuning of the curve parameters and, hence, easily constrain the path to stay within a predefined sequence of connected polytopes.

4. B-spline and Equivalent Bézier Curves

This section presents the notions of B-spline curves and their equivalent Bézier representations. The focus is on their definitions, transformations, and further geometrical properties, while more details on both types of curves could be found in the literature [11,12,15,18,20,21]. The same notations as in some of the previous work [8,9] is used intentionally, for easy reference.

4.1. Definition of B-spline Curves

A *clamped uniform* B-spline curve $z(t) : [t_s, t_f] \to \mathbb{R}^m$ of degree d is defined with n control points $P_i \in \mathbb{R}^m$ ($i \in \{1, \ldots, n\}, n \geq d+1$) as

$$z(t) = \sum_{i=1}^{n} P_i B_{i,d,\xi}(t) = \boldsymbol{P}\boldsymbol{B}_{d,\xi}(t),\ t \in [t_s, t_f], \tag{10}$$

with $\boldsymbol{P} \triangleq [P_1 \ \cdots \ P_n] \in \mathbb{R}^{m \times n}$ gathering the control points that control the shape of the curve and needs to be defined in the path planning problem. The vector $\boldsymbol{B}_{d,\xi}(t) \triangleq [B_{1,d,\xi}(t) \ \ldots \ B_{n,d,\xi}(t)]^\top : \mathbb{R} \to \mathbb{R}^n$ contains the B-spline basis functions of the degree d, whose recursive definition is given by [15,16,24]

$$B_{i,0,\xi}(t) = \begin{cases} 1, \text{ for } \tau_i \leq t \leq \tau_{i+1}, \\ 0, \text{ otherwise}, \end{cases} \quad \forall i \in \{1, \ldots, n+d\}, \tag{11}$$

$$B_{i,d,\xi}(t) = \frac{t - \tau_i}{\tau_{i+d} - \tau_i} B_{i,d-1,\xi}(t) + \frac{\tau_{i+d+1} - t}{\tau_{i+d+1} - \tau_{i+1}} B_{i+1,d-1,\xi}(t), \forall d \geq 1. \tag{12}$$

Here, the time instances τ_j are *clamped* and *uniformly distributed* in a knot vector ξ:

$$\xi = \{\tau_1 \leq \tau_2 \leq \cdots \leq \tau_{n+d+1}\}, \quad (13)$$

$$\tau_j = \begin{cases} t_s, & 1 \leq j \leq d, \\ t_s + (j - d - 1)\Delta, & d+1 \leq j \leq n+1, \\ t_f, & n+2 \leq j \leq n+d+1, \end{cases} \quad (14)$$

with $\Delta = (t_f - t_s)/(n-d)$. The *clamped* and *uniform* B-spline curve $z(t)$ from (10) has exactly $(n-d)$ consecutive intervals equally distributed within $[t_s, t_f]$. The partial curve within the jth interval ($j \in \{1, \ldots, n-d\}$) is given by

$$z(j,t) \triangleq z(t),\ t \in [t_s + (j-1)\Delta, t_s + j\Delta). \quad (15)$$

The B-spline curve $z(t)$ as defined in (10)–(14) possesses the following properties:

(P1) The jth interval $z(j,t)$ of the curve as in (15) only depends on its $(d+1)$ neighbor control points. More specifically, $z(j,t)$ stays within their convex hull:

$$z(j,t) = \sum_{i=j}^{j+d} P_i B_{i,d,\xi}(t) \in \text{Conv}\{\boldsymbol{P}_j\}, \quad (16)$$

with $\boldsymbol{P}_j \triangleq [P_j \cdots P_{j+d}]$ containing $(d+1)$ consecutive control points from (10).

(P2) The first and last control points P_1 and P_n from (10) are also the starting and ending points of the curve $z(t)$:

$$z(t_s) = P_1, \quad z(t_f) = P_n. \quad (17)$$

(P3) Derivatives of B-spline basis functions can be expressed as a linear combination of B-spline basis functions:

$$\frac{\partial \boldsymbol{B}_{d,\xi}(t)}{\partial t} = M_{d,d-1} L_{d,d-1} \boldsymbol{B}_{d,\xi}(t), \quad (18)$$

with $\boldsymbol{B}_{d,\xi}$ as in (10). The two matrices $M_{d,d-1} \in \mathbb{R}^{n \times (n-1)}$ and $L_{d,d-1} \in \mathbb{R}^{(n-1) \times n}$ are given in Theorems 4.1–4.3 of reference [16].

Various works in the literature have employed the aforementioned properties to adapt the B-spline framework to the problems of path/trajectory planning with obstacle avoidance and waypoint constraints. For example, in [11,14], the authors use B-splines to generate trajectories for a quadcopter system with waypoint constraints. In [13,15,18], B-spline is introduced as a general framework for obstacle and collision avoidance for more aerial vehicles. However, the local B-spline control boundary of each interval Conv$\{\boldsymbol{P}_j\}$ as in (16) is relatively large in comparison with the curve interval $z(j,t)$ itself (c.f. Figure 1) [8,20], which causes unnecessary extra conservativeness to the motion planning problems. This problem is solved in the next section with the introduction of the equivalent Bézier representation of the B-spline curve, which provides us with a tighter local control boundary for each section of the curve.

4.2. Local Equivalent Bézier Representation

As proven in various works from the CAD (computer-aided design) community [20,21], any interval of a B-spline curve, as defined in (10), e.g., $z(j,t)$ from (15), is also a Bézier curve of the same degree:

$$z(j,t) = \sum_{i=1}^{d+1} \overline{P}_{(j-1)d+i} B_{i,d,\overline{\xi}_j}(t). \quad (19)$$

Here, $\overline{P}_k$ is the Bézier control point ($k \in \{(j-1)d+1,\ldots,jd+1\}$). The formulation uses the same basis function $B_{i,d,\overline{\xi}_j}$ as defined in (11) and (12), but with a new knot vector $\overline{\xi}_j$ constructed by repeating the start and end of the interval

$$\overline{\xi}_j = \{\underbrace{t_s+(j-1)\Delta,\ldots,t_s+(j-1)\Delta}_{d+1 \text{ knots}},\underbrace{t_s+j\Delta,\ldots,t_s+j\Delta}_{d+1 \text{ knots}}\}, \tag{20}$$

with $(t_s+(j-1)\Delta, t_s+j\Delta)$ as in (15). The Bézier control points $\overline{P}_k$ ($k \in \{(j-1)d+1,\ldots,jd+1\}$) as used in (19) can be calculated from the $(d+1)$ original B-spline control points $\{P_j,\ldots,P_{j+d}\}$ by using the following matrix transformation:

$$\overline{\boldsymbol{P}}_j = \boldsymbol{P}_j A(d,n,j). \tag{21}$$

Here, $\overline{\boldsymbol{P}}_j \triangleq [\overline{P}_{(j-1)d+1}\cdots\overline{P}_{jd+1}]$ and $\boldsymbol{P}_j \triangleq [P_j\cdots P_{j+d}]$ consist of $(d+1)$ Bézier and B-spline control points, respectively. The B-spline-to-Bézier conversion matrix $A(d,n,j) \in \mathbb{R}^{(d+1)\times(d+1)}$ is recursively defined in [20], while a new calculation method for the matrix is proposed in the next section. More interestingly, every Bézier control point is a convex combination of the B-spline control points [20]. This means that every column in the matrix $A(d,n,j)$ adds up to 1. Since the total number of intervals is fixed at $(n-d)$ from (15), it is possible to calculate $A(d,n,j)$ for all $j \in [1,\ldots,n-d]$, then reformulate the transformation between the Bézier and B-spline control points as follows:

$$\overline{\boldsymbol{P}} = \boldsymbol{P}\overline{A}(d,n), \tag{22}$$

with $\overline{\boldsymbol{P}} \triangleq [\overline{P}_1\cdots\overline{P}_{\overline{n}}]$ consisting of all the Bézier control points and $\boldsymbol{P}$ as in (10). The total number of Bézier control points needed to express the whole B-spline curve of degree d is

$$\overline{n} = (n-d)d+1, \tag{23}$$

where n is the number of B-spline control points from (10).

As a Bézier curve is also a B-spline curve, the same properties of a local convex hull container (16) and endpoint interpolation (17) are applied to any interval of the curve. This helps to extend the *geometrical* properties (16)–(17) of the B-spline curve $z(t)$ by applying (16) and (17) to each jth interval $z(j,t)$ as in (19) of the curve for all $j \in \{1,\ldots,n-d\}$:

(P1*) The jth interval $z(j,t)$ stays within the convex hull of its $(d+1)$ Bézier control points,

$$z(j,t) \in \text{Conv}\{\overline{\boldsymbol{P}}_j\} \subset \text{Conv}\{\boldsymbol{P}_j\}, \tag{24}$$

with $\boldsymbol{P}_j, \overline{\boldsymbol{P}}_j$ being the B-spline and equivalent Bézier control points from (21). The convexity property (24) is significantly tighter than the standard one in (16), as proven in [21] and illustrated hereinafter.

(P2*) The B-spline curve $z(t)$ passes through $(n-d+1)$ waypoints, which can be determined by using only the B-spline control points (including the first and last control points as two endpoints):

$$z(t_s+(j-1)\Delta) = \overline{P}_{(j-1)d+1}, \tag{25}$$

for all $j \in \{1,\ldots,n-d+1\}$. The Bézier control points $\overline{P}_{(j-1)d+1}$ are actually expressed in terms of the B-spline control points (22). The proof is straightforward as $(\overline{P}_{(j-1)d+1}, \overline{P}_{(j-1)d+1})$ are the two Bézier control points, which start and end the jth interval, respectively. Hence, they belong to the curve according to the property **P2** (17).

The next section introduces the new algorithm used for calculating the local transformation matrix $A(d, n, j)$ from (21) for the jth interval and the complete matrix $\overline{A}(d, n)$ as in (22) for the whole B-spline curve.

4.3. Calculation of B-spline-to-Bézier Conversion Matrix

The core idea of the proposed algorithms is to consider the matrix $\overline{A}(d, n, j)$ as a variable to solve for in (21). For the predefined j^{th} interval of a B-spline curve (i.e., of the degree d and having n control points), a sufficient number of sets consisting of randomly generated B-spline control points is collected together with their equivalent Bézier control points. Then, $\overline{A}(d, n, j)$ is solved by using the linear Equation (21). The process is repeated for all $j \in \{1, \ldots, n-d\}$, except for some special circumstances (i.e., the repetition of values of some middle matrices as discussed in Section 4.3.3); the results are gathered into the complete transformation matrix $\overline{A}(d, n)$, as in (22). Note that the B-spline curve is formulated in an m-dimensional space in (10), but only 1D control points are needed to calculate the matrices. Therefore, this section is restricted to 1D points $\boldsymbol{P} \triangleq [P_1 \ \cdots \ P_n] \in \mathbb{R}^{1\times n}$ and 1D function $z(t)$ as in (10).

4.3.1. Equivalent Bézier Control Points of One Interval

It is possible to solve the equivalent Bézier control points $\overline{\boldsymbol{P}}_j$ of the jth interval of the B-spline curve $z(t)$ from (10) given the specific values of the B-spline control points $\boldsymbol{P}$ and the degree d. The idea is to uniformly sample the jth time interval $[t_s + (j-1)\Delta, t_s + j\Delta)$ into $(d+1)$ instants: $\{{}^{(1)}t_j, \ldots, {}^{(d+1)}t_j\}$ (e.g., by using *linespace*) and solve the following linear equation for $\overline{\boldsymbol{P}}_j$:

$$\boldsymbol{B}_{d,j}\overline{\boldsymbol{P}}_j = \begin{bmatrix} z\left({}^{(1)}t_j\right) \\ \vdots \\ z\left({}^{(d+1)}t_j\right) \end{bmatrix}, \tag{26}$$

with the square matrix $\boldsymbol{B}_{d,j} \in \mathbb{R}^{(d+1)\times(d+1)}$ defined as:

$$\boldsymbol{B}_{d,j} = \begin{bmatrix} B_{1,d,\overline{\xi}_j}\left({}^{(1)}t_j\right) & \cdots & B_{d+1,d,\overline{\xi}_j}\left({}^{(1)}t_j\right) \\ \vdots & \ddots & \vdots \\ B_{1,d,\overline{\xi}_j}\left({}^{(d+1)}t_j\right) & \cdots & B_{d+1,d,\overline{\xi}_j}\left({}^{(d+1)}t_j\right) \end{bmatrix}.$$

4.3.2. Conversion Matrix of One Interval

Next, $(d+1)$ sets of n control points are randomly selected and denoted as ${}^{(1)}\boldsymbol{P}, \ldots, {}^{(d+1)}\boldsymbol{P}$. We further define ${}^{(1)}\boldsymbol{P}_j, \ldots, {}^{(d+1)}\boldsymbol{P}_j$ as the control points of the jth interval taken from ${}^{(1)}\boldsymbol{P}, \ldots, {}^{(d+1)}\boldsymbol{P}$, respectively. Since the conversion matrix $A(d, n, j)$ remains the same for different values of the control points (i.e., but not for different numbers of control points), the following equation holds true:

$$\begin{bmatrix} {}^{(1)}\boldsymbol{P}_j \\ \vdots \\ {}^{(d+1)}\boldsymbol{P}_j \end{bmatrix} A(d, n, j) = \begin{bmatrix} {}^{(1)}\overline{\boldsymbol{P}}_j \\ \vdots \\ {}^{(d+1)}\overline{\boldsymbol{P}}_j \end{bmatrix}, \tag{27}$$

in which ${}^{(i)}\overline{\boldsymbol{P}}_j$ is calculated by using (26). Solving (27) provides the conversion matrix $A(d, n, j)$ for the jth interval.

Remark 1. *In [20], the conversion matrices are calculated by using a recursive definition and not by directly solving as proposed in* (27). *Furthermore, the calculation in [20] treats the matrix $A(d, n, j)$ the same for all the intervals and for all control point numbers (i.e., $A(d, n, j)$ is simplified to $A(d)$ in [20]), which is not true. The order of j, with respect to the total number of intervals*

$(n-d)$, plays an important role in the calculation; hence, the matrix needs to be considered as $A(d,n,j)$ as in our work. For more details, the algorithm given in [20] calculates the value of $A(d,n,j)$ only for $j \in \{d,\ldots,n-2d+1\}$ and $n \geq 3d-1$. It is just a subset of our general consideration of $(n-d)$ intervals, i.e., $j \in \{1,\ldots,n-d\}$ and $n \geq d+1$, which are due to the natural definition of a B-spline curve (10).

4.3.3. Conversion Matrix of the Whole Curve

Theoretically, one can repeat solving (27) for all $j \in \{1,\ldots,n-d\}$ with the same sets of control points ${}^{(1)}\boldsymbol{P},\ldots,{}^{(d+1)}\boldsymbol{P}$ in order to obtain $(n-d)$ conversion matrices $A(d,n,j) \in \mathbb{R}^{(d+1)\times(d+1)}$ for $(n-d)$ intervals. However, according to the analysis in [20], the values of the conversion matrix $A(d,n,j)$ remain the same for $j \in \{d,\ldots,n-2d+1\}$ when $n \geq 3d-1$ (i.e., the domain in which the algorithm given in [20] is validated). Therefore, if our algorithm runs into these distinguished cases, it does not repeat the computation but makes use of the previously stored values.

These matrices are then stacked into the complete matrix $\overline{A}(d,n) \in \mathbb{R}^{n\times\overline{n}}$ with $\overline{n}$ number of equivalent Bézier control points (23). The ending point of an interval is also the starting point of the next one and these points should not be repeated in the complete conversion matrix. The reader is referred to Figure 3 in reference [20] for an illustrative example of how to stack these matrices.

4.3.4. Evaluation of the B-spline-to-Bézier Conversion Algorithm

Figure 5 shows the calculation time (in milliseconds) of our proposed algorithm (implemented in Python on a normal personal computer). The complete conversion matrix $\overline{A}(d,n)$ from (22) is computed with the curve degrees $d \in \{2,3,4\}$ and with the number of control points n up to 50. It can be observed that a higher degree requires more computation time. More interestingly, when increasing the number of control points n, the computation time grows at the beginning but then seems to be steady. The reason is due to the special case of $n \geq 3d-1$ in which the algorithm can make use of the repeated value of the conversion matrix without a recalculation, as mentioned in Section 4.3.3. Even though the calculation time is only up to a maximum of $\{20,10,4\}$ milliseconds for the $\{4,3,2\}$-degree cases, respectively, in practice, it is recommended to calculate these conversion matrices beforehand for a predicted range of n (e.g., up to hundreds) and a specific value of degree d, and then store them for online usage with real-time applications. During the online process, the calculation of $\overline{A}(d,n)$ is only performed when a new value is needed, and the result can be stored in a bank for future usage.

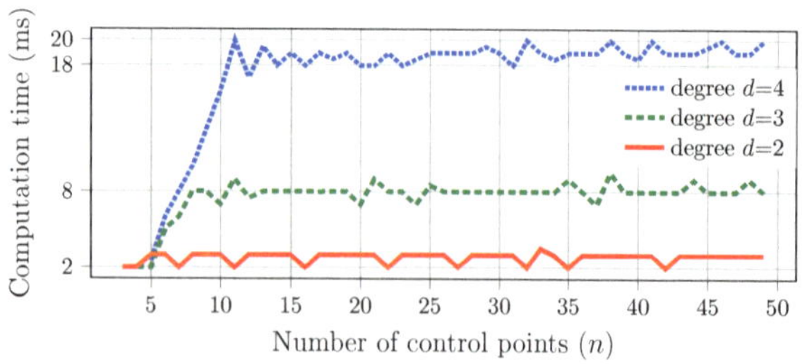

Figure 5. Computation time of the B-spline-to-Bézier conversion algorithm with respect to the number of control points and the curve's degree.

4.4. Application of B-spline-to-Bézier Conversion on 2D Path Planning

This section presents an application of the B-spline-to-Bézier conversion on 2D path planning for mobile robots in simulation. A case study of path planning in a polytopic corridor with waypoint constraints is further discussed, in which the advantages of using the B-spline-to-Bézier conversion are clearly demonstrated.

Figure 6 presents the path planning result with the B-spline-to-Bézier conversion method proposed in Section 4.3. The reference B-spline path (plotted in a solid green line) is required to stay entirely within the polytopic corridor and pass through three waypoints $\mathbb{W} = \{(0,3),(2,3),(0.5,2)\}$, which are intentionally chosen to be inside the corridor.

The first step is to convert the B-spline control points (10) to the equivalent Bézier points (22) and then apply Variant 1 introduced in Section 5.1 for placing the Bézier points, such that the curve stays inside the connected polytopes. It includes the ending point constraints, i.e., the first and last control points equal to the first and last waypoints, respectively. Finally, the extended property **P2*** (25) is applied to constrain the 9th Bézier control point to be the middle waypoint $(2,3)$.

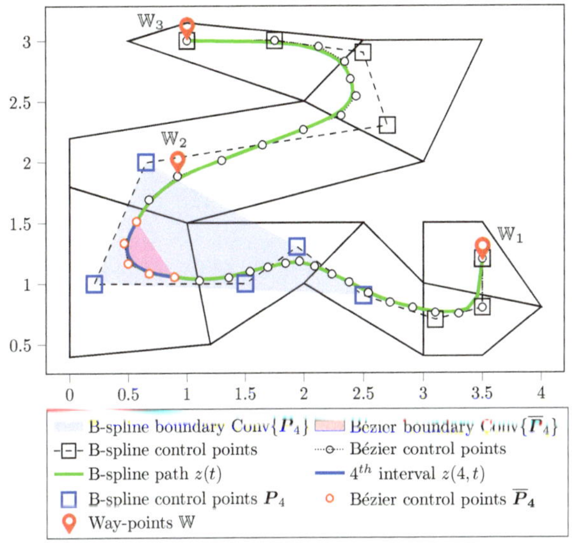

Figure 6. Path planning results in a polytopic corridor using fourth-order B-spline curves and a comparison between B-spline versus Bézier boundaries.

In Figure 6, the B-spline control points are plotted with square marks while the equivalent Bézier control points are plotted with circle marks. The fourth interval and its B-spline control points are highlighted in blue, which shows that the B-spline control boundary is relatively large in comparison with the curve itself and violates the safety constraint. On the other hand, the Bézier control boundary of this interval (plotted with red circle marks and filled with pink) is significantly smaller and completely stays inside the corridor. The 9th Bézier control point (marked with a red flag) is placed exactly at $(2,3)$, which enforces the path to pass through this waypoint and, hence, satisfies all of the requirements.

The next section introduces the constraint formulations that make use of the equivalent Bézier representation for solving the path planning problem in a sequence of polytopes. The benefits of using the Bézier format over the original B-spline formulation are also highlighted via two variants of constraints.

5. B-spline Path Planning Algorithms in a Sequence of Connected Polytopes

This section presents our approaches for optimally placing the control points $\{P_1,\ldots,P_n\}$, such that the B-spline curve $z(t)$, as defined in (10), satisfies all of the requirements of our path generation problem (4)–(6) and has a minimal length. For a quick summary, the curve needs to start at a point P_s, end at another point P_f, and completely stay inside the safe region $\mathcal{S}$:

$$z(t_s) = P_s, \quad z(t_f) = P_f, \quad z(t) \in \mathcal{S}, \tag{28}$$

with $\mathcal{S} = S_1 \cup S_2 \cup \cdots \cup S_q$ ($q \geq 2$) from (6) and $\{P_s, P_f\}$ the start and end poses from (17). Hereinafter, two different formulations of constraints are introduced in order to tackle the aforementioned problems, one using a minimal number of control points and the other requiring more control points but guaranteeing the existence of a solution.

5.1. Variant 1: Constraint Formulation with a Minimal Number of Control Points

Proposition 1. *The requirements* (28) *are satisfied if the following conditions are guaranteed:*

(C1) *Number of control points:*

$$n = q + d. \tag{29}$$

(C2) *Start and end points:*

$$P_1 = P_s, \quad P_{q+d} = P_f. \tag{30}$$

(C3) *All the Bézier control points in one interval belong to one extended polytope* (8)*:*

$$\overline{P}_k \in S_{j,j+1}, \; \forall \, \overline{P}_k \in \overline{\mathbf{P}}_j \text{ and } \forall \, j \in \{1, \ldots, q\}, \tag{31}$$

with $\overline{\mathbf{P}}_j$ consisting of $(d+1)$ Bézier control points given in terms of $(d+1)$ B-spline control points $\mathbf{P}_j$ as in (21)*; $S_{j,j+1}$ is the extended polytope as in* (8)*.*

Proof. At first, the starting and ending constraints from (28) are satisfied by condition C2 (37) due to the endpoint interpolation property (17) of the B-spline curves.

Next, by using $n = d + q$ control points as in (29), the curve $z(t)$ from (10) has q intervals. Within each interval j, $j \in \{1, \ldots, q\}$, the following equation holds:

$$z(j, t) \in \text{Conv}\{\overline{\mathbf{P}}_j\} \subseteq S_{j,j+1}, \tag{32}$$

in which the convexity property is given in (24) and the latter is due to the fact that all $(d+1)$ points in $\overline{\mathbf{P}}_j$ stay inside $S_{j,j+1}$ as constrained by (31). The result in (32) leads to:

$$z(t) \in \bigcup_{j=1}^{q} S_{j,j+1} \equiv \mathcal{S}, \; t \in [t_s, t_f]. \tag{33}$$

This completes the proof. □

Remark 2. *Using the Bézier representation* (19) *allows us to formulate the constraint* (31)*, such that it is possible to enforce "each interval $z(j,t)$ to be inside each extended polytope" as proven in* (32)*. This cannot be done if the original B-spline convexity property* (16) *is employed instead. The reason is that two consecutive B-spline boundaries share d common points (e.g., $\mathbf{P}_j = [P_j \cdots P_{j+d}]$ and $\mathbf{P}_{j+1} = [P_{j+1} \cdots P_{j+d+1}]$ from* (16)*). This leads to the fact that if the B-spline control points are employed in condition C3* (31)*, i.e., $P_k \in S_{j,j+1}, \forall P_k \in \mathbf{P}_j, \forall j \in \{1, \ldots, q\}$, then, the following necessary condition is required:*

$$\bigcap_{i=j}^{j+d} S_{i,i+1} \neq \varnothing, \forall j \in \{1, \ldots, q\}, \tag{34}$$

which is clearly not guaranteed for the extended polytopes defined in (8)*.*

On the other hand, there is only one common point for the Bézier representation (24) *(e.g., $\overline{\mathbf{P}}_j$ and $\overline{\mathbf{P}}_{j+1}$ share one common point $\overline{P}_{jd+1}$). Therefore, the necessary condition for the solution of* (31) *to exist is already satisfied, i.e.,*

$$S_{j,j+1} \cap S_{j+1,j+2} \neq \varnothing, \forall j \in \{1, \ldots, q-1\}, \tag{35}$$

with $S_{j,j+1} \cap S_{j+1,j+2} = T_j$ as defined in (7) *and* (8)*.*

This approach exploits the property of the equivalent Bézier representation, which allows formulating the control points for each interval independently, as in (26). Therefore, it is possible to impose the constraint of "each interval within each extended polytope", which appears to be the choice with the minimum number of control points in our analysis. However, the existence of the solution for the set of constraints (29)–(31) is not always guaranteed. This issue, unfortunately, may cause bugs and become stuck during online deployment. Therefore, we introduce another approach that requires more control points (i.e., more decision variables and heavier computation) but provides a guaranteed solution.

5.2. Variant 2: Constraint Formulation with Guaranteed Solution

Proposition 2. *The requirements* (28) *are satisfied if the B-spline control points of $z(t)$ are chosen according to the following conditions:*

*(**C1**) Number of B-spline control points:*

$$n = d(q-1) + 2, \tag{36}$$

which allows the curve to have $d(q-2)+2$ intervals as given in (15).

*(**C2**) Start and end points:*

$$P_1 = P_s, \quad P_n = P_f. \tag{37}$$

with n as in (36).

*(**C3**) First and last intervals stay in the first and last (i.e., $S_{q,q+1} \equiv S_q$) extended polytopes* (8), *respectively:*

$$\overline{\boldsymbol{P}}_1 \in S_{1,2}, \ \overline{\boldsymbol{P}}_{d(q-1)+2} \in S_{q,q+1}, \tag{38}$$

*(**C4**) and every other extended polytope contains d consecutive intervals:*

$$\overline{\boldsymbol{P}}_j \in S_{k,k+1}, \forall j \in \{d(k-2)+2,\ldots,d(k-1)+1\}, \\ \forall k \in \{2,\ldots,q-1\}, \tag{39}$$

with $\overline{\boldsymbol{P}}_j$ consisting of $(d+1)$ Bézier control points (which control the jth interval) given in terms of $(d+1)$ B-spline control points $\boldsymbol{P}_j$ as in (21) *and $S_{k,k+1}$ as the extended polytope in* (8).

Proof. The proof is similar to the one in Proposition 1, except for the existence of a feasible solution. Therefore, only its sketch is presented hereinafter:

(1) Condition **C2** (37) helps to ensure the start and end points of the path.
(2) Condition **C1** provides a sufficient number of intervals of the curve for the existence of a feasible solution. All of the intervals are then constrained to stay within the safe region $\mathcal{S}$ by two conditions **C3–C4** because each Bézier control boundary (i.e., the convex hull of the corresponding $(d+1)$ Bézier control points (24)) is inside one extended polytope.
(3) Solutions for the complete problem (36)–(39) always exist. One can be found by placing the $d(q-1)+2$ original B-spline control points according to two conditions:
    (i) The first and last points chosen according to (37).
    (ii) Having d points in every transition zone T_k:

$$\{P_{(k-1)d+2},\ldots,P_{kd+1}\} \in T_k, \forall k \in \{1,\ldots,q-1\}, \tag{40}$$

which is feasible since the transition T_k is not empty for all $k \in \{1,\ldots,q-1\}$ as defined in (7). The next step is to prove that condition (40) ensures the satisfaction of the two conditions **C3–C4** (38)–(39) on the Bézier control points.

For **C3**, regarding the first interval of the curve, it is true that $P_1 = P_s \in S_{1,2}$ from (37) and $\{P_2,\ldots,P_{d+1}\} \in T_1 \subseteq S_{1,2}$ which ensure $\overline{\boldsymbol{P}}_1 \in S_{1,2}$ as $\text{Conv}\{\overline{\boldsymbol{P}}_1\} \subset \text{Conv}\{\boldsymbol{P}_1\}$ from

(21). A similar argument is applied to the last interval of the curve; together, they lead to the satisfaction of (37).

Regarding **C4**, every extended polytope $S_{k,k+1}$ with $k \in \{2, \dots, q-1\}$ contains $2d$ B-spline control points:

$$\{P_{(k-2)d+2}, \dots, P_{kd+1}\} \in S_{k,k+1}, \tag{41}$$

which is due to (40) and the fact that both $T_k \subset S_{k,k+1}$ and $T_{k+1} \subset S_{k,k+1}$ (7)–(9). Then, for d consecutive intervals $(k-2)d+2, \dots, (k-1)d+1$, their Bézier control points satisfy:

with $\overline{\boldsymbol{P}}_i$, $\boldsymbol{P}_i$ as in (21). Note that we have $\mathrm{Conv}\{\boldsymbol{P}_i\} \subset S_{k,k+1}$ since $\boldsymbol{P}_i \in S_{k,k+1}$, $\forall i \in [(k-2)d+2, \dots, (k-1)d+1]$ due to (41). Finally, the condition (39) is ensured and the resulting B-spline path satisfies all of the requirements (28). This also completes the proof. □

Remark 3. *In comparison with our approach of using a "minimal number of control points" as in* (29)–(31)*, Proposition 2 adds d intervals to a middle polytope instead of using only one interval as in* (31)*. It allows controlling the curve's shape within each polytope completely and independently and, thus, always guarantee the existence of the solution. Furthermore, the feasible solution* (40) *is built upon the B-spline format. It is obviously only a subset of the Bézier constraints* (39)*, while both of them can ensure the path planning requirements* (28)*. As a result, we actually gain more feasibility and flexibility when switching to the equivalent Bézier format.*

5.3. Path Generation Problem with Minimal Length

This section presents the complete optimization problem used to solve the B-spline reference path satisfying the constraints (28) and minimizing the curve's length. The property **P3** in (18) of the B-spline curve $z(t)$ from (10) is exploited in order to formulate the length cost into a quadratic function of the control points $P_i \in \mathbb{R}^2$, $i \in \{1, \dots, n\}$ (with n, the number of control points, chosen as in (29) or (36)). By denoting $\boldsymbol{P} \triangleq [P_1 \cdots P_n]$ from (10), the optimization problem is given by:

$$\boldsymbol{P}^* = \arg\min_{\boldsymbol{P}} \int_{t_s}^{t_f} \|\dot{z}(t)\|^2 dt, \tag{42}$$

subject to constraints (29)–(31) or (36)–(39) depending on the variants.

The property (18) of the B-spline curve leads to:

$$\dot{z}(t) = \boldsymbol{P} M_{d,d-1} L_{d,d-1} \boldsymbol{B}_{d,\xi}(t) = \sum_{i=1}^{n} Q_i B_{i,d,\xi}(t), \tag{43}$$

with $Q_i \in \mathbb{R}^2$ being the ith column of $Q = \boldsymbol{P} M_{d,d-1} L_{d,d-1} \in \mathbb{R}^{2\times 2q}$. Therefore, the optimization problem (42) is reformulated into:

$$\boldsymbol{P}^* = \arg\min_{\boldsymbol{P}} \sum_{i=1}^{n}\sum_{j=1}^{n} Q_i^\top Q_j \int_{t_s}^{t_f} B_{i,d,\xi}(t) B_{j,d,\xi}(t) dt, \tag{44}$$

subject to constraints (29)–(31) or (36)–(39) depending on the variants.
which clearly has a quadratic cost function since the integral terms are independent of the decision variables $\boldsymbol{P} = [P_1 \cdots P_n]$.

Finally, the reference path $p(t)$ as required in (4)–(6) is taken as:

$$p(t) = \boldsymbol{P}^* \boldsymbol{B}_{d,\xi}(t), \tag{45}$$

in which the optimal control points $\boldsymbol{P}^*$ are obtained from solving the optimization problem (44) and the B-spline basis functions $\boldsymbol{B}_{d,\xi}$ as used in (10) is defined with $[t_s, t_f] = [0, 1]$.

The theoretical background of our path planning algorithms using B-spline parametrization is complete. The next section will introduce the public repository containing the implementation of the whole path planning process and its usage guidelines.

6. Navigation with Polytopes Toolbox

The algorithms discussed throughout this paper were implemented in Python, published and maintained as the *Navigation with Polytopes* (https://gitlab.rob.uni-luebeck.de/robPublic/navigation_with_polytopes, accessed on 19 February 2023) toolbox. The toolbox can be used either as stand-alone scripts for research purposes or as a global path planner that is compatible with ROS (robot operating system) navigation tools. It provides a framework for the construction of a polytope map from a standard occupancy grid map, searching for an appropriate sequence of polytopes and planning a minimal-length path with different options on the B-spline or Bézier characterizations.

6.1. Introduction to the Toolbox

The repository of the toolbox (https://gitlab.rob.uni-luebeck.de/robPublic/navigation_with_polytopes, accessed on 19 February 2023) is organized in the following structure:

- `navigation_with_polytopes`—toolbox with source code.
- `navigation_with_polytope_ros`—integration of the toolbox into ROS.
- `Examples`–sample python scripts for the illustration of the toolbox.

It provides three main features:

- Constructs a polytope map from a grid map.
- Finds an appropriate sequence of polytopes.
- Plans a B-spline path with different algorithms.

The outcomes of each task can be seen in Figure 7 for a given grid map. Within the scope of this paper, more details on the feature of planning the reference B-spline path using the equivalent Bézier representation are presented hereinafter. The toolbox provides the function *bspline_path_planner_polytope_map*, which receives five parameters: the starting and ending points, the polytope map, the degree d of the curve, and the *method*. Three options for *method* have been implemented as follows:

(1) *bezier_min* calls the Variant 1 algorithm given in Section 5.1, which uses the proposed B-spline to Bézier conversion method with a minimal number of control points [8];
(2) *bezier_guarantee* (default option) uses the Variant 2 algorithm given in Section 5.2 with a guaranteed solution [9];
(3) *bspline_guarantee* returns the algebraic solution (40) of Proposition 2. The whole calculation is done with the original B-spline format and with a guaranteed solution.

The function returns both the path defined as a list of points, and the B-spline control points $\boldsymbol{P}$ for constructing the analytical formulation $z(t)$ (10) of the path if needed. The optimal path planning problem is implemented in Pyomo [25], Python 3, and with the solver IPOPT [26]. For ease of use, two interfaces are provided: stand-alone scripts for quick tests and easy modifications as well as a global planner package in ROS for practical usages.

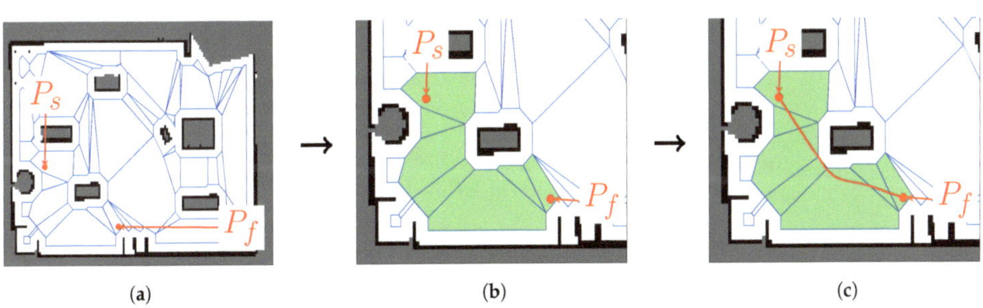

Figure 7. Main tasks of the *Navigation with Polytopes* toolbox. (**a**) Polytope map from a grid map; (**b**) finding a sequence of polytopes; (**c**) planning a B-spline reference path.

6.2. ROS Integration

As part of the toolbox, a ROS1 package that may be used as a global path planner is also provided for convenience and integration into current projects. Two ROS nodes can be found in the ROS package *navigation with polytopes*:

- *poly_map_construct*—creates a polytope map from a given grid map by using the procedure outlined in Section 3.1.
- *bspline_path_planner_node*—given the current pose and goal, the node performs the whole path planning process. In addition to performing the tasks as the first node, it searches for the ideal sequence of polytopes and publishes the B-spline path as mentioned in Section 4.

The *bspline_path_planner_node* takes several parameters for the creation of a polytope map-like robot footprint (offset o_p), RDP inner and outer epsilons, path planner parameters, such as the B-spline degree, method, etc., and parameters, such as the map frame, base frame of the robot, etc. The package provides a sample launch file, which contains all of the necessary parameters for the node. The results of the toolbox's ROS integration are illustrated in Figure 8. The ROS package is validated using a sample environment from Gazebo, as shown in Figure 8a, and the results of the path planning algorithms from the toolbox are visualized in Rviz, as shown in Figure 8b. The polytopes (plotted in blue) are visualized in Rviz using the *jsk_ visualization* (https://github.com/jsk-ros-pkg/jsk_visualization, accessed on 19 February 2023) package. Both ROS nodes mentioned above will publish the polytope map and sequence as a msg type *jsk_ recognition_ msgs/PolygonArray* for visualization purposes in Rviz. Usage instruction and structure information (subscribed and published topics of nodes) are available in the repository.

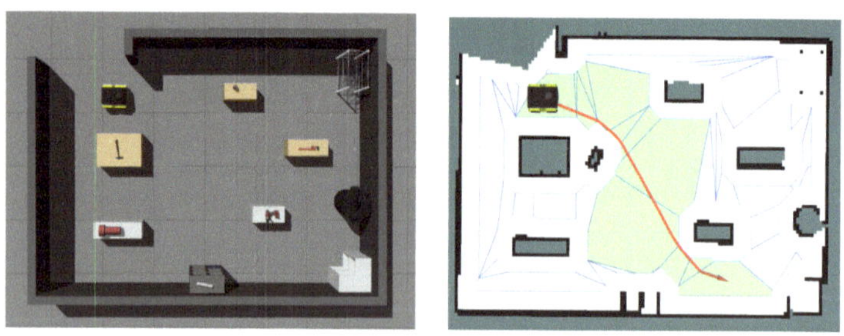

(**a**) Gazebo simulation world. (**b**) Path planning result in Rviz.

Figure 8. Toolbox's result on a grid map obtained from a Gazebo simulation.

7. Validation Results

The validation results of the *Navigation with Polytopes* toolbox are illustrated in this section. It firstly presents an exploration strategy for mapping an unknown environment given a top-down view figure (e.g., satellite Google Earth image). Then, the proposed path planning algorithm as well as other methods are validated in different grid maps, which are collected from realistic simulations and an actual environment. The evaluation of the toolbox when being used with ROS in various Gazebo simulation environments is presented, together with the comparisons with the default path-planning methods employed by the ROS navigation stack.

7.1. Exploration Strategy for Creating the Occupancy Grid Map

The proposed path planning process requires constructing a polytope map from a standard grid map. One simple method for creating such grid maps is to use the laser-based SLAM package *gmapping* while driving the robot around either manually or autonomously. We implemented an exploration program, which receives a top-down image of the environment around the robot, then allows the user to select points that will be connected as the exploration path for the robot to follow afterwards, autonomously, by using the *move_base* function in ROS (c.f. Figure 9a). Note that a simple sketch of the environment is sufficient for the program, but a screenshot of the simulation from a top-down view or satellite map image of the field is better. Moreover, the size of the map must be specified in meters and the starting position and orientation of the robot has to be specified. The exploration and mapping process is visualized in real-time in the program window as shown in Figure 9a. The dashed white line indicates the sequence in which the exploration path passes through the predefined points. Boundaries of mapped obstacles are highlighted in red, unexplored parts of the map are grayed-out and explored areas are colored. The robot's position is marked by the robot symbol. After obtaining the grid map, one can apply the *Navigation with Polytopes* toolbox to construct the polytope map and plan a B-spline reference path (given by solid red line) as shown in Figure 9b. A comparison with the standard path planner *Navfn* in ROS (with its result plotted in a solid green line) is also presented there, which shows similarities and comparable performances (e.g., smoothness, shape, length) of the two methods. More details on the applications of the toolbox and comparisons with the *Navfn* method will be discussed in the next section.

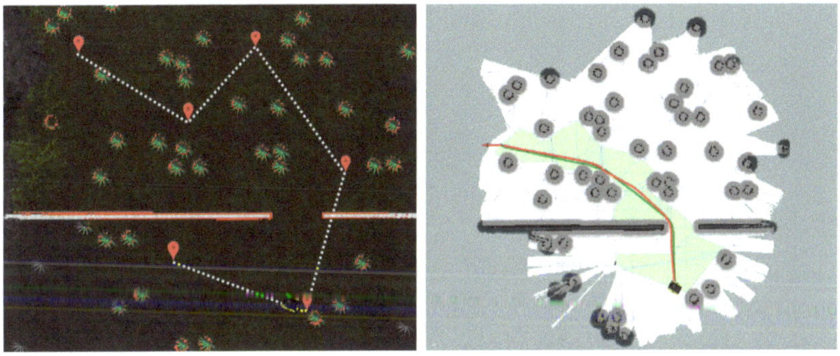

(a) (b)

Figure 9. Illustration of the exploration strategy using a top-down figure of the environment and the corresponding path planning results (reference paths obtained from the *Navigation with Polytopes* toolbox and from the standard *Navfn* planner of ROS plotted in red and green lines, respectively). (**a**) Exploration program window; (**b**) path planning results after the exploration.

7.2. Simulation Results

Figure 10 shows the results of the whole path planning process performed by the toolbox on different grid maps. Figures 10a–c were captured from an agricultural field, while Figure 10d resulted from an indoor scenario after an earthquake with scattered furniture. All were simulated with high fidelity in Gazebo. Note that more results from different aspects of the aforementioned two scenarios, as well as another laboratory and office maps, are provided in Appendix A. In all scenarios, the toolbox performs the sequential steps as described throughout the paper: (i) constructing a polytope map, (ii) finding the sequence of polytopes that leads from start to goal points, and (iii) planning a reference path using one of the three methods available in the toolbox. The polytope maps are bordered in blue, the sequence of polytopes allowing safe travel from P_s to P_f is filled with green, and the B-spline reference path is plotted in red. The computation time for constructing the polytope map for the complex agricultural field (c.f. Figure 10a–c) is around 500 ms, while for the indoor scenario after the earthquake shown in Figure 10d, it takes up to 900 ms due to numerous small and cluttered obstacles. Next, the optimal path planning process consumes 1225, 1080, and 1460 ms for the three scenarios shown in Figure 10b, 10c, and 10d, respectively. Figures 9b and 10b–d also present the comparison of the toolbox with the path planning results of the standard *Navfn* function of ROS. The average computation time of the *Navfn* planner is around 100 ms, which is much less than the proposed toolbox. It is understandable as the *Navfn* planner is basically a modified version of the A* algorithm [5,27], which is well-known for its fast searching capability. However, the *Navigation with Polytopes* toolbox serves as a global path planner, which only runs once at the beginning of the navigation task in order to find a safe and optimal path; spending several seconds is an acceptable trade-off with numerous advantages that the B-spline path provides in comparison with the A* path. For more details, in Figures 10d and A1a, the *Navfn* path planner plans the paths through the unknown gray area, which are shorter than the B-spline paths but possibly unsafe. The reason is due to the high safety demand of the *Navigation with Polytopes* toolbox, it only considers the explored and free regions when constructing the polytope map while the *Navfn* path planner allows the movements inside the unexplored area. Furthermore, the paths resulting from the *Navfn* planner (plotted in solid green) are not as smooth as the B-spline path (in red) and are longer in most of the cases. These prove the effectiveness of the proposed minimal-length path planning algorithms in (44). Another important property of the planned paths is their smoothness. As shown in Figure 10b as well as Figure A1b–d in Appendix A, the B-spline paths (plotted in solid red lines) are significantly smoother than the results from the *Navfn* path planner (plotted in solid green lines) despite the usage of the *Savitzky–Golay* path smoother, which is already implemented in the *Navfn* planner. After extensive simulation and experimental trials, the *Navigation with Polytopes* toolbox was tested carefully with various maps of different environments and of various sizes to validate its scalability and robustness, as shown in Figures 10 and A1.

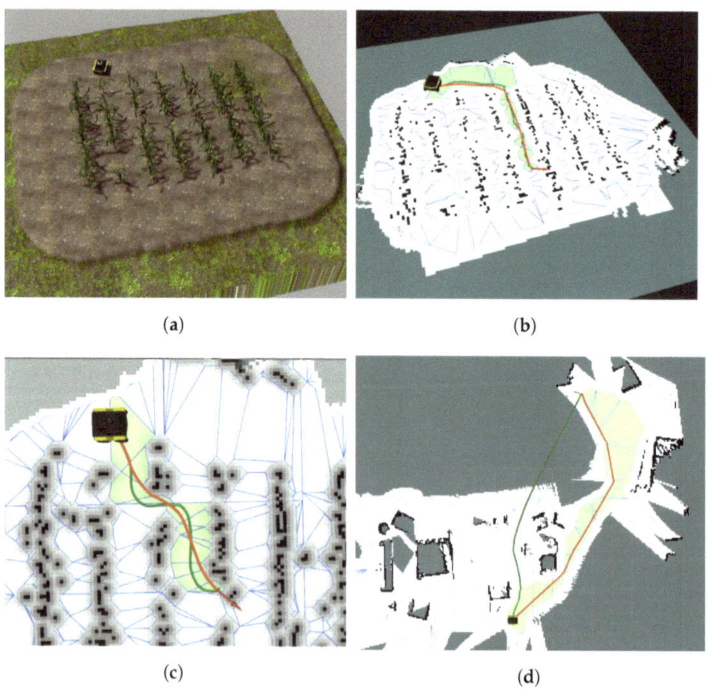

Figure 10. Comparisons of different path planning methods in occupancy grid maps: *Navigation with Polytopes* toolbox (red lines) versus the standard *Navfn* of ROS (green lines). (**a**) Agriculture field in the Gazebo simulation; (**b**) results in the field map: test case 1; (**c**) results in the field map: test case 2; (**d**) results in an earthquake-affected house.

8. Discussion

The *Navigation with Polytopes* toolbox and its theoretical background were introduced in this paper. The toolbox currently serves as a global path planner and is compatible with the ROS navigation package. It takes the standard occupancy grid map of the environment, the current robot's position, and the selected goal as inputs and provides a safe and smooth reference path with an optimal length to the goal. Major distinctions (with respect to existing works in the literature) are the construction of a polytope map from the original grid map, the usage of the B-spline curve via its equivalent Bézier representation on the constrained path planning problem, and the conversion from B-spline to Bézier control points of the curve. As an unavoidable consequence of the optimization usage, the computation time is higher than any standard path planning methods using the grid and graph search strategies (e.g., Dijkstra, RRT [2–4]). However, the results obtained with the toolbox show advantages over the standard methods employed in the current path planner of the ROS navigation package, such as a shorter length, a smoother profile, and enhanced safety. Another advantage of the toolbox is that optimization can be reformulated with much flexibility. In the case of the agricultural field map (c.f. Figure 10a), the optimization problem (44) can be enforced to plan the path along the center lines of the rows and not pass through them by adding an additional constraint on the connectivity of the polytope map, i.e., by not considering two polytopes to be connected if they cut through the predefined rows. We emphasize that the use of a polytope map allows for the integration of various optimal control methods to solve different navigation problems, in addition to the primary goal of serving as a global path planner. By constructing the polytope map, the toolbox can become a framework for easily integrating existing optimal control techniques into realistic grid map data. The problems to be tackled are not limited to path/trajectory planning, but also navigation, motion control, localization, etc. The transformation of the obstacle-free space (i.e., non-occupied cells in a grid map) into a polytope map, as well as finding an appropriate polytope sequence, allows simplifying and representing a safe environment with only linear constraints (i.e., polytopic constraints). They have been employed in various optimal control applications, such as MPC (model predictive control) and mixed-integer-programming [15,28–30], e.g., in [28], the authors introduce an MPC controller for the safe navigation of a mobile robot within a polytope, which can push the system far away from the selected boundaries, such as walls. This controller will be added to the toolbox as the local navigation controller in the near future.

Another worthy extension would be to improve the technique of finding an appropriate sequence of polytopes by taking into account the narrowness of the corridor (i.e., only distance is counted for now). It is needed to evaluate a trade-off between a short but narrow corridor and a long but spacious one. More kinetic constraints will be taken into account, such as turning the radius and speed into the path planning algorithms, considering their effects on the solvability of the final optimization problem.

9. Conclusions

This paper presents the process to solve the path planning problem for a mobile robot given a standard grid map of the surrounding environment. It first constructs a polytope map of the free space and then seeks a sequence of connected polytopes leading to the goal with minimal distance. Next, a B-spline path is planned within the sequence and connects the two end points. Specifically, the B-spline path is converted into its equivalent Bézier representation in order to reduce the conservativeness of the path planning problem. Another contribution is the new technique to calculate the B-spline-to-Bézier conversion matrix, which covers all partitions of the curve. Two variants of constraints that enforce the B-spline path to stay within the aforementioned sequence of polytopes are presented with proofs. The whole procedure is implemented in Python and is publicly available as the *Navigation with Polytopes* (https://gitlab.rob.uni-luebeck.de/robPublic/navigation_with_polytopes, accessed on 19 February 2023) toolbox, which is ready to use and compatible with ROS as a global path planner for navigation.

Author Contributions: Conceptualization, N.T.N. and G.S.; methodology, N.T.N.; software, N.T.N. and P.T.G.; validation, N.T.N., P.T.G. and N.F.K.; formal analysis, N.T.N.; investigation, N.T.N. and P.T.G.; resources, F.E.; data curation, N.T.N. and P.T.G.; writing—original draft preparation, N.T.N., P.T.G. and N.F.K.; writing—review and editing, all authors; visualization, N.T.N., P.T.G. and N.F.K.;

supervision, G.S. and F.E.; project administration, F.E.; funding acquisition, F.E. All authors have read and agreed to the published version of the manuscript.

Funding: This research was funded by the German Ministry of Food and Agriculture (BMEL), project no. 28DK133A20.

Institutional Review Board Statement: Not applicable.

Informed Consent Statement: Not applicable.

Data Availability Statement: Not applicable.

Acknowledgments: We would like to thank our former and present colleagues at the University of Lübeck, especially Heiko Hamann, Lars Schilling, Michael Sebastian Angern, and Arne Sahrhage, for the fruitful discussions and technical support during the implementation of this work.

Conflicts of Interest: The authors declare no conflicts of interest. The funders had no role in the design of the study; in the collection, analyses, or interpretation of data; in the writing of the manuscript; or in the decision to publish the results.

Abbreviations

The following abbreviations are used in this manuscript:

ROS	robot operating system
SLAM	simultaneous localization and mapping
CAD	computer-aided design
RDP	Ramer–Douglas–Peucker
MPC	model predictive control

Appendix A. Illustration of the Path Planning Results from the Navigation with Polytopes Toolbox

This section provides additional examples of the path planning results obtained from the *Navigation with Polytopes* toolbox compared to the standard *Navfn* path planner in ROS. The paths from the toolbox are plotted in solid red lines, while those obtained from *Navfn* are plotted in solid green lines for four different environments: (i) an indoor scenario after an earthquake in Figure A1a, (ii) an agricultural field (cf. Figure 8a) in Figure A1b, (iii) a laboratory in Figure A1c, and (iv) an office floor in Figure A1d.

(**a**) Earthquake-affected house.

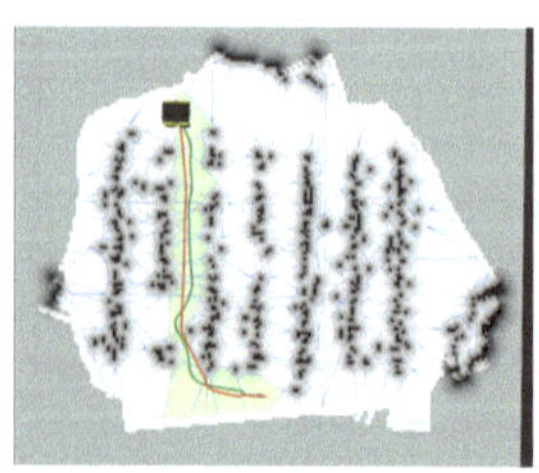

(**b**) Agricultural field.

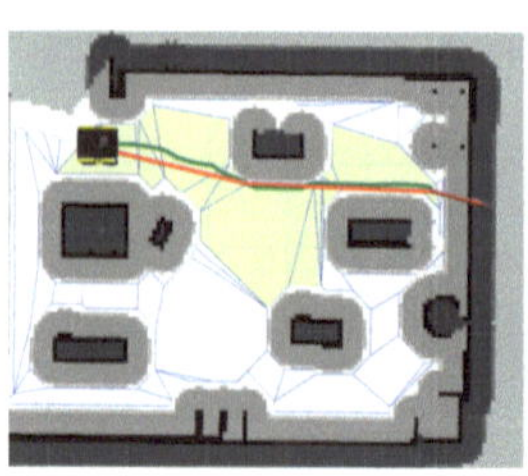

(**c**) Laboratory.

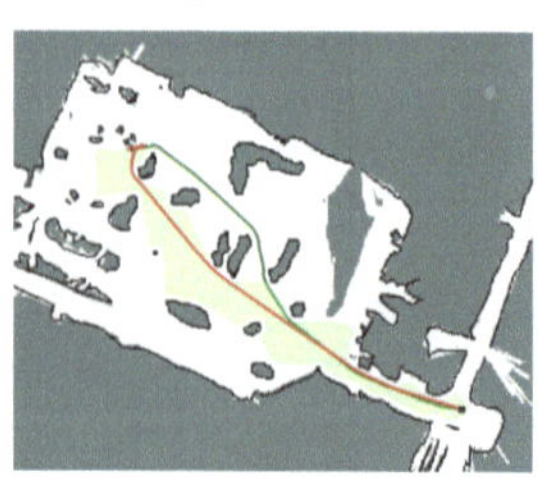

(**d**) Office floor

Figure A1. Path planning results for different occupancy grid maps; compares the performance of the *Navigation with Polytopes* toolbox (red lines) versus the standard *Navfn* in ROS (green lines).

References

1. LaValle, S.M. *Planning Algorithms;* Cambridge University Press: New York, NY, USA, 2006.
2. Ab Wahab, M.N.; Nefti-Meziani, S.; Atyabi, A. A comparative review on mobile robot path planning: Classical or meta-heuristic methods? *Annu. Rev. Control* **2020**, *50*, 233–252. [CrossRef]
3. Zhang, H.Y.; Lin, W.M.; Chen, A.X. Path planning for the mobile robot: A review. *Symmetry* **2018**, *10*, 450. [CrossRef]
4. Sánchez-Ibáñez, J.R.; Pérez-del Pulgar, C.J.; García-Cerezo, A. Path Planning for Autonomous Mobile Robots: A Review. *Sensors* **2021**, *21*, 7898. [CrossRef] [PubMed]
5. Kim, C.; Suh, J.; Han, J.H. Development of a hybrid path planning algorithm and a bio-inspired control for an omni-wheel mobile robot. *Sensors* **2020**, *20*, 4258. [CrossRef] [PubMed]
6. Schildbach, G.; Borrelli, F. A dynamic programming approach for nonholonomic vehicle maneuvering in tight environments. In Proceedings of the 2016 IEEE Intelligent Vehicles Symposium (IV), Gothenburg, Sweden, 19–22 June 2016; pp. 151–156.
7. Nguyen, N.T.; Prodan, I. Stabilizing a multicopter using an NMPC design with a relaxed terminal region. *IFAC-PapersOnLine* **2021**, *54*, 126–132. [CrossRef]
8. Nguyen, N.T.; Schilling, L.; Angern, M.S.; Hamann, H.; Ernst, F.; Schildbach, G. B-spline path planner for safe navigation of mobile robots. In Proceedings of the 2021 IEEE/RSJ International Conference on Intelligent Robots and Systems (IROS), Prague, Czech Republic, 27 September–1 October 2021; pp. 339–345.
9. Nguyen, N.T.; Gangavarapu, P.T.; Sahrhage, A.; Schildbach, G.; Ernst, F. Navigation with polytopes and B-spline path planner. In Proceedings of the 2023 IEEE International Conference on Robotics and Automation (ICRA), London, UK, 29 May–2 June 2023 .
10. Grisetti, G.; Stachniss, C.; Burgard, W. Improved Techniques for Grid Mapping With Rao-Blackwellized Particle Filters. *IEEE Trans. Robot.* **2007**, *23*, 34–46. [CrossRef]
11. Nguyen, N.T.; Prodan, I.; Lefèvre, L. Flat trajectory design and tracking with saturation guarantees: A nano-drone application. *Int. J. Control* **2020**, *93*, 1266–1279. [CrossRef]
12. Manyam, S.G.; Casbeer, D.W.; Weintraub, I.E.; Taylor, C. Trajectory Optimization For Rendezvous Planning Using Quadratic Bézier Curves. In Proceedings of the 2021 IEEE/RSJ International Conference on Intelligent Robots and Systems (IROS), Prague, Czech Republic, 27 September–1 October 2021; pp. 1405–1412.
13. Stoican, F.; Ivănuçcă, V.M.; Prodan, I.; Popescu, D. Obstacle avoidance via B-spline parametrizations of flat trajectories. In Proceedings of the 24th Mediterranean Conference on Control and Automation (MED'16), Athens, Greece, 21–24 June 2016; pp. 1002–1007.
14. Stoican, F.; Prodan, I.; Grøtli, E.I.; Nguyen, N.T. Inspection Trajectory Planning for 3D Structures under a Mixed-Integer Framework. In Proceedings of the 2019 IEEE International Conference on Control & Automation (ICCA'19), Edinburgh, UK, 16–19 July 2019; pp. 1349–1354.
15. Prodan, I.; Stoican, F.; Louembet, C. Necessary and sufficient LMI conditions for constraints satisfaction within a B-spline framework. In Proceedings of the 2019 IEEE 58th Conference on Decision and Control (CDC), Nice, France, 11–13 December 2019; pp. 8061–8066.
16. Suryawan, F.; De Doná, J.; Seron, M. Splines and polynomial tools for flatness-based constrained motion planning. *Int. J. Syst. Sci.* **2012**, *43*, 1396–1411. [CrossRef]
17. Berglund, T.; Brodnik, A.; Jonsson, H.; Staffanson, M.; Soderkvist, I. Planning smooth and obstacle-avoiding B-spline paths for autonomous mining vehicles. *IEEE Trans. Autom. Sci. Eng.* **2009**, *7*, 167–172. [CrossRef]
18. Zhang, X.; Wang, C.; Chui, K.T.; Liu, R.W. A real-time collision avoidance framework of MASS based on B-spline and optimal decoupling control. *Sensors* **2021**, *21*, 4911. [CrossRef] [PubMed]
19. Maekawa, T.; Noda, T.; Tamura, S.; Ozaki, T.; Machida, K.I. Curvature continuous path generation for autonomous vehicle using B-spline curves. *Comput.-Aided Des.* **2010**, *42*, 350–359. [CrossRef]
20. Romani, L.; Sabin, M.A. The conversion matrix between uniform B-spline and Bézier representations. *Comput. Aided Geom. Des.* **2004**, *21*, 549–560. [CrossRef]
21. Böhm, W. Generating the Bézier points of B-spline curves and surfaces. *Comput.-Aided Des.* **1981**, *13*, 365–366.
22. Amsters, R.; Slaets, P. Turtlebot 3 as a robotics education platform. In *Robotics in Education: Current Research and Innovations 10;* Springer: Berlin/Heidelberg, Germany, 2020; pp. 170–181.
23. Douglas, D.H.; Peucker, T.K. Algorithms for the reduction of the number of points required to represent a digitized line or its caricature. *Cartogr. Int. J. Geogr. Inf. Geovisualizat.* **1973**, *10*, 112–122.
24. Piegl, L.; Tiller, W. B-spline Curves and Surfaces. In *The NURBS Book;* Springer: Berlin/Heidelberg, Germany, 1995; pp. 81–116.
25. Hart, W.E.; Watson, J.P.; Woodruff, D.L. Pyomo: Modeling and solving mathematical programs in Python. *Math. Program. Comput.* **2011**, *3*, 219–260.
26. Wächter, A.; Biegler, L.T. On the implementation of an interior-point filter line-search algorithm for large-scale nonlinear programming. *Math. Program.* **2006**, *106*, 25–57.
27. Raja, P.; Pugazhenthi, S. Optimal path planning of mobile robots: A review. *Int. J. Phys. Sci.* **2012**, *7*, 1314–1320. [CrossRef]
28. Nguyen, N.T.; Schildbach, G. Tightening polytopic constraint in MPC designs for mobile robot navigation. In Proceedings of the 2021 25th International Conference on System Theory, Control and Computing (ICSTCC), Iasi, Romania, 20–23 October 2021; pp. 407–412.

29. Caregnato-Neto, A.; Maximo, M.R.; Afonso, R.J. Real-time motion planning and decision-making for a group of differential drive robots under connectivity constraints using robust MPC and mixed-integer programming. *Adv. Robot.* **2022**, *37*, 356–379. [CrossRef]
30. Nezami, M.; Nguyen, N.T.; Männel, G.; Abbas, H.S.; Schildbach, G. A Safe Control Architecture Based on Robust Model Predictive Control for Autonomous Driving. In Proceedings of the 2022 American Control Conference (ACC), Atlanta, GA, USA, 8–10 June 2022; pp. 914–919.

sensors

MDPI

Article

Improving Mobile Robot Maneuver Performance Using Fractional-Order Controller

Daniel Acosta, Bibiana Fariña, Jonay Toledo and Leopoldo Acosta *

Computer Science and System Department, Universidad de La Laguna, 38200 Canary Island, Spain; alu0100908480@ull.edu.es (D.A.); bfarinaj@ull.edu.es (B.F.); jttoledo@ull.edu.es (J.T.)
* Correspondence: lacosta@ull.edu.es

Abstract: In this paper, the low-level velocity controller of an autonomous vehicle is studied. The performance of the traditional controller used in this kind of system, a PID, is analyzed. This kind of controller cannot follow ramp references without error, so when the reference implies a change in the speed, the vehicle cannot follow the proposed reference, and there is a significant difference between the actual and desired vehicle behaviors. A fractional controller is proposed which changes the ordinary dynamics allowing faster responses for small times, at the cost of slower responses for large times. The idea is to take advantage of this fact to follow fast setpoint changes with a smaller error than that obtained with a classic non-fractional PI controller. Using this controller, the vehicle can follow variable speed references with zero stationary error, significantly reducing the difference between reference and actual vehicle behavior. The paper presents the fractional controller, studies its stability in function of the fractional parameters, designs the controller, and tests its stability. The designed controller is tested on a real prototype, and its behavior is compared to a standard PID controller. The designed fractional PID controller overcomes the results of the standard PID controller.

Keywords: fractional control; autonomous vehicle; robotics

Citation: Acosta, D.; Fariña, B.; Toledo, J.; Acosta, L. Improving Mobile Robot Maneuver Performance Using Fractional-Order Controller. *Sensors* **2023**, *23*, 3191. http://doi.org/10.3390/s23063191

Academic Editors: David Cheneler and Stephen Monk

Received: 8 February 2023
Revised: 12 March 2023
Accepted: 14 March 2023
Published: 16 March 2023

1. Introduction

Mobile robotics is a very active research area. This includes the design and implementation of autonomous robots. These robots are capable of making intelligent decisions based on localization, path planning, obstacle detection and avoidance, and environment analysis modules. One of the key parameters for the success of a mobile robot is robot control. The robot must obey the decisions made by higher control layers in the most precise way. Any variation between the maneuver received and the actual maneuver executed can result in final application failure and more complicated high-level control.

Our research group has been working for some time with a low-cost electric golf cart [1]. The objective is to turn a standard golf cart into an autonomous vehicle so that some mechanical and electric modifications were made on it. The drive that generates traction is a direct current motor and a drive by wire steering system that coexists with manual steering is included. The prototype includes an on board computer, sensors and software that turn it into an autonomous robot capable of transporting two passengers in non-structured environments. The vehicle localize itself [2,3], makes navigation decisions [4,5], detects obstacles [6], avoids them [7,8], and plans the best route in real time [9]. The robot applies this plan using a steering and velocity control, and the quality of the control limits the final performance of the vehicle. This turns this vehicle into a good framework to test different self-driven vehicle strategies.

The sensor set includes two encoders attached to each rear wheel to obtain odometric information, an IMU, a centimeter GPS, three Lidars, and a stereo vision system. The software is developed on Robotic Operative System (ROS). The software is structured in layers, from the low level where there are sensors and actuators to the high level where there are planning layers able to make intelligent decisions based on the environment. Figure 1 shows the prototype, a fully electric two-seat golf cart.

Figure 1. The sensors system to measure cart speed.

The measured speed for the control is obtained from the odometric system of the prototype. The odometric system is based on encoders coupled to rear wheels, as shown in Figure 2. Each encoder provides 1024 pulses per revolution and each revolution of the wheel generates a revolution of the encoder (1:1 coupling). Wheel rotation is transferred to the encoders through a flexible mechanical transmission system that goes from the center of each wheel to the encoder placed on the side of the vehicle (see Figure 2). Encoder output is connected to an ad hoc electronics that samples the encoders signal every 0.5 ms. The electronics is designed to measure and integrate the encoder signals and the output is transmitted to the on-board computer at every integration period of 20 ms. The integration is made in the microcontroller installed in the ad hoc electronics, based on Euler integration, and collecting encoder increments for the integration time.

Figure 2. The odometric sensor coupled to the rear wheels.

This paper focuses on the lower software level of the vehicle, the motor traction control when a trajectory is being tracked. The traditional way of approaching the control of a system is using a PID controller [10]. It is a convenient and easy way to apply a solution for controlling any system, but usually the performance of the controller system is not the best. To improve the control quality of the final system, some alternatives are available in the literature. Specifically, the classic PI controller used in the first designs was replaced by a new fractional control in order to achieve better maneuverability under certain conditions.

In particular, the control engineering benefited from the advantages of adding fractional operators to controllers. Incorporating integral and derivative fractional parts into a controller makes it possible to have two additional parameters to tune compared to the integer versions. These two parameters are the corresponding fractional orders. The objective is precisely to take advantage of the fractional controllers to obtain a better performance in the maneuverability of the prototype.

In this paper, a new fractional controller is used as the speed control for the autonomous vehicle. This controller allows it to follow the applied commands in a more precise way. Specifically, if the command sent by the high-level control is a speed increase or reduction, a traditional PID is not able to follow the trajectory without a stationary error, so the actual speed is different from the desired speed, as shown in Section 5. The fractional controller proposed is able to follow these variable speed commands with zero stationary error, and the error between reference and command is reduced as Section 7 results shows. This advantage allows a more precise and accurate movement of the autonomous cart.

Figure 3 shows the implementation of the low-level controller that is made in the autonomous vehicle, where a standard PI controller is substituted by a fractional PI.

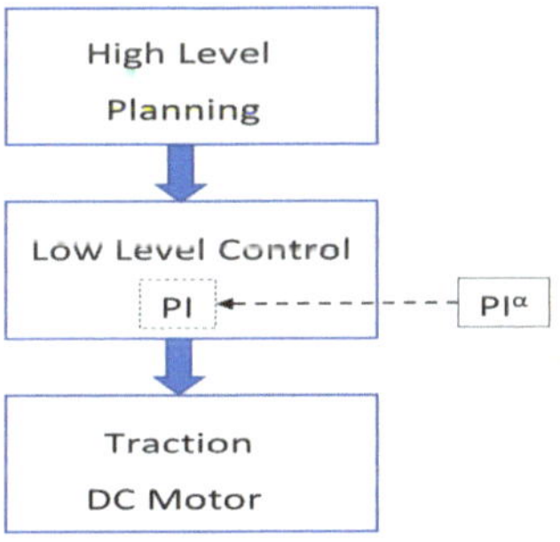

Figure 3. Overall block diagram of the autonomous cart.

2. Previous Work

Fractional calculus studies the generalization of integer-order derivatives and integrals to a fractional-order derivatives and integrals. This means that traditional calculus use integer indices in its derivatives and integrals, however, the fractional calculus allows to use fractional derivatives and integrals describing a more complex function. The fractional calculus may be considered an old and yet novel topic. It is an old topic because, starting from some speculations of G.W. Leibniz in 1695 and L. Euler in 1730 , it has been progressively developed up to now. However, it may also be considered a novel topic because its applications began in recent decades. A complete description of the fractional mathematics can be found in [11]. In [12], a survey with the advances in fractional calculus since the 1970s is shown, which includes numerical applications to implement the actual fractional systems that can work in real time. In [13], a survey of many applications of fractional calculus, examples, and possible implementations are presented. It also contains a separate chapter of fractional order control systems, which opens new perspectives in control theory.

Fractional models have been applied to different problems to characterize the dynamics of processes with complex behaviors, as in [14], where fractional kinetic equations of the diffusion are presented as a useful approach for the description of transport dynamics in complex systems which are governed by anomalous diffusion and non-exponential relaxation patterns. Methods to find the solution are introduced, and for some special cases, exact solutions are calculated. This report demonstrates that fractional equations have come of age as a complementary tool in the description of anomalous transport processes.

In [15], fractional calculus is applied to the control of a revolute planar robotic manipulator. The fractional derivatives required by the control can be obtained by adopting numerical real-time signal processing. Numerical experiments illustrated the feasibility and effectiveness of the approach. Ref. [16] presents the possibilities of fractional calculus applied to system identification and control engineering, but also into sensing and filtering domains. The fractional-order electronic component has led to the possibility of analog filtering techniques from a practical perspective, enlarging the horizon to a wider frequency range, with increased robustness to component variation, stability, and noise reduction. Fractional-order digital filters have developed to provide an alternative solution to higher-order integer-order filters, with increased design flexibility and better performance.

The control of autonomous vehicles includes multiple steps, including route planning, behavioral decision-making, motion planning, and vehicle control [17]. The last step, vehicle control is usually made with a standard PID controller. A survey of the different strategies applied in the low-level control of the vehicles can be found in chapter 2 of reference [18], where the authors distinguish between model-based and model-free controllers. Model-based controllers are more complicated to implement, and when the system changes, as for example for battery discharge, its performance is reduced. However, model-free controllers are more difficult to adjust, but more robust to changes in the model. The fractional controller presented in this paper can be classified as model-free, but with better performance than standard PID controllers. In [19], the longitudinal control task is addressed by implementing adaptive PID control using two different approaches: genetic algorithms (GA-PID) and then neural networks (NN-PID), respectively, adapting the controller to the non-linearities and the change in system characteristics. In [?], a control schema to manage low-level vehicle actuators (steering throttle and brake) based on fuzzy logic, an artificial intelligence technique that is able to mimic human procedural behavior is presented, in this case, when performing the driving task.

In this paper, a new approach to controlling the speed of an autonomous robot is presented, where the fractional-order controller is used to improve the performance in reference tracking. The advantages of this kind of controller include the fact that it allows to obtain a better performance in robot tracking the following sections will show.

3. Fractional Integral and Derivative

Given a real function dependent on time f(t), its fractional integral $\left(I^{\alpha}_{0^+} f\right)(t)$ of order α is defined as Equation (1).

$$\left(I^{\alpha}_{0^+} f\right)(t) \triangleq \frac{1}{\Gamma(\alpha)} \int_0^t \frac{f(\tau)}{(t-\tau)^{1-\alpha}} d\tau \tag{1}$$

where α is the real positive integration order and $\Gamma(\alpha)$ is the Gamma function. The Laplace transform of this integral equation can be defined as Equation 2.

$$L\{I^{\alpha} f(t)\} = \int_0^{\infty} e^{-st} \left(\frac{1}{\Gamma(\alpha)} \int_0^t \frac{f(\tau)}{(t-\tau)^{1-\alpha}} d\tau \right) dt = \frac{1}{s^{\alpha}} F(s) \tag{2}$$

with $I^{\alpha}_{0^+} \equiv I^{\alpha}$ and zero initial conditions.

The definition of the fractional integral is unique. However, for the definition of the fractional derivative, there are various proposals.

The Lagrange's rule for differential operators is used to define the Riemann–Liouville fractional derivative ${}^{RL}D^{\beta}_{0^+} f(t)$ of order β for a function $f(t)$. Given $n \in \mathbb{N}$ such that

$n-1<\beta\leq n$, the Riemann–Liouville derivative is obtained, computing the n-th order derivative over the integral of order $(n-\beta)$ which is defined in Equation (3).

$$ {}^{RL}D^{\beta}_{0+}f(t)\triangleq D^{n}\left(I^{n-\beta}_{0+}f\right)(t) \quad (3)$$

where $D\equiv\frac{d}{dt}$ and ${}^{RL}D^{\beta}_{0+}\equiv D^{\beta}$ is used.

In a very similar way to the previous definition, changing the order of the derivative and the integral, it is possible to define the Caputo fractional derivative ${}^{c}D^{\beta}_{0+}f(t)$ of order β in Equation (4).

$$ {}^{c}D^{\beta}_{0+}f(t)\triangleq\left(I^{n-\beta}_{0+}(D^{n}f)\right)(t) \quad (4)$$

The advantage of the Caputo derivative over the Riemann Liouville derivative, Equation (4), is that it is not necessary to define the fractional-order initial conditions when solving differential equations.

Another alternative definition for the fractional derivative is that of Grünwald–Letnikov ${}^{GL}D^{\beta}_{0+}f(t)$ (Equation (5)).

$$ {}^{GL}D^{\beta}_{0+}f(t)=\lim_{h\to 0}\frac{1}{h^{\beta}}\sum_{j=0}^{\infty}(-1)^{j}\binom{\beta}{j}f(t+(\beta-j)h) \quad (5)$$

where $\binom{\beta}{j}$ is defined in Equation (6).

$$\binom{\beta}{j}=\frac{\Gamma(\beta+1)}{\Gamma(j+1)\Gamma(\beta-j+1)} \quad (6)$$

It can be shown that the above definitions of the fractional derivative are equivalent for a wide class of functions [13].

The Laplace transform of the fractional derivative D^{β} is given in Equation (7).

$$L\left\{D^{\beta}f(t)\right\}=s^{\beta}F(s) \quad (7)$$

when $n-1<\beta\leq n$ and $f(0)=f'(0)=\cdots=f^{(n-1)}(0)=0$.

It is important to note that the classical derivative of a function at an instant t is a local operator. However, the fractional derivative of a function at time t depends on past values, and it is therefore an operator with memory.

3.1. Fractional Systems

A non-integer linear time-invariant system with input $u(t)$ and output $y(t)$ can be represented in Equation (8).

$$\sum_{k=0}^{n}a_{k}D^{\alpha_{k}}y(t)=\sum_{k=0}^{m}b_{k}D^{\beta_{k}}u(t) \quad (8)$$

where $\alpha_k,\beta_k\in\mathbb{R}$ and $n\geq m$.

If the orders of derivation α_k and β_k can be represented as a term $k\alpha$, with k = 0, 1, 2,... the system is said to be of commensurate order Equation (9)

$$\sum_{k=0}^{n}a_{k}D^{k\alpha}y(t)=\sum_{k=0}^{m}b_{k}D^{k\alpha}u(t) \quad (9)$$

and its transfer function is defined in Equation (10)

$$G(s)=\frac{Y(s)}{U(s)}=\sum_{k=0}^{m}b_{k}(s^{\alpha})^{k}/\sum_{k=0}^{n}a_{k}(s^{\alpha})^{k} \quad (10)$$

It should be noted that a complex variable function such as Equation (11) is multi–valued. Its domain is a Riemann surface, with a finite number of sheets when $\forall k,\ \alpha_k \in \mathbb{Q}^+$. The q sheets of the Riemann surface, with $\alpha = 1/q$, are determined by

$$F(s) = \sum_{k=0}^{n} a_k s^{\alpha_k} \\ s = |s|e^{j\phi},\quad (2k+1)\pi < \phi < (2k+3)\pi \tag{11}$$

where $k = -1, 0, \ldots, q-2$. Note that only the roots of the principal sheet are meaningful [21].

The stability study of this type of control system is the key of its applicability. The stability analysis is performed, finding an integer index m such as $m\alpha_k$ which is an integer for $k = 0, 1, \ldots, n$. Then, it is possible to define a transformation between the complex plane s and a new complex plane v, where $s = v^m$.

Figure 4 shows that the first Riemann sheet is a slice of the complex plane v, which is limited for a θ range of $\left(-\frac{\pi}{m}, \frac{\pi}{m}\right)$. The line with $\theta = \frac{\pi}{2m}$ splits the first Riemann sheet into two zones. This line is the stability boundary and the zone above the stability boundary is the stability region [21–23].

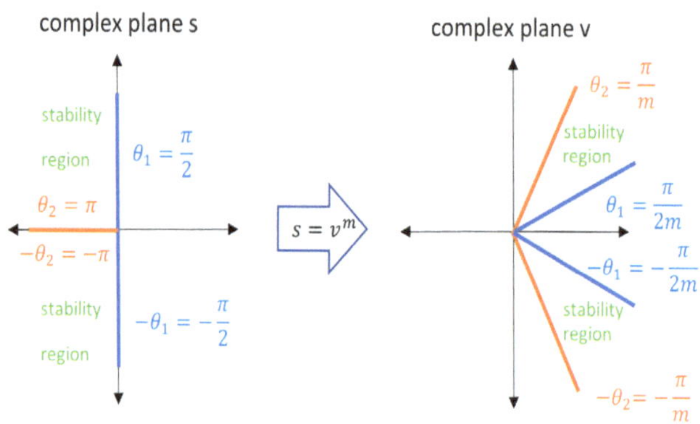

Figure 4. Transformation of the stability region from plane s to plane v.

3.2. $PI^{\alpha}D^{\beta}$ Fractional Controller

In the control theory, the classical PID has been modified by replacing the ordinary integral term for a fractional integral of order α, and by replacing the ordinary derivative term for a fractional derivative of order β. Indeed, Podlubny [24] proposed a generalization of the classical PID controller known as $\mathrm{PI}^{\alpha}D^{\beta}$, with $0 < \alpha, \beta < 1$. The fractional PID has two new tuning parameters (the fractional order of the integral and derivative actions) and it has shown a better performance in both time and frequency domains than its classical counterpart on some applications [25,26].

The $\mathrm{PI}^{\alpha}D^{\beta}$ controller expression in the time domain is shown in Equation (12) where $e(t)$ is the error and $u(t)$ the control input.

$$u(t) = k_p e(t) + k_i D^{-\alpha} e(t) + k_D D^{\beta} e(t) \tag{12}$$

The transfer function of the $\mathrm{PI}^{\alpha}D^{\beta}$ controller is described in Equation (13).

$$C(s) = K_p + \frac{K_i}{s^{\alpha}} + K_d s^{\beta}\,,\quad \alpha,\beta > 0 \tag{13}$$

4. Prototype Description

In order to carry out the controller design, a model of the traction response of the vehicle must be obtained first. For this, a constant voltage has been used as an open-loop input and the readings from the optical encoder coupled to the rear wheels are measured. This relates motor inputs with velocity output in open loop.

With the measured response, model adjustment has been made. Figure 5 shows the measured and model output for the same input. The right part of the figure corresponds to the part in which traction is not exerted and the vehicle stops due to the friction of the wheels with the ground.

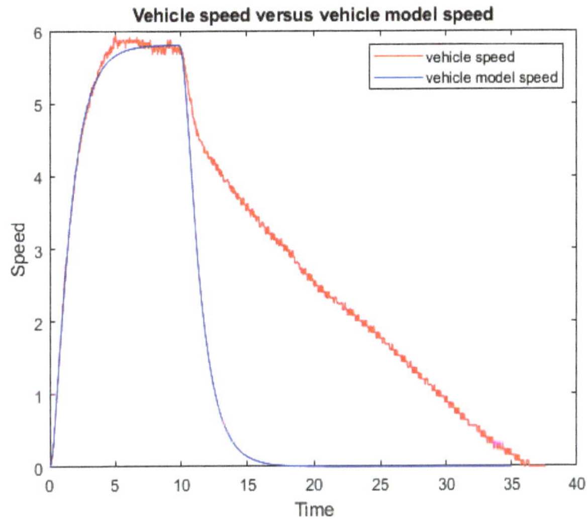

Figure 5. Step response for the electric vehicle.

The adjusted model cart is represented by the state variables described in Equation (14).

$$A = \begin{bmatrix} 0.00 & 1.00 \\ -1.85 & -3.05 \end{bmatrix}; B = \begin{bmatrix} 0.00 \\ 1.85 \end{bmatrix}; C = \begin{bmatrix} 1.00 & 0.00 \end{bmatrix}; D - 0.00 \tag{14}$$

The prototype can also be described by the transfer function of Equation (15).

$$G(s) = \frac{K}{(\tau_1 s + 1)(\tau_2 s + 1)} \tag{15}$$

with $K = 1$, $\tau_1 = 1.20$, $\tau_2 = 0.45$.

5. Fractional Control Application

It should be highlighted that the introduction of fractional terms means that the dynamics of the closed-loop system does not depend on exponentials but on Mittag–Leffler functions described in Equation (16).

$$E_\alpha(z) = \sum_{r=0}^{\infty} \frac{z^r}{\Gamma(1 + \alpha r)}, \quad \alpha > 0 \quad and \quad z \epsilon \mathbb{C} \tag{16}$$

where Γ is the Gamma function. When $\alpha = 1$, the exponential is obtained as a particular case $E_1(z) = e^z$.

An important fact is that, unlike what happens with the product of two exponentials (Mittag–Leffler functions with $\alpha = 1$) which is another exponential function, the product of two Mittag–Leffler functions with $\alpha \neq 1$ is not a Mittag–Leffler function but is obtained by Equation (17).

$$\begin{aligned} E_\alpha(ax)E_\alpha(ay) &= \sum_{r=0}^{\infty} \frac{a^r g_\alpha(r;x,y)}{\Gamma(1+\alpha r)} \\ g_\alpha(r;x,y) &= \sum_{i=0}^{r} \binom{r}{i}_\alpha x^{r-i} y^i \\ \binom{r}{i}_\alpha &= \frac{\Gamma(1+\alpha r)}{\Gamma(1+\alpha i)\Gamma(1+\alpha(r-i))} \end{aligned} \tag{17}$$

It should be noted that, if $\alpha = 1$, this expression reduces to a binomial and the classical expression for the product of exponential is obtained. This effect has multiple consequences, but in this paper, the change in the time scale produced by the Mittag–Leffler functions is particularly interested. Thus, for rapid change signals, the dynamics are much faster than for an exponential, while for slow change signals, the opposite occurs, that is, the dynamics given by the Mittag–Leffler function is much slower than that of an exponential. To show this behavior in a simple way, the Mittag–Leffler functions for the simplest situation, represented in the fractional differential Equation (18), has been chosen.

$$\frac{dy(t)}{dt} + c_1 \frac{d^{(\beta} y(t)}{dt^{(\lambda}} + c_2 I^{(\alpha} y(t) = -y(t) \tag{18}$$

Note that the differential Equation (18) corresponds to a system with no input, and to observe the dynamics, an initial condition other than zero must be chosen. Thus, it was considered $y(0) = 2$.

Figure 6a shows the behavior when the values $c_1 = 1$; $c_2 = 0$ were chosen. Only the dynamics generated by the fractional derivative term is present. Figure 6b shows the dynamic when $c_1 = 0$; $c_2 = 1$ have been chosen as parameters, so only the dynamics generated by the fractional integral term is present. The bandwidth of the controller can be adapted in function of the coefficients β in Figure 6a and α in Figure 6b, although the fractional controller gives more degrees of freedom to adjust the system behavior, changing the time response for the derivative and integral part.

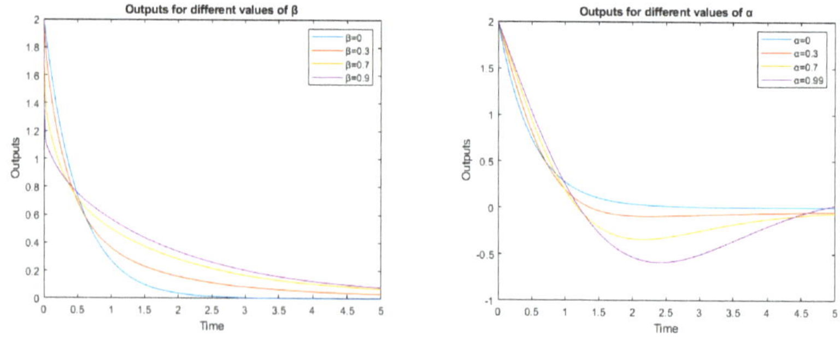

(**a**) Comparison of classical dynamics (β = 0) with fractional derivative dynamics for different values of β.

(**b**) Comparison of classical dynamics (α = 0) with fractional integral dynamics for different values of α.

Figure 6. Behavior of the fractional terms with different parameters.

The standard closed-loop transfer function of error versus reference is shown in Equation (19).

$$\frac{E(s)}{R(s)} = \frac{1}{1 + G(s)C(s)} \tag{19}$$

A controller $C(s)$ a PI$^{\alpha}$ shown in Equation (20) is proposed.

$$C(s) = \left(\frac{K_p s^{\alpha} + K_i}{s^{\alpha}}\right) \tag{20}$$

so the controller system transfer function is shown in Equation (21).

$$\frac{E(s)}{R(s)} = \frac{s^{\alpha}(\tau_1 s + 1)(\tau_2 s + 1)}{s^{\alpha}((\tau_1 s + 1)(\tau_2 s + 1) + KK_p) + KK_i} \tag{21}$$

The objective is to control a golf cart, so the possible commands that the path planning layer can send to the controller are a constant speed reference, and a speed change reference. Step $(l = 1)$ for constant speed and ramp $(l = 2)$ for change in the speed are considered as the possible inputs for the controller systems. The possible input references for the controller are shown in Equation (22).

$$R(s) = \frac{r}{s^l} \tag{22}$$

As is well known, to calculate the stationary error, the final value theorem is applied in Equation (23).

$$e_{stat} = \lim_{s \to 0} sE(s) = \lim_{s \to 0} \frac{rs^{\alpha+1}(\tau_1 s + 1)(\tau_2 s + 1)}{s^{\alpha+l}((\tau_1 s + 1)(\tau_2 s + 1) + KK_p) + KK_i s^l} \tag{23}$$

If the reference is for the step type ($l = 1$), the limit of Equation (23) is shown in Equation (24).

$$e_{stat} = \lim_{s \to 0} \frac{rs^{\alpha}(\tau_1 s + 1)(\tau_2 s + 1)}{s^{\alpha}((\tau_1 s + 1)(\tau_2 s + 1) + KK_p) + KK_i} \tag{24}$$

so, the final stationary error depends on α as shown in Equation (25)

$$\begin{cases} \alpha = 0; & e_{stat} = \frac{r}{K(K_p + K_i)} \\ \alpha > 0; & e_{stat} = 0 \end{cases} \tag{25}$$

For this kind of reference, the classic PI can be used where $\alpha = 1$ and with zero error in the stationary. However, if the reference is ramp type ($l = 2$), where the speed change from an initial value to a final one, the tracking stationary error can be calculated as Equation (26).

$$e_{stat} = \lim_{s \to 0} \frac{r(\tau_1 s + 1)(\tau_2 s + 1)}{s((\tau_1 s + 1)(\tau_2 s + 1) + KK_p) + KK_i s^{1-\alpha}} \tag{26}$$

the final stationary error depends on α as shown in Equation (27).

$$\begin{cases} c0 \leq \alpha < 1; & e_{stat} = \infty \\ \alpha = 1; & e_{stat} = \frac{r}{KK_i} \\ \alpha > 1; & e_{stat} = 0 \end{cases} \tag{27}$$

In this case, the classical integer solution with $\alpha = 2$ obtains a zero stationary error, but it can make the closed-loop system unstable. For this reason, a fractional controller is used to achieve a zero stationary error, and it is necessary to carry out a stability analysis to assure stability. For this, it is considered as a final control transfer function Equation (28).

$$G(s)C(s) = \frac{K}{(\tau_1 s + 1)(\tau_2 s + 1)}\left(\frac{K_p s^{\alpha} + K_i}{s^{\alpha}}\right) \tag{28}$$

and the frequency response must be calculated according to Equation (29).

$$
\begin{aligned}
mag &= 20log_{10}\left(\frac{\sqrt{K_p^2\omega^{2\alpha} + K_i^2 + 2K_pK_i\omega^\alpha cos\left(\frac{\pi}{2}\alpha\right)}}{\omega^\alpha}\right) \\
&+ 20log_{10}\left(\frac{K}{\sqrt{(\tau_1\omega)^2 + 1}\sqrt{(\tau_2\omega)^2 + 1}}\right) \\
fase &= arctag\left(\frac{K_p\omega^\alpha sin\left(\frac{\pi}{2}\alpha\right)}{K_p\omega^\alpha cos\left(\frac{\pi}{2}\alpha\right) + K_i}\right) - \frac{\pi}{2}\alpha - arctag(\tau_1\omega) - arctag(\tau_2\omega)
\end{aligned}
\tag{29}
$$

To also guarantee stability and robustness, the hypotheses described in [21,27] will be used. Phase margin φ_m has typically been used as a measure of stability and robustness. Thus, the phase margin φ_m will be considered to define the desired nominal damping of the system. On the other hand, the crossover frequency ω_{cg} that fixes the desired nominal speed of the response of the system will also be used.

In order to calculate the gain crossover frequency ω_{cg} , the equality defined in Equation (30) must be verified.

$$
|C(j\omega)G(j\omega)|(K_p, K_i, \alpha)|_{\omega=\omega_{cg}} = 1 \tag{30}
$$

This value will depend on the parameters that characterize the controller, that is K_p, K_i, α. At the frequency ω_{cg}, the phase margin φ_m is calculated according to Equation (31).

$$
arg[C(j\omega)G(j\omega)](K_p, K_i, \alpha)|_{\omega=\omega_{cg}} = -\pi + \varphi_m \tag{31}
$$

The two previous conditions by imposing values for ω_{cg} and φ_m are established. Thus, the three parameters of the controller K_p, K_i, α are set as unknowns, a third condition that sets the phase of the open-loop system to be flat at ω_{cg} and consequently to be approximately constant in an interval around ω_{cg} according to Equation (32) is defined. The value obtained for α is fixed greater than 1, a condition which has been previously seen as necessary to achieve zero steady-state error when faced with ramp-type references.

$$
\frac{d(arg[C(j\omega)G(j\omega)])}{d\omega}(K_p, K_i, \alpha)\Big|_{\omega=\omega_{cg}} = 0 \tag{32}
$$

The third condition establishes robustness against gain variations which guarantees robustness locally. The gain range depends on the frequency range at approximately ω_{cg} for which the phase keeps flat. This frequency range will be longer or shorter depending on the controller and the process.

6. Methods Discussion

The path-planning algorithm for the autonomous vehicle is based on a search in a space of the possible movements for the robot [28,29]. The path is divided in primitives; small actions can combine to make complex robot movements. The primitives of the cart include, different steering wheel angles and different displacement speeds. The combination of these primitives can compose any desired movement, and the path-planning algorithm joins the primitives looking for the best path.

The position of the steering wheel can be set accurately using a standard PID controller; however, a standard controller cannot accurately track the desired translation speed generated by the primitive. Focusing on cart movement primitives, 3 different primitives can be can highlighted.

- The cart keeps the actual speed, which is equivalent to a step reference ($l = 1$);
- The cart increases its speed, which is equivalent to a ramp reference ($l = 2$);
- The cart reduces its speed, which is equivalent to a ramp with negative slope ($l = 2$).

Constant speed can be kept by a standard controller with zero stationary error, so it can be assumed that this primitive is correctly followed. However, the primitives of increasing or decreasing speeds are different; this kind of command involves a ramp command, so the speed increases or decreases from one starting speed to a final one. These primitives are very difficult to follow by a standard controller, and the tracking error for this kind of command can be high. If the primitive is not followed correctly, the final cart control will be poor, and the cart performance can be limited.

The fractional control proposed in this paper is a practical solution for the cart speed control. This implementation improves the performance of the whole system, so the primitives are correctly followed, and the movement of the robot is similar to that planned by the path-planning algorithm.

7. Results

As mentioned, the design process consists of setting the values of the crossover frequency ω_{cg} and the phase margin φ_m. Figure 7 shows the results for two values of the crossover frequency ω_{cg}, and the effect it produces on the Bode diagram. In both cases, it can be observed how for the value of the crossover frequency that ω_{cg} the phase reaches a maximum, and therefore, the derivative is zero. This fact corresponds to the robustness condition imposed. However, the overall width of the maximum in the phase diagram decreases as the crossover frequency ω_{cg} increases, and therefore, the overall robustness decreases.

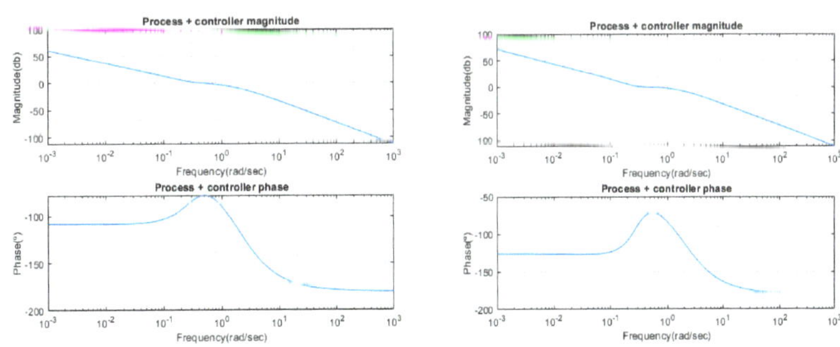

(**a**) Bode diagram for $K_p = 1.2$; $K_i = 0.3$; $\alpha = 1.2$; for which φ_m is 105° and ω_{cg} is 0.5 rad/s. (**b**) Bode diagram for $K_p = 1.4$; $K_i = 0.25$; $\alpha = 1.4$; for which φ_m is 105° and ω_{cg} is 0.4 rad/s.

Figure 7. Bode plot of the system with different parameters.

Figure 8 presents temporal simulations that show how the error is reduced when the values of the K_p and K_i parameters are increased. Note that the vertical scale on which the error is represented changes. On the other hand, as the α value increases, the response becomes faster, but also more oscillating. Furthermore, for all values of K_p and K_i, the closed-loop system becomes unstable when $\alpha = 2$, as shown in Figure 8d.

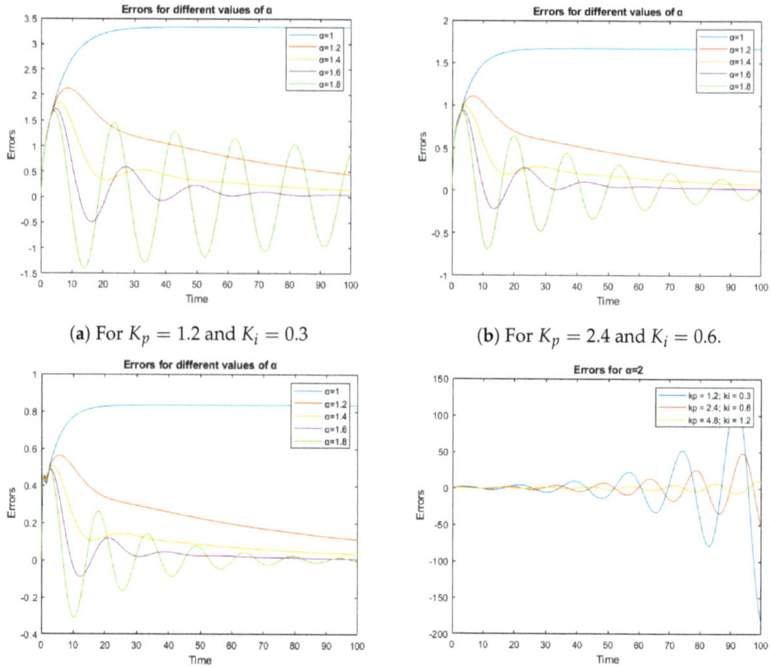

(**a**) For $K_p = 1.2$ and $K_i = 0.3$ (**b**) For $K_p = 2.4$ and $K_i = 0.6$.

(**c**) For $K_p = 4.8$ and $K_i = 1.2$. (**d**) For all values of K_p, K_i above and $\alpha = 2$.

Figure 8. Tracking errors are shown for a ramp reference in different situations.

Table 1 shows a stability analysis for some representative cases presented in Figure 8.

Table 1. Poles in the first sheet of the Riemann surface for some representative cases presented in the temporal simulations. The case of the last row corresponds to a situation where $\alpha = 2.2$, which is included to illustrate the presence of poles in the instability region.

Parameters	m	Poles in Stability Region	Poles in Instability Region
$K_p = 1.2$ $K_i = 0.3$; $\alpha = 1.2$	5	1.0059 + 0.5396i	———
		1.0059 − 0.5396i	———
		0.6407 + 0.3570i	———
		0.6407 − 0.3570i	———
$K_p = 2.4$; $K_i = 0.6$; $\alpha = 1.4$	5	1.0768 + 0.5192i	———
		1.0768 − 0.5192i	———
		0.7177 + 0.3305i	———
		0.7177 − 0.3305i	———
$K_p = 4.8$; $K_i = 1.2$; $\alpha = 1.8$	5	1.1590 + 0.5089i	———
		1.1590 − 0.5089i	———
		0.7945 + 0.2773i	———
		0.7945 − 0.2773i	———
$K_p = 4.8$; $K_i = 1.2$; $\alpha = 2$	1	−1.5566 + 2.8745i	0.0302 + 0.4543i
		−1.5566 − 2.8745i	0.0302 − 0.4543i
$K_p = 1.2$; $K_i = 0.3$; $\alpha = 2.2$	5	1.0213 + 0.5399i	0.8001 + 0.2129i
		1.0213 − 0.5399i	0.8001 − 0.2129i

In order to evaluate the proposed fractional PI controller, a series of experiments were conducted involving an electrical vehicle following different movement primitives. The goal of the vehicle was to maintain the desired speed with the smallest error. To facilitate this, the vehicle had to control its own power according to the path. The command can change a lot for the same speed, depending on the slope of the road, the pavement roughness, the battery level etc. The experiments were conducted using Simulink with the Real-Time Workshop toolbox, and the vehicle was is obtained from the odometer sensor. The set point for the Simulink model, which included the fractional controller, was the movement primitive generated by the path planning module, and the control action was transmitted to the vehicle's control hardware via a serial protocol. The tests were carried out under different slope, pavement, and battery conditions. The objective of this paper was to improve the longitudinal controller for an autonomous vehicle. To measure the performance of the reference tracking, the difference between the reference velocity and the actual velocity of the prototype is used as a metric. If the reference tracking is good, the vehicle will be able to better follow the high-level primitives. This means that the maneuverability will increase, reducing the tracking error. High level layers will correct the control error introduced by system control, but if we reduce this error, the performance of the whole prototype will increase.

Figure 9 shows the cart speed error during two experiments following different primitives. From 0 to 10 s, the cart receives a movement primitive of acceleration, and should follow a speed ramp. The standard PID controller cannot follow the reference and it maintains a constant error, however, the fractional controller reduces the error over time. From 10 to 25 s, a constant speed primitive is set. The traditional integer controller significantly reduces the error, but the fractional controller reduces the error almost to 0. For the two real tests presented, the values K_p and K_i are maintained as fixed, while the value of alpha has been changed. The results obtained correspond to what was predicted by the simulations, where in Figure 9b, the error is reduced when α is increased and the system remains stable.

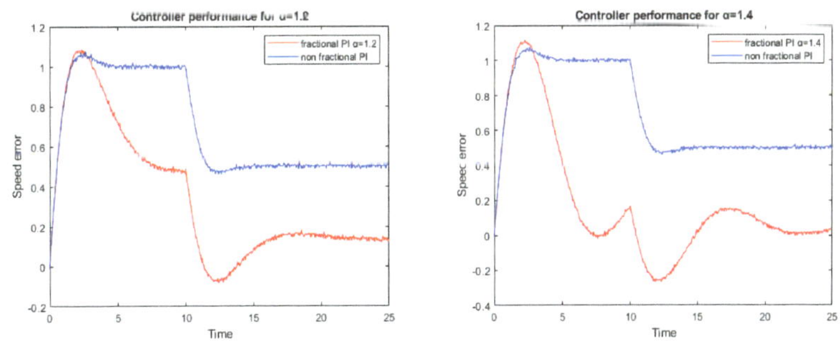

(**a**) (**b**)

Figure 9. Tracking errors for a fixed reference of 2.5 m with $K_p = 1.2$; $K_i = 1$. (**a**) For a PI versus $PI^{1.2}$; and (**b**) For a PI versus $PI^{1.4}$.

Figure 10 shows the ratio between the control command effort of a fractional strategy versus integer strategy. The command is bigger for the fractional controller, and when α is increased, the ratio also grows. This is the expected behavior, so the error using a fractional controllers is also smaller.

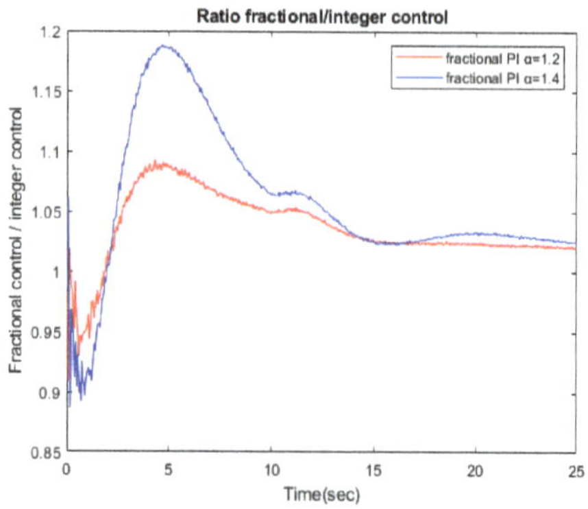

Figure 10. Ratio fractional control/integer control for $PI^{1.2}$ versus $PI^{1.4}$.

8. Implementation of Fractional Module s^α

The values that have been used for α as a fractional order coefficient in the real tests have been $\alpha = 1.2$ and $\alpha = 1.4$. In the actual implementation, two fractional modules have been used with $\alpha = 0.5$ and $\alpha = 0.7$. To obtain the value $\alpha = 1.2$, a 0.5 module and a 0.7 module were connected in series, while two 0.7 modules were connected in series to obtain the value $\alpha = 1.4$.

The Matsuda approximation has been used to obtain the two modules. First, a frequency range is chosen between a lower frequency ω_l and a higher frequency ω_h where the approximation is valid. It is also necessary to give the degree n of the approximation, which will determine $N = 2n$. Then, $N + 1$ logarithmically distributed frequencies are calculated in the range of $[\omega_l, \omega_h]$, Equation (33).

$$\omega_k = \omega_l \left(\frac{\omega_h}{\omega_l} \right)^{\frac{k}{N}} \quad k = 0, \ldots, N \tag{33}$$

and $N + 1$ coefficients are defined for each frequency ω_k which we will call $d_i(\omega_k)$ Equation (34).

$$d_i(\omega_k) = \begin{cases} (\omega_k)^\alpha & i = 0; k = 0, \ldots, N \\ \frac{\omega_k - \omega_{i-1}}{d_{i-1}(\omega_k) - d_{i-1}(\omega_{i-1})} & i = 1, \ldots, N; k = 0, \ldots, N \end{cases} \tag{34}$$

Note that these coefficients must be calculated recursively. From the $d_i(\omega_k)$, we will define c_k as Equation (35) shows.

$$c_k = d_k(\omega_k) = \begin{cases} \omega_0^\alpha & k = 0 \\ \frac{\omega_k - \omega_{k-1}}{d_{k-1}(\omega_k) - d_{k-1}(\omega_{k-1})} & k = 1, \ldots, N \end{cases} \tag{35}$$

With the c_k values, it is possible to write the following truncated continued fraction expansion that approximates s^α, as in Equation (36).

$$s^\alpha \cong c_0 + \cfrac{s - \omega_0}{c_1 + \cfrac{s - \omega_1}{c_2 + \cfrac{s - \omega_2}{c_3 + \cfrac{s - \omega_3}{c_4 + \cdots}}}} \tag{36}$$

It is usual to write Equation (36) in a compact way by using the following notation, Equation (37).

$$s^\alpha \cong c_0 + \frac{s - \omega_0}{c_1 +} \frac{s - \omega_1}{c_2 +} \cdots \frac{s - \omega_{N-1}}{c_N} \tag{37}$$

Note that since N is even, the degree of the numerator and denominator coincide. If N is odd, the degree of the numerator is one greater than the degree of the denominator.

For this reason, $N = 2n$ was chosen. Table 2 shows the values of c_k for the ninth-order approximation used for the modules with $\alpha = 0.5$ and with $\alpha = 0.7$.

Performing operations on the above equation can be easily reduced to a quotient of polynomials in s as Equation shown in (38).

$$s^{\alpha} \cong \frac{N(s)}{D(s)} = \frac{\sum_{J=0}^{n} b_j s^j}{\sum_{J=0}^{n} a_j s^j} \tag{38}$$

In this case, a ninth-order approximation, i.e., $n = 9$ is chosen. Table 3 shows the values of a_k and b_k for the ninth-order approximation used for the modules with $\alpha = 0.5$ and with $\alpha = 0.7$. In Figure 11, the frequency representations of Matsuda approximations of different orders are shown, proving that the ninth order is a good approximation.

Table 2. Coefficients of the continued fraction expansion for the ninth-order Matsuda approximation.

A	Coefficients C
0.5	$C_0 = 10^{-3}$; $C_1 = 2.5647^{-3}$; $C_2 = 4.0132^{-3}$ $C_3 = 6.2796^{-3}$; $C_4 = 9.8260^{-3}$; $C_5 = 1.5375^{-2}$ $C_6 = 2.4058^{-2}$; $C_7 = 3.7645^{-2}$; $C_8 = 5.8905^{-2}$ $C_9 = 9.2172^{-2}$; $C_{10} = 1.4423^{-1}$; $C_{11} = 2.2568^{-1}$ $C_{12} = 3.5313^{-1}$; $C_{13} = 5.5256^{-1}$; $C_{14} = 8.6461^{-1}$ $C_{15} = 1.3529$; $C_{16} = 2.1170$; $C_{17} = 3.3125$ $C_{18} = 5.1832$
0.7	$C_0 = 6.3096^{-5}$; $C_1 = 2.6337^{-2}$; $C_2 = 6.7040^{-4}$ $C_3 = 2.9510^{-2}$; $C_4 = 2.6694^{-3}$; $C_5 = 4.6540^{-2}$ $C_6 = 9.7623^{-3}$; $C_7 = 7.7435^{-2}$; $C_8 = 3.4763^{-2}$ $C_9 = 1.3109^{-1}$; $C_{10} = 1.2257^{-1}$; $C_{11} = 2.2337^{-1}$ $C_{12} = 4.3051^{-1}$; $C_{13} = 3.8159^{-1}$; $C_{14} = 1.5097$ $C_{15} = 6.5256^{-1}$; $C_{16} = 5.2909$; $C_{17} = 1.1164$ $C_{18} = 18.537$

Table 3. Coefficients of numerator and denominator of the ninth-order Matsuda approximation.

α	Numerator $N(s)$	Denominator $D(s)$
0.5	$b_0 = 8.76$ $b_1 = 52.260$ $b_2 = 30.508$ $b_3 = 2.4739$ $b_4 = 3.1015^{-2}$ $b_5 = 6.2017^{-5}$ $b_6 = 1.9589^{-8}$ $b_7 = 9.1993^{-13}$ $b_8 = 5.3200^{-18}$ $b_9 = 1.7783^{-24}$	$a_0 = 1$ $a_1 = 29.916$ $a_2 = 51.732$ $a_3 = 11.016$ $a_4 = 3.4874^{-1}$ $a_5 = 1.7441^{-3}$ $a_6 = 1.3912^{-6}$ $a_7 = 1.7156^{-1}$ $a_8 = 2.9388^{-15}$ $a_9 = 4.9261^{-21}$
0.7	$b_0 = 25.939$ $b_1 = 1.2228^2$ $b_2 = 58.519$ $b_3 = 3.9356$ $b_4 = 4.1069^{-2}$ $b_5 = 6.8328^{-5}$ $b_6 = 1.7874^{-8}$ $b_7 = 6.8520^{-13}$ $b_8 = 3.0830^{-18}$ $b_9 = 5.6234^{-25}$	$a_0 = 1$; $a_1 = 54.825$ $a_2 = 1.2185^2$ $a_3 = 31.784$ $a_4 = 1.2151$ $a_5 = 7.3032^{-3}$ $a_6 = 6.9986^{-6}$ $a_7 = 1.0406^{-9}$ $a_8 = 2.1745^{-14}$ $a_9 = 4.6127^{-20}$

To obtain the discrete version, we used Tustin's discretization, as in Equation (39).

$$s \cong \frac{2}{T}\left(\frac{1-z^{-1}}{1+z^{-1}}\right) \tag{39}$$

where z^{-1} is the delay operator.

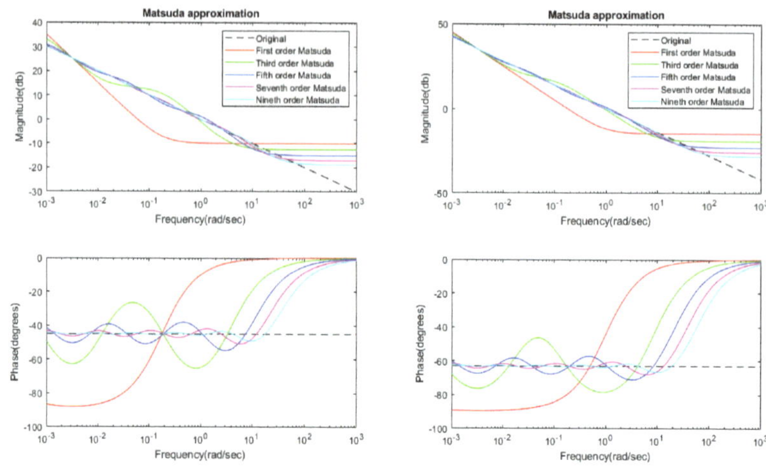

(a) (b)

Figure 11. Magnitude and phase responses of Matsuda approximation of different orders: (**a**) for $s^{0.5}$; and (**b**) for $s^{0.7}$.

The approximation between the actual module in function of the N coefficient is shown in Figure 11, where both modules with $\alpha = 0.5$ and $\alpha = 0.7$ and its adjustment in function of the approximation degree N are shown. The ninth-order approximation follows in the frequency range the behavior of the fractional order controller with a negligible error. The computation cost of the implementation of these modules is also very small.

9. Discussion

As the results section shows, the use of fractional-order controllers represents a clear improvement in system control. When tracking control primitives for an autonomous vehicle, it is able to track them with less error than traditional controllers. Specifically, when the received command is a ramp, which is equivalent to a speed change in a certain slope, the fractional controller is capable of following it with an error in the stationary state of 0.

To achieve an equivalent performance using traditional non-fractional systems, it is necessary to use a double PID, however, this compromises the stability of the system. The use, as has been demonstrated in previous sections, of a PID with integral index $\alpha > 1$ allows obtaining a stationary error 0, but guarantees stability.

The tests carried out in simulation demonstrate that bandwidth and gain adjustment can be carried out with this type of controller. We also check how the index of the integral part of the PID affects the stability of the system, ensuring a stable value with correct tracking and a low stationary error with a coefficient of $\alpha = 1.4$. This demonstrates the better performance of this type of controller compared to the traditional ones.

The tests carried out on the real prototype confirm these results, with a much lower primitive tracking error than the previously used PID controller. The difference in computation time and complexity are clearly compensated thanks to the better performance of the overall system.

Using this type of controller, a more reliable autonomous vehicle system is obtained, capable of better following trajectories and performing more precise maneuvers, thus facilitates the control of high-level systems.

10. Conclusions

The low-level controller of an autonomous vehicle can make the difference in the performance of its activity. In this case, the analysis and implementation of the traction motor control for an autonomous cart is presented. A traditional PID control generates stationary output errors in the controller variable, but it is not valid for tracking speed changes, so a solution looking for a better tracking performance is presented.

In this article, a fractional PI^{α} controller, with a parameter $\alpha > 1$ has been proposed for the speed ramp tracking problem of an electric car. It must be taken into account that the approach normally used in the literature considers fractional orders within the interval $(0, 1]$.

Several simulations have been carried out that allowed demonstrating the better performance of the PI^{α} controller, as well as an implementation in the electric vehicle that showed a remarkable reduction in the error.

The controller was applied to an autonomous electric cart, improving the low-level control performance and obtaining a better path tracking. The ability to follow more closely follow the trajectory facilitates the high-level tasks. This controller facilitates navigation in narrow areas and with multiple obstacles.

Author Contributions: Methodology, J.T.; Software, D.A.; Validation, D.A.; Investigation, D.A., L.A., and B.F.; Writing—original draft, J.T. and L.A.; Writing—review and editing, B.F. and J.T.; Supervision, J.T. All authors have read and agreed to the published version of the manuscript.

Funding: Spanish Ministry of Science and Technology under the SIRTAPE project DPI2017-90002-R.

Data Availability Statement: Not applicable.

Acknowledgments: The authors gratefully acknowledge the contribution from the Spanish Ministry of Science and Technology under the SIRTAPE project DPI2017-90002-R. Foundation.

Conflicts of Interest: The authors declare no conflict of interest.

References

1. Toledo, J.; Piñeiro, J.D.; Arnay, R.; Acosta, D.; Acosta, L. Improving Odometric Accuracy for an Autonomous Electric Cart. *Sensors* **2018**, *18*, 200. [CrossRef]
2. Perea-Strom, D.; Morell, A.; Toledo, J.; Acosta, L. GNSS Integration in the Localization System of an Autonomous Vehicle Based on Particle Weighting. *IEEE Sen. J.* **2020**, *20*, 3314–3323. [CrossRef]
3. Perea, D.; Hernández-Aceituno, J.; Morell, A.; Toledo, J.; Hamilton, A.; Acosta, L. MCL with sensor fusion based on a weighting mechanism versus a particle generation approach. In Proceedings of the 16th International IEEE Conference on Intelligent Transportation Systems (ITSC 2013), The Hague, The Netherlands, 6–9 October 2013 ; pp. 166–171. [CrossRef]
4. Morales, N.; Toledo, J.; Acosta, L. Generating automatic road network definition files for unstructured areas using a multiclass support vector machine. *Inf. Sci.* **2016**, *329*, 105–124. [CrossRef]
5. Morales, N.; Toledo, J.; Acosta, L. Path planning using a Multiclass Support Vector Machine. *Appl. Soft Comput.* **2016**, *43*, 498–509. [CrossRef]
6. Morales, N.; Toledo, J.T.; Acosta, L.; Arnay, R. Real-time adaptive obstacle detection based on an image database. *Comput. Vis. Image Underst.* **2011**, *115*, 1273–1287. [CrossRef]
7. Morales, N.; Toledo, J.; Acosta, L.; Sánchez-Medina, J. A Combined Voxel and Particle Filter-Based Approach for Fast Obstacle Detection and Tracking in Automotive Applications. *IEEE Trans. Intell. Transp. Syst.* **2017**, *18*, 1824–1834. [CrossRef]
8. Morales, N.; Morell, A.; Toledo, J.; Acosta, L. Fast Object Motion Estimation Based on Dynamic Stixels. *Sensors* **2016**, *16*, 1182. [CrossRef]
9. Hernández-Aceituno, J.; Arnay, R.; Toledo, J.; Acosta, L. Using Kinect on an Autonomous Vehicle for Outdoors Obstacle Detection. *IEEE Sen. J.* **2016**, *16*, 3603–3610. [CrossRef]
10. Barbosa, R.S.; Jesus, I.S. Special Issue on Algorithms for PID Controllers 2021. *Algorithms* **2023**, *16*, 35. [CrossRef]
11. Kilbas, A.; Srivastava, H.; Trujillo, J. *Theory and Applications of Fractinal Differential Equations*; Elsevier: Amsterdam, The Netherlands, 2006; Volume 204. [CrossRef]

12. Machado, J.T.; Kiryakova, V.; Mainardi, F. Recent history of fractional calculus. *Commun. Nonlinear Sci. Numer. Simul.* **2011**, *16*, 1140–1153. [CrossRef]
13. Podlubny, I. *Fractional Differential Equations: An Introduction to Fractional Derivatives, Fractional Differential Equations, to Methods of Their Solution and Some of Their Applications*; Mathematics in Science and Engineering; Academic Press: London, UK, 1999.
14. Metzler, R.; Klafter, J. The random walk's guide to anomalous diffusion: A fractional dynamics approach. *Phys. Rep.* **2000**, *339*, 1–77. [CrossRef]
15. Lopes, A.M.; Tenreiro Machado, J.A. Fractional-Order Sensing and Control: Embedding the Nonlinear Dynamics of Robot Manipulators into the Multidimensional Scaling Method. *Sensors* **2021**, *21*, 7736. [CrossRef] [PubMed]
16. Muresan, C.I.; Birs, I.R.; Dulf, E.H.; Copot, D.; Miclea, L. A Review of Recent Advances in Fractional-Order Sensing and Filtering Techniques. *Sensors* **2021**, *21*, 5920. [CrossRef] [PubMed]
17. Paden, B.; Cáp, M.; Yong, S.Z.; Yershov, D.S.; Frazzoli, E. A Survey of Motion Planning and Control Techniques for Self-Driving Urban Vehicles. *IEEE Trans. Intell. Veh.* **2016**, *1*, 33–55. [CrossRef]
18. Joseph, L.M.A. *Autonomous Driving and Advanced Driver-Assistance Systems (ADAS): Applications, Development, Legal Issues, and Testing*; CRC Press: Boca Raton, FL, USA, 2021. [CrossRef]
19. Kebbati, Y.; Ait-Oufroukh, N.; Vigneron, V.; Ichalal, D.; Gruyer, D. Optimized self-adaptive PID speed control for autonomous vehicles. In Proceedings of the 2021 26th International Conference on Automation and Computing (ICAC), Portsmouth, UK, 2–4 September 2021; pp. 1–6. [CrossRef]
20. Naranjo, J.; Jiménez, F.; Gómez, O.; Zato, J. Low Level Control Layer Definition for Autonomous Vehicles Based on Fuzzy Logic. *Intell. Autom. Soft Comput.* **2012**, *18*, 333–348. [CrossRef]
21. Monje, C.; Chen, Y.; Vinagre, B.; Xue, D.; Feliu, V. *Fractional Order Systems and Control—Fundamentals and Applications*; Springer: Cham, Switzerland, 2010. [CrossRef]
22. Alagoz, B.B. Hurwitz stability analysis of fractional order LTI systems according to principal characteristic equations. *ISA Trans.* **2017**, *70*, 7–15. [CrossRef]
23. Senol, B.; Ates, A.; Baykant Alagoz, B.; Yeroglu, C. A numerical investigation for robust stability of fractional-order uncertain systems. *ISA Trans.* **2014**, *53*, 189–198. [CrossRef]
24. Podlubny, I. Fractional-order systems and PI/sup /spl lambda//D/sup /spl mu//-controllers. *IEEE Trans. Autom. Control* **1999**, *44*, 208–214. [CrossRef]
25. Chen, Y.; Petras, I.; Xue, D. Fractional order control—A tutorial. In Proceedings of the 2009 American Control Conference, St. Louis, MO, USA, 10–12 June 2009; pp. 1397–1411. [CrossRef]
26. Wang, N.; Wang, J.; Li, Z.; Tang, X.; Hou, D. Fractional-Order PID Control Strategy on Hydraulic-Loading System of Typical Electromechanical Platform. *Sensors* **2018**, *18*, 3024. [CrossRef]
27. Monje, C.A.; Vinagre, B.M.; Feliu, V.; Chen, Y.Q. Tuning and auto-tuning of fractional order controllers for industry applications. *Control Eng. Pract.* **2008**, *16*, 798–812. [CrossRef]
28. Arnay, R.; Morales, N.; Morell, A.; Hernandez-Aceituno, J.; Perea, D.; Toledo, J.T.; Hamilton, A.; Sanchez-Medina, J.J.; Acosta, L. Safe and Reliable Path Planning for the Autonomous Vehicle Verdino. *IEEE Intell. Transp. Syst. Mag.* **2016**, *8*, 22–32. [CrossRef]
29. Morales, N.; Arnay, R.; Toledo, J.; Morell, A.; Acosta, L. Safe and reliable navigation in crowded unstructured pedestrian areas. *Eng. Appl. Artif. Intell.* **2016**, *49*, 74–87. [CrossRef]

sensors

MDPI

Article

Self Calibration of a Sonar–Vision System for Underwater Vehicles: A New Method and a Dataset

Nicolas Pecheux, Vincent Creuze *, Frédéric Comby and Olivier Tempier

LIRMM, Univ. Montpellier, CNRS, Montpellier, France
* Correspondence: vincent.creuze@lirmm.fr

Abstract: Monocular cameras and multibeam imaging sonars are common sensors of Unmanned Underwater Vehicles (UUV). In this paper, we propose a new method for calibrating a hybrid sonar–vision system. This method is based on motion comparisons between both images and allows us to compute the transformation matrix between the camera and the sonar and to estimate the camera's focal length. The main advantage of our method lies in performing the calibration without any specific calibration pattern, while most other existing methods use physical targets. In this paper, we also propose a new sonar–vision dataset and use it to prove the validity of our calibration method.

Keywords: calibration; multibeam imaging sonar; monocular camera; dataset

Citation: Pecheux, N.; Creuze, V.; Comby, F.; Tempier, O. Self Calibration of a Sonar–Vision System for Underwater Vehicles: A New Method and a Dataset. *Sensors* **2023**, *23*, 1700. https://doi.org/10.3390/s23031700

Academic Editors: David Cheneler and Stephen Monk

Received: 10 January 2023
Revised: 31 January 2023
Accepted: 2 February 2023
Published: 3 February 2023

1. Introduction

Remotely Operated Vehicles (ROVs) are used for a wide range of underwater operations either physically impossible or technically complicated for divers, from inspections of industrial offshore structures to scientific deep-sea explorations. Usually, ROVs are equipped with at least one monocular video camera to pilot the ROV and to observe its surroundings. For more autonomous robots, this camera can be used for navigation by determining the robot's position from the observed objects and features, for obstacle avoidance by tracking objects in the camera and determining the risk and time for the robot to encounter them, or even for autonomous docking using visual targets. Another example of such applications is station-keeping, which gives the ROV increased stability when standing still during inspections. This can be achieved by using homography to estimate the movement of the robot and then compensate for it [1]. Furthermore, object detection algorithms can help guide the pilot to its goal. This can be achieved using object segmentation, as presented in [2], by combining multiple visual cues (gradient, colour disparity, pixel intensity, etc.). However, all these methods are limited by optical cameras' sensitivity to low-light conditions, colour degradation, turbidity, and noise. To cope with these problems, many techniques have been proposed to enhance underwater images, as presented in the survey [3,4]. There are also solutions to denoise underwater images using a variation of the wavelet transform [5]. Some of these algorithms are quite simple and can even be used for low-power platforms [6], such as for Autonomous Underwater Vehicles (AUVs).

In addition to the camera, an imaging sonar may be added for specific operations (inspections of underwater structures, target localisation, etc.). An example of an ROV equipped with such sensors is shown in Figure 1. The imaging sonar allows to detect objects at a larger range or under poor visibility conditions. Moreover, sonars allow to obtain information regarding dimension and distances, which is not the case of monocular cameras. These advantages of the sonars over the cameras are counterbalanced by two limitations: a slower frame rate, due to the sound propagation; a poorer resolution, due to the limited number of acoustic beams and the quite low frequency of the emitted acoustic waves (typically less than 1.2 MHz). There are several classes of sonars. In this paper, we will only consider multibeam imaging sonars, often called "acoustic cameras". Unlike single-beam scanning sonars, multibeam imaging sonars use several beams at the same time (typically 256), allowing a much higher update rate (typically 10 to 30 fps, depending on the range). The acoustic beams have a quite large vertical aperture (typically 20°) while having a narrow horizontal width (less than 1°). Thanks to their long range and their ability to work in turbid waters, sonars are very useful for underwater object or landmark detection and recognition. In [7], this is achieved by processing the beams composing the sonar image and by looking for combined bright spots and acoustic shadows in the acoustic image; then, comparing the sizes of the detected bright and shadow zones to a known template of the landmark leads to its recognition. These landmarks are then used for the localisation of Autonomous Underwater Vehicles (AUVs). Another use of sonar imaging is marine life detection for ecological surveys [8] using machine learning algorithms such as k-nearest neighbours, support vector machines, and random forests. To classify them, the detected targets are described using many parameters, such as their size, intensity, speed, time in the image, or time of the observation. Sonars can also be used to detect dangerous objects. For example, in [9], the authors used a CNN-based approach to identify underwater mines lying on the seafloor. In pipeline following and inspection, sonars are also often employed. A recent approach used a constant false alarm algorithm to extract the pipeline in spite of the noise in the sonar image [10].

Figure 1. The Hilarion ROV of the DRASSM equipped with an acoustic camera (Oculus 1200M multibeam sonar from BluePrint Subsea) and a monocular video camera (Sony ER8530).

Combining the sonar with a monocular camera allows to benefit from both sensors' advantages: long range sensing, distance and dimension measurements, robustness to turbidity in the sonar image, easier identification of objects in the optical images, etc. Figure 2 shows acquisitions of the same scene by a video camera and sonar. However, this requires knowledge of the transformation matrix between the two sensors, thus allowing to match pixels of the sonar image with pixels of the optical image. Furthermore, the knowledge of this matrix allows to improve piloting experience. Indeed, areas of the optical image can be highlighted where obstacles or objects of interest are detected by the sonar.

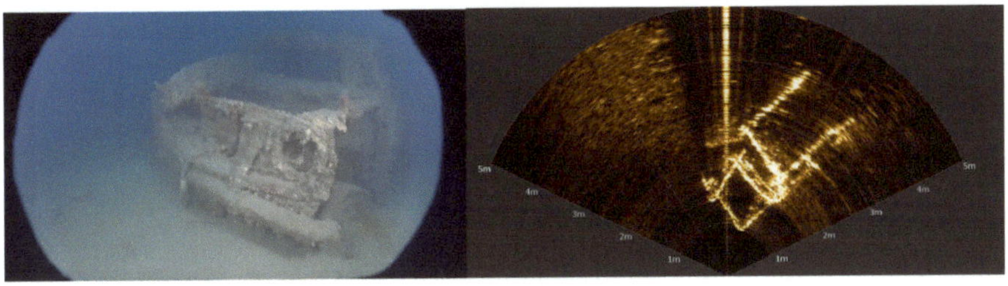

Figure 2. Optical image of a car acquired by a UHD (4K) camera (on the left) and the corresponding acoustic image obtained by a multibeam imaging sonar (on the right). One can observe the bright lines corresponding to the edges of the wreck.

In this paper, we propose to study an acquisition system associating a monocular camera and a multibeam imaging sonar. As mentioned above, to adequately exploit such a system, it is necessary to perform a calibration, i.e., to determine the existing transformation between the two sensors. Most existing calibration methods rely on purpose-made physical calibration patterns, which contain both optical patterns (such as checkers or aruco markers) and acoustically detectable patterns (made of materials with different textures or different backscattering properties). For example, in [11,12], the authors use a grid where the edges create bright lines, which intersect at corners, creating eligible feature points in both acoustic and optical images. Corners in both images are associated to their known positions in the grid. With enough points, it is then possible to find the transformation matrix linking the two sets of points by using the Levenberg–Marquardt algorithm. This process is quite similar to the one used for the calibration of standard optical stereovision systems. More recently, a paper proposed to use patterns such as aruco markers with metal rods [13] or bolts [14], allowing differences in sound reflection. These differences lead to bright spots where the material is highly reflective and dark spots where it is not. This creates patterns visible in both the optical and the acoustic images. Another approach consists in using a known 3D object, including an optical pattern such as a chessboard pattern [15]. By comparing the acoustic view of the object and the image of the optical pattern, it is possible to find the transformation between the two sensors.

There are other hybrid sensors' associations for underwater perception. One of these methods uses a stereo camera placed alongside a sonar [16]. This adds the distance information to the visual data, thus allowing to match them with the distances from the sonar image. Another method, combining a monocular camera and an acoustic sensor, uses an echosounder instead of a sonar [17]. While not giving an acoustic image of the scene, this gives a distance map that can be overlapped with the optical image. Additionally, an original idea came from using a multidirectional microphone array [18]. This kind of sensor proposes the idea of using multiple microphones placed at various positions. This could be advantageous when the payload of the vehicle is limited.

As seen previously, most calibration methods between a sonar and an optical camera rely on a specific calibration object with features that can be detected and matched in both the acoustic and the optical images. These approaches are efficient, but their use at sea may be limited by some difficulties, such as the sea state or the requirement of divers and the time needed to immerse the object and to calibrate the system, especially from large vessels or offshore structures. A pre-calibration in a pool or in a harbour is not always enough, as the ROV maintenance teams often modify the system on the field to adapt it to various types of missions (pipeline inspection, hull inspection, manipulation, etc.) or simply because the maintenance implies frequent disassembly and reassembly of the robot, thus inducing small changes in the relative positions of the sensors. In this context and at the moment, we have found only one team who proposed a targetless calibration method. This approach is based on natural contours [19] and uses the fact that not only can edges be easily detected in optical images but they also create detectable bright lines in the acoustic images. Using these contours, the article proposes to match segmented images of the two sensors in order to perform the calibration. As for target-based approaches, this method may be limited by the field constraints because many underwater environments do not offer the adequate natural shapes and textures (i.e., allowing easy matching of optical and acoustic contours).

In this paper, we propose a new calibration method, using only very common underwater elements (rock, underwater structures, wrecks, etc.) without requiring any specific shape. Thus, our self-calibration technique is dedicated to hybrid sensing systems composed of a monocular camera and a multibeam imaging sonar. Unlike most existing methods, this technique does not require any artificial calibration pattern, and uses only elements of the observed scene without necessitating any knowledge about them. Our method first extracts acoustic feature points in the sonar image and tracks them with optical flow to compute their motion in two consecutive sonar frames. Then, a comprehensive search algorithm estimates the best transformation matrix by projecting these motions onto the optical image and by comparing the motions predicted from the acoustic image with the motion actually observed in the optical images. The proposed method also allows to estimate the focal length of the optical camera and, thus, does not require any prior knowledge of its intrinsic matrix. This method is validated by experiments on field data gathered during archaeological surveys. The results presented highlight the ability of the method to estimate the focal length of the monocular camera, as well as the transformation matrix between the two sensors. Another contribution of this paper is the introduction of a dataset. This dataset includes combined optical and sonar images acquired on archaeological underwater sites in the Mediterranean sea. The paper is organised as follows. In Section 2, we introduce the sensors' models and the notations. Section 3 presents the calibration method. Then, the experimental performances of our algorithm are evaluated on field data and the results are presented and analysed in Section 4. This chapter also presents the content of the public dataset accompanying this paper. The conclusion gives some perspectives on future works and usage of this method.

2. Problem Statement, Notations and Models

2.1. Problem Statement

We consider two sensors: one monocular optical camera and one acoustic camera. Each variable associated with the monocular camera (respectively, the acoustic camera) will be referenced with a subscript o (respectively, s). Let us define $\mathfrak{R}_s$ as the frame associated to the sonar and $\mathfrak{R}_o$ as the frame associated to the optical camera as shown in Figure 3. Then, a 3D point is denoted $P_s : (X_s, Y_s, Z_s)^\intercal$ in the sonar frame, while the same point is denoted $P_o : (X_o, Y_o, Z_o)^\intercal$ in the optical frame. The transformation between the two frames $\mathfrak{R}_o$ and $\mathfrak{R}_s$ is composed of a 3D rotation matrix R_s^o and a translation matrix T_s^o.

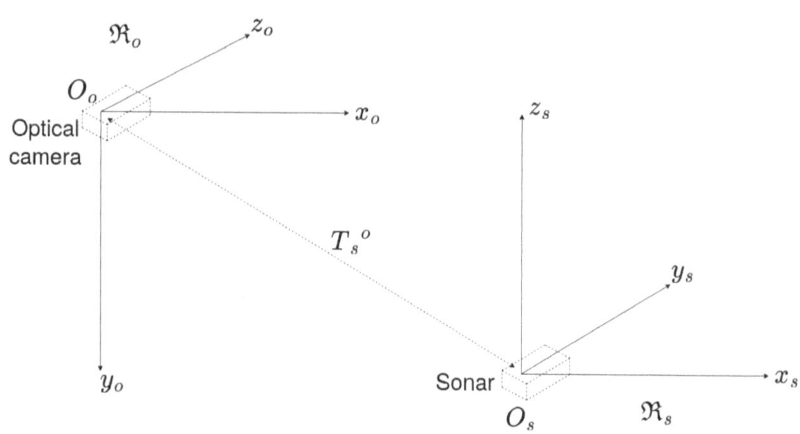

Figure 3. The camera frame $\mathfrak{R}_o$, the sonar frame $\mathfrak{R}_s$, and the translation vector T_s^o.

The rotation matrix R_s^o is defined by three angles α, β, and γ around the axes x_s, y_s, and z_s, respectively. Using the Euler angles with the (z, y, x) convention, the rotation matrix is defined by Equation (1).

$$R_s^o = R_x(\alpha)R_y(\beta)R_z(\gamma) = \begin{pmatrix} 1 & 0 & 0 \\ 0 & cos\alpha & -sin\alpha \\ 0 & sin\alpha & cos\alpha \end{pmatrix} \begin{pmatrix} cos\beta & 0 & sin\beta \\ 0 & 1 & 0 \\ -sin\beta & 0 & cos\beta \end{pmatrix} \begin{pmatrix} cos\gamma & -sin\gamma & 0 \\ sin\gamma & cos\gamma & 0 \\ 0 & 0 & 1 \end{pmatrix} \quad (1)$$

The translation vector $T_s^o = (t_x, t_y, t_z)^\intercal$ has three components, one for each translation along the axes of the sonar frame. Then, a 3D point $P_s : (X_s, Y_s, Z_s)^\intercal$ in the sonar frame can be expressed in the camera frame using Equation (2).

$$P_o = R_s^o P_s + T_s^o \quad (2)$$

where $P_o : (X_o, Y_o, Z_o)^\intercal$ are the coordinates of the 3D point P_o in $\Re_o$, and R_s^o and T_s^o have been defined above and are the elements that we want to estimate through our calibration method.

2.2. Monocular Camera's Model

This section details the camera model used to project a 3D point expressed in R_o into the 2D image frame. Using the well-known pin-hole model, the projection is expressed in Equation (3).

$$p_o = \frac{1}{Z_o} K P_o \quad (3)$$

where P_o is a 3D point expressed in the camera frame; $p_o : (u, v, 1)^\intercal$ is the corresponding pixel in the optical image; and K is the intrinsic matrix of the camera, defined by Equation (4).

$$\mathbf{K} = \begin{pmatrix} f_x & s & c_u \\ 0 & f_y & c_v \\ 0 & 0 & 1 \end{pmatrix} \quad (4)$$

where (f_x, f_y) are the focal length in pixel/m along the two axes, s is the skew parameter describing the non-orthogonality of pixels, and (c_u, c_v) are the coordinates of the optical centre of the camera expressed in pixels. For our method, we assume that the skew parameter s is equal to zero since it is now the case for most cameras thanks to modern manufacturing techniques (as said in [20]), and we also assume that coordinates (c_u, c_v) correspond to the middle of our image. Only the focal length remains unknown, with the assumption that f_x and f_y have the same value, noted f. Even though f can be obtained by a classic intrinsic calibration, we decided to include it in our calibration method to simplify as much as possible the calibration process to the ROV's operator.

2.3. Sonar's Projection Model

In this paper, we consider the case of a multibeam imaging sonar, which processes the echoes received along multiple beams to create an image. The principle of multibeam sonar imaging is illustrated in Figure 4.

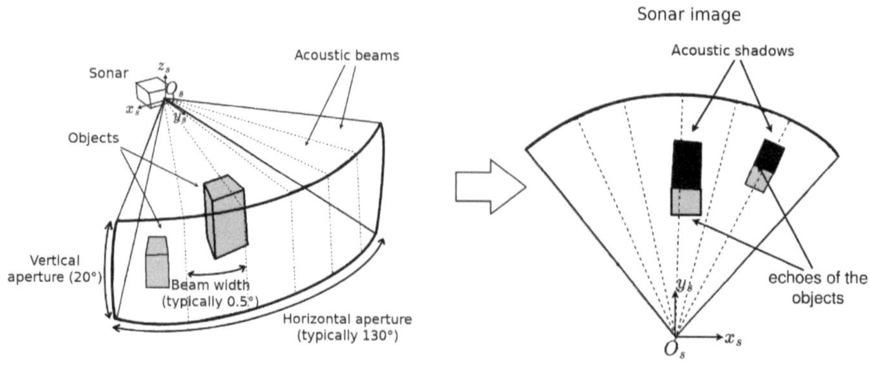

Figure 4. Illustration of the principle of a multibeam imaging sonar. The sensor produces wide acoustic beams, which are reflected by the objects they reach. The echoes are then received by an array of transducers forming many beams. The echoes create bright points in the acoustic image. The areas located behind the objects do not receive any sound, thus creating dark zones, corresponding to the acoustic shadows. The length of the shadow generally depends on the object's height.

In what follows, p_S: $(\rho, \theta)^\top$ will be the polar coordinates of p_S—the projection of the 3D point P_S in the 2D sonar image I_S; ρ is the distance in meters between the sonar frame's origin O_S and the point P_S; while θ is the horizontal azimuth angle with respect to the central line of the sonar image (Figure 5).

As one will remark, p_S has only two coordinates, ρ and θ, while the elevation angle ϕ does not appear. This is because the sonar cannot discriminate the echoes from points having the same horizontal azimuth and the same distance but different elevations. So, every 3D point in spherical coordinates P_{S_k}: $(\rho, \theta, \phi)^\top$ with the same distance ρ and azimuth θ will be projected on the same point p_S: $(\rho, \theta)^\top$ of the sonar image as long as their elevation ϕ is within the range of the vertical aperture of the sonar. Figure 5 illustrates this.

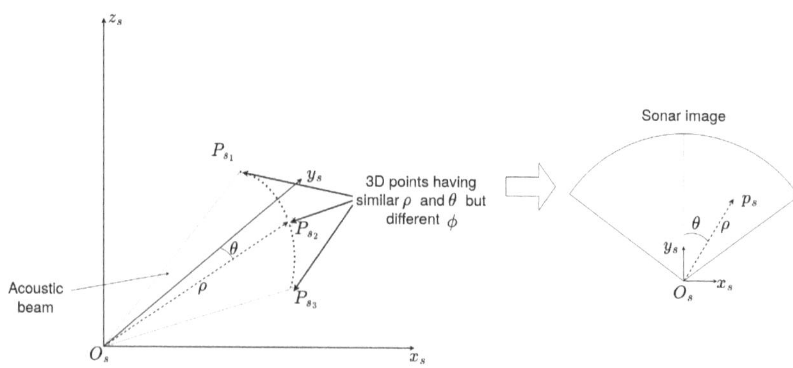

Figure 5. Illustration of the sonar elevation incertitude effect. In the figure, we can see that the three 3D points P_{S_1}, P_{S_2}, and P_{S_3}—having the same azimuth angles θ and the same range ρ but different elevation angles ϕ along the doted arc—will be projected on the same point p_S in the sonar image (on the right).

This inability to discriminate the elevation angle has been studied in works concerning 3D reconstruction from sonar images. To deal with this, the existing methods either rely on a single sonar or on adding an additional sonar placed orthogonally [21] to compute the elevation angles by using the azimuth angles observed in the images from the second sonar. Another approach [22] consists in using multiple views by moving the sonar up and down and then in tracking points in these views to determine their elevation from their displacement along an acute angle of the object. The limitation of this method as it is presented by the author is that in case of smooth objects, extracting and following a feature can be difficult and can lead to errors in the elevation's estimation. Another possibility is to consider the intensity of the points as an image of the elevation [23]. Even though the intensity of a pixel in the sonar image is linked to the echoes of each point of the arc of the acoustic beam, this method only works when used close to the ocean floor and with objects with a similar composition since backscattered intensity varies depending on the material. One last solution is to track the bright spot of an object and its shadow [24]. By combining the robot position and considering the evolution of the object's position in time, particularly the moment when it leaves the image, this method allows to determine the elevation of certain points and the height of objects.

In our case, because of the possible absence of targets and the complexity of the environment, these method will not be used. Instead, we use an interval of elevation values $[\phi]$, where ϕ_{min} and ϕ_{max} (the minimum and maximum values of the interval) are defined by the sonar's vertical aperture. Using this interval, we can find the interval of 3D points $[P_s]$ corresponding to each sonar point p_s by using Equation (5).

$$[P_s] = \begin{cases} X_s = \rho sin(\theta)cos([\phi]) \\ Y_s = \rho cos(\theta)cos([\phi]) \\ Z_s = \rho sin([\phi]) \end{cases} \tag{5}$$

where the values of ϕ belong to the interval $[\phi_{min}, \phi_{max}]$. Using this method means that each point in the sonar image may come from an arc of 3D points.

2.4. Frame Transformation

As stated before, the calibration consists in finding the parameters to go from a pixel p_s of the sonar image to its corresponding pixel p_o in the optical image. This transformation relies on the sensors' models and the transformation between the sensors' frames. First, starting from the sonar image point p_s, its corresponding sets of points $[P_s]$ can be obtained from Equation (5). Then, for the set of points $[P_s]$, a corresponding set of points $[P_o]$ is found in the optical camera frame by applying Equation (2) on each points of $[P_s]$. Finally, from $[P_o]$ and using Equation (3), the corresponding set of points $[p_o]$ in the optical image can be found. In summary, for a point in the sonar image p_s with an azimuth angle θ and a range ρ, as well as a value of ϕ in the interval $[\phi_{min}, \phi_{max}]$, we obtain a corresponding set $[p_o]$ in the optical image. This transformation is summarised by Equation (6):

$$[p_o] = \frac{1}{Z_o} K(R_s^o \begin{pmatrix} \rho sin(\theta)cos(\phi) \\ \rho cos(\theta)cos(\phi) \\ \rho sin(\phi) \end{pmatrix} + T_s^o) \tag{6}$$

where ρ and θ are the coordinates of p_s in the sonar image, ϕ is within $[\phi_{min}, \phi_{max}]$, and the other variables have been introduced in previous sections. In Equation (6), we need to estimate the translation vector T_s^o, the rotation matrix R_s^o, as well as the focal length f of the camera (inside the camera's intrinsic matrix K). In order to find these parameters, we introduce a new calibration algorithm in the following section.

3. Calibration Method

3.1. Selection of a Set of Feature Points in the Sonar Images

To compute K, R_s^o, T_s^o, and f (i.e., to calibrate the optical–acoustic system), similarly to stereovision calibration, we need to select a set of corresponding feature points in both the sonar image and the optical image.

To associate points between a camera image and a sonar image, a recent method proposes to use feature matching (SuperGlue, a feature matching method based on graph neural networks) between the optical and the style-transferred sonar image (CNN-based style transfer) [25,26]. Another optical–acoustic matching is proposed by the same research team, based on the Dense Adaptive Self-Correlation Descriptor (DASC), which provides better results than other descriptor techniques such as Scale-Invariant Feature Transform (SIFT), Binary Robust Invariant Scalable Keypoints (BRISK), and Accelerated-KAZE (A-KAZE) [27]. The goal of the authors was not to calibrate the opti-acoustic system, and one notes that rotation, translation, and scale differences between two images were corrected prior to the images' preprocessing, thanks to the knowledge of the relative sensor's transformation. Even if the results obtained by the matching process in [27] are impressive and very relevant, the method requires that the calibration parameters of the opti-acoustic system are known. This method also necessitates that, after style transfer, the acoustic image contains patterns relatively similar to the ones of the optical image. Although in many situations the calibration of the opti-acoustic system can be performed before the mission, for the reasons given in the introduction, we propose today a method for automatic calibration. Moreover, in natural underwater environments, it may happen that the acoustic image bears no resemblance to the optical, as depicted in Figure 2, thus reducing the effectiveness of the descriptor-based methods. For this reason, in this paper, we propose a motion-based method aiming at performing the calibration of the opti-acoustic system. Relying on the comparison of the local motion in both images (optical flow), our method does not rely on the visual similarity of the images; thus, it can work in any type of environment (except completely flat bottoms) and we do not need any artificial pattern or calibration target. This method is described below.

Before selecting the points, we need to process the sonar images in order to reduce the background noise and other disturbances such as schools of fish that would appear as multiple clustered spots in the images. To suppress these, we apply a low-pass filter on the sonar image using its Fourier transform, results are shown in Figure 6. In the denoised image, denoted I_{s_i}, we select n feature points using the Shi-Tomasi algorithm. On sonar images, the Shi-Tomasi detector offers the advantage of selecting less outliers than the Harris detector would.

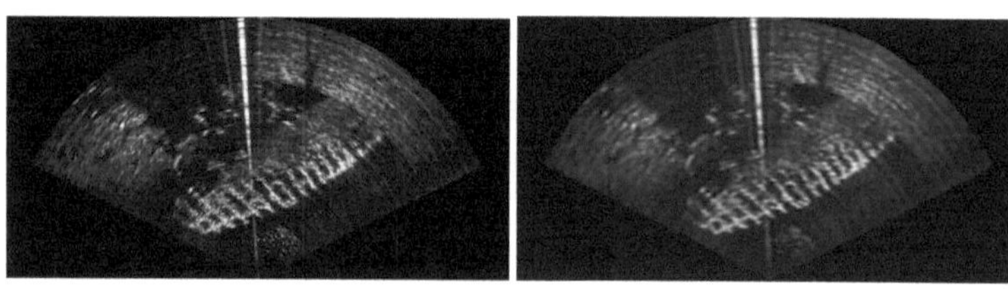

(a) **(b)**

Figure 6. (**a**) The unfiltered sonar image. (**b**) The sonar image after using a low-pass filter.

Among the n selected feature points in the sonar image, the ones located farther than an adjustable range ρ_{max} are discarded, since they may not be visible in the optical camera due to turbidity or the lack of light (deep sea). The value of ρ_{max} is set depending on the water's turbidity and the lighting capabilities of the robot. In what follows, we will use $\rho_{max} = 2$ m. We also discard the points that may be occluded by other selected points located closer on the same acoustic beam. The final set of selected points is denoted $\{p_{s_i}\}$ in the following. Figure 7 summarises the selection process, while Figure 8 gives an example on real sonar images. It is important to note that if the number of points in $\{p_{s_i}\}$ is below a certain threshold n_{min}, the image is discarded and the algorithm will go on to the next image. For a correct behaviour of the calibration process, experiments have demonstrated that n_{min} should be equal to at least 10 points. Once a large enough set of points $\{p_{s_i}\}$ has been selected in the current sonar image I_{si}, these points are tracked in the next sonar image $I_{s_{i+1}}$ using the Lucas–Kanade tracking algorithm [28]. Thus, we obtain the set of sonar points $\{p_{s_{i+1}}\}$ corresponding to the tracked positions of $\{p_{s_i}\}$ in the second image $I_{s_{i+1}}$.

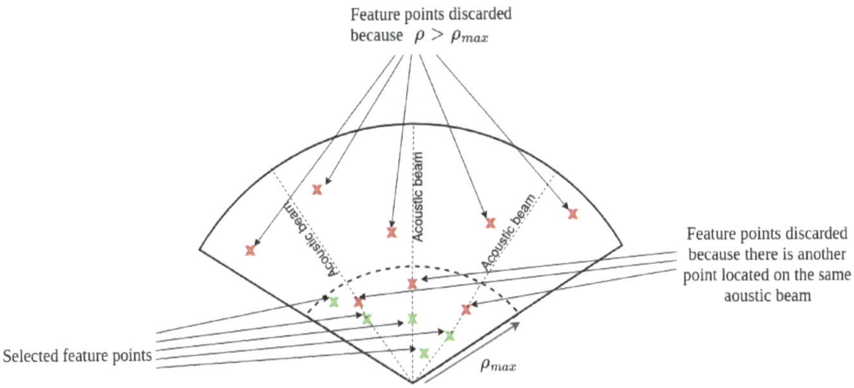

Figure 7. Illustration of the points' selection process in the sonar image.

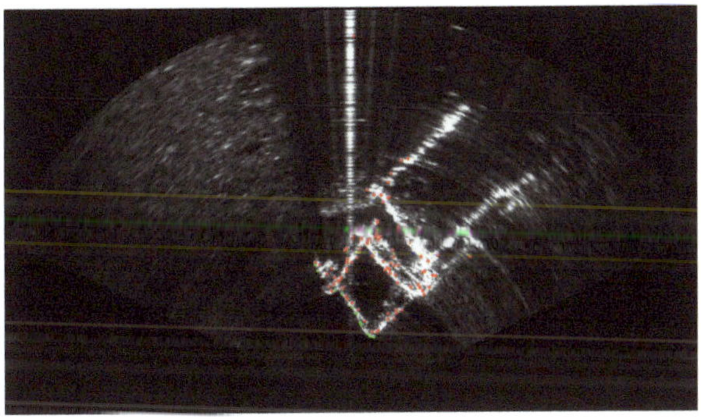

Figure 8. Selection process of the sonar feature points. Similarly to Figure 7, green points are the remaining points p_{s_i} after the suppression of the red points located further than ρ_{max} range or occluded by a closer point located on the same acoustic beam.

3.2. Projection and Evaluation

First, we consider an arbitrary initial value for T_i, R_i, and f_i, the sought-after parameters. Using Equations (5) and (6) presented in Section 2, we can project each starting point p_{s_i} and the corresponding end point $p_{s_{i+1}}$ into the optical images I_{o_i} and $I_{o_{i+1}}$ acquired at the same times t_i and t_{i+1}, thus obtaining the corresponding sets of optical starting points noted p_{o_i} and end points $p_{o_{i+1}}$. As stated before, these optical points represent an arc of points for each of the selected sonar points. We then use the Lucas–Kanade optical flow to estimate end points in the next optical image based on the optical movement of the starting points p_{o_i}, thus obtaining the estimated end points $\{\hat{p}_{o_{i+1}}\}$. An example of this projection process is shown in Figure 9 for a single sonar point.

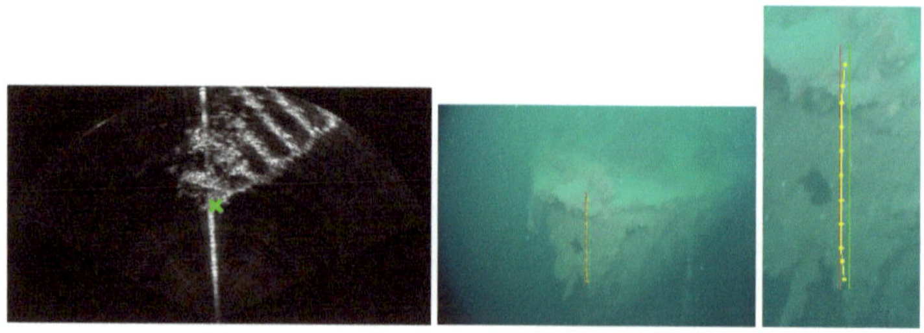

Figure 9. (**a**) The selected sonar point in green. (**b**) The corresponding projected points in the optical image, with the starting (green), ending (red), and estimated points (yellow); a zoomed image of the projected points is proposed in (**c**).

So, for each selected point in the sonar image, we have a starting arc of optical points, an arc of optical end points corresponding to its tracked counterpart, and an arc of estimated points from the movement in the optical image. Using these, we can compute the projection score (i.e., a proximity score between the points computed from the optical movement and the end points obtained by projection of the tracked acoustic points).

The relative score for the *j-th* point is defined in Equation (7), where d is the minimal distance between the estimated points and the end points, and d_{max} is the distance between the starting and ending points. This is illustrated in Figure 10.

This score is calculated by considering the estimated end points with the biggest displacement with respect to the starting points and their distance to the end points, noted d, as well as the distance between the arc of end points and the arc of starting points, noted d_{max}. The score for the *j-th* point among the n selected points is calculated by Equation (7) and described by Figure 10. We decided to represent the score with a distance ratio to mitigate the effect of parameters that could act as scale factors. We call scale factors the parameters such as focal length that will impact the scale of the projection, thus changing the spacing of the point by themselves.

$$score_j = \frac{abs(d_{max} - d)}{d_{max}} \tag{7}$$

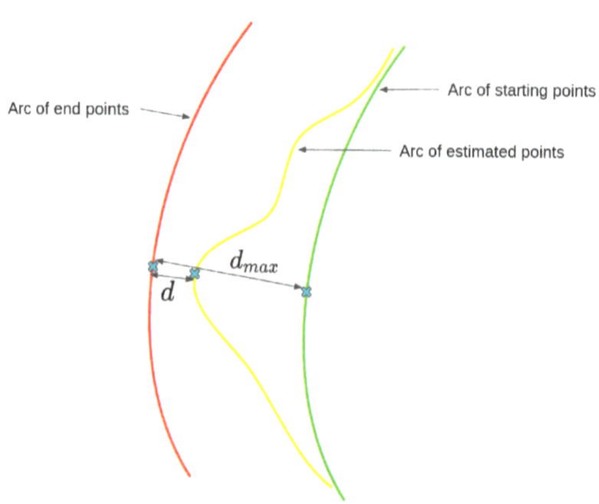

Figure 10. Description of the score calculation for one projected point, with the starting points (green), ending points (red), and estimated points (yellow) from the optical flow, as well as the distances used to compute the score.

Then, by taking the mean score of every projected point, we obtain the score for one group of images (two consecutive sonar images and their corresponding optical images).

3.3. Estimation of the Calibration Parameters

To compute the projection parameters (i.e., calibration parameters), we iterate through all the parameters, realising an exhaustive search in a parameter space whose limits can be either defined by the dimensions of the robot or chosen by the operator according to the rough knowledge of the robot's configuration if it is available (note that the method will work even without any prior knowledge about the geometric configuration of the setup). Since an exhaustive search can take a long time, we use an adaptive search, starting with a coarser step, and then using a finer step to find the calibration parameters. In addition, we also need to use multiple image pairs to obtain a finer estimation.

To conclude this section, all the steps of the calibration algorithm are represented in Algorithm 1.

Algorithm 1 Research of the calibration algorithm on one set of camera and sonar image pairs.

$I_{s_i} \leftarrow getNextSonarImage()$
$I_{o_i} \leftarrow getNextCameraImage()$
$I_{s_{i+1}} \leftarrow getNextSonarImage()$
$I_{o_{i+1}} \leftarrow getNextCameraImage()$
$p_{s_i} \leftarrow selectFeaturePoints(I_{s_i})$
$p_{s_{i+1}} \leftarrow LucasKannade(I_{s_i}, I_{s_{i+1}}, p_{s_i})$
$scoreMin \leftarrow +\infty$
for all R_s^o, T_s^o and f **do**
  $[p_{o_i}, p_{o_{i+1}}] \leftarrow projectPointsSonarToCamera(R_s^o, T_s^o, f, p_{s_i}, p_{s_{i+1}})$
  $\hat{p}_{o_{i+1}} \leftarrow LucasKannade(I_{o_i}, I_{o_{i+1}}, p_{o_i})$
  $projectionScore \leftarrow computeScore(p_{o_i}, p_{o_{i+1}}, \hat{p}_{o_{i+1}})$
  if $score < scoreMin$ **then**
    $scoreMin \leftarrow score$
    $[T_{min}, R_{min}, f_{min}] \leftarrow [R_s^o, T_s^o, f]$
  end if
end for
Return : $[T_{min}, R_{min}, f_{min}]$

4. Experimental Validation and Dataset

4.1. Experimental Setup

To test our calibration algorithm, we performed two campaigns at sea with two different ROVs. These tests were performed on wrecks under the supervision of the Department of Underwater Archaeological Research (DRASSM) of the French ministry of culture. The first set of tests were performed with the *Hilarion* ROV equipped with a Sony 4K ER8530 optical camera and an Oculus 1200M multibeam imaging sonar (Figure 1).

Hilarion inspected underwater car wrecks located in the Mediterranean Sea, 60 m deep. Such wrecks are interesting for these experiments since they present sharp angles, thus facilitating the detection of feature points thanks to the bright echoes they create in the sonar images. The second set of tests were performed with the *Basile* ROV, equipped with the same Oculus 1200M multibeam imaging sonar and a monocular imaging camera, both mounted on a mechanical frame, allowing to accurately change the geometric parameters (e.g., distance and orientation of the camera with respect to the sonar) and thus allowing us to control the ground truth of the extrinsic calibration parameters, as shown in Figure 11. During this second mission, the ROV observed various wrecks (cars, barges, boats, etc.) located around 60 m deep. We created a software allowing synchronisation of the images from the two sensors, as well as the IMU of the robot.

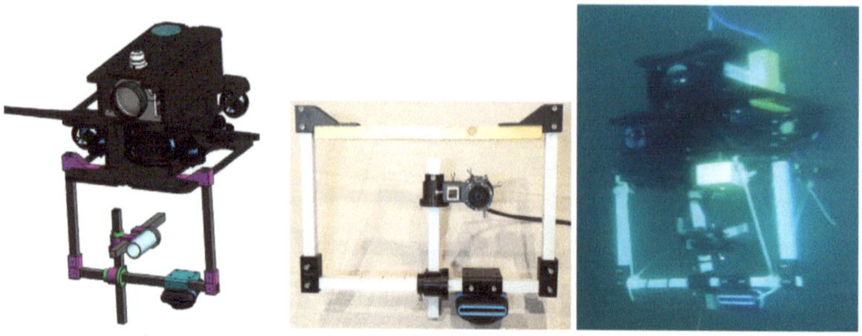

(a) (b) (c)

Figure 11. (**a**) The CAD model of the Basile ROV and its frame, allowing to modify the relative positions of the camera and the sonar. (**b**) The mechanical frame with the optical camera and the sonar. (**c**) The frame attached to the *Basile* ROV during a dive in Marseille.

4.2. Dataset

The dataset we created contains 17572 monocular images and the 8577 corresponding sonar images. We also added the IMU data of the ROV during the mission, despite them not being useful for our calibration method. We named this dataset the "shipwreck sensing dataset" and it is publicly available here https://www.lirmm.fr/shipwreck-dataset/ (accessed on 1 February 2023).

Details on the nature of the data and their acquisition are presented in Table 1. In order to see if our algorithm works for various positions of the sensors, we acquired images with different configurations, as presented in Table 2. The choice of these ground truth configurations was made to try parameters independently, the first one serving as a reference and the two others introducing variation on a single parameter. A representation of each of the extrinsic parameters is shown in Figure 12.

Table 1. Technical data about the sensors of the Basile ROV.

Monocular Video Camera	
Camera model	Optovision HD mini IP camera
Image size	720 × 480 pixels
Frame rate	30 fps
Sonar	
Sonar model	Oculus 1200 M
Image size	1024 × 507 pixels
Frame rate	10 fps
Horizontal aperture	130°
Vertical aperture	20°
Angular resolution	0.5°
IMU data frequency	20 Hz

Table 2. The three geometric configurations of the sensors available in the dataset.

	T_x (cm)	T_y (cm)	T_z (cm)	α (°)	β (°)	γ (°)	f (pixel/m)
Configuration I	0	5	0	0	0	0	600
Configuration II	0	15	0	0	0	0	600
Configuration III	10	5	0	0	0	0	600

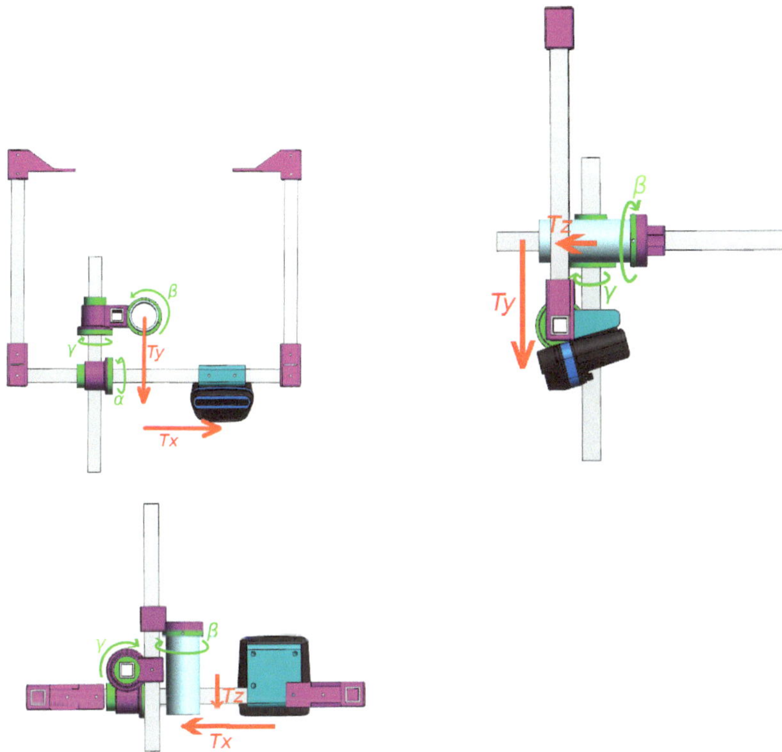

Figure 12. Representation of the extrinsic parameters as part of the frame used to set them during the experiments at sea. It is important to note that the rotations are expressed along the sonar frame $\mathfrak{R}_s$ while the translations are expressed along the optical camera frame $\mathfrak{R}_o$, as defined in Figure 3.

4.3. Experimental Evaluation of the Calibration Algorithm

Taking sonar and camera image pairs from this dataset, we tested our algorithm using the steps described in Section 3. The code was made in C++ with the OpenCV library and executed on a Dell precision 5520 with an Intel Xeon E3-1505M v6 3.00 GHz processor.

First, we tested our algorithm on an increasing number of image pairs to show the evolution of the error. The error is the absolute value of the difference between the parameters obtained with the algorithm and the ground truth (relative positions of the two sensors on the frame, and focal length computed from a standard optical calibration of the camera). The results are presented in Figure 13. One observes that the algorithm converges very fast (5 to 6 pairs of images) to errors smaller than 1 cm and 1 degree.

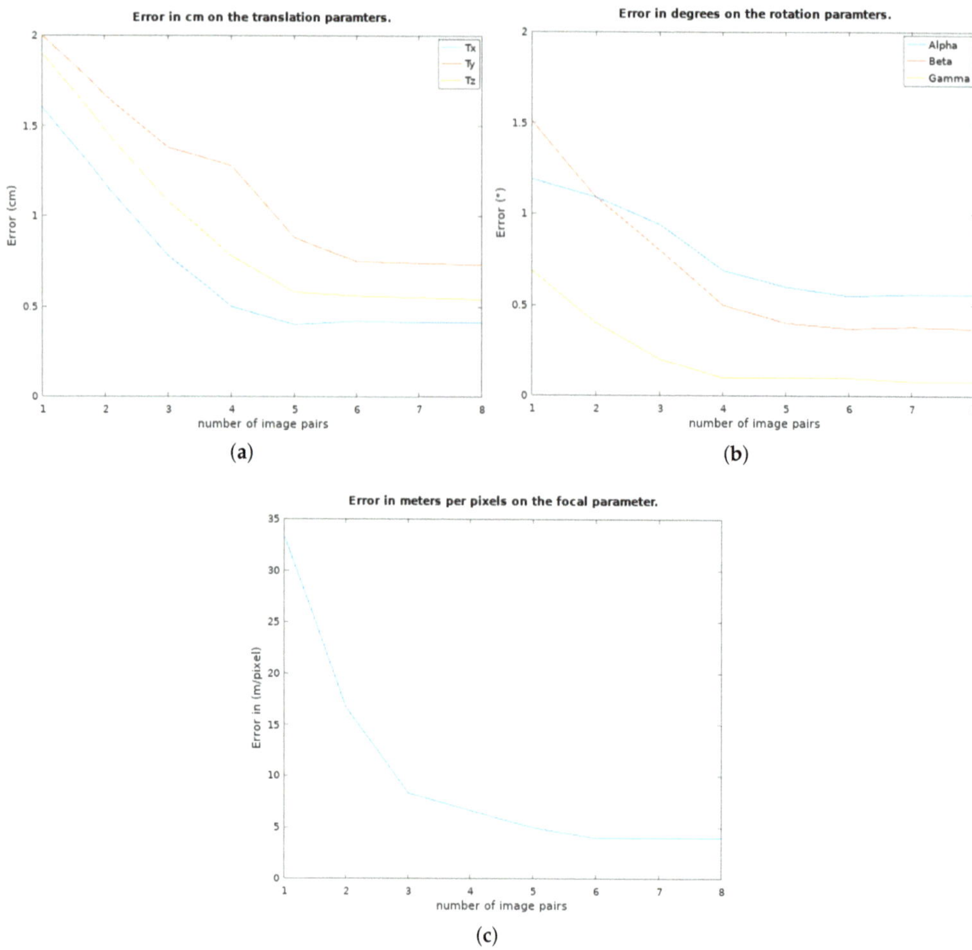

Figure 13. (**a**) The evolution of the mean error on the calibration parameters over the number of image pairs used. In (**a**), the translation parameters (T_s, T_y, and T_z) are presented in meters, (**b**) the rotation parameters (α, β, and γ) in degrees, and (**c**) is the focal parameter.

As we could expect, the results yield a bigger error on the β and T_z parameters because of the elevation uncertainty in the sonar images, creating a larger vertical zone where the projection can match the movement. Similarly with the error on the parameters shown in Figure 13, the evolution of the reprojection error in pixels is shown in Figure 14. This reprojection error is defined by the minimal distance between the projected arcs and the known position where they should be. An example of points projected with the found calibration parameters in comparison to their goal is shown in Figure 15.

The achieved results allow to accurately convey information (the position of an object from one to the other, for example) from the sonar to the camera and vice-versa, notably for the position of objects seen by the sonar from further away. Even though a greater number of images yields a lower error, it is at the cost of the time required to obtain the results. Since this is an exhaustive search without any optimisation, the time increases with the number of image pairs (around 4 h per pair), requiring several hours to compute the calibration parameters, despite using a coarser step to reduce search time (typically searching by 5 cm/° every iteration, then reducing the step to 3, and then 1).

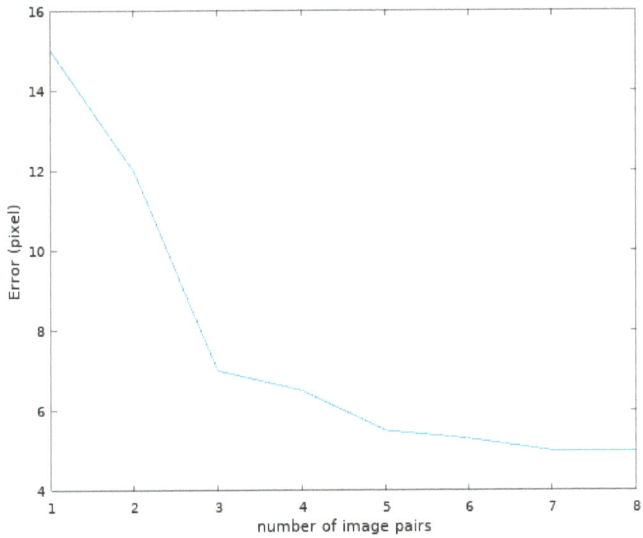

Figure 14. The mean reprojection error (in pixels) depending on the number of image pairs used for the calibration. As a reminder, the images have dimensions of 720 × 480 pixels.

This steep increase of the required time can be explained by the sequential implementation of this algorithm (no parallelisation). An improvement on that matter could be a subject of future work. The purpose of this brute force approach was to validate the algorithm before improving its time of execution. To end this section, Table 3 summarises the results yielded on all the configurations available in the dataset.

These results show that we are able to achieve a precise estimation of all the parameters despite the differences in configuration. Even though an error still persists, we consider it sufficiently low for applications making these two sensors work together. For example, with such precision we could highlight in the optical image the position of a distant object visible only in the sonar image.

Table 3. Results obtained with our method for the three geometric configurations.

	T_x (cm)	T_y (cm)	T_z (cm)	α (°)	β (°)	γ (°)	Focal
Configuration I ground truth	0	5	0	0	0	0	600
Configuration I estimated	1.2	3.8	0.9	0.7	1.0	0.1	570
Configuration II ground truth	0	15	0	0	0	0	600
Configuration II estimated	0.5	14.2	0.8	0.3	1.1	0.4	610
Configuration III ground truth	10	5	0	0	0	0	600
Configuration III estimated	8.7	4.0	0.8	0.7	1.0	0.1	570

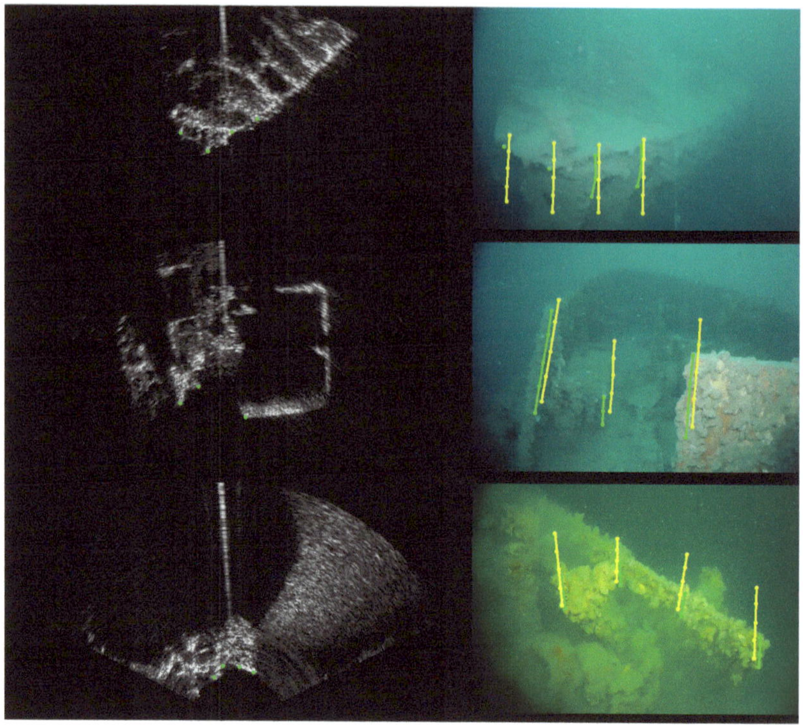

Figure 15. Examples of reprojection once the calibration parameters have been obtained. In green are the selected points in the sonar image and their corresponding area of presence in the camera image. In yellow are the arcs of points obtained using the found calibration parameters.

Table 4 presents a comparison with results from the literature. One can observe that we obtain better performances for translation estimation and we obtain 0.5 degree less accurate results for rotation estimations. This shows that using movement is an effective way to compute the calibration parameters.

Table 4. Comparison between existing methods and our algorithm.

Algorithm	Error on T_x (m)	Error on T_y (m)	Error on T_z (m)	Error on α (°)	Error on β (°)	Error on γ (°)
[12]	0.02	0.05	0.1	0.1	1.0	0.003
[14]	0.0	0.05	0.1	1.0	5.0	0.0
Our algorithm	0.01	0.015	0.05	1.0	1.5	0.5

The main limitation of our method in its current form is the important time required to estimate the calibration parameters. This makes our method unusable for short missions; however, it could still be of used for long-term missions. This drawback is counterbalanced by the fact that our method does not require any specific calibration pattern and can be performed in any natural environment. The computation of the parameters relies on brute force; thus, it is likely to be optimised in the future in several ways. As gradient-based techniques are likely to fail with such a problem, we will consider other approaches in the coming months, such as genetic algorithms. In addition to this, although it is not required for the convergence of the algorithm, a rough measurement of the relative positions of the two sensors with a very reasonable accuracy of several centimetres and several degrees would drastically reduce the search space and, thus, will help them to converge much faster.

5. Conclusions

In this article, we presented a new targetless calibration method for a system combining an acoustic camera (i.e., multibeam imaging sonar) and an optical monocular camera. This method uses the pixels' motion in the images of the two sensors. After a presentation of the model of each sensor, we showed that we could project the movement of feature points of the sonar image into the optical image. Using the optical flow of the optical image to obtain an estimate of the movement of projected points in the optical image, a distance score was calculated, allowing us to compute the calibration parameters through an exhaustive search. The important upside of this method is that it does not require a calibration pattern. This will help for robotic operations at sea, which may require frequent recalibration due to changes in the sensors' positions and orientations. The obtained level of accuracy is sufficient to merge the data acquired by the two sensors and is close to the one obtained by existing calibration methods based on a target. Future works will consist in optimising the algorithm to improve the search speed, with the goal of reaching a far better execution time, preferably below an hour, while keeping the same precision. For now, plans for this method are to use it to highlight in the optical image the distant structures (objects, rocks, pipelines, etc.) that are visible only to the sonar, in order to give better indication to the ROV's operator.

Author Contributions: Conceptualization, N.P., V.C., F.C. and O.T.; methodology, N.P., V.C. and F.C.; software, N.P. and V.C.; validation, N.P. and V.C.; formal analysis, N.P., V.C. and F.C.; investigation, N.P., V.C., F.C. and O.T.; resources, N.P., V.C. and F.C.; data curation, N.P.; writing—original draft preparation, N.P.; writing—review and editing, N.P., V.C. and F.C.; visualization, N.P., V.C. and F.C.; supervision, V.C. and F.C.; project administration, V.C. and F.C.; funding acquisition, V.C. and F.C. All authors have read and agreed to the published version of the manuscript.

Funding: This work was partly funded by a doctoral grant from the Occitanie Regional Council (ALDOCT 000-941).

Institutional Review Board Statement: Not applicable.

Informed Consent Statement: Not applicable.

Data Availability Statement: As stated in Section 4.2, all the data presented in this article and acquired in order to test this method are available at https://www.lirmm.fr/shipwreck-dataset/ (accessed on 1 February 2023). There, both optical and sonar images as well as the IMU data during the mission can be downloaded.

Acknowledgments: For this article, we would like to thank the teams of the DRASSM (Denis Dégez, Marine Sadania, and the crew of the Alfred Merlin research vessel) for their valuable help to perform the experiments at sea.

Conflicts of Interest: The authors declare no conflict of interest.

References

1. Nguyen, L.H.; Hua, M.D.; Allibert, G.; Hamel, T. A Homography-Based Dynamic Control Approach Applied to Station Keeping of Autonomous Underwater Vehicles Without Linear Velocity Measurements. *IEEE Trans. Control. Syst. Technol.* **2021**, *29*, 2065–2078. [CrossRef]
2. Chen, Z.; Zhang, Z.; Bu, Y.; Dai, F.; Fan, T.; Wang, H. Underwater object segmentation based on optical features. *Sensors* **2018**, *18*, 196. [CrossRef] [PubMed]
3. Moghimi, M.K.; Mohanna, F. Real-time underwater image enhancement: A systematic review. *J. -Real-Time Image Process.* **2021**, *18*, 1509–1525. [CrossRef]
4. Zhuang, P.; Li, C.; Wu, J. Bayesian retinex underwater image enhancement. *Eng. Appl. Artif. Intell.* **2021**, *101*, 104171. [CrossRef]
5. Li, C.; Lian, S.; Niu, J.; Wang, C.; Zhou, X. Research on Underwater Image Denoising Based on Wavelet Threshold Method. In Proceedings of the 2022 7th International Conference on Intelligent Computing and Signal Processing (ICSP), Xi'an, China, 15–17 April 2022; pp. 1941–1947.
6. Kis, A.; Balta, H.; Ancuti, C. Underwater Image Enhancement on Low-Cost Hardware Platform. In Proceedings of the 2021 International Symposium ELMAR, Zadar, Croatia, 13–15 September 2021; pp. 97–100.
7. Pyo, J.; Cho, H.; Yu, S.C. Beam slice-based recognition method for acoustic landmark with multi-beam forward looking sonar. *IEEE Sens. J.* **2017**, *17*, 7074–7085. [CrossRef]

8. Cotter, E.; Polagye, B. Automatic classification of biological targets in a tidal channel using a multibeam sonar. *J. Atmos. Ocean. Technol.* **2020**, *37*, 1437–1455. [CrossRef]
9. Palomeras, N.; Furfaro, T.; Williams, D.P.; Carreras, M.; Dugelay, S. Automatic Target Recognition for Mine Countermeasure Missions Using Forward-Looking Sonar Data. *IEEE J. Ocean. Eng.* **2021**, *47*, 141–161. [CrossRef]
10. Tulsook, S.; Kasetkasem, T.; Tipsuwan, Y.; Sugino, N.; Chanwimaluang, T.; Hoonsuwan, P. A Pipeline Extraction on Forward-Looking Sonar Images Using the Self-Organizing Map. In Proceedings of the 2018 15th International Conference on Electrical Engineering/Electronics, Computer, Telecommunications and Information Technology (ECTI-CON), Chiang Rai, Thailand, 18–21 July 2018; pp. 584–587. [CrossRef]
11. Negahdaripour, S. Calibration of DIDSON forward-scan acoustic video camera. In Proceedings of the OCEANS 2005 MTS/IEEE, Washington, DC, USA, 17–23 September 2005; pp. 1287–1294.
12. Negahdaripour, S.; Sekkati, H.; Pirsiavash, H. Opti-acoustic stereo imaging: On system calibration and 3-D target reconstruction. *IEEE Trans. Image Process.* **2009**, *18*, 1203–1214. [CrossRef] [PubMed]
13. Yang, D.; He, B.; Zhu, M.; Liu, J. An extrinsic calibration method with closed-form solution for underwater opti-acoustic imaging system. *IEEE Trans. Instrum. Meas.* **2020**, *69*, 6828–6842. [CrossRef]
14. Lindzey, L.; Marburg, A. Extrinsic Calibration between an Optical Camera and an Imaging Sonar. In Proceedings of the OCEANS 2021: San Diego–Porto, San Diego, CA, USA, 20–23 September 2021; pp. 1–8.
15. Hurtos, N.; Cufi, X.; Salvi, J. Calibration of optical camera coupled to acoustic multibeam for underwater 3D scene reconstruction. In Proceedings of the OCEANS'10 IEEE SYDNEY, Sydney, NSW, Australia, 24–27 May 2010; pp. 1–7.
16. Lagudi, A.; Bianco, G.; Muzzupappa, M.; Bruno, F. An alignment method for the integration of underwater 3D data captured by a stereovision system and an acoustic camera. *Sensors* **2016**, *16*, 536. [CrossRef] [PubMed]
17. Roznere, M.; Li, A.Q. Underwater monocular image depth estimation using single-beam echosounder. In Proceedings of the 2020 IEEE/RSJ International Conference on Intelligent Robots and Systems (IROS), Las Vegas, NV, USA, 24 October 2020–24 January 2021; pp. 1785–1790.
18. Legg, M.; Bradley, S. A combined microphone and camera calibration technique with application to acoustic imaging. *IEEE Trans. Image Process.* **2013**, *22*, 4028–4039. [CrossRef] [PubMed]
19. Barat, C.; Rendas, M.J. Exploiting natural contours for automatic sonar-to-video calibration. In Proceedings of the Europe Oceans 2005, Brest, France, 20–23 June 2005; Volume 1, pp. 271–275.
20. Szeliski, R. *Computer Vision: Algorithms and Applications*; Springer Nature: Berlin/Heidelberg, Germany, 2022; p. 46.
21. McConnell, J.; Martin, J.D.; Englot, B. Fusing Concurrent Orthogonal Wide-aperture Sonar Images for Dense Underwater 3D Reconstruction. In Proceedings of the 2020 IEEE/RSJ International Conference on Intelligent Robots and Systems (IROS), Las Vegas, NV, USA, 24 October 2020–24 January 2021; pp. 1653–1660.
22. Ji, Y.; Kwak, S.; Yamashita, A.; Asama, H. Acoustic camera-based 3D measurement of underwater objects through automated extraction and association of feature points. In Proceedings of the 2016 IEEE International Conference on Multisensor Fusion and Integration for Intelligent Systems (MFI), Baden, Germany, 19–21 September 2016; pp. 224–230.
23. Aykin, M.; Negahdaripour, S. On feature extraction and region matching for forward scan sonar imaging. In Proceedings of the 2012 Oceans, Hampton Roads, VA, USA, 14–19 October 2012; pp. 1–9.
24. Cho, H.; Kim, B.; Yu, S.C. AUV-based underwater 3-D point cloud generation using acoustic lens-based multibeam sonar. *IEEE J. Ocean. Eng.* **2017**, *43*, 856–872. [CrossRef]
25. Jang, H.; Kim, G.; Lee, Y.; Kim, A. CNN-based approach for opti-acoustic reciprocal feature matching. In Proceedings of the IEEE International Conference on Robotics and Automation Workshop, Montreal, QC, Canada, 20–24 May 2019.
26. Jang, H.; Yoon, S.; Kim, A. Multi-session underwater pose-graph slam using inter-session opti-acoustic two-view factor. In Proceedings of the 2021 IEEE International Conference on Robotics and Automation (ICRA), Xi'an, China, 30 May–5 June 2021; pp. 11668–11674.
27. Gwon, D.H.; Shin, Y.S.; Kim, Y.; Kim, A.; Lee, Y.; Choi, H.T. Nontemporal relative pose estimation for opti-acoustic bundle adjustment. In Proceedings of the OCEANS 2016 MTS/IEEE Monterey, Monterey, CA, USA, 19–23 September 2016; pp. 1–5.
28. Bouguet, J.Y. Pyramidal implementation of the affine lucas kanade feature tracker description of the algorithm. *Intel Corp.* **2001**, *5*, 4.

sensors

MDPI

Article

A Strategy for Controlling Motions Related to Sensory Information in a Walking Robot Big Foot

Ivan Chavdarov [1,2,*], Kaloyan Yovchev [1,2], Lyubomira Miteva [2], Aleksander Stefanov [2,3] and Dimitar Nedanovski [2,4]

1 Institute of Robotics, Bulgarian Academy of Sciences, Acad. G. Bonchev St, Bl. 1, 1113 Sofia, Bulgaria
2 Faculty of Mathematics and Informatics, University of Sofia "St. Kliment Ohridski", 1504 Sofia, Bulgaria
3 Institute of Mathematics, Bulgarian Academy of Sciences, Acad. G. Bonchev Str., Bl. 8, 1113 Sofia, Bulgaria
4 Institute for Nuclear Research and Nuclear Energy, 72 Tsarigradsko Shosse Blvd., 1784 Sofia, Bulgaria
* Correspondence: ivannc@uni-sofia.bg

Abstract: Acquiring adequate sensory information and using it to provide motor control are important issues in the process of creating walking robots. The objective of this article is to present control algorithms for the optimization of the walking cycle of an innovative walking robot named "Big Foot". The construction of the robot is based on minimalist design principles—only two motors are used, with which Big Foot can walk and even overcome obstacles. It is equipped with different types of sensors, with some of them providing information necessary for the realization of an optimized walk cycle. We examine two laws of motion—sinusoidal and polynomial—where we compare the results with constant angular velocity motion. Both proposed laws try to find balance between minimizing shock loads and maximizing walking speed for a given motor power. Experimental results are derived with the help of a 3D-printed working prototype of the robot, with the correct realization of the laws of motion being ensured by the use of a PD controller receiving data from motor encoders and tactile sensors. The experimental results validate the proposed laws of motion and the results can be applied to other walking robots with similar construction.

Keywords: walking robot; motor control; movement of sensors

Citation: Chavdarov, I.; Yovchev, K.; Miteva, L.; Stefanov, A.; Nedanovski, D. A Strategy for Controlling Motions Related to Sensory Information in a Walking Robot Big Foot. *Sensors* **2023**, *23*, 1506. https://doi.org/10.3390/s23031506

Academic Editors: David Cheneler and Stephen Monk

Received: 30 December 2022
Revised: 23 January 2023
Accepted: 24 January 2023
Published: 29 January 2023

1. Introduction

Mobile robots are designed to function in a complex environment, which requires that they have to possess specific capabilities, such as: climbing stairs [1–3]; avoiding obstacles and moving on uneven terrain [4,5]. Many of their applications involve them covering a given area, while bypassing obstacles within it. Some of the tasks that mobile robots are usually designed for include: cleaning [6,7], grass cutting [8], agricultural applications [9], and underwater operations [10]. They are capable of moving in an unstructured and uneven environment [1] and can take part in rescue missions or research inspections. Thus, the use of suitable sensors becomes necessary.

In general, providing adequate sensor input is quite complicated. For example, in [11], an integrated laser-based perception is applied for planning the steps and control of a walking humanoid robot in an unknown rugged terrain. A perception system determines the surrounding environment with an accuracy of several centimeters and the robot's movements are realized based on input data obtained by scanning with a laser sensor. The authors of [4] consider the issue of perception of an uneven terrain and its mapping with a walking robot equipped with low-cost optical range sensors providing only 2D information. A Hokuyo URG-04LX light sensor and laser scanner are used. The mapping algorithm and methods to remove artifacts that lead to quality errors in the map are applied. Article [12] describes the design and testing of a bipedal robot. Each of its legs is equipped with six servo motors. A gyroscope and an accelerometer are used to measure the current position of the robot's structure in the space. Control algorithm stabilizes the robot in an upright position. Potentiometers placed in the axes allow measuring of the angular positions of the individual servomechanisms during movement.

An important problem in the control of mobile robots is the preservation of stability. Recovering from a fall is usually hard (often impossible without outside assistance), so measures must be taken to avoid it. Stability is divided into two groups: dynamic and static. Static stability means that the robot maintains balance without constantly making adjustments to its steering. In this case, the projection of the robot's center of gravity always lies in the support polygon defined by its legs (or wheels). Static stability implies that the robot can stop at any time in its walking cycle without losing balance [13]. To maintain dynamic stability, the robot must actively balance its body. This requires much more complex control algorithms and also usually the robot has a large number of degrees of freedom. The forces and moments in the robot's legs are an important factor when one investigates dynamical stability [14].

The planning of the gait for walking robots is a complicated task, which needs to be consistent with the terrain [11]. Wayfinding methods proposed in the literature can be divided into two categories: offline and online planning strategies. Offline strategies use a previously known map of the region where the robot performs a certain task, using different path-planning approaches: genetic algorithms [15], cellular decomposition [16], neural networks [17], etc. In the online strategies, obstacles are detected in real time using various sensors [18]. Articles [6–10] consider coverage path planning and obstacle-avoiding algorithms. The goal is to find a path that covers all points in a given region [7]. One of the widely used methods is the Boustrophedon cell decomposition [19]. In [20], the bipedal walking of a robot is realized by a method of control based on information received from various sensors. The control of the walking function uses separation of the movements in the sagittal and lateral planes. The effectiveness of the proposed method is investigated with a walking robot, "BLR-G2", equipped with pressure sensors in the feet. These sensors provide information about the state of contact with the floor. This contributes to a realization of a smooth walk with a good grip on the surface. Article [21] presents a hexapod robot walking on uneven terrain. A trajectory generator is used for precise control of its legs. Trajectories are further shaped by sensory information. Thus, the robot passes through obstacles of different sizes and rough surfaces. The results show that by integrating the trajectory generator on foot, the sensor-driven six-legged robot can achieve better terrain adaptability and better walking performance. The bipedal robot "Johnny" is designed to achieve a dynamically stable gait, which enables high walking speeds [22]. Very accurate and fast sensors have been developed for this purpose. The design uses six component force-torque sensors. The control scheme is based on the information from these sensors and the robot can walk on an unstructured terrain.

Compared with the wheeled robots, the walking robots have much more complex structure, have more motors, and are slower. However, they are able to overcome higher and more complicated obstacles. There are studies that aim to reduce the complexity of the walking robots while maintaining their advantages. The authors of [23] present a conceptual design of a new minimalist biped walking robot with four degrees of freedom. The proposed mechanism combines sensing the stability and balancing of the robot during the steps. The sensor mechanism uses an additional flexible ankle joint that is able to provide a measurement of the instability of the biped robot. A balance mass and control algorithm are used to maintain the lateral stability of the robot. The authors of [24] propose a concept for a bipedal robot with vertical stabilization of the robot base and minimization of its lateral oscillations. This robot uses only six actuators and has a good energy balance compared with purely articulated biped robots.

The above literature review for the walking robots could be summarized as follows:

- Since their primary function is to work in an undefined and complex environment, they have to perform complicated spatial movements, which usually requires a sophisticated mechanical design;

- In order to obtain appropriate information for the environment, they have to be equipped with a sufficient number of different types of sensors;
- The control system has to be able to handle the processing of a large amount of sensory information including the motion planning algorithms and changing conditions of the unstructured environment.

Mechanical designs of robots with a large number of degrees of freedom, sensors, and complex control algorithms are often used to solve these problems. However, such an approach leads to significant disadvantages: low reliability, need for more energy, and high production and maintenance cost.

We ask the following questions:

- Could we use the robot's movements in order to obtain more information from its sensors?
- What is the minimal number of degrees of freedom which allows for a creation of a walking robot with good functional capabilities?
- In which way can sensory information be used to improve the walking performance of the robot?

We propose an innovative design with only two degrees of freedom called "Big Foot". For this design, the first question is examined in article [25]. The answers to the remaining questions depend on details that must be further specified, i.e., what is the expected walking environment, expected capabilities of the robot, cost, etc. Our proposed design is capable of walking on even or uneven surfaces using only two motors and having a low overall cost. The aim of this paper is to try to optimize the motion of the robot for the case of walking on a plane (or a surface that is close to a plane) by examining two types of laws of motion: polynomial and sinusoidal. The overall goal is to find a compromise that achieves sufficient walking speed, while keeping impact shocks low, and is compatible with motor power constraints. The results are verified experimentally by using a 3D-printed prototype. The realization of the desired laws of motion is ensured by the use of a PD controller reading data from the motor encoders and the tactile sensors on the robot's feet.

This paper is structured as follows: Section 2 examines the overall structure of the robot and some of the previous work on the subject; Section 3 presents in detail the structure of the walking mechanism and its kinematics. The used laws of motion are also located here. This section also contains the experimental setup and details on the 3D-printed prototype and motors and sensors used; Section 4 contains the experimental results and their comparison with the theoretical laws of motion; Section 5 is a short discussion; Section 6 contains some concluding remarks; and Section 7 provides the patents, Supplementary Materials, and other information.

2. Background and Related Work

The development of the robot in question went through a few iterations, with the original idea first presented in articles [26,27] and patent [28]. The robot has only two degrees of freedom. It consists of a base (base 1), on which the body (body 2) of the robot is connected by means of a vertical rotary joint (with axis R1), in which a shaft (shaft 3) is mounted, which drives the symmetrical arms (legs) 4R and 4L of the robot. Shaft 3 is perpendicular to axis R1. In Figure 1a, the principle scheme of the robot is given; 1b is a 3D model; and 1c is a photo of a 3D-printed robot prototype.

In the two symmetrical arms (legs) 4L and 4R, the robot feet 5L and 5R are bearing mounted. The rotations R1 and R2 are driven by DC motors and transmission mechanisms. The parts identified as 6R and 6L are two belts or gears that provide parallel movement of the feet 5L and 5R relative to base 1.

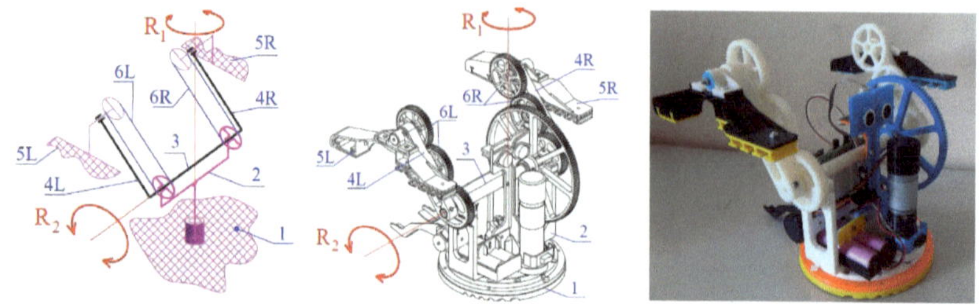

(a) (b) (c)

Figure 1. The robot Big Foot: principle scheme (**a**); 3D model (**b**); photo of a 3D-printed prototype (**c**).

Although the robot has only two motors and a relatively simple design, it moves by walking and rotates on 360[°], avoids obstacles, and even climbs stairs suitable to its size. In [25], the main kinematic dependencies of the robot are presented and a design based on a proportional distribution of the potential energy during the movement of the robot is proposed. A simulation of its movements while climbing an obstacle is given. The robot's ability to passively adapt to high obstacles in order to overcome them is discussed. The authors of [29] present numerous experiments with 3D-printed models of the robot with different shapes and materials of the feet, which lead to an increase in its movement capabilities in a complicated environment. The dynamics of the robot is developed in [30].

Although the mechanics of the robot is relatively well studied both theoretically and experimentally, its sensor systems and possibilities for exploring the surrounding environment are discussed in only one article [25]. The 3D printed model of the "Big Foot" robot has five tactile sensors and one force sensor attached to the bottom of the circular base 1 (see Figure 2a). When the robot moves, it steps on the sensors and activates the tactile buttons. As there may be bumps on the surface the robot is moving on, one or more of the buttons may not be pressed and activated. Thus, the walking robot can be used to detect surface irregularities or to "read" inscriptions or drawings that are embossed or indented in the surface (Figure 2b,c).

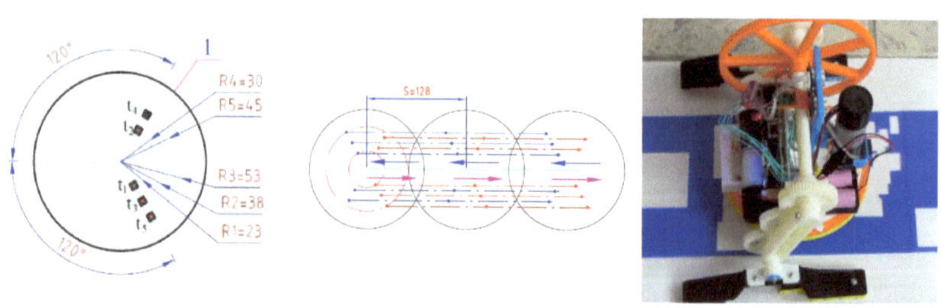

(a) (b) (c)

Figure 2. Location of the tactile sensors (t_i; $i = 1 - 5$) on the base of the robot "Big Foot" (**a**); exemplary trajectory for scanning of bumps (**b**); experiment using the robot for scanning light unevenness on the surface (**c**).

The location of the tactile sensors is chosen in such a way that the robot could measure up to five different points on the surface each time it touches it. The skillful combination of the two rotational movements of the robot (R1 and R1) with the sensors at its base are used to enrich the received sensory information. In [25], such a strategy for the study of irregularities with tactile sensors is considered, and a video with the programmed movements can be seen in the following link: https://youtu.be/RYRJZcYdIX0 (accessed on 20 January 2023).

The robot is also equipped with other sensors: magnetic encoders of the motors, a gyroscope, a magnetometer, and an accelerometer located in body 2.

Here, we further develop the idea presented in [25] to show that sensory signals combined with the movements of a robot based on a minimalist design are useful both for receiving external information and for control of the robot's walking movements.

A word on notations: we will use degrees for angles where possible (for example in graphics) but will switch to radians when needed.

3. Materials and Methods

The main methods, which we apply to develop and test the strategy for managing the information received from the walking robot's sensors are as follows: application of kinematics to determine the necessary motions, velocities, and laws for motor control; mathematical analysis for defining a suitable time dependence of the velocity, which ensures smooth robot movement; synthesis of the control algorithm based on sensor information; design, 3D modeling, and printing of a prototype for experiments; and experimental validation.

3.1. Kinematics of Walking

In [27], we consider the kinematics of our walking robot. There are two phases of the walking function (see Table 1). The walking mechanism occurs only in the motor which turns shaft 3 (Figure 1). For one revolution of shaft 3, body 1 of the robot is successively moved along an arc corresponding to the angle φ_B and feet 5 along the arc φ_S (Figure 3). These angles are defined as a function of the step length S of the robot:

$$\varphi_B = 2arctan\left(\frac{S}{2(L_2 - L_4)}\right), \tag{1}$$

$$\varphi_S = 2\pi - \varphi_B. \tag{2}$$

Table 1. Processes during the different phases of motion.

Phase	Motionless Parts	Instantaneous Center of Velocity for Arm 4	Time Interval	Movement of the Robot	Active Sensor
I	Feet 5L and 5R	Point A (Figure 3)	$T_1 = t_1 - t_0$	Yes	Encoder
II	Base 1 and body 2	Point B (Figure 3)	$T_2 = t_2 - t_1$	No	Tactile sensors, encoder
Transitional process	Undefined	Undefined (instantaneously changes from A to B)	Undefined short time	Undefined	Accelerometer

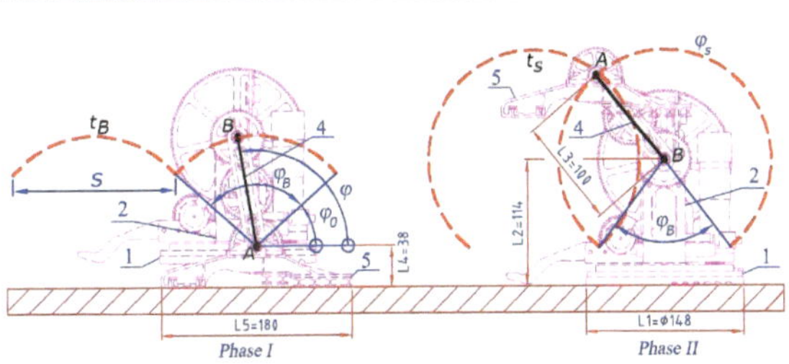

Figure 3. Walking phases (on even horizontal surface) for the robot "Big Foot". The trajectory of the base (robot's body) is t_B while t_s is the trajectory of the robot's feet. The dimensions $L_j = 1 \div 5$, are in millimeters.

L_2 and L_4 are the distances shown in Figure 3 and step S is defined as [26]:

$$s = 2\sqrt{L_3^2 - (L_2 - L_4)^2} \tag{3}$$

As a result of Equations (1)–(3), for both phases of walking, the angles φ_B and φ_S (rotation angle of link 2) depend only on the dimensions L_2, L_3, and L_4 of the links 1, 2, 4, and 5.

The movements of the robot are carried out successively as follows. In the first phase of walking the robot rests on feet 5 and in the second phase it rests on base 1. During the transition between the first and second phase, the instantaneous center of velocity of arm 4 changes with a jump from point B's instantaneous center of velocity to point A. Thus, the elements of the robot undergo shock loads during this transition.

If we assume that the motor maintains a constant speed of shaft 3, which drives link 4, then for the robot's velocity v_x on a horizontal plane, we have:

$$v_x = \dot{x}_{c1} = \omega L_3 \sin(\varphi). \tag{4}$$

Note that there is forward movement only during the first phase of walking (when the feet are on the ground). During this phase, we have $\varphi = \omega t + \varphi_0$. The robot's velocity–time graph is provided in Figure 4.

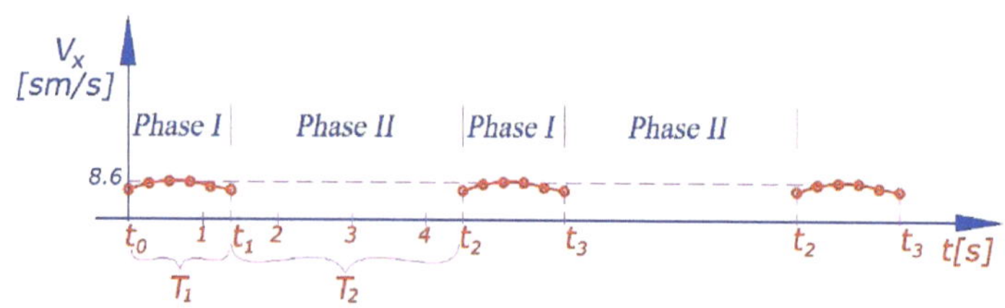

Figure 4. Change in the robot's linear velocity v_x over time obtained according to Figure 3 and Equation (4) in the case of the uniform rotation of shaft 3 with the angular velocity $\omega = 50$ [deg/s].

The robot's velocity v_x is a periodic function of time with the period $T = T_1 + T_2 = t_2 - t_0$ (see Figure 4). This period is divided into two parts. In the first part, $T_1 = t_1 - t_0$, the robot's feet 5 are on the ground and the robot is moving. In the second part, $T_2 = t_2 - t_1$, the robot's base/body (links 1 and 2) are on the ground and arm 4 and feet 5 are rotating (see Figure 3). Obviously the time, $t_2 - t_1$, in which feet do not touch the ground should be minimized. A generalized overview of the robot's movement is provided in Table 1.

3.2. Law of Motion Synthesis

In order to find an appropriate control law for the motor, which drives the walking mechanism, we are guided by the following ideas/aims: the robot's movement should be as fast as possible; impact loads in the transition between the two walking phases must be minimal; and the movement should be smooth and the available sensors should be used in the control process. The shock phenomenon is observed when a sudden (instantaneous) change in the speed of a body is caused by the action of instantaneous forces. The impact force reaches large magnitudes during the collision process. The momentum of the impact leads to a step change in the velocity of the body [31]:

$$J = m(v - v_0) = \lim_{\tau \to 0} \int_{t_0}^{t_0+\tau} F d\tau, \tag{5}$$

where J is the impulse of the force F, v is the velocity at a moment of time $t_0 + \tau$, which is very close to the moment of time t_0 at which the contact between the two bodies occurs, m is the mass of the body, and τ is a short interval of time. In our case, at the moment of contact, $v_0 = 0$, since the body becomes immobile. If we reduce the velocity v at the time $t_0 + \tau$, which is very close to the contact time, we will reduce the impact load. Moreover, if the velocity changes smoothly, we will have small inertial forces. For these reasons, we require the following initial conditions for the angle φ_1, the angular velocity $\dot{\varphi}_1$, and the angular acceleration $\ddot{\varphi}_1$ (the dots denote the time derivative) for phase I (feet are on the ground):

$$\varphi_1(t_0) = \varphi_1(0) = \varphi_0, \quad \varphi_1(t_1) = \varphi_1(T_1) = \varphi_B + \varphi_0, \tag{6}$$

$$\dot{\varphi}_1(t_0) = \dot{\varphi}_1(0) = 0, \quad \dot{\varphi}_1(t_1) = \dot{\varphi}_1(T_1) = 0, \tag{7}$$

$$\ddot{\varphi}_1(t_0) = \ddot{\varphi}_1(0) = 0, \quad \ddot{\varphi}_1(t_1) = \ddot{\varphi}_1(T_1) = 0. \tag{8}$$

Figures 3 and 4 explain the parameters in Equations (6)–(11). In a similar way, for the phase II robot's base on the ground we have:

$$\varphi_2(t_1) = \varphi_2(0) = \varphi_B + \varphi_0, \quad \varphi_2(t_2) = \varphi_2(T_1 + T_2) = \varphi_B + \varphi_S + \varphi_0, \tag{9}$$

$$\dot{\varphi}_2(t_1) = 0, \quad \dot{\varphi}_2(t_2) = \dot{\varphi}_2(T_1 + T_2) = 0, \tag{10}$$

$$\ddot{\varphi}_2(t_1) = 0, \quad \ddot{\varphi}_2(t_2) = \ddot{\varphi}_2(T_1 + T_2) = 0 \tag{11}$$

The motor's limitations and load should also be taken into account. During each phase, the motor can achieve angular accelerations in the interval $0 \leq |\dot{\varphi}(t)| \leq \omega_{max}$ and angular accelerations in the interval $0 \leq |\ddot{\varphi}(t)| \leq \varepsilon_{max}$.The maximal values are determined by the power of the motor and the moments of inertia of the corresponding links.

We consider two types of time dependence for the angular velocity which meet the conditions stated above: sinusoidal and polynomial.

3.2.1. Sinusoidal Dependence

The trigonometric functions sine and cosine are suitable for constructing a law of motion, which smoothly varies the velocity from zero to maximum and then decreases it again to zero. A smooth increase in the angular velocity ω when starting from rest and a smooth stop can be ensured if we use a function of the form:

$$\omega(t) = \dot{\varphi}(t) = A - Acos(kt). \tag{12}$$

Here, A is the amplitude of the angular velocity and k defines the frequency. For the separate phases I and II of motion we are only interested in one period of the function in Equation (12). After integration, for the law of motion of link (arm) 4 we obtain:

$$\varphi(t) = At - A\frac{1}{k}sin(kt) + C. \tag{13}$$

The constant C sets the initial angle of rotation for the phases I and II. For the angular acceleration ε of link 4 we have:

$$\varepsilon(t) = \ddot{\varphi}(t) = Aksin(kt). \tag{14}$$

During phase I time is in the interval $t \in [0, T_1]$. The coefficient k is determined from Equations (7) and (12), $k = \frac{2\pi}{T_1}$. The constant C is determined by the first condition in Equation (6). We obtain $C = \varphi_0$ if the robot's base 1 is on the ground and the motor rotates the links (4L and 4R). The angle φ_0 corresponds to the moment when the phase of movement changes, which is read by the tactile sensors (see Figure 3). From the second condition in Equations (6) and (13) we obtain: $A = \frac{\varphi_B}{T_1}$. Thus, we could write Equations (12)–(14) for phase I in the form:

$$\varphi_1(t) = \frac{\varphi_B}{T_1}\left[t - \frac{T_1}{2\pi}sin\left(\frac{2\pi}{T_1}t\right)\right] + \varphi_0, \tag{15}$$

$$\dot{\varphi}_1(t) = \frac{\varphi_B}{T_1}\left[1 - cos\left(\frac{2\pi}{T_1}t\right)\right], \tag{16}$$

$$\ddot{\varphi}_1(t) = \frac{2\pi\varphi_B}{T_1^2}sin\left(\frac{2\pi}{T_1}t\right). \tag{17}$$

During this phase, the maximal angular velocity is $\dot{\varphi}_{1max} = 2\frac{\varphi_B}{T_1}$ and is reached at time $t = \frac{T_1}{2}$. The maximal angular acceleration $\ddot{\varphi}_{1max} = \frac{\varphi_B}{T_1^2}2\pi$ is reached at $t = \frac{T_1}{4}$, and with the opposite sign at $t = \frac{3T_1}{4}$. If the maximal angular velocity and acceleration are known, then one could determine the least possible time, T_{1min}, for the execution of phase I:

$$T_{1min} = \min\left[\frac{2\varphi_B}{\dot{\varphi}_{1max}}, \sqrt{\frac{2\pi\varphi_B}{\ddot{\varphi}_{1max}}}\right]. \tag{18}$$

Since the angle φ_B is significantly smaller than φ_S and $\varphi_B + \varphi_S = 2\pi$, usually $T_{1min} = \left|\sqrt{\frac{2\pi\varphi_B}{\ddot{\varphi}_{1max}}}\right|$.

In a similar way, using Equations (9)–(14) for phase II, corresponding to time $t \in [T_1, T_1 + T_2]$, we obtain: $k = \frac{2\pi}{T_2}$, $C = \varphi_0 + \varphi_B$, $A = \frac{\varphi_s}{T_2}$. Reaching the angle $\varphi_0 + \varphi_B$ is confirmed by the tactile sensors in the robot's base 1. Equations (12)–(14) for phase II are modified as follows:

$$\varphi_2(t) = \frac{\varphi_S}{T_2}\left[(t - T_1) - \frac{T_2}{2\pi}sin\left(\frac{2\pi}{T_2}(t - T_1)\right)\right] + \varphi_0 + \varphi_B, \tag{19}$$

$$\dot{\varphi}_2(t) = \frac{\varphi_S}{T_2}\left[1 - cos\left(\frac{2\pi}{T_2}(t - T_1)\right)\right], \tag{20}$$

$$\ddot{\varphi}_2(t) = \frac{2\pi\varphi_S}{T_2^2}sin\left(\frac{2\pi}{T_2}(t - T_1)\right). \tag{21}$$

Here, the maximal angular velocity is $\dot{\varphi}_{2max} = 2\frac{\varphi_S}{T_2}$ and is reached at time $t = T_1 + \frac{T_2}{2}$. The maximal angular acceleration, $\ddot{\varphi}_{2max} = \frac{\varphi_S}{T_2^2}2\pi$, is reached at $t = T_1 + \frac{T_2}{4}$, and with the opposite sign at $t = T_1 + \frac{3T_2}{4}$. Again, if the maximal angular velocity and acceleration are known, then one could determine the least possible time, T_{2min}, for the execution of phase II:

$$T_{2min} = \min\left[\frac{2\varphi_S}{\dot{\varphi}_{2max}}, \sqrt{\frac{2\pi\varphi_S}{\ddot{\varphi}_{2max}}}\right]. \tag{22}$$

Since $\varphi_S > \varphi_B$, this phase is performed in a longer time compared with phase I and it is expected that the motor will reach its maximal angular velocity, which corresponds to $T_{2min} = \frac{2\varphi_S}{\dot{\varphi}_{2max}}$.

3.2.2. Polynomial Dependence

Another suitable function for a smooth variation in the angular velocity ω during the change in the phases of movement is a polynomial of degree four:

$$\omega(t) = \dot{\varphi}(t) = a_1t^4 + a_2t^3 + a_3t^2 + a_4t + a_5. \tag{23}$$

Such a polynomial has at most 3 extreme points. The analysis is similar to that of the sinusoidal law. After integration, we obtain for the law of motion:

$$\varphi(t) = \frac{a_1t^5}{5} + \frac{a_2t^4}{4} + \frac{a_3t^3}{3} + a_5t + C, \tag{24}$$

Here, C is a constant of integration which again sets the initial angle of rotation for the phases I and II. After differentiating (23), for the angular acceleration ε of link 4 we have:

$$\varepsilon(t) = \ddot{\varphi}(t) = 4a_1t^3 + 3a_2t^2 + 2a_3t + a_4, \tag{25}$$

We determine the coefficients a_i and the constants of integration from Equations (6)–(11). Again, we will review each phase separately, and reaching the angle φ_0 corresponds to the moment when the phases of movement change.

During phase I time is in the interval $t \in [0, T_1]$. We have 6 coefficients $a_1, \ldots, a_5$, C and 6 Equations (6)–(8), and solving them leads to:

$$\varphi_1(t) = \varphi_B\left(\frac{10t^3}{{T_1}^3} + \frac{-15t^4}{{T_1}^4} + \frac{6t^5}{{T_1}^5}\right) + \varphi_0, \tag{26}$$

$$\dot{\varphi}_1(t) = \frac{\varphi_B}{T_1}\left(\frac{30t^2}{{T_1}^2} + \frac{-60t^3}{{T_1}^3} + \frac{30t^4}{{T_1}^4}\right), \tag{27}$$

$$\ddot{\varphi}_1(0) = \frac{\varphi_B}{{T_1}^2}\left(\frac{60t}{T_1} + \frac{-180t^2}{{T_1}^2} + \frac{120t^3}{{T_1}^3}\right). \tag{28}$$

During this phase, the maximal angular velocity is $\dot{\varphi}_{1max} = \frac{15}{8}\frac{\varphi_B}{T_1}$ and is reached at time $t = \frac{T_1}{2}$. The maximal angular acceleration, $\ddot{\varphi}_{1max} = \frac{10}{\sqrt{3}}\frac{\varphi_B}{T_1^2}$, is reached at $t = \left(\frac{1}{2} - \frac{\sqrt{3}}{6}\right)T_1$, and with the opposite sign at $t = \left(\frac{1}{2} + \frac{\sqrt{3}}{6}\right)T_1$. Again, if the maximal angular velocity and acceleration are known, then one could determine the least possible time, T_{1min}, for the execution of phase I:

$$T_{1min} = \min\left[\frac{15}{8}\frac{\varphi_B}{\dot{\varphi}_{1max}},\quad \sqrt{\frac{10}{\sqrt{3}}\frac{\varphi_B}{\ddot{\varphi}_{1max}}}\right] \tag{29}$$

This was the case for the sinusoidal time dependence since the angle φ_B is significantly smaller than φ_S and $\varphi_B + \varphi_S = 2\pi$, usually $T_{1min} = \sqrt{\frac{10}{\sqrt{3}}\frac{\varphi_B}{\ddot{\varphi}_{1max}}}$.

In a similar way, using Equations (9)–(11) and (23)–(25) for phase II, corresponding to time $t \in [T_1, T_1 + T_2]$, we obtain:

$$\varphi_2(t) = \varphi_S\left(\frac{10(t-T_1)^3}{{T_2}^3} + \frac{-15(t-T_1)^4}{{T_2}^4} + \frac{6(t-T_1)^5}{{T_2}^5}\right) + \varphi_0 + \varphi_B, \tag{30}$$

$$\dot{\varphi}_2(t) = \frac{\varphi_S}{T_2}\left(\frac{30(t-T_1)^2}{{T_2}^2} + \frac{-60(t-T_1)^3}{{T_2}^3} + \frac{30(t-T_1)^4}{{T_2}^4}\right), \tag{31}$$

$$\ddot{\varphi}_2(0) = \frac{\varphi_S}{{T_2}^2}\left(\frac{60(t-T_1)}{T_2} + \frac{-180(t-T_1)^2}{{T_2}^2} + \frac{120(t-T_1)^3}{{T_2}^3}\right). \tag{32}$$

The maximal angular velocity is $\dot{\varphi}_{2max} = \frac{15}{8}\frac{\varphi_S}{T_2}$ and is reached at time $t = T_1 + \frac{T_2}{2}$. The maximal angular acceleration, $\ddot{\varphi}_{2max} = \frac{10}{\sqrt{3}}\frac{\varphi_B}{T_2^2}$, is reached at $t = T_1 + \left(\frac{1}{2} - \frac{\sqrt{3}}{6}\right)T_2$, and with the opposite sign at $t = T_1 + \left(\frac{1}{2} + \frac{\sqrt{3}}{6}\right)T_2$. For the minimal time for execution we have:

$$T_{2min} = \min\left[\frac{15}{8}\frac{\varphi_B}{\dot{\varphi}_{2max}},\quad \sqrt{\frac{10}{\sqrt{3}}\frac{\varphi_B}{\ddot{\varphi}_{2max}}}\right]. \tag{33}$$

As in the sinusoidal case, this phase is performed in a longer time compared with phase I and it is expected that the motor will reach its maximal angular velocity, which corresponds to $T_{2min} = \frac{15}{8}\frac{\varphi_B}{\dot{\varphi}_{2max}}$.

3.2.3. Experiment

The considered construction of the "Big foot" robot uses two motor reducers of type FIT0277 (12 V-Motor: DC; with encoder, with gearbox; 12VDC; 230 mA; 146 rpm; 51:1) with magnet encoders. The output shaft revolutions are 146 RPM. From the transmission's (see Figure 5) gear ratio, $i = \frac{z_2}{z_1} = \frac{124}{40} = 3.1$, of the motor's parameters, we obtain the maximal value for the angular velocity $\omega_{max} = \dot{\varphi}_{max} = 150[^\circ/s]$ and angular acceleration $\varepsilon_{max} = \ddot{\varphi}_{max} = 130\left[^\circ/s^2\right]$ for the angular acceleration of the links 4L and 4R. Thus, we are able to determine the least possible durations, T_{1min}, T_{2min}, for the phases I and II.

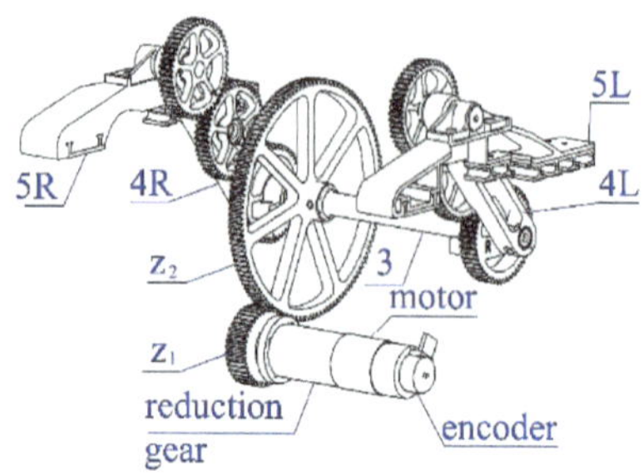

Figure 5. Walking mechanism.

Angular velocity control is realized by feedback with a PD-type controller. This controller receives as input the error between the set angular velocity and the current angular velocity, measured in number of encoder readings. The output of the PD controller is the necessary correction of the signal supplied to the motor driver. The motor driver receives as input an integer from 0 to 255. The controller parameters are experimentally set to P = 0.05 D = 0.00025. P is the proportional term, D is the derivative gain. The proportional term produces an output value that is proportional to the current error value. The derivative of the process error is calculated by determining the slope of the error over time and multiplying this rate of change by the derivative gain. In the transition between the two stages of the movement, the shock load on the robot structure is maximal and there is the greatest need for correction of the input value to the motor driver.

The robot is equipped with tactile sensors in the base (Figure 1a), which allow the moment of contact of the base with the surface to be accurately registered and to determine the phase of the movement. A sensor for measuring acceleration (accelerometer) is also installed on the robot. This sensor allows us to read the acceleration along the vertical z axis that acts on the structure when the transition between the two phases takes place. The sensor is set to read values between ±2 g, where $g = 9.81\left[m/s^2\right]$ is the gravitational acceleration. When the robot is at rest, the sensor reads that the gravitational acceleration and its readings are equal to 1 g, respectively.

Two types of experiments were conducted. In the first type, the constant angular velocity of the motor is set, in which arm 4 of the robot has the angular velocities: $\omega_1 = 118[^\circ/s]$ and $\omega_2 = 59[^\circ/s]$.

The second type are the experiments with angular velocity control according to Equations (15)–(17) and (19)–(21), subject to restrictions (6–11) and the maximal allowed angular velocity and acceleration for link 4.

4. Results

From Equation (3) for the robot's step we obtain $S = 128[mm]$ and the rotation range of the base in phase I is determined by Equation (2). Thus, the maximal angles are $\varphi_B = 80.5[^\circ]$ and $\varphi_S = 279.5[^\circ]$. These are results calculated theoretically using the designed dimensions of the robot. In order to specify these parameters, measurements have been made based on the information from the motor's encoder. Experimentally, we have found that the encoder takes 4575 readings per full revolution of $360[^\circ]$ of arm 4 and feet 5. Thus, one encoder reading equals 0.0787 degrees. Maximal angular velocity (when the feet are in the air) of 1650 readings per second has been experimentally confirmed, which corresponds to $130[^\circ/s]$. The angle φ_B of phase I (when the feet are on the ground and the base is in the air) is 975 encoder readings, i.e., $\varphi_B = 77[^\circ]$. The angle φ_S of phase II is 3600 encoder readings, i.e., $\varphi_s = 283[^\circ]$.

Equations (18), (22), (29), and (33) determine the times T_{1min}, T_{2min}, and the periods in Equations (12) and (23). The results are presented in Table 2.

Table 2. Minimal duration of phases I and II for the considered control laws.

Law	Minimal Duration of the Phase [s]		Period T [s]
	T_{1min} for Phase I	T_{2min} for Phase II	
Polynomial	1.84	3.54	5.38
Sinusoidal	1.92	3.78	5.70

Thus, Equations (15)–(22) set the sinusoidal motor control laws, while Equations (26)–(33) set the polynomial motor control laws. Figure 6a–c shows a comparison of the angular position, velocity, and acceleration assignments during the entire motion for one period, T, under the two control laws.

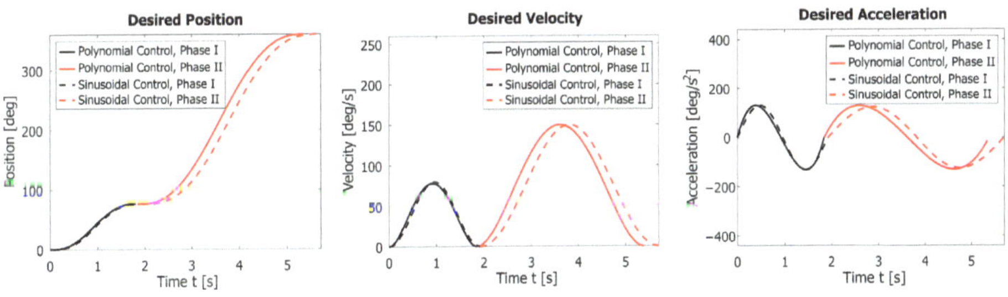

Figure 6. Comparison of the assignment angle positions φ for the two control laws (**a**). Comparison of the assigned angle velocities $\dot{\varphi}$ (**b**). Comparison of the assigned angle accelerations $\ddot{\varphi}$ (**c**).

The following figures present raw unfiltered data from the accelerometer and encoder. Figure 7 contains the results of performing two rotations of the robot's feet at a set constant angular velocity $\omega_1 = 118[^\circ/s]$, which is close to the maximal one. At this rate, the average execution times for phases I and II are $T_1 = 0.80[s]$ and $T_2 = 2.56[s]$, respectively. It takes an average of $T = 3.36[s]$ for a full walk cycle. During the movement, the robot experiences the following minimal and maximal acceleration values along the z axis (the axis normal to the walking plane): -0.26 g and 1.99 g, reported by the accelerometer (Figure 7). These values subject the robot to a strong external load and are not suitable when it is used for a long time.

Next, we reduced the constant angular velocity by half to $\omega_1 = 59[^\circ/s]$. Figure 8 presents the results of two complete rotations of the robot's legs. The average execution times of phases I and II are respectively $T_1 = 2.00[s]$ and $T_2 = 5.06[s]$. It takes an average of $T = 7.06[s]$ for one full walk cycle. During the movement, the robot experiences the following minimal and maximal acceleration along the z axis: 0.34 g and 1.47 g.

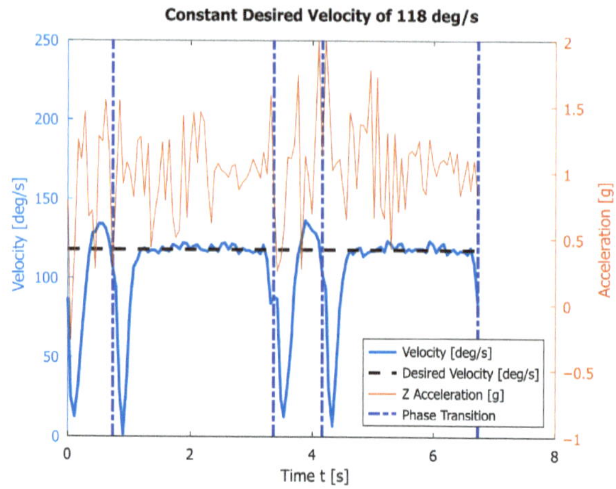

Figure 7. The angular velocity read by the encoder and the vertical acceleration obtained by the accelerometer during the motion of link 4 with a constant angular velocity y $\omega_1 = 118[^\circ/s]$.

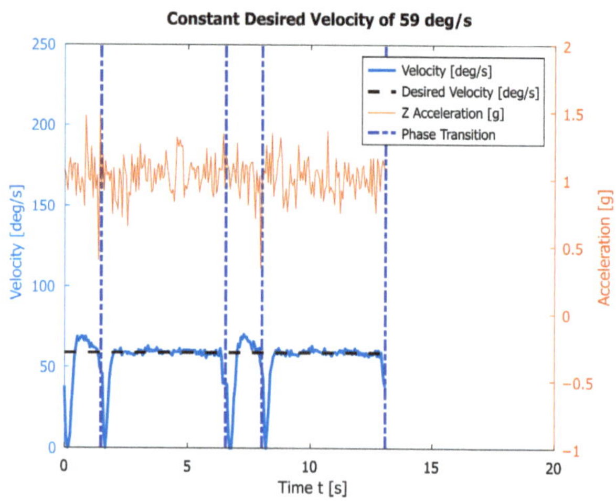

Figure 8. The angular velocity read by the encoder and the vertical acceleration obtained by the accelerometer during the motion of link 4 with a constant angular velocity $\omega_2 = 59[^\circ/s]$.

The third experiment uses the polynomial control law. Now velocities are set in a way ensuring that at the start and at the end of both phase I and phase II the velocities and accelerations are zero. The results of two complete rotations of the robot's legs are shown in Figure 9. With the motion planned in this way, the average execution times of phases I and II are $T_1 = 2.11[s]$ and $T_2 = 3.88[s]$, respectively. It takes an average of $T = 5.99[s]$ for one full walk cycle. During the movement, the robot experiences the following minimal and maximal accelerations along the z axis: 0.34 g and 1.57 g. The achieved maximal angular velocity is $\omega_{max} = 149[^\circ/s]$.

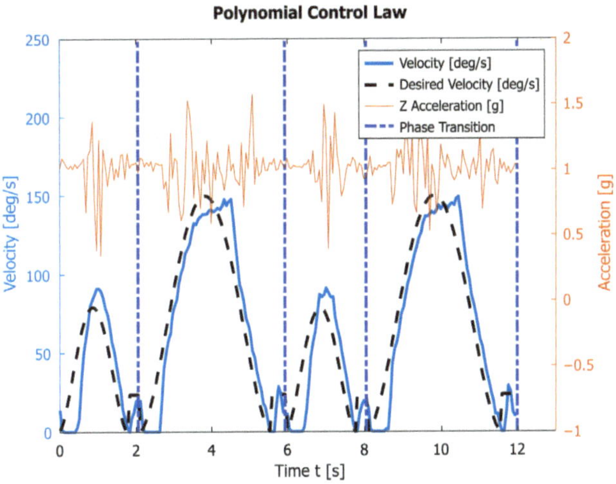

Figure 9. The angular velocity read by the encoder and the vertical acceleration obtained by the accelerometer during the motion of link 4 subjected to the polynomial control law.

The fourth experiment uses the sinusoidal control law. Again, the velocities are set in a way ensuring that at the start and at the end of both phase I and phase II the velocities and accelerations are zero. The results of two complete rotations of the robot's legs are shown in Figure 10. The average execution times of phases I and II are $T_1 = 2.17[s]$ and $T_1 = 4.16[s]$, respectively. It takes an average of $T = 6.32[s]$ for one full rotation. During the movement, the robot experiences the following minimal and maximal accelerations along the z axis: 0.22 g and 1.94 g. The achieved maximal angular velocity is $\omega_{max} = 150[^\circ/s]$.

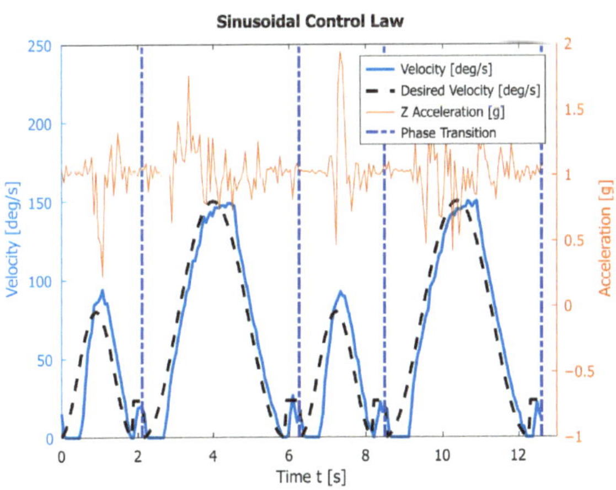

Figure 10. The angular velocity read by the encoder and the vertical acceleration obtained by the accelerometer during the motion of link 4 subjected to sinusoidal control law.

In the third and fourth experiments, the robot experiences lower acceleration along the z axis during the entire motion compared with the motion at velocity $\omega_1 = 118[^\circ/s]$. In real conditions, as seen in Figures 9 and 10, the robot needs a minimal additional time of about 0.30 s on average for transition between the two phases. This is due to the use of a PD-type controller, as well as the physical characteristics of the electric motors and the mechanics of the robot itself. During this time, the motor passes through the moment of zero acceleration. Furthermore, it starts from zero acceleration, while lifting the robot's body, i.e., it overcomes the weight of the structure. This delay can be eliminated if the two phases are planed with a time overlap.

A comparison of the angle change φ (legs' positions) when using polynomial and sinusoidal law is shown in Figure 11.

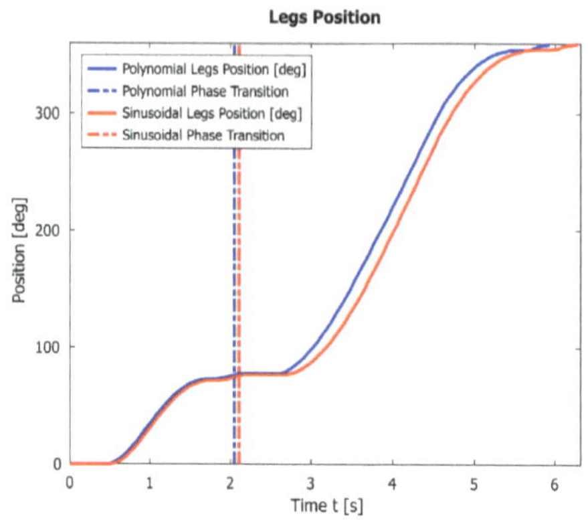

Figure 11. A comparison of the legs' positions when using polynomial and sinusoidal law.

A video with some of the experiments is available at the following link: https://youtu.be/zo1276JLs0k (accessed on 20 January 2023).

5. Discussion

The robot "Big Foot" is an innovative design with minimal degrees of freedom and as such it is difficult to make a direct comparison with other designs. For example, ref. [32] deals with similar issues when trying to develop control algorithms for the joints of bipedal walking robots. They approach the problems in three steps: planning method, mathematical modeling (dynamics), and control algorithms. While we also have a dynamical model of the design [30], the simplicity and static stability allows for a purely kinematic approach (with adequate support from sensory input), with the only restrictions being motor loads and impact shocks. Note that in [32], the author tried to solve similar problems (minimizing impact shocks) using similar methods (a PD controller) with a key difference being that they include force control methods.

In article [33], the authors conduct a simulation of a walking robot with a similar analysis. They present angles and angular velocities with and without impact, and their effect on walking speed. They, however, do not consider accelerations and impact shocks.

The presented theoretical and experimental results are in good agreement but there are some differences. A difference is observed between the calculated rotation angles φ_B and φ_s for the two phases and the experimentally measured ones from the motor encoder. In reality, the dimensions have inaccuracies as there are slacks in the joints as well as elasticities, which lead to a deviation of the actual values for the angles for the two phases. This experiment is important for accurate determination of the coefficients in the control laws.

The theoretically calculated intervals for the two phases of motion provided in Table 2 differ from the experimentally obtained results presented in Figures 9 and 10. This is due to an inaccurate determination of the actual coefficients and the fact that the proposed model does not take into account the dynamics of the process. However, since the velocities are low, the inaccuracies from the dynamics are insignificant. The experimentally obtained values for T_i are bigger than the theoretical ones.

Graphs in Figure 6 show that both proposed laws provide a smooth increase in velocity and acceleration and satisfy the initial conditions. However, the polynomial law completes one period for $T = 5.38[s]$, which is $0.32[s]$ faster than the sinusoidal law. This is also confirmed by the result given in Figure 11.

The experimental results given in Figures 7 and 8 show that under a motion with a constant angular speed $\omega_1 = 118[^\circ/s]$ of the link 4, which is close to the maximum permissible one for the motor, the accelerometer reports very high acceleration fluctuations during the transition process. This leads to significant loads on the robot structure, which are also visible in the attached video. Decreasing the speed leads to a reduction in accelerations and shock loads, but at a constant angular velocity it significantly reduces the speed of the robot.

The experimental graphs in Figures 9 and 10 show the deviation of the angular velocity from the theoretical one. These deviations are largest at the transition points between the phases and at the maxima of the functions. The polynomial law executes one period in $T = 5.99[s]$, which is $0.33[s]$ faster than the sinusoidal law. This difference is very close to the theoretically obtained value. Therefore, the polynomial law can be used to make the robot move faster.

Figures 7–10 also contain the accelerations normal to the walking surface (labeled z-axis) read by the accelerometer. One could notice that the application of polynomial and sinusoidal control laws (Figures 9 and 10) lead to much lower values than the accelerations obtained with the maximum angular velocity given in Figure 7. These accelerations are close to those obtained in Figure 8 at the average angular velocity. An important advantage of both laws is that they significantly shorten the execution time of each period while keeping low values of accelerations along the Z-axis. This corresponds to small dynamic loads.

Figures 9 and 10 also show some disadvantages of applying the sinusoidal and polynomial control laws. In the transition between the two phases, there is a delay, which is a result of two things: the motor needs to overcome a significant torque at low speeds, which is difficult for the DC motor to do; the PD controller tries to ensure the correct motor angle, with close to zero angular velocity. When the angular velocity increases sufficiently, the PD controller tries to "catch up", as is evident by the blue line in Figures 9 and 10. This leads to another difference between experimental and theoretical results, located around the maximal values of the angular velocity. The controller is trying to compensate the difference between the real and expected velocities, which leads to overshoot when the expected velocity rapidly changes at the maximum. Note that the difference is more dire for shorter periods. The first issue could be solved by using more powerful motors, but this necessitates changes to the mechanical construction and electronics of the robot. Both issues can probably be solved by the implementation of a more sophisticated controller (for example a full PID controller). The delay can also be eliminated if the two phases are planed with a time overlap. Improvements in those directions are planned for future work.

6. Conclusions

We present a theoretical and experimental approach for the control of an innovative design of a walking robot with only two degrees of freedom, named "Big Foot". Our approach aims to reduce shock loads while trying to maximize walking speed over a flat surface. The proposed algorithm utilizes a PD controller using the robot's tactile sensors and encoders to determine the transition between the phases of walking, and the motor's angular velocity. Three different laws of motion were compared: constant angular velocity, polynomial, and sinusoidal. Theoretically and experimentally, it is shown that the polynomial law leads to higher walk speed compared with the other laws, while maintaining low motor loads and low impact shocks.

The flaws in experimental realization could be eliminated by using a more complicated control algorithm (for example a full PID controller), more powerful motors, or more sophisticated laws of motion (time overlap between different phases of walking). The proposed scheme can be generalized in two ways: by considering collision and obstacle avoidance; and by walking in an uneven and/or unstructured environment. The proposed approach may be applicable to the control of the walking mechanisms of similar mobile robots.

7. Patents

Chavdarov I, Tanev T, and Pavlov V. Walking robot. Patent application № 111362. Published summary—Bulletin № 6, 30 June 2014, p. 11, in Bulgarian.

Supplementary Materials: The following supporting information can be downloaded at: https://youtu.be/RYRJZcYdIX0—Video S1 (Figure 2); https://youtu.be/zo1276JLs0k—Video S2 (Figure 11).

Author Contributions: Conceptualization, I.C.; methodology, I.C. and K.Y.; software, K.Y. and L.M.; validation, I.C., L.M. and K.Y.; formal analysis, I.C., A.S. and K.Y.; investigation, I.C., K.Y., L.M., D.N. and A.S.; resources, I.C.; data curation, K.Y.; writing—original draft preparation, A.S., L.M. and D.N.; writing—review and editing, D.N., L.M., I.C. and K.Y; visualization, K.Y. and I.C.; supervision, I.C.; project administration, I.C.; funding acquisition, I.C. All authors have read and agreed to the published version of the manuscript.

Funding: This work was supported by the European Regional Development Fund within the OP "Science and Education for Smart Growth 2014–2020" and Project CoC "Smart Mechatronic, Eco And Energy Saving Systems And Technologies", № BG05M2OP001-1.002-0023.

Institutional Review Board Statement: Not applicable.

Informed Consent Statement: Not applicable.

Data Availability Statement: Not applicable.

Conflicts of Interest: The authors declare no conflict of interest.

References

1. Ben-Tzvi, P.; Ito, S.; Goldenberg, A.A. A mobile robot with autonomous climbing and descending of stairs. *Robotica* **2009**, *27*, 171–188. [CrossRef]
2. Figliolini, G.; Ceccarelli, M. Climbing stairs with EP-WAR2 biped robot. In Proceedings of the 2001 IEEE International Conference on Robotics & Automation, Seoul, Republic of Korea, 21–26 May 2001; pp. 4116–4121.
3. Jeon, B.S.; Yeo, S.J.; Jeong, Y.C.; Kwak, K.W.; Kim, S.H. Bio-Mimetic Articulated Mobile Robot overcoming stairs by using a slinky moving mechanism. In Proceedings of the ICAD2009, The Fifth International Conference on Axiomatic Design, Caparica, Portugal, 25–27 March 2009; pp. 173–179.
4. Grzelczyk, D.; Awrejcewicz, J. On the Controlling of Multi-Legged Walking Robots on Stable and Unstable Ground. In *Dynamical Systems Theory*; IntechOpen: Rijeka, Croatia, 2019.
5. Mikolajczyk, T.; Mikołajewska, E.; Al-Shuka, H.F.N.; Malinowski, T.; Kłodowski, A.; Pimenov, D.Y.; Paczkowski, T.; Hu, F.; Giasin, K.; Mikołajewski, D.; et al. Recent Advances in Bipedal Walking Robots: Review of Gait, Drive, Sensors and Control Systems. *Sensors* **2022**, *22*, 4440. [CrossRef] [PubMed]
6. Yasutomi, F.; Yamada, M.; Tsukamoto, K. Cleaning robot control. In Proceedings of the 1988 IEEE International Conference on Robotics and Automation, Philadelphia, PA, USA, 24–29 April 1988; Volume 3, pp. 1839–1841.
7. Bormann, R.; Hampp, J.; Hagele, M. New brooms sweep clean—An autonomous robotic cleaning assistant for professional office cleaning. In Proceedings of the IEEE International Conference on Robotics and Automation, Seattle, WA, USA, 26–30 May 2015.
8. Bosse, M.; Nourani-Vatani, N.; Roberts, J. Coverage Algorithms for an Under-actuated Car-Like Vehicle in an Uncertain Environment. In Proceedings of the IEEE International Conference on Robotics and Automation, Rome, Italy, 10–14 April 2007; pp. 698–703.
9. Hameed, I.; Bochtis, D.; Sørensen, C.L. An Optimized Field Coverage Planning Approach for Navigation of Agricultural Robots in Fields Involving Obstacle Areas. *Int. J. Adv. Robot. Syst.* **2013**, *10*, 231. [CrossRef]
10. Galceran, E.; Carreras, M. Efficient seabed coverage path planning for ASVs and AUVs. In Proceedings of the 2012 IEEE/RSJ International Conference on Intelligent Robots and Systems, Vilamoura-Algarve, Vilamoura, Portugal, 7–12 October 2012; pp. 88–93.
11. Nishiwaki, K.; Chestnutt, J.; Kagami, S. Autonomous navigation of a humanoid robot over unknown rough terrain using a laser range sensor. *Int. J. Robot. Res.* **2012**, *31*, 1251–1262. [CrossRef]
12. Nowak, P.; Milecki, A.; Białek, M. Construction and Control of the Bipedal Walking Robot. In *MATEC Web of Conferences*; EDP Sciences: Grenoble, France, 2019; Volume 252, p. 02009.
13. Siegwart, R.; Illah, R. *Nourbakhsh, Introduction to Autonomous Mobile Robots, A Bradford Book*; The MIT Press Cambridge: London, UK, 2004.
14. Rodriguez, N.E.N.; Carbone, G.; Ceccarelli, M. Design Evolution of low-cost humanoid robot CALUMA. In Proceedings of the 12th IFToMM World Congress, Besanon, France, 18–21 June 2007.
15. Hameed, I.; Bochtis, D.; Sørensen, C.L. Driving Angle and Track Sequence Optimization for Operational Path Planning Using Genetic Algorithms. *Appl. Eng. Agric.* **2011**, *27*, 1077–1086. [CrossRef]

16. Li, Y.; Chen, H.; Joo, M.E.; Wang, X. Coverage path planning for UAVs based on enhanced exact cellular decomposition method. *Mechatronics* **2011**, *21*, 876–885. [CrossRef]
17. Lakshmanan, A.K.; Mohan, R.E.; Ramalingam, B.; Le, A.V.; Veerajagadeshwar, P.; Tiwari, K.; Ilyas, M. Complete coverage path planning using reinforcement learning for Tetromino based cleaning and maintenance robot. *Autom. Constr.* **2020**, *112*, 103078. [CrossRef]
18. Yazici, A.; Kirlik, G.; Parlaktuna, O.; Sipahioglu, A. A Dynamic Path Planning Approach for Multirobot Sensor-Based Coverage Considering Energy Constraints. In *IEEE Transactions on Cybernetics*; IEEE: Denver, CO, USA, 2014; Volume 44, pp. 305–314.
19. Galceran, E.; Carreras, M. A survey on coverage path planning for robotics. *Robot. Auton. Syst.* **2013**, *61*, 1258–1276. [CrossRef]
20. Furusho, J.; Sano, A. Sensor-Based Control of a Nine-Link Biped. *Int. J. Robot. Res.* **1990**, *9*, 83–98. [CrossRef]
21. Chen, W.; Liu, T.; Li, W.; Wang, J.; Wu, X.; Liu, D. Locomotion control with sensor-driven reflex for a hexapod robot walking on uneven terrain. *Trans. Inst. Meas. Control.* **2016**, *38*, 956–970. [CrossRef]
22. Loffler, K.; Gienger, M.; Pfeiffer, F.; Ulbrich, H. Sensors and Control Concept of a Biped Robot. *IEEE Trans. Ind. Electron.* **2004**, *51*, 972–980. [CrossRef]
23. Jo, H.S.; Mir-Nasiri, N. Stability Control of Minimalist Bipedal Robot in Single Support Phase. *Procedia Eng.* **2012**, *41*, 113–119. [CrossRef]
24. Virgala, I.; Miková, L'.; Kelemenová, T.; Varga, M.; Rákay, R.; Vagaš, M.; Semjon, J.; Jánoš, R.; Sukop, M.; Marcinko, P.; et al. A Non-Anthropomorphic Bipedal Walking Robot with a Vertically Stabilized Base. *Appl. Sci.* **2022**, *12*, 4108. [CrossRef]
25. Chavdarov, I. Walking robot realized through 3D printing. *Comptes Rendus L'acad Emie Bulg. Sci.* **2016**, *69*, 1069–1076.
26. Chavdarov, I.; Naydenov, B. Design and kinematics of a 3-D printed walking robot "Big Foot", overcoming obstacles. *Int. J. Adv. Robot. Syst.* **2019**, *16*, 1729881419891329. [CrossRef]
27. Chavdarov, I.; Tanev, T.; Pavlov, V. Walking Robot. Patent application No 111362, 30 June 2014. (in Bulgarian)
28. Chavdarov, I.; Krastev, A.; Naydenov, B.; Pavlova, G. Analysis and experiments with a 3D printed walking robot to improve climbing obstacle. *Int. J. Adv. Robot. Syst.* **2020**, *17*, 1729881420925282. [CrossRef]
29. Stefanov, A.; Chavdarov, I.; Nedanovski, D.; Boiadjhiev, G. Dynamics and Control of a 3D Printed Walking Robot. In Proceedings of the 2019 International Conference on Software, Telecommunications and Computer Networks (SoftCOM), Split, Croatia, 19–21 September 2019; pp. 1–5. [CrossRef]
30. Miteva, L.; Chavdarov, I.; Yovchev, K.; Naydenov, B. Design of a Sensor System for a Minimalistic Walking Robot with Two Degrees of Freedom. In Proceedings of the 2021 International Conference on Software, Telecommunications and Computer Networks (SoftCOM), Split, Croatia, 23–25 September 2021; pp. 1–5. [CrossRef]
31. Awrejcewicz, J. Theory of Impact. In *Classical Mechanics. Advances in Mechanics and Mathematics*; Springer: New York, NY, USA, 2012; Volume 29. [CrossRef]
32. Leng, X.; Piao, S.; Chang, L.; He, Z.; Zhu, Z. Universal Walking Control Framework of Biped Robot Based on Dynamic Model and Quadratic Programming. *Complexity* **2020**, *2020*, 2789039. [CrossRef]
33. Jánoš, R.; Sukop, M.; Semjon, J.; Tuleja, P.; Marcinko, P.; Kočan, M.; Grytsiv, M.; Vagaš, M.; Miková, L'.; Kelemenová, T. Stability and Dynamic Walk Control of Humanoid Robot for Robot Soccer Player. *Machines* **2022**, *10*, 463. [CrossRef]

sensors

Article

Semantic-Structure-Aware Multi-Level Information Fusion for Robust Global Orientation Optimization of Autonomous Mobile Robots

Guofei Xiang [1,2], Songyi Dian [1,*], Ning Zhao [2] and Guodong Wang [2]

1 College of Electrical Engineering, Sichuan University, Chengdu 610065, China
2 National Key Laboratory of Special Vehicle Design and Manufacturing Integration Technology, Baotou 014031, China
* Correspondence: scudiansy@scu.edu.cn

Abstract: Multi-camera-based simultaneous localization and mapping (SLAM) has been widely applied in various mobile robots under uncertain or unknown environments to accomplish tasks autonomously. However, the conventional purely data-driven feature extraction methods cannot utilize the rich semantic information in the environment, which leads to the performance of the SLAM system being susceptible to various interferences. In this work, we present a semantic-aware multi-level information fusion scheme for robust global orientation estimation. Specifically, a visual semantic perception system based on the synthesized surround view image is proposed for the multi-eye surround vision system widely used in mobile robots, which is used to obtain the visual semantic information required for SLAM tasks. The original multi-eye image was first transformed to the synthesized surround view image, and the passable space was extracted with the help of the semantic segmentation network model as a mask for feature extraction; moreover, the hybrid edge information was extracted to effectively eliminate the distorted edges by further using the distortion characteristics of the reverse perspective projection process. Then, the hybrid semantic information was used for robust global orientation estimation; thus, better localization performance was obtained. The experiments on an intelligent vehicle, which was used for automated valet parking both in indoor and outdoor scenes, showed that the proposed hybrid multi-level information fusion method achieved at least a 10-percent improvement in comparison with other edge segmentation methods, the average orientation estimation error being between 1 and 2 degrees, much smaller than other methods, and the trajectory drift value of the proposed method was much smaller than that of other methods.

Keywords: simultaneous localization and mapping (SLAM); semantic; information fusion; orientation estimation; mobile robots

Citation: Xiang, G.; Dian, S.; Zhao, N.; Wang, G. Semantic-Structure-Aware Multi-Level Information Fusion for Robust Global Orientation Optimization of Autonomous Mobile Robots. *Sensors* **2023**, *23*, 1125. https://doi.org/10.3390/s23031125

Academic Editors: David Cheneler and Stephen Monk

Received: 26 December 2022
Revised: 12 January 2023
Accepted: 13 January 2023
Published: 18 January 2023

1. Introduction

With the rapid development of sensing, computation, manufacturing, and control technologies in recent years, various kinds of robots have been coming into our lives and work gradually, such as unmanned aerial vehicles, robot vacuum cleaners, intelligent vehicles, autonomous disinfection robots, logistics delivery robots, and so on. These different kinds of robots have been transforming our social lives ever-increasingly [1,2]. For example, during the current COVID-19 pandemic, we always expect to use robots to replace humans to complete the disinfection work in public places, due to different regions in the environment having different risks. Thus, in these environments, one of the most-important prerequisites for the robots to accomplish the task safely, autonomously, and efficiently is that the robot should know its own location relative to the environment. Therefore, it is of great importance to endow the robot with the ability of autonomous navigation. Simultaneous localization and mapping (SLAM) [3] is such a technique that can build the environment map and compute the location in the map simultaneously. With the help of SLAM, when a robot enters an uncertain or even unknown environment, it can make use of the structured environment information to determine its location on the map; it can also reconstruct its surroundings by only relying on its own sensors. Thus, the robot can move and complete specific tasks in a prior unknown and unstructured environment autonomously [4,5].

SLAM has been widely applied in various applications. Since the robot has to build the environment map and locate its positions by the sensors carried by it, the most-fundamental problem is how to accurately and robustly extract information that can be used for map building and localization. Conventionally, the information extraction techniques have been utilized to obtainfeatures directly from the available sensors by some manual rule-based methods, then these features will be used for map building and localization. With the rapid development of deep-neural-network-based learning methods, we can obtain much high-level semantic information from the environment by the nature of the multi-level information processing mechanism. Thus, the accuracy and robustness of the SLAM system can be improved further by this high-level semantic information [3,6,7].

Semantic information plays a positive role in the SLAM system of mobile robots. The semantic information can enhance the robustness against those disturbances in the environment and systems [8–10]. During the computing of semantic information, the effective information generally flows from low-level, low-accuracy to high-level, high-accuracy by the nature of the hierarchical structure of the deep neural network; thus, this high-level information can adapt to various variations in the environment [11–14]. Some works showed that the stability of the features can be enhanced by filtering out those features of dynamic or potentially moving objects. Some works showed that this high-level semantic information lowers the sensitivity to sensor noise to some extent, such as straight lines, triangles, and planes, which are more robust than the original sensor data [15,16]. Some work showed that semantic information can improve the accuracy and reliability of data association by taking advantage of relevance, heterogeneity, and invariance in semantics [17,18]. On the other hand, semantic information can support the accomplishment of various high-level tasks to a great extent, since these high-level tasks are usually encoded by some natural-language-like semantics, consisting of some specific, abstract symbols, logic, and temporal operators, such as linear temporal logic, signal temporal logic, metric temporal logic, and so on [19–22]. However, though semantic SLAM has achieved state-of-the-art performance in various practical applications, the purely data-driven-based neural network model has also shown many disadvantages, such as the non-explainability issue and the huge amount of labeled data requirement, both of which limit its adaptability in different environments, especially the human beings involved and safety-critical scenarios [23,24]. Therefore, in this work, we hope to resolve the aforementioned problem partially by embedding rule-based and explainable structures, allowing the robot to be able to adapt to the new environment efficiently.

In practical applications, in order to be able to achieve the all-around perceptual coverage of objects in any direction in the task scene, mobile robots often need to be equipped with multiple vision sensors to meet this demand. In smart cars, for example, the vision system on board often consists of multiple high-resolution cameras with different viewing angles to cover a 360-degree field of view around the vehicle. Having a wider field of view means that the robots can observe more information about the environment, but it also means that the amount of computation required to process this information is greatly increased. In addition, under the current situation that the perception method based on the deep neural network model has become the mainstream, more raw perception data input also require corresponding orders of magnitude labeled data, which undoubtedly greatly increases the cost and time of the perception model when adapting to new scenarios, which is not conducive to rapid deployment and application in new scenarios.

In summary, the main contributions of this article are summarized as follows:

- We propose a multi-level semantic-structure-aware global orientation estimation framework, which consists of a semantic information extraction module and a global orientation estimation module.
- In the semantic information extraction stage, we attempted to process the surround view synthesized images obtained after the inverse perspective mapping (IPM) of the original image and use the passable area segmentation mode, which more easily obtains the annotation data, fully combines the potential prior information in the task, and obtains the boundary including the passable area. The semantic information including visual feature points and ground marking edges in the passable area can effectively reduce the requirements of the semantic perception model for labeled data and, finally, meet the needs for tasks such as mapping and positioning.
- In the orientation estimation stage, we designed a segmentation method for marker lines based on the structural rules of the Manhattan world, which can be used to obtain from the image a collection of line segments that conform to the structural assumptions of Manhattan and the dominant orientation of these lines; thus, we can distinguish the marker lines from the noise lines.
- We validated the effectiveness of the proposed scheme by taking the semantic perception task of intelligent vehicles equipped with multi-vision systems in the automatic valet parking task as an example.

The remainder of this work is organized as follows: In Section 2, some related works are described. In Section 3, we describe our system architecture, followed by semantic information extraction in Section 4. After that, we present the semantic-aware global orientation estimation method in Section 5, with the experiments presented in Section 6 and the conclusion in Section 7.

2. Related Works

2.1. Multi-Camera SLAM System

In general, increasing the number of cameras in a mobile robot system can effectively improve its perception range, thereby increasing the potential improvement of its perception ability [25,26]. Therefore, researchers are paying more and more attention to how to model multi-vision systems and use the characteristics of multi-vision systems to improve the accuracy and robustness of the system, and the relevant research results have been widely used in mobile robot SLAM tasks [27–29].

Although direct processing of multi-camera images maximizes the use of the information in the original image, it also means that sufficient computing resources are required to enable fast processing, so it may not be suitable for tasks that require high real-time performance. As an alternative scheme, some recent research has also been performed to use the image after surround view synthesizing as the input, which can greatly improve the processing efficiency of the system while achieving satisfactory accuracy.

2.2. Feature Extraction Techniques for SLAM

The environmental feature information utilized in traditional visual SLAM frameworks typically includes sparse corner features (e.g., SIFT, ORB) in indirect methods and pixel luminance information in direct methods [30–32]. On top of this, there are many ways to further improve SLAM by detecting features such as geometric elements, such as line segments and planes, in the scene reliability of the system [12,33,34].

With the rapid development of deep learning in recent years, more and more methods based on deep neural networks have been integrated into visual SLAM systems to improve the system's perception and utilization of environmental information. As the main means of semantic perception, semantic segmentation and object detection networks are widely used in the acquisition of pixel-level and object-level semantic information, respectively, and bring additional semantic and geometric constraints to the SLAM system, especially in scenarios with many dynamic objects, which significantly improves the stability and accuracy of the SLAM system. Semantic SLAM has been widely studied and implemented in various kinds of robots by the substantial progress of deep learning techniques, especially alongside high-performance computing machines, such as graphics processing units (GPUs). For example, the image-segmentation-based neural network models include the FCN [35], SegNet [36], PSPNet [37], ICNet [38], DeepLab [39], MobileNet [40], and their extensions; the image-based object detection network models include RCNN [41], YOLO [42], SSD [43], and their extensions; point cloud segmentation network models include PointNet [44] and its extensions; point cloud-based object detection network modes include VoxelNet [45], PointPillars [46], and their extensions. These semantic segmentation and object detection networks have been applied to extract object-level and point-level semantics, which bring prior structures into the SLAM system, and more accurate and robust performances are obtained. However, these deep-network-based methods need huge amounts of labeled data, which are usually not affordable for practical applications. Therefore, we propose a hybrid and multi-level information fusion scheme to deal with this problem.

2.3. Semantic Information with Synthesized Surround View Image

The homogeneous transform or IPM relied on by surround view synthesizing techniques is a very classic image processing method [47,48]. With the rapid development of vision systems in recent years, this technology has also been widely used, such as the reverse assistance system in cars, which usually uses this technology to enable the driver to easily observe the situation of surrounding objects and the distance from the robot. Similarly, when mobile robots complete tasks such as mapping, positioning, and navigation, researchers also focus on how to extract effective semantic information from the synthesized surround view image, such as various pavement markings, obstacles, and passable areas, and describe them in different forms of representation such as point clouds and occupied grids to help complete related tasks [49,50].

However, because models based on deep neural networks often require a large amount of manually labeled data for training, they are difficult to quickly scale and apply to new scenarios. Therefore, this work considered only the rough passable space segmentation results to assist in extracting the semantic information in the surround view synthesized image and obtain rich and effective semantic information such as passable area boundaries, pavement sparse feature points, and pavement marking edges, which not only improves the effective information quality of the input SLAM system, but also greatly reduces the requirements for annotated data, so it can be applied to mapping positioning and navigation tasks in new scenes more quickly than the previous methods.

2.4. Semantic-Feature-Based Global Localization

Due to the high ability of convolutional neural networks to discover complex patterns, NetVLAD uses the network to generate global descriptors directly end-to-end [51]. LoST uses the network to learn local key points, global descriptors, and semantic information to complete VPR tasks with extreme changes in view and appearance [52]. In addition, some methods even use the network directly to give the results of pose estimation end-to-end [53–55]. Although their results show extremely high accuracy and robustness to noise in the dataset, they are invariably dependent and data-dependent, so the generalization performance of these methods is not satisfactory.

As a high-level feature, semantic information is compact and has good stability, so it is suitable as a reference for visual localization. Some research work used specific landmarks as masks to avoid extracting feature information in some dynamic regions [56]. VLASE uses the semantic edges extracted by the network to characterize the image and, then, achieve localization. Some methods use columnar landmarks as a special location reference to improve the positioning accuracy of robots [57]. The aforementioned methods only consider the information of the episemantic class. However, the spatial relationship between semantics also implies information about the place. Therefore, the method proposed in this work uses both episemantic and spatial distribution information to pursue a complete description of the scene. Some studies use graph models to encode scenes and their topologies, but building graphs from images is a difficult problem, especially in sparse landmark scenes [58,59]. In this work, semantic information is explicitly encoded as semantic vectors, and spatial geometric relationships are represented in the form of histograms. The descriptor constructed by this encoding method has a compact structure and good interpretability, which is commonly used in various scenarios.

3. Structure-Aware Global Orientation Estimation System

In this work, we built a SLAM framework based on semantic-aware structural information for estimating the robot's global orientation state, as shown in Figure 1. The framework takes a synthesized surround view image as the input and outputs a drift-free trajectory and a map consisting of marker lines. As we can see, the proposed SLAM system consists mainly of two procedures, structured semantic information extraction and orientation estimation. During the structured semantic information extraction stage, the passable space in the image is firstly extracted with the help of the semantic segmentation network model as a mask for feature extraction; thus, we can improve the computational efficiency of the system while retaining most of the effective information. Since the boundary of the passable area contains geometric information about the environment, it could also be converted into a point cloud representation by LiDAR, which effectively complements the visual feature point information. In addition, since the passable space contains rich pavement marking edges, by further using the distortion characteristics of the reverse-perspective projection process, the distortion edges could be effectively eliminated, and the hybrid edge information could be used for mapping and localization tasks. During the orientation estimation stage, based on the dominant orientation information, the global orientation of each frame can be preliminarily estimated without drift. In order to further improve the anti-interference performance, we built a local map and designed it to optimize the state factor in the local map using the global orientation error constraint. Finally, the estimation of the global orientation is obtained, and a line map that reflects the structure of the real scene can be reconstructed.

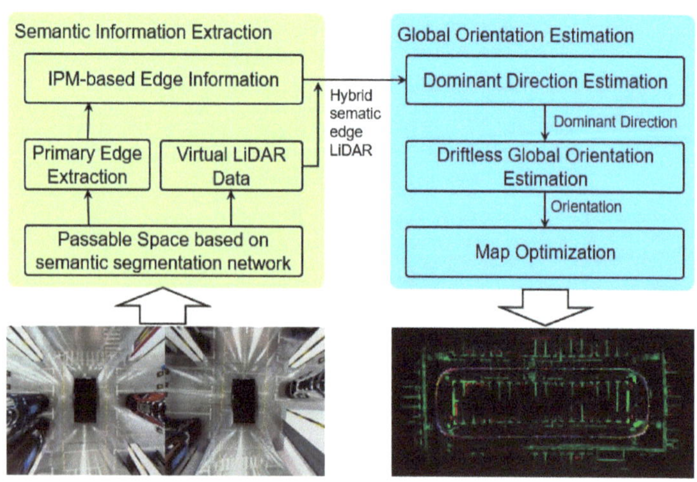

Figure 1. Semantic-structure-aware global orientation estimation framework.

4. Semantic Information Extraction

4.1. Virtual LiDAR Data Generation

Conventionally, LiDAR can accurately measure the distance information of objects in the environment relative to the robot itself, and the virtual radar proposed in this work obtains similar distance measurement information through the secondary processing of semantic segmentation results, simulating the detection results of LiDAR sensors. Since most of the area in the original image is the ground and objects on the ground, we considered using IPM to convert the original image to a top view and synthesized it to obtain a surveillance-synthesized image. The process of IPM transformation can be described as

$$\begin{bmatrix} x_{veh} \\ y_{veh} \\ 1 \end{bmatrix} = \lambda \begin{bmatrix} \mathbf{R}_c & \mathbf{t}_c \end{bmatrix}^{-1}_{\text{col}:1,2,4} \pi_c^{-1}\left(\begin{bmatrix} u_{fish} \\ v_{fish} \\ 1 \end{bmatrix}\right), \quad (1)$$

where $\left[u_{fish}, v_{fish}\right]^T$ denotes the pixel coordinates of the fish-eye camera image, $\pi_c(\cdot)$ denotes the projection model with its inverse projection model as $\pi_c^{-1}(\cdot)$, $\begin{bmatrix} \mathbf{R}_c & \mathbf{t}_c \end{bmatrix}$ denotes the external parameters of the camera, i.e., the homogeneous transformation matrix from the robot coordinate system to the camera coordinate system, and $[]_{\text{col}:k}$ denotes the k-th column of the matrix. λ denotes the scaling factor, which can be obtained by calibrating the correspondence between the transformed image and the actual scene size, and $\begin{bmatrix} x_{\text{veh}} & y_{veh} \end{bmatrix}^T$ denotes the position of the point in the final robot coordinate system.

After IPM transformation, a virtual camera directly above the robot and shot vertically downward can be further constructed, which can map the points under the obtained robot coordinate system to obtain the final synthesized surround view image as

$$\begin{bmatrix} u_{ipm} \\ v_{ipm} \\ 1 \end{bmatrix} = \mathbf{K}_{ipm} \begin{bmatrix} x_{veh} \\ y_{veh} \\ 1 \end{bmatrix}, \quad (2)$$

where $\begin{bmatrix} u_{ipm} & v_{ipm} \end{bmatrix}^T$ denotes the pixel coordinate of the synthesized surround view image and $\mathbf{K}_{ipm}$ denotes the internal parameter matrix for the virtual camera.

To obtain the segmentation results of free spaces, this paper trained a semantic segmentation network model to distinguish between passable space and non-passable space. Then, through the morphological processing at the image level, the segmentation results can be further modified to obtain a more ideal segmentation effect, and the boundaries of the passable area can be further extracted. Since the scale transformation coefficient between the pixel distance of the image and the actual distance can also be obtained at the same time during the calibration process of surround view synthesizing, the pixel distance between the point on the passable area boundary in the image and the pixel distance of the image center can be directly converted into the actual distance under the robot coordinate system or virtual radar sensor coordinate system. Then, according to the scanning method using LiDAR, all boundary points are sampled at fixed angular intervals, and the boundary points in the same angle window can be represented by the closest point so that the final virtual radar measurement data can be obtained. The process of virtual LiDAR data generation is shown in Figure 2.

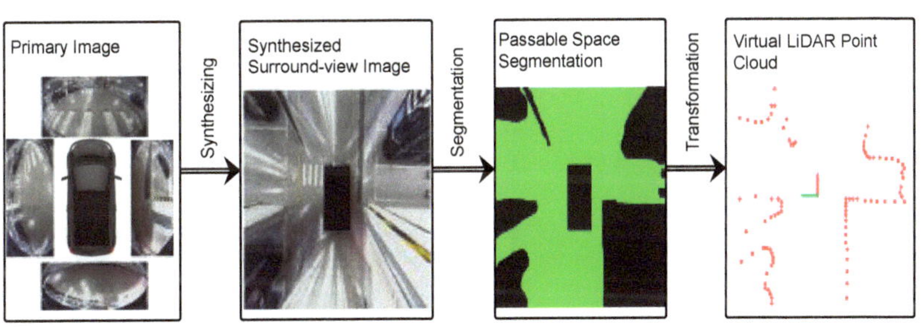

Figure 2. The process of virtual LiDAR data generation.

4.2. IPM-Based Hybrid Semantic Edge Extraction

The original synthesized surround view image often contains a large number of ground markings, which can be fed into SLAM systems as high-quality road sign information. However, the information of the synthesized surround view image is also disturbed by a large number of ground spots. At the same time, the reverse-perspective projection transformation during the synthesizing process will distort objects with a certain height on the ground. Therefore, an edge segmentation module needs to be designed to reject the above interference edges, so that the high-quality effective edges on the ground can be retained. The process of hybrid semantic edge extraction is shown in Figure 3.

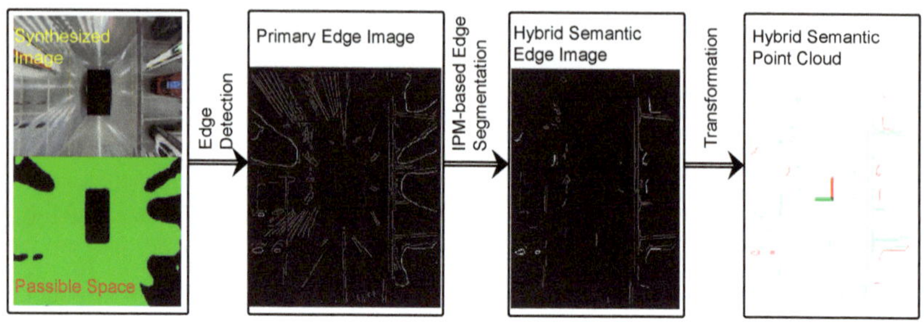

Figure 3. The procedure of hybrid semantic edges extraction.

Primary edge extraction: The edges on the input surround view synthesized image can be extracted by traditional edge detection methods, such as the Canny edge detector and LSD detector. On the one hand, the edge of the ground mark and the projected edge of the object can be well extracted; on the other hand, traditional edge detectors will also detect those invalid edges, such as the edges of surrounding robots, pillars, and light spots, so the original edges need to be processed to some extent. Due to the segmentation of the passable space, edges that are located on objects above the ground level can be easily removed. However, there is still a considerable part of the spot edge that cannot be removed in this way, and the part of the passable space boundary that is affected by distortion requires additional processing; otherwise, it will not be able to be entered into the subsequent SLAM system and achieve the ideal positioning and mapping results.

IPM-based edge segmentation: Considering that the distortion edges are mainly centered on the photocenter of each camera and distributed in the direction of the rays, an intuitive idea is to build a polar coordinate system with its photocenter as the origin for each camera and count the number of edge points in each direction. However, the segmentation method based on ray accumulation has problems such as erroneous removal of dense small edges, such as the edge of zebra crossings, threshold coupling of angle parameters and the number of edge points in the fan, and inaccurate results. In order to consider the geometric distribution of edges, further attempts can be made to detect line segments in the edge image. Specifically, traditional linear detectors, such as those based on the Hough transform, can extract segments of a length from the original edge image, then calculate the distance between the line in which each segment is located and the camera center of its field of view. Finally, those segments that are small enough away are marked as distorted edge areas. However, the segmentation method based on segment detection has problems such as the erroneous exclusion of some unconnected edge points, and the distance between the line segment and the optical center will also affect the selection of the final distance threshold.

To further consider the fine structure of the edge, different consecutive edges can first be distinguished before edge segmentation. Then, for each edge, the Douglas–Peucker algorithm is used to approximate the edges. The line segments obtained by approximating the contour of each edge are connected to the center of the camera photocenter in the field of view, and then, the angle between the line and the line segment is calculated to remove those parts whose angles are less than a certain threshold. The main advantage of the segmentation method based on polyline approximation is that it simplifies the operation of segment detection and limits segment approximation to the interior of each edge. At the same time, for edges at different distances, the way of evaluating the angle is more stable and consistent than the previous way of evaluating the distance, especially for those edges that are very far from the robot. The algorithm of the polyline-approximation-based edge segmentation method is shown in Algorithm 1.

Algorithm 1 Polyline-approximation-based edge segmentation algorithm.

Input: Synthesized surround view image within the passable area $\mathbf{I}_{edge}$.
Output: Valid edge image after segmentation $\mathbf{I}_{seg}$.
Initializing: The maximum allowable error of the polyline approximation D_{max}, the maximum allowable angle threshold for the line and polyline between the midpoint of the effective polyline and the camera optical center θ_{max}, reject range r.
Extracting the contours of all edges in $\mathbf{I}_{edge}$ to form the edge profile set $\mathcal{C}$;
for $c \in \mathcal{C}$ **do**
  Computing the polyline set $\mathcal{L}$ by taking advantage of Douglas–Peucker operators and D_{max} to approximate the contours with polylines;
  for $l \in \mathcal{L}$ **do**
    Computing the camera center note position C of the field of view based on the endpoint of the polyline l;
    Computing the connecting line l_C between the middle point of l and the camera optical center C;
    Computing the angle θ between l and l_C;
    if $\theta < \theta_{\max}$ **then**
      Removing all the edge points covered by l along with reject range r from $\mathbf{I}_{seg}$;
    end if
  end for
end for

5. Semantic-Aware Global Orientation Estimation

5.1. Local Dominant Direction Estimation

In a real navigation environment, there is a large number of road markings on the ground, which is a good reference landmark for SLAM systems. However, there are also various disturbing noises on the ground, such as glare or water stains. Therefore, it is necessary to distinguish the marker lines from the noise lines. By observation, it can be seen that most of the marker lines are parallel or perpendicular to each other, while the noise lines are disordered. Inspired by this phenomenon, this section designs a segmentation method for marker lines based on the structural rules of the Manhattan world, which can be used to obtain from the image a collection of line segments that conform to the structural assumptions of Manhattan and the dominant orientation of these lines, as shown in Figure 4. The left (a) is the raw input image. The middle (b) is the extracted raw lines, and the right (c) is the line segmentation result. Green lines are the preserved marking lines, and red lines are the noisy lines. The bottom row shows three consecutive images. Green lines are marking lines; the orange arrow is the local dominant direction of the current frame, and the blue arrow is the global dominant direction.

In the initial frame, the line with the highest number of perpendicular and parallel to the other lines is considered to be the initial dominant direction x_d, and the vertical or parallel between two lines can be evaluated by computing,

$$\theta - \arccos\left(\frac{x_i \cdot x_j}{|x_i||x_j|}\right) < \delta, \tag{3}$$

where δ denotes the tolerance and θ can be set as 0 for the parallel lines' evaluation and $\frac{\pi}{2}$ for the perpendicular lines' evaluation. Thus, we can build two sets for these two types of lines as $L_{//}, L_{\perp}$ for parallel lines and perpendicular lines, respectively. Then, we optimize the dominant direction:

$$\min_{a,b} \sum_{l_i \in L_{//}} |x'_d \cdot l_i| + \sum_{l_j \in L_{\perp}} |x_d \cdot l_j|. \tag{4}$$

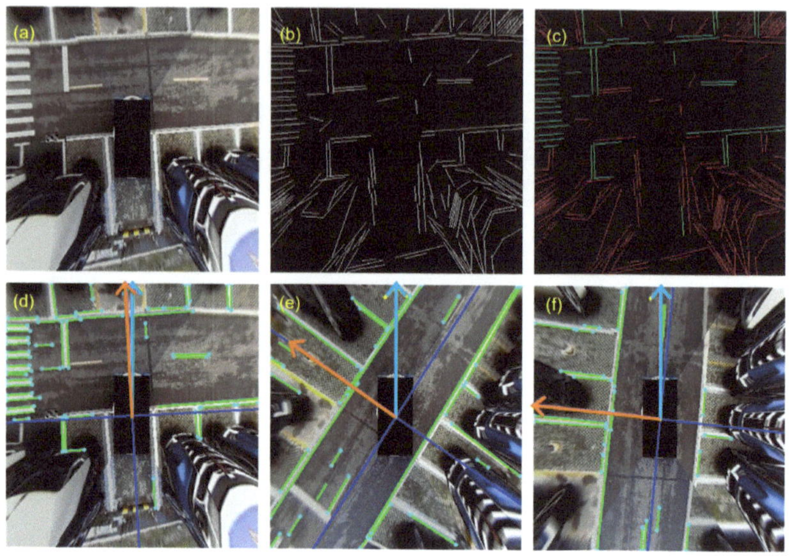

Figure 4. The three subfigures in the upper row illustrate the procedure of structure information extraction: (**a**) denotes the raw input synthesized image, (**b**) the extracted raw lines, and (**c**) the segmentation results, in which green lines denote the preserved marking lines and red lines denote the noisy lines. (**d–f**) denote the relationship between the local dominant direction and global orientation, in which the orange arrow denotes the local dominant direction and the blue arrow denotes the global dominant direction for the current frame, respectively, while the green lines denote the marking lines.

The first term denotes that the parallel line intersects with the dominant direction at point $x'_d = [b, -a, 0]$, and the second term means the perpendicular line intersects with the dominant direction at point $x_d = [a, b, 0]$.

Due to the robot's dynamic constraints, the maximum directional change in adjacent frames is assumed to be η. After successful initialization, in order to reduce the amount of calculation, subsequent frames will only look for a new dominant direction and the corresponding set of structure lines from candidate line features in the interval with the dominant direction of the previous frame less than η.

5.2. Global Orientation Optimization

When a new image appears, first extract the marker line collection and determine the local dominant direction according to the method described in the subsection before. As shown in Figure 4, the orientation of the current frame equals that angle between the global orientation and the dominant direction of the current frame; thus, it can be computed as

$$\theta_z = \arccos\left(\frac{I_g \cdot I_c}{|I_g||I_c|}\right), \tag{5}$$

where I_g denotes the global orientation and I_c denotes the dominant direction of the current frame.

A collection of marker lines may contains some noise lines. However, these noise lines also satisfy the geometric rules of the Manhattan world assumption, so they have little effect on the orientation estimates. This way of estimating orientation is independent of the orientation estimation of adjacent frames, so there is no cumulative error. When the orientation is known, only the translation term remains to be estimated. In this way, the originally nonlinear pose estimation problem is transformed into a linear least-squares problem.

Due to the existence of occlusion, visual blurring, and other factors, the consistency of line segment detection is poor, and a line segment is often split into two independent line segments, or the length of the line segment will change. This results in the line feature not binding enough on the amount of translation in the extension direction and may even introduce incorrect constraints. Although it is also possible to extract the features of points separately to constrain the amount of translation, this will cost additional computational resources. As shown in Figure 4, the endpoints of many segments are also corner points. Therefore, in order to solve the above problem, this paper takes the endpoints of the line segment as point features and uses bidirectional optical flow to trace these endpoints to establish data associations between endpoints, rather than between line features. More specifically, mapping the feature points from the reference frame to the current frame, denoted as P_r, re-mapping the feature points from the current frame to the reference frame, denoted as P_c, only if enough point pairs are obtained, we can compute the translation term as

$$^{W}\mathrm{t}_{B_i} = {}^{W}\mathbf{R}_{B_j} \cdot \bar{\mathrm{p}}_{B_j} + {}^{W}\mathrm{t}_{B_j} - {}^{W}\mathbf{R}_{B_i} \cdot \bar{\mathrm{p}}_{B_i}, \tag{6}$$

where $\bar{p}$ denotes the mean position of the point sets. Then, estimate the translation of the current frame via the reprojection errors.

If the current frame is more than 10 frames away from the previous keyframe and the required number of features is met, the current frame is selected as the keyframe and added to the local map. Then, optimize the local map using an objective function that minimizes the global orientation residuals and reprojection errors as

$$\min_{^{W}\mathbf{T}_{B_i}} w \sum_i \left(1 - \frac{I_i \cdot ({}^{B_i}\mathbf{R}_W I_g)}{|I_i||{}^{B_i}\mathbf{R}_W I_g|}\right) + \sum_i \left\| \mathrm{p}_i - {}^{W}\mathbf{T}_{B_i} \pi_s^{-1}(\mathrm{u}_i) \right\|_2, \tag{7}$$

where the former term means the global orientation residual errors since the angle between I_i and $^{B_i}\mathbf{R}_W I_g$ should be zero. $^{B_i}\mathbf{R}_W$ denotes the transformation of x_g from the world coordinates to the robot frame B. The latter term means the reprojection errors. $^{W}\mathbf{T}_{B_i} = [{}^{W}\mathbf{R}_{B_i}, {}^{W}\mathrm{t}_{B_i}] \in \mathrm{SE}(2)$, and π_s^{-1} transform the pixel u_i to the world coordinates W. Here, the end-to-line reprojection error commonly used by other methods is not used because the point-to-point reprojection error is more constrained and accurate in the amount of translation.

After local map optimization, the global orientation can be found, which is not disturbed by the accumulated errors. At the same time, the position of the line in the world coordinate system is also determined. Because line features are primarily marked lines on the pavement, maps reconstructed from line features reflect the structure of the road and the markings on the roads.

6. Experimental Results

6.1. Experiments' Configurations

In this experiment, an automotive platform equipped with four fisheye cameras was mounted for parking assistance purposes, as shown in Figure 5. The experimental data were recorded by a car and wheel speedometer equipped with four fisheye cameras. All cameras have a 190-degree viewing angle, take images at a frequency of 25 Hz, and have an image resolution of 1920*1208. Each camera was connected to the in-vehicle computing platform NVIDIA Drive PX2 via a Gigabit multimedia serial interface. Timestamp synchronization is guaranteed by the hardware. The composite ring view has a resolution of 384*384 and covers a range of 15.3 m by 15.3 m around the vehicle circumference. The measured values of the wheel speedometer and IMU are obtained via the CAN bus. The experimental computing platform has a configuration of 3.3 GHz Intel i5-4590 CPU and 16 GB memory. In the indoor scenes with weak GPS signals, the true value of panning is provided by the fusion of wheel odometry and high-precision IMU measurements. In the outdoor scenes, the true value of planning is provided by the high-precision differential GPS.

Figure 5. (**a**) The experimental platform with four fisheye cameras. (**b**) The outdoor and (**c**) indoor parking scenes.

6.2. Evaluation of Structural Information

Figure 6 shows some sample results of different edge segmentation methods at the same level of recall. For the original edge picture corresponding to each test sample, the truth value of the edge segmentation result can be given by manual annotation, and then, the difference between the segmentation result of each method and the true value can be compared, including the part that was correctly retained, the part that was mistakenly rejected, and the part that was incorrectly retained. As we can see, the splitting method based on ray accumulation retains too many wrong spot edges, usually because the same long edge is incorrectly divided into different sector areas, and the length of each segment does not reach the set threshold, so it cannot be correctly rejected. The segmentation method based on line segment detection can detect most of the spot edges, but for the less straight edges in the distance, especially the boundaries of the passable area in the distance, the set distance threshold cannot be reached, so it will be incorrectly retained. The segmentation method based on polyline approximation can successfully remove most of the distorted edges, while only a small number of effective edges are erroneously rejected because they are exactly in the direction of the camera's field of view rays.

More specifically, as shown in Table 1, when the recall of all methods is controlled at about 0.73, the segmentation method based on polyline approximation can achieve the best segmentation effect, which is 24.3 percent higher than the segmentation method based on ray accumulation and 11.9 percent higher than the segmentation method based on line segment detection.

Table 1. Comparisons of the precision and recall rates among different edge segmentation methods.

Methods	Precision	Recall
Ray-accumulation-based segmentation	0.621	0.731
Line-segment-detection-based segmentation	0.745	0.729
Polyline-approximation-based segmentation	0.864	0.730

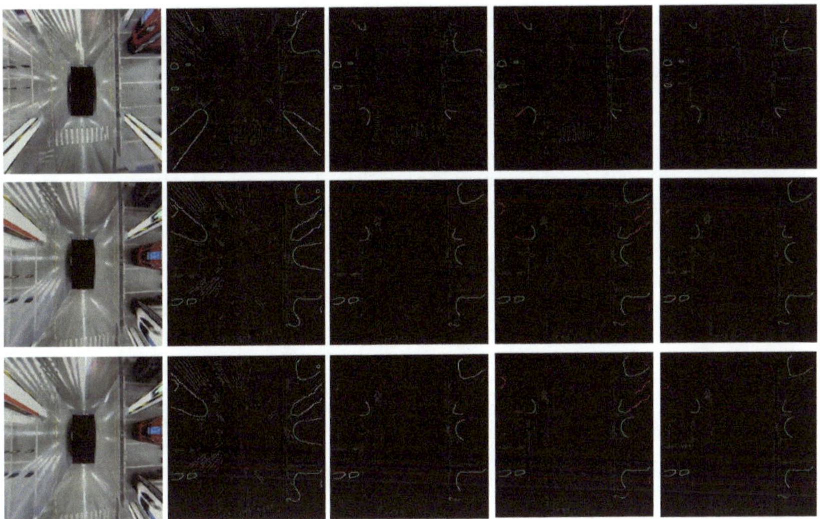

Figure 6. The comparison of different edge segmentation methods. Each row represents the results for different structure extraction methods for the same synthesized image, which is illustrated in the first column, and each column represents the results for different synthesized images with the same structure information extraction method, which is described at the bottom of each column. The first column denotes the three input images, the second column the ground truth of edge information, the third column the edge information extracted by the ray accumulation method, the fourth column the edge information extracted by the line segment detection method, and the fifth column the edge information extracted by polyline approximation method. The manually labeled ground truth edges are drawn in green. For each column of segmented edges, the green lines denote that the edges are correctly preserved, while the red lines denote that the edges are not correctly preserved, and the white lines denote those missed for the specific method. Moreover, The color intensity means whether the edges are inside the free space (brighter) or on the contour of the free space (darker).

6.3. Global Orientation Optimization

We compared our semantic-structure-based method (semantic-aware) with three other methods, ORB feature-point-based method (ORB-based), primary-edge-based method (primary-edge-based), and wheel-speedometer-based method (wheel-speedometer-based). In Figure 7, we show the comparisons of the orientation estimation errors for the different methods. As the trajectory becomes longer, the orientation angle error of the three comparison methods gradually increases. The main reason is that errors accumulate over time. However, our method achieves a global estimation of the change in direction and avoids the occurrence of accumulated errors. Therefore, the orientation error of this algorithm is independent of the length of the trajectory. When the vehicle turns, the orientation error of all methods increases significantly. This is mainly caused by blurry images caused by fast rotation. Table 2 shows a comparison of the mean orientation estimation errors of the different methods. The average orientation estimation error of the methods presented in this work is stable between 1 and 2 degrees, well below other methods, both indoors and outdoors. This also verifies that, among the different types of feature information, the structure line information is the most robust to the interference factors, and the orientation results estimated based on it are the most accurate.

The comparisons of the trajectories for the different methods are shown in Figure 8. Since the proposed method has the most-accurate orientation estimation, the trajectory drift value of the proposed method is much smaller than that of other methods. The trajectory estimated using other information has a significant drift. At the same time, it can be seen from the trajectory comparison chart that, although the trajectory using structural information is more consistent with the true value in the upward direction, there are also jagged oscillations and even mutations in some local areas. This is actually due to environmental factors: when there are few marked lines on the road surface in the environment and when the line features are not accurate enough due to blurry images, the estimated dominant direction will oscillate, rather than be smooth enough. Therefore, although the positioning strategy based on structural information proposed in this paper performs well in orientation estimation, the overall trajectory still has errors compared with the truth trajectory.

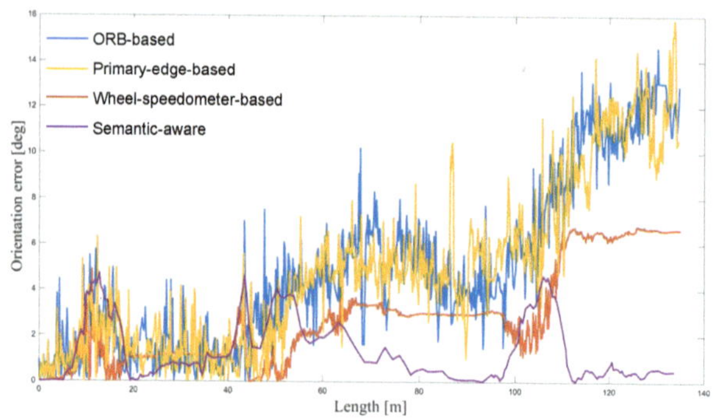

Figure 7. The comparisons of the orientation estimation errors of different methods.

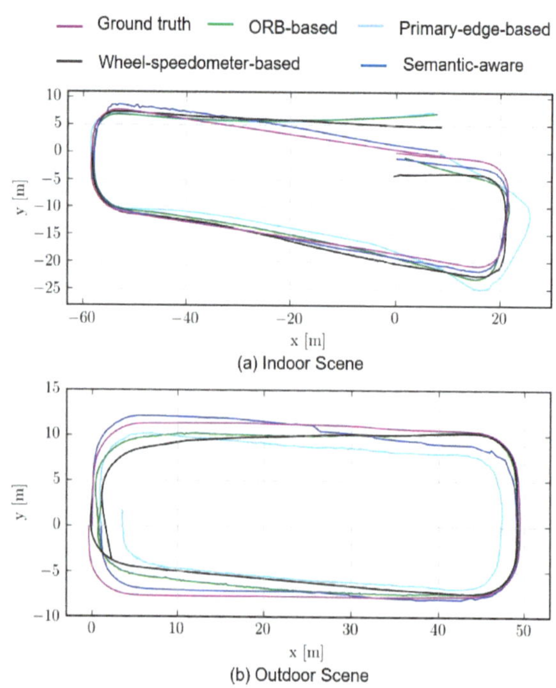

Figure 8. The comparisons of the trajectory estimation for different methods. (**a**) Indoor Scene, (**b**) Outdoor Scene..

Table 2. Comparison of the orientation estimation errors of different methods.

Methods	ORB-Based (°)	Primary-Edge-Based (°)	Wheel-Speedometer-Based (°)	Semantic-Aware (°)
Outdoor-navigation	6.426	5.885	3.815	1.491
Indoor-navigation	4.839	4.342	3.095	1.897

7. Conclusions

This paper used high-level semantic information and spatial distribution information to assist the visual location recognition task and used the semantic structure information to realize the global orientation estimation method without drift interference. In this paper, a visual semantic perception system based on the synthesized surround view image was proposed for the multi-eye surround vision system, which was used to obtain the visual semantic information required for SLAM tasks. Different from the traditional method of obtaining feature information from the original multi-eye image, the original multi-eye image was transformed and synthesized by reverse-perspective projection to obtain a synthesized surround view image that could describe the scene in an all-round way, to improve the computational efficiency of the system while retaining most of the effective information. To retain the effective information in the synthesized surround view image and remove the features that had been distorted during the synthesizing process or located on the dynamic object, this paper extracted the passable space in the image with the help of the semantic segmentation network model as a mask for feature extraction. Since the boundary of the passable area contained geometric information about the environment, it could also be converted into a point cloud representation by LiDAR, which effectively complements the visual feature point information. In addition, since the passable space contained rich pavement marking edges, by further using the distortion characteristics of the reverse perspective projection process, the distortion edges could be effectively eliminated, and the hybrid edge information could be used for mapping and localization tasks. The experiments based on the indoor and outdoor automated valet parking verify that the proposed scheme can achieve more precise edge segmentation results, much smaller orientation estimation error, and better trajectory estimation. In the future, we hope to investigate the generalization of our algorithm in more diverse scenarios, such as light changes and more dynamic tasks.

Author Contributions: Conceptualization, G.X. and S.D.; methodology, writing, visualization, G.X. and N.Z.; investigation, N.Z. and G.W.; validation, G.W., supervision, S.D. All authors have read and agreed to the published version of the manuscript.

Funding: This work was partially supported by the Natural Science Foundation of Sichuan Province under Grants 2023NSFSC0475 and 2023NSFSC1441, the Fundamental Research Funds for the Central Universities under Grant 2022SCU12004, and the Funds for National Key Laboratory of Special Vehicle Design and Manufacturing Integration Technology under Grant GZ2022KF007.

Institutional Review Board Statement: Not applicable.

Informed Consent Statement: Not applicable.

Data Availability Statement: Not applicable.

Conflicts of Interest: The authors declare no conflict of interest.

References

1. Thrun, S. Probabilistic robotics. *Commun. ACM* **2002**, *45*, 52–57. [CrossRef]
2. Correll, N.; Hayes, B.; Heckman, C.; Roncone, A. *Introduction to Autonomous Robots: Mechanisms, Sensors, Actuators, and Algorithms*; MIT Press: Cambridge, MA, USA, 2022.
3. Cadena, C.; Carlone, L.; Carrillo, H.; Latif, Y.; Scaramuzza, D.; Neira, J.; Reid, I.; Leonard, J.J. Past, present, and future of simultaneous localization and mapping: Toward the robust-perception age. *IEEE Trans. Robot.* **2016**, *32*, 1309–1332. [CrossRef]
4. Qin, T.; Li, P.; Shen, S. Vins-mono: A robust and versatile monocular visual-inertial state estimator. *IEEE Trans. Robot.* **2018**, *34*, 1004–1020. [CrossRef]

5. Park, J.; Shin, U.; Shim, G.; Joo, K.; Rameau, F.; Kim, J.; Choi, D.-G.; Kweon, I.S. Vehicular multi-camera sensor system for automated visual inspection of electric power distribution equipment. In Proceedings of the 2019 IEEE/RSJ International Conference on Intelligent Robots and Systems (IROS), Macau, China, 3–8 November 2019; pp. 281–288.
6. Garg, S.; Sunderhauf, N.; Dayoub, F.; Morrison, D.; Cosgun, A.; Carneiro, G.; Wu, Q.; Chin, T.-J.; Reid, I.; Gould, S.; et al. Semantics for robotic mapping, perception and interaction: A survey. *Found. Trends Robot.* **2020**, *8*, 1–224. [CrossRef]
7. Kostavelis, I.; Gasteratos, A. Semantic mapping for mobile robotics tasks: A survey. *Robot. Auton. Syst.* **2015**, *66*, 86–103. [CrossRef]
8. Yu, C.; Liu, Z.; Liu, X.-J.; Xie, F.; Yang, Y.; Wei, Q.; Fei, Q. Ds-slam: A semantic visual slam towards dynamic environments. In Proceedings of the 2018 IEEE/RSJ International Conference on Intelligent Robots and Systems (IROS), Madrid, Spain, 1–5 October 2018; pp. 1168–1174.
9. Bescos, B.; Fácil, J.M.; Civera, J.; Neira, J. Dynaslam: Tracking, mapping, and inpainting in dynamic scenes. *IEEE Robot. Autom. Lett.* **2018**, *3*, 4076–4083. [CrossRef]
10. Xiao, L.; Wang, J.; Qiu, X.; Rong, Z.; Zou, X. Dynamic-slam: Semantic monocular visual localization and mapping based on deep learning in dynamic environment. *Robot. Auton. Syst.* **2019**, *117*, 1–16. [CrossRef]
11. Zhou, H.; Zou, D.; Pei, L.; Ying, R.; Liu, P.; Yu, W. Structslam: Visual slam with building structure lines. *IEEE Trans. Veh. Technol.* **2015**, *64*, 1364–1375. [CrossRef]
12. Zuo, X.; Xie, X.; Liu, Y.; Huang, G. Robust visual slam with point and line features. In Proceedings of the 2017 IEEE/RSJ International Conference on Intelligent Robots and Systems (IROS), Vancouver, BC, Canada, 24–28 September 2017; pp. 1775–1782.
13. Im, J.-H.; Im, S.-H.; Jee, G.-I. Extended line map-based precise vehicle localization using 3d lidar. *Sensors* **2018**, *18*, 3179. [CrossRef] [PubMed]
14. Bao, S.Y.; Bagra, M.; Chao, Y.-W.; Savarese, S. Semantic structure from motion with points, regions, and objects. In Proceedings of the 2012 IEEE Conference on Computer Vision and Pattern Recognition, Providence, RI, USA, 16–21 June 2012; pp. 2703–2710.
15. Yang, S.; Scherer, S. Cubeslam: Monocular 3-d object slam. *IEEE Trans. Robot.* **2019**, *35*, 925–938. [CrossRef]
16. Chen, X.; Milioto, A.; Palazzolo, E.; Giguere, P.; Behley, J.; Stachniss, C. Suma++: Efficient lidar-based semantic slam. In Proceedings of the 2019 IEEE/RSJ International Conference on Intelligent Robots and Systems (IROS), Macau, China, 3–8 November 2019; pp. 4530–4537.
17. Lianos, K.-N.; Schonberger, J.L.; Pollefeys, M.; Sattler, T. Vso: Visual semantic odometry. In Proceedings of the European Conference on Computer Vision (ECCV), Munich, Germany, 8–14 September 2018; pp. 234–250.
18. Li, J.; Koreitem, K.; Meger, D.; Dudek, G. View-invariant loop closure with oriented semantic landmarks. In Proceedings of the 2020 IEEE International Conference on Robotics and Automation (ICRA), Paris, France, 31 May–31 August 2020; pp. 7943–7949.
19. Deng, W.; Huang, K.; Chen, X.; Zhou, Z.; Shi, C.; Guo, R.; Zhang, H. Semantic rgb-d slam for rescue robot navigation. *IEEE Access* **2020**, *8*, 221320–221329. [CrossRef]
20. Vasilopoulos, V.; Kantaros, Y.; Pappas, G.J.; Koditschek, D.E. Reactive planning for mobile manipulation tasks in unexplored semantic environments. In Proceedings of the 2021 IEEE International Conference on Robotics and Automation (ICRA), Xi'an, China, 30 May–5 June 2021; pp. 6385–6392.
21. Kantaros, Y.; Kalluraya, S.; Jin, Q.; Pappas, G.J. Perception-based temporal logic planning in uncertain semantic maps. *IEEE Trans. Robot.* **2022**, *38*, 2536–2556. [CrossRef]
22. Li, S.; Park, D.; Sung, Y.; Shah, J.A.; Roy, N. Reactive task and motion planning under temporal logic specifications. In Proceedings of the 2021 IEEE International Conference on Robotics and Automation (ICRA), Xi'an, China, 30 May–5 June 2021; pp. 12618–12624.
23. Charalampous, K.; Kostavelis, I.; Gasteratos, A. Recent trends in social aware robot navigation: A survey. *Robot. Auton. Syst.* **2017**, *93*, 85–104. [CrossRef]
24. Qi, X.; Wang, W.; Yuan, M.; Wang, Y.; Li, M.; Xue, L.; Sun, Y. Building semantic grid maps for domestic robot navigation. *Int. J. Adv. Robot. Syst.* **2020**, *17*, 1729881419900066. [CrossRef]
25. Zhang, Z.; Rebecq, H.; Forster, C.; Scaramuzza, D. Benefit of large field-of-view cameras for visual odometry. In Proceedings of the 2016 IEEE International Conference on Robotics and Automation (ICRA), Stockholm, Sweden, 16–21 May 2016; pp. 801–808.
26. Houben, S.; Quenzel, J.; Krombach, N.; Behnke, S. Efficient multi-camera visual-inertial slam for micro aerial vehicles. In Proceedings of the 2016 IEEE/RSJ International Conference on Intelligent Robots and Systems (IROS), Daejeon, Korea, 9–14 October 2016; pp. 1616–1622.
27. Aguilar, W.G.; Manosalvas, J.F.; Guillén, J.A.; Collaguazo, B. Robust motion estimation based on multiple monocular camera for indoor autonomous navigation of micro aerial vehicle. In Proceedings of the International Conference on Augmented Reality, Virtual Reality and Computer Graphics, Otranto, Italy, 24–27 June 2018; Springer: Cham, Switzerland, 2018; pp. 547–561.
28. Heng, L.; Choi, B.; Cui, Z.; Geppert, M.; Hu, S.; Kuan, B.; Liu, P.; Nguyen, R.; Yeo, Y.C.; Geiger, A.; et al. Project autovision: Localization and 3d scene perception for an autonomous vehicle with a multi-camera system. In Proceedings of the 2019 International Conference on Robotics and Automation (ICRA), Montreal, QC, Canada, 20–24 May 2019; pp. 4695–4702.
29. Hu, J.; Yang, M.; Xu, H.; He, Y.; Wang, C. Mapping and localization using semantic road marking with centimeter-level accuracy in indoor parking lots. In Proceedings of the 2019 IEEE Intelligent Transportation Systems Conference (ITSC), Auckland, New Zealand, 27–30 October 2019; pp. 4068–4073.

30. Campos, C.; Elvira, R.; Rodríguez, J.J.G.; Montiel, J.M.; Tardós, J.D. Orb-slam3: An accurate open-source library for visual, visual–inertial, and multimap slam. *IEEE Trans. Robot.* **2021**, *37*, 1874–1890. [CrossRef]
31. Engel, J.; Schöps, T.; Cremers, D. Lsd-slam: Large-scale direct monocular slam. In Proceedings of the European Conference on Computer Vision, Zurich, Switzerland, 6–12 September 2014; Springer: Cham, Switzerland, 2014; pp. 834–849.
32. Wang, R.; Schworer, M.; Cremers, D. Stereo dso: Large-scale direct sparse visual odometry with stereo cameras. In Proceedings of the IEEE International Conference on Computer Vision, Venice, Italy, 22–29 October 2017; pp. 3903–3911.
33. Hartmann, J.; Klüssendorff, J.H.; Maehle, E. A comparison of feature descriptors for visual slam. In Proceedings of the 2013 European Conference on Mobile Robots, Barcelona, Spain, 25–27 September 2013; pp. 56–61.
34. Yang, S.; Song, Y.; Kaess, M.; Scherer, S. Pop-up slam: Semantic monocular plane slam for low-texture environments. In Proceedings of the 2016 IEEE/RSJ International Conference on Intelligent Robots and Systems (IROS), Daejeon, Korea, 9–14 October 2016; pp. 1222–1229.
35. Long, J.; Shelhamer, E.; Darrell, T. Fully convolutional networks for semantic segmentation. In Proceedings of the IEEE Conference on Computer Vision and Pattern Recognition, Boston, MA, USA, 7–12 June 2015; pp. 3431–3440.
36. Badrinarayanan, V.; Kendall, A.; Cipolla, R. Segnet: A deep convolutional encoder-decoder architecture for image segmentation. *IEEE Trans. Pattern Anal. Mach. Intell.* **2017**, *39*, 2481–2495. [CrossRef] [PubMed]
37. Zhao, H.; Shi, J.; Qi, X.; Wang, X.; Jia, J. Pyramid scene parsing network. In Proceedings of the IEEE Conference on Computer Vision and Pattern Recognition, Honolulu, HI, USA, 21–26 July 2017; pp. 2881–2890.
38. Zhao, H.; Qi, X.; Shen, X.; Shi, J.; Jia, J. Icnet for real-time semantic segmentation on high-resolution images. In Proceedings of the European Conference on Computer Vision (ECCV), Munich, Germany, 8–14 September 2018; pp. 405–420.
39. Chen, L.-C.; Papandreou, G.; Kokkinos, I.; Murphy, K.; Yuille, A.L. Deeplab: Semantic image segmentation with deep convolutional nets, atrous convolution, and fully connected crfs. *IEEE Trans. Pattern Anal. Mach. Intell.* **2017**, *40*, 834–848. [CrossRef] [PubMed]
40. Howard, A.; Sandler, M.; Chu, G.; Chen, L.-C.; Chen, B.; Tan, M.; Wang, W.; Zhu, Y.; Pang, R.; Vasudevan, V.; et al. Searching for mobilenetv3. In Proceedings of the IEEE/CVF International Conference on Computer Vision, Seoul, Republic of Korea, 27 October–2 November 2019; pp. 1314–1324.
41. Ren, S.; He, K.; Girshick, R.; Sun, J. Faster r-cnn: Towards real-time object detection with region proposal networks. In Proceedings of the 28th International Conference on Neural Information Processing Systems, Montreal, QC, Canada, 7–12 December 2015.
42. Redmon, J.; Divvala, S.; Girshick, R.; Farhadi, A. You only look once: Unified, real-time object detection. In Proceedings of the IEEE Conference on Computer Vision and Pattern Recognition, Las Vegas, NV, USA, 27–30 June 2016; pp. 779–788.
43. Liu, W.; Anguelov, D.; Erhan, D.; Szegedy, C.; Reed, S.; Fu, C.-Y.; Berg, A.C. Ssd: Single shot multibox detector. In Proceedings of the European Conference on Computer Vision, Amsterdam, The Netherlands, 11–14 October 2016; Springer: Cham, Switzerland, 2016; pp. 21–37.
44. Qi, C.R.; Su, H.; Mo, K.; Guibas, L.J. Pointnet: Deep learning on point sets for 3d classification and segmentation. In Proceedings of the IEEE Conference on Computer Vision and Pattern Recognition, Honolulu, HI, USA, 21–26 July 2017; pp. 652–660.
45. Zhou, Y.; Tuzel, O. Voxelnet: End-to-End Learning for Point Cloud Based 3d Object Detection. In Proceedings of the IEEE Conference on Computer Vision and Pattern Recognition, Salt Lake City, UT, USA, 18–23 June 2018; pp. 4490–4499.
46. Lang, A.H.; Vora, S.; Caesar, H.; Zhou, L.; Yang, J.; Beijbom, O. Pointpillars: Fast encoders for object detection from point clouds. In Proceedings of the IEEE/CVF Conference on Computer Vision and Pattern Recognition, Long Beach, CA, USA, 15–20 June 2019; pp. 12697–12705.
47. Szeliski, R. *Computer Vision: Algorithms and Applications*; Springer: Cham, Switzerland, 2022.
48. Mallot, H.A.; Bülthoff, H.H.; Little, J.; Bohrer, S. Inverse perspective mapping simplifies optical flow computation and obstacle detection. *Biol. Cybern.* **1991**, *64*, 177–185. [CrossRef] [PubMed]
49. Roddick, T.; Cipolla, R. Predicting semantic map representations from images using pyramid occupancy networks. In Proceedings of the IEEE/CVF Conference on Computer Vision and Pattern Recognition, Seattle, WA, USA, 13–19 June 2020; pp. 11138–11147.
50. Pan, B.; Sun, J.; Leung, H.Y.T.; Andonian, A.; Zhou, B. Cross-view semantic segmentation for sensing surroundings. *IEEE Robot. Autom. Lett.* **2020**, *5*, 4867–4873. [CrossRef]
51. Arandjelovic, R.; Gronat, P.; Torii, A.; Pajdla, T.; Sivic, J. Netvlad: Cnn architecture for weakly supervised place recognition. In Proceedings of the IEEE Conference on Computer Vision and Pattern Recognition, Las Vegas, NV, USA, 27–30 June 2016; pp. 5297–5307.
52. Garg, S.; Suenderhauf, N.; Milford, M. Semantic–geometric visual place recognition: A new perspective for reconciling opposing views. *Int. J. Robot. Res.* **2022**, *41*, 573–598. [CrossRef]
53. Kendall, A.; Grimes, M.; Cipolla, R. Posenet: A convolutional network for real-time 6-dof camera relocalization. In Proceedings of the IEEE International Conference on Computer Vision, Santiago, Chile, 7–13 December 2015; pp. 2938–2946.
54. Sarlin, P.-E.; Cadena, C.; Siegwart, R.; Dymczyk, M. From coarse to fine: Robust hierarchical localization at large scale. In Proceedings of the IEEE/CVF Conference on Computer Vision and Pattern Recognition, Long Beach, CA, USA, 15–20 June 2019; pp. 12716–12725.
55. Dusmanu, M.; Rocco, I.; Pajdla, T.; Pollefeys, M.; Sivic, J.; Torii, A.; Sattler, T. D2-net: A trainable cnn for joint description and detection of local features. In Proceedings of the IEEE/CVF Conference on Computer Vision and Pattern Recognition, Long Beach, CA, USA, 15–20 June 2019; pp. 8092–8101.

56. Poggenhans, F.; Salscheider, N.O.; Stiller, C. Precise localization in high-definition road maps for urban regions. In Proceedings of the 2018 IEEE/RSJ International Conference on Intelligent Robots and Systems (IROS), Madrid, Spain, 1–5 October 2018; pp. 2167–2174.
57. Yu, X.; Chaturvedi, S.; Feng, C.; Taguchi, Y.; Lee, T.-Y.; Fernandes, C.; Ramalingam, S. Vlase: Vehicle localization by aggregating semantic edges. In Proceedings of the 2018 IEEE/RSJ International Conference on Intelligent Robots and Systems (IROS), Madrid, Spain, 1–5 October 2018; pp. 3196–3203.
58. Li, K.; Zhang, X.; Kun, L.; Zhang, S. Vision global localization with semantic segmentation and interest feature points. In Proceedings of the 2020 IEEE/RSJ International Conference on Intelligent Robots and Systems (IROS), Las Vegas, NV, USA, 24 October 2020–24 January 2020; pp. 4581–4587.
59. Liu, Y.; Petillot, Y.; Lane, D.; Wang, S. Global localization with object-level semantics and topology. In Proceedings of the 2019 International Conference on Robotics and Automation (ICRA), Montreal, QC, Canada, 20–24 May 2019; pp. 4909–4915.

sensors

MDPI

Article

Dynamic Balance Control of Double Gyros Unicycle Robot Based on Sliding Mode Controller

Yang Zhang, Hongzhe Jin * and Jie Zhao

State Key Laboratory and System, Harbin Institute of Technology, Harbin 150001, China
* Correspondence: hongzhejin@hit.edu.cn

Abstract: This paper presents a doublegyroscope unicycle robot, which is dynamically balanced by sliding mode controller and PD controller based on its dynamics. This double−gyroscope robot uses the precession effect of the double gyro system to achieve its lateral balance. The two gyroscopes are at the same speed and in reverse direction so as to ensure that the precession torque of the gyroscopes does not interfere with the longitudinal direction of the unicycle robot. The lateral controller of the unicycle robot is a sliding mode controller. It not only maintains the balance ability of the unicycle robot, but also improves its robustness. The longitudinal controller of the unicycle robot is a PD controller, and its input variables are pitch angle and pitch angular velocity. In order to track the set speed, the speed of the unicycle robot is brought into the longitudinal controller to facilitate the speed control. The dynamic balance of the designed double gyro unicycle robot is verified by simulation and experiment results. At the same time, the anti−interference ability of the designed controller is verified by interference simulation and experiment.

Keywords: unicycle robot; sliding mode control; dynamics; mobile robot

Citation: Zhang, Y.; Jin, H.; Zhao, J. Dynamic Balance Control of Double Gyros Unicycle Robot Based on Sliding Mode Controller. *Sensors* **2023**, *23*, 1064. https://doi.org/10.3390/s23031064

Academic Editors: David Cheneler and Stephen Monk

Received: 10 December 2022
Revised: 8 January 2023
Accepted: 12 January 2023
Published: 17 January 2023

1. Introduction

As a wheeled mobile robot, the unicycle robot has high flexibility due to its contact with the ground being a single point of contact. However, the characteristics of the unicycle robot, such as difficult balance control, high coupling, weak anti−interference ability, also bring great challenges to the research into unicycle robots. There are many researchers studying unicycle robots. According to the classification of lateral balance of unicycle robots, there are the following: Stanford and MIT use horizontal rotors to achieve the balance control of the unicycle robot [1,2]; The gyrover unicycle robot designed by Carnegie Mellon University and Yangsheng Xu team controls the robot's lateral balance according to the high−speed flywheel [3–6]; Pusan National University of Korea uses vertical rotor structure to achieve the balance of the unicycle robot [7]; Chiba University of Japan [8] and the Asian Institute of Technology in Thailand [9] use the precession effect of double gyros to achieve the balance control of the unicycle robot; J. Shen and D. Hong applied universal wheel to realize the balance control of the designed unicycle robot [10,11]. Many studies on the unicycle robot focus on its static balance control. However, as a mobile robot, its mobility and anti−interference ability are also worth studying. Kwok Wai Au used the state feedback controller to realize the displacement tracking of the Grover unicycle robot [12]. Umashankar Nagarajan et al. used PID controller to realize the dynamic balance control of the designed unicycle robot through off−line motion trajectory [13]. It can be seen from previous studies that the research on dynamic balance control of unicycle robot is also important. Due to the point contact movement of the unicycle robot, its anti−interference ability is worse than other mobile robots. However, many previous studies on the controller of the unicycle robot focus on the balance control of the unicycle robot, while neglecting to enhance its anti−interference ability. Therefore, the dynamic anti−interference balance control of the unicycle robot is studied in this paper.

As a highly complex structure of the unicycle robot, the double gyros unicycle robot has the characteristics of control difficulty, serious self vibration and interference. Compared with other unicycle robots, the double gyros has better balance characteristics due to its gyro effect. Therefore, the dynamic balance control and anti−interference of the designed unicycle robot are studied in this paper. The dynamics of the double gyros unicycle robot has been completed in previous research [14]. In this paper, according to the dynamic model of the unicycle robot, sliding mode controller is used to control its lateral dynamic balance. The sliding mode controller has anti−interference ability to ensure that the unicycle robot can still maintain balance when facing lateral interference under the dynamic motion. Its dynamic anti−interference ability has been verified in simulation and experiment. Longitudinal balance takes pitch angle, pitch angular speed and the speed of unicycle robot as controller inputs. The PD controller is used to ensure the dynamic speed tracking and longitudinal balance of the unicycle robot. According to the corresponding relationship between the pitch angle and the bottom wheel speed in the longitudinal dynamics, at the target speed, the corresponding target pitch angle is given to ensure balance.

The rest of this paper is as follows. Section 2 is the model of the designed unicycle robot, which includes lateral dynamics model, longitudinal dynamics model and experimental platform model. Section 3 is the controller designed according to the dynamics of the unicycle robot. Section 4 provides the simulation results. Section 5 provides the experimental results. Section 6 presents the conclusion and future work.

2. Model

In this section, based on the Lagrangian dynamics method, the dynamic model of the designed double gyros unicycle robot is decomposed into two parts: lateral dynamics and longitudinal dynamics. The dynamics of the double gyros unicycle robot are shown in previous work [14]. Its Lagrangian dynamics are as follows:

$$M(q)\ddot{q} + N(\dot{q}, q) = Q. \tag{1}$$

$M(q) \in \mathbb{R}^{6\times6}$ and $N(\dot{q}, q) \in \mathbb{R}^{6\times1}$ the inertia matrix and nonlinear term, respectively, where

$$M(q) = \begin{pmatrix} m_{11} & m_{12} & m_{13} & m_{14} & 0 & m_{16} \\ m_{21} & m_{22} & m_{23} & 0 & 0 & m_{26} \\ m_{31} & m_{32} & m_{33} & m_{34} & 0 & 0 \\ m_{41} & 0 & m_{43} & m_{44} & 0 & 0 \\ 0 & 0 & 0 & 0 & m_{55} & 0 \\ m_{61} & m_{62} & 0 & 0 & 0 & m_{66} \end{pmatrix}$$

$$N(\dot{q}, q) = (n_1\ n_2\ n_3\ n_4\ n_5\ n_6)^T.$$

This paper only considers the linear dynamics of the unicycle robot, so the influence of yaw angle on the unicycle robot is ignored. For the system, $\boldsymbol{q}$ and Q are

$$q = (\varphi, \delta, \theta, \omega, \alpha, \beta)^T,\ Q = (0, 0, 0, \tau_\omega, \tau_\alpha, \tau_\beta)^T$$

where τ_ω, τ_α and τ_β are torque of the bottom, torque of precession system and torque of gyro system, respectively. Next, the dynamics are decomposed into lateral dynamics and longitudinal dynamics.

2.1. Lateral Dynamics

Figure 1 shows the lateral model of the unicycle robot. It consists of double gyro rotation and precession system. The two gyroscopes have the same speed and opposite direction. The precession angular speed of these two gyroscopes is the same, and the direction is opposite. This can ensure that the gyroscopic moment generated by the precession of the gyroscope can offset the longitudinal interference. In the figure, δ Is the roll angle of the unicycle robot, α_1, α_2 are the precession angles of the left and right sides of the unicycle robot. β_1, β_2 are the rotation angles of the left and right gyroscopes, respectively.

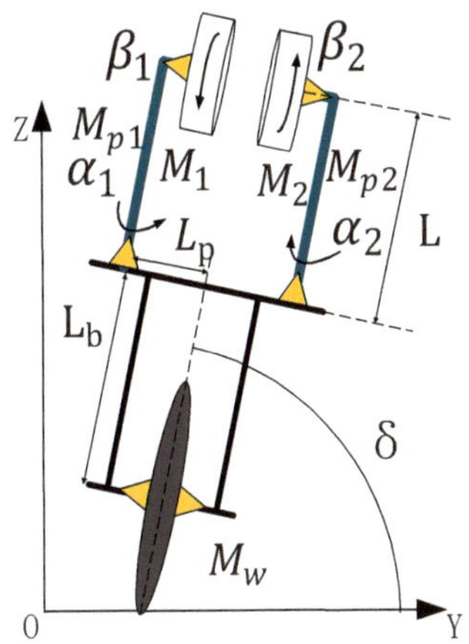

Figure 1. Lateral model of the unicycle robot.

According to the dynamic Equation (1), the gyro rotation is assumed to be constant, and the influence of yaw angle on the side direction of the unicycle robot is ignored. The lateral dynamic is as follows:

$$m_{22}\ddot{\delta} + m_{23}\ddot{\theta} + n_2 = 0 \tag{2}$$

When the unicycle robot is dynamically balanced, its lateral direction is regarded as an inverted pendulum model, so the influence of pitch angle and bottom wheel speed can be ignored. Where parameter $I_{1z} - I_{1x}$ is small, so the lateral dynamic equation can be solved as follows:

$$m_{22}\ddot{\delta} + 2\dot{\beta}\dot{\alpha}c\alpha I_{1z} - G_{roll}\delta = 0 \tag{3}$$

In order to simplify the dynamic equations, cx and sx mean $cos(x)$ and $sin(x)$, respectively. Where I_{1z} is the moment of inertia of gyroscopes' center of gravity to Z axe for left, G_{roll} is the lateral gravity component, and its equation is as follows:

$$G_{roll} = M_w g R_W + M_b g(L_b + R_W) + 2M_{p1} g(L + R_W) + 2M_1 g(L + R_W) \tag{4}$$

where M_w, M_b, M_{p1} and M_1 are mass of the wheel, mass of the frame, mass of the precession frame and mass of the gyroscope, respectively. R_W, L_b and L are radius of the wheel, distance of frame's center of gravity from center of wheel and distance of gyro's center of gravity from precession frame's center of gravity for left. g is the unit of gravity. It can be seen from the lateral dynamic equation that the roll angle is mainly affected by the rotation speed and precession angular speed of the double gyros. If precession angle α is too large, the lateral precession torque will be smaller. Therefore, in the lateral controller, the precession angular velocity should be used to control the lateral balance of the unicycle robot when the gyros' rotation speed is constant. At the same time, in order to keep the precession angle as small as possible to ensure lateral controllability, the precession angle is regarded as the input variable in the lateral controller.

2.2. Longitudinal Dynamics

Figure 2 shows the longitudinal model of the unicycle robot. It is composed of the bottom wheel and frame. In the figure, θ is the pitch angle of the unicycle robot and ω is the rotation angle of the bottom wheel of the unicycle robot.

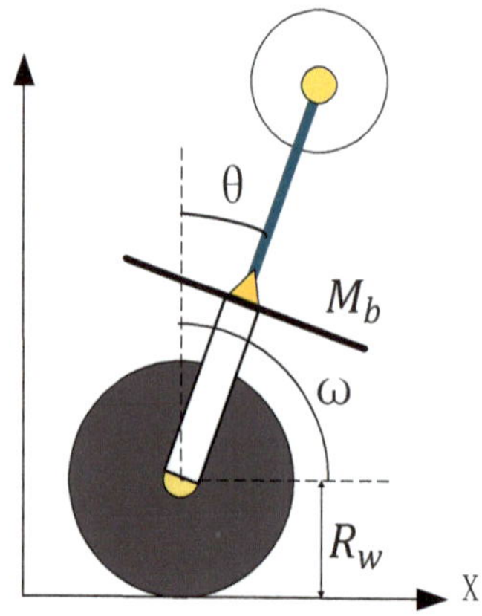

Figure 2. Longitudinal model of the unicycle robot.

According to the dynamic Equation (1), ignore the influence of yaw angle on the longitudinal direction of the unicycle robot. The longitudinal dynamic is as follows:

$$m_{32}\ddot{\delta} + m_{33}\ddot{\theta} + m_{34}\ddot{\omega} + n_3 = 0 \quad (5)$$

When the unicycle robot is dynamically balanced, its longitudinal direction is regarded as an inverted pendulum model, so the influence of roll angle can be ignored. Where parameter $I_{1z} - I_{1x}$ is small, so the longitudinal dynamic equation can be solved as follows:

$$m_{33}\ddot{\theta} + m_{34}\ddot{\omega} - G_{pitch}\theta = 0 \quad (6)$$

where G_{pitch} is the longitudinal gravity component, and its equation is as follows:

$$G_{pitch} = M_b g L_b + 2M_{p1} g L + 2M_1 g L \quad (7)$$

It can be seen from the longitudinal dynamic equation that its longitudinal balance is mainly affected by the bottom wheel. Although the precession angle rotation of the double gyros also affects the longitudinal direction, $(4s\alpha c\alpha I_{1x} - 4s\alpha c\alpha I_{1z})$ is small, and its impact can be ignored. Therefore, in the longitudinal control, the torque of the bottom wheel is taken as the longitudinal control variable, and in order to ensure that the bottom wheel can track the set speed, the speed of the unicycle is added to the controller.

2.3. Experimental Platform Model

As shown in Figure 3, it is the CAD model of the designed unicycle robot. It is mainly composed of three parts: gyro system, bottom wheel system and frame system. In the gyro system, the gyro is driven by a brushless DC motor, and precession rotation is achieved by gear transmission. The left and right gyroscopes are the same. In the frame system, the symmetrical structure is used to distribute the drive motor to ensure the balance of the unicycle robot, and the belt drive is used to drive the gear to realize the precession rotation. The left and right precession rotation is driven by the same gear input shaft to ensure the same precession speed. Moreover, its driving motor is a DC servo motor on the frame. The bottom wheel is also driven by the DC servo motor on the frame, and its transmission mode is belt. The left and right sides of the bottom pulley belt drive have the same structure to ensure the symmetry of the unicycle robot in the structure. In the experiment section, the experiments are completed according to the designed unicycle robot platform.

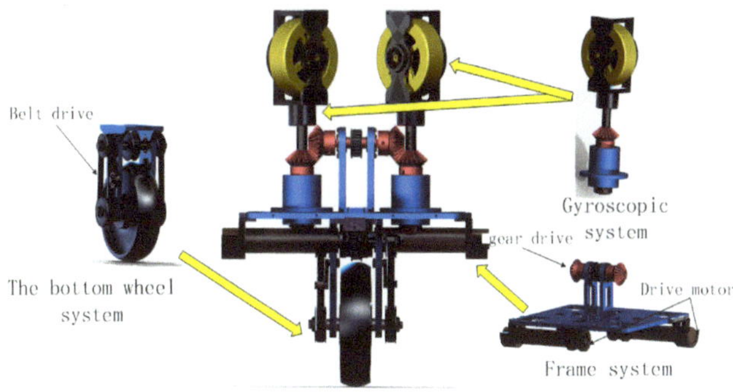

Figure 3. CAD model of the unicycle robot.

3. Design of the Controllers

In this section, the lateral and longitudinal controllers are designed according to the dynamic model of the double gyros unicycle robot. The control block diagram of the unicycle robot is shown in Figure 4. According to the dynamic model, the lateral controller is a sliding mode controller to ensure the dynamic balance of the unicycle robot and increase its anti−interference ability. The longitudinal controller is a PD controller, in which a speed tracker is added to ensure the dynamic speed tracking of the unicycle robot. The sliding mode controller has excellent anti−interference ability. In order to enhance the anti−interference ability of the unicycle robot, the lateral controller is designed as a sliding mode controller. At the same time, it helps to improve the lateral balance ability. The longitudinal controller is used to ensure the longitudinal balance of the unicycle robot while tracking the set speed. The longitudinal control state variable has not only pitch angle, but also speed value. In order to simplify the lateral controller and let the longitudinal balance track the set speed at the same time, the longitudinal controller is designed as a PD controller.

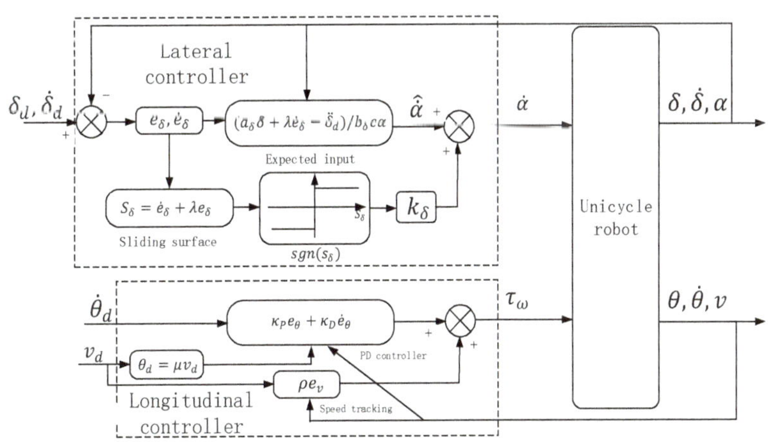

Figure 4. The control block diagram of the unicycle robot.

3.1. Lateral Controller

The lateral dynamic equation is shown in Equation (3). In the dynamic controller of the unicycle robot, the rotation angle speed of the gyroscopes is constant ($\ddot{\beta} = 0$). The variable $\dot{\beta}$ in dynamics can be regarded as a constant value. Therefore, the lateral dynamic equation can be changed as follows:

$$\ddot{\delta} = a_\delta \delta - b_\delta \dot{\alpha} c\alpha \tag{8}$$

where a_δ and b_δ are lateral dynamic parameters, both of which are constant values. The roll angle error equation is as follows:

$$e_\delta = \delta - \delta_d \tag{9}$$

where δ_d is the desired roll angle. According to the roll angle error value e_δ, its lateral sliding surface is as follows:

$$s_\delta = \dot{e}_\delta + \lambda e_\delta \tag{10}$$

where λ is a constant parameter. The derivative of the lateral sliding surface can be obtained as follows:

$$\dot{s}_\delta = \ddot{e}_\delta + \lambda \dot{e}_\delta \tag{11}$$

Bring Equation (8) into Equation (11) and take precession angular velocity $\dot{\alpha}$ as the output value of lateral balance controller. For lateral balance, if $\dot{s}_\delta \to 0$, the expected input value $\hat{\dot{\alpha}}$ is as follows:

$$\hat{\dot{\alpha}} = \left(a_\delta \delta + \lambda \dot{e}_\delta - \ddot{\delta}_d\right) / b_\delta c\alpha \tag{12}$$

Since the parameter values a_δ and b_δ are estimated values, they are not accurate. Due to the error between the actual precession angular velocity $\dot{\alpha}$ and expected precession angular velocity $\hat{\dot{\alpha}}$, in order to reduce the interference caused by the error to the unicycle robot and increase the anti−interference of the controller, a discontinuous term is added between the actual and expected precession angular velocity to make it swing in the designed sliding surface s_δ. The precession angular velocity can be obtained as follows:

$$\dot{\alpha} = \hat{\dot{\alpha}} - k_\delta sgn(s_\delta) \tag{13}$$

k_δ is the coefficient of discontinuous function $sgn(s_\delta)$. Where function $sgn(s_\delta)$ is a discontinuous function, the equation is as follows:

$$sgn(s_\delta) = \begin{cases} +1 & s_\delta > 0 \\ -1 & s_\delta < 0 \end{cases} \tag{14}$$

Thus, a lateral sliding mode controller can be obtained.

3.2. Longitudinal Controller

The longitudinal dynamic equation is shown in Equation (6). It can be seen from the dynamic equation that the longitudinal balance is affected by the bottom wheel. If the longitudinal direction of the unicycle robot is regarded as the inverted pendulum model, the torque of the bottom wheel is as follows:

$$\tau_\omega = m_\theta \ddot{\omega} \tag{15}$$

Where m_θ is the longitudinal component of the mass of the unicycle robot. Therefore, the bottom wheel torque τ_ω can be regarded as the output value of the longitudinal controller. The longitudinal balance controller is a PD controller, and its control equation is as follows:

$$\tau_\omega = \kappa_P \theta + \kappa_D \dot{\theta} \tag{16}$$

κ_P and κ_D are the pitch angle and pitch angle velocity coefficients in the longitudinal PD controller, both of which are constant. In order to track the set bottom wheel speed, the bottom wheel speed tracking is added on the basis of PD controller. The control equation can be obtained as follows:

$$\tau_\omega = \kappa_P\theta + \kappa_D\dot{\theta} + \rho_P\dot{e}_\omega \tag{17}$$

where ρ_P is the constant coefficient of bottom wheel speed error and $\dot{e}_\omega$ is the bottom wheel speed error, and its equation is as follows:

$$\dot{e}_\omega = \dot{\omega} - \dot{\omega}_d \tag{18}$$

ω_d is the desired bottom wheel speed. The unicycle robot does not move at a constant speed in the process of speed tracking. When the bottom wheel has acceleration, the pitch angle must be disturbed. In this case, if the target pitch angle set by the longitudinal PD control is 0, it is bound to cause certain interference to the pitch angle. In the longitudinal balance controller, the corresponding equation is designed for the target forward speed value and the target pitch angle value. The equation is as follows:

$$\theta_d = \mu_\omega\dot{\omega}_d \tag{19}$$

In order to control the speed of the unicycle robot, the angular velocity of the bottom wheel is converted into the speed of the robot. Since the relationship between the robot speed and the bottom wheel angular velocity is $v = R_w\dot{\omega}$, Equation (19) can be transformed as follows:

$$\theta_d = \mu v_d \tag{20}$$

v_d is the target speed of the unicycle robot. μ Is the constant coefficient of Equation (20). Therefore, the longitudinal controller changes as follows:

$$\tau_\omega = \kappa_P e_\theta + \kappa_D\dot{e}_\theta + \rho e_v \tag{21}$$

where e_v is the error of the speed of the unicycle robot. It is as follows:

$$e_v = v - v_d \tag{22}$$

The error value of pitch angle is as follows:

$$e_\theta = \theta - \theta_d \tag{23}$$

4. Simulation

In order to test the balance ability and anti−interference ability of the designed controller on the double gyros unicycle robot, it is verified in the three−dimensional simulation environment. The simulation software is Vrep. The designed CAD model is brought into the simulation software, and the connection pair and model parameters are set to ensure the authenticity of the simulation. The rotation speeds of the double gyros are set to 7000 rpm and the direction is opposite. The precession angular velocities on the left and right sides are the same and in opposite directions. The precession angular velocity is taken as the lateral balance control output, and the bottom wheel torque is taken as the longitudinal balance control output. The set tracking speed of the bottom wheel is 0.08 m/s. In the simulation, a 3N pulse interference is generated on the side of the unicycle robot at 10 s, and the anti−interference ability of the designed controller is tested. The initial coefficients of the controller are given according to the estimated parameters in the dynamics of the designed unicycle robot, and then the performance of the unicycle robot is achieved through minor adjustment. The sliding mode controller can enhance the balance ability and anti−interference ability of the unicycle robot by adjusting the coefficient of discontinuous function k_δ. The simulation curves are shown in Figure 5. Figure 5a shows the roll angle in the simulation. It can be seen that the roll angle of the unicycle robot fluctuates slightly and keeps balance. After interference, the roll angle has a large swing and immediately returns to a stable state. Figure 5b shows the pitch angle in the simulation. Due to the unicycle robot displaying a dynamic movement in the process of moving forward, it has a certain value in the pitch angle to ensure the longitudinal balance of the unicycle robot. After the lateral interference has a small interference to the longitudinal, the pitch angle returns to the equilibrium state immediately. Figure 5c shows the precession angle in the simulation. After the lateral interference, the precession angle returns to the balance state immediately after the large swing to maintain the dynamic balance of the unicycle robot. Figure 5d shows the speed curve of the unicycle robot in the simulation. It can be seen that the unicycle robot can track the set moving speed of 0.08 m/s with minimal gap. According to the simulation, it can be seen that the designed controller has the ability to maintain balance with large disturbances, and the unicycle robot can also track the set speed.

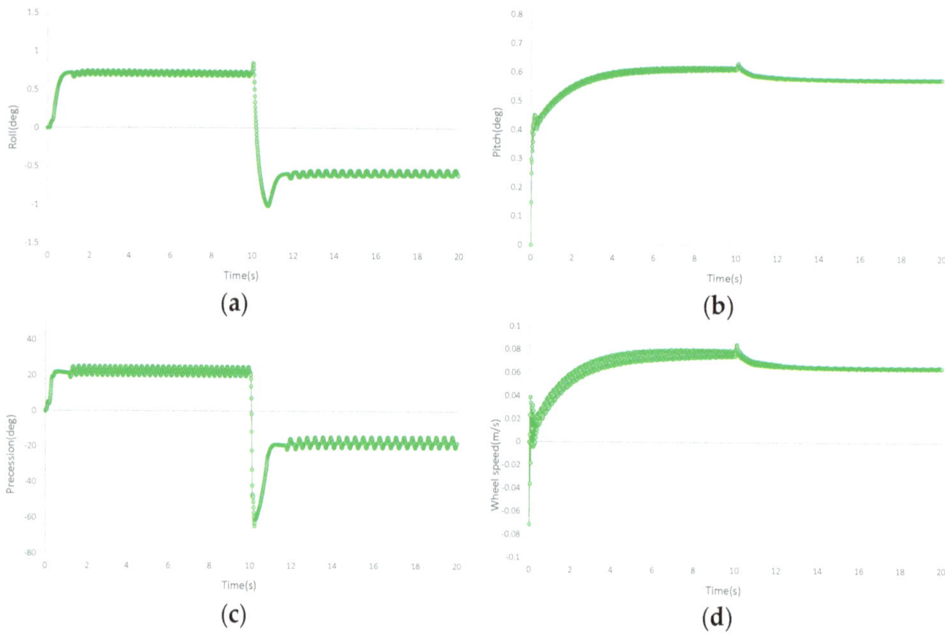

Figure 5. Curves of the unicycle robot in simulation. (**a**) The roll angle curve in the simulation. (**b**) The pitch angle curve in the simulation. (**c**) The precession angle curve in the simulation. (**d**) The speed curve in the simulation.

In the simulation, continuous stochastic interference is added for comparison. After the unicycle robot is balanced for 4 s, a stochastic pulse interference within 1–2N is added every 2 s. PD controller is added as comparison controller to the simulation. The longitudinal controller of PD controller is the same as the designed controller. The lateral controller is a PD controller composed of roll angle and roll angular velocity. However, in the dynamic balance simulation of the unicycle robot under 3N pulse disturbance, PD controller is difficult to keep the balance. Therefore, the 3N pulse interference curve only has the designed controller curves. The curves are shown in Figure 6. Figure 6a–d are the comparison curves of roll angle, pitch angle, precession angle and speed, respectively. In the case of continuous stochastic interference, the pitch angle and speed of PD controller and designed controller have little influence. In the roll angle, the designed controller has less vibration and is more stable than the PD controller in the continuous stochastic interference. In the precession angle, the designed controller has smaller amplitude and faster in convergence speed than PD controller. It can be seen that the designed controller has better anti−interference ability and is more stable than the PD controller in the case of dynamic balance of the unicycle robot in the continuous stochastic interference.

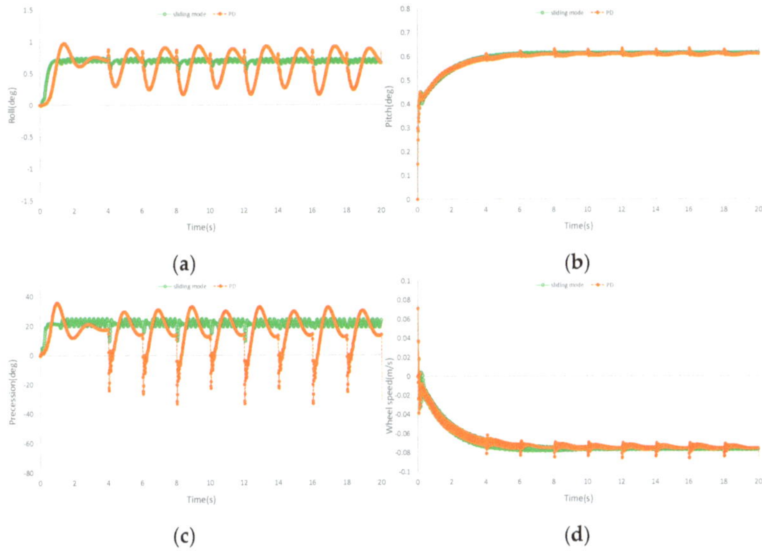

Figure 6. Curves of stochastic pulse interference in simulation. (**a**) The roll angle curve of stochastic pulse interference in the simulation. (**b**) The pitch angle curve of stochastic pulse interference in the simulation. (**c**) The precession angle curve of stochastic pulse interference in the simulation. (**d**) The speed curve of stochastic pulse interference in the simulation.

5. Experimental

The dynamic balance ability and stability of the controller to the robot are verified by using the designed experiment platform of the double gyros unicycle robot. The roll angle and pitch angle of the unicycle robot are obtained by the gyroscope sensor mpu9250 installed in the center of the unicycle robot. The precession angle and the angle of the bottom wheel are calculated according to the encoder on the precession drive motor and bottom wheel drive motor, respectively. The gyroscopes have constant speed and the same left, right and opposite direction. The experiment mainly includes dynamic balance experiment and dynamic interference experiment. The dynamic interference experiment is to verify the dynamic anti−interference ability with a 0.18 kg mass block by placing it in the lateral, middle and longitudinal directions when the unicycle robot is moving forward dynamically.

Figure 7 shows the dynamic balance experiment curve of the unicycle robot. Figure 7a–d show the roll angle curve, the pitch angle curve, the precession angle curve and the speed curve of the unicycle robot in the dynamic balance experiment, respectively. It can be seen from the figures that with the designed controller, the roll angle of the double gyros unicycle robot is stable near the zero position and fluctuates slightly. As the longitudinal dynamic balance controller is designed, the pitch angle has a corresponding tilt angle while the speed is tracked. In the experiment, the set speed is 0.2 m/s. According to Equation (18), the corresponding pitch angle is shown in Figure 7b. Although the speed fluctuates greatly due to the influence of the bottom wheel transmission mode, the speed can basically track the set speed value. The PD controller also can keep the unicycle robot balanced and stable.

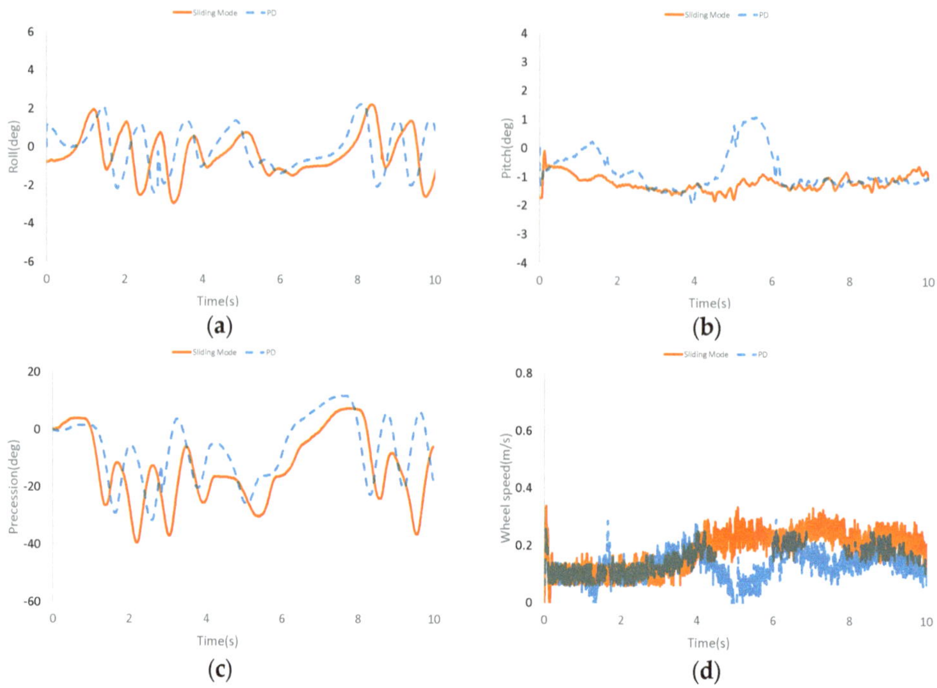

Figure 7. Dynamic balance curves. (**a**) The roll angle curve in dynamic balance experiment. (**b**) The pitch angle curve in dynamic balance experiment. (**c**) The precession angle curve in dynamic balance experiment. (**d**) The speed curve in dynamic balance experiment.

Figure 8 shows the curves of the left side interference dynamic experiment of the unicycle robot. Place the 0.18 kg mass on the left side of the unicycle robot (90 cm from the centre) when the unicycle robot is balanced for 10 s. In the dynamic balance experiment with 0.18 kg mass block placed, the PD controller is difficult to keep the unicycle robot balanced, so the step interference dynamic balance experiment of the PD controller is not shown here. The unicycle robot is affected by the step interference caused by small gravel falling on it. The mass block is placed at the unicycle robot when it is moving forward dynamically to simulate the anti−interference ability of the unicycle robot in the face of step interference. The 0.18 kg mass block is used to simulate the balance ability of the unicycle robot under the influence of large step interference. Figure 8a–d show the roll angle curve, the pitch angle curve, the precession angle curve and the speed curve of the dynamic interference balance of the unicycle robot, respectively. It can be seen that the roll angle and precession angle are greatly affected by interference. After interference, the roll angle keeps balance after fluctuation, and the precession angle keeps balance after large swing. The lateral interference has little effect on the pitch angle, and the speed has slight interference, but the speed remains stable after balance.

Figure 9 shows the curves of the middle interference dynamic experiment of the unicycle robot. Place the 0.18 kg mass on the middle of the unicycle robot when the unicycle robot is balanced for 10 s. Figure 9a–d show the roll angle curve, the pitch angle curve, the precession angle curve and the speed curve of the dynamic interference balance of the unicycle robot, respectively. It can be seen from the curves that when the mass block is placed in the middle of the unicycle robot, the longitudinal and lateral effects are very small.

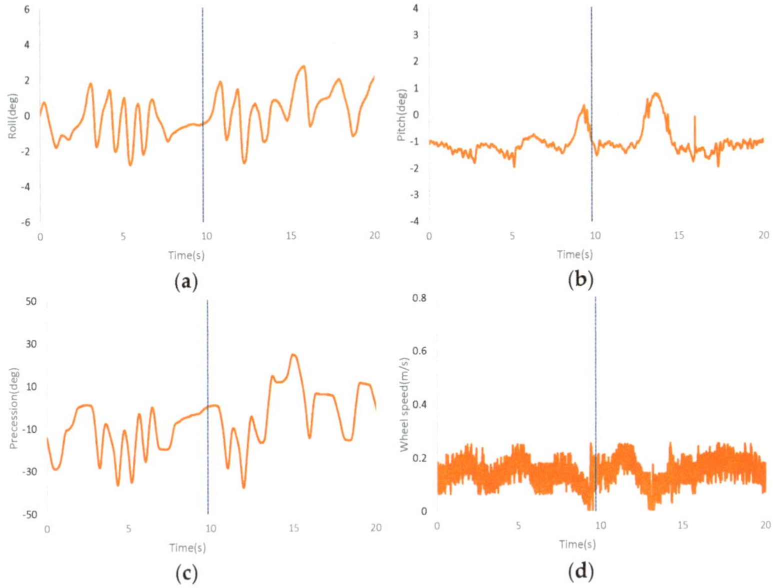

Figure 8. Dynamic balance curves after placing a mass block laterally. (**a**) The roll angle curve of lateral placement of material block experiment. (**b**) The pitch angle curve of lateral placement of material block experiment. (**c**) The precession angle curve of lateral placement of material block experiment (**d**) The speed curve of lateral placement of material block experiment.

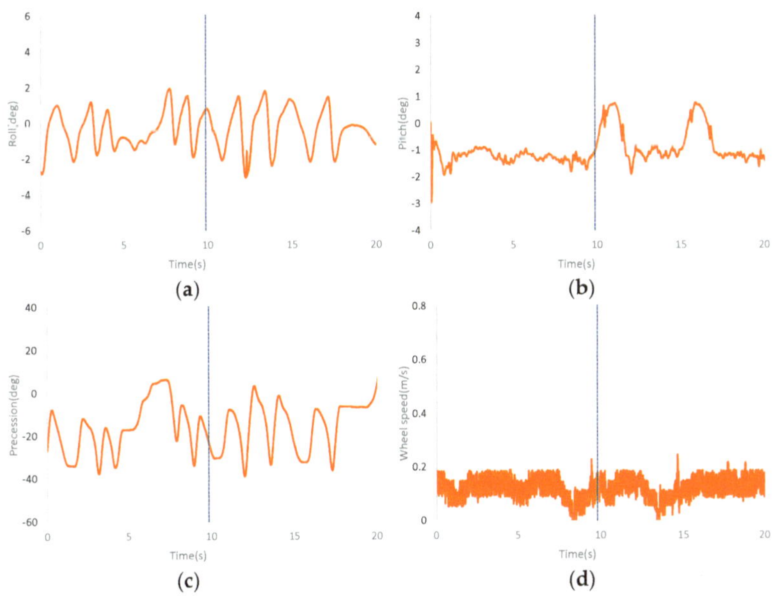

Figure 9. Dynamic balance curves after placing a mass block in middle. (**a**) The roll angle curve of placing material block in the middle. (**b**) The pitch angle curve of placing material block in the middle. (**c**) The precession angle curve of placing material block in the middle. (**d**) The speed curve of placing material block in the middle.

Figure 10 shows the curves of the interference dynamic experiment on the rear of the unicycle robot. Place the 0.18 kg mass on the rear of the unicycle robot (43 cm from the centre) when the unicycle robot is balanced for 12 s. Figure 10a–d show the roll angle curve, the pitch angle curve, the precession angle curve and the speed curve of the dynamic interference balance of the unicycle robot, respectively. It can be seen from the curves that the interference on the rear has little influence on the lateral of the unicycle robot, and the roll angle and precession angle fluctuate slightly. Rear interference has great influence on the longitudinal direction. After the interference, the pitch angle becomes larger and has a large swing. After the interference, the pitch angle has a large swing and is balanced at a larger angle value. The speed is greatly increased after interference, but it is still within the set speed range.

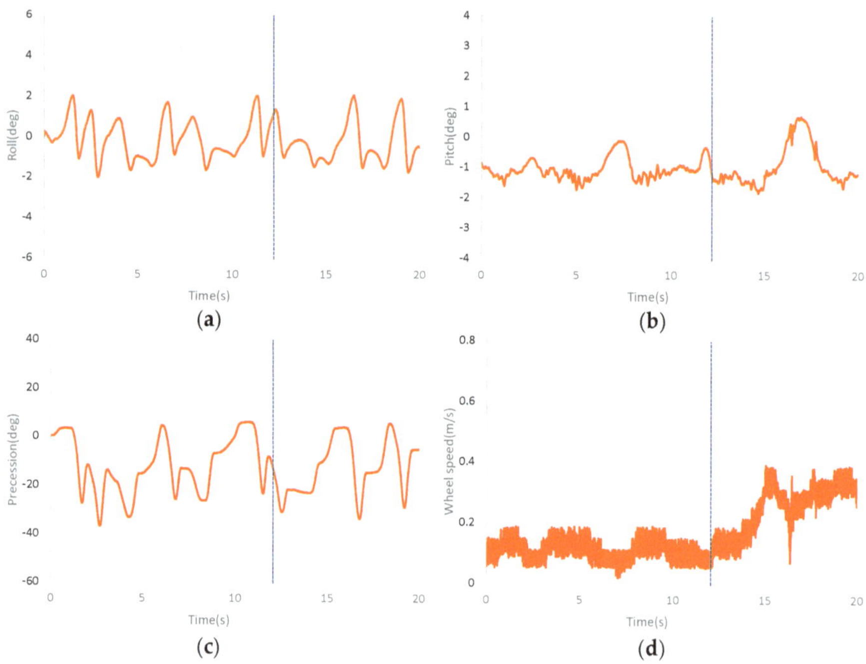

Figure 10. Dynamic balance curves after placing a mass block longitudinally. (**a**) The roll angle curve of placing material block longitudinally. (**b**) The pitch angle curve of placing material block longitudinally. (**c**) The precession angle curve of placing material block longitudinally. (**d**) The speed curve of placing material block longitudinally.

The dynamic balance experiment curves with continuous stochastic interference are shown in Figure 11. Figure 11a–d are the comparison curves of roll angle, pitch angle, precession angle and speed, respectively. It can be seen from the curves that continuous stochastic interference has a greater impact on roll angle and precession angle. However, after the interference, the roll angle and precession angle can still quickly recover to balance. Compared with PD controller, the designed controller has smaller amplitude in roll angle and precession angle. The interference also affects the pitch angle and speed. The pitching angle can still recover the balance state after interference. Although there is a large deviation in speed after interference, it can slowly recover to the tracking speed. Compared with PD controller, the designed controller has smaller oscillation and faster recovery when the pitch angle is subject to continuous stochastic interference and the speed fluctuates less.

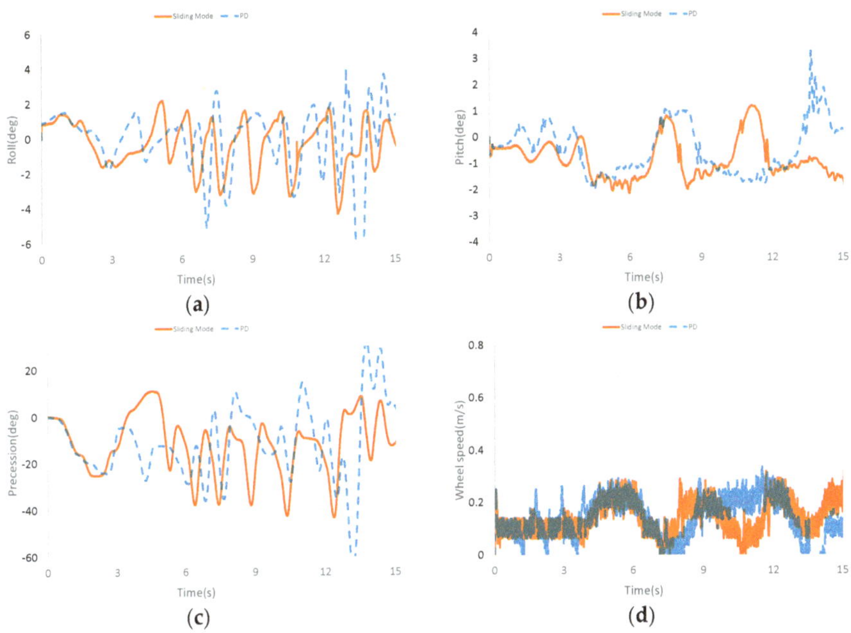

Figure 11. Curves of stochastic pulse interference experiment. (**a**) The roll angle curve of stochastic pulse interference experiment. (**b**) The pitch angle curve of stochastic pulse interference experiment. (**c**) The precession angle curve of stochastic pulse interference experiment. (**d**) The speed curve of stochastic pulse interference experiment.

According to the dynamic test and interference experiment of the designed controller on the double gyros unicycle robot, it can be seen that the designed controller can track the set speed of the unicycle robot and has the ability of anti−interference. The interference on the rear has a strong interference with the longitudinal direction of the unicycle robot, but the robot can keep balance and track the set speed after the interference. It can be seen that the double gyros unicycle robot can maintain dynamic balance, have anti−interference ability and track the set speed by using the designed controller.

6. Conclusions

In this paper, the dynamic balance controller is designed according to the dynamic model of the double gyros unicycle robot. The lateral controller is a sliding mode controller, which can improve the anti−interference ability of the unicycle robot. In the longitudinal controller, PD controller is used to balance. At the same time, due to the relationship between the pitch angle and moving speed, the corresponding equation is designed, and the designed speed tracking equation is taken into the longitudinal controller to ensure that the speed can track the designed speed. In the simulation, the dynamic balance and anti−interference ability of the designed controller are verified. According to the dynamic balance experiment, lateral, middle and longitudinal interference experiments, the dynamic balance and anti−interference ability of the designed controller on the double gyros unicycle robot are verified. The contribution of this paper is to design the lateral sliding film controller and the longitudinal speed tracking controller for the double gyros unicycle robot to achieve its dynamic balance ability and dynamic anti−interference ability. In the latter research, we will study the yaw angle control and autonomous motion of the double gyros unicycle robot.

Author Contributions: Conceptualization, Y.Z. and H.J.; methodology, Y.Z.; software, Y.Z.; validation, Y.Z.; formal analysis, Y.Z.; investigation, Y.Z.; resources, Y.Z.; data curation, Y.Z.; writing—original draft preparation, Y.Z.; writing—review and editing, Y.Z.; visualization, Y.Z.; supervision, H.J. and J.Z.; project administration, H.J. and J.Z.; funding acquisition, H.J. and J.Z. All authors have read and agreed to the published version of the manuscript.

Funding: This work was supported in part by the STI 2030—Major Project 2021ZD0201400, the National Natural Science Foundation of China under Grants 92048301 and 61473102.

Data Availability Statement: The data is unavailable due to privacy.

Conflicts of Interest: No conflict of interest exists in the submission of this manuscript, and the manuscript is approved by all authors for publication. I would like to declare on behalf of my co−authors that the work described was original research that has not been published previously, and not under consideration for publication elsewhere, in whole or in part. All the authors listed have approved the manuscript that is enclosed.

References

1. Schoonwinkel, A. Design and Test of a Computer-Stabilized Unicycle. Ph.D. Thesis, Stanford University, Stanford, CA, USA, 1987.
2. Vos, D.; Von Flotow, A.H. Dynamics and nonlinear adaptive control of an autonomous unicycle: Theory and Experiment. In Proceedings of the 29th IEEE Conference on Decision and Control, Honolulu, HI, USA, 5–7 December 1990; pp. 182–187.
3. Benjamin Brown, H., Jr.; Xu, Y. A Single-Wheel, Gyroscopically Stabilized Robot. In Proceedings of the IEEE International Conference on Robotics and Automation, Minneapolis, MN, USA, 22–28 April 1996; pp. 3658–3663.
4. Brown, H.B.; Xu, Y. A single wheel, gyroscopically stabilized robot. *IEEE Robot. Autom. Mag. A Publ. IEEE Robot. Autom. Soc.* **1997**, *3*, 4.
5. Xu, Y.; Au, K.W.; Nandy, G.C.; Brown, H.B. Analysis of actuation and dynamic balancing for a single-wheel robot. In Proceedings of the 1998 IEEE/RSJ International Conference on Intelligent Robots and Systems. Innovations in Theory, Practice and Applications (Cat. No.98CH36190) IEEE, Victoria, BC, Canada, 17 October 1998.
6. Zhu, Z.; Naing, M.P.; Al-Mamun, A. A 3-D simulator using ADAMS for design of an autonomous gyroscopically stabilized single wheel robot. In Proceedings of the 2009 IEEE International Conference on Systems, Man and Cybernetics IEEE, San Antonio, TX, USA, 11–14 October 2009.
7. Lee, J.; Han, S.; Lee, J. Decoupled Dynamic Control for Pitch and Roll Axes of the Unicycle Robot. *IEEE Trans. Ind. Electron.* **2013**, *60*, 3814–3822. [CrossRef]
8. Dao, M.Q.; Liu, K.Z. Gain-Scheduled Stabilization Control of a Unicycle Robot. *JSME Int. J. Ser. C Mech. Syst. Mach. Elem. Manuf.* **2005**, *48*, 649–656. [CrossRef]
9. Chantarachit, S.; Parnichkun, M. Development and control of a unicycle robot with double flywheels. *Mechatronics* **2016**, *40*, 28–40. [CrossRef]
10. Shen, J.; Hong, D. OmBURo: A Novel Unicycle Robot with Active Omnidirectional Wheel. In Proceedings of the 2020 IEEE International Conference on Robotics and Automation (ICRA), Paris, France, 31 May–31 August 2020; pp. 8237–8243.
11. Ahn, J.S.; Hong, D. Dynamic Analysis and Steering Control of a Novel Unicycle Robot with Active Omnidirectional Wheel. In Proceedings of the 2021 18th International Conference on Ubiquitous Robots (UR), Gangneung, Korea, 12–14 July 2021; pp. 149–155. [CrossRef]
12. Xu, Y.; Au, S.W. Stabilization and Path Following of a Single Wheel Robot. *IEEE/ASME Trans. Mechatron.* **2004**, *9*, 407–419. [CrossRef]
13. Nagarajan, U.; Kantor, G.; Hollis, R.L. Trajectory Planning and Control of an Underactuated Dynamically Stable Single Spherical Wheeled Mobile Robot. In Proceedings of the 2009 IEEE International Conference on Robotics and Automation, Kobe International Conference Center, Kobe, Japan, 12–17 May 2009; pp. 3743–3748.
14. Jin, H.; Zhang, Y.; Zhang, H.; Liu, Z.; Liu, Y.; Zhu, Y.; Zhao, J. Steering control method for an underactuated unicycle robot based on dynamic model. *Math. Probl. Eng.* **2018**, *2018*, 5240594. [CrossRef]

sensors MDPI

Article

Improving Tracking of Trajectories through Tracking Rate Regulation: Application to UAVs

Fernando Diaz-del-Rio, Pablo Sanchez-Cuevas, Pablo Iñigo-Blasco and J. L. Sevillano-Ramos *

ETS Ingeniería Informática, Universidad de Sevilla, Av. Reina Mercedes s/n, 41012 Sevilla, Spain
* Correspondence: jlsevillano@us.es

Abstract: The tracking problem (that is, how to follow a previously memorized path) is one of the most important problems in mobile robots. Several methods can be formulated depending on the way the robot state is related to the path. "Trajectory tracking" is the most common method, with the controller aiming to move the robot toward a moving target point, like in a real-time servosystem. In the case of complex systems or systems under perturbations or unmodeled effects, such as UAVs (Unmanned Aerial Vehicles), other tracking methods can offer additional benefits. In this paper, methods that consider the dynamics of the path's descriptor parameter (which can be called "error adaptive tracking") are contrasted with trajectory tracking. A formal description of tracking methods is first presented, showing that two types of error adaptive tracking can be used with the same controller in any system. Then, it is shown that the selection of an appropriate tracking rate improves error convergence and robustness for a UAV system, which is illustrated by simulation experiments. It is concluded that error adaptive tracking methods outperform trajectory tracking ones, producing a faster and more robust convergence tracking, while preserving, if required, the same tracking rate when convergence is achieved.

Keywords: UAV; mobile robots; path following; trajectory tracking; error adaptive tracking; Lyapunov stability theory

Citation: Diaz-del-Rio, F.; Sanchez-Cuevas, P.; Iñigo-Blasco, P.; Sevillano-Ramos, J.L. Improving Tracking of Trajectories through Tracking Rate Regulation: Application to UAVs. *Sensors* **2022**, *22*, 9795. https://doi.org/10.3390/s22249795

Academic Editors: David Cheneler and Stephen Monk

Received: 10 November 2022
Accepted: 8 December 2022
Published: 13 December 2022

1. Introduction

In a state space system, all the possible internal states of the system can be represented as a vector of variables. Typical control engineering problems in these systems are the stabilization problem, i.e., how to take the system to a fixed point in its state space, and the *tracking* problem, i.e., how to follow a desired trajectory or path. This tracking problem has been profusely studied in the area of motion control of mobile robots and autonomous vehicles, where the desired path is either memorized or previously generated [1,2].

In the particular case of UAVs (Unmanned Aerial Vehicles), it must be remarked that paths are usually defined as a set of straight lines and circular-orbit paths connecting several waypoints. This means that these paths usually contain singular points in the intersections of these lines, that is, they are not feasible trajectories for the UAV itself, but imprecise paths that the UAV cannot accurately track. Nonetheless, a convenient interpolation can convert this piecewise path into a smooth UAV trajectory passing over the desired waypoints, which should fulfill its own state equations. Note that having well-defined feasible desired trajectories is important when using UAVs safely (e.g., avoiding collisions) in many applications, such as multi-UAV systems, cluttered urban environments, etc. Navigation sensors are usually integrated into the robot in order to determine its current position and, thus, calculate the errors between the desired and actual trajectory.

The most common tracking method is called "*trajectory tracking*" (TT) or "*reference tracking*" and it explicitly considers time in the tracking [1]. In this case, the controller aims to bring the robot as near as possible to a moving target (or *reference*) point (Figure 1, top right). It is like servosystems (Figure 1, top left) where it must be guaranteed that the system will approach the desired point in a deterministic time. Examples of this kind of tracking can be found in most industrial robot applications (due to their strict real time characteristics). In mobile robots, pursuing a real moving objective (such as an antimissile system) is an example of a task that needs time determinism.

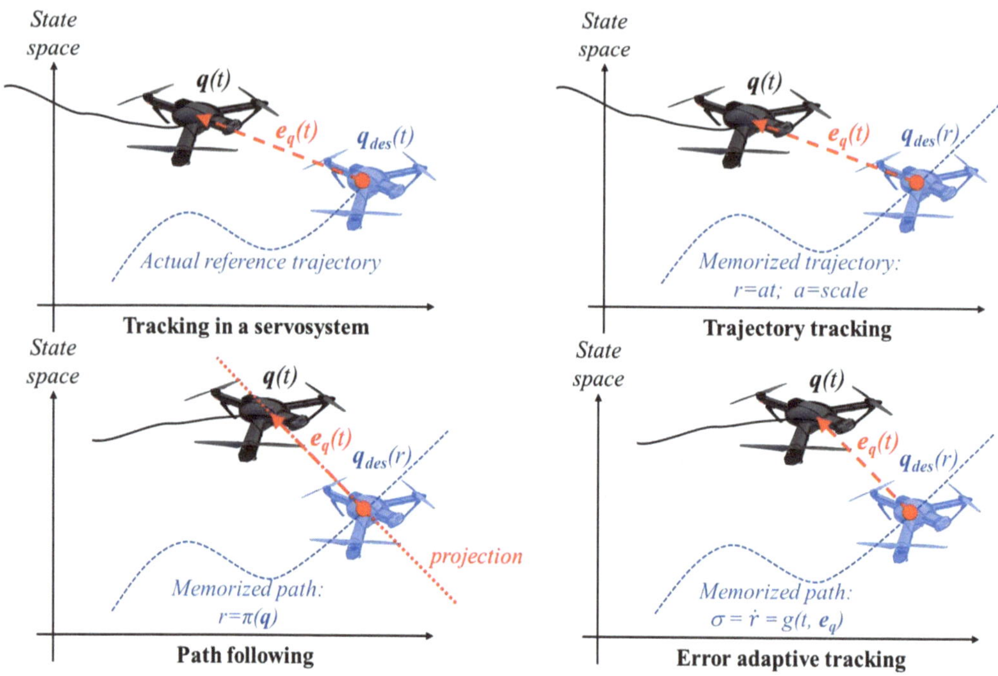

Figure 1. Classification of tracking methods regarding the descriptor parameter.

The second group, usually called "*path following*" (PF) (Figure 1, bottom left), does not consider timing requirements and simply tries to converge to a path. A reference point on the path must be selected at each instant according to some relation between actual robot state and path shape, e.g., the "closest" point to the robot's position. Consequently, a notable PF inconvenience is assuring the projection uniqueness for all possible paths.

A common example of PF is car driving, which can be extended to most Intelligent Transportation Systems (ITS) applications. For instance, in cars, usual control approaches select a point at a look-ahead distance on the road and the vehicle is driven to that point. Linear speed is preset, while orientation (or steering) is the single control variable used to perform the convergence.

However, there is confusion in the literature regarding the terminology used for these methods. For instance, a tracking rate that adapts to system errors has been used in [3] to improve the TT guidance results for underactuated vehicles in the presence of parametric modeling uncertainties, although these authors use the term 'path following' to refer to their implementation. There are other approaches, such as the one inspired by the Dynamic Time Warping (DTW) algorithm (studied extensively in the automatic speech recognition literature) in [4], where a strictly increasing rate of progression ($\dot{r} > 0$) is selected by minimizing a cost function for finite-duration movements.

Although we can find several excellent compendia of both methods in some classic books [1], the question of which tracking method is the most adequate for a given application is an active research area and many papers choose to implement a TT or a PF controller for UAVs and other mobile robots, depending on the application or with the purpose of easing the finding of a suitable controller (see more references in context in the rest of this section). This question has been elucidated for some simple paths and specific systems: in [5], TT and PF controllers were investigated for a linear time-invariant system with unstable zero dynamics, and it was demonstrated (for the simple PF task of moving the vehicle along a straight line) that there is a fundamental performance limitation for TT, which does not appear for PF method.

In the field of UAVs, many variants have appeared, which are called "guidance laws", and are actually based on TT or PF methods. Many of them use a virtual target point (VTP) on the path, which is selected through some projection, such as the line-of-sight (LOS) point situated at a certain look-ahead distance from the nearest path point to the robot. The selection of this point implies that they are a PF variant. Among these methods, we find a first set that uses simple and intuitive methods, such as the classic carrot-chasing algorithms and the Pure Pursuit. The number of works that have used these simple methods for UAVs is considerable, with [6,7] being perhaps the first ones.

There is another set of methods that select the projection point using a pair of circles that intersect with the desired path, which have been named "nonlinear guidance laws" (NLGL). These methods were used many years ago [8], and are still very common in recent years [9].

An alternative to guidance laws, which appeared 15 years ago, are those based on vector fields (VFs). A VF is built for each position in the state space and as a function of each specific path, that is, it is a geometric approach that computes a special projection that returns a vector. This vector defines some of the desired variables that the system must follow. Thus, according to our classification, VF are also a type of PF. It is worth mentioning that, up to date, not all state variables are determined by the VF, and the rest of the variables that remain free must be calculated by the controller. In this respect, a Lyapunov-based controller can be simplified because some of the desired states are predefined by the VF [10,11].

To sum up, designing VFs in 3D is not simple, and requires significant work [12]. Maybe the first proposal of a VF-based PF algorithm was developed in [13], as an intuitive and easy way to compute the desired heading angle for simple paths, such as straight lines and circles. Many other VFs for specific paths have followed since then, such as [14–16]. No VF has been implemented yet for any generic path; hence, this method should evidently come across the same drawbacks as PF. It is not guaranteed that the virtual field exists for a generic path, even for a simple one, such as a pure rotation around the robot center of mass.

In TT, time is the usual descriptor parameter of a path. Since time is an intuitive parameter, TT seems to be the most straightforward method. However, for the rest of the approaches, other path-descriptor parameters are possible. For example, in differential geometry, the natural arc parameter, which makes the linear speed equal to one, is generally preferred. In this paper, for the sake of generality, the descriptor parameter is denoted by r. Therefore, other groups of tracking methods can be defined to explicitly control the progression rate of a moving virtual target to be tracked; i.e., they impose a pace for r or a value for $\dot{r}$ (derivative with respect to time). Equivalently, in these methods, the real robot is forced to follow a virtual robot (also called "reference" robot) that goes along the reference path at a variable pace, which may be null when necessary; i.e., the reference robot can "wait" for the real one [17,18]. This pace can be selected with several purposes (as shown below).

Some scattered examples can be found in the literature, where the explicit control of progression of the "virtual target" (that is, the VTP) helps design a control law. For instance, in [18] a complete practical application, where the motion of the descriptor parameter was governed by a differential equation depending on the errors' and path's shape, was developed. In [19], a term related to the curvature (called "curvature effort" in that work) was defined, and a penalty factor based on the curvature effort was introduced in the dynamics of the path's description parameter to prevent the performance degradation of the tracking when the dynamic and kinematic constraints are exceeded. In [20], the target progression was tailored to design a nonlinear adaptive control law, which yields the convergence of the (closed-loop system) error trajectories to zero in the presence of parametric modelling uncertainties.

This family of methods can be considered a different path-tracking method that can be named *error adaptive tracking* (EAT) [13] (Figure 1, bottom right). Furthermore, EAT methods can be divided into two categories, depending on whether time deterministic following is expected or not. Basic EAT variants can be named "non-deterministic" EAT (NDEAT) because no aspect of time determinism is pursued. On the other hand, tracking rate adaptation to system errors can be combined with convergence of r to time (that is, convergence to the TT method). In this case, the rate of r can be extended to include the "inaccuracy in the deterministic tracking", i.e., the difference between the descriptor parameter r and time t. For this reason, this variant can be named "soft" deterministic error adaptive tracking (SDEAT) [21]. A tailored control law that includes a variable tracking (similar to that of SDEAT) of the virtual target that helps design the control law was exploited in [22] (these authors called it path tracking).

The aim of this paper is to provide a formal description and generalization of the EAT tracking method (which was used in particular cases of terrestrial and underwater vehicles [21]), and to show how it can be used *in any system using the same control law*, with the additional advantage of improving error convergence and robustness. Afterwards, EAT method is applied to a UAV model to show its benefits. We must emphasize that this paper is not focused on the design of new control laws. The selection of a tracking method, or more specifically, a proper form for $\dot{r}$ when using EAT, has been exploited in some systems [3,20,22] with the aim of finding a stabilizing control law (mainly via second Lyapunov method). On the contrary, the present study is focused on how the EAT method (instead of TT) can be applied using the same controller just by selecting a proper pace for $\dot{r}$, having the additional benefits that error convergence and robustness are improved. As a result, the burden of finding a new controller or a special path projection (see VF, carrot-chasing and so on in the Introduction section) will not be necessary. An additional benefit is that the switching between these two tracking methods (TT and EAT) can be performed smoothly, since the controller is exactly the same. Our approach is here particularized for a UAV, but it is worth mentioning that the same procedure can be applied to other non-linear systems. In fact, this is the first time (to the best of our knowledge) that an EAT method has been applied to a UAV, despite the fact that many tracking methods have been implemented for UAVs (see Introduction).

In order to understand properly the different tracking methods and the notation used in this work, we present first the simple case of a one-state system: $\dot{x} = u$, where x is the coordinate and u is the input. The goal is to follow a reference x_{des}={$x_{des}(r)$} made by a virtual system, which must fulfill $x'_{des} = u_{des}(r)$, where (′) holds for derivative with respect to r. Let us suppose that the whole reference path is known (memorized) and r is extended all over real line R. Thus, the following relations for r hold: $u_{des}(t) = \dot{r}\, u_{des}(r)$, and $\dot{x}_{des} = u_{des}(t) = \dot{r}\, u_{des}(r)$.

This one-coordinate system will permit us to easily extract and analyze the differential equations implicated in this proposal and to clearly see EAT running. Moreover, we will be able to gain insight of the influence of tracking rate election over system behavior. Note that the concept of trajectory does not exist for a coordinate only, but our analysis and conclusions can be extended to a system with several coordinates.

For a fair comparison between TT and EAT, we will use the same control law for both methods. We briefly present a problem that affects the model considered by the controller (which are usual for UAVs). Evidently, with no perturbation, big errors or unmodeled effects stationary tracking would be perfect and, consequently, there would be no need for studying improvements introduced via tracking rate changes. The main idea is that, using the EAT method, problems that affect the tracking will be partially "absorbed" by the tracking rate in order to reduce tracking errors and to improve convergence.

If TT were applied, the system error would be merely the difference: $e(t) = x(t) - x_{des}(t)$, and the state equation for the error would be: $\dot{e}(t) = u(t) - u_{des}(t)$;

It is clear that a simple convergent control law is: $u(t) = u_{des}(t) - K_p e$, $K_p > 0$

This yields the TT error equation: $\dot{e}(t) = -K_p e$.

The solution of the previous equation is: $e(t) = e(0)\, exp(-K_p t)$, where $e(0)$ is the initial error. Therefore, exponential convergence is ensured, with a characteristic time constant $t = (K_p)^{-1}$.

Figure 2 depicts the role that the tracking methods play in the feedback control of a simple one-state plant. Common blocks for the plant, controller and sensors work as usual. However, the desired trajectory sent to the controller is computed through the product of the desired path profile $u_{des}(r)$ and the $\dot{r}$ evolution, which is selected by the desired tracking method.

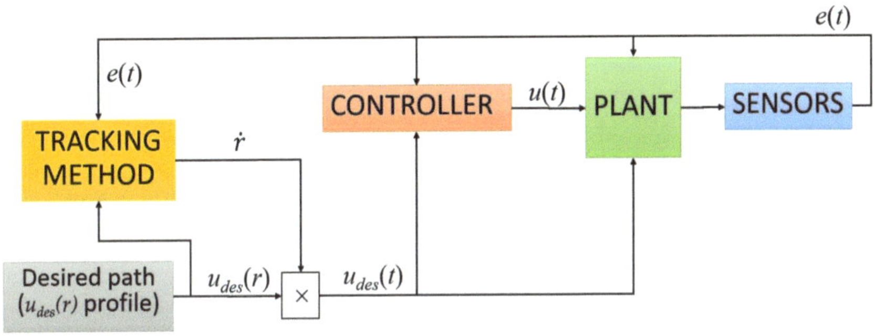

Figure 2. Diagram block of the control of a plant that considers the descriptor parameter evolution.

On the other hand, if EAT were to be applied, the system error (a superscript r is added to clearly distinguish this error definition from that of TT) would be: $e^r(t) = x(t) - x_{des}(r(t))$, and the error equation: $\dot{e}^r(t) = u(t) - \dot{r} u_{des}(r)$. Now the same simple control law yields:

$$\dot{e}^r = -K_p e^r + \dot{r} u_{des}(r) - u_{des}(r);$$

An intuitive proposal for a NDEAT tracking rate can be (this intuitive form fulfills completely the mathematical condition given by the Lemma in the next section):

$$\dot{r} = 1 + \frac{K_r}{u_{des}} e^r \; ; \; K_r > 0$$

This yields to the EAT error equation: $\dot{e}^r(t) = -K_p e^r - K_r e^r$. The error evolution includes a new parameter K_r that considers the rate of tracking, which is: $e(t) = e(0)\, exp((-K_p - K_r)t)$, where $e(0)$ is the initial error. Exponential convergence for EAT method has now the characteristic time constant $t = (K_p + K_r)^{-1}$, which is faster than that of TT because error decreasing is produced by two causes: the control law and the tracking rate selection.

Likewise, if SDEAT were to be applied, system error could be defined as: $e^{rt}(t) = x(t) - x_{des}(r(t)) + A_{rt}(t - r)$, $A_{rt} > 0$. Therefore, an intuitive proposal for SDEAT tracking rate could be:

$$\dot{r} = 1 + \frac{K_{rt}}{A_{rt} + u_{des}} e^{rt} \; ; \; K_r t > 0$$

Transient behavior is more interesting with respect to the advancement along the desired path r because this will be the main objective when following a memorized path, for example, in mobile robot applications. Previous equations were simulated using MATLAB, with the following conditions: simulation time = 10 s, $e(0) = -1.5$ m, $K_p = 0.5\ s^{-1}$, $K_r = 0.5\ s^{-1}$, and reference path defined by $u_{des}(r) = 1$ m/s. In Figure 3a we represent TT and NDEAT error behaviors as a function of parameter r (being $r = t$ for TT). Likewise, $\dot{r}$ evolution is shown in Figure 3b.

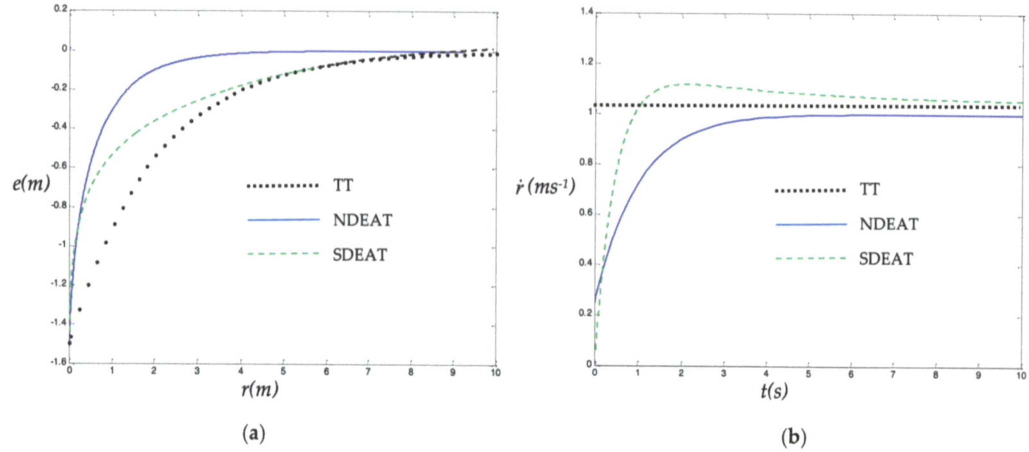

Figure 3. (**a**) TT, NDEAT and SDEAT transient behaviors for a big initial error. (**b**) $\dot{r}$ evolution.

The most interesting fact in Figure 3b happens during the first transitory moments for EAT methods. Here, $\dot{r}$ gets low values, thus, the desired point $x_{des}(r)$ "waits" for the robot to approach it. For a system with several coordinates, this means that a faster convergence to the desired path can be reached with the EAT method. At the extreme case, if errors were big, this approximation would become a straight line to the reference $x_{des}(r(t))$. This desirable behavior clearly resembles that of the PF method [1,5].

On the contrary, for TT we find the usual behavior of a tracking system; the objective advances continuously pulling the system ahead. In this way, in a system with several coordinates, this attraction will prevent the system from approaching the path, and so the system convergence will be delayed in relation to the path parameter r. Note also that the final value of r for NDEAT in Figure 3b is lower than the final value of r (or t) for TT because $\dot{r}$ takes small values for the first seconds.

In the case of SDEAT, it can be seen that it behaves similar to NDEAT during the first moments when errors are big (Figure 3a). That is, $\dot{r}$ stays low (Figure 3b) because the system intends to approach point $x_{des}(r)$ (r is almost constant). Therefore, we will achieve a fast convergence to the path (in case of several coordinates). However, when the system is approaching convergence, $\dot{r}$ grows above one, and the system tries to be deterministic by tracking the reference point $x_{des}(t)$, that is, the system recovers real-time conditions. In this last part, the system velocity continues to be slightly bigger than the reference velocity $u_{des}(r)$ to reduce the difference $r-t$. SDEAT evolution of Figure 3 has been reproduced with exactly the same conditions and control law. Constants for SDEAT were chosen as $A_{rt} = 2.0\ \text{ms}^{-1}$, $K_{rt} = 2.0\ \text{s}^{-1}$.

In addition, let us analyze the existence of a parametric modeling uncertainty. Imagine that the real system equation has a δ deviation from the ideal model, that is, the real system behaves actually as $\dot{x} = (1+\delta)u$. Applying the same control law, the error equation becomes now:

$$\dot{e}(t) = -K_p(1+\delta)e + (d + 1 - \dot{r})\, u_{des}(r)$$

Therefore, a steady error e_{ss} is unavoidably present. In the case of TT ($\dot{r} = 1$), when the stationary state is reached ($\dot{e} \rightarrow 0$), we arrive at

$$e_{ss,TT} = \frac{\delta u_{des}}{(1+\delta)K_p}$$

However, this error would be scaled down by $\dot{r}$ if NDEAT were used. Applying the NDEAT proposal for $\dot{r}$, the stationary state can be easily found as:

$$e_{ss,NDEAT} = \frac{\delta u_{des}}{(1+\delta)K_p + K_r}$$

It is clear that $e_{ss,NDEAT}$ becomes smaller than $e_{ss,TT}$, as the reduction in tracking rate (that is, $\dot{r} < 1$ for NDEAT) is absorbing the unmodeled behavior. More exactly for constants $K_p = K_r = 0.5\ s^{-1}$, the stationary error is divided by almost 50%.

To sum up, two important considerations must be taken into account:

- As the control law and the system become more complex, the design of tracking rate $\dot{r}$ should contemplate more circumstances as it is discussed below for the case of UAVs.
- For complex systems, it is obviously more difficult to find robust control laws that behave well enough under several problems (such as unmodeled behaviors, motor delay responses and so on). When this happens, the EAT method may provide a new form of avoiding oscillations, divergences, error enlargements, etc. This will be presented at the following sections, where a generic method is formulated to extend the EAT for any controller that is based on a Lyapunov function.

The paper is organized as follows. Basic formulation and the lemmas to use EAT methods from a Lyapunov-based controller are stated in Section 2. In Section 3, an application case is discussed in depth, a UAV model. Its asymptotically stable control law is considered, and its validity for various EAT methods is demonstrated. Evaluation of these methods through simulation is discussed to illustrate the EAT benefits in Section 4. Finally, conclusions are presented in the last section.

2. Conversion of Trajectory Tracking into Error Adaptive Tracking for Lyapunov-Based Controllers

In this section, we present a theoretical formulation of the tracking equations and some Lemmas for using EAT straightforwardly from a Lyapunov-based controller. Let us consider the state space representation of a dynamic system model: $\dot{q} = f_q(q, u, t)$, where q is the state vector of dimension m, u is the input vector of dimension n, and t is time. The initial conditions are given by $q(0)$. A *memorized, reference or desired* path or trajectory (or merely *path*) $q_{des}(r)$ can be described by a single-descriptor parameter, namely r. This path should be covered by the system, that is, it must fulfill the system model: $q'_{des} = f_q(q_{des}, u_{des}, r)$, where ($'$) denotes derivative with respect to r. As we are interested in convergence to a path, we assume in this work that the desired trajectory has no end and it never stops, that is, r (-∞, ∞), $q'_{des} \neq 0, \forall r$ (because if it ended at a certain point, this would be a problem of stabilization instead of tracking).

To study the tracking, an error state vector should be defined through a diffeomorphism $e = h_e(q, r, t)$, such that $e = 0$ if and only if $q(t) = q_{des}(r(t))$. Note that r dependence can now be introduced in the definition of error e. Likewise, the input vector may be redefined as $v = h_v(q, r, u, t)$ in order to express the dynamics in a more convenient form. In practical cases, the dependence of error and input vectors on r are usually through $q_{des}(r)$ and $u_{des}(r)$, which are merely functions of r. Therefore, a new state variable r appears, whose state equation can be freely defined ("modeled") in general as: $\dot{r} = \sigma(e, r, t)$, where σ can be considered a new input, so dependence on $\mathbf{v}$ is prevented in σ. Thus, the system error model is now $\dot{e} = f(e, r, v, \sigma, t)$, and the initial conditions are given by $e(0)$, $r(0)$.

Minimal tracking control objectives for these state variables $\{e, r\}$ can be set to

$$e \rightarrow 0 \text{ when } t \rightarrow \infty \tag{1a}$$

$$|\dot{r}| \text{ bounded } \forall t, r \rightarrow \infty \text{ when } t \rightarrow \infty \tag{1b}$$

Objective (1b) is needed to prevent r from jumping suddenly, so tracking is performed smoothly. These objectives are to be satisfied through the proper selection of control laws:

$$v = c_v(e, r, t) \tag{2a}$$

$$\sigma = g(e, r, t). \tag{2b}$$

We want to point out that the expression $g(e, r, t)$ is just a control law for r, and the behavior of r (or its rate of progression) can be designed for any application—still preserving the objectives stated in (1b).

A trivial rate of progression can be performed just by identifying parameter r with time, that is, $r(t) = t$ or $\sigma = 1$, which yield the simplest TT. This would mean that $q_{des}(r)$ advanced continuously pulling the system forward. In this case, error coordinates can be simply defined as $e_q(t) = q(t)\text{-}q_{des}(t)$. Nevertheless, TT can be extended with a more general assumption: let parameter r be a strict increasing function of time to fulfill objectives in (1b), for example, $r = at$, $a > 0$. Therefore, error coordinates can be $e_q(t) = q(t)\text{-}q_{des}(r(t))$.

However, if time was not critical, the tracking methodology could be freely designed, as the whole trajectory is known a priori. The most common alternative in the mobile robotics literature is *path following*. This is based on some relation between the actual system state and the whole path. This relation or *projection* will give us the desired point $q_{des}(r)$, i.e., the descriptor parameter r as a function of the actual position and the path. Differently from the TT case, here the real system must aim to follow this point. For example, the desired point is usually selected to be the point on the path that is "closest" to the actual robot's position [23,24]. The main drawback of PF is that (to the authors' knowledge) projection uniqueness has not been guaranteed for all possible paths $q_{des}(r)$. This problem may occur when the projection is fulfilled for an interval of r [23], which means that (1b) cannot be satisfied (r is undetermined in this interval). Finally, we have the EAT method, which has very interesting properties: it can be applied to every tracking system and to all sets of paths (because it does not depend on any projection); it can also consider timing requirements, and, as it will be discussed in the next section, its controller design can be performed in a way similar to that of TT, but achieving faster convergence and higher robustness. In the approach presented here, *the same control law* can be used for TT and EAT methods, thus, facilitating the design of the controller and allowing to choose between TT and EAT when needed.

It can be noted that PF on the one hand, and TT or EAT on the other, must define completely dissimilar control laws. For example, in a system with two error variables and two inputs, the PF projection causes a constraint that eliminates one of the variables [23], which implies that PF control law is applied only to one variable. A common approach consists of setting one of the inputs to be constant, thus, forcing the system to move. Another approach consists of selecting a relation for both inputs that implies the overall input to never be null. This "motion exigency" is not present in EAT (or TT). Depending on the regulations performed for both tracking methods, one method will present potential benefits over the other or vice versa. On the other hand, the control law will be identical when applying NDEAT or SDEAT (see Lemmas below), and no design or tuning of a new controller is needed.

Moreover, when using EAT, the convergence will be, at least, as fast as that obtained with TT. In the next sections, this general procedure is illustrated for a PVTOL (planar vertical take-off and landing aircraft). The subsequent tests will demonstrate that *a)* EAT presents a faster convergence than TT, and *b)* it is valid even for those paths where PF cannot be applied. Nonetheless, as mentioned before, the procedure shown here can be applied to any other system that uses a Lyapunov-based control law.

Before presenting the Lemmas, it should be pointed out that their objective is to change the tracking from TT to EAT due to the good properties of the latter. As a direct consequence of these Lemmas, when using EAT the question would be: what tracking rate must be selected for the EAT method so that the same controller presents a faster convergence to the desired posture? In this respect, finding an appropriate tracking rate is crucial for nonholonomic system controllers. This is a direct consequence of the Brockett's Theorem [25], which prevents the existence of a smooth feedback stabilization control law. We mean that a smooth controller may fail if the tracking rate is different from $\sigma = 1$ (this is evident if the rate becomes near $\sigma = 0$, that is, if the tracking tends to stabilization). Another consequence of nonholonomic constraints is that derivatives of a Lyapunov function V cannot be a negative definite function but only a negative semidefinite function. This can be observed for those postures where only an error is not null and the movement that should reduce this error is prohibited, such as a lateral displacement in a car.

The next Lemmas allow selecting a valid rate that additionally improves initial TT convergence and introduces an interesting relation with the PF method (see Remark 3).

Lemma 1. *Conversion of TT into NDEAT.*

Let $\dot{q} = f_q(q, u, t)$ be the model of a system that must follow a smooth desired path given by $q_{des}(r)$, $r(-\infty,\infty)$, which fulfills $q'_{des} = f_q(q_{des}, u_{des}, r)$, with $u_{des}(r) \neq 0\, \forall r$. Let $e^t = h_e(q, q_{des}(t))$ be the definition of TT errors, and $e^r = h_e(q, q_{des}(r(t)))$ those of EAT errors, being h_e also smooth. Let us suppose that there exists a positive definite Lyapunov function $V(\mathbf{e}^t)$, with the intention that a smooth control law $v = c_v(e^t, t)$ makes $\dot{V}$ be negative semidefinite and uniformly continuous, so it can be proved that e^t=0 is a global asymptotically stable equilibrium point for the path $q_{des}(t)$.

With these assumptions and for the same control law evaluated on r: $v = c_v(e^r, r)$, we have that $e^r = h_e(q, q_{des}(r(t))) = 0$ is also a global asymptotically stable equilibrium point, if a uniformly continuous NDEAT rate $\dot{r} = \sigma(e^r, r) = 1 + e_\sigma(e^r, r)$ is chosen, so that $e_\sigma \left\langle \frac{\partial V}{\partial e}\Big|_{e=e^r}, \frac{\partial h_e}{\partial q_{des}} f_q(q_{des}, u_{des}, r) \right\rangle$ is negative semidefinite with respect to e. Here, $\langle x, y \rangle$ represents the dot product of x, y and $\frac{\partial}{\partial x}$ stands for the gradient for a scalar function, or the Jacobian for a vector function.

Proof of Lemma 1. Using the chain law, $\dot{\mathbf{q}}_{\mathbf{des}} = \sigma\, \mathbf{q}'_{\mathbf{des}} = \sigma\, \mathrm{f_q}(\mathbf{q}_{\mathbf{des}}, \mathbf{u}_{\mathbf{des}}, \mathrm{r}) = \sigma\, \mathrm{f}_{\mathrm{q,des}}$, where it has been called $\mathrm{f}_{\mathrm{q,des}} = \mathrm{f_q}(\mathbf{q}_{\mathbf{des}}, \mathbf{u}_{\mathbf{des}}, \mathrm{r})$ for clarity purposes. Deriving $\mathbf{e}^{\mathbf{r}} = \mathrm{h_e}(\mathbf{q}, \mathbf{q}_{\mathbf{des}}(\mathrm{r(t)}))$ and using the change $\sigma = 1 + \mathrm{e}_\sigma$,

$$\dot{e}^r = \frac{\partial h_e}{\partial q}\dot{q} + \frac{\partial h_e}{\partial q_{des}}\dot{q}_{des} = \frac{\partial h_e}{\partial q}\dot{q} + \frac{\partial h_e}{\partial q_{des}} f_{q,des} + e_\sigma \frac{\partial h_e}{\partial q_{des}} f_{q,des}$$

The previous derivative can be expressed as:

$$\dot{e}^r = \dot{e}^t_{t=r} + e_\sigma \frac{\partial h_e}{\partial q_{des}} f_{q,des}\,,$$

where $\dot{e}^t_{t=r} = \dot{q} - q'_{des}(t=r)$ represents the tracking error rate for the TT case when $t = r$. By computing the derivative of $V(\mathbf{e}^\mathbf{r})$, we obtain

$$\dot{V}(e^r) = \left\langle \frac{\partial V}{\partial e}\Big|_{e=e^r}, \dot{e}^r \right\rangle = \left\langle \frac{\partial V}{\partial e}\Big|_{t=r}, \dot{e}^r \right\rangle = \left\langle \frac{\partial V}{\partial e}\Big|_{t=r}, \dot{e}^t_{t=r} \right\rangle + e_\sigma \left\langle \frac{\partial V}{\partial e}\Big|_{e=e^r}, \frac{\partial h_e}{\partial q_{des}} f_{q,des} \right\rangle$$

Using the hypothesis, it holds that $\dot{V}$ is negative (at least) semidefinite and uniformly continuous in time. Therefore, the resulting NDEAT tracking will also make $\mathbf{e}^\mathbf{r}$ be a globally asymptotically stable equilibrium point for the path $q_{des}(r)$, as was e^t. □

Remark 1. *Note that the convergence of the NDEAT is, at least, as fast as that of the TT because the term* $e_\sigma \left\langle \frac{\partial V}{\partial e}\Big|_{e=e^r}, \frac{\partial h_e}{\partial q_{des}} f_{q,des} \right\rangle$ *decreases or maintains the temporal rate of V.*

Remark 2. *Note that a simple election like* $e_\sigma = -K_\sigma \left\langle \frac{\partial V}{\partial e}\Big|_{e=e^r}, \frac{\partial h_e}{\partial q_{des}} f_{q,des} \right\rangle$; $K_\sigma > 0$ *implies that* $e_\sigma \to 0$; $\sigma \to 1$ *, i.e., the same desired tracking rate is preserved when the convergence is achieved.*

Remark 3. *Path following controllers usually project real system posture over the reference path by choosing that reference point that minimizes some kind of distance. A common and sensible distance is given by the proper Lyapunov function* [1]. *In fact, this function gives an idea of the amount of error, so the point on the path with minimum errors is chosen. In this case, the projecting point looks for the value of r that makes the derivative of V null for a fixed system state, that is,*

$$\left.\frac{\partial V}{\partial r}\right|_{q=constant}=0\,;\ \left.\frac{\partial V}{\partial e}\right|_{e=e^r}\left.\frac{\partial e}{\partial r}\right|_{q=constant}=\left\langle\left.\frac{\partial V}{\partial e}\right|_{e=e^r},\frac{\partial h_e}{\partial q_{des}}f_{q,des}\right\rangle=0\,;$$

The previous equation zeroes the same term that multiplies to e_σ in $\dot{V}(\mathbf{e^r})$, which implies that the chosen e_σ of Remark 2 makes EAT tracking tend to that of PF. Thereby, EAT conserves most of the advantages of the PF method [1,5], while avoiding its main obstacle: the non-uniqueness of the selection of a point in the path when the robot is far from it (in other words, the need for the robot to stay in a tube around the path).

If the SDEAT method were to be applied, a way to obtain the proper function $\sigma=\sigma(e^r,r,t)$ is to select the next variant of the Lyapunov function: $V_2(\mathbf{e^r},t)=V(\mathbf{e^r})+\frac{1}{2}A_{rt}(r-t)^2$, A_{rt}>0. Proceeding correspondingly,

$$\dot{V}_2(\mathbf{e^r},t)=\dot{V}+e_\sigma A_{rt}(r-t)=\left\langle\left.\frac{\partial V}{\partial e}\right|_{t=r},\dot{e}^t_{t=r}\right\rangle+e_\sigma\left(\left\langle\left.\frac{\partial V}{\partial e}\right|_{e=e^r},\frac{\partial h_e}{\partial q_{des}}f_{q,des}\right\rangle+A_{rt}(r-t)\right)$$

An evident SDEAT proposal that keeps $\dot{V}_2$ uniformly continuous in time and negative definite (other σ are possible), is making the last term quadratic by carrying out

$$\sigma=1-K_\sigma\left(\left\langle\left.\frac{\partial V}{\partial e}\right|_{e=e^r},\frac{\partial h_e}{\partial q_{des}}f_{q,des}\right\rangle+A_{rt}(r-t)\right),\ K_\sigma>0$$

This allows us to enunciate the following Lemma.

Lemma 2. *Conversion of TT into SDEAT.*

For the same conditions of Lemma 1, $\mathbf{e}=h_e(\mathbf{q},\mathbf{q_{des}}(r(t)))=0$ is a global asymptotically stable equilibrium point, if a uniformly continuous SDEAT rate $\dot{r}=\sigma(\mathbf{e^r},r)=1+e_\sigma(\mathbf{e^r},r,t)$ is chosen, so that $e_\sigma\left(\left\langle\left.\frac{\partial V}{\partial e}\right|_{e=e^r},\frac{\partial h_e}{\partial q_{des}}f_{q,des}\right\rangle+A_{rt}(r-t)\right)$ is negative semidefinite with respect to $\mathbf{e}$ and to $(r-t)$, with $A_{rt}>0$.

Proof. The proof can be guided in a way similar to that of Lemma 1, using the given form of $\dot{V}_2(\mathbf{e^r},t)$.□

Remark 4. *Note that r tends to t, so TT tracking rate can be achieved in the end.*

Straightforward case uses can be easily obtained. For example, using Remark 2, a uniformly continuous NDEAT rate for $\dot{r}=1+e_\sigma$ can be found for the TT controller of the WMR presented in Section 34.4.2 of [1]. There, the authors use the kinematic model of a unicycle robot:

$$\begin{cases}\dot{x}=u_1\cos\theta\\ \dot{y}=u_1\sin\theta\\ \dot{\theta}=u_2\end{cases}\quad\begin{cases}\dot{x}_{des}=u_{1,des}\cos\theta_{des}\\ \dot{y}_{des}=u_{1,des}\sin\theta_{des}\\ \dot{\theta}_{des}=u_{2,des}\end{cases}$$

where state $\mathbf{q}=(x,y,\theta)$ represents the Cartesian coordinates of the driven wheel middle point and the orientation with respect to a fixed frame, and (u_1, u_2) the linear and angular speed of this point. Subscript 'des' is used for the virtual robot (the desired trajectory). They introduce the TT error definition $\mathbf{z^t} = h_e(\mathbf{q},\mathbf{q_{des}}(t)) = (z_1, z_2, z_3)$ valid for $\theta - \theta_{des} \neq \pm\pi/2$ and given by

$$\begin{cases} \dot{z}_1 = (x - x_{des})\cos\theta_{des} + (y - y_{des})\sin\theta_{des} \\ \dot{z}_2 = -(x - x_{des})\sin\theta_{des} + (y - y_{des})\cos\theta_{des} \\ \dot{z}_3 = \tan(\theta - \theta_{des}) \end{cases}$$

where dependence on *(t)* has been suppressed for desired coordinates for clarity reasons.

Additionally, through the definition of the Lyapunov function $V = \frac{1}{2}\left(z_1^2 + z_2^2 + \frac{1}{k_2}z_3^2\right)$; $k_2 > 0$, they propose a globally asymptotically stable TT control law that makes $\dot{V}$ negative semidefinite and uniformly continuous, provided that $|u_{1,des}|$ is uniformly continuous and does not tend to zero.

In order to apply NDEAT, gaining the benefits described previously, the necessary terms for Lemma 1 are calculated:

$$\left.\frac{\partial V}{\partial e}\right|_{e=e^r} = \left(z_1, z_2, \frac{1}{k_2}z_3\right), \ \frac{\partial h_e}{\partial \mathbf{q_{des}}} f_{q,des} = \left(-u_{1,des} + z_2u_{2,des}, z_1u_{2,des}, -u_{2,des}\left(1 + z_3^2\right)\right)$$

Finally, using Remark 2, an expression for the NDEAT tracking rate that preserves the validity of the same controller is obtained:

$$e_\sigma = -k_\sigma\left(\left\langle \left.\frac{\partial V}{\partial e}\right|_{e=e^r}, \frac{\partial h_e}{\partial \mathbf{q_{des}}} f_{q,des} \right\rangle\right) = k_\sigma\left(z_1u_{1,des} + \frac{1}{k_2}z_3u_{2,des}\left(1 + z_3^2\right)\right); k_\sigma > 0$$

With this e_σ, Lemma 1 ensures that the same control law is globally asymptotically stable for errors $\mathbf{z^r}=h_e(\mathbf{q},\mathbf{q_{des}}(r))$, and the convergence of the NDEAT tracking is, at least, as fast as that of TT. Note that if timing requirements were needed, the SDEAT method can be found in a similar manner.

Although several benefits of NDEAT tracking can be revealed for simple models, these benefits can be better observed for more complex robots including non-linearities. In the next section, EAT is compared with TT for a UAV, whose inputs saturate when their values surpass a certain bound.

3. Application of EAT for the PVTOL

Mobile robots, and more specifically UAVs, are a traditional application example when studying and selecting tracking methods because most of them do not need strict timing requirements. The interesting control problems associated with the vertical/short takeoff and landing aircraft has turned PVTOL into one of the most studied benchmarks for controller design. More concretely, the fact that PVTOL has non-minimum phase zero dynamics associated with its center of mass suggested that path following controllers could be more appropriate than tracking controllers [3].

Firstly, we recall in the next paragraphs the main equations for the PVTOL according to [26]. We refer the reader to this classic paper for further details on this system. Afterwards, the tracking method is transformed from TT to NDEAT and SDEAT by using the previous Lemmas. Finally, in the next section, several results are shown to demonstrate the benefits of using EAT methods instead of TT.

A simplified model for PVTOL is given by [26]:

$$\begin{aligned} m\ddot{x} &= -\sin\theta\, T \\ m\ddot{y} &= -mg + \cos\theta\, T \\ \ddot{\theta} &= -\omega_n^2\sin\theta + k_sT\sigma_2(u_2) \\ \dot{T} &= -k_t(T - \sigma_1(T_d)) \\ \dot{T}_d &= u_1 \end{aligned} \quad (3)$$

where x, y are the lateral and vertical positions, θ is the pitch angle, T is the actual propeller thrust in Newtons (which is controlled through a second order dynamics by the input u_1, being T_d the desired thrust), and input u_2 is the stabilator input used to generate a pitching movement. Constants in (3) are: $m = 2.15$ kg (mass of the aircraft), $g = 4.98$ ms^{-2} (effective gravity), $\omega_n = \sqrt{33}\text{s}^{-1}$ (natural frequency in pitch), $k_s = 5.4$ kg^{-1}m^{-1} (stabilator constant), and $k_t = 4$ s^{-1} (thrust constant). Functions σ are saturations for the thrust and the stabilator inputs, with the following upper and lower limits: $max(\sigma_1) = 16$, $min(\sigma_1) = 0$, $max(\sigma_2) = 1$, $min(\sigma_2) = -1$.

It is well known that linearized forms of systems with nonholonomic constraints, such as WMRs (Wheeled Mobile Robots) and UAVs, are not controllable [27]. For this reason, other alternatives, such as feedback linearization have been profusely studied for these systems. In [26] it is shown that system (3) is feedback linearizable provided that saturations are not active. Using the linearizing coordinate change

$$\begin{cases} \boldsymbol{x} = \left(x, \dot{x}, y, \dot{y}, \theta, \dot{\theta}, T, T_d\right) \mapsto \boldsymbol{z} \equiv \left(x, \dot{x}, \ddot{x}, \dddot{x}, y, \dot{y}, \ddot{y}, \dddot{y}\right) \\ \boldsymbol{u} = (u_1, u_2) \mapsto v \equiv \left(x^{(4)}, y^{(4)}\right) \end{cases} \tag{4}$$

the linearized dynamics (valid outside the saturation) result in

$$\begin{array}{l} \boldsymbol{A_o} = \begin{bmatrix} 0 & 1 & 0 & 0 \\ 0 & 0 & 1 & 0 \\ 0 & 0 & 0 & 1 \\ 0 & 0 & 0 & 0 \end{bmatrix}; \mathbf{B}_o = \begin{bmatrix} 0 \\ 0 \\ 0 \\ 1 \end{bmatrix} \\ \boldsymbol{A} = \text{block diagonal}\,\{\mathbf{A}_o, \boldsymbol{A_o}\} \\ \boldsymbol{B} = \text{block diagonal}\,\{\mathbf{B}_o, \boldsymbol{B_o}\} \\ \dot{\boldsymbol{z}} = \boldsymbol{Az} + \boldsymbol{Bv} \end{array} \tag{5}$$

A reference or desired path to be followed can be a feasible trajectory that fulfills (3). Therefore, it can be expressed as a function of a descriptor parameter r: $\mathbf{x_{des}}(r)$, with desired inputs $\mathbf{u}_{\text{des}}(r)$. Alternatively, for the linearized system, the desired path $\{\mathbf{z_{des}}(r), \mathbf{v}_{\text{des}}(r)\}$ must fulfill $z'_{des} = \boldsymbol{A}z_{des} + \boldsymbol{B}v_{des}$.

Now we recall the control law for the linearized system (5) presented in [26], and we show that the same law can be used for EAT using the appropriate tracking rate (see Lemmas 1 and 2). Therefore, no design or tuning of a new controller is needed. A linear control law for system (5) is:

$$v = v_{des} + \boldsymbol{K}(z - z_{des}) \tag{6}$$

where $\boldsymbol{A_c} = \boldsymbol{A} + \boldsymbol{BK}$ is Hurwitz, which provides local stable trajectory tracking (global in (z, v) if there are no restrictions in these coordinates). Linear dynamics result in $\dot{e}_z = A_c e_z$, being $\boldsymbol{e_z} = z\text{-}z_{des}$. According to [26], the experience with the actual PVTOL shows that $\boldsymbol{K}$ can be decoupled for lateral and vertical modes and that values for $\boldsymbol{K}$: $K_o = \begin{bmatrix} -3604 & -2328 & -509.25 & -39 \end{bmatrix}$, $\boldsymbol{K} = \text{block diagonal}\{\mathbf{K}_o, \boldsymbol{K_o}\}$ make the system perform properly. System input $\boldsymbol{u}$ can be calculated from $\boldsymbol{v}$, through (4) and (3). Thus, provided that $\{\mathbf{x}(t), \mathbf{u}(t)\}$ stays within the valid region, global exponential convergence to the desired trajectory (that is, to $\boldsymbol{e_z} = 0$) is guaranteed. Consequently, given a positive definite symmetric matrix $\boldsymbol{Q}$ ($\boldsymbol{Q} > 0$), there exists a unique positive definite symmetric matrix, $\boldsymbol{P} > 0$, that fulfills the Lyapunov equation $\boldsymbol{A_c^T P} + \boldsymbol{PA_c} + \boldsymbol{Q} = 0$.

Let us first analyze the case of NDEAT. Using Lemma 1, it is evident that: if (*a*) the same control (6) is applied, being $V = \boldsymbol{e_z^T P e_z}$ a Lyapunov candidate (from now on the superscript ^{r} of the error is omitted for simplicity reasons), and (*b*) the chosen uniformly continuous NDEAT rate $\sigma = 1 + e_\sigma(\boldsymbol{e_z}, r)$ makes $-2e_\sigma\left(\boldsymbol{z'^T_{des} P e_z}\right)$ negative semidefinite with respect to $\boldsymbol{e_z}$, then $\boldsymbol{e_z} = z\text{-}z_{des}(r(t)) = 0$ is a globally asymptotically stable equilibrium point (if there are no restrictions in $\boldsymbol{z}$, $\boldsymbol{v}$ coordinates).

The proof can be guided by using the change $\sigma = 1 + e_\sigma$, as in Lemma 1. In this case, the new linear dynamics for the same control law (6) is $\dot{e}_z = \dot{z} - \sigma\, z'_{des} = A_c e_z - e_\sigma z'_{des}$. By computing the derivative of V, and using the new linear dynamics, we obtain

$$\dot{V} = e_z^T \left(A_c^T P + P A_c \right) e_z - 2e_\sigma \left(z'^T_{des} P e_z \right) \quad (7)$$

Using the hypothesis and the Lyapunov equation, it holds that $\dot{V} = -e_z^T Q e_z - 2e_\sigma \left(z'^T_{des} P e_z \right)$ is negative definite and uniformly continuous in time.

An evident proposal that makes the last term of (7) quadratic is:

$$\sigma = 1 + 2K_\sigma \left(z'^T_{des} P e_z \right) = 1 + 2K_\sigma z'_{des}, e_z P,\ K_\sigma > 0 \quad (8)$$

but other functional proposals for σ are possible (depending on the application tracking requirements).

Remark 5. *The proposed form (8) of σ includes a dot product, which gives an idea of the relative posture (the "sign" of the errors) of the real and virtual robots when the robot is not too "far" (according to the distance function given by matrix* **P***) from the reference point. When the robot is "ahead" with respect to $z_{des}(r)$ (along the direction specified by z ¢$_{des}(r)$), then this dot product will be positive, but when the robot is delayed, it will be negative. If this product is zero, the errors are perpendicular to the desired velocities, and the robot is neither ahead nor delayed. Therefore, the tracking rate can be greater or less than 1. When the robot is ahead (according to the previous dot product criterion), it is intuitive that a faster rate will get the reference point closer to the robot. On the other hand, for a delayed robot, the lower value of σ means that the desired point will "wait" for the robot.*

If the SDEAT method were to be applied, another Lyapunov function can be defined according to Lemma 2. Using $V_2 = e_z^T P e_z + \frac{1}{2} A_{rt} (r - t)^2$ and proceeding correspondingly, (7) results in:

$$\dot{V}_2 = e_z^T \left(A_c^T P + P A_c \right) e_z - e_\sigma \left(2z'^T_{des} P e_z - A_{rt}(r - t) \right)$$

An evident SDEAT proposal that keeps $\dot{V}_2$ uniformly continuous in time and negative definite (other σ are possible), is to make the last term quadratic by carrying out

$$\sigma = 1 + K_\sigma \left(2z'_{des}, e_z P - A_{rt}(r - t) \right),\ K_\sigma > 0 \quad (9)$$

Therefore, if *a)* the same control (6) is applied and *b)* the uniformly continuous EAT rate $\sigma = 1 + e_\sigma$ makes $-2e_\sigma \left(z'^T_{des} P e_z \right) - A_{rt}(r - t)$ negative definite, then $e_z = 0$ is a global asymptotically stable equilibrium point. The proof can be guided in a way similar to that of the proof of Lemma 2.

4. Simulation Results

In this section, simulation results comparing the TT and EAT behaviors for the previous UAV system are presented and discussed. Although the results are shown for this system, these analyses and conclusions can be extended to other systems for which a classic control law was previously obtained. Advancing one of the conclusions of our results, using numerically the same control law for EAT leads to a faster and much more robust convergence as Remark 1 points out. The reason is that errors that affect the tracking are partially "absorbed" by the tracking rate $\dot{r}$. Therefore, a final thought is worth mentioning: for complex systems, it is more difficult to find robust control laws that behave well enough under several problems. When this happens, the EAT method may provide a form of avoiding oscillations, divergences, error enlargements, etc.

Two tests are to be analyzed using a SIMULINK/MATLAB model [28]: (a) the diverse initial extreme conditions, and (b) the convergence to a desired path that cannot be executed by the PF presented in [26] because the whole path fulfills the projection used there. The first test is focused on big errors because in the case of small errors the simulated system's behavior will be similar to that of an exponential convergent system. The second test is intended to confirm the advantage of EAT versus PF, which is the validity of EAT for all kinds of paths.

The chosen form of σ for NDEAT is that of (8) with $K_s = 0.0010\ \mathrm{m}^{-1}\mathrm{s}^{-1}$. To conduct a fair comparison between NDEAT and SDEAT methods, the constant K_s of SDEAT is the same as that of NDEAT, with $A_{rt} = 400\ \mathrm{m}^2/\mathrm{s}^2$. Greater values of A_{rt} will bring the tracking closer to that of TT, while lower ones will bring it closer to that of NDEAT. An important property of σ is the linearity around $\boldsymbol{e_z} = 0$.

As our intention is to show the benefits of EAT against TT for a same controller, we have used exactly the same controller of [26]. Therefore, the Lyapunov equation matrices used are $\boldsymbol{Q} = \boldsymbol{I}$, and by solving Lyapunov equation: $\boldsymbol{P}$ = block diagonal $\{\boldsymbol{P_o}, \boldsymbol{P_o}\}$, with

$$P_0 = \begin{bmatrix} 436.3905\ 281.1773\ 53.4184\ 0.0001 \\ 281.1773\ 189.7791\ 39.2207\ 0.1210 \\ 53.4184\ 39.2207\ 10.4683\ 0.0780 \\ 0.0001\ 0.1210\ 0.0780\ 0.0148 \end{bmatrix}$$

The first test (Figures 4–6) analyzes the tracking of a periodic lateral motion $x_{des}(r) = A\ sin(w_{ref}\ r)$ with constant $y_{des}(r) = 0$, where $w_{ref} = 2\pi/5\ \mathrm{s}^{-1}$, and $A = 1.857\ \pi/2$ m. The initial desired state and value of parameter r are: $\mathbf{z_{des}} = 0$; $r = 0$, for all tests. Two considerable initial position errors in x and y are tested in order to compare EAT with TT (see Table 1, where the corresponding figures are also shown). The rest of the real initial states are the same as those desired.

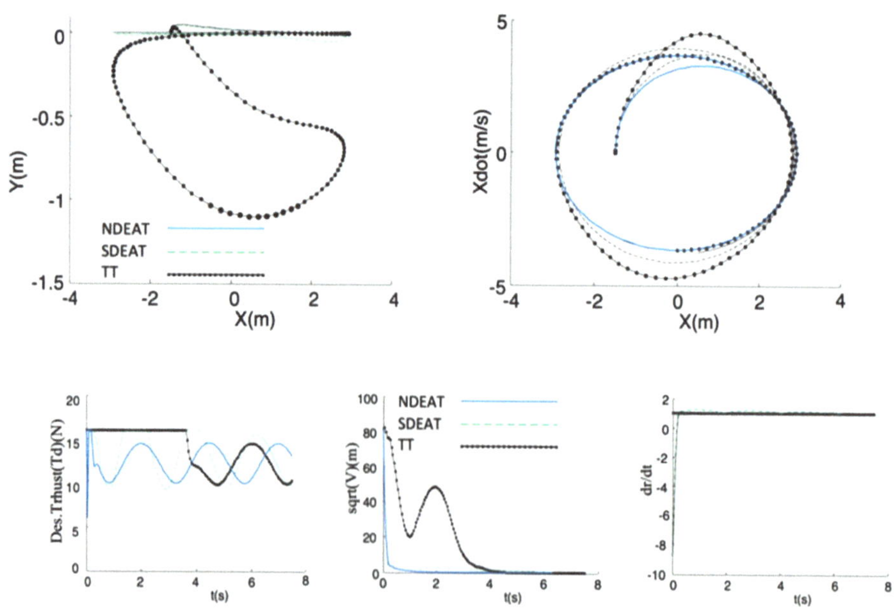

Figure 4. Comparison of NDEAT, SDEAT, and TT when the robot is delayed (desired path consists on a periodic horizontal motion $x_{des}(r) = A\ sin(w_{ref}\ r)$, $y_{des}(r) = constant$). **Up**: **Left**, Y/X trajectories; **Right**: evolution of $\dot{X}/X$ **Bottom**: **Left**, Desired Thrust T_d (with saturation). **Middle**: Lyapunov function versus time. **Right**: evolution of $\sigma = dr/dt$ with respect to time.

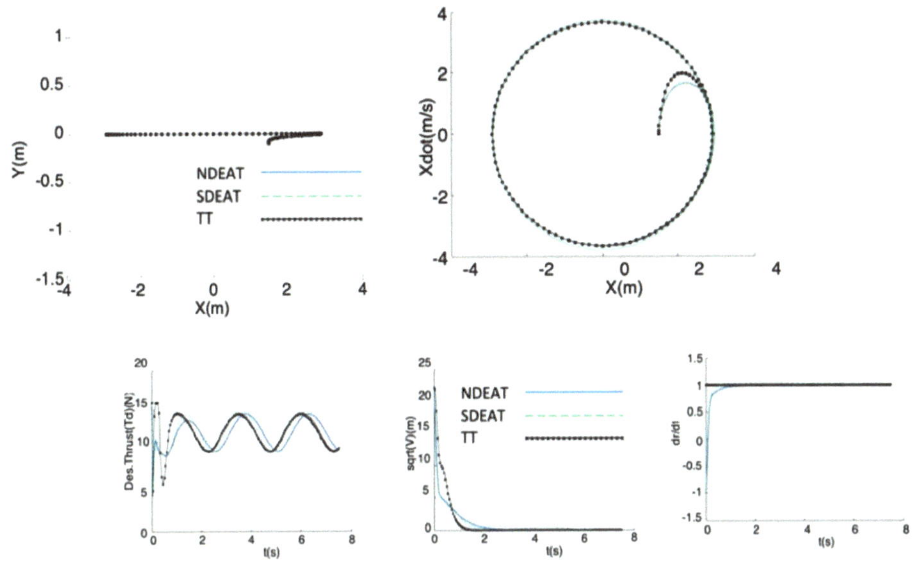

Figure 5. Comparison of NDEAT, SDEAT and TT when the robot is ahead (same desired path as previous Figure 4). **Up**: **Left**, Y/X trajectories; **Right**: evolution of $\dot{X}/X$ **Bottom**: **Left,** Desired Thrust Td (with saturation). **Middle**: Lyapunov function versus time. **Right**: evolution of $\sigma = dr/dt$ with respect to time.

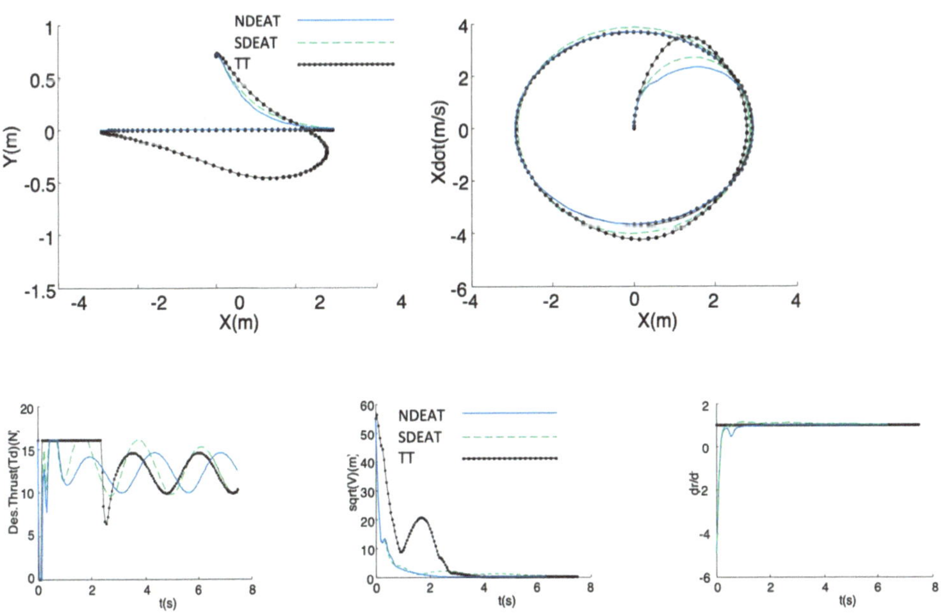

Figure 6. Comparison of NDEAT, SDEAT and TT when the robot is neither delayed nor ahead. **Up**: **Left**, Y/X trajectories (same desired path as previous Figure 4); **Right**: evolution of $\dot{X}/X$ **Bottom**: **Left**, Desired Thrust Td (with saturation). **Middle**: Lyapunov function versus time. **Right**: evolution of $\sigma = dr/dt$ with respect to time.

Table 1. Initial robot positions and their corresponding figures.

Figure	*x-xdes* (m)	*y-ydes* (m)	Case
Figure 4	−1.5	0	robot is delayed
Figure 5	1.5	−0.1	robot is ahead
Figure 6	0	0.7	robot is neither ahead nor delayed

The experiment is delayed for only 7.5 s, in order to see the transients more clearly. For all figures, NDEAT curves are drawn with solid lines, SDEAT with dashed lines and TT with dash-dotted lines.

For all the tests, it can be observed that the method with the fastest and best convergence to the path is NDEAT. According to Remark 1, NDEAT and SDEAT convergences are obviously faster than that of TT. This fact can be also verified by several reasons. First, Lyapunov function diminution is very much faster for EAT than for TT (especially when the robot is delayed). Moreover, the Lyapunov function is not decreasing when the system comes into the saturation zone ($T_d > 16\ N$), which occurs in many occasions and for long periods when using TT. This zone should be avoided because when the system enters this zone, the feedback linearization control is not valid. As a consequence, PVTOL control is lost and in some occasions coordinate Y goes so low that the actual system may collide with the ground (a value below −0.5 m is not possible for the real PVTOL). In any case, the evolution of s is the expected one: during the first transients, it reaches values that are far from one, in order to "look for" the best desired reference (which fulfills that z'_{des}, e_{zP} is near zero). Afterwards, it remains near one for NDEAT but greater than one for SDEAT, which is necessary in order to reduce the difference between r and t. After 7 s, parameter r has almost reached t in SDEAT; on the contrary, a gap remains when NDEAT is applied.

Other interesting points are the following: when the robot is ahead, all tracking methods behave satisfactorily. Since the actual system response is slow (mainly due to the second order dynamics of input u_1) this case is not as critical as the delay in the robot's posture. Nevertheless, TT is the only method that comes into saturation in the first moments of this test. Finally, TT shows also significant problems for the last trial (the robot is neither ahead nor delayed). Although coordinate X presents a small error, Y falls considerably. This is because the TT reference posture is continuously increasing, which implies high input values that take the system out of the linear zone. In conclusion, one of the strongest points of NDEAT is that its behavior is almost the same as the PF approach found in [26]. This can be observed very patently when the robot is delayed.

Moreover, and due to its large errors, the real speeds demanded by TT are greater than those of NDEAT (Figure 7). It is obvious that input limitations will further degrade TT's response. In the end, the TT method introduces more oscillations than EAT, and it has a transient response that separates the robot from the desired path. This is a well-known advantage of PF that EAT retains [1]. Concretely, for the PVTOL system the states variables barely enter the saturation zone when applying NDEAT. Moreover, if time determinism were needed, SDEAT would be a possible option, which avoids TT drawbacks and maintains the system's response near to that of NDEAT.

In the second test, the segment that fulfills the PF projection used in [26] is previously computed to serve as the path for the tracking (thick line in Figure 8). This path is composed of the points that are equidistant from the origin according to the projection. The initial posture of the robot is the origin ($\mathbf{x} = 0$), and the desired initial position is $(X,Y) = (-1.5,0)$. As can be seen, there is no racking problem for any EAT method. Moreover, as expected, NDEAT converges slightly better than SDEAT, and SDEAT converges better than TT. Note that no comparison with PF is possible because the projection is not defined for it. Moreover, and due to its large errors, the real speeds demanded by TT are greater than those of NDEAT. It is obvious that speed limitations will degrade TT's response even more. In the end, the TT method introduces more oscillations than EAT, and it has a transient response that separates the robot from the desired path. This is also another advantage of PF [1] that EAT also retains.

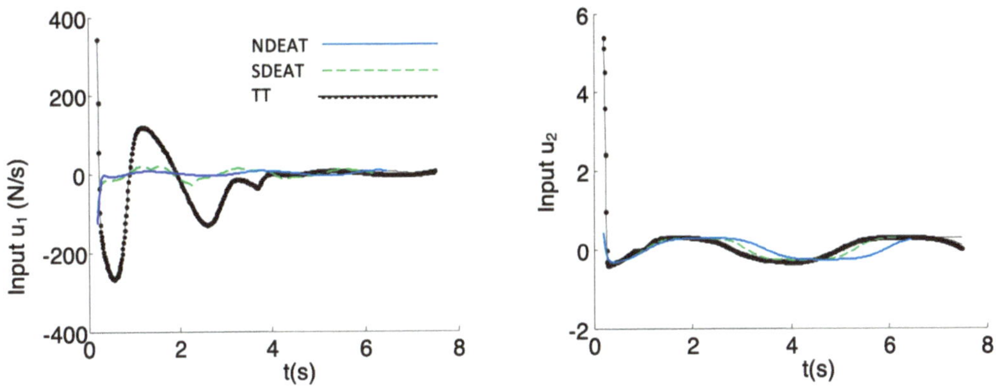

Figure 7. Inputs u when the robot is delayed. During transients, NDEAT demands less input than SDEAT, and SDEAT demands much less than TT. The first instants are not depicted because TT inputs rise to very high values, which would reduce the scale of the plot too much.

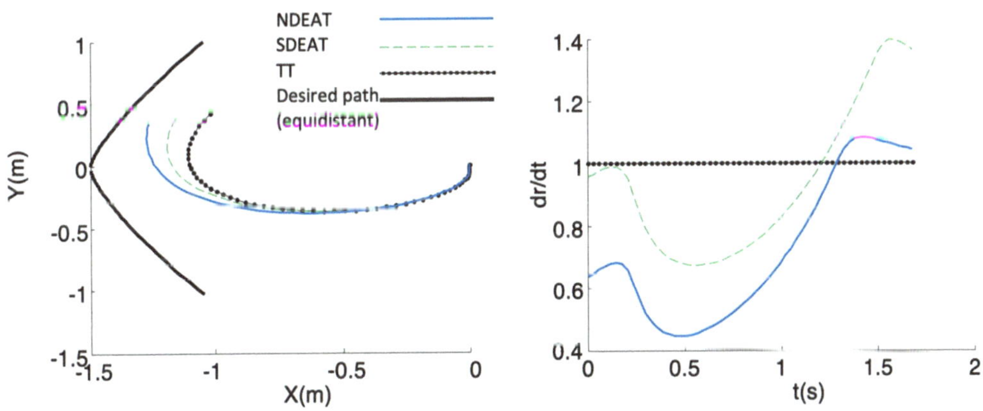

Figure 8. NDEAT, SDEAT and TT when the path is equidistant to the origin. **Left**: XY desired and real trajectories. **Right**: evolution of $\sigma = dr/dt$ with respect to time.

Similar comparisons can be made for other perturbations, such as saturation of inputs, path's curvature discontinuities, introduction of a scale between the inputs demanded by the control and the real ones, etc. On the whole, this is because the inputs demanded by TT are usually greater than those asked by EAT (see Figure 7), so TT experiences more problems in the tracking. Thus, the EAT method is more robust than TT against perturbations or unmodeled dynamics because the adaptive variation of r facilitates robustness. Finally, it is important to observe that the qualitative behavior of EAT is similar to that of PF, i.e., $\dot{r}$ is reduced in the presence of large errors until the system approaches the path. Obviously, both methods are constructed in a very different way, so it is not easy to make a quantitative comparison.

5. Conclusions

In this work, it is illustrated how a Lyapunov-based trajectory tracking control law can be used for error adaptive tracking methods for any system. This is carried out by selecting the proper rate for the progression of the descriptor parameter of the reference curve. This way, the burden of finding a new controller is not necessary. When EAT method is applied to a UAV (PVTOL), it is shown, through several tests, that error adaptive tracking methods outperform trajectory tracking ones using exactly the same controller (with identical parameters). We can conclude that the behavior of the several alternatives of error adaptive tracking is much better than that of a trajectory tracking under large errors, disturbances, unmodeled parameters or delayed response. This is because in error adaptive tracking, pace adapts to system errors. Two additional advantages of error adaptive tracking are also presented: a) it conserves most of the advantages of the path following method, and b) it avoids one of the main drawbacks of the path following method, that is, it is valid for all feasible trajectories. Finally, the benefits of the variant here called "soft deterministic error adaptive tracking" are also illustrated. This alternative presents almost the same excellent behavior as any error adaptive tracking when errors are large because temporal determinism is ignored in these situations. However, once errors have decreased, it gains the additional benefit of taking into account timing determinism as in classic trajectory tracking.

Author Contributions: Conceptualization, F.D.-d.-R., P.S.-C., P.I.-B. and J.L.S.-R.; methodology, F.D.-d.-R., P.S.-C., P.I.-B. and J.L.S.-R.; software, F.D.-d.-R. and P.S.-C.; validation, F.D.-d.-R. and P.I.-B.; formal analysis, F.D.-d.-R., P.S.-C., P.I.-B. and J.L.S.-R.; investigation, F.D.-d.-R., P.S.-C., P.I.-B. and J.L.S.-R.; writing—review and editing, F.D.-d.-R., P.S.-C., P.I.-B. and J.L.S.-R.; project administration, F.D.-d.-R. and J.L.S.-R. All authors have read and agreed to the published version of the manuscript.

Funding: This work has been supported by the Spanish project (with support from the European Regional Development Fund) PID2019-110455GB-I00 (AEI/FEDER, EU) and by the Andalusian Regional Excellence Research Project (with support from the European Regional Development Fund) project US-1381077 (JJAA/FEDER, UE).

Conflicts of Interest: The authors declare no conflict of interest.

References

1. Siciliano, B.; Khatib, O. (Eds.) *Springer Handbook of Robotics*; Springer Handbooks; Springer International Publishing: Cham, Switzerland, 2016; ISBN 978-3-319-32550-7.
2. Palacín, J.; Rubies, E.; Clotet, E.; Martínez, D. Evaluation of the Path-Tracking Accuracy of a Three-Wheeled Omnidirectional Mobile Robot Designed as a Personal Assistant. *Sensors* **2021**, *21*, 7216. [CrossRef] [PubMed]
3. Aguiar, A.P.; Hespanha, J.P. Trajectory-Tracking and Path-Following of Underactuated Autonomous Vehicles With Parametric Modeling Uncertainty. *IEEE Trans. Autom. Control.* **2007**, *52*, 1362–1379. [CrossRef]
4. Kingston, P.; Egerstedt, M. Time and Output Warping of Control Systems: Comparing and Imitating Motions. *Automatica* **2011**, *47*, 1580–1588. [CrossRef]
5. Aguiar, A.P.; Dačić, D.B.; Hespanha, J.P.; Kokotović, P. Path-Following or Reference Tracking? *IFAC Proc. Vol.* **2004**, *37*, 167–172. [CrossRef]
6. Breivik, M.; Fossen, T.I. Principles of Guidance-Based Path Following in 2D and 3D. In Proceedings of the 44th IEEE Conference on Decision and Control, Seville, Spain, 12–15 December 2005; pp. 627–634.
7. Conte, G.; Duranti, S.; Merz, T. Dynamic 3D Path Following for an Autonomous Helicopter. *IFAC Proc. Vol.* **2004**, *37*, 472–477. [CrossRef]
8. Park, S.; Deyst, J.; How, J.P. Performance and Lyapunov Stability of a Nonlinear Path Following Guidance Method. *J. Guid. Control. Dyn.* **2007**, *30*, 1718–1728. [CrossRef]
9. Xavier, D.M.; Natassya, B.F.S.; Kalinka, R.L.J.C.B. Path-Following Algorithms Comparison Using Software-in-the-Loop Simulations for UAVs. In Proceedings of the 2019 IEEE Symposium on Computers and Communications (ISCC), Barcelona, Spain, 29 June–3 July 2019; pp. 1216–1221.
10. Lawrence, D.A.; Frew, E.W.; Pisano, W.J. Lyapunov Vector Fields for Autonomous Unmanned Aircraft Flight Control. *J. Guid. Control. Dyn.* **2008**, *31*, 1220–1229. [CrossRef]
11. Chen, H.; Chang, K.; Agate, C.S. Tracking with UAV Using Tangent-plus-Lyapunov Vector Field Guidance. In Proceedings of the 2009 12th International Conference on Information Fusion, Seattle, WA, USA, 6–9 July 2009; pp. 363–372.
12. Unmanned Aerial Vehicle Path Following: A Survey and Analysis of Algorithms for Fixed-Wing Unmanned Aerial Vehicless. *IEEE Control. Syst.* **2014**, *34*, 42–59. [CrossRef]

13. del Rio, F.D.; Jimenez, G.; Sevillano, J.L.; Amaya, C.; Balcells, A.C. Error Adaptive Tracking for Mobile Robots. In Proceedings of the IEEE 2002 28th Annual Conference of the Industrial Electronics Society, IECON 02, Sevilla, Spain, 5–8 November 2002; Volume 3, pp. 2415–2420.
14. Nelson, D.R.; Barber, D.B.; McLain, T.W.; Beard, R.W. Vector Field Path Following for Small Unmanned Air Vehicles. In Proceedings of the 2006 American Control Conference, Minneapolis, MN, USA, 14–16 June 2006; p. 7.
15. Frew, E.W.; Lawrence, D. Tracking Expanding Star Curves Using Guidance Vector Fields. In Proceedings of the 2012 American Control Conference (ACC), Montreal, QC, Canada, 27–29 June 2012; pp. 1749–1754.
16. Fari, S.; Wang, X.; Roy, S.; Baldi, S. Addressing Unmodeled Path-Following Dynamics via Adaptive Vector Field: A UAV Test Case. *IEEE Trans. Aerosp. Electron. Syst.* **2020**, *56*, 1613–1622. [CrossRef]
17. Salazar, S.; González-Hernández, I.; Lopez, R.; Lozano, R. Simulation and Robust Trajectory-Tracking for a Quadrotor UAV. In Proceedings of the 2014 International Conference on Unmanned Aircraft Systems (ICUAS), Orlando, FL, USA, 27–30 May 2014; pp. 1167–1174.
18. Egerstedt, M.; Hu, X.; Stotsky, A. Control of Mobile Platforms Using a Virtual Vehicle Approach. *IEEE Trans. Autom. Control.* **2001**, *46*, 1777–1782. [CrossRef]
19. Maček, K.; Petrović, I.; Siegwart, R. A Control Method for Stable and Smooth Path Following of Mobile Robots. In Proceedings of the 2nd European Conference on Mobile Robots (ECMR), Ancona, Italy, 31 August–3 September 2005. [CrossRef]
20. Lapierre, L.; Soetanto, D.; Pascoal, A. Nonsingular Path Following Control of a Unicycle in the Presence of Parametric Modelling Uncertainties. *Int. J. Robust Nonlinear Control.* **2006**, *16*, 485–503. [CrossRef]
21. Basañez, L. Asociación Española de Robótica y Automatización Tecnologías de la Producción, International Federation of Robotics. In Proceedings of the 40th International Symposium on Robotics Programme - AER-ATP, Barcelona, Spain, 10–13 March 2009.
22. Xiang, X.; Lapierre, L.; Liu, C.; Jouvencel, B. Path Tracking: Combined Path Following and Trajectory Tracking for Autonomous Underwater Vehicles. In Proceedings of the 2011 IEEE/RSJ International Conference on Intelligent Robots and Systems, San Francisco, CA, USA, 25–30 September 2011; IEEE: San Francisco, CA, USA, 2011; pp. 3558–3563.
23. Díaz del Río, F.; Jiménez, G.; Sevillano, J.L.; Vicente, S.; Civit Balcells, A. A Path Following Control for Unicycle Robots. *J. Robot. Syst.* **2001**, *18*, 325–342. [CrossRef]
24. Encarnacao, P.; Pascoal, A. Combined Trajectory Tracking and Path Following: An Application to the Coordinated Control of Autonomous Marine Craft. In Proceedings of the 40th IEEE Conference on Decision and Control (Cat. No.01CH37228), Orlando, FL, USA, 4–7 December 2001; IEEE: Orlando, FL, USA, 2001; Volume 1, pp. 964–969.
25. Brockett, R.W. Asymptotic Stability and Feedback Stabilization. In *Differential Geometric Control Theory*; Birkhauser: Boston, MA, USA, 1983; pp. 181–208.
26. Hauser, J.; Hindman, R. Maneuver Regulation from Trajectory Tracking: Feedback Linearizable Systems. *IFAC Proc. Vol.* **1995**, *28*, 595–600. [CrossRef]
27. De Luca, A.; Oriolo, G.; Vendittelli, M. Control of Wheeled Mobile Robots: An Experimental Overview. In *Ramsete*; Nicosia, S., Siciliano, B., Bicchi, A., Valigi, P., Eds.; Lecture Notes in Control and Information Sciences; Springer Berlin Heidelberg: Berlin, Heidelberg, 2001; Volume 270, pp. 181–226. ISBN 978-3-540-42090-3.
28. Fernando Diaz Del Rio, ETSII—Universidad de Sevilla. PVTOL_tracking_methods.Zip. *MATLAB Central File Exchange*. Available online: Https://www.Mathworks.Com/Matlabcentral/Fileexchange/46938-Pvtol_tracking_methods-Zip (accessed on 9 November 2022).

sensors

MDPI

Article

OCTUNE: Optimal Control Tuning Using Real-Time Data with Algorithm and Experimental Results

Mohamed Abdelkader [1,*], Mohamed Mabrok [2] and Anis Koubaa [1]

[1] College of Computer & Information Sciences, Robotics & Internet of Things Laboratory, Prince Sultan University, P.O. Box 66833, Riyadh 11586, Saudi Arabia

[2] Mathematics Program, Department of Mathematics, Statistics and Physics, College of Arts and Sciences, Qatar University, Doha P.O. Box 2713, Qatar

* Correspondence: mabdelkader@psu.edu.sa

Abstract: Autonomous robots require control tuning to optimize their performance, such as optimal trajectory tracking. Controllers, such as the Proportional–Integral–Derivative (PID) controller, which are commonly used in robots, are usually tuned by a cumbersome manual process or offline data-driven methods. Both approaches must be repeated if the system configuration changes or becomes exposed to new environmental conditions. In this work, we propose a novel algorithm that can perform online optimal control tuning (OCTUNE) of a discrete linear time-invariant (LTI) controller in a classical feedback system without the knowledge of the plant dynamics. The OCTUNE algorithm uses the backpropagation optimization technique to optimize the controller parameters. Furthermore, convergence guarantees are derived using the Lyapunov stability theory to ensure stable iterative tuning using real-time data. We validate the algorithm in realistic simulations of a quadcopter model with PID controllers using the known Gazebo simulator and a real quadcopter platform. Simulations and actual experiment results show that OCTUNE can be effectively used to automatically tune the UAV PID controllers in real-time, with guaranteed convergence. Finally, we provide an open-source implementation of the OCTUNE algorithm, which can be adapted for different applications.

Keywords: robotics; unmanned aerial vehicles; control tuning; open-source

Citation: Abdelkader, M.; Mabrok, M.; Koubaa, A. OCTUNE: Optimal Control Tuning Using Real-Time Data with Algorithm and Experimental Results. *Sensors* **2022**, 22, 9240. https://doi.org/10.3390/s22239240

Academic Editors: Stephen Monk and David Cheneler

Received: 31 October 2022
Accepted: 21 November 2022
Published: 28 November 2022

1. Introduction

Control tuning is a fundamental concept in any control system's design cycle; see, for instance, Refs. [1,2] and the references therein. In particular, robotic systems require control tuning to perform different levels of autonomous tasks with the desired performance. In these systems, conventional controllers, such as the Proportional–Integral–Derivative (PID) controller, are usually tuned using an iterative manual process or offline data-driven methods.

For instance, in quadrotor control, known open-source autopilots, such as PX4 [3] and Ardupilot [4], use either manual tuning or non-optimal auto-tuning methods. Furthermore, many dynamical systems exhibit complex characteristics, such as non-linearity, time-varying parameters, and time delay, which leads to different operating conditions and/or disturbances, leading to poor control.

Generally speaking, control tuning methodologies can be classified as offline methods and online methods. In the offline methods, such as linear–quadratic Gaussian control (LQG) [5,6] and H-infinity control [7], the controller requires an accurate model of the system dynamics under control. The controller is designed for the model of the system and is tuned before the implementation stage. These controllers work well with systems that have an accurate linear model. However, they give poor performance otherwise. On the other hand, in the online adaptive methods, the model is often required as well. However, the controller can adapt to the un-modeled system dynamics, which is well-known under the adaptive control theory [8,9], which is well-developed and established in linear and nonlinear control systems.

Online model-free methods also exist in several studies [10–12]. Another class of gradient-descent-based algorithms also exist in the literature. For instance, a control method based on an adaptive PID neural network and particle swarm optimization (PSO) algorithm was developed in [13]. In [14], the investigation of adaptive learning control for underwater vehicles (AUVs) with unknown uncertainties using gradient descent algorithm is presented, where the unknown nonlinear functions in the system are approximated by radial basis function neural networks.

In [15], another adaptive gradient descent algorithm combined with a fuzzy system was developed to improve the attitude estimation accuracy and adaptability of unmanned underwater vehicles under various ocean environments. Many attempts have been made to build auto-tuned PID controllers using different adaptive learning techniques [16]. For instance, the authors in [17,18] used genetic algorithms to tune a PID. Furthermore, the use of a neural network to tune a PID controller through extensive training was discussed in [19]. However, these techniques have several drawbacks, such as a lack of stability guarantees, slow convergence, or implementation constraints.

In this paper, we develop an online learning algorithm based on the backpropagation learning technique to learn a controller for a dynamical system without knowing the system model. Our control-learning algorithm is based on the work presented in [20,21], where the backpropagation learning technique is used in system identification for linear dynamical models. The use of backpropagation learning techniques in training systems is becoming the norm due to the extensive use of backpropagation algorithms in the modern machine-learning domain. The accessibility of the backpropagation algorithms in several software packages, such as TensorFlow [22] and PyTorch [23], has made them more attractive and easy to use.

The backpropagation learning technique was used in several attempts in PID tuning. For instance, in [24], a fuzzy PID controller, which is a combination of a fuzzy controller with a PID neural network (PIDNN), was proposed. In [25], a conventional Neuro PID controller for linear or nonlinear systems that was unaffected by the unpredictability of the system's parameters and disturbances, such as noise, was developed. However, again, these methods require a model for the controlled system as they lack stability guarantees.

This paper proposes a novel, implementable, and fast algorithm that can perform online optimal control tuning (OCTUNE) of a discrete linear time-invariant (LTI) controller in a classical feedback system, only using real-time system signals, i.e., no model required for the system under control. The OCTUNE algorithm uses the backpropagation optimization technique to optimize a performance function (squared error between the desired and actual signals) in real-time. Furthermore, convergence guarantees are derived using the Lyapunov stability theory to ensure stable tuning using online real-time data.

We demonstrate the effectiveness and practicality of the OCTUNE algorithm by applying it to the tuning of a discrete PID controller (a particular case of an LTI controller) that is used to stabilize the angular rates of a quadrotor unmanned aerial vehicle (UAV). The demonstration is performed in a realistic simulation environment using the Gazebo simulator and the robot operating system (ROS). The simulation results show how OCTUNE can be effectively used to automatically tune the UAV angular rate PID controllers using real-time signals in a fraction of a minute with a low number of online iterations. Finally, an open-source implementation of the OCTUNE algorithm is provided, which can be adapted for different applications.

The contributions of this work are summarized as follows.

- An online and model-free optimal auto-tuning algorithm for a generic LTI controller is developed, called OCTUNE, which is demonstrated using realistic simulations of a quadrotor system.
- Convergence proof of the OCTUNE algorithm is derived in order to guarantee safe control learning/tuning.

- We provide our implementation as an open-source software package of OCTUNE to facilitate the use and adaptation of the presented algorithm for different applications. The links to the open-source software is provided in the Supplementary Materials section.

The remainder of the paper is organized as follows. The problem statement is presented in Section 2. The optimal tuning algorithm is derived in Section 3, followed by convergence analysis in Section 4. Realistic simulation results of the OCTUNE algorithms for a quadrotor tuning application are discussed in Section 5. Finally, our conclusions and future work are discussed in Section 6.

2. Problem Statement

This section defines the controller architecture to be optimized using the OCTUNE algorithm described in Section 3 to improve the reference tracking in a classical feedback system. In this work, we assume a standard feedback system as shown in Figure 1, where the system is represented by an unknown discrete-time plant, $P(z)$. The controller is assumed to be a discrete-time linear time-invariant (LTI) system of the following transfer function.

$$C(z) = \frac{B(z)}{A(z)} = \frac{b_0 + b_1 z^{-1} + \cdots + b_{n_b} z^{-n_b}}{1 + a_1 z^{-1} + \cdots + a_{n_a} z^{-n_a}}, \tag{1}$$

where a's and b's are the controller's denominator and numerator coefficients, respectively. The system signals, reference $r(k)$, controller output $u(k)$, and output $y(k)$ for time instances $k = 0, 1, 2, \ldots$ are all assumed to be measurable in real-time.

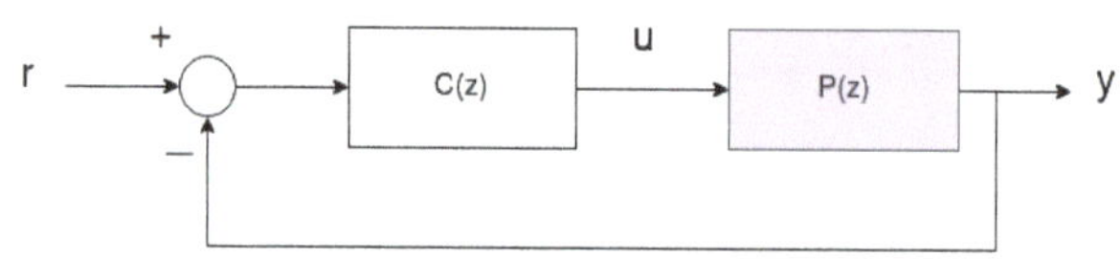

Figure 1. A feedback system with an unknown discrete-time plant, $P(z)$, and a discrete-time LTI controller, $C(z)$.

It is assumed that the controller in Equation (1) is initially stabilizing the feedback system in Figure 1. However, the performance defined later in Equation (2) may not be initially optimal. In other words, the system output $y(k)$ is poorly tracking the reference $r(k)$.

The objective of this work is to find a controller structure of the form $C(z)$, which optimizes the performance of the system response—defined later in Equation (2). To achieve this objective, we propose the OCTUNE algorithm, which updates the controller's parameters a and b in real-time as shown in Figure 2. The OCTUNE block shown in Figure 2 receives $r(k)$, $y(k)$, and $u(k)$ signals in real-time and computes the updated controller parameters in order to minimize the error between the reference signal $r(k)$ and the actual output signal $y(k)$. In addition, the OCTUNE algorithm updates the controller parameters while guaranteeing stable convergence to the minimum error using only real-time signals.

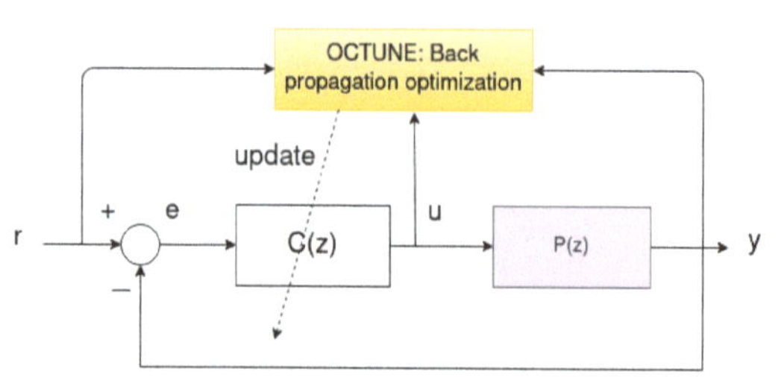

Figure 2. A feedback system with controller $C(z)$ coefficients updated by the OCTUNE algorithm. The OCTUNE algorithm receives the reference, actual, and controller output signals and performs update operations to update the controller parameters.

3. Control Tuning Algorithm

The objective of the OCTUNE is for the system output $y(k)$ to track the desired reference signal $r(k)$ as accurately as possible. Therefore, we define the objective function L that is to be optimized as follows.

$$L(k) = \frac{1}{2} \sum_{i=k-N}^{k} e^2(i) \tag{2}$$

where N is the number of available data samples, and the error $e(k)$ at time instant k is defined as the difference between the desired reference signal $r(k)$ and the corresponding output signal $y(k)$,

$$e(k) = r(k) - y(k) \tag{3}$$

The objective function L can be written in a compact form as follows.

$$L = \frac{1}{2}||E||^2 \tag{4}$$

where $|| \cdot ||$ is the Euclidean norm, and

$$E = [e(k-N), e(k-N+1), \ldots, e(k)]^T \tag{5}$$

In the following section, an algorithm based on the backpropagation method is developed to compute the controller parameters in Equation (1) that minimize the objective function defined in Equation (4) given the system signals $r(k), y(k)$, and $u(k)$.

Optimization Using Backpropagation

Backpropagation (BP) is a widely used algorithm in machine learning for efficiently training artificial neural networks (ANNs) [26]. BP computes the gradient (partial derivatives) of the loss function with respect to the weights of the neural network. The partial derivatives are then used to update the weight values. This process is repeated until convergence is achieved. The objective of this work is to compute the controller parameters (analogous to weights in ANNs) that minimize the loss function in Equation (2).

As depicted in the computational graph in Figure 3, the backpropagation operations (represented using dashed lines) use the chain rule to compute the partial derivatives $\frac{\partial L}{\partial a}, \frac{\partial L}{\partial b}$ through the intermediate partial derivatives $\frac{\partial L}{\partial y}, \frac{\partial L}{\partial u}$. Then, the computed partial derivatives $\frac{\partial L}{\partial a}, \frac{\partial L}{\partial b}$ are used to compute the new controller parameters a, b using the delta rule (gradient descent), Equation (16).

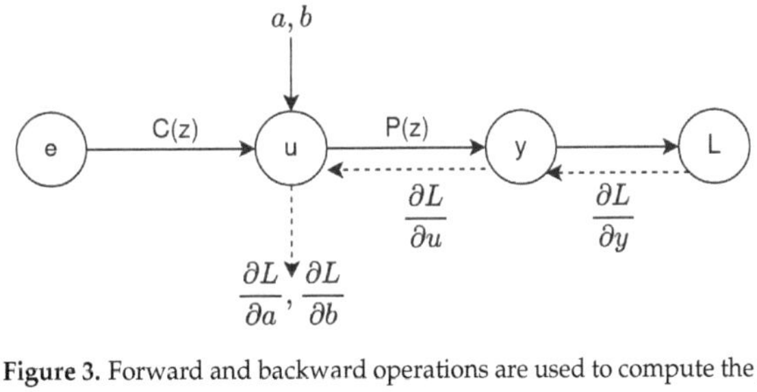

Figure 3. Forward and backward operations are used to compute the partial derivatives. Solid arrows represent forward propagation, and backpropagation is represented by dashed arrows.

The optimization problem is defined as follows,

$$\min_{a_i, b_i} L = \frac{1}{2} \sum_{i=k-N}^{k} e^2(i) \tag{6}$$

In order to use the backpropagation algorithm [26] to solve (6), the partial derivatives of the objective function L with respect to the controller coefficients a_i, b_i need to be calculated, which is described as follows.

The objective L is directly a function of the error e; hence, the partial derivative of L with respect to the error e_k at time k, and the error vector E at all N samples, are the first derivatives that need to be computed as follows.

$$\begin{aligned} \frac{\partial L}{\partial e_k} &= e(k) \\ \frac{\partial L}{\partial E} &= E \end{aligned} \tag{7}$$

Next, going backward in the chain, the partial derivative of L with respect to the output y_k at time k (and y for all N samples) is defined as follows,

$$\begin{aligned} \frac{\partial L}{\partial y_k} &= \frac{\partial L}{\partial e_k}\frac{\partial e_k}{\partial y_k} = -e(k) \\ \frac{\partial L}{\partial y} &= -E \end{aligned} \tag{8}$$

Next, using the chain rule, the change of L with respect to the controller denominator coefficients a_i is

$$\begin{aligned} \frac{\partial L}{\partial a_i} &= \sum_{t=k-N}^{k} \frac{\partial L}{\partial e_t}\frac{\partial e_t}{\partial a_i} \\ &= \sum_{t=k-N}^{k} \frac{\partial L}{\partial e_t}\frac{\partial e_t}{\partial y_t}\frac{\partial y_t}{\partial u_t}\frac{\partial u_t}{\partial a_i} \\ &= \sum_{t=k-N}^{k} e(t)(-1)\frac{\partial y_t}{\partial u_t}\frac{\partial u_t}{\partial a_i} \\ i &= 1, \ldots, n_a \end{aligned} \tag{9}$$

Equation (9) can be written in a compact vector form as follows.

$$\begin{aligned} \frac{\partial L}{\partial a} &= -J_a E \; \in R^{n_a} \\ J_a &= \begin{bmatrix} \frac{\partial y_{k-N}}{\partial u_{k-N}}\frac{\partial u_{k-N}}{\partial a_1} & \cdots & \frac{\partial y_k}{\partial u_k}\frac{\partial u_k}{\partial a_1} \\ \vdots & \vdots & \vdots \\ \frac{\partial y_{k-N}}{\partial u_{k-N}}\frac{\partial u_{k-N}}{\partial a_{n_a}} & \cdots & \frac{\partial y_k}{\partial u_k}\frac{\partial u_k}{\partial a_{n_a}} \end{bmatrix} \in R^{n_a \times (N+1)} \end{aligned} \tag{10}$$

Similarly, the change of L with respect to the controller's numerator coefficients b_j can be calculated as follows.

$$\begin{aligned} \frac{\partial L}{\partial b_j} &= \sum_{t=k-N}^{k} \frac{\partial L}{\partial e_t}\frac{\partial e_t}{\partial b_j} \\ &= \sum_{t=k-N}^{k} \frac{\partial L}{\partial e_t}\frac{\partial e_t}{\partial y_t}\frac{\partial y_t}{\partial u_t}\frac{\partial u_t}{\partial b_j} \\ &= \sum_{t=k-N}^{k} e(t)(-1)\frac{\partial y_t}{\partial u_t}\frac{\partial u_t}{\partial b_j} \\ j &= 0, \ldots, n_b \end{aligned} \tag{11}$$

Equation (11) can also be written in a compact vector form.

$$\frac{\partial L}{\partial b} = -J_b E \ \in R^{n_b}$$
$$J_b = \begin{bmatrix} \frac{\partial y_{k-N}}{\partial u_{k-N}} \frac{\partial u_{k-N}}{\partial b_0} & \cdots & \frac{\partial y_k}{\partial u_k} \frac{\partial u_k}{\partial b_0} \\ \vdots & \vdots & \vdots \\ \frac{\partial y_{k-N}}{\partial u_{k-N}} \frac{\partial u_{k-N}}{\partial a_{n_b}} & \cdots & \frac{\partial y_k}{\partial u_k} \frac{\partial u_k}{\partial a_{n_b}} \end{bmatrix} \in R^{n_b \times (N+1)} \tag{12}$$

For compactness, let us define the following quantities, W is the controller parameter vector, J is the Jacobean matrix of all intermediate, data-driven, partial derivatives, and $\frac{\partial L}{\partial W}$ is the gradient vector of L with respect to the controller parameters W.

$$W = [a_1, \ldots, a_{n_a}, b_0, \ldots, b_{n_b}]^T \in R^{n_a+n_b} \tag{13}$$

$$J = \begin{bmatrix} J_a \\ J_b \end{bmatrix} \in R^{(n_a+n_b)\times(N+1)} \tag{14}$$

$$\frac{\partial L}{\partial W} = \begin{bmatrix} \frac{\partial L}{\partial a} \\ \frac{\partial L}{\partial b} \end{bmatrix} = \frac{\partial E}{\partial W} \frac{\partial L}{\partial E} = -J \cdot E \in R^{n_a+n_b} \tag{15}$$

Using Equation (15), the update rule of the controller coefficients a_i, b_i can be written as follows.

$$\begin{aligned} W &:= W + \Delta W = W - \alpha \frac{\partial L}{\partial W} \\ &= W + \alpha J \cdot E \end{aligned} \tag{16}$$

The calculations of $\frac{\partial u}{\partial a_i}$ and $\frac{\partial u}{\partial b_i}$ are performed similar to the calculations of $\frac{\partial y}{\partial b_j}$ and $\frac{\partial y}{\partial a_j}$ in [20], and omitted here for brevity. In comparison to [20], in this work, we are identifying the coefficients of the controller's transfer function instead of the plant's.

In [20], the calculation of $\frac{\partial y}{\partial u}$, which is needed in Equations (9) and (11), was performed using the known linear structure of the plant $P(z)$. However, in this work, this cannot be conducted in the same way as $P(z)$ is assumed to be unknown. Instead, we assume that the system signals y and u are sampled fast enough, and the following first-order approximation is used.

$$\frac{\partial y}{\partial u} \approx \frac{\Delta y}{\Delta u} = \frac{y(k) - y(k-1)}{u(k) - u(k-1)} \tag{17}$$

Algorithm 1 presents a pseudo code of the OCTUNE algorithm.

Algorithm 1: Pseudo code of the OCTUNE algorithm

Data: N_{itr}: Maximum number of iterations, N_t: Maximum optimization time
Data: W_0: Initial controller parameters
Result: $W = W^*$, optimal controller parameters
$W \leftarrow W_0$, Equation (13)
while *Not converged* **do**
  Compute error vector E, Equation (5) ;
  Compute $\frac{\partial L}{\partial E}$, Equation (7) ;
  Compute J_a, Equation (10);
  Compute J_b, Equation (12);
  Compute J, Equation (14);
  Compute $\frac{\partial L}{\partial W}$, Equation (15);
  Compute $|\lambda_{\min}|$, the absolute value of the smallest eigenvalue of $(-J \cdot J^T)$;
  Compute optimal learning rate, $\alpha^* = \frac{2}{|\lambda_{\min}|}$;
  Update W using Equation (16);
end

4. Convergence Analysis

The convergence of the controller coefficients a_i, b_i (or W) depends on the choice of the training rate α in Equation (16). High values of α can diverge the controller coefficients, while overly small values can guarantee convergence but with a slow training speed, which might not be practical for real-time applications. In this section, the procedure of selecting the proper values of α is developed.

Let $V(k)$ be a discrete Lyapunov function [27] that is defined as follows.

$$V(E) = \frac{1}{2}||E||^2 \tag{18}$$

where $||\cdot||$ is the 2-norm. The Lyapunov function $V(E) = 0$ only when $E = 0$. The change of V, ΔV is defined as follows.

$$\begin{aligned} \Delta V &= V(E_{k+1}) - V(E_k) \\ &= \frac{1}{2}\left[||E_{k+1}||^2 - ||E_k||^2\right] \\ &= \Delta E^T \left[E_k + \frac{\Delta E}{2}\right] \end{aligned} \tag{19}$$

The error difference ΔE can be written as follows.

$$E_{k+1} = E_k + \Delta E = E_k + \left(\frac{\partial E_k}{\partial W_k}\right)^T \cdot \Delta W_k \tag{20}$$

Using Equations (15) and (16),

$$E_{k+1} = E_k - \alpha J^T J E_k \tag{21}$$

Therefore, ΔE can be defined as follows

$$\Delta E = -\alpha J^T J E \tag{22}$$

Theorem 1. *Let α be the learning rate used in the backpropagation algorithm in Equation (16) and $|\lambda_{min}|$ be the absolute value of the smallest eigenvalue of $(-J \cdot J^T)$, where J is defined in Equation (14). Then, the convergence of the controller coefficients W is guaranteed if α is chosen such that it satisfies the following relationship.*

$$0 < \alpha < \frac{2}{|\lambda_{min}|} \tag{23}$$

Proof. Plugging Equation (22) into Equation (19) yields

$$\begin{aligned}\Delta V &= (-\alpha J^T JE)^T \left[E + \frac{1}{2}(-\alpha J^T JE)\right] \\ &= \frac{-\alpha^2}{2} E^T J^T \left[\frac{2}{\alpha} I - JJ^T\right] JE\end{aligned} \tag{24}$$

From Equation (24), $\Delta V < 0$ if $\alpha > 0$ and $\frac{2}{\alpha} I - JJ^T$ is positive definite, which can be achieved by choosing α as in (23). With $V(E) > 0$ for $E \neq 0$ and $\Delta V < 0$ satisfied by Equation (23), the convergence of W in Equation (16) is guaranteed. The optimal convergence can be achieved by $\alpha^* = \frac{2}{|\lambda_{\min}|}$. □

5. Validation: Quadrotor Tuning

This section presents realistic simulation results of the proposed OCTUNE algorithm applied to a practical use case of tuning a quadrotor's PID angular rate control loops in real-time during flight. The angular rate stabilization is the innermost control loop and is the most critical one, which affects all the other higher control loops, such as the attitude, linear velocity, and position. For example, refer to the control architecture of the PX4 open-source autopilot PX4 control architecture [28].

Many UAVs use open-source autopilots, such as ArduPilot [4] and PX4 [3], in custom UAV research and development works. Usually, the custom-built UAVs that use off-the-shelf autopilots with open-source software, such as PX4 require iterative tuning of the PID control loops, which is generally performed manually before further development and flight testing. This manual process is essential to have a desirable flight performance. However, it can be cumbersome and time-consuming, as it requires manually performing flight tests, collecting data, manually analyzing them, and finally tuning the PID gains.

This manual tuning is conducted for each degree of freedom, i.e., three rotational (roll, pitch, and yaw) and three translational (x, y, and z) degrees, repeated many times until the desired control performance is achieved. In addition, a re-tuning process is needed if the UAV configuration is changed, for example, by adding or removing a payload. Moreover, the PID control loops might be tuned to work in specific environmental conditions, such as low wind speed. Therefore, it will need to be re-tuned to perform well against different disturbance sources and levels. An algorithm that can automatically and reliably tune controllers in such situations and in real-time is greatly beneficial as it saves time and optimizes performance.

The OCTUNE algorithm presented in this work effectively and practically addresses the above mentioned issues in real-time with no manual iterations or interventions. The OCTUNE algorithm is demonstrated with realistic quadrotor simulations in the following sections. A link to the video of the simulation experiments is provided in the Supplementary Materials section.

5.1. Simulation Setup

The quadrotor simulation setup consists of the four main components depicted in Figure 4 and described as follows.

- Gazebo simulator : An open-source robot simulator [29] that accurately and efficiently simulates several types of robots in complex indoor and outdoor environments with multiple options of robust physics engines. It also has strong integration with the robot operating system (ROS) to facilitate software development and integration. The robot model simulated in this work is an actual quadrotor UAV called Iris; see Figure 5. The quadrotor model has several plugins to simulate the onboard sensors (Inertial measurement unit, GPS, and magnetometer) and the propulsion system. The model also models the mechanical structure of the drone with its mass and inertial characteristics.

- PX4 autopilot: This is the autopilot firmware that interfaces with Gazebo to receive the simulated sensors readings, to perform control and operations, and to send motor commands to the motor plugins of the Gazebo quadrotor model. The PX4 autopilot firmware implements the PID control loops tuned using the OCTUNE algorithm. The autopilot firmware in simulation (called software in the loop, SITL) is the same as the one on actual autopilot hardware, except that the actual sensors and motors are replaced with simulated ones.
- MAVROS: MAVROS is a software package that interfaces between the PX4 autopilot and the robot operating system (ROS) [30]. Interfacing PX4 with ROS makes the software development and integration extremely streamlined and can be easily deployed on actual hardware with almost no modifications to the software used in the simulation. The MAVROS communicates the required signals (target, controller output, and actual), and the PID controller gains between the OCTUNE application and the PX4 autopilot.
- OCTUNE: This is the implementation of the OCTUNE algorithm as a ROS software package (node in ROS terminology) for real-time tuning. The OCTUNE node receives the quadrotor signals (target, actual, and controller output), and the PID gains from the MAVROS node in real-time. After the signals and the current gains are used to compute the updated PID gains by the OCTUNE node, the new gains are sent to the PX4 autopilot via the MAVORS node.

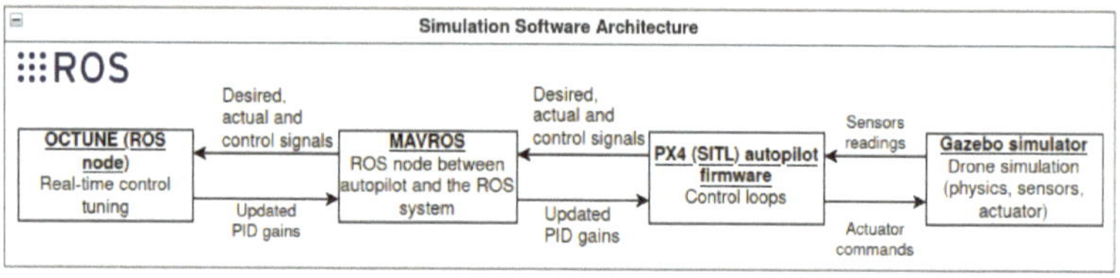

Figure 4. Abstract of the software architecture used in the simulations. Bold and underlined text represent software packages that are explained in Section 5.1.

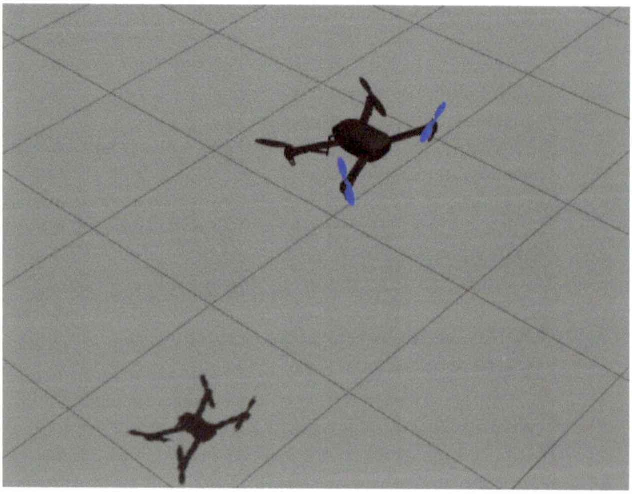

Figure 5. A snapshot of the Iris quadrotor model flying in the Gazebo simulator.

5.2. Controller

The OCTUNE algorithm requires the definition of the controller's transfer function as defined in Equation (1)—namely, the numerator coefficients b_i and denominator coefficients a_i. An angular rate PID controller can be represented as a discrete-time transfer function [31] as follows.

$$
\begin{aligned}
C(z) = \frac{U(z)}{E(z)} &= \frac{b_0 + b_1 z^{-1} + b_2 z^{-2}}{1 - z^{-1}} \\
b_0 &= K_p + K_d/T + K_i T \\
b_1 &= -2K_d/T - K_p \\
b_2 &= K_d/T
\end{aligned} \tag{25}
$$

where b_0, b_1, and b_2 are the controller's transfer function numerator's coefficients; K_p, K_i, and K_d are the proportional, integral, and differential PID gains; and T is the sampling time in seconds. With some algebraic manipulations, the PID gains can be calculated from the controller's coefficients.

$$
\begin{aligned}
K_p &= -2b_2 - b_1 \\
K_i &= (b_0 + b_1 + b_2)/(T) \\
K_d &= b_2/T
\end{aligned} \tag{26}
$$

5.3. Algorithm Implementation and State Machine

For the real-time safe implementation and execution of the OCTUNE algorithm, a state machine was designed to control the transitions between the different stages of the tuning process. The four primary states are depicted in detail in Figure 6 and described as follows.

- IDLE state: In the IDLE state, the tuning application waits for the operator to send a start signal. Upon receiving the start signal, the application transitions to the next state—the Get Data State .
- Initial Gain State: In this state, the initial (current) PID gains are requested from the autopilot. If there are no failures in receiving the initial gains, the application transitions to the next state, the Get Data Sate. Otherwise, it returns to the IDLE state.
- Get Data State: In this state, the required data for the tuning process, such as target, actual, and control output signals, are stored in buffers in real-time, over a predefined time period or number of samples. Once sufficient data samples are received, they are post-processed to align the data samples according to their time stamps and up-sampled to reduce the high-frequency noise in the acquired signals. If data post-processing is successfully performed, the application transitions to the next state—the Optimization state. Otherwise, the tuning process is stopped, and the application transitions to the IDLE state.
- Optimization state: In this state, an update step of the OCTUNE algorithm, Equation (16) is performed using the data collected and prepared in the Get Data State. The optimal learning rate α^* in Equation (23) is also computed in this state. If the update step is completed successfully, the application transitions back to the Get Data State to prepare a new set of signals for a new update iteration. If a termination condition is reached, such as the maximum optimization iteration or maximum optimization time, the state-machine is terminated, and the application transitions to the IDLE state to be ready for a new tuning cycle.

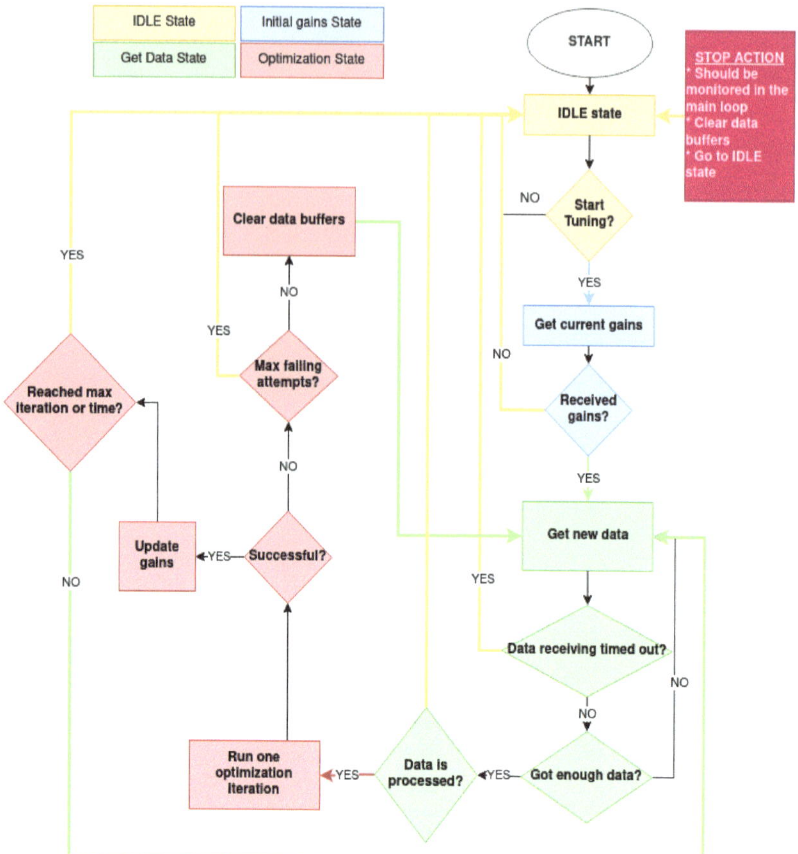

Figure 6. The auto-tuning state machine.

The aforementioned state-machine implementation is used to run multiple simulations of real-time tuning processes for the quadrotor system, which is discussed in the following sections.

5.4. Simulation Results with a Static Learning Rate, α

As mentioned, a primary contribution of this work is to guarantee stability during the tuning process in real-time, which is proved using the condition on the learning rate α, Equation (23). To demonstrate this, we compare the effect of executing the OCTUNE algorithm with a fixed learning rate α and with the optimal one in Equation (23) in simulations.

A simulation run was performed with a fixed learning rate $\alpha = 0.001$ for the angular rates of the roll and pitch PID control loops. In this simulation, the following steps were followed.

1 The quadrotor was commanded to take off in position stabilization mode and hover at 2 m above the ground. The quadrotor was initially stable.
2 The proportional gain of the pitch rate PID control loop was increased from 0.2 to 0.6 in order to introduce high-frequency oscillations.
3 The OCTUNE algorithm was started to tune the PID gains.
4 At the end of the tuning process, the tuning performance was shown using different plots as shown in Figure 7.

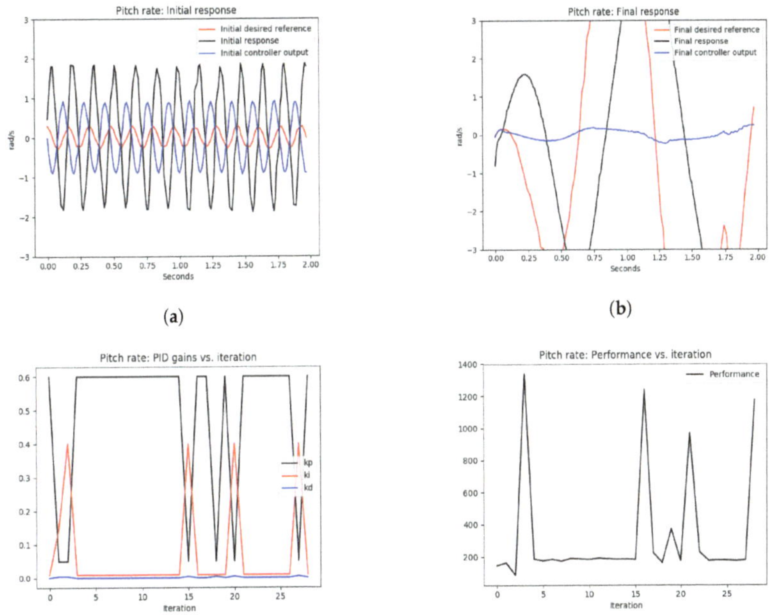

(**a**) (**b**) (**c**) (**d**)

Figure 7. The pitch-rate tuning process during hovering. A fixed learning rate, $\alpha = 0.001$, was used. The quadrotor started with an oscillatory angular pitch response and ended with a worse response after tuning due to the use of a non-optimal fixed learning rate. (**a**) signals before tuning, (**b**) signals after tuning, (**c**) pitch rate PID gains, (**d**) performance error $V(E)$ over iterations.

As shown in Figure 7a, the quadrotor initially had an oscillatory response in the angular rate control of the pitch axis due to high proportional gain; see the $K_p = 0.6$ value at iteration 1 in Figure 7c. Since the learning rate $\alpha = 0.001$ was fixed over the entire optimization iterations, it resulted in the divergence of the system output in Figure 7b, the oscillatory behavior of the PID gains in Figure 7c, and non-diminishing performance error in Figure 7d. Therefore, using a fixed value of the learning rate α can be dangerous to the system tuning process as this cannot guarantee convergence.

In the next section, the simulations are performed with the optimal condition on the learning rate α^* to guarantee stability during the tuning process.

5.5. Simulation Results with an Optimal Learning Rate, α^*

To guarantee the convergence of the performance metric L in Equation (4) of the angular rate control loops of the quadrotor system, the learning rate α was computed at each iteration, according to Equation (23), using the absolute value of the minimum eigenvalue of the Jacobean matrix J in Equation (14), which was constructed using the real-time signals, $r(t), y(t), u(t)$.

Similar simulation steps were followed as in the static learning rate case, starting with the quadrotor in a hover state, increasing the proportional gain of both roll and pitch angular rate PID controllers to obtain high frequency oscillations, and finally starting the OCTUNE algorithm to tune the PID gains in using real-time simulated signals.

As shown in Figures 8 and 9, the initial roll and pitch angular rate responses showed high-frequency oscillations due to the high proportional gains. After tuning the control loops using the OCTUNE algorithm over 28 iterations for 60 s, the control loops were stabilized as shown in Figures 8b and 9b. In addition, the performance error L eventually diminished as shown in Figures 8c and 9c.

In Figures 8d and 9d, we can see that the learning rate α changes over iterations to guarantee that the performance error eventually converges. In comparison to the oscillating gains in Figure 7c, the gains in Figures 8e and 9e are not oscillating and are tuned to reduce the performance error, which results in stable tracking of the angular rates as shown in Figures 8b and 9b. The proportional gains are lowered to reduce the oscillations, and the integral gains are increased to reduce the steady-state error. The differential gains, however, have a minimal change, which is reasonable as high differential gains can cause system instability.

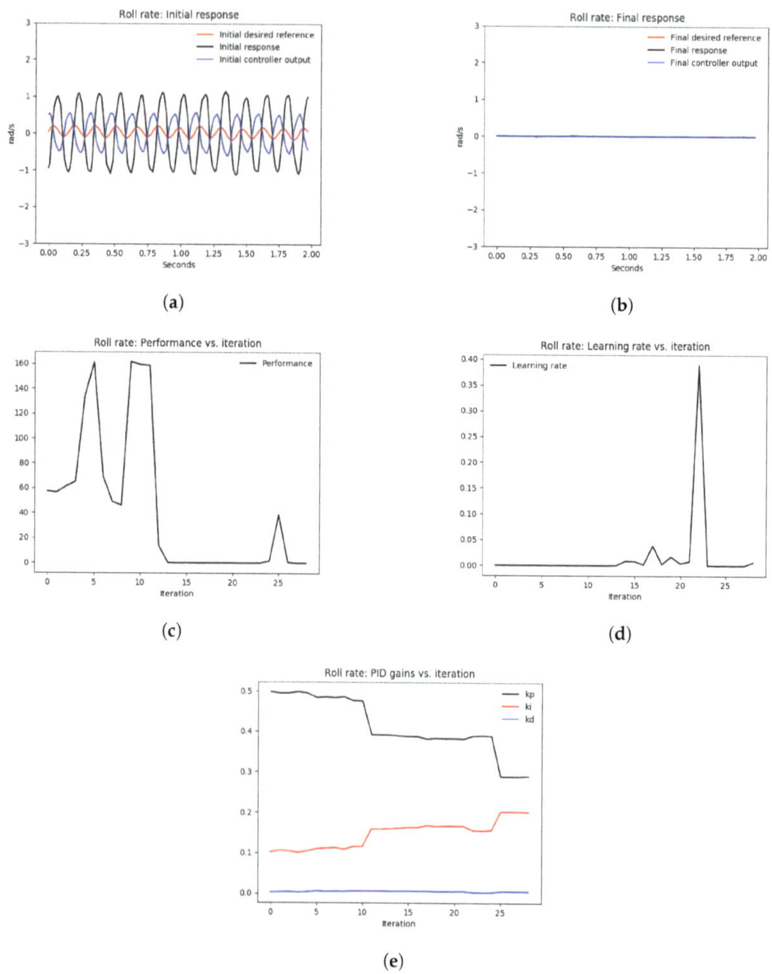

Figure 8. The tuning process of the roll-rate PID controller during hovering. The quadrotor starts with an oscillating behavior due to poorly tuned PID gains. Eventually, the angular rate loops are stabilized after the real-time tuning process. (**a**) signals before tuning, (**b**) signals after tuning, (**c**) performance error $V(E)$ over iterations, (**d**) learning rate α over tuning iterations, (**e**) PID gains.

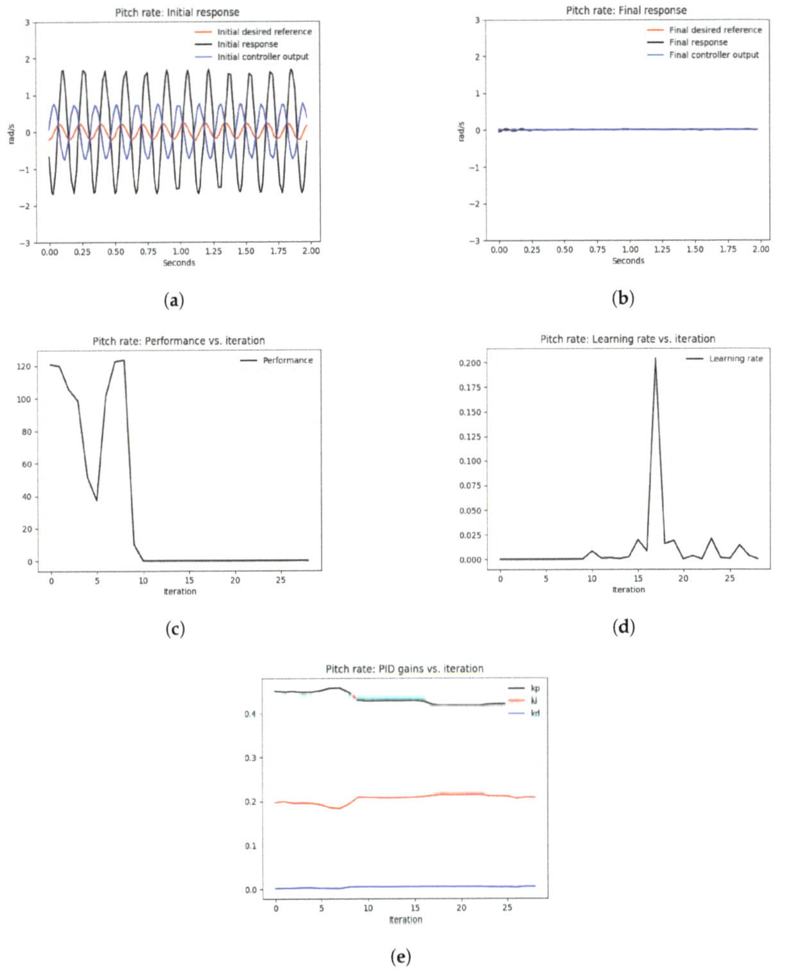

Figure 9. The tuning process of the pitch rate PID controller during hovering. The quadrotor starts with an oscillating behavior due to poorly tuned PID gains. Eventually, the angular rate loops are stabilized after the real-time tuning process. (**a**) signals before tuning, (**b**) signals after tuning, (**c**) performance error $V(E)$ over iterations, (**d**) learning rate α over tuning iterations, (**e**) PID gains.

To provide numerical assessment of the tuning performance, we computed the mean squared error $\text{MSE} = \frac{1}{n}\sum_{i=1}^{n}(r(i) - y(i))^2$ before and after tuning. The number of data samples is constant in all experiments $n = 200$, with time length $T = 2$ seconds and the sampling rate $dt = 0.01$.

Table 1 provides the mean squared error (MSE) of the simulation experiments of the pitch and roll-rate PID controllers, with an optimal learning rate α^* as depicted in Figures 8 and 9. As shown in Table 1, the MSE for the roll rate after tuning is 5% of the MSE before tuning. Similarly, for the pitch rate control, the MSE after tuning is 0.82% of the MSE before tuning. This shows a significant improvement in the reference tracking of the rate PID control loops after the tuning process.

Table 1. The mean squared error (MSE) for the simulation results with the optimal learning rate, α^*.

Experiment	MSE before Tuning	MSE after Tuning
Roll rate tuning	0.59	0.03
Pitch rate tuning	1.21	0.01

5.6. Hardware Experiments

This subsection provides validation results of the OCTUNE algorithm on a real quadcopter platform. The quadcopter used in the presented experiments is depicted in Figure 10. Three experiments were conducted in order to evaluate the OCTUNE performance under different initial PID gains. The PID controllers that were tuned in the hardware experiments were the same as the ones performed in simulation, which control the roll and pitch rates. The experiments were conducted in an indoor environment, and the quadcopter was controlled by a pilot. A link to the video of the hardware experiments are provided in the Supplementary Materials section. Each experiment's design and results are presented as follows.

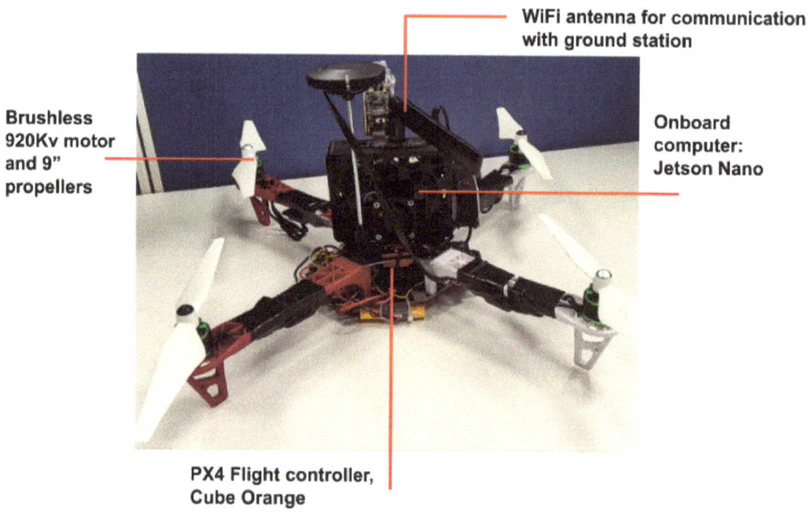

Figure 10. F450 quadcopter platform used in the OCTUNE hardware experiments.

5.6.1. Experiment 1

In this experiment, the PID gains of the roll and pitch rates were left at their default values, and the OCTUNE algorithm was executed during flight. The experiment steps are described as follows.

1 The drone is started on the ground with disarmed motors. The PID gains of the roll/pitch speed control loops are left at their default values ($P = 0.15, I = 0.2$, $D = 0.003$).
2 The pilot flies the quadcopter to a hover position.
3 The OCTUNE process is started.
4 The pilot performs some maneuvers with the quadcopter in order to excite the system.
5 The OCTUNE process is stopped automatically after the indicated maximum optimization time, 120 s, is reached, and the logs and plots are saved.

As can be seen from Figure 11, the initial response as shown in Figure 11a and the final response as shown in Figure 11b show similar tracking performance. However, Figure 11e shows an increase in the P gain to have relatively faster tracking. Furthermore, the performance error in Figure 11c is small (≤ 3), which indicates acceptable tracking of the actual pitch rate signal to the desired one. This experiment demonstrates that starting for good PID gains that stabilize the system with good performance error, the OCTUNE algorithm does not drive the control system to an unstable state. It just improves its performance or at least maintains the current low-error performance. A similar observation of the roll axis can be seen in Figure 12.

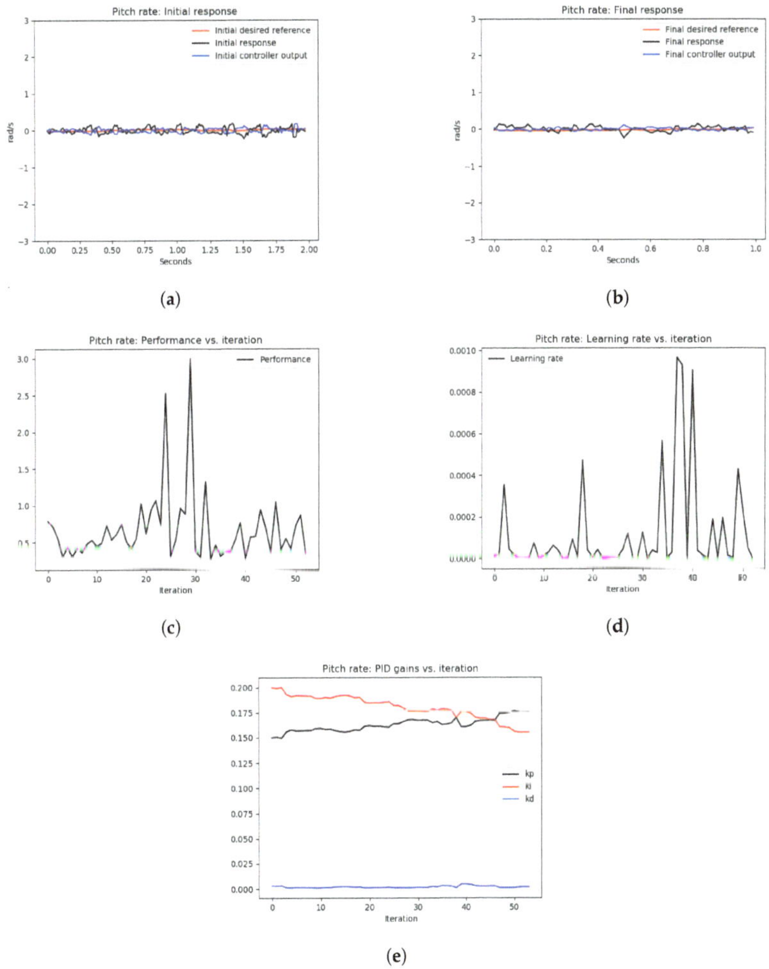

Figure 11. The results of the tuning process of the pitch rate PID controller in Experiment 1. (**a**) signals before tuning, (**b**) signals after tuning, (**c**) performance error $V(E)$ over iterations, (**d**) learning rate α over tuning iterations, (**e**) PID gains.

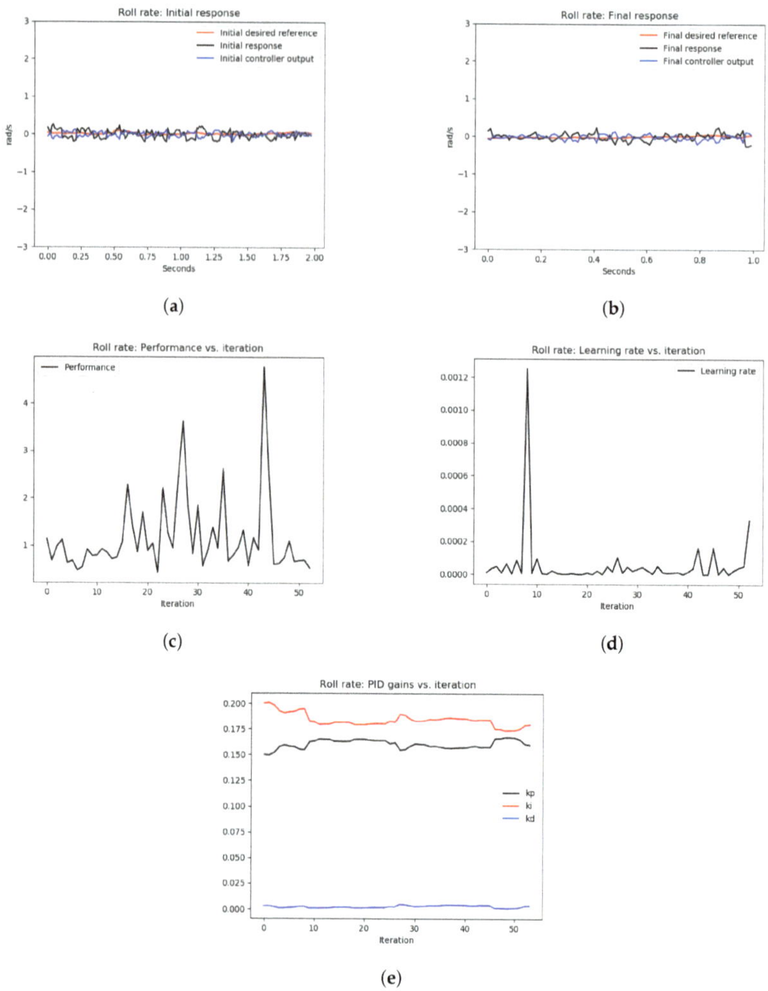

Figure 12. The results of the tuning process of the roll-rate PID controller in Experiment 1. (**a**) signals before tuning, (**b**) signals after tuning, (**c**) performance error $V(E)$ over iterations, (**d**) learning rate α over tuning iterations, (**e**) PID gains.

5.6.2. Experiment 2

In this experiment, the proportional gain (P) of the roll and pitch rates was increased dramatically, four times more than the default values (from 0.15 to 0.6, and, in order to introduce high-frequency oscillations, the OCTUNE algorithm was executed during flight, which should eventually tune the controllers to obtain rid of the oscillations. The experiment steps are described as follows.

1 Initially, the drone is on the ground, and the motors are disarmed. The PID gains of the roll/pitch speed control loops are left at their default values ($p = 0.15$, $I = 0.2$, $D = 0.003$).
2 The pilot flies the quadcopter to a hover position.
3 The P gains of the roll/pitch speed control loops are set to high values (from 0.15 to 0.6) to introduce high-frequency oscillations.
4 The OCTUNE process is started during the flight

5 The pilot tries to keep the quadcopter in hover position while tuning is running.
6 After the quadcopter stabilizes, the pilot performs some maneuvers with the quadcopter in order to excite the system and make sure the system is tuned well.
7 The OCTUNE process is stopped automatically after the indicated maximum optimization time, in the table below, is reached, and the logs and plots are saved.

The tuning results of Experiment 2 are depicted in Figures 13 and 14, for pitch and roll axes, respectively. As can be seen from Figures 13a and 14a, the initial response of the pitch and roll rates, respectively, show high-frequency oscillations as expected because the P gain of both controllers was increased dramatically. After executing the OCTUNE algorithm for 100 iterations (approximately 2 min), the performance error eventually decreased (see Figures 13c and 14c) and the proportional gains were decreased as well (see Figures 13e and 14e). As a result, the reference tracking is improved as shown in Figures 13b and 14b.

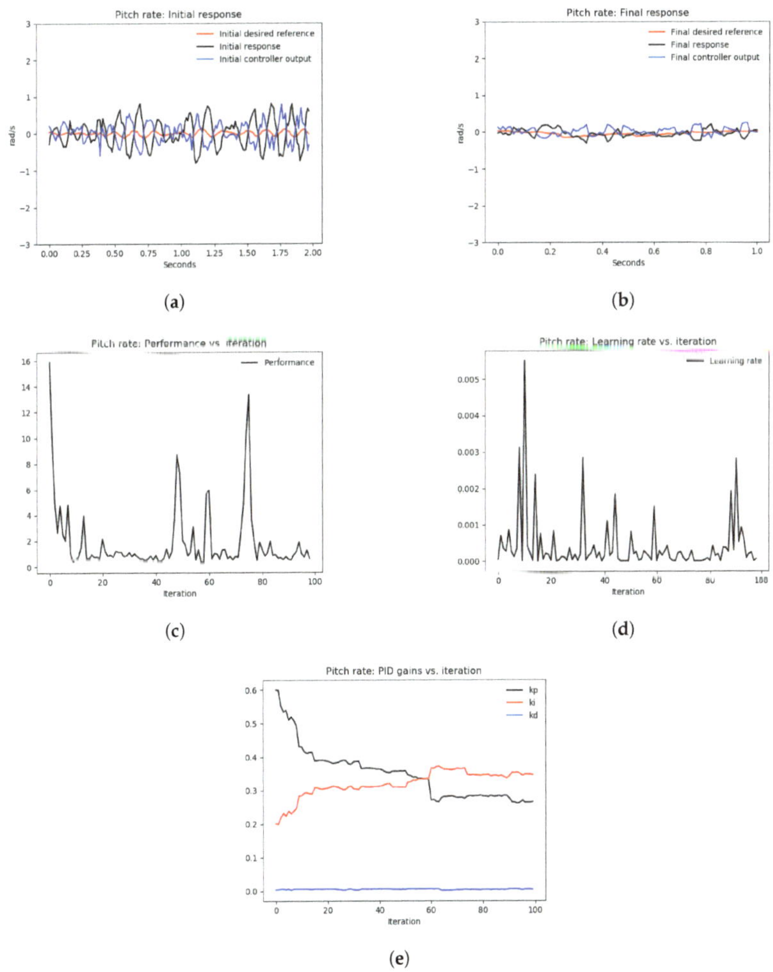

Figure 13. The results of the tuning process of the pitch rate PID controller in Experiment 2. (**a**) signals before tuning, (**b**) signals after tuning, (**c**) performance error $V(E)$ over iterations, (**d**) learning rate α over tuning iterations, (**e**) PID gains.

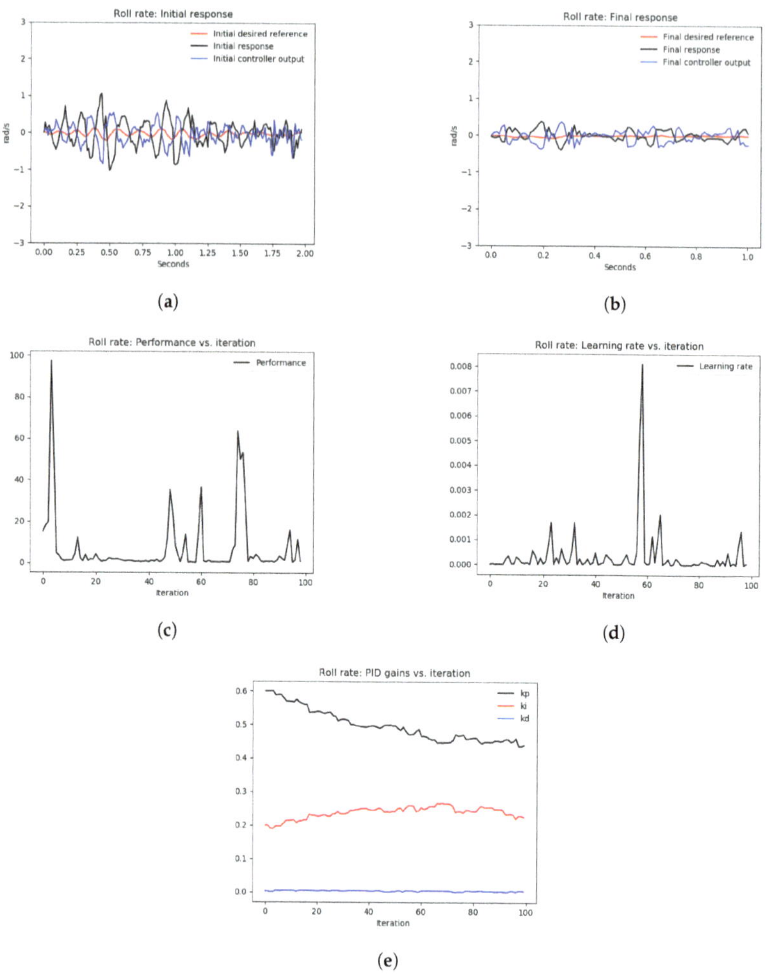

Figure 14. The results of the tuning process of the roll-rate PID controller in Experiment 2. (**a**) signals before tuning, (**b**) signals after tuning, (**c**) performance error $V(E)$ over iterations, (**d**) learning rate α over tuning iterations, (**e**) PID gains.

Similar to the simulation experiments, we computed the MSE of the error signal in the hardware experiments before and after tuning. The MSE results are shown in Table 2. As shown in Table 2, the MSE of teh tracking error of the roll-rate PID controller after tuning is 5.9% of the MSE before tuning. Similarly, the MSE of tracking error of the pitch rate PID controller after tuning is 4.4% of the MSE before tuning. This shows significant improvement of the tracking performance after the tuning process in real-time.

Table 2. The mean squared error (MSE) for hardware Experiment 2.

Experiment	MSE before Tuning	MSE after Tuning
Roll rate tuning	0.17	0.01
Pitch rate tuning	0.16	0.007

6. Conclusions

In this paper, we presented the OCTUNE algorithm, which can be used for the optimal control tuning of an LTI controller (such as a PID) in a classical feedback system without the knowledge of the plant model and using only real-time signals.

The OCTUNE algorithm was validated in realistic simulations of a quadrotor UAV model and on a real quadrotor platform, in which the angular rates of PID controllers were stabilized in a fraction of a minute. The OCTUNE algorithm can run in real-time and continuously tune the controllers to account for any changes in the physical system (e.g., a change of payload) or environment (e.g., wind conditions), with proven convergence. In addition, an open-source implementation of the OCTUNE is available to facilitate the adaptation of the algorithm in different applications.

In future works, it would be interesting to generalize the OCTUNE algorithm to some nonlinear controllers with guaranteed convergence. Furthermore, the trade-off between robustness and optimality in real-time data-driven tuning is an exciting property to address.

Supplementary Materials: Supporting videos of simulations and hardware experiments, as well as the open-source codes, can be found in the following links: Video of the OCTUNE algorithm in quadcopter simulations: https://youtu.be/OY9XY9CdGhA; Video of the OCTUNE algorithm with actual quadcopter experiments: https://youtu.be/a3mrDvK2b-c; Open-source code of the OCTUNE algorithm: https://github.com/mzahana/octune; A ROS wrapper package to use OCTUNE with the PX4 flight controller: https://github.com/mzahana/px4_octune_ros. These links are accessed on 30 October 2022.

Author Contributions: Conceptualization, M.M. and M.A.; methodology, Mohamed Mabrok and M.A.; software, M.A.; validation, M.A.; formal analysis, M.A. and M.M.; data curation, M.A.; writing—original draft preparation, M.A. and M.M.; writing—review and editing, M.A., M.M. and A.K.; visualization, M.A.; supervision, M.A. and A.K.; project administration, M.A. All authors have read and agreed to the published version of the manuscript.

Funding: This research was funded by Prince Sultan University grant number SEED-2021-CCIS-90.

Institutional Review Board Statement: Not applicable.

Data Availability Statement: Not applicable.

Acknowledgments: The authors would like to thank Prince Sultan University for their support in providing the required equipment required for conducting the work described in this paper.

Conflicts of Interest: The authors declare no conflict of interest.

Abbreviations

The following abbreviations are used in this manuscript:

OCTUNE	Optimal control tuning algorithm
PID	Proportional, integra, and derivative controller
UAV	Unmanned aerial vehicle
ROS	Robot operating system

References

1. Garriga, J.L.; Soroush, M. Model predictive control tuning methods: A review. *Ind. Eng. Chem. Res.* **2010**, *49*, 3505–3515. [CrossRef]
2. Borase, R.P.; Maghade, D.; Sondkar, S.; Pawar, S. A review of PID control, tuning methods and applications. *Int. J. Dyn. Control* **2021**, *9*, 818–827. [CrossRef]
3. Meier, L.; Honegger, D.; Pollefeys, M. PX4: A node-based multithreaded open source robotics framework for deeply embedded platforms. *Proc. IEEE Int. Conf. Robot. Autom.* **2015**, *2015*, 6235–6240. [CrossRef]
4. ArduPilot Open Source Autopilot System. Available online: https://ardupilot.org/ (accessed on 15 March 2022).
5. Lindquist, A. On feedback control of linear stochastic systems. *SIAM J. Control* **1973**, *11*, 323–343. [CrossRef]
6. Whittle, P. Risk-sensitive linear/quadratic/Gaussian control. *Adv. Appl. Probab.* **1981**, *13*, 764–777. [CrossRef]
7. Bansal, A.; Sharma, V. Design and analysis of robust H-infinity controller. *Control. Theory Inform.* **2013**, *3*, 7–14.
8. Åström, K.J. Theory and applications of adaptive control—A survey. *Automatica* **1983**, *19*, 471–486. [CrossRef]

9. Tao, G. *Adaptive Control Design and Analysis*; John Wiley & Sons: Hoboken, NJ, USA, 2003; Volume 37.
10. Fliess, M.; Join, C. Model-free control. *Int. J. Control* **2013**, *86*, 2228–2252. [CrossRef]
11. Xu, D.; Jiang, B.; Shi, P. A novel model-free adaptive control design for multivariable industrial processes. *IEEE Trans. Ind. Electron.* **2014**, *61*, 6391–6398. [CrossRef]
12. Hou, Z.; Jin, S. *Model Free Adaptive Control*; CRC Press: Boca Raton, FL, USA, 2013.
13. Kang, J.; Meng, W.; Abraham, A.; Liu, H. An adaptive PID neural network for complex nonlinear system control. *Neurocomputing* **2014**, *135*, 79–85. [CrossRef]
14. Qiu, J.; Ma, M.; Wang, T.; Gao, H. Gradient descent-based adaptive learning control for autonomous underwater vehicles with unknown uncertainties. *IEEE Trans. Neural Netw. Learn. Syst.* **2021**, *32*, 5266–5273. [CrossRef] [PubMed]
15. Lyu, F.; Xu, X.; Zha, X. An adaptive gradient descent attitude estimation algorithm based on a fuzzy system for UUVs. *Ocean. Eng.* **2022**, *266*, 113025. [CrossRef]
16. Yu, C.C. *Autotuning of PID Controllers: A Relay Feedback Approach*; Springer Science & Business Media: Berlin/Heidelberg, Germany, 2006.
17. Uren, K.; van Schoor, G. Genetic Algorithm based PID Tuning for Optimal Power Control of a Three-shaft Brayton Cycle based Power Conversion Unit. *IFAC Proc. Vol.* **2012**, *45*, 685–690. [CrossRef]
18. Maddi, D.; Sheta, A.; Davineni, D.; Al-Hiary, H. Optimization of PID Controller Gain Using Evolutionary Algorithm and Swarm Intelligence. In Proceedings of the 2019 tenth International Conference on Information and Communication Systems (ICICS), Irbid, Jordan, 11–13 June 2019; pp. 199–204. [CrossRef]
19. Narendra, K.S.; Parthasarathy, K. Neural networks and dynamical systems. *Int. J. Approx. Reason.* **1992**, *6*, 109–131. [CrossRef]
20. Forgione, M.; Piga, D. *dynoNet*: A neural network architecture for learning dynamical systems. *Int. J. Adapt. Control. Signal Process.* **2021**, *35*, 612–626. [CrossRef]
21. Peng, J.; Dubay, R. Identification and adaptive neural network control of a DC motor system with dead-zone characteristics. *ISA Trans.* **2011**, *50*, 588–598. [CrossRef] [PubMed]
22. Abadi, M.; Agarwal, A.; Barham, P.; Brevdo, E.; Chen, Z.; Citro, C.; Corrado, G.S.; Davis, A.; Dean, J.; Devin, M.; et al. TensorFlow: Large-Scale Machine Learning on Heterogeneous Systems, 2015. Available online: tensorflow.org (accessed on 30 October 2022).
23. Paszke, A.; Gross, S.; Massa, F.; Lerer, A.; Bradbury, J.; Chanan, G.; Killeen, T.; Lin, Z.; Gimelshein, N.; Antiga, L.; et al. PyTorch: An Imperative Style, High-Performance Deep Learning Library. In *Advances in Neural Information Processing Systems 32*; Wallach, H., Larochelle, H., Beygelzimer, A., d'Alché-Buc, F., Fox, E., Garnett, R., Eds.; Curran Associates, Inc.: La Jolla, CA, USA, 2019; pp. 8024–8035.
24. Yongquan, Y.; Ying, H.; Bi, Z. A PID neural network controller. In Proceedings of the International Joint Conference on Neural Networks, Istanbul, Turkey, 26–29 June 2003; Volume 3, pp. 1933–1938.
25. Patel, R.; Kumar, V. Multilayer neuro PID controller based on back propagation algorithm. *Procedia Comput. Sci.* **2015**, *54*, 207–214. [CrossRef]
26. Goodfellow, I.; Bengio, Y.; Courville, A. *Deep Learning*; MIT Press: Cambridge, MA, USA 2016. Available online: http://www.deeplearningbook.org (accessed on 30 October 2022).
27. Khalil, H. *Nonlinear Systems*; Pearson Education, Prentice Hall: Hoboken, NJ, USA, 2002.
28. PX4 Control Architecture. Available online: http://docs.px4.io/master/en/flight_stack/controller_diagrams.html (accessed on 7 April 2022).
29. Koenig, N.; Howard, A. Design and Use Paradigms for Gazebo, An Open-Source Multi-Robot Simulator. In Proceedings of the IEEE/RSJ International Conference on Intelligent Robots and Systems, Sendai, Japan, 28 September–2 October 2004; pp. 2149–2154.
30. Quigley, M.; Conley, K.; Gerkey, B.; Faust, J.; Foote, T.; Leibs, J.; Wheeler, R.; Ng, A. ROS: An open-source Robot Operating System. *ICRA Workshop Open Source Softw.* **2009**, *3*, 5.
31. Åström, K.J.; Wittenmark, B. *Computer-Controlled Systems: Theory and Design*, 2nd ed.; Prentice-Hall, Inc.: Hoboken, NJ, USA, 1990.

sensors

Article

Neural Network-Based Autonomous Search Model with Undulatory Locomotion Inspired by *Caenorhabditis Elegans*

Mohan Chen, Dazheng Feng *, Hongtao Su, Meng Wang and Tingting Su

National Laboratory of Radar Signal Processing, Xidian University, Xi'an 710071, China
* Correspondence: dzfeng@xidian.edu.cn

Abstract: *Caenorhabditis elegans* (*C. elegans*) exhibits sophisticated chemotaxis behavior with a unique locomotion pattern using a simple nervous system only and is, therefore, well suited to inspire simple, cost-effective robotic navigation schemes. Chemotaxis in *C. elegans* involves two complementary strategies: klinokinesis, which allows reorientation by sharp turns when moving away from targets; and klinotaxis, which gradually adjusts the direction of motion toward the preferred side throughout the movement. In this study, we developed an autonomous search model with undulatory locomotion that combines these two *C. elegans* chemotaxis strategies with its body undulatory locomotion. To search for peaks in environmental variables such as chemical concentrations and radiation in directions close to the steepest gradients, only one sensor is needed. To develop our model, we first evolved a central pattern generator and designed a minimal network unit with proprioceptive feedback to encode and propagate rhythmic signals; hence, we realized realistic undulatory locomotion. We then constructed adaptive sensory neuron models following real electrophysiological characteristics and incorporated a state-dependent gating mechanism, enabling the model to execute the two orientation strategies simultaneously according to information from a single sensor. Simulation results verified the effectiveness, superiority, and realness of the model. Our simply structured model exploits multiple biological mechanisms to search for the shortest path concentration peak over a wide range of gradients and can serve as a theoretical prototype for worm-like navigation robots.

Keywords: bio inspired model; autonomous search; undulatory locomotion; *Caenorhabditis elegans*; chemotaxis

Citation: Chen, M.; Feng, D.; Su, H.; Wang, M.; Su, T. Neural Network-Based Autonomous Search Model with Undulatory Locomotion Inspired by *Caenorhabditis Elegans*. *Sensors* **2022**, *22*, 8825. https://doi.org/10.3390/s22228825

Academic Editors: David Cheneler and Stephen Monk

Received: 18 October 2022
Accepted: 12 November 2022
Published: 15 November 2022

1. Introduction

Biological systems are important inspirational resources for mobile-robot control research. Even the simplest organisms have unique locomotion patterns and remarkable spatial orientation abilities, which depend on their powerful nervous systems. The nematode *Caenorhabditis elegans* (*C. elegans*) has a small, compact anatomy, a fully mapped nervous system comprising only 302 neurons [1,2], and a rich behavioral repertoire; thus, it is an ideal organism for linking neural activity to behavior. '*C. elegans* moves in an undulatory fashion by generating sinusoidal dorsoventral bends that propagate from anterior to posterior; the locomotion is involved in most, if not all, of its behavior. Furthermore, *C. elegans* exhibits chemotaxis toward numerous environmental cues, including salt; chemotaxis is the ability to move up (or down) a concentration gradient of a chemical attractant (or repellent). In *C. elegans*, chemotaxis is performed using two parallel strategies [3]: klinokinesis and klinotaxis. Klinokinesis [4] is a biased random walk in which sharp turns occur more frequently in response to a declining (or rising) concentration gradient. The klinotaxis strategy [3] gradually adjusts the movement direction toward the line of steepest ascent (or descent) within the gradient. These behaviors can also be important functions for mobile robots. Chemotaxis-inspired navigation methods can control robots to perform specific tasks, such as chemical leak location [5,6]; radiation measurement [7]; and environment monitoring [8]. Worm-like undulation robots can be deployed in certain special scenarios, such as in pipelines [9] and complex terrain [10,11].

From an engineering perspective, the *C. elegans* chemotaxis behavior is attractive for robotic navigation control. First, this is because the two chemotaxis strategies serve complementary roles to ensure a short search path. Klinokinesis allows the robot to use temporal gradients of environmental variables to quickly correct its direction away from the target, while klinotaxis allows the robot to gradually optimize the path using spatial gradients to align its movement with the steepest gradient direction. Second, *C. elegans* performs chemotaxis only by sensing concentration changes at a single point in its head, suggesting that a robot could mimic the two chemotaxis strategies with a single sensor. This is reasonable, especially for small or resource-constrained mobile robots, as it enables them to make precise steering decisions according to temporal and spatial gradients while carrying a single sensor. Moreover, the clearly delineated nervous system of *C. elegans* provides researchers with the opportunity to study the functional neural circuits [12–14] and mechanisms underlying these behaviors [15–18]. Therefore, these behaviors can be replicated using simple models with efficient biological neural mechanisms, and such models could potentially incorporate these biological methods into robot control applications.

Several studies have explored navigation models inspired by chemotaxis locomotion in *C. elegans*. In early works, researchers [19–21] simulated chemotaxis behavior using recurrent networks; hence, they explored computational rules and behavioral strategies for chemotaxis in *C. elegans*. Additionally, Morse et al. [22] designed an autonomous robot to perform chemotaxis-like phototaxis behavior under the control of a simulated neural network. Xu et al. [23] trained dynamic neural networks with single or dual sensory neurons to perform navigation tasks featuring salt attraction and toxin avoidance, mimicking the klinokinesis or klinotaxis strategy; subsequently, those researchers added a speed regulation mechanism to their model [24]. Some studies focused on implementing *C. elegans* chemotaxis-inspired contour tracking by designing spiking neural networks [25,26] and utilizing a neuromorphic processor [27]. However, in the above works, each model was regarded as a point, and the whole-body movement of *C. elegans* was ignored. Deng et al. [28] incorporated the body undulatory locomotion of *C. elegans* into a navigation model emulating chemotaxis for the first time; however, the wave propagation was modeled using an added phase lag term, which was unrealistic, and navigation-induced head deflections could not be propagated. A follow-up study [29] further incorporated the proprioception mechanism [15,30], which is a biological mechanism responsible for the propagation of undulatory waves along the body in *C. elegans*. Demin et al. [31] and Costalago-Meruelo et al. [32] both trained neural circuit models and combined physics engines to simulate chemotaxis and body locomotion in *C. elegans*.

Existing models typically leverage one strategy only. Most (i.e., [19–29,31]) imitate klinokinesis; the model decides to turn or continue straight based on the temporal gradients of the environmental variables only and, therefore, a short search path cannot be guaranteed. Other models (i.e., [23,24,32]) mimic klinotaxis exclusively; they steer in the correct direction depending on the spatial gradient perpendicular to their current path, but usually under the assumption that two sensors are spaced at a certain distance to directly determine the spatial gradient. However, biological studies [14,33] have found that *ASEL* and *ASER* neurons in *C. elegans* are responsible for sensing salt, and function by sensing concentration changes at a single point in the head due to proximity. The state-dependent gating neural mechanism [34,35] indicates that the klinotaxis steering response of *C. elegans* relies on the internal state of the nervous system at the time of the sensory stimulus during undulatory locomotion. Our previous work [17] further identified this mechanism in a neural model based on the *C. elegans* connectome.

Thus, owing to the omission of one chemotaxis strategy, current *C. elegans*-inspired navigation models are often unable to use the temporal and spatial gradient information simultaneously to rapidly search for a gradient source in the environment. In particular, current models typically lack the ability to capture spatial gradient information with a single sensor. To address this problem, we designed a bio-inspired network model that integrates the two strategies (i.e., klinokinesis and klinotaxis) and body undulatory locomotion of *C. elegans* in the context of one sensor. The model exploits the advantages of the complete *C. elegans* chemotaxis behavior, aiming to provide an efficient robotic autonomous search scheme and afford a simple network control prototype for worm-like navigation robots. The proposed model is a multi-joint rigid link system with a network circuit that controls the joint angles. To implement complex behavior in this simple model, we simplified the network circuit as much as possible based on the functional neural circuits of *C. elegans* and incorporated multiple biological mechanisms. The main contributions of this study are as follows.

First, a central pattern generator (CPG) was evolved using a genetic algorithm (GA) to spontaneously generate rhythmic oscillatory signals. Furthermore, a repeating minimal network unit with proprioceptive feedback was designed, which is sufficient to propagate the undulatory wave from anterior to posterior. As such, the model can perform realistic body undulatory locomotion during both forward and navigation-induced steering movements; this behavior was verified through simulation experiments.

Second, dynamic adaptive sensory neuron models were constructed based on the electrophysiological characteristics of the salt sensory neurons in *C. elegans*, which convert information from the sensor into the sensory inputs of the network circuit. Moreover, the state-dependent gating mechanism was incorporated into the model, thereby realizing klinotaxis behavior in the context of a single sensor. Klinokinesis behavior is implemented by fitting a logic function. The model can make steering decisions leveraging the klinotaxis and klinokinesis strategies simultaneously; klinokinesis allows for rapid steering away from the target, while klinotaxis continuously adjusts the movement direction toward the side with the higher concentration. The model was tested in simulations, demonstrating its stable search for environmental variable peaks in directions close to the steepest gradients over a wide range of gradients. The search path was significantly shorter than those of the models employing a single strategy. In addition, the quantitative analysis verified the realness of the two strategies implemented by the proposed model.

The remainder of this paper is organized as follows. Section 2 details the proposed methodology; the overall model architecture and biological basis are first introduced and a detailed description of the sub-network circuits controlling the undulatory locomotion and navigation behavior is then provided. Section 3 presents the experiment results and corresponding analyses, and comparative discussion. Section 4 summarizes the study and suggests future research.

2. Methodology

2.1. Overall Model Architecture and Biological Basis

The overall model architecture consists of a multi-joint rigid link system and a simple network circuit based on the anatomical structure and functional neural circuits of *C. elegans*, as shown in Figure 1. *C. elegans* is approximately 1 mm long and elliptically cylindrical in shape. It moves on its side on agar and bends in the dorsoventral plane [36], with this movement being mediated by a neuromuscular circuit. Its 95 body wall muscles are arranged along the four body quadrants [37]: the dorsal left, dorsal right, ventral left, and ventral right (DL, DR, VL, and VR, respectively). Each quadrant contains 24 or 23 muscles, which are staggered in pairs. Based on its muscle structure, *C. elegans* was typically modeled with 11 or 12 segments in previous works [28,29]. Similarly, our model contains 12 rigid rods of length l, with 11 rotatable joints and 13 nodes, as shown in Figure 1a. Joint 1 controls the head orientation and Node 1 corresponds to the head tip.

The *C. elegans* neural circuits for navigation locomotion are well understood. The motoneurons located in the head and along the ventral neural cord (VNC) drive the dorsoventral muscles to contract and relax rhythmically, generating undulatory locomotion with sinusoidal body waves. Laser ablation studies [13] have shown that, among the five types of VNC motoneurons, the cholinergic B-type excitatory motoneurons are essential for forward locomotion; these include 11 ventral B-type motoneurons (VB neurons) and seven dorsal B-type motoneurons (DB neurons) that form neuromuscular junctions with ventral and dorsal muscle cells, respectively. The four SMB-class and four SMD-class head motoneurons [12] play important roles in regulating the undulatory amplitude and sharp turns, respectively, and innervate the head and neck muscles. In the salt-sensorimotor pathway of *C. elegans*, asymmetric chemosensory neurons (ASEL/R) [14,33] located in the head sense salt stimuli in the environment and communicate with motoneurons through interneurons, inducing head steering. Based on this understanding, we designed a simplified network circuit, the outputs of which control the joint angles of the rigid link system. As shown in Figure 1b, the circuit consists of a head circuit responsible for generating undulations and making navigation decisions, and 10 repeating minimal VNC units responsible for wave propagation. Because this model is two-dimensional, the muscles degenerate to the dorsal and ventral muscles (DMs and VMs, respectively). In addition, the head circuit contains a CPG; however, it is still unclear which neurons belong to the CPG in the head of *C. elegans*.

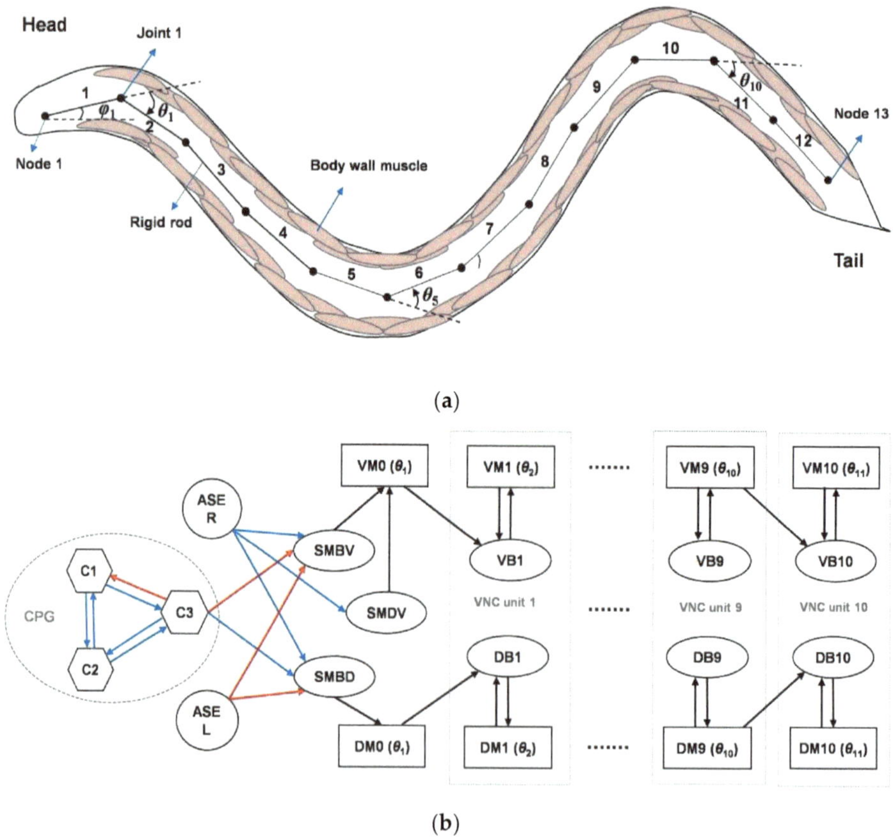

Figure 1. Overall model architecture: (**a**) multi-joint rigid link system abstracted from *Caenorhabditis elegans* (*C. elegans*) structure; and (**b**) network circuit. The circles, hexagons, ovals, and rectangles represent sensory neurons, oscillatory neurons, motoneurons, and muscle cells, respectively. The blue and red lines represent excitatory and inhibitory neural connections, respectively, where the central pattern generator (CPG) connection polarities were determined through evolution and the rest were specifically designed. The black lines represent neuromuscular connections.

2.2. Definitions

For clarity, some definitions of the kinematics-related terms used in this study are shown in Figure 2. Unless otherwise specified, the position and locomotion trajectory of the model default to those of the head tip (i.e., Node 1 of the rigid link system). Because the model performs undulatory locomotion, the translation direction is the forward direction of the model during movement for one cycle. The normal direction is that 90° counterclockwise to the translation direction. The instantaneous locomotion direction can be decomposed into a translation component and a perpendicular component (the same or opposite to the normal direction). The turning bias is the angle between the current translation direction and the translation direction after one cycle. Sides D and V correspond to the left and right sides, respectively, and a left/right sweep corresponds to half a cycle of movement toward the right/left.

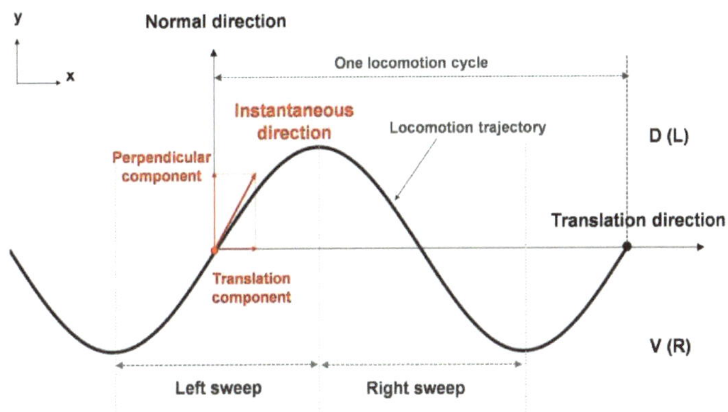

Figure 2. Definitions of kinematics-related terms.

2.3. Undulatory Control Circuit

The key to realizing undulatory locomotion is the generation and propagation of an undulatory wave, which requires encoding of rhythmic oscillatory signals in both temporal sequences and spatial patterns. For simplicity, we designed an undulatory control circuit in accordance with the biological view [15,38] that a single CPG produces head-bending waves in *C. elegans*, and that these waves propagate through body from anterior to posterior via proprioceptive feedback. Note, however, that the existence of multiple oscillators in the mid-body VNC motor circuit of *C. elegans* has been suggested [16,39].

2.3.1. CPG

CPGs are neuronal circuits capable of producing rhythmic outputs without rhythmic inputs, typically as a result of reciprocal inhibitory interactions between neurons. Biomimetic CPG controllers are often used in rhythmic motion robots [40,41]. In the proposed model, the CPG consists of three interconnected dynamic neurons (i.e., C1, C2, and C3). In previous models [28,29], a sinusoidal voltage was preassigned to a neuron (i.e., as an oscillator) and the CPG neuron phases were regulated through neuronal interactions. In contrast, we made no explicit *a priori* assumption regarding the way neurons generate oscillations and evolved them to spontaneously generate oscillatory voltages. This mechanism is more in line with biological reality, and from an engineering perspective, the design of a specialized neuron with sinusoidal oscillations is not required.

The neurons in the CPG have first-order nonlinear dynamics, which are expressed by the following first-order ordinary differential equation (ODE):

$$\tau_i \cdot \frac{dV_i(t)}{dt} = -(V_i - E_i^{rest}) + \sum_j w_{i,j} \cdot f(V_j + b_j), \quad (1)$$

where $V_i(t)$ denotes the voltage of neuron i at time t, τ is the time constant, and E^{rest} is the resting potential. The second term on the right is the input current from the other neurons, where $w_{i,j}$ represents the connection weight from neuron j to i, b is the constant bias, and $f(\cdot)$ is a sigmoidal function that expresses the nonlinear transmission from the presynaptic neurons to the postsynaptic neurons.

In the proposed model, a simple evolutionary algorithm known as the genetic algorithm (GA) [42] evolves the CPG parameters. There are 15 parameters to be determined: τ_i [0.05, 2]; $w_{i,j}$ [−10, 10]; E_i^{rest} [−10, 10]; b_i [−10, 10] (the values in square brackets are the ranges). The parameters are encoded as a real-valued vector with a range of [−1, 1], which corresponds to one individual. The initial population is composed of 500 random individuals and evolves into a new population generation through crossover, mutation, and selection operations. During crossover, two individuals are randomly selected as parents and a new individual, the child, is obtained through two-point recombination. The child is then mutated by adding Gaussian noise with mean zero and a standard deviation (s.d.) of 0.2 to each element of the vector. During selection, the parent with the lower fitness is selected and replaced with the child. If 300 generations are obtained, or the fitness of the best individual exceeds the threshold, the iteration ends.

The goal of this evolution is for the CPG output neuron C3 to generate a sinusoidal oscillatory voltage with a cycle of $T_{osc} = 4$ s (approximating the *C. elegans* cycle recorded in biological data [35]). Therefore, the model employs the fitness function F, which is the product of multiple components: $F = F_1 \cdot F_2 \cdot F_3$, in terms of $V_{C3}(t)$, and

$$F_1 = 1 - \left| \frac{1}{T} \cdot \int_0^T V_{C3}(t)dt \right|, \quad (2)$$

$$F_2 = f_N(\max(V_{C3}(t)), -\min(V_{C3}(t))), \quad (3)$$

$$F_3 = f_N\left(\frac{1}{T} \cdot \int_0^T \left| \frac{dV_{C3}(t)}{dt} \right| dt,\ \frac{4 \cdot \max(V_{C3}(t))}{T_{osc}} \right), \quad (4)$$

where T is the time length and $f_N(x, x_0) = (x/x_0) \cdot \exp(1 - x/x_0)$ is a normalization function that reaches a maximum of 1 at $x = x_0$. Here, F_1 is combined with F_2 to reward voltage dynamics with zero mean and with the same positive and negative amplitudes. Hence, the same left–right sweep amplitude is ensured for the undulatory locomotion; that is, the translation direction is straight. F_3 is an oscillatory criterion that encourages the voltage change rate to match that of the ideal sinusoidal function, with reference to the criterion used in [43]. The negative values for these functions are set to zero.

The specific amplitude of the voltage is not specified in the evolution, because the output strength can be tuned by connection weights. In addition, some evolved solutions may exhibit damped oscillations that must be re-evaluated over a longer period. Finally, the best individual with stable oscillation is selected from the multiple solutions evolved by the GA to form the CPG.

2.3.2. Undulation Generation and Propagation

As shown in Figure 1b, the *SMBD* and *SMBV* motoneurons receive antiphase oscillatory inputs from neuron C3, with the same connection weight strengths but opposite algebraic signs. For simplicity, *SMBD* and *SMBV* are modeled as neurons with a linear synaptic transfer function, and their voltage dynamics are expressed by the following ODE:

$$\begin{aligned} \tau_{SMB} \cdot \frac{dV_{SMBD/V}(t)}{dt} = -\left(V_{SMBD/V}(t) - E_{SMB}^{rest}\right) + w_{SMBD/V,C3} \cdot V_{C3}(t) + \\ w_{SMB,ASEL} \cdot V_{ASEL}(t) + w_{SMB,ASER} \cdot V_{ASER}(t) \end{aligned} \quad (5)$$

The symbols have the same meanings as in Equation (1), except that the subscripts refer to different neurons; thus, the definitions are not repeated here. $w_{SMBD,C3} = -w_{SMBV,C3} > 0$, and the other *SMBD* and *SMBV* parameters are set to the same values. Here, we default to zero sensory input from ASEL/R; this setting is discussed below.

When affected by oscillatory inputs, *SMBD* and *SMBV* generate oscillatory voltages with opposite phases, which are in turn fed to muscles *DM0* and *VM0*, respectively, via excitatory neuromuscular junctions. The *DM0* and *VM0* activation states are expressed as follows:

$$\tau_{A_0} \cdot \frac{dA_{DM0}(t)}{dt} = -A_{DM0}(t) + w^{m}_{M0,SMB} \cdot V_{SMBD}(t), \tag{6}$$

$$\tau_{A_0} \cdot \frac{dA_{VM0}(t)}{dt} = -A_{VM0}(t) + w^{m}_{M0,SMB} \cdot V_{SMBV}(t) + w^{m}_{M0,SMD} \cdot V_{SMDV}(t), \tag{7}$$

where $A_i(t)$ denotes the activation state of muscle i at time t and $w^{m}_{M0,SMB} > 0$ is the neuromuscular junction weight from *SMBD/V* to *D/VM0*. In addition, *VM0* also receives input from *SMDV* with $w^{m}_{M0,SMD} > 0$; this mechanism is related to the navigation function, as discussed below.

The angle of Joint 1 is controlled by the difference between the nonlinear outputs of *DM0* and *VM0*:

$$O_{D/VM0}(t) = f(A_{D/VM0}(t) + b^{m}_{0}), \tag{8}$$

$$\theta_1(t) = \omega_0 \cdot (O_{DM0}(t) - O_{VM0}(t)), \tag{9}$$

where $O_i(t)$ and $\theta_1(t)$ denote the output of muscle i and the angle of Joint 1 at time t, respectively; and b^{m}_{0} and ω_0 are the output bias and steering coefficient of *D/VM0*, respectively. In this manner, Joint 1 exhibits rhythmic rotation over time under the control of the network outputs. An increase/decrease in $\theta_1(t)$ means that Rod 1 is turning left/right relative to Rod 2 (i.e., counterclockwise/clockwise).

Biologically, local proprioceptive coupling of adjacent body parts in *C. elegans* converts rhythmic motion near the head into bending waves along the body through a cascade of muscle-to-neuron feedback, with B-type motoneurons transducing the proprioceptive signals [15]. Therefore, in the proposed model, Joints 2–11 are controlled by 10 repeating minimal VNC units, each containing a pair of B-type motoneurons and a pair of muscles, where neurons *DBi* and *VBi* ($i = 1, 2, \ldots, 10$) receive feedback currents from the muscles of their own units and the anterior adjacent units. The voltage dynamics of DB and VB are as follows:

$$\tau_B \cdot \frac{dV_{D/VBi}(t)}{dt} = -(V_{D/VBi}(t) - E^{rest}_{B}) - I^{shape}_{D/VBi}(t). \tag{10}$$

The parameters of all units are the same. Further, $I^{shape}_{DBi}(t)$ and $I^{shape}_{VBi}(t)$ are the proprioceptive feedback currents of *DBi* and *VBi*, respectively, and are expressed as follows:

$$I^{shape}_{DBi}(t) = \sum_{j=0,1} w^{shape}_{j} \cdot f(p_j \cdot \theta_{i-j+1}(t)) \cdot \left(V_{DBi}(t) - E^{shape}_{j}\right), \tag{11}$$

$$I^{shape}_{VBi}(t) = \sum_{j=0.1} w^{shape}_{j} \cdot f(-p_j \cdot \theta_{i-j+1}(t)) \cdot \left(V_{VBi}(t) - E^{shape}_{j}\right), \tag{12}$$

where the terms with $j = 0, 1$ represent the feedback currents from the considered unit and the anterior adjacent unit, respectively; w^{shape} is the proprioceptive feedback weight; E^{shape} is the reversal potential of the B-type neurons to the feedback inputs; and p is a constant coefficient.

Through substitution of Equation (9), Equations (11) and (12) can be rewritten as

$$I^{shape}_{DBi}(t) = \sum_{j=0,1} w^{shape}_{j} \cdot f\left(q_{i,j} \cdot \left(O_{DM(i-j)}(t) - O_{VM(i-j)}(t)\right)\right) \cdot \left(V_{Di}(t) - E^{shape}_{j}\right), \tag{13}$$

$$I^{shape}_{VBi}(t) = \sum_{j=0,1} w^{shape}_{j} \cdot f\left(q_{i,j} \cdot \left(O_{VM(i-j)}(t) - O_{DM(i-j)}(t)\right)\right) \cdot \left(V_{Vi}(t) - E^{shape}_{j}\right), \tag{14}$$

where $q_{i,j} = p_j \cdot \omega_0$ for $i = 1$ and $q_{i,j} = p_j \cdot \omega_1$ for $i > 1$. Here, ω_1 is the steering coefficient of *DMi* and *VMi* ($i > 1$).

Similar to Equations (6), (8) and (9), the voltages of DBi and VBi ($i = 1, 2, \ldots, 10$) affect the activation states of DMi and VMi, respectively, and their output differences control the angle $\theta_{i+1}(t)$. Thus,

$$\tau_{Am_i} \cdot \frac{dAm_{D/VMi}(t)}{dt} = -Am_{D/VMi}(t) + wm_{Am_i,B} \cdot V_{D/VBi}(t), \tag{15}$$

$$O_{D/VMi}(t) = f(A_{D/VMi}(t) + b_1^m), \tag{16}$$

$$\theta_{i+1}(t) = \omega_1 \cdot (O_{DMi}(t) - O_{VMi}(t)). \tag{17}$$

The angles of all joints have been determined at this stage. The Joint 1 angle is determined by the CPG rhythm, and the angles of the other joints are determined by the outputs of the corresponding VNC units. Influenced by the feedback current, the joint-angle oscillation controlled by the output of each VNC unit is the same as that of its anterior joint angle, but with a certain phase lag; therefore, soon after the anterior rod turns, the posterior rod is forced to turn in the same direction. In this manner, the rigid link system has a sinusoidal wave shape, and the shape changes with time.

2.4. Navigation Control Circuit

The proposed model simulates chemotaxis behavior in *C. elegans*, utilizing parallel strategies (i.e., klinokinesis and klinotaxis) to navigate based on information from a single sensor. To develop our model, we first constructed sensory neuron models to process the sensor concentration information. We then designed the navigation control circuit (i.e., the head circuit) to make steering decisions in accordance with a concentration peak search. In the designed circuit, the sensory neurons communicate directly with SMBD, SMBV, and SMDV, which in turn connect to *DM*0 and *VM*0. Hereafter, the chemical concentration is taken as a representative environmental variable.

2.4.1. Adaptive Sensory Neuron Models

Electrophysiological recordings [33] have shown the following characteristics of two salt sensory neurons in *C. elegans*: (S1) *ASEL* and *ASER* in *C. elegans* act as ON and OFF cells, respectively, in response to increases and decreases in salt concentration, respectively. (S2) The *ASEL/R* voltage response peak increases with an increase of the up/down step amplitude of concentration and tends to saturation. Following this characteristic, we constructed *ASEL* and *ASER* models using a simple conductance-based approach.

The *ASEL/R* dynamics is expressed by the following ODE:

$$\tau_{ASE} \cdot \frac{dV_{ASEL/R}(t)}{dt} = -(V_{ASEL/R}(t) - E_{ASE}^{rest}) - g_{ASEL/R}(t) \cdot (V_{ASEL/R}(t) - E_{ASE}^{ext}), \tag{18}$$

where E_{ASE}^{ext} is the *ASEL/R* reversal potential to the external input. Further, $g_{ASEL/R}(t)$ denotes the *ASEL/R* conductance at time t, which is determined by the concentration information. All constant parameters of the *ASEL* and *ASER* neurons are equal.

For *ASEL*, $g_{ASEL}(t)$ is derived as follows:

$$\begin{cases} g_{ASEL}(t) = g_{\max} \cdot \tanh\left(\frac{a \cdot |\Delta C(t)|}{1 + b \cdot C_N(t)}\right), & if\ \Delta C(t) > 0 \\ \tau_g \cdot \frac{dg_{ASEL}(t)}{dt} = -g_{ASEL}(t), & otherwise, \end{cases} \tag{19}$$

where $g_{\max}$ is the maximum conductance; τ_g is the delay time constant of the conductance; $a, b > 0$ are two constant coefficients; and $\Delta C(t) = C(t) - C(t - \Delta t)$ represents the concentration difference between two consecutive samples, where $C(t)$ represents the instantaneous concentration detected at time t and Δt represents the sampling interval. The hyperbolic tangent function ($\tanh(\cdot)$) ensures conductance saturation.

The sensory neurons of *C. elegans* can change their sensitivities to adapt to an environment in which they have lived for some time [44]; however, this characteristic has not been incorporated into previous navigation models inspired by *C. elegans*. In the present model, $C_N(t)$, a term representing the historical average absolute concentration difference over the past N s is introduced, which is expressed as follows:

$$C_N(t) = \frac{1}{N} \cdot \sum_{t'=t-N}^{t} |\Delta C(t')|. \tag{20}$$

This term allows sensory neurons to have an adaptive characteristic: (S3) sensory neurons can adaptively modulate their response sensitivities to gradients according to the local gradient amplitude detected over a recent period; the sensitivity is high when the amplitudes of the recently detected gradient are small; and decreases when the recent gradient amplitudes are large. Thus, the model can operate effectively over a wide range of gradients.

Consequently, when there is no concentration gradient, the *ASEL* conductance is zero by default. When a positive gradient is detected, the conductance becomes positive, activating ASEL; after the gradient disappears, the conductance value gradually returns to zero and the *ASEL* voltage gradually returns to the resting potential ($E^{ext}_{ASE} = 0$ mV here).

As regards *ASER*, the dynamic equations of its conductance $g_{ASER}(t)$ obey a set of similar equations to Equations (19) and (20), with the replacements $L \rightarrow R$ and $\Delta C(t) > 0 \rightarrow \Delta C(t) < 0$; hence, *ASER* responds only to concentration decreases.

2.4.2. Klinotaxis Control Circuit

According to the state-dependent gating mechanism [17,34,35], klinotaxis in *C. elegans* may depend on the alternating sensitivities to sensory inputs of neural outputs responsible for controlling dorsal and ventral turns during sinusoidal locomotion. We designed the klinotaxis control circuit based on this neural mechanism, in which *SMBD* and *SMBV* receive sensory inputs and coordinate with *DM*0 and *VM*0 to regulate steering.

A negative bias b_{ij}^{m} with a large absolute value is set for *DM*0 and *VM*0, and shifts their outputs toward the lower saturation region owing to the nonlinear sigmoidal function (see Equation (8)). Figure 3 shows the dynamics of the klinotaxis control circuit in the absence of a sensory input. The antiphase voltages of *SMBD* and *SMBV* drive the activation states of *DM*0 and *VM*0 to undergo antiphase oscillations, and the outputs of *DM*0 and *VM*0 alternately saturate. During the half period when the SMBD/V voltage is small, the D/*VM*0 output is saturated to zero, where the output is insensitive to the input. As such, *DM*0 and *VM*0 form a state-dependent gating.

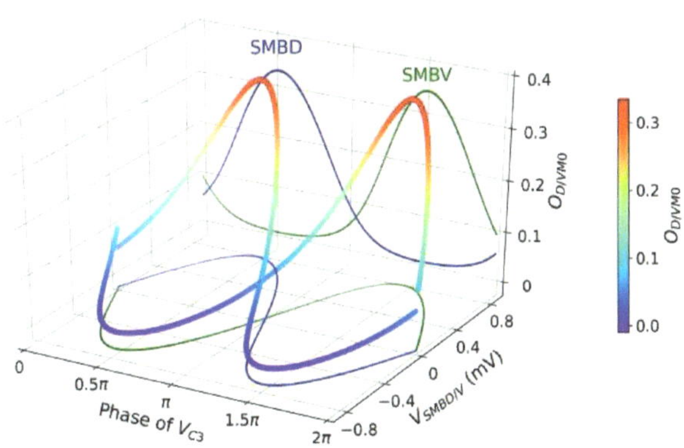

Figure 3. Dynamics of *SMBD/V* motoneuron and *D/VM0* muscle in the absence of sensory input. The solid-colored curves are the projections of three-dimensional (3D) curves onto each plane.

In the presence of sensory input and when a concentration increase/decrease is detected during undulatory locomotion, the perpendicular component direction relative to the instantaneous locomotion direction is the side with higher/lower concentration.

Therefore, the model should steer in the same/opposite direction to the perpendicular component of the instantaneous direction. This suggests that *ASEL* and *ASER* should have antagonistic effects on the neural output; therefore, we set $w_{SMB,ASEL} < 0$ and $w_{SMB,ASER} > 0$.

The above two designs ensure two key features of the klinotaxis behavior, respectively: (f1) When the same concentration change is detected during the right and left sweeps, opposite steering is induced (i.e., state dependence), and (f2) when concentration changes with opposite polarity are detected at the same locomotion phase, opposite steering is induced (i.e., sensory dependence).

Figure 4 shows how the control circuit yields a steering decision corresponding to klinotaxis behavior when a concentration increase or decrease (an up or down step, respectively) is detected at two different locomotion phases. The up/down step evokes a continuous voltage response from *ASEL/R* over a certain timescale, which is transmitted by motoneurons; this response yields a subsequent significant decrease/increase in the output of the sensitive muscle and an almost constant output of the saturated muscle. Therefore, the difference between the dorsal and ventral output decreases/increases, causing a decrease/increase in the bending angle of the subsequent undulatory locomotion, which has a duration of approximately half a cycle. As a result, the trajectories are biased toward the same/opposite side of the perpendicular component. Figure 5 shows the cumulative changes in the subsequent outputs of *DM*0 and *VM*0 induced by applying an up or down step at each locomotion phase. When a concentration step is detected at a phase with a larger perpendicular component (such as phases *c* and *e*), the difference between the resulting cumulative differences of the dorsal and ventral muscle outputs is larger, indicating that the steering amplitude is larger. For phases where the perpendicular component is zero (phases *d* and *f*), the difference between the resulting cumulative differences of the dorsal and ventral muscle outputs is approximately zero; therefore, almost no steering is induced. Furthermore, for the right sweep (phases between 0.5π and 1.5π), the change in amplitude of O_{DM0} exceeds that of O_{VM0}, and the reverse for the left sweep; that is, the steering is reversed for the right and left sweeps. Consequently, the circuit can correct the model direction according to gradients in the normal direction throughout the locomotion; hence, the model steers to the side with the higher concentration.

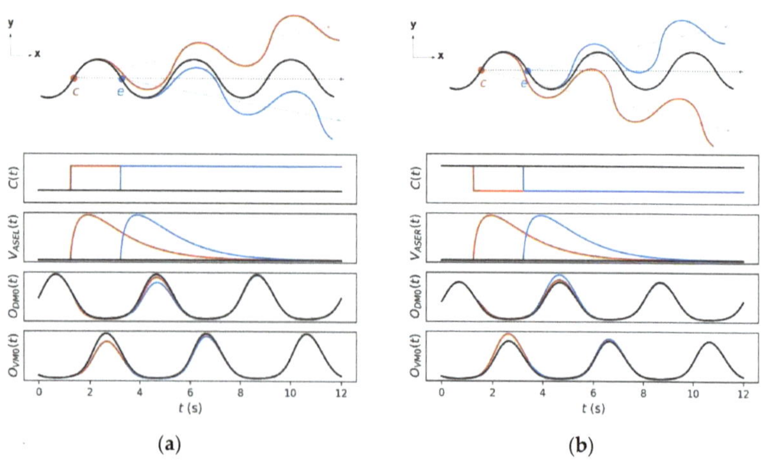

Figure 4. Changes in sensory neuron voltages and muscle outputs (bottom) as well as steering responses (top) caused by application of an up (**a**); or down (**b**) steps at two different locomotion phases. Black, orange, and blue represent the cases where no sensory stimulus is applied and where the sensory stimulus is applied at phase points *c* and *e*, respectively.

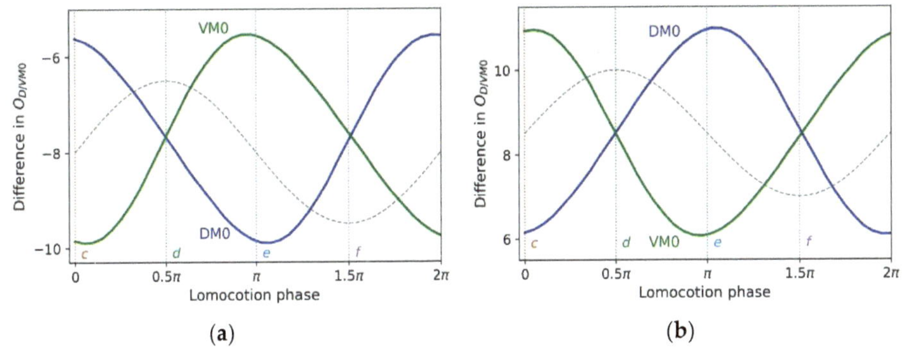

Figure 5. Cumulative differences in *D/VM0* output induced by application of an up (**a**); or down (**b**) step at each locomotion phase, where phase points *c* and *e*, respectively, correspond to phases *c* and *e* in Figure 4. The black dashed lines represent the shape of the locomotion trajectory.

2.4.3. Klinokinesis Control Circuit

In the proposed model, steering mimicking klinokinesis behavior is mediated by an *SMDV* motoneuron, the dynamic equation of which is shown below.

$$\tau_{SMDV} \cdot \frac{dV_{SMDV}(t)}{dt} = -(V_{SMDV}(t) - E^{rest}_{SMDV}) - w_{SMDV,ASER} \cdot f(V_{ASER}(t)) \cdot (V_{SMDV} - E_{SMDV,ASER}), \quad (21)$$

where $w_{SMDV,ASER}$ and $E_{SMDV,ASER}$ are the connection weight from *ASER* to *SMDV* and the corresponding reversal potential, respectively.

In this study, the *SMDV* response voltage was designed to fit the logic function shown in Figure 6 through the parameter settings. This approach is similar to that in the literature [29]. The klinokinesis-related steering depends on the negative temporal gradient; therefore, the *SMDV* response voltage is a function of the *ASER* voltage. When a positive gradient is detected, *ASER* has no sensory response ($V_{ASER} = 0$ mV). This implies that the model is moving toward the increased concentrations and the model therefore does not need to turn. Therefore, *SMDV* does not generate a voltage response ($V_{SMDV} = 0$ mV) and does not transmit a turning signal to *VM0*. However, when a negative gradient is detected, the *ASER* response voltage activates *SMDV*, which in turn increases the activation state and output of *VM0* (see Equation (7)); thus, the model turns right. The greater the deviation between the movement direction and direction of the concentration peak, the larger would be the amplitude of the negative gradient detected and the stronger would be the extent to which the model should correct the direction; accordingly, the *SMDV* voltage response should be large to send a large turning signal. Therefore, the *SMDV* voltage amplitude is proportional to that of the *ASER* voltage and tends to be saturated, enabling the model to correct the direction according to the deviations from the peak direction while preventing oversteering.

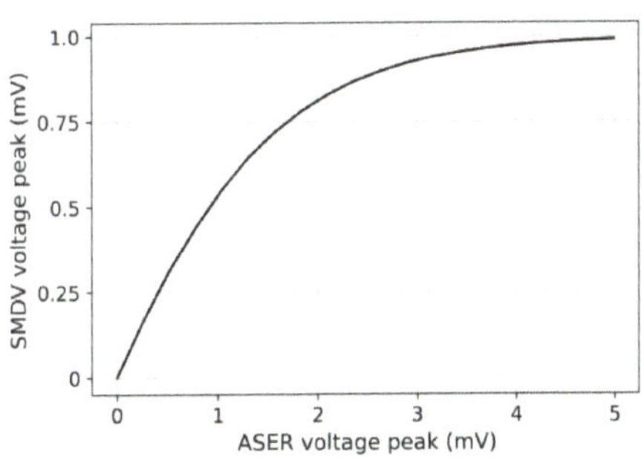

Figure 6. Logic function for mimicking klinokinesis behavior.

3. Simulation Results and Discussion

To verify the effectiveness of the proposed model, we tested its undulatory locomotion behavior without sensory input and its autonomous search behavior under simulated scenarios with concentration gradients. Then, we observed the shape of the link system during the steering behavior and quantitatively analyzed the search trajectories. In the simulation experiments, the model was assumed to have $l = 0.1$ mm and a constant-velocity $v = 0.25$ mm/s based on data for real *C. elegans*. Except for the CPG parameters, the parameters in the network circuit were determined by trial and error. All experiments were conducted using Python 3. 8, and the ODEs were solved by Euler integration with a 0.01-s step.

3.1. Rhythmic Patterns of CPG and Joint Angles

Figure 7 shows the voltage dynamics of the CPG neurons, obtained from our simulations. Through the neuron interactions, the C3 neuron, as the CPG output neuron, generated an approximate sinusoidal oscillatory voltage with a 4-s period and the same positive and negative amplitudes. In the absence of sensory input, the oscillatory voltage of C3 (processed and transmitted by the head network circuit) caused the Joint-1 angle to vary periodically, being centered at zero with the same period, as shown in Figure 8a. Figure 8b shows the changes in all joint angles over time. All joint angles exhibited the same rhythmic oscillation pattern, but the oscillation of each joint (except Joint 1) had a time lag of approximately $T_{osc}/10$ compared with that of its anterior adjacent joint. Through backward propagation, the oscillation of the Joint-11 angle was almost synchronized with that of Joint 1.

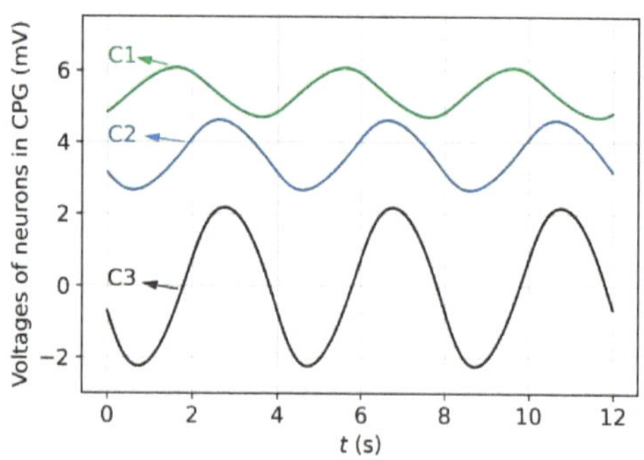

Figure 7. Voltage dynamics of CPG neurons.

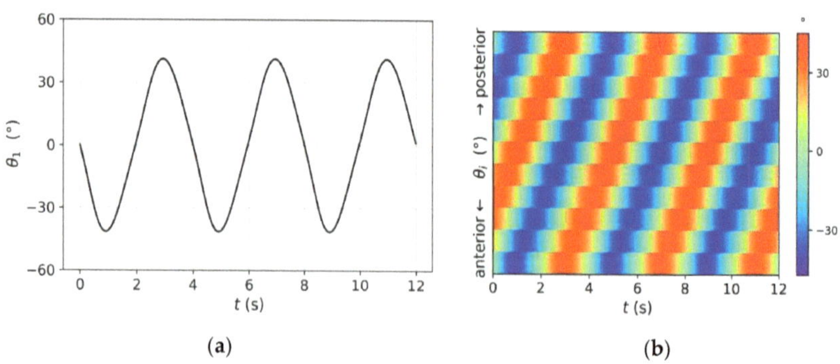

Figure 8. (**a**) periodic changes in Joint-1 angle; and (**b**) kymogram depicting changes in all joint angles over time.

3.2. Forward Undulatory Locomotion Behavior of Model

We observed the behavior of the model under the rhythmic pattern shown in Figure 8. Figure 9a shows the positions and shapes of the rigid link system at 1-s intervals during the first period. We initialized the $t = 0$ s position as $(0, 0)$ and $\varphi_1(0) = 150°$. In terms of spatial mode, $\theta_1(0) = 0°$ and the joint angles from front to back increased, decreased, and increased again, thereby shaping the link system as a sinusoidal wave, similar to the real-world behavior of *C. elegans*. As time progressed, Joint 1 rotated periodically and the model moved forward in a fluctuating pattern, forming a sinusoidal trajectory. The posterior joints and, hence, the shape of the link system, varied accordingly. For example, the model was transformed from the left sweep at $t = 0$ s to the right sweep at $t = 1$ s and 2 s. After one period (at $t = 4$ s), the trajectory drew a sine curve with a complete period and the system shape returned to that at $t = 0$ s. The system wavelength during locomotion was approximately 0.83 times the system length (measurements of real *C. elegans* on agar fall in the range of 0.4 to 0.9 [43]). Additionally, from the trajectory in Figure 9b, the model traveled straight with an undulating pattern in the absence of sensory input, because the positive and negative swing amplitudes of θ_1 were consistent (Figure 8a).

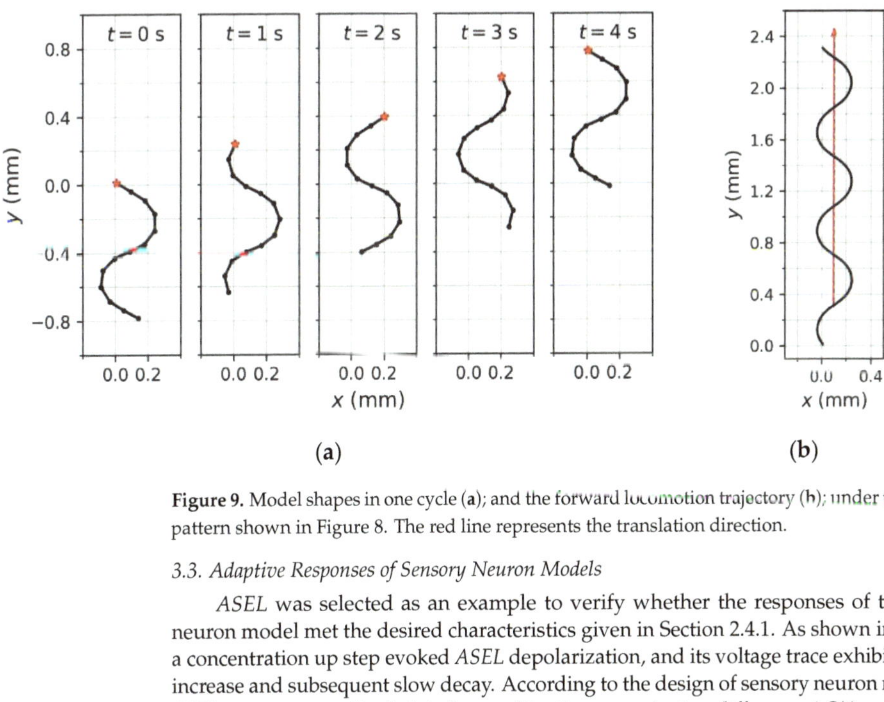

Figure 9. Model shapes in one cycle (**a**); and the forward locomotion trajectory (**b**); under the rhythmic pattern shown in Figure 8. The red line represents the translation direction.

3.3. Adaptive Responses of Sensory Neuron Models

ASEL was selected as an example to verify whether the responses of the sensory neuron model met the desired characteristics given in Section 2.4.1. As shown in Figure 4a, a concentration up step evoked *ASEL* depolarization, and its voltage trace exhibited a rapid increase and subsequent slow decay. According to the design of sensory neuron models, the *ASEL* response amplitude is influenced by the concentration difference $\Delta C(t)$ and historical average absolute concentration difference $C_N(t)$ (refer to Equations (19) and (20)). Figure 10 shows the *ASEL* response voltage peak as a function of $\Delta C(t)$ and $C_N(t)$, from which the following observations can be drawn:

- *ASEL* responded to concentration increases ($\Delta C(t) > 0$) only, as required in S1;
- When positive $\Delta C(t)$ was small, the *ASEL* response amplitude was small. The *ASEL* voltage peak increased with the increase in $\Delta C(t)$ for any given $C_N(t)$ (Figure 10b,c). This indicates that for the local scenario or the scenario with small gradient differences (i.e., the model did not need to re-adapt to the new gradient range), the *ASEL* response amplitude was proportional to the detected temporal gradient, as required in S2. On this basis, the model can control the steering amplitude by comparing magnitudes of the same-polarity gradients;

- When $C_N(t)$ was small, the *ASEL* response amplitude was large, implying that *ASEL* was highly sensitive to the gradient. The *ASEL* voltage peak decreased with the increase in $C_N(t)$ for any positive $\Delta C(t)$ (Figure 10b,d). This indicates that the *ASEL* response was adaptive to the local environment. When exposed to large concentration gradients for a period, the sensory neurons became less sensitive to the gradients. Thus, the characteristic in S3 was satisfied, allowing the model to operate effectively across a wide range of gradients.

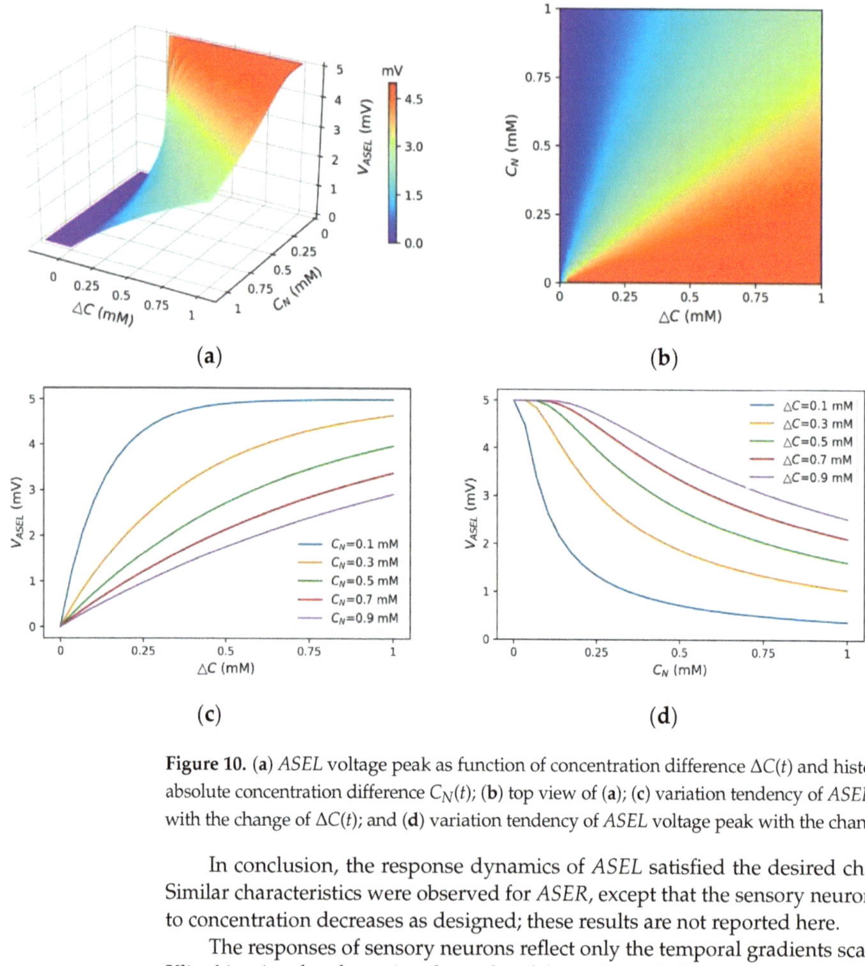

Figure 10. (**a**) *ASEL* voltage peak as function of concentration difference $\Delta C(t)$ and historical average absolute concentration difference $C_N(t)$; (**b**) top view of (**a**); (**c**) variation tendency of *ASEL* voltage peak with the change of $\Delta C(t)$; and (**d**) variation tendency of *ASEL* voltage peak with the change of $C_N(t)$.

In conclusion, the response dynamics of *ASEL* satisfied the desired characteristics. Similar characteristics were observed for *ASER*, except that the sensory neuron responded to concentration decreases as designed; these results are not reported here.

The responses of sensory neurons reflect only the temporal gradients scaled by $C_N(t)$. Klinokinesis-related steering depends solely on the scaled temporal gradient, according to the *ASER* response. Meanwhile, klinotaxis-related steering depends on the normal gradient, i.e., gradients in the normal direction (Figure 2), according to the *ASEL* and *ASER* responses. This is achieved via the state-dependent gating mechanism, that is, the klinotaxis control circuit combines the network internal state with the sensory neuron responses to implicitly extract the normal gradient information corresponding to the current locomotion phase (see Section 2.4.2 for details).

3.4. Autonomous Search Behavior of Model

To verify that the proposed model can search for a concentration peak in a direction close to the steepest gradient, we performed simulations with concentrations having Gaussian distributions.

First, we observed the model trajectories in a scenario where the concentration peak position was (0, 0) and the maximum concentration was 50 mM. The model began moving from different initial positions and with different initial orientations. For comparison, we eliminated one strategy (klinokinesis or klinotaxis) from the proposed model, and the same experiments were also performed for models utilizing a single strategy. The navigation method of the model with klinotaxis eliminated is similar to that of the previous models [29]. Figure 11 shows several trajectories for the three models (two parallel strategies, klinokinesis only, and klinotaxis only) obtained for the same initial conditions. Comparison of these trajectories yielded the following observations:

- For various initial conditions, the three models moved forward in a sinusoidal fashion and successfully reached the concentration peak; however, their search trajectories varied;
- As regards the search trajectories of the klinokinesis-only model (Figure 11b), for the positive gradient direction, the model did not steer, even if there was a deviation from the peak direction. The model corrected the locomotion direction by right turns only when negative gradients were encountered; this behavior is consistent with that of the previous models [29]. In other words, the model could only ensure that it was moving close to the peak, but not that it was using a short search path. In such cases, a large locomotion undulation is advantageous, because a large swing facilitates detection of a negative gradient and adjustment to a more favorable direction. However, a large swing also yields a longer path and may increase the search time;
- As regards the search trajectories of the klinotaxis-only model (Figure 11c), this model continuously and gradually veered toward the side with higher concentration throughout the undulatory locomotion. The orientation adjustment of the model was slightly slower than that of the klinokinesis-only model for negative gradients (compare the beginnings of trajectories whose beginning position are (20, −10) in Figure 11b,c). Because the *ASEL* responses reduced the Joint-1 oscillation amplitude when the model moved toward positive gradients, the model swing amplitude decreased; hence, the search paths were shortened;
- Our model, which integrates both strategies, yielded significantly shortened search paths (Figure 11a). Regardless of the initial position and orientation, the model reached a peak in a direction close to the steepest gradient. If the search began in a direction away from the peak, the klinokinesis strategy allowed the model to make sharp turns to rapidly correct the direction. Meanwhile, the klinotaxis strategy allowed the model to gradually optimize the locomotion direction according to the deviation between the instantaneous and peak directions.

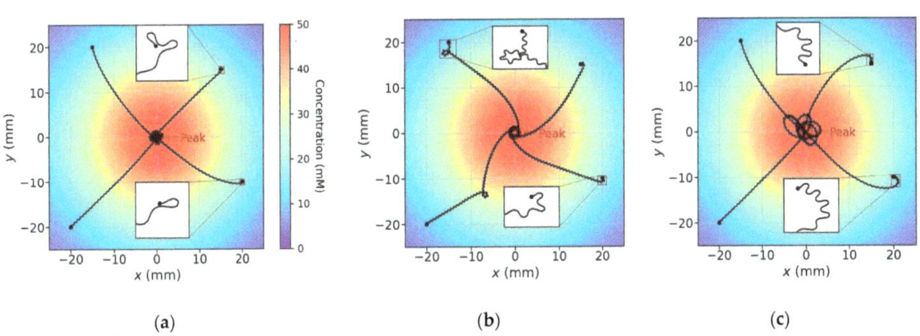

Figure 11. Search trajectories of three models in a given scenario: (**a**) the model exploiting parallel (klinokinesis and klinotaxis) strategies; (**b**) the klinokinesis-only model; and (**c**) the klinotaxis-only model. The four trajectories of each model have different initial positions and orientations, and the initial conditions were the same for all models. The color of each position in the scenario represents the concentration of that position (marked by the color bar in (**a**)).

Two metrics were used to measure the model search performance: the *arrival rate* and *SSR*. The *arrival rate* is the ratio of the number of times the model reached the concentration peak to the total number of experiments, where the peak point is defined as a circular area within 1 mm of the maximum concentration point in the simulations. The *SSR* is the ratio of the model search time to reach the concentration peak to that corresponding to the shortest path (i.e., the linear distance from the initial position to the concentration peak position).

We then tested the three models in 10 scenarios with different gradient ranges. The maximum concentration for each scenario varied from 50 to 1400 mM. In each scenario, the models began moving from four different initial positions with 10 different initial orientations; that is, each model was run 40 times per scenario for a total of 400 times across all scenarios. Figure 12 shows the average search performance results obtained for the three models in each scenario, and Table 1 lists the average results for the models across all experiments. The following conclusions were drawn:

- The *arrival rates* of all three models were 1, indicating that the models reached the concentration peaks in all experiments regardless of whether they used single or parallel strategies. In addition, the standard deviations of the *SSR*s for all models were very small, indicating that the models had robust search performance for different scenarios and initial conditions;
- Compared with the models using a single strategy, the average *SSR* of the model using parallel strategies was the smallest and close to 1, indicating that the search paths were effectively shortened by simultaneous use of the two complementary strategies, and that the search paths were approximated to the optimal paths although the model moved along a waveform path instead of a straight line;
- The average *SSR* of the klinokinesis-only model was much larger; this was because it did not optimize the path in the direction of the positive gradients and the swing amplitude of the undulatory locomotion was large (see Figure 11b).

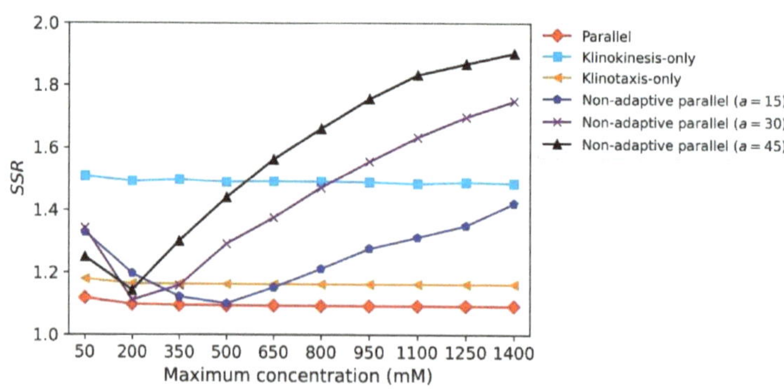

Figure 12. Search performance of all models in scenarios with different concentration gradient ranges. Each point represents the average *SSR* for the model searches, where the model started from different initial positions and orientations.

Table 1. Search performance results for different models across all experiments (±standard deviation).

Strategies		*Arrival Rate*	Average *SSR*
Klinokinesis only		1.0	1.4922 ± 0.07309
Klinotaxis only		1.0	1.1642 ± 0.09253
Parallel (klinokinesis & klinotaxis)	Adaptive	1.0	1.0964 ± 0.05162
	Non-adaptive ($a = 15$)	0.8	1.2474 ± 0.32701
	Non-adaptive ($a = 30$)	1.0	1.4388 ± 0.36184
	Non-adaptive ($a = 45$)	1.0	1.5720 ± 0.43820

Additionally, to verify the role of dynamic adaptation in the sensory neuron models, we conducted the same experiments on a non-adaptive model. In that case, the sensory neuron models did not contain the normalization term with respect to $C_N(t)$. That is, the *ASEL* conductance activation term in Equation (19) was modified to $g_{ASEL}(t) = g_{\max} \cdot \tanh(a \cdot |\Delta C(t)|)$, and the corresponding *ASER* term was similarly modified. The coefficient, a, took on three different values. Comparison of the results presented in Figure 12 and Table 1 reveals that the non-adaptive models had comparable search performance to that of the adaptive model within a narrow gradient range, and that the search performance deteriorated seriously beyond this range. For small a in particular, the response amplitudes of the sensory neurons in the small-gradient scenario were too small, which may have caused a model search failure (i.e., the *arrival rate* was not 1). In contrast, the adaptive sensory neuron models could dynamically adjust their response sensitivities to the gradient according to the gradient amplitude in the preceding time period, so that the model maintained stable search performance in all scenarios.

3.5. Analysis of Search Behavior

The shape of the rigid link system during the steering induced by the navigation behavior was observed. Figure 13 shows shapes at five different stages as the model underwent an Ω turn (a sharp turn performed by *C. elegans*). When the head turned, the rigid link system bent accordingly (i.e., at times $t2$, $t3$, and $t4$). The link system shape at each stage was close to the corresponding trajectory shape. After the head returned to a normal swing, the entire link system returned to the normal sinusoidal undulation (i.e., at time $t5$). This suggests that the rigid link system could be shaped similarly to the body of a real worm during steering.

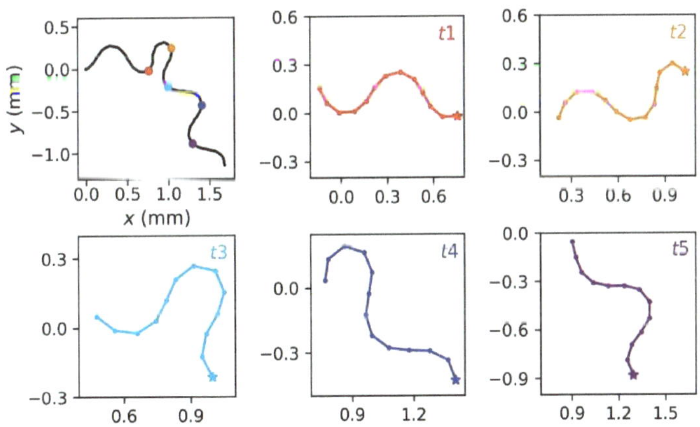

Figure 13. Shape of rigid link system during steering. The upper left diagram shows the model trajectory, and the shapes of the rigid link system at the position indicated by the five colored points are shown in the similarly colored sub-graphs. The pentagram represents the head tip (i.e., Node 1 of the system).

Additionally, to verify that the navigation decisions generated by the model conformed to the two chemotaxis strategies of *C. elegans*, we performed a quantitative analysis of all trajectories of the models using klinokinesis only and klinotaxis only. First, we calculated and statistically analyzed the relationship between the turning biases and normal concentration gradients in the klinotaxis-only trajectories; the result are shown in Figure 14a. The average turning bias was positively correlated with the normal gradient, which is consistent with the statistical results of biological experiments [3]. This suggests that the proposed model mimicked the klinotaxis behavior of real-world *C. elegans*, steering to the side with the higher concentration. We then calculated the relationship between the turning bias and the temporal gradient of the concentration in the klinokinesis-only trajectories, as shown in Figure 14b. For positive gradients, the turning bias was almost zero. For negative gradients, the turning bias was negative (i.e., the model turned right) and the amplitude was proportional to the gradient amplitude. These results have the same trend as biological results [3,4]. In addition, comparing Figure 14a,b, the steering amplitude obtained for the klinokinesis behavior was greater than that for the klinotaxis behavior; this is also consistent with biological findings [3].

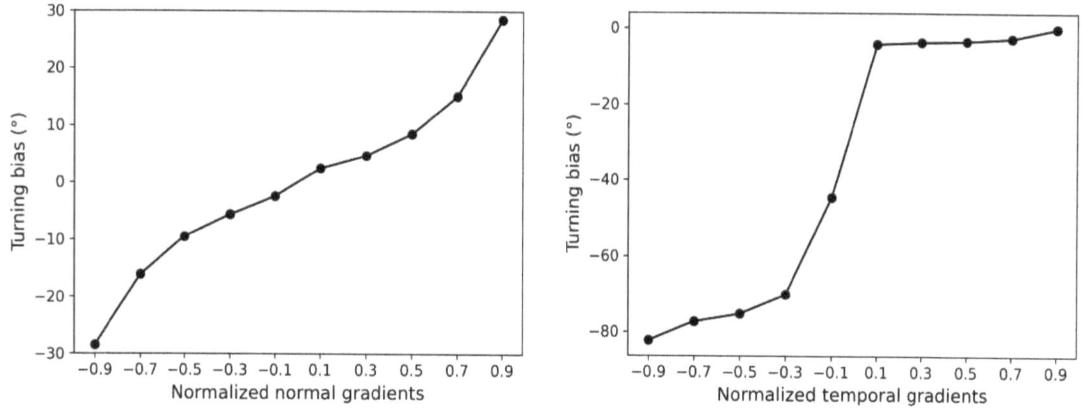

(a) (b)

Figure 14. Quantitative analysis of search behavior strategies: (**a**) turning bias vs. Normal concentration gradient of klinotaxis trajectories; and (**b**) turning bias vs. temporal gradient of klinokinesis trajectories. All gradients were linearly normalized to between −1 and 1.

3.6. Discussion

This section highlights the differences and advantages of our study compared with previous related research. Most existing navigation models inspired by *C. elegans* chemotaxis aim to realize the chemotaxis behavior and, on this premise, explore the underlying mechanisms from the biological perspective; therefore, less attention is paid on the search performance of models from an engineering perspective. In contrast, the purpose of this study is to replicate the complete chemotaxis behavior of *C. elegans*, including parallel chemotaxis strategies and body movement, in the context of one sensor, and to provide an easy-to-implement and good-performance model for worm-like robot navigation control. Table 2 lists the capabilities and properties of related navigation models in the literature and those of the proposed model, which are summarized as follows:

(1) Previous models typically adopted one strategy to perform navigation tasks. In contrast, our model combines two strategies to improve the search performance;
(2) Our model can realize the klinotaxis behavior with a single sensor by incorporating the state-dependent gating mechanism, whereas previous models that mimic klinotaxis typically require two sensors to obtain the required spatial gradient;
(3) Our model can realize body undulatory locomotion during steering by incorporating a proprioceptive mechanism, similar to the model in [29]. However, the structure of our model is simpler;
(4) Our model exhibits adaptive sensitivity to the concentration gradient to cope with scenarios with various gradient ranges, a function which is absent in the previous models.

Table 2. Comparison of capabilities and properties of models in the literature and in our study (✔ represents the presence of this capability or property in the model).

Articles		Capabilities			Properties	
		Klinokinesis	Klinotaxis	Body Undulatory Locomotion	Single Sensor	Adaptive Sensitivity to Gradients
Xu et al. [23,24]	Model 1	✔			✔	
	Model 2		✔			
Santurkar et al. [25]		✔			✔	
Shukla et al. [26]		✔			✔	
Kishore et al. [27]		✔			✔	
Costalago-Meruelo et al. [32]			✔	✔		
Deng et al. [28]		✔		✔	✔	
Deng et al. [29]		✔		✔	✔	
Our study		✔	✔	✔	✔	✔

Additionally, in simulations, the proposed model exhibited realistic undulatory locomotion and a stable search for the shortest-path concentration peak over a wide range of gradients. Moreover, the simulation results have demonstrated that the search performance of our model combining two strategies is significantly better than that of models using a single strategy.

For further comparison, we conducted a comparative experiment between our model and the model in [29]; this model also uses a single sensor and incorporates undulatory locomotion of the body, and the implementation approach of klinokinesis in our model is similar to that of this model. We reproduced the model in [29] using the original logic function and parameters. We tested our model against this model according to the scenario used in [29]; the peak concentration and diffusion range in this scenario are very small. Figure 15a,b show the search trajectories of our model and the model from [29], respectively. The trajectory patterns were consistent with those shown in Figure 11a,b, respectively. Multiple experiments were conducted with four different initial positions and 10 different initial orientations; the average *SSR*s were 1.961 for the reproduced model and 1.486 for our model. The ratio of average *SSR* of the reproduced model to that of our model was 1.32, which was close to the results in Section 3.4; as shown in Table 1, the ratio of average *SSR* of the klinokinesis-only model to that of our model was 1.36. In addition, the reproduced model needed to normalize the concentration gradient in the scenario in advance, while, in practice, it is difficult to obtain *a priori* knowledge of the gradient range of an unknown scenario.

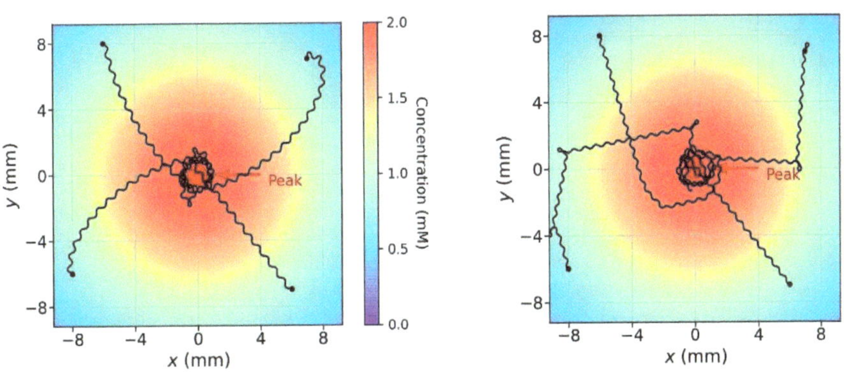

(**a**) (**b**)

Figure 15. Search trajectories of our model (a); and the reproduced model (b) in the scenario used in [29].

4. Conclusions

To incorporate new biological methods into mobile-robot control, a neural network-based autonomous search model with undulatory locomotion inspired by *C. elegans* was proposed in this study. The developed model is the first to simultaneously mimic the body locomotion and the klinokinesis and klinotaxis strategies of *C. elegans* with a single sensor, to search for environmental variable peaks in directions close to the steepest gradients. Multiple biological outcomes are incorporated into the model so that the simple structure is sufficient to achieve complex *C. elegans*-like behavior. Specifically, the CPG and proprioceptive mechanism of *C. elegans* are incorporated in the model to achieve undulatory locomotion, as well as the electrophysiological characteristics of salt-sensory neurons and the state-dependent gating mechanism; the latter is included to achieve klinotaxis behavior in the case of a single sensor. In addition, klinokinesis behavior is realized by fitting a logic function. In this study, the effectiveness and realness of the proposed model were demonstrated through simulation experiments. The model exhibited stable search performance across a wide range of gradients and outperformed models using a single strategy, while exhibiting realistic body undulation.

In summary, the developed model constitutes a simple bio-inspired network control prototype for worm-like navigation robots. In future work, extending the model by including more navigation strategies will contribute to addressing possible problems in complex scenarios, such as lack of gradient or local extremum. Additionally, the developed model will be considered for application in actual robots.

Author Contributions: Conceptualization, M.C. and D.F.; methodology, M.C., D.F. and H.S.; software, M.C.; validation, M.C., M.W. and T.S.; resources, D.F. and H.S.; writing—original draft preparation, M.C.; writing—review and editing, M.C., M.W. and T.S.; supervision, D.F. and H.S. All authors have read and agreed to the published version of the manuscript.

Funding: This research was funded by the National Natural Science Foundation of China [grant number 61971470].

Institutional Review Board Statement: Not applicable.

Informed Consent Statement: Not applicable.

Data Availability Statement: Not applicable.

Conflicts of Interest: The authors declare no conflict of interest. The funders had no role in the design of the study; in the collection, analyses, or interpretation of data; in the writing of the manuscript; or in the decision to publish the results.

References

1. White, J.G.; Southgate, E.; Tomson, J.N.; Brenner, S. The structure of the nervous system of the nematode *Caenorhabditis elegans*. *Philos. Trans. R. Soc. Lond. B Biol. Sci.* **1986**, *314*, 1–340. [PubMed]
2. Cook, S.J.; Jarrell, T.A.; Brittin, C.A.; Wang, Y.; Bloniarz, A.E.; Yakovlev, M.A.; Nguyen, K.; Tang, L.T.; Bayer, E.A.; Duerr, J.S. Whole-animal connectomes of both *Caenorhabditis elegans* sexes. *Nature* **2019**, *571*, 63–71. [CrossRef] [PubMed]
3. Iino, Y.; Yoshida, K. Parallel use of two behavioral mechanisms for chemotaxis in *Caenorhabditis elegans*. *J. Neurosci.* **2009**, *29*, 5370–5380. [CrossRef] [PubMed]
4. Pierce-Shimomura, J.T.; Morse, T.M.; Lockery, S.R. The fundamental role of pirouettes in *Caenorhabditis elegans* chemotaxis. *J. Neurosci.* **1999**, *19*, 9557–9569. [CrossRef] [PubMed]
5. Macedo, J.A.; Marques, L.; Costa, E. A comparative study of bio-inspired odour source localisation strategies from the state-actionperspective. *Sensors* **2019**, *19*, 2231. [CrossRef] [PubMed]
6. Li, J.-G.; Cao, M.-L.; Meng, Q.-H. Chemical source searching by controlling a wheeled mobile robot to follow an online planned route in outdoor field environments. *Sensors* **2019**, *19*, 426. [CrossRef]
7. Huo, J.; Liu, M.; Neusypin, K.A.; Liu, H.; Guo, M.; Xiao, Y. Autonomous search of radioactive sources through mobile robots. *Sensors* **2020**, *20*, 3461. [CrossRef]
8. Bayat, B.; Crasta, N.; Crespi, A.; Pascoal, A.M.; Ijspeert, A. Environmental monitoring using autonomous vehicles: A survey of recent searching techniques. *Curr. Opin. Biotech.* **2017**, *45*, 76–84. [CrossRef]
9. Webster, R.J.; Romano, J.M.; Cowan, N.J. Mechanics of precurved-tube continuum robots. *IEEE Trans. Robot.* **2009**, *25*, 67–78. [CrossRef]

10. Iguchi, Y.; Nakajima, M.; Ariizumi, R.; Tanaka, M. Step climbing control of snake robot with prismatic joints. *Sensors* **2022**, *22*, 4920. [CrossRef]
11. Zhao, X.; Dou, L.; Su, Z.; Liu, N. Study of the navigation method for a snake robot based on the kinematics model with MEMS IMU. *Sensors* **2018**, *18*, 879. [CrossRef] [PubMed]
12. Gray, J.M.; Hill, J.J.; Bargmann, C.I. Inaugural Article: A circuit for navigation in *Caenorhabditis elegans*. *Proc. Natl. Acad. Sci. USA* **2005**, *102*, 3184–3191. [CrossRef] [PubMed]
13. Chalfie, M.; Sulston, J.E.; White, J.G.; Southgate, E.; Thomson, J.N.; Brenner, S. The neural circuit for touch sensitivity in *Caenorhabditis elegans*. *J. Neurosci.* **1985**, *5*, 956–964. [CrossRef] [PubMed]
14. Bargmann, C.I.; Horvitz, H.R. Chemosensory neurons with overlapping functions direct chemotaxis to multiple chemicals in *C. elegans*. *Neuron* **1991**, *7*, 729–742. [CrossRef]
15. Wen, Q.; Po, M.D.; Hulme, E.; Chen, S.; Liu, X.; Kwok, S.W.; Gershow, M.; Leifer, A.M.; Butler, V.; Fang-Yen, C. Proprioceptive coupling within motor neurons drives *C. elegans* forward locomotion. *Neuron* **2012**, *76*, 750–761. [CrossRef]
16. Fouad, A.D.; Teng, S.; Mark, J.R.; Liu, A.; Alvarez-Illera, P.; Ji, H.; Du, A.; Bhirgoo, P.D.; Cornblath, E.; Guan, S.A. Distributed rhythm generators underlie *Caenorhabditis elegans* forward locomotion. *eLife* **2018**, *7*, e29913. [CrossRef]
17. Chen, M.; Feng, D.; Su, H.; Su, T.; Wang, M. Neural model generating klinotaxis behavior accompanied by a random walk based on *C. elegans* connectome. *Sci. Rep.* **2022**, *12*, 3043. [CrossRef]
18. Yu, Y.V.; Xue, W.; Chen, Y. Multisensory integration in *Caenorhabditis elegans* in comparison to mammals. *Brain Sci.* **2022**, *12*, 1368. [CrossRef]
19. Ferrée, T.C.; Marcotte, B.A.; Lockery, S.R. Neural network models of chemotaxis in the nematode *Caenorhabditis elegans*. *Adv. Neural Inf. Process. Syst.* **1996**, *9*, 55–61.
20. Ferrée, T.C.; Lockery, S.R. Computational rules for chemotaxis in the nematode *C. elegans*. *J. Comput. Neurosci.* **1999**, *6*, 263–277. [CrossRef]
21. Dunn, N.A.; Lockery, S.R.; Pierce-Shimomura, J.T.; Conery, J.S. A neural network model of chemotaxis predicts functions of synaptic connections in the nematode *Caenorhabditis elegans*. *J. Comput. Neurosci.* **2004**, *17*, 137–147. [CrossRef] [PubMed]
22. Morse, T.M.; Ferree, T.C.; Lockery, S.R. Robust spatial navigation in a robot inspired by chemotaxis in *Caenorhabditis elegans*. *Adapt Behav.* **1998**, *6*, 393–410. [CrossRef]
23. Xu, J.X.; Deng, X.; Ji, D. Biological neural network based chemotaxis behaviors modeling of *C. elegans*. In Proceedings of the International Joint Conference on Neural Networks, Barcelona, Spain, 18–23 July 2010; pp. 1–6.
24. Xu, J.X.; Deng, X. Biological modeling of complex chemotaxis behaviors for *C. elegans* under speed regulation—A dynamic neural networks approach. *J. Comput. Neurosci.* **2013**, *35*, 19–37. [CrossRef] [PubMed]
25. Santurkar, S.; Rajendran, B. *C. elegans* chemotaxis inspired neuromorphic circuit for contour tracking and obstacle avoidance. In Proceedings of the IEEE International Joint Conference on Neural Networks, Killarney, Ireland, 12–17 July 2015; pp. 1–8.
26. Shukla, S.; Dutta, S.; Ganguly, U. Design of spiking rate coded logic gates for *C. elegans* inspired contour tracking. In Proceedings of the 27th International Conference on Artificial Neural Networks, Rhodes, Greece, 4–7 October 2018; pp. 273–283.
27. Kishore, A., Saraswat, V.; Ganguly, U. Simplified klinokinesis using spiking neural networks for resource-constrained navigation on the neuromorphic processor Loihi. In Proceedings of the International Joint Conference on Neural Networks, Shenzhen, China, 18–22 July 2021; pp. 1–8.
28. Deng, X.; Xu, J.X. A 3D undulatory locomotion model inspired by *C. elegans* through DNN approach. *Neurocomputing* **2014**, *131*, 248–264. [CrossRef]
29. Deng, X.; Xu, J.X.; Wang, J.; Wang, G.; Chen, Q. Biological modeling the undulatory locomotion of *C. elegans* using dynamic neural network approach. *Neurocomputing* **2016**, *186*, 207–217. [CrossRef]
30. Krieg, M.; Pidde, A.; Das, R. Mechanosensitive body-brain interactions in *Caenorhabditis elegans*. *Curr. Opin. Neurobiol.* **2022**, *75*, 102574. [CrossRef]
31. Demin, A.V.; Vityaev, E.E. Learning in a virtual model of the *C. elegans* nematode for locomotion and chemotaxis. *Biol. Inspired Cogn. Archit.* **2014**, *7*, 9–14. [CrossRef]
32. Costalago-Meruelo, A.; Machado, P.; Appiah, K.; Mujika, A.; Leskovsky, P.; Alvarez, R.; Epelde, G.; McGinnity, T.M. Emulation of chemical stimulus triggered head movement in the *C. elegans* nematode. *Neurocomputing* **2018**, *290*, 60–73. [CrossRef]
33. Suzuki, H.; Thiele, T.R.; Faumont, S.; Ezcurra, M.; Lockery, S.R.; Schafer, W.R. Functional asymmetry in *Caenorhabditis elegans* taste neurons and its computational role in chemotaxis. *Nature* **2008**, *454*, 114–117. [CrossRef]
34. Izquierdo, E.J.; Lockery, S.R. Evolution and analysis of minimal neural circuits for klinotaxis in *Caenorhabditis elegans*. *J. Neurosci.* **2010**, *30*, 12908–12917. [CrossRef]
35. Izquierdo, E.J.; Beer, R.D. Connecting a connectome to behavior: An ensemble of neuroanatomical models of *C. elegans* klinotaxis. *PLoS Comput. Biol.* **2013**, *9*, e1002890. [CrossRef] [PubMed]
36. Niebur, E.; Erdös, P. Theory of the locomotion of Nematodes: Dynamics of undulatory progression on a surface. *Biophys. J.* **1991**, *60*, 1132–1146. [CrossRef]
37. Waterston, R. *The Nematode C. elegans*; Wood, W., Ed.; Cold Spring Harbor Laboratory Press: New York, NY, USA, 1988; pp. 281–335.
38. Erdös, P.; Niebur, E. The neural basis of the locomotion of nematodes. *Lect. Notes Phys.* **1990**, *368*, 253–267.

39. Xu, T.; Huo, J.; Shao, S.; Po, M.; Kawano, T.; Lu, Y.; Wu, M.; Zhen, M.; Wen, Q. Descending pathway facilitates undulatory wave propagation in *Caenorhabditis elegans* through gap junctions. *Proc. Natl. Acad. Sci. USA* **2018**, *115*, E4493–E4502.
40. Mikolajczyk, T.; Mikołajewska, E.; Al-Shuka, H.F.N.; Malinowski, T.; Kłodowski, A.; Pimenov, D.Y.; Paczkowski, T.; Hu, F.; Giasin, K.; Mikołajewski, D.; et al. Recent advances in bipedal walking robots: Review of gait, drive, sensors and control systems. *Sensors* **2022**, *22*, 4440. [CrossRef]
41. Han, Q.; Cao, F.; Yi, P.; Li, T. Motion control of a gecko-like robot based on a central pattern generator. *Sensors* **2021**, *21*, 6045. [CrossRef]
42. Back, T. *Evolutionary Algorithm in Theory and Practice*; Oxford University Press: New York, NY, USA, 1996; pp. 106–130.
43. Olivares, E.; Izquierdo, E.J.; Beer, R. A neuromechanical model of multiple network rhythmic pattern generators for forward locomotion in *C. elegans*. *Front. Comput. Neurosci.* **2021**, *15*, 572339. [CrossRef]
44. Maruyama, I.N. Receptor Guanylyl Cyclases in Sensory Processing. *Front. Endocrinol.* **2017**, *7*, 173.

sensors

MDPI

Review

Virtual Obstacles for Sensors Incapacitation in Robot Navigation: A Systematic Review of 2D Path Planning

Thabang Ngwenya, Michael Ayomoh * and Sarma Yadavalli

Department of Industrial and Systems Engineering, University of Pretoria, Hatfield, Pretoria 0028, South Africa
* Correspondence: michael.ayomoh@up.ac.za; Tel.: +27-12-420-2832; Fax: +27-12-362-5103

Abstract: The field of mobile robot (MR) navigation with obstacle avoidance has largely focused on real, physical obstacles as the sole external causative agent for navigation impediment. This paper has explored the possible option of virtual obstacles (VOs) dominance in robot navigation impediment in certain navigation environments as a MR move from one point in the workspace to a desired target point. The systematically explored literature presented reviews mostly between the years 2000 and 2021; however, some outlier reviews from earlier years were also covered. An exploratory review approach was deployed to itemise and discuss different navigation environments and how VOs can impact the efficacy of both algorithms and sensors on a robotic vehicle. The associated limitations and the specific problem types addressed in the different literature sources were highlighted including whether or not a VO was considered in the path planning simulation or experiment. The discussion and conclusive sections further recommended some solutions as a measure towards addressing sensor performance incapacitation in a robot vehicle navigation problem.

Keywords: mobile robot navigation; virtual obstacles; sensor incapacitation; environmental conditions

Citation: Ngwenya, T.; Ayomoh, M.; Yadavalli, S. Virtual Obstacles for Sensors Incapacitation in Robot Navigation: A Systematic Review of 2D Path Planning. *Sensors* **2022**, *22*, 6943. https://doi.org/10.3390/s22186943

Academic Editors: Stephen Monk and David Cheneler

Received: 31 July 2022
Accepted: 6 September 2022
Published: 14 September 2022

1. Introduction

Over the years, mobile robots (MRs) have been deployed to smartly assist humans in routines requiring navigation intelligence in the work environment [1,2]. This has partly been facilitated through the use of sensors. Sensor technology types coupled with guidance, control, and navigation decision-making algorithms are primarily responsible for the intelligence in MR path planning. The appearance of intelligence associated with MRs is capable of failing when they are exposed to certain environmental conditions capable of causing sensor malfunction. A few of these malfunctions can be seen in extreme weather conditions such as overheating temperatures with extreme heat emissions (e.g., emissions from groundwater in mines and gas leaks underground in the case of intelligent underground mine rovers) [3] and in clustered domains such as collapsed buildings, cave-ins, or fire outbreak in buildings in the case of search and rescue robots, among other scenarios. According to [3], excessive heat, wind, and obstacles can hinder the functional ability of the sensors hence, causing MRs to react abruptly. The literature has also confirmed that an inertial measurement sensor is prone to failure when there is an electromagnetic interference with its signal emissions [4]. The malfunctioning of a sensing device often results in the display of false data, hence impacting the overall accuracy and behavioural intelligence of a MR [5]. Another shortcoming associated with sensors in MRs is the little scope of significant distance estimation and blind areas [6] which can be orchestrated in the environmental domain. In the earlier review exercises [7–10] on 2D robot navigation (RN), efficacy was mostly measured and assessed based on algorithmic strength.

As a result, the current review is focused on investigating RN incapacitation based on environmental conditions that can impede the performance of sensors. The review is anchored on the fact that sensory incapacitation, as with algorithmic ineffectiveness, can hinder a robot from successfully navigating to a desired target point (TP). Sensory incapacitators in the context of this review are not visible, physical objects, but rather invisible or unseen, virtual phenomena as earlier discussed. Potential sensory incapacitation environments for robotic vehicles are often facilitated by magnetic fields, electric fields, clustered and dark environments, environments infiltrated with nuclear radiations and harmful gases, among others. Sensors that are negatively impacted in these environments include but are not limited to LiDAR, ultrasonic, radar, GPS, and infrared sensors [11]. In underground mining for instance, poor conditions such as dangle nano-size dust particles and unclear lighting can significantly limit the performance of a vision sensor [12]. Sensory incapacitation herein is not about mechanical or electrical fault on a mounted sensor, but rather the impact of invisible, unseen, external environmental influences. These invisible sensory incapacitating phenomena are referred to as virtual obstacles (VOs) in the context of this research. VOs are neither visible to the human eyes nor the mounted sensory devices on a MR; rather, they remain invisible and can affect the navigation of a robot towards its desired TP by interfering with the transductive effectiveness of the sensor, hence resulting in wrong metric output. There is a gap in exploring challenges associated with functionality of sensors when MRs are deployed in environments containing VOs.

The review exercise herein is aimed at systematically analysing research works in the field of 2D MR navigation with a view towards exploring how much attention was attributed to understanding VOs as possible causes of sensing incapacitation which can result in poor path planning (PP) as much as an ineffective algorithm. The review aims to understand sensors, sensing incapacitation domains, and how these can influence 2D domain navigating robot in navigation environments such as the underground domains for mining activities, cluttered domains, harmful gaseous environments, and others. The rest of the paper is divided into two additional sections, viz., Section 2, which addresses a review of specific research works and their algorithmic and sensing incapacitation considerations for effectiveness in robot PP, and Section 3, which focuses on discussions, findings, recommendations, and future work.

2. Most Commonly Used Navigation Methodologies for MRs

Researchers have applied several methodologies in addressing the 2D RN problem amidst workspace obstacles with considerations given mostly to validating the algorithmic efficacy or inefficacy in the deployed workspaces. In these research works, some papers only discussed the use of an algorithm without a mention of hardware utilisation, especially when the validation process is simulation based. This review has focused on the most commonly used classical and heuristic approaches in 2D robot path planning in both algorithmic control and sensory incapacitation discussions regarding the efficacy or inefficacy of PP in diverse navigation environments. In the reviewed papers [13–93], a noticeable gap can be observed in the literature regarding information and discussions about the likely environments where sensors may fail in the case of experimental validations which can lead to the malfunctioning or inefficacy in the algorithmic outputs based on environmental influences on the mounted sensors. The researchers very often, present discussions on the efficiency of the deployed algorithms without referring to sensory incapacitation even in a possible medium to high-risk experimental environment. Mostly, these algorithms were validated on simulation platforms, with a few validated experimentally using real dynamic obstacles (DOs) and real static obstacles (SOs) environment, with a mention of VOs which can act as sensor incapacitators.

2.1. An Overview of Algorithms and Navigation Approaches

The following Sections 2.1.1 and 2.1.2 have presented a review of various path planning research works in a bid to investigate if any of these algorithms were deployed to address possible VOs as much as the real obstacles in a robot navigation workspace. In addition, a mild classification was carried out to be sure which algorithms were deployed in a strictly simulation environment, an experimental environment, or a hybrid environment. Very often, researchers in this problem domain are mostly concerned about the efficacy of their algorithms in the presence of physical workspace obstacles. Hence, this section is not seeking to compare the efficacy of algorithms or their degree of sensorial independence in a robot navigation mission (i.e., if an algorithm can be used independent of a sensor or not).

2.1.1. Classical Approaches

The following classical techniques have been reviewed in this section: Simultaneous Localisation and Mapping (SLAM), as well as some commonly used algorithmic solutions that can either be fused into the concept of SLAM to facilitate its localization and mapping features or be independently deployed and used directly as standalone control algorithms in the navigation and control of 2D robotic vehicles. These include Light Detection and Ranging (LiDAR), Vector Field Histogram (VFH) and the Artificial Potential Field (APF)/Virtual Force Field (VFF).

Simultaneous Localisation and Mapping (SLAM)

SLAM is one of the most used methodologies that addresses the path planning problem via the construction of a workplace map with no prior knowledge of the environment by a navigating robot. It further localises the MR within the map without any human involvement, as discussed by Taheri and Xia [13]. It was further asserted by Taheri and Xia [13] that using low-level sensors makes utilising SLAM technique difficult. As a result, SLAM is associated with observation errors associated with sensors and caused by the changes in physical factors of the environment. Moreover, other researchers [14–21] conducted research works on SLAM for RN with real obstacles; however, there was no mention of VOs in their research. A need to investigate and explore possible navigation inefficiencies or inaccuracies from the point of view of sensor impairment cannot be overemphasised. In [12], the author implemented SLAM in underground mining and found that the challenges were directly linked to VOs such as dust and illumination challenges for the sensors. However, there is less literature that addresses SLAM efficiency where the different types of sensors are capable of malfunctioning due to VOs orchestrated by environmental influences. Despite SLAM being an effective path planning technique, the thinking of the future in experimental path planning is in both algorithmic and sensory assessment. For instance, will an incorrect localisation of the navigating robot always be linked to algorithmic deficiency? Can there be some temporary or permanent perceptory conditions resulting from the environment hence contributing to the poor signal prompting and incorrect readings? The same thinking applies to mapping. Could it be that some areas within the workspace are not accessible by the sensors due to unseen influences? Inaccuracies resulting from virtual conditions can be a source of navigation blind spots and are seen as posing critical challenges to a navigating robot especially when all attention is on the physical obstacles and visual environment.

Light Detection and Ranging (LiDAR)

LiDAR is an eminent dynamic distance detecting path planning sensor system which is utilised as a range estimation sensor which consistently sends a beam of light and utilises pivoting radiations at a steady rate. It also registers the distance between the object and itself with high precision. LiDAR improves outcomes when combined with different sensors [22]. In [12], the author highlighted that vision sensors are greatly restricted by VOs, which may affect the success of LiDAR beams in the underground mines. Over the years, researchers have explored this technique to make more useful improvements. Ghorpade et al. [23] proposed an efficient OA model using the 2D LiDAR for an MR to accomplish proper constant execution and improve the precision of OA focused on independent mechanical frameworks intended for military applications. However, the paper does not address the limitations that the sensors used have in this environment.

Madhavan and Adharsh [24] used a deliberate methodology to dodge impediments on most minor expense work guidelines, which are limited to simulation environment with static obstacles. Additionally, Baras et al. [25] used LiDAR and Raspberry Pi to address navigation problems while the autonomous vehicle avoided impediments. The results show that the approach can navigate safely in less luminous environments. Future adjustments might anticipate impediments to movement and may more efficiently explore in a dynamic workspace. Similarly, Dong et al. [26] and Ren Yee et al. [27] used real-time experiments in the presence of static and dynamic obstacles. However, it could be assumed that the experiments were conducted in a workspace that is void of VOs capable of resulting in the failure of the mounted sensors, as the algorithm was effective without failing. However, as a remote sensing device that uses laser pulse-like lighting beams for high resolution maps in surveillance, among other uses, deploying the LiDAR system in, for instance, an underground facility for autonomous path planning in a mining environment may result in visibility related challenges. LiDAR sensors facilitate robotic vehicles to visualise the ambient environment by generating and measuring several data points, and then creating a dynamic navigation map of the static or changing environment. The LiDAR sensor has a deficiency in measuring distances through interceptions such as heavy rain, snow, and fog. In addition, the LiDAR sensor measurement capability can be adversely impacted by contamination from sunlight during the day as pointed out in Atmospheric Chemistry and Physics, European Geosciences Union [28]. When LiDAR receives scattered radiations from the Sun, they easily become saturated, following that the solar radiation has so much influence on a diverse set of wavelengths. In general, the performance of LiDAR depreciates as the weather conditions deteriorates. How all of these culminate into a negative impact on a robot and to what degree remains an open investigation to be carried out.

Vector Field Histogram (VFH)

The literature from [29–36] highlight that there is a gap in applying this technique in environments with VOs, because even in the real world it is not considered that experiments involving sensors can malfunction and cause wrong algorithmic outputs. The VFH was pioneered by Borenstein and Koren [29]. The technique was very robust and efficient. Ulrich and Borenstein improved the VFH in 1998 [30] and 2000 [31]. The technique was developed to diminish the restriction of potential-field strategies (i.e., robot motions while dodging the obstacles) [32]. Yim and Park [33] used VFH in RN with SOs. Kumar and Kaleeswari [34] implemented the VFH in a robot with DOs and SOs. Future work will consider the use of a potential field strategy. Alagic et al. [35] proposed a modified VFH technique in an MR framework. Their VFH calculation gave local movement arranging and obstruction evasion dependent on ready sensor estimations. Results demonstrated the VFH calculation's capability to explore RN prior to the TP evading impediments. The disadvantage with this technique is that it gives mediocre results for local PP regarding travel time and distance covered. Diaz and Marin [36] improved on the algorithm proposed by [30].

Artificial Potential Field (APF)/Virtual Force Field (VFF)

In [37–51], the application of the VFF technique even when mixed with other approaches was limited to user-friendly environments. Hence, there is still insufficient literature where the efficiency and effectiveness of this approach is tested in the presence of VOs. The APF PP innovation previously proposed by Khatib is on a fundamental level appropriate for constant control [37]. APF is also known as VFF, which Borenstein and Koren [38] pioneered. The drawback is that it falls into the local minima trap (LMT) and neglects the TP. The essential thought of APF is to make the robot move using forces such that obstacles produce a repulsive force (RF), and TP delivers an attractive force (AF) on a robot. The paper [39] provides crucial functions in understanding APF. Chiang et al. [40] used APF-stochastic reachable strategy for PP in complex workplaces. Extended work by Malone et al. [41] for PP was in a highly intricate and dynamic workplace with impediments. Sudhakara et al. [42] investigated OA and navigation of a wheeled robot using amended APF in unstructured environments. Results showed that the enhanced APF may be adequately used in the direction arranging of wheeled robots and can be applied progressively in real-time situations. The improved APF calculation adjusts well in specific and complex conditions with a short travel time. Lu et al. [43] and Lin et al. [44] algorithm dependent on the improved APF to tackle the issue of local optimum.

A discrete artificial potential field (DAPF) for robot PP introduced by Lazarowska [45] utilises the idea of an APF and alters it for use in a discrete setup space. Results showed that the DAPF calculation is fit for finding a crash-free way for a robot in dynamic and static conditions. The advantage of this is the close ongoing activity, which makes it helpful for practical applications. Moreover, a new pattern in RN research is the use of a hybrid approach (HA) to accomplish better outcomes. Shin and Kim [46] pioneered a HA that combines positioning risk (PR) and the APF. They designed a flowchart that mapped out the methodology premised on the use of a temporary goal (TG). e The algorithm is triggered when the MR does not reach its TP because of LMT caused by obstacles. Results from their paper showed that the proposed PR-APF generated more than 90% success paths while the APF failed to generate up to 50% success paths which constitutes a significant limitation for the unenhanced APF method. Another HA is the hybrid virtual force field (HVFF) approach. This approach integrates the virtual force field (VFF), virtual obstacle concept (VOC), and the virtual goal concept (VGC). The HVFF flowchart as presented in their paper showed a few navigations rules. One of these is such that if a MR is obstructed by either a lengthy or concave shaped obstacle, the VFF, VOC and VGC should be triggered otherwise implement VFF and VGC else, implement the VFF procedure only. Olunloyo and Ayomoh [47] proposed the HVFF approach to take care of PP in both static and dynamic obstacles scenario [48–51]. The methodology endeavours to impersonate human knowledge by recognising a nearby local minimal trap causative obstacle as an entity, while continuing away from the trap towards the target point. Despite advances with this technique, there is still limited investigation on this algorithm in environments that can impact on the functionability of sensors. Moreover, an outlook where magnetic field forces can interfere negatively with this algorithm has not been explored especially in underground mines, as the APF group of methods are directly linked to attraction and repulsion of forces from the workspace objects.

2.1.2. Heuristic Approaches

This section presents a review of the following techniques considering their deployment in a VO navigation environment: Fuzzy Logic (FL), Neural Network (NN), Particle Swarm Optimisation (PSO), Genetic Algorithm (GA), Ant Colony Optimisation (ACO), and Firefly Algorithm (FA). In [52–93], these algorithms were effectively deployed for RN in presence of DOs and SOs; however, they never experimented for environments, as discussed in Section 1.

Fuzzy Logic (FL)

FL was introduced by Zadeh [52] and extensively used in robotics engineering to guide robots. FL control is appropriate for minimal effort robots that do not need highly complex routes and are motivated by human thinking. The behaviour-based FL by Qing-yong et al. [53] includes OA. Jaradat et al. [54] investigated RN in a dynamic environment where they integrated FL with APF. The disadvantage is the LMT, where the robot becomes caught while sitting tight for an obstacle. Pandey et al. [55] developed an FL for taking care of the PP issue in the presence of various states of SOs to discover crash freeway. Outcomes showed that the technique empowers the MR to arrive at the objective without impacting. In the future, an improvement will be by streamlining with the assistance of optimisation algorithms. In [56,57], improvements have been made on FL, but the experiments do not consider environments with VOs, but rather only DOs and SOs. Batti et al. [58] extended the use of FL for OA in labyrinth workspace. Similarly, Mohanty et al. [59] proposed a new model called Takagi-Sugeno FL developed to address PP via an enhanced wall following approach. However, in their paper, the approach was limited to the static obstacles problem. Moreover, in [60] an improvement of FL was proposed by applying it in a complex environment that involves more than two DOs. Moreover, in discussions of the papers reviewed for FL, there are insufficient details on the application of this method in underground mines. The capability of FL in MR is still limited to environments that do not consider toxic environments for sensors (i.e., VOs). An exploration of the effectiveness of this algorithm in different environments remains to be investigated.

Neural Network (NN)

NN is a considerable plan of equivalent spread planning segments related to graph geology. It has filled quickly in recognising objects and obstacle discovery in a picture. Recently, the issue of recognising obstacles in the RN system has been essential [61]. The popular methodology used to solve this problem in the past years has been convolutional neural networks (CNN) [62,63]. Chi and Lee [64] used various principles that were actualised for the control technique to keep away from the obstacle effectively. The proposed framework with the NN control approach has illustrated the adequacy of dodging the obstacles. It needs further exploratory examination in other environments with VOs such as underground mining. Moreover, in [65–68], researchers continued to explore the NN technique with environments consisting of DOs and SOs. Wei and Ye [69] proposed an obstacle avoidance (OA) framework dependent on GA-supported OIF-Elman NN. In their paper, they showed three layers that make up the design of an OIF-Elman network structure. The layers include the input, hidden and output layers. A context layer is also included in the hidden layer. The context layer inside the hidden layer is the primary feedback mechanism of Elman The framework presents a versatile navigation procedure for robots to evade impediments in a workspace. Results presented, showed that the OIF-Elman network is quite successful with OA. This approach was applied in an indoor environment in order to avoid the effect of illumination. Zhang et al. [70] focused on improving NN for RN in complex environments. The simulation results showed considerable efficiency and effectiveness despite the change of different conditions such as weather conditions and road changes. However, the application is not tested in extreme weather conditions where sensors can malfunction.

Particle Swarm Optimisation (PSO)

PSO is broadly utilised in versatile RRs tending to RR planning and confinement issues in the obscure workplace [71]. Examination of different methodologies and results showed that the FL matched with PSO provides the ideal outcomes in separation voyages [72]. Atyabi et al. [73] introduced an extension of the PSO in robotics to improve performance. The research considered the environment with SOs and DOs. Results showed potential under the conditions considered; however, the method cannot work fully in robotics. This is very limiting to further investigation for environments with VOs, as they can negatively impact sensors. Future work will examine the effectiveness of this method under real world applications. Another technique named the PSO-IAC [74] is used to determine the objective of approaching the OA issue for a 6° of freedom controller of the home assistance robot. Simulation outcomes demonstrated that the PSO-IAC calculation gives the quickest combination capacity. The suggested control plan can cause the controller of the home assistance robot to show up at the objective situation with and without impediments. However, home environments do not consist mainly of VOs, as it is an environment that is safe for humans compared to underground mining environment.

Meerza et al. [75] built up a PSO-based robot PP calculation that impacts shirking capacity for SOs and DOs. They will test their calculation in a certifiable workplace in the future. Alaliyat et al. [76] proposed powerful PP calculations dependent on PSO to manage the complex dynamic workplace. Outcomes indicated that without any earlier information about a workplace, the robot could accomplish its objective of evading SOs and DOs. In the future, they plan to acquire a super robot that can learn and retain the circumstances during its navigation. It's not clear if VOs would form part of the future consideration for their proposed real experiment. Tian et al. [77] deployed the use of remote sensing in finding multiple robots and impediments, which utilised an improved counterfeit clever calculation. One limitation to the method is the calculation of the union speed to improve the worldwide pursuit execution and failure to manage the circumstance that numerous robots may collide. In the future, a hypothetical exploration of PSO calculation and obstruction evasion calculation to manage different testing improvement issues will be studied. The noticeable gap in discussion of PSO means that there are insufficient details when it comes to its effectiveness in environments with obstacles that makes sensors malfunction. Underground mining is an environments where this technique still needs to be explored for effectiveness and efficacy.

Genetic Algorithm (GA)

GA is a known technique-based enhancement instrument that follows the guideline of hereditary qualities and joint determination [78]. Application of this technique to software engineering was introduced first in 1975 [79]. The utilisation of GA for the versatile RR issue is significant in the static workplace. Reproduction results introduce the investigation as they were within sight of a polygonal impediment. Xiao et al. [80] embraced the strategy to accomplish the objective of the route. Many scientists have given less attention to the sight of a DO in an uncertain workplace [81]. To improve results in robot PP, numerous scientists have joined in on using GA and other techniques to obtain a HA [7]. Patle et al. [82] state that in the future, the work may stretch to cause the crossbreed regulator for the ongoing open-air workplace usage. Germi et al. [83] tended to alter the first potential field calculation to better the exhibition of the calculation in dynamic conditions. Choueiry et al. [84] introduced a survey of the PP enhancement issue and a calculation for robot PP in a static environment using GA as a device to discover number of steps while staying away from obstacles. The designed flowchart of the proposed approach took into consideration, the workspace grid size, initial and target positions and obstacles distribution all serving as inputs. If the MR does not reach the TP the number of steps in the GA algorithm are increased. Lopez-Gonzalez et al. [85] utilised GA to accomplish distance-based development by using two unique sorts of chromosomes. Aghda and Mirfakhrae [86] consolidated the GA-FL to improve directing. In the quest to improve GA, this approach is incompetent in dynamic environments [50], but produces good outcomes in simulation. However, as this approach has not been tested in environments with VOs, this is open for future investigation.

Ant Colony Optimisation (ACO)

ACO applies in the robot system field, particularly the PP issue [87]. ACO resolves this issue to determine the mechanical flying-vehicle course for a war zone [88]. Zhangqi et al. [89] proposed improvement measures and applied GA to the advancement and arrangement boundaries of the essential ACO. Simulation outcomes showed that the improved ideal path length is essentially not exactly the fundamental ACO. Wang et al. [90] improved APF first, implementing a strategy for a piecewise capacity of fascination potential that suggests that the robot can, without much of a stretch, slam into the obstruction. The limitation of this approach is that the model contains numerous boundaries which makes it difficult to tune. Even the flowchart design depicts this limitation by having a lot of decision blocks. Researchers will discover the relationships between these boundaries in the future. Yi et al. [91] produced dynamic change data as indicated by the contrast between the best way of the past age and the current ACO cycle. Ma et al. [92] addressed the automated submerged vehicle two-dimensional independent PP issue in the climate influenced by sea momentum and obstacles. Results showed that this calculation could rapidly locate the ideal global arrangement where the unpredictable workplace is. However, there is no clear indication on the impact of these conditions on the sensors. Zhao [93] proposed the ideal way of anticipating whether robots are dependent on ACO by contemplating the connected writing and effective methods of robot PP. Results showed that the model could wisely pick a robot with DO evasion efficiently.

3. Discussion

It is clear that there is a noticeable gap in the literature in respect of VOs, as there is insufficient consideration, information, or discussions about such environments where sensors can fail, or inefficacy of an algorithmic output based on environmental influences on mounted sensors. The researchers very often present discussions on the efficiency of the deployed algorithms without referring to sensory incapacitation even in a possible medium- to high-risk experimental environment. Mostly, these algorithms were validated on simulation platforms with a few validated experimentally using real dynamic obstacles (DOs) and/or real static obstacles (SOs) environment.

In as much as some measures appear to be in place regarding the combating of VOs in both open and obscured environments for MRs, for instance LiDAR generally uses various filtering methods to filter dust while real industrial robots generally have redundant sensors to process information to ensure their stable operation under VO conditions etc., this paper recommends a concept premised on holistic path planning. Holistic path planning should integrate VOs thinking as much as real obstacles thinking in robot navigation problems and solution proffering. While purposeful experiments on robot navigation to examine the efficacy of general sensorial incapacitation due to extreme or obscured environmental factors are still lacking in the literature (see introductory sections), future research will present robot vehicle navigation limitations based on sensorial incapacitated experiments. The experiments will utilise the same set of robotic vehicles in two different navigation environments depicting different (i.e., normal and extreme) environments over different trials with the conduct of statistical significance of the difference in navigation output over time. In addition, it is recommended that in a traditional robot navigation task, when obstacle avoidance and goal reachability becomes challenging, a robotist should verify the functionality of the onboard sensors, power unit, and actuators. If all are in a good operational condition, VOs capable of incapacitating the onboard sensor types may likely be in play and should be verified. This troubleshooting recipe can be of a greater assistance in extreme or obscured environments involving robot vehicle navigation.

Furthermore, following that VOs can interfere with or influence both the internal and external workings of nearly all sensor devices through the interception of both receptive and emitted sensorial signals, leading to wrong computation and misleading robot navigation decision, an additional measure of solution to address a possible external influence can revolve around the integration of a machine learning assisted algorithm for sensors response accuracy and interpretation of propagated signals. Based on this proposed solution, each time a sensor emits and receives a feedback signal from the external environment based on the obstacles along its navigation path, the machine learning algorithm should be able to compare the most recent and similar emitted signal from its historic emissions and see if the disparities between the receptive signals for the same or similar emitted signals are significantly different. In the case of a significant difference, the robot can send out a beep sound, which is an indication of a possible external interference to its sensorial computation. However, regarding internal distortions caused by VOs, a sensor-proof capability, which would protect the limitations as explained in previous sections, can go a long way in securing the hardware.

Key Findings from This Research Are as Presented Below

Few papers have addressed the RN problem in the presence of VOs. Virtual obstacles are not visible to both the mounted sensors and the human eyes. However, these can affect the navigation of a robot towards the desired TP by interfering with the operations of the sensors, resulting in wrong sensorial output metrics. Examples of experimentally unverified VOs include magnetic field influence on sensors (infrared sensors), extreme temperature effect on sensors (freezing temperature, boiling temperature), as well as infrared sensors, electric field effect on sensors, and so forth.

Occasionally, the effect of frictionless navigation environment on the navigating wheels of a robot can also impede the display of intelligence in target point attainment. Even though frictionlessness is not a VO in the context of an obstruction, it serves as a virtual impedance to a MR in its bid to accurately reach and stop at a desired target point. Also, in a noisy, clustered environment, the performance of a sonar sensor can be subject to some form of impedance. Furthermore, a vision system-controlled navigation can be influenced by the degree of illumination a robot is exposed to.

Furthermore, additional significant limitations with some of the methodologies presented in the literature is the processing speed and performance in complex nonconventional navigation environments as a result of certain environmental impediments. Future work in this field will present specific sensory experimental quantitative information covering different VO prone environments as earlier presented.

4. Conclusions

This paper has explored the problem of VOs in robot navigation obstruction in certain extreme or obscured navigation environments as a robot travels from one point to another within the workspace. Based on this, the current review has investigated robot navigation incapacitation resulting from environmental conditions that can hamper the performance of a sensor. The review is premised on the fact that sensory incapacitation, just as with algorithmic ineffectiveness, can hinder a robot from successfully navigating to a desired position in a given workspace. Sensory incapacitators in the context of this review are not orchestrated by visible, physical objects, but rather by invisible, virtual phenomena as earlier presented. Furthermore, based on the possible influence of VOs on the navigation intelligence of robots due to sensory incapacitation, the robust and all-encompassing concept of SLAM, as previously discussed in this review, is considered to be more skewed towards algorithm effectiveness in the control of a robot than the tracking of a robot's hardware incapacitation, nevertheless with a generic consideration given to onboard hardware units. It is quite obvious that there are not any categoric considerations for sensors incapacitation based on VOs (see Durrant-Whyte and Bailey [17]; Taheri and Xia [13]). Based on the above, it is suggested that the broad concept of SLAM be extended or modified to address both "algorithm effectiveness and sensors signal" (emission and reception) monitoring and evaluation, especially when a robot is navigating in an obscured environment. This can be achieved by deploying a modified concept of SLAM with the acronym "SLAAAM", representing "Simultaneous Localisation Assessment Adaptation and Mapping". The assessment component in "SLAAAM" which is the first "A", would address the disruption in sensory signal emission and reception and prepare the robot for "adaptation" which is the second "A". The assessment would be carried out by way of a swift analysis and evaluation of receptive signals. The deployment of the assessment process will require an onboard vision sensor with both (obstacle proximity response measurement and imaging capability) and a non-vision sensor with (obstacle proximity response measurement capability).

Usually in an operational environment, a vision sensor will scan the ambient environment to generate images of physical obstacles while also keeping record of the measured obstacle's distance during the simultaneous mapping process. Similarly, the non-vision sensors (e.g., infrared or sonar sensors) would intermittently send out and receive sensory signals for proximity distance measurement from obstacles in the ambient environment. This assessment process is such that when the processed receptive signal by the vision and non-vision sensors are somewhat misaligned, not necessarily with each other but with their default sensing attributes when they sense obstacles (for instance, a vision sensor capturing no obstacle image yet exhibiting some sensory receptive features in response to a non-existent obstacle), may arguably signify the presence of a sensory interceptive medium which is obviously a virtual obstructive medium.

Even though different sensors are expected to react to different VOs based on their operational mode, their respective incapacitative response would remain the same for every VO they are prone to. For instance, a vision sensor will often not be able to produce any captured image when a virtual rather than a physical obstacle is present within its sensing zone. However, the non-vision sensors such as infrared and sonar will have their emissions intercepted and a false receptive response signal propagated. Finally, the adaptation component of "SLAAAM" would prompt the robot to respond to an unusual obstacle scenario as depicted by the assessment process above, hence causing the affected sensing devices to be triggered off intermittently as the robot withdraws from the affected sensory incapacitated mapped region to avoid a partial or absolute damage of the incapacitated sensing device. The sensors are left in the normally "on" status and immediately the robot is out of the mapped incapacitated region.

Table 1 presents a summary of pathplanning methodologies discussed above and their taxonomy covering: Types of obstacles, obstacle geometry, approach used, results, year, TP and number of robot(s) deployed. Additionally, the taxonomy breakdown covers references where the algorithms were tested for effectiveness and deployment ommited the ones used just for the literature.

Table 1. Analysis of various path planning and navigation algorithms amidst obstacles.

Ref No	Techniques	Environment Consists of			Technique Used as		Result		Year	Obstacle(s) Shape		Target Point (TP)		Robot	
		SOs	DOs	VOs	Stand Alone	Hybrid	SR	RTR		Cv	Cx	Single TP	Multi TP	Single Robot	Multi Robot
Classical Approach															
[12]	SLAM	Yes	No	Yes	Yes	No	No	Yes	2019	Yes	Yes	Yes	No	Yes	No
[13]		No	No	No	Yes	No	No	No	2021	No	No	No	No	No	No
[14]		Yes	No	No	Yes	No	No	Yes	2014	Yes	Yes	Yes	No	Yes	No
[15]		No	Yes	No	No	Yes	Yes	No	2014	No	Yes	Yes	No	Yes	No
[16]		Yes	Yes	No	Yes	No	No	Yes	2016	Yes	Yes	Yes	No	Yes	No
[17]		Yes	No	No	Yes	No	Yes	Yes	2006	Yes	Yes	Yes	No	Yes	No
[18]		Yes	Yes	No	No	Yes	No	Yes	2018	Yes	Yes	Yes	No	Yes	No
[19]		Yes	Yes	No	No	Yes	Yes	No	2021	Yes	Yes	Yes	No	Yes	No
[20]		Yes	No	No	Yes	No	Yes	Yes	2021	No	Yes	Yes	No	Yes	No
[21]		Yes	No	No	Yes	No	Yes	No	2021	No	Yes	No	No	Yes	No
[23]	Light Detection and Ranging (LiDAR)	Yes	No	No	Yes	No	Yes	No	2017	No	Yes	Yes	No	Yes	No
[24]		Yes	No	No	Yes	No	Yes	No	2019	No	Yes	Yes	No	Yes	No
[25]		Yes	No	No	Yes	No	No	Yes	2019	Yes	Yes	Yes	No	Yes	No
[26]		Yes	Yes	No	Yes	No	Yes	Yes	2020	Yes	Yes	Yes	No	Yes	No
[27]		Yes	Yes	No	Yes	Yes	No	Yes	2020	Yes	Yes	Yes	No	Yes	No
[29]	Vector Field Histogram (VFH)	Yes	No	No	Yes	No	No	Yes	1991	No	Yes	Yes	No	Yes	No
[30]		Yes	No	No	Yes	No	No	Yes	1998	No	Yes	Yes	No	Yes	No
[31]		Yes	No	No	Yes	No	Yes	Yes	2000	No	Yes	Yes	No	Yes	No
[32]		No	Yes	No	Yes	No	No	Yes	2012	No	Yes	Yes	No	Yes	No
[33]		Yes	No	No	Yes	No	Yes	No	2014	No	Yes	Yes	No	Yes	No
[34]		Yes	Yes	No	Yes	No	No	Yes	2016	Yes	Yes	Yes	No	Yes	No
[35]		Yes	Yes	No	Yes	No	Yes	No	2019	No	Yes	Yes	No	Yes	No
[36]		Yes	Yes	No	Yes	No	No	Yes	2020	No	Yes	Yes	No	No	Yes
[37]	Artificial Potential Field (APF)/ Virtual Force Field (VFF)	Yes	No	No	Yes	No	Yes	No	2014	No	Yes	Yes	No	Yes	No
[38]		Yes	No	No	Yes	No	No	Yes	1989	Yes	No	Yes	No	Yes	No
[39]		Yes	No	No	Yes	No	No	Yes	1985	Yes	Yes	Yes	No	Yes	No
[40]		Yes	Yes	No	No	Yes	Yes	No	2015	Yes	Yes	Yes	No	Yes	No
[41]		No	Yes	No	No	Yes	Yes	No	2017	Yes	Yes	Yes	No	Yes	No
[42]		Yes	No	No	Yes	No	Yes	No	2018	Yes	No	Yes	No	Yes	No
[43]		Yes	No	No	Yes	No	Yes	No	2020	No	Yes	Yes	No	Yes	No
[44]		Yes	No	No	Yes	No	Yes	No	2020	Yes	Yes	Yes	No	Yes	No
[45]		Yes	Yes	No	Yes	No	Yes	Yes	2019	No	Yes	Yes	No	Yes	No
[46]		Yes	No	No	No	Yes	Yes	Yes	2021	Yes	Yes	Yes	No	Yes	No

Table 1. *Cont.*

Ref No	Techniques	Environment Consists of			Technique Used as		Result		Year	Obstacle(s) Shape		Target Point (TP)		Robot	
		SOs	DOs	VOs	Stand Alone	Hybrid	SR	RTR		Cv	Cx	Single TP	Multi TP	Single Robot	Multi Robot
[49]		Yes	No	Yes	No	Yes	Yes	No	2009	Yes	Yes	Yes	No	Yes	No
[48]		No	Yes	Yes	No	Yes	Yes	No	2009	Yes	Yes	Yes	No	Yes	No
[49]		No	Yes	No	No	Yes	Yes	No	2010	Yes	Yes	Yes	No	Yes	No
[50]		Yes	No	No	No	Yes	Yes	No	2011	Yes	Yes	Yes	No	Yes	No
[51]		Yes	No	No	No	Yes	Yes	No	2021	Yes	Yes	Yes	Yes	Yes	No
Heuristic Approach															
[53]	Fuzzy Logic (FL)	Yes	No	No	Yes	No	Yes	No	2009	Yes	Yes	Yes	No	Yes	No
[54]		Yes	Yes	No	No	Yes	Yes	No	2012	No	Yes	Yes	No	Yes	No
[55]		Yes	Yes	No	Yes	No	Yes	No	2014	No	Yes	Yes	No	Yes	No
[56]		Yes	Yes	No	Yes	No	Yes	Yes	2016	Yes	Yes	Yes	No	No	Yes
[57]		Yes	No	No	Yes	No	Yes	Yes	2018	No	Yes	Yes	No	Yes	No
[58]		Yes	No	No	Yes	No	Yes	No	2019	Yes	Yes	Yes	No	Yes	No
[59]		Yes	No	No	Yes	No	Yes	Yes	2020	Yes	Yes	Yes	No	Yes	No
[60]		Yes	Yes	No	Yes	No	Yes	No	2021	No	Yes	Yes	Yes	Yes	Yes
[64]	Neural Network (NN)	Yes	No	No	Yes	No	No	Yes	2011	No	Yes	Yes	No	Yes	No
[65]		Yes	No	No	Yes	No	Yes	No	2014	No	Yes	Yes	No	Yes	No
[66]		Yes	No	No	Yes	No	Yes	No	2004	Yes	Yes	Yes	No	Yes	No
[67]		Yes	Yes	No	Yes	No	Yes	No	2019	No	Yes	Yes	No	No	Yes
[68]		Yes	No	No	Yes	No	No	Yes	2020	Yes	Yes	Yes	No	Yes	No
[69]		Yes	No	No	No	Yes	Yes	No	2020	Yes	Yes	Yes	No	Yes	No
[70]		Yes	Yes	No	Yes	No	Yes	No	2020	Yes	Yes	Yes	No	Yes	No
[73]	Particle Swarm Optimisation (PSO)	Yes	Yes	No	Yes	No	Yes	No	2010	No	Yes	Yes	No	Yes	No
[74]		Yes	Yes	No	Yes	No	Yes	No	2016	No	Yes	Yes	No	Yes	No
[75]		Yes	Yes	No	Yes	No	Yes	No	2018	Yes	Yes	Yes	No	Yes	No
[76]		Yes	Yes	No	Yes	No	Yes	No	2019	Yes	Yes	Yes	No	Yes	No
[77]		Yes	Yes	No	Yes	No	Yes	No	2021	Yes	Yes	No	Yes	No	Yes
[83]	Genetic Algorithm (GA)	Yes	Yes	No	Yes	No	Yes	Yes	2018	No	Yes	Yes	No	Yes	No
[84]		Yes	No	No	Yes	No	Yes	No	2019	Yes	Yes	Yes	No	Yes	No
[85]		No	Yes	No	Yes	No	Yes	Yes	2020	No	Yes	No	Yes	No	Yes
[86]		Yes	Yes	No	No	Yes	Yes	No	2020	No	Yes	Yes	No	Yes	No
[89]	Ant Colony Optimisation (ACO)	Yes	No	No	Yes	No	Yes	No	2011	Yes	Yes	Yes	No	Yes	No
[90]		Yes	No	No	No	Yes	Yes	No	2018	Yes	Yes	Yes	No	Yes	No
[91]		Yes	No	No	Yes	No	Yes	No	2019	Yes	Yes	Yes	No	Yes	No
[92]		Yes	Yes	No	No	Yes	Yes	No	2020	No	Yes	Yes	No	Yes	No
[93]		Yes	Yes	No	Yes	No	Yes	No	2020	Yes	Yes	Yes	No	Yes	No

Cv = Concave, Cx = Convex, SR = Simulation Result, RTR = Real Time Result.

Author Contributions: Writing—original draft, T.N.; Conceptualization, M.A.; Supervision, M.A.; Writing—review & editing, M.A. and S.Y. All authors have read and agreed to the published version of the manuscript.

Funding: Funding support for this research was received from the Department of Industrial and Systems Engineering, University of Pretoria.

Institutional Review Board Statement: Not applicable.

Informed Consent Statement: Not applicable.

Data Availability Statement: Not applicable.

Conflicts of Interest: The authors have no conflict of interest.

References

1. Li, H.; Savkin, A.V. An algorithm for safe navigation of mobile robots by a sensor network in dynamic cluttered industrial environments. *Robot. Comput. Integr. Manuf.* **2018**, *54*, 65–82. [CrossRef]
2. Abiyev, R.; Ibrahim, D.; Erin, B. Navigation of mobile robots in the presence of obstacles. *Adv. Eng. Softw.* **2010**, *41*, 1179–1186. [CrossRef]
3. Fraiwan, M.; Alsaleem, A.; Abandeh, H.; Aljarrah, O. Obstacle avoidance and navigation in robotic systems: A land and aerial robots study. In Proceedings of the 2014 5th International Conference on Information and Communication Systems (ICICS), Irbid, Jordan, 1–3 April 2014; pp. 1–5.
4. Hur, H.; Ahn, H.-S. Unknown Input H\bm∞ Observer-Based Localization of a Mobile Robot with Sensor Failure. *IEEE ASME Trans. Mechatron.* **2014**, *19*, 1830–1838. [CrossRef]
5. Fourlas, G.K.; Karras, G.C.; Kyriakopoulos, K.J. Sensors fault diagnosis in autonomous mobile robots using observer-based technique. In Proceedings of the 2015 International Conference on Control, Automation and Robotics, Singapore, 20–22 May 2015; pp. 49–54.
6. Wang, R.; Chen, L.; Wang, J.; Zhang, P.; Tan, Q.; Pan, D. Research on autonomous navigation of mobile robot based on multi ultrasonic sensor fusion. In Proceedings of the 2018 IEEE 4th Information Technology and Mechatronics Engineering Conference (ITOEC), Chongqing, China, 14–16 December 2018; pp. 720–725.
7. Patle, B.; Pandey, A.; Parhi, D.; Jagadeesh, A. A review: On path planning strategies for navigation of mobile robot. *Def. Technol.* **2019**, *15*, 582–606. [CrossRef]
8. Pandey, A.; Pandey, S.; Parhi, D.R. Mobile robot navigation and obstacle avoidance techniques: A review. *Int. Rob. Auto J.* **2017**, *2*, 96–105. [CrossRef]
9. Hoy, M.; Matveev, A.; Savkin, A. *Algorithms for Collision-Free Navigation of Mobile Robots in Complex Cluttered Environments: A Survey*; Cambridge University Press: Cambridge, UK, 2014.
10. Niloy, A.K.; Shama, A.; Chakrabortty, R.K.; Ryan, M.J.; Badal, F.R.; Tasneem, Z.; Ahamed, H.; Moyeen, S.I.; Das, S.K.; Ali, F.; et al. Critical Design and Control Issues of Indoor Autonomous Mobile Robots: A Review. *IEEE Access* **2021**, *9*, 35338–35370. [CrossRef]
11. Guo, P.; Kim, H.; Virani, N.; Xu, J.; Zhu, M.; Liu, P. RoboADS: Anomaly Detection Against Sensor and Actuator Misbehaviors in Mobile Robots. In Proceedings of the 2018 48th Annual IEEE/IFIP International Conference on Dependable Systems and Networks (DSN), Louxembourg, 25–28 June 2018; pp. 574–585. [CrossRef]
12. Androulakis, V.; Sottile, J.; Schafrik, S.; Agioutantis, Z. Elements of Autonomous Shuttle Car Operation in Underground Coal Mines. In Proceedings of the 2019 IEEE Industry Applications Society Annual Meeting, Baltimore, MD, USA, 29 September–3 October 2019; pp. 1–7. [CrossRef]
13. Taheri, H.; Xia, Z.C. SLAM; Definition and evolution. *Eng. Appl. Artif. Intell.* **2021**, *97*, 104032. [CrossRef]
14. Moreno-Armendariz, M.A.; Calvo, H. Visual SLAM and Obstacle Avoidance in Real Time for Mobile Robots Navigation. In Proceedings of the International Conference on Mechatronics, Electronics and Automotive Engineering, Cuernavaca, Mexico, 18–21 November 2014; pp. 44–49.
15. Iizuka, S.; Nakamura, T.; Suzuki, S. Robot navigation in dynamic environment using Navigation function APF with SLAM. In Proceedings of the 10th France-Japan/8th Europe-Asia Congress on Mecatronics (MECATRONICS2014- Tokyo), Tokyo, Japan, 27–29 November 2014; pp. 89–92.
16. Sqalli, M.; Tatsuno, K.; Kurabe, K.; Ando, H.; Obitsu, H.; Itakura, R.; Aoto, T.; Yoshino, K. Improvement of a tele-presence robot autonomous navigation Using SLAM algorithm. In Proceedings of the International Symposium on Micro-Nano Mechatronics and Human Science (MHS), Nagoya, Japan, 28–30 November 2016; pp. 1–7.
17. Durrant-Whyte, H.; Bailey, T. Simultaneous localization and mapping: Part I. *IEEE Robot. Autom. Mag.* **2006**, *13*, 99–110. [CrossRef]
18. Kim, P.; Chen, J.; Cho, Y.K. SLAM-driven robotic mapping and registration of 3D point clouds. *Autom. Constr.* **2018**, *89*, 38–48. [CrossRef]
19. Liu, Z. Implementation of SLAM and path planning for mobile robots under ROS framework. In Proceedings of the 6th International Conference on Intelligent Computing and Signal Processing (ICSP), Xi'an, China, 9–11 April 2021; pp. 1096–1100.
20. Gobhinath, S.; Anandapoorani, K.; Anitha, K.; Sri, D.D.; DivyaDharshini, R. Simultaneous Localization and Mapping [SLAM] of Robotic Operating System for Mobile Robots. In Proceedings of the 2021 7th International Conference on Advanced Computing and Communication Systems (ICACCS), Coimbatore, India, 19–20 March 2021; pp. 577–580. [CrossRef]
21. An, J.; Mou, H.; Lu, R.; Zhou, L. A Study on SLAM Based on Probabilistic Motion Model of Mobile Robot. In Proceedings of the 2021 5th International Conference on Robotics and Automation Sciences (ICRAS), Wuhan, China, 11–13 June 2021; pp. 56–59. [CrossRef]
22. Gul, F.; Rahiman, W.; Alhady, S.S.N.; Chen, K. A comprehensive study for robot navigation techniques. *Cogent Eng.* **2019**, *6*, 1632046. [CrossRef]
23. Ghorpade, D.; Thakare, A.D.; Doiphode, S. Obstacle Detection and Avoidance Algorithm for Autonomous Mobile Robot using 2D LiDAR. In Proceedings of the International Conference on Computing, Communication, Control and Automation (ICCUBEA), Pune, India, 17–18 August 2017; pp. 1–6.
24. Madhavan, T.R.; Adharsh, M. Obstacle Detection and Obstacle Avoidance Algorithm based on 2-D RPLiDAR. In Proceedings of the International Conference on Computer Communication and Informatics (ICCCI), Coimbatore, India, 23–25 January 2019; pp. 1–4.

25. Baras, N.; Nantzios, G.; Ziouzios, D.; Dasygenis, M. Autonomous Obstacle Avoidance Vehicle Using LIDAR and an Embedded System. In Proceedings of the 8th International Conference on Modern Circuits and Systems Technologies (MOCAST), Thessaloniki, Greece, 13–15 May 2019; pp. 1–4.
26. Dong, H.; Weng, C.-Y.; Guo, C.; Yu, H.; Chen, I.-M. Real-Time Avoidance Strategy of Dynamic Obstacles via Half Model-Free Detection and Tracking With 2D Lidar for Mobile Robots. *IEEE ASME Trans. Mechatron.* **2020**, *26*, 2215–2225. [CrossRef]
27. Ren Yee, P.D.; Pinrath, N.; Matsuhira, N. Autonomous Mobile Robot Navigation Using 2D LiDAR and Inclined Laser Rangefinder to Avoid a Lower Object. In Proceedings of the 59th Annual Conference of the Society of Instrument and Control Engineers (SICE), Chiang Mai, Thailand, 23–26 September 2020; pp. 1404–1409.
28. Wu, D.L.; Chae, J.H.; Lambert, A.; Zhang, F.F. Characteristics of CALIOP attenuated backscatter noise: Implication for cloud/aerosol detection. *Atmos. Chem. Phys.* **2011**, *11*, 2641–2654. [CrossRef]
29. Borenstein, J.; Koren, Y. The vector field histogram–fast obstacle avoidance for mobile robots. *IEEE Trans. Robot. Autom.* **1991**, *7*, 278–288. [CrossRef]
30. Ulrich, I.; Borenstein, J. VFH+: Reliable obstacle avoidance for fast mobile robots. In Proceedings of the 1998 IEEE International Conference on Robotics and Automation, Leuven, Belgium, 20 May 1998; pp. 1572–1577.
31. Ulrich, I.; Borenstein, J. VFH/sup */: Local obstacle avoidance with look-ahead verification. In Proceedings of the IEEE International Conference on Robotics and Automation, San Francisco, CA, USA, 24–28 April 2000; pp. 2505–2511.
32. Babinec, A.; Dekan, M.; Duchoň, F.; Vitko, A. Modifications of VFH Navigation Methods for Mobile Robots. *Procedia Eng.* **2012**, *48*, 10–14. [CrossRef]
33. Yim, W.J.; Park, J.B. Analysis of mobile robot navigation using vector field histogram according to the number of sectors, the robot speed and the width of the path. In Proceedings of the 14th International Conference on Control, Automation and Systems (ICCAS 2014), Gyeonggi-do, Korea, 22–25 October 2014; pp. 1037–1040.
34. Kumar, J.S.; Kaleeswari, R. Implementation of Vector Field Histogram based obstacle avoidance wheeled robot. In Proceedings of the International Conference on Green Engineering and Technologies (IC-GET), Coimbatore, India, 19 November 2016; pp. 1–6.
35. Alagic, E.; Velagic, J.; Osmanovic, A. Design of Mobile Robot Motion Framework based on Modified Vector Field Histogram. In Proceedings of the International Symposium ELMAR, Zadar, Croatia, 23–25 September 2019; pp. 135–138.
36. Diaz, D.; Marin, L. VFH+D: An Improvement on the VFH+ Algorithm for Dynamic Obstacle Avoidance and Local Planning. *IFAC Pap.* **2020**, *53*, 9590–9595. [CrossRef]
37. Zhou, L.; Li, W. Adaptive Artificial Potential Field Approach for Obstacle Avoidance Path Planning. In Proceedings of the Seventh International Symposium on Computational Intelligence and Design, Hangzhou, China, 13–14 December 2014; pp. 429–432.
38. Borenstein, J.; Koren, Y. Real-time Obstacle Avoidance for Fast Mobile Robots. *IEEE Trans. Syst. Man Cybern.* **1989**, *19*, 1179–1187. [CrossRef]
39. Khatib, O. Real-time Obstacle Avoidance for Manipulators and Mobile Robots. *Int. J. Robot. Res.* **1985**, *1*, 90–98.
40. Chiang, H.-T.; Malone, N.; Lesser, K.; Oishi, M.; Tapia, L. Path-guided artificial potential fields with stochastic reachable sets for motion planning in highly dynamic environments. In Proceedings of the IEEE International Conference on Robotics and Automation (ICRA), Seattle, WA, USA, 26–30 May 2015; pp. 2347–2354.
41. Malone, N.; Chiang, H.-T.; Lesser, K.; Oishi, M.; Tapia, L. Hybrid Dynamic Moving Obstacle Avoidance Using a Stochastic Reachable Set-Based Potential Field. *IEEE Trans. Robot.* **2017**, *33*, 1124–1138. [CrossRef]
42. Sudhakara, P.; Ganapathy, V.; Priyadharshini, B.; Sundaran, K. Obstacle Avoidance and Navigation Planning of a Wheeled Mobile Robot using Amended Artificial Potential Field Method. *Procedia Comput. Sci.* **2018**, *133*, 998–1004. [CrossRef]
43. Lu, X.S.; Li, E.; Guo, R. An Obstacles Avoidance Algorithm Based on Improved Artificial Potential Field. In Proceedings of the IEEE International Conference on Mechatronics and Automation (ICMA), Beijing, China, 13–16 October 2020; pp. 425–430.
44. Lin, X.; Wang, Z.; Chen, X. Path Planning with Improved Artificial Potential Field Method Based on Decision Tree. In Proceedings of the 27th Saint Petersburg International Conference on Integrated Navigation Systems (ICINS), St. Petersburg, Russia, 25–27 May 2020; pp. 1–5.
45. Lazarowska, A. Discrete Artificial Potential Field Approach to Mobile Robot Path Planning. *IFAC PapersOnLine* **2019**, *52*, 277–282. [CrossRef]
46. Shin, Y.; Kim, E. Hybrid path planning using positioning risk and artificial potential fields. *Aerosp. Sci. Technol.* **2021**, *112*, 106–640. [CrossRef]
47. Olunloyo, V.O.S.; Ayomoh, M.K.O. Autonomous Mobile Robot Navigation Using Hybrid Virtual Force Field Concept. *Eur. J. Sci. Res. EJSR* **2009**, *31*, 204–228.
48. Olunloyo, V.O.S.; Ayomoh, M.K.O.; Ibidapo-Obe, O. A path planning model for an autonomous vehicle in an unstructured obstacle domain. In Proceedings of the 14th IASTED International Conference, Cambridge, MA, USA, 2–4 November 2009; pp. 180–187.
49. Olunloyo, V.O.S.; Ayomoh, M.K.O. A Hybrid Path Planning Model for Autonomous Mobile Vehicle Navigation. In Proceedings of the 25th International Conference of CAD/CAM, Robotics & Factories of the Future Conference, Pretoria, South Africa, 13–16 July 2010; pp. 1–12.
50. Olunloyo, V.O.S.; Ayomoh, M.K.O. An Efficient Path Planning Model in an Unstructured Obstacle Domain. In Proceedings of the IASTED International Conference Robotics and Applications, Vancouver, BC, Canada, 1–3 June 2011; pp. 38–45.

51. Ngwenya, T. Mobile Robot Optimum Trajectory Development using a Hybrid Reactive Navigation Model. Master Dissertation, University of Pretoria, Pretoria, South Africa, 2021.
52. Zadeh, L.A. The concept of a linguistic variable and its application to approximate reasoning—I. *Inf. Sci.* **1975**, *8*, 199–249. [CrossRef]
53. Qing-Yong, B.A.O.; Shun-Ming, L.I.; Wei-Yan, S.; Mu-Jin, A.N. A Fuzzy Behavior–Based Architecture for Mobile Robot Navigation in Unknown Environments. In Proceedings of the International Conference on Artificial Intelligence and Computational Intelligence, Shanghai, China, 7–8 November 2009; pp. 257–261.
54. Jaradat, M.; Garibeh, M.; Feilat, E.A. Autonomous mobile robot planning using hybrid fuzzy potential field. *Soft Comput.* **2012**, *15*, 153–164. [CrossRef]
55. Pandey, R.K.; Sonkar, K.K.; Parhi, D.R. Path planning navigation of mobile robot with obstacles avoidance using fuzzy logic controller. In Proceedings of the IEEE 8th International Conference on Intelligent Systems and Control (ISCO), Coimbatore, India, 10–11 January 2014; pp. 39–41.
56. Almasri, M.M.; Elleithy, K.M.; Alajlan, A.M. Development of efficient obstacle avoidance and line following mobile robot with the integration of fuzzy logic system in static and dynamic environments. In Proceedings of the IEEE Long Island Systems, Applications and Technology Conference (LISAT), Farmingdale, NY, USA, 29–29 April 2016; pp. 1–6.
57. Singh, N.H.; Thongam, K. Mobile Robot Navigation Using Fuzzy Logic in Static Environments. *Procedia Comput. Sci.* **2018**, *125*, 11–17. [CrossRef]
58. Batti, H.; Jabeur, C.B.; Seddik, H. Mobile Robot Obstacle Avoidance in labyrinth Environment Using Fuzzy Logic Approach. In Proceedings of the International Conference on Control, Automation and Diagnosis (ICCAD), Grenoble, France, 2–4 July 2019; pp. 1–5.
59. Mohanty, P.K.; Kundu, S.; Srivastava, S.; Dash, R.N. A New T-S Model Based Fuzzy Logic Approach for Mobile Robots Path Planning. In Proceedings of the IEEE International Women in Engineering (WIE) Conference on Electrical and Computer Engineering (WIECON-ECE), Bhubaneswar, India, 26–27 December 2020; pp. 476–480.
60. Oleiwi, B.K.; Mahfuz, A.; Roth, H. Application of Fuzzy Logic for Collision Avoidance of Mobile Robots in Dynamic-Indoor Environments. In Proceedings of the 2nd International Conference on Robotics, Electrical and Signal Processing Techniques (ICREST), Dhaka, Bangladesh, 5–7 January 2021; pp. 131–136.
61. Verbitsky, N.S.; Chepin, E.V.; Gridnev, A.A. Experimental studies of a convolutional neural network for application in the navigation system of a mobile robot. *Procedia Comput. Sci.* **2018**, *145*, 611–616. [CrossRef]
62. Ribeiro, D.; Mateus, A.; Miraldo, P.; Nascimento, J.C. A real-time deep learning pedestrian detector for robot navigation. In Proceedings of the Autonomous Robot Systems and Competitions (ICARSC), Coimbra, Portugal, 26–28 April 2017; pp. 165–171.
63. Pershina, Z.; Kazdorf, S.; Abrosimov, V. Application of algorithms for object recognition based on deep convolutional neural networks for visual navigation of a mobile robot. In Proceedings of the 25th Saint Petersburg International Conference on Integrated Navigation Systems (ICINS), Saint Petersburg, Russia, 28–30 May 2018; pp. 1–2.
64. Chi, K.; Lee, M.R. Obstacle avoidance in mobile robot using Neural Network. In Proceedings of the International Conference on Consumer Electronics, Communications and Networks (CECNet), Xianning, China, 16–18 April 2011; pp. 5082–5085.
65. Motlagh, O.; Nakhaeinia, D.; Tang, S.H.; Karasfi, B.; Khaksar, W. Automatic navigation of mobile robots in unknown environments. *Neural Comput. Appl.* **2014**, *24*, 1569–1581. [CrossRef]
66. Janglova, D. Neural networks in mobile robot motion. *Int. J. Adv. Robot. Syst.* **2004**, *1*, 15–22. [CrossRef]
67. Yu, J.; Ji, J.; Miao, Z.; Zhou, J. Neural network-based region reaching formation control for multi-robot systems in obstacle environment. *Neurocomputing* **2019**, *333*, 11–21. [CrossRef]
68. Saleem, K.A.; Jabri, A.A.; Maashri, W.A.; Maawali, W.; Mesbah, M. Obstacle-Avoidance Algorithm Using Deep Learning Based on RGBD Images and Robot Orientation. In Proceedings of the 7th International Conference on Electrical and Electronics Engineering (ICEEE), Antalya, Turkey, 14–16 April 2020; pp. 268–272.
69. Wei, H.; Ye, Q. Mobile Robot Obstacle Avoidance System Based on GA-Aided OIF-Elman Network. In Proceedings of the 4th International Conference on Robotics and Automation Sciences (ICRAS), Wuhan, China, 12–14 June 2020; pp. 6–10.
70. Zhang, Y.; Ge, R.; Lyu, L.; Zhang, J.; Lyu, C.; Yang, X. A Virtual End-to-End Learning System for Robot Navigation Based on Temporal Dependencies. *IEEE Access* **2020**, *8*, 134111–134123. [CrossRef]
71. Tang, X.-L.; Li, L.-M.; Jiang, B.-J. Mobile robot SLAM method based on multi-agent particle swarm optimized particle filter. *J. China Univ. Posts Telecommun.* **2014**, *21*, 78–86. [CrossRef]
72. Algabri, M.; Hassan, M.; Hedjar, R.; Alsulaiman, M. Comparative study of soft computing technique for mobile robot navigation in an environment. *Comput. Hum. Behav.* **2015**, *50*, 42–56. [CrossRef]
73. Atyabi, A.; Phon-Amnuaisuk, S.; Ho, C.K. Applying Area Extension PSO in Robotic Swarm. *J. Intell. Robot. Syst.* **2009**, *58*, 253–285. [CrossRef]
74. Lin, C.-J.; Li, T.-H.S.; Kuo, P.-H.; Wang, Y.-H. Integrated particle swarm optimization algorithm based obstacle avoidance control design for home service robot. *Comput. Electr. Eng.* **2016**, *56*, 748–762. [CrossRef]
75. Meerza, S.I.A.; Islam, M.; Uzzal, M.M. Optimal Path Planning Algorithm for Swarm of Robots Using Particle Swarm Optimization Technique. In Proceedings of the 3rd International Conference on Information Technology, Information System and Electrical Engineering (ICITISEE), Yogyakarta, Indonesia, 13–14 November 2018; pp. 330–334.

76. Alaliyat, S.; Oucheikh, R.; Hameed, I. Path Planning in Dynamic Environment Using Particle Swarm Optimization Algorithm. In Proceedings of the 8th International Conference on Modeling Simulation and Applied Optimization (ICMSAO), Manama, Bahrain, 15–17 April 2019; pp. 1–5.
77. Tian, S.; Li, Y.; Kang, Y.; Xia, J. Multi-robot path planning in wireless sensor networks based on jump mechanism PSO and safety gap obstacle avoidance. *Future Gener. Comput. Syst.* **2021**, *118*, 37–47. [CrossRef]
78. Bremermann, H.J. *The Evolution of Intelligence. The Nervous System as a Model of Its Environment*; University of Washington: Washington, DC, USA, 1958.
79. Holland, J.H. *Adaptation in Natural and Artificial Systems*, 1st ed.; University of Michigan Press: Ann Arbor, MI, USA, 1975.
80. Xia, J.; Michalewicz, Z.; Zhang, L.; Trojanowski, K. Adaptive evolutionary planner/navigator for mobile robot. *Trans. Evol. Comput.* **1997**, *1*, 18–28. [CrossRef]
81. Shi, P.; Cui, Y. Dynamic path planning for mobile robot based on genetic algorithm in unknown environment. In Proceedings of the Chinese Control and Decision Conference, Xuzhou, China, 26–28 May 2010.
82. Patle, B.K.; Parhi, D.R.; Jagadeesh, A.; Kashyap, S.K. Matrix-binary codes based genetic algorithm for path planning of mobile robot. *Comput. Electr. Eng.* **2018**, *67*, 708–728. [CrossRef]
83. Germi, S.B.; Khosravi, M.A.; Fard, R.F. Adaptive GA-based Potential Field Algorithm for Collision-free Path Planning of Mobile Robots in Dynamic Environments. In Proceedings of the 6th RSI International Conference on Robotics and Mechatronics (IcRoM), Tehran, Iran, 23–25 October 2018; pp. 28–33.
84. Choueiry, S.; Owayjan, M.; Diab, H.; Achkar, R. Mobile Robot Path Planning Using Genetic Algorithm in a Static Environment. In Proceedings of the Fourth International Conference on Advances in Computational Tools for Engineering Applications (ACTEA), Beirut, Lebanon, 3–5 July 2019; pp. 1–6.
85. López-González, A.; Campaña, J.M.; Martínez, E.H.; Contro, P.P. Multi robot distance based formation using Parallel Genetic Algorithm. *Appl. Soft Comput.* **2020**, *86*, 105929. [CrossRef]
86. Aghda, S.A.F.; Mirfakhraei, M. Improved routing in dynamic environments with moving obstacles using a hybrid Fuzzy-Genetic algorithm. *Future Gener. Comput. Syst.* **2020**, *112*, 250–257. [CrossRef]
87. Guan-Zheng, T.; Huan, H.E.; Aaron, S. Ant colony system algorithm for real time globally optimal path planning of mobile robots. *Acta Autom. Sin.* **2007**, *33*, 279–285.
88. Chen, Y.; Su, F.; Shen, L.C. Improved ant colony algorithm based on PRM for UAV route planning. *J. Syst. Simul.* **2009**, *21*, 1658–1666.
89. Zhangqi, W.; Xiaoguang, Z.; Qingyao, H. Mobile Robot Path Planning based on Parameter Optimization Ant Colony Algorithm. *Procedia Eng.* **2011**, *15*, 2738–2741. [CrossRef]
90. Wang, H.; Wang, Z.A.; Yu, L.; Wang, X.; Liu, C. Ant Colony Optimization with Improved Potential Field Heuristic for Robot Path Planning. In Proceedings of the 37th Chinese Control Conference (CCC), Wuhan, China, 25–27 July 2018; pp. 5317–5321.
91. Yi, Z.; Yanan, Z.; Xiangde, L. Path Planning of Multiple Industrial Mobile Robots Based on Ant Colony Algorithm. In Proceedings of the 16th International Computer Conference on Wavelet Active Media Technology and Information Processing, Chengdu, China, 14–15 December 2019; pp. 406–409.
92. Ma, Y.; Mao, Z.; Wang, T.; Qin, J.; Ding, J.; Meng, X. Obstacle avoidance path planning of unmanned submarine vehicle in ocean current environment based on improved firework-ant colony algorithm. *Comput. Electr. Eng.* **2020**, *87*, 106–773. [CrossRef]
93. Zhao, H. Optimal Path Planning for Robot Based on Ant Colony Algorithm. In Proceedings of the International Wireless Communications and Mobile Computing (IWCMC), Limassol, Cyprus, 15–29 June 2020; pp. 671–675.

sensors

MDPI

Article

Intelligent Smart Marine Autonomous Surface Ship Decision System Based on Improved PPO Algorithm

Wei Guan *, **Zhewen Cui** and **Xianku Zhang**

Navigation College, Dalian Maritime University, Dalian 116026, China; cuizhewen123@dlmu.edu.cn (Z.C.); zhangxk@dlmu.edu.cn (X.Z.)
* Correspondence: gwwtxdy@dlmu.edu.cn

Abstract: With the development of artificial intelligence technology, the behavior decision-making of an intelligent smart marine autonomous surface ship (SMASS) has become particularly important. This research proposed local path planning and a behavior decision-making approach based on improved Proximal Policy Optimization (PPO), which could drive an unmanned SMASS to the target without requiring any human experiences. In addition, a generalized advantage estimation was added to the loss function of the PPO algorithm, which allowed baselines in PPO algorithms to be self-adjusted. At first, the SMASS was modeled with the Nomoto model in a simulation waterway. Then, distances, obstacles, and prohibited areas were regularized as rewards or punishments, which were used to judge the performance and manipulation decisions of the vessel Subsequently, improved PPO was introduced to learn the action–reward model, and the neural network model after training was used to manipulate the SMASS's movement. To achieve higher reward values, the SMASS could find an appropriate path or navigation strategy by itself. After a sufficient number of rounds of training, a convincing path and manipulation strategies would likely be produced. Compared with the proposed approach of the existing methods, this approach is more effective in self-learning and continuous optimization and thus closer to human manipulation.

Keywords: decision-making; deep reinforcement learning; Nomoto; PPO; SMASS

Citation: Guan, W.; Cui, Z.; Zhang, X. Intelligent Smart Marine Autonomous Surface Ship Decision System Based on Improved PPO Algorithm. *Sensors* **2022**, *22*, 5732. https://doi.org/10.3390/s22155732

Academic Editors: David Cheneler and Stephen Monk

Received: 21 June 2022
Accepted: 28 July 2022
Published: 31 July 2022

1. Introduction and Background

Since the 1970s, the combination of robot technologies and vehicles has led to the emergence of drones, unmanned vehicles, and unmanned ships [1]. Among them, a ship sailing on the sea is seriously affected by wind and surges. The decision-making and path planning of intelligent ships have been considered significant academic problems. Ships are generally under-actuated due to their large tonnage, slow speed, and relatively weak power. The autonomous navigation of ships has to meet huge inertia and complex navigation rules; therefore, the requirements for smart ships are much higher than those for unmanned vehicles. A ship operator faces many challenges, including those associated with the dynamic environment, insufficient power, and uncertainties in perception. According to the report of the International Maritime Organization, more than 80 percent of maritime accidents are caused by misunderstandings of the situation and by human error in decision-making resulting from failure to comply with the International Regulations for Preventing Collisions at Sea (COLREGs). Therefore, artificial intelligence for ship navigation is considered very difficult, and its core functions are path planning and intelligent decision-making.

Ship intelligent decision-making can be divided into two types: path planning and obstacle avoidance. One is the traditional model-based obstacle avoidance algorithm. For many years, the A* algorithm was the dominant approach in relevant research. A Swiss boat named Avalon was capable of generating a persuasive path to a given destination and avoiding both static and dynamic obstacles based on the A* algorithm [2]. Several heuristic function values of the current path grid are compared by the A* algorithm to gradually determine the next path grid, which can accurately avoid obstacles. However, when there were multiple minimum values, the optimal path could not be searched by the A* algorithm. Sudden obstacles would make the ship fall into the local optimum. Zhang et al. improved the Rapidly Exploring Random Tree (RRT) algorithm so that the convergence rate of the algorithm was significantly improved [3]. However, the path was randomly selected by the RRT algorithm, and the probability of encountering narrow channels was low. It was not appropriate to navigate in a narrow channel or to face multiple static obstacles. An ant colony optimization (ACO) and a clustering-based algorithm were proposed to settle the path planning of the USV by Liu et al. [4]. The improved ant colony optimization was used to adaptively select the appropriate search range, and the smoothing mechanism was used to adjust the path to achieve global path planning. An improved artificial potential field method (APF) was proposed by Shaorong Xie et al. The problem of USV falling into the unreachable local optimal target could be improved by this method, but there were still problems such as the poor accuracy of the algorithm and falling into local optimum in complex environments [5]. The gravitational field and repulsion field functions were required to be set separately; thus, this method does not apply to any environment. A new artificial potential field (APF) method was improved by Hongguang Lv et al. Different from the method proposed by Shaorong Xie et al., the new modified repulsive potential field function and the corresponding virtual force were introduced in the algorithm [6]. Appropriate functionality and security requirements were added to the corresponding virtual force to ensure compliance with the International Regulations for Preventing Collisions at Sea (COLREGs). However, with the complexity of modern maritime systems, a complete collision avoidance model is difficult to establish in many path planning and navigation problems. In most model-based algorithms, uncertainty is difficult to predict in practical applications.

Another is a model-free reinforcement learning algorithm that learns optimal strategies by interacting with the environment. At present, the development of artificial intelligence technology, especially reinforcement learning, provides a new possibility to satisfy the requirements of the path planning of intelligent ships. Reinforcement learning has attracted extensive attention in recent years, which emphasizes the learning of agents from the environment to behavior mapping and seeks the most correct or best action decision by maximizing value functions. A ship path planning algorithm based on Q learning was proposed by Chen C. et al. Combined with the ship mathematical model, the USV could obtain a higher reward value by learning the action-value function [7]. However, the reinforcement learning algorithm had an insufficient perception of the external environment, and the action state information was difficult to be searched. In addition, the experimental environment was too simple, without considering the decision problem of complex multi-obstacles. An algorithm was proposed by Everett et al. that generates an appropriate collision-free path even when the number of dynamic obstacles is changed by using Long short-term memory (LSTM) [8]. A Deep-Q-Learning (DQN) algorithm linking perception and decision-making was proposed by Jingwei Zhang et al. The algorithm could acquire external images by depth camera information and extracts image features as inputs of DQN [9]. Decision problems can be solved by this algorithm, but the use of a depth camera and convolution network makes the calculation huge. Moreover, when sailing in harsh sea conditions, the depth camera could not be effectively put into use, and the method would be not convincing. A DQN-based path planning obstacle avoidance algorithm was proposed by Haiqing Shen et al. The algorithm could be successfully simulated with human experience and International Regulations for Preventing Collisions at Sea [10]. However, the DQN algorithm has an overestimation problem, and an unmanned surface vessel (USV) is prone to action selection error in a more complex environment. An algorithm combining Deep-Q-Learning (DQN) and the artificial potential field (APF) was proposed by Lingyu Li et al., which was used for USV path planning [11]. This algorithm made deep reinforcement learning more purposeful in the early stage of training and had a faster convergence effect. However, the method based on Q-learning seemed inadequate in solving the problem of continuous action. A Multi-Experience Library Framework was proposed by Zijian Hu et al. for Unmanned Aerial Vehicle (UAV) autonomous motion planning. The algorithm generated expert experience by model predictive control and simulated annealing [12]. When applying this algorithm to a complex unknown simulation environment constructed based on the parameters of the real UAV, the training experiment results showed that the novel Deep reinforcement learning (DRL) algorithm led to a performance improvement exceeding twenty percent, as compared to the state-of-the-art Deep Deterministic Policy Gradient (DDPG). DDPG has a slightly better decision-making effect than value-based learning algorithms in complex environments. Choosing the maximum probability action in each step under continuous action can make calculation much simpler. A new quantitative risk assessment method was proposed by Do-Hyun Chun et al. In the calculation of collision risk (CR), the distance closest point of approach (DCPA) and time closest point of approach (TCPA) were determined by ship length and ship speed [13]. This algorithm could take the collision risk assessment CR as one of the inputs of the neural network, but the experiment was too simple to generalize. An obstacle avoidance method based on the combination of PPO and the Line of Sight (LOS) guidance system was proposed by Luman Zhao et al. This algorithm could ensure that the ship moves along the predetermined path and avoids collision with the moving ship. Due to the limitation of the LOS algorithm, this experiment cannot avoid collision in a complex environment [14]. An improved DQN algorithm was proposed by Xinli Xu et al. The network weight was set by them to slowly approach the current value; in other words, the target network would approach the evaluation network gradually [15]. It could reduce the correlation between the current value function and the target value function to some extent. In addition, the reward function of the algorithm made the USV alter different angles, and the reward value was also different. However, the algorithm based on value learning was still overestimated, and the problems of static obstacles and generalization were not considered in the experiment. A distributed sensor-level collision avoidance policy for multi-robot systems was proposed by Pinxin Long et al., which could directly map raw sensor measurements to an agent's steering commands in terms of movement velocity [16]. This experiment verified the learned sensor-level collision avoidance strategy in various simulation scenarios and conducted a comprehensive performance evaluation. This experiment also demonstrated that the learned policy could be well generalized to new scenarios that did not appear in the entire training period, including navigating a heterogeneous group of robots and a large-scale scenario with 100 robots. Pinxin Long et al.'s experiment had a strong generalization ability, which is worth learning.

Based on the above research, an intelligent smart marine autonomous surface ship (SMASS) decision system based on an improved PPO algorithm was proposed in this paper. The main contributions of this article were as follows:

- An intelligent SMASS decision-making system based on the Proximal Policy Optimization (PPO) algorithm was proposed in this paper, which could make the critic network and action network converge faster.
- Through the Gazebo simulation environment, sensors such as laser radar and navigation radar were used to obtain external environmental information. Intelligent SMASS could make complex path planning decisions in different environments. After the training, if unknown obstacles are placed on the map, the intelligent ship could still successfully avoid obstacles.
- The Nomoto model was brought into the training of this experiment. Training the model could meet the needs of practical engineering.

The rest of this paper is organized as follows. Section 2 introduces the composition of the intelligent SMASS system, ship mathematical model, and COLREGs. Section 3 introduces a deep reinforcement learning algorithm and improved Proximal Policy Optimization (PPO) algorithm. Section 4 mainly introduces the reward function setting and network setting. Section 5 mainly introduces the design of the Gazebo simulation and the analysis of experimental results. Section 6 is the summary of this paper and the future research planning.

2. Intelligent Ship Decision System and Ship Mathematical Model

In building a complete set of the intelligent smart marine autonomous surface ship (SMASS) decision-making systems, it was necessary to clarify the components of the system, the functions of each part, and the relationship between different parts [17]. There are three parts included in the intelligent smart marine autonomous surface ship (SMASS) decision-making system, namely, the sensing part, the decision-making part, and the control part, as shown in Figure 1.

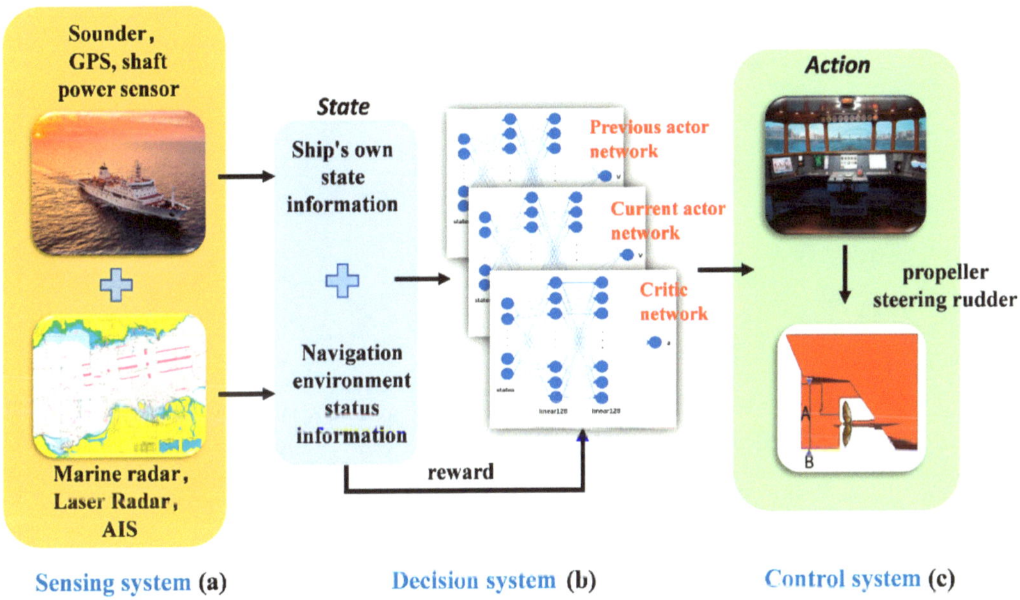

Figure 1. Intelligent ship navigation system: (**a**) the sensing part, (**b**) the decision part, and (**c**) the control part.

2.1. Intelligent Ship Decision System

The sensing part is divided into the SMASS's own state information and navigation environment information. The sensing part is mainly composed of navigation radar, laser radar, GPS, shaft power sensor, bathymeter, speed sensor, and AIS. The SMASS's own state information includes the SMASS's course, speed, position, oil consumption remaining, propeller speed, and hull structure strength [18]. Navigation environment information includes other ship heading speed TCPA, DCPA, hydrological information, velocity, channel depth, meteorological information (temperature, humidity, wind direction, wind speed), electronic chart information, navigation mark distribution, etc. In this paper, laser radar and positioning systems were used in the environmental perception of intelligent SMASS. The decision part includes path planning before sailing and obstacle avoidance during self-service navigation. In this paper, improved PPO algorithms were used for SMASS path planning and obstacle avoidance. The algorithm has the following advantages:

- With autonomous learning ability, the convergence rate was faster than the common calculation method.
- The trained intelligent SMASS navigation system could obtain strong generalization, which would solve different scene problems. For example, it can solve the problem

of path planning for SMASS sailing in broad waters, narrow waters, and restricted waters. In local path planning, it could successfully avoid unknown obstacles that do not appear on the electronic chart.

- The path planning problem and SMASS decision problem could be solved simultaneously. SMASS could find the optimal path to the target point through known obstacle information. Under the unknown environment, the SMASS could detect the position of obstacles by laser radar and accurately avoid the obstacles.

2.2. Ship Mathematical Model

The mathematical model of ship motion is significant for ship motion simulation. The ship motion model can be divided into the linear model and the nonlinear model. The linear model is mainly used to optimize or train the control simulator, neural network decision-making, and controller design [19]. To describe the motion of a ship, a ship motion coordinate system was established, as shown in Figure 2.

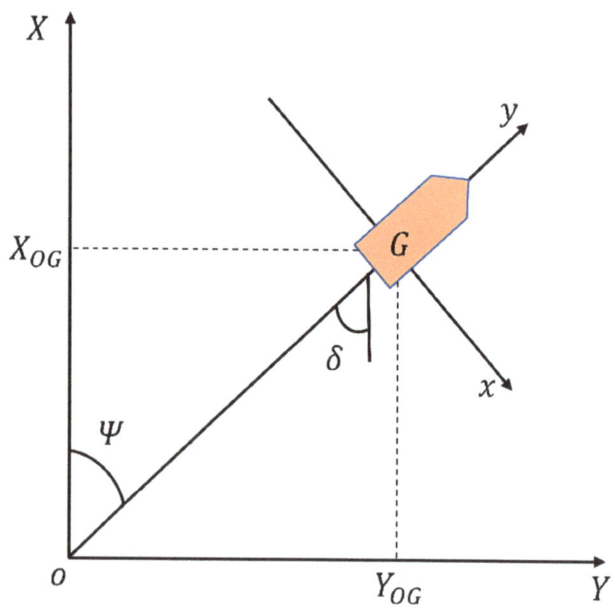

Figure 2. Ship motion mathematical model.

In this figure, G represents the position of the center of gravity of the ship, XOY indicates the hydrostatic water plane, O is the origin, X_{OG} represents the projection of the center of ship gravity on the X and Y axes, respectively, ψ is the heading of the ship, and δ indicates the ship rudder angle. Considering only the ship lateral drift velocity v and yaw angular velocity r, the ship motion mathematic model could be expressed as:

$$\begin{bmatrix} \dot{v} \\ \dot{r} \end{bmatrix} = \begin{bmatrix} a_{11} & a_{12} \\ a_{21} & a_{22} \end{bmatrix} \begin{bmatrix} v \\ r \end{bmatrix} + \begin{bmatrix} b_{11} \\ b_{21} \end{bmatrix} \delta \tag{1}$$

where $a_{11}, a_{12}, a_{21}, a_{22}, b_{11}$, and b_{21} are the ship maneuverability parameters [20]. Ignoring the lateral drift velocity v in Equation (1), the response equation of the ship steering rudder to yaw motion can be written as:

$$T_1 T_2 \ddot{r} + (T_1 + T_2)\dot{r} + r = K_0 \delta + K_0 T_3 \dot{\delta} \tag{2}$$

where T_1, T_2, T_3, and K_0 are maneuverability indexes. Their values could be estimated by $T_1T_2 = 1/(a_{11}a_{22} - a_{12}a_{21})$, $T_1 + T_2 = (a_{11} + a_{22})/(a_{12}a_{21} - a_{11}a_{22})$, $T_3 = b_{21}/(a_{21}b_{11} - a_{11}b_{21})$, and $K_0 = (a_{21}b_{11} - a_{11}b_{21})/(a_{11}a_{22} - a_{12}a_{21})$. Then, the Laplace transform of Equation (2) could be carried out to obtain the transfer function of the ship steering control system, as shown in Equation (3):

$$G(s) = \frac{\psi(s)}{\delta(s)} = \frac{K_0(1 + T_3s)}{s(1 + T_1s)(1 + T_2s)} \tag{3}$$

For ships with large inertia, the dynamic characteristics are the most important in the low-frequency range [21]. Thus, let the following formula show that $s = jw \rightarrow 0$, and, ignoring the second and third-order small quantities, the Nomoto model can be obtained by:

$$G_{\varphi\delta}(s) = \frac{\psi}{\delta} = \frac{K_0}{s(T_0s + 1)} \tag{4}$$

The differential equation form of the Nomoto model is written as shown in Equation (5):

$$T\ddot{\psi} + \dot{\psi} = K\delta \tag{5}$$

The value T represents the coefficient ratio of the inertia moment to the damping moment [22]. A large T value indicates a large inertia moment and a small damping moment during ship motion. The value K actually refers to the angular velocity value of yaw motion by each rudder angle. The large K means a large yaw moment and a small damping moment produced by the rudder.

Taking the ship as a rigid body, when the ship steers at any rudder angle δ, the yaw rate is r. The above formula can be seen as the yaw motion equation of the ship when it steers. When the ship turns, altering her course, at any rudder angle, assuming that the initial conditions are $t = 0$, $\delta = \delta_0$, and $r = 0$, the yaw angle at any time can be calculated by KT Equation (6):

$$r = K\delta(1 - e^{-t/T}) \tag{6}$$

Ship yaw angle r is the derivative of ψ with respect to time. As shown in Equation (7).

$$\psi = K\delta_0(t - T + T \cdot e^{-t/T}) \tag{7}$$

There are two advantages of using the Nomoto model in this experiment:

- In the low-frequency range, the spectrum of the Nomoto model is very close to that of the high order model.
- The designed controller has low order and is easy to implement.

2.3. COLREGs

To solve SMASS path planning and obstacle avoidance problems based on DRL, maritime collision avoidance rules should be considered. COLREGS is a maritime traffic rule that is stipulated in the high seas and all navigational waters connected to the high seas to ensure the safety of ship navigation, preventing ship collision. Therefore, intelligent ship decision-making systems should also act in accordance with COLREG to ensure the safety of maritime navigation [23]. According to the COLREGS, the relative position of the two ships is divided into four obstacle avoidance strategy regions, such as in Figure 3.

The four collision avoidance rules involved in COLREGS Chapter 2 Regulation 13 to 17 are as follows. The corresponding collision avoidance actions are displayed in Figure 3.

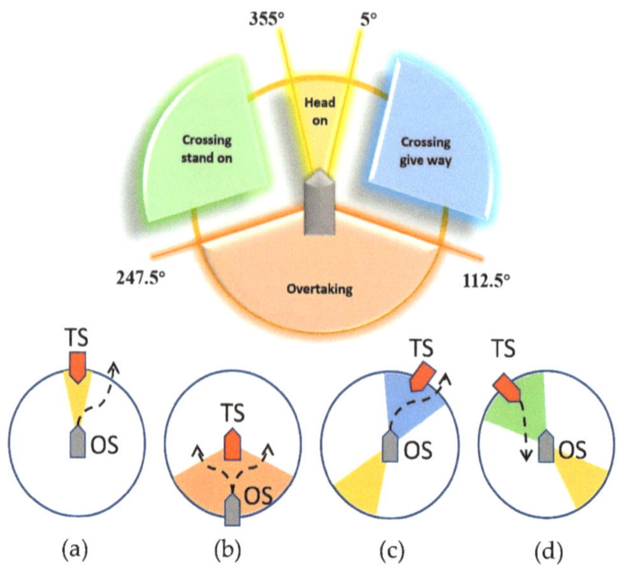

Figure 3. Encounter situations defined by COLREGs: (**a**) Head-on, (**b**) Overtaking, (**c**) Crossing give-way, and (**d**) Crossing stand-on.

(1) Head-on

The encounter situation refers to the opposite or nearly opposite course (where the course usually refers to the bow direction of the ship rather than the track direction) of the two mobile ships under the condition of mutual seeing, and there is a risk of collision. The opposite direction means the relative azimuth of the target ship (TS) and own ship (OS) is in $[355^\circ, 360^\circ]$ or $[0^\circ, 5^\circ]$. Two ships should alter course port through passing the starboard side of another ship. The head-on situation is displayed in Figure 3a.

(2) Overtaking

The overtaking situation means that the speed of the rear ship is greater than that of the front ship. When the own ship chases the target ship in a certain direction 22.5 degrees behind the target ship, the target ship is a stand-on ship, and the own ship should give way to the target ship. The overtaking situation is displayed in Figure 3b.

(3) Crossing give-way

When two ships meet and there is a risk of collision, the relative position of the target ship and the own ship is in $[5^\circ, 112.5^\circ]$. In this case, the own ship should give way to the target ship. According to COLREGs, the own ship should alter her course to starboard to avoid a collision. The crossing give-way situation is displayed in Figure 3c.

(4) Crossing stand-on

When two ships meet and the relative position of the target ship and the own ship is in $[247.5^\circ, 355^\circ]$, there is a risk of collision. In this case, the ship is stand-on, and the target ship should give way to the own ship. If the target ship does not take avoidance action timely, the own ship should take action to avoid the collision. The crossing stand-on situation is displayed in Figure 3d.

3. Improved PPO Algorithm

3.1. Deep Reinforcement Learning

At present, artificial intelligence technologies have developed rapidly; especially after AlphaGo defeated Lee Sedol, the nine-stage chess player, reinforcement learning has risen rapidly to provide new possibilities for intelligent ship path planning. The Q-learning algorithm could obtain the best behavior decision-making through the optimal action-value function [24]. However, the marine environment is too complex, and a ship sailing on the sea faces many uncertainties. The Q-table could seem inadequate in solving complex problems.

The development of deep reinforcement learning is greatly accelerated by neural networks [25]. With the change of the agent's external environment, through the backpropagation of neural networks, the weights of neural networks could be updated to simulate complex functions. Deep reinforcement learning algorithms are divided into two categories: value learning and strategy learning.

Reinforcement learning based on value function is represented by Deep Q-Learning (DQN), and the problem of correlation and non-static distribution could be solved by the experience replay method. The current Q value is generated by the evaluation network, and the target Q value is generated by the target network [26]. The experience replay stores the transfer samples (s_t, a_t, r_t, s_{t+1}) from each time step agent that interacts with the environment into the replay memory unit. Then, small-batch data in the memory library are selected for training, but the DQN algorithm is not accurate in estimating the action value Q, so there are some errors. Suppose DQN's estimate of the real action is unbiased, then the error is noise with an average of 0. $q = \max_a Q(s, a; \omega)$ is maximized based on DQN action a and used to compute TD_{target}. Adding noise to the action-value function will make $q \geq \max_a(S_{t+1}, a, \omega)$. Obtaining the Q value at the next moment is an overestimation. Although noise does not change the mean value, it will make the maximum value of Q greater than the maximum value of x. Expectations for the maximum of Q will also be greater than the maximum value of x. Updating DQN estimates at time t with TD_{target} also means updating itself with itself. Uniform overestimation does not make DQN a problem with action selection because each action overestimation is the same agent and will still choose to score high action. However, non-uniform overestimation will make DQN have problems in the action selection. Double Deep Q-Learning (Double DQN) was proposed by Google DeepMind to solve the overestimation problem of DQN [27]. Although the estimation made by Double Deep Q-Learning is relatively small, its overestimation of the maximum value cannot be solved fundamentally. This is why reinforcement learning based on value learning was abandoned in this paper.

The Actor-Critic (AC) algorithm is representative of strategy learning. There are two neural networks that exist in the AC algorithm. One is used to interact with the environment to select actions, and the other is used to evaluate the quality of actions, and the network parameters are updated by gradient descent. The AC algorithm is good but difficult to converge. Compared with random strategies, deterministic strategies adopt different action probabilities at the same state when solving continuous action problems, but the maximum probability is only one. Double actor neural networks and double critic neural networks were used in the Deep Deterministic Policy Gradient (DDPG) algorithm to improve the convergence of neural networks [28]. The algorithm can only take action with the maximum probability; however, by removing the probability distribution, the algorithm will be much simpler. In 2017, a Proximal Policy Optimization (PPO) algorithm was proposed by OpenAI [29]. The Policy Gradient algorithm is very sensitive to the step size, but it is difficult to select the appropriate step size. If the difference between the new and old strategies is too large, it is not conducive to learning. The problem of uncertain learning rates in the Policy Gradient algorithm could be solved by the PPO algorithm; if the learning rate is too large, then the learned strategy is not easy to converge. On the contrary, if the learning rate is too small, it will take a long time. The proportion of current and previous strategies could be used in the PPO algorithm, which would limit the update range of the current strategy, so that the Policy Gradient algorithm would not be so sensitive to a slightly larger learning rate.

3.2. Improved PPO Algorithm

The current and previous strategy networks were used by the traditional PPO algorithm to improve the uncertainty of the learning rate, but they still had a large variance. A generalized advantage estimate was proposed by John Schulman et al. to improve the TRPO algorithm [30], which can also be used to improve the PPO algorithm.

First, the application of baseline in strategy learning should be understood. The baseline could be regarded as a function b independent of action a.

$$\begin{aligned} E_{A\sim\pi}\left[b\cdot\frac{\partial \ln \pi(A|S;\theta)}{\partial\theta}\right] &= b\cdot E_{A\sim\pi}\left[\frac{\partial \ln \pi(A|S;\theta)}{\partial\theta}\right] \\ &= b\cdot \textstyle\sum_a \pi(a|S;\theta)\cdot\frac{\partial \ln \pi(a|s;\theta)}{\partial\theta} \\ &= b\cdot \textstyle\sum_a \pi(a|s;\theta)\cdot\frac{1}{\pi(a|s;\theta)}\cdot\frac{\partial \pi(a|s;\theta)}{\partial\theta} \\ &= b\cdot \textstyle\sum_a \frac{\partial \pi(a|s;\theta)}{\partial\theta} \\ &= b\cdot \frac{\partial \sum_a \pi(a|s;\theta)}{\partial\theta} \\ &= 0 \end{aligned} \tag{8}$$

where a is the action taken for the agent, s is the current state, and θ is the network parameter. The essence of the policy function is the probability density function. Taking Equation (9) to the equality of policy gradient update will obtain the advantage function.

$$\begin{aligned} \frac{\partial V_\pi(S)}{\partial\theta} &= E_{A\sim\pi}\left[\frac{\partial \ln \pi(A|S;\theta)}{\partial\theta}\cdot Q_\pi(S,A)\right] \\ &= E_{A\sim\pi}\left[\frac{\partial \ln \pi(A|S;\theta)}{\partial\theta}\cdot (Q_\pi(S,A)-b)\right] \end{aligned} \tag{9}$$

Although the gradient is not affected by the value of b, it affects the Monte Carlo approximation. When b approaches Q_π, the variance of the Monte Carlo approximation will decrease, and the convergence rate will improve. The value of b is $V_\pi(S_t)$, where $V_\pi(S_t)$ is independent of action a, and then the advantage function is obtained. The action value function can be seen as the conditional expectation of the return value U_t to s_t, a_t, and the state value function can be seen as the conditional expectation of the action value function to s_t; thus, the equation can be obtained:

$$\begin{aligned} Q_\pi(s_t,a_t) &= E_{S_{t+1},A_{t+1}}[R_t+\gamma Q_\pi(s_{t+1},a_{t+1})] \\ &= E_{S_{t+1}}\left[R_t+\gamma E_{A_{t+1}}(Q_\pi(s_{t+1},a_{t+1}))\right] \\ &= E_{S_{t+1}}[R_t+\gamma V_\pi(S_{t+1})] \end{aligned} \tag{10}$$

$$\begin{aligned} V_\pi(S_t) &= E_{A_t}[Q_\pi(S_t,A_t)] \\ &= E_{A_t}\left[E_{S_{t+1}}[R_t+\gamma V_\pi(S_{t+1})]\right] \\ &= E_{A_t,S_{t+1}}[R_t+\gamma V_\pi(S_{t+1})] \end{aligned} \tag{11}$$

At this time, the Monte Carlo approximation of Q_π and V_π can be obtained:

$$Q_\pi(s_t,a_t)\approx r_t+\gamma V_\pi(s_{t+1}) \tag{12}$$

$$V_\pi(s_t)\approx r_t+\gamma V_\pi(s_{t+1}) \tag{13}$$

Because the value b in the dominant function is $V_\pi(s_t)$, it is a definite value, so it is not necessary to use the Monte Carlo approximation. The unbiased estimation of the strategy gradient can be expressed as:

$$\begin{aligned} g(a_t) &= \frac{\partial \ln \pi(a_t|s_t;\theta)}{\partial\theta}(Q_\pi(s_t,a_t)-V_\pi(s_t)) \\ &\approx \frac{\partial \ln \pi(a_t|s_t;\theta)}{\partial\theta}\cdot(r_t+\gamma V_\pi(s_{t+1})-V_\pi(s_t)) \\ &\approx \frac{\partial \ln \pi(a_t|s_t;\theta)}{\partial\theta}\cdot(r_t+\gamma V_\pi(s_{t+1};\omega)-V_\pi(s_t;\omega)) \end{aligned} \tag{14}$$

We can define $\eta_t^V = r_t + \gamma V_\pi(s_{t+1}) - V_\pi(s_t)$ and subtract the $K - step$ advantage from the baseline function, then we can obtain the following equation:

$$\hat{G}_t^{(\infty)} = \sum_{k=0}^{\infty} \gamma^k \eta_{t+1}^V = -V(s_t) + \sum_{k=0}^{\infty} \gamma^k r_{t+1} \tag{15}$$

Therefore, the generalized advantage estimation can be obtained. The formula is as follows:

$$\begin{aligned} \hat{G}_t &= (1-\lambda)(\hat{G}_t^{(1)} + \lambda \hat{G}_t^{(2)} + \lambda^2 \hat{G}_t^{(3)} + \ldots) \\ &= (1-\lambda)(\eta_t^V + \lambda(\eta_t^V + \gamma\eta_{t+1}^V) + \lambda^2(\eta_t^V + \gamma\eta_{t+1}^V + \gamma^2\eta_{t+2}^V) + \ldots) \\ &= (1-\lambda)\left(\eta_t^V\left(\tfrac{1}{1-\lambda}\right) + \gamma\eta_{t+1}^V\left(\tfrac{\lambda}{1-\lambda}\right) + \gamma^2\eta_{t+1}^V\left(\tfrac{\lambda^2}{1-\lambda}\right) + \ldots\right) \\ &= \sum_{k=0}^{\infty} (\gamma\lambda)^k \eta_{t+1}^V \end{aligned} \tag{16}$$

The loss function of the PPO algorithm is:

$$L^{PPO}(\theta) = E\left[\min(\mu_\theta^t G^t, clip(\mu_\theta^t, 1-\varepsilon, 1+\varepsilon))G^t\right] \tag{17}$$

$$\mu_\theta^t = \frac{\pi(a_t|s_t)}{\pi_{old}(a_t|s_t)} \tag{18}$$

In this equation, μ_θ^t is the ratio of probability. The ratio of the probability is that the strategy before updating takes a specific operation in a specific state to the probability that the current strategy takes the same operation in the same state. The ratio is between $1-\varepsilon$ and $1+\varepsilon$ according to the range of the super parameter ε. Therefore, there is a great change between the previous strategy and the current strategy. The PPO loss iteration is shown in Figure 4.

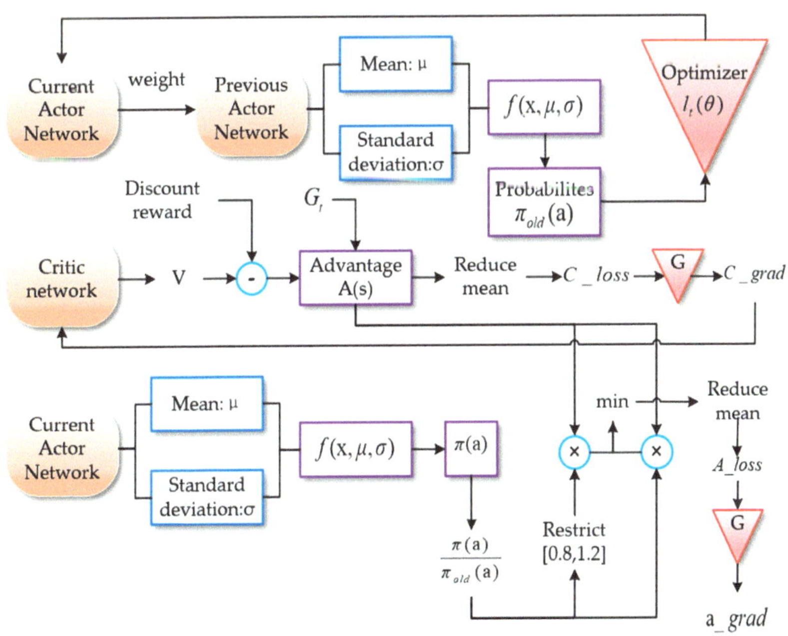

Figure 4. Loss function structure of PPO algorithm.

4. Neural Network Design and Reward Function

4.1. Network Construction and Input and Output Information

State information is input by the actor network and critic network in the PPO algorithm. Two-dimensional plane coordinates of the ship (x_p, y_p); rudder angle and rudder angular velocity of the operating system (δ, δ_1); and 24 laser radar vector lines $(\chi_1, \chi_2, \chi_3 \cdots \chi_{24})$ were used as the state information of the environment.

To avoid collisions with other ships, the navigator should adjust the direction of their own ship to ensure the navigation safety of ships in designated waters. The collision avoidance method of an autonomous ship can be created through a sufficient learning process by simulating the appropriate decision-making skills that the navigator could acquire over a long period of experience [31]. In this experiment, the output data are the rudder angle of the SMASS. The course and path of the SMASS would be affected by the change of rudder angle. The altering course to port is defined as negative, and altering course to starboard is defined as positive. The action space of this experiment is $[-45°, -25°, 0°, 25°, 45°]$. The obstacle avoidance process of the SMASS is shown in Figure 5.

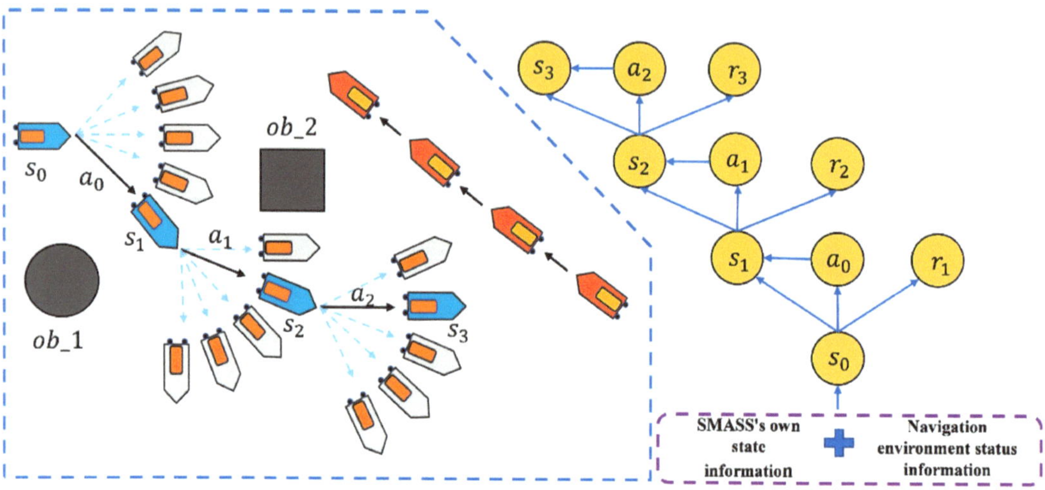

Figure 5. The update process of the SMASS collision avoidance algorithm.

In this paper, the deep neural network was used to fit the policy function π. Among them, the actor network adopted a two-layer full connection layer with 128 neurons. The Relu activation function was used and the network input was state S. The obtained expectation and standard deviation were put into the Gaussian distribution, the probability density function was obtained using the strategy distribution, and the probability corresponding to different action a was the output. The critic network adopted two fully connected layers with 128 neurons and the Relu activation function. The network input was state S, the output of the actor selected action score. The PPO algorithm and environment interaction process are shown in Figure 6.

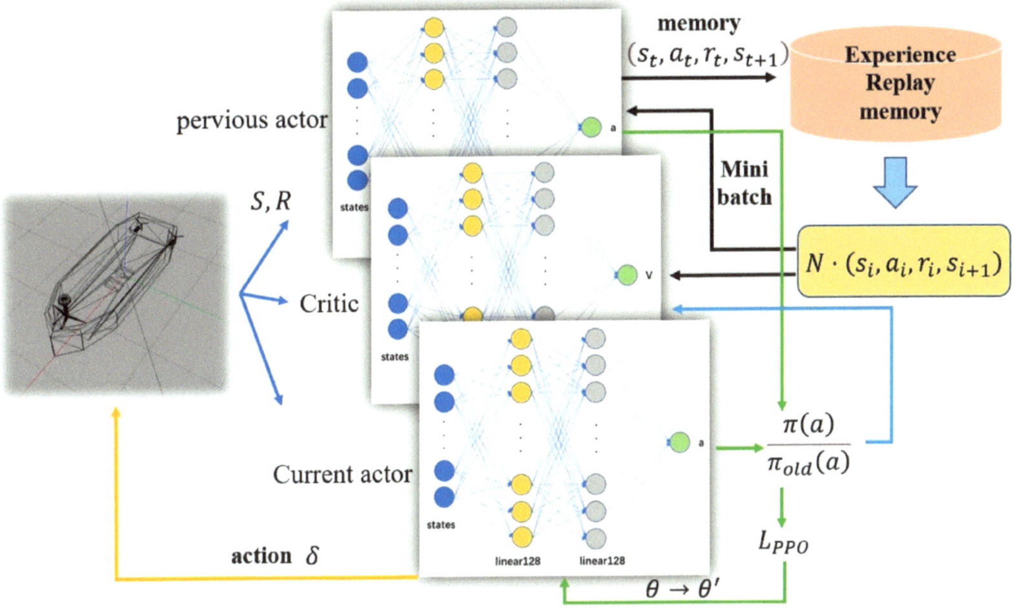

Figure 6. Flow chart of the PPO algorithm and environment interaction. Input is state vector S, output is ship steering angle δ. Both critic network and actor network are connected by a linear layer with 128 neurons using the Relu activation function.

The probability obtained by the previous strategy was optimized with other relevant parameters, and the difference in the *new_Actor* network was obtained. The obtained difference was put into the *new_Actor*, so that the strategy of the global network is new, and the strategy of the regional network is old. The critic network output is the value V, using discount reward, value subtraction, and generalized advantage estimate optimization to obtain the advantage function. Then, the gradient descent algorithm was used to calculate the error and update the network parameters. The proportion of the current and previous strategies was multiplied by the advantage function. One part was directly multiplied, and the other part was multiplied after $1 - \varepsilon$ and $1 + \varepsilon$, according to the range of the super parameter ε. The minimum value of the two was taken, and then the error was calculated.

To break the correlation of data and ensure the convergence of policy functions, an empirical playback memory can be set to store the historical motion state. Under each time step t, the intelligent ship entered a new state after interacting with the environment, and the updated state was put into the memory. In the process of the neural network training, a small batch of state samples were extracted from the memory to ensure the stability of the training.

4.2. Reward Function

According to the task of SMASS path planning and obstacle avoidance, the reward function was set to the following five parts: goal approach reward, yaw angle reward, target point reward, obstacle avoidance reward, and COLREGs reward as shown in the Figure 7.

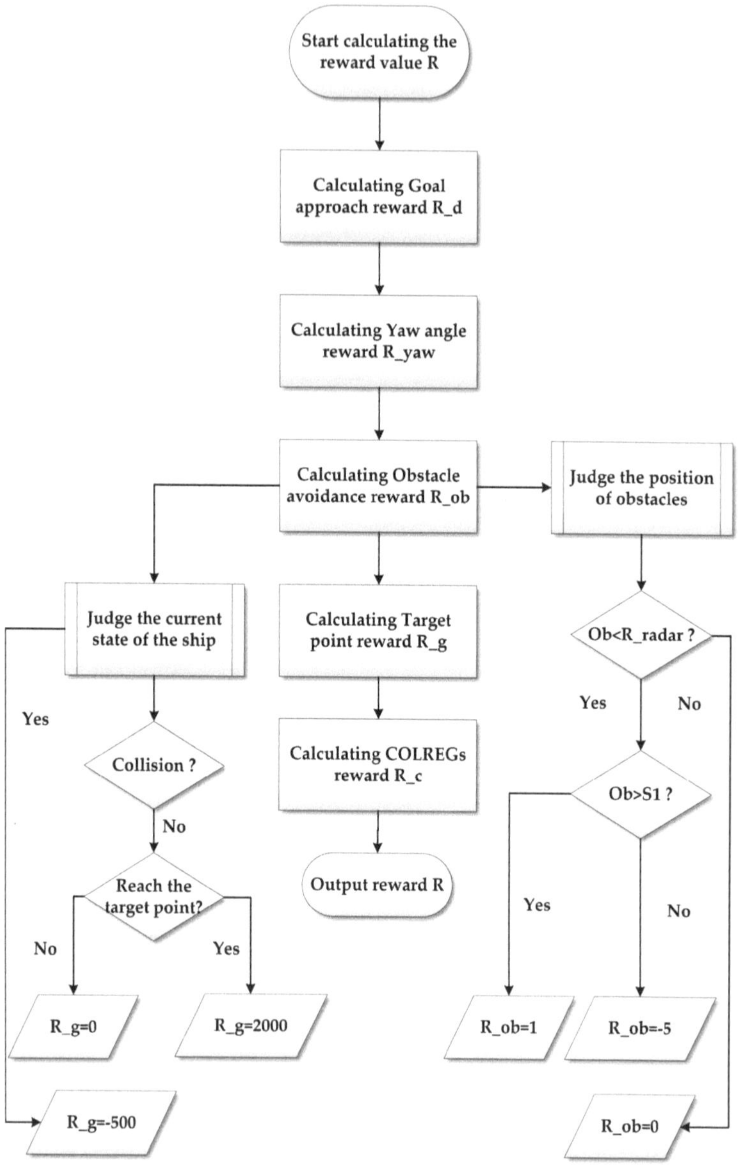

Figure 7. Calculation process of the reward function.

(1) Goal approach reward

The primary task to solve the intelligent SMASS path planning was to make the SMASS reach the target position. The goal approach reward value was set as follows:

$$R\_d = -\lambda_g \cdot \sqrt{(x_p - x_g)^2 + (y_p - y_g)^2} \tag{19}$$

where x_p and y_p are the coordinates of the current position of the ship, x_g and y_g are the coordinates of the target point, and λ_g is the weight of the target proximity reward.

(2) Yaw angle reward

When the SMASS is planning the path, the heading angle should be taken as an important indicator. As shown in Figure 8, the connection between the current position of the ship and the position of the target point should be regarded as the shortest distance, and the SMASS motion direction should be along this direction as far as possible. The Yaw angle reward function is set as follows:

$$R\_yaw = tr \cdot \lambda_a \cdot 2^{(\varepsilon yaw)^2}\sqrt{(x_p-x_g)^2+(y_p-y_g)^2} \tag{20}$$

where yaw is the yaw angle between the SMASS and the target point; tr is the reward coefficient of the yaw angle, which indicates that the reward values obtained from different angles are different; λ_a is the weight of the yaw angle reward; and ε is the adjustment parameter of the reward value and distance.

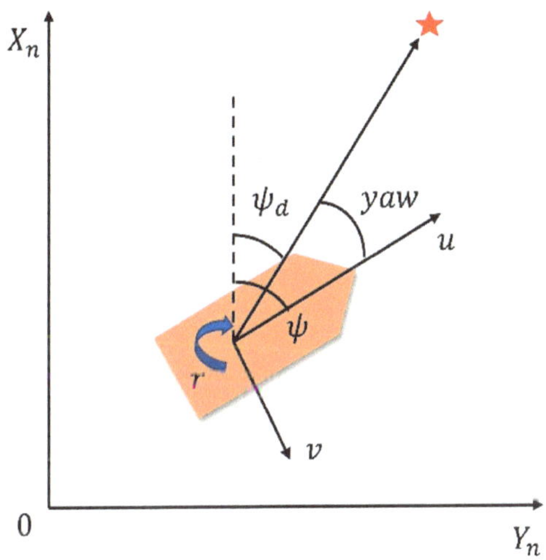

Figure 8. Definition of heading angle error.

(3) Target point reward

In order to get the SMASS to the target point, it is necessary to set a reward at the target point position. At the same time, the SMASS should also receive a negative reward when it collides with obstacles during navigation. The reward value is set as follows:

$$R\_g = \begin{cases} -500\ collision \\ 2000\ goal \end{cases} \tag{21}$$

(4) Obstacle avoidance reward

The laser radar detection range of the SMASS is a circle, launching 24 detection lines from the center of the circle; R_radar is the radar radius, and the reward is 0 when the static obstacle is outside the radar radius. As shown in Figure 9, S_1 is set as the safe distance between the SMASS and the obstacle. When the distance between the SMASS and the obstacle is less than S_1, a negative reward will be obtained. The reward value is set as follows:

$$R\_ob = \begin{cases} 0\ ob > R\_radar \\ -5\ ob < R\_radar, ob < S_1 \\ 1\ ob < R\_radar, ob > S_1 \end{cases} \tag{22}$$

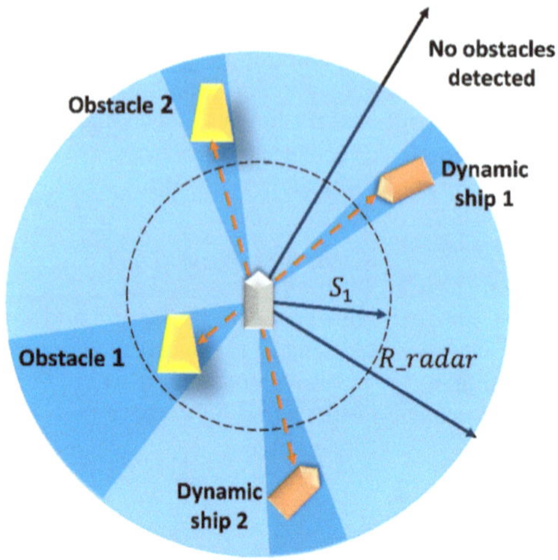

Figure 9. The process of obstacle detection by SMASS laser radar.

(5) COLREGs reward

In order to make the trained SMASS behavior satisfy COLREGs, a COLREGs reward function was introduced. The distance between SMASS and the target point was designed in the COLREGs reward.

While SMASS needs to keep heading, the rudder angle should be 0. In addition, when SMASS needs to avoid obstacles or target ships, she should alter her course to starboard. These are defined as satisfying COLREGs. Otherwise, SMASS should alter her course to port or hold heading after encountering obstacles or target ships, which is considered to be a violation of COLREGs. When the SMASS operations comply with COLREGs, the SMASS would obtain positive rewards. However, when SMASS violates COLREGs, it will be punished. Hence, the reward function can be set as follows:

$$R\_c = \begin{cases} 0 & contrary\ to\ COLREGs, \\ \lambda_c \cdot \sqrt{(x_p - x_g)^2 + (y_p - y_g)^2} & else. \end{cases} \tag{23}$$

where λ_c is the weight of the COLREGs reward function.

Therefore, the calculation process of the total reward function is shown in Figure 7 and is expressed as follows:

$$R = R\_d + R\_yaw + R\_g + R\_ob + R\_c \tag{24}$$

5. Simulation

5.1. Design of Simulation

The training environment is necessary for the intelligent SMASS deep reinforcement learning. A designed unmanned ship training environment can quickly test algorithms [32]. Hence, multiple simulation scenarios were set up to train mobile SMASS for path planning. Based on the improved PPO algorithm proposed above and the construction of the neural network framework, the neural network was trained. The computer configuration was as follows: Intel Core i9-11900K, NVIDIA GTX3090, 24 G video memory, 32 G main memory, and 512 G SSD storage. Gazebo and VScode were used for joint simulation and established a three-dimensional navigation environment in Gazebo to simulate different waters and build a SMASS model, as shown in Figure 10.

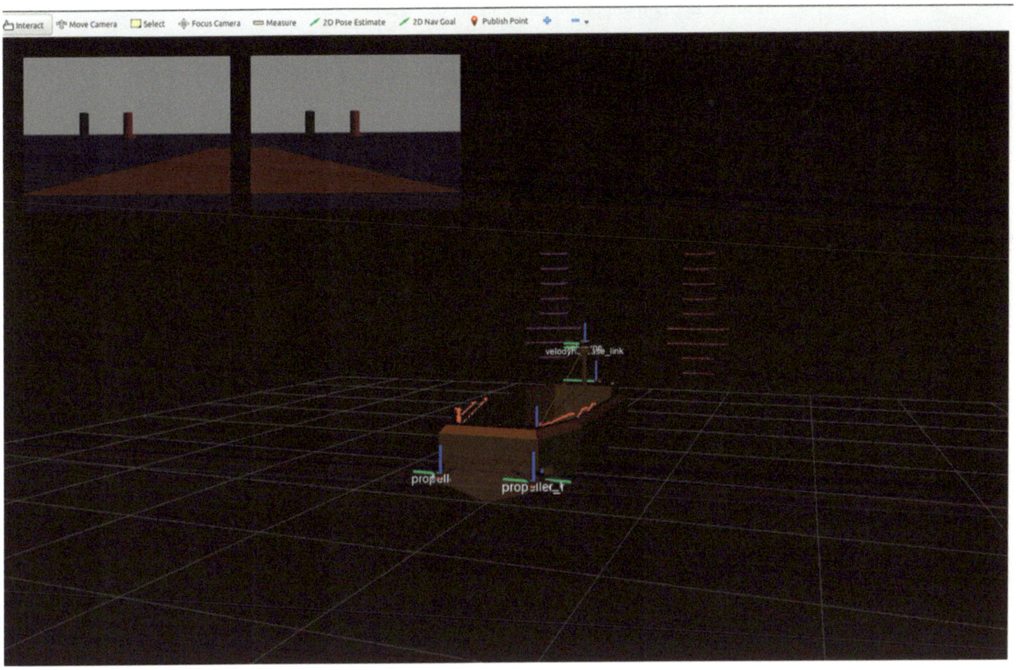

Figure 10. Ship model built in Gazebo simulation environment.

Some restrictions were attached to the SMASS model. SMASS cannot slow down her speed and can only alert her course during the voyage. The SMASS inertia was appropriately increased to simulate the real motion state of the SMASS. In the SMASS steering phase, with the increase of the rudder angle, the rudder transverse force and rudder force turn the SMASS moment. In the transition stage, the transverse velocity and angular velocity were generated under the action of transverse force and rudder force transfer torque, and the increasingly obvious oblique shipping motion made the ship enter the accelerated rotation state. When the SMASS moved in a fixed-length cycle, the steering force transfer torque, drift angle hydrodynamic transfer torque, and resistance transfer torque were balanced. The acceleration of the rotational angular is zero, and the rotational angular velocity was the largest and most stable at this value. This experiment assumed that the SMASS navigated in still water.

5.2. Network Training Process

Experimental parameter settings are shown in Table 1. The Gazebo environment platform module is responsible for generating a navigation environment and simulating SMASS simulation. The environment module could generate and calculate SMASS position and SMASS movement information. When the SMASS reached the target, the training task was over, and entered the next training. When the SMASS encountered obstacles, it stopped training immediately and was placed in the initial position for the next training. The SMASS obstacle avoidance decision training process was divided into two environments, environment one (Env1) and environment two (Env2).

Table 1. Experiment parameter information table.

Experimental Parameters	Symbol	Value
Discounted rate	γ	0.95
Lambda	λ	0.99
Clipping hyperparameter	ε	0.20
Target reward weight	λ_g	10.0
Reward coefficient	tr	1.00
Yaw angle reward weight	λ_a	0.30
COLREGs reward weight	λ_c	1.20
Safe distance	S_1	0.50
Radar radius	R_radar	4.50

In the experiment, the initial position of the SMASS in the simulation environment was (0,0). There were six static obstacles in the simulation environment, and the coordinates of these six static obstacles were (0.46, 1.78), (−0.57, −1.75), (1.68, 3.78), (0.62, −4.44), (0.13, 6.08), and (−1.15, −6.18). There were two target points, and the coordinates were (1.00, −7.00) and (2.00, 7.00), as shown in Figure 11. In the early stage of environmental interaction, ships extremely lacked driving experience and collision avoidance experience. The trained SMASS could not navigate towards the target and avoid a collision.

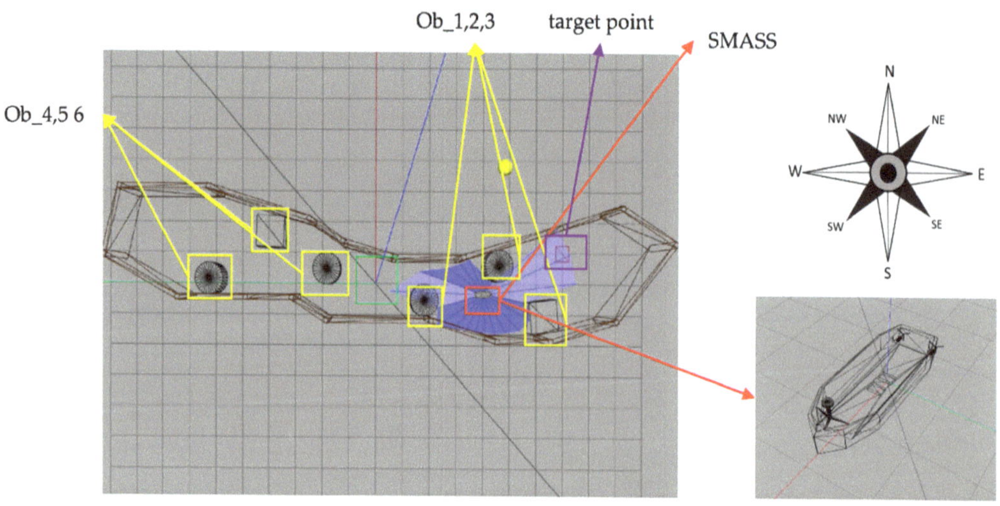

Figure 11. Gazebo simulation environment (Env1). From right to left are six obstacles (ob1, ob2, ob3, ob4, 0b5, and ob6), the blue part is the laser radar range, and the blue line is the laser radar detection line. The left purple box is the target point.

After 1000 training times, SMASS could avoid obstacle 1 and obstacle 2. When the SMASS sailed on the port side of obstacle 2, the course remained unchanged. When encountering obstacle 2, the SMASS took two consecutive port alters of 25° and moved towards the upper right under obstacle 2. When it was 0.6 miles from obstacle 1, her course to port was altered to 45°, along with obstacle 1 upward obliquely. The SMASS continuously steered port and starboard and changed course during movement, but the SMASS could not reach the target point and collided with the environmental framework during the wandering process. SMASS collision avoidance obstacles 1, 2, and 3 are shown in Figure 12.

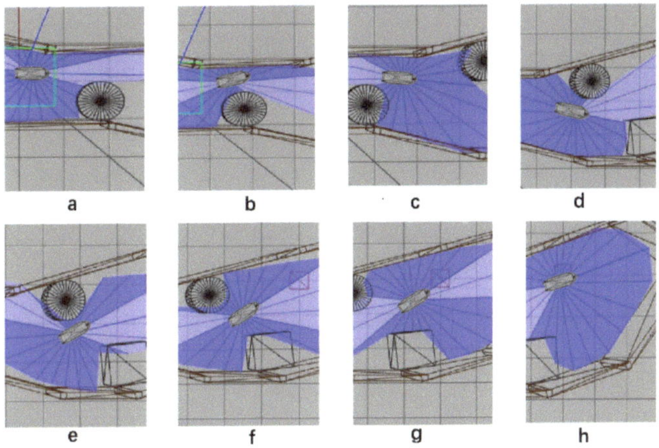

Figure 12. The processes of SMASS avoiding obstacle in Env1 after 1200 training times. Subfigures (**a–c**) show the process of SMASS avoiding ob3. Subfigures (**d–f**) show the process of SMASS avoiding ob2. Subfigures (**g**,**h**) show the process of SMASS avoiding ob1.

After training about 1200 times, the SMASS successfully reached the first target point. Subsequently, the SMASS continuously altered her course to port 45° and sailed to the target point 2. When the SMASS passed under obstacle 3 and navigated towards obstacle 4, her course was altered to starboard 25°, then port and starboard rudder were altered continuously to ensure heading stability.

After training 1500 times, the SMASS could maintain her course and sail to the target point. The SMASS first altered her course to port 25° close to the upper starboard of obstacle 6, and then turned starboard by 25° twice in succession, passing over Obstacle 6, successfully reaching the target point 2. The collision avoidance process is shown in Figure 13. In training environment one, the SMASS successfully avoided six obstacles. In the process of SMASS obstacle avoidance, the change curve of the SMASS steering angle with time is shown in Figure 14.

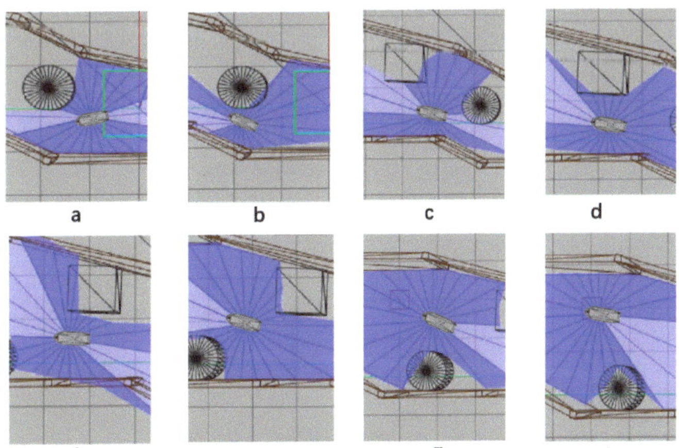

Figure 13. The processes of SMASS avoiding obstacle in Env1 after 1500 training times. Subfigures (**a–c**) show the process of SMASS avoiding ob4. Subfigures (**d–f**) show the process of SMASS avoiding ob5. Subfigures (**g**,**h**) show the process of SMASS avoiding ob6.

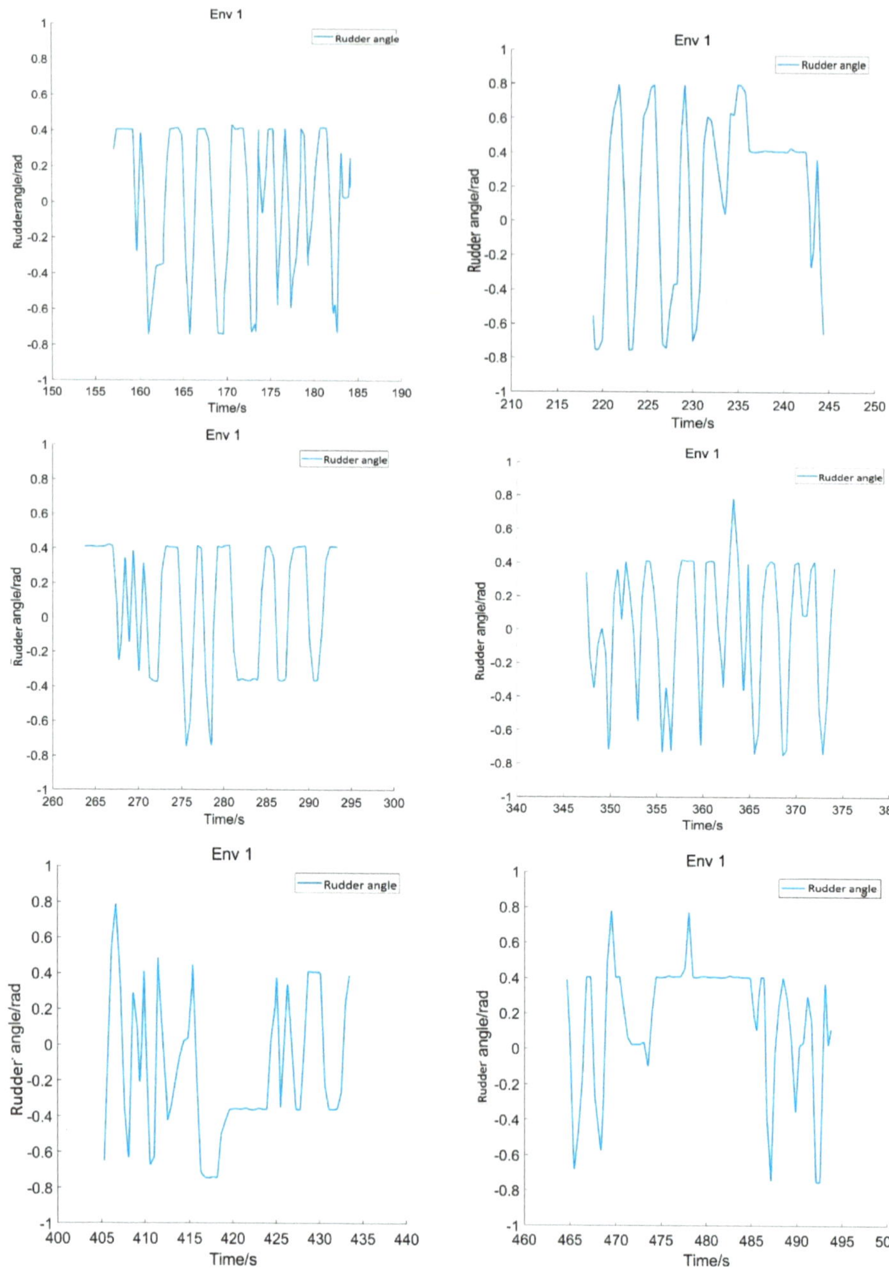

Figure 14. Rudder angle changes of the SMASS sailing in environment 1.

There were five obstacles in the second simulation environment, and the coordinates of these five obstacles were (−1.7, 3.2), (−1.6, −0.5), (2.7, −2.0), (5.4, −1.6), and (−3.8, 1.6). The coordinates of the two target points were (6.0, −3.0) and (−4.5, 3.0), as shown in Figure 15.

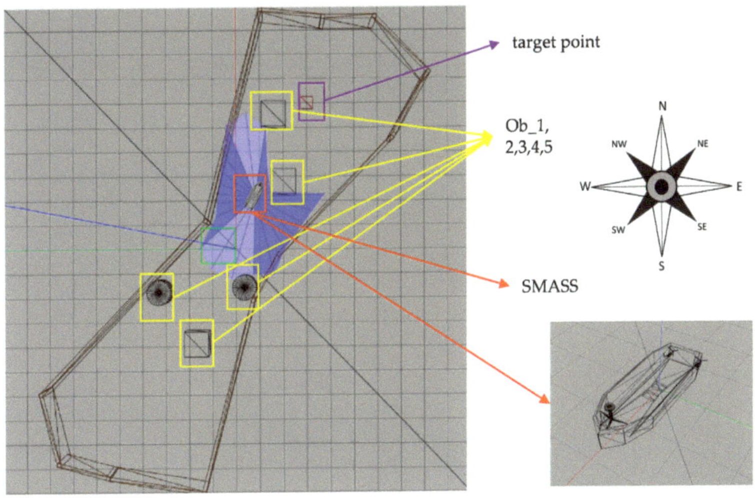

Figure 15. Gazebo simulation environment (Env2). From up to down, there are five obstacles (ob1, ob2, ob3, ob4, and ob5); the blue part is the laser radar range, and the blue line is the laser radar detection line. The purple box is the target point.

After training about 1200 times, the SMASS frequently operated the rudder and reached the first target point. In the process of sailing to the second target point, the SMASS chose to sail around obstacle 1 from above, as shown in Figure 16b. After reaching the target point, the SMASS chose to alter course to port 45° to sail a distance on the left upper side, and then frequently operated the rudder. When the SMASS reached the top left of obstacle 1, the SMASS chose to alter the course to port 45° to drive down.

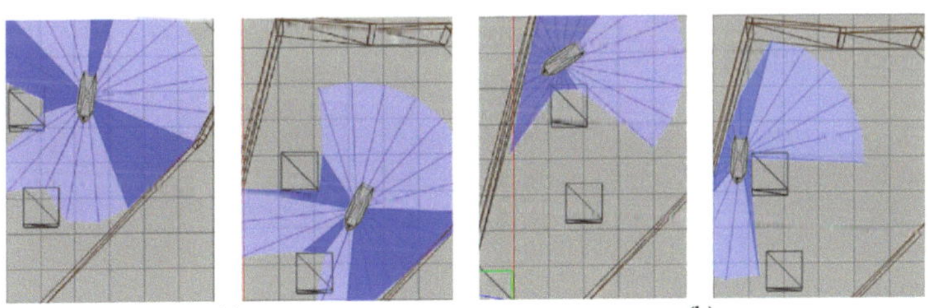

Figure 16. SMASS avoids Obstacle 1 in Env 2. Figure (**a**) shows the obstacle avoidance process of the ship after 1400 training times. Figure (**b**) shows the obstacle avoidance process of the ship after 1200 training times.

After training 1400 times, the SMASS almost did not collide with five obstacles or enter the minimum distance S_1 between the SMASS and the static obstacle. The reward value obtained by the SMASS crossing between obstacle 1 and obstacle 2 was greater than that obtained by the SMASS bypassing above obstacle 1. As shown in Figure 16a, when the distance between the SMASS and obstacle 1 was greater than 0.5 miles, the SMASS altered her course to starboard 25°. When the SMASS was 0.4 miles away from obstacle 2, the SMASS chose to alter her course to starboard 25° and moved forward 0.5 miles. Subsequently, the SMASS altered her course to port and avoided obstacle 2. At the same time, when the SMASS arrived at target 2 and got ready to return to target 1, the reward value obtained by the SMASS passing through the left side of obstacle 5 was larger than that passing through the right side. After passing obstacle 5, the SMASS chose to alter her course to port by 25°. The SMASS altered her course to starboard 45° after passing through obstacle 5. The process of SMASS obstacle avoidance is shown in Figure 17. In the process of SMASS obstacle avoidance, the change of the SMASS steering angle is shown in Figure 18.

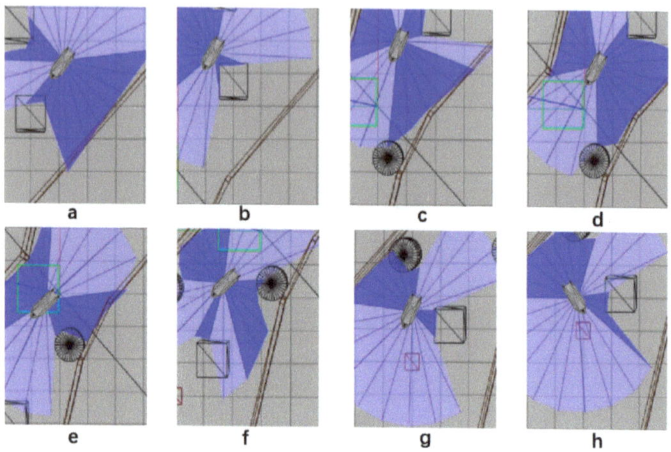

Figure 17. The processes of SMASS avoiding obstacle in Env2. Subfigures (**a**–**c**) show the process of SMASS avoiding ob1 and ob2. Subfigures (**d**–**f**) show the process of SMASS avoiding ob3. Subfigures (**g**,**h**) show the process of SMASS avoiding ob4 and ob5.

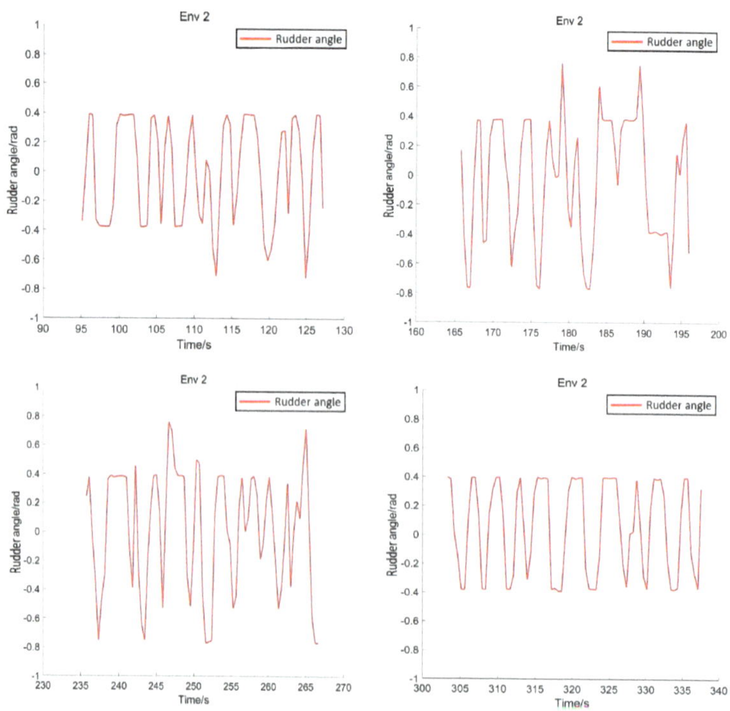

Figure 18. Rudder angle changes of the SMASS sailing in environment 2.

5.3. Comparison Experiment

To verify the effectiveness of the improved PPO algorithm, this paper compared the improved PPO algorithm with the other classic strategy-based reinforcement learning algorithms (such as the AC algorithm, DDPG algorithm, and traditional PPO algorithm). As shown in Figure 19, after training 20,000 times, the actor-network in the AC algorithm converged after training 11,000 times, and the critic network converged after training 10,000 times. The results showed that the convergence rate of the AC algorithm was not satisfied, and the loss value was high. While the DDPG algorithm converged after about training 10,000 times, the algorithm still had the problem of high loss value. When solving SMASS decision-making problems, the traditional PPO algorithm converged after 8000 training times, which was better than the AC algorithm and DDPG algorithm. However, the improved PPO algorithm converged after 6000 training times; the convergence rate was significantly better than the traditional PPO algorithm, and the loss was greatly improved. Hence, it can be found that the convergence rate of the improved PPO algorithm could increase by about 25% compared to the traditional PPO algorithm. Compared with the traditional DDPG and AC algorithms, the convergence rate of the improved PPO algorithm could increase by about 50%.

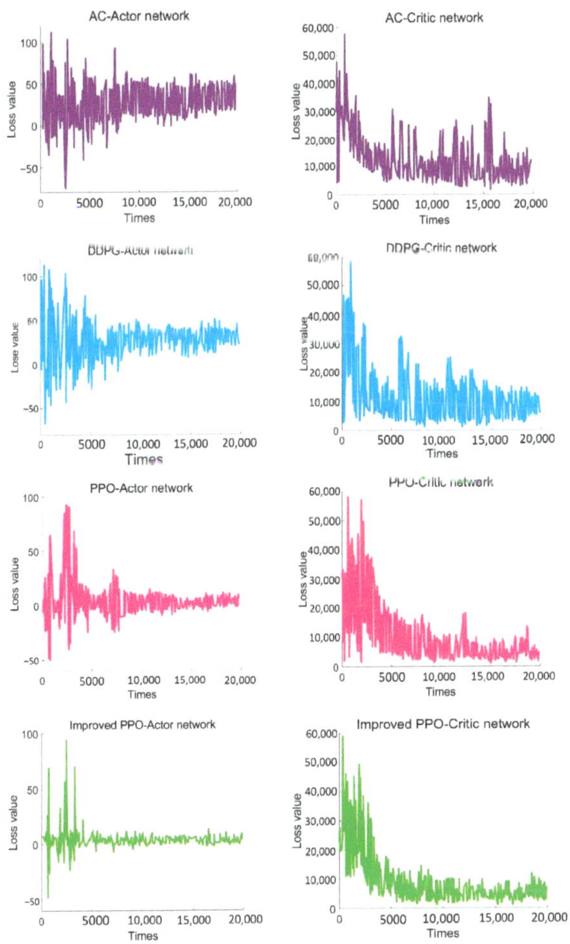

Figure 19. The comparative experiment of the convergence curve.

The Generalized Advantage Estimation Algorithm directly affects the convergence speed and convergence quality of the PPO algorithm. In this experiment, four groups of comparison experiments were conducted to prove the influence of differences in the generalized advantage estimation on the PPO algorithm. Taking the training environment as an example, four λ values were selected for comparative experiments, which were 0.8, 0.9, 0.95, and 0.99, respectively.

The convergence of actor and critic networks when λ was 0.8 is shown in Figures 20 and 21. The convergence of the actor network was not obvious, and the critic network was not converged obviously after 24,000 training sessions.

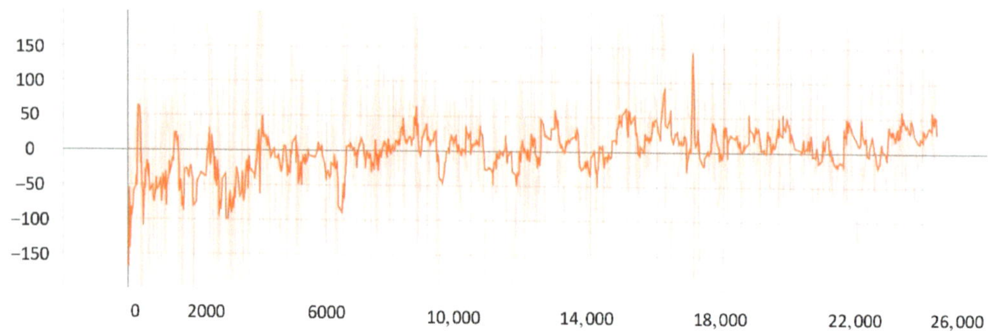

Figure 20. The loss function value change with episodes of actor network when λ was 0.8.

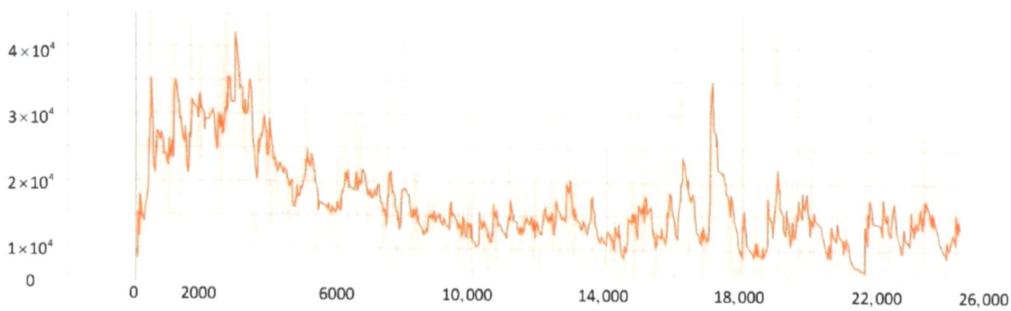

Figure 21. The loss function value change with episodes of critic network when λ was 0.8.

The convergence of actor and critic networks when λ was 0.9 is shown in Figures 22 and 23. Compared with the actor network convergence curve when λ was 0.8, the actor network convergence was better, but the critic network still did not converge after 22,000 training sessions.

The convergence of actor and critic networks when λ was 0.95 is shown in Figures 24 and 25. The convergence rate of the actor network was faster than when λ was 0.9 in the early convergence effect, and the overall convergence trend was shown. In addition, the convergence effect of the critic network was significantly better than when λ was 0.9.

The convergence of actor and critic networks when λ was 0.99 is shown in Figures 26 and 27. The convergence rate of the actor network was much faster than that of the curve when λ was 0.95. In addition, when λ was 0.99, the convergence quality and stability of the actor network and critic network were better than the curve when λ was 0.95.

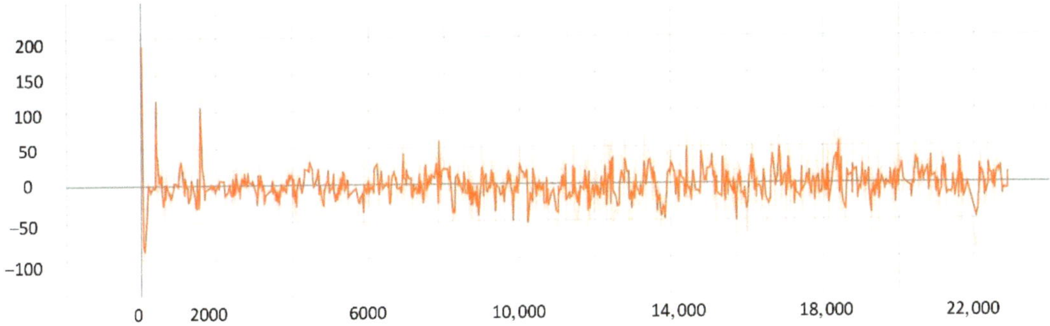

Figure 22. The loss function value change with episodes of actor network when λ was 0.9.

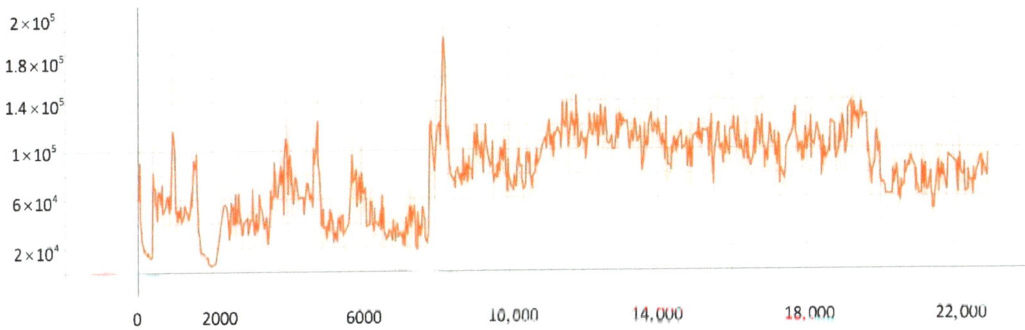

Figure 23. The loss function value change with episodes of critic network when λ was 0.9.

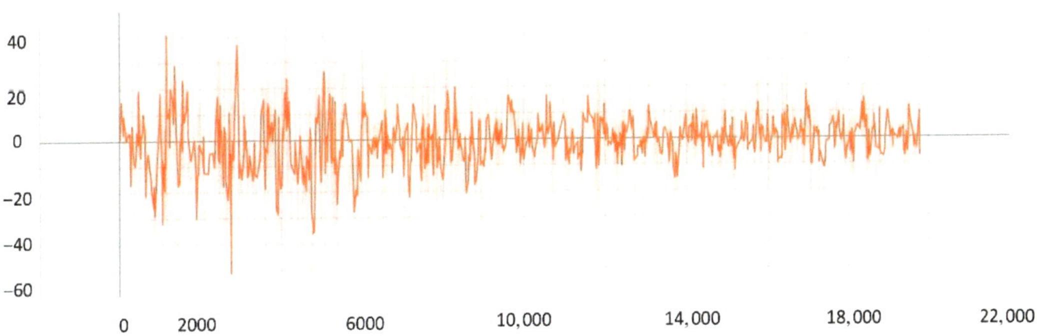

Figure 24. The loss function value change with episodes of actor network when λ was 0.95.

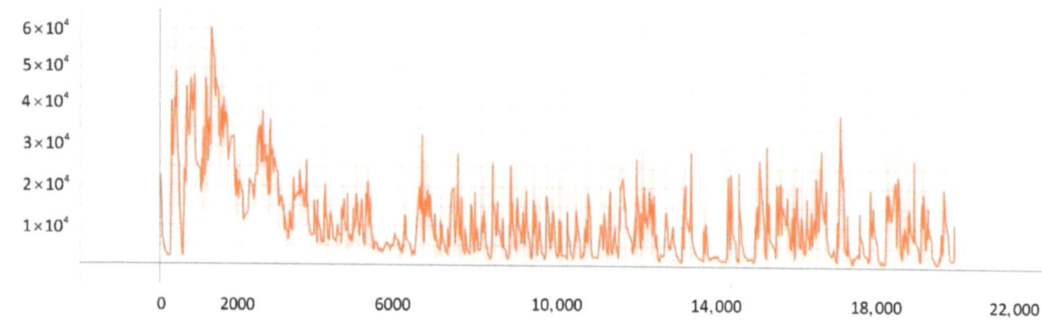

Figure 25. The loss function value change with episodes of critic network when λ was 0.95.

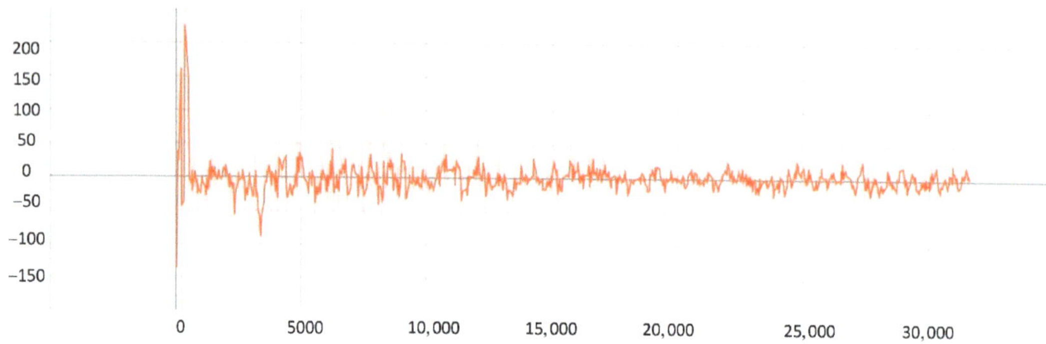

Figure 26. The loss function value change with episodes of actor network when λ was 0.99.

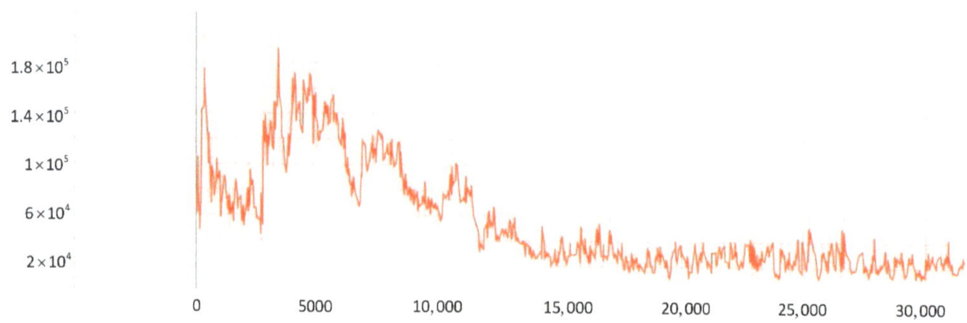

Figure 27. The loss function value change with episodes of critic network when λ was 0.99.

5.4. Verification Simulation

Generalizability refers to the ability of trained models to apply to new data and make accurate predictions. When the training is insufficient, the fitting ability of the decision-making system is not obvious. The disturbance of training data is insufficient to make the decision-making system change significantly. With the increase of training times, the fitting ability of the decision-making system is gradually enhanced. The disturbance can be detected by the decision-making system. A model is often trained too well on training data, that is, overfitting, so that it cannot be generalized. In order to prove the generalization of the proposed SMASS intelligent obstacle avoidance model in this paper, several different simulation environments were constructed to verify the generalizability of the trained SMASS obstacle avoidance network.

The eight representative simulation environments were extracted and displayed as shown in Figure 28. The initial and end positions of each environment were shown in Table 2. There were five obstacles in environment 3. Environments 4, 5, and 6 were used to simulate the navigation of SMASS in relatively narrow waters. The number of obstacles in environment 7 was not too much, but the environment was more complex. There were only two obstacles in environment 8, but the navigable waters were very narrow to simulate the SMASS obstacle avoidance in narrow waters. Environment 9 was relatively open, but there were multiple obstacles located along a line. The environment was used to test whether the SMASS could find the optimal path when there were multiple obstacles in the environment. In environment 10, the navigation area with more obstacles was very narrow, which could be used to simulate the SMASS complex obstacle avoidance navigation in complex narrow waters. In each environment, the collision avoidance processes from the starting position to the end position were described by six graphs (as shown in Figures 29 and 30). Moreover, the SMASS steering rudder angle of collision avoidance processes in each environment are shown in Figures 31–33.

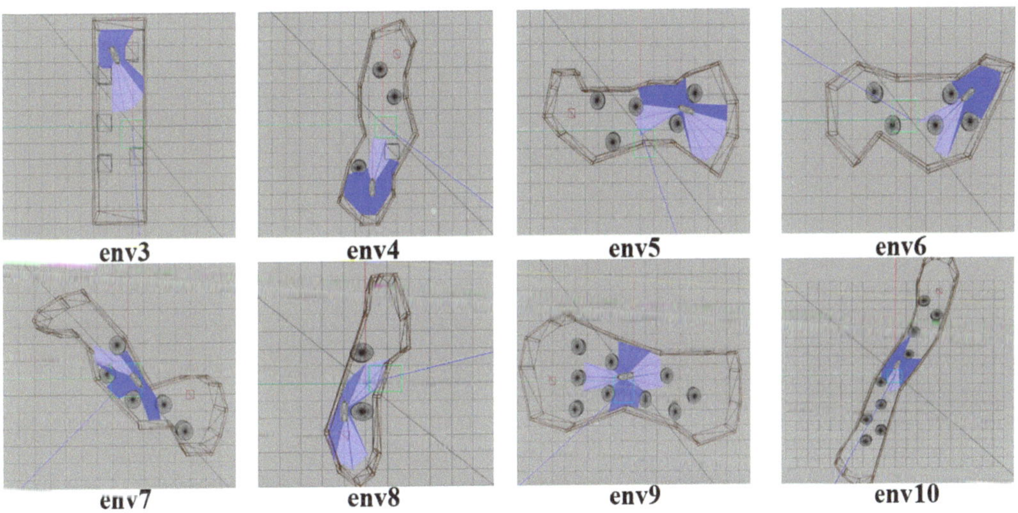

Figure 28. Experimental environment to verify the universality of the SMASS decision-making network.

Table 2. Training environment and verification environment target point coordinates.

Gazebo Environment	Initial Position	End Position
Env 1(Train)	(1.0, −7.0)	(2.0, 7.0)
Env 2(Train)	(6.0, −3.0)	(−4.5, 3.0)
Env 3(Verification)	(5.0, 2.0)	(−4.0, 1.0)
Env 4(Verification)	(5.0, −1.0)	(−5.0, 1.0)
Env 5(Verification)	(1.0, 5.0)	(1.0, −5.0)
Env 6(Verification)	(1.0, 4.0)	(2.0, −5.0)
Env 7(Verification)	(3.0, 2.0)	(−1.0, −4.0)
Env 8(Verification)	(−3.0, 1.0)	(4.0, −1.0)
Env 9(Verification)	(0.0, 7.0)	(1.0, −6.0)
Env 10(Verification)	(9.0, −4.0)	(−9.0, 3.0)

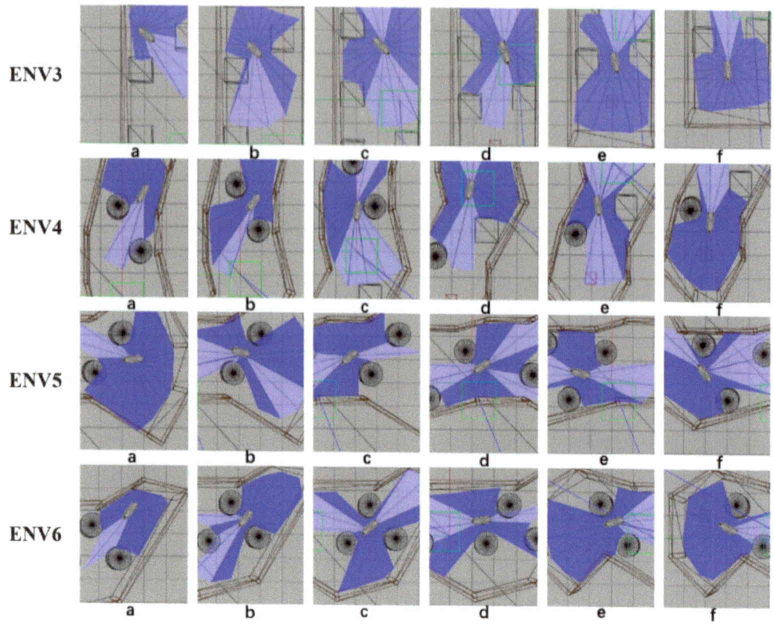

Figure 29. Obstacle avoidance process from environment 3 to environment 6. The SMASS collision avoidance process of each environment is shown by six subgraphs (**a**)–(**f**).

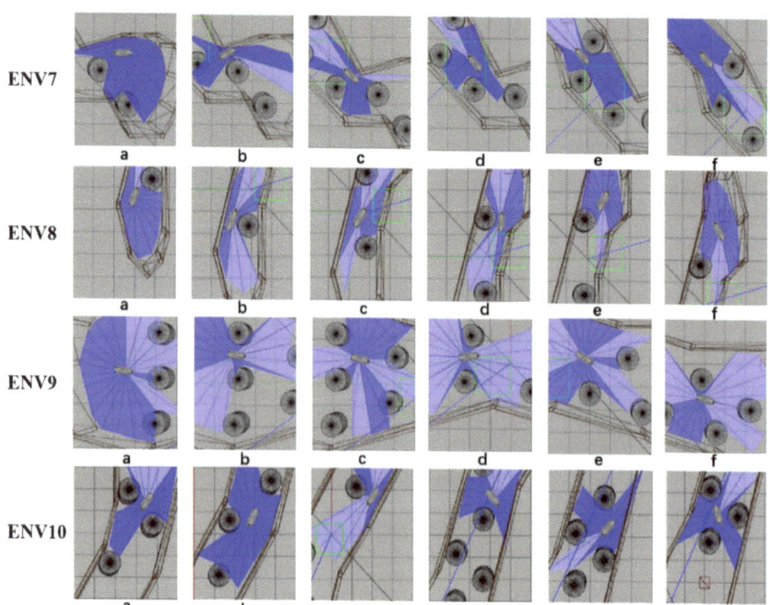

Figure 30. Obstacle avoidance process from environment 7 to environment 10. The SMASS collision avoidance process of each environment is shown by six subgraphs (**a**)–(**f**).

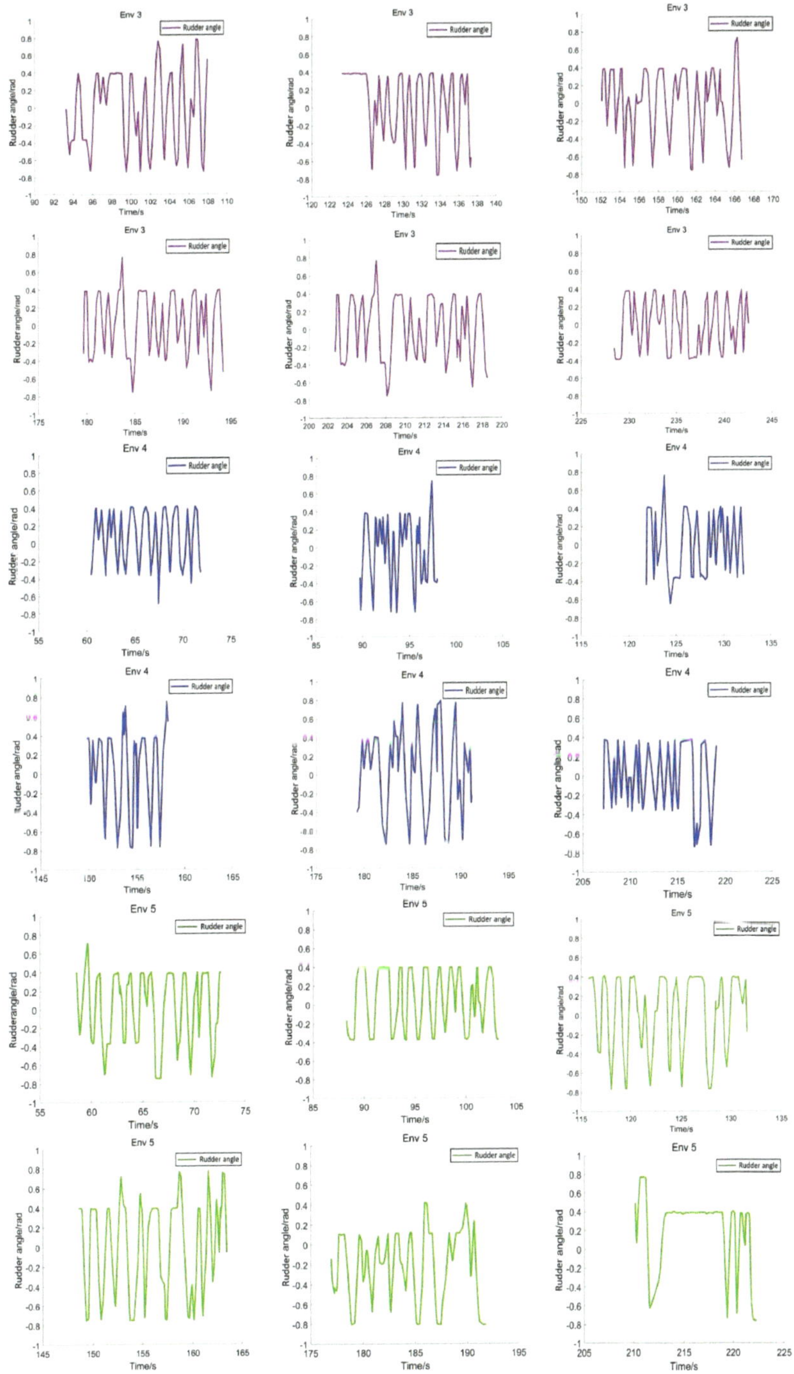

Figure 31. Rudder angle changes of the SMASS sailing in environment 3, environment 4, and environment 5.

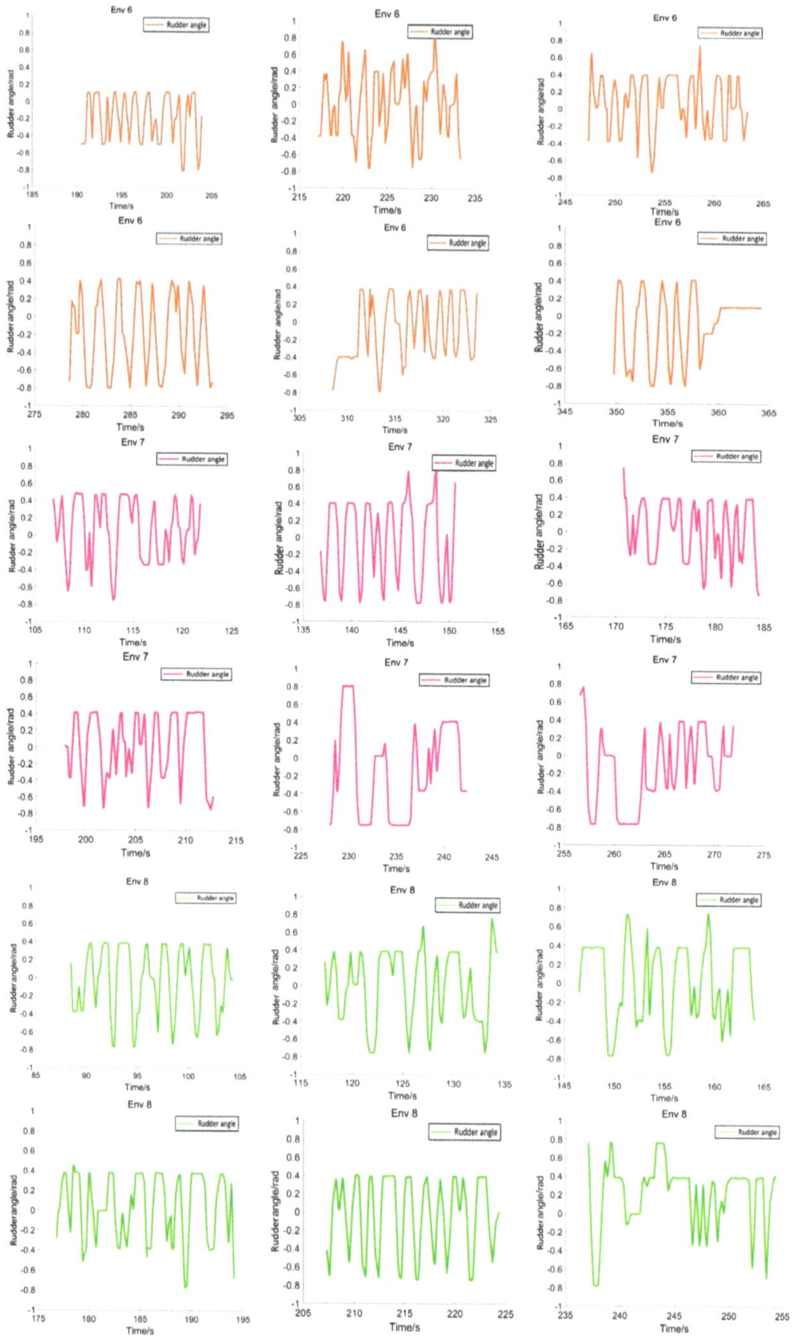

Figure 32. Rudder angle changes of the SMASS sailing in environment 6, environment 7, and environment 8.

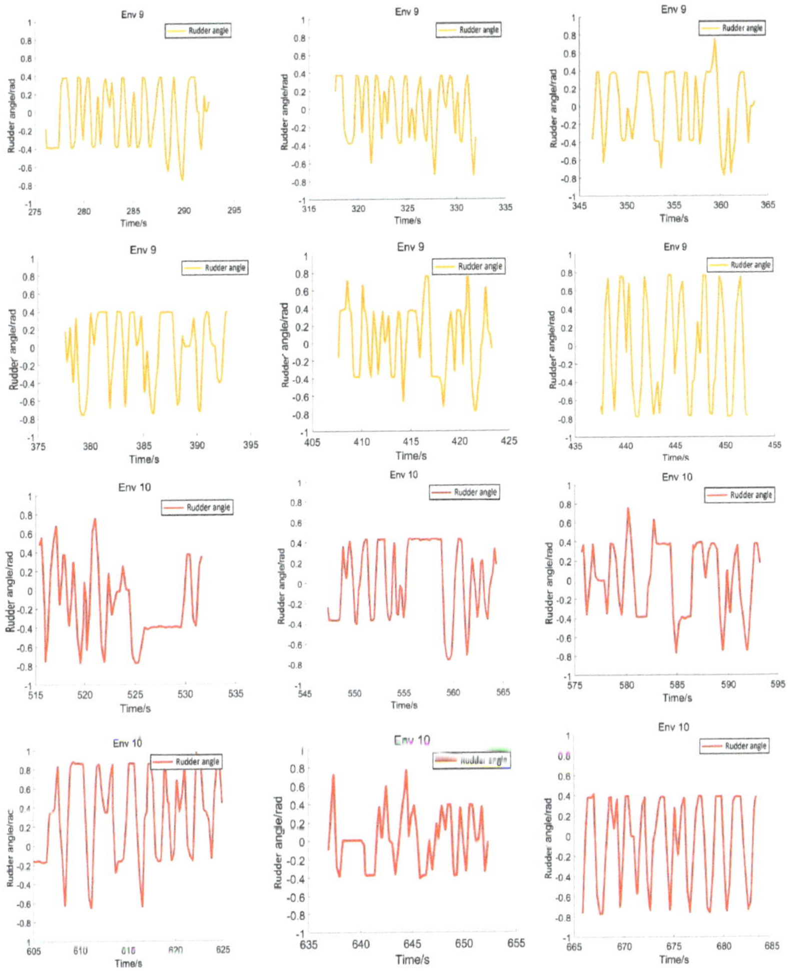

Figure 33. Rudder angle changes of the SMASS sailing in environment 9 and environment 10.

In addition, the avoidance simulations of sailing target ships were carried out to verify the trained SMASS obstacle avoidance capability. Taking the No. 9 environment as an example, these sailing target ships met the trained SMASS under the different collision encounter situations, and the trained SMASS could avoid them accurately and safely according to COLREGs.

As shown in Figure 34, the left side of the figure is the sailing path of the SMASS and three target ships, and the right side is the SMASS avoidance process in the simulation environment. The first target ship (TS01) and the SMASS formed a crossing give-way situation, and the SMASS altered her course to starboard to avoid the first target ship. When the SMASS met the second target ship (TS02), the two ships are formed a crossing stand-on situation. Then, the SMASS kept her course and altered starboard to avoid the second target ship. When the SMASS passed through the middle position, the third target ship (TS03) and the SMASS formed the head-on situation. Then, the SMASS altered course to starboard to avoid the third target ship.

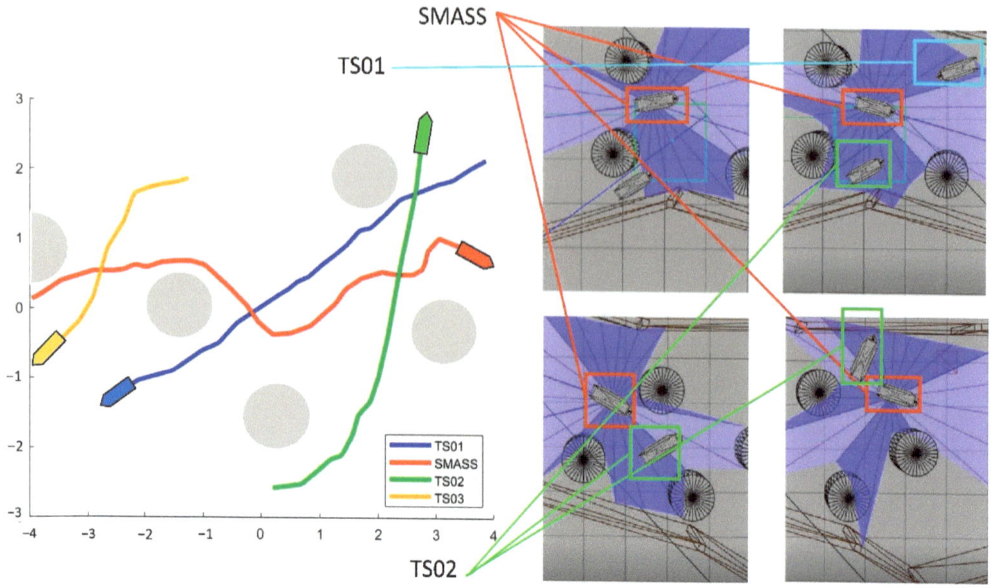

Figure 34. Trained SMASS avoids other ships in environment 9.

6. Conclusions

An improved PPO algorithm for path planning and obstacle avoidance in different complex waters was presented in this paper. SMASS can perform complex local path planning and obstacle avoidance operations when external information is not fully accepted. In this experiment, five factors were considered in the design of the reward function, namely, the relationship between target position, angle, and distance, COLREGs, the reward for safety obstacle avoidance, and whether to reach the target point. This algorithm also performed well in complex waters composed of different numbers of obstacles. The contributions of this experiment are as follows:

- The improved PPO algorithm is superior to other traditional model-free reinforcement learning algorithms based on strategy learning in solving ship decision-making and local path planning problems. The improved PPO algorithm has the advantages of fast convergence and low loss value.
- The improved PPO algorithm has a strong self-learning ability and strong generalization, which could be used to solve the SMASS local path planning and collision avoidance decision-making simultaneously in different complex navigation environments.

Some works should be explored in the future. In the experiment, there are some limitations in setting obstacles into cylinders and squares. Actual obstacles such as islands and navigable areas are not suitable to be set into base shapes. The design of complex obstacles is one of the directions in the future study. In addition, the rudder angle output in this study was the command rudder angle, which has a certain deviation from the execution rudder angle. This is also an important factor to be considered in future studies.

Author Contributions: Conceptualization, W.G. and Z.C.; methodology, W.G. and Z.C.; software, Z.C.; writing—original draft preparation, Z.C.; writing—review and editing, W.G. and Z.C.; resources, X.Z. All authors have read and agreed to the published version of the manuscript.

Funding: The paper was partially supported by the National Natural Science Foundation of China (No. 51409033, 52171342), the Fundamental Research Funds for the Central Universities (No. 3132019343, 3132022126), and Dalian Innovation Team Support Plan in the Key Research Field (2020RT08). The authors would like to thank the reviewers for their valuable comments.

Institutional Review Board Statement: Not applicable.

Informed Consent Statement: Not applicable.

Data Availability Statement: Not applicable.

Conflicts of Interest: The authors declare no conflict of interest.

References

1. Seuwou, P.; Banissi, E.; Ubakanma, G.; Sharif, M.S.; Healey, A. Actor-Network Theory as a Framework to Analyse Technology Acceptance Model's External Variables: The Case of Autonomous Vehicles. In *International Conference on Global Security, Safety, and Sustainability*; Springer: Berlin/Heidelberg, Germany, 2017; pp. 305–320.
2. Erckens, H.; Busser, G.A.; Pradalier, C.; Siegwart, R.Y. Avalon Navigation Strategy and Trajectory Following Controller for an Autonomous Sailing Vessel. *IEEE Robot. Autom. Mag.* **2010**, *17*, 45–54. [CrossRef]
3. Zhang, Z.; Wu, D.F.; Gu, J.D.; Li, F.S. A Path-Planning Strategy for Unmanned Surface Vehicles Based on an Adaptive Hybrid Dynamic Stepsize and Target Attractive Force-RRT Algorithm. *J. Mar. Sci. Eng.* **2019**, *7*, 132. [CrossRef]
4. Liu, X.Y.; Li, Y.; Zhang, J.; Zheng, J.; Yang, C.X. Self-Adaptive Dynamic Obstacle Avoidance and Path Planning for USV Under Complex Maritime Environment. *IEEE Access* **2019**, *7*, 114945–114954. [CrossRef]
5. Xie, S.R.; Wu, P.; Peng, Y.; Luo, J.; Qu, D.; Li, Q.M.; Gu, J. The Obstacle Avoidance Planning of USV Based on Improved Artificial Potential Field. In Proceedings of the IEEE International Conference on Information and Automation (ICIA), Hailar, China, 28–30 July 2014; pp. 746–751.
6. Lyu, H.; Yin, Y. COLREGS-Constrained Real-time Path Planning for Autonomous Ships Using Modified Artificial Potential Fields. *J. Navig.* **2019**, *72*, 588–608. [CrossRef]
7. Chen, C.; Chen, X.Q.; Ma, F.; Zeng, X.J.; Wang, J. A knowledge-free path planning approach for smart ships based on reinforcement learning. *Ocean. Eng.* **2019**, *189*, 106299. [CrossRef]
8. Everett, M.; Chen, Y.F.; How, J.P. Motion Planning Among Dynamic, Decision-Making Agents with Deep Reinforcement Learning. In Proceedings of the 25th IEEE/RSJ International Conference on Intelligent Robots and Systems (IROS), Madrid, Spain, 1–5 October 2018; pp. 3052–3059.
9. Zhang, J.; Springenberg, J.T.; Boedecker, J.; Burgard, W. Deep Reinforcement Learning with Successor Features for Navigation across Similar Environments. In Proceedings of the 2017 IEEE/RSJ International Conference on Intelligent Robots and Systems (IROS), Vancouver, BC, Canada, 24–28 September 2017.
10. Shen, H.; Hashimoto, H.; Matsuda, A.; Taniguchi, Y.; Terada, D.; Guo, C. Automatic collision avoidance of multiple ships based on deep Q-learning. *Appl. Ocean. Res.* **2019**, *86*, 268–288. [CrossRef]
11. Li, L.; Wu, D.; Huang, Y.; Yuan, Z.-M. A path planning strategy unified with a COLREGS collision avoidance function based on deep reinforcement learning and artificial potential field. *Appl. Ocean. Res.* **2021**, *113*, 102759. [CrossRef]
12. Hu, Z.; Wan, K.; Gao, X.; Zhai, Y.; Wang, Q. Deep Reinforcement Learning Approach with Multiple Experience Pools for UAV's Autonomous Motion Planning in Complex Unknown Environments. *Sensors* **2020**, *20*, 1890. [CrossRef]
13. Chun, D.-H.; Roh, M.-I.; Lee, H.-W.; Ha, J.; Yu, D. Deep reinforcement learning-based collision avoidance for an autonomous ship. *Ocean. Eng.* **2021**, *234*, 109216. [CrossRef]
14. Zhao, L.; Roh, M.-I.; Lee, S.-J. Control method for path following and collision avoidance of autonomous ship based on deep reinforcement learning. *J. Mar. Sci. Technol.-Taiwan* **2019**, *27*, 293–310.
15. Xu, X.L.; Lu, Y.; Liu, X.C.; Zhang, W.D. Intelligent collision avoidance algorithms for USVs via deep reinforcement learning under COLREGs. *Ocean. Eng.* **2020**, *217*, 107704. [CrossRef]
16. Long, P.; Fan, T.; Liao, X.; Liu, W.; Zhang, H.; Pan, J. Towards Optimally Decentralized Multi-Robot Collision Avoidance via Deep Reinforcement Learning. In Proceedings of the IEEE International Conference on Robotics and Automation (ICRA), Brisbane, QLD, Australia, 21–25 May 2018; pp. 6252–6259.
17. Guan, W.; Peng, H.W.; Zhang, X.K.; Sun, H. Ship Steering Adaptive CGS Control Based on EKF Identification Method. *J. Mar. Sci. Eng.* **2022**, *10*, 294. [CrossRef]
18. Guan, W.; Zhou, H.T.; Su, Z.J.; Zhang, X.K.; Zhao, C. Ship Steering Control Based on Quantum Neural Network. *Complexity* **2019**, *2019*, 3821048. [CrossRef]
19. Zhang, X.-K.; Han, X.; Guan, W.; Zhang, G.-Q. Improvement of integrator backstepping control for ships with concise robust control and nonlinear decoration. *Ocean. Eng.* **2019**, *189*, 106349. [CrossRef]
20. Perera, L.P.; Oliveira, P.; Guedes Soares, C. System Identification of Nonlinear Vessel Steering. *J. Offshore Mech. Arct. Eng.* **2015**, *137*, 031302. [CrossRef]
21. Nomoto, K.; Taguchi, T.; Honda, K.; Hirano, S. On the steering qualities of ships. *Int. Shipbuild. Prog.* **1957**, *4*, 354–370. [CrossRef]
22. Zhang, J.; Yan, X.; Chen, X.; Sang, L.; Zhang, D. A novel approach for assistance with anti-collision decision making based on the International Regulations for Preventing Collisions at Sea. *Proc. Inst. Mech. Eng. Part M J. Eng. Marit. Environ.* **2012**, *226*, 250–259. [CrossRef]
23. Vagale, A.; Oucheikh, R.; Bye, R.T.; Osen, O.L.; Fossen, T.I. Path planning and collision avoidance for autonomous surface vehicles I: A review. *J. Mar. Sci. Technol.* **2021**, *26*, 1292–1306. [CrossRef]

24. Dearden, R. Bayesian Q-learning. In Proceedings of the Fifteenth National/tenth Conference on Artificial Intelligence/innovative Applications of Artificial Intelligence, Madison, WI, USA, 26–30 July 1998.
25. Rumelhart, D.; Hinton, G.E.; Williams, R.J. Learning Representations by Back Propagating Errors. *Nature* **1986**, *323*, 533–536. [CrossRef]
26. Mnih, V.; Kavukcuoglu, K.; Silver, D.; Graves, A.; Antonoglou, I.; Wierstra, D.; Riedmiller, M. Playing Atari with Deep Reinforcement Learning. *arXiv* **2013**, arXiv:1312.5602.
27. Hasselt, H.V.; Guez, A.; Silver, D. Deep Reinforcement Learning with Double Q-learning. In Proceedings of the Thirtieth AAAI Conference on Artificial Intelligence, Phoenix, AZ, USA, 12–17 February 2016; Volume 30. [CrossRef]
28. Lillicrap, T.P.; Hunt, J.J.; Pritzel, A.; Heess, N.; Erez, T.; Tassa, Y.; Silver, D.; Wierstra, D. Continuous control with deep reinforcement learning. *arXiv* **2015**, arXiv:1509.02971.
29. Schulman, J.; Wolski, F.; Dhariwal, P.; Radford, A.; Klimov, O. Proximal Policy Optimization Algorithms. *arXiv* **2017**, arXiv:1707.06347.
30. Schulman, J.; Moritz, P.; Levine, S.; Jordan, M.; Abbeel, P. High-Dimensional Continuous Control Using Generalized Advantage Estimation. *arXiv* **2015**, arXiv:1506.02438.
31. Fan, Y.; Sun, Z.; Wang, G. A Novel Reinforcement Learning Collision Avoidance Algorithm for USVs Based on Maneuvering Characteristics and COLREGs. *Sensors* **2022**, *22*, 2099. [CrossRef] [PubMed]
32. Duguleana, M.; Mogan, G. Neural networks based reinforcement learning for mobile robots obstacle avoidance. *Expert Syst. Appl.* **2016**, *62*, 104–115. [CrossRef]

MDPI AG
Grosspeteranlage 5
4052 Basel
Switzerland
Tel.: +41 61 683 77 34

Sensors Editorial Office
E-mail: sensors@mdpi.com
www.mdpi.com/journal/sensors

www.ingramcontent.com/pod-product-compliance
Lightning Source LLC
Chambersburg PA
CBHW041433210926
51850CB00013B/3417
9783725833887